Model Emergent Dynamics in Complex Systems

Mathematical Modeling and Computation

About the Series

The SIAM series on Mathematical Modeling and Computation draws attention to the wide range of important problems in the physical and life sciences and engineering that are addressed by mathematical modeling and computation; promotes the interdisciplinary culture required to meet these large-scale challenges; and encourages the education of the next generation of applied and computational mathematicians, physical and life scientists, and engineers.

The books cover analytical and computational techniques, describe significant mathematical developments, and introduce modern scientific and engineering applications. The series will publish lecture notes and texts for advanced undergraduate- or graduate-level courses in physical applied mathematics, biomathematics, and mathematical modeling, and volumes of interest to a wide segment of the community of applied mathematicians, computational scientists, and engineers.

Appropriate subject areas for future books in the series include fluids, dynamical systems and chaos, mathematical biology, neuroscience, mathematical physiology, epidemiology, morphogenesis, biomedical engineering, reaction-diffusion in chemistry, nonlinear science, interfacial problems, solidification, combustion, transport theory, solid mechanics, nonlinear vibrations, electromagnetic theory, nonlinear optics, wave propagation, coherent structures, scattering theory, earth science, solid-state physics, and plasma physics.

A. J. Roberts, *Model Emergent Dynamics in Complex Systems*

John B. Drake, *Climate Modeling for Scientists and Engineers*

Erik M. Bollt and Naratip Santitissadeekorn, *Applied and Computational Measurable Dynamics*

Daniela Calvetti and Erkki Somersalo, *Computational Mathematical Modeling: An Integrated Approach Across Scales*

Jianke Yang, *Nonlinear Waves in Integrable and Nonintegrable Systems*

A. J. Roberts, *Elementary Calculus of Financial Mathematics*

James D. Meiss, *Differential Dynamical Systems*

E. van Groesen and Jaap Molenaar, *Continuum Modeling in the Physical Sciences*

Gerda de Vries, Thomas Hillen, Mark Lewis, Johannes Müller, and Birgitt Schönfisch, *A Course in Mathematical Biology: Quantitative Modeling with Mathematical and Computational Methods*

Ivan Markovsky, Jan C. Willems, Sabine Van Huffel, and Bart De Moor, *Exact and Approximate Modeling of Linear Systems: A Behavioral Approach*

R. M. M. Mattheij, S. W. Rienstra, and J. H. M. ten Thije Boonkkamp, *Partial Differential Equations: Modeling, Analysis, Computation*

Johnny T. Ottesen, Mette S. Olufsen, and Jesper K. Larsen, *Applied Mathematical Models in Human Physiology*

Ingemar Kaj, *Stochastic Modeling in Broadband Communications Systems*

Peter Salamon, Paolo Sibani, and Richard Frost, *Facts, Conjectures, and Improvements for Simulated Annealing*

Lyn C. Thomas, David B. Edelman, and Jonathan N. Crook, *Credit Scoring and Its Applications*

Frank Natterer and Frank Wübbeling, *Mathematical Methods in Image Reconstruction*

Per Christian Hansen, *Rank-Deficient and Discrete Ill-Posed Problems: Numerical Aspects of Linear Inversion*

Michael Griebel, Thomas Dornseifer, and Tilman Neunhoeffer, *Numerical Simulation in Fluid Dynamics: A Practical Introduction*

Khosrow Chadan, David Colton, Lassi Päivärinta, and William Rundell, *An Introduction to Inverse Scattering and Inverse Spectral Problems*

Charles K. Chui, *Wavelets: A Mathematical Tool for Signal Analysis*

Model Emergent Dynamics in Complex Systems

A. J. Roberts

University of Adelaide
Adelaide, South Australia
Australia

siam.
Society for Industrial and Applied Mathematics
Philadelphia

10 9 8 7 6 5 4 3 2 1

Library of Congress Cataloging-in-Publication Data

Roberts, A. J. (Anthony John), 1957- author.
Model emergent dynamics in complex systems / A. J. Roberts, University of Adelaide, Adelaide, South Australia, Australia.
pages cm. – (Mathematical modeling and computation)
Includes bibliographical references and index.
ISBN 978-1-611973-55-6
1. Dynamics–Mathematical models. 2. Computational complexity. 3. Differential equations–Asymptotic theory. I. Title.
QA845.R62 2014
515′.39–dc23

2014024843

For Barbara

and all her support

over the decades

taken to develop the ideas

and connections in this book

Contents

Preface

This book explores how to derive relatively simple dynamical equations that model complex physical interactions. The book arises out of the growing interest in and applications of modern dynamical systems theory. Due to my background, fluid flows and other continuum dynamics form many of the applications we investigate. The triple aim of the book is to use sound theory to explore algebraic techniques, develop interesting applications, and discover general modeling principles.

Mentor and colleague Prof. E. O. Tuck was discussing the undergraduate curriculum with another who was espousing the need for courses on elementary mathematics with an advanced approach. Prof. Tuck's riposte was "but that is exactly the opposite of what I want; I want advanced mathematics from an elementary approach." Similarly, this book aims to develop advanced mathematical modeling methods and discuss their subtleties with as elementary mathematics as possible. The assumed background knowledge is common undergraduate linear algebra, calculus, and differential equations, but there is no need for functional analysis, advanced differential geometry, or even complex analysis.

The basis for the methodology is both the theory and the geometric picture of both coordinate transforms and invariant manifolds in dynamical systems: in particular, we heavily use center and slow manifolds. The wonderful aspect of this approach is the range of geometric interpretations of the modeling process. Simple geometric pictures inspire sound methods of analysis and construction. Further, the pictures that we draw of state spaces also provide a route for better assessing limitations and strengths in a model. Geometry and algebra form a powerful partnership.

> ... duality between algebra and geometry was discovered by René Descartes: every geometric object has an algebraic description, every algebraic formula determines a geometric object. Humans tend to use the algebraic version for calculation, and the geometric one for imagination.
>
> Fearful symmetry, *Stewart and Golubitsky.*

The theme of this book is that coordinate transforms and center manifolds provide a powerfully enhanced and unified view of a swath of other complex system modeling methodologies, such as averaging, homogenization, multiple scales, singular perturbations, two timing, and WKB theory.

One main reason complex systems are complex is that there are many interacting components. We generally posit some network of interacting "agents" or "particles." The simplest such complex networks arise through the highly regular structure of space where nearest neighbor interactions dominate. The domain of spatiotemporal dynamics provides us with many examples. Under time evolution, coherent patterns (stripes on a tiger) or incoherent patterns (turbulence) emerge over time. We seek to find ways

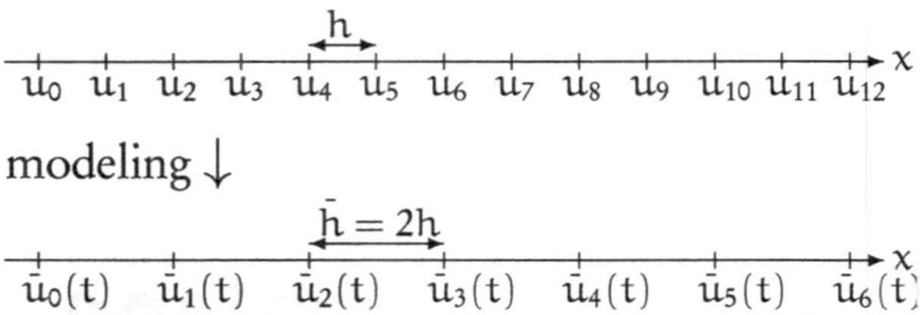

Figure 1. *Our modeling analyzes the dynamics* (0.1) *of* $u_i(t)$ *on the (upper) fine-scale lattice, spacing* h, *and systematically maps it into the dynamics* (0.2) *of* $\bar{u}_j(t)$ *on the (lower) coarse grid, spacing* $2h$.

to characterize, to model, the coherent or incoherent behavior that we see. What is the aggregate behavior? How can the whole appear to be more than the sum of its parts?

Example 0.1 (dynamics on a 1D lattice). Let us loosely overview one application in multiscale modeling. Distribute "particles" at the grid points of a 1D lattice (with spacing h, say), as shown in the upper part of Figure 1. Each of these particles has some property called u, perhaps temperature, that evolves in time so the ith particle has property $u_i(t)$. For simplicity, restrict attention to linear dynamics. Then the generic, spatially discrete, only nearest neighbor interactions system that preserves total u-stuff is the advection-diffusion equations

$$\frac{du_i}{dt} = -c\tfrac{1}{2}(u_{i+1} - u_{i-1}) + d(u_{i+1} - 2u_i + u_{i-1}) \tag{0.1}$$

for some constants c and d. This equation moves u-stuff around with "advection speed" ch and spreads u-stuff with "diffusion" dh^2.

Suppose we seek to model the dynamics (0.1) on a grid coarser by a factor of two, spacing $\bar{h} = 2h$, as in the lower part of Figure 1. Grid point j on the coarser grid would correspond to fine grid point $i = 2j$. Let $\bar{u}_j(t)$ be the coarse grid values of the u-stuff. Techniques we develop (Roberts, 2009c, e.g.) justifiably model the fine-grid dynamics by the coarse-grid equation

$$\frac{d\bar{u}_j}{dt} \approx -\bar{c}\tfrac{1}{2}(\bar{u}_{j+1} - \bar{u}_{j-1}) + \bar{d}(\bar{u}_{j+1} - 2\bar{u}_j + \bar{u}_{j-1}),$$
$$\text{where} \quad \bar{c} = \tfrac{1}{2}c \quad \text{and} \quad \bar{d} = \tfrac{1}{4}d + \frac{c^2}{16d}. \tag{0.2}$$

This is another advection-diffusion equation for the movement and spread of u-stuff but with appropriately renormalized coefficients to suit the coarser grid.

The renormalization of the coefficients in (0.2) has components $\tfrac{1}{2}c$ and $\tfrac{1}{4}d$ that all traditional linear modeling techniques would derive. Our more careful techniques show that the coarse "diffusion" should be enhanced in proportion to c^2/d. This last correction helps ensure that the coarse-scale model preserves stability. It arises through more carefully resolving the consequences of the microscale dynamics on the fine grid.

Having generated a mapping of dynamics from one grid to a coarser grid, we may repeat the mapping across a whole hierarchy of grids, as indicated schematically by Figure 2. At each level of the grid the dynamics would be an advection-diffusion

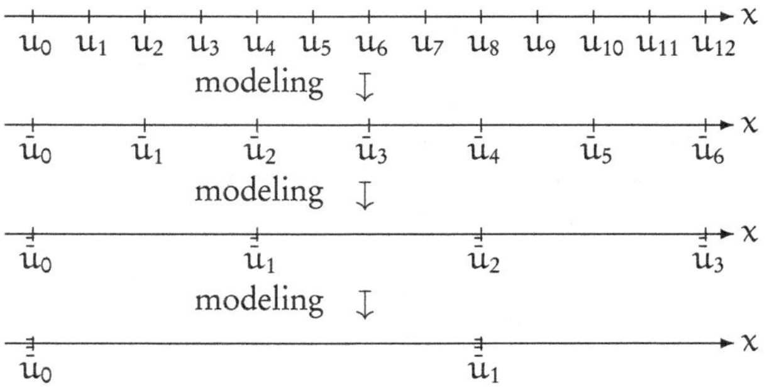

Figure 2. *The modeling* (0.1)$\mapsto$(0.2) *from a fine lattice to a coarser lattice may be repeated across a hierarchy of lattices to understand dynamics on many length scales.*

equation. Although each of the models (0.1) and (0.2) expresses simple linear dynamics, the transformation of the model from one grid scale to another is *nonlinear* through the nonlinear dependence of $\bar{d}$ upon c and d. Repeating the nonlinear transformation (0.1)$\mapsto$(0.2) across many grid scales morphs diffusion-dominated microscale dynamics into an upwind, advection-dominated, macroscale model. Analogous nonlinear transformations of other systems have the potential to demonstrate and illuminate emergence of qualitatively new dynamical rules on different scales of a complex system.

Example 0.1 links many mathematical and computational techniques, such as multigrid algorithms, wavelets, cellular automata, lattice dynamics, and renormalization. Our main tool to tease out such emergent dynamics will be judicious coordinate transforms and their analogues in invariant manifolds. Where possible our analysis will be rigorous. However, rigor rapidly disintegrates in the face of practical realities: in many of the applications discussed, rigorous support is lacking, but let's not let that lack prevent us from making progress in understanding the modeling of emergent dynamics.

> Mathematics is not a careful march down a well-cleared highway, but a journey into a strange wilderness, where explorers often get lost. Rigour should be a signal to the historian that the maps have been made, and the real explorers have gone elsewhere. *W.S. Anglin*

I emphasize that modeling from one scale to another is a nonlinear process. Example 0.1 shows this nonlinearity in the nonlinear mapping of coefficients from one scale to the next. Why emphasize this nonlinearity? Because most people mostly use linear arguments in modeling. For such linear derivations the dynamics on one scale looks much the same as that on another, due to the linearity; thus they view the whole as always the sum of its parts. But because we here recognize the inherent nonlinearity in modeling, we see that the whole is more than the sum of its parts.[1]

Another major reason complex systems are complex is the interaction of many physical processes: "For example, a complete computational model of a large-scale fusion device

[1] Huge discussions take place over "emergence" and "strong emergence" in complex systems. The mapping of systems from microscale to macroscale, being nonlinear, need not be uniquely invertible. Because we recognize this possibility of multiple causations, the nonlinear nature of modeling from one scale to another supports the "strong emergence" tenet that macroscale phenomena need not be always traceable to the microscale.

is a complex system involving issues of fluid dynamics, deformation of solid materials, thermal effects, ablation, fracture, corrosion and aging of materials, radiation and many other phenomena" (Brown et al., 2008, §2.1.1, e.g.). Most systematic approaches to modeling require multiphysical processes to all interact at the same "order" in the modeling. In contrast, techniques developed herein flexibly allow different physical effects to be of different orders of magnitudes, even in different locations in space-time.

Example 0.2 (thin fluid flow on a fiber). Even the relatively simple flow of a thin film of fluid along a cylindrical fiber is physically complicated (Roberts and Li, 2006, e.g.). Gravity drains the fluid along the fiber but also pulls fluid tangentially around the fiber in some places and normally develops a "hydrostatic" pressure in other places with gradients that drive the flow in other directions. Surface tension then tends to form drops of shape induced by curvature of the cylindrical fiber and the curvature of the drop. The flow of the fluid in the drops is then affected by inertia, whereas the thinner fluid outside the drops is little influenced by inertia. All these processes occur in the flow, with different balances occurring in different locations and times. Sound modeling has to cope with such multiphysics.

Find more information Developments and applications further to those developed herein are documented on my web page for this book.[2] My website also provides services to construct slow and center manifold models of autonomous or nonautonomous differential equation systems that you might enter.

[2] http://www.siam.org/books/mm20

Part I

Asymptotic methods solve algebraic and differential equations

Part I Contents

This part of the book introduces important algebraic techniques for approximating solutions to algebraic and differential equations. These form a basis for developing techniques for understanding the technically challenging partial differential equations typically found in applications.

Our approach is generally perturbative. When analyzing a complex problem of interest, we find a similar, but easier, problem and then seek the solution of the interesting problem as a perturbation of the easier problem. Such perturbations naturally generate power series solutions. Chapter 1 thus explores some of perhaps the simplest nonlinear problems—perturbed algebraic equations. These algebraic equations illuminate the solution and interpretation of power series solutions of nonlinear problems. Although these algebraic problems are almost ridiculously basic—we start with quadratics—nonetheless the general framework established here later empowers you to solve complicated modeling problems of real interest.

Generally we invoke an iterative approach driven by the residual of the governing algebraic equation. We explore how to easily and flexibly implement such iterations in computer algebra. Iteration is ideal for computer algebra: we want computers to do the tedious repetitive tasks for us—those for which it is worth investing our time, making sure the computer is doing what we want. Throughout the book, computer algebra handles some of the incredible details in similarly solving complicated modeling problems of real interest.

Although many people prefer to phrase problems as so-called singular perturbations, section 1.2 argues that phrasing problems as regular perturbations is more powerful.

Section 1.4 introduces the key features of the computer algebra package Reduce that we use. Look to this section for help getting started with Reduce and for translations to other computer algebra languages.

Constructing power series solutions of differential equations is an ideal application of computer algebra. It leads immediately to techniques to solve vastly more complicated problems. The aim of Chapter 2 is to use simple iteration to develop power series solutions of linear and nonlinear differential equations. As done for the algebraic equations of Chapter 1, Chapter 2 makes iteration flexible by basing iteration upon the residual of the governing equations.

It eventuates that many of the series solutions we construct are not convergent—they are divergent. Section 2.3 shows that with care we can still manipulate and use divergent series to make useful predictions.

Amazingly, nonlinear coordinate transforms empower us to derive simple models of nonlinear oscillations—detailed in Chapter 3; the model describes just the amplitude and frequency of the oscillations. Other extant methods, such as that of multiple scales or averaging, also model nonlinear oscillations, but Takens (1974) recognized that the so-called normal-form transformation is a much sounder basis. Such coordinate transforms extend to modeling oscillations with complex exponentials $e^{\pm i\omega t}$. The evolution of their complex amplitudes provides a form that we readily interpret to discover overall properties of the oscillations.

An analogy is the following. Imagine standing at the side of a two-way highway; the cars whiz by so fast you cannot see any details inside. Now get in a car driving one way on the highway: the cars driving the other way are still a mystery, but the cars driving the same direction can now have their inside details viewed at leisure. That is, some things become clear by shifting to a moving coordinate system, while others stay mysterious. The same clarification appears in the use of time-dependent coordinate transforms, as shown in sections 3.6–3.7.

Chapter 1

Perturbed algebraic equations solved iteratively

Contents

We start modeling the dynamics of complex systems by first revisiting algebraic equations. The reason is that in modeling we need to solve the polynomial equations that arise for eigenvalues.

Example 1.1. Consider a toy system of two coupled differential equations for two attributes, $u_1(t)$ and $u_2(t)$, interacting together:

$$\dot{u}_1 = \frac{du_1}{dt} = -u_1 + u_2 \quad \text{and} \quad \dot{u}_2 = \frac{du_2}{dt} = +u_1 - u_2 \,. \tag{1.1}$$

In the classic algebra of differential equations we would solve this equation analytically to write a general solution as $u_1 = c_1 + c_2 e^{-2t}$ and $u_2 = c_1 - c_2 e^{-2t}$, for some integration constants c_i. We do *not* follow this algebraic route in this book, as realistic nonlinear differential equations almost never have analytic solutions.

Alternatively, we could numerically solve the differential equations (1.1) and then plot the solutions against time. Figure 1.1 shows the results. The problem is that it

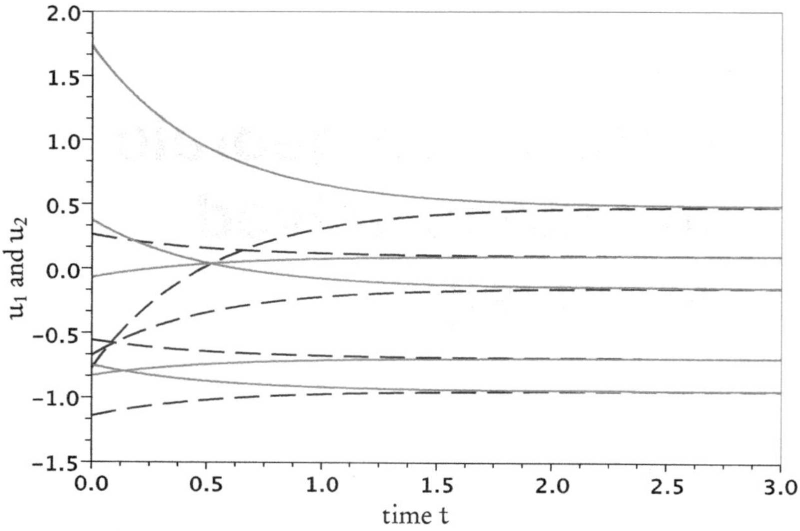

Figure 1.1. *Values of solutions of* (1.1)—$u_1(t)$ *(dashed) and* $u_2(t)$ *(solid)—plotted against time. The structure of the long-time dynamics is hard to see.*

is hard to see the cohesive structure that emerges over long times. Generally we do not plot solutions against time.

Instead we generally plot what happens in state space, here the u_1u_2-plane. Such state space plots show the structure of emergent states. Figure 1.2 shows the evolution from 50 different initial conditions. All solutions collapse onto the line $u_1 = u_2$. The solutions then are constant. This state of constant $u_1 = u_2$ is the emergent dynamics of this simple system.

We understand these emergent dynamics by the linear algebra of matrices. The eigenvalues and eigenvectors of the matrix implicitly on the right-hand side of (1.1) lead to a linear transform to mean and difference variables $u_1 = v(t) - w(t)$ and $u_2 = v(t) + w(t)$: substitute this transformation, and then a little algebra shows that the system (1.1) becomes the decoupled pair $\dot{v} = 0$ and $\dot{w} = -2w$. We predict from the algebraic form of these two equations that w decays quickly to zero to leave constant v as the long-time dynamics. Later sections use analogous coordinate transforms to construct emergent dynamics.

But perturbations and change are generic. What if the system is not precisely (1.1)? Perhaps $\dot{u}_2 = 0.9u_1 - u_2$. The linear algebra still applies but is perturbed. The key initial step of finding eigenvalues is then to solve a perturbed polynomial equation. We need techniques to solve such perturbed equations.

Algebraic polynomial equations introduce some of the basic ideas we use to solve and analyze complex problems. This section explores some simple algebraic equations along the lines of Hinch (1991, Chap. 1) and Bender and Orszag (1981, §7.1–2). Computer algebra empowers our analysis.

> On two occasions I have been asked [by members of Parliament!], "Pray, Mr. Babbage, if you put into the machine wrong figures, will the right answers come out?"

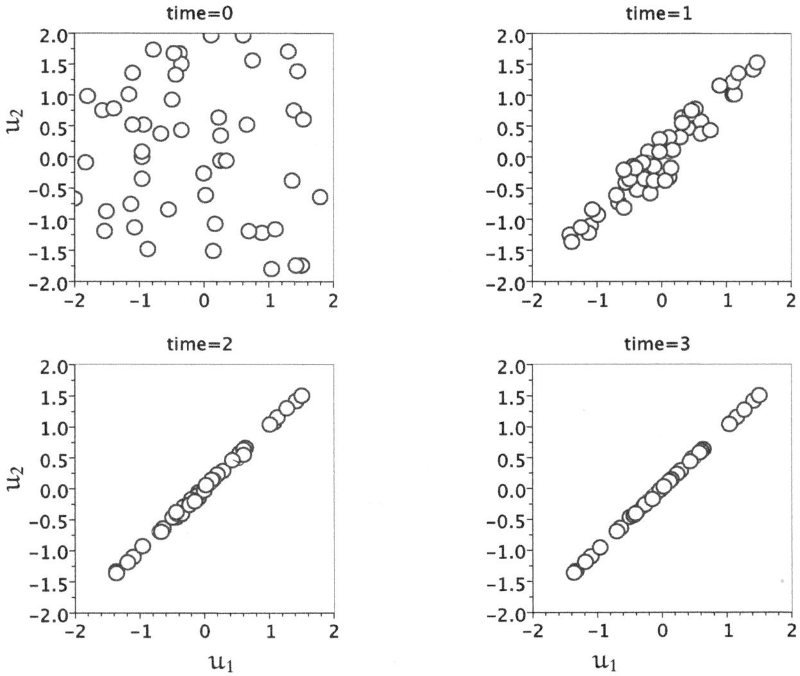

Figure 1.2. *Fifty initial conditions in the* u_1u_2*-plane at time* $t = 0$ *evolve according to* (1.1). *Their locations at later times demonstrate that all solutions collapse onto the line* $u_1 = u_2$.

> I am not able rightly to apprehend the kind of confusion of ideas that could provoke such a question. *Charles Babbage*

The irony in the above quote is that this is exactly what we do: we supply the computer with a wrong answer, albeit roughly correct, and then iterate to improve the answer until it becomes right![3]

This book adopts iteration throughout.

1.1 ▪ Base iteration upon the residual

Let us start with this simple quadratic equation for an unknown x:

$$x^2 + 0.1\,x - 1 = 0\,. \tag{1.2}$$

Suppose we did not know the following famous formula for solving such a nonlinear equation: $x = \frac{1}{2}\left(-0.1 \pm \sqrt{0.1^2 + 4}\right) = 0.95125\,, -1.05125$. Forget that you know how to solve quadratics like this. Instead we argue that the quadratic (1.2) is nearly the same as the simpler quadratic

$$x^2 - 1 = 0\,. \tag{1.3}$$

[3]However, if one interprets Babbage's term "figures" to mean the combination of "data+program" together, then there is no irony. Our iteration uses appropriate "wrong" data together with the "right" program to form a correct algorithm that gives the right answers.

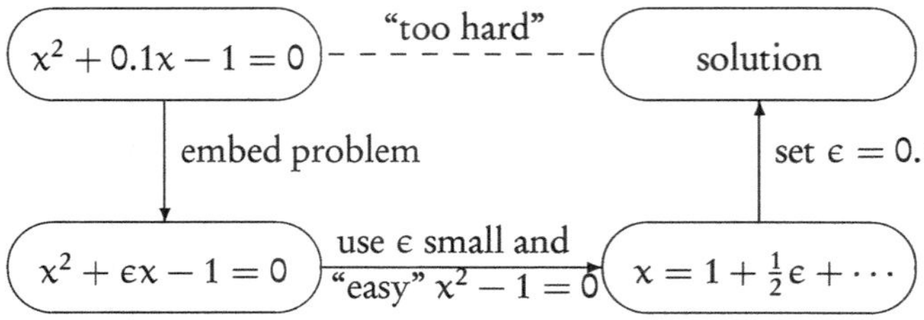

Figure 1.3. *Schematically shows our common approach of embedding a "hard" problem in a class of problems, solving the class approximately for a small parameter, and then evaluating to approximate the solution of the original.*

The simpler quadratic (1.3) has easier solution $x = \pm 1$, which for the purposes of argument suppose we know. In a manner to become familiar, we use the simpler problem (1.3), and its solution, to perturbatively solve the more complicated problem (1.2). Figure 1.3 shows the approach that we typically use in difficult problems.

Connect the simpler quadratic (1.3) to the original quadratic (1.2) as an instance of the more general quadratic problem

$$x^2 + \epsilon x - 1 = 0. \tag{1.4}$$

Analogous to what we do in much more interesting difficult problems, and as shown in Figure 1.3, introduce the parameter ϵ so that the general quadratic (1.4) encompasses both the simpler quadratic (1.3), the case $\epsilon = 0$, and the original quadratic (1.2), the case $\epsilon = 0.1$. This approach of embedding the particular problem (1.2) in a more general problem (1.4) is immensely useful.

Now we proceed to expand the solution in powers of "small" parameter ϵ (Figure 1.3). Our aim is to use simple iteration to recover the Taylor series

$$x = 1 - \tfrac{1}{2}\epsilon + \tfrac{1}{8}\epsilon^2 - \tfrac{1}{128}\epsilon^4 + \cdots \tag{1.5}$$

of the supposedly unknown, exact, positive root of the quadratic (1.4), namely, $x = -\epsilon/2 + \sqrt{1 + \epsilon^2/4}$.

1.1.1 ▪ Use the residual to guide improvements

Let's solve this quadratic problem (1.4) in a manner analogous to how we later straightforwardly solve incredibly difficult, multiscale, multiphysics problems. We seek power series solutions to the perturbed quadratic (1.4) *using the residual* of the quadratic equation.

Define the residual

$$\operatorname{Res}_{(1.4)}(x) := x^2 + \epsilon x - 1. \tag{1.6}$$

The residual is a function of the variable x (and the parameter ϵ). Our aim is to find the variable x as a function of parameter ϵ for which the residual is zero and hence to have solved the quadratic (1.4).

Start from the simple case of $\epsilon = 0$, for which we know the positive solution $x = 1$. Correspondingly, the residual $\operatorname{Res}_{(1.4)}(1) = 1^2 + 0 \cdot 1 - 1 = 0$. Thus set the initial approximation $x_0 = 1$.

- Now seek an improvement appropriate for nonzero parameter ϵ. Seek $x_1 = x_0 + \hat{x}_0 = 1 + \hat{x}_0$, where $\hat{x}_0$ is a *small* correction to the approximate initial solution x_0. The "smallness" of corrections is crucial.

Substitute $x = x_1 = 1 + \hat{x}_0$ into the quadratic (1.4):

$$\begin{aligned} &(1+\hat{x}_0)^2 + \epsilon(1+\hat{x}_0) - 1 = 0 \\ \iff\ &2\hat{x}_0 + \hat{x}_0^2 + \epsilon + \epsilon\hat{x}_0 = 0\,. \end{aligned}$$

Recall that both parameter ϵ and the correction $\hat{x}_0$ are small: this means we drop any products of these terms, such as $\hat{x}_0^2$ and $\epsilon\hat{x}_0$ above. The iteration is then to solve the straightforward linear equation

$$2\hat{x}_0 + \epsilon = 0\,.$$

This gives correction $\hat{x}_0 = -\frac{1}{2}\epsilon$, and consequently the next approximation is $x_1 = 1 - \frac{1}{2}\epsilon$. Indeed these are the first two terms in the Taylor series (1.5) of the positive root of the quadratic (1.4).

Now consider the ϵ term in the previous linear equation that drives the correction $\hat{x}_0$: from (1.6) recognize it is simply the residual $\operatorname{Res}_{(1.4)}(x_0)$ of the quadratic (1.4) evaluated at the initial approximation; for brevity denote this residual by $\operatorname{Res}_0 := \operatorname{Res}_{(1.4)}(x_0)$, which here is equal to ϵ.

- Second, seek a further improvement. Seek $x_2 = x_1 + \hat{x}_1 = 1 - \frac{1}{2}\epsilon + \hat{x}_1$, where $\hat{x}_1$ is a small correction to the approximate solution x_1.

  Substitute $x = x_2 = 1 - \frac{1}{2}\epsilon + \hat{x}_1$ into the quadratic (1.4):

$$\begin{aligned} &(1 - \tfrac{1}{2}\epsilon + \hat{x}_1)^2 + \epsilon(1 - \tfrac{1}{2}\epsilon + \hat{x}_1) - 1 = 0 \\ \iff\ &-\tfrac{1}{4}\epsilon^2 + 2\hat{x}_1 + \hat{x}_1^2 = 0\,. \end{aligned}$$

  Recall that the correction $\hat{x}_1$ is small: thus drop any product such as $\hat{x}_1^2$ above. Although parameter ϵ is also smallish, we actually want to evaluate solutions at finite ϵ; thus we do *not* neglect products involving *only* the parameter ϵ, such as the ϵ^2 terms above—it is retained; we only drop ϵ products when *also* involving corrections, such as the $\epsilon\hat{x}_0$ terms in the previous iteration. The iteration is then to solve the straightforward linear equation

$$2\hat{x}_1 - \tfrac{1}{4}\epsilon^2 = 0\,.$$

  This gives correction $\hat{x}_1 = \frac{1}{8}\epsilon^2$, and consequently the next approximation is $x_2 = 1 - \frac{1}{2}\epsilon + \frac{1}{8}\epsilon^2$. Indeed these are the first three terms in the Taylor series (1.5) of the positive root of the quadratic (1.4).

  Again, look at the ϵ^2 term in the linear equation above that drives the correction: recognize from (1.6) that it is simply the residual $\operatorname{Res}_{(1.4)}(x_1)$ of the quadratic (1.4) evaluated at the first nontrivial approximation, namely $\operatorname{Res}_1 := \operatorname{Res}_{(1.4)}(x_1) = -\frac{1}{4}\epsilon^2$.

The next subsection continues with further iteration aided by computer algebra. The key is that the corrections are guided by the residual of the quadratic equation (1.4).

1.1.2 ▪ Computer algebra makes iteration easy

I wish to God these calculations had been executed by steam.

Charles Babbage

Algorithm 1.1 *Reduce code to solve the quadratic* (1.4) *by iteration using successive corrections* (1.8).

```
x:=1;                       % set initial approx to root
let eps^10=>0;              % truncate power series
for iter:=1:99 do begin% repeat iteration
    res:=x^2+eps*x-1;   % current residual of quadratic
    write x:=x-res/2;   % correct approx using residual
    if res=0 then iter:=1000000;% until residual approx 0
end;
```

Reexamine solving the quadratic (1.4) by iteration. Suppose that at the jth iterate we have an approximation x_j, expressed as a truncated power series in the parameter ϵ. Seek a small correction $\hat{x}_j$ so that $x_{j+1} = x_j + \hat{x}_j$ is a better approximation. How can it be better? By reducing the residual (1.6) of the quadratic equation (1.4). So substitute $x = x_j + \hat{x}_j$ into the quadratic (1.4) to see that

$$\begin{aligned} &(x_j + \hat{x}_j)^2 + \epsilon(x_j + \hat{x}_1) - 1 = 0 \\ \iff\ & x_j^2 + \epsilon x_j - 1 + (2x_j + \epsilon)\hat{x}_j + \hat{x}_j^2 = 0 . \end{aligned}$$

That is, we seek to use the equation

$$\operatorname{Res}_j + (2x_j + \epsilon)\hat{x}_j + \hat{x}_j^2 = 0 \tag{1.7}$$

to guide our choice of correction $\hat{x}_j$. Note that $\operatorname{Res}_j := \operatorname{Res}_{(1.4)}(x_j) = x_j^2 + \epsilon x_j - 1$ is the residual (1.6) of the quadratic (1.4) at the current jth iteration. Invoke three general principles:

- we cannot expect to solve (1.7) exactly, as that would be tantamount to solving the original quadratic (1.4) exactly;
- instead, omit all products of corrections such as $\hat{x}_j^2$; and
- approximate all coefficients of linear terms in the correction $\hat{x}_j$ by the unperturbed value, here setting $x_j \approx 1$ and $\epsilon \approx 0$ so that $(2x_j + \epsilon)\hat{x}_j \approx 2\hat{x}_j$.

Then the equation (1.7) is approximately

$$\operatorname{Res}_j + 2\hat{x}_j = 0 \quad \Longrightarrow \quad \text{correction } \hat{x}_j = -\tfrac{1}{2}\operatorname{Res}_j . \tag{1.8}$$

Recognize that this is effectively what we did by hand in the previous subsection. Now proceed to iterate with the power and reliability of computer algebra.

Look at the computer algebra code of Algorithm 1.1. To execute this code, I recommend you type it into a text file for Reduce to execute. That way you will be able to easily repair errors in typing or conception. Type Algorithm 1.1 into a text file, say named `quad`. Then it is *most important* to surround the code by the two lines

```
on div; off allfac; on revpri;
...
end;
```

- The first line tells Reduce to print algebraic expressions formatted appropriately for perturbation expansions—*always use this first line for all Reduce code in this book.*

Dialogue 1.1 *Output using Algorithm* 1.1.

```
[aroberts@wilton Asymptotic]$ reduce
Loading image file :/usr/local/reduce/lisp/psl/linux/red/reduce.img
REDUCE 3.7, 15-Apr-1999 ...

1: in quad$

x := 1
         1
x := 1 - ---*eps
         2
         1         1  2
x := 1 - ---*eps + ---*eps
         2         8
         1         1  2       1     4
x := 1 - ---*eps + ---*eps  - -----*eps
         2         8          128
         1         1  2       1     4      1      6
x := 1 - ---*eps + ---*eps  - -----*eps  + ------*eps
         2         8          128          1024
          1      8
      - -------*eps
        32768
         1         1  2       1     4      1      6
x := 1 - ---*eps + ---*eps  - -----*eps  + ------*eps
         2         8          128          1024
          5      8
      - -------*eps
        32768
         1         1  2       1     4      1      6
x := 1 - ---*eps + ---*eps  - -----*eps  + ------*eps
         2         8          128          1024
          5      8
      - -------*eps
        32768
2:
```

- The three dots denote the Reduce code of Algorithm 1.1 and the Reduce code of all other algorithms later explored.
- The last of the above lines tells Reduce to stop scanning the file for input.

To get the Reduce commands on a file executed, use the `in` command: here `in quad$`, at the prompt `1:`, where the termination by the `$` means that the commands are not echo-printed.

The Reduce dialogue might then be as in Dialogue 1.1. The last printed line is the power series expansion for a solution of the quadratic equation (1.4). The code of Algorithm 1.1 follows the iteration we have developed:

- make the initial approximation $x_0 = 1$;
- in Reduce we efficiently truncate power series solutions by telling Reduce to discard—for example, with `let eps^10=>0;`—all terms in parameter ϵ with exponent of 10 or higher;
- then the iteration is to compute the residual with the current approximation followed by updating the approximation using this residual—safe iteration is done in a `for-do` loop;

- the iteration continues until the residual is zero, to the order of truncation in powers of parameter ϵ specified earlier—we exit the loop by setting the loop counter to a value higher than the maximum specified in the `for-do` statement.

This simple computer algebra gives us the first six nonzero terms in the Taylor series (1.5) in parameter ϵ for solutions of the quadratic (1.4). Retaining higher powers in the iteration will give us correspondingly more terms in the Taylor series (1.5).

1.1.3 ▪ There are many advantages to this iterative approach

> The whole of the developments and operations of analysis are now capable of being executed by machinery.... As soon as an Analytical Engine exists, it will necessarily guide the future course of science. *Charles Babbage,* 1864

The approach is relatively easy The computer algebra code is simpler than other approaches.

The approach is reliable The iteration is self-checking in that the crucial requirement for getting the correct power series is that you code the computation of the residual correctly—as the iteration only terminates when the residual is zero. All other parts of the coding only affect how long you have to wait for the answer to emerge.

Infinite loops? Waiting an infinite time for an infinite loop to terminate is useless. An infinite loop indicates that you must improve how a correction is computed from the residual.

You can easily generate high orders Computing higher-order terms in the power series is, until you run out of time or memory, simply a matter of increasing the order of truncation of the power series and rerunning the iterative code.

A perturbation need not be small In the above example power series solution of the quadratic equation (1.4), although not immediately obvious, the power series converges for $|\epsilon| < 2$. Thus this approach finds the solution to problems quite a large distance from the base, here $x^2 - 1 = 0$. For example, the quadratic $x^2 + x - 1 = 0$ has positive solution given by the $\epsilon = 1$ case of our power series in the supposedly small parameter ϵ.

The approach is flexible Any problem in a wide class may be solved simply by changing the computation of the residual. For example, to solve the cubic $\epsilon x^3 + x^2 + \epsilon x - 1 = 0$, just change the computation of the residual accordingly. Simply changing the computation of the residual will work provided the initial approximation is unchanged and the rule for using the residual to give a correction remains unchanged.

1.2 ▪ Rescale singular perturbations

Now let's move on to consider the innocuous looking quadratic

$$\epsilon x^2 + x - 1 = 0\,. \tag{1.9}$$

As for all quadratics, this has two solutions. But when the parameter $\epsilon = 0$, the equation reduces to $x - 1 = 0$ and *only one solution is apparent.* This is an example of a *singular perturbation*. Singular perturbations arise when the basic case, here $\epsilon = 0$, differs qualitatively from the general cases, here $\epsilon \neq 0$. Interesting physical problems often appear singular.

Perturbation problems that are not singular are termed *regular perturbations*.

Here we resolve the apparent contradiction between $\epsilon = 0$ and $\epsilon \neq 0$ by finding the exact roots of the quadratic (1.9) and exploring their behavior as $\epsilon \to 0$. The two roots are, from $x = (-1 \pm \sqrt{1+4\epsilon})/(2\epsilon)$,

$$x^{(1)} = \quad 1 - \epsilon + 2\epsilon^2 - 5\epsilon^3 + \cdots,$$
$$x^{(2)} = -\frac{1}{\epsilon} - 1 + \epsilon - 2\epsilon^2 + 5\epsilon^3 - \cdots.$$

The root $x^{(1)}$ is near 1 and is well behaved. The other root, $x^{(2)}$, disappears to $-\infty$ as $\epsilon \to 0$; hence $x^{(2)}$ is not apparent when we simply set $\epsilon = 0$.

1.2.1 ▪ Iteration finds the well-behaved root

As before, use iteration to find the root $x^{(1)}$.

Suppose you know an approximation x_j to this root, such as the initial approximation $x_0 = 1$. Seek a small correction $\hat{x}_j$ so that $x_{j+1} = x_j + \hat{x}_j$ better satisfies the singular quadratic (1.9). Substitute $x = x_{j+1} = x_j + \hat{x}_j$ into the quadratic (1.9) to find

$$\begin{aligned} &\epsilon(x_j + \hat{x}_j)^2 + (x_j + \hat{x}_j) - 1 = 0 \\ \iff\ &\epsilon x_j^2 + 2\epsilon x_j \hat{x}_j + \epsilon \hat{x}_j^2 + x_j + \hat{x}_j - 1 = 0 \\ \iff\ &\text{Res}_j + (2\epsilon x_j + 1)\hat{x}_j + \epsilon \hat{x}_j^2 = 0, \end{aligned} \tag{1.10}$$

where $\text{Res}_j := \text{Res}_{(1.9)}(x_j) = \epsilon x_j^2 + x_j - 1$ is the residual of the quadratic (1.9) for the currently known approximation. As before, neglect products of small corrections such as $\hat{x}_j^2$, and replace coefficients of terms linear in $\hat{x}_j$ by their $\epsilon \to 0$ limit, here $(2\epsilon x_j + 1)\hat{x}_j \approx +\hat{x}_j$. Thus approximate equation (1.10) for the small corrections by

$$\text{Res}_j + \hat{x}_j = 0 \quad \Longrightarrow \quad \text{correction } \hat{x}_j = -\text{Res}_j. \tag{1.11}$$

Use this correction to update the approximation to the root.

For example, here

$$\begin{aligned} &x_0 = 1 \implies \text{Res}_0 = \epsilon \implies \\ &x_1 = 1 - \epsilon \implies \text{Res}_1 = -2\epsilon^2 + \epsilon^3 \implies \\ &x_2 = 1 - \epsilon + 2\epsilon^2 - \epsilon^3 \implies \cdots. \end{aligned}$$

Thus two iterates find the first few terms in the Taylor series of the well-behaved root of the singular quadratic (1.9); the ϵ^3 term in x_2 will be corrected in the next iteration.

1.2.2 ▪ Iteration also finds the singular root

Now adapt the iteration to find the singular root $x^{(2)}$.

Again suppose you know an approximation x_j to this root, such as the initial approximation $x_0 = -1/\epsilon$. Seek a small correction $\hat{x}_j$ so that $x_{j+1} = x_j + \hat{x}_j$ better satisfies the singular quadratic (1.9). Substitute $x = x_{j+1} = x_j + \hat{x}_j$ into the quadratic (1.9) to find (1.10) as before. Also as before, neglect products of small corrections, such as $\hat{x}_j^2$, and replace coefficients of terms linear in $\hat{x}_j$ by their $\epsilon \to 0$ limit. *The difference here* is that $x_j \approx -1/\epsilon$ and so $(2\epsilon x_j + 1)\hat{x}_j \approx -\hat{x}_j$. Thus approximate equation (1.10) for the small corrections by

$$\operatorname{Res}_j - \hat{x}_j = 0 \quad \Longrightarrow \quad \text{correction } \hat{x}_j = +\operatorname{Res}_j. \tag{1.12}$$

Use this correction to update the approximation to the singular root.

For example, here

$$\begin{aligned} x_0 &= -\frac{1}{\epsilon} \Longrightarrow \operatorname{Res}_0 = -1 \Longrightarrow \\ x_1 &= -\frac{1}{\epsilon} - 1 \Longrightarrow \operatorname{Res}_1 = \epsilon \Longrightarrow \\ x_2 &= -\frac{1}{\epsilon} - 1 + \epsilon \Longrightarrow \cdots . \end{aligned}$$

Thus two iterations find the first few terms in the power series of the singular root of the singular quadratic (1.9).

Deduce the initial approximation But how do we know to start with $x_0 = -1/\epsilon$? This x_0 follows because *nontrivial solutions must come from a balance of at least two dominant terms in the governing equation.* Consider the possibilities exhaustively.

- Suppose the last two terms of the quadratic (1.9) are the dominant balance. That is, $x - 1 = 0$ is the approximate equation. Then the initial approximation will be $x \approx 1$. The neglected term $\epsilon x^2 \approx \epsilon$ is then smaller than the two terms forming the balance, namely the x and 1, and so the initial approximation is valid. We saw that this led us to the well-behaved root.

- Suppose the first two terms of the quadratic (1.9) form the dominant balance. That is, $\epsilon x^2 + x = 0$ is the approximate equation. Then the initial approximation will be solutions of $\epsilon x^2 + x = 0$, namely the equation $x(\epsilon x + 1) = 0$, giving solutions $x \approx 0$ and $x \approx -1/\epsilon$.

  - The $x \approx 0$ case is not relevant because then the neglected term -1 is larger than the retained terms in the quadratic (1.9).

  - However, for the $x \approx -1/\epsilon$ case, the neglected term -1 is smaller than the two terms forming the balance, both of large size $\pm 1/\epsilon$, and so the initial approximation is valid. We saw that this case led us to find the singular root.

- Last, suppose the first and the last terms of the quadratic (1.9) are the dominant balance. That is, $\epsilon x^2 - 1 = 0$ is the approximate equation. Then the initial approximation could be either of $x \approx \pm 1/\sqrt{\epsilon}$. The neglected term $x \approx \pm 1/\sqrt{\epsilon}$ is then larger than both the two terms forming the balance, and so neither of these initial approximations can be valid. This case cannot lead to a root of the singular quadratic (1.9).

Hinch (1991, §1.2) and Bender and Orszag (1981, §7.2) further explore this "dominant balance" sort of reasoning.

Algorithm 1.2 *Reduce code to solve the rescaled quadratic* (1.13) *by iteration. Here find the root* $X \approx -1$, *that is,* $x \approx -1/\epsilon$.

```
x:=-1;
let eps^7=>0;
for iter:=1:99 do begin
    res:=x^2+x-eps;
    write x:=x+res;
    if res=0 then iter:=1000000;
end;
```

Dialogue 1.2 *Output of Algorithm* 1.2.

```
x := -1

x :=  - 1 - eps
                   2
x :=  - 1 - eps + eps
                   2          3          4
x :=  - 1 - eps + eps   - 2*eps   + eps
                   2          3          4          5          6
x :=  - 1 - eps + eps   - 2*eps   + 5*eps   - 6*eps   + 6*eps
                   2          3          4           5           6
x :=  - 1 - eps + eps   - 2*eps   + 5*eps   - 14*eps   + 26*eps
                   2          3          4           5           6
x :=  - 1 - eps + eps   - 2*eps   + 5*eps   - 14*eps   + 42*eps
                   2          3          4           5           6
x :=  - 1 - eps + eps   - 2*eps   + 5*eps   - 14*eps   + 42*eps
```

1.2.3 ▪ Rescale to remove the singularity

Working with quantities that go to infinity is awkward. Instead a better strategy is to rescale singular problems so quantities stay finite. We generally do this, avoiding singular perturbations, because high-order systematic analysis of regular perturbations is easier.[4]

Consider the singular quadratic (1.9). Scale the variable $x = X/\epsilon$ in terms of a new variable X. Substitute into the singular quadratic (1.9) to find the corresponding quadratic for the new variable X:

$$\epsilon\frac{X^2}{\epsilon^2} + \frac{X}{\epsilon} - 1 = 0 \quad \Longleftrightarrow \quad X^2 + X - \epsilon = 0\,. \tag{1.13}$$

Now solve this corresponding quadratic (1.13) for roots X using the regular methods. This works because the rescaling of the singular quadratic (1.9) produces the regular quadratic (1.13).

For example, use computer algebra. The Reduce code in Algorithm 1.2 finds the root

$$X = -1 - \epsilon + \epsilon^2 - 2\epsilon^3 + 5\epsilon^4 + \cdots\,,$$

which corresponds to the singular root $x^{(2)} = X/\epsilon$. See the output from the Reduce code in Dialogue 1.2. Remember to surround the code of Algorithm 1.2 by

[4]There is so much theory developed for singular perturbations that many people scale all problems into singular perturbations. I aim to convince you that there are many important modeling issues that become clear when we instead scale problems as regular perturbations.

Algorithm 1.3 *Reduce code to solve the rescaled quadratic* (1.13) *by iteration. Here find the root* $X \approx 0$, *that is,* $x \approx 1$.

```
x:=0;
let eps^7=>0;
for iter:=1:99 do begin
    res:=x^2+x-eps;
    write x:=x-res;
    if res=0 then iter:=1000000;
end;
```

```
on div; off allfac; on revpri;
...
end;
```

in a text file, and then execute the commands in the file using the `in textfile$` command.

Example 1.2. Estimate the large negative root of the quadratic $\frac{1}{5}x^2 + x - 1 = 0$.

Solution: This quadratic is an instance of the singular quadratic (1.9) with parameter $\epsilon = \frac{1}{5}$. Recognize that the finite number $\frac{1}{5}$ perhaps may be treated as small, and hence analysis based upon small ϵ is valid. Consequently, substitute $\epsilon = \frac{1}{5}$ into our above series to estimate

$$X \approx -1 - 0.2 + 0.04 - 0.016 + 0.008 - 0.0045 + 0.0027 = -1.170$$

to three decimal places. Each term in the sum is roughly half that of the preceding term, so expect the error in this estimate to be roughly half the size of the last term, and hence no more than three decimal places are justified. Thus the large negative root of the quadratic $\frac{1}{5}x^2 + x - 1 = 0$ is

$$x = 5X \approx -5.849.$$

This compares very well with the exact root -5.854.

Find the other root, $x^{(1)} \approx 1$, using the same program, but start from $X_0 = 0$ (or $X_0 = \epsilon$) corresponding to $x_0 = 1$, *and* change the update to X from the correction. See these two changes from Algorithm 1.2 in the Reduce code of Algorithm 1.3. Dialogue 1.3 lists the output from the Reduce code. Hence the other root $x^{(1)} = X/\epsilon = 1 - \epsilon + 2\epsilon^2 + \cdots$, as before.

Avoid singular perturbations Dynamical systems often have a combination of very fast damped modes and much slower modes. Others phrase their modeling analysis as a "singular perturbation problem" by regarding the damping as so fast that the damping occurs in infinitesimal time. However, in most instances the "singular" nature is removed simply by rescaling time. We always do so. Our approach is one of regular perturbations: we avoid singular perturbations by rescaling time, and because of the well-posed nature of our dynamical modeling methodology.

Dialogue 1.3 *Output from Algorithm* 1.3.

```
x := 0

x := eps
             2
x := eps - eps
             2         3       4
x := eps - eps  + 2*eps   - eps
             2         3         4         5         6
x := eps - eps  + 2*eps   - 5*eps  + 6*eps   - 6*eps
             2         3         4          5          6
x := eps - eps  + 2*eps   - 5*eps  + 14*eps   - 26*eps
             2         3         4          5          6
x := eps - eps  + 2*eps   - 5*eps  + 14*eps   - 42*eps
             2         3         4          5          6
x := eps - eps  + 2*eps   - 5*eps  + 14*eps   - 42*eps
```

1.3 ▪ Undetermined coefficients are quicker than algebra

Shall I refuse my dinner because I do not fully understand the process of digestion? *O. Heaviside*

How many of the previous computer algebra exercises did you do? If you did more than a couple, I bet that you ran an infinite loop in at least one of the exercises.

Why? Because it is nontrivial to deduce the update rule, such as (1.8) or (1.11), for correcting an approximation based upon a residual. The algebraic derivation contains enough detail—apparently uninteresting detail—that people make little errors. At least I generally do.

This section explores a simpler but reliable alternative: namely a variant of *the method of undetermined coefficients*. The method recognizes that *generally* the desired correction depends linearly upon the residual. Consequently, let the coefficient of the linear dependence be some as yet unknown parameter, and then find the unknown parameter via performing two iterations.

Example 1.3. Reconsider finding the root $x \approx 1$ of the quadratic equation (1.4) using computer algebra. Suppose we do not know that the correction to an approximation x_j is $\hat{x}_j = -\frac{1}{2}\operatorname{Res}_j$, (1.8).

However, we do know that *generally there is a linear dependence*, a proportionality, between residual and correction. Thus, propose that the coefficient of proportionality is some constant, say a, that we need to determine. For this problem, execute *precisely two iterations* of the proposed computer algebra code; see Algorithm 1.4. Dialogue 1.4 lists the Reduce output.

Note in these first two iterations that $\operatorname{Res}_1 = \epsilon$ and $\operatorname{Res}_2 = (1+2a)\epsilon + a(1+a)\epsilon^2$. For general coefficient a, both the first and second residuals are of size ϵ. However, the second residual Res_2 will be smaller, namely of size ϵ^2, when we choose the parameter a so that the coefficient of the ϵ term in Res_2 becomes zero. That is, choose parameter $a = -\frac{1}{2}$ so that $1+2a = 0$. Then the rule for correcting an approximation from a residual is precisely (1.8), which we deduced before with algebra.

Algorithm 1.4 *Precisely two iterations of this Reduce code tell us how to use the residuals to update the corrections. Simply choose the parameter* a *so that the second residual is "better" than the first.*

```
factor eps;
x:=1;
let eps^10=>0;
for iter:=1:2 do begin
    write res:=x^2+eps*x-1;
    write x:=x+a*res;
end;
```

Dialogue 1.4 *Output of Algorithm* 1.4.

```
x := 1

res := eps

x := 1 + eps*a
                           2       2
res := eps*(1 + 2*a) + eps *(a + a )
                       2       2   2    3
x := 1 + eps*(2*a + 2*a ) + eps *(a   + a )
```

The next step would then be to edit the proposed Algorithm 1.4 by setting the parameter a to its useful value of $a = -\frac{1}{2}$ and then performing as many iterations as necessary to obtain the desired power series approximation, as in Algorithm 1.1.

Recognize the enormously efficient use of your time in this method. By parametrizing the linear dependence of a correction upon a residual, you write the computer algebra code you would use anyway if you knew the parameter, execute two iterations to find the useful value, and then quickly edit the code to set the value of the parameter. The only crucial part of the computer algebra code is the computation of the residual: if that computation is correctly coded, then in using this method you will construct a correct perturbation power series solution.

Let's see this method of undetermined coefficients in action again in a second straightforward example.

Example 1.4. Reconsider the task of finding the negative root $X \approx -1$ of the quadratic equation (1.13) using computer algebra. Again suppose we do not know the correction for any given residual.

However, we do know that *generally the correction is linearly proportional to the residual.* Thus propose that the coefficient of proportionality is some constant, say a, to be determined. Then for this problem, execute *two iterations* of the proposed computer algebra code; see Algorithm 1.5. Dialogue 1.5 lists the Reduce output.

Note that $\text{Res}_1 = -\epsilon$ and that $\text{Res}_2 = (a-1)\epsilon + a^2\epsilon^2$. The second residual, Res_2, will be smaller than the first, Res_1, only when we choose the parameter a so that the coefficient of the ϵ term in Res_2 is set to zero. That is, choose parameter $a = 1$ so that $a - 1 = 0$. Then the rule for correcting an approximation from a residual becomes precisely that implemented before in Algorithm 1.2.

Algorithm 1.5 *Precisely two iterations of this Reduce code tell us how to use the residuals to update the corrections for quadratic* (1.13). *Simply choose the parameter* a *so that the second residual is "better" than the first.*

```
factor eps;
x:=-1;
let eps^10=>0;
for iter:=1:2 do begin
    write res:=x^2+x-eps;
    write x:=x+a*res;
end;
```

Dialogue 1.5 *Output of Algorithm* 1.5.

```
x := -1

res :=  - eps

x :=  - 1 - eps*a
                             2  2
res := eps*( - 1 + a) + eps *a
                             2       2  3
x :=  - 1 + eps*( - 2*a + a ) + eps *a
```

This method of undetermined coefficients finds a critical parameter for an iteration to converge. In more complicated problems, where we solve systems of equations, the linear relation between residuals and corrections is, in general, a finite-dimensional linear transformation. Thus in more complicated problems we have to determine an unknown matrix of coefficients. But nonetheless, a couple of iterations of the iteration scheme are usually enough to determine the coefficients.

1.4 ▪ Introducing Reduce computer algebra

In the previous sections I assumed you could already use the computer algebra package Reduce. This section provides an introduction to Reduce for those that need more help getting started.

One reason to use Reduce is that it is perhaps the fastest general purpose computer algebra system (Fateman, 2003, e.g.). Another reason is that Reduce is freely available.[5] Check that you can start and run Reduce;[6] it should open up a window saying something like[7]

```
Reduce (Free CSL version), 17-Feb-10 ...

1:
```

The "`1:`" is a prompt for a command: to finish Reduce, type `quit;` followed by pressing the return or enter key. If this works, you can run Reduce.

[5] `http://reduce-algebra.sourceforge.net/` provided the software at the time of writing (September 28, 2014).

[6] Perhaps by typing into a terminal window the command `reduce` or `reduce -w`.

[7] I generally use such a teletype font for computer instructions and dialogue.

1.4.1 ▪ Explore a little Reduce

- Start Reduce by typing `reduce` in a command or terminal window, or perhaps clicking on its icon. To exit from Reduce type the command `quit;` followed by pressing the enter key.
- All Reduce statements must be terminated with a semicolon. Do not forget. They are subsequently executed by pressing the enter key.
- Reduce uses exact arithmetic by default: for example, to find 100! in full gory detail type `factorial(100);` and then press enter (I will not mention the enter key again unless necessary).
- Identifiers denote either variables or expressions: in `f:=2*x^2+3*x-5;` the identifier `x` is a variable, whereas `f`, after the assignment with `:=`, contains the above expression; similarly, after `g:=x^2-x-6;`, `g` contains an algebraic expression. Often we use single-letter identifiers for brevity, although we can use up to about 20 letters and digits to form an identifier.
- Expressions may be

  added with `f+g;`

  subtracted with `f-g;`

  multiplied with `f*g;`

  divided with `f/g;`

  exponentiated with `f^3;`

  and so on.
- Straightforward equations may be solved (by default equal to zero), such as `solve(x^2-x-6,x);` or via an expression previously found, such as `solve(f,x);`.

  Systems of equations may be solved by giving a list (enclosed in braces) of equations and a list of variables to be determined. For example, `solve({x-y=2,a*x+y=0},{x,y});` returns the solution parametrized by a.
- Basic calculus is a snap:

  differentiation uses the function `df` as in `df(f,x);` to find the first derivative, `df(g,x,x);` for the second, or `df(sin(x*y),x,y);` for a mixed derivative.

  The product rule for differentiation is *verified* for the above two functions by `df(f*g,x)-df(f,x)*g-f*df(g,x);` reducing to zero.

  integration is similar, `int(f,x);` giving the integral of the polynomial in `f`, *without an integration constant,* but perhaps more impressive is the almost instant integration of `int(x^5*cos(x^2),x);`.

  Repeated integration must be done by repeated invocations of `int`, not by further arguments, as for `df`. Instead, for example, `int(f,x,0,2);` will give you the definite integral from 0 to 2.
- One can substitute an expression for a variable in another expression. For example, the composition $f(g(x))$ is computed by `sub(x=g,f);`.

- Reduce allows you to use many lines for the one command: a command is not terminated until the semicolon is typed. Reduce alerts you to the fact that you are still entering the one command by displaying the prompt again. Thus if you forget the semicolon, just type a semicolon at the new prompt and then press the enter key to execute what you had typed on the previous lines.
- Reduce is case insensitive: lowercase variables/names and uppercase variables/names both refer to the same things.
- If Reduce displays the message `Declare xxx operator ?`, then you have probably mistyped something and the best answer is to type `N` and then press enter.

1.4.2 ▪ Use operators to help pattern matching

A subtle but immensely powerful feature of computer algebra, especially Reduce, is its flexible pattern matching and replacement. This section introduces useful pattern matching and replacement with the aid of operators.

An operator for double integration Suppose that in some problem you need to compute a lot of double integrals, $\iint \cdots dx\,dx$. Of course you could use Reduce's native integrator twice: `int(int(...,x),x)`. But such double invocation is inefficient. Instead, define an operator, here called `iint`, that does the double integration in one step. For the moment, tell Reduce that `iint` is an operator, and tell Reduce what to do with some powers of x:

```
operator iint;
let {iint(1)=>x^2/2,
     iint(x)=>x^3/6,
     iint(x^2)=>x^4/12,
     iint(x^3)=>x^5/20};
```

Try it. It works: wherever you type one of the above patterns, such as `iint(x)`, you get the right-hand side. But ... try, say, `iint(2*x+5*x^2)`: we just get `iint(5*x^2  + 2*x)` back as the answer. That is no use.

Linear operators distribute We need to tell Reduce that the operator `iint` *distributes* over addition and constant multiplication so that, for example, Reduce will know that `iint(2*x+5*x^2)` is the same as the distributed `2*iint(x)+5*iint(x^2)`. The Reduce command `linear` does precisely this distribution provided we also specify that x is the variable of interest. Replace the above definitions with

```
operator iint;
linear iint;
let {iint(1,x)=>x^2/2,
     iint(x,x)=>x^3/6,
     iint(x^2,x)=>x^4/12,
     iint(x^3,x)=>x^5/20};
```

Now execute the commands in Dialogue 1.6. Why do we need the second argument x for `iint`? So that Reduce knows that other variables are constants as far as `iint` is concerned. Thus we get the correct answer from `iint(3*(a+x)^3,x)` because Reduce knows to factor the constant `a` outside of `iint`. But ... what if we try

Dialogue 1.6 *Linear operators distribute over addition and constant multiplication.*

```
3: iint(2*x+5*x^2,x);
  3
 x *(5*x + 4)
--------------
      12

4: iint(3*(a+x)^3,x);
    2      3         2          2    3
 3*x *(10*a  + 10*a *x + 5*a*x  + x )
-------------------------------------
                   20

5: iint(c*x^5,x);
        5
c*iint(x ,x)
```

`iint(c*x^5,x)`? The useless answer is `c*iint(x^5,x)` because we have not told Reduce what to do with high powers of x.

Flexible patterns Reduce empowers us to match and transform general patterns, not just specific. Here we continue the example by showing how to transform general powers of x, not just the specific powers coded above. That is, we want `iint` to know that $\iint x^n \, dx \, dx = \frac{1}{(n+1)(n+2)} x^{n+2}$ *in general.* Do this by introducing a tilde variable on the left-hand side of the `let` transformations. Here code

```
operator iint;
linear iint;
let {iint(1,x)=>x^2/2,
     iint(x,x)=>x^3/6,
     iint(x^~n,x)=>x^(n+2)/(n+1)/(n+2) };
```

Now see it at work with something more complicated, such as

```
6: iint((a+x)^5,x);
  2       5         4          3  2          2  3          4     5
 x *(21*a  + 35*a *x + 35*a *x  + 21*a *x  + 7*a*x  + x )
-------------------------------------------------------------
                              42
```

It works! Try it for any polynomial in x, with any number of other constants. It works because `iint(x^~n,x)` matches any power of x; the `~n` tells Reduce that anything can appear as the exponent, and when it does, replace the n on the right-hand side by whatever is the exponent on the left.

Why do we have to separately code transformations for `iint(1,x)` and `iint(x,x)`? Because these are *not* in the form of x raised to some power. *We* know that mathematically $1 = x^0$ and $x = x^1$, but Reduce does not assume these *in its pattern matching*. Reduce knows to replace `x^0` by `1` and `x^1` by `x`, but in its pattern matching it will not assume the reverse. Thus we must code the transformation of such special powers separately from the general case.

Build in subsidiary conditions In the above example of double integration we ignored any integration constants. Our code above always provides an answer for which the value and its derivative at $x = 0$ are both zero. That is, our above code *implicitly* assumed the answer has to be such that its value and derivative at $x = 0$ have to be zero. What if instead we wanted an answer that was zero at $x = 0$ and zero at $x = 1$, and there is no constraint on the derivative? Answer: build such subsidiary *homogeneous* conditions into the operator `iint`.

A little algebra gives the general double integral

$$\iint x^n\,dx\,dx = \frac{1}{(n+1)(n+2)}x^{n+2} + ax + b\,.$$

Make this zero at $x = 0$ by choosing $b = 0$, as we did implicitly above. Make the answer zero at $x = 1$ by choosing $a = -1/[(n+1)(n+2)]$. That is, the general particular double integral we need is $\iint x^n\,dx\,dx = \frac{1}{(n+1)(n+2)}(x^{n+2} - x)$. Code this rule instead.

```
operator iint;
linear iint;
let {iint(1,x)=>(x^2-x)/2,
     iint(x,x)=>(x^3-x)/6,
     iint(x^~n,x)=>(x^(n+2)-x)/(n+1)/(n+2) };
```

Then see that it works on some examples:

```
6: y:=iint((b*x+x^2)^2,x);
                 2  3       2          4                5
       x*(5*b *x   - 5*b  + 6*b*x   - 6*b + 2*x   - 2)
y := ----------------------------------------------
                                60
7: sub(x=1,y);
0
```

In essence the operator `iint` solves the second-order differential equation $d^2y/dx^2 = f$ with Dirichlet boundary conditions that $y(0) = y(1) = 0$. We just type `y:=iint(f,x)`, provided the right-hand side f is a polynomial in x.

If the right-hand side is nonpolynomial, then either code more suitably general rules or resort to the powerful general Reduce integration operator `int`. *But remember that a specifically coded operator, like* `iint`, *is very much quicker for Reduce to execute.*

1.4.3 ▪ Summary of some Reduce commands

... the different branches of Arithmetic—Ambition, Distraction, Uglification and Derision.

the Mock Turtle in Alice in Wonderland *by Lewis Carroll*

- Download and install Reduce from (as at September 28, 2014) `http://reduce-algebra.sourceforge.net/`.
- Reduce instructions must be terminated and separated by a semicolon or a dollar character (a new line does not suffice).
- `quit;` or `bye;` terminates Reduce execution.
- Use `on div;`, `off allfac;` and `on revpri;` to improve the printing of power series.

- `:=` is the assignment operator.
- The normal arithmetic operators are `+`, `-`, `*`, `/`, and `^` for addition, subtraction, multiplication, division, and exponentiation, respectively.
- `write` displays the result of an expression, although Reduce automatically displays the results of each command that is not in a loop.
- `int(y,x)` provides *an* integral of the expression in `y` with respect to the variable `x`, provided Reduce can actually do the integral.
- `df(y,x)` returns the derivative of the expression in `y` with respect to the variable `x`; `df(y,x,z)` will return the second derivative of `y` with respect to `x` and `z`.
- Often we use a symbol to denote a function; for example, we might want to use u to denote the function $u(x)$ or perhaps $u(x,t)$. In these two cases, respectively, use the statements `depend u,x;` and `depend u,x,t;`. Then, for example, Reduce transforms `df(a*u^2,x)` to the expression `2*a*u*df(u,x)`.
- `factorial(n)` returns the value of $n!$.
- The function `trigsimp()` helps manipulate trigonometric functions: for example, `trigsimp(sin(x)^2*cos(x),combine)` transforms to $(\cos x - \cos 3x)/4$; whereas `trigsimp(cos(3*x),expand)` transforms to $(1 - 4\sin^2 x)\cos x$.
- `for n:=2:5 do`, for example, will repeat whatever statement follows for values of the variable used, here n, over the range specified in the command, here from 2 to 5 inclusive.

  Prematurely terminate a `for-do` loop by setting, somewhere within the loop, the loop variable to a high value. For the above example, executing within the loop the statement `n:=10;` will terminate the loop at the end of the current pass.
- `begin...end` is used to group statements into one—often needed for loops.
- The `let` statement does pattern matching and replacement; for example `let x^15=>0;` tells Reduce to subsequently discard any term involving x to the power 15 or more.

  A tilde variable on the left-hand side represents any expression: for example, `let e^(i*~q)=>cos(q)+i*sin(q)` replaces all complex exponentials by the trigonometric functions; for example, `e^(pi*i*x)` would transform to `cos(pi*x)+i*sin(pi*x)`.

  Make a `let` rule conditional with `when`: for example,

  ```
  let linv(cos(~n*q),q)=>cos(n*q)/(4-n^2) when n neq 2;
  ```

  helps solve the ODE $d^2u/dq^2 + 4u = \sum_n c_n \cos(nq)$ as it helps form $u = \sum_{n\neq 2} c_n \frac{\cos(nq)}{4-n^2}$ without the possibility of dividing by zero.
- `repeat...until...` will repeatedly execute a statement until the given condition is true (but prefer the safer for-loop).

- `end;` is also used to terminate reading in a file of Reduce commands.
- `in "..."$` tells Reduce to execute the commands contained in the specified file.

  For those who appreciate style, the 2012 version of Reduce introduced the command `in_tex "filename.tex"$` that inputs a LaTeX file and only interprets as Reduce commands material between `\begin{reduce}` and `\end{reduce}` pairs. This empowers you to easily produce beautiful documentation of computer algebra.[8]
- `operator ...` defines the name of an "operator" that we use to symbolically operate on its arguments—its actions are typically defined by some `let` rules.
- `linear ...` declares that the named operator is to expand *linearly* its first argument with respect to its second argument.
- Possibly use operators to represent subscripted variables, as in `operator x;` and then `y:=x(1)^2*sin(x(2));`. However, one can also use the `mkid` function: for example, `y:=mkid(x,1)^2*sin(mkid(x,2));` is identical to `y:=x1^2*sin(x2);`, whereas the loop `for j:=1:5 do set(mkid(y,j),x^j);` sets `y1,...,y5` to respective powers of `x`.
- `out "filename"$` directs all output to the file `filename` instead of the interactive console window. Use this command to save the results of an executed file for later reference. To restore output to the console window, and to flush all output to the file, execute the companion command `shut "filename"$`.
- `solve(exps,vars)` attempts to find values of the variables in the list `vars` such that the expressions in the list `exps` are all zero.

  For example, `solve({x-y/2,y+a*x-1},{x,y});` finds the solution `{{x=1/(a+2),y=2/(x+2)}}` of the linear equations $x - y/2 = 0$ and $ax + y = 1$.
- `rederr "message";` terminates a loop with some message and asks whether you want to continue with commands after the loop; it also interrupts execution of the `in` file and asks whether you want to continue the `in` file.
- Reduce does have a matrix/vector facility, via the `mat()` function, but I mostly find it inconvenient. For many purposes, synthesize vectors using an operator to represent basis vectors.
  - Define `operator ee;`. Then, for example, execute the statement `r:=x*ee(1)+y*ee(2)+z*ee(3)` to represent the position vector $\mathbf{r} = (x, y, z)$.
  - Define an infix operator to access individual elements of a vector, perhaps

    ```
    infix #;
    precedence #,^;
    let ~a#~b=>coeffn(a,ee(b),1);
    ```

    Then, for example, $\texttt{r\#1} = x$, $\texttt{r\#2\^{}2} = y^2$, and $\texttt{sin(r\#3)} = \sin z$.

[8] `http://www.maths.adelaide.edu.au/anthony.roberts/exampleReduce.tex` [September 28, 2014].

- Depending upon requirements, you could define a product of vectors to be either an inner product or a component by component product:
  * `let{ee(~i)^2=>1,ee(~i)*ee(~j)=>0 when i neq j};` gives the dot product, for example, simplifying `r*r` $= x^2 + y^2 + z^2$;
  * `let{ee(~i)^2=>ee(i),ee(~i)*ee(~j)=>0 when i neq j};` gives a component by component product, for example, `r*r` $= (x^2, y^2, z^2)$.
- Linear operators provide a useful mechanism for defining matrix multiplication by their actions on basis vectors, especially for matrices with patterns such as Toeplitz matrices. For example, multiplication by the Toeplitz second-order finite difference matrix

$$D = \begin{bmatrix} -2 & 1 & 0 & \cdots & 0 \\ 1 & -2 & 1 & & 0 \\ 0 & 1 & -2 & \ddots & \vdots \\ \vdots & & \ddots & \ddots & 1 \\ 0 & \cdots & 0 & 1 & -2 \end{bmatrix}$$

  is effected by `dd(r,ee)` after defining

```
operator dd; linear dd;
let dd(ee(~i),ee)=>-2*ee(i)
    +(if i=1 then 0 else ee(i-1))
    +(if i=n then 0 else ee(i+1));
```

- However, to use the inbuilt matrix facility, note the following.
  - `matrix` declares some variables as matrices and initializes them to zeros. For example, `matrix a(3,2),b(2,2);` creates a 3×2 and a 2×2 matrix of zeros.
  - `mat(...)` creates a matrix according to its argument, which must be a comma-separated list of rows, each row being parentheses around a comma-separated list of entries. For example,

$$\texttt{z:=mat((1,-x,y),(x,0,y))} \quad \text{represents} \quad \begin{bmatrix} 1 & -x & y \\ x & 0 & y \end{bmatrix}$$

    and stores the result in `z`. Then access individual parts of matrix `z` with subscripts in parentheses; for example, `z(1,2)`, which in this case is the value `-x`.
  - `tp(...)` transposes its matrix argument; e.g., `tp mat((6,2,-3));` forms the 3×1 matrix $[6, 3, -3]^T$.
  - Essential is the `map(...)` function that applies a specified operation (first argument) to all the entries in the specified matrix (second argument). For example, older versions of Reduce needed `map(df(~a,t),g)` to form the matrix of `t` derivatives of each entry in matrix `g`.
  - Compute the inverse of a matrix, say `aa`, with simply `1/aa`. For example, the vector `x:=1/aa*b` solves $A\mathbf{x} = \mathbf{b}$; you then need to access individual components of `x` as `x(j,1)`.

1.4.4 ▪ Translate between computer algebra languages

Tables 1.1–1.3 list corresponding commands in some computer algebra languages. The aim of the tables is to help you implement given algorithms in other computer algebra languages.

1.5 ▪ Exercises

Exercise 1.1. Change the computation of the residual, but *not* the update rule, in the computer algebra code of Algorithm 1.1 to find power series solutions to the following algebraic equations for the root $x \approx 1$:

1. $\epsilon x^3 + x^2 + \epsilon x - 1 = 0$.
2. $(1-\epsilon)x^2 + 2\epsilon x - 1 = 0$.
3. $x^3 - \frac{1}{2}x^2 + \epsilon x - \frac{1}{2} = 0$.

For each of these algebraic equations, estimate the root for $\epsilon = 0.1, 0.5$, and 1. Discuss how accurate, or even valid, you consider your estimates.

Answer:

1. $1 - \epsilon + \frac{3}{2}\epsilon^2 - 3\epsilon^3 + \frac{55}{8}\epsilon^4 - 17\epsilon^5 + \cdots$.
2. $1 - \frac{1}{2}\epsilon - \frac{1}{8}\epsilon^2 + \frac{1}{16}\epsilon^3 + \frac{11}{128}\epsilon^4 + \frac{7}{256}\epsilon^5 + \cdots$.
3. $1 - \frac{1}{2}\epsilon - \frac{1}{16}\epsilon^2 + \frac{1}{64}\epsilon^3 + \frac{31}{1024}\epsilon^4 + \frac{91}{4096}\epsilon^5 + \cdots$.

Exercise 1.2. Change the computation of the residual *and* the rule for updating corrections in the computer algebra code of Algorithm 1.1 to find power series solutions to the following algebraic equations for the root $x \approx 1$:

1. $x^3 + x^2 + \epsilon x - 2 = 0$.
2. $(1-\epsilon)x^2 + (1+2\epsilon)x - 2 = 0$.
3. $x^4 - \epsilon x^2 - \epsilon x - 1 = 0$.

Use one of the series to estimate a root of $x^3 + x^2 + x - 2 = 0$; discuss the accuracy of the estimate.

Answer:

1. $1 - \frac{1}{5}\epsilon + \frac{1}{125}\epsilon^2 + \frac{8}{3125}\epsilon^3 + \frac{1}{15625}\epsilon^4 - \frac{154}{1953125}\epsilon^5 - \frac{598}{48828125}\epsilon^6 + \cdots$.
2. $1 - \frac{1}{3}\epsilon - \frac{1}{27}\epsilon^2 + \frac{7}{243}\epsilon^3 + \frac{31}{2187}\epsilon^4 - \frac{41}{19683}\epsilon^5 - \frac{251}{59049}\epsilon^6 + \cdots$.
3. $1 + \frac{1}{2}\epsilon - \frac{1}{16}\epsilon^3 + \frac{1}{32}\epsilon^4 + \frac{1}{128}\epsilon^5 - \frac{5}{256}\epsilon^6 + \cdots$.

Exercise 1.3. In complicated problems we often need power series expansions for elementary functions such as $\sqrt{1+\epsilon}$, $1/\sqrt{1+\epsilon^2}$, and $1/(1+\epsilon)^2$. Of course we could explicitly code these elementary functions using the binomial expansion. But it is just as easy to use the same iteration to create the power series expansions of these functions simultaneously with the iteration to solve the complicated problem. To do this, just recast such elementary functions as the solution of an algebraic equation and apply the techniques we have just explored.

Table 1.1. *Translate between different languages a lot of the computer algebra code used herein. Blank cells represent things that are unknown to me. This table gives some elementary translations; the next two tables are more advanced.*

Description	Reduce	Maple	*Mathematica*
Letter case	`insensitive`	`sensitive`	`sensitive`
Terminate statements	`;` or `$`	`;`	`;`
Assignment	`:=`	`:=`	`=`
Arithmetic operators	`+ - * / ^`	`+ - * / ^`	`+ - * / ^`
Logical operators	`= neq`	`= <>`	`== =!=`
Constants i, π, ∞	`i, pi, infinity`	`I, Pi`	`I, Pi, Infinity`
Functions sin, $\log_e$, e.g.	`sin(...), log(...)`	`sin(...), log(...)`	`Sin[...], Log[...]`
Strings	`"..."`		`"..."`
Printing	`write ...`	`print(...)`	`Print[...]`
Value of previous computation	`ws`	`%`	`%`
Exit	`quit;`	`quit;`	`Exit[];`

Table 1.2. *Translate between different languages a lot of the computer algebra code used herein: mainly mathematical operations. Blank cells represent things that are unknown to me.*

Description	Reduce	Maple	*Mathematica*
$n!$	`factorial(n)`	`factorial(n)`	`n!`
df/dx	`df(f,x)`	`diff(f,x)`	`D[f,x]`
$\partial^2 f/\partial x\partial y$	`df(f,x,y)`	`diff(f,x,y)`	
$\partial^n f/\partial x^n$	`df(f,x,n)`		`D[f,{x,n}]`
$\int f\,dx$	`int(f,x)`	`int(f,x)`	`Integrate[f,x]`
$\int_a^b f\,dx$	`int(f,x,a,b)`	`int(f,x=a..b)`	`Integrate[x,{x,a,b}]`
Substitution: if $f(x)$, g then $f(g)$	`sub(x=g,f)`	`subs(x=g,f)`	`f/.x->g`
Simplification		`expand(...)` or `simplify(...)`	`Expand[...]` or `Simplify[...]`
Trigonometric simplification	`trigsimp(..., combine)`	`combine(..., trig)`	
Extract coefficient of x	`coeffn(f,x,1)`	`coeff(f,x)`	`Coefficient[f,x]`
Local pattern replacement $a \mapsto b$	`(... where a=>b)`	`algsubs(a=b, ...)`	

Table 1.3. *Translate between different languages a lot of the computer algebra code used herein: programming structures and things. Blank cells represent things that are unknown to me.*

Description	Reduce	Maple	*Mathematica*
Comment lines	% ...	# ...	
For-loop $n = a, a+1, a+2, \ldots, b$	for n:=a:b do statement	for n from a to b do statements od;	For[n=a; n<b, n++, statements];
Group statements together	begin statements end or << statements >>		
Repeat loop	repeat begin statements end until condition;	while condition do statements od;	While[condition, statements];
Execute commands from file	in file$	read file;	«file
Depend r upon t	depend r,t;	alias(r=r(t));	
Global pattern replacement $a \mapsto b$	let a=>b;		
Truncate power series to error $\mathcal{O}(x^9)$	let x^9=>0;	algsubs(x^9=0, expand(...))	x^n_ ^:=0 /; n>8;

1. For example, define $x = \sqrt{1+\epsilon}$; then square this definition to the algebraic equation $x^2 = 1+\epsilon$, and seek x as a function of parameter ϵ using iteration guided by the residual of this algebraic equation.
2. Similarly find the power series of $1/\sqrt{1+\epsilon^2}$.
3. Similarly find the power series of $1/(1+\epsilon)^2$.

Answer:

1. $1 + \frac{1}{2}\epsilon - \frac{1}{8}\epsilon^2 + \frac{1}{16}\epsilon^3 - \frac{5}{128}\epsilon^4 + \frac{7}{256}\epsilon^5 - \frac{21}{1024}\epsilon^6 + \cdots$.
2. $1 - \frac{1}{2}\epsilon^2 + \frac{3}{8}\epsilon^4 - \frac{5}{16}\epsilon^6 + \cdots$.
3. $1 - 2\epsilon + 3\epsilon^2 - 4\epsilon^3 + 5\epsilon^4 - 6\epsilon^5 + 7\epsilon^6 + \cdots$.

Exercise 1.4. Explore all three roots of the singular algebraic equation $\epsilon x^3 + x^2 + \epsilon x - 1 = 0$.

Exercise 1.5. Rescale x so that the singular cubic equation $\epsilon x^3 + x^2 + \epsilon x - 1 = 0$ becomes regular. Then write, test, and run computer algebra code to generate power series approximations for all three roots of the cubic. Hence estimate the three roots of $\frac{1}{4}x^3 + x^2 + \frac{1}{4}x - 1 = 0$; discuss the accuracy of the estimates.

Exercise 1.6. Rescale x to find approximations, as power series in parameter ϵ, to the two real roots of $\epsilon^2 x^6 - \epsilon x^4 - x^3 + 8 = 0$. Exhaust all scaling possibilities.

Answer: One real root comes from assuming x is of size 1, another real root comes from the scaling $x = X/\epsilon^{2/3}$ (Bender and Orszag, 1981, §7.2).

Exercise 1.7. Use the method of undetermined coefficients to find the linear relation between residual and correction for some of the earlier exercises. Especially explore any exercises you have not yet done.

Chapter 2

Power series solve ordinary differential equations

Contents

Example 2.1 (power series emerge). Recall the linear dynamics of Example 1.1: a linear transform separated the emergent dynamics from ignorable transients. But most scientific dynamics are nonlinear. So now consider the toy nonlinear dynamics of

$$\dot{u}_1 = -u_1 + u_2 - u_1^2 \quad \text{and} \quad \dot{u}_2 = +u_1 - u_2 + u_2^2, \tag{2.1}$$

where, as always, overdots denote d/dt. The numerical solutions plotted in Figure 2.1 demonstrate that all solutions quickly collapse onto a curve in the u_1u_2-plane. Thus the long-time emergent dynamics lies on this curve. But over long times do the solutions stay within the picture? Or do they "explode" to infinity? Linearly, as in Example 1.1, the long-time dynamics is $u_1 = u_2$. In the case when $u_1 = u_2$ precisely, then the quadratic reactions remove as much as they generate, so simple arguments, such as linear transforms or averaging, cannot predict the effective long-time influence of the nonlinear terms.

However, consider the system in new variables V and W, whose nonlinear coordinate curves are shown in Figure 2.2. The corresponding algebraic coordinate transform, that $u_1 = v - w$ and $u_2 = v + w$, where in turn

$$v = V - VW + \frac{3}{4}VW^2 + \cdots \quad \text{and} \quad w = W + \frac{1}{2}V^2 - \frac{1}{2}W^2 + \frac{1}{4}W^3 + \cdots, \tag{2.2}$$

leads to the dynamics of V and W being

$$\dot{V} = V^3 + \cdots \quad \text{and} \quad \dot{W} = (-2 - V^2 + \cdots)W \tag{2.3}$$

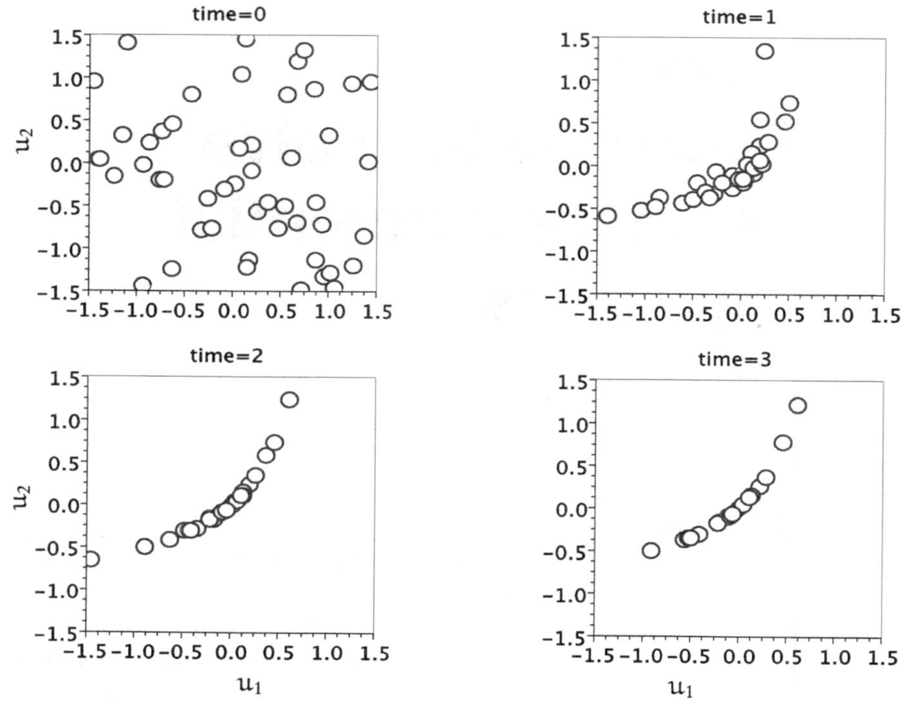

Figure 2.1. *Fifty initial conditions in the* u_1u_2*-plane at time* $t = 0$ *evolve according to* (2.1). *Their locations at later times demonstrate that all solutions collapse onto a curve in the* u_1u_2*-plane.*

One may check this transform by substitution and some straightforward algebra (albeit tedious). Now, $W \to 0$ exponentially quickly from the coefficient $(-2-V^2)$ in (2.3) and as seen in the collapse to a curve in Figure 2.1. Thus the emergent dynamics is that V evolves as $\dot{V} \approx V^3$, which eventually explodes like $1/\sqrt{c-t}$. That is, these nearly balanced quadratic reactions are destabilizing.

Power series approximations (2.2) of nonlinear coordinate transforms do the job of finding the emergent dynamics. The approximations (2.2)–(2.3) arise by solving differential equations associated with the system (2.1), but the algebra is awful. Thus the task of this chapter is to develop computer algebra to construct power series solutions of differential equations. The techniques introduced in this chapter are prototypes for all the subsequent asymptotic and perturbative analysis of dynamical systems. The techniques build on from those introduced in Chapter 1 for solving algebraic equations. Discover much more in the approximate solution of differential equations by, for example, reading Chapter 3 in the book by Bender and Orszag (1981). Here we limit attention to introducing those techniques later pertinent in modeling complex dynamical systems.

2.1 ▪ Picard iteration is straightforward

Computers are extremely good are repeating the same thing many times over. We use this aspect to find power series solutions of some basic differential equations and then move on to solve some "horrible" nonlinear differential equations. Examples develop the approach.

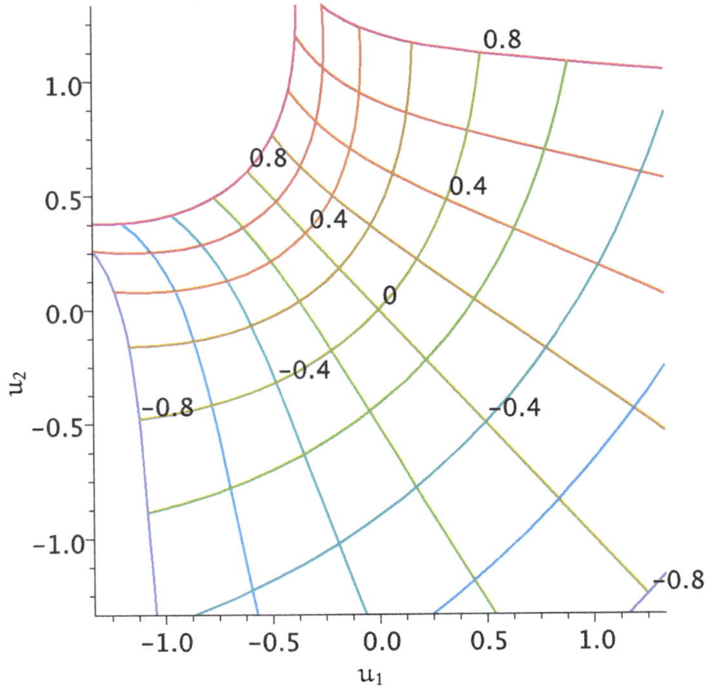

Figure 2.2. *V and W coordinate curves of the coordinate transform* (2.2). *In these coordinates, equation* (2.3) *describes the dynamics of system* (2.1).

Example 2.2. The solution to $y'' + y = 0$, $y(0) = 1$, and $y'(0) = 0$ is $y = \cos x$. Find the Taylor series solution by iteration: first by hand and second using computer algebra.

Solution: Rearrange this ODE to $y'' = -y$ and then formally integrate twice to $y = -\iint y\,dx\,dx$. These integrals on the right-hand side are indefinite integrals, so constants of integration, say $a + bx$, should appear on the right-hand side. But we know that the cosine solution to $y'' + y = 0$ has $y(0) = 1$ and $y'(0) = 0$, so surely we should set $a = 1$ and $b = 0$ to account for these initial conditions. Thus

$$y = 1 - \iint y\,dx\,dx, \tag{2.4}$$

where *here* the integrals are implicitly the definite integral from 0 to x. This rearrangement incorporates the information of the ODE and its initial conditions.

In the form (2.4) we readily find its power series solution by iteration: given an approximation $y_n(x)$, obtain a new approximation by evaluating

$$y_{n+1} = 1 - \iint y_n\,dx\,dx\,.$$

First try by hand starting from $y_0 = 1$:

- $y_1 = 1 - \iint 1\,dx\,dx = 1 - \frac{1}{2}x^2$;
- $y_2 = 1 - \iint 1 - \frac{1}{2}x^2\,dx\,dx = 1 - \frac{1}{2}x^2 + \frac{1}{24}x^4$.

Dialogue 2.1 *Reduce for Example 2.2.*

```
1: on div;
2: off allfac;
3: on revpri;
4: y:=1;
y := 1
5: y:=1-int(int(y,x),x);
           1  2
y := 1 - ---*x
           2
6: y:=1-int(int(y,x),x);
           1  2     1   4
y := 1 - ---*x  + ----*x
           2        24
8: for n:=3:8 do y:=1-int(int(y,x),x);
9: y;
       1  2     1   4      1    6       1      8       1        10
  1 - ---*x  + ----*x  - -----*x  + -------*x  - ---------*x
       2        24        720        40320        3628800
            1          12         1             14        1              16
     + -----------*x      - -------------*x      + -----------------*x
        479001600            87178291200            20922789888000
```

These are the first few terms in the Taylor series for $\cos x$. Now try using Reduce to do the algebra:

- first type the three commands

  `on div;`, `off allfac;`, and `on revpri;`

  (remember the semicolon to logically terminate each command and the return or enter key to get Reduce to execute the line you have typed)—these commands tell Reduce to format its output in a nice way for power series;
- set variable `y` to the first approximation by typing `y:=1;`;
- type `y:=1-int(int(y,x),x);` to assign the first approximation, $y_1 = 1 - x^2/2$, to the variable `y`—`int(y,x)` computes *an* integral with respect to `x` of whatever is in `y`; fortunately for us, for polynomial `y` it computes the integral that is zero at $x = 0$;
- type `y:=1-int(int(y,x),x);` again to compute y_2;
- iterative loops are standard in computer languages, and computer algebra is no exception, so type

  `for n:=3:8 do y:=1-int(int(y,x),x);`

  to compute further iterations. But nothing was printed, so finally type `y;` to see the resulting power series for $\cos x$.

The entire Reduce dialogue should look like Dialogue 2.1.

Example 2.3. Find the general Taylor series solution to $y'' + y = 0$ using computer algebra (Reduce).

Algorithm 2.1 *Use iteration to find the general solution of the* ODE *in Example* 2.3.

```
    factor a,b;
    y:=a+b*x;
    for n:=1:4 do write
        y:=a+b*x-int(int(y,x),x);
```

Dialogue 2.2 *Reduce output of Algorithm* 2.1.

```
1: in textfile$
y := b*x + a
              1  3             1  2
y := b*(x - ---*x ) + a*(1 - ---*x )
              6                2
              1  3       1   5               1  2      1   4
y := b*(x - ---*x  + -----*x ) + a*(1 - ---*x  + ----*x )
              6         120                   2        24
              1  3       1   5       1     7
y := b*(x - ---*x  + -----*x  - ------*x )
              6         120        5040
                 1  2      1   4      1    6
   + a*(1 - ---*x  + ----*x  - -----*x )
                 2        24        720
              1  3       1   5       1     7        1       9
y := b*(x - ---*x  + -----*x  - ------*x  + --------*x )
              6         120        5040        362880
                 1  2      1   4      1    6        1      8
   + a*(1 - ---*x  + ----*x  - -----*x  + -------*x )
                 2        24        720        40320
```

Solution: In the previous example we built in the specific initial conditions appropriate to $y = \cos x$, namely $y(0) = 1$ and $y'(0) = 0$. By making the integration constants arbitrary, via iterating

$$y = a + bx - \iint y \, dx \, dx \,,$$

we then recover the general solution parametrized by a and b, where $y(0) = a$ and $y'(0) = b$. Let's do it. Type the commands in Algorithm 2.1 into a text file (remember to start with `on div;`, `off allfac;`, and `on revpri;` and to finish with `end;`):

- type `factor a,b;` to get Reduce to group all terms in a and all terms in b;
- set the initial value to something simple satisfying the initial conditions `y:=a+b*x;`;
- iterate four times with

  `for n:=1:4 do write y:=a+b*x-int(int(y,x),x);`

  using the `write` command to print each iterate.

Dialogue 2.2 lists the Reduce output. See how easily this generates the Taylor series for $y = a \cos x + b \sin x$.

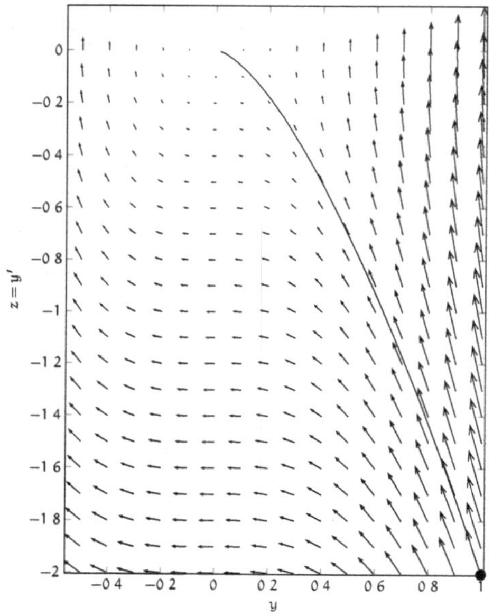

Figure 2.3. *Vector field and trajectory of solution for Example* 2.4.

Picard iteration Theoretical properties of solutions to differential equations are often established by writing the differential equation as an integral equation. For example, the differential equation $y' = f(x, y)$ such that $y(0) = a$ may be equivalently written as the integral equation $y = a + \int f(x, y)\,dx$ (the integrals in this paragraph are over 0 to x). Similarly, as seen above, the second-order differential equation $y'' = f(x, y, y')$ such that $y(0) = a$ and $y'(0) = b$ is equivalent to the integral equation $y = a + bx + \iint f(x, y, y')\,dx\,dx$. The previous example also shows that such an integral equation formulation may form the basis for an iterative solution: given an approximate solution $y_n(x)$, we generally expect

$$y_{n+1}(x) = a + bx + \iint f(x, y_n, y_n')\,dx\,dx$$

to be a better solution of the differential equation. This is Picard iteration.

Now let us try a rather hard differential equation—one almost impossible to quantitatively solve except via power series methods. We now use precisely the same Picard iteration to solve a *nonlinear* ODE.

Example 2.4. Find the Taylor series solution to the nonlinear ODE $y'' = 6y^2$ with initial conditions $y(0) = 1$ and $y'(0) = -2$.

Before solving this ODE as a power series (by design its exact solution just happens to be $y = 1/(1 + x)^2$), investigate it qualitatively using techniques developed in courses on systems of first-order differential equations. Introduce $z(x) = y'$; then the equivalent system is

$$y' = z, \quad z' = 6y^2 .$$

Hence the evolution in the phase plane is dictated by the arrows shown in Figure 2.3 with the particular trajectory starting from the initial condition $(1, -2)$.

Algorithm 2.2 *Use Reduce to find the power series solution to the nonlinear* ODE *of Example* 2.4 *via its integral representation* (2.5).

```
1   let x^8=>0;
2   y:=1-2*x;
3   for n:=1:5 do write
4       y:=1-2*x+6*int(int(y^2,x),x);
```

Solution: Now we find its power series solution! As before, recast the ODE in the following form, which also incorporates the initial conditions, by formally integrating the ODE twice:

$$y = 1 - 2x + 6\iint y^2\,dx\,dx\,, \tag{2.5}$$

where again the repeated x integral is assumed done so that each integral is zero at $x = 0$. Then iterate, starting from $y_0 := 1 - 2x$, say:

$$\begin{aligned}
y_1 &:= 1 - 2x + 6\iint 1 - 4x + 4x^2\,dx\,dx \\
&= 1 - 2x + 3x^2 - 4x^3 + 2x^4\,; \\
y_2 &:= 1 - 2x + 6\iint \left(1 - 2x + 3x^2 - 4x^3 + 2x^4\right)^2 dx\,dx \\
&= 1 - 2x + 6\iint 1 - 4x + 10x^2 - 20x^3 + 29x^4 - 32x^5 \\
&\qquad + 28x^6 - 16x^7 + 4x^8\,dx\,dx \\
&= 1 - 2x + 3x^2 - 4x^3 + 5x^4 - 6x^5 \\
&\qquad + \frac{29}{5}x^6 - \frac{32}{7}x^7 + 3x^8 - \frac{4}{3}x^9 + \frac{4}{15}x^{10}\,.
\end{aligned}$$

The length of these approximations is quickly becoming horrible. But that is just why computers are made. Before rushing in to use Reduce, observe that here the quadratic nonlinearity y^2 is going to generate very high powers of x, most of which we do not want. For example, in y_2 the terms up to x^5 are correct, but all the higher powers are as yet wrong.[9] Another iteration would generate a 22nd-order polynomial for y_3, of which only the first eight coefficients are correct; the rest are rubbish. In Reduce, discard such high-order terms in a power series by using, for example, the command `let x^8=>0;`, which tells Reduce to discard, set to zero, or otherwise ignore all terms with a power of x of eight or more. This is just what we want; thus the first line of Algorithm 2.2 sets the order of truncation of the power series. The second line sets the initial approximation that $y \approx 1 - 2x$. The third and fourth lines code a loop to iterate (2.5). Put these commands in a text file, say named `node`, surrounded by the standard two lines

```
on div; off allfac; on revpri;
...
end;
```

[9]The quadratic nonlinearity y^2 rapidly generates high powers of x in the expressions. However, the iteration plods along only getting one or two orders of x more accurate each iteration.

Dialogue 2.3 *Reduce output of Algorithm 2.2.*

```
1: in node$

y := 1 - 2*x
                   2        3        4
y := 1 - 2*x + 3*x   - 4*x    + 2*x
                   2        3        4        5      29   6      32   7
y := 1 - 2*x + 3*x   - 4*x    + 5*x    - 6*x    + ----*x    - ----*x
                                                  5            7
                   2        3        4        5        6        7
y := 1 - 2*x + 3*x   - 4*x    + 5*x    - 6*x    + 7*x    - 8*x
                   2        3        4        5        6        7
y := 1 - 2*x + 3*x   - 4*x    + 5*x    - 6*x    + 7*x    - 8*x
```

Then execute the Reduce file using the `in` command to see the output in Dialogue 2.3. The iteration settles on the correct power series, but all terms with powers of eight or higher in x are neglected. Thus triumphantly write the solution of this nonlinear ODE as

$$y = 1 - 2x + 3x^2 - 4x^3 + 5x^4 - 6x^5 + 7x^6 - 8x^7 + \mathcal{O}(x^8),$$

where $\mathcal{O}(x^8)$ (read "*order of* x^8," as Definition 2.2 later defines precisely) tells us that the error in the power series, the neglected terms, is x^8 or higher powers.

In the above three examples we developed the Taylor series about $x = 0$. To find Taylor series about any point $x = c$ it is simply a matter of changing the independent variable to, for example, $t = x - c$ and then finding the Taylor series in t. We continue to find only Taylor series about $x = 0$ because that is (nearly) all we need to also find other power series solutions.

2.2 ▪ Iteration is very flexible

So far we have simply rearranged an ODE in order to derive an iteration that generates the desired power series solution.[10] This section discusses why this strategy works at all, and what extension we need to solve a very wide range of differential equations.

The iteration works because integration is basically a smoothing operation. This smoothing by integration tends to reduce errors in a power series. For example, suppose an error was $\mathcal{O}(x^3)$ (see Definition 2.2), so that the error is roughly about 10^{-3} when $x = 0.1$, say. Then integrating the error twice will lead to an error $\mathcal{O}(x^5)$ in the integral, which is much smaller in magnitude, roughly 10^{-5} when $x = 0.1$. Conversely differentiation magnifies errors: two derivatives of an error $\mathcal{O}(x^3)$ become an error $\mathcal{O}(x)$, which, at roughly 10^{-1} when $x = 0.1$, is much larger. To make errors smaller, equivalently to push them to higher powers in x, we generally need to integrate. Thus an integral reformulation of an ODE is the basis for a successful iterative solution.

How do we know how many iterations should be performed? One answer is simple: keep iterating until there is no more change to the solution. One consequence of the answer though is that we have to keep track of the change in the approximations.

[10] What we have done is rather remarkable. Recall learning about fixed point iteration as a method of solving linear and nonlinear equations. Fixed point iteration is also employed here. The remarkable difference is that previously you simply found the numerical value that satisfies a given equation; here you found the function, via its power series, that satisfies the given differential equation. This is iteration in a function space—a much more difficult task. Nonetheless the strategy of appropriately rearranging the equation and iterating works.

A good way to find the change in an approximation is to solve for it explicitly. That is, in analogy with the solution of algebraic equations in section 1.1, we solve for corrections to each approximation. But first we have to find an equation for the small corrections to an approximate solution at each iteration. This leads us to a powerful iterative framework, based upon the residual of the ODE, which we develop further and explore by example.

Example 2.5 (Legendre functions). Use iteration to find the general Taylor series solutions to Legendre's equation

$$(1-x^2)y''-2xy'+ky=0 \quad \text{for } k=m(m+1)$$

to an error $\mathcal{O}(x^{10})$ for initial conditions $y(0)=1$ and $y'(0)=0$.

Solution: Immediately an initial approximation is

$$y_0=1\,,$$

as this satisfies the initial conditions. The iterative challenge is, given a ***known*** approximation y_n, find an improved solution

$$y_{n+1}(x)=y_n(x)+\hat{y}_n(x)\,,$$

where $\hat{y}_n$ is the as yet unknown correction to the approximation that we have to find. Now substitute this form for y_{n+1} into the ODE and rearrange to put all the known terms on the right-hand side and all the unknown on the left:

$$-(1-x^2)\hat{y}_n''+2x\hat{y}_n'-k\hat{y}_n=(1-x^2)y_n''-2xy_n'+ky_n\,.$$

This looks like a differential equation for the as yet unknown correction $\hat{y}_n$ forced by the known right-hand side, the residual of Legendre's equation evaluated at the current approximation, $\text{Res}_n := (1-x^2)y_n''-2xy_n'+ky_n$. For example, the first residual from $y_0=1$ is $\text{Res}_0=k$. But this ODE for the correction $\hat{y}_n$ is far too complicated—indeed if we could solve it exactly then the problem would be over immediately. Instead, seek a simplification to make the ODE for $\hat{y}_n$ tractable while still useful. The general principles of the simplification are that in any terms involving $\hat{y}_n$:

- near the point of expansion $x=0$, x is much smaller than 1 and x^2 is even smaller still; thus neglect higher powers of x relative to lower powers—so in this example we replace the $(1-x^2)$ factor by 1 because the x^2 is negligible in comparison to 1 for the small x near the point of expansion;
- because differentiation increases errors—since differentiation by x corresponds roughly to lowering the power of x by 1 (equivalently it roughly corresponds to dividing by x)—neglect low-order derivatives of $\hat{y}_n$ provided they are not also divided by x. (In this example $x\hat{y}_n'$ is roughly of the same "size" as $\hat{y}_n$ because the derivative makes it larger but the multiplication by x cancels this effect, but both of these terms are smaller than $\hat{y}_n''$, which is roughly $1/x^2$ times as large.)

After this simplification, the ODE for a correction then reduces to

$$-\hat{y}_n''=\text{Res}_n(x)=(1-x^2)y_n''-2xy_n'+ky_n\,.$$

Algorithm 2.3 *Reduce code to construct solutions of Legendre's equation in Example* 2.5.

```
factor x;
y:=1;
let x^10=>0;
for iter:=1:99 do begin
    res:=(1-x^2)*df(y,x,2)-2*x*df(y,x)+k*y;
    res:=-int(int(res,x),x);
    write y:=y+res;
    if res=0 then iter:=1000000;
end;
```

In the first iteration, as $\text{Res}_0 = k$, this ODE is $-\hat{y}_0'' = k$, which upon integrating twice leads to the requisite correction being $\hat{y}_0 = -kx^2/2$.

But what about the constants of integration? In this approach the initial approximation satisfies the initial conditions $y(0) = 1$ and $y'(0) = 0$. We ensure these initial conditions are satisfied by all approximations through ensuring all the corrections $\hat{y}_n$ satisfy the corresponding *homogeneous* initial conditions $\hat{y}_n(0) = \hat{y}_n'(0) = 0$. Thus, for example, the change $\hat{y}_0$ above is indeed correct. Hence the next approximation is $y_1 = 1 - kx^2/2$.

We could continue doing this by hand, but the plan is to use computer algebra to do the tediously repetitious iteration. Algorithm 2.3 codes the following steps.

- Set the initial approximation $y := 1$.
- Discard any powers generated of $\mathcal{O}(x^{10})$ by `let x^10=>0;`.
- Iterate until the correction is negligible using a safe for-do loop.
- Inside the loop:
  - compute residual, `res:=(1-x^2)*df(y,x,x)-2*x*df(y,x)+k*y;`;
  - compute correction by two integrations, `res:=-int (int(res,x),x);`;
  - and lastly update the approximation, `write y:=y+res;`.
- Group terms in powers of variable x by `factor x;` at the start of the code.

The Reduce output might be as listed in Dialogue 2.4. This Reduce output gives the desired Taylor series to be

$$
\begin{aligned}
y = 1 - \frac{k}{2}x^2 + \left(\frac{1}{24}k^2 - \frac{1}{4}k\right)x^4 - \left(\frac{1}{720}k^3 - \frac{13}{360}k^2 + \frac{1}{6}k\right)x^6 \\
+ \left(\frac{1}{40320}k^4 - \frac{17}{10080}k^3 + \frac{101}{3360}k^2 - \frac{1}{8}k\right)x^8 + \mathcal{O}(x^{10}).
\end{aligned}
$$

Example 2.6. Find the Taylor series solution to errors $\mathcal{O}(x^8)$ to the nonlinear ODE $y'' + (1+x)y' - 6y^2 = 0$ such that $y(0) = 1$ and $y'(0) = -1$.

Solution: Again immediately write down an initial approximation consistent with the initial conditions: namely $y_0 = 1 - x$. Then, given a known approximation, say $y_n(x)$, seek an improved approximation $y_{n+1}(x) = y_n(x) + \hat{y}_n(x)$,

Dialogue 2.4 *Reduce output of Algorithm* 2.3.

```
1: in leg$
y := 1
            1  2
y := 1 - ---*x *k
          2
          1  2      4      1      1  2
y := 1 - ---*x *k + x *( - ---*k + ----*k )
          2                  4      24
y :=
    1  2      4      1      1  2      6      1      13   2      1  3
1 - ---*x *k + x *( - ---*k + ----*k ) + x *( - ---*k + -----*k  - -----*k )
    2                 4      24                 6       360        720
          1  2      4      1      1  2
y := 1 - ---*x *k + x *( - ---*k + ----*k )
          2                  4      24
        6      1      13   2      1  3
   + x *( - ---*k + -----*k  - -----*k )
              6      360        720
        8      1      101   2      17    3      1     4
   + x *( - ---*k + ------*k  - -------*k  + -------*k )
              8      3360        10080        40320
          1  2      4      1      1  2
y := 1 - ---*x *k + x *( - ---*k + ----*k )
          2                  4      24
        6      1      13   2      1  3
   + x *( - ---*k + -----*k  - -----*k )
              6      360        720
        8      1      101   2      17    3      1     4
   + x *( - ---*k + ------*k  - -------*k  + -------*k )
              8      3360        10080        40320
```

where $\hat{y}_n(x)$ is some as yet unknown correction. Substitute into the differential equation and rearrange to deduce the following ODE for the correction:

$$-\hat{y}_n'' - (1+x)\hat{y}_n' + 6\hat{y}_n^2 + 12y_n\hat{y}_n = \operatorname{Res}_n = y_n'' + (1+x)y_n' - 6y_n^2 ,$$

where here, as always, $\operatorname{Res}_n(x)$ is the known residual evaluated for the current approximation. Now simplify the left-hand side:

- since x is "small" (in the power series expansion), $1 + x \approx 1$ and similarly $y_n \approx 1$ from the initial condition $y(0) = 1$, so the left-hand side first simplifies to
$$-\hat{y}_n'' - \hat{y}_n' + 6\hat{y}_n^2 + 12\hat{y}_n ;$$
- but also the correction $\hat{y}_n$ must be small (as each $\hat{y}_n$ is to make a small improvement in the solution), and so $\hat{y}_n^2$ must be much smaller still and should be neglected—for example, we typically expect the first correction $\hat{y}_0$ to be $\mathcal{O}(x^2)$ whence $\hat{y}_0^2 = \mathcal{O}(x^4)$, which is much smaller and negligible in the first iteration—hence the left-hand side simplifies further to
$$-\hat{y}_n'' - \hat{y}_n' + 12\hat{y}_n ;$$
- last, differentiation effectively decreases the order of any term so that the second-derivative term dominates the others above, and so the ODE for the change becomes simply
$$-\hat{y}_n'' = \operatorname{Res}_n = y_n'' + (1+x)y_n' - 6y_n^2 .$$

Algorithm 2.4 *Reduce code to solve the nonlinear* ODE *in Example 2.6. Note: It is safest to specify that the integrals are from* 0 *so that we know the corrections satisfy* $\hat{y}(0) = \hat{y}'(0) = 0$.

```
y:=1-x;
let x^8=>0;
for iter:=1:99 do begin
    res:=df(y,x,2)+(1+x)*df(y,x)-6*y^2;
    res:=-int(int(res,x),x);
    write y:=y+res;
    if res=0 then iter:=1000000;
end;
```

For example, begin the first iteration by computing the residual

$$\text{Res}_0 = 0 + (1+x)(-1) - 6(1-x)^2 = -7 + 11x - 6x^2 .$$

Then changing sign and integrating twice gives the first correction

$$\hat{y}_0 = -\iint \text{Res}_n \, dx\, dx = \frac{7}{2}x^2 - \frac{11}{6}x^3 + \frac{1}{2}x^4 ,$$

after recalling that we need to satisfy homogeneous initial conditions $\hat{y}_n'(0) = \hat{y}_n(0) = 0$ for the corrections in order to ensure that the solution satisfies the specified initial conditions. Thus the first corrected approximation is

$$y_1 = 1 - x + \frac{7}{2}x^2 - \frac{11}{6}x^3 + \frac{1}{2}x^4 .$$

Now explore further with the computer algebra code in Algorithm 2.4. See in the output of Dialogue 2.5 how the nonlinearity generates a lot of high-order rubbish, but the iteration soon corrects it all. Thus conclude that the Taylor series solution is

$$y = 1 - x + \frac{7}{2}x^2 - 3x^3 + \frac{25}{6}x^4 - \frac{257}{60}x^5 + \frac{219}{40}x^6 - \frac{1433}{252}x^7 + \mathcal{O}(x^8) .$$

The general principles Note the following general principles in this iterative approach to finding power series solutions to linear and nonlinear ODEs.

- Make an initial approximation consistent with the initial conditions of the ODE.
- Seek a simple ODE for successive corrections by substituting $y_{n+1} = y_n + \hat{y}_n$ into the differential equation, grouping all the known terms into the residual Res_n, and then neglecting all but the dominant terms involving the change $\hat{y}_n$:
  - neglect all nonlinear terms in the small change $\hat{y}_n$;
  - approximate all coefficient factors of $\hat{y}_n$ by the lowest-order term in x;
  - and, counting each derivative with respect to x as equivalent to a division by x, keep only those terms of lowest order in x.

Dialogue 2.5 *Reduce output of Algorithm* 2.4.

```
2: in nod_r$
y := 1 - x
          7  2      11  3     1  4
y := 1 - x + ---*x  - ----*x  + ---*x
          2          6          2
          7  2        3    31  4     121  5     199  6     83  7
y := 1 - x + ---*x  - 3*x  + ----*x  - -----*x  + -----*x  - ----*x
          2                8         40         60         42
          7  2        3    25  4     169  5     83  6     8543  7
y := 1 - x + ---*x  - 3*x  + ----*x  - -----*x  + ----*x  - ------*x
          2                6         40         16        1680
          7  2        3    25  4     257  5     787  6     789  7
y := 1 - x + ---*x  - 3*x  + ----*x  - -----*x  + -----*x  - -----*x
          2                6         60         144        140
          7  2        3    25  4     257  5     219  6     9551  7
y := 1 - x + ---*x  - 3*x  + ----*x  - -----*x  + -----*x  - ------*x
          2                6         60         40         1680
          7  2        3    25  4     257  5     219  6     1433  7
y := 1 - x + ---*x  - 3*x  + ----*x  - -----*x  + -----*x  - ------*x
          2                6         60         40         252
          7  2        3    25  4     257  5     219  6     1433  7
y := 1 - x + ---*x  - 3*x  + ----*x  - -----*x  + -----*x  - ------*x
          2                6         60         40         252
```

This process is close kin to *linearization*.

- Iteratively make changes as guided by the residuals until the changes are zero to some order of error in x. Utilize computer algebra.

Applying these principles becomes more involved when we apply them in developing power series about a singular point of an ODE. Let's investigate a couple of examples.

Example 2.7 (Bessel function of order 0). Find the power series solution of $x^2y'' + xy' + x^2y = 0$ that is well behaved at $x = 0$ to an error $\mathcal{O}(x^{10})$—namely find the low orders of a power series proportional to the Bessel function $J_0(x)$.

Solution: First, find and solve the indicial equation by substituting $y = x^r + \mathcal{O}(x^{r+1})$. Here the ODE becomes[11]

$$\begin{aligned} x^2y'' + xy' + x^2y &= r(r-1)x^r + rx^r + x^{r+2} + \mathcal{O}(x^{r+1}) \\ &= r^2x^r + \mathcal{O}(x^{r+1}) \,. \end{aligned}$$

The only way this can be zero for all small x is to satisfy the indicial equation $r^2 = 0$. This indicial equation leads, as discussed by Kreyszig (2011, §4.4) and Bender and Orszag (1981, §3.2–3), to the homogeneous solutions of the ODE being approximately $y \approx a + b\log x$. The logarithm is *not* well behaved as $x \to 0$; hence we set $b = 0$ and just seek solutions that tend to a constant as $x \to 0$. Without loss of generality, because we can multiply by a constant later, we choose to find solutions such that $y(0) = 1$.

Second, make an initial approximation to the solution. After the above discussion of the indicial equation, choose $y_0 = 1$.

[11]As x^{r+2} is absorbed into the error term $\mathcal{O}(x^{r+1})$.

Third, given a known approximation $y_n(x)$, seek an improved approximation $y_{n+1}(x) = y_n(x) + \hat{y}_n(x)$, where $\hat{y}_n(x)$ is some small correction. Substitute this into the ODE, neglect $x^2\hat{y}_n$ because it is two orders of x smaller than either $x^2\hat{y}_n''$ or $x\hat{y}_n'$, and deduce that the correction $\hat{y}_n$ should satisfy

$$-x^2\hat{y}_n'' - x\hat{y}_n' = \text{Res}_n = x^2 y_n'' + xy_n' + x^2 y_n\,. \tag{2.6}$$

Solving this for the correction $\hat{y}_n$ is no longer simply a matter of integrating twice. However, rearranging the form of the ODE (2.6) we again express the solution in terms of two integrations. Notice that the left-hand side is identical to $-x(x\hat{y}_n')'$, whence

$$\begin{aligned} & -x(x\hat{y}_n')' = \text{Res}_n \\ \iff\; & x\hat{y}_n' = -\int \frac{\text{Res}_n}{x}\,dx \\ \iff\; & \hat{y}_n = -\int \frac{1}{x}\int \frac{\text{Res}_n}{x}\,dx\,dx\,. \end{aligned}$$

Apply this iteration here.

1. In the first iteration, $y_0 = 1$, so the residual $\text{Res}_0 = x^2$. Thus

$$\begin{aligned} \hat{y}_0 &= -\int \frac{1}{x}\int \frac{x^2}{x}\,dx\,dx \\ &= -\int \frac{1}{x}\left(\tfrac{1}{2}x^2 + b\right)\,dx \\ &= -\frac{1}{4}x^2 - b\log x + a \end{aligned}$$

for integration constants a and b.

These integration constants include $a - b\log x$ in $\hat{y}_0$, but we cannot tolerate any component in $\log x$, as it behaves badly at $x = 0$, so $b = 0$, and a has to be chosen zero in order to ensure $y_n(0) = 1$. This consideration applies at all iterations. Hence $y_1 = 1 - x^2/4$.

2. In the second iteration compute $\text{Res}_1 = -x^4/4$. Thus, setting the integration constants to zero as before,

$$\begin{aligned} \hat{y}_1 &= -\int \frac{1}{x}\int \frac{-x^4/4}{x}\,dx\,dx \\ &= -\int \frac{1}{x}\left(-\frac{x^4}{16}\right)\,dx \\ &= \frac{x^4}{64}\,. \end{aligned}$$

Hence $y_2 = 1 - x^2/4 + x^4/64$.

Continuing the iteration by the computer algebra of Algorithm 2.5, find the output listed in Dialogue 2.6. Thus the Taylor series of the Bessel function $J_0(x)$ is

$$y = J_0(x) = 1 - \frac{1}{4}x^2 + \frac{1}{64}x^4 - \frac{1}{2304}x^6 + \frac{1}{147456}x^8 + \mathcal{O}(x^{10})\,.$$

Algorithm 2.5 *Reduce code to find the well-behaved Bessel function of order* 0. *Example 2.7 shows how iteration constructs the Taylor series solution.*

```
y:=1;
let x^10=>0;
for iter:=1:99 do begin
    res:=x^2*df(y,x,2)+x*df(y,x)+x^2*y;
    res:=-int(int(res/x,x)/x,x);
    write y:=y+res;
 if res=0 then iter:=1000000;
end;
```

Dialogue 2.6 *Reduce output of Algorithm* 2.5.

```
2: in besso$
y := 1
        1  2
y := 1 - ---*x
        4
        1  2     1   4
y := 1 - ---*x  + ----*x
        4        64
        1  2     1   4       1    6
y := 1 - ---*x  + ----*x  - ------*x
        4        64         2304
        1  2     1   4       1    6        1     8
y := 1 - ---*x  + ----*x  - ------*x  + --------*x
        4        64         2304         147456
        1  2     1   4       1    6        1     8
y := 1 - ---*x  + ----*x  - ------*x  + --------*x
        4        64         2304         147456
```

Example 2.8 (Bessel functions of order 0). Find the power series expansion about $x = 0$, to errors $\mathcal{O}(x^{10})$, of the *general solution* to Bessel's ODE with $\nu = 0$, namely $x^2y'' + xy' + x^2y = 0$.

Solution: The indicial equation shows that in general the dominant component in the solution is $a + b\log x$ for any a and b. (These were also naturally obtained in the integration constants of the previous example.) Use this as the first approximation y_0 and see what ensues. The derivation of the equation for the iteration changes; equation (2.6) remains the same.

Code the iteration as in Algorithm 2.6. Dialogue 2.7 lists the output. That is, as Kreyszig (2011, p. 213) assures us, for double roots of the indicial equation, the general solution is of the form $y = ay_1(x) + by_2(x)$, where

$$y_1 = 1 - \frac{1}{4}x^2 + \frac{1}{64}x^4 - \frac{1}{2304}x^6 + \frac{1}{147456}x^8 + \mathcal{O}(x^{10}),$$
$$y_2 = y_1(x)\log x + \frac{1}{4}x^2 - \frac{3}{128}x^4 + \frac{11}{13824}x^6 - \frac{25}{1769472}x^8 + \mathcal{O}(x^{10}).$$

Use residuals to improve approximate solutions Getting computers to do tedious algebra may be adapted to a wide variety of problems. An iteration will improve an

Algorithm 2.6 *Reduce code to find the general series solution of Bessel's* ODE *of order* 0. *The* `factor` *command improves the appearance of the printed output.*

```
factor a,b,log;
y:=a+b*log(x);
let x^10=>0;
for iter:=1:99 do begin
    res:=x^2*df(y,x,2)+x*df(y,x)+x^2*y;
    res:=-int(int(res/x,x)/x,x);
    write y:=y+res;
    if res=0 then iter:=1000000;
end;
```

approximation provided corrections deduced from the residuals are appropriate. The key is to deduce a simple and sensible approximation to the equation for the corrections. But the correctness of the ultimate result depends only upon being able to evaluate the residuals correctly. That we drive the residuals to zero to some level of accuracy in a finite number of iterations ensures that the procedure terminates. But if it terminates, the zero residual ensures that we constructed a solution.

2.3 ▪ Divergent series are useful

> Divergent series are the invention of the devil, and it is shameful to base on them any demonstration whatsoever. *Niels Hendrik Abel,* 1828

All the series constructed so far are convergent. However, in modeling complex systems mostly we construct divergent series. Somehow we must make sense of such divergent series. This section explores the concept of asymptotic series and how they form useful approximations even if divergent.

Let's start the exploration with an innocuous-looking example that almost immediately presents new challenges.

Example 2.9 (Stieltjes series). Find the Taylor series, about $x = 0$, of the solution to the linear Stieltjes ODE

$$x^2y'' + (1+3x)y' + y = 0 \quad \text{such that} \quad y(0) = 1. \tag{2.7}$$

Why is there only one initial condition, $y(0) = 1$? Because this is a singular perturbation problem as the coefficient of the highest derivative vanishes at the point of expansion $x = 0$. Unlike the singular algebraic equations, which could be rescaled into regularity, the singular nature of the ODE (2.7) cannot be removed. Indeed the singular nature is fundamental to the interest of this Stieltjes ODE.

Solution: Solve the ODE (2.7) by iteration. The initial approximation is $y_0 = 1$ from the given initial condition. As before, suppose $y_n(x)$ is an approximate solution to the ODE; seek a correction $\hat{y}_n(x)$ so that $y_{n+1} = y_n(x) + \hat{y}_n(x)$ is a better approximation. Substitute y_{n+1} into the ODE (2.7) to deduce

$$x^2\hat{y}_n'' + (1+3x)\hat{y}_n' + \hat{y}_n + \text{Res}_{(2.7)} = 0,$$

Dialogue 2.7 *Reduce output of Algorithm 2.6.*

```
6: in bessg_r$
y := a + log(x)*b
      1     2              1   2                        1   2
y := ---*b*x   + a*(1 - ---*x ) + log(x)*b*(1 - ---*x )
      4                    4                            4
        1   2       3    4              1   2      1    4
y := b*(---*x   - -----*x ) + a*(1 - ---*x   + ----*x )
        4          128                   4         64
                                1   2      1    4
      + log(x)*b*(1 - ---*x   + ----*x )
                                4          64
        1   2       3    4       11       6
y := b*(---*x   - -----*x   + -------*x )
        4          128         13824
              1   2     1    4      1       6
      + a*(1 - ---*x   + ----*x   - ------*x )
              4         64        2304
                                1   2      1    4      1       6
      + log(x)*b*(1 - ---*x   + ----*x   - ------*x )
                                4          64         2304
        1   2       3    4       11       6       25        8
y := b*(---*x   - -----*x   + -------*x   - ---------*x )
        4          128         13824        1769472
              1   2     1    4      1       6       1        8
      + a*(1 - ---*x   + ----*x   - ------*x   + --------*x )
              4         64        2304         147456
                                1   2      1    4      1       6       1        8
      + log(x)*b*(1 - ---*x   + ----*x   - ------*x   + --------*x )
                                4          64         2304         147456
        1   2       3    4       11       6       25        8
y := b*(---*x   - -----*x   + -------*x   - ---------*x )
        4          128         13824        1769472
              1   2     1    4      1       6       1        8
      + a*(1 - ---*x   + ----*x   - ------*x   + --------*x )
              4         64        2304         147456
                                1   2      1    4      1       6       1        8
      + log(x)*b*(1 - ---*x   + ----*x   - ------*x   + --------*x )
                                4          64         2304         147456
```

where $\text{Res}_{(2.7)}$ is the residual of the ODE (2.7) at the current approximation. Now approximate the operator acting on the correction $\hat{y}_n$ to give the equation

$$\hat{y}'_n = -\text{Res}_{(2.7)}:$$

- $3x \ll 1$ for x near zero and so is omitted;
- $x^2\hat{y}''_n$ is of size $\hat{y}_n$, both are smaller than $\hat{y}'_n$, and so both are omitted.

Use Algorithm 2.7 to find the power series solution to the Stieltjes ODE (2.7). Dialogue 2.8 lists the Reduce output, now without trace printing in the loop. These are the first nine terms in the so-called Stieltjes series

$$y = \sum_{n=0}^{\infty} (-1)^n n! x^n . \tag{2.8}$$

Exercise 2.16 asks you to confirm that this infinite sum satisfies the Stieltjes ODE (2.7).

Algorithm 2.7 *Use Reduce to find the power series solution to the Stieltjes ODE (2.7).*

```
let x^9=>0;
y:=1;
for iter:=1:99 do begin
    res:=x^2*df(y,x,2)+(1+3*x)*df(y,x)+y;
    res:=-int(res,x);
    y:=y+res;
    if res=0 then iter:=1000000;
end;
write y;
```

Dialogue 2.8 *Reduce output of Algorithm 2.7.*

```
4: in stieltjes$
y := 1
             2       3        4         5         6          7           8
1 - x + 2*x  - 6*x  + 24*x  - 120*x  + 720*x  - 5040*x  + 40320*x
```

We encountered no difficulty constructing the power series (2.8) solution. But look closely. The coefficients in the series grow more and more rapidly. The ratio test informs us of convergence, or otherwise: the nth term in the series is $a_n = (-1)^n n! x^n$; the ratio of consecutive terms $|a_n/a_{n-1}| = nx$ by cancellation; for $x \neq 0$ this ratio $|a_n/a_{n-1}| = nx \to \infty$ as $n \to \infty$. This power series diverges for all $x \neq 0$. The radius of convergence of this power series is zero!

Such divergence almost everywhere reflects the irredeemably singular nature of the Stieltjes ODE (2.7) at $x = 0$.

Divergent power series are common Later examples of power series indicate not only that such divergent series occur in the modeling of nonlinear dynamical systems, but also that divergent power series are the norm. Indeed, be surprised when we find convergent power series solutions! Mostly we only find a few terms in a power series solution. Even though we generally will not generate enough information to indicate whether the power series converges or diverges, expect the power series to be divergent. *The outstanding issue for us is to make sense of the use of such divergent power series.*

Example 2.10 (error summing the Stieltjes series). Exercise 2.16 asks you to show that the Stieltjes integral

$$y(x) = \int_0^\infty \frac{e^{-t}}{1+xt}\,dt \tag{2.9}$$

solves the Stieltjes ODE (2.7).[12]

[12] The Stieltjes integral (2.9) illuminates the singularity at $x = 0$ that is the root cause of the divergence of any power series solution to the Stieltjes ODE. The Stieltjes integral is clearly an analytic function of x. But see in the integral (2.9) that there is a pole at $t = -1/x$. For negative x this pole lies on the path of integration. To navigate around this pole, the contour of integration may be moved to one side or the other in the complex t-plane. The change in the integral, depending upon the deformation of the integration contour, represents a branch cut in the analytic dependence—a branch cut that exists for all negative x, including those arbitrarily close to $x = 0$. Hence $x = 0$ is a singular point and any power series representation should diverge. However, the strength of the branch cut is very small as $|x| \to 0$. The strength of the branch cut is the jump in the function value, which is just $2\pi i$ times the strength of the pole at $t = -1/x$. Rewriting the integrand as

Now use this solution to illuminate the power series. An integration by parts of the Stieltjes integral (2.9) leads to

$$\begin{aligned} y(x) &= \left[(1+xt)^{-1}(-e^{-t})\right]_{t=0}^{t=\infty} - \int_0^\infty x(1+xt)^{-2}e^{-t}\,dt \\ &= 1 - x\int_0^\infty (1+xt)^{-2}e^{-t}\,dt\,. \end{aligned}$$

This integration by parts extracts the first term in the Stieltjes series (2.8) from the integral. Further integration by parts extracts more terms in the Stieltjes series:

$$\begin{aligned} y(x) &= 1 - x\int_0^\infty (1+xt)^{-2}e^{-t}\,dt \\ &= 1 - x + 2x^2\int_0^\infty (1+xt)^{-3}e^{-t}\,dt \\ &= 1 - x + 2x^2 - 6x^3\int_0^\infty (1+xt)^{-4}e^{-t}\,dt \\ &\;\;\vdots \\ &= 1 - x + 2!x^2 - 3!x^3 + \cdots + (-1)^N N!x^N + \varepsilon_N(x)\,, \end{aligned}$$

where the remainder term

$$\varepsilon_N(x) = (-1)^{N+1}(N+1)!x^{N+1}\int_0^\infty (1+xt)^{-N-2}e^{-t}\,dt\,.$$

The Stieltjes series appears, but, importantly, the series appears with a definite expression for the error $\varepsilon_N(x)$ in the Nth partial sum.

The error is bounded since $1+xt \geqslant 1$ for $x, t \geqslant 0$; thus $(1+xt)^{-N-2} \leqslant 1$ and hence

$$|\varepsilon_N(x)| \leqslant (N+1)!x^{N+1}\int_0^\infty e^{-t}\,dt = (N+1)!x^{N+1}\,.$$

Although the series is divergent, this bound on the error of some partial sum may be small enough for practical purposes. For example, the four-term partial sum $1 - x + 2x^2 - 6x^3$ approximates the Stieltjes integral to two decimal places when $24|x|^4 < 0.01$, that is, for $|x| < 0.14$.

Observe that the error $\varepsilon_N(x)$ in the partial sums of the Stieltjes series is bounded by the magnitude of the next term in the series. Experience suggests that errors of partial sums are often about the size of the first, lowest-order, neglected term in the series. Occasionally we invoke the following conjecture.

Conjecture 2.1. *As a gross but practical generalization, the error in an asymptotic series may be estimated by the first neglected term in the series.*

$e^{-t}/[x(t+1/x)]$, the strength of the pole at $t = -1/x$ is thus $e^{1/x}/x$. Consequently, the strength of the branch cut is *exponentially small*, $2\pi i e^{1/x}/x$, as $x \to 0-$. Being exponentially small, the Stieltjes series in powers of x cannot discern the branch cut other than being divergent.

2.3.1 ▪ Asymptotic power series and their properties

> Divergent series converge faster than convergent series because they don't have to converge.
> *George F. Carrier*

Because divergent series are endemic in our modeling, we henceforth phrase analysis in terms of asymptotic power series rather than Taylor series. The difference in algebraic manipulations is usually negligible. The difference primarily lies in how the series are interpreted. For example, we do not necessarily expect a partial sum to improve as more terms are taken. Instead, and this is the core idea of asymptotic series, we just demand that any specific partial sum improve as the parameter gets smaller.

In this section we formalize the basis and properties of asymptotic power series along the lines of Hinch (1991, Chap. 2), and Bender and Orszag (1981, §3.8).

Definition 2.2 (basic notation). *We use the following notation (Bender and Orszag,* 1981, *§3.4):*

- *"$f(x)$ is much smaller than $g(x)$ as $x \to x_0$" is written as*

$$f(x) \ll g(x) \text{ as } x \to x_0\,, \quad \textit{meaning} \quad \lim_{x\to x_0} \frac{f(x)}{g(x)} = 0\,,$$

and is alternatively sometimes written as $f(x) = o(g(x))$ as $x \to x_0$;

- *"$f(x)$ is of the* **order of** *$g(x)$ as $x \to x_0$" is written as*

$$f(x) = \mathcal{O}(g(x)) \text{ as } x \to x_0\,, \quad \textit{meaning} \quad \frac{f(x)}{g(x)} \textit{ is bounded as } x \to x_0\,.$$

In such limits, the limit point x_0 may be 0, ∞, or a one-sided limit such as $+0$. We also write $f(x) = h(x) + \mathcal{O}(g(x))$ to mean $f - h = \mathcal{O}(g)$, and similarly for $o()$.

Example 2.11 (some asymptotic relations).

1. $\cos x = 1 + \mathcal{O}(x^2)$ since $1 - \cos x = \mathcal{O}(x^2)$ as $x \to 0$. Consider

$$\begin{aligned}\lim_{x\to 0} \frac{1-\cos x}{x^2} &= \lim_{x\to 0} \frac{\sin x}{2x} && \text{by l'Hôpital}\\ &= \lim_{x\to 0} \frac{-\cos x}{2} && \text{by l'Hôpital}\\ &= \tfrac{1}{2} \text{ and hence bounded.}\end{aligned}$$

2. $\sin x = o(\sqrt{x})$ as $x \to 0$. Consider

$$\begin{aligned}\lim_{x\to +0} \frac{\sin x}{\sqrt{x}} &= \lim_{x\to +0} \frac{\cos x}{\frac{1}{2}x^{-1/2}} && \text{by l'Hôpital}\\ &= \lim_{x\to +0} 2\sqrt{x}\cos x\\ &= 0.\end{aligned}$$

3. $x^p = \mathcal{O}(x^q)$ as $x \to 0$ only for $p \geqslant q$. Consider

$$\begin{aligned}\lim_{x\to 0}\frac{x^p}{x^q} &= \lim_{x\to 0} x^{p-q}\\ &= \begin{cases} 0 & \text{if } p-q>0,\\ 1 & \text{if } p=q,\\ \infty & \text{if } p-q<0.\end{cases}\end{aligned}$$

Therefore the ratio is bounded only when $p \geqslant q$.

4. $x^n \ll e^x$ as $x \to +\infty$. Let's prove by induction: for the case $n = 0$, the limit $\lim_{x\to+\infty} x^0/e^x = \lim_{x\to+\infty} e^{-x} = 0$; thus $x^0 = 1 \ll e^x$, whereas for general exponent n,

$$\begin{aligned}\lim_{x\to+\infty}\frac{x^n}{e^x} &= \lim_{x\to+\infty}\frac{nx^{n-1}}{e^x} \quad \text{by l'Hôpital}\\ &= n\lim_{x\to+\infty}\frac{x^{n-1}}{e^x}\\ &= n\times 0 \quad \text{by case } n-1\\ &= 0;\end{aligned}$$

thus by induction $x^n \ll e^x$, as $x \to +\infty$ for all n.

5. $e^{-1/x} = o(x^n)$ as $x \to +0$ for any power n. Substitute $1/x$ for x and proceed as for the previous example.

Beware that some people occasionally misuse the symbols $\ll$, $\mathcal{O}()$, and $o()$: they use the symbols without an appropriate accompanying limit. An implicit limit is all right and often done: I often just write "$\mathcal{O}(x^3)$" because the context supplies the limit "as $x \to 0$." But statements involving phrases such as "$1 \ll 10$" are not strictly correct because there is no limit. In such misuse, people appeal to the qualitative idea, not the strict meaning.

Now let us proceed to understand how to sensibly write and manipulate infinite power series *irrespective of whether they converge or not.*

Definition 2.3 (asymptotic power series). *The infinite sum* $\sum_{n=0}^{\infty} a_n x^n$ *is said to be an* **asymptotic power series** *to a function* $f(x)$ *as* $x \to 0$, *written* $f(x) \sim \sum_{n=0}^{\infty} a_n x^n$, *when the error in every partial sum is always much smaller than the size of the last term included:*

$$f(x) - \sum_{n=0}^{m} a_n x^n \ll x^m \quad \text{as } x \to 0 \quad \text{for all } m. \tag{2.10}$$

Sometimes the limits are restricted to be one sided. A partial sum of an asymptotic power series is usually called an **asymptotic approximation.**

The combination of symbols "$\sum_{n=0}^{\infty} a_n x^n$" does not represent a series that we may sum at any x. Instead the combination of symbols as a whole represents a process as the variable $x \to 0$. All this definition requires is that we can add a finite number of terms in the sum, and that such partial sums behave themselves as $x \to 0$.

Example 2.12 (the Stieltjes power series is asymptotic). The Stieltjes series (2.8) is written in powers of x. Example 2.10 derived that the error in a partial sum of the Stieltjes series

$$|\varepsilon_m(x)| \leqslant (m+1)!\,x^{m+1} \ll x^m \quad \text{as } x \to +0.$$

Hence the Stieltjes series is an asymptotic power series as $x \to +0$.

Example 2.13 (Taylor series are asymptotic). Every convergent Taylor series is also an asymptotic power series. Let the function $f(x) = \sum_{n=0}^{\infty} f^{(n)}(0)x^n/n!$ converge in some neighborhood of $x = 0$. Then for any truncation m consider

$$\begin{aligned}
&\lim_{x\to 0} \frac{f(x) - \sum_{n=0}^{m} f^{(n)}(0)x^n/n!}{x^m} \\
&= \lim_{x\to 0} \frac{f^{(m)}(x) - f^{(m)}(0)}{m!} \quad \text{by l'Hôpital } m \text{ times} \\
&= 0, \quad \text{as } f \text{ is arbitrarily differentiable.}
\end{aligned}$$

That is, $f(x) - \sum_{n=0}^{m} a_n x^n \ll x^m$ as $x \to 0$ for all m. Hence a Taylor series is an asymptotic power series.

2.3.2 ▪ Properties of asymptotic power series

Uniqueness If a function possesses an asymptotic power series, then the series is *unique*. The coefficients of the asymptotic power series are uniquely and recursively determined by the limits

$$a_m = \lim_{x\to 0} \frac{f(x) - \sum_{n=0}^{m-1} a_n x^n}{x^m} .$$

If these limits exist, then the limits are by definition unique and hence our coefficients are unique.

Nonuniqueness There are many functions with the same asymptotic power series. For example, consider

$$\frac{1}{1-x} \sim \sum_{n=0}^{\infty} x^n \quad \text{as } x \to +0,$$

$$\text{but also } \frac{1}{1-x} + e^{-1/x} \sim \sum_{n=0}^{\infty} x^n \quad \text{as } x \to +0 .$$

Two functions that differ by something with a zero asymptotic power series, namely $\sum_{n=0}^{\infty} 0x^n$, such as the function $e^{-1/x}$ used above, have the same asymptotic power series. However, the difference between members in the class of functions is much smaller than any power of x.

Equating coefficients We may occasionally write

$$\sum_{n=0}^{\infty} a_n x^n \sim \sum_{n=0}^{\infty} b_n x^n \quad \text{as } x \to 0 .$$

This asymptotic notation means that the two asymptotic power series $\sum_{n=0}^{\infty} a_n x^n$ and $\sum_{n=0}^{\infty} b_n x^n$ are asymptotic to the same class of functions as $x \to 0$. By the uniqueness of asymptotic coefficients, it follows that $a_n = b_n$ for all n. That is, *we may equate coefficients of asymptotic power series irrespective of convergence or divergence.*

Arithmetic operations follow naturally A further consequence of the uniqueness of coefficients is that asymptotic power series may be summed, subtracted, multiplied, or divided in the usual manner (Bender and Orszag, 1981, p. 125).

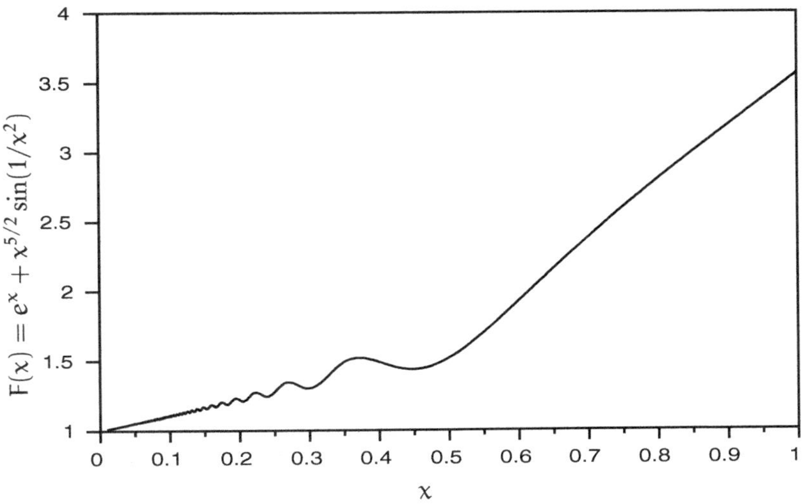

Figure 2.4. *An example function with a limited asymptotic power series.*

Calculus is trickier An asymptotic power series of a function $f(x)$ can be integrated term by term near the origin (the point of expansion).

However, asymptotic power series cannot always be differentiated. The problem is that very small terms may nonetheless oscillate wildly. For example, consider two functions $f(x)$ and $g(x)$ differing by a negligible oscillatory amount:

$$g(x) = f(x) + e^{-1/x}\sin(e^{1/x}).$$

As $x \to +0$, f and g have the same power series because $e^{-1/x}$ has the zero power series. Yet their derivatives differ markedly:

$$g'(x) = f'(x) + x^{-2}e^{-1/x}\sin(e^{1/x}) - x^{-2}\cos(e^{1/x}).$$

The last term on the right-hand side, which grows like x^{-2}, shows the enormous difference between the derivatives as $x \to +0$.

In practice we proceed by assuming that the highest derivative we need does have a valid asymptotic power series. Then asymptotic power series exist for all the lower-order derivatives by integration.

In practice, also, the above notion of asymptotic power series is so useful that we apply it even when the formal infinite series does not exist.

Example 2.14. Let the function $F(x) = e^x + x^{5/2}\sin(1/x^2)$, for $x \geqslant 0$, as plotted in Figure 2.4. We start computing its asymptotic series.

1. For $m = 0$,

$$a_0 = \lim_{x\to 0} \frac{F(x) - 0}{x^0} = \lim_{x\to 0} \frac{e^x + x^{5/2}\sin(1/x^2)}{1} = 1 + 0 = 1.$$

By subtracting 1 from each part, the first limit then also shows $F(x) = 1 + o(1)$.

2. For $m = 1$, and using l'Hôpital in the middle,

$$\begin{aligned} a_1 &= \lim_{x\to 0} \frac{F(x) - 1}{x^1} \\ &= \lim_{x\to 0} \frac{e^x - 1 + x^{5/2}\sin(1/x^2)}{x} \\ &= \lim_{x\to 0} \frac{e^x - 1}{x} + \lim_{x\to 0} \frac{x^{5/2}\sin(1/x^2)}{x} \\ &= \lim_{x\to 0} \frac{e^x}{1} + \lim_{x\to 0} x^{3/2}\sin(1/x^2) \\ &= 1 + 0\,. \end{aligned}$$

By subtracting 1 from each part, the first limit then also shows $F(x) = 1 + x + o(x)$.

3. For $m = 2$, and using l'Hôpital twice in the middle,

$$\begin{aligned} a_2 &= \lim_{x\to 0} \frac{F(x) - 1 - x}{x^2} \\ &= \lim_{x\to 0} \frac{e^x - 1 - x + x^{5/2}\sin(1/x^2)}{x^2} \\ &= \lim_{x\to 0} \frac{e^x - 1 - x}{x^2} + \lim_{x\to 0} \frac{x^{5/2}\sin(1/x^2)}{x^2} \\ &= \lim_{x\to 0} \frac{e^x}{2} + \lim_{x\to 0} x^{1/2}\sin(1/x^2) \\ &= \tfrac{1}{2} + 0\,. \end{aligned}$$

By subtracting $\frac{1}{2}$ from each part, the first limit then also shows $F(x) = 1 + x + \frac{1}{2}x^2 + o(x^2)$.

4. For $m \geqslant 3$ the process stops, as the limits no longer exist due to the fractional power.

We are restricted to the statement that $F(x) = 1 + x + \frac{1}{2}x^2 + o(x^2)$, although a little more work proves $F(x) = 1 + x + \frac{1}{2}x^2 + \mathcal{O}(x^{5/2})$.

Example 2.15 (bad derivative). Consider the function $F(x) = e^x + x^{5/2}\sin(1/x^2)$ given in Example 2.14. It has the beginnings of an asymptotic power series, $F(x) = 1 + x + \frac{1}{2}x^2 + \mathcal{O}(x^{5/2})$, so we would be tempted to assert that its derivative $F'(x) = 1 + x + \mathcal{O}(x^{3/2})$. But alas no.

The derivative $F'(x) = e^x + \frac{5}{2}x^{3/2}\sin(1/x^2) + 2x^{-1/2}\cos(1/x^2)$, which, because of the last term, has no limit as $x \to 0$. See Figure 2.5. Hence the derivative does not even have a leading term in an asymptotic power series. Be careful with differentiation.

2.3.3 ▪ General asymptotic series

Power series are most common in our applications, so you may skip the rest of this section. But recall the Bessel functions of Example 2.7; sometimes the logarithm appears in an asymptotic analysis. Sometimes powers of logarithms appear. Thus we define asymptotic series more generally than just for powers of a small parameter.

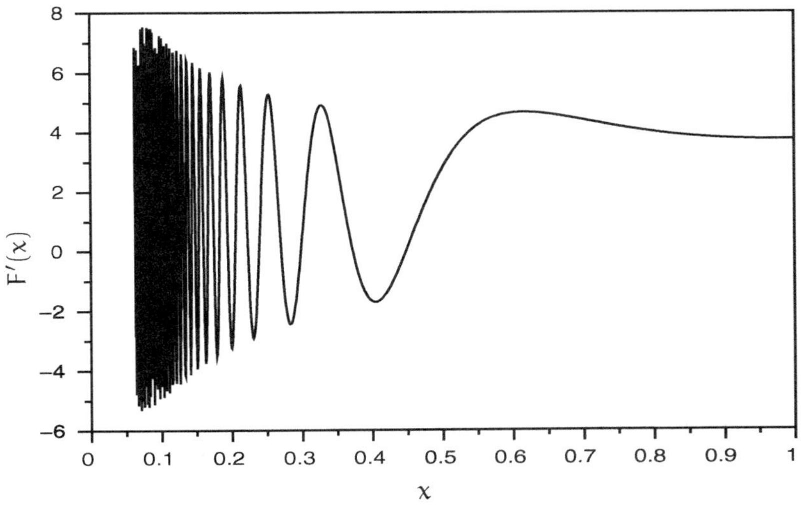

Figure 2.5. *The derivative of the function given in Example* 2.14.

Definition 2.4 (more notation). *The notation (Bender and Orszag,* 1981, §3.4*) for*

- "$f(x)$ *is* **asymptotic** *to* $g(x)$ *as* $x \to x_0$" *is*

$$f(x) \sim g(x) \text{ as } x \to x_0\,, \quad \textit{meaning} \quad \lim_{x\to x_0} \frac{f(x)}{g(x)} = 1\,.$$

In such limits, the limit point x_0 *may be* 0, ∞, *or a one sided limit such as* $0+$.

Example 2.16 (asymptotic relations).

1. $x \ll \log x$ as $x \to 0$;
2. $e^{-1/x} \ll x^2$ as $x \to 0+$ (but *not* as $x \to 0-$);
3. $(\log x)^3 \ll x^{1/5}$ as $x \to +\infty$;
4. $e^x \sim e^x + x$ as $x \to +\infty$;
5. $\sin x \sim x$ as $x \to 0$.

A critical property of a power series is that it is the sum of smaller and smaller terms. To generalize such a sum to an asymptotic series, we introduce an infinite sequence of "smaller and smaller" functions. Let $\phi_n(x)$, $n = 0, 1, 2, \ldots$, be an infinite sequence of functions such that each is much smaller than its predecessor; that is, $\phi_n(x) \ll \phi_{n-1}(x)$ as $x \to x_0$ for $n = 1, 2, \ldots$. For example:

- Taylor series arise from choosing $\phi_n(x) = (x - x_0)^n/n!$ as $1 \gg (x - x_0) \gg (x - x_0)^2/2 \gg \cdots$ as $x \to x_0$.
- Laurent series arise from choosing $\phi_n(x) = 1/x^{n+1}$ as $x \to \infty$.
- We could choose the sequence of fractional powers $x^{-2/3} \gg x^{-1/6} \gg x^{1/3} \gg x^{5/6} \gg x^{4/3} \gg \cdots$ as $x \to 0+$.

- Perhaps involve the logarithm as in the sequence $1 \gg x \log x \gg x \gg x^2 \log^2 x \gg x^2 \log x \gg x^2 \gg x^3 \log^3 x \gg \cdots$ as $x \to 0$.

Definition 2.5 (asymptotic series). *The infinite sum* $\sum_{n=0}^{\infty} a_n \phi_n(x)$ *is said to be an* **asymptotic series** *to a function* $f(x)$ *as* $x \to x_0$, *written* $f(x) \sim \sum_{n=0}^{\infty} a_n \phi_n(x)$, *when the error in every partial sum is always much smaller than the size of the last term included:*

$$f(x) - \sum_{n=0}^{m} a_n \phi_n(x) \ll \phi_m(x) \quad \text{as } x \to x_0 \quad \text{for all } m. \tag{2.11}$$

A partial sum of an asymptotic series is usually called an **asymptotic approximation.**

2.3.4 ▪ Properties of asymptotic series

Uniqueness If a function possesses an asymptotic series, then the series is *unique* for that sequence of $\phi_n(x)$. In principle, the coefficients of the asymptotic series are uniquely and recursively determined by the limits

$$a_m = \lim_{x \to x_0} \frac{f(x) - \sum_{n=0}^{m-1} a_n \phi_n(x)}{\phi_m(x)}.$$

However, the uniqueness is for one given sequence of $\phi_n(x)$. For example, the coefficients differ in the two asymptotic approximations, as $x \to 0$, of

$$\begin{aligned} \tan x &\sim x + \tfrac{1}{3}x^3 + \tfrac{2}{15}x^5 + \cdots \\ &\sim \sin x + \tfrac{1}{2}\sin^3 x + \tfrac{3}{8}\sin^5 x + \cdots . \end{aligned}$$

Nonuniqueness There are many functions with the same asymptotic series. Two functions that differ by something with a zero asymptotic series, $\sum_{n=0}^{\infty} 0\phi_n(x)$, have the same asymptotic series. That is, any given asymptotic series represents a whole class of functions. However, the difference between members in the class of functions is much smaller than all elements in the sequence $\phi_n(x)$.

Equating coefficients We may occasionally write

$$\sum_{n=0}^{\infty} a_n \phi_n(x) \sim \sum_{n=0}^{\infty} b_n \phi_n(x) \quad \text{as } x \to x_0,$$

even though the definition only specifies asymptoticity between a function and a series. We take this notation to mean that the two series $\sum_{n=0}^{\infty} a_n \phi_n(x)$ and $\sum_{n=0}^{\infty} b_n \phi_n(x)$ are asymptotic to the same class of functions as $x \to x_0$. By the uniqueness of asymptotic coefficients, it follows that $a_n = b_n$ for all n. That is, *we may equate coefficients of asymptotic series.*

Arithmetic operations follow naturally A further consequence of the uniqueness of coefficients is that asymptotic series may be summed, subtracted, multiplied, or divided in the usual manner (Bender and Orszag, 1981, p. 125).

Calculus is trickier An asymptotic series of a function $f(x)$ can be integrated term by term when the function $f(x)$ is integrable near x_0 (and the sequence functions $\phi_n(x)$ are all positive).

However, asymptotic series cannot always be differentiated. As before, the problem is that very small terms may nonetheless oscillate wildly.

In summary, when you know that the function and the requisite derivatives all have an asymptotic series, then the uniqueness of asymptotic series ensures that formal manipulations of the infinite sums of the series yield correct answers (Bender and Orszag, 1981, p. 127).

2.4 ▪ Exercises

Exercise 2.1. Modify the iteration of Example 2.2 to find the Taylor series solution to the ODE $y'' - 2y = 0$ such that $y(0) = 1$ and $y'(0) = 0$ using Reduce and to errors $\mathcal{O}(x^{10})$.

Answer: $y = 1 + x^2 + \frac{1}{6}x^4 + \frac{1}{90}x^6 + \frac{1}{2520}x^8 + \mathcal{O}(x^{10})$.

Exercise 2.2. Similarly use Reduce to find the Taylor series solution to errors $\mathcal{O}(x^{15})$ to the ODE $y'' + xy = 0$ such that $y(0) = a$ and $y'(0) = b$. The Taylor series multiplied by a and b are those of two linearly independent solutions to *Airy's equation*, mentioned by Kreyszig (2011, p. 198, pp. 958–60).

Answer: $y = b(x - \frac{1}{12}x^4 + \frac{1}{504}x^7 - \frac{1}{45360}x^{10} + \frac{1}{7076160}x^{13}) + a(1 - \frac{1}{6}x^3 + \frac{1}{180}x^6 - \frac{1}{12960}x^9 + \frac{1}{1710720}x^{12}) + \mathcal{O}(x^{15})$.

Exercise 2.3. Use Reduce to find the Taylor series of the solution to $y' = \cos(x)y$ such that $y(0) = 1$ to errors $\mathcal{O}(x^{10})$. Hint: Replace $\cos x$ in the code by its Taylor series; you may use `factorial(n)` to compute $n!$. Compare your answer to that of the exact analytic solution obtained by recognizing that the ODE is separable.

Answer: $y = 1 + x + \frac{1}{2}x^2 - \frac{1}{8}x^4 - \frac{1}{15}x^5 - \frac{1}{240}x^6 + \frac{1}{90}x^7 + \frac{31}{5760}x^8 + \frac{1}{5670}x^9 + \mathcal{O}(x^{10})$.

Exercise 2.4. Modify the analysis of Example 2.4 to use Reduce to find the Taylor series solution to errors $\mathcal{O}(x^{10})$ of the nonlinear ODE $y'' = 6y^2$ such that $y(0) = 1$ and $y'(0) = b$, where b is some arbitrary constant. Note: Because this is a nonlinear ODE, the solution depends nonlinearly upon b, in contrast to linear ODEs, which would show a linear dependence only.

Answer: $y = 1 + 3x^2 + 3x^4 + 3x^6 + \frac{18}{7}x^8 + b(x + 2x^3 + 3x^5 + \frac{24}{7}x^7 + \frac{25}{7}x^9) + b^2(\frac{1}{2}x^4 + x^6 + \frac{45}{28}x^8) + b^3(\frac{1}{7}x^7 + \frac{5}{14}x^9) + \mathcal{O}(x^{10})$.

Exercise 2.5. Use Reduce to find the Taylor series solution of the nonlinear ODE $y'' = (1 + x)y^3$ to errors $\mathcal{O}(x^{10})$ such that $y(0) = 2$ and $y'(0) = -3$.

Answer: $y = 2 - 3x + 4x^2 - \frac{14}{3}x^3 + \frac{11}{2}x^4 - \frac{25}{4}x^5 + \frac{211}{30}x^6 - \frac{47}{6}x^7 + \frac{2081}{240}x^8 - \frac{41243}{4320}x^9 + \mathcal{O}(x^{10})$.

Exercise 2.6. Modify the Reduce computer algebra of Example 2.5 to find the Taylor series of the general solution to Legendre's equation in the specific case $k = 3$ to an error $\mathcal{O}(x^{10})$.

Answer: $y = a(1 - \frac{3}{2}x^2 - \frac{3}{8}x^4 - \frac{17}{80}x^6 - \frac{663}{4480}x^8) + b(x - \frac{1}{6}x^3 - \frac{3}{40}x^5 - \frac{27}{560x}x^7 - \frac{159}{4480}x^9) + \mathcal{O}(x^{10})$.

Exercise 2.7. Modify the arguments and the Reduce computer algebra of Example 2.5 to find the Taylor series, to an error $\mathcal{O}(x^{10})$, of a *general* solution to the following three ODEs:

1. $(x-2)y' = xy$,
2. $(1-x^2)y' = 2xy$,
3. $y'' - 4xy' + (4x^2-2)y = 0$.

Exercise 2.8. (Laguerre polynomials). Use Reduce to iteratively construct the power series solution to the Laguerre equation $xy'' + (1-x)y' + ny = 0$ with initial condition $y(0) = 1$ (only one condition is needed, as other solutions are unbounded at $x = 0$). First, for fixed parameter $n = 2$ find $y = 1 - 2x + \frac{1}{2}x^2$. Second, for general n factorize the coefficients, at least as far as computed, and observe that the power series truncates for integer n to give so-called Laguerre polynomials. Hint: $xy'' + y' = (xy')'$.

Exercise 2.9. Code an operator that constructs particular solutions of the differential equation $x^2y'' + 2y = f$ for polynomial right-hand sides f.

Exercise 2.10. Code an operator that constructs solutions of the differential equation $x^2y'' + xy' - 4y = f$ for polynomial right-hand sides f. The subsidiary conditions are that solutions must be well behaved near $x = 0$, whereas $y = 0$ at $x = 1$. Assume that the polynomial f never has a quadratic, x^2, component.

Exercise 2.11. Modify the computer algebra code for Example 2.7 to find the Taylor series, to errors $\mathcal{O}(x^{10})$, of the well-behaved solution of the nonlinear ODE $x^2y'' + x^2y' + xy^3 = 0$ such that $y(0) = 2$.

Answer: $y = 2 - 4x + 4x^2 - \frac{32}{9}x^3 + \frac{26}{9}x^4 - \frac{56}{25}x^5 + \frac{3404}{2025}x^6 - \frac{832}{675}x^7 + \frac{1199}{1350}x^8 - \frac{4142}{6561}x^9 + \mathcal{O}(x^{10})$.

Exercise 2.12. Use Reduce to help you find the power series about $x = 0$, to errors $\mathcal{O}(x^{10})$, of the well-behaved solutions of the linear ODE $x^2y'' + x^3y' + (x^2-2)y = 0$. Hint: $x^2y'' - 2y = (x^4(y/x^2)')'$. Then modify your Reduce code to find the power series of the one-parameter family of well-behaved solutions to the nonlinear ODE $x^2y'' + x^3y' + (x^2-2)y + y^2 = 0$.

Answer: Well-behaved solutions of the linear ODE are $y = a(x^2 - \frac{3}{10}x^4 + \frac{3}{56}x^6 - \frac{1}{144}x^8) + \mathcal{O}(x^{10})$ for any multiplicative constant a. Nonlinear solutions, parametrized by the same a, are $y = a(x^2 - \frac{3}{10}x^4 + \frac{3}{56}x^6 - \frac{1}{144}x^8) + a^2(-\frac{1}{10}x^4 + \frac{11}{280}x^6 - \frac{661}{75600}x^8) + a^3(\frac{1}{140}x^6 - \frac{11}{3150}x^8) - \frac{17}{37800}a^4x^8 + \mathcal{O}(x^{10})$.

Exercise 2.13. Use Reduce to help find the power series about $x = 0$, to errors $\mathcal{O}(x^{20})$, of the well-behaved solutions of the ODE $xy'' + 3y' + 3x^2y = 0$. Hint: $xy'' + 3y' = (x^3y')'/x^2$.

Answer: Well-behaved solutions are proportional to $y = 1 - \frac{1}{5}x^3 + \frac{1}{80}x^6 - \frac{1}{2640}x^9 + \frac{1}{147840}x^{12} - \frac{1}{12566400}x^{15} + \frac{1}{1507968000}x^{18} + \mathcal{O}(x^{20})$.

Algorithm 2.8 *Reduce code for Exercise* 2.15.

```
y:=2*x;
let x^20=>0;
for iter:=1:99 do begin
    res:=(1-x^3)*df(y,x,2)-(y^2-x^2)*df(y,x);
    res:=-int(int(res,x),x);
    write y:=y+res;
    if res=0 then iter:=1000000;
end;
```

Exercise 2.14. Find the power series expansions about $x = 0$, to errors $\mathcal{O}(x^{10})$, for the two-parameter general solution to the linear ODE $x^2y'' - \sin(x)y' + y = 0$, with the aid of computer algebra. Hint: Expand $\sin x$ in a Taylor series and write $x^2y'' - xy'$ in the form $x^{2-p}(x^p y')'$.

Answer: $y = (a + b\log x)(x - \frac{1}{24}x^3 + \frac{7}{3840}x^5 - \frac{89}{1161216}x^7 + \frac{6721}{2229534720}x^9) + b(\frac{1}{23040}x^5 + \frac{11}{11612160}x^7 - \frac{5951}{44590694400}x^9) + \mathcal{O}(x^{10})$.

Exercise 2.15. Algorithm 2.8 lists some Reduce code to iteratively find a power series solution to an ODE. What is the differential equation it purports to solve? Its initial conditions? What is the value of y after the first iteration of the loop? What is the order of error in the computed power series after the loop terminates?

Answer: $(1-x^3)y''-(y^2-x^2)y' = 0$, such that $y(0) = 0$ and $y'(0) = 2$. $y^{(1)} = 2x+\frac{1}{2}x^4$. The ultimate error is $\mathcal{O}(x^{20})$.

Exercise 2.16. Use algebra to show that the divergent Stieltjes series (2.8) formally solves the Stieltjes ODE (2.7). By considering $d/dt\left[te^{-t}/(1+xt)^2\right]$, show that the Stieltjes integral (2.9) satisfies the Stieltjes ODE (2.7). Argue that the eight-term partial sum of the Stieltjes series (2.8) approximates the Stieltjes integral to four decimal places for $|x| < 0.08$.

Exercise 2.17. For any specified positive x, estimate the number of terms in the Stieltjes series that will give the smallest error bound—the optimal truncation. Hence, and using Stirling's formula for a factorial, deduce that the Stieltjes series can be summed to an error $|\varepsilon_{\text{optimal}}| \approx \sqrt{\frac{2\pi}{x}}e^{-1/x}$.

Exercise 2.18. Argue that (Bender and Orszag, 1981, Ex. 3.64)

$$\int_0^\infty \frac{e^{-t}}{1+xt^2}\,dt \sim \sum_{n=0}^{\infty}(-1)^n(2n)!\,x^n \quad x \to 0+.$$

Exercise 2.19. Let the sequence $\phi_n(x)$, $n = 0, 1, 2, \dots$, satisfy $\phi_n \ll \phi_{n-1}$ as $x \to x_0$ for all n.

1. Show that if $f(x) \sim \sum_{n=0}^{\infty} a_n\phi_n(x)$ and $g(x) \sim \sum_{n=0}^{\infty} b_n\phi_n(x)$, then $\alpha f + \beta g \sim \sum_{n=0}^{\infty}(\alpha a_n + \beta b_n)\phi_n(x)$ as $x \to x_0$.
2. Let $\psi_n(x) = \int_{x_0}^{x} \phi_n(t)\,dt$ and suppose all the functions $\phi_n(x)$ are positive near x_0. Deduce that $f(x) \sim \sum_{n=0}^{\infty} a_n\phi_n(x)$ as $x \to x_0$ implies $\int_{x_0}^{x} f(t)\,dt \sim \sum_{n=0}^{\infty} a_n\psi_n(x)$ as $x \to x_0$.

Chapter 3

A normal form of oscillations illuminates their character

Contents

Rapid oscillations are a simple form of dynamics that we understand. Typically we want to know the long-term evolution of the oscillations: do they decay in amplitude? Do they persist for all time? Do they blow up? Does their frequency evolve over long times? For example, a wave approaching a beach is recognizably a wave up to the moment it breaks into tumultuous surf. Yet as waves propagate toward the beach they slow down, change direction, and simultaneously increase in amplitude (like depth$^{-1/4}$) to breaking, yet retain the same frequency. Normal-form coordinate transformations help us to uncover such behavior.

The fundamental modeling question here is: how do we "look beyond" the rapid oscillations to the relatively slow long-term dynamics of the amplitude? How do we resolve the rapid oscillations, yet describe the long term? The answer developed in this chapter is to choose a coordinate transform that separates the rapid from the slow.

Classic nonlinear oscillators yield their secrets with these normal-form coordinate transform methods. This chapter investigates the frequency shift in the Duffing oscillator (Hinch, 1991, §6.2), the subtleties of Morrison's problem; the appearance of a limit cycle in the van der Pol oscillator (Hinch, 1991, §7.1), and the stability of forced oscillations in Mathieu's ODE (Hinch, 1991, §7.2). Section 3.1 starts with a basic system.

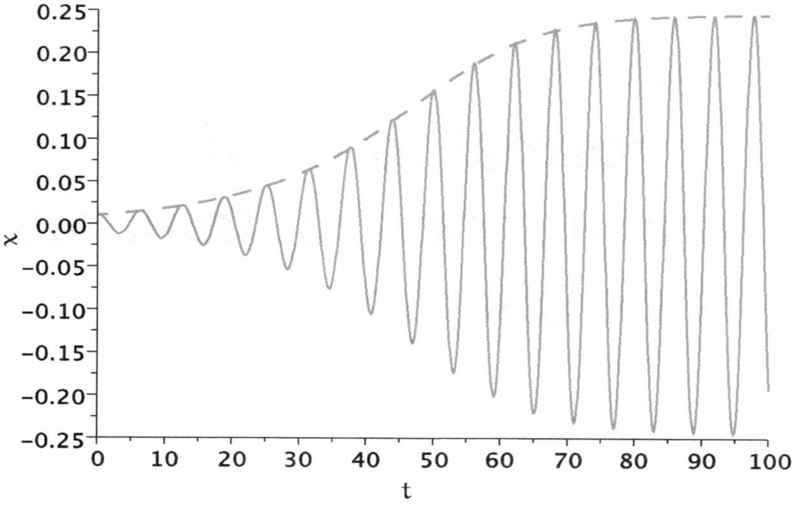

Figure 3.1. *A solution* $x(t)$ *of the simple nonlinear oscillator* (3.1) *for parameter* $\epsilon = 0.06$. *The oscillation amplitude grows until it reaches about* 0.25. *This growth matches the amplitude* $r(t)$ *of the oscillations predicted by the model* (3.3).

3.1 ▪ Simple evolving nonlinear oscillations

Consider the oscillations described by the coupled pair of nonlinear ODEs

$$\dot{x} = \epsilon x - y - (x^2 + y^2)(x + y), \tag{3.1a}$$
$$\dot{y} = x + \epsilon y + (x^2 + y^2)(x - y), \tag{3.1b}$$

where ϵ is some small positive parameter. See the plot of an example solution $x(t)$ in Figure 3.1.

First, the amplitude and frequency may be extracted from the linear dynamics. Neglect the products and cross-products of x, y, and the small ϵ to describe the linear part of the dynamics:

$$\dot{x} = -y \quad \text{and} \quad \dot{y} = x\,.$$

These describe simple harmonic oscillations. The general solution is $x = A\cos(t + B)$ and $y = A\sin(t + B)$ for any constants A and B. The key step is to alternatively describe these oscillations in terms of their amplitude r and phase θ as

$$x = r\cos\theta \quad \text{and} \quad y = r\sin\theta\,, \tag{3.2a}$$
$$\text{where } \dot{r} = 0 \quad \text{and} \quad \dot{\theta} = 1\,. \tag{3.2b}$$

This is completely equivalent to the general solution because $r = A$ solves $\dot{r} = 0$, whereas $\theta = t + B$ solves $\dot{\theta} = 1$. The point of rewriting the dynamics in terms of amplitude r and phase θ is that we separate the shape of the oscillations, the shapes $x = r\cos\theta$ and $y = r\sin\theta$, from the evolution of the characteristics of the oscillations, namely $\dot{r} = 0$ and the frequency $\dot{\theta} = 1$. The aim of the subsequent sections is to form this useful view without actually constructing, nor necessarily knowing, the details of such solutions.

Second, explore the evolution of the amplitude r and phase θ of the nonlinear oscillations. These amplitude and phase variables are polar coordinates for the xy-plane,

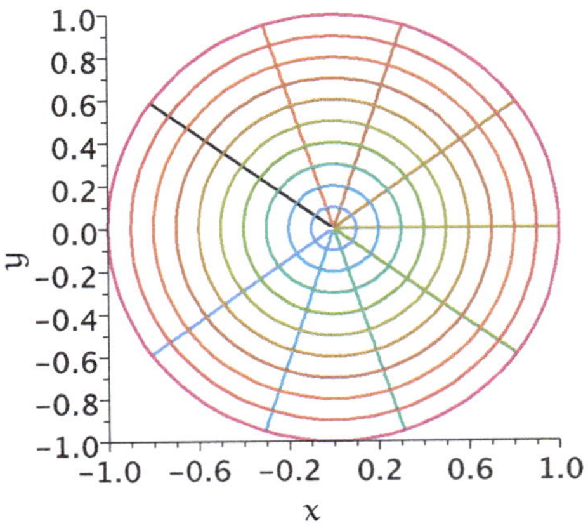

Figure 3.2. *Coordinate curves of the polar coordinates* (r, θ), *equation* (3.2a), *that simplify the fast dynamics of the oscillator* (3.1a)–(3.1a) *to the simple, slow, amplitude-phase dynamics* (3.3).

as plotted in Figure 3.2. Change coordinates by rewriting the nonlinear ODEs (3.1a) and (3.1b) in terms of amplitude r and phase angle θ variables instead of x and y. The aim is to find the evolution of the amplitude that we readily see in Figure 3.1 without explicitly constructing a solution. That is, substitute $x = r\cos\theta$ and $y = r\sin\theta$, where both amplitude r and phase angle θ vary with time. Exercise 3.1 asks you to do the straightforward but slightly messy algebra to discover the ODEs

$$\dot{r} = \epsilon r - r^3 \quad \text{and} \quad \dot{\theta} = 1 + r^2 . \tag{3.3}$$

From the amplitude equation $\dot{r} = \epsilon r - r^3$ we can readily see that there are two equilibria in r, the fixed points $r = 0$ and $r = \sqrt{\epsilon}$. Deduce from the sign of the right-hand side of $\dot{r} = \epsilon r - r^3$ that all solutions evolve to the stable $r = \sqrt{\epsilon}$—note this evolution in the smooth curve of Figure 3.1. That is, all oscillations eventually have amplitude $\sqrt{\epsilon}$. But the frequency $\dot{\theta} = 1 + r^2$ depends upon the amplitude. Thus the oscillations change their frequency in time but eventually settle on the frequency $\dot{\theta} = 1 + \epsilon$ when the amplitude $r = \sqrt{\epsilon}$. Nonlinear oscillations generally have a frequency correction.

Third, the transform from the original x and y variables to new r and θ variables empowers us to easily interpret the long-term evolution of the oscillations. In this sense, this transformation is to a normal form of the differential equations (Takens, 1974, e.g.).

The reason we make progress in this straightforward example is that the shape of the oscillations does not change with amplitude. Figure 3.3 shows that the oscillations are perfectly circular, $x = r\cos\theta$ and $y = r\sin\theta$, except for the slow evolution of amplitude r. In general, nonlinear oscillations are deformed circles. This deformation is a major complication, which we explore and overcome in the next example of the Duffing oscillator.

3.2 ▪ Duffing oscillator has a nonlinear frequency correction

Figure 3.4 shows some examples of nonlinear oscillations governed by Duffing's ODE (Bender and Orszag, 1981, §11.1–2)

$$\ddot{x} + x - x^3 = 0 . \tag{3.4}$$

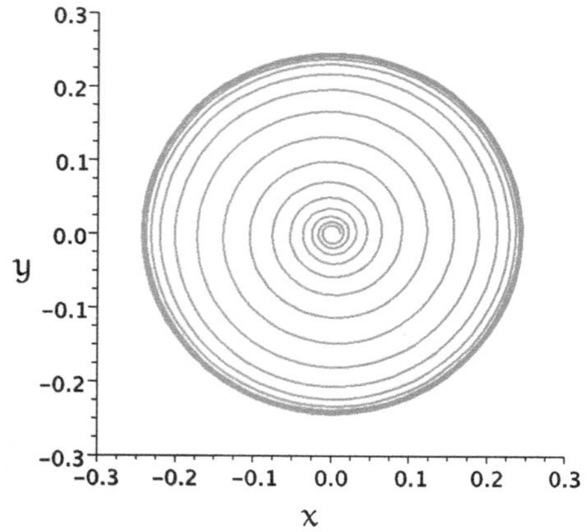

Figure 3.3. *A trajectory in the* xy*-plane of the simple nonlinear oscillator* (3.1) *for parameter* $\epsilon = 0.06$. *The oscillation spirals to a perfect circle.*

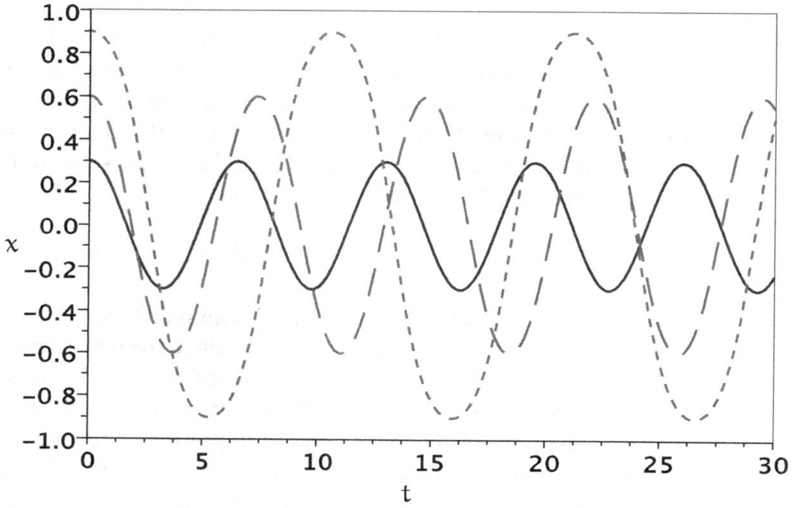

Figure 3.4. *Three solutions* $x(t)$ *of Duffing's ODE* (3.4) *show the perpetual oscillations for each of the three amplitudes.*

When the amplitude of the oscillations is small enough, the ODE reduces to the classic linear oscillator $\ddot{x} + x = 0$. This has general solution $x(t) = A\cos(t + B)$ for arbitrary constants A and B. What happens when we take into account the effects of the nonlinear terms x^3?

The answer here is that the frequency of the oscillations changes a little and the shape of the oscillations changes a bit as you can see in Figure 3.5. To see this algebraically, transform the Duffing equation to radius and phase variables. *Such a transform of the variables is an example of the very general and powerful normal-form transformation.*

As a preliminary step, the general linear solution $x(t) = A\cos(t+B)$ may be written as $x(t) = r(t)\cos\theta(t)$, where $\dot{r} = 0$ and $\dot{\theta} = 1$. The relation $x = r\cos\theta$ is a coordinate transform from variable x (and $\dot{x}$) to amplitude (radius) and phase angle coordinates r and θ, respectively. Under this transformation the linear oscillation ODE $\ddot{x} + x = 0$

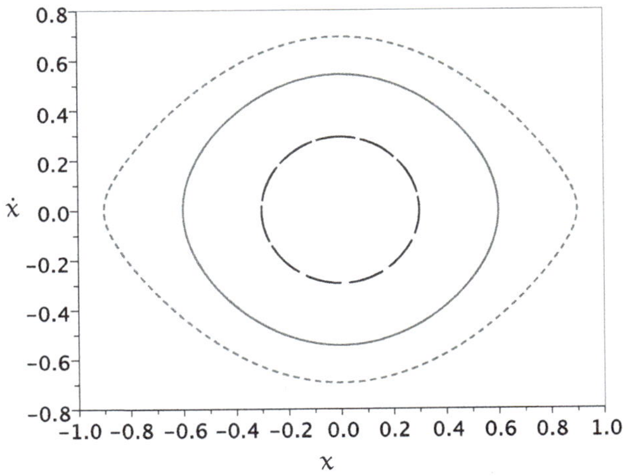

Figure 3.5. *Three trajectories in the* $x\dot{x}$*-plane of Duffing's ODE* (3.4)*. Note in the largest oscillation the noncircular shape that is crucial to the nonlinear dynamics.*

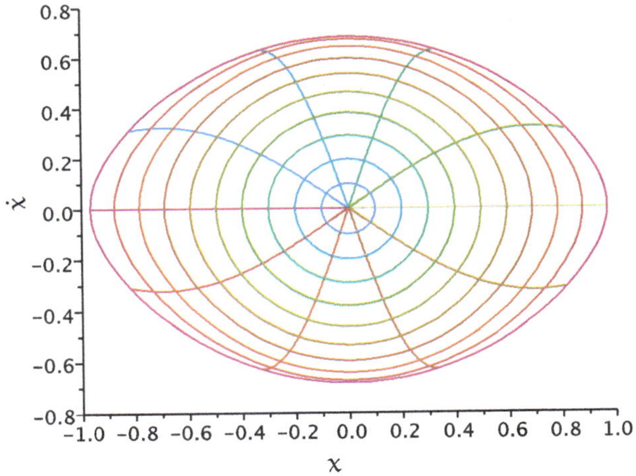

Figure 3.6. *Coordinate curves of the amplitude-phase coordinates* (r, θ)*, equation* (3.8a)*, that simplify the fast dynamics of the Duffing oscillator* (3.4) *to the simple, slow, amplitude-phase dynamics* (3.8b)*.*

becomes the simple pair of ODEs $\dot{r} = 0$ and $\dot{\theta} = 1$. From these last two ODEs, read off that the linear oscillator has constant amplitude r and a fixed frequency $\omega = \dot{\theta} = 1$. In the amplitude-phase variables it is easy to interpret the solution of the ODE without actually solving the ODE; we only transform the ODE.

Now explore the complicated algebra of the similar transform of the nonlinear Duffing ODE (3.4). Assume that the amplitude r of the oscillations is small enough so that in this initial analysis we neglect terms quartic or higher in r.

Transform to amplitude-phase variables We seek the new coordinate system, plotted in Figure 3.6, that simplifies our description of the Duffing oscillations. The fundamental transformation to amplitude-phase variables seeks solutions of Duffing's ODE (3.4) in the form of a correction to linear oscillations $x \approx r\cos t$; namely we seek

$$x = r\cos\theta + \hat{x}(r,\theta) \tag{3.5a}$$
$$\text{such that } \dot{r} = \hat{g}(r) \quad\text{and}\quad \dot{\theta} = 1 + \hat{h}(r)\,, \tag{3.5b}$$

where the corrections, denoted by a hat, are small—they are quadratic in amplitude r *relative* to the leading terms. This generalizes the transform of the linear oscillator: we allow

- the nonlinear terms to change the shape of the oscillations through $\hat{x}(r,\theta)$, but the change is only by a small amount;
- the amplitude r to evolve slowly in time through $\hat{g}(r)$; and
- the phase to evolve at a slightly different rate through $\hat{h}(r)$, that is, the frequency may change a little with amplitude r.

Although we only use the dynamic variable x here, recall that there is a companion variable $y = \dot{x} = -r\sin\theta + \cdots$, plotted as the vertical axis in Figure 3.6, to make the conversion to (r,θ) variables a genuine coordinate transform from (x,y) variables. This coordinate transformation insists that the right-hand sides of (3.5b) be only a function of the slowly varying amplitude r. Also the $\hat{x}$ of (3.5a) depends upon phase angle θ *only* through trigonometric (periodic) functions. In this way the dynamics of Duffing's ODE (3.4) will be *approximated in a manner that is both easy to interpret (slowly varying) and uniformly valid in time ($\hat{x}$ is bounded).*

Substitute the transformation (3.5a) into Duffing's ODE (3.4) using the evolution equations (3.5b) to replace time derivatives. Since the corrections we seek are small, omit products of corrections, as such products will be very small. The most complicated parts are the time derivatives:

$$\begin{aligned}\dot{x} &= \dot{r}\cos\theta - r\sin\theta\,\dot{\theta} + \hat{x}_r\dot{r} + \hat{x}_\theta\dot{\theta}\\ &= \hat{g}\cos\theta - r\sin\theta - r\sin\theta\,\hat{h} + \hat{x}_r\hat{g} + \hat{x}_\theta + \hat{x}_\theta\hat{h}\\ &\approx \hat{g}\cos\theta - r\sin\theta - r\sin\theta\,\hat{h} + \hat{x}_\theta\,,\end{aligned} \tag{3.6}$$

upon neglecting the very small products of the small corrections. Differentiate this again to approximate

$$\begin{aligned}\ddot{x} &= \hat{g}_r\dot{r}\cos\theta - \hat{g}\sin\theta\dot{\theta} - \dot{r}\sin\theta - r\cos\theta\,\dot{\theta} - \dot{r}\sin\theta\,\hat{h}\\ &\quad - r\cos\theta\,\dot{\theta}\hat{h} - r\sin\theta\,\hat{h}_r\dot{r} + \hat{x}_{r\theta}\dot{r} + \hat{x}_{\theta\theta}\dot{\theta}\\ &= \hat{g}_r\hat{g}\cos\theta - \hat{g}\sin\theta\,(1+\hat{h}) - \hat{g}\sin\theta - r\cos\theta\,(1+\hat{h})\\ &\quad - \hat{g}\sin\theta\,\hat{h} - r\cos\theta\,(1+\hat{h})\hat{h} - r\sin\theta\,\hat{h}_r\hat{g} + \hat{x}_{r\theta}\hat{g}\\ &\quad + \hat{x}_{\theta\theta}(1+\hat{h})\\ &\approx -2\hat{g}\sin\theta - r\cos\theta - 2r\cos\theta\,\hat{h} + \hat{x}_{\theta\theta}\,,\end{aligned}$$

upon neglecting the very small products of corrections. Then Duffing's ODE (3.4) becomes approximately

$$\begin{aligned}&-2\hat{g}\sin\theta - r\cos\theta - 2r\cos\theta\,\hat{h} + \hat{x}_{\theta\theta} + r\cos\theta + \hat{x}\\ &-\tfrac{1}{4}r^3(\cos 3\theta + 3\cos\theta) = 0\,,\end{aligned}$$

upon also omitting products of small corrections $\hat{x}$ and the relatively small amplitude r (as in the omission of the term $3r^2\cos^2\theta\,\hat{x}$). The $r\cos\theta$ terms cancel, as these satisfy the linear oscillator part of Duffing's ODE (3.4):

$$-2\hat{g}\sin\theta - 2r\cos\theta\,\hat{h} + \hat{x}_{\theta\theta} + \hat{x} - \tfrac{1}{4}r^3(\cos 3\theta + 3\cos\theta) = 0\,. \qquad (3.7)$$

This equation governs how to account for the nonlinearity x^3 in Duffing's ODE (3.4).

Aim for simple evolution Recall that $\hat{g}$ and $\hat{h}$ modify the evolution of the amplitude r and the phase θ: we aim to keep these as simple as possible in order to keep the new "coarse-scale" description of the evolution as simple as possible. Thus we aim to choose the shape correction $\hat{x}$ to cancel with as many other terms in equation (3.7) as possible.

For example, to cancel the term $-\frac{1}{4}r^3\cos 3\theta$ we choose $-\frac{1}{32}r^3\cos 3\theta$ to be a component of $\hat{x}$. This choice modifies the shape of the oscillations.

But we cannot choose $\hat{x}$ to eliminate the term $-\frac{3}{4}r^3\cos\theta$ in (3.7). The reason is that any general solution of $\hat{x}_{\theta\theta} + \hat{x} - \frac{3}{4}r^3\cos\theta = 0$ involves the particular solution $\hat{x} = \frac{3}{8}r^3\theta\sin\theta$. Such a term is not allowed in $\hat{x}$ because it contains a bare θ factor, that is, one not inside a trigonometric function. We cannot tolerate such a so-called secular term in $\hat{x}$ because it would make the supposedly small correction $\hat{x}$ large for large phase angle θ—the term $\theta\sin\theta$ grows without bound. We have no choice; we must instead cancel the $-\frac{3}{4}r^3\cos\theta$ term in (3.7) through modifying the evolution of the amplitude or phase by choosing $\hat{g}$ or $\hat{h}$ appropriately. The corrections $\hat{g}$ and $\hat{h}$ must not have a phase angle θ dependence, as otherwise r and $\dot{\theta}$ would evolve quickly through a dependence upon phase θ. Such a dependence would make interpretation of the transformed equations impossible. Thus the component in $\cos\theta$ in equation (3.7) must be canceled by a component in the correction $\hat{h}$ to the phase evolution. Consequently choose corrections $\hat{h} = -\frac{3}{8}r^2$ and $\hat{g} = 0$.

Interpret the amplitude-phase form Our nonlinear coordinate transform, plotted in Figure 3.6, is then

$$x \approx r\cos\theta - \tfrac{1}{32}r^3\cos 3\theta \qquad (3.8a)$$

$$\text{and } y = \dot{x} \approx -(r - \tfrac{3}{8}r^3)\sin\theta + \tfrac{3}{32}r^3\sin 3\theta$$

$$\text{such that } \dot{r} \approx 0 \quad \text{and} \quad \dot{\theta} \approx 1 - \tfrac{3}{8}r^2\,. \qquad (3.8b)$$

The nonlinearity in Duffing's ODE (3.4) changes the shape of the oscillations through equation (3.8a) (see Figure 3.5); the amplitude of the oscillations persists unchanged for all time as $\dot{r} = 0$,[13] but the frequency is slightly modified through (3.8b)—Figure 3.4 shows the noticeable decrease in frequency for larger-amplitude oscillations.

Phrasing modeling as a coordinate transform has the following important consequence. Here view the coordinate transform (3.8a) as an exact transform, not an approximation—with coordinate curves plotted in Figure 3.6. In this coordinate system the exact Duffing dynamics evolves r and θ. So far there is no approximation, just a reparametrization of

[13]That the Duffing oscillations have constant amplitude connects to conservation inherent in the Duffing oscillator. Multiply Duffing's ODE (3.4) by $\dot{x}$ and integrate in time to deduce that solutions must satisfy $\frac{1}{2}\dot{x}^2 + \frac{1}{2}x^2 - \frac{1}{4}x^4 = E$, where E is an integration constant. Since $\frac{1}{2}\dot{x}^2$ looks like a kinetic energy term, and $V(x) = \frac{1}{2}x^2 - \frac{1}{4}x^4$ could be a potential energy, we identify E as the conserved total energy. That we can integrate Duffing's ODE (3.4) in this way shows that "energy" is conserved, and hence we expect the oscillations to have constant amplitude.

the dynamics. The *only* approximation is when we truncate the description of $\dot r$ and $\dot\theta$ to equation (3.8b). As the only place an approximation takes place, it is easy to control errors in modeling by using such coordinate transforms.

This example introduces how a normal-form coordinate transformation empowers us to accurately interpret the long-term evolution of nonlinear oscillations.

The following theorem (Takens, 1974), based upon the construction we employ for Duffing's ODE (3.4), assures us that the amplitude and frequency model for nonlinear oscillators is generally applicable.

Theorem 3.1 (polar normal form). *Autonomous, second-order, nonlinear oscillators satisfying ODEs of the form*

$$\ddot x + \omega^2 x + f(x, \dot x) = 0 \tag{3.9}$$

for smooth, C^p, strictly nonlinear functions f[14] *can always be put into the normal form of amplitude and phase angle variables, r and θ, respectively:*

$$\begin{aligned} x &= r\cos\theta + X(r,\theta) + o(r^p) \quad \textit{such that} \\ \dot r &= g(r) + o(r^p) \quad \textit{and} \quad \dot\theta = \omega + h(r) + o(r^{p-1}), \end{aligned} \tag{3.10}$$

where X, g, and h are some strictly nonlinear functions of amplitude r and X is 2π-periodic in angle θ.

Proof. With the slight change that the frequency $\dot\theta = \omega + \cdots$, the method of construction of the normal form for Duffing's ODE (3.4) generalizes to any smooth nonlinear function f.

Use induction. Suppose we have found a coordinate transform $x = r\cos\theta + X(r,\theta)$ such that $\dot r = g(r)$ and $\dot\theta = \omega + h(r)$ satisfy the ODE (3.9) to a residual Res $= \mathcal{O}(r^k)$, $k \leqslant p$. Seek a correction to the coordinate transform to $x = r\cos\theta + X(r,\theta) + \hat x(r,\theta)$ such that $\dot r = g(r) + \hat g(r)$ and $\dot\theta = \omega + h(r) + \hat h(r)$ that reduces the residual, that is, increases the order of the residual. Substitute into the ODE (3.9), and linearize in the corrections to derive

$$-2\omega\hat g\sin\theta - 2\omega r\cos\theta\,\hat h + \omega^2(\hat x_{\theta\theta} + \hat x) + \text{Res} = 0\,.$$

Since the nonlinear function f is smooth, the residual is smooth, at least at order k in amplitude r, and so has a Fourier series: Res $= \sum_n \left[a_n\cos(n\theta) + b_n\sin(n\theta)\right]$. Choose $\hat g = b_1/(2\omega)$, $\hat h = a_1/(2\omega r)$, and $\hat x = \sum_{n\neq 1}\left[a_n\cos(n\theta) + b_n\sin(n\theta)\right]/(n^2-1)/\omega^2$ to ensure that the new coordinate transform has residual Res $= o\left(r^k\right)$. This proceeds for as long as the nonlinearity f is smooth, $k \leqslant p$, so by induction the theorem holds. □

3.3 ▪ Iteration easily constructs the normal form

The major workload in the normal-form transformation to amplitude-angle variables is the transformation of the time derivatives. That the time derivatives are the major workload is common to many systematic modeling procedures. This section uses the power of computer algebra to greatly aid the normal-form transformation of Duffing's ODE (3.4).

[14] If the nonlinearity f is infinitely differentiable, then the order of error in radius r in the coordinate transform is not generally zero. Instead the order of error in (3.10) is just smaller than any power of r, for example, $\mathcal{O}(\exp[-1/r^2])$.

Algorithm 3.1 *Reduce code to deduce the normal-form transform of Duffing's* ODE (3.4) *to amplitude-angle variables* r *and* θ, *respectively denoted by* `r` *and* `q`.

```
factor r;
% new amplitude and phase angle variables
depend r,t;
depend q,t;
let { df(r,t)=>g, df(q,t)=>h };
% to solve x_qq+x+res=0
operator linv; linear linv;
let { linv(1,q)=>-1
    , linv(cos(~n*q),q)=>cos(n*q)/(n^2-1) };
% initial linear approximation
x:=r*cos(q);
g:=0;
h:=1;
% iterate to make residual zero
let r^7=>0;
for iter:=1:99 do begin
   res:=trigsimp( df(x,t,t)+x-x^3 ,combine);
   g:=g+(gd:=coeffn(res,sin(q),1)/2);
   h:=h+(hd:=coeffn(res,cos(q),1)/2/r);
   x:=x+linv(res-2*gd*sin(q)-2*hd*r*cos(q),q);
   if res=0 then iter:=1000000;
end;
```

Before proceeding to explain the procedure, look at the computer algebra results. Algorithm 3.1 determines that the nonlinear term x^3 modifies the shape of the oscillations (3.8a) by the next-order terms to

$$x = r\cos\theta - \tfrac{1}{32}r^3\cos 3\theta + \tfrac{1}{1024}r^5(-21\cos 3\theta + \cos 5\theta) + \mathcal{O}(r^7), \tag{3.11}$$

together with

$$\begin{aligned} y = \dot{x} &= -r\sin\theta + \tfrac{3}{32}r^3(4\sin\theta + \sin 3\theta) \\ &\quad + \tfrac{1}{1024}r^5(60\sin\theta + 27\sin 3\theta - 5\sin 5\theta) + \mathcal{O}(r^7). \end{aligned}$$

More important, from the computer algebra output, the normal form for the dynamics of the amplitude and phase of the oscillations of Duffing's ODE (3.4) is

$$\dot{r} = 0 \quad \text{and} \quad \dot{\theta} = 1 - \tfrac{3}{8}r^2 - \tfrac{15}{256}r^4 - \tfrac{123}{8192}r^6 + \mathcal{O}(r^8)\,. \tag{3.12}$$

That is, the amplitude of the Duffing oscillations is conserved, $\dot{r} = 0$; the nonlinear term x^3 in Duffing's ODE (3.4) only modifies the frequency $\dot{\theta}$ by an amplitude-dependent amount.

You might regard the program of Algorithm 3.1 as rather long for such a simple problem. However, *a wide variety of oscillation equations may be modeled with this one program simply by modifying the one-line computation of the residual.* Is that not marvelous? Try it in the exercises.[15]

[15] Furthermore, other algorithms for normal forms, such as that proposed by Rand and Armbruster's (1987, Chap. 3), are considerably more involved to apply.

Dialogue 3.1 *Output of Algorithm* 3.1.

```
1: in duffing$
x := cos(q)*r
g := 0
h := 1
2: x:=x;
                  1   3
x := r*cos(q)  - ----*r  *cos(3*q)
                  32
         5        21                  1
     + r *( - ------*cos(3*q) + ------*cos(5*q))
                1024                1024
         7    417                 43                   1
     - r *(-------*cos(3*q) - -------*cos(5*q) + -------*cos(7*q))
              32768               32768                32768
3: drdt:=g;
drdt := 0
4: dqdt:=h;
               3   2     15    4     123    6
dqdt := 1 - ---*r   - -----*r   - ------*r
               8        256        8192
```

Algorithm 3.1 implements the iterative procedure to the Reduce output in Dialogue 3.1. The four main parts of Algorithm 3.1 are the following.

- Define that variable names `r` and `q` depend implicitly upon time. Also use a `let` statement to tell Reduce to replace time derivatives of these variables by whatever asymptotic approximation we have found to date.
- The discussion after equation (3.7) identified that we need to solve the ODE $\hat{x}_{\theta\theta} + \hat{x} + \text{Res} = 0$ for the correction $\hat{x}$ forced by the current residual Res. Here the residual will be of the form of a sum of $\cos n\theta$ terms. For each such term in the residual, we introduce into the correction $\hat{x}$ the component $\cos n\theta/(n^2-1)$ to satisfy this ODE.

  The linear operator `linv` defines this process for us: being linear, the operator distributes over addition and multiplication by any factor independent of its second argument `q`, which denotes the phase angle θ. Then the pattern matching of the `let` statement maps residuals to the corresponding corrections.
- Set the linear approximation to the oscillations: $x = r\cos\theta$ such that $g = \dot{r} = 0$ and $h = \dot{\theta} = 1$.
- Iterate until the computed residual is zero to the specified $\mathcal{O}(r^8)$ error.
  - Compute the residual of Duffing's ODE (3.4), ensuring all products of sines and cosines are combined.
  - Extract any component in the residual in $\cos\theta$ and $\sin\theta$, as these would otherwise generate unallowable secular terms in the correction, and use these components to update the evolution of $\dot{r}$ and $\dot{\theta}$.
  - Last, use the pattern matching of `linv` to correct the shape of the oscillations.

The algorithm only terminates when the residual is zero to the specified error. Thus the computation of the residual is the critical part of the algorithm to ensure a correct normal-form transformation.

3.4 ▪ Carefully define emergent variables

We might want to predict the largest oscillation in the nonlinear Duffing ODE (3.4). We know the largest oscillation is the so-called heteroclinic orbit connecting equilibria at $x = \pm 1$. This heteroclinic orbit has infinite period, that is, zero frequency. Thus we could predict it by solving $\dot\theta = h(r) = 0$ from $h(r)$ given by equation (3.12). But after finding such a "radius" r such that $\dot\theta = 0$, what is r? We do not know. We have not defined exactly what the variable r is to be. We need to do so. The need for a careful definition of emergent variables is common in asymptotic analysis, but few people do define their emergent variables carefully.

One can decide the new variables to suit any of a variety of needs. For example, here let us say that we want the new variable r to be identical to x whenever the angle $\theta = 0$. What do we do? First we choose the initial linear approximation to satisfy the condition, which $x = r\cos\theta$ does (fortuitously). Then we ensure that all updates to the coordinate transform $x = r\cos\theta + \hat{x}(r,\theta)$ satisfy $\hat{x}(r,0) = 0$ so that when $\theta = 0$ we will have $x = r$. We can always do this by adding the appropriate amount of the homogeneous solution, $\cos\theta$, to all corrections $\hat{x}$.

Similarly, we tighten the meaning of the phase θ by requiring $\dot{x} = 0$ when $\theta = 0$. This particular requirement easily holds for Duffing's ODE (3.4) as we simply never introduce any component of $\sin\theta$ into x.

Often, but not always, the easiest way to ensure that the new variables have the correct definition is to automatically build the definition into the operator `linv`: when we generate a component in the coordinate transform of $\cos n\theta$ we also immediately subtract the corresponding component in $\cos\theta$. That is, we redefine the pattern matching:

```
let { linv(1,q)=>-1+cos(q)
    , linv(cos(~n*q),q)=>(cos(n*q)-cos(q))/(n^2-1) };
```

Executing the modified code, we find the frequency

$$\dot\theta = h = 1 - \frac{3}{8}r^2 - \frac{21}{256}r^4 - \frac{81}{2048}r^6 + \mathcal{O}(r^8).$$

The difference with (3.12) is now that here we know that $x = r$ when the phase $\theta = 0$. That is, when we now find that the above $h = 0$ for $r = \pm 1.2765$ we then predict that the homoclinic orbit goes through $x = \pm 1.2765$. Although 27% in error (due to the finite-amplitude effects), at least, by defining the variable r, now we are empowered to make definite predictions.

Tiger stripes or leopard spots Transforming to radius-angle variables leads to models for pattern formation such as the stripes on a tiger or the spots on a leopard. The formation of such complex patterns in space is described as the evolution in time of "spatial oscillations." The phase $\theta(x,y,t)$ then varies significantly in space as it evolves slowly in time. The radius variable r measures the amplitude of the spatial oscillations at any location in (x,y,t), but, as in the introductory example, we still get an equation roughly of the form $\partial_t r = r - r^3 + \cdots$ so that generically the spatial pattern evolves to be the same finite amplitude across all space. However, the local phase θ is not "tied" to any particular value and instead drifts around slowly under the influence of variations in the pattern and the boundary conditions: it evolves according to a nonlinear diffusion equation, $\partial_t\theta = \nabla^2\theta + \cdots$. It is this phase evolution that we see in the complex evolution of many spatial patterns, such as the formation of stripes on a tiger. This is a topic for further study.

Interestingly, in such pattern formation there are generally three time scales. The fast scale is that of local transients toward a nonequilibrium pattern. The medium time scale is that of the local amplitude of the pattern evolving to its local quasi-equilibrium. The slowest time scale is that of the global evolution of the phase giving rise to the global pattern.

3.5 ▪ Extend to nonlinear electromagnetic waves

The main theme—developing coordinate transformations to understand oscillations—continues in section 3.6. This section diverts us to an interesting application of the approach to electromagnetic wave propagation.

Kirkinis (2012) uses the mysterious methodology of the "renormalization group" to predict electromagnetic wave reflection from a change in the medium. This section analyzes the same problem using our transformation to a polar coordinates normal form.

First recall that a wave traveling over time t to the right in space x might look like the field $u(x,t) = \cos(x-t)$. Conversely, a wave to the left might be $u(x,t) = \cos(x+t)$. When the wave to the right has amplitude r and the wave to the left has amplitude s, we superpose the waves as in the field $u(x,t) = r\cos(x-t) + s\cos(x+t)$. But often we prefer the equivalent complex form $u(x,t) = \Re\{re^{i(x-t)} + se^{-i(x+t)}\}$ because we like to factor out the periodicity in time via the complex factor e^{-it}; that is, superposed waves are to be written $u(x,t) = \Re\{(re^{ix} + se^{-ix})e^{-it}\}$. But these are only pure sinusoidal waves: general waves in a nonlinear media will be nonsinusoidal. In such a case let's write an electromagnetic field as $u(x,t) = \Re\{\psi(x)e^{-it}\}$ for some complex field $\psi(x)$ that describes the spatial structure of the wave. A component in the complex field $\psi(x)$ of the form e^{ix} represents a right-traveling wave, whereas a component of the form e^{-ix} represents a left-traveling wave.

Having factored out the time dependence, our interest lies in the spatial "evolution" of the complex field $\psi(x)$.

Second, Kirkinis (2012) asserts that the complex field $\psi(x)$ satisfies the nonlinear complex differential equation

$$\frac{d^2\psi}{dx^2} + \psi + \epsilon|\psi|^2\psi = 0 \tag{3.13}$$

for some small nonlinearity parameter ϵ. Neglecting the nonlinearity, the linear version of this complex ODE is $d^2\psi/dx^2 + \psi \approx 0$, which has linearly independent solutions $\psi = e^{\pm ix}$ representing right- and left-traveling waves. We seek a description of the nonlinear dynamics of such waves when they interact through the nonlinearity of the medium as expressed in the small term $\epsilon|\psi|^2\psi$.

Unfortunately, Theorem 3.1 does not directly apply despite the ODE (3.13) looking very much like the ODE (3.9). Crucially here we deal with a complex field $\psi(x)$ and we deal with a nonlinearity in the ODE (3.13) that involves the nonanalytic function $|\psi|$. Nonetheless, the coordinate transformation approach and spirit of section 3.2 simplify our view of the ODE (3.13).

Superpose linear waves Third, note that a second-order differential equation, such as the ODE (3.13), is equivalent to two first-order ODEs. But the ODE (3.13) is complex so it is equivalent to four real first-order ODEs. Thus to fully characterize the solutions of the ODE (3.13) we must transform to a description involving four real variables. Let's choose real variables analogous to polar coordinates, but adapted to the complex field $\psi(x)$: we write solutions $\psi(x)$ as

$$\psi \approx re^{i\theta} + se^{-i\vartheta},$$

where the four real variables r, θ, s, and ϑ evolve in space x according to a set of four real ODEs. Physically, and using primes to denote derivatives d/dx, r denotes the amplitude of the right-traveling wave, evolving slowly, $r' \approx 0$; θ denotes the phase of the right-traveling wave, evolving near linearly, $\theta' \approx 1$; s denotes the amplitude of the left-traveling wave, evolving slowly, $s' \approx 0$; and ϑ denotes the phase of the right-traveling wave, evolving near linearly, $\vartheta' \approx 1$. This completes a linear description of the electromagnetic field in these new complex polar variables.

3.5.1 ▪ Cubic interaction gives frequency shifts

Now let's find a first correction to the linear superposition. Using our four new real variables we seek a complex field

$$\psi \approx re^{i\theta} + se^{-i\vartheta} + \hat{\psi}(r, s, \theta, \vartheta)$$

for small corrections $\hat{\psi}$. Correspondingly we also seek small corrections to the evolution, their ODEs, of the four new real variables:

$$r' = \hat{r}(r, s, \theta, \vartheta), \qquad s' = \hat{s}(r, s, \theta, \vartheta),$$
$$\theta' = 1 + \hat{\theta}(r, s, \theta, \vartheta), \qquad \vartheta' = 1 + \hat{\vartheta}(r, s, \theta, \vartheta),$$

where the hats denote the small corrections to the ODEs. Now substitute into the ODE, but the derivatives get complicated. Simplify by neglecting products of small quantities (hatted quantities):

$$\begin{aligned}\frac{d\psi}{dx} &\approx r'e^{i\theta} + ire^{i\theta}\theta' + s'e^{-i\vartheta} - ise^{-i\vartheta}\vartheta' + \frac{\partial\hat{\psi}}{\partial r}r' + \frac{\partial\hat{\psi}}{\partial\theta}\theta' + \frac{\partial\hat{\psi}}{\partial s}s' + \frac{\partial\hat{\psi}}{\partial\vartheta}\vartheta' \\ &\approx \hat{r}e^{i\theta} + ire^{i\theta}(1+\hat{\theta}) + \hat{s}e^{-i\vartheta} - ise^{-i\vartheta}(1+\hat{\vartheta}) + \frac{\partial\hat{\psi}}{\partial\theta} + \frac{\partial\hat{\psi}}{\partial\vartheta} \\ &\approx ire^{i\theta} - ise^{-i\vartheta} + (\hat{r} + ir\hat{\theta})e^{i\theta} + (\hat{s} - is\hat{\vartheta})e^{-i\vartheta} + \frac{\partial\hat{\psi}}{\partial\theta} + \frac{\partial\hat{\psi}}{\partial\vartheta}.\end{aligned}$$

This expression for ψ' is directly analogous to the expression (3.6) derived for $\dot{x}$ in the polar coordinate version of Duffing's ODE (3.4). Differentiate once more, again neglecting products of small quantities, and omitting the intermediate steps, derive

$$\begin{aligned}\frac{d^2\psi}{dx^2} &\approx \cdots \\ &\approx -re^{i\theta} - se^{-i\vartheta} + 2i(\hat{r} + ir\hat{\theta})e^{i\theta} - 2i(\hat{s} - is\hat{\vartheta})e^{-i\vartheta} + (\partial_\theta + \partial_\vartheta)^2\hat{\psi}.\end{aligned}$$

This expression for ψ'' is directly analogous to the expression (3.7) derived earlier for $\ddot{x}$ in the polar coordinate version of Duffing's ODE (3.4). Consequently, we deduce that

$$\frac{d^2\psi}{dx^2} + \psi \approx +2i(\hat{r} + ir\hat{\theta})e^{i\theta} - 2i(\hat{s} - is\hat{\vartheta})e^{-i\vartheta} + (\partial_\theta + \partial_\vartheta)^2\hat{\psi} + \hat{\psi}.$$

The rest of the ODE (3.13) is the cubic nonlinearity, which, remembering that variables r, s, θ, and ϑ are real, is

$$\begin{aligned}
\epsilon|\psi|^2\psi &= \epsilon\bar{\psi}\psi^2 \\
&= \epsilon\left(re^{-i\theta} + se^{+i\vartheta} + \bar{\hat{\psi}}\right)\left(re^{i\theta} + se^{-i\vartheta} + \hat{\psi}\right)^2 \\
&= \epsilon\left[(r^3 + 2rs^2)e^{i\theta} + (s^3 + 2r^2 s)e^{-i\vartheta} + r^2 s e^{i(2\theta+\vartheta)} + rs^2 e^{-i(\theta+2\vartheta)}\right] \\
&\quad + \mathcal{O}(\epsilon^2, \epsilon\hat{\psi})
\end{aligned}$$

Adding the last two expressions, we thus deduce that to make progress in satisfying the nonlinear complex ODE (3.13) we must solve

$$\begin{aligned}
&2(i\hat{r} - r\hat{\theta})e^{i\theta} + 2(-i\hat{s} - s\hat{\vartheta})e^{-i\vartheta} + (\partial_\theta + \partial_\vartheta)^2\hat{\psi} + \hat{\psi} \\
&+ \epsilon\left[(r^3 + 2rs^2)e^{i\theta} + (s^3 + 2r^2 s)e^{-i\vartheta} + r^2 s e^{i(2\theta+\vartheta)} + rs^2 e^{-i(\theta+2\vartheta)}\right] \approx 0. \qquad (3.14)
\end{aligned}$$

This equation for small corrections (hatted quantities) is analogous to equation (3.7) for Duffing's ODE (3.4).

Simplify evolution as far as possible The first line of equation (3.14) is the corrections that must cancel the "forcing" terms in the second line of (3.14). Consider the forcing terms one by one, starting from the last.

- To cancel with the last term $\epsilon rs^2 e^{-i(\theta+2\vartheta)}$ we seek a component in $\hat{\psi} \propto e^{-i(\theta+2\vartheta)}$. Straightforward undetermined coefficients then determine that the component must be $\frac{1}{8}\epsilon rs^2 e^{-i(\theta+2\vartheta)}$.

- To cancel with the penultimate term $\epsilon r^2 s e^{i(2\theta+\vartheta)}$ we seek a component in $\hat{\psi} \propto e^{i(2\theta+\vartheta)}$. Straightforward undetermined coefficients then determine that the component must be $\frac{1}{8}\epsilon r^2 s e^{i(2\theta+\vartheta)}$.

- Now consider the term $\epsilon(s^3 + 2r^2 s)e^{-i\vartheta}$. We cannot put a corresponding term into $\hat{\psi}$ because $\hat{\psi} = e^{-i\vartheta}$ is a homogeneous solution of $(\partial_\theta + \partial_\vartheta)^2\hat{\psi} + \hat{\psi} = 0$. Instead we must cancel the term $\epsilon(s^3 + 2r^2 s)e^{-i\vartheta}$ using the term $2(-i\hat{s} - s\hat{\vartheta})e^{-i\vartheta}$ by correcting the evolution of $\hat{s}$ and $\hat{\vartheta}$. By inspection, choosing $\hat{s} = 0$ and $\hat{\vartheta} = \epsilon(\frac{1}{2}s^2 + r^2)$ will suffice.

- Correspondingly, since $e^{i\theta}$ is also a homogeneous solution of $(\partial_\theta + \partial_\vartheta)^2\hat{\psi} + \hat{\psi} = 0$, we must correct the evolution by $\hat{r} = 0$ and $\hat{\theta} = \epsilon(\frac{1}{2}r^2 + s^2)$ to cancel with the first forcing term $\epsilon(r^3 + 2rs^2)e^{i\theta}$.

Putting the above together, in analogy to the polar form (3.8) of Duffing's ODE (3.4), here we transform the nonlinear complex ODE (3.13) to the complex polar form

$$\psi \approx re^{i\theta} + se^{-i\vartheta} + \tfrac{1}{8}\epsilon r^2 s e^{i(2\theta+\vartheta)} + \tfrac{1}{8}\epsilon rs^2 e^{-i(\theta+2\vartheta)}, \qquad (3.15a)$$

$$\begin{aligned}
\text{such that } r' &\approx 0, \quad \theta' \approx 1 + \epsilon(\tfrac{1}{2}r^2 + s^2), \\
s' &\approx 0, \quad \vartheta' \approx 1 + \epsilon(r^2 + \tfrac{1}{2}s^2). \qquad (3.15b)
\end{aligned}$$

Remember to view equation (3.15a) as a coordinate transform of the original ODE (3.13) to new variables $(r, s, \theta, \vartheta)$ and their ODEs (3.15b). This transform brings out that the main effect of wave-wave interactions in this nonlinear medium is to modify the spatial wavenumber of each wave as seen in the corrections to θ' and ϑ'.

3.5.2 ▪ Waves reflect at and transmit through an interface

Analogous to the discussion by Kirkinis (2012) we use this coordinate transform to predict wave transmission and reflection in the wave propagation from a linear to a nonlinear medium. Suppose there is a boundary between a linear medium, $\epsilon = 0$, to the left and a nonlinear medium, finite ϵ, to the right. Recall that $re^{i\theta}$ represents a right-propagating wave, whereas $se^{-i\vartheta}$ represents a left-propagating wave.

- In the nonlinear medium on the right there will be no left-propagating wave, $s = 0$. Despite the finite ϵ, because $s = 0$ the wave shape from (3.15a) is simply $\psi = r_t e^{i\theta}$ with gradient affected by the change in wavenumber to $\psi' = i(r_t + \frac{1}{2}\epsilon r_t^3)e^{i\theta}$, where r_t denotes the amplitude of the transmitted wave.
- Without loss of generality, define the wave phases θ and ϑ so that the interface between linear and nonlinear media is at $\theta = \vartheta = 0$. Then the wave field on the right has, at the interface,
$$\psi_0 = r_t \quad \text{and} \quad \psi_0' = i(r_t + \tfrac{1}{2}\epsilon r_t^3).$$
- To the left of the interface the medium is linear, $\epsilon = 0$, so the wave shape is the simple superposition of incident and reflected waves, $\psi = r_i e^{i\theta} + s_r e^{-i\vartheta}$, where r_i denotes the given amplitude of the incident wave and s_r is the amplitude of the reflected wave. Because $\epsilon = 0$, the gradient field is simply $\psi' = ir_i e^{i\theta} - is_r e^{-i\vartheta}$. Evaluate at the interface to find that the field and its gradient just to the left must be
$$\psi_0 = r_i + s_r \quad \text{and} \quad \psi_0' = ir_i - is_r\,.$$
- The field and its gradient must be continuous across the interface, so we equate the two expressions for each of ψ_0 and ψ_0'. That is,
$$r_i + s_r = r_t \quad \text{and} \quad ir_i - is_r = i(r_t + \tfrac{1}{2}\epsilon r_t^3).$$
Eliminating the reflected amplitude s_r from the above two equations gives
$$r_i = r_t + \tfrac{1}{4}\epsilon r_t^3 \quad \Longrightarrow \quad r_t = r_i - \tfrac{1}{4}\epsilon r_i^3 + \mathcal{O}(\epsilon^2).$$
Since the errors in the coordinate transform are $\mathcal{O}(\epsilon^2)$, this is an acceptable approximation to the transmission amplitude. Correspondingly, substituting into $r_i + s_r = r_t$, we then immediately find that the reflected wave amplitude is $s_r = -\frac{1}{4}\epsilon r_i^3 + \mathcal{O}(\epsilon^2)$.[16]

The coordinate transform readily determines wave reflection and transmission across such an interface.

3.5.3 ▪ Iteration constructs the normal form

As in section 3.3 for Duffing's ODE (3.4), the computer algebra of Algorithm 3.2 readily constructs higher-order versions of the nonlinear coordinate transform. Moreover, our approach caters for many other variations of similar complex wave equations.

The basis has already been established for the computer algebra iteration of Algorithm 3.2. We transform the differential equation (3.13) to new amplitude-phase

[16] Since the amplitude s_r is negative we should really have incorporated a phase shift of π into ϑ. But it is simpler not to.

Algorithm 3.2 *Reduce code to deduce the normal-form coordinate transform of the nonlinear complex* ODE (3.13) *to complex polar variables* r, s, θ, *and* ϑ, *respectively denoted by* `r`, `s`, `q`, *and* `p`.

```
% amplitude-phase variables
depend r,x; depend q,x,pq;
depend s,x; depend p,x,pq;
let { df(r,x)=>drdx, df(q,x)=>dqdx
    , df(s,x)=>dsdx, df(p,x)=>dpdx };
% initial linear approximation
psi:=r*(eiq:=cos(q)+i*sin(q))+s*(eip:=cos(p)-i*sin(p));
drdx:=dsdx:=0; dqdx:=dpdx:=1;
% to solve [(d/dq+d/dp)^2+1]psi+res=0
operator linv; linear linv;
let { linv(1,pq)=>-1
    , linv(cos(~~m*q+~~n*p),pq)=>cos(m*q+n*p)/((m+n)^2-1)
      when (m+n)^2 neq 1
    , linv(sin(~~m*q+~~n*p),pq)=>sin(m*q+n*p)/((m+n)^2-1)
      when (m+n)^2 neq 1 };
operator mean; linear mean;
let { mean(1,pq)=>1
    , mean(cos(~~m*q+~~n*p),pq)=>0 when m+n neq 0
    , mean(sin(~~m*q+~~n*p),pq)=>0 when m+n neq 0 };
% iterate corrections to specified error
let eps^3=>0;
for iter:=1:9 do begin write
    res:=trigsimp( df(psi,x,x)+psi+eps*sub(i=-i,psi)*psi^2
        ,combine);
    ceq:=mean(trigsimp(res*sub(i=-i,eiq),combine),pq);
    drdx:=drdx-coeffn(ceq,i,1)/2;
    dqdx:=dqdx+coeffn(ceq,i,0)/2/r;
    cep:=mean(trigsimp(res*sub(i=-i,eip),combine),pq);
    dsdx:=dsdx+coeffn(cep,i,1)/2;
    dpdx:=dpdx+coeffn(cep,i,0)/2/s;
    psi:=psi+linv(res-ceq*eiq-cep*eip,pq);
    if res=0 then write iter:=1000000+iter;
end;
```

variables r, s, θ, and ϑ; linearly, $\psi \approx re^{i\theta} + se^{-i\vartheta}$ such that $r' \approx s' \approx 0$ and $\theta \approx \vartheta \approx 1$. Equation (3.14) then determines that in general we need to solve the following equation for corrections to the coordinate transform:

$$2(i\hat{r} - r\hat{\theta})e^{i\theta} + 2(-i\hat{s} - s\hat{\vartheta})e^{-i\vartheta} + (\partial_\theta + \partial_\vartheta)^2\hat{\psi} + \hat{\psi} + \mathrm{Res} = 0\,, \qquad (3.16)$$

where Res denotes the residual of the ODE (3.13) at the current approximation. Let's describe how Algorithm 3.2 implements this iteration to find (3.15) and higher-order terms. The algorithm adapts Algorithm 3.1 to this problem.

- The dynamics is to be expressed in new variables r, s, θ, and ϑ denoted in Algorithm 3.2 by variables `r`, `s`, `q`, and `p`. These variables must depend upon the independent variable $x = x$. But also we need to be able to distinguish between the slow modulation of amplitudes r and s and the fast oscillations through θ and ϑ, so we additionally let q and p depend upon pq.

The new variables have x-derivatives stored in variables `drdx`, `dsdx`, `dqdx`, and `dpdx`. The iteration updates these stored expressions.

- The linear approximation is then coded: $\psi \approx re^{i\theta} + se^{-i\vartheta}$ such that $r' \approx s' \approx 0$ and $\theta \approx \vartheta \approx 1$. Curiously, here it is more convenient to use $e^{i\theta}$ in its trigonometric form $\cos\theta + i\sin\theta$, and similarly for other complex exponentials.[17] So for $e^{i\theta}$ define `eiq` $= \cos\theta + i\sin\theta$ and for $e^{-i\vartheta}$ define `eip` $= \cos\vartheta - i\sin\vartheta$.

- The operators `mean` and `linv` help determine corrections and so are described in the context of the iteration.

- Compute expressions to some set error; here errors are $\mathcal{O}(\epsilon^3)$. Then iterate in a for-loop until the residual, computed in `res`, is zero (to the specified error).

- Within the loop first calculate the residual of the differential equation (3.13), using `trigsimp` to simplify the products of sine and cosines. Compute the nonlinear term $|\psi^2|\psi$ as $\bar{\psi}\psi^2$, where simply changing $i \mapsto -i$ obtains the complex conjugate $\bar{\psi}$ (as all other variables are real).

- Then we use part of this residual to update the evolution, the derivatives, of r and θ. Multiplying (3.16) by $e^{-i\theta}$ leads to $2i\hat{r} - 2r\hat{\theta} + \cdots + \text{Res}\, e^{-i\theta} = 0$, so we identify the updates as coming from the part of $(\text{Res}\, e^{-i\theta})$ that is independent of the oscillations. That is, we extract the mean of $(\text{Res}\, e^{-i\theta})$ to update r' and θ'. Operator `mean` does this. Then the imaginary part determines the correction to r', obtained by `coeffn`, and the real part determines the correction to θ', obtained by `coeffn`.

- Determine corrections to the derivatives s' and ϑ' via (3.16) by similarly extracting and using the mean of $(\text{Res}\, e^{+i\vartheta})$.

- Last, corrections $\hat{\psi}$ to the field satisfy the ODE $(\partial_\theta + \partial_\vartheta)^2\hat{\psi} + \hat{\psi} + \text{Res} = 0$, where this new residual accounts for the terms just used in correcting the evolution. Linear operator `linv` codes the solution to this ODE as, for example,

$$\left[(\partial_\theta + \partial_\vartheta)^2 + 1\right]\cos(m\theta + n\vartheta) = -[(m+n)^2 - 1]\cos(m\theta + n\vartheta).$$

Such updates are always computable provided $|m + n| \neq 1$, that is, provided we have appropriately determined the corrections to the evolution and removed such troublesome terms.

[17] The main reason that we use trigonometric functions rather than the complex exponential form arises in solving (3.16) for corrections. We have to use the real part of subexpressions to correct phase derivatives θ' and ϑ' and the imaginary part of subexpressions to correct amplitude derivatives r' and s'. For more general nonlinearities than that in ODE (3.13) these must be expressed in terms of sine and cosines. Hence, so far, it is simpler that every expression use such trigonometric functions. Section 3.6 shows how complex-amplitude variables can restore the appeal of complex exponentials in complicated oscillation problems.

Executing Algorithm 3.2 verifies not only the derivation of the transformation (3.15), but also higher-order versions such as

$$\begin{aligned}
\psi &\approx re^{i\theta} + se^{-i\vartheta} + \epsilon\tfrac{1}{8}\left[r^2 s e^{i(2\theta+\vartheta)} + rs^2 e^{-i(\theta+2\vartheta)}\right] \\
&\quad + \epsilon^2\tfrac{1}{64}\left[r^3 s^2 e^{i(3\theta+2\vartheta)} + r^2 s^3 e^{-i(2\theta+3\vartheta)} - (11r^2s^3 + 10r^4 s)e^{i(2\theta+\vartheta)}\right. \\
&\qquad \left. - (11r^3s^2 + 10rs^4)e^{-i(\theta+2\vartheta)}\right] + \mathcal{O}(\epsilon^3), \\
r' &= s' = \mathcal{O}(\epsilon^3), \\
\theta' &\approx 1 + \epsilon(\tfrac{1}{2}r^2 + s^2) - \epsilon^2\tfrac{1}{16}(2r^4 + 6r^2s^2 + 7s^4) + \mathcal{O}(\epsilon^3), \\
\vartheta' &\approx 1 + \epsilon(r^2 + \tfrac{1}{2}s^2) - \epsilon^2\tfrac{1}{16}(7r^4 + 6r^2s^2 + 2s^4) + \mathcal{O}(\epsilon^3).
\end{aligned} \tag{3.17}$$

We are definite about the errors because the iteration only terminates when the ODE (3.13) is satisfied to residuals of this order.

Recall that we view (3.17) as a coordinate transform from one second-order ODE (3.13) for one complex ψ to an equivalent system of first-order ODEs for four real variables. In this new form (3.17) we readily discern key features: the wave shape ψ is modified by the wave-wave interaction, but the wave amplitudes are conserved as r' and s' are zero (to this order), and the nonlinearity modifies the effective wavenumber through modification of θ' and ϑ'.

3.6 ▪ Oscillations have simple complex modulation

Eppur si muove. [And yet it moves.] — *Galileo,* 1633

Expressing oscillations and waves in terms of amplitude and phase angle variables is a powerful aid in interpreting their dynamics. However, when problems involve time-dependent coefficients or forcing, such as the classic Mathieu's ODE, which we examine shortly (subsection 3.7.1), or stochastic oscillations, or problems with inhomogeneous physics (which normally invoke the WKB approximation), then amplitude and phase variables become too difficult. In such cases we appeal to the even more powerful and more flexible method of expressing oscillations in terms of complex exponentials such as $\exp(i\omega t)$ and $\exp(-i\omega t)$.

The complex amplitudes of these complex exponentials contain the amplitude and phase angle information obtained in sections 3.1–3.4.

Example 3.1. Consider the pair of ODEs (3.1) introduced in section 3.1. There we wrote, as in equation (3.3),

$$x = r\cos\theta \quad \text{such that} \quad \dot r = \epsilon r - r^3 \quad \text{and} \quad \dot\theta = 1 + r^2 .$$

Thus, for example, from $\dot r = \epsilon r - r^3$ we readily predict persistent oscillations of amplitude $r = \sqrt{\epsilon}$. Now we rewrite the above form of oscillations in terms of complex exponentials $\exp(i\omega t)$ and $\exp(-i\omega t)$.

First, set the angle $\theta = t + \phi(t)$, where the phase shift ϕ evolves according to $\dot\phi = r^2$. Crucially, for small-amplitude r the phase shift ϕ evolves slowly, as its time derivative $\dot\phi = r^2$ is very small.

Second, expand the cosine in terms of complex exponentials:

$$\begin{aligned} x &= r\cos\theta \\ &= r\cos(t+\phi) \\ &= r\tfrac{1}{2}\left[e^{i(t+\phi)} + e^{-i(t+\phi)}\right] \\ &= \tfrac{1}{2}re^{i\phi}e^{it} + \tfrac{1}{2}re^{-i\phi}e^{-it} \\ &= ae^{it} + \bar{a}e^{-it}, \end{aligned}$$

where the complex amplitude $a = \frac{1}{2}re^{i\phi}$, and its complex conjugate $\bar{a} = \frac{1}{2}re^{-i\phi}$. This writes the solution x in terms of the two complex exponential oscillating solutions $e^{\pm it}$ of the linearized version of the ODEs, namely $\dot{x} = -y$ and $\dot{y} = x$.

Third, the complex amplitude a evolves in time with r and ϕ. From $a = \frac{1}{2}re^{i\phi}$ we deduce

$$\begin{aligned} \dot{a} &= \tfrac{1}{2}\dot{r}e^{i\phi} + \tfrac{1}{2}re^{i\phi}i\dot{\phi} \\ &= \tfrac{1}{2}(\epsilon r - r^3)e^{i\phi} + \tfrac{1}{2}re^{i\phi}ir^2 \\ &= \tfrac{1}{2}re^{i\phi}\left[\epsilon + (-1+i)r^2\right] \\ &= a[\epsilon + (-1+i)4|a|^2], \end{aligned}$$

using $r^2 = 4a\bar{a} = 4|a|^2$. In summary, we may choose to write the solution of ODEs (3.1) as

$$x = ae^{it} + \bar{a}e^{-it} \quad \text{such that} \quad \dot{a} = a[\epsilon + (-1+i)4|a|^2].$$

Now interpret this form:

- For small complex amplitude a, $\dot{a} \approx \epsilon a$, with solution $a \propto e^{\epsilon t}$, so the oscillations $ae^{it} + \bar{a}e^{-it}$ grow in time like $e^{\epsilon t}$.
- The oscillations grow until $|a| = \frac{1}{2}\sqrt{\epsilon}$ when the right-hand side factor $\epsilon + (-1+i)4|a|^2$ becomes pure imaginary, namely $i\epsilon$, and thereafter the complex amplitude a just rotates in the complex plane like $e^{i\epsilon t}$—this is the nonlinear frequency shift.

All features of the nonlinear oscillation of section 3.1 appear here using complex amplitudes and complex exponentials.

Again, this analysis of Example 3.1 is straightforward because the shape of the oscillations is essentially circular—see the trajectories in Figure 3.3. But any analysis is significantly more complicated for typical nonlinear oscillations because their shape may be far from circular. Let's revisit the nonlinear oscillations of ODE (3.18) below in Exercise 3.3.

Furthermore, let's immediately turn to the computer to do the tedious algebra involved.

Example 3.2 (modify Duffing). Consider the second-order ODE

$$\ddot{x} + x + \dot{x}^3 = 0. \tag{3.18}$$

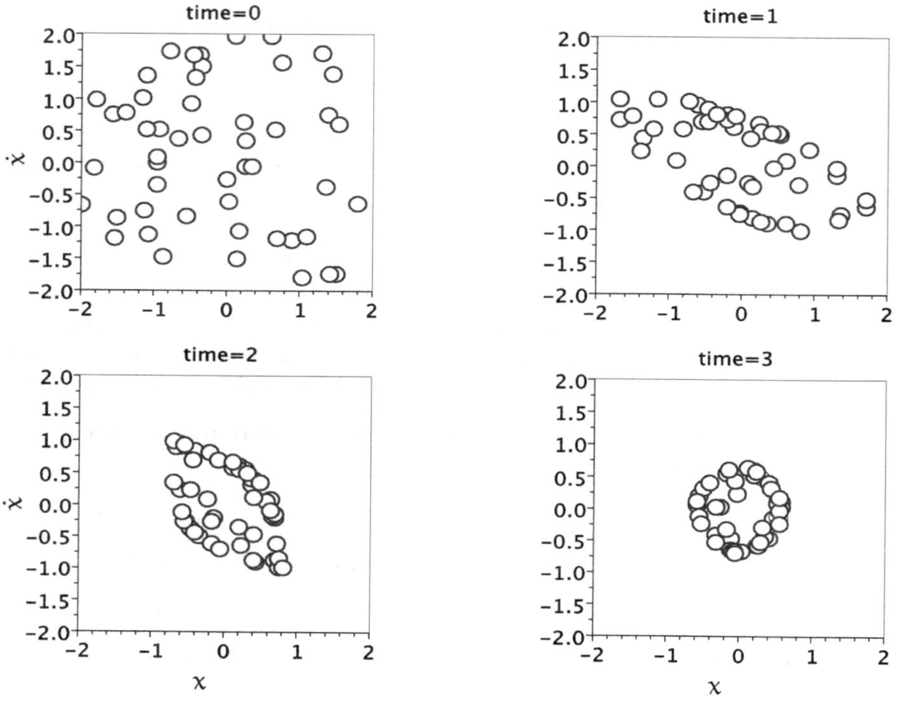

Figure 3.7. *Evolution of* (3.18) *from* 50 *random initial conditions: all solutions collapse toward the origin; thereafter they rotate about the origin and very slowly reduce in amplitude.*

Figure 3.7 shows some snapshots of the evolution (also see Exercise 3.3). The eventual model is that the nonlinear oscillations take the form

$$\begin{aligned} x &= ae^{it} + \bar{a}e^{-it} + i\tfrac{1}{8}\big(-a^3e^{i3t} + \bar{a}^3e^{-i3t}\big) \\ &\quad + \tfrac{3}{64}\big(-a^5e^{i5t} + 9a^4\bar{a}e^{i3t} + 9a\bar{a}^4e^{-i3t} - \bar{a}^5e^{-i5t}\big) \\ &\quad + \mathcal{O}\big(|a|^6\big), \end{aligned} \tag{3.19a}$$

where the complex amplitude evolves according to

$$\dot{a} = -\tfrac{3}{2}|a|^2a + i\tfrac{9}{16}|a|^4a + \mathcal{O}\big(|a|^6\big)\,. \tag{3.19b}$$

View (3.19a), together with $y = \dot{x}$, as a time-dependent coordinate transform of the $x\dot{x}$ phase plane. Figure 3.8 shows four snapshots of the coordinate transform. In principle, such time-dependent coordinate systems capture *all* the dynamics of the nonlinear oscillator. The value of this particular coordinate transform is that its time dependence, a deformed rotation, is especially chosen so that the small-amplitude dynamics are *simply* described by the model (3.19b).

Before proceeding to derive this model, note that the real part of the right-hand side, $-\frac{3}{2}|a|^2a$, controls the slow decay of the oscillation to zero amplitude, as suggested by Figure 3.7. In contrast, the imaginary part of the right-hand side, $i\frac{9}{16}|a|^4a$, governs the nonlinear frequency correction, which is essentially unobservable in simulations. To see these two roles, simply substitute $a = \frac{1}{2}r(t)\exp[i\phi(t)]$ into the complex-amplitude model (3.19b), which then becomes

$$\tfrac{1}{2}\dot{r}e^{i\phi} + i\tfrac{1}{2}r\dot{\phi}e^{i\phi} = -\tfrac{3}{16}r^3e^{i\phi} + i\tfrac{9}{512}r^5e^{i\phi} + \mathcal{O}\big(r^6\big)\,.$$

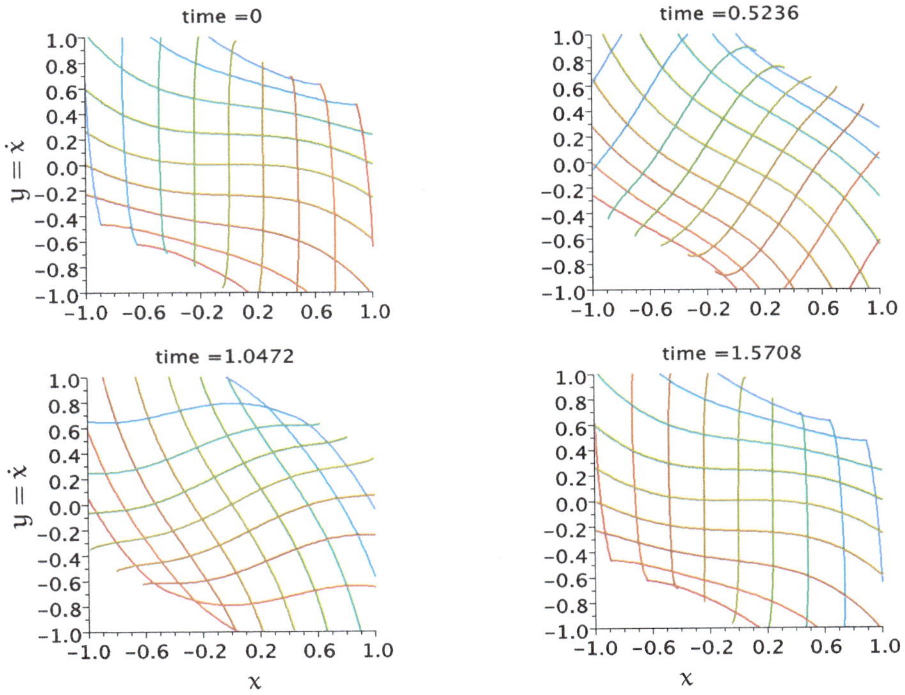

Figure 3.8. *The form* (3.19a) *is a time-dependent rotating coordinate transform: set complex amplitude* $a = \frac{1}{2}(A + iB)$ *for real* A *and* B*; then plot "coordinate curves" in the* $x\dot{x}$*-plane of constant* A *and* B *at different times over a quarter period. These coordinate curves deform as they rotate to suit the dynamics of the nonlinear oscillator* (3.18).

Divide by $\exp[i\phi]$ and equate real and imaginary parts to deduce

$$\dot{r} = -\tfrac{3}{8}r^3 + \mathcal{O}(r^6) \quad \text{and} \quad \dot{\phi} = \tfrac{9}{256}r^4 + \mathcal{O}(r^5)\,.$$

After this division by $\exp[i\phi]$, the real part of the left-hand side is $\frac{1}{2}\dot{r}$ and the imaginary part of the left-hand side is $\frac{1}{2}r\dot{\phi}$; consequently the real and imaginary parts of the right-hand side control the amplitude and phase evolution, respectively ($\dot{r}$ and $\dot{\phi}$). These are the typical roles for the real and imaginary parts of $\dot{a}$.

We use the residual of the ODE (3.18), $\ddot{x} + x + \dot{x}^3 = 0$, to drive corrections to the complex-amplitude normal form of the ODE. Algorithm 3.3 obtains the above normal form for the ODE. Let's explore the code in blocks.

- Reduce inconveniently knows too much about the exponential function. Thus we define a function `cis`, implemented as an `operator`, so that $\operatorname{cis} u = e^{iu}$ with only the properties that we need. These properties are defined by the pattern matching of the first `let` statement in the algorithm, namely its derivative, combining products and squares, and $\operatorname{cis} 0 = e^{i0} = 1$.
- Define the operator `linv` to find corrections to the solution x for any given residual: the task is to solve $\hat{x}_{tt} + \hat{x} + \text{Res} = 0$. This task is greatly simplified because we recognize that the time dependence in a correction $\hat{x}$ is dominated by the oscillations through $\operatorname{cis} nt = e^{int}$. The reason is that the

Algorithm 3.3 *Reduce code to find the complex modulation of the nonlinear oscillations of the* ODE (3.18).

```
factor i,small;
% complex exponential cis(u)=exp(i*u)
operator cis;
let { df(cis(~v),~u)=>i*cis(v)*df(v,u)
    , cis(~u)*cis(~v)=>cis(u+v)
    , cis(~u)^2=>cis(2*u)
    , cis(0)=>1 };
% to solve x_tt+x+res=0
operator linv; linear linv;
let { linv(1,cis)=>-1
    , linv(cis(~n*t),cis)=>cis(n*t)/(n^2-1) };
% complex-amplitude variables
depend a,t;
depend b,t;
let { df(a,t)=>ga, df(b,t)=>gb };
% initial linear approximation
x:=small*(a*cis(t)+b*cis(-t));
ga:=gb:=0;
% iterate to nonlinear transformation
let small^6=>0;
for iter:=1:99 do begin
   res:=df(x,t,t)+x+df(x,t)^3;
   ga:=ga+i/2/small*(ca:=coeffn(res,cis(+t),1));
   gb:=gb-i/2/small*(cb:=coeffn(res,cis(-t),1));
   x:=x+linv(res-ca*cis(t)-cb*cis(-t),cis);
   if res=0 then iter:=1000000;
end;
```

time dependence in the evolution of the complex amplitude a is much slower, and hence the time derivatives of a are negligible when finding an approximate correction.[18] Consequently, a term in the residual of the form e^{int} is approximately, but effectively, accounted for by a correction to the solution x of $e^{int}/(n^2-1)$. The pattern matching of `linv` implements this.

- Express the solution x in terms of the complex amplitude a, which depends upon time t, and its complex conjugate b. Use a separate independent variable for the complex conjugate, as it is easiest to perform the algebra. Indeed the analysis then holds for fully complex solutions of the ODE; the symmetry that maintains $b = \bar{a}$ follows automatically.

  We then tell Reduce that time derivatives of these two complex amplitudes are to be found in `ga` and `gb`.

- The linear approximation is then $x \approx ae^{it} + be^{-it}$, where $\dot{a} = g_a \approx 0$ and $\dot{b} = g_b \approx 0$.

[18] In principle, we could try to account for the evolution of the complex amplitude a in seeking a correction for any given residual, but this would be a very complicated task. Instead, we prefer to correct the transformation in many approximate steps, rather than fewer accurate steps.

Dialogue 3.2 *Output of Algorithm* 3.3.

```
1: in oscex$
x := small*(cis(t)*a + cis( - t)*b)
ga := gb := 0
2: x;
                                          5   27              4
small*(cis(t)*a + cis( - t)*b) + small *(----*cis(3*t)*a *b
                                              64
       3              5   27               4
   - ----*cis(5*t)*a  + ----*cis( - 3*t)*a*b
       64                 64
       3               5
   - ----*cis( - 5*t)*b )
       64
          3      1           3    1               3
 + i*small *( - ---*cis(3*t)*a  + ---*cis( - 3*t)*b )
                 8                 8
3: ga;
   3       2  2      9        4  3  2
 - ---*small *a *b + ----*i*small *a *b
   2                 16
4: gb;
   3       2    2    9        4  2  3
 - ---*small *a*b  - ----*i*small *a *b
   2                 16
```

But I introduced a strange variable `small`; what is its role? The variable `small` simply counts the number of complex amplitudes a and b in any given term. As there will always be a `small` carried with each a and b in the computation, the nonlinearities will multiply the `small`'s as they do the amplitudes. We need to count the amplitudes, so we truncate the asymptotic series *consistently* by, for example, here setting `small^6=>0`: this discards all terms in the residual with six or more amplitude factors. Introducing the dummy counting variable `small` is the simplest way to truncate such multivariate asymptotic series.

- Finally, the loop computes the residual and derives corrections to the shape of the oscillations and the evolution of the complex amplitudes a and b.

  To easily determine the coefficients $\pm i/2$ in the correction of the amplitudes' evolution, put some variable name there, iterate the loop twice only, and see what value of the variable makes progress in reducing the residual.

  The division by the amplitude-counting variable `small` accounts for the fact that when $\dot{a}$ is replaced by g_a, there is already one `small` factor in the variable a of $\dot{a}$, and so we always need to carry one less `small` factor in the evolution expressions g_a and g_b.

Execute this program and see the Reduce output in Dialogue 3.2. Ignore the counting variable `small`, although it is useful for organizing the printed output, and note both the shape (3.19a) of the oscillations in the expression for x and the evolution of the complex amplitude (3.19b) in the expression in `ga`.

Such Reduce code needs only minor modifications to generate complex-amplitude normal forms for other oscillatory ODEs; perhaps replace the cubic nonlinearity by the quadratic $\dot{x}^2$ for another example. These normal forms extract from the form of the equations governing fast nonlinear oscillations the slow evolution of their amplitude and phase. In this normal form we readily interpret the behavior of the oscillations.

Alternatively, artificially parametrize nonlinearity Some may not like using variables such as `small` to count the nonlinear order of a term. A simpler and equivalent alternative is to artificially parametrize the nonlinearity. For example, to solve $\ddot{x}+x+\dot{x}^3=0$ we embed it as the $\epsilon=1$ case of the family of ODEs $\ddot{x}+x+\epsilon\dot{x}^3=0$. Analyze this family by treating the parameter ϵ as small. For Example 3.2 one obtains exactly the same description but with parameter ϵ appearing—compare the following with (3.19):

$$\begin{aligned} x &= ae^{it}+\bar{a}e^{-it}+\epsilon i\tfrac{1}{8}\big(-a^3e^{i3t}+\bar{a}^3e^{-i3t}\big) \\ &\quad +\epsilon^2\tfrac{3}{64}\big(-a^5e^{i5t}+9a^4\bar{a}e^{i3t}+9a\bar{a}^4e^{-i3t}-\bar{a}^5e^{-i5t}\big) \\ &\quad +\mathcal{O}(\epsilon^3), \end{aligned} \tag{3.20}$$

where the complex amplitude evolves according to

$$\dot{a}=-\epsilon\tfrac{3}{2}|a|^2a+\epsilon^2 i\tfrac{9}{16}|a|^4a+\mathcal{O}(\epsilon^3)\,. \tag{3.21}$$

Then substitute $\epsilon=1$ to obtain precisely the same modeling—apart from the error term, which has to be inferred.

Interpret the complex exponential normal form In essence the complex exponential normal form invokes a time-dependent coordinate transform, such as the one shown in Figure 3.8. The new complex-amplitude coordinates a and b parametrize location in complex x and $y=\dot{x}$ by the time-periodic complex exponentials $e^{\pm it}$. A tragedy of our three-dimensional world is that we cannot visualize the full wonder of this four-dimensional complex coordinate transform (two real dimensions for each of the two complex variables x and $\dot{x}$). Perhaps the easiest way to appreciate the time dependence in the coordinate transform is the following argument for *real* solutions when $b=\bar{a}$. Recall that

$$x(t)=ae^{it}+\bar{a}e^{-it} \quad\text{and}\quad y(t)=\dot{x}\approx iae^{it}-i\bar{a}e^{-it}\,.$$

Use A and B to parametrize the complex subspace $a=\frac{1}{2}(A+iB)$ and $b=\bar{a}=\frac{1}{2}(A-iB)$; then substituting, separating real and imaginary parts, rearrange the above to the vector identity

$$(x,y)=A(\cos t,-\sin t)+B(-\sin t,-\cos t)\,.$$

That is, write locations in the xy-plane (the phase plane) as a linear combination of the two orthogonal vectors $(\cos t,-\sin t)$ and $(-\sin t,-\cos t)$—consider these two vectors as unit basis vectors of a coordinate system. The coefficients A and B of the linear combination are the coordinates in this rotating coordinate system. The novel feature is simply that the unit basis vectors of the coordinate system rotate (clockwise) in time as seen clearly near the origin in Figure 3.8. Such uniform rotation of the basis unit vectors empowers the coordinate system to simply describe oscillations of nearly the same frequency. The complex exponential normal form is so powerful because the beautiful properties of the exponential function easily allow variations in amplitude and phase of the complex exponentials to be encompassed by variations in the complex amplitudes a and b.

Figure 3.8 also demonstrates how the complex exponential form also adapts the coordinate transform to cater for nonlinear deformations of the trajectories of a nonlinear oscillator at finite amplitude. The underlying coordinate curves "breathe" with the oscillation to empower us to simply describe the evolution of the oscillations' complex amplitude.

Example 3.3 (electromagnetic waves). Recall the electromagnetic wave system of section 3.5. The spatial structure is governed by equation (3.13), namely

$$\frac{d^2\psi}{dx^2} + \psi + \epsilon|\psi|^2\psi = 0$$

for some complex field $\psi(x)$. In the absence of nonlinearity, parameter $\epsilon = 0$, the general solution is $\psi = ae^{ix} + be^{-ix}$, representing right- and left-traveling waves, respectively.[19]

Here we let amplitudes a and b be complex-valued functions of position x. We aim to transform the one second-order differential equation for complex-valued ψ into two first-order complex differential equations for the two complex-valued amplitudes a and b. In particular, we aim this transform to be valid in the parameter domain where the complex amplitudes are "slowly" evolving.

In a first iterative step, let's seek a field $\psi = ae^{ix} + be^{-ix} + \hat{\psi}(a, b, x)$, for small $\hat{\psi}$, such that the x-derivatives a' and b' are small. The derivative $\psi' \approx (a' + ia)e^{ix} + (b' - ib)e^{-ix} + \hat{\psi}'$, upon neglecting $(\partial\hat{\psi}/\partial a)a'$ and $(\partial\hat{\psi}/\partial b)b'$ as being small. Similarly, the second derivative $\psi'' \approx (2ia' - a)e^{ix} + (-2ib' - b)e^{-ix} + \hat{\psi}''$, upon also neglecting smaller a'' and b''. The nonlinearity $|\psi|^2\psi = \bar{\psi}\psi^2$, upon using overbars for complex conjugation, so the governing differential equation (3.13) becomes approximately

$$\begin{aligned} 0 &= 2ia'e^{ix} - 2ib'e^{-ix} + \hat{\psi}'' + \hat{\psi} \\ &\quad + a^2\bar{b}e^{i3x} + (\bar{a}a^2 + 2a\bar{b}b)e^{ix} + (2\bar{a}ab + \bar{b}b^2)e^{-ix} + \bar{a}b^2e^{-i3x}. \end{aligned}$$

Summing coefficients of $e^{\pm ix}$ to zero leads to the approximate evolution

$$\begin{aligned} a' &\approx +i(\tfrac{1}{2}\bar{a}a^2 + a\bar{b}b) = +i(\tfrac{1}{2}|a|^2 + |b|^2)a\,, \\ b' &\approx -i(\tfrac{1}{2}\bar{b}b^2 + b\bar{a}a) = -i(\tfrac{1}{2}|b|^2 + |a|^2)b\,. \end{aligned}$$

Then incorporate the $e^{\pm i3x}$ terms into the field through $\hat{\psi} = \frac{1}{8}(a^2\bar{b}e^{i3x} + \bar{a}b^2e^{-i3x})$; that is, the complex field is

$$\psi \approx ae^{ix} + be^{-ix} + \tfrac{1}{8}(a^2\bar{b}e^{i3x} + \bar{a}b^2e^{-i3x}).$$

This complex-amplitude transformation of the dynamics is precisely equivalent to the complex polar transformation (3.15) found in section 3.5. It empowers us to describe and understand oscillations where the properties of the oscillations vary relatively slowly.

[19]Here there is no notion that complex amplitudes a and b need to be complex conjugates of each other. In this application, a and b are independently complex.

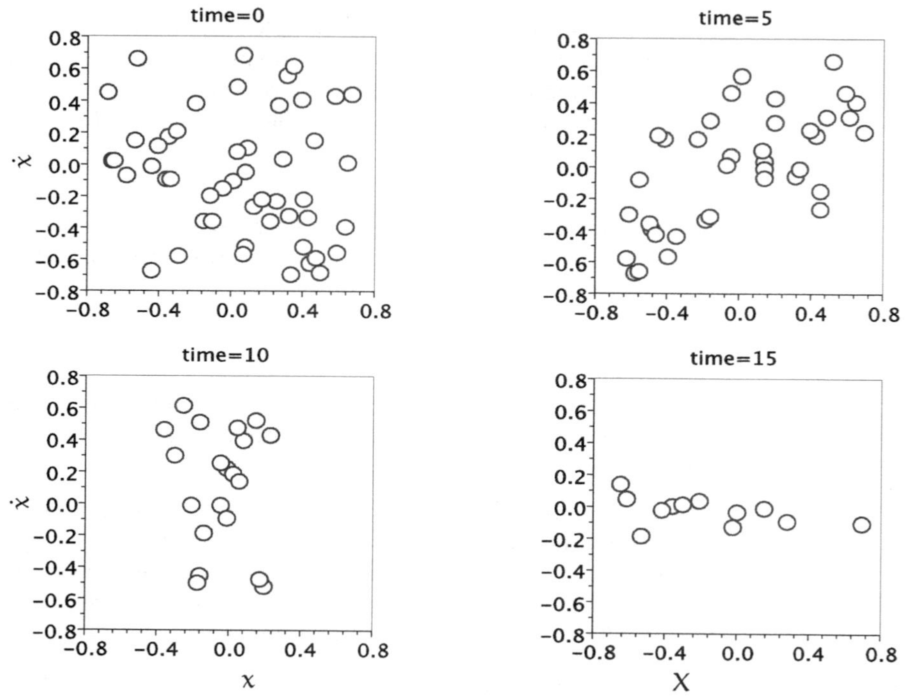

Figure 3.9. *Fifty initial conditions in the phase plane evolve according to Mathieu's ODE* (3.22) *with* $\omega = 2$ *and* $\epsilon = 0.2$. *They rotate with frequency one. Over long times solutions collapse to a rotating line and then grow along that line.*

3.7 ▪ Forcing or parameter variation also modulates oscillations

So far we have discussed how nonlinearities affect the amplitude and frequency of oscillations. Periodic forcing also modifies oscillatory dynamics in marvelous ways. In the case of Mathieu's ODE, periodic forcing generates growing instabilities in certain parameter regions. Similarly, oscillations involving slow variations in properties, such as slowly changing the length of a pendulum, also modify oscillations. The complex-amplitude normal form empowers us to analyze both of these situations.

Similar but generalized analysis applies to traveling waves and other more complicated oscillatory dynamics. But here we restrict attention to simple oscillations as an introduction to the methodology.

3.7.1 ▪ Mathieu's ODE displays resonant instabilities

The linear Mathieu's ODE,

$$\ddot{x} + (1 + 2\epsilon \cos \omega t)x = 0, \tag{3.22}$$

is an example of an ODE whose coefficients are periodic; here the parameter ϵ gives the size of the periodic variations, and ω is their frequency relative to the natural frequency of the base oscillator. Although linear, Mathieu's ODE (3.22) has complications caused by the varying coefficient $\cos \omega t$: Figure 3.9 demonstrates the solutions growing exponentially in time for frequencies near "resonance." A normal-form transformation to complex-amplitude variables shows the regions of stability and instability.

Algorithm 3.4 *Reduce code fragment to transform Mathieu's* ODE (3.22) *into a normal form of complex amplitudes of complex exponentials. This code follows the code from Algorithm* 3.3 *for complex exponentials* `cis`, *the linear operator* `linv`, *and the complex-amplitude variables.*

```
% frequency of variations
w:=2;
coswt:=(cis(w*t)+cis(-w*t))/2;
% initial linear approximation
x:=a*cis(t)+b*cis(-t);
ga:=gb:=0;
% iterate to nonlinear transformation
let eps^3=>0;
for iter:=1:99 do begin
   res:=df(x,t,t)+(1+2*eps*coswt)*x;
   ga:=ga+i/2*(ca:=coeffn(res,cis(+t),1));
   gb:=gb-i/2*(cb:=coeffn(res,cis(-t),1));
   x:=x+linv(res-ca*cis(t)-cb*cis(-t),cis);
   if res=0 then iter:=1000000;
end;
```

Let's investigate two cases of different frequencies in the varying coefficient: $\omega = 2$ gives the strongest response of the solution to the fluctuations in the coefficient, and $\omega = 1$ when the natural frequency is the same as that of the fluctuations.

Harmonic fluctuations excite instability To discover the behavior of the oscillations seen in Figure 3.9 we transform to complex-amplitude variables. The Reduce code to do this transformation is a variation of that already discussed for the nonlinear oscillations of Example 3.2. Copy some parts of Algorithm 3.3 and modify to form Algorithm 3.4. The modifications

- represent the periodic fluctuations $\cos\omega t$ in the coefficient of Mathieu's ODE (3.22) by `coswt`, which is written in terms of complex exponentials $e^{\pm i\omega t}$ for some set frequency ω;
- change the residual computation to that for Mathieu's ODE (3.22);
- discard the dummy counting variable `small`, as in this linear problem we have no need to count the amplitude factors;
- instead, truncate the asymptotic series in the size of the fluctuations; here `eps^3=>0` gives expressions to errors $\mathcal{O}(\epsilon^3)$.

Execute Algorithm 3.4 to see the Reduce output in Dialogue 3.3. Thus deduce that the oscillations of Mathieu's ODE (3.22) with $\omega = 2$ have shape

$$x = ae^{it} + be^{-it} + \frac{\epsilon}{8}\left(ae^{i3t} + be^{-i3t}\right) + \frac{\epsilon^2}{192}\left(ae^{i5t} - 9be^{i3t} + 9ae^{-i3t} + be^{-i5t}\right) + \mathcal{O}(\epsilon^3). \tag{3.23}$$

Dialogue 3.3 *Output of Algorithm* 3.4.

```
1: in mathieu$
w := 2
            1              1
coswt := ---*cis(2*t) + ---*cis( - 2*t)
            2              2
x := cis(t)*a + cis( - t)*b
ga := gb := 0
2: x;

                                          1                     1
cis(t)*a + cis( - t)*b + eps*(---*cis(3*t)*a + ---*cis( - 3*t)*b)
                                          8                     8
         2      3                   1
  + eps  *( - ----*cis(3*t)*b + -----*cis(5*t)*a
                64                 192
                3                     1
            - ----*cis( - 3*t)*a + -----*cis( - 5*t)*b)
                64                   192
3: ga;
 1            3     2
---*eps*i*b + ----*eps  *i*a
 2            16
4: gb;
   1             3     2
 - ---*eps*i*a - ----*eps  *i*b
   2             16
```

Alternatively, view (3.23) as, together with $y = \dot{x}$, a time-dependent coordinate transform such as the snapshots shown in Figure 3.10. Any dynamics may be captured in these coordinates. But this particular coordinate transform is chosen to simplify our description of the dynamics for small parameter ϵ.

The important outcome of the transform is that the complex amplitude of the oscillations evolves according to the normal form

$$\dot{a} = i\tfrac{1}{2}\epsilon b + \mathcal{O}(\epsilon^2) \quad \text{and} \quad \dot{b} = -i\tfrac{1}{2}\epsilon a + \mathcal{O}(\epsilon^2)\,. \tag{3.24}$$

Having constant coefficients, this pair of ODEs for the complex amplitude is much simpler than Mathieu's ODE (3.22). To find the instability, seek solutions to this pair of coupled ODEs proportional to $e^{\lambda t}$: deduce the characteristic equation $\lambda^2 - \frac{1}{4}\epsilon^2 = 0$, and hence the complex amplitudes are a linear combination of the two components $\exp(\pm\frac{1}{2}\epsilon t)$. Consequently we predict an instability on a time scale of $1/\epsilon$ from the growing component $\exp(\frac{1}{2}\epsilon t)$.

Synchronous fluctuations also excite instability What happens when the forcing is $\cos t$ rather than $\cos 2t$? Find out simply by setting `w:=1` in Algorithm 3.4; then execute the Reduce code to see the output in Dialogue 3.4. Thus deduce that the oscillations of Mathieu's ODE (3.22) with $\omega = 1$ have shape

$$x = ae^{it} + be^{-it} + \epsilon\left[-(a+b) + \tfrac{1}{3}ae^{i2t} + \tfrac{1}{3}b^{-i2t}\right] + \frac{\epsilon^2}{24}\left(ae^{i3t} + be^{-i3t}\right) + \mathcal{O}(\epsilon^3)\,. \tag{3.25}$$

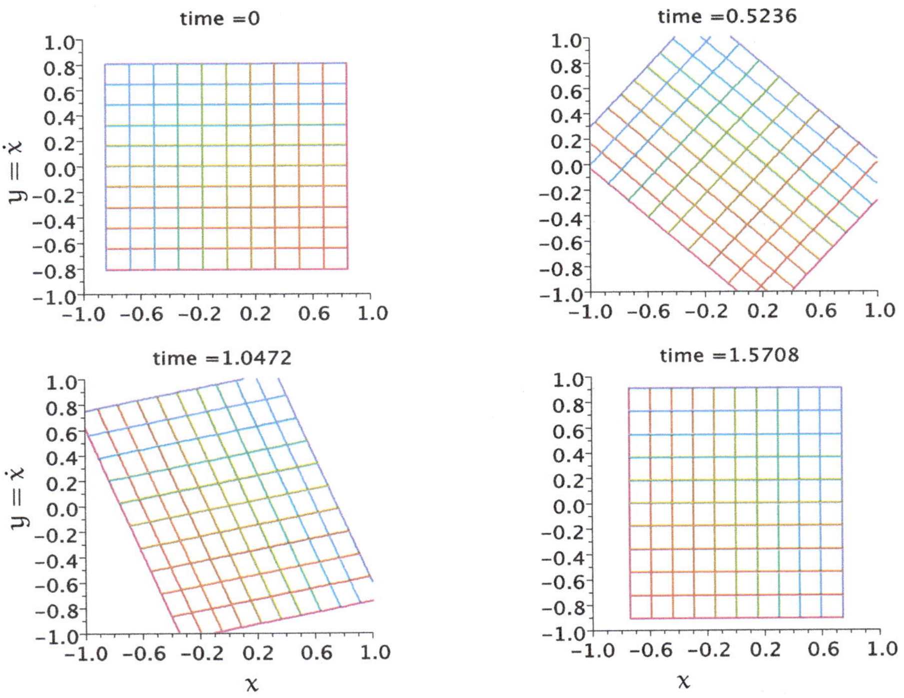

Figure 3.10. *The form* (3.23) *is a time-dependent, rotating coordinate transform: set complex amplitude* $a = \frac{1}{2}(A + iB)$ *for real* A *and* B*; then plot coordinate curves in the* $x\dot{x}$*-plane of constant* A *and* B *at different times over a quarter period for parameter* $\epsilon = \frac{1}{2}$. *These coordinate curves deform as they rotate to suit the dynamics of the forced linear Mathieu's ODE* (3.22).

Dialogue 3.4 *Output of Algorithm* 3.4 *with forcing frequency changed to* $\omega = 1$.

```
1: in mathieu$
w := 1
           1           1
coswt := ---*cis(t) + ---*cis( - t)
           2           2
x := cis(t)*a + cis( - t)*b
ga := gb := 0
2: x;
cis(t)*a + cis( - t)*b
                        1                   1
 + eps*( - b - a + ---*cis(2*t)*a + ---*cis( - 2*t)*b)
                        3                   3
      2   1                  1
 + eps *(----*cis(3*t)*a + ----*cis( - 3*t)*b)
          24                 24
3: ga;
   2       1       1
eps *i*( - ---*b - ---*a)
           2       3
4: gb;
   2     1       1
eps *i*(---*b + ---*a)
         3       2
```

More important, the complex amplitude of the oscillations evolves according to the normal form

$$\dot{a} = -i\epsilon^2(\tfrac{1}{3}a + \tfrac{1}{2}b) + \mathcal{O}(\epsilon^3) \quad \text{and} \quad \dot{b} = +i\epsilon^2(\tfrac{1}{2}a + \tfrac{1}{3}b) + \mathcal{O}(\epsilon^3)\,. \tag{3.26}$$

Here the coefficients in the ODEs are of $\mathcal{O}(\epsilon^2)$, whereas in the $\omega = 2$ case they are of $\mathcal{O}(\epsilon)$: any instability is weaker in this case of $\omega = 1$. To find the instability, seek solutions to this pair of coupled ODEs proportional to $e^{\lambda t}$ and deduce the characteristic equation $\lambda^2 - \frac{5}{36}\epsilon^4 = 0$; hence the complex amplitudes are a linear combination of $\exp(\pm\sqrt{5}\epsilon^2 t/6)$. Consequently we will see an instability on a long time scale of $1/\epsilon^2$ from the growing mode $\exp(\sqrt{5}\epsilon^2 t/6)$.

Where are the stability boundaries? The previous analysis shows that exactly at these critical frequencies, the response of Mathieu's oscillator is that of growing oscillations. But such growth only occurs close enough to the critical cases. Let's discover how close by detuning the natural oscillations and seeing how the growth rate depends upon the detuning parameter.

Consider the $\omega = 2$ case and modify Mathieu's ODE (3.22) to

$$\ddot{x} + (1 + \delta + 2\epsilon \cos 2t)x = 0$$

for a small detuning parameter δ. The natural oscillations of the ODE, without any imposed fluctuations $\epsilon = 0$, then have frequency $\sqrt{1+\delta}$. This natural frequency is a little different from the subharmonic of the fluctuations $\cos 2t$: the parameter δ controls the amount of the difference. Correspondingly modify the Reduce code of Algorithm 3.4:

- change the residual computation to

```
res:=df(x,t,t)+(1+delta+2*eps*coswt)*x;
```

- and ensure that parameter δ is treated as small by truncating the asymptotic series in it as well as ϵ:

```
let { eps^2=>0, delta^2=>0 };
```

Then execute the Reduce code to find the output of Dialogue 3.5. That is, the complex amplitude of the oscillations of Mathieu's ODE (3.22) is governed by[20]

$$\dot{a} = i\tfrac{1}{2}\delta a + i\tfrac{1}{2}\epsilon b + \mathcal{O}(\epsilon^2, \delta^2) \quad \text{and} \quad \dot{b} = -i\tfrac{1}{2}\epsilon a - i\tfrac{1}{2}\delta b + \mathcal{O}(\epsilon^2, \delta^2)\,. \tag{3.27}$$

Again, to find any instability, seek solutions to this pair of coupled ODEs proportional to $e^{\lambda t}$ and deduce the characteristic equation $\lambda^2 + \frac{1}{4}\delta^2 - \frac{1}{4}\epsilon^2 = 0$; hence the growth rates are $\lambda = \pm\frac{1}{2}\sqrt{\epsilon^2 - \delta^2}$. Only for $|\epsilon| > |\delta|$ will there be instability. The boundaries of the instability are $\delta = \pm\epsilon$.

These boundaries are only approximate. Retaining higher-order terms in parameters ϵ and δ gives a characteristic equation that predicts more accurate stability boundaries.

[20]Equation (3.27) is an example of a multivariable asymptotic expansion. These will occur more often as we consider physical problems with more competing physical processes that need parametrization. We have not established the properties for the manipulation of multivariable asymptotic series; there are some subtleties that we blithely ignore for now.

Dialogue 3.5 *Output of Algorithm 3.4 but with the natural frequency detuned by a small amount.*

```
1: in mathieu$
w := 2
             1              1
coswt := ---*cis(2*t) + ---*cis( - 2*t)
             2              2
x := cis(t)*a + cis( - t)*b
ga := gb := 0
2: ga;
 1              1
---*i*a*delta + ---*eps*i*b
 2              2
3: gb;
    1                1
 - ---*i*b*delta - ---*eps*i*a
    2                2
```

3.7.2 ▪ WKB theory is also a normal form

The periodic fluctuations in Mathieu's ODE (3.22) are easily catered for because, for small parameter ϵ, the coefficient of the x term in the ODE was never far from some fixed constant, namely the constant 1. But when water waves propagate from the deep ocean to the shallow coast, there is an enormous change in depth, causing major changes to the wave's properties. Similarly, a quantum particle is trapped by a continuously varying potential well that varies through the energy of the particle and hence changes the quantum probability wave so much that it is reflected. We proceed to extend the analyzes and algebra of the preceding sections to this situation of widely varying coefficients in an ODE that supports oscillatory dynamics. Our analysis parallels and reproduces the results of the well-known WKB theory (Bender and Orszag, 1981, §11.1, e.g.).[21]

The canonical problem with varying coefficients is the linear oscillator with time-varying frequency $\omega(t)$:

$$\ddot{x} + \omega^2 x = 0\,. \tag{3.28}$$

For example, in the potential well of a quantum particle, with t representing the spatial coordinate, ω^2 is the difference between the energy of the particle and the potential energy at any position t. Now, if ω is roughly constant near any time, then surely the solution x should locally be a linear combination of $\exp(\pm i\omega t)$. Actually this is not quite correct. A better approximation is that the solution x is a linear combination of $\exp(\pm i \int \omega\, dt)$: the phase angle $\theta(t) = \int \omega\, dt$ does indeed change by 2π over a local period $\Delta t \approx 2\pi/\omega$, and so $\exp(\pm i\theta(t))$ are global smooth oscillations that locally have the correct frequency.

Thus we approach WKB solutions to the varying oscillator (3.28) by posing solutions of the form

$$x = a e^{i\theta} + b e^{-i\theta} \quad \text{such that} \quad \dot{a} = g_a(\omega, a) \quad \text{and} \quad \dot{b} = g_b(\omega, b)\,, \tag{3.29}$$

where the complex amplitudes a and b evolve relatively slowly and similar to the rate at which the natural frequency ω varies. For real solutions x, the complex amplitudes a and b will be complex conjugates; if derived correctly, the evolution (3.29) will maintain this complex conjugate symmetry.

[21] Hinch (1991, §7.5) identifies 12 people who each played an important part in the development of WKB theory: Liouville, Green, Horn, Rayleigh, Gans, Jeffrey, Wentzel, Kramers, Brillouin, Langer, Olver, and Meyer.

Algorithm 3.5 *Reduce code to construct the* WKB *approximation to the slowly varying linear oscillator* (3.28). *Use definitions of the complex exponential and the complex-amplitude variables of Algorithm* 3.3 *followed by this code.*

```
% phase angle and omega=w has small derivatives
depend q,t;
depend w,tt; depend tt,t;
let { df(q,t)=>w, df(w,t)=>eps*df(w,tt) };
% initial linear approximation
x:=a*cis(q)+b*cis(-q);
ga:=gb:=0;
% iterate to normal-form transformation
let eps^3=>0;
for iter:=1:99 do begin
    res:=df(x,t,t)+w^2*x;
    ga:=ga+i/2/w*(ca:=coeffn(res,cis(+q),1));
    gb:=gb-i/2/w*(cb:=coeffn(res,cis(-q),1));
    showtime;
    if res=0 then iter:=1000000;
end;
```

Computer algebra derives the WKB solution by adapting ideas in Algorithms 3.1 and 3.3. Algorithm 3.5 is the result.

- The first innovation is to write the oscillations in terms of the phase θ (represented by `q` in the Reduce code) such that $d\theta/dt = \omega$, as implemented by the `let` rule `df(q,t)=>w`.

- But we need to account for derivatives of the varying frequency ω. Use the dummy parameter `eps` to count the number of time derivatives in each term. Insert the count by splitting the time derivative of the frequency `w` into two via the two commands `depend w,tt;` and `depend tt,t;`. Then count with the pattern match and replacement `df(w,t)=>eps*df(w,tt)`.[22]

  To form the asymptotic expansion, assume that the variations of the frequency ω are slow enough so that terms with n derivatives of ω are asymptotically smaller than those with $n-1$ derivatives and hence we may discard terms with more than a set number of derivatives. As coded in Algorithm 3.5, we work to errors $\mathcal{O}(\epsilon^3)$, that is, we retain only terms with zero, one, or two derivatives of frequency ω. This limit may of course be increased as you wish.

- Finally, in this linear problem there is no need to correct the form of the oscillations; hence we do not need to change `x` and neither do we need the pattern matching of `linv`. For nonlinear ODEs we would need `linv`.

Execute the above Reduce code to get the output of Dialogue 3.6. That is, the WKB normal form for the linear varying oscillator (3.28) is that $x = a e^{i\theta} + b e^{-i\theta}$ for phase

[22] Those embedded in the methodology of multiple scales will want to call `tt` $= T = \epsilon t$ a "slow time." Resist the temptation. In our normal-form approach, the dummy parameter ϵ is only there to count the number of derivatives of frequency ω. The superficial similarity between $T = \epsilon t$ and a slow time is because in multiple scales the parameter ϵ is in essence doing the same job of counting derivatives.

Dialogue 3.6 *Output of Algorithm* 3.5.

```
1: in "wkb.red"$
x := b*cis( - q) + a*cis(q)
ga := gb := 0
2: ga;
   1                    -1         2
 - ---*a*eps*df(w,tt)*w     + a*eps *i
   2
   3           2  -3     1              -2
*(---*df(w,tt)  *w     - ---*df(w,tt,2)*w   )
   8                     4
3: gb;
   1                    -1         2
 - ---*b*eps*df(w,tt)*w     + b*eps *i
   2
     3           2  -3     1              -2
*( - ---*df(w,tt)  *w    + ---*df(w,tt,2)*w   )
     8                     4
```

$\theta = \int \omega \, dt$, where the complex amplitudes evolve according to

$$\dot{a} = -\frac{\dot{\omega}}{2\omega} a + i\left(\frac{3\dot{\omega}^2}{8\omega^3} - \frac{\ddot{\omega}}{4\omega^2}\right) a + \mathcal{O}(\partial_t^3 \omega), \tag{3.30a}$$

$$\dot{b} = -\frac{\dot{\omega}}{2\omega} b - i\left(\frac{3\dot{\omega}^2}{8\omega^3} - \frac{\ddot{\omega}}{4\omega^2}\right) b + \mathcal{O}(\partial_t^3 \omega), \tag{3.30b}$$

and where the error $\mathcal{O}(\partial_t^3 \omega)$ encompasses all terms with three or more derivatives of the slowly varying frequency ω.

For example, we use this model to predict the conservation of action. Consider the leading-order model for the complex amplitude a: $\dot{a} + \frac{\dot{\omega}}{2\omega} a = 0$. Multiply by $2\omega a$ and integrate to obtain $a^2\omega = $ constant, that is, $a^2\omega$ is conserved. Since the energy of an oscillation is proportional to $a^2\omega^2$, this conservation rule is that the action = energy/ω = constant as the oscillations' natural frequency changes.

Example 3.4 (estimate eigenvalues with the WKB normal form). Use the WKB normal form that $x = ae^{i\theta} + \bar{a}e^{-i\theta}$ for phase $\theta = \int \omega \, dt$ and the complex amplitudes evolve according to (3.30a) to estimate the eigenvalues E of[23]

$$\ddot{x} + E(t+\pi)^4 x = 0 \quad \text{such that} \quad x(0) = x(\pi) = 0. \tag{3.31}$$

From the earlier analysis, here I set $b = \bar{a}$ to ensure real solutions.

Solution: This ODE is of the form of the linear oscillator (3.28) with varying frequency $\omega = \sqrt{E}(t+\pi)^2$. To be definite, define the phase

$$\theta = \int_0^t \omega(\tau) \, d\tau = \sqrt{E} \int_0^t (\tau + \pi)^2 \, d\tau = \tfrac{1}{3}\sqrt{E}[(t+\pi)^3 - \pi^3].$$

To satisfy the boundary condition that $x(0) = 0$, since $\theta(0) = 0$, $e^{\pm i\theta(0)} = 1$, so $x(0) = a + \bar{a} = 0$, requiring a to be pure imaginary. As a first approximation, truncate the normal form to $\dot{a} = -\dot{\omega}a/(2\omega)$; then throughout the domain of

[23]For example, this ODE might describe the oscillations of a quantum particle in a quantum potential well, in which case the eigenvalues E represent the discrete energy levels of the quantum dynamics.

integration the complex amplitude a remains pure imaginary. Thus the only way to ensure $x(\pi) = 0$ is to require $\theta(\pi) = n\pi$ for some integral n. This equation estimates the eigenvalues:

$$\tfrac{1}{3}\sqrt{E_n}7\pi^3 \approx n\pi \quad \Longrightarrow \quad E_n \approx \frac{9n^2}{49\pi^4}\,.$$

Bender and Orszag (1981, p. 492) show that this approximation for the eigenvalues is amazingly good. We expect it to be good for large n as then the variations in natural frequency ω are relatively small per oscillation: indeed E_{10} is accurate to a relative error of 0.13%. But the above formula estimates even the fundamental eigenvalue E_1 to a relative error of no more than 8.1%.

Bender and Orszag (1981, Chap. 10) discuss many more aspects of the WKB approximation, including conditions of its validity. The normal form derived above is equivalent to the WKB approximation they discuss. However, our approach empowers us to analyze much more difficult problems, such as nonlinear oscillators with large variations in coefficients provided the variations take place slowly. For example, Exercise 3.22 asks you to analyze a nonlinear Duffing oscillator with a varying frequency. The necessary modifications to the computer algebra are straightforward. *This one normal-form approach fantastically allows us to solve straightforwardly a wide range of difficult problems.*

Our approach here is also an alternative to the so-called Magnus expansion for the approximation of rapidly oscillating solutions to differential equations with varying coefficients (Blanes et al., 2009, e.g.).

Furthermore, our normal-form approach also closely parallels aspects of homogenization methods and theory; for example, Bensoussan et al. (1978) or Rosencrans (1997) may introduce you to aspects of homogenization. Many practical situations involve dispersion or other macroscopic transport through a material with complicated microscopic structure—groundwater flow through soil, for example. Homogenization analyzes the microstructure and the transport through the microstructure to derive an equation for the effective macroscopic transport. Here we similarly analyze the microscopic shape of oscillations to derive equations governing the evolution of the macroscopic amplitude and phase of the oscillations over long macroscopic times. Our analysis is simply based upon a normal-form change in the coordinates. *The normal-form coordinate transform approach subsumes or illuminates many other modeling methods.*

3.8 ▪ Exercises

Exercise 3.1 (simple nonlinear oscillator). Consider the oscillator of section 3.1 described by the ODEs (3.1). Change from Cartesian variables to amplitude $r(t)$ and phase angle $\theta(t)$ variables,

$$x = r\cos\theta \quad \text{and} \quad y = r\sin\theta\,,$$

to discover the ODEs (3.3) governing the evolution of amplitude and angle. Hint: Differentiate $r^2 = x^2 + y^2$ to determine $\dot r$.

Exercise 3.2 (Stuart–Landau oscillator). Consider the oscillations $\mathbf{x}(t) = (x(t), y(t))$ governed by the nonlinear system (Takeshita and Feres, 2010, e.g.)

$$\dot{\mathbf{x}} = \begin{bmatrix} 1 & -a \\ a & 1 \end{bmatrix}\mathbf{x} - \begin{bmatrix} 1 & -b \\ b & 1 \end{bmatrix}\rho(|\mathbf{x}|)\mathbf{x},$$

where a and b are constants and $\rho(r)$ is any smooth function, such as $\rho(r) = r^2$, for which $\rho(1) = 1$ and $\rho'(1) > 0$. Transform to amplitude $r(t)$ and phase angle $\theta(t)$ variables (polar coordinates),

$$x = r\cos\theta \quad \text{and} \quad y = r\sin\theta\,,$$

to discover the normal form

$$\dot{r} = [1 - \rho(r)]r \quad \text{and} \quad \dot{\theta} = a - b\rho(r)\,.$$

Deduce that there exists a stable so-called limit cycle at $r = 1$ of period $2\pi/(a-b)$.

Further, find that exponential spirals are the so-called isochrons for the limit cycle. Isochrons are those curves that all have the same ultimate evolution; thus we seek curves that map into a shifted version of themselves into the future. Since the evolution on the limit cycle is $\dot{\theta} = a - b$, seek invariant curves, the isochrons, of the form $\theta = f(r) + (a - b)t$. Differentiate, rearrange, and cancel to derive $df/dr = b/r$ and thus proceed to deduce that the isochrons are exponential spirals.

Exercise 3.3 (modify Duffing). Modify the amplitude-angle transformation of Duffing's ODE (3.4) to interpret the long-term evolution of the nonlinear oscillations (Bender and Orszag, 1981, pp. 552–3, e.g.) governed by the ODE (3.18), $\ddot{x} + x + \dot{x}^3 = 0$. Hint: The algebra of the expansion of the second time derivative $\ddot{x}$ is identical.

Answer: $\dot{r} \approx -\frac{3}{8}r^3$ and $\dot{\theta} \approx 1$: the oscillations decay slowly to zero amplitude while their frequency remains approximately constant, as shown by Figure 3.7 (but see Exercise 3.5).

Exercise 3.4 (damped Duffing). Modify the computer algebra of Algorithm 3.1 to interpret the long-term evolution of the nonlinear oscillations governed by the damped Duffing equation (O'Malley, Jr., and Kirkinis, 2010, eq. (15))

$$\ddot{x} + 2\epsilon\dot{x} + x - x^3 = 0 \quad \text{for small damping } \epsilon > 0\,.$$

Hint: Include a pattern-matching rule for the terms that arise in the residual involving $\sin n\theta$.

Exercise 3.5 (modify Duffing). Modify the computer algebra of Algorithm 3.1 to interpret the long-term evolution of the nonlinear oscillations governed by (3.18). Hint: Include a pattern-matching rule for the terms that arise in the residual involving $\sin n\theta$.

Answer: $\dot{r} \approx -\frac{3}{8}r^3 - \frac{351}{8192}r^7$ and $\dot{\theta} \approx 1 + \frac{9}{256}r^4 + \frac{1755}{262144}r^8$. Oscillations decay slowly to zero amplitude, roughly $r \propto t^{-1/2}$.

Exercise 3.6 (nonlinear pendulum). Modify the computer algebra of Algorithm 3.1 to find the frequency change in the nonlinear pendulum governed by the ODE

$$\ddot{x} + \sin x = 0\,, \tag{3.32}$$

where x is the angle of the pendulum from the vertical. Hint: When computing the residual for any given approximation to the pendulum's angle $x(r,\theta)$, use an appropriate Taylor series of $\sin x$ rather than the sine function as is.

Answer: $x \approx r\cos\theta - \frac{1}{192}r^3\cos 3\theta$ such that $\dot r = 0$ and frequency $\dot\theta \approx 1 - \frac{1}{16}r^2 + \frac{1}{1024}r^4$.

Exercise 3.7 (nonlinear pendulum extension). Modify your solution for Exercise 3.6. Recall that the pendulum equation is $\ddot x + (g/\ell)\sin x = 0$, where g is gravity and ℓ is the length of the pendulum; use your computer algebra to find how one should, in principle, vary the length ℓ with amplitude r of the oscillation so that the frequency $\dot\theta$ remains precisely 1 (to some order of error in amplitude r). Hint: Seek $g/\ell = k(r)$, where $k \approx 1$ and you update an approximation to k instead of updating $\dot\theta$.

Further extension: How can you vary the length ℓ with pendulum angle x so that the frequency remains precisely 1.

Answer: $\ell \approx g(1 - \frac{1}{8}r^2 + \frac{3}{512}r^4)$ and $\ell = g\sin x/x$.

Exercise 3.8 (van der Pol cycles). Use a normal-form transformation to amplitude-phase variables in order to explore the behavior of the van der Pol equation (Kreyszig, 2011, §3.5, e.g.)

$$\ddot x + x + \epsilon(x^2 - 1)\dot x = 0 \tag{3.33}$$

for small values of the nonlinearity parameter ϵ. Describe the qualitative behavior of its solutions. Hint: Instead of discarding high powers of amplitude r, discard high powers of parameter ϵ.

Answer: $\dot r \approx \epsilon(\frac{1}{2}r - \frac{1}{8}r^3)$ and $\dot\theta \approx 1 + \frac{1}{8}\epsilon^2(-1 + r^2 - \frac{7}{32}r^4)$. Oscillations tend to a limit cycle of radius 2 with a slightly lower frequency than that of the linear dynamics.

Exercise 3.9 (Morrison's oscillations). Describe the qualitative behavior of solutions to Morrison's ODE

$$\ddot x + x + \epsilon\dot x^3 - 3\epsilon^2\dot x = 0, \tag{3.34}$$

using its normal-form transformation to amplitude and phase variables. Why do you need to know at least the terms in ϵ^2?

Answer: $\dot r \approx -\frac{3}{8}\epsilon r^3 + \frac{3}{2}\epsilon^2 r$ and $\dot\theta \approx 1 + \frac{9}{256}\epsilon^2 r^4$. Solutions tend to a limit cycle of radius 2ϵ with a very slightly higher frequency than that of the linear oscillations. The ϵ^2 term in $\dot r$ destabilizes the origin to form the limit cycles.

Exercise 3.10 (Chiba's example). Adapt the amplitude-phase transformation to the case of oscillations arising in the system of two first-order interacting differential equations (O'Malley, Jr., and Kirkinis, 2010, eq. (23))

$$\dot x = y - y^2 \quad\text{and}\quad \dot y = -x - \epsilon y \tag{3.35}$$

for small $\epsilon > 0$ and small-amplitude oscillations. Confirm that the linear term $-\epsilon y$ is stabilizing for small parameter ϵ. What is the stability influence of the quadratic nonlinear term $-y^2$?

Exercise 3.11 (electromagnetic waves). Modify just one line of Algorithm 3.2 to find the amplitude-phase dynamics of the spatial waves in the complex van der Pol ODE

$$\frac{d^2\psi}{dx^2} + \psi + \epsilon\left(|\psi|^2 - 1\right)\frac{d\psi}{dx} = 0 .$$

Comment on the evolution of the waves in view of the amplitude-phase dynamics.

Exercise 3.12 (general electromagnetic waves). General nonlinearities in the complex wave equation of Exercise 3.11 cause more difficulties. The reason is that the evolution of the amplitude-phase variables r, s, θ, and ϑ generally involves $\cos(\theta - \vartheta)$, $\sin(\theta - \vartheta)$, and harmonics of these: although θ and ϑ both evolve rapidly, their difference evolves slowly and so can appear in the evolution of the amplitudes r and s; for example, we find $r' \approx -\epsilon^2\frac{1}{3}r^2 s\sin(\theta - \vartheta)$. In this example, seek a simplifying coordinate transform to amplitude-phase variables of the complex ODE

$$\frac{d^2\psi}{dx^2} + \psi + \epsilon|\psi|^2 = 0 .$$

Modify Algorithm 3.2 as follows:

- change the computation of the residual appropriately;
- add rules for the operator `mean` to zero the mean of trigonometric functions with just one phase angle argument and to pass on to the evolution trigonometric functions with argument $m\theta + n\vartheta$, where $m + n = 0$;
- add rules for the operator `linv` to cater for trigonometric functions with just one phase angle argument; and
- linearize the products of trigonometric functions in `ceq*eiq+cep*eip`.

Execute your new algorithm to find that

$$\begin{aligned}
\psi &\approx re^{i\theta} + se^{-i\vartheta} + \epsilon\left[i(s^2 - r^2) - \tfrac{2}{3}\sin(\theta + \vartheta)\right],\\
r' &\approx -\epsilon^2\tfrac{1}{3}r^2 s\sin(\theta - \vartheta),\\
\theta' &\approx 1 + \epsilon^2\left[\tfrac{2}{3}s^2 - \tfrac{1}{2}r^2 - \tfrac{1}{3}rs\cos(\theta - \vartheta)\right].
\end{aligned}$$

Modify the computation of the residual to find amplitude-phase versions of other complex wave ODEs.

Exercise 3.13 (suite of oscillators). Modify the code of Algorithm 3.3 to both analyze and interpret the nonlinear oscillations of the following oscillators using the complex-amplitude normal form:

1. Duffing's ODE (3.4);
2. the nonlinear pendulum (3.32);
3. the van der Pol equation (3.33);
4. Morrison's ODE (3.34).

Confirm that the interpretation of the complex-amplitude normal form agrees with that of the normal form in amplitude-angle variables.

Exercise 3.14 (web service oscillators). A web service to construct center manifolds (Roberts, 2012) also empowers us to construct complex-amplitude normal forms. Use this web service to construct complex-amplitude normal forms of the systems mentioned in Exercise 3.13. Compare the web service results with those you obtain in Exercise 3.13.

For example Consider the ODE (3.18), $\ddot{x}+x+\dot{x}^3=0$. Write as a system of two first-order ODEs using $u_1(t)=x(t)$ and $u_2(t)=\dot{x}(t)$:

$$\dot{u}=\begin{bmatrix}\dot{u}_1\\ \dot{u}_2\end{bmatrix}=\begin{bmatrix}u_2\\ -u_1-u_2^3\end{bmatrix}=\underbrace{\begin{bmatrix}0 & 1\\ -1 & 0\end{bmatrix}}_{\mathcal{L}}u+\underbrace{\begin{bmatrix}0\\ -u_2^3\end{bmatrix}}_{f(u)}.$$

Consequently, input the following into the web service.[24]

- For the *linear matrix*, input the two rows of the above matrix $\mathcal{L}$ as `(0,1),(-1,0)`.
- For the *nonlinearity*, input the column vector **f** as `(0,-u(2)^3)`.
- For the *center subspace* basis, we need the eigenvectors of the matrix: since the characteristic equation $\det(\lambda I-\mathcal{L})=\lambda^2+1=0$, eigenvalues are $\lambda=\pm i$; then find that the corresponding complex eigenvectors of the oscillations of interest are $\mathbf{v}\propto(1,\pm i)$, so enter the basis as `(1,i),(1,-i)`.
- For the *adjoint subspace* basis find that the corresponding eigenvectors of the transpose, $\mathcal{L}^T$, are $\mathbf{z}=(1,\mp i)$, so enter this basis as `(1,-i),(1,i)`.
- Enter the magic word, click `Submit`, and wait a few seconds for the result.

The web service solves the system as $\dot{u}_1=u_2$ and $\dot{u}_2=-\varepsilon u_2^3-u_1$, where the introduced parameter ε labels and controls nonlinearity terms. It then reports that the center manifold is the following, which we compare with (3.19a):

$$\begin{aligned}u_1&=+e^{-ti}s_2+e^{ti}s_1\\&\quad+\varepsilon\left(\tfrac{1}{8}e^{-3ti}s_2^3 i+\tfrac{3}{4}e^{-ti}s_2^2 s_1 i-\tfrac{1}{8}e^{3ti}s_1^3 i-\tfrac{3}{4}e^{ti}s_2 s_1^2 i\right),\\u_2&=-e^{-ti}s_2 i+e^{ti}s_1 i\\&\quad+\varepsilon\left(\tfrac{3}{8}e^{-3ti}s_2^3-\tfrac{3}{4}e^{-ti}s_2^2 s_1+\tfrac{3}{8}e^{3ti}s_1^3-\tfrac{3}{4}e^{ti}s_2 s_1^2\right).\end{aligned}$$

The expression for u_1 is almost the same as (3.19a) upon identifying $s_1\approx a$ and $s_2\approx\bar{a}$ (and upon ignoring the labeling ε). The only difference is the cubic terms involving $\varepsilon e^{\pm it}$, indicating that the web service uses a slightly different definition of the amplitude, that is, $a=s_1+\mathcal{O}(|s|^3)$ and $\bar{a}=s_2+\mathcal{O}(|s|^3)$. This provides a slightly different time-dependent basis for describing the oscillations.

The web service also reports that in this time-dependent basis, the amplitude variables s_1 and s_2 evolve according to $\dot{s}_1=-\frac{3}{2}\varepsilon s_2 s_1^2-\frac{27}{16}\varepsilon^2 s_2^2 s_1^3 i$ and its complex conjugate $\dot{s}_2=-\frac{3}{2}\varepsilon s_2^2 s_1+\frac{27}{16}\varepsilon^2 s_2^3 s_1^2 i$. The first term of these is effectively identical to the earlier (3.19b) and predicts slow decay of the oscillations. The second term in

[24] http://www.maths.adelaide.edu.au/anthony.roberts/gencm.php.

these predicts a nonlinear frequency shift that is apparently different from (3.19b), but this difference is accounted for by the small difference between the amplitudes a and $\bar{a}$ of (3.19a) and the current amplitudes s_1 and s_2.

Similarly, use this web service to construct complex-amplitude normal forms of the systems mentioned in Exercise 3.13.

Exercise 3.15 (complex electromagnetic waves). Recall from Example 3.3 the coordinate transform to complex-amplitude form of the electromagnetic wave ODE. Example 3.3 deduced the leading-order nonlinear transformation. In this exercise, adapt the computer algebra of Algorithm 3.3 to derive this and higher-order transformations.

Changing the independent variable from time t to space x is a straightforward part of this exercise. The main challenge is to deal with the complex conjugation in the nonlinearity in the ODE (3.13): $|\psi|^2\psi = \bar{\psi}\psi^2$. I suggest that a good way to proceed is to define a complex conjugation operator that has all the requisite properties. Add to Algorithm 3.3 a complex conjugation operator, perhaps cc, with the following properties:

$$\overline{uv} \mapsto \bar{u}\bar{v}, \quad \overline{u/v} \mapsto \bar{u}/\bar{v}, \quad \overline{u+v} \mapsto \bar{u}+\bar{v}, \quad \overline{u^p} \mapsto \bar{u}^p,$$
$$\bar{\bar{u}} \mapsto u, \quad \frac{d\bar{u}}{dp} \mapsto \overline{\frac{du}{dp}}, \quad \overline{e^{iu}} \mapsto e^{-iu},$$
$$\bar{i} \mapsto -i, \quad \overline{-i} \mapsto i, \quad \bar{n} \mapsto n, \quad \text{and also } \bar{\epsilon} \mapsto \epsilon,$$

for any variables u and v, real variable p, and real number n. Execute the algorithm and verify Example 3.3. Compute to higher order, say errors $\mathcal{O}(\epsilon^5)$, and interpret the results.

Modify the nonlinearity in the differential equation and execute the algorithm to confirm that the approach works immediately for a range of differential equations. In particular, compare the transformation with that for Exercises 3.11 and/or 3.12.

Exercise 3.16 (FitzHugh–Nagumo). The FitzHugh–Nagumo system of ODEs models neurone spiking in brains: the dynamic variables are $v(t)$ and $w(t)$, representing electrical dynamics in the neurone; physical parameters are I, α, γ, and ϵ; the ODE system is

$$\dot{v} = v(v-\alpha)(1-v) - w + I \quad \text{and} \quad \dot{w} = \epsilon(v - \gamma w).$$

Physical interest lies in small ϵ, which gives rise to relaxation oscillations that look like a train of spikes. But suppose we wish to view such nonlinear oscillations via a Hopf bifurcation to nonlinear oscillations.

1. Our first task is to identify a combination of parameters from which we can most easily view the emerging oscillations.

   We most comfortably deal with oscillators in the form $\ddot{x} + x =$ nonlinear. Here let $c(v) = v(v-\alpha)(1-v)$ and eliminate w from equations by differentiating the v ODE to derive that the following ODE is equivalent to the FitzHugh–Nagumo system:

$$\ddot{v} + \epsilon v = c'(v)\dot{v} + \epsilon\gamma[I - \dot{v} + c(v)].$$

2. This is in the form of a linear oscillator with frequency $\sqrt{\epsilon}$ modified by nonlinearities on the right-hand side. But our analysis is easier if we fix the frequency. So scale time to $\tau = \sqrt{\epsilon}t$ and derive

$$\ddot{v} + v = c'(v)\dot{v}/\sqrt{\epsilon} + \gamma[I - \sqrt{\epsilon}\dot{v} + c(v)] \tag{3.36}$$

in terms of derivatives with respect to the new time.

3. When parameters γ and α are both small (but parameters I and ϵ are finite), deduce that the right-hand side can be viewed as a perturbation to small oscillations of the left-hand side.

4. Hence modify Algorithm 3.3 to derive complex-amplitude models of the small-amplitude oscillations in the FitzHugh–Nagumo system, for example,

$$\sqrt{\epsilon}\dot{a} \approx (\gamma I - \tfrac{1}{2}\alpha - \tfrac{1}{2}\epsilon\gamma)a - (\tfrac{3}{2} + i\tfrac{2}{3}/\sqrt{\epsilon})|a|^2 a.$$

5. Looking at the evolution equation for the complex amplitude, approximately for what combination of parameters is the zero-amplitude solution linearly neutral in its stability?

Exercise 3.17 (FitzHugh–Nagumo extended). Reconsider the Hopf bifurcations in the regularized FitzHugh–Nagumo system (3.36). Consider constant V.

- For which parameters α and I will $v = V$ be an equilibrium of the oscillator (3.36)?
- Argue that the FitzHugh–Nagumo system has nonlinearly perturbed, small-amplitude oscillations about $v(t) = V$ for the parameter regime of small parameter γ for nearby α and I.
- Hence modify Algorithm 3.3 to derive complex-amplitude models of the small-amplitude oscillations in the FitzHugh–Nagumo system.

Exercise 3.18 (couple Duffing oscillators). Consider the pair of coupled Duffing oscillators for signal $x_1(t)$ and $x_2(t)$,

$$\ddot{x}_j + x_j + x_j^3 - x_{3-j}^2 x_j = 0, \quad j = 1, 2. \tag{3.37}$$

The $x_{3-j}^2 x_j$ terms couple the dynamics of these two oscillators. Model the system's long-term evolution.

1. Observe that linearized solutions of (3.37) are $x_j = a_j e^{it} + b_j e^{-it}$ for $j = 1, 2$. Adapt the Reduce code of Algorithm 3.3 to derive the evolution of the four complex amplitudes a_j and b_j under the nonlinear interactions: define separate amplitudes for the two $j = 1, 2$; define separate evolution expressions for $\dot{a}_j$ and $\dot{b}_j$; and create separate but coupled coordinate transforms x_j via computing separate residuals of (3.37) for each $j = 1, 2$.

   Fortunately, the update rules for corrections have the same form as in Algorithm 3.3. If we were to use amplitude-angle variables to model the coupled system (3.37), then we would have to deduce and implement a whole new suite of update rules. A major simplification of the complex exponential approach is that the update rules remain the same for such coupled problems.

2. Use the cubic model to verify the set of "equilibria" $a_2 = a_1$ and $b_2 = b_1$—such equilibria correspond to phase-locked solutions of the system (3.37) oscillating with constant amplitude and frequency. Derive that these are the only real equilibria, perhaps via the substitution $a_j = r_j e^{i\phi_j}$.
3. Linearize the cubic model about these equilibria to deduce that perturbations to perfect phase-locking are either neutral, with eigenvalues $\lambda = 0$ (twice), or oscillate slowly about the phase locking, with eigenvalues $\lambda = \pm i 2\sqrt{3}|a_1|^2$.

Answer: The cubic interaction model is $\dot{a}_j = i(\frac{3}{2} b_j a_j^2 - a_j a_{3-j} b_{3-j} - \frac{1}{2} a_{3-j}^2 b_j) + \cdots$, and similarly for b_j.

Exercise 3.19 (Stoleriu's coupled oscillators). Iulian Stoleriu (2012) considered the following pair of coupled oscillators with parameter σ for $x(t)$ and $y(t)$:

$$\ddot{x} + \sin x = \sigma y \quad \text{and} \quad \ddot{y} + y \cos x = 0\,. \tag{3.38}$$

1. Rewrite as a system of four coupled first-order ODEs.
2. Linearize about the origin and find that the eigenvalues are $\pm i$, each of multiplicity two, for all σ.
3. Find the corresponding eigenvectors and generalized eigenvectors for all σ.
4. As in Exercise 3.18, develop a nonlinear time-dependent coordinate transform to complex-amplitude variables for the two oscillating and interacting modes. Expand the trigonometric functions in (3.38) in an appropriate low-order Taylor series. Do *not* assume small σ.
5. Interpret the evolution equations for the complex amplitudes.

Exercise 3.20 (high-order stability boundaries). Determine higher-order approximations, $\delta = \pm\epsilon + \cdots$, to the stability boundaries for Mathieu's ODE (3.22) with frequency $\omega = 2$. Hint: Iteratively seek δ as a power series expansion in ϵ in the *same* iteration as the construction of the complex exponential normal form. Recognize that the stability boundary corresponds to eigenvalue $\lambda = 0$, and so the equation to determine δ is the algebraic equation that the determinant of (3.27) must be zero. A difficulty is determining the necessary correction to δ that drives the determinant to zero.

Exercise 3.21 (eigenvalues). Use all the terms found in the normal form (3.30a) for the linear varying oscillator (3.28) to deduce more accurate estimates for the eigenvalues E_n of the ODE (3.31).

Exercise 3.22 (multiple physical effects). Construct a WKB model for the slowly varying version of Duffing's oscillator

$$\ddot{x} + \omega^2 x + \epsilon x^3 = 0\,.$$

For simplicity let the parameter ϵ measure the magnitude of the slow variations in time of the frequency ω *and* the magnitude of the nonlinearity x^3. Use Algorithm 3.5 as a base, adapting and incorporating the `linv` operator of Algorithm 3.3 to update the shape x of the oscillations. Interpret your results.

Answer: The shape of the oscillations is $x = ae^{i\theta}+be^{-i\theta}+\epsilon\big[\frac{1}{8\omega^2}a^3e^{i3\theta}+\frac{1}{8\omega^2}b^3e^{-i3\theta}\big]+\mathcal{O}(\epsilon^2)$, where the complex amplitude evolves according to $\dot a = -\frac{\epsilon\dot\omega}{2\omega}a+\frac{i3\epsilon}{2\omega}a^2b+\frac{9\epsilon^2\dot\omega}{4\omega^3}a^2b+\frac{i\epsilon^2}{\omega^3}\big[-\frac{15}{16}a^3b^2+\frac{3}{8}\dot\omega^2a-\frac{1}{4}\omega\ddot\omega a\big]+\mathcal{O}(\epsilon^3)$, and similarly the complex conjugate for $\dot b$.

Exercise 3.23 (delayed feedback grows oscillations). There is much interest in delay differential equations (DDEs). Explore the modeling of the following oscillator with a term having a delay of τ (Kirkinis and O'Malley Jr., 2011, §2.2, e.g.):

$$\ddot x + x + \epsilon x(t-\tau) = 0\,. \tag{3.39}$$

1. Make a coordinate transform to complex-amplitude variables $x = a(t)e^{it} + b(t)e^{-it}$ to find the equivalent-"amplitude" ODE $\ddot a+2i\dot a+\epsilon e^{-i\tau}a(t-\tau) = 0$, and similarly for $b(t)$.
2. So far no approximation has been made, but now seek slowly varying oscillations when the strength of the feedback is small. Rearrange the amplitude ODE, and assuming small parameter ϵ, argue that
$$\dot a \approx \tfrac{1}{2}i\epsilon e^{-i\tau}a(t-\tau) - \tfrac{1}{8}i\epsilon^2e^{-i2\tau}a(t-2\tau) + \cdots\,.$$
What do you expect the form would be if you iterated further?
3. Now, because the amplitude $a(t)$ is slowly varying, we approximate $a(t-\tau)$ by a Taylor series in delay τ: show that $a(t-\tau) \approx a(t) - \frac{1}{2}i\tau\epsilon e^{-i\tau}a$. Hence deduce that the amplitude ODE may be approximated by $\dot a \approx \frac{1}{2}i\epsilon e^{-i\tau}a$.
4. Explain why expanding the complex amplitude $a(t-\tau)$ in a Taylor series in τ has a much wider range of validity than expanding the solution $x(t-\tau)$.
5. Deduce that the approximate growth rate and frequency change of the oscillations due to the small delayed feedback are $\frac{1}{2}\epsilon\sin\tau$ and $\frac{1}{2}\epsilon\cos\tau$, respectively. For what range of ϵ and τ do you predict the oscillations will be stabilized? Destabilized?
6. What are the $\mathcal{O}(\epsilon^2)$ corrections to the growth rate and frequency change?

Exercise 3.24 (difficult delay). Write, debug, and run a Reduce script to construct the amplitude model of the delay feedback oscillations of Exercise 3.23. Base your code on Algorithm 3.3: there seem to be many ways to modify the algorithm, but the following notes may help.

- The ODE residual will have amplitudes evaluated at $t-\tau$, so you will need to code for the Taylor series of such expressions. In order to recognize when they occur, perhaps explicitly include the time argument in the amplitude variables.
- The operator `cis` involves both `t` and `tau`; perhaps use the time derivative of its argument to find the coefficient of the time dependence.
- Extracting the $e^{\pm it}$ component in the residual is more complicated, as the exponential's argument also has delays in it, and similarly for solving for the modifications to x.
- You might truncate expansions in ϵ by scaling the delay term coefficient with `small`.

When you get code working for the linear ODE in Exercise 3.23, also apply your code to the nonlinear delay ODE

$$\ddot{x} + x + x^2\dot{x} + \epsilon x(t-\tau) = 0\,.$$

Explain your algorithm, and report and interpret the results.

Exercise 3.25 (drilling vibrations). Stone and Askari (2002) derived the following nondimensional DDE model of the vibrations that occur in drilling:

$$\ddot{x} + \delta\dot{x} + x = \beta\{1 - \mu[x - x(t-\tau)]\}(p_0 + p_1\dot{x} + p_2\dot{x}^2).$$

The delay time τ is proportional to the period of the rotation of the drill, and the parameter β relates to the depth of a cut: these two parameters can be varied in operation. The other parameters, μ and p_j, characterize fixed physical properties of the drill. Campbell (2009) invoked center manifold theory to predict the onset and evolution of vibrations about the equilibrium solution $x = \beta p_0$, which represents the desired steady drilling. Here adapt the approach of this section, and the previous two exercises in particular, to model vibrations about this state of steady drilling.

- Treat parameters δ and β as small and the oscillation amplitude as arbitrary (not necessarily small).
- Adapt the previous computer algebra code to derive the leading-order (cubic in amplitude) nonlinear model for the vibrations.
- Interpret this leading-order model and determine parameter regimes in τ and β of stability and instability.
- Confirm that your code generates higher-order models without difficulty.

Part I Summary

- Solving algebraic equations in Chapter 1 illustrates some of the key features seen in solving complicated problems:
  - solve a difficult problem by writing it as a perturbation of a solvable problem;
  - use the residual to iteratively construct power series solutions;
  - use computer algebra to handle the messy details.
- Rescaling variables so that infinities become finite (see section 1.2) transforms some apparently singular problems into regular problems.
- Easily determine the critical coefficient of proportionality between residual and correction using a few iterations of your proposed computer algebra code; see section 1.3.
- Power series solutions of differential equations empower us to later tackle vastly more complicated problems. Iteration is a straightforward and flexible approach to constructing power series solutions (section 2.2).
- Divergent power series are the norm in our applications (section 2.3). Hence we invoke the methodology of asymptotic series. One crudely estimates the error of a partial sum of an asymptotic series by the size of the first neglected term. Asymptotic series are unique. This uniqueness ensures that we can formally operate on asymptotic series (with care).
- For oscillatory dynamics, the change of variables to a normal form in amplitude-angle variables, polar coordinates, introduced in Chapter 3, provides a powerful and simple model of the oscillations.

  This normal-form approach is significantly easier to implement and interpret than other methodologies, such as multiple scales or averaging.
- In modeling dynamics we must interpret the algebra: the mapping from an ODE in x, say, to amplitude-angle variables (r, θ) is a coordinate transform; the evolution of the amplitude r determines the modulation of oscillations; and the evolution of the angle θ determines the frequency.
- Complex exponentials provide an even more flexible and powerful base for describing and interpreting nonlinear, forced, and/or varying oscillations. The approach is to use complex exponentials as a time-dependent basis for the dynamics.

Part II

Center manifolds underpin accurate modeling

Part II Contents

This part develops the basic concepts of center manifolds and their use in accurately modeling dynamical systems. Subsequent parts explore important classes of applications as well as further illuminating and extending the theory.

The challenge is to transform a description of detailed dynamics into an accurate model of coarse interesting quantities. For two examples: How can a description of millions of interacting particles/agents be summed up by a simple population model?

How can the infinite resolution of a PDE be described by a spatial discretization? Our quest is to do such modeling transformations accurately.

The flexible power of the center manifold approach to creating dynamical models comes from three theorems introduced in Chapter 4. The three theorems assure us of three key aspects:

- a model *exists*, depending upon a linear picture of the dynamics;
- the model *emerges and is relevant* because all nearby dynamics are attracted to those of the model;
- we may *construct* the model to a controlled order of error.

The key is to find a domain where there is a separation of time scales between long-lasting modes of interest and rapidly decaying modes of relatively uninteresting details.

Section 4.6 discusses how other sound methodologies produce equivalent models. I contend that the center manifold approach is the clearest and most powerful methodology. The approach also supports both local and global models (section 4.7) and resolves controversies over the correct structure functions (section 4.8).

Many people consider that a linear change of basis (coordinate system) is trivial (undergraduate mathematics) and so assume that it is sufficient to study systems where the slow and fast (critical and decaying) variables are linearly separated (by diagonalization, or similar, of the linear operator). Thus the theory of Chapter 4 only addresses systems with a linearly separated form: the emergent center/slow variables $\mathbf{x}$ are separate from the uninteresting stable/unstable variables $\mathbf{y}$ in the linearized system. The linear operators of PDEs can also be separated into slow and fast modes, say by diagonalizing the linear operator through taking the Fourier transform (when applicable). But it is far easier to interpret mathematical formulas in the original physical variables rather than in transformed variables. Furthermore, it is far less error prone to code the computation of the residuals of original physical equations, rather than transforming equations and then coding the transformed system. These two reasons imply that we need the general algorithm of Chapter 5 (Coullet and Spiegel, 1983) because it works with the physically meaningful original variables and physically meaningful original equations.

Importantly, the construction algorithm of section 5.3 is coordinate free: it can be applied to the form of equations preferred by any user.

Center manifold theory also underpins global models (section 5.5.1). Although introduced as being local to an equilibrium, the theory also applies when there is a parametrized family of equilibria, a subspace of equilibria for example: the theory then supports a dynamic model that applies across the whole family to generate a model that is local in some variables but global in other variables. One application is to chemical reactions. Other applications include many singularly perturbed systems.

Our theory also applies in a finite domain and so rigorously supports modeling for finite-sized parameters and variables. Thus to estimate the model accurately at finite values of parameters and variables, we need to compute asymptotic approximations to different orders in the parameters and variables. The Newton diagrams introduced in section 5.5.2 empower us to very flexibly create and use models.

Chapter 4

The center manifold emerges

Contents

Remove all redundancies from the problem and reduce it to the simplest elements.
Rene Descartes

Here we develop a rigorous view to justify low-dimensional approximations to the emergent evolution of dynamical systems. The treatment is based on that expounded by Carr (1981). However, this and other related literature was primarily developed to answer questions about stability; we have a different purpose and so differ in emphasis. We aim to develop *low-dimensional and accurate models of dynamics*. These models are justified by an *exponentially quick collapse* of all solutions into a low-dimensional set of states;

the later emergent long-term evolution therein forming a reduced low-dimensional dynamical system modeling the original.

Our whole discussion develops from basic knowledge of nonlinear dynamical systems as described in many textbooks, such as Chapter 4 in the textbook by Kreyszig (2011). The following definition reminds us of some key concepts.

Definition 4.1 (key concepts).

- *We explore the evolution in time* t *of a vector of variables* $u(t)$ *in* n *dimensions,* $\mathbb{R}^n$*. The* **state space** *(*$\mathbb{R}^n$ *or some subset) is the set of all valid values of* u *in the application.*
- *In our* **dynamical system** *the state vector* $u(t)$ *evolves according to a set of coupled* ODEs *expressed as* $du/dt = \dot{u} = f(t, u)$*. Mostly we restrict attention to* **autonomous** ODEs *where the right-hand side has no explicit time:* $\dot{u} = f(u)$.
- *An* **equilibrium, fixed point,** *or* **critical point** *of such an autonomous dynamical system is a point* u_* *such that* $f(u_*) = 0$ *so that* $u(t) = u_*$*, constant, is a solution for all time. Typically we move the origin to be located at an important equilibrium if the origin is not already an important equilibrium.*
- *A* **trajectory** *of a dynamical system is the path of a solution* $u(t)$ *of the* ODEs *drawn or imagined in the state space* $\mathbb{R}^n$.

4.1 ▪ Introduce some basic examples

Consider the dynamical system

$$\dot{x} = ax^3 \quad \text{and} \quad \dot{y} = -y + y^2, \tag{4.1}$$

where a is some constant (say negative) and an overdot denotes d/dt. Figure 4.1 shows some snapshots of numerical solutions in the xy state space: all solutions in the shown domain collapse quickly onto the line $y = 0$; thereafter all solutions evolve slowly to the origin.

Analytic solutions confirm this picture. The two equations of the ODE (4.1) are uncoupled and so each may be solved independently of the other by separation of variables. Given the initial condition that $x = x_0$ and $y = y_0$ at $t = 0$, the solution is

$$x(t) = \frac{x_0}{\sqrt{1 - 2ax_0^2 t}} \quad \text{and} \quad y(t) = \frac{1}{1 + \left(\frac{1}{y_0} - 1\right)\exp(t)}. \tag{4.2}$$

For the above simple dynamical system, observe, in both the algebraic solutions and in the trajectories, that provided initial $y_0 < 1$ the system $u(t)$ tends exponentially quickly to the line $y = 0$, labeled $\mathcal{M}$ in the trajectories of Figure 4.2. Ignoring this rapidly decaying transient behavior, the behavior that emerges from the whole system is that of one degree of freedom, evolving according to $\dot{x} = ax^3$ and all the while staying on $\mathcal{M}$. This long-term evolution is generally much slower than exponential,[25] namely the algebraic evolution $x = (c_1 - 2at)^{-1/2}$. This exponential collapse of the dimensionality of the dynamics from the full 2D state space to the 1D $\mathcal{M}$ really does occur very quickly: within a time $\Delta t \approx 6$, all initial conditions in the picture have evolved to be indistinguishable from being on $\mathcal{M}$, whereas in the slow algebraic evolution, it takes $\Delta t \approx 10000/a$ to reach the origin to within $\Delta x = 0.01$.

[25] Near the origin at least.

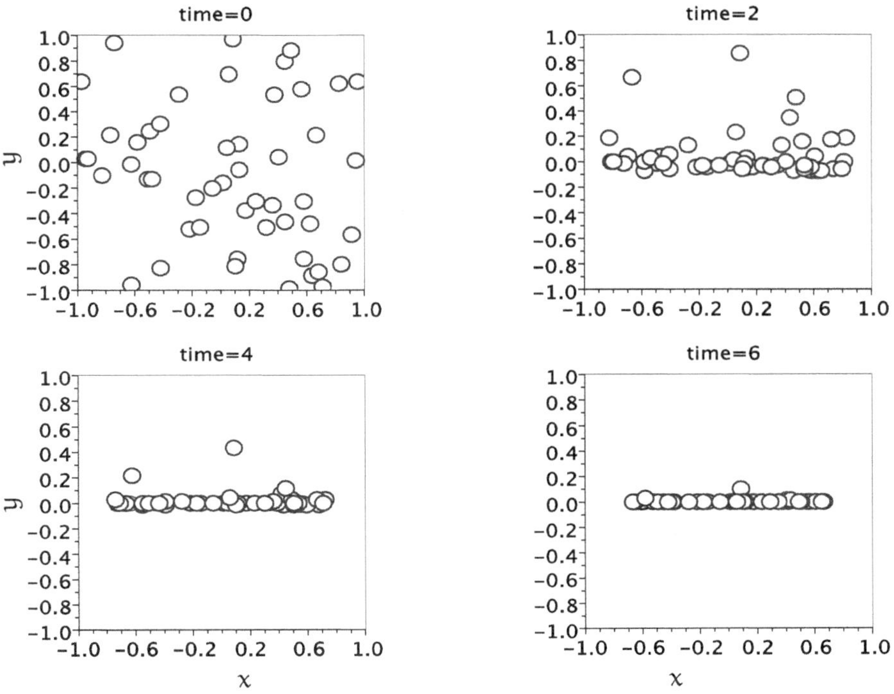

Figure 4.1. *Locations at four times in the* xy*-plane of solutions to the ODE* (4.1) *starting from* 50 *random initial conditions: the slow center manifold* $y = 0$ *emerges quickly.*

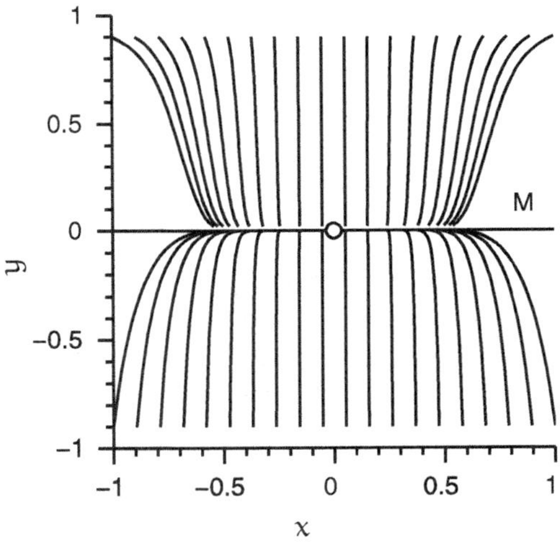

Figure 4.2. *Trajectories of the dynamical system* (4.1) *showing the rapid approach to* $\mathcal{M}$, $y = 0$.

Usually we concentrate our interest upon what happens near the origin in the state space of the dynamical system. Requiring nearness to the origin implies that we perform what is called a *local analysis*. However, many locales are big enough to provide interesting and rich results. Being based on the origin is no constraint because we move the origin to any reference equilibrium of interest. *In application, we just need some equilibrium, or fixed point, on which to anchor the analysis.*

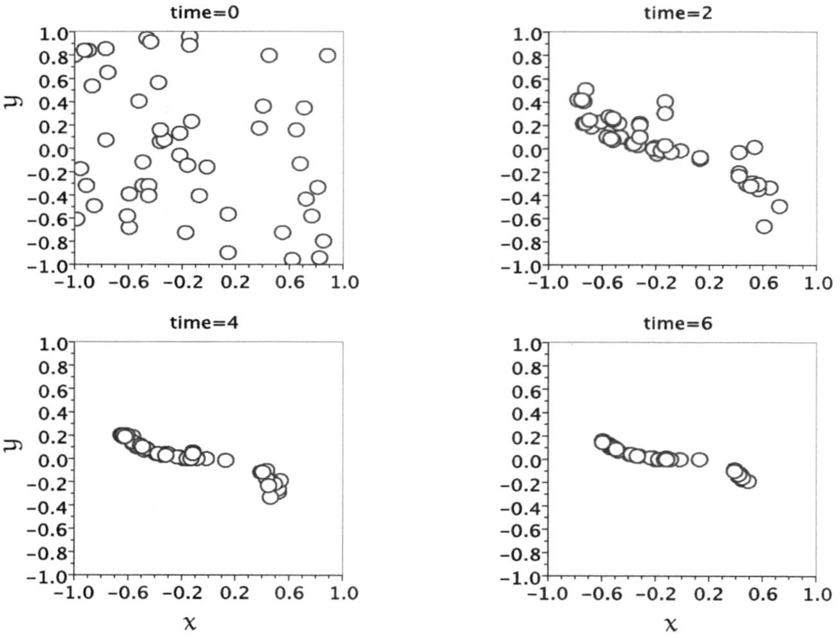

Figure 4.3. *Locations in the* xy*-plane at four times of solutions to the ODE* (4.3) *starting from* 50 *random initial conditions: the curved slow center manifold emerges quickly.*

4.1.1 ▪ Variables are usually coupled

Suppose now that the pair of equations are coupled, viz.

$$\dot{x} = ax^3 + x^2y \quad \text{and} \quad \dot{y} = -y + y^2 + xy - x^3 . \tag{4.3}$$

Figure 4.3 shows snapshots of solutions illustrating the emergence of reduced dynamics on a curve near the origin. Algebra cannot so easily determine the long-term behavior near the origin of this dynamical system. *Linearizing* about the origin, we find $\dot{x} = 0$ and $\dot{y} = -y$, and so we see that y decays exponentially quickly (approximately) and that x does not evolve. But this picture is not satisfactory because, as in the system (4.1), x is not constant but evolves algebraically. *The nonlinear terms in the equations must modify the too simple picture obtained from linearization to that of Figure* 4.3.

A coordinate transform simplifies Waving an algebraic magic wand (to be explained subsequently, in Chapter 11), I introduce a coordinate transform from x and y variables to new X and Y variables—Figure 4.4 plots the coordinate curves—such that

$$x = X - X^2Y + \cdots \quad \text{and} \quad y = Y - Y^2 - X^3 + \cdots .$$

This coordinate transform is cunningly crafted to transform the nonlinear dynamical system (4.3) into

$$\begin{aligned} \dot{X} &= aX^3 - X^5 + \cdots , \\ \dot{Y} &= (-1 + X - 2X^3 + 2X^4 + \cdots)Y . \end{aligned} \tag{4.4}$$

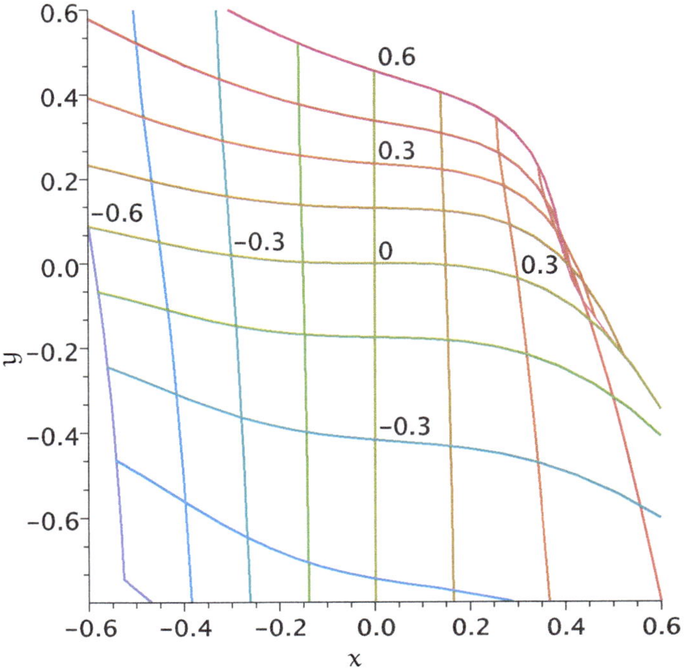

Figure 4.4. *Coordinate curves of the new* (X, Y) *coordinate system that transforms the dynamical system* (4.3) *into the decoupled system* (4.4). *Labels are corresponding coordinate values of some of the coordinate curves.*

These new variables highlight important dynamics. First, the new variable Y must decay exponentially quickly to zero from all initial conditions in some finite domain around $X = 0$ as, by continuity, the rate $-1+X-2X^3+2X^4+\cdots$ must be negative in some finite domain around $X = 0$. Thus, given its exponentially quick decay, Y does not appear in the long-term dynamics; only the dynamics of X emerges in the long term. Second, after $Y \to 0$ we see that $x = X$, so from the X-dynamics the emergent long-term dynamics are $\dot{x} \approx ax^3 - x^5$. Last, these dynamics occur for original variable $y \approx -X^3 \approx -x^3$, which is the curved attractor seen in the numerical simulations of Figure 4.3.

Problem Such beautiful complete coordinate transforms are impractical in most applications. The reason for their impracticality is that in applications there are usually a large number, often infinite, of exponentially decaying transients. Correspondingly, there are usually only relatively few modes that emerge to describe the long-term dynamics. It is completely impractical to transform a large/infinite number of transient modes that we ultimately just set to zero, when all we want is the relatively few emergent modes. Center manifold theory provides an equivalent, alternative, but economical method to extract the emergent dynamics.

Focus only on the emergent dynamics The center manifold theory that we now turn to asserts that (4.3) has an exponentially attractive curve $\mathcal{M}$ on which all the long-term evolution takes place ($Y = 0$ in Figure 4.4). The curve $\mathcal{M}$ may be described by $y = h(x)$. In essence, the theory says that the linear picture is largely correct except that it is "bent" by nonlinear terms, as seen in Figure 4.5.

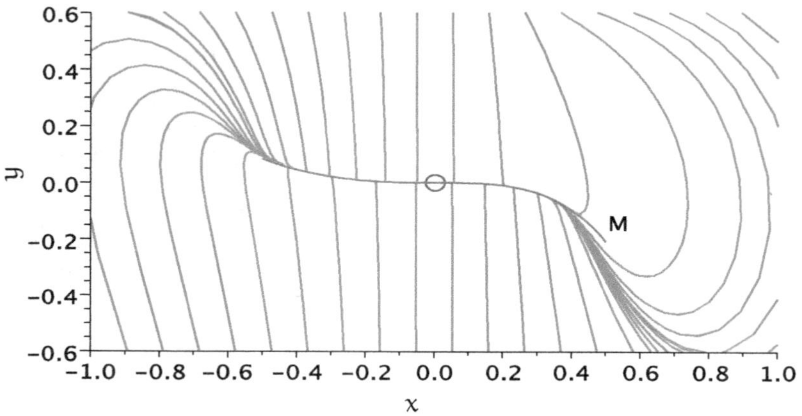

Figure 4.5. *Trajectories of the dynamical system* (4.3) *showing the rapid approach to the curved center manifold* $\mathcal{M}$ (4.5).

Direct calculation shows that here the center manifold $\mathcal{M}$ and the evolution thereon are

$$\begin{aligned} y &= h(x) \sim -x^3 - x^4 \quad \text{as } x \to 0\,, \\ \text{where} \quad \dot{x} &= ax^3 + x^2 h(x) \sim ax^3 - x^5\,. \end{aligned} \tag{4.5}$$

Furthermore, the theory asserts something stronger than mere attraction to $\mathcal{M}$. Theory asserts that the *evolution* on $\mathcal{M}$ will emerge; for all solutions $(x(t), y(t))$ of the full system (4.3), with initial point (x_0, y_0) sufficiently near to the origin, there exists a solution, say $x = s(t)$, of (4.5) such that[26]

$$\left.\begin{aligned} x(t) &= s(t) + \mathcal{O}(e^{-t}) \\ y(t) &= h(s(t)) + \mathcal{O}(e^{-t}) \end{aligned}\right\} \quad \text{as } t \to \infty\,.$$

That is, ODE (4.5) becomes valid exponentially quickly as a long-term, smaller degrees of freedom description of the full system (4.3). Thus the ODE (4.5) forms a model of the original system; with fewer degrees of freedom it is a simpler, more tractable system to handle. The ODE (4.5) describes the emergent dynamics.

These are the same conclusions as deduced from the normal-form coordinate transform (4.4). The difference is that we economically do not resolve the transients that are irrelevant to the long-term model.

4.2 ▪ Eigenvalues imply existence

For the generic theory of this section we move from the earlier examples in the two variables x and y to consider dynamical systems in n variables. Let the system variables be grouped into the vector $\mathbf{u} \in \mathbb{R}^n$. We base analysis about an equilibrium, so set the $\mathbf{u}$-origin to be at the equilibrium. Then the differential equations for the evolution are written as

$$\dot{\mathbf{u}} = \mathcal{L}\mathbf{u} + \mathbf{f}(\mathbf{u}), \tag{4.6}$$

[26] We often say that "the system settles onto $\mathcal{M}$ exponentially quickly," that is, the center manifold $\mathcal{M}$ emerges quickly.

Table 4.1. *Matrix eigenvalue/eigenvector problems have the following important properties.*

The standard eigenproblem for eigenvalues λ and eigenvectors $\mathbf{e}$ consists of solving $\mathcal{L}\mathbf{e} = \lambda\mathbf{e}$ for some real $n \times n$ matrix $\mathcal{L}$ (analogous properties hold for complex matrices).

1. There are n eigenvalues, $\lambda_1, \dots, \lambda_n$, when counted and listed according to their multiplicity.
2. When the matrix $\mathcal{L}$ is real, any complex eigenvalue occurs as one of a complex conjugate pair.
3. Corresponding to each eigenvalue λ_j, there exists (at least) an eigenvector $\mathbf{e}_j \neq 0$ (and all their linear combinations).
4. If the matrix $\mathcal{L}$ is symmetric, then
   (a) all eigenvalues λ_j are real;
   (b) eigenvectors $\mathbf{e}_j$ corresponding to different eigenvalues are orthogonal;
   (c) the *Rayleigh quotient* gives the eigenvalue $\lambda_j = \mathbf{e}_j \cdot (\mathcal{L}\mathbf{e}_j)$ and is typically used to approximate well the eigenvalue given only a rough approximation $\mathbf{e}_j$ to the eigenvector.
5. Generalized eigenvectors often arise in physical applications to the motion of bodies and particles but are only potentially possible when the matrix $\mathcal{L}$ is not symmetric.
   - When a given eigenvalue has multiplicity two, say λ_1, but there is only one linearly independent eigenvector, say $\mathbf{e}_1$, then we find a *generalized eigenvector* $\mathbf{e}_2$ by solving $\mathcal{L}\mathbf{e}_2 = \lambda\mathbf{e}_2 + \mathbf{e}_1$.
   - When a given eigenvalue has multiplicity larger than two, say λ_1 has multiplicity $k > 2$, then we iterate as follows. Let $\{\mathbf{e}_1, \dots, \mathbf{e}_j\}$ be the currently known linearly independent set of eigenvectors and generalized eigenvectors, $j < k$. Find a basis (preferably orthogonal) for the null space of the matrix

$$\mathcal{L}' := \begin{bmatrix} \mathcal{L} - \lambda_1 I & \mathbf{e}_1 & \cdots & \mathbf{e}_j \\ \mathbf{e}_1^T & 0 & \cdots & 0 \\ \vdots & \vdots & & \vdots \\ \mathbf{e}_j^T & 0 & \cdots & 0 \end{bmatrix}.$$

   For each (linearly independent) vector in the basis, its first n components are a new *generalized eigenvector* to add to the growing set. Stop iterating when $j = k$.
6. The set of eigenvectors and generalized eigenvectors is complete; they span $\mathbb{R}^n$.

where $\mathcal{L}$ $(n \times n)$ is a constant matrix (or linear operator in general) and $\mathbf{f}$ is a strictly nonlinear function. Further, the linear operator $\mathcal{L}$ must have certain additional properties:[27]

- no eigenvalues with a positive real part;
- m eigenvalues with a zero real part, $\lambda_1, \dots, \lambda_m$, and corresponding linearly independent eigenvectors $\mathbf{e}_1, \dots, \mathbf{e}_m$ (repeat eigenvalues according to their multiplicity, and include any generalized eigenvectors[28]);

[27]These properties ensure that there is a linear coordinate transform, $u \to (x, y)$, such that the linear system, $\dot{u} = \mathcal{L}u$, separates into the center modes $\dot{x} = Ax$ and the exponentially decaying (stable) modes $\dot{y} = By$.

[28]Generalized eigenvectors often arise in physical applications to the motion of bodies and particles; see Exercise 4.16 for example. There may be a zero eigenvalue of multiplicity two associated with mean position and mean momentum. The corresponding eigenvector gives only the position subspace. A corresponding generalized eigenvector is needed to span the momentum subspace as well. See Table 4.1.

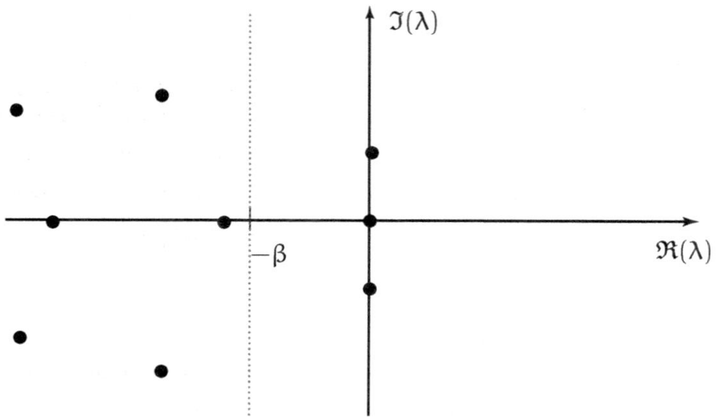

Figure 4.6. *A schematic spectrum of the linearized version of the dynamical system* (4.6).

- the remaining eigenvalues, $\lambda_{m+1}, \dots, \lambda_n$, have negative real part; the real parts are bounded away from 0 by some bound $-\beta$, that is, $\Re\lambda_j < -\beta < 0$ for $j = m + 1, \dots, n$.

Figure 4.6 shows a schematic picture of the spectrum of the class of linearized problems.

We normally assume that the function $\mathbf{f}$ is smooth, at least twice differentiable at the origin, whence $\mathbf{f}$ is a strictly nonlinear function if at the origin $f_i = \partial f_i/\partial u_j = 0$ for all i, j.

Definition 4.2. *A set $\mathcal{M} \subset \mathbb{R}^n$ is said to be an* **invariant manifold** *of the dynamical system* (4.6) *if for any $\mathbf{u}_0 \in \mathcal{M}$ the solution of* (4.6) *with $\mathbf{u}(0) = \mathbf{u}_0$ stays in $\mathcal{M}$ for some finite time, that is, $\mathbf{u}(t) \in \mathcal{M}$ for $0 \leqslant t < T$.*[29]

One algebraic way to check that a manifold $\mathcal{M}$ is invariant is the following. Suppose $\mathcal{M}$ is defined by some function $\mathbf{h}(\mathbf{u}) = \mathbf{0}$. If $d\mathbf{h}/dt = \mathbf{0}$ under the dynamics of the system, then $\mathcal{M}$ is invariant. Another way is to show that $\dot{\mathbf{u}}$ is tangent to the manifold $\mathcal{M}$. These two ways are equivalent since $dh/dt = \dot{\mathbf{u}} \cdot \nabla \mathbf{h}$; since $\nabla \mathbf{h}$ is normal to $\mathbf{h} = 0$, then the dot product being zero ensures that $\dot{\mathbf{u}}$ is tangent to $\mathbf{h} = 0$, that is, tangent to the manifold.

Example 4.1. Consider the dynamical system

$$\dot{x} = -y + x^2 - 1 \quad \text{and} \quad \dot{y} = x + xy\,.$$

Show that both the line $y = -1$ and the unit circle are invariant.

- Test the line by setting $h(x, y) = y + 1$. Then $dh/dt = \dot{y} = x + xy = xh$ by the ODEs. But on the line, $h = 0$, so that $dh/dt = xh = 0$.
- Test the circle by setting $h(x, y) = x^2 + y^2 - 1$. Then $dh/dt = 2x\dot{x} + 2y\dot{y} = 2x(-y + x^2 - 1) + 2y(x + xy) = 2x(x^2 + y^2 - 1) = 2xh$. But on the circle, $h = 0$, so that $dh/dt = 2xh = 0$.

[29] Some people require the trajectory to stay in $\mathcal{M}$ for all time. But this requirement is overly restrictive in practice.

Example 4.2 (invariant subspaces). Consider the linear system $\dot{u} = \mathcal{L}u$. Suppose operator $\mathcal{L}$ has eigenvalues λ_j and corresponding eigenvectors e_j. Then for any (finite) subset of eigenvectors, say $e_1, \ldots, e_k$, the subspace $\mathbb{E} = \operatorname{span}\{e_1, \ldots, e_k\}$ is an invariant manifold of the linear dynamics.

Any $u \in \mathbb{E}$ may be written as the linear combination $u = \sum_{j=1}^{k} c_j e_j$,. Hence $\dot{u} = \mathcal{L}u = \sum_{j=1}^{k} c_j \mathcal{L} e_j = \sum_{j=1}^{k} c_j \lambda_j e_j = \sum_{j=1}^{k} c_j' e_j \in \mathbb{E}$, as it is another linear combination of the basis eigenvectors. Since $\dot{u} \in \mathbb{E}$, it is tangent to $\mathbb{E}$ and hence the subspace $\mathbb{E}$ is an invariant manifold.

In its most abstract, Definition 4.2 asserts that the invariant manifold $\mathcal{M}$ is just the union of a set of trajectories of the system (4.6). But the term "invariant manifold" is usually used with some *implicit* extra properties, such as smoothness, completeness, or analyticity near the origin or other equilibrium.

For example, if $f = 0$, so that the dynamical system (4.6) is linear, then there exist two invariant manifolds of prime interest to us:

- the *center subspace* $\mathbb{E}_c := \operatorname{span}\{e_1, \ldots, e_m\}$, consisting of all the nondecaying modes and often parametrized by "center variables" x; and
- the *stable subspace* $\mathbb{E}_s := \operatorname{span}\{e_{m+1}, \ldots, e_n\}$, consisting of all the exponentially decaying modes and often parametrized by "stable variables" y.

The center subspace interests us because all solutions to the linear problem $\dot{u} = \mathcal{L}u$, starting from any initial condition, tend exponentially quickly, at least as fast as $\exp(-\beta t)$, to the center subspace $\mathbb{E}_c$. Once on $\mathbb{E}_c$, solutions evolve according to some equation $\dot{x} = \mathcal{G}x$, where $\mathcal{G}$ is the restriction of $\mathcal{L}$ to $\mathbb{E}_c$. That is, the behavior on the low-dimensional (m-dimensional) center subspace captures the long-term dynamics of the original linear system.

As discussed informally in the previous section, analogous results hold, such as the following, for fully nonlinear systems such as (4.6).

Theorem 4.3 (existence). *Given the more relaxed condition on the eigenvalues of $\mathcal{L}$ that $|\Re\lambda_j| > \beta$ for $j = m+1, \ldots, n$, there exists, in a neighborhood $\mathcal{D}$ of the origin, an m-dimensional* invariant manifold $\mathcal{M}$ *for (4.6), with tangent space $\mathbb{E}_c$ at the origin, in the form $u = v(s)$ (that is, locations on $\mathcal{M}$ are parametrized by $s \in \mathbb{R}^m$). The flow on $\mathcal{M}$ is governed by the m-dimensional dynamical system*

$$\dot{s} = g(s) = \mathcal{G}s + \mathcal{O}(|s|^2), \tag{4.7}$$

where the nonlinear function g may be determined from v, and $\mathcal{G}$ is the restriction of $\mathcal{L}$ to $\mathbb{E}_c$. Because of the nature of the eigenvalues of $\mathcal{L}$ on $\mathbb{E}_c$, this invariant manifold is called a **center manifold** *of the system. All trajectories that stay in the neighborhood $\mathcal{D}$ for all time belong to a center manifold.*

Haragus and Iooss (2011), for example, give a version of this theorem for infinite-dimensional dynamics. This book discusses only finite-dimensional theory, even though we do apply the methodology to infinite-dimensional problems.

Definition 4.4 (slow manifold). *If, as commonly occurs, the m eigenvalues of the center subspace $\mathbb{E}_c$ are precisely zero (not just with zero real part), then the center manifold $\mathcal{M}$ is more informatively called a* **slow manifold** *(and $\mathbb{E}_c$ is called the* **slow subspace** *$\mathbb{E}_0$).*

Example 4.3 (an exact slow manifold). Consider the dynamical system $\dot{x} = -xy$ and $\dot{y} = -y + x^2 - 2y^2$. The origin is an equilibrium. Linearized about the origin, the dynamics are $\dot{x} \approx 0$ and $\dot{y} \approx -y$, and hence the linear operator has eigenvalues 0 and -1. By the existence theorem, the system has a center manifold (slow manifold) through the origin.

Since the center (slow) subspace $\mathbb{E}_c$ is the x-axis, as $\dot{x} \approx 0$, the slow manifold may be parametrized by x. In this particularly basic example we confirm that a slow manifold is $y = x^2$ by showing that this parabola is invariant under the dynamics. Consider the time derivative $d(y - x^2)/dt = \dot{y} - 2x\dot{x} = -y + x^2 - 2y^2 + 2x^2y = (y - x^2)(-1 - 2y)$. Thus, somewhat incestuously, this time derivative is zero on $y - x^2 = 0$ and so for any initial condition on this parabola, the system stays on the parabola. The parabola is an invariant manifold. Being tangent to the slow subspace at the origin, the parabola must be a slow manifold.

We aim to model large-scale systems with many interacting modes. The best-understood large-scale systems are partial differential equations.

Example 4.4 (well-mixed reaction). Consider some field $u(x, t)$ governed by the reaction-diffusion PDE $\partial u/\partial t = \partial^2 u/\partial x^2 + f(u)$ for some strictly nonlinear reaction $f(u)$, that is, $f(0) = f'(0) = 0$. The diffusion term $\partial^2 u/\partial x^2$ may model some given spatial mixing of the field $u(x, t)$. Suppose, for definiteness, that this reaction-diffusion takes place on a finite interval, say the nondimensional interval $[0, \pi]$, with insulating boundary conditions $\partial u/\partial x = 0$ at $x = 0, \pi$.

Since $f(0) = 0$, the origin $u = 0$ is an equilibrium. Since the reaction is strictly nonlinear, the PDE linearizes about the origin to the diffusion PDE $\partial u/\partial t = \partial^2 u/\partial x^2$ with insulating boundary conditions $\partial u/\partial x = 0$ at $x = 0, \pi$. Standard separation of variables shows that this linear PDE has linearly independent solutions $u_k = e^{\lambda_k t} \cos kx$ for integer $k = 0, 1, 2, \ldots$ and corresponding eigenvalues $\lambda_k = -k^2$. Since there is one zero eigenvalue, there exists a slow manifold tangent to the slow subspace of spatially constant solutions $\propto \cos 0x$.

Consider the special case when the reaction f has no explicit x dependence. Then solutions constant in x are invariant in the dynamics. Thus the slow manifold that emerges is $u = s(t)$ (constant in x), on which the evolution is $\dot{s} = f(s)$. This evolution has just one degree of freedom, $s(t)$, instead of the infinite number of modes inherent in the original PDE. This argument justifies modeling well-mixed reactors by the overall parameter $s(t)$ rather than the detailed field $u(x, t)$.

Example 4.5 (parametrized reaction-diffusion). What if we relax the reaction constraint that $f'(0) = 0$? Also, what if we change the boundary conditions to those of fixed value (temperature/concentration) of $u = 0$ at $x = 0, \pi$.

There is still an equilibrium at the origin $u = 0$. Linearized about the origin, the dynamics is $\partial u/\partial t = \partial^2 u/\partial x^2 + f_0' u$, where $f_0' = f'(0)$ is some number. With the zero-value boundary conditions, separation of variables gives linearly independent modes $u_k = e^{\lambda_k t} \sin kx$ for integer $k = 1, 2, 3, \ldots$ and eigenvalues $\lambda_k = -k^2 + f_0'$. Several cases arise depending upon f_0':

- if $f_0' < 1$, then all eigenvalues are negative and there is no center manifold;

- if $f_0' = 1$, then all eigenvalues are negative ($\leqslant -3$) except for $\lambda_1 = 0$ and so there is a one-dimensional slow manifold tangent to the slow subspace $\propto \sin x$;[30]
- if $f_0' > 1$ and is not a perfect square, then most eigenvalues are negative, a few are positive, and there is no center manifold;
- if $f_0' > 1$ and is a perfect square, then most eigenvalues are negative, and one is zero so there is a center manifold, but the system does not satisfy our requirements, as a few eigenvalues are positive.

Example 4.6 (infinite domain diffusion). Consider diffusion $\partial u/\partial t = \partial^2 u/\partial x^2$ on an infinite domain with "boundary conditions" that the field u be bounded. Separation of variables leads to a Fourier integral decomposition that independent modes are $e^{\lambda t + ikx}$ for eigenvalue $\lambda = -k^2$ for, now, all real wavenumbers k. Although there is a zero eigenvalue, corresponding to wavenumber $k = 0$, there is no center manifold apparent as there are other modes with eigenvalues, albeit all negative, arbitrarily close to zero. Thus there is no clear separation of a slow center manifold from the other decaying dynamics in the system.[31]

Example 4.7 (nonlinear mixing). Consider the nonlinear diffusion PDE $\partial u/\partial t = \partial/\partial x(u\, \partial u/\partial x)$ with insulating boundary conditions $\partial u/\partial x = 0$ at $x = 0, \pi$. Such nonlinear diffusion is weakly analogous to the nonlinear mixing, of turbulence: local eddies generate the local mixing, but the energy in the local eddies is proportional to the local flow velocity, here denoted by u.

Here there is a whole subspace of equilibria: $u(x,t) = U =$ constant. Consider perturbations $\hat{u}$ to each equilibrium: $u(x,t) = U + \hat{u}(x,t)$. Then the PDE, $\partial(U+\hat{u})/\partial t = \partial/\partial x[(U+\hat{u})\partial/\partial x(U+\hat{u})]$, can be linearized to become $\partial\hat{u}/\partial t = U\partial^2\hat{u}/\partial x^2$ such that $\partial\hat{u}/\partial x = 0$ at $x = 0, \pi$. Standard separation of variables gives modes $\hat{u}_k = e^{\lambda_k t}\cos kx$ for integer wavenumber $k = 0, 1, 2, \ldots$ and eigenvalues $\lambda_k = -Uk^2$. Since the eigenvalues are $0, -U, -4U, -9U, \ldots$, for equilibria $U > 0$ there is a center (slow) manifold about each and every equilibrium.

Necessarily, the equilibria with $U > 0$ form the slow manifold, as shown schematically by Figure 4.7. The local domains of applicability overlap and merge to a large "global" domain of theoretical support. This is a specific example of the general rule that a connected subspace/manifold of equilibria will always form at least part of a slow manifold of each of those equilibria.

Example 4.8 (toy atmosphere). Lorenz (1986) proposed a system of five coupled equations to model important characteristics of atmospheric dynamics. Nonlinear oscillations in the variables u, v, and w represent schematically the slowly traveling waves of high- and low-pressure systems in the atmosphere. Oscillations in

---

[30] Because the eigenvector $\sin x$ of the slow subspace has nontrivial spatial structure, expect that models based upon averaging inside the domain will be at least quantitatively in error. The spatial structure is not the same as a constant structure; hence, although averaging may be qualitatively acceptable, averaging is not quantitatively correct.

[31] A marvelous transformation used by Wayne (1997), transforming to log-time and straining space by $\tau = \log t$ and $\xi = x/\sqrt{t}$, discretizes the spectrum of this continuum diffusion. Then center manifold theory applies to nonlinear variations of this basic diffusion to assure us of the existence and relevance of similarity solutions.

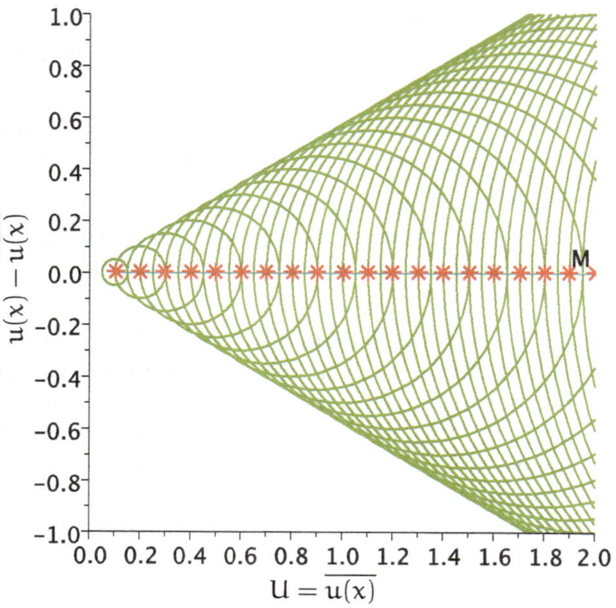

Figure 4.7. *Schematic diagram of the neighborhoods of validity (green) for the nonlinear mixing PDE of Example* 4.7. *The vertical axis represents the infinite-dimensional space of mean zero fields* $u(x)$. *The local neighborhoods (green) about each point on the slow manifold (red) overlap and merge to give an enlarged global domain of validity.*

variables x and z represent schematically the relatively fast "inertial waves" that live on the density stratification in the atmosphere. The Lorenz system of ODEs is

$$\begin{aligned}
\dot{u} &= -vw + bvz,\\
\dot{v} &= \quad uw - buz,\\
\dot{w} &= -uv,\\
\dot{x} &= -z,\\
\dot{z} &= \quad x + buv.
\end{aligned}$$

Linearizing, note that only the x and z equations have any linear part, so there are three zero eigenvalues corresponding to the u, v, and w variables. The x- and z-dynamics linearize to $\dot{x} = -z$ and $\dot{z} = x$, which have eigenvalues $\lambda = \pm i$. That is, all five eigenvalues have zero real part. The entire state space is the center manifold.[32]

Proof of Theorem 4.3 The proof of this Existence Theorem 4.3, and indeed of the next two theorems, usually involves the fixed point theorem and may be found in books by Carr (1981), Iooss and Adelmeyer (1992), and Shilnikov et al. (1998). Section 11.3 describes alternative corresponding theorems with proofs based upon near-identity nonlinear coordinate transformations.

[32]The interest in this example lies in the 3D slow manifold of u, v, and w dynamics among the relatively rapid oscillations of the x and z variables. Analysis of this slow manifold among rapid oscillations is a story for another chapter.

Smoothness The center manifold is, in some neighborhood of the origin, at least as differentiable as the nonlinear terms $\mathbf{f}$. However, it may not be analytic even though $\mathbf{f}$ is; this nonanalyticity appears in some of the examples discussed later.

The neighborhood of existence The Existence Theorem 4.3 asserts that a center manifold exists in some neighborhood $\mathcal{D}$ of an equilibrium but leaves wide open the question of how big that neighborhood might be. Capiński and Roldán (2012) recently derived methods for placing some quantitative bounds on the minimum size of the domain $\mathcal{D}$ of validity and applied the methods to the restricted three-body problem.[33] However, Capiński and Roldán (2012) appear to place more restrictions on the system (4.6), such as a restriction that the linear operator of the stable modes is negative definite. The recent work of Lamarque et al. (2012) views the issue of the domain of validity in the framework of a nonlinear coordinate transform and thus establishes limits based upon the invertibility of the coordinate transform. Guckenheimer et al. (2012) also recently addressed the issue of providing rigorous bounds on a center manifold, but only for systems with one stable mode and two slow modes. As yet, estimating the finite extent of the existence and emergence of center manifold dynamics is open to further research.

A center manifold need not be unique Nonuniqueness of a center manifold disturbs some people (Verhulst, 2005, Ex. 8.7, e.g.). However, upon investigating examples, it is readily apparent that the distinction between the various center manifolds *is of the same order as the difference between trajectories with different initial conditions, which are nonetheless approaching each other exponentially quickly*. Thus when we view the center manifold as forming a low-dimensional model of the long-term dynamics of the full system, the differences between the possible center manifolds are of the same size as those differences between the exact solutions, which we chose to ignore in the first place. Such nonuniqueness, when it arises, is irrelevant to our modeling.

Example 4.9 (nonuniqueness). Consider the system

$$\dot{x} = -x^2 \quad \text{and} \quad \dot{y} = -y\,. \tag{4.8}$$

Here *a* center manifold (slow manifold) is just $y = 0$. However, solve the system (4.8) exactly to find invariant manifolds $y = C\exp(-1/x)$ for $x > 0$ and for any C. Any of these will do just as well for a slow manifold. The reason is that $\exp(-1/x)$ is extremely flat at the origin so it is tangent to the center subspace. But on each of these slow manifolds, the y variable must still be decaying exponentially quickly. Such decay we set out to ignore in the modeling. To be consistent we must ignore the corresponding differences between the different slow manifolds. That is, the nonuniqueness is irrelevant to modeling.

Unique power series In Example 4.9 observe that the power series of the various slow manifolds are all the same; in the example the power series is the trivial $C\exp(-1/x) \sim 0 + 0x + 0x^2 + \cdots$. So the power series of the center manifold *is* unique—this uniqueness is generally true. We generally seek to find the unique power series of center manifolds that may or may not be unique themselves.

[33] The restricted three-body problem is the gravitational motion of a near massless third body in the gravitational field of two relatively massive bodies that are in elliptical orbits.

Inertial manifolds If any attractor of the dynamical system lies inside the neighborhood $\mathcal{D}$, then the Existence Theorem 4.3 assures us that the attractor must lie within a center manifold. This holds whether the attractor is an equilibrium, a limit cycle, a strange attractor, or anything else. Together with the Emergence Theorem 4.5 of the next section, this case ensures that a center manifold also serves as a so-called inertial manifold (Temam, 1990; Foias et al., 1988; Chueshov and Scheutzow, 2001, e.g.).

4.3 ▪ Center manifolds emerge

A theory has only the alternative of being wrong. A model has a third possibility—it might be right but irrelevant. *Manfred Eigen (1927–)*

Based on the rationale of neglecting rapidly decaying transients, our aim is to consider (4.7) as a simple model system for the full dynamical system (4.6). It is simpler in two senses: simpler because it has fewer variables, m instead of n, and simpler because the rate of evolution is much slower. This simplification will be considerable if n is large compared with m, as often happens in practice. However, *we must be assured that solutions of the model* (4.7) *do indeed correspond accurately to solutions of the full system* (4.6); *moreover, they must correspond to solutions that emerge in the long term.*

Some researchers, such as Robinson (1996) and Fenichel (1974), use the term "asymptotic completeness" or "asymptotic phase" to address this issue of the relevance of a model.

Example 4.10 (initial slip). For a near trivial example, consider the system

$$\dot{x} = xy \quad \text{and} \quad \dot{y} = -y\,, \tag{4.9}$$

which has slow manifold $y = 0\,, x = s$ on which the evolution is just $\dot{s} = 0$ (as $\dot{s} = \dot{x} = xy|_{y=0} = 0$). All solutions of the model are constant, $s(t) = s_0$ for some s_0, whereas the full system has solution

$$y(t) = y_0 e^{-t} \quad \text{and} \quad x(t) = x_0 \exp\left[y_0(1 - e^{-t})\right], \tag{4.10}$$

for which $x(t) \to x_0 e^{y_0}$ (a constant)[34] exponentially quickly as time $t \to \infty$ (like e^{-t}). As is generally the case, every solution of the full system approaches a solution of the model, here a constant, exponentially quickly.

There is a difference between solutions approaching a manifold and approaching a solution on the manifold. The latter is what we require: as in the previous example, solutions of the full system should approach a solution of the model. However, the next example shows that just approaching a low-dimensional manifold is *not* sufficient by itself to ensure approach to solutions on that manifold.

Example 4.11 (approach need not be to solutions). Modify the system in Example 4.10 to

$$\dot{x} = x + xy \quad \text{and} \quad \dot{y} = -y\,,$$

which again has an exponentially attractive invariant manifold $y = 0\,, x = s$.[35] On this invariant manifold, $\dot{s} = s$ with solution $s = s_0 e^t$. One might be tempted to assert that this is the long-term model.

[34] That the long-term solution $s(t) = x_0 e^{y_0}$ is not equal to x_0 is sometime called "initial slip" in physics (van Kampen, 1985).

[35] $y = 0$ is called the unstable manifold.

However, the solution of the full system is $y(t) = y_0 e^{-t}$ and

$$x(t) = x_0 \exp\left[t + y_0\left(1 - e^{-t}\right)\right] \sim \left(x_0 e^{y_0}\right) e^t - \left(x_0 y_0 e^{y_0}\right) \quad \text{as } t \to \infty .$$

Although the exponential term matches the model, with $s_0 = x_0 e^{y_0}$, the additional constant term in this long-time asymptote ensures that there is generally an unavoidable $\mathcal{O}(1)$ discrepancy between the full system and the low-dimensional model.

The discrepancy in the evolution in the previous example occurs despite the exponential attraction to $y = 0$. However, the following theorem (Iooss and Adelmeyer, 1992, p. 7, e.g.) guarantees the emergence and long-term accuracy of models based upon center manifolds.

Theorem 4.5 (emergence). *The neighborhood $\mathcal{D}$ may be chosen so that all solutions of* (4.6) *staying in $\mathcal{D}$ tend exponentially quickly to some solution of* (4.7). *That is, for all solutions $\mathbf{u}(t) \in \mathcal{D}$ for all $t \geqslant 0$ there exists a solution $\mathbf{s}(t)$ of the model* (4.7) *such that*

$$\mathbf{u}(t) = \mathbf{v}\left(\mathbf{s}(t)\right) + \mathcal{O}\left(e^{-\beta' t}\right) \quad \text{as } t \to \infty \tag{4.11}$$

for some $0 > -\beta' \geqslant -\beta$, the upper bound on the negative eigenvalues of the linear operator $\mathcal{L}$.

This theorem is crucial; it asserts that for a wide variety of initial conditions the solutions of the full system decay exponentially quickly to a solution predicted by the low degrees of freedom model. That is, the evolution on the center manifold model emerges.

For practical purposes, the rate of attraction β' may be *estimated* by β. The rate of attraction may differ from β due to the size of the domain $\mathcal{D}$ and the nature of the nonlinearities, and so the theorem is conservative.

The trajectories $\mathbf{u}(t)$ have to be sufficiently small for this theorem to apply, but in practice sufficiently small can be quite generous. For example, consider the system (4.1): $\dot{x} = ax^3$ and $\dot{y} = -y + y^2$. When the parameter $a \leqslant 0$, then the exact solution (4.2) shows that all trajectories with initial $y(0) < 1$ asymptote exponentially quickly to the slow manifold $y = 0$. Thus the entire half-plane $y < 1$ comprises initial conditions to which the conclusions of this theorem apply, that is, the neighborhood $\mathcal{D}$ is the entire half-plane; the conclusions *need not* just apply to some small neighborhood of the origin. Every solution in this large finite domain is exponentially quickly modeled.[36]

Another restriction on the application of the theorem is that it only applies to trajectories that stay within the neighborhood of the origin—perhaps because the origin is stable. However, it is clear from the system (4.1) and its solution (4.2) that even when the origin is unstable, when the parameter $a > 0$, the center manifold $y = 0$ is approached exponentially quickly by a large variety of trajectories; it is just that the evolution on the center manifold pushes the solution a long way away in a finite time, here to infinity.

[36]In contrast, the theory of singular perturbations provides significantly less assurance. Let the (small) parameter ϵ measure the ratio in time scales of the decay of the fast variables ($\mathbf{y}$) and the decay of the slow variables ($\mathbf{x}$). Tikhonov's theorem (Verhulst, 2005, p. 99, e.g.) asserts that *in the limit as $\epsilon \to 0$* the solution of the full dynamics tends to the solution of the low-dimensional model *in a finite time interval*, whereas the center manifold Emergence Theorem 4.5 asserts the exponential attractiveness on a *specific time scale* of the low-dimensional model for *finite* ϵ (characteristic of the size of the neighborhood $\mathcal{D}$), and also asserts for *all time* that the dynamics stays within $\mathcal{D}$. I contend that for many problems, center manifold methods are more powerful than singular perturbation methods.

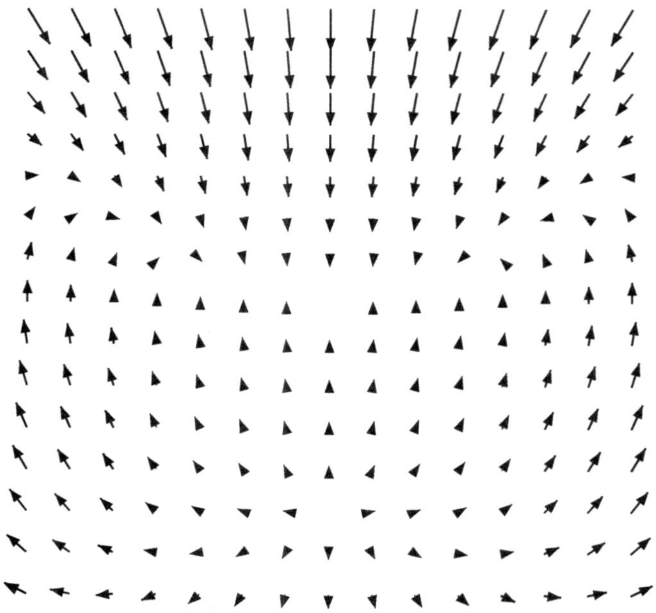

Figure 4.8. *Vector field* $\mathsf{f}(\mathsf{u})$ *of some dynamical system* $\dot{\mathsf{u}} = \mathsf{f}(\mathsf{u})$.

I contend that even if the origin is unstable, in many cases the model is still relevant, albeit only for a finite time.

An older version (Carr, 1981, e.g.) of the emergence theorem appears as the following two corollaries.

Corollary 4.6. *If the origin of* (4.6) *is stable, then for all* $\mathsf{u}(0)$ *sufficiently small there exists a solution* $\mathsf{s}(t)$ *of the model* (4.7) *such that the exponential attraction* (4.11) *holds.*

This follows immediately from Theorem 4.5 because if the origin is stable, then all sufficiently small $\mathsf{u}(0)$ lead to trajectories that stay in $\mathcal{D}$, hence satisfying the requirements of the theorem, and so must exponentially quickly approach a solution of the model.

Corollary 4.7. *The origin of* (4.6) *is stable if and only if the origin of* (4.7) *is stable.*

This corollary requires more work to prove, as detailed by Carr (1981), and so the proof is omitted here.

Example 4.12. Figure 4.8 draws the vector field $\mathsf{f}(\mathsf{u})$ of some dynamical system $\dot{\mathsf{u}} = \mathsf{f}(\mathsf{u})$: for each location u in the plane, the figure plots a small vector proportional to $\mathsf{f}(\mathsf{u})$. On this figure,

- draw the location of any equilibria;
- sketch any slow manifold you can see (to the extent that the slow manifold is reasonably clear);

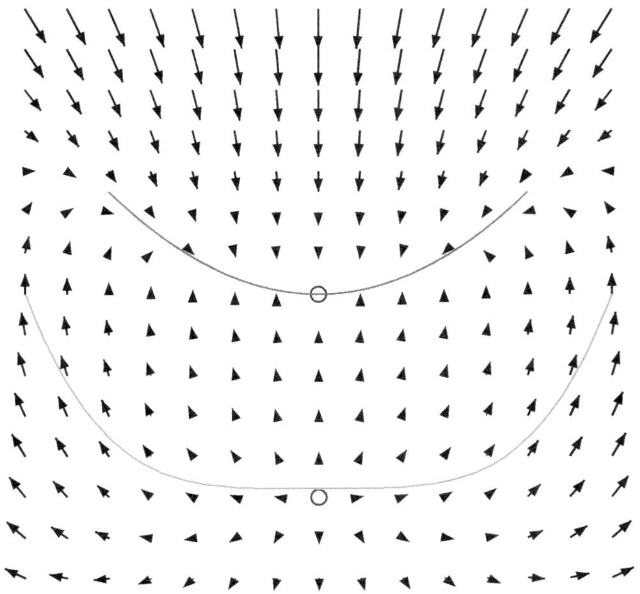

Figure 4.9. *Vector field* $f(u)$ *of the dynamical system* $\dot{u} = f(u)$ *in Example* 4.12*: the two equilibria are marked by small circles; the slow manifold is the curve through the middle of the picture although one could imagine it extending further; the domain of emergence appears to be at least all the shown domain above the lower curve.*

- outline a finite domain in which you reasonably expect the slow manifold existence and emergence theorems to apply.

Solution: Figure 4.9 shows a solution.

Many people want to create and use models of *slow* dynamics by neglecting *fast* oscillations, rather than fast decay. Lorenz (1986) introduced the toy "atmospheric" system in Example 4.8 to illustrate such modeling. It is straightforward to create such models using the methods described in section 4.4. The fundamental stumbling block in such important applications is that there is no strong emergence theorem to support the relevance of such a slow model among fast oscillations.

Example 4.13 (cannot ignore fast oscillations). Consider the system

$$\dot{x} = -x^3 + xy^2 + xz^2\,, \quad \dot{y} = -z\,, \quad \dot{z} = y\,.$$

Linearizing about the equilibrium at the origin, x is a slow variable (corresponding to eigenvalue zero), and y and z undergo fast oscillations (corresponding to eigenvalues $\pm i$).

The slow manifold, because the fast variables are here independent of the slow x, is simply to set the fast modes to zero, $y = z = 0$. Hence the slow model is $\dot{x} = -x^3$. This slow model predicts algebraic decay of x to zero, $x \propto 1/\sqrt{t}$.

However, such decay is *not* representative for nearby solutions. Convert y and z to polar coordinates by invoking radius and angle variables $r(t)$ and $\theta(t)$, $y = r\cos\theta$ and $z = r\sin\theta$; then the system becomes

$$\dot{x} = -x^3 + xr^2, \quad \dot{r} = 0, \quad \dot{\theta} = 1.$$

The slow manifold $y = z = 0$ corresponds to $r = 0$.

Solutions in the *neighborhood* ($r \neq 0$) of the slow manifold are not quantitatively modeled, nor even qualitatively modeled, by the slow manifold solutions ($r = 0$): because $\dot{x} = x(-x^2 + r^2)$, nearby solutions are instead attracted, like e^{-rt}, to one of the two finite-amplitude equilibria at $x = \pm r$. As seen here, the presence of fast oscillations of amplitude $r \neq 0$ generally causes neighboring solutions to drift away from predictions of the slow model.

The issue of how such fast oscillations generally affect long-term modeling is taken up again in detail in Chapter 13.

A physical analogue of the previous example is that water waves generate macroscale currents, drift, that would not exist if the water waves were not present. Such wave-induced currents are most noticeable along a shoreline in rips and alongshore currents.

The strong support of the Emergence Theorem 4.5 applies only to the emergence of the center manifold dynamics among decaying, dissipating modes.

4.4 ▪ Approximately construct a slow manifold

Essentially, all models are wrong, but some are useful.

George E. P. Box, 1987

We need to find an equation whose solution gives the center manifold $\mathcal{M}$. Recall that the center manifold, being invariant under the dynamics, is composed of trajectories of the original system of equations. Thus we must solve some version of the original equations.

Example 4.14. Let us see how an example satisfies the dynamical equations. For the example, recall that for the system (4.3), namely

$$\dot{x} = ax^3 + x^2y \quad \text{and} \quad \dot{y} = -y + y^2 + xy - x^3,$$

we claimed that (4.5) forms a slow manifold model. To higher order the model is

$$y \approx -x^3 - x^4 + (3a-1)x^5 \quad \text{such that} \quad \dot{x} \approx ax^3 - x^5.$$

To see that this is a slow invariant manifold, just substitute into the full differential equations: first, the RHS of the original $\dot{x}$ equation becomes

$$\text{RHS} = ax^3 + x^2y = ax^3 + x^2[-x^3 - x^4 - (3a-1)x^5] = ax^3 - x^5 + \mathcal{O}(x^6),$$

which matches the claimed evolution, $\dot{x}$, to error $\mathcal{O}(x^6)$. Second, the original $\dot{y}$ equation has both sides the same to the above order of error:

$$\text{LHS} = \dot{y} = \frac{dy}{dx}\dot{x} = (-3x^2 - 4x^3)(ax^3 - x^5) = -3ax^5 + \mathcal{O}(x^6)$$

matches its

$$\begin{aligned}\text{RHS} &= x^3 + x^4 - (3a-1)x^5 + (-x^3 - x^4 + (3a-1)x^5)^2 \\ &\quad + x(-x^3 - x^4 + (3a-1)x^5) - x^3 \\ &= -3ax^5 + \mathcal{O}(x^6).\end{aligned}$$

Thus the original equations are satisfied to residuals of $\mathcal{O}(x^6)$ or better. Surely the error in the slow manifold and its evolution should be of the same order. In general it is.

We now deduce a general equation to be solved for the center manifold. Obtain an equation straightforwardly by substituting the assumed relations, that $u(t) = v(s(t))$ where $\dot{s} = g(s) \approx \mathcal{G}s$, into the general evolution equation (4.6) to find

$$\begin{aligned}\dot{u} &= \frac{\partial v}{\partial s}\dot{s} \quad \text{by the chain rule where} \quad \frac{\partial v}{\partial s} = \left[\frac{\partial v_i}{\partial s_j}\right] \\ &= \frac{\partial v}{\partial s} g(s) \quad \text{by (4.7)},\end{aligned}$$

and thus the center manifold, $u = v(s)$, and the evolution thereon, $\dot{s} = g(s)$, satisfy

$$\mathcal{L}v(s) + f(v(s)) = \frac{\partial v}{\partial s} g(s). \tag{4.12}$$

This is the equation to be solved for a center manifold $\mathcal{M}$.

But equation (4.12) is not yet well posed. A vital extra condition is that the tangent space of $\mathcal{M}$ is the center subspace $\mathbb{E}_c$ at the origin:

$$\mathbb{E}_c = \mathcal{T}v(0),$$

where $\mathcal{T}v(s) = \operatorname{span}\{\partial v/\partial s_j\}$ is the tangent space of v at the point parametrized by s. More crudely, this requires that v be quadratically near $\mathbb{E}_c$: $v \in \mathbb{E}_c + \mathcal{O}(|s|^2)$ as $s \to 0$. This tangency condition ensures that the constructed center manifold truly contains the whole of the center modes, and nothing but the center modes.

Without the tangency condition, the solution of (4.12) could be based on an almost arbitrary mixture of linear modes. Indeed, other invariant manifolds of note satisfy (4.12) but are tangent to different vector subspaces as $s \to 0$. For example, the *stable manifold* has tangent space $\mathbb{E}_s$; the *center-unstable manifold* has tangent space $\mathbb{E}_c \oplus \mathbb{E}_u$; and lastly the *slow manifold* has tangent space $\mathbb{E}_0 = \operatorname{span}\{e_j^0 \mid \lambda_j = 0\}$, as the slow manifold is composed of all modes corresponding to precisely zero eigenvalues, not the real part zero. Most center manifolds that we explore will be slow manifolds.

It is typically impossible to find exact solutions to (4.12), as finding exact solutions is tantamount to solving the original system (4.6). However, in applications we approximate $\mathcal{M}$ to any desired accuracy by asymptotically solving (4.12).[37]

The errors in the center manifold $\mathcal{M}$ are in proportion to the residual of the governing equations. For functions $v^{(k)} : \mathbb{R}^m \to \mathbb{R}^n$ (imagine that $v^{(k)}$ approximates the shape of the center manifold v) and $g^{(k)} : \mathbb{R}^m \to \mathbb{R}^m$ (imagine that $g^{(k)}$ approximates the right-hand side $g(s)$ of the evolution) define the *residual*

$$\operatorname{Res}(v^{(k)}, g^{(k)}) = \mathcal{L}v^{(k)}(s) + f\left(v^{(k)}(s)\right) - \frac{\partial v^{(k)}}{\partial s} g^{(k)}(s), \tag{4.13}$$

and observe that $\mathcal{M}$ satisfies $\operatorname{Res}(v(s), g(s)) = 0$.

[37]Recall that although a center manifold need not be unique, its asymptotic power series is unique.

Theorem 4.8 (approximation). *If the tangent space* $\mathcal{T}\mathbf{v}^{(k)}(0) = \mathbb{E}_c$ *and the residual* $\operatorname{Res}(\mathbf{v}^{(k)}, \mathbf{g}^{(k)}) = \mathcal{O}(s^p)$ *as* $s \to 0$ *for some* $p > 1$, *where* s *denotes* $|\mathbf{s}|$, *then* $\mathbf{v}(\mathbf{s}) = \mathbf{v}^{(k)}(\mathbf{s}) + \mathcal{O}(s^p)$ *and* $\mathbf{g}(\mathbf{s}) = \mathbf{g}^{(k)}(\mathbf{s}) + \mathcal{O}(s^p)$ *as* $s \to 0$.

That is, if we can satisfy the equations (4.12) of the governing system to some order of accuracy, then the center manifold is given to the same order of accuracy. Alternatively, the *errors* in the center manifold shape and evolution are of the *same order* as the residual of the governing equation (4.12).

In almost all cases we solve (4.12) by iteration.

Example 4.15 (iteration approximates). Consider (4.3) and seek the slow manifold parametrized by x; that is, seek $y = h(x)$, which is equivalent to the parametrization $x = s$ and $y = h(s)$. We require the function h to be at least quadratic in x so that the constructed center manifold is tangent to $\mathbb{E}_c$, the x-axis, at the origin. Then the x equation asserts that the evolution will be $\dot{x} = ax^3 + x^2h$, whence the y equation becomes

$$\frac{\partial h}{\partial x}\left[ax^3 + x^2h\right] = -h + h^2 + xh - x^3.$$

This equation is equivalent to (4.12) with $\mathbf{v}(s) = (s, h(s))$ and $\dot{s} = \dot{x}$.

Solve this equation by iteration. Since h is quadratic in x, the dominant term is the term linear in h on the right-hand side. Rearrange so that the dominant linear term in h is by itself on the left-hand side:

$$h = -\frac{\partial h}{\partial x}\left[ax^3 + x^2h\right] + h^2 + xh - x^3. \tag{4.14}$$

Iteration starting with $h^{(0)} = 0$ in the right-hand side leads to

$$\begin{aligned}
h^{(1)} &= -x^3,\\
h^{(2)} &= -x^3 - x^4 + 3ax^5 + x^6 - 3x^7,\\
h^{(3)} &= -x^3 - x^4 + (3a-1)x^5 + (1+7a)x^6 + \cdots - 63x^{15},
\end{aligned}$$

and so on; part of $h^{(3)}$ was mentioned earlier. This iteration appears to be converging, as an asymptotic expansion, to some description of the slow manifold.

Corollary 4.9 (below) of the approximation Theorem 4.8 assures us that

- $h^{(1)}$ describes the slow manifold $\mathcal{M}$ with errors $\mathcal{O}(h^{(2)} - h^{(1)}) = \mathcal{O}(x^4)$, whereas
- $h^{(2)}$ has errors $\mathcal{O}(h^{(3)} - h^{(2)}) = \mathcal{O}(x^5)$.

Such simple iteration works for dynamical systems in the linearly separated form (4.15). We now deduce that the error in any iterate is of the order of the difference with the next iterate.

Almost all previously documented statements of the existence, emergence, and approximation theorems assume that the description of the physical dynamical system $\dot{u} = \mathcal{L}u + f(u)$ has been transformed by a linear change of variables to the linearly separated form

$$\dot{\mathbf{x}} = A\mathbf{x} + \mathbf{f}_1(\mathbf{x}, \mathbf{y}) \quad \text{and} \quad \dot{\mathbf{y}} = B\mathbf{y} + \mathbf{f}_2(\mathbf{x}, \mathbf{y}), \tag{4.15}$$

where $\mathbf{x}$ contains all the critical center modes (eigenvalues of matrix A all have zero real part) and $\mathbf{y}$ contains all the stable modes (eigenvalues of matrix B all have negative real part).[38] Consequently, almost everyone assumes that the parametrization of the center manifold is $\mathbf{x} = \mathbf{s}$. As long as we are allowed quite general smooth, albeit nonlinear, changes of coordinates, then the statements of the theorems given here are exactly equivalent to what others previously documented. I state the theorems in terms of $\mathbf{u}$ because then the theorems are closer to what we need in applications where we want to deal with physically meaningful variables.

Nonetheless, many people limit themselves to systems of the separated form (4.15). Because of the simplicity, we also derive an approximation corollary for this case. We seek a center manifold in the form

$$\mathbf{y} = \mathbf{h}(\mathbf{x}) \quad \text{such that} \quad \dot{\mathbf{x}} = A\mathbf{x} + \mathbf{f}_1(\mathbf{x}, \mathbf{h}(\mathbf{x})). \tag{4.16}$$

The leading approximation to the center manifold is $\mathbf{h}^{(0)} = \mathbf{0}$, as the center subspace is $\mathbf{y} = \mathbf{0}$. Then we construct successive approximations to the center manifold via the iteration

$$-B\mathbf{h}^{(k+1)} = \mathbf{f}_2(\mathbf{x}, \mathbf{h}^{(k)}) - \frac{\partial \mathbf{h}^{(k)}}{\partial \mathbf{x}} \left[A\mathbf{x} + \mathbf{f}_1(\mathbf{x}, \mathbf{h}^{(k)})\right]. \tag{4.17}$$

This iteration converges only if matrix A has eigenvalues that are purely zero, the slow manifold case; if not, another approach is needed. Nonetheless, whether the iteration converges or not, the following corollary applies.

Corollary 4.9 (approximation). *Applying the iteration* (4.17) *to approximate a slow manifold of* (4.15), *the error in* $\mathbf{h}^{(k)}$ *is* $\mathcal{O}(\mathbf{h}^{(k+1)} - \mathbf{h}^{(k)})$ *as* $\mathbf{x} \to \mathbf{0}$.

Proof. The $\mathbf{x}$ equation of system (4.15) is satisfied identically by $\dot{\mathbf{x}} = A\mathbf{x} + \mathbf{f}_1(\mathbf{x}, \mathbf{h}^{(k)})$. The residual in approximating the system (4.15) comes solely from the $\mathbf{y}$ equation. Using the iteration (4.17), consider

$$\begin{aligned}
B\mathbf{h}^{(k+1)} - B\mathbf{h}^{(k)} &= -B\mathbf{h}^{(k)} - \mathbf{f}_2(\mathbf{x}, \mathbf{h}^{(k)}) + \frac{\partial \mathbf{h}^{(k)}}{\partial \mathbf{x}} \left[A\mathbf{x} + \mathbf{f}_1(\mathbf{x}, \mathbf{h}^{(k)})\right] \\
&= -B\mathbf{h}^{(k)} - \mathbf{f}_2(\mathbf{x}, \mathbf{h}^{(k)}) + \frac{\partial \mathbf{h}^{(k)}}{\partial \mathbf{x}} \dot{\mathbf{x}} \\
&= -B\mathbf{h}^{(k)} - \mathbf{f}_2(\mathbf{x}, \mathbf{h}^{(k)}) + \dot{\mathbf{y}} \\
&= \operatorname{Res}_{(4.15)}(\mathbf{h}^{(k)}).
\end{aligned}$$

Using the Approximation Theorem 4.8, the error in the kth iterate $\mathbf{h}^{(k)}$ is then determined as $\mathcal{O}(\operatorname{Res}_{(4.15)}(\mathbf{h}^{(k)})) = \mathcal{O}(B\mathbf{h}^{(k+1)} - B\mathbf{h}^{(k)}) = \mathcal{O}(\mathbf{h}^{(k+1)} - \mathbf{h}^{(k)})$ as $\mathbf{x} \to \mathbf{0}$. □

As seen in Example 4.15, often the iteration (4.17) obtains corrections accurate to one higher order in $\mathbf{x}$ in each iteration. With certain symmetries we sometimes obtain two orders of accuracy each iteration. Sometimes, associated with generalized eigenvectors in the center subspace, we need two or more iterations for each order improvement in the accuracy. No matter how many iterates are actually performed, or even if the iteration converges, the Approximation Corollary 4.9 determines the order of error in the result.

The evolution on the center manifold in Example 4.15 is then

$$\dot{x} = ax^3 + x^2\left[h^{(1)} + \mathcal{O}(x^4)\right] = x^3\left(a - x^2\right) + \mathcal{O}(x^6).$$

[38] For example, the variables $\mathbf{x}$ and $\mathbf{y}$ correspond to the master and slaved variables of Haken's synergetics (Haken, 1983, 1996, e.g.).

Observe from this equation that the origin is unstable for $a > 0$ and we expect a slow evolution over long times to either of two equilibria at approximately $s = \pm\sqrt{a}$. Conversely, for $a \leqslant 0$ solutions slowly decay, like $t^{-1/2}$, to the stable origin. Such slow evolution to whichever attractor is the ultimate destination forms the low-dimensional model of (4.3).

But, you may comment, if the parameter $a > 0$, then the origin is unstable and the Emergence Theorem 4.5 need not apply. However, if the parameter a is small enough so that the equilibrium attractors at $\pm\sqrt{a}$ lie in the neighborhood $\mathcal{D}$, then solutions will indeed stay within $\mathcal{D}$ and the emergence theorem applies to guarantee relevance of the model. Section 4.5 similarly rescues the center manifold theorems in important applications by a neat and extraordinarily powerful trick.

Example 4.16 (multiple variables). Let us arbitrarily damp the fast waves in the toy Lorenz system of Example 4.8:

$$\begin{aligned}
\dot{u} &= -vw + bvz\,, \\
\dot{v} &= uw - buz\,, \\
\dot{w} &= -uv\,, \\
\dot{x} &= -x - z\,, \\
\dot{z} &= x - z + buv\,.
\end{aligned}$$

This system is in the separated form (4.15) upon identifying

$$\mathbf{x} = (u, v, w), \quad \mathbf{y} = (x, z), \quad A = 0\,, \quad B = \begin{bmatrix} -1 & -1 \\ 1 & -1 \end{bmatrix}.$$

As the eigenvalues of matrix A are all precisely zero, and the eigenvalues of B are $-1 \pm i$, there exists an emergent slow manifold parametrized by the slow variables $\mathbf{x} = (u, v, w)$. The iteration scheme (4.17) becomes

$$-B\mathbf{h}^{(k+1)} = \begin{bmatrix} 0 \\ buv \end{bmatrix} - \begin{bmatrix} \dot{x} \\ \dot{z} \end{bmatrix}$$

and starts from the tangent space approximation that $\mathbf{h}^{(0)} = \mathbf{0}$.

- Substituting $h^{(0)} = 0$ into the right-hand side gives right-hand side $(0, buv)$. As the inverse
$$-B^{-1} = \tfrac{1}{2}\begin{bmatrix} 1 & -1 \\ 1 & 1 \end{bmatrix},$$
the first nontrivial approximation to the shape of the slow manifold is $\mathbf{y} = (x, z) \approx \mathbf{h}^{(1)} = \tfrac{1}{2}(-1, 1)buv$.
- Consequently $\dot{x} \approx -\tfrac{1}{2}b(\dot{u}v + u\dot{v}) = \cdots = \tfrac{1}{2}b(u^2 - v^2)(w - \tfrac{1}{2}b^2uv)$, and similarly for $\dot{z}$. Thus substituting $h^{(1)} = \tfrac{1}{2}(-1, 1)buv$ into the right-hand side gives $(0, buv) - \tfrac{1}{2}(1, -1)b(u^2 - v^2)(w - \tfrac{1}{2}b^2uv)$. Using the inverse B^{-1}, the second nontrivial approximation to the shape of the slow manifold is $\mathbf{y} = (x, z) \approx \mathbf{h}^{(2)} = \tfrac{1}{2}(-1, 1)buv + \tfrac{1}{2}(1, 0)b(u^2 - v^2)(w - \tfrac{1}{2}b^2uv)$.

Since the lowest-order terms in $\mathbf{h}^{(2)} - \mathbf{h}^{(1)}$ are cubic terms in variables $\mathbf{x} = (u, v, w)$, then we know that the slow manifold is $\mathbf{y} = \mathbf{h}(\mathbf{x}) = \mathbf{h}^{(1)}(\mathbf{x}) + \mathcal{O}(|\mathbf{x}|^3) = \tfrac{1}{2}(-1, 1)buv + \mathcal{O}(|\mathbf{x}|^3)$.

4.5 ▪ Introducing parameters widens applications

What's the use of a newborn baby?

anonymous, possibly Benjamin Franklin

If it were not for one generalization, then the theory of the previous sections would be extremely academic. After all, how many systems of interest are exactly at critical, that is, with the exactly required spectrum? The answer is almost none. At first sight, center manifold theory cannot be applied usefully to realistic physical applications.

In applications the requirement that the real parts of the eigenvalues be precisely zero is far too restrictive. Instead, surely eigenvalues near zero—characterizing modes that decay at a small rate—will be important over long times. Fortunately, center manifold theory adapts to cater for this situation. Furthermore, the same adaptation also caters for growing modes provided that they grow slowly. The trick is to treat the slow growth or decay as if it really did correspond to eigenvalues of zero real part but was just perturbed by a small amount. The pitchfork bifurcation is a widespread realization of such a perturbation.

4.5.1 ▪ Case study: A simple pitchfork bifurcation

We are often forced to introduce a perturbation parameter when there is no other analytical way to make progress *(Bender and Orszag,* 1981, *p.* 351*)*

Suppose we wanted to model the definite system

$$\dot{x} = 0.1\,x - xy \quad \text{and} \quad \dot{y} = -y + x^2. \tag{4.18}$$

This system has an equilibrium at the origin. Linearize to $\dot{x} \approx 0.1\,x$ and $\dot{y} \approx -y$ with eigenvalues 0.1 and −1. As just mentioned, center manifold theory does not directly apply as there are no eigenvalues with zero real part. But simulations, snapshots shown in Figure 4.10, demonstrate that there is still strong attraction to a low-dimensional manifold, and then subsequent slow evolution; these are exactly the characteristics of a slow manifold model. To apply our theory, the trick is to write this definite system as one in a wider family of systems that do have a slow manifold.

Generalize the previous system by introducing a parameter a that measures small effects. Here, embed the previous system as the $a = 0.1$ case of this prototype bifurcation system,

$$\dot{x} = ax - xy \quad \text{and} \quad \dot{y} = -y + x^2, \tag{4.19}$$

where a is a parameter that in application will be some value such as $a = 0.1$.[39]

The linearized version of the system (4.19) is $\dot{x} = ax$ and $\dot{y} = -y$, which has eigenvalues of a and −1. Since none of the eigenvalues are zero (except for the special case $a = 0$), we apparently cannot apply center manifold theory.

For this prototype system (4.19) the mathematical trick to rescue the application of center manifold theory is the following. Consider the three-variable dynamical system

$$\dot{a} = 0, \quad \dot{x} = ax - xy, \quad \text{and} \quad \dot{y} = -y + x^2. \tag{4.20}$$

[39]One might find the equilibria of the dynamical system by solving $\dot{x} = ax - xy = 0$ and $\dot{y} = -y + x^2 = 0$. This pair of simultaneous equations has equilibria $x = 0$ and $y = 0$ for all a, and when parameter $a > 0$, $x = \pm\sqrt{a}$ and $y = a$. These are the equilibria, but what about the dynamics—stability, for instance? Of especial interest is when the parameter a crosses 0, as then the number of equilibria changes, that is, a bifurcation occurs.

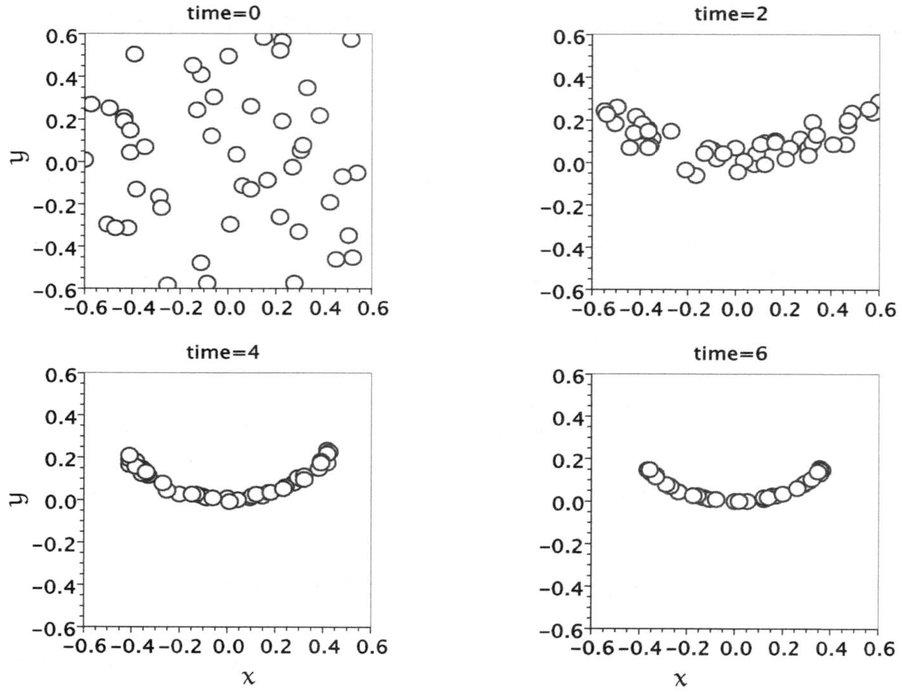

Figure 4.10. *Locations in the* xy*-plane of solutions to the ODE* (4.18) *starting from* 50 *random initial conditions at four times: the near-parabolic slow manifold emerges quickly.*

The equation $\dot{a} = 0$ implies that a is a fixed constant parameter and so this system (4.20) describes *exactly* the same dynamics as the previous system (4.19). However, the linearized system is $\dot{a} = 0$, $\dot{x} = 0$, and $\dot{y} = -y$; the ax term in the original $\dot{x}$ equation becomes a nonlinear term by this trick! This system of linear equations has two zero eigenvalues and one negative eigenvalue of -1; thus we may and do apply center manifold theory to prove the existence and emergence of a slow manifold of the nonlinear system (4.20). Since their solutions are the same, the modeling of (4.20) then provides a model of the prototype system (4.19). Since the center manifold exists and emerges in a finite neighborhood of $a = x = y = 0$, solutions at finite parameter a are encompassed, which empowers us to model the original physical system (4.18).[40]

Iteration constructs the slow manifold Let us seek a slow center manifold for system (4.20), equivalently (4.19), of the form $(a, x, y) = (a, x, h(a, x))$. The slow manifold is a two-dimensional surface in the axy space, as shown in Figure 4.11, because there are two zero eigenvalues in the linearized equations. Substituting this form, and rearranging

[40] A more sophisticated version of this trick was used by Arneodo et al. (1985) to *prove* the existence of chaos in a real fluid system. Knowing that chaotic dynamics require three-dimensional state space, they sought a physical system with a slow manifold with three dimensions. They thus imagined a fluid system with three material parameters representing three competing physical instabilities. At a critical value of these three parameters there are three critical modes in the slow manifold (together with the three dimensions of the parameter space, making a six-dimensional slow manifold, but only three are of dynamical modes; the other dimensions are of the three parameters). Using the three parameters to unfold the dynamics, they were able to show that chaotic dynamics exists arbitrarily close to the critical point. Thus the Emergence Theorem 4.5 applies and guarantees the existence of such chaos in the fluid flow.

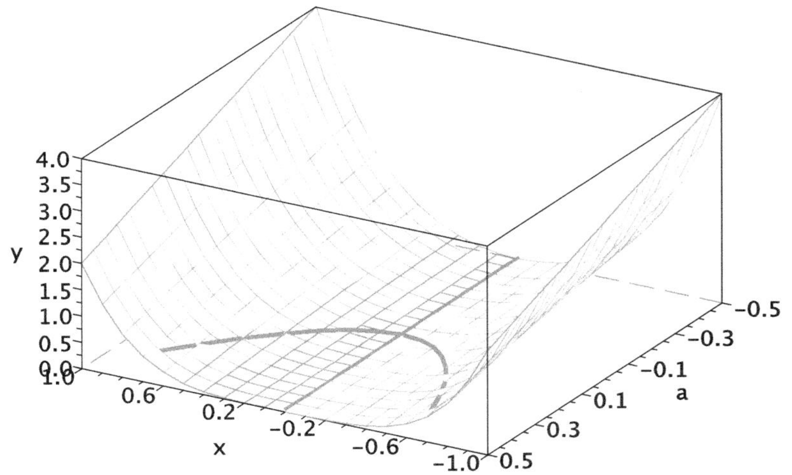

Figure 4.11. *Approximate center manifold of* (4.19) *in* axy*-space. Also superimposed is the pitchfork of equilibria.*

to put the dominant term on the left-hand side, we deduce that h must satisfy

$$h = x^2 - \frac{\partial h}{\partial x} x(a - h).$$

Solving this equation iteratively, for simplicity, leads to approximations

$$\begin{aligned} h^{(0)} &= 0, \\ h^{(1)} &= x^2, \\ h^{(2)} &= x^2 - 2ax^2 + 2x^4. \end{aligned}$$

Now $h^{(2)} - h^{(1)} = -2ax^2 + 2x^4 = \mathcal{O}(|(a,x)|^3) = \mathcal{O}((a^2+x^2)^{3/2})$ as $(a,x) \to 0$, and therefore by the Approximation Corollary 4.9, the slow manifold $y = h(a,x) = x^2 + \mathcal{O}(|(a,x)|^3)$. Similarly, $h^{(3)} - h^{(2)} = \mathcal{O}(|(a,x)|^4) = \mathcal{O}((a^2+x^2)^2)$, and so the slow manifold is better described by

$$y = h(a,x) = (1-2a)x^2 + \mathcal{O}(|(a,x)|^4),$$

as shown in Figure 4.11.[41]

The slow manifold is local to the origin The slow manifold is *local* to the origin in the axy-space: the equilibrium at the basis of the theorems is $a = x = y = 0$, and the asymptotic errors are expressed $\mathcal{O}(|(a,x)|^p)$. In particular, the slow manifold is only valid for small enough parameter a, that is, near enough to the bifurcation. Figure 4.13 (left) (later in the chapter) shows the bifurcation of the equilibria in the ax-plane and a schematic of the possible domain of validity about the origin in the ax-plane.[42]

[41]Those familiar with bifurcations may want to treat the parameter a as second order. However, the center manifold theorems presented so far have no concept of variables of different orders of magnitude: the orders of errors are only expressed in terms of vector length $|\mathbf{s}|$ or $|\mathbf{x}|$. Further, requiring "parameter a to be second order" is putting the cart before the horse: supposing that it "should be" second order comes from knowing phenomena in the long-term dynamics, but at this stage we are only justifying and creating the low-dimensional model; we do not yet know the long-term phenomena. Thus we shun potentially arbitrary a priori assumptions.

[42]The finite neighborhood $\mathcal{D}$ of attraction to the slow manifold extends out into the y direction, but for simplicity Figure 4.13 does not attempt to plot the y structure.

Algorithm 4.1 *Outline of the general iteration to construct a center manifold model by iteration.*

```
1: preliminaries;
2: initial linear approximation;
3: repeat
4:    compute residual;
5:    correct the center manifold;
6: until residual is small enough.
```

Interpret the slow manifold dynamics This slow manifold emerges exponentially quickly, roughly like $\exp(-t)$, as this is the linear decay of the fast y mode; once on the slow manifold the system evolves according to

$$\dot{a} = 0,$$
$$\dot{x} = ax - xh \approx ax - (1-2a)x^3.$$

The parameter a remains as a constant. Thus from this *Landau equation* for x we readily discern that if parameter $a \leqslant 0$ then the origin $(x, y) = (0, 0)$ is stable and hence is the attractor for large time. Contrariwise, if parameter $a > 0$, then the origin is unstable, and the solutions are attracted to one of the finite-amplitude equilibria located at $x \approx \pm\sqrt{a}$.

One important subtlety is that although the origin in the xy-plane may not be stable, depending upon the parameter a, the analysis was actually done in the axy-space. Further, the origin in the axy-space *is* stable; a trajectory started near the origin in the axy-space stays near the origin for all time—if it starts within a distance δ of the origin it will stay within a distance of $\sqrt{\delta}$. Despite the possible lack of stability of the origin in the xy-plane, the Emergence Theorem 4.5 still applies and our preceding claims that the slow manifold model emerges exponentially quickly is indeed a rigorous result. We are assured that the dynamics near the bifurcation is correctly modeled.

4.5.2 ▪ Computer algebra iteration

The approximate construction of a center manifold model is significantly easier for a system of the separated form (4.15). Indeed all the example toy dynamical systems introduced so far have been in the separated form (4.15). As a prelude to the general algorithm of the next chapter—where such separation is either inconvenient, awkward, or impractical—here we explore constructing slow manifold models for the separated system (4.15).

Computer algebra readily implements iterative algorithms such as the previous. Let's develop such an algorithm here for the specific example bifurcation system (4.19). Algorithm 4.1 outlines the generic iteration for dynamical systems in the special separated form of (4.15).

Complete details of a Reduce program for the particular example system (4.19) follow.

1. The preliminaries mainly involve telling Reduce that the variable called x is to depend upon time (`t`) and that whenever Reduce sees the derivative of x with respect to time (`df(x,t)`), it is to replace it with the corresponding right-hand side of the differential equation. Thus

```
depend x,t;
let df(x,t)=>a*x-x*y;
```

Algorithm 4.2 *Iteration finds the slow manifold of the simple bifurcation problem* (4.19).

```
depend x,t;
let df(x,t)=>a*x-x*y;
y:=0;
let {a^2=>0, x^4=>0};
for it:=1:99 do begin
   res:=-df(y,t)-y+x^2;
   y:=y+res;
   if res=0 then it:=10000;
end;
```

2. The usual initial linear approximation to the shape of the slow manifold is simply the tangent approximation at the origin, namely that

```
y:=0;
```

3. Now iterate in a loop. Control the truncation of the asymptotic approximation by telling Reduce to discard any factor in a^2 or higher and any factor in x^4 or higher; thus all expressions are computed to an error of $\mathcal{O}(a^2, x^4)$:[43]

```
let {a^2=>0, x^4=>0};
for it:=1:99 do begin
   . . .
   if res=0 then it:=10000;
end;
```

I recommend implementing the repeat-until loop of Algorithm 4.1 as the for-loop above for safety (to avoid the possibility of infinite loops). Within the above loop:

(a) The $\dot{y}$ equation must be satisfied by driving its residual to zero. For whatever approximation y contains, since Reduce knows how x depends upon time, compute the residual as

```
res:=-df(y,t)-y+x^2;
```

(b) Then, because of the simplicity of this example, the correction to the shape of the slow manifold is simply

```
y:=y+res;
```

In general we would have to solve a linear equation for the update: from iteration (4.17) the general update is of the form $\mathbf{h}^{(k+1)} = \mathbf{h}^{(k)} - B^{-1}\,\mathrm{Res}$.

Putting these code fragments together gives Algorithm 4.2. Executing Algorithm 4.2 recomputes the results determined earlier by hand.

Of course with computer algebra one may partake in an orgy of computation, just for the fun of it. Here we compute the asymptotic expressions to excruciatingly high order

[43] Reduce is *precisely literal* in how it matches its patterns except for this one case of the pattern of a simple variable raised to some power being replaced by zero. In this case it also replaces all higher powers by zero.

by changing the order at which terms are discarded. Using `let {a=>0,x^20=>0};`, we get the computer to tell us

$$y = x^2 + 2x^4 + 12x^6 + 112x^8 + 1,360x^{10} + 19,872x^{12} + 335,104x^{14} + 6,359,040x^{16} + 133,560,576x^{18} + \mathcal{O}(x^{20}, a).$$

Evidently, by the rapid growth in the coefficients, this appears to be a *divergent* power series in x.[44] Such divergence is typical. Thus we have to be very careful in the use of these low-dimensional models. Typically, the best model is some low-order truncation of such asymptotic series. Although *formally not* the most accurate, a low-order truncation is generally more widely applicable. Due to such divergence, seeking higher accuracy often leads to a smaller domain of validity of the power series (distinct from the neighborhood of validity appearing in the center manifold existence and emergence theorems, which is ultimately determined by the dynamics of the system).

Alternatively we may compute the asymptotic expressions to excruciatingly high order in the parameter a by invoking `let {a^8=>0,x^4=>0};`; we get the computer to tell us

$$y = x^2(1 - 2a + 4a^2 - 8a^3 + 16a^4 - 32a^5 + 64a^6 - 128a^7) + \mathcal{O}(x^4, a^8).$$

The coefficient of x^2 in this approximation appears to be a *convergent* power series in a.[45] This example shows what experience indicates in general: *expect multivariable series approximations to slow manifolds to be divergent in dynamic variables like* x, *but the coefficients to be convergent in fixed parameters like* a.

In applications we sometimes do want to compute approximations to high orders in some parameters, but only low orders in dynamic variables. Theorem 4.10, below, justifies the required flexibility.

4.5.3 ▪ Approximation may be more flexible

Parameters are extremely important in application of center manifold theory. A useful approximation theorem is specialized to this case. Consider the general dynamical system in $\ell + n$ variables,

$$\dot{\epsilon} = 0 \quad \text{and} \quad \dot{u} = \mathcal{L}u + f(\epsilon, u), \tag{4.21}$$

where

- ϵ is a vector of ℓ parameters;
- $\mathcal{L}$ is independent of both the parameters ϵ and the variables u, and $\mathcal{L}$ has the same eigenstructure as before;
- the nonlinear terms f are quadratic at the *origin of the parameter-variable space, namely at* $(\epsilon, u) = 0$.

[44] Where does this divergence come from? An answer *here* is that the slow manifold has a weak, high-order singularity at the finite-amplitude equilibria $(x, y) = (\pm\sqrt{a}, a)$—along the parabola in Figure 4.11. Linearization shows that trajectories very close to these equilibria have the approximate shape of the singular $y \propto |x \mp \sqrt{a}|^{1/(2a)}$: the slow manifold, albeit infinitely differentiable at the origin, will generally be one of these trajectories through the finite-amplitude equilibria and thus only finitely differentiable. Now let parameter $a \to 0$ (along the parabola in Figure 4.11), with the singularities at the finite equilibria coming arbitrarily close to the origin to cause the divergence we see at $a = 0$, albeit at an order becoming infinitely high. Expect analogous singular behavior to be hidden in general dynamical modeling.

[45] Whether the pattern continues or not, whether convergent or not, we are justified in writing $y = x^2/(1 + 2a) + \mathcal{O}(x^4, a^8)$, because it has the same power series to the known terms, and then using $y \approx x^2/(1 + 2a)$ for practical prediction if we prefer.

The extended system (4.21) satisfies the requirements for the existence of a center manifold. Seeking a center manifold in the form

$$u = v(\epsilon, s), \quad \text{where} \quad \dot{s} = g(\epsilon, s),$$

we use the following generalized theorem to test the accuracy of approximations $v^{(k)}(\epsilon, s)$ and $g^{(k)}(\epsilon, s)$ (Li, 1999, Chap. 2).

Theorem 4.10 (flexible approximation). *If the tangent space* $\mathcal{T}v^{(k)}(0,0) = \mathbb{R}^\ell \times \mathbb{E}_c$ *and the residual* $\operatorname{Res}(v^{(k)}, g^{(k)}) = \mathcal{O}(\epsilon^q + s^p)$ *as* $(\epsilon, s) \to 0$ *for some* $p, q > 1$ *(where* Res *is as before and* ϵ *and* s *denote* $|\epsilon|$ *and* $|s|$ *respectively), then* $v(\epsilon, s) = v^{(k)}(\epsilon, s) + \mathcal{O}(\epsilon^q + s^p)$ *and* $g(\epsilon, s) = g^{(k)}(\epsilon, s) + \mathcal{O}(\epsilon^q + s^p)$ *as* $(\epsilon, s) \to 0$.

The equivalent statement for errors $\mathcal{O}(\epsilon^q, s^p)$ *is also true.*

The perhaps new use of the "order of" notation has the following meaning.

- A term $\epsilon^m s^n = \mathcal{O}(\epsilon^q, s^p)$ if and only if either $m \geqslant q$ or $n \geqslant p$. In other words, $\mathcal{O}(\epsilon^q, s^p)$ is a shorthand for terms $\mathcal{O}(\epsilon^q) + \mathcal{O}(s^p)$.
- A term $\epsilon^m s^n = \mathcal{O}(\epsilon^q + s^p)$ if and only if $m/q + n/p \geqslant 1$. This comes from the definition of $\mathcal{O}()$ by considering when $\epsilon^m s^n/(\epsilon^q + s^p)$ is bounded as $(\epsilon, s) \to 0$.

  For example, terms a^2, $a^2 s$, as^2, as^3, and s^4 are all $\mathcal{O}(a^2 + s^4)$.

Example 4.17 (pitchfork bifurcation continued). In the example system (4.19), the residual $\operatorname{Res}(h^{(1)}) = h^{(2)} - h^{(1)} = -2ax^2 + 2x^4$.

- Because the ax^2 term in the residual is cubic, we earlier concluded that the error in $h^{(1)}$ approximating the slow manifold $\mathcal{M}$ is $\mathcal{O}(|(a, x)|^3)$.
- Alternatively, we may also observe that the two terms in the residual are first $\mathcal{O}(a)$ for fixed x and second $\mathcal{O}(x^2)$ for fixed a, and hence this theorem now asserts that the error in $h^{(1)}$ approximating the slow manifold $\mathcal{M}$ is $\mathcal{O}(a, x^2)$.
- In yet another alternative, we may also observe that the two terms in the residual are $\mathcal{O}(a^2 + x^4)$, and hence this theorem also asserts that the error in $h^{(1)}$ approximating the slow manifold $\mathcal{M}$ is correspondingly $\mathcal{O}(a^2 + x^4)$.

  Consequently, for example, this last alternative shows that there can be no $a^0 x^3$ term in $\mathcal{M}$—a stronger result than claimable in other alternatives.

All of the above are valid error statements. Choose whichever one you like for your purposes.

In using computer algebra we often find it most convenient to work to residuals and errors $\mathcal{O}(\epsilon^p, s^q)$. Nonetheless, we may report other errors by subsequently truncating the multivariate asymptotic expansions in other ways.

4.6 ▪ Alternative sound methods are equivalent

The local analysis of the pitchfork bifurcation is the best one can do: limits of validity are determined by the characteristics of the solutions. No matter how one may twist and turn with various mathematical analyses, the predictions and domain of validity are

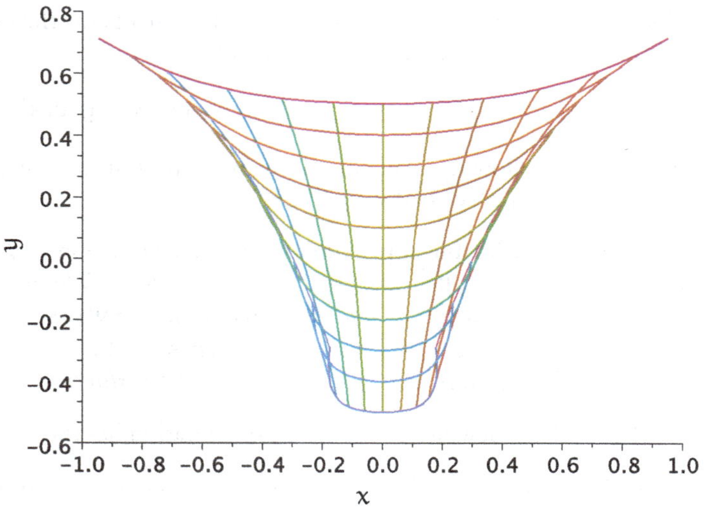

Figure 4.12. X *and* Y *coordinate curves in the* xy*-plane that simplify the dynamics of the example system* (4.19) *for parameter* $a = 0.1$. *The particular coordinate curves plotted are* $-0.5 : 0.1 : 0.5$.

essentially the same for any sound modeling. The reason is that the underlying dynamics we are modeling is the same, and any sound modeling has to respect that dynamics.

Let's use the simple pitchfork bifurcation of the system (4.19) as an example to illustrate the extent of slow manifold modeling and then the equivalence with alternative methods.

4.6.1 ▪ A coordinate transform displays the domain of attraction

For simple systems like (4.19), normal-form coordinate transforms illustrate more details. By magical arguments developed elsewhere (Chapter 11), consider the system (4.19) in new variables X and Y where, in an asymptotic expansion,

$$\begin{aligned} x &= X + XY + \tfrac{1}{2}XY^2 + \tfrac{1}{6}XY^3 + 2X^3Y + \cdots, \\ y &= Y + (1 - 2a + 4a^2)X^2 - 2X^2Y^2 + 2X^4 + \cdots. \end{aligned} \tag{4.22}$$

Figure 4.12 plots the coordinate curves of this XY system. Some tedious algebra developed elsewhere shows that in these new variables the system (4.19) becomes

$$\begin{aligned} \dot{X} &= aX - (1 - 2a + 4a^2)X^3 - 2X^5 + \cdots, \\ \dot{Y} &= (-1 + 2X^2 + 4X^4 + \cdots)Y. \end{aligned} \tag{4.23}$$

I emphasize that because this is the result of the coordinate transform in Figure 4.12, we have not lost or changed any dynamics: all dynamics of the original system (4.19) near the origin are encompassed by (4.23). Now, what does system (4.23) tell us? Note the magical result that the X equation in (4.23) is independent of Y and so gives the evolution of the slow variable X for all time. The Y equation in (4.23) then predicts the exponentially quick decay to the emergent slow manifold $Y = 0$. However, this decay only occurs for slow variable X such that the rate $-1 + 2X^2 + 4X^4 + \cdots$ is negative. We cannot be definite because we do not know this asymptotic expansion exactly, but say this rate is negative for $|X| < 1/2$ as then at least the known part $-1 + 2X^2 + 4X^4 < -1/4 < 0$. Figure 4.12

plots the coordinate transform (4.22) for $|X| < 1/2$ so we are reasonably assured that any initial conditions within the plotted region are exponentially quickly attracted to the slow manifold.[46] The plot is also limited to $|Y| < 1/2$, as the asymptotics are local to the origin and the coordinate transform appears to degenerate for larger Y, as well as larger X. This special coordinate transform provides good evidence of the finite extent of the neighborhood $\mathcal{D}$ referred to in the center manifold theorems.

4.6.2 ▪ An equivalent global manifold

As in many other asymptotic methods, one might choose to scale the dependent variables, here x and y, to obtain a global model. Let's try such scaling, as others do, focusing on just $a \geqslant 0$ for simplicity.

Here one would argue that the natural scaling is to change variables to $X(t)$ and $Y(t)$, where $x = \sqrt{a}X(t)$ and $y = aY(t)$. Substitute into (4.19) and cancel factors of $\sqrt{a}$ and a to derive

$$\dot{X} = a(X - XY) \quad \text{and} \quad \dot{Y} = -Y + X^2. \tag{4.24}$$

Now the parameter a is small near the bifurcation, so adjoin the trivial ODE $\dot{a} = 0$. Then consider this system in the extended (a, X, Y)-space by the recipe of finding the equilibria, linearizing, and then appealing to center manifold theory. There is a set of equilibria: $X = s$, $Y = s^2$, and $a = 0$, that is, the set $(0, s, s^2)$.[47] Linearize about each of these equilibria by, for example, substituting $(a, X, Y) = (0 + \epsilon(t), s + \xi(t), s^2 + \eta(t))$ and neglecting products of small quantities ϵ, ξ, and η to find that the Jacobian is the linear operator (matrix)

$$\mathcal{L} = \begin{bmatrix} 0 & 0 & 0 \\ s - s^3 & 0 & 0 \\ 0 & 2s & -1 \end{bmatrix}.$$

Being a (lower) triangular matrix, the eigenvalues are along the diagonal, namely 0 (twice) and -1: that is, at each equilibrium there is a slow manifold. The finite neighborhoods of validity along the set of equilibria overlap and merge to form a *globally valid slow manifold* in X; the finite neighborhoods in the ϵ direction are still just *local* in a. Being global in X sounds wonderful, so surely we have gained something. But alas no.

The slow manifold will be $Y = X^2 + \mathcal{O}(a)$, so that the evolution is $\dot{X} = a(X - X^3) + \mathcal{O}(a^2)$, which as before predicts the bifurcation to finite-amplitude equilibria. The domain of validity is global in X, which nicely encompasses the predicted finite-amplitude equilibria, but typically the width of the domain decreases as X increases, as shown schematically in the right panel of Figure 4.13. The shaded domain is local in parameter a and global in X, as required. But this domain is *precisely the same* domain as the local domain shaded in the left panel of the straightforward unscaled analysis. The two approaches, scaled and unscaled, are equivalent.

4.6.3 ▪ A singular perturbation is also equivalent

In many applications one recognizes that some physical processes are much more rapid than others; for example, some chemical reactions may take place in microseconds,

---

[46] Initial conditions a long way outside this region are also attracted to the slow manifold. However, their transient dynamics are not simple decay; they are complicated. Thus their attraction to the slow manifold lies outside this theory.

[47] There are also the equilibria $(a, 0, 0)$, but these have eigenvalues $\{0, a, -1\}$, which only have a nontrivial slow manifold in the case $a = 0$ that is encompassed by the body of this example.

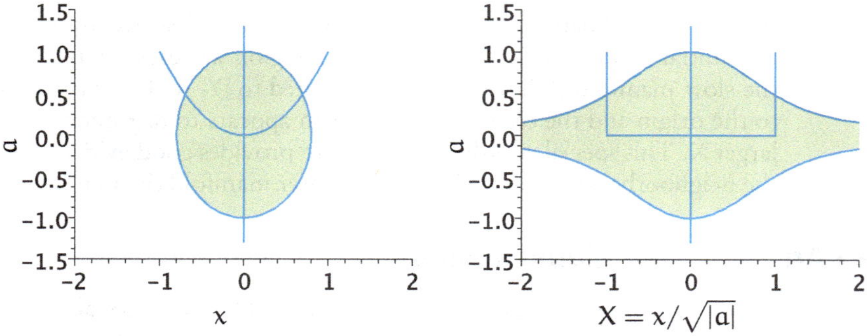

Figure 4.13. *Schematic diagram of the finite domain* $\mathbb{U}$ *of validity of the slow manifold model of the bifurcation in the system* (4.19)*: left, shows the* local *domain of the pitchfork appropriate to straightforward analysis; right, shows the same* global *domain of the pitchfork when scaled as in system* (4.24).

whereas other interesting reactions take place in milliseconds. The singular perturbation approach is to replace the large rates of the fast dynamics by a parameter such as $1/\epsilon$. Then one deduces a model in the limit of small parameter ϵ. Let's do the equivalent here.

Here the fast rates are the order-1 rates of the y equation in the system (4.19). So the singular perturbation form of (4.19) is

$$\dot{x} = ax - xy \quad \text{and} \quad \dot{y} = \frac{1}{\epsilon}(-y + x^2), \tag{4.25}$$

to be analyzed in the limit of small ϵ. The physically relevant value is $\epsilon = 1$. Here we perform an equivalent analysis via regularizing the system by changing time to τ, where $t = \epsilon\tau$. Then the above singular perturbation system (4.25) is identical to the regularized system

$$\frac{d\epsilon}{d\tau} = 0, \quad \frac{dx}{d\tau} = \epsilon(ax - xy), \quad \text{and} \quad \frac{dy}{d\tau} = -y + x^2.$$

Analogous to the previous example, this system has a manifold of equilibria $(\epsilon, x, y) = (0, s, s^2)$. Almost as before, about each of these equilibria the Jacobian is the linear operator (matrix)

$$\mathcal{L} = \begin{bmatrix} 0 & 0 & 0 \\ as - s^3 & 0 & 0 \\ 0 & 2s & -1 \end{bmatrix}.$$

The same conclusions follow: there exists a relevant slow manifold, approximately $y \approx x^2$, globally valid in x and local in ϵ. A difference here is that the existence and relevance are also global in parameter a: is this globality a great boon? Again, alas no.

Figure 4.14 schematically shows the global domain of validity of this singular perturbation model. But the physically relevant parameter is $\epsilon = 1$, shown by the flat plane in Figure 4.14: the intersection of this physical plane with the global domain of singular perturbations results in the same finite domain of physical validity in the ax-plane as determined earlier and plotted in Figure 4.13.

The three different methods of analyzing this one example show one case of how the different methods obtain equivalent models and equivalent domains of validity.

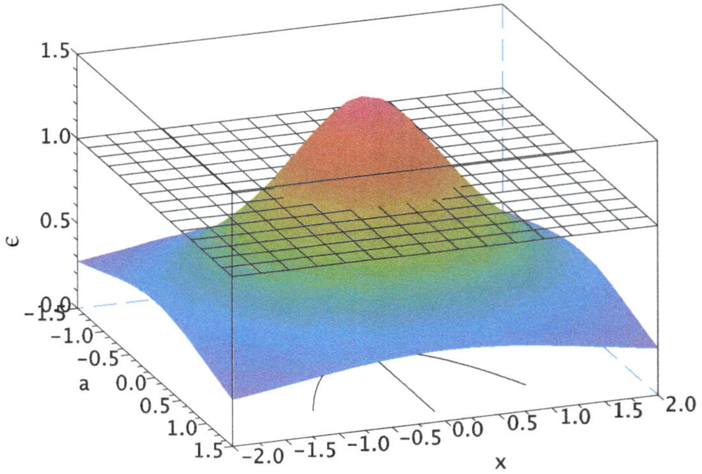

Figure 4.14. *Schematic diagram of the finite domain* $\mathbb{U}$, *below the curved surface, of validity of the singularly perturbed version of the bifurcation in* (4.19). *The physically relevant* $\epsilon = 1$ *(shown) results in a finite domain of physical validity in the* ax*-plane. For reference, the pitchfork bifurcation is shown in the* ax*-plane.*

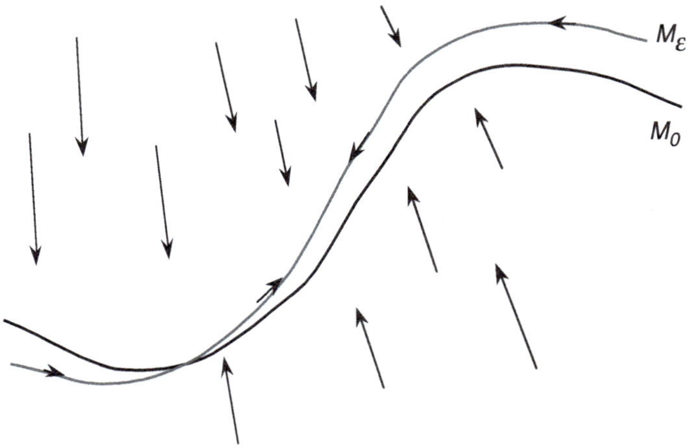

Figure 4.15. *Schematic diagram of a manifold of equilibria,* $\mathcal{M}_0$ *when* $\epsilon = 0$, *approximating a global slow manifold,* $\mathcal{M}_\epsilon$ *for* $\epsilon \neq 0$.

4.7 ▪ Manifold equilibria: Chemical kinetics

As indicated earlier, sometimes instead of having an isolated equilibrium of the dynamical system, there exists a whole continuous class—a manifold $\mathcal{M}_0$—of equilibria. We could construct a slow manifold model local to each of these equilibria as each such equilibrium is exponentially attractive except for the tangent, which is neutral. However, it is generally better to construct a slow manifold that is *global* along $\mathcal{M}_0$ but *local* to $\mathcal{M}_0$ as a whole; this is shown schematically in Figure 4.15.

Consider the following basic system typical of the kinetics of enzyme reactions:

$$\dot{x} = -x + (x + c)z \quad \text{and} \quad \epsilon\dot{z} = x - (x + 1)z,$$

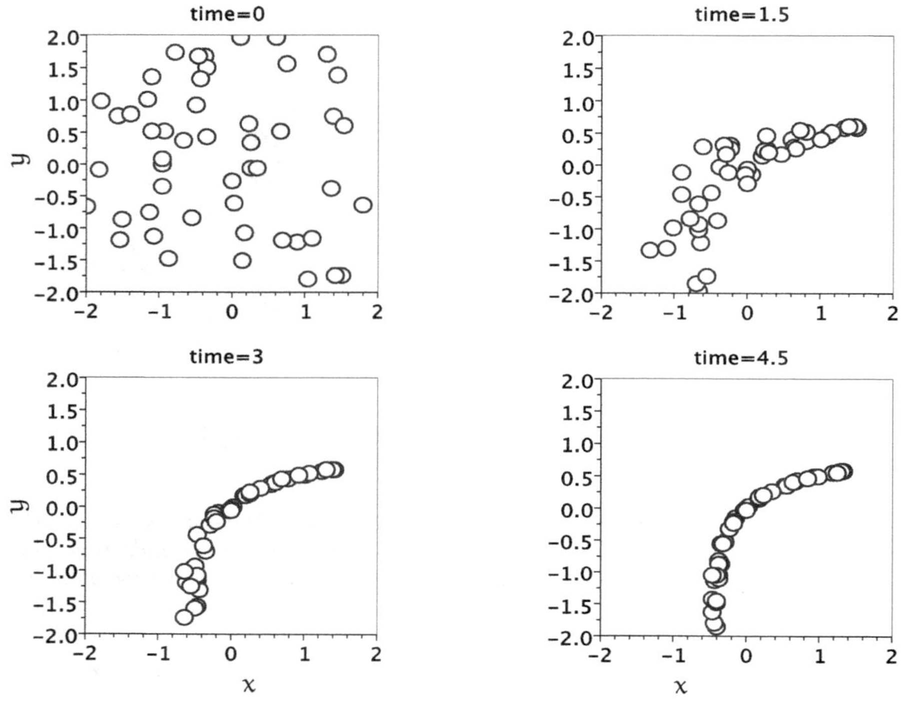

Figure 4.16. *Locations at four times in the xz-plane of solutions to the ODE* (4.26) *starting from* 50 *random initial conditions: the slow center manifold $z = x/(1+x)$ emerges quickly.*

where $\epsilon > 0$ is small and c is a constant $0 < c < 1$. Figure 4.16 shows simulations of the system collapsing onto a curving slow manifold. Heuristically, since ϵ is small, z evolves very quickly to a state where $0 \approx x - (x+1)z$, that is, $z \approx x/(1+x)$. Substituting this into the first of the pair of governing equations gives that x evolves over ordinary time scales according to $\dot{x} \approx -(1-c)x/(1+x)$. We now show how center manifold theory puts this heuristic argument on a sound basis.

The first task is to stretch time so that the time scales fit within the framework of the theory. What is important is the *relative separation* between the two time scales of the ignorable transients and the interesting dynamics: we maintain the relative separation. Stretch time by letting $t = \epsilon\tau$ so that ordinary times for t correspond to long times in τ and the very rapid transients occurring in a t-time of order ϵ occur on a τ-time of order 1; note that $d/dt = \frac{1}{\epsilon} d/d\tau$. The enzyme kinetics ODEs then become (as simulated in Figure 4.16)

$$\frac{dx}{d\tau} = \epsilon\,[-x + (x+c)z] \quad \text{and} \quad \frac{dz}{d\tau} = x - (x+1)z. \tag{4.26}$$

When $\epsilon = 0$, the curve $z = x/(1+x)$ is a manifold of equilibria, $\mathcal{M}_0$, with exponential attraction in the z direction, at an x-dependent rate $\lambda = -(1+x)$. For ϵ small but nonzero, formally done by adjoining the trivial evolution equation $d\epsilon/d\tau = 0$ and considering the dynamics local to $\mathcal{M}_0$ in ϵxz-space, the manifold of quasi-equilibrium is perturbed to some manifold, $\mathcal{M}$, and on $\mathcal{M}$ the evolution is slow. By basing the analysis on the whole of $\mathcal{M}_0$ we obtain an approximate model that is globally valid along the whole of $\mathcal{M}$, just provided ϵ is small enough.

Second, we seek the departure from $\mathcal{M}_0$. Introducing y to measure the departure, we substitute $z = x/(1+x) + y$. The governing ODEs (4.26) may then be written as

$$\frac{dx}{d\tau} = \epsilon\left[-\frac{(1-c)x}{1+x} + (x+c)y\right],$$
$$\frac{dy}{d\tau} = -(1+x)y - \frac{\epsilon}{(1+x)^2}\left[-\frac{(1-c)x}{1+x} + (x+c)y\right].$$

Rearranging the last to

$$(1+x)y = -\frac{\epsilon}{(1+x)^2}\left[-\frac{(1-c)x}{1+x} + (x+c)y\right] - \frac{dy}{d\tau},$$

one iteration then straightforwardly derives that there exists a slow manifold

$$y = h(x,\epsilon) = \epsilon\frac{(1-c)x}{(1+x)^4} + \mathcal{O}(\epsilon^2),$$

global in x, on which the system evolves according to

$$\frac{dx}{d\tau} = -\epsilon\frac{(1-c)x}{1+x}\left[1 - \epsilon\frac{x+c}{(1+x)^3}\right] + \mathcal{O}(\epsilon^3).$$

Furthermore, this slow manifold would be reached very quickly; the transient would decay something like $\exp[-(1+x)\tau] = \exp[-(1+x)t/\epsilon]$. Because the error factors are (largely) independent of x, the approximation theorem asserts that the accuracy of the slow manifold model is independent of x and hence is globally valid.

4.8 ▪ Case study: Melting is a simple Stefan problem

The choice of approximating function is a constant source of debate.

Mitchell and Myers (2010)

Consider a one-dimensional bar on the point of melting (such as ice at 0°C). Heat one end of the bar: how does the bar melt into a liquid? Figure 4.17 illustrates the situation. Such problems with a moving interface, here between melt and solid, are called Stefan problems.

First, mathematically formulate perhaps the simplest such problem (Mitchell and Myers, 2010, e.g.). For simplicity write all quantities as nondimensional. Let the bar and melt stretch from $x = 0$ to $x \to \infty$ with temperature field $u(x,t)$. Initially set the bar to be precisely at its melting temperature $u(x,0) = 0$. Heat the end at $x = 0$ to maintain the fixed-temperature boundary condition $u(0,t) = 1$ (nondimensionally). As shown in Figure 4.17, at any time the bar will have melted to a position $x = s(t)$ so that temperature $u(x,t) > 0$ for the melt $0 \leqslant x < s$, whereas temperature $u = 0$ for the solid $s < x < \infty$. Assume continuity of temperature at the melt-solid interface so that $u(s,t) = 0$. Heat diffuses through the melt according to the diffusion PDE

$$\frac{\partial u}{\partial t} = \frac{\partial^2 u}{\partial x^2}. \tag{4.27}$$

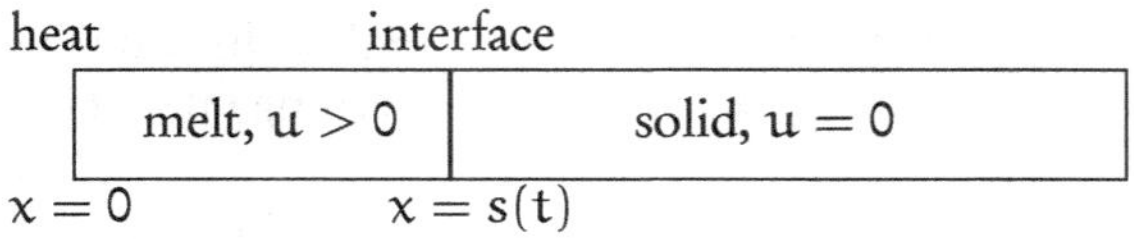

Figure 4.17. *Schematic diagram of a melting bar after heat is applied to the left end* $x = 0$*: the interface between melt and solid moves to be at* $x = s(t)$ *at time* t.

The interesting interface, $x = s(t)$ between the melt and the solid, moves in proportion to the flux of heat diffusing from the melt into the interface. As the flux is proportional to $-\partial u/\partial x$, the movement of the interface is

$$\frac{ds}{dt} = \dot{s} = -k\frac{\partial u}{\partial x} \quad \text{at } x = s^-\,, \tag{4.28}$$

where k is some physical constant depending upon density, latent heat of melting, and thermal conductivities. This completes the mathematical model, which is expressed in terms of the dynamics of the temperature field $u(x,t)$ and the melt-solid interface location $x = s(t)$.

Second, center manifold theory supports the model of the movement of the interface so that $s \propto \sqrt{t}$. Where is this slow manifold?[48]

1. As always, first find an equilibrium. For the temperature field to be in equilibrium, $\partial u/\partial t = 0$. From the differential equation this implies $\partial^2 u/\partial x^2 = 0$, and so the field has to be linear in space: $u = 1 - x/s$ satisfies the boundary conditions at the heated end and the interface. But then the interface must be evolving as $ds/dt = -k(\partial u/\partial x) = k/s$, and hence there is no equilibrium.

   That is, the interface must be evolving unless we artificially set the physical parameter $k = 0$. As in the bifurcation problem of section 4.5, adjoin the trivial $dk/dt = 0$ to the dynamics so that we consider parameter k as a dynamical variable. In essence, we embed the physical problem with some specific physical value for parameter k, in the family of problems that occur for small parameter k.

   Consequently, a set of equilibria are that the interface $s =$ constant, temperature field $u = 1 - x/s$, and parameter $k = 0$. We are in the fortunate position of deriving a model global in interface location s because of this one-parameter set of equilibria.

2. Second, linearize about each of these equilibria to find that the spectrum supports a slow manifold. Consider parameter k as small and that there are small variations to the linear temperature distribution.[49] Seek heat fields $u = 1 - x/s + \hat{u}(x,t)$ where $\hat{u}$ is small: the linear diffusion PDE (4.27) and boundary conditions lead to

$$\frac{\partial \hat{u}}{\partial t} = \frac{\partial^2 \hat{u}}{\partial x^2} \quad \text{such that} \quad \hat{u} = 0 \text{ at } x = 0, s\,.$$

[48] Among other matters, Mitchell and Myers (2010) review how various modelers assume different spatial structures for the heat in the melt and then deduce quantitatively different models for the movement of the interface. Such discrepancies between modelers are common. In conventional modeling there is often no good rationale to distinguish between the various assumptions on the spatial structures. In contrast, center manifold theory uses the governing physical equations themselves to determine spatial structures. This systematic framework provides a definite sequence of models of increasing refinement.

[49] The analysis here is a little cavalier. An exercise for you, the reader, is to fill in details in the analysis to find the same conclusion, but for completely correct reasons.

Since this diffusion PDE is constant coefficient, solutions must be exponential and/or trigonometric: from the boundary conditions solutions must have spatial structure $\sin(\ell\pi x/s)$ for integer $\ell = 1, 2, 3, \ldots$. Seeking temperature perturbations in the form $\hat{u}_\ell = e^{\lambda_\ell t}\sin(\ell\pi x/s)$, the linearized diffusion PDE determines the growth rates (eigenvalues) $\lambda_\ell = -\pi^2\ell^2/s^2$. These are all negative. There is no slow subspace, and hence no slow manifold, associated *directly* with the temperature field.

Additionally, there is a zero eigenvalue associated with the location of the interface. The interface evolution equation (4.28) reduces to $ds/dt = 0$ when $k = 0$, as occurs at the identified equilibria. Thus the spectrum of the linearized system is the set $\{0, -\pi^2/s^2, -4\pi^2/s^2, \ldots\}$.

3. From the pattern of eigenvalues in this spectrum, there exists an emergent slow manifold for all s. Because a slow manifold exists for all s, the slow manifold is global in s. Further, because the slow manifolds are based around parameter $k = 0$, the slow manifolds exist for at least some finite domain of small parameter k.

   From the leading negative eigenvalue, we predict that transients will decay with a rate roughly $-\pi^2/s^2$. Interestingly, the slow manifold is extremely strongly attractive at the onset of heating: when the interface s is close to zero, the decay rates are extremely large (negative). The slow manifold model is only weakly attractive at large times, when s is large, but by then the transients will have decayed anyway.[50]

4. Now proceed to construct approximations to the slow manifold. Observe that the Stefan problem (4.27)–(4.28) is in the separated form (4.15): the "fast" modes of $u(x,t)$ are governed by (4.27), whereas the "slow" mode $s(t)$ is governed by (4.28). Hence we apply the iteration (4.17).

   The first approximation to the slow manifold is that the temperature field is linear: $u \approx u^{(1)} = 1 - x/s$. Evaluating the interface condition (4.28) gives the slow manifold evolution $\dot{s} \approx -k\partial u^{(1)}/\partial x = k/s$. What does this model predict? One would solve it by separation of variables to predict the interface $s \approx \sqrt{2kt}$.

   What is the second approximation? The refinement is driven by the heat equation in the melt: the linear distribution $u = 1 - x/s$ is an equilibrium, but because the interface s is moving, the temperature field must be out of equilibrium. This out-of-equilibrium nature shows up in the time derivative $\partial u/\partial t$, which by the chain rule becomes $\partial u/\partial t \approx \partial u^{(1)}/\partial t = x\dot{s}/s^2 \approx kx/s^3$. Such out-of-equilibrium "forcing" of the heat equation engenders the next correction.

   The essence of the iteration (4.17) is to put the dominant linear terms on the left-hand side, and everything else on the right-hand side (as everything else is small). For the heat equation (4.27), the dominant linear term is $\partial^2 u/\partial x^2$ and the "everything else" is just the time derivative. Thus the iteration (4.17) is to solve $\partial^2 u^{(2)}/\partial x^2 = \partial u^{(1)}/\partial t = kx/s^3$. Integrate twice with boundary conditions $u(0,t) = 1$ and $u(s,t) = 0$ to give the second approximation $u^{(2)} = 1 - x/s + (k/6)(x^3/s^3 - x/s)$. This predicted heat field is cubic in x.

[50] Some of you may know that there is a nice self-similarity solution for this particular Stefan problem. Deriving the self-similarity solution, by itself, only assures you of self-consistency, not emergence. This is one aspect that our slow manifold tells us that the exact self-similarity solution does not: center manifold theory guarantees that the self-similar solution will emerge in the dynamics (Suslov and Roberts, 1998, e.g.).

Algorithm 4.3 *Iteration finds the slow manifold of melting in the Stefan problem of section* 4.8.

```
on div; off allfac; on revpri;
let k^4=>0;
depend s,t;
let df(s,t)=>-k*sub(x=s,df(u,x));
operator linv; linear linv;
let { linv(x^~p,x)=>(x^(p+2)-x*s^(p+1))/(p+1)/(p+2)
    , linv(x,x)=>(x^3-s^2*x)/6
    , linv(1,x)=>(x^2-s*x)/2 };
u:=1-x/s;
for iter:=1:99 do begin
    res:=df(u,t)-df(u,x,2);
    u:=u+linv(res,x);
    if res=0 then iter:=1000000;
end;
sdsdt:=s*df(s,t);
end;
```

Evaluating the interface condition (4.28) then gives the refined slow manifold evolution $\dot{s} \approx -k\partial u^{(2)}/\partial x = (k - \frac{1}{3}k^2)/s$.

Further approximations may be constructed analogously.

5. Computer algebra readily constructs further approximations to the slow manifold. Algorithm 4.3 uses the residuals of the governing equations to derive that the heat field and interface evolution are

$$u = \left(1 - \frac{x}{s}\right) - \frac{k}{6}\left(\frac{x}{s} - \frac{x^3}{s^3}\right) + \frac{k^2}{360}\left(19\frac{x}{s} - 10\frac{x^3}{s^3} - 9\frac{x^5}{s^5}\right) + \mathcal{O}(k^3),$$

$$\dot{s} = \frac{1}{s}\left[k - \tfrac{1}{3}k^2 + \tfrac{7}{45}k^3\right] + \mathcal{O}(k^4).$$

Evidently, the slow manifold model is $\dot{s} = f(k)/s$ for some function $f(k) \approx k - \frac{1}{3}k^2 + \frac{7}{45}k^3$.[51] Elementary calculus then gives solutions $s = \sqrt{2f(k)t}$ to predict the classic square-root time growth of a diffusion-limited process.

How does Algorithm 4.3 work?

- The slow manifold is local in parameter k so we must truncate the analysis in some power of k: `let k^4=>0` implies the residuals are $\mathcal{O}(k^4)$.
- The slow manifold is also parametrized by the evolving variable s (interface location), so we tell Reduce that it depends upon time t: `depend s,t;`. Further, the interface condition $ds/dt = -k(\partial u/\partial x)$ tells us how variable s evolves, so code that as well:

  ```
  let df(s,t)=>-k*sub(x=s,df(u,x));
  ```

- Given any residual of the diffusion equation (4.27), arising through $\partial u/\partial t$ from the out-of-equilibrium dynamics, we need to solve $\partial^2\hat{u}/\partial x^2 = \text{Res}_{(4.27)}$

[51] High-order computations strongly suggest that the power series for $f(k)$ converges with an unphysical, convergence-limiting singularity lying at $k^* \approx -1.29$.

such that $\hat{u} = 0$ at $x = 0, s$. The linear operator `linv` does this by coding the particular solutions that arise for any component in the residual of the form x^p.

- The iterative improvement must start with an initial guess, here say $u = 1 - x/s$, so code `u:=1-x/s;`.
- Use a `for` loop to iterate until the computed residual `res` is zero to the specified order of error: exit out of the iterative `for` loop by setting the iteration count to a ridiculously high value.
- Within the iteration, compute the residual of the heat equation and then use it to update the temperature field:

```
res:=df(u,t)-df(u,x,x);
u:=u+linv(res,x);
```

This simple algorithm works because the equation (4.28) for the slow mode $s(t)$ is linearly separated from the heat equation (4.27) for the fast modes in the field $u(x, t)$.

Try it. Have some fun with an orgy of algebra: truncate to high order in parameter k; what if there is a small radiative loss from the melt, say $-\epsilon u$? What if there is a weak reaction within the melt, say ϵu^2?

4.9 ▪ Further theory underpins our range of applications

Some will want to explore more details of the theoretical support. Chapter 11 introduces general, finite-dimensional, nonlinear coordinate transforms that immediately prove the existence and emergence of center manifold dynamics as a long-term model of the system dynamics. Chapter 19 extends this approach to nonautonomous systems. This section instead points to some of the published research that underpins the theory used in this book.

Perhaps the first systematic treatment of invariant manifolds was by Kelley (1967b), who proved existence and smoothness in a neighborhood of an equilibrium, periodic orbit, or periodic surface. Knobloch and Aulbach (1982) traced key ideas back to Bohl (1904), who established that center manifolds contain all solutions that are sufficiently small.

Possibly the best place to start exploring details of the underlying theory is the book by Carr (1981). This classic book gathered together proofs of existence, emergence, and approximation theorems for finite-dimensional autonomous systems with center manifolds. The last chapter proves analogous results for a wide class of infinite-dimensional systems (such as partial differential equations in Banach spaces) that have a finite-dimensional center manifold. The article by Knobloch and Aulbach (1982) is also a good starting point in that it overviews the state of theory at the time, including nonautonomous systems, but does not give proofs.

The next major step in theoretical development was a series of three papers by Aulbach and Wanner (1996,9, 2000). Not only did their research address systems in infinite-dimensional Banach spaces, but the systems could have center manifolds of infinite dimensions (as needed in Part III on macroscale spatial modeling). Also, the theory addressed nonautonomous systems in Banach spaces (as needed to underpin Chapter 19). Further, they extended the theory to cover cases where the center eigenvalues have real part near zero, not just precisely zero: such an extension is very useful in applications (as developed by Chapter 11). Aulbach and Wanner (2000) proved that the center, stable, and unstable modes can be nonlinearly separated and hence established the existence and emergence of center manifolds.[52]

Aulbach and Wanner (1996, 1999, 2000) place major restrictions on the nonlinearity in the system—namely it is to be Lipschitz and bounded. The advantage of these restrictions is that the theory proves *global* existence and emergence. One disadvantage is that Lipschitz and bounded nonlinearities are rare in applications. To apply their theory one typically needs to introduce cut-off functions that mollify the nonlinearity, and then the resultant invariant manifolds are only assured to exist and emerge locally. Haragus and Iooss (2011), for example, show how to use such cut-off functions to be able to prove local existence and emergence for systems with more general nonlinearity.

One aspect missing from the last two paragraphs is any mention of an approximation theorem. Fortunately, Potzsche and Rasmussen (2006) provide us with such a theorem for infinite-dimensional nonautonomous systems: their Proposition 3.6 establishes that if we satisfy the governing equations to some asymptotic order of error, then the approximation of invariant manifolds is asymptotically correct to the same order of error.

Although there are many other published center manifold theorems, the ones mentioned here appear to me to be the crucial major extant theorems to support our practical modeling.

However, all such theorems are "forward" theorems; that is, they address the issues for a precisely given dynamical system. There appears to be scope for research into "backward" theorems analogous to those that have been found to be so useful in numerical analysis (Grcar, 2011, e.g.). Indeed, Chapter 11 suggests that such backward theorems could be based upon normal-form coordinate transforms and could establish significant estimates for the size of the domain of validity of the theoretical support.

4.9.1 ▪ Alternative approximation methods

Throughout, this book constructs models by invoking algebraic asymptotic approximations of center manifolds. This subsection mentions a few alternatives.

There are a variety of methods for computing numerically stable or unstable manifolds. Krauskopf et al. (2005) usefully reviewed several different approaches. However, one limitation is that these approaches only seem suitable for (un)stable manifolds of low dimension: the most common examples are just 2D. More important, the numerical methods appear to rely on the uniqueness of (un)stable manifolds.

But center manifolds are not unique, although the uniqueness should be irrelevant for the purposes of the chosen modeling (Example 4.9). Any methodology for construction of center manifolds has to cope with this nonuniqueness. Algebraic asymptotic expansions cope with nonuniqueness by forming a "smoothest" approximation: all other possible center manifolds differ by an exponentially small amount. Numerical methods of construction have to cope with the finite size of such exponentially small amounts when at a finite distance from the base equilibrium.

Nonetheless, Jolly and Rosa (2005) approximated center manifolds by adapting an iterative algorithm, and applied the method to two test problems. Again the test problems had center manifolds of low dimension. In contrast, this book establishes a framework

[52]Many people maintain, for good reasons, a distinction between invariant manifolds of autonomous systems and the corresponding notion of integral manifolds of nonautonomous systems. In this book I do not make the distinction and so loosely term them all invariant manifolds.

for modeling systems when the emergent dynamics may have tens to thousands of dimensions. Further, it is unclear how Jolly and Rosa (2005) addressed the nonuniqueness.

The pitchfork bifurcation example of section 4.5.1 illuminates the uniqueness problem. Recall that $y \approx x^2 - 2ax^2 + 2x^4$ (Figure 4.11) approximates a center manifold of the system (4.20): $\dot{a} = 0$, $\dot{x} = ax - xy$, and $\dot{y} = -y + x^2$. The center manifold appears smooth, but hidden are singularities in the center manifold shape at the finite-amplitude equilibria $x = \pm\sqrt{a}$. Because the Jacobian of the system at the equilibria is $\begin{bmatrix} 0 & \mp\sqrt{a} \\ \pm 2\sqrt{a} & -1 \end{bmatrix}$, with eigenvalues $-1, -2a$, the trajectories about the equilibria have shape like $|x \mp \sqrt{a}|^{1/(2a)}$. Consequently, the center manifold is unique in joining $x = y = 0$ to the attractors $(x, y) = (\pm\sqrt{a}, a)$, but must generally they have a singularity of order of the exponent $1/(2a)$. But there is no unique way to continue the center manifold outside of the attractors $(x, y) = (\pm\sqrt{a}, a)$. The crucial difficulty is that any decision a numerical method takes locally to resolve the local singularity and local nonuniqueness needs to be coordinated globally to ensure that the center manifold construction remains coherent.

The algebraic asymptotic construction employed herein deals with the issue by finding that the problems are exponentially small as $(a, x) \to 0$ and hence negligible. Alternatively, Chapter 11 argues that there is a dynamical system close to the one specified that has a beautifully smooth center manifold.

On-the-fly construction Another class of methods to approximate slow manifolds contains those obtained directly and locally from numerical simulation. These often involve repeated integration forward and backward in time until the simulation relaxes to a state of quasi-equilibrium. Maas and Pope (1992) introduced the method of *intrinsic low-dimensional manifolds* for chemical combustion problems (see also Bykov et al., 2006). *Computational singular perturbations* is another method that analyzes local dynamics to approximately find a slow manifold (Valorani et al., 2005, e.g.) and has the possible advantage that it additionally identifies the local fast directions. *Projective integration* is a component of "equation-free" methods that evolves a simulation to the local slow manifold and then estimates the slow manifold later in time via a macroscale projective step (Kevrekidis and Samaey, 2009, e.g.). But finding such slow manifolds by simulation for nonautonomous or stochastic systems is much more involved (Kan et al., 2013, e.g.). These methods primarily provide a localized estimate of the slow manifold; for a global picture it appears that we need algebraic asymptotic approximations.

4.10 ▪ Summary

Current center manifold theorems support low-dimensional models of dynamical systems. The existence of a model is based upon the structure of the eigenvalues (the spectrum) of an equilibrium, or a subspace of equilibria. The emergence theorem ensures that the model is relevant apart from initial transients. Approximations can be found so straightforwardly that we program computers to handle the messy details. Table 4.2 gives the typical framework for the approach.

One apparent restriction is that the theorems are local to the anchoring equilibrium. However, in applications the "sufficiently small" locale may be generous in size—sometimes the entire state space. Importantly, the theory guarantees the existence and emergence of the model for all time in a *finite-sized* domain; in contrast, other methodologies typically assert properties for a finite time, often only as a parameter tends to zero, and often only with some uncontrollable error.

Table 4.2. *These general steps invoke center manifold theory to model emergent dynamics in complicated physical systems.*

1. Embed the physical problem in a useful family of cognate problems.
2. Equilibria: Find the equilibria upon which one might anchor an analysis; a subspace of equilibria empowers global approximations.
3. Linearization: The spectrum of the dynamics linearized about the equilibria identifies center/slow modes among stable/fast modes.
4. Theorems: Invoke existence and emergence theorems, as far as possible, to assure us of the relevance of the modeling.
5. Construction: By hand or with computer algebra, construct approximations to the center/slow manifold model based upon driving the residuals of the equations to zero to some order of error in small parameters. Invoke theory to assert that the model is then accurate to the same order.
6. Convergence: Where possible, find evidence that the approximations are sufficiently accurate at parameter values of interest.
7. Interpretation: Identify how components in the model relate to the original physical processes.
8. Regularization: Consider modifying the model—with terms of order of the error—to improve, in some sense, the finite-amplitude behavior; usually one seeks to remove an unphysical instability.

The finite domain of validity of the center manifold theorems arises from the intrinsic dynamics of the system. Thus expect that the differing incantations of various methodologies are not going to magically change the validity of the modeling when evaluated at corresponding physical parameters—provided the methodologies are sound.

Remember that in the context of modeling dynamics, time derivatives du/dt are *not* linear (at best the time derivative is quasi-linear). This nonlinearity follows from the chain rule: in modeling we seek the physical field $u = v(s)$, where the modeling variables s evolve according to $ds/dt = g(s)$. By the chain rule $du/dt = (\partial v/\partial s)ds/dt = (\partial v/\partial s)g$. Hence du/dt hides the nonlinear product of the modeling unknowns v and g. Many people are misled by the apparent linearity of du/dt to propose and use modeling methodologies that do not respect this hidden nonlinearity. Such modeling methods inevitably incur errors. Sometimes such methods lead to "paradoxes" that cannot be resolved without realizing the hidden nonlinearity in du/dt. Be aware of this nonlinear nature of time derivatives when *modeling* dynamics.

4.11 ▪ Exercises

Exercise 4.1 (graphical slow manifolds). Figures 4.18–4.21 draw the vector fields $f(u)$ of some dynamical systems $\dot{u} = f(u)$: for each location u in the plane, the figures plot a small vector proportional to $f(u)$. On these figures,

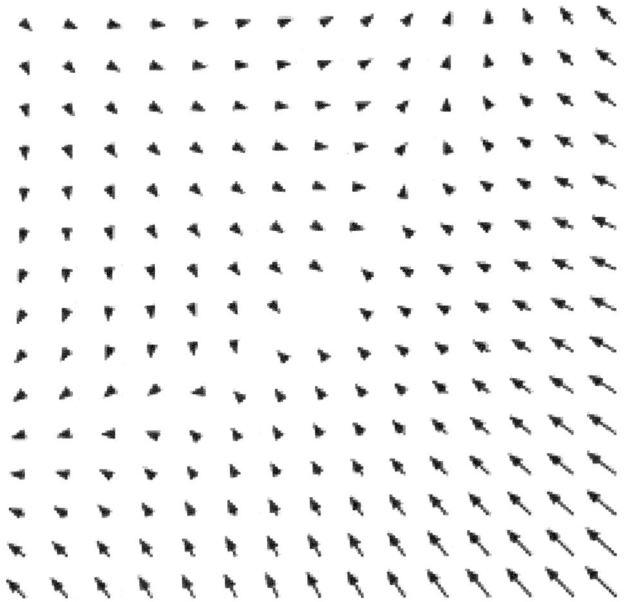

Figure 4.18. *Vector field* $\mathsf{f}(\mathsf{u})$ *of some dynamical system* $\dot{\mathsf{u}} = \mathsf{f}(\mathsf{u})$ *for Exercise* 4.1.

- draw the location of any equilibria;
- sketch any slow manifold you can see (to the extent that the slow manifold is reasonably clear);
- if applicable, outline a finite domain inside which you reasonably expect the slow manifold existence and emergence theorems to apply.

Exercise 4.2 (a slow manifold). Recalling Example 4.16, consider

$$\begin{aligned} \dot{x} &= -xy\,, \\ \dot{y} &= -2y + z + x^2\,, \\ \dot{z} &= y - z + x^2\,. \end{aligned}$$

Start by following the usual procedure: find a convenient equilibrium, linearize the dynamics, find eigenvalues and eigenvectors, identify a slow subspace, and modify the slow subspace into a slow manifold.

- Deduce that there exists a one-dimensional slow manifold: $y \approx 2x^2$ and $z \approx 3x^2$.
- What are the next-order corrections to the shape of the slow manifold?
- What is the corresponding evolution on the slow manifold? Does theory guarantee that this 1D model is valid exponentially quickly and for all time?

Exercise 4.3 (uncover the instability). Consider the dynamical system

$$\dot{x} = xy - x^3 \quad \text{and} \quad \dot{y} = -y + 2x^2.$$

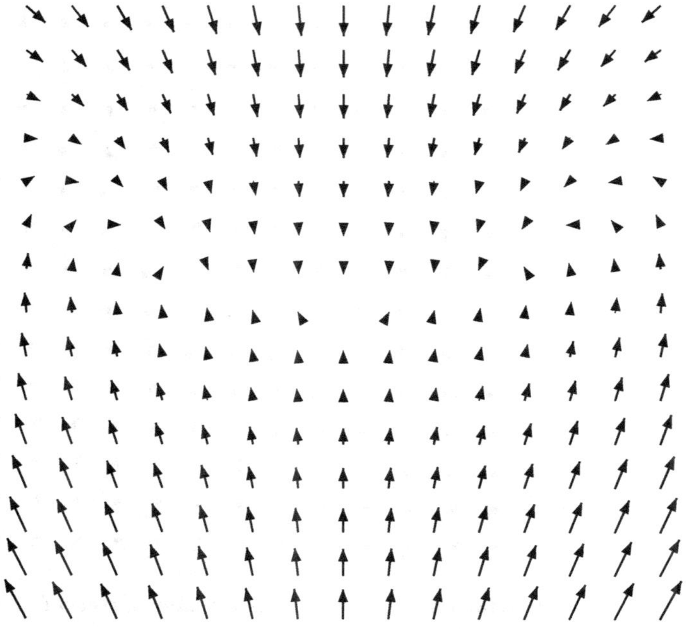

Figure 4.19. *Vector field* $\mathbf{f}(\mathbf{u})$ *of some dynamical system* $\dot{\mathbf{u}} = \mathbf{f}(\mathbf{u})$ *for Exercise* 4.1.

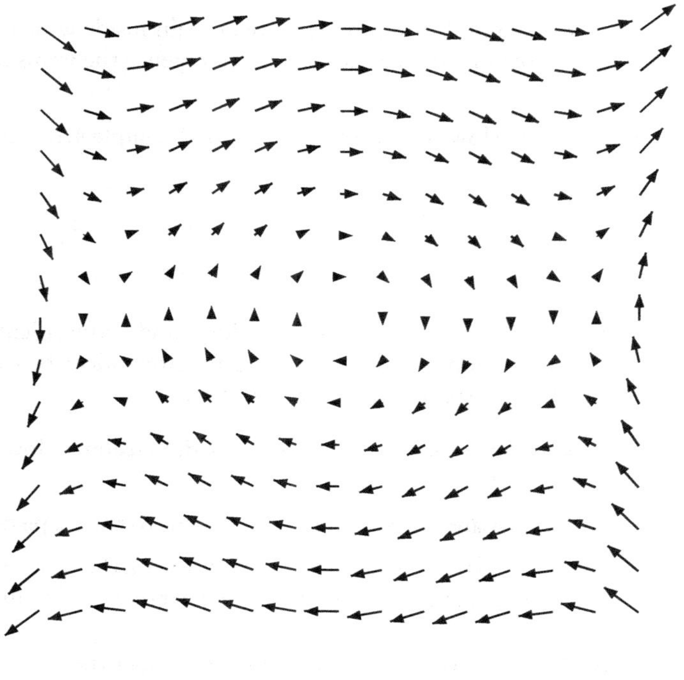

Figure 4.20. *Vector field* $\mathbf{f}(\mathbf{u})$ *of some dynamical system* $\dot{\mathbf{u}} = \mathbf{f}(\mathbf{u})$ *for Exercise* 4.1.

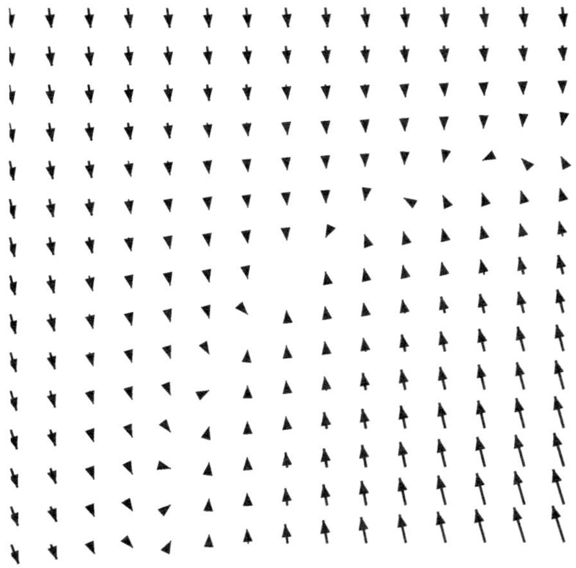

Figure 4.21. *Vector field* $f(u)$ *of some dynamical system* $\dot{u} = f(u)$ *for Exercise* 4.1.

Start by following the usual procedure: find a convenient equilibrium, linearize the dynamics, find eigenvalues and eigenvectors, and identify a slow subspace.

1. Argue for the existence of a slow manifold tangent to $y = 0$.
2. Determine the approximate shape of the slow manifold, and justify the accuracy of your approximation.
3. Hence determine a one-dimensional model in x for the dynamics near the origin. Does emergence theory justify the use of your model?
4. To model the dynamics badly, one might argue that simply substituting $y = 0$ into the x equation would suffice as a model for x-evolution. If you were to do this, what qualitatively wrong prediction would you make about the dynamics near the origin?

Exercise 4.4 (uncover the instability II). Repeat Exercise 4.3 for the dynamical system (due to Peter Brenner)

$$\dot{x} = x^2 y - x^5 \quad \text{and} \quad \dot{y} = -y + x^2.$$

Exercise 4.5 (equilibrium elsewhere). Consider the system (Chicone, 2006, p. 347)

$$\dot{p} = -\tfrac{1}{5}p(3p^5 - 5q^4p^2 + 13q^2p - 3),$$
$$\dot{q} = -\tfrac{1}{5}q(p^5 + q^2p - 1);$$

it has an equilibrium at $(p, q) = (1, 0)$. Deduce the existence, emergence, and approximate shape of the slow manifold of this equilibrium. Is the equilibrium stable or unstable?

Answer: $p \approx 1 - \frac{13}{15}q^2$ such that $\dot{q} \approx \frac{2}{3}q^3$.

Exercise 4.6 (a slow subspace). In problems that are linear in the evolving variables, the slow manifold is only curved in its parameter dependence. The slow manifold is a linear subspace in the dynamic variables. Consider the toy (linear) system

$$\dot{x} = -ax \quad \text{and} \quad \dot{y} = -2y + ax$$

for some fixed parameter a.

1. In classic linear analysis, seek solutions proportional to $e^{\lambda t}$ and find that eigenvalues $\lambda = -1 \pm \sqrt{1-a^2}$.
2. Alternatively, use our theorems to deduce that there exists a slow manifold based around the parameter value $a = 0$.
3. Construct approximations to the evolution on the slow manifold. Confirm that the slow manifold is linear in the xy-plane but nonlinear in a. Compare the evolution for the slow x-dynamics with the eigenvalues of part 1.
4. Discuss why the slow manifold model must break down for parameter a where $|a|$ is comparable to one.

Exercise 4.7 (explore theory). The purpose of this exercise is to explore how the center manifold theorems give different support to low-dimensional models depending upon how you establish the linear basis of the modeling. Consider the two equivalent dynamical systems

$$\dot{x} = ax^3 + x^2y\,,$$
$$\dot{y} = -y + y^2 + xy - x^3$$

and

$$\dot{a} = 0\,,$$
$$\dot{x} = ax^3 + x^2y\,,$$
$$\dot{y} = -y + y^2 + xy - x^3.$$

They both have an equilibrium at the origin: $(0,0)$ for the first, and $(0,0,0)$ for the second. Write and run a Reduce program to construct approximations to the slow manifolds, based at the origin, of these two systems. *Compare and contrast* the assurances that the first three center manifold Theorems 4.3, 4.5, and 4.8 give to the two slow manifold models.

Exercise 4.8 (rate of attraction). The Emergence Theorem 4.5 guarantees exponentially quick convergence to a center manifold in some finite domain. Explore some examples. The following systems all have a slow manifold through an equilibrium at the origin. Find or estimate the domain of initial conditions in which solutions are attracted to the slow manifold at a rate $\mathcal{O}(e^{-t/2})$ as $t \to \infty$ (the second and third parts are more subtle than they might appear).

1. $\dot{x} = 0$ and $\dot{y} = -(1-x^2)y$ with slow manifold $y = 0$.
2. $\dot{x} = -x^3$ and $\dot{y} = -(1-x^2)y$ with slow manifold $y = 0$.
3. $\dot{\epsilon} = 0$, $\dot{x} = \epsilon x$, $\dot{y} = -(1-x^2)y$ with slow manifold $y = 0$.

Exercise 4.9 (domain of attraction). Consider the simple system $\dot{x} = y$ and $\dot{y} = -xy$. (Riaza (2012) attributes this system to Fiedler et al. (2000).)

1. Deduce that there is a whole subspace of equilibria.
2. Compute the Jacobian, its eigenvalues, and maybe its eigenvectors, to deduce that the subspace of equilibria is an attractive slow manifold only for $x > 0$.
3. Recall that trajectories in such 2D systems are sometimes found by solving the differential equation
$$\frac{dy}{dx} = \frac{dy/dt}{dx/dt}.$$
Here find that the trajectories are parabolas in the xy-plane, and hence deduce that all initial conditions in the half-plane $y > 0$ together with all initial conditions $y > -x^2/2$ for $x > 0$ end up attracted to the slow manifold.

Exercise 4.10 (many valid error statements). This exercise demonstrates that the Approximation Theorem 4.8 justifies different orders of errors depending upon how one introduces parameters and variables. Consider the prototype bifurcation system (4.19), but now set the parameter $a = \alpha^2$ and adjoin $\dot{\alpha} = 0$ to the system. That is, consider the equivalent system
$$\dot{\alpha} = 0, \quad \dot{x} = \alpha^2 x - xy, \quad \text{and} \quad \dot{y} = -y + x^2.$$

1. Show that through the origin, and in its neighborhood, there exists a two-dimensional emergent slow manifold, exactly as before.
2. Construct approximations iteratively and observe that the iterates are exactly as before but with $a = \alpha^2$.
3. Recognize that the only difference is in the order of error of each iterate. For example, here the order of error in $h^{(1)}$ is $\mathcal{O}(h^{(2)} - h^{(1)}) = \mathcal{O}(|(\alpha, x)|^4)$, which is equivalent to $\mathcal{O}(a^2 + x^4)$.
4. Conversely, reconsider the prototype bifurcation system (4.19) with variable $x = \xi^2$. Write and consider the system (4.19) as a system in (a, ξ, y). Repeat the above steps.
5. Compare and contrast the statements on the order of error, and their relation to the Flexible Approximation Theorem 4.10.

Exercise 4.11 (center manifolds triumph). Consider the bifurcation that occurs at the origin as parameter ϵ increases through 0 to small positive values in the following dynamical system:
$$\dot{x} = \epsilon x + x^3 - xy,$$
$$\dot{y} = -y + x^2 + y^2.$$

Then argue the following. Start by following the usual procedure: find a convenient equilibrium, linearize the dynamics, find eigenvalues and eigenvectors, and identify a slow subspace.

1. Deduce that there exists a slow manifold on which the evolution is $\dot{x} = \epsilon x - x^5 + \mathcal{O}(\epsilon^2 + x^6)$. Hence argue that theory guarantees the predictions of this model, that is, there exist stable equilibria at $x \approx \pm\epsilon^{1/4}$ for small enough ϵ.

2. Argue that simple projection onto the slow subspace (analogous to cross-sectional averaging) fails because substituting $y = 0$ into the x equation predicts unbounded explosive growth.
3. Argue that the method of multiple scales fails: substitute the appropriately scaled variables, $x = \sqrt{\epsilon}X(T)$, $y = \epsilon Y(T)$, with $T = \epsilon t$ the slow time to obtain ODEs for X and Y; in the Y equation neglect all but the (two) leading-order terms; then deduce the multiple-scale model $dX/dT \approx X$, that is, $\dot{x} \approx \epsilon x$, which incorrectly predicts unbounded exponential growth.[53]

Exercise 4.12 (instability at higher order). Now explore an example that shows that the systematic nature of center manifold theory is essential. Consider the three-dimensional system

$$\begin{aligned} \dot{x} &= y\,, \\ \dot{y} &= -xz\,, \\ \dot{z} &= -z + x^2 + xy\,. \end{aligned}$$

1. Linearize the system and deduce that it has a two-dimensional slow manifold, with z being a quickly decaying variable.
2. The adiabatic approximation is simply to assume one can set $\dot{z} = 0$ and hence deduce $z \approx x^2 + xy$. Always be wary of the adiabatic approximation.[54] Let us see why: define the energy functional $E = x^4 + 2y^2$; then its rate of change in time is
$$\frac{dE}{dt} = 4x^3\dot{x} + 4y\dot{y} = 4x^3y - 4xyz\,,$$
which upon invoking the adiabatic approximation becomes
$$\frac{dE}{dt} \approx 4x^3y - 4xy(x^2 + xy) = -4x^2y^2 < 0$$
all the time except for the isolated times when $x = 0$ or $y = 0$. Hence the adiabatic approximation predicts that the energy E always decreases and thus the origin must be an attracting stable equilibrium. But this is not so.
3. Use iteration to find that the slow manifold is
$$z = x^2 - xy + \mathcal{O}(x^4 + y^2)\,.$$
Consequently deduce that the energy E actually increases, at least for small x and y, and so the origin must be unstable.

---

[53] This failure can be avoided, but one has to be a mathematical contortionist in order to do so. One might view multiple scales as analogous to the epicycle theory of planetary motion: it works, but one needs ever more "cycles" as demands for accuracy are increased; conversely, the center manifold approach is analogous to Kepler's recognition of elliptical orbits.

[54] Unfortunately and confusingly, different people adopt different naming conventions. The adiabatic approximation is also known as the quasi-steady state approximation. For another example, Verhulst (2005, §8.6) uses the term "slow manifold" to mean what we call the "adiabatic manifold" obtained by simply setting to zero the derivatives of the rapidly decaying variables. Consequently, all "slow manifold" models of Verhulst (2005) have a finite error, $\mathcal{O}(\epsilon)$, which is hard to reduce because of the base adiabatic approximation. Be wary of the possibility of confusion between "slow manifolds" and "adiabatic manifolds."

Exercise 4.13 (Davis–Skodje problem). Valorani et al. (2005) used the Davis–Skodje problem, the system

$$\frac{dx}{dt} = -x\,, \quad \frac{dy}{dt} = \frac{1}{\epsilon}\left(-y + \frac{x}{1+x}\right) - \frac{x}{(1+x)^2}\,,$$

to illustrate the method of computational singular perturbations.

1. Rescale time to transform this system from a singular to a regular perturbation for small parameter ϵ.
2. Determine the set of equilibria for parameter $\epsilon = 0$.
3. Linearize about each equilibrium and argue that a slow manifold exists globally in x and locally in ϵ. What is the extent of the global domain in x?
4. Show that $y = x/(1+x)$ is an exact emergent slow manifold of the Davis–Skodje problem for all parameters ϵ significantly less than one.

Exercise 4.14 (heterogeneous catalysis—two global slow manifolds). Bykov and Gol'dshtein (2013) explored the nature of fast-slow modeling with an example of heterogeneous catalysis (attributed to Yablonsky et al. (2003)). In a relevant parameter regime this example possesses two slow manifolds, which are both global and which interestingly intersect. The Langmuir–Hinshelwood mechanism for CO oxidation on platinum is written as

$$\begin{aligned} 2Z + O_2 &\xrightarrow{k_1} 2ZO\,, \\ Z + CO &\xleftrightarrow{k_2} ZCO\,, \\ ZO + ZCO &\xrightarrow{k_3} 2Z + CO_2\,. \end{aligned}$$

Assume that there is plentiful O_2, CO, and CO_2 so that we are only interested in the dynamics on the catalysis: Z denotes the free catalyst; whereas ZO and ZCO denote species adsorbed on the catalyst surface.

1. Let $x(t)$ and $y(t)$ denote the *fractions* of two of the catalyst variables; physical values are then $x + y \leqslant 1$ and $x, y \geqslant 0$. Argue that the kinetic model may be written in the form

$$\frac{dx}{dt} = 2k_1(1-x-y)^2 - k_3xy\,, \tag{4.29a}$$

$$\frac{dy}{dt} = k_2(1-x-2y) - k_3xy\,. \tag{4.29b}$$

   Which species do $x(t)$ and $y(t)$ measure?
2. Figure 4.22 plots some trajectories of the system (4.29a)–(4.29b). To confirm physical relevance, prove that for rates $k_j \geqslant 0$ solutions starting in the physical domain $x+y \leqslant 1$ and $x, y \geqslant 0$ stay in this domain for all time: consider $\dot{y}$ on $y = 0$, $\dot{x}$ on $x = 0$, and $d(x+y)/dt$ on $x+y = 1$.
3. Physically it is known that the third reaction is fast; that is, the rate constant $k_3 \gg k_1, k_2$ (as chosen for Figure 4.22). Thus, for simplicity assume we have chosen the time scale so that $k_3 = 1$, and then the modeling is to assume that rates k_1 and k_2 are small. In the limit of vanishing rates k_1 and k_2 argue that $x = 0$ and $y = 0$ are two slow subspaces of equilibria.

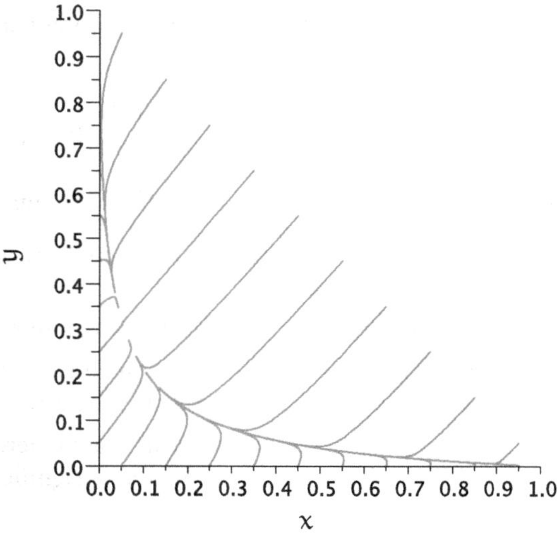

Figure 4.22. *Example trajectories of the heterogeneous catalysis system* (4.29) *for parameters* $k_1 = 0.02$, $k_2 = 0.05$, *and* $k_3 = 1$.

4. Find the nonuniform rate of attraction to the slow subspaces of equilibria. Have an existential argument with a colleague over the intersection of the two slow subspaces. Reason that there should be some sort of slow manifold(s) near these slow subspaces for small rates k_1 and k_2.

5. Approximate the slow manifold(s) separately from the two slow subspaces. How does the nonuniform rate of attraction affect the quality of the following approximations?
   (a) Argue that a first approximation is $y \approx k_2(1-x)/x$, on which $\dot{x} \approx 2k_1(1-x)^2 - k_2(1-x)$. Predict two equilibria on the slow manifold.
   (b) Argue that a first approximation is alternatively $x \approx 2k_1(1-y)^2/y$, on which $\dot{y} \approx k_2(1-2y) - 2k_1(1-y)^2$. Predict an equilibrium on the slow manifold.

Exercise 4.15 (varying parameter). In experiments one often slowly changes a parameter to explore a range of parameter values. We also model the effects of such time-varying parameters. Consider the simple bifurcation problem (4.19) but now suppose the parameter a varies slowly in time as prescribed externally. We aim to construct the modified slow manifold[55]

$$y \approx (1 - 2a + 2\dot{a} + 4a^2 - 12a\dot{a} + 12\dot{a}^2)x^2, \tag{4.30}$$

on which the evolution is the modified

$$\dot{x} \approx ax - (1 - 2a + 2\dot{a} + 4a^2 - 12a\dot{a} + 12\dot{a}^2)x^3. \tag{4.31}$$

[55] There are two interesting ways to regard the approximation (4.30)–(4.31). One way is to adjoin dynamical equations for the evolution of parameter a, such as $\dot{a} = b$ where b is a small constant, and then this approximation is a standard slow manifold in space of a dimension or two higher. The second way is to view the approximation (4.30)–(4.31) as the first example of a time-varying nonautonomous slow manifold for which I have not yet presented any supporting theory.

1. Substitute this approximation (4.30)–(4.31) into the governing equations (4.19) and deduce that the approximation is correct to errors $\mathcal{O}(a^3, x^4, \ddot{a})$, in some sense.
   The presence of the higher time derivative $\ddot{a}$ in the residual indicates that the approximation is also asymptotic in time derivatives of parameter a. That is, assume that parameter a varies slowly enough in time that higher-order time derivatives are negligible.
2. Use computer algebra to construct the approximation (4.30)–(4.31) by modifying Algorithm 4.2.
   (a) Recognize that parameter a varies with time by adding into the preamble `depend a,t;`.
   (b) But truncate the order of time derivatives by the pattern matching, where `~p` matches any expression and the condition `when p>1` restricts the matching,
   `let df(a,t,~p)=>0 when p>1;`
   (c) Now comes a complication: we must neglect high powers of a whether they are "bare," as in a^3, or hidden within a time derivative, as in $a\dot{a}^2$. The latter is represented in Reduce as `a*df(a,t)^2` and so is not neglected by the `let` statement in Algorithm 4.2. Instead we cunningly introduce an artificial "bookkeeping" parameter, say `eps`, that simply counts the number of factors of parameter a in any term: change the differential equations to `eps*a` wherever the parameter a occurs, here in the statement `let df(x,t)=>eps*a*x-x*y;`. Then neglect a term when the count is high enough via
   `let eps^3=>0;`
   (d) Execute the modified code to find the approximation (4.30)–(4.31).
3. Use your computer algebra to compute the approximation to second-order time derivatives of parameter a.

Answer: $y \approx (1 - 2a + 2\dot{a} - 2\ddot{a} + 4a^2 - 12a\dot{a} + 12\dot{a}^2 + 16a\ddot{a} - 40\dot{a}\ddot{a} + 40\ddot{a}^2)x^2$.

Exercise 4.16 (two bound particles). Two particles with mass m move in one dimension with positions x_1 and x_2 and momentum p_1 and p_2. The two particles move under the influence of some force field $K(x)$ and are bound to each other by a strong "spring" with force $\pm(x_2 - x_1)/\epsilon$. Thus the equations of motion are (van Kampen, 1985, p. 92)

$$\dot{x}_1 = \frac{p_1}{m}, \qquad \dot{p}_1 = K(x_1) + \frac{1}{\epsilon}(x_2 - x_1),$$
$$\dot{x}_2 = \frac{p_2}{m}, \qquad \dot{p}_2 = K(x_2) + \frac{1}{\epsilon}(x_1 - x_2).$$

For small parameter ϵ the strong spring dominates, so expect $x_1 \approx x_2$. Departures from this approximate equality will generate fast dynamics on a time scale proportional to some power of ϵ, so there should exist some slow dynamics. Let's try to find emergent slow dynamics.

1. Rescale variables: $t^* = t/\sqrt{m\epsilon}$, $x_i^* = x_i$, and "velocity" $u_i^* = p_i\sqrt{\epsilon/m}$. Dropping the stars, deduce

$$\dot{x}_i = u_i, \quad \dot{u}_i = x_{3-i} - x_i + \epsilon K(x_i).$$

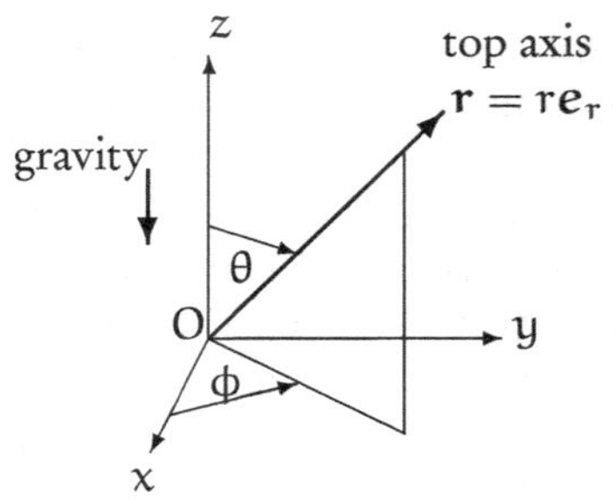

Figure 4.23. *A top spins about an axis* $\mathbf{r}$ *with its tip at the origin O. Downward gravity causes the axis to precess around the vertical z-axis: the azimuthal angle* ϕ *varies linearly at a rate that depends upon the polar inclination* θ.

2. Formally adjoin the trivial $\dot\epsilon = 0$ and deduce that equilibria are $x_1 = x_2$ and $u_i = \epsilon = 0$.
3. Linearize about these equilibria, ignoring the trivial $\dot\epsilon = 0$, and deduce that the characteristic equation is $\lambda^2(\lambda^2 + 2) = 0$.
4. Hence argue that there may be slow dynamics with $x_1 \approx x_2$ and $v_1 \approx v_2$ (from a generalized eigenvector), flagged by $\lambda = 0$, but that the surrounding dynamics are oscillations, λ pure imaginary, so these slow dynamics are not attractive/emergent.

This nonemergence in classical nondissipative dynamics is typical and is explored elsewhere (Chapter 13) with the aid of normal-form coordinate transforms.

Exercise 4.17 (spinning top). Consider a spinning top on a table. Depending upon the angle of its inclination to the vertical, the axis of the top will precess. Berry and Shukla (2011) interpreted this precession as a slow manifold. Deduce that this precession may be a slow manifold, but it is *not* an emergent center manifold.

Define Cartesian (x, y, z) and spherical (r, θ, ϕ) coordinates as in Figure 4.23, with the axis of the spinning top being aligned along the radial unit vector $\mathbf{e}_r$. By equating the rate of change of angular momentum to the applied torque, Berry and Shukla (2011, eq. (3.6)) derived a vector differential equation for the top's orientation vector $\mathbf{r}$. Their vector differential equation contains one large parameter G measuring the strength of the rotation, which tries to keep the top's orientation fixed, relative to the strength of gravity, which tries to tip the top.

Rescaling time replaces the large parameter G by a small parameter $\epsilon = 1/G^2$. Then let's introduce an orientation "velocity" $\mathbf{u}(t)$ so that the system becomes the regular perturbation system

$$\dot{\mathbf{r}} = \mathbf{u} \quad \text{and} \quad \mathbf{r} \times \dot{\mathbf{u}} + \mathbf{u} + \epsilon \mathbf{r} \times \mathbf{e}_z = \mathbf{0}. \tag{4.32}$$

1. Set parameter $\epsilon = 0$, corresponding to extremely fast rotation or negligible gravity, and deduce the family of equilibria of the top oriented in any direction: $\mathbf{r} = \mathbf{R}$ and $\mathbf{u} = \mathbf{0}$.
2. Linearize the dynamics (4.32) about each of these equilibria: seek $\mathbf{r}(t) = \mathbf{R} + \hat{\mathbf{r}}(t)$ and $\mathbf{u}(t) = \mathbf{0} + \hat{\mathbf{u}}(t)$ for hatted quantities small.

3. To find the spectrum, seek solutions proportional to $e^{\lambda t}$ in Cartesian coordinates. Deduce that eigenvalues are $\lambda = 0$ (three times) and $\lambda = \pm i/|\mathbf{R}|$.
4. Hence argue that there may be slow dynamics, flagged by $\lambda = 0$, but that the surrounding dynamics are oscillations, λ pure imaginary, so the slow dynamics are not attractive/emergent.

This nonemergence in classical nondissipative dynamics is typical and is explored elsewhere with the aid of normal-form coordinate transforms.

Exercise 4.18 (fixed flux Stefan melting). Repeat, modified, much of the analysis of section 4.8 for the Stefan problem of a melting bar in the case where the heating at the end $x = 0$ is one of fixed heat flux instead of fixed temperature. Instead of $u(0,t) = 1$, apply the boundary condition $\partial u/\partial x = -1$ at $x = 0$. Justify the existence of and approximately construct the slow manifold to find that the long-term evolution thereon takes the form $\dot s = kf(ks)$; comment on the power series for f.

Answer: $u = (s - x) + k\left(\frac{1}{2}x^2 - \frac{1}{2}s^2\right) + k^2\left(s^3 - sx^2\right) + \mathcal{O}(k^3)$ such that $\dot s = k - k^2 s + 2k^3 s^2 + \mathcal{O}(k^4)$.

Exercise 4.19 (radial Stefan melting). Repeat, modified, much of the analysis of section 4.8 for the Stefan problem of a melting plate. The heating is to be applied at a tiny region around the origin (as if a Bunsen burner heats the origin). Define polar coordinates on the plate and seek radial solutions only by assuming no angular dependence, $\partial/\partial\theta = 0$. The heat diffusion in 2D for the field $u(r,t)$, instead of PDE (4.27), is then the radial

$$\frac{\partial u}{\partial t} = \frac{1}{r}\frac{\partial}{\partial r}\left(r\frac{\partial u}{\partial r}\right).$$

Consider the case where the heating at the origin $r \approx 0$ is one of fixed heat flux; that is, apply the boundary condition $r\partial u/\partial r \to -1$ as $r \to 0$. Justify the existence of and approximately construct an emergent slow manifold to find that the long-term evolution thereon takes the form $\dot s = f(k)/s$; comment on the power series for f and the long-term dynamics.[56]

Answer: $u = \log(s/r) - \frac{k}{4}(1 - r^2/s^2) + \mathcal{O}(k^2)$ such that $s\,ds/dt = k - k^2/2 + \mathcal{O}(k^3)$.

Exercise 4.20 (water infiltration). Consider surface water infiltrating from above into dry subsoil. This is like the earlier Stefan problems in that there is a moving boundary between the wet soil and the dry soil. A difference is that gravity pulls the water downward so the physical transport involves advection, not just diffusion.

Adapt section 4.8. Let the x coordinate measure distance from the soil surface into the subsoil, and let the water density in the soil be $u(x,t)$. Suppose at the soil surface the water is pooled, so that the surface density $u(0,t) = 1$ for all time. Beneath the soil, as well as diffusion of water, gravity pulls the water downward according

[56] Yes, there does exist a self-similar solution for this problem (which as an extension you could find). The point of the exercise is to find that it emerges.

to Darcy's law that the flux is proportional to gravity and hence conservation of water requires $u(x,t)$ to satisfy the PDE[57]

$$\frac{\partial u}{\partial t} = \frac{\partial^2 u}{\partial x^2} - c\frac{\partial u}{\partial x}, \quad 0 < x < s(t),$$

for some constant c proportional to the gravitational forcing. The wet region of the soil is $0 < x < s(t)$. At the interface $x = s(t)$ with the dry soil we must have the water density $u(s,t) = 0$ and we assume the derivative condition that

$$\frac{ds}{dt} = \dot{s} = -k\frac{\partial u}{\partial x} \quad \text{at } x = s\,.$$

Invoke a slow manifold as in section 4.8 to argue that for small k the water density

$$u \approx \frac{1 - e^{c(x-s)}}{1 - e^{-cs}},$$

where the interface location evolves like $\dot{s} \approx kc/(1 - e^{-cs})$.

What is the next correction to this model?

Exercise 4.21 (three time scales). Consider for small parameter ϵ the singular differential equation (O'Malley, Jr., and Kirkinis, 2010, eq. (26))

$$\epsilon^3\frac{d^3u}{dt^3} + \epsilon\frac{d^2u}{dt^2} + \frac{du}{dt} + u = 0$$

for the function $u(t)$. Let us discover, by modeling the dynamics on each scale, that solutions evolve on three time scales. Because the ODE is linear, the algebra is straightforward, even though the context of the task is nontrivial.

1. (a) Regularize the problem by changing to the fast time scale. From the ratio of coefficients of the two highest derivatives, change to fast time $\tau = t/\epsilon^2$ so that the slow-time derivative $d/dt = \epsilon^{-2}d/d\tau$. Using overdots to denote fast-time derivatives, show that the ODE becomes the regular (not singular) ODE $\dddot{u} + \ddot{u} + \epsilon\dot{u} + \epsilon^3 u = 0$.
   (b) Change to the system of first-order ODEs in $x = \epsilon u$, $y = \dot{u}$, and $z = \ddot{u}$: $\dot{x} = \epsilon y$, $\dot{y} = z$, and $\dot{z} = -z - \epsilon y - \epsilon^2 x$.
2. (a) Deduce that the fast-time slow manifold is $z \approx -(\epsilon^2 + \epsilon^3)x - (\epsilon + \epsilon^2 + \epsilon^3)y$.
   (b) Observe that the evolution on this fast-time slow manifold is $\mathcal{O}(\epsilon)$, so rescale time to the intermediate time scale $\hat{\tau} = \epsilon\tau = t/\epsilon$ to obtain, using primes to denote $d/d\hat{\tau}$, the intermediate model $x' = y$ and $y' \approx -(1 + \epsilon + \epsilon^2)y - (\epsilon + \epsilon^2)x$.
3. (a) Deduce that the intermediate-time slow manifold is $y \approx -(\epsilon + \epsilon^2)x$.
   (b) Return to the original time to find that the evolution on this intermediate-time slow manifold is the model on the original time of $dx/dt \approx -(1 + \epsilon)x$.

Which is the right model to use? The original fast, the intermediate, or the slow? The answer depends upon how you will use the model.

[57] Here the diffusion of water is most likely due to the stochastic nature of the soil structure causing different droplets of water to take different-length "wiggly" paths through the soil. Such different-length paths result in such an effective diffusion.

Exercise 4.22 (microscale nonlinear diffusion). Consider a microscale lattice, indexed by j, with some evolving field $u_j(t)$ at each lattice point. Suppose they interact by the nonlinear variant of discrete diffusion

$$\dot{u}_j = u_j(u_{j-1} - 2u_j + u_{j+1}) \quad \text{for } j = 1, \ldots, n. \tag{4.33}$$

At the extremes of the lattice invoke the "Robin" boundary conditions, parametrized by b, normally $0 \leqslant b \leqslant 1$,

$$b(u_0 + u_1) + (1 - b)(u_0 - u_1) = 0$$
$$\text{and } b(u_{n+1} + u_n) + (1 - b)(u_{n+1} - u_n) = 0.$$

Analyze the dynamics for the tiny case of $n = 2$ dynamic lattice points.

1. Transform to new, mean and difference, dynamic variables $a(t) = (u_1+u_2)/2$ and $c(t) = (u_2 - u_1)/2$ to find that the dynamics are
$$\dot{a} = -2a^2 b - 2(1 + b)c^2 \quad \text{and} \quad \dot{c} = -2(1 + 2b)ac. \tag{4.34}$$
2. First find that a slow manifold model exists that explains the collective behavior of the dynamics.
   (a) Show that $b = c = 0$ are equilibria for all a, attractive for $a > 0$.
   (b) Hence deduce that there exists an emergent slow manifold in some neighborhood of $b = c = 0$ and $a > 0$.
   (c) Confirm that the slow manifold is $c = 0$, on which the amplitude a evolves according to $\dot{a} = -2a^2 b$.
   (d) Interpret these results in terms of the two lattice variables $u_1(t)$ and $u_2(t)$.
3. Second, reason that the slow manifold model extends over a large range of the parameter space.
   (a) Returning to the full dynamics (4.34), show that $c = 0$ is invariant for all a and b.
   (b) Show that $c = 0$ is attractive for $a > 0$ and $b > -1/2$.
   (c) Hence deduce an emergent dynamic model for the mean amplitude variable a.
   (d) Confirm this emergence for $b = 1$ by analytically solving the equations in the case when the difference variable c is small enough so that c^2 is negligible: find that $a \approx a_0/(1 + 2a_0 t)$ and $c \approx c_0/(1 + 2a_0 t)^3$.
   (e) Interpret these results in terms of the two lattice variables $u_1(t)$ and $u_2(t)$.

Exercise 4.23 (weakness in formal solutions). Some modelers rearrange equations with a small parameter multiplying a time derivative; then iteration generates higher and higher order time derivatives in their expansions (Roberts, 1985, e.g.). The results are sound in the appropriate parameter regime, but some modelers are tempted too far. Explore the concept and pitfalls in the example system, linear for simplicity,

$$\dot{x} = -y \quad \text{and} \quad \dot{y} = -\frac{1}{p}y + x. \tag{4.35}$$

The small parameter p characterizes the separation of time scales of the evolution of the slow mode x and the fast mode y. Physically, x represents interesting macroscopic "mean" variables, and y represents uninteresting microscale structures.

1. Rearrange the fast equation to the iteration $y = px - p\dot{y}$; then iterate formally starting from $y^{(0)} = 0$ and recognizing that variable x depends upon time t. Find formally that $y = px - p^2\dot{x} + p^3\ddot{x} - p^4\dddot{x} + \cdots$.
2. Substitute this formal expansion into the slow equation to find the model, remembering that parameter p is small,

$$\dot{x} = -px + p^2\dot{x} - p^3\ddot{x} + p^4\dddot{x} - \cdots .$$

3. Depending upon the truncation of the formal exact expansion on the right-hand side, show that the model predicts solutions $x \propto \exp(\lambda t)$ for eigenvalues
   (a) $\lambda = -p$,
   (b) $\lambda = -p/(1 - p^2)$,
   (c) $\lambda \approx -p, -1/p^3$, and
   (d) $\lambda \approx -p, \pm 1/p^2$.
4. Discuss the strengths and weaknesses of these model truncations—compare to the exact eigenvalue of the two modes, $\lambda = -(1 \pm \sqrt{1 - 4p^2})/(2p)$.
5. Find that the exact slow manifold of the original system may be expressed as $y = x(1 - \sqrt{1 - 4p^2})/(2p)$. Compare the slow manifold model with the above.

Chapter 5

Construct slow center manifolds iteratively

Contents

In previous examples of the application of center manifold theory it is straightforward to organize an iterative procedure to approximate the center manifold. In practice, the approximation process is more involved, and various researchers have devised a variety of schemes, many based on asymptotic power series. I recommend an iterative approach based upon reducing the residuals of the governing differential equations. As instigated by Coullet and Spiegel (1983), *the approach is coordinate free: in any application the analysis reflects the dynamics, not any particular basis we may choose to describe this dynamics.*

This chapter develops a flexible iterative algorithm, eminently suitable for computer algebra and based upon the center manifold theorems, for deriving low-dimensional models of general dynamical systems. First, section 5.1 investigates the modeling of a relatively simple dynamical system, a Burgers-like PDE, and introduces the basic concepts of the iteration scheme. A computer algebra implementation follows in section 5.2. Section 5.3 details the approach for dynamical systems of the general form (4.6). The examples of subsections 5.3.5–5.3.8 and 5.4 illustrate applications.

I restrict the scope of this section to slow manifolds, as the more general case of a center manifold with some pure imaginary eigenvalues is significantly more complicated, as discussed in Part VI.

5.1 ▪ Forced Burgers pitchfork bifurcation

Consider the following variation of Burgers' PDE for some field $u(x,t)$ and featuring growth at some controlled rate, ru; nonlinearity, uu_x; and dissipation, u_{xx}:

$$\frac{\partial u}{\partial t} = -u\frac{\partial u}{\partial x} + ru + \frac{\partial^2 u}{\partial x^2}, \quad u(0) = u(\pi) = 0. \tag{5.1}$$

View this as an infinite-dimensional dynamical system, the state space being the set of all smooth functions $u(x)$ on $[0,\pi]$.

Equilibrium For all values of the control parameter r there is an equilibrium at the origin, that is, a trivial equilibrium state is $u = 0$.

Linearization shows existence and emergence The linearization of the PDE (5.1) about this equilibrium, namely $u_t = ru + u_{xx}$, has constant coefficients in x and t. Seeking $u = e^{\lambda t}v(x)$, we need to solve the eigenproblem $v'' + rv = \lambda v$. This has the trigonometric modes $v_k(x) = \sin kx$ with associated eigenvalues $\lambda_k = r - k^2$ for wavenumbers $k = 1, 2, 3, \ldots$.[58] Thus the origin $u = 0$ becomes unstable as r crosses 1, because the $k = 1$ mode, $v_1(x) = \sin x$, begins to grow exponentially quickly, and the system undergoes a bifurcation.

To find the details of this pitchfork bifurcation is a straightforward task for center manifold theory. From the spectrum λ_k there is a slow manifold for control parameter $r = 1$. Thus explore dynamics near this critical parameter by setting $r = 1 + \epsilon$ and considering the parameter ϵ to be small (by notionally adjoining $\dot\epsilon = 0$). Linearly, exactly at critical, $\epsilon = 0$, all modes decay exponentially quickly except for the critical mode $\sin x$; it has a zero decay rate and therefore is long lasting; by the Existence Theorem 4.3 we are assured that there exists a slow manifold. Nonlinearly, and for ϵ and $u(x)$ near 0, all modes decay exponentially except for the critical mode, which slowly evolves. Thus, neglecting the exponentially decaying transients, we accurately model the dynamics solely in terms of the evolution of the amplitude of the $\sin x$ mode; define $a(t)$ to be its amplitude. By the Emergence Theorem 4.5, expect the evolution in time of amplitude a to form an accurate one-dimensional model of the original infinite-dimensional dynamical system (5.1).

The tangent space is the first approximation The first stage is to identify a linear approximation to the slow manifold and the evolution thereon. The critical eigenvector (mode) corresponding to the zero eigenvalue gives

$$u(x,t) \approx a \sin x \quad \text{such that} \quad \dot a \approx 0, \tag{5.2}$$

where a measures the amplitude of the mode. Equation (5.2) gives the tangent plane approximation to the slow manifold as the slow subspace $\mathbb{E}_0 = \operatorname{span}\{\sin x\}$. Figure 5.1 shows a representation of this slow subspace as a sequence of fields for different parameters $a(t)$.

[58] Note how these eigenvalues and eigenvectors satisfy the properties of a regular Sturm–Liouville eigenproblem: Table 5.1 list the properties. Remember these Sturm–Liouville properties, as they provide an enormously useful structure for the eigenvalues and eigenvectors for many systems of interest.

Table 5.1. *Regular Sturm–Liouville eigenvalue problems (Haberman, 1987, §5.3). Many of these properties also apply to important nonregular problems.*

The regular Sturm–Liouville eigenvalue problem for $u(x)$ consists of the ODE

$$\frac{d}{dx}\left(p(x)\frac{du}{dx}\right) + q(x)u = \lambda r(x)u\,, \quad a < x < b\,,$$

subject to Robin/Dirichlet/Neumann boundary conditions of the form $\alpha_1 u(a)+\alpha_2 u'(a) = 0$ and $\beta_1 u(b) + \beta_2 u'(b) = 0$. The coefficient functions are all real, and $p(x), r(x) > 0$ for $a \leqslant x \leqslant b$. Such eigenvalue problems have marvelous properties.

1. All eigenvalues λ are real.
2. There is a countably infinite number of eigenvalues $\lambda_1 > \lambda_2 > \lambda_3 > \cdots$:
   (a) there is a largest (most positive) eigenvalue λ_1;
   (b) there is no smallest (most negative) eigenvalue; further, $\lambda_n \to -\infty$ as $n \to \infty$.
3. Corresponding to each eigenvalue λ_n, there is an eigenfunction $v_n(x)$ (unique to within scaling) that has exactly $n-1$ zeros for $a < x < b$.
4. The set of eigenfunctions $\{v_n(x)\}$ is complete: any piecewise smooth function $f(x)$ can be written as a generalized Fourier series, the convergent sum
$$f(x) = \sum_{n=1}^{\infty} a_n v_n(x) \quad \left(= \tfrac{1}{2}[f(x-) + f(x+)]\right).$$
5. Eigenfunctions corresponding to different eigenvalues are orthogonal relative to the *weight function* $r(x)$:
$$\int_a^b v_n(x)v_m(x)r(x)\,dx = 0 \quad \text{if } m \neq n\,.$$
6. The *Rayleigh quotient* gives the eigenvalue
$$\lambda = \frac{[pvv']_a^b + \int_a^b \left[qv^2 - pv'^2\right]dx}{\int_a^b v^2 r\,dx}$$
and is typically used to approximate well the eigenvalue given only a rough approximation v to the eigenfunction.

Iteration improves the approximation The second stage is to seek iterative improvements to the description of the slow manifold and the low-dimensional evolution thereon. Figure 5.2 represents the slow manifold that we seek: the fields $u(x)$ for different amplitudes a are distorted versions of the linear sine curve shown in the subspace of Figure 5.1. Suppose that at one stage of the iteration we have the approximate model

$$u \approx v(x, a, \epsilon) \quad \text{such that} \quad \dot{a} \approx g(a, \epsilon)\,;$$

it is approximate because the residual of the governing differential equation (5.1) is

$$\begin{aligned}\operatorname{Res}(v, g, \epsilon) &:= \frac{\partial u}{\partial t} + u\frac{\partial u}{\partial x} - ru - \frac{\partial^2 u}{\partial x^2} \\ &= \frac{\partial v}{\partial a}g - (1+\epsilon)v + v\frac{\partial v}{\partial x} - \frac{\partial^2 v}{\partial x^2}\,.\end{aligned} \tag{5.3}$$

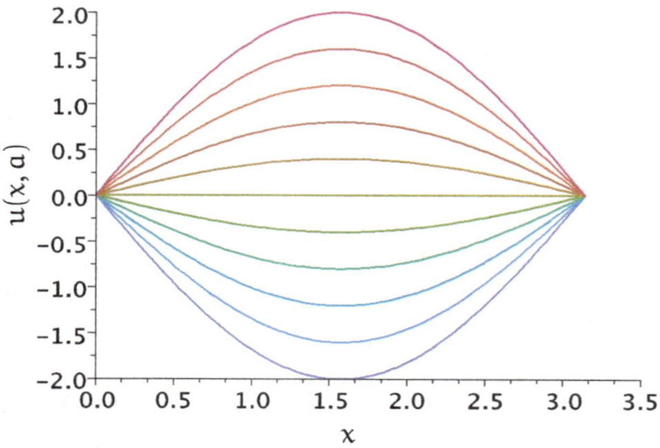

Figure 5.1. *Representation of the slow subspace* $\mathbb{E}_0$ *for Burgers' PDE* (5.1) *exactly at critical* $r = 1$. *Plots the field* $u(x)$ *as a function of the amplitude* $a \in [-2, 2]$*: being a subspace, the field is proportional to* $\sin x$.

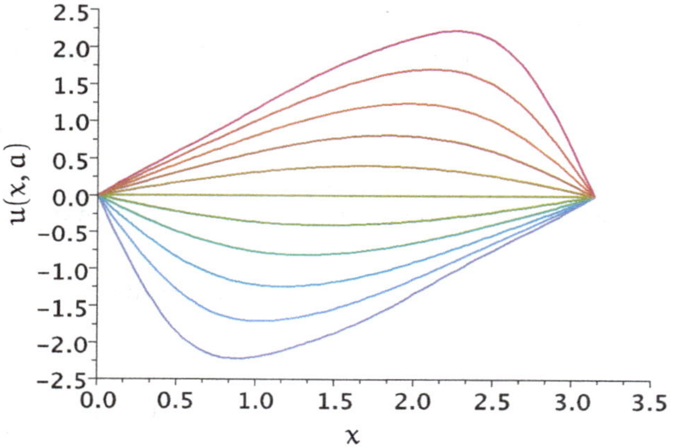

Figure 5.2. *Representation of the slow manifold for Burgers' PDE* (5.1) *exactly at critical* $r = 1$. *Plots the field* $u(x)$ *as a function of the amplitude* $a \in [-2, 2]$*: the field is a distorted sine curve.*

For example, in the first iteration, starting from the initial approximation (5.2), the residual $\text{Res} = -\epsilon a \sin x - a^2 \sin x \cos x$. In any iteration the residual is of some magnitude, say $\text{Res} = \mathcal{O}(\epsilon^p + a^q)$, for some exponents p and q. The aim of each iteration is to improve the residual (increase p and q) so that, by the Flexible Approximation Theorem 4.10, we improve the asymptotic accuracy of the model.

We seek to find small corrections, indicated by hats, so that

$$u \approx v(x, a, \epsilon) + \hat{v}(x, a, \epsilon) \quad \text{such that} \quad \dot{a} \approx g(a, \epsilon) + \hat{g}(a, \epsilon)$$

is a better approximation to the slow manifold and the evolution thereon. Substituting the above into the governing PDE (5.1), using the chain rule for time derivatives, leads to

$$\begin{aligned}&\frac{\partial v}{\partial a} g + \frac{\partial v}{\partial a}\hat{g} + \frac{\partial \hat{v}}{\partial a} g + \frac{\partial \hat{v}}{\partial a}\hat{g}\\ &= (1+\epsilon)v - v\frac{\partial v}{\partial x} + \frac{\partial^2 v}{\partial x^2} + (1+\epsilon)\hat{v} - \hat{v}\frac{\partial v}{\partial x} - v\frac{\partial \hat{v}}{\partial x} - \hat{v}\frac{\partial \hat{v}}{\partial x} + \frac{\partial^2 \hat{v}}{\partial x^2},\end{aligned}$$

that is,

$$\operatorname{Res}(v, g, \epsilon) + \frac{\partial v}{\partial a}\hat{g} + \frac{\partial \hat{v}}{\partial a}g + \frac{\partial \hat{v}}{\partial a}\hat{g} = (1+\epsilon)\hat{v} - \hat{v}\frac{\partial v}{\partial x} - v\frac{\partial \hat{v}}{\partial x} - \hat{v}\frac{\partial \hat{v}}{\partial x} + \frac{\partial^2 \hat{v}}{\partial x^2}.$$

It is impossible to solve this for the perfect corrections in one step. Instead we seek an approximate equation for the corrections of $\mathcal{O}(\epsilon^p + a^q)$ by the following.

1. Ignore products of corrections (hatted quantities) because they will be much smaller, $\mathcal{O}(\epsilon^{2p} + a^{2q})$, than the sought corrections. Then, for example, in the first iteration we would like to solve

$$\operatorname{Res} + \sin x\, \hat{g} = (1+\epsilon)\hat{v} - a\cos x\, \hat{v} - a\sin x\, \frac{\partial \hat{v}}{\partial x} + \frac{\partial^2 \hat{v}}{\partial x^2}.$$

   But such an equation is still too difficult for analytic solution in general as it involves the variables a and ϵ in its coefficients of the unknown corrections $\hat{v}$.

2. Also recognize that near the "anchoring" equilibrium at the origin, both a and ϵ are small. Thus terms such as $\epsilon\hat{v}$ and $a\cos x\, \hat{v}$ are very small, small in comparison to $\hat{v}$ terms, and may be neglected to lead to

$$\hat{v} + \frac{\partial^2 \hat{v}}{\partial x^2} = \sin x\, \hat{g} + \operatorname{Res}.$$

   More generally, wherever quantities multiply a correction factor, then replace the quantities by their zeroth-order approximation, $\partial v/\partial a \approx \sin x$, $v \approx 0$, and $g \approx 0$ (this introduces errors $\mathcal{O}(\epsilon^{p+1} + a^{q+1})$).

In a first iteration from the linear approximation (5.2) this is

$$\hat{v} + \frac{\partial^2 \hat{v}}{\partial x^2} = \hat{g}\sin x - \epsilon a \sin x + \frac{1}{2}a^2 \sin 2x.$$

We wish to solve such equations for $\hat{v}$. However, $\hat{v} + \hat{v}_{xx}$, with boundary conditions $\hat{v}(0) = \hat{v}(\pi) = 0$, is singular, as the critical mode, $\sin x$, is always a homogeneous solution. Thus we have to put the right-hand side into the range of the operator $\mathcal{L} = 1 + \partial_{xx}$ before we can find $\hat{v}$. Here this is easily done by choosing $\hat{g} = \epsilon a$, called the *solvability condition*. Having made that choice, then the ODE $\hat{v} + \hat{v}_{xx} = \frac{1}{2}a^2 \sin 2x$ has as a solution $\hat{v} = -\frac{1}{6}a^2 \sin 2x$ obtained by the method of undetermined coefficients (just trying $\hat{v} = A\sin 2x$). Thus this first iteration deduces $u \approx a\sin x - \frac{1}{6}a^2 \sin 2x$, which shows the nonlinear steepening/flattening of negative/positive slopes, and also shows that the evolution is $\dot{a} \approx \epsilon a$, which exhibits the loss of stability of the equilibrium $a = 0$ when ϵ becomes positive.

Further iterations in this example (see section 5.2) lead to the slow manifold being

$$u = a\sin x - \frac{1-\epsilon/3}{6}a^2 \sin 2x + \frac{1}{32}a^3 \sin 3x + \mathcal{O}(\epsilon^2 + a^4), \tag{5.4}$$

as shown schematically in Figure 5.2, on which the system evolves according to (a Landau equation)

$$\dot{a} = \epsilon a - \frac{1}{12}a^3 + \mathcal{O}(\epsilon^2 + a^4). \tag{5.5}$$

Algorithm 5.1 *Outline of the general iteration to construct a slow manifold model.*

```
preliminaries;
initial linear approximation;
repeat
    compute residual;
    find solvability condition;
    compute correction to the slow manifold;
    update approximations;
until residual is small enough.
```

Algorithm 5.2 *Iteration finds the slow manifold of the forced Burgers* PDE (5.1).

```
    operator linv;  linear linv;
    let linv(sin(~k*x),x)=>sin(k*x)/(1-k^2);
    depend a,t;  let df(a,t)=>g;
    u:=a*sin(x);  g:=0;  r:=1+eps;
    let {eps^2=>0, a^4=>0};
    for iter:=1:99 do begin
        resb:=df(u,t)-r*u+u*df(u,x)-df(u,x,2);
        write resb:=trigsimp(resb,combine);
        g:=g+(gd:=-coeffn(resb,sin(x),1));
        u:=u+linv(resb+gd*sin(x),x);
        if resb=0 then iter:=1000000;
    end;
```

The Emergence Theorem 4.5 assures us (as $\epsilon = a = 0$ is a stable equilibrium) that this one-dimensional model of the original infinite-dimensional dynamical system (5.1) is valid exponentially quickly in time. From the model (5.5), for example, we deduce the quantitative shape of the pitchfork bifurcation: there are stable equilibria at $a \approx \pm 2\sqrt{3\epsilon}$. Physically, these equilibria represent a balance between the growth ru, the nonlinear steepening of the uu_x term, and the dissipation of u_{xx}.

5.2 ▪ Computer algebra implementation

> It is unworthy of excellent persons to lose hours like slaves in the labour of calculation.
> *Gottfried Wilhelm von Leibniz*

A principal reason for adopting an iterative approach is that it is simply and reliably implemented in computer algebra. Based upon the above derivation, Algorithm 5.1 gives the general outline of the requisite iteration.

Algorithm 5.2 details a Reduce program for the particular example PDE (5.1). A reason for using Reduce is that it has excellent pattern-matching and replacement capabilities through its `operator` and `let` statements.

1. The preliminaries are the following.

    - Define the operator `linv` to act as the inverse of $\mathcal{L}$:

```
operator linv; linear linv;
let linv(sin(~k*x),x)  => sin(k*x)/(1-k^2);
```

- declaring the operator `linear` tells Reduce to expand sums and products in the first argument and to only leave functions of the second argument inside the operator, for example, the expression $\mathtt{linv(}\epsilon a \sin x + 2a^2 \sin 2x\mathtt{,x)}$ expands to $\epsilon a\mathtt{linv(}\sin x\mathtt{,x)+}2a^2\mathtt{linv(}\sin 2x\mathtt{,x);}$
- the `let` statement defines the action of the operator as the solution to $\hat{v} + \hat{v}_{xx} = \sin kx$, namely $\hat{v} = \frac{1}{1-k^2} \sin kx$, with the tilde before the `k` on the left-hand side matching any pattern (no action is defined for the singular case $k = 1$ because the pattern `sin(~k*x)` does not match `sin(x)`—any appearance of `linv(sin(x),x)` usefully signals an error).

- Establish first that the parametric variable `a` is to depend upon time, as we use a as the time-dependent amplitude in the model, and second that time derivatives of amplitude a, `df(a,t)`, are to be replaced by the value of `g`, at the time of replacement, as `g` is to store the current approximate model evolution equation, such as (5.5):

```
depend a,t;   let df(a,t)=>g;
```

2. Assign the linear approximation (5.2) of the slow manifold to be the initial value of the variables `u` and `g`:

```
u:=a*sin(x);   g:=0;   r:=1+eps;
```

3. Perform the iterations in a "safe" for-loop. Control the truncation of the asymptotic approximation by telling Reduce to discard any factor in ϵ^2 or higher and any factor in a^4 or higher; thus all expressions are computed to an error of $\mathcal{O}(\epsilon^2, a^4)$. Terminate the for-loop in Reduce by setting the loop counter to a large value.

```
let {eps^2=>0, a^4=>0};
for iter:=1:99 do begin
    . . .
    if res=0 then iter:=1000000;
end;
```

Within the above loop:

(a) Compute the residual:

```
res:=df(u,t)-r*u+u*df(u,x)-df(u,x,x);
write res:=trigsimp(res,combine);
```

The code is a very direct translation of the governing PDE (5.1) into Reduce symbols—using this approach you only implement the governing equations, and all the messy details of the asymptotic expansions are dealt with by the computer algebra engine.

Like most computer algebra engines, the products of trigonometric terms are *not* by default linearized, so here explicitly combine products of trigonometric terms using the `trigsimp` function as shown.

(b) The solvability condition eliminates any component in $\sin x$ in the right-hand side by choosing the correction to `g`:

```
g:=g+(gd:=-coeffn(res,sin(x),1));
```

(c) Solve for the correction to u and update the current approximation:

```
write u:=u+linv(res+sin(x)*gd,x);
```

The program derives high-order effects in amplitude a or parameters ϵ simply by increasing the order of the factors that are discarded. Computing to high order in ϵ, by letting `eps^10=>0`, for example, indicates that

$$\dot{a} = \epsilon a - \frac{a^3}{12(1+\epsilon/3)} + \mathcal{O}(a^5)\,.$$

Quite different dynamical systems with the same linear structure may be analyzed simply by changing the computation of the residual.[59]

5.3 ▪ Iteration generally constructs slow manifold models

Now consider a general dynamical system in the form

$$\dot{\mathbf{u}} = \mathcal{L}\mathbf{u} + \mathbf{f}(\mathbf{u}, \boldsymbol{\epsilon})\,, \tag{5.6}$$

where, as usual, $\mathbf{u}(t)$ is the evolving state vector; $\mathcal{L}$ is a linear operator whose spectrum, as required by center manifold theory, is discrete and separates into m eigenvalues of zero real part, the critical eigenvalues, and eigenvalues with strictly negative real part; $\boldsymbol{\epsilon} \in \mathbb{R}^\ell$ is a vector of ℓ parameters; and $\mathbf{f}$ is a function that is strictly nonlinear when considered as a function of $\mathbf{u}$ and $\boldsymbol{\epsilon}$ together. The aim is to find a low-dimensional model $\dot{\mathbf{s}} = \mathbf{g}(\mathbf{s}, \boldsymbol{\epsilon})$ for the evolution of the m amplitudes $\mathbf{s}$ of the critical modes.

The parameters $\mathbf{s}$ of the center manifold can be any reasonable measure of the amplitude of the critical modes; you choose the physical meaning of $\mathbf{s}$ as best suits you and the system. This low-dimensional dynamics occurs on the exponentially attractive center manifold described parametrically as $\mathbf{u} = \mathbf{v}(\mathbf{s}, \boldsymbol{\epsilon})$.

However, in this chapter we do not consider dynamical systems with pure imaginary eigenvalues, $i\omega \neq 0$, as the analysis is significantly more complicated. *Here we restrict our attention to slow manifolds based on modes with purely zero eigenvalues.*

5.3.1 ▪ Identify a suitable equilibrium

We need an equilibrium on which to base the modeling analysis and construction (or a family of equilibria for a global model). Here the system (5.6) is defined such that the origin is the equilibrium on which analysis is anchored.

Often we first embed the problem of interest in a family of related problems. For example, suppose we are interested in the specific system

$$\dot{x} = 0.1x - xy\,, \quad \dot{y} = -2y + x^2.$$

Then the origin $x = y = 0$ is an equilibrium, but it is apparently unsuitable for the next step because there are no critical modes, that is, no linearization with zero eigenvalues. However, embed this particular system as the $a = 0.1$ case of the family

$$\dot{a} = 0\,, \quad \dot{x} = ax - xy\,, \quad \dot{y} = -2y + x^2.$$

[59] By way of comparison, Rand and Armbruster's (1987, pp. 27–34) MACSYMA code for constructing the slow manifold of a finite-dimensional dynamical system, based upon power series manipulation, uses 53 lines of active code (although some are for input of the dynamical equations), whereas the above algorithm has only 14 lines of active code.

Then a suitable equilibrium is $a = x = y = 0$, as it has critical modes, and we should be able to access the case of interest, $a = 0.1$, because of the finite domain assured by the center manifold theorems.

5.3.2 ▪ Identify the m critical modes

The m critical modes (eigenvectors) are those associated with the zero eigenvalues. These critical modes are necessary in order to project the linear dynamics and nonlinear perturbations onto the slow modes of interest. Find these modes from the nontrivial solutions, e_j, of $\mathcal{L}e_j = 0$, but in general we need the critical slow eigenspace, $\mathbb{E}_0$, and so we also need to find all the generalized eigenvectors.[60] Then, in terms of modal amplitudes s_j, a linear approximation to the slow manifold and the evolution thereon is the general linear combination

$$u(t) \approx \sum_j e_j s_j = \mathcal{E}s \quad \text{such that} \quad \dot{s} \approx \mathcal{G}s\,, \tag{5.7}$$

where the columns of matrix $\mathcal{E} := [e_j]$ span the slow eigenspace $\mathbb{E}_0$, and where $\mathcal{G}$ may be chosen in Jordan form in the case of generalized eigenvectors (Table 4.1); $\mathcal{G}$ is zero if there are no generalized eigenvectors (in this case of a slow manifold). In order to model the nonlinear dynamics the linear approximation (5.7) needs to be modified by the nonlinear effects.

5.3.3 ▪ The eigenvectors may not match your amplitudes

You must *choose* a meaning for the amplitudes s, either explicitly or implicitly. Generally we choose a linear definition (but not always). For example, for Burgers' PDE (5.1) we implicitly chose amplitude a to be the component of $\sin x$ in the solution field: $a := \frac{2}{\pi}\int_0^\pi \sin x\, u\, dx$. Let's consider the cases where we *define* the amplitudes linearly: $s_j := \langle r_j, u\rangle = r_j^T u$ for some chosen vectors r_j. That is, upon forming the restriction matrix $\mathcal{R} := [r_j]$, we define the amplitude vector $s := \mathcal{R}^T u$.

Figure 5.3 shows the idea. We *choose* to parametrize positions on the slow manifold $\mathcal{M}$ by the distance of points on $\mathcal{M}$ in the direction of some vector r. Sometimes we may choose r to be in the same direction as the basis vector e, sometimes not. The choice is yours, either implicitly or explicitly.

The issue is that the eigenvectors obtained in the previous step may not match these definitions of the amplitudes; instead we should use some other linear combination. Denote the (generalized) eigenvectors found previously by e'_j and the matrix $\mathcal{E}' := [e'_j]$. We want new (generalized) eigenvectors as a linear combination of these: $e_j = \sum_i e'_i c_{ij}$, that is, $\mathcal{E} = \mathcal{E}'C$. Now, using the above,

$$s = \mathcal{R}^T u = \mathcal{R}^T\mathcal{E}s = \mathcal{R}^T\mathcal{E}'Cs\,.$$

This has to be true for all amplitudes s, so we must have $\mathcal{R}^T\mathcal{E}'C = I$, the identity. Hence the linear combination matrix $C = (\mathcal{R}^T\mathcal{E}')^{-1}$. The basis eigenvectors for our chosen amplitudes, and the leading approximation (5.7), are then the columns of $\mathcal{E} = \mathcal{E}'C$.

[60] Table 4.1 describes how to find generalized eigenvectors. When finding generalized eigenvectors, solve $\mathcal{L}e_{j+1} = \sum_{i=1}^{j} g_{ij}\, e_i$ for the generalized eigenvector e_{j+1} and coefficients g_{ij}. Repeat. The coefficients g_{ij} are the entries in $\mathcal{G}$. Such generalized eigenvectors are common in mechanical problems where mean position and mean velocity are slow variables; then the linear equations for the slow variables are $\dot{x} = v$ and $\dot{v} = 0$. That is, the matrix $\mathcal{G}$ is nonzero and hence there must be generalized eigenvectors in the basis of $\mathbb{E}_0$. The ordinary eigenvector gives the position dimension, and the generalized eigenvector gives the velocity dimension.

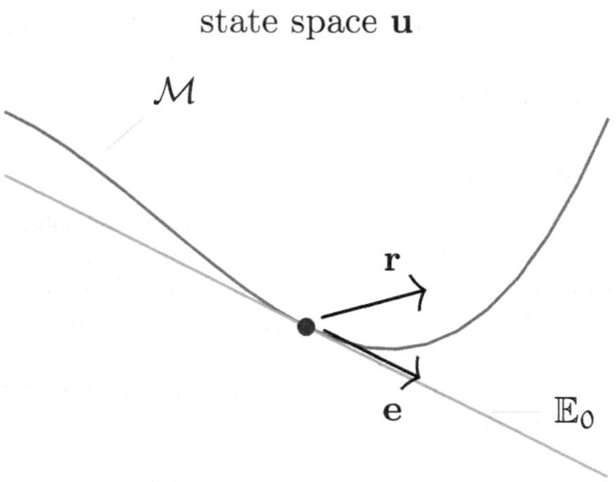

Figure 5.3. *Schematic diagram of an equilibrium (small disc) with slow subspace $\mathbb{E}_0$ spanned by* $\mathbf{e}$. *The slow manifold $\mathcal{M}$ is tangent to $\mathbb{E}_0$ at the equilibrium. To measure positions on the slow manifold we may use amplitudes proportional to distances in the direction* $\mathbf{r}$, *that is, amplitude* $s = \langle \mathbf{r}, \mathbf{u} \rangle$.

Lemma 5.1 (linear problems). *In modeling a (parametrized) linear system, the definition of the amplitudes is almost immaterial.*

Proof. Suppose we wish to model the dynamics of the linear system $\dot{\mathbf{u}} = \mathcal{L}(\boldsymbol{\epsilon})\mathbf{u}$. The dependence upon the parameters $\boldsymbol{\epsilon}$ is irrelevant to the issue of alternative amplitudes, so I omit $\boldsymbol{\epsilon}$ hereafter. Suppose the system has an m-dimensional slow manifold. Being a linear system, the slow manifold is linearly parametrized as $\mathbf{u} = \mathcal{V}\mathbf{s}$, on which $\dot{\mathbf{s}} = \mathcal{G}\mathbf{s}$ for some amplitudes defined by the linear $\mathbf{s} = \mathcal{R}^\mathsf{T}\mathbf{u}$: since $\dot{\mathbf{u}} = \mathcal{L}\mathbf{u}$ becomes $\mathcal{V}\mathcal{G}\mathbf{s} = \mathcal{L}\mathcal{V}\mathbf{s}$ then $\mathcal{V}\mathcal{G} = \mathcal{L}\mathcal{V}$ follows. Now suppose we wish to use another definition of the amplitude, namely $\mathbf{s}' = \mathcal{R}'^\mathsf{T}\mathbf{u}$.

First, there exists a linear transform T such that $\mathbf{s} = \mathsf{T}\mathbf{s}'$, since $\mathbf{s}' = \mathcal{R}'^\mathsf{T}\mathbf{u} = \mathcal{R}'^\mathsf{T}\mathcal{V}\mathbf{s} = \mathcal{R}'^\mathsf{T}\mathcal{V}\mathsf{T}\mathbf{s}'$ is true for all $\mathbf{s}'$ by choosing transform $\mathsf{T} = (\mathcal{R}'^\mathsf{T}\mathcal{V})^{-1}$, which exists for equivalent parametrizations.

Second, confirm that $\mathbf{u} = (\mathcal{V}\mathsf{T})\mathbf{s}'$ such that $\dot{\mathbf{s}}' = \mathsf{T}^{-1}\mathcal{G}\mathsf{T}\mathbf{s}'$ is the same slow manifold by confirming that the residual $\dot{\mathbf{u}} - \mathcal{L}\mathbf{u} = \mathcal{V}\mathsf{T}\dot{\mathbf{s}}' - \mathcal{L}\mathcal{V}\mathsf{T}\mathbf{s}' = \mathcal{V}\mathsf{T}\mathsf{T}^{-1}\mathcal{G}\mathsf{T}\mathbf{s}' - \mathcal{V}\mathcal{G}\mathsf{T}\mathbf{s}' = \mathcal{V}\mathcal{G}\mathsf{T}\mathbf{s}' - \mathcal{V}\mathcal{G}\mathsf{T}\mathbf{s}' = \mathtt{0}$. Since the residual is zero, the new system is also an invariant manifold model. Since $\dot{\mathbf{s}}' = (\mathsf{T}^{-1}\mathcal{G}\mathsf{T})\mathbf{s}'$ is the corresponding transformation of $\dot{\mathbf{s}} = \mathcal{G}\mathbf{s}$, the two models, with the two different definitions of amplitudes, describe exactly the same dynamics. □

Further, if the evolution operator $\mathcal{G}$ and the transformation T commute, then not only is the evolution equation in the new variables dynamically the same, it also has exactly the same algebraic form. If they commute, then $\dot{\mathbf{s}}' = (\mathsf{T}^{-1}\mathcal{G}\mathsf{T})\mathbf{s}' = (\mathsf{T}^{-1}\mathsf{T}\mathcal{G})\mathbf{s}' = \mathcal{G}\mathbf{s}'$.

Amplitudes for a normal form In applications where the center/slow manifold is of small dimensions, say one- or two-dimensional, people often choose to parametrize the center manifold to make it easier to predict the characteristics of solutions (Arnold and Imkeller, 1998; Takens and Vanderbauwhede, 2010; Haragus and Iooss, 2011, e.g.). The parametriztion of the center manifold is chosen so that the model, $\dot{\mathbf{s}} = \mathbf{g}(\mathbf{s})$, is in a normal form that has been tabulated and characterized by Guckenheimer and Holmes (1983), for

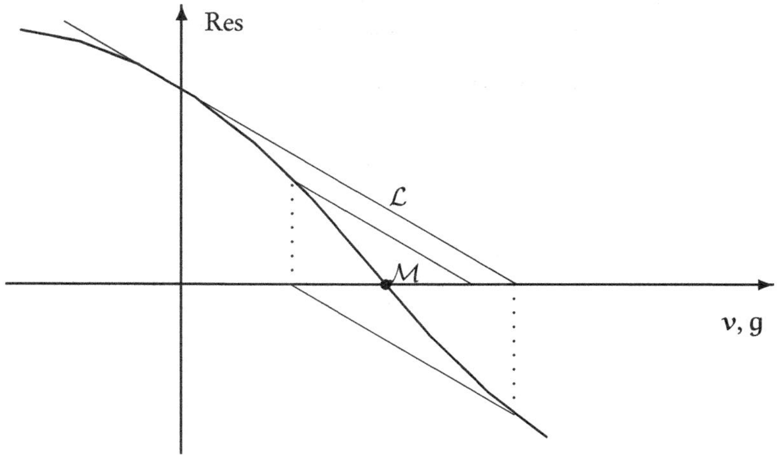

Figure 5.4. *Schematic diagram of the iteration to approximate the shape of a slow manifold, represented by $\mathcal{M}$. The straight-line extrapolation represents the linear operator $\mathcal{L}$ acting upon the current residuals* Res. *Each iteration typically improves the accuracy by a fixed amount, usually an extra order in a multinomial asymptotic expansion.*

example, for typically one- and two-dimensional center manifolds. Such a focus on the solution structure greatly enhances the nature of predictions one may make, at the cost of complicating its relation to the original physical variables. You have this choice in systems with center manifolds of small dimension.

5.3.4 ▪ Seek iterative improvements

Figure 5.4 shows the concept of the iterative scheme. Evaluating the residual characterizes how far a given approximation, v and g, is from $\mathcal{M}$. Then solving $\mathcal{L}\hat{v} = \text{Res}$ for a correction $\hat{v}$ in essence uses the linear dynamics about the equilibrium to determine a correction in order to get closer to $\mathcal{M}$. When the linear dynamics about the equilibrium is close enough to the dynamics about any point on $\mathcal{M}$, then the iteration should converge. In the schematic Figure 5.4, this is equivalent to requiring that the slope of the residual curve near $\mathcal{M}$ be roughly that near $v = 0$.

Given some approximate description of the slow manifold and the low-dimensional evolution thereon we seek an improvement. The aim is to find a low-dimensional description that satisfies the nonlinear dynamical equation (5.6). As in iterative methods for finding the zero of a function, we use the residual of the governing equations to guide corrections. The iteration scheme is successful as long as it ultimately drives the residual to zero to the desired order of error—see the Flexible Approximation Theorem 4.10. Suppose that at any one stage of the iteration we have the approximate model

$$u \approx v(s, \epsilon) \quad \text{such that} \quad \dot{s} \approx g(s, \epsilon),$$

which is approximate because the residual of the governing differential equation (5.6) is

$$\text{Res}\,(v, g, \epsilon) := -\dot{u} + \mathcal{L}u + f(u, \epsilon) = -\frac{\partial v}{\partial s}g + \mathcal{L}v + f(v, \epsilon) = \mathcal{O}\big(\epsilon^p + s^q\big) \qquad (5.8)$$

for some order of error, p and q, and where s and ϵ denote $|\mathbf{s}|$ and $|\epsilon|$, respectively. Seek small corrections, indicated by hats, to the approximation so that

$$\mathbf{u} \approx \mathbf{v}(\mathbf{s}, \epsilon) + \hat{\mathbf{v}}(\mathbf{s}, \epsilon) \quad \text{such that} \quad \dot{\mathbf{s}} \approx \mathbf{g}(\mathbf{s}, \epsilon) + \hat{\mathbf{g}}(\mathbf{s}, \epsilon)$$

is a better approximation to the slow manifold and the evolution thereon. Substituting into the governing differential equation (5.6), and using the chain rule for time derivatives, leads to

$$\left(\frac{\partial \mathbf{v}}{\partial \mathbf{s}} + \frac{\partial \hat{\mathbf{v}}}{\partial \mathbf{s}}\right)(\mathbf{g} + \hat{\mathbf{g}}) = \mathcal{L}\mathbf{v} + \mathcal{L}\hat{\mathbf{v}} + \mathbf{f}(\mathbf{v} + \hat{\mathbf{v}}, \epsilon).$$

Given that it is impossible to solve this for the perfect corrections in one step, seek an approximate equation for the corrections of $\mathcal{O}(\epsilon^p + s^q)$ by

- ignoring products of corrections (hatted quantities) because they will be small, $\mathcal{O}(\epsilon^{2p} + s^{2q})$, compared with the dominant effect of the linear correction terms, to give
$$\frac{\partial \mathbf{v}}{\partial \mathbf{s}}\mathbf{g} + \frac{\partial \mathbf{v}}{\partial \mathbf{s}}\hat{\mathbf{g}} + \frac{\partial \hat{\mathbf{v}}}{\partial \mathbf{s}}\mathbf{g} = \mathcal{L}\mathbf{v} + \mathcal{L}\hat{\mathbf{v}} + \mathbf{f}(\mathbf{v}, \epsilon) + \left.\frac{\partial \mathbf{f}}{\partial \mathbf{u}}\right|_{\mathbf{v}} \hat{\mathbf{v}},$$
- and, wherever they are multiplied by a correction factor, replacing quantities by their zeroth-order approximation in s and ϵ. This replacement introduces errors $\mathcal{O}(\epsilon^{p+1} + s^{q+1})$.[61]

Thus we wish to solve

$$\frac{\partial \mathbf{v}}{\partial \mathbf{s}}\mathbf{g} + \mathcal{E}\hat{\mathbf{g}} + \frac{\partial \hat{\mathbf{v}}}{\partial \mathbf{s}}\mathcal{G}\mathbf{s} = \mathcal{L}\mathbf{v} + \mathcal{L}\hat{\mathbf{v}} + \mathbf{f}(\mathbf{v}, \epsilon).$$

In general, the term $\partial\hat{\mathbf{v}}/\partial\mathbf{s}\,\mathcal{G}\mathbf{s}$ must be retained because the operator $\partial/\partial\mathbf{s}\,\mathcal{G}\mathbf{s}$ is of zeroth order in s—multiplication by $\mathbf{s}$ raises the order by one, but the gradient $\partial/\partial\mathbf{s}$ reduces the order by one. Thus, rearranging and recognizing that $\partial\mathbf{v}/\partial\mathbf{s}\mathbf{g} = \partial\mathbf{v}/\partial t$ by the chain rule, we aim to solve the *homological equation*

$$-\mathcal{L}\hat{\mathbf{v}} + \frac{\partial \hat{\mathbf{v}}}{\partial \mathbf{s}}\mathcal{G}\mathbf{s} + \mathcal{E}\hat{\mathbf{g}} = \operatorname{Res}(\mathbf{v}, \mathbf{g}, \epsilon) := -\frac{\partial \mathbf{v}}{\partial t} + \mathcal{L}\mathbf{v} + \mathbf{f}(\mathbf{v}, \epsilon) \tag{5.9}$$

for the hatted correction quantities. The great advantage of this iterative approach is that the right-hand side, which drives the corrections, is simply the residual of the governing system (5.6) evaluated at the current approximation. Thus at any iteration we just deal with physically meaningful expressions; all the detailed algebraic machinations of asymptotic expansions in other procedures are absent.

The case when the m critical eigenvalues have no imaginary component may be simplified. Arrange the variables so that $\mathcal{G}$ is upper/lower triangular, such as in the often physically appealing Jordan form. Then significantly simplify the algorithm by neglecting the term $\partial\hat{\mathbf{v}}/\partial\mathbf{s}\,\mathcal{G}\mathbf{s}$. The cost is that of increasing the number of iterations needed by a factor of no more than m, the multiplicity of the zero eigenvalue of $\mathcal{L}$. In the remaining analysis, assume this neglect for simplicity. The more general case is deferred until Part VI on Hopf bifurcations.

[61]The approximation of replacing quantities by their zeroth order slows the iteration convergence to linear. The iteration could converge quadratically if we retained the current approximation of the quantities. Such an iteration would be a version of Newton's method for solving equations. However, I know of no practical problem where this quadratic convergence can be practically realized for slow manifolds.

The main detail is then to solve homological equations of the restricted form

$$-\mathcal{L}\hat{\mathbf{v}} + \mathcal{E}\hat{\mathbf{g}} = \text{Res} \tag{5.10}$$

for some computed residual Res. Recognize that there are more unknowns than components in this equation; its solution is not unique. Freedom comes from the freedom to parametrize the slow manifold via the amplitudes $\mathbf{s}$ in an almost arbitrary manner. The freedom is only resolved by defining a precise meaning for the m amplitudes $\mathbf{s}$; for example, as described in the previous subsection, often one *chooses* a linear definition $\mathbf{s} = \mathcal{R}^T\mathbf{u}$. Then in an update we require the additional equation $0 = \mathcal{R}^T\hat{\mathbf{v}}$ in order to preserve the meaning of $\mathbf{s}$.

You may decide to choose other definitions for your amplitudes $\mathbf{s}$—such as an energy-related amplitude—such alternatives give rise to similar considerations to those that follow.

There are two approaches to solving (5.10) for corrections.

1. Numerically, it is easiest to adjoin the amplitude condition to the equation and solve

$$\begin{bmatrix} -\mathcal{L} & \mathcal{E} \\ \mathcal{R}^T & 0 \end{bmatrix} \begin{bmatrix} \hat{\mathbf{v}} \\ \hat{\mathbf{g}} \end{bmatrix} = \begin{bmatrix} \text{Res} \\ 0 \end{bmatrix}. \tag{5.11}$$

   This approach is also convenient for an algebraic approach to small systems.

2. However, in algebraic analysis of complex systems it is usually more convenient to adopt the following procedure, which we used for Burgers' PDE (and which may be familiar to you as part of other asymptotic methods). Rewrite (5.10) as $-\mathcal{L}\hat{\mathbf{v}} = \text{Res} - \mathcal{E}\hat{\mathbf{g}}$, and recognize that $\mathcal{L}$ is singular due to the zero eigenvalue of multiplicity m. Then choose the m components of $\hat{\mathbf{g}}$ to place the right-hand side in the range of $\mathcal{L}$; achieve this by taking the inner product of the equation with a basis of (adjoint) left eigenvectors $\mathbf{z}_j$: find these by solving $\mathcal{L}^T\mathbf{z}_j = 0$.[62] The inner products give a set of m *solvability conditions*[63], which upon forming matrix $\mathcal{Z} = [\mathbf{z}_j]$ are

$$(\mathcal{Z}^T\mathcal{E})\hat{\mathbf{g}} = \mathcal{Z}^T\,\text{Res}, \quad \text{equivalently} \quad \langle\mathcal{Z},\mathcal{E}\rangle\hat{\mathbf{g}} = \langle\mathcal{Z},\text{Res}\rangle. \tag{5.12}$$

   Having put the right-hand side in the range of $\mathcal{L}$, solve $-\mathcal{L}\hat{\mathbf{v}} = \text{Res} - \mathcal{E}\hat{\mathbf{g}}$ for $\hat{\mathbf{v}}$ in any convenient way, making the solution unique by accounting for the definition of the amplitudes $\mathbf{s}$.

   For example, in matrix algebra, one convenient way for solving $-\mathcal{L}\hat{\mathbf{v}} = \mathbf{R}$ is the following (once you know $\mathbf{R} = \text{Res} - \mathcal{E}\hat{\mathbf{g}}$ is in the range of $\mathcal{L}$); see Algorithm 5.3, line 8, below. First, normalize the left eigenvectors $\mathbf{z}_j$ to be orthogonal to the (right) eigenvectors $\mathbf{e}_j$ so that $\mathcal{Z}^T\mathcal{E} = I$ (this can always be done). Second, the amplitude requirement is in general that $\mathcal{R}^T\hat{\mathbf{v}} = \mathbf{r}$ for some $\mathbf{r}$: often $\mathbf{r} = 0$, and often $\mathcal{R} = \mathcal{Z}$. Then meet the required $-\mathcal{L}\hat{\mathbf{v}} = \mathbf{R}$ and $\mathcal{R}^T\hat{\mathbf{v}} = \mathbf{r}$ by solving, for any $c \neq 0$,

$$(-\mathcal{L} + c\mathcal{E}\mathcal{R}^T)\hat{\mathbf{v}} = \mathbf{R} + c\mathcal{E}\mathbf{r}; \tag{5.13}$$

   that is, $\hat{\mathbf{v}} = (-\mathcal{L} + c\mathcal{E}\mathcal{R}^T)^{-1}(\mathbf{R} + c\mathcal{E}\mathbf{r})$.

[62] For reference, Table 5.2 summarizes essential information about inner products, adjoints, and left eigenvectors.

[63] Often one chooses amplitudes $\mathbf{s}$ so that $\mathcal{R} = \mathcal{Z}$ and then $\mathcal{Z}^T\mathcal{E} = \langle\mathcal{Z},\mathcal{E}\rangle = I$, the identity. Further, the matrix appearing in (5.11) is then symmetric when $\mathcal{L}$ is symmetric.

Table 5.2. *Summary of essential linear algebra information on inner products, adjoints, left eigenvectors, and self-adjoint operators. Generalization of this summary to complex-valued vectors involves introducing complex conjugates.*

Consider vectors $\mathbf{u}, \mathbf{v}$ in finite dimensions, $\mathbf{u}, \mathbf{v} \in \mathbb{R}^n$; smooth functions $u(x), v(x)$ on some domain $a \leqslant x \leqslant b$; or potentially smooth functions on a multidimensional domain D. For simplicity, call all of these "vectors."

- Start by defining an *inner product* of two vectors $\mathbf{u}$ and $\mathbf{v}$, denoted $\langle \mathbf{u}, \mathbf{v} \rangle$, which must satisfy symmetry, $\langle \mathbf{u}, \mathbf{v} \rangle = \langle \mathbf{v}, \mathbf{u} \rangle$; linearity, $\langle a\mathbf{u} + b\mathbf{v}, \mathbf{w} \rangle = a \langle \mathbf{u}, \mathbf{w} \rangle + b \langle \mathbf{v}, \mathbf{w} \rangle$; and positivity, $\langle \mathbf{u}, \mathbf{u} \rangle \geqslant 0$, with equality only if $\mathbf{u} = \mathbf{0}$.

  Two examples are the usual dot product in finite dimensions or, for functions, and for a fixed specified weight function $r(x)$, the weighted integral $\langle u, v \rangle = \int_a^b u(x)v(x)r(x)\,dx$. This last turns the Sturm–Liouville integrals of Table 5.1 into inner products.
- The *length* $|\mathbf{u}|$ of a vector $\mathbf{u}$, or 2-norm, is defined by $|\mathbf{u}|^2 = \langle \mathbf{u}, \mathbf{u} \rangle$.
- Two (nonzero) vectors are *orthogonal* when the inner product $\langle \mathbf{u}, \mathbf{v} \rangle = 0$.
- Generalizing the concept of the transpose of a matrix, define the *adjoint* of a matrix/operator $\mathcal{L}$ to be the matrix/operator $\mathcal{L}^T$ such that for all vectors $\mathbf{u}$ and $\mathbf{v}$ the inner product $\langle \mathcal{L}\mathbf{u}, \mathbf{v} \rangle = \langle \mathbf{u}, \mathcal{L}^T\mathbf{v} \rangle$.

  For example, using the weight function $r(x) = 1$, integration by parts shows that the *adjoint* of the derivative operator ∂_x is its negative: $\partial_x^T = -\partial_x$. But such statements *must* be accompanied with statements about boundary conditions, such as "the operator ∂_x *with* boundary condition that the function is zero at $x = a$ has adjoint operator $-\partial_x$ *with* boundary condition that the function is zero at $x = b$."
- Then define the *left eigenvectors* $\mathbf{z}$, also called *adjoint eigenvectors*, of an operator $\mathcal{L}$ as eigenvectors of the adjoint operator: $\mathcal{L}^T\mathbf{z} = \lambda\mathbf{z}$ for eigenvalue λ. Importantly for deriving (5.12), left eigenvectors $\mathbf{z}_i$ are orthogonal to eigenvectors $\mathbf{v}_j$ of distinct eigenvalues: $\langle \mathbf{z}_i, \mathbf{v}_j \rangle = 0$ when eigenvalues $\lambda_i \neq \lambda_j$.
- Generalizing the concept of a symmetric matrix, a *self-adjoint* operator $\mathcal{L}$ is one that equals its adjoint: $\mathcal{L}^T = \mathcal{L}$. Consequently, for a self-adjoint operator the various eigenvectors and left eigenvectors are identical.

  For example, the Sturm–Liouville operators of Table 5.1 are self-adjoint.
- Eigenvectors and eigenvalues of a self-adjoint operator have all the nice properties of eigenvectors and eigenvalues of a symmetric matrix.

  For example, all the properties of the Sturm–Liouville operators of Table 5.1 match properties of symmetric matrices.

- Although $\mathcal{L}$ is singular, $(-\mathcal{L} + c\mathcal{E}\mathcal{R}^T)$ is invertible for general $c \neq 0$.
- Further, premultiplying equation (5.13) (left) by $\mathcal{Z}^T$ leads to

$$\begin{aligned}
\text{LHS} &= (-\mathcal{Z}^T\mathcal{L} + c\mathcal{Z}^T\mathcal{E}\mathcal{R}^T)\hat{\mathbf{v}} = (0 + cI\mathcal{R}^T)\hat{\mathbf{v}} = c\mathcal{R}^T\hat{\mathbf{v}}, \\
\text{RHS} &= \mathcal{Z}^T\mathbf{R} + c\mathcal{Z}^T\mathcal{E}\mathbf{r} = 0 + cI\mathbf{r} = c\mathbf{r}.
\end{aligned}$$

  Hence the amplitude requirement is met.
- Finally, because solutions must satisfy $\mathcal{R}^T\hat{\mathbf{v}} = \mathbf{r}$, then the two components $c\mathcal{E}\mathcal{R}^T\hat{\mathbf{v}}$ and $c\mathcal{E}\mathbf{r}$ cancel on the two sides of (5.13), so that the solution also satisfies the required $\mathcal{L}\hat{\mathbf{v}} = \mathbf{R}$.

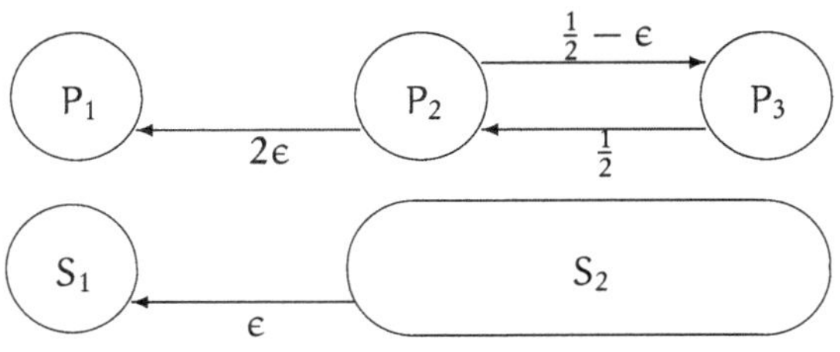

Figure 5.5. *Top is the three state continuous-time Markov chain of the example of section* 5.3.6. *The rates of change of state are shown on the arrows. Bottom is the two-state, quasi-stationary slow manifold model of the long-term evolution.*

The last step of each iteration is to update the approximations for the slow manifold shape and the evolution thereon.

5.3.5 ▪ Example: Burgers-like dynamics

The analysis of the system (5.1) in section 5.1 is an example of the general scheme. However, instead of the state vector $\mathbf{u}(t)$ being in n dimensions, the state vector is the function $u(x,t)$ in an infinite-dimensional Banach space of functions of x. The linear operator $\mathcal{L} = 1+\partial^2/\partial x^2$, with zero boundary conditions at $x = 0, \pi$. The slow subspace is spanned by the eigenvector $e_1(x) = \sin x$, which, by the self-adjoint nature of $\mathcal{L}$ (Table 5.2) is also the left eigenvector $z_1(x) = \sin x$, and which we implicitly chose to define the amplitude. The solvability condition in the update equation, $\mathcal{L}\hat{v} - \hat{g}\sin x = \text{Res}$, could be computed via the inner product $\langle \sin x, \text{Res}\rangle := \int_0^\pi \sin x \text{ Res } dx$; however, simply extracting the coefficient of $\sin x$ on the right-hand side is more efficient. In this example, the right-hand side of $\mathcal{L}\hat{v} = \hat{g}\sin x + \text{Res}$ is always expressible as a linear combination of sine functions; hence *an* inverse of operator $\mathcal{L}$ was straightforward. In general, such as the nonlinearity u_x^2, the residual is not immediately expressible as a linear combination of sine functions and one then has to work considerably harder to solve for $\hat{v}$. We avoid such hard work in this first example.

5.3.6 ▪ Example: Quasi-stationary probability distribution

Consider the continuous-time Markov chain, shown at the top of Figure 5.5, with three states, labeled P_1, P_2, and P_3. Let $p_i(t)$ denote the probability that the system is in state P_i at time t. The following set of linear ordinary differential equations governs the dynamics:[64]

$$\begin{aligned}
\dot{p}_1 &= 2\epsilon p_2\,,\\
\dot{p}_2 &= -\left(\tfrac{1}{2}+\epsilon\right)p_2 + \tfrac{1}{2}p_3\,,\\
\dot{p}_3 &= \left(\tfrac{1}{2}-\epsilon\right)p_2 - \tfrac{1}{2}p_3\,.
\end{aligned}$$

We should explicitly adjoin $\dot{\epsilon} = 0$ and write in matrix-vector form for a four-dimensional vector of variables $(\epsilon, p_1, p_2, p_3)$. However, treating fixed parameters explicitly as an extra variable is always a bit painful. Usually, as here, we adopt the convention of staying with the original three physical variables (p_1, p_2, p_3) and just treating parameter ϵ as

[64]Such systems are often called *master equations*. The analogue in systems of continuous states is called a Fokker–Planck equation or Kolmogorov forward equation.

small. That is, put ϵ terms in the nonlinearity $\mathbf{f}$. Thus write the system in matrix-vector form

$$\dot{\mathbf{p}} = \mathcal{L}\mathbf{p} + \mathbf{f}, \tag{5.14}$$

where

$$\mathbf{p} = \begin{bmatrix} p_1 \\ p_2 \\ p_3 \end{bmatrix}, \quad \mathcal{L} = \begin{bmatrix} 0 & 0 & 0 \\ 0 & -\frac{1}{2} & \frac{1}{2} \\ 0 & \frac{1}{2} & -\frac{1}{2} \end{bmatrix}, \quad \text{and } \mathbf{f} = \begin{bmatrix} 2 \\ -1 \\ -1 \end{bmatrix} \epsilon p_2 .$$

Equilibria? The origin, $\epsilon = p_1 = p_2 = p_3 = 0$, is the equilibrium about which we model the long-term dynamics.[65]

Linearization's leading approximation The linear operator $\mathcal{L}$ has a two-dimensional slow eigenspace corresponding to eigenvalue zero and has one eigenvalue of -1. As the nonlinearities are smooth, theory implies that there exists an emergent 2D slow manifold tangent to the slow subspace.

The slow subspace is spanned by, for example, the two eigenvectors $\mathbf{e}_1' = (1, 0, 0)$ and $\mathbf{e}_2' = (0, 1, 1)$. And if we were explicitly representing parameter ϵ as a dynamical variable, then we would also have a third eigenvector corresponding to ϵ pointing into a fourth dimension—but we do not do this explicitly. We fix the basis eigenvectors using the definition of the amplitudes s_1 and s_2. As drawn schematically at the bottom of Figure 5.5, we *choose* to define

- $s_1 = \Pr(\text{being in state } 1) = p_1$, and
- $s_2 = \Pr(\text{being in states 2 or } 3) = p_2 + p_3$.

That is, we choose

$$\mathbf{s} = \mathcal{R}^\mathsf{T}\mathbf{p} \quad \text{for the restriction} \quad \mathcal{R} = \begin{bmatrix} 1 & 0 \\ 0 & 1 \\ 0 & 1 \end{bmatrix}.$$

Then the appropriate basis vectors for our chosen amplitudes come from the columns of

$$\mathcal{E} = \mathcal{E}'(\mathcal{R}^\mathsf{T}\mathcal{E}')^{-1} = \mathcal{E}' \begin{bmatrix} 1 & 0 \\ 0 & 2 \end{bmatrix}^{-1} = \begin{bmatrix} 1 & 0 \\ 0 & 1/2 \\ 0 & 1/2 \end{bmatrix}.$$

Consequently, our chosen definition of the amplitudes forces our basis eigenvectors of the slow subspace $\mathbb{E}_0$ to be

$$\mathbf{e}_1 = (1, 0, 0) \quad \text{and} \quad \mathbf{e}_2 = \left(0, \tfrac{1}{2}, \tfrac{1}{2}\right).$$

As $\mathcal{L}$ is symmetric, the (adjoint) left eigenvectors can be the same as the above right eigenvectors: $\mathbf{z}_j = \mathbf{e}_j$. Let's make this choice (although other choices for $\mathbf{z}_j$ would also serve).

The first linear approximation is then that the slow manifold is the slow subspace

$$\mathbf{p} = \mathbf{e}_1 s_1 + \mathbf{e}_2 s_2 = \left(s_1, \tfrac{1}{2}s_2, \tfrac{1}{2}s_2\right) \quad \text{such that} \quad \dot{s}_1 = \dot{s}_2 = 0 .$$

[65]There are lots of other equilibria, a subspace, but this system is linear in the probabilities so all equilibria are equivalent. In a nonlinear system we generally like to take advantage of a subspace of equilibria, but in a linear system we gain nothing at the cost of adding complication.

This first approximation predicts that if the system has not yet been absorbed, $s_1 = 0$, then the system is in the quasi-stationary state of approximately $\mathbf{e}_2 = (0, \frac{1}{2}, \frac{1}{2})$.

Iteration improves the model Second, with this approximation, the residual of equation (5.14) is

$$\operatorname{Res}_{(5.14)} := -\dot{\mathbf{p}} + \mathcal{L}\mathbf{p} + \mathbf{f} = \left(1, -\tfrac{1}{2}, -\tfrac{1}{2}\right)\epsilon s_2 .$$

To solve $-\mathcal{L}\hat{\mathbf{p}} = \operatorname{Res}_{(5.14)} - \mathcal{E}\hat{\mathbf{g}}$, choose the evolution correction $\hat{\mathbf{g}}$ so the right-hand side is in the range of $\mathcal{L}$:

- the inner product of this equation with $\mathbf{z}_1 = \mathbf{e}_1$ determines the correction $\hat{g}_1 = \epsilon s_2$, and
- the inner product of this equation with $\mathbf{z}_2 = \mathbf{e}_2$ determines the correction $\hat{g}_2 = -\epsilon s_2$.

Upon setting $\hat{\mathbf{g}}$, the right-hand side vanishes, so there is no change to the slow manifold, $\hat{\mathbf{p}} = \mathbf{0}$, at least to terms linear in parameter ϵ. Consequently, center manifold theory assures us that exponentially quickly the system settles onto the slow manifold

$$\mathbf{p} = (p_1, p_2, p_3) = (1, 0, 0)s_1 + \left(0, \tfrac{1}{2}, \tfrac{1}{2}\right)s_2 + \mathcal{O}(\epsilon^2),$$

where the quasi-probabilities evolve slowly according to

$$\dot{s}_1 = \epsilon s_2 + \mathcal{O}(\epsilon^2) \quad \text{and} \quad \dot{s}_2 = -\epsilon s_2 + \mathcal{O}(\epsilon^2).$$

That is, the system rapidly reaches a quasi-stationary balance between states P_2 and P_3, with probability then "leaking" from states P_2 and P_3 into the absorbing state P_1 over long time scales of order $1/\epsilon$, as shown schematically in Figure 5.5.[66]

Quasi-stationary distributions identify ocean collectors You may have heard of the Sargasso Sea. Owing to surface currents, the Sargasso accumulates a high concentration of waste. Similar garbage collection areas occur in most oceans. The coherent oceanic gyres and eddies house low-dispersion regions where waste or biomass is trapped over long periods. These persistent slow dispersive regions are crucial, as they act as transport barriers. Such regions may be found as a quasi-stationary state of the dynamics of ocean transport, both in two dimensions (Froyland et al., 2007, e.g.) and in three dimensions (Dellnitz et al., 2009, e.g.).

5.3.7 ▪ Example: Computer algebra I

Now reconsider the two-variable system of Example 2.1:

$$\dot{u}_1 = -u_1 + u_2 - u_1^2 \quad \text{and} \quad \dot{u}_2 = +u_1 - u_2 + u_2^2. \tag{5.15}$$

[66]Interpreting the parameters s_i as probabilities leads to physical meaning for negative probabilities! Consider the evolution of the Markov chain when known to start from the state P_3, that is, the initial $\mathbf{p}(0) = (0, 0, 1)$, compared to the evolution starting from the mixed quasi-stationary state S_2, that is, $\mathbf{p}(0) \approx (0, \frac{1}{2}, \frac{1}{2})$. In the latter case the initial values for the model probabilities are $\mathbf{s} = (0, 1)$, as the system is on the slow manifold in state S_2. However, in the first case there is a delay in the transient settling on the slow manifold; during that delay the system is dominantly in state P_3 and so unable to "leak" probability into $S_1 = P_1$. There must be an effective delay in the realization of the long-term evolution $\dot{s}_1 \approx \epsilon s_2$. Such a delay must be reflected in an initial condition for the model "probabilities" $\mathbf{s}(0)$ such that $s_1(0)$ is negative! Exercise 5.2 helps you through the corresponding algebra.

Algorithm 5.3 *Iteration finds the slow manifold of the system* (5.15) *using the* `solve` *function.*

```
on div; off allfac; on revpri; factor s;
depend s,t; let df(s,t)=>g;
u1:=u2:=s; g:=0;
let s^5=>0;
for iter:=1:99 do begin
    res1:=-df(u1,t)-u1+u2-u1^2;
    res2:=-df(u2,t)+u1-u2+u2^2;
    soln:=solve({du1+du2=0
        ,+du1-du2+dg=res1
        ,-du1+du2+dg=res2},{dg,du1,du2});
    u1:=u1+sub(soln,du1);
    u2:=u2+sub(soln,du2);
    g:=g+sub(soln,dg);
    if {res1,res2}={0,0} then iter:=1000000;
end; end;
```

Figure 2.1 clearly shows that numerical simulations of the dynamics rapidly collapse onto a curve in the u_1u_2-plane. We now find the emergent long-term dynamics after this rapid collapse.

Equilibria? The origin is an equilibrium. In this system there is also an equilibrium at $\mathbf{u} = (-2, 2)$, which appears irrelevant, as it does not have a slow subspace.

Linearization determines existence Near the origin the linearized dynamics is

$$\dot{\mathbf{u}} \approx \mathcal{L}\mathbf{u} = \begin{bmatrix} -1 & 1 \\ 1 & -1 \end{bmatrix} \mathbf{u}\,.$$

Matrix $\mathcal{L}$ has eigenvalues $\lambda = 0, -2$ and, since the nonlinearities are smooth, the system has an emergent slow manifold (with transients roughly like e^{-2t}, as explicitly shown by the W ODE in (2.3) after a coordinate transform). The eigenvector corresponding to the zero eigenvalue is $\mathbf{e}_1 = (1, 1)$, so a linear approximation to the slow manifold is the slow subspace $\mathbf{u} = \mathbf{e}_1 s = (1, 1)s$.

First correction The initial approximation to the slow manifold is the slow subspace $\mathbf{u} = \mathbf{e}_1 s = (1, 1)s$ such that $\dot{s} = g(s) = 0$. Evaluating (5.15) for this approximation, we find the residuals $\text{Res}_1 = -s^2$ and $\text{Res}_2 = +s^2$. The equation for updates, $-\mathcal{L}\hat{\mathbf{u}} + \mathcal{E}\hat{g} = \text{Res}$, here becomes

$$+\hat{u}_1 - \hat{u}_2 + \hat{g} = -s^2, \quad -\hat{u}_1 + \hat{u}_2 + \hat{g} = +s^2, \quad \hat{u}_1 + \hat{u}_2 = 0\,.$$

The third equation comes from the parametrization requirement that we choose $\langle \mathbf{e}_1, \mathbf{u} \rangle = u_1 + u_2 = 2s$; that is, we choose $\mathbf{r} = \mathbf{e}_1/2$. One readily sees that $\hat{g} = 0$ and $\hat{u}_1 = -\hat{u}_2 = -s^2/2$ satisfy the three required equations.

Iterate to construct Algorithm 5.3 constructs the slow manifold model to $\mathcal{O}(s^5)$. Variable s parametrizes the slow manifold so evolution thereon is $\dot{s} = g$. The initial linear approximation, $\mathbf{u} = (1,1)s$, is set with the two components `u1:=u2:=s;`. The core of the algorithm is the evaluation of the residuals for each of the two equations in the system (5.15). Then the `solve` operator finds corrections denoted by the prefix `d`; the last two of the equations to solve are the two components of (5.10); the first equation in the list is the amplitude requirement that the dot product $(1,1)\cdot\hat{\mathbf{u}} = \hat{u}_1 + \hat{u}_2 = 0$, coming from the chosen amplitude definition $(1,1)\cdot\mathbf{u} = 2s$. The algorithm then adds these corrections to the slow manifold model. Finally, exit the loop when the residuals are both zero. Algorithm 5.3 finds that the curved slow manifold apparent in Figure 2.1 is

$$u_1 = s - \tfrac{1}{2}s^2 + \tfrac{3}{8}s^4 + \mathcal{O}(s^5), \quad u_2 = s + \tfrac{1}{2}s^2 - \tfrac{3}{8}s^4 + \mathcal{O}(s^5),$$

where $\dot{s} = s^3 + \mathcal{O}(s^5)$. This agrees with the long-term nonlinear growth seen in Example 2.1.

Other parametrizations But this is only one possible description among infinitely many. We could instead choose to parametrize the slow manifold and the evolution thereon by $u_1 = s$—for some reason we choose $\mathbf{r}_1 = (1,0)$. The linear approximation to the slow manifold is the same in this case: $\mathbf{u} \approx (1,1)s$. The only difference lies in computing the updates: instead of requiring $\hat{u}_1 + \hat{u}_2 = 0$ we here require $\hat{u}_1 = 0$ so that $u_1 = s$ is maintained throughout the iteration. Change the algorithm, execute, and find that the slow manifold could also be described as

$$u_1 = s, \quad u_2 = s + s^2 + s^3 + \tfrac{1}{2}s^4 + \mathcal{O}(s^5), \quad \text{where } \dot{s} = s^3 + \tfrac{1}{2}s^4 + \mathcal{O}(s^5).$$

This is the same manifold, the same evolution, it is just a different parametrization.

Or perhaps you might want the parameter s to be proportional to the signed distance from the origin: $s^2 = (u_1^2 + u_2^2)/2$. The linear approximation $u_1 = u_2 = s$ satisfies this condition. Seek corrections that maintain the condition by also solving $s^2 = [(u_1 + \hat{u}_1)^2 + (u_2 + \hat{u}_2)^2]/2 \approx (u_1^2 + u_2^2)/2 + u_1\hat{u}_1 + u_2\hat{u}_2 \approx (u_1^2 + u_2^2)/2 + s\hat{u}_1 + s\hat{u}_2$. That is, in the solve step, change the amplitude equation to $\hat{u}_1 + \hat{u}_2 = [s^2 - (u_1^2 + u_2^2)/2]/s$. Find that the slow manifold is equivalently

$$u_1 = s - \tfrac{1}{2}s^2 - \tfrac{1}{8}s^3 + \tfrac{1}{2}s^4 + \mathcal{O}(s^5), \quad u_2 = s + \tfrac{1}{2}s^2 - \tfrac{1}{8}s^3 - \tfrac{1}{2}s^4 + \mathcal{O}(s^5),$$

where $\dot{s} = s^3 + \mathcal{O}(s^5)$.

5.3.8 ▪ Example: Computer algebra II

Algorithm 5.4 also constructs the slow manifold of the system (5.15). However, it uses the solvability condition (5.12), the solution equation (5.13), and the matrix facility in Reduce. Given the vector of residuals, $\text{Res} = (\text{Res}_1, \text{Res}_2)$, here the correction to the evolution is $\hat{g} = \mathbf{z}_1 \cdot \text{Res} = (\tfrac{1}{2}, \tfrac{1}{2}) \cdot \text{Res}$—the left eigenvector $\mathbf{z}_1 \propto \mathbf{e}_1$, as the matrix $\mathcal{L}$ is symmetric. Then knowing we require $\hat{u}_1 + \hat{u}_2 = 0$ to maintain the definition of s, namely $\mathbf{z}_1^T\hat{\mathbf{v}} = 0$, we invert the modified matrix to find the update. Algorithm 5.4 generates the same model as before.

5.4 ▪ Case study: Small patches model macroscales

Kevrekidis and colleagues (Kevrekidis and Samaey, 2009, e.g.) are developing multiscale analysis to empower microscopic computational simulations to address macroscopic issues of interest. The generic endeavor is termed "equation-free" because the methods use

Algorithm 5.4 *Iteration finds the slow manifold of the system* (5.15) *using the solvability condition* (5.12) *and the solution equation* (5.13).

```
on div; off allfac; on revpri; factor s;
ll:=mat((-1,1),(1,-1));
e1:=tp mat((1,1)); z1:=e1/2;
depend s,t; let df(s,t)=>g;
u:=e1*s; g:=0;
let s^5=>0;
for iter:=1:9 do begin
    resd:=-df(u,t)+ll*u-tp mat((u(1,1)^2,-u(2,1)^2));
    gd:=(tp z1)*resd; g:=g+gd(1,1);
    u:=u+1/(-ll+e1*(tp z1))*(resd-e1*gd);
    if resd=tp mat((0,0)) then iter:=1000000;
end; end;
```

only computational simulations of the physics of the application: the methods avoid having to know a macroscale algebraic model. The gap-tooth scheme empowers simulations of detailed microscopic dynamics on small patches of space to be coupled together over large intervening spatial gaps to predict the macroscale dynamics over large spatial regions. Here we demonstrate one way that center manifold analysis certifies the efficacy of the gap-tooth scheme.

Scenario Imagine that you have a highly detailed computational simulation of some complicated spatiotemporal system. But the computational simulation is far too slow to simulate over the spatial domain of your interest. The gap-tooth scheme simulates the system on small patches of space, with appropriate coupling across unsimulated space. For simplicity here, suppose that the spatial domain is 1D, say $-1 \leqslant x \leqslant 1$, and that there is only one evolving field of macroscopic interest, $u(x,t)$, say, as shown in Figure 5.6. For extreme simplicity here, let's use just one small patch, say $-r \leqslant x \leqslant r$ for small r (Figure 5.6 uses the case $r = 0.5$ for clarity). In implementation we would only compute on the small fraction r of the domain; on most of the domain we compute nothing—just interpolate. The gap-tooth scheme poses that it is a "good thing" to couple this small patch of simulation to the distant boundaries at $x = \pm 1$ with classic Lagrange interpolation (Roberts and Kevrekidis, 2007).

Challenge Approximate the slow manifold of macroscopic dynamics to establish that such cross-space coupling is appropriate.

How do we analyze the computational simulation? We do not know what the simulation is, but let's suppose that it is in the class of reaction-diffusion equations. Then analysis of a class of reaction-diffusion equations provides answers that apply to all simulations in that class. Here consider all simulations that are effectively in the class of the reaction-diffusion PDE

$$\frac{\partial u}{\partial t} = K\frac{\partial^2 u}{\partial x^2} + f(u) \tag{5.16}$$

for some diffusion constant K and some reaction f. (Figure 5.6 simulated this system with diffusivity $K = 0.02$ and reaction $f = 0.7(u - u^3)$.) Further, suppose the simulation requires us to provide Dirichlet boundary conditions on the edge of its simulation domain.

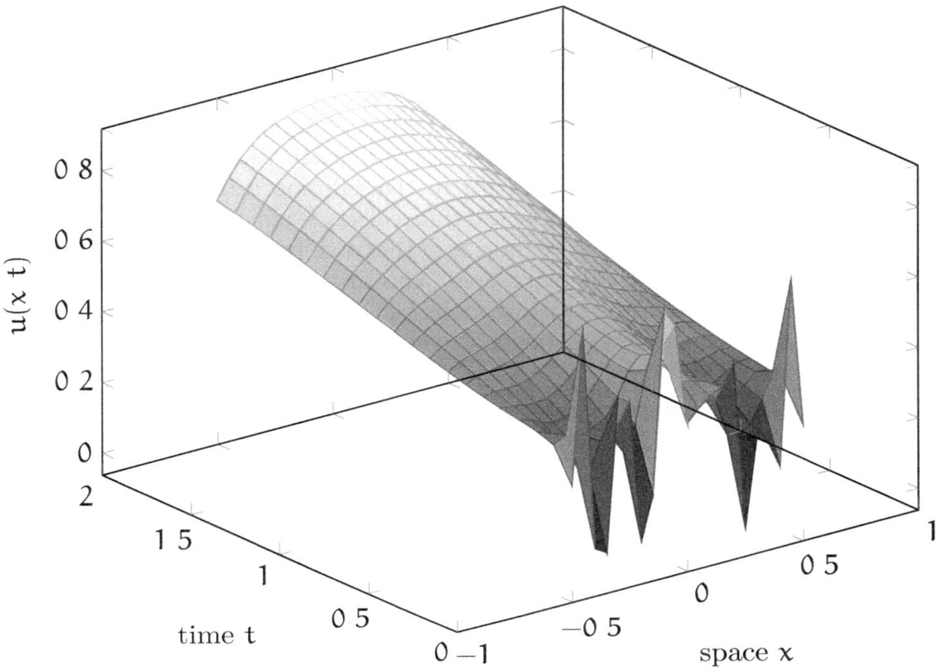

Figure 5.6. *A space-time simulation for a field* $u(x,t)$ *of a reaction-diffusion PDE on one small patch of space,* $-0.5 < x < 0.5$, *coupled to domain boundary conditions of* $u(\pm 1, t) = 0$. *"Random" initial conditions rapidly smooth in the microscale dynamics until for* $t \geqslant 1$ *the field in the patch is a smooth sample of the macroscale field.*

Here the simulation is only on the small patch $-r \leqslant x \leqslant r$, so we would need to provide values for $u(\pm r, t)$.

Interpolation determines the macroscopic field The macroscale has fixed field at the macroscale boundaries $x = \pm 1$ of, say, zero. The macroscale field also has to "pass through" the field u in the patch: define the macroscale patch value

$$U(t) := u(0,t), \tag{5.17}$$

that is, U is the patch field sampled at its midpoint. Then the gap-tooth scheme interpolates through the three points $(\pm 1, 0)$ and $(0, U)$ with the parabola $u(x,t) = (1-x^2)U(t)$.

Effective patch boundary conditions The parabolic interpolation provides boundary conditions for the simulation (as used for Figure 5.6). The parabola estimates the field on the edge of the patch as $u(\pm r, t) = (1 - r^2)U(t)$. But equation (5.17) defines $U(t) = u(0,t)$, so the parabolic interpolation of the gap-tooth scheme effectively provides boundary conditions to the simulator of

$$u(\pm r, t) = (1 - r^2)u(0,t). \tag{5.18}$$

Unusually, these are nonlocal boundary conditions: they relate the value of the field at the patch edges to that in the middle. It is the gap-tooth scheme that effectively provides these conditions. Our challenge is to understand the nature of solutions of generic reaction-diffusion PDEs (5.16) when supplied with the boundary conditions (5.18). Then we predict the nature of the gap-tooth scheme when used for this class of simulators.

Where is the slow manifold? Let's think about the dynamics of the reaction-diffusion PDE (5.16) on a small patch. Being a small patch, the diffusion will rapidly mix or smooth the field $u(x,t)$ across the patch: the diffusion is the dominant microscale dynamics. That is, for small patches the diffusive decay is rapid, as seen in the decaying transients of Figure 5.6. We measure the size of the patch by r: as the patch gets smaller, $r \to 0$, the decay gets more and more rapid. This smacks of a singular perturbation problem. Our approach to singular perturbations is to rescale the problem to make it regular. Only then will we clearly extract the slow manifold dynamics, if any.

Rescale space and time The patch size is proportional to r, so adopt the natural microscale space scaling $x = r\xi$, so that $\partial/\partial x = (1/r)\partial/\partial\xi$. The cross-patch diffusion time is proportional to r^2, so adopt the natural, fast, microscale time scale of $\tau = r^2 t$ so that $\partial/\partial t = (1/r^2)\partial/\partial\tau$. Multiplying by r^2, the reaction-diffusion PDE (5.16) and boundary conditions (5.18) then become

$$\frac{\partial u}{\partial \tau} = K\frac{\partial^2 u}{\partial \xi^2} + r^2 f(u), \quad u(\pm 1, t) = (1 - r^2)u(0,t), \tag{5.19}$$

for the microscale field $u(\xi, \tau)$. In this form we readily find a slow manifold model for small patch size r. The model predicts the emergent dynamics of the gap-tooth scheme.

Equilibria? One set of equilibria in the microscale system (5.19) is clearly the field $u =$ constant when $r = 0$. We use this subspace of equilibria as the base of the slow manifold modeling. The definition (5.17) of the macroscale variable, that $U = u(0,t)$, implies we parametrize the equilibria by U, that is, as $u(\xi,\tau) = U$ and $r = 0$. The resultant slow manifold model, being based upon this subspace of equilibria, is usefully *global* in the macroscale variable U.

Linearization establishes existence Straightforwardly linearize the microscale problem (5.19) by substituting $u = U + \hat{u}(\xi,\tau)$ for small $\hat{u}$ and r. Omit products of small quantities immediately to derive

$$\frac{\partial \hat{u}}{\partial \tau} = K\frac{\partial^2 \hat{u}}{\partial \xi^2}, \quad \hat{u}(\pm 1, \tau) = \hat{u}(0,\tau). \tag{5.20}$$

Seek solutions $\hat{u} = e^{\lambda K \tau} v(\xi)$ for eigenvalues λ and eigenfunctions $v(\xi)$. Substituting and factoring out the exponential leads to the eigenproblem

$$\frac{d^2 v}{d\xi^2} - \lambda v = 0, \quad v(\pm 1) = v(0). \tag{5.21}$$

First observe that for eigenvalue $\lambda = 0$, eigenfunction v must be linear in ξ, but the boundary conditions then force v to be constant. An eigenvalue is $\lambda = 0$ and the corresponding eigenfunction is $v = 1$. This eigensolution corresponds to the subspace of equilibria.

Second, we expect spatially trigonometric solutions of the constant-coefficient eigenproblem (5.20), so set eigenvalue $\lambda = -k^2 < 0$, whence a general solution of the ODE is $v = A\cos k\xi + B\sin k\xi$, where we see that k denotes a microscale wavenumber. The patch boundary conditions in (5.20) then require $A\cos k \pm B\sin k = A$. Two cases arise.

- Subtracting the two boundary conditions leads to $2B\sin k = 0$ with nontrivial solutions ($B \neq 0$) when wavenumber $k = \pi, 2\pi, 3\pi, 4\pi, \ldots$. That is, microscale eigenvalues are $\lambda_n = -n^2\pi^2$ with corresponding eigenfunction $\sin n\pi\xi$ for $n = 1, 2, 3, 4, \ldots$.
- Adding the two boundary conditions leads to $2A(\cos k - 1) = 0$ with nontrivial solutions ($A \neq 0$) when wavenumber $k = 2\pi, 4\pi, 6\pi, \ldots$. That is, the microscale eigenvalues $\lambda_n = -n^2\pi^2$ for even n also have corresponding eigenfunction $\cos n\pi\xi$.[67]

Evidently, the spectrum of the linearized problem (5.20) is of one zero eigenvalue, and all the rest negative ($\lambda \leqslant -\pi^2$). Theorems 4.3 and 4.5 then establish the existence and emergence of a slow manifold for the reaction-diffusion PDE (5.19). Further,

- when $r = 0$, the slow manifold is tangent to the field $u =$ constant;
- the slow manifold is global in the field amplitude U and local in the small patch size r; and
- as the slowest decay is $e^{-K\lambda_1\tau} = e^{-K\pi^2\tau}$, the slow manifold emerges on a microscale time $1/(K\pi^2)$, which in the original time is emergence on the fast time scale $r^2/(K\pi^2)$.

A first approximation proves effective Let's find the first nontrivial approximation to the slow manifold of the microscale reaction-diffusion problem (5.19). Seek microscale field $u(\xi,\tau) = U + \hat{u}(\xi, U, r)$ such that $dU/d\tau = \hat{g}(U,r)$ for small $\hat{u}$ and $\hat{g}$, and where we treat the patch size r as also small. Substitute this ansatz into the system (5.19), expand reaction f in a Taylor series about $u = U$, and omit products of small quantities (except keeping r^2 but not higher powers), leading to

$$\hat{g} = K\frac{\partial^2\hat{u}}{\partial\xi^2} + r^2 f(U), \quad \hat{u}(\pm 1) = \hat{u}(0) - r^2 U\,.$$

This ODE and boundary conditions determine the first nontrivial approximation to the slow manifold. But we also must recognize our amplitude definition, that $U = u(0,\tau)$, which additionally requires $0 = \hat{u}$ at $\xi = 0$.

Obtain a general solution of the ODE by integrating twice to find that the microscale field may be

$$\hat{u} = \frac{1}{2K}\left[\hat{g} - r^2 f(U)\right]\xi^2 + A\xi + B\,.$$

The amplitude condition determines $B = 0$. The difference of the two boundary conditions determines $A = 0$. The sum of the two boundary conditions leads to $\frac{1}{K}[\hat{g} - r^2 f(U)] = -2r^2 U$, which determines the slow manifold evolution correction $\hat{g} = r^2 f(U) - 2Kr^2 U$. That is, the first nontrivial approximation to the slow manifold and

[67] Because the equation $\cos k - 1 = 0$ has double roots at $k = n\pi$ for even n there should be another eigenfunction other than $\cos n\pi\xi$. There is: it is the *generalized* eigenfunction $\xi\sin n\pi\xi$. The existence of such generalized eigenfunctions brings out that the eigenproblem (5.20) is not self-adjoint. Thus there may well be other eigenvalues and eigenfunctions other than those identified here. Inform me if you find any.

the evolution thereon is that the microscale field

$$u \approx U - r^2\xi^2 U \quad \text{such that} \quad \frac{dU}{d\tau} \approx -2Kr^2U + r^2 f(U). \tag{5.22}$$

In terms of macroscale variables, the field $u \approx (1-x^2)U$, which matches the interpolation from the distant boundaries at $x = \pm 1$. The evolution on the slow macroscale time is correspondingly $dU/dt \approx -2KU + f(U)$, which contains the effects of the reaction $f(U)$ and a decay due to diffusion to the distant boundary values $u(\pm 1, t) = 0$.

Evidently the gap-tooth scheme behaves appropriately when applied to simulate systems in the class of systems that manifest themselves on the macroscale as the reaction-diffusion PDE (5.16). This conclusion applies no matter how small the patch size r: in principle, we are empowered to simulate the system by computing only over an arbitrarily small part of the domain and still make reasonable macroscale predictions.

Our analysis of this case study only invoked one patch in one spatial dimension—for simplicity. In applications, the gap-tooth scheme would involve many small patches spread out over the domain, in a multidimensional spatial domain. Analysis should be correspondingly generalized (Roberts et al., 2013, e.g.). We leave such general analysis to another chapter.

Higher-order corrections determine patch effects With the computer algebra iteration of Algorithm 5.1 we straightforwardly compute the slow manifold to higher order in patch size r. Such higher-order computation shows the leading-order errors in the gap-tooth scheme with parabolic interpolation.

Algorithm 5.5 details a Reduce program that constructs the slow manifold model for the general advection-diffusion PDE (5.1).

1. The preliminaries are the following.

    - Improve printing by factoring in terms of the small parameter, the small patch size r.
    - Define the operator `linv` to act as the inverse of $\tilde{\mathcal{L}}v = d^2v/d\xi^2$ with boundary conditions $v(0) = 0$ (to preserve the midpatch value) and $v(+1) = v(-1)$ (part of the homogeneous patch boundary conditions). Operator $\tilde{\mathcal{L}}$ is a desingularized modification of the singular operator $\mathcal{L}v = d^2v/d\xi^2$ with patch homogeneous boundary conditions $v(\pm 1) = v(0)$: the modification is to apply only the difference of the two patch boundary conditions, and to adjoin the amplitude condition. The inverses of $\mathcal{L}$ and $\tilde{\mathcal{L}}$ are identical when applied to right-hand sides in the range of $\mathcal{L}$, as we do.
        - Recall that declaring the operator `linear` tells Reduce to expand sums and products in the first argument and to only leave functions of the second argument inside the operator.
        - The `let` statement defines the solution for any power of ξ in the residual of the PDE (5.19): the double tilde on exponent `~~p` matches exponents *including* the trivial exponent of one; however, it does not match the exponent of zero, so additionally define an action for ξ^0.
    - Similarly define the linear operator `solg` to generate appropriate updates to the evolution for any computed residual of the PDE (5.19).

    The formula for `solg` is mysterious. It comes from the surprising adjoint of the linear operator $\mathcal{L}$. The surprise is in the boundary conditions:

Algorithm 5.5 *Iteration finds the slow manifold of the macroscopic dynamics of the general reaction-diffusion* PDE (5.19) *on a small patch, size* r, *coupled to "distant" boundaries.*

```
factor r;  % improve printing
% solves v''=RHS s.t. v(0)=0 and v(+1)=v(-1)
operator linv;  linear linv;
let {linv(xi^~~p,xi)=>(xi^(p+2)-(1-(-1)^p)*xi/2)/(p+1)/(p+2)
    , linv(1,xi)=>(xi^2)/2 };
% solvability condition operator
operator solg;  linear solg;
let {solg(xi^~~p,xi)=>(1+(-1)^p)/(p+2)/(p+1)
    , solg(1,xi)=>1 };
% parametrize slow manifold with uu
depend uu,tau;  let df(uu,tau)=>g;
% Taylor series of reaction f(u) about u=U
operator f;  depend f,uu;
let df(f(~k),~a)=>f(k+1)*df(uu,a);
fu:=f(0)+for k:=1:3 sum f(k)*(u-uu)^k/factorial(k);
% linear approximation is constant field in patch
u:=uu;  g:=0;
% truncate asymptotic expansion in patch size r
let r^5=>0;
% iterate until equations satisfied
for iter:=1:9 do begin
  respde:=-df(u,tau)+kk*df(u,xi,2)+r^2*fu;
  resrbc:=sub(xi=+1,u)-(1-r^2)*sub(xi=0,u);
  reslbc:=sub(xi=-1,u)-(1-r^2)*sub(xi=0,u);
  resamp:=sub(xi=0,u)-uu;
  g:=g+(gd:=-kk*(resrbc+reslbc)+solg(respde,xi));
  u:=u+linv(-respde+gd,xi)/kk+xi*(reslbc-resrbc)/2;
  if {respde,resrbc,reslbc,resamp}={0,0,0,0}
  then write iter:=10000+iter;
end;
```

$\mathcal{L}^T z = d^2z/d\xi^2$ such that $z(\pm 1) = 0$, $z(\xi)$ is continuous at $\xi = 0$, but the derivative $dz/d\xi$ jumps at $\xi = 0$. Consequently the left eigenvector z of the zero eigenvalue is $z = 1-|\xi|$. Then `solg`$(\xi^p, \xi) = \langle z, \xi^p \rangle = 2/(p+1)/(p+2)$ for even exponents p, and zero for odd.

- Establish first that the variable `uu` is to depend upon time τ, as we choose U as the time-dependent amplitude in the modeling, and second that time derivatives, `df(uu,tau)`, are to be replaced by the value of `g`, as `g` is to store the current approximate model evolution.

2. The nonlinear reaction in the reaction-diffusion PDE (5.19) is to be general $f(u)$. But we do know that $u \approx U$, so let's code into `fu` the Taylor series that $f(u) = f(U)+f'(U)(u-U)+f''(U)(u-U)^2/2+\cdots$. In the computer algebra we choose to represent the derivative $f^{(k)}(U)$ by the symbol `f(k)`. Reduce must then be told that these derivatives depend upon $U =$ `uu`, and it must be told the chain rule for any derivatives.

3. Assign the linear approximation of the slow manifold to be the initial value of the variables u and g.

4. Perform the iterations in a safe for-loop. Control the truncation of the asymptotic approximation by telling Reduce to discard any factor in r^5 so that all expressions are computed to an error of $\mathcal{O}(r^5)$.

   (a) Compute the residuals of the required equations as a very direct translation of the algebraic form of the reaction-diffusion PDE (5.19), the patch boundary conditions (5.19), and the "amplitude" condition (5.17).

   (b) Choose the correction to the evolution g: I found the coefficients coded here by the quick and dirty method of undetermined coefficients of section 1.3.

   (c) Solve for the correction to the microscale field u and update the current approximation: again I used the method of undetermined coefficients to finalize the code.

   (d) Terminate the iterative for-loop when all four residuals are zero to the specified order of error.

The computer algebra of Algorithm 5.5 successfully terminates in three iterations. It predicts that the microscale field in the patch is

$$u = (1 - r^2\xi^2)U + r^4\tfrac{1}{12}\xi^2(1-\xi^2)\left[2KU - f(U) + Uf'(U)\right] + \mathcal{O}(r^5). \tag{5.23}$$

The first term is the parabolic interpolation, modified by the second term, which captures some subpatch interactions between reaction and diffusion. The corresponding evolution on the slow manifold, upon restoring the original time $t = r^2\tau$, predicts the dynamics of the gap-tooth model to be

$$\frac{dU}{dt} = \left[-2KU + f(U)\right] + \tfrac{1}{6}r^2\left[-2KU + f(U) - Uf'(U)\right] + \mathcal{O}(r^3). \tag{5.24}$$

This governs the emergent dynamics of the gap-tooth scheme on any PDE in the class of reaction-diffusion PDEs (5.16). The leading-order appropriate model is modified by finite-sized patch effects, the $\mathcal{O}(r^2)$ terms: such modifications would improve accuracy if we improved the interpolation (Roberts and Kevrekidis, 2007).

Very high orders indicate convergence One limit of interest is when the patch covers the entire domain. When the patch size $r \to 1$ the patch boundary conditions (5.18), based upon parabolic interpolation, reduce to the exact physical boundary conditions that $u = 0$ at $x = \pm 1$. Thus our slow manifold (5.23)–(5.24) should reproduce the exact dynamics upon setting the patch size to one.

Setting the patch size $r = 1$ is problematic because the slow manifold is based upon small patch size r. Nonetheless center manifold theory supports the existence of the slow manifold model at finite r. The question is: does this domain of existence extend to $r = 1$? We find evidence for convergence by looking at high-order expansions.

Computer algebra readily computes to high order. However, the nonlinear reaction $f(u)$ generates a combinatorial explosion in terms. To eliminate such an explosion, neglect reaction and restrict attention to the linear diffusion PDE (5.19) with $f = 0$. Algorithm 5.5, with high-order truncation in patch size r, then computes the following power series for the evolution on the slow manifold:

$$\frac{dU}{dt} = -KU\left[2 + \tfrac{1}{3}r^2 + \tfrac{4}{45}r^4 + \tfrac{1}{35}r^6 + \tfrac{16}{1575}r^8 + \tfrac{8}{2079}r^{10} + \tfrac{32}{21021}r^{12} + \mathcal{O}(r^{13})\right].$$

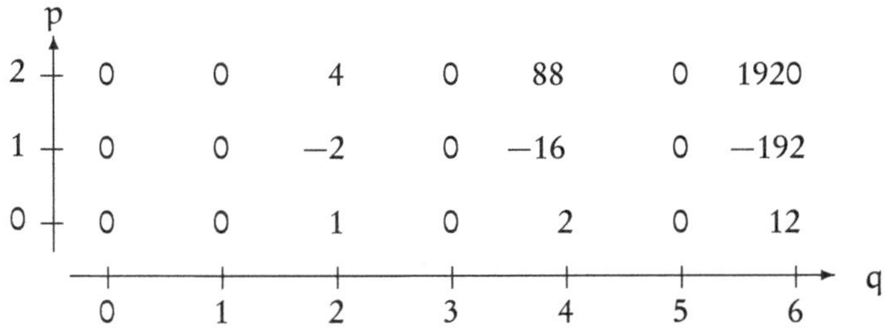

Figure 5.7. *Lay out the multinomial coefficients of approximations to the slow manifold $\mathcal{M}$ on a grid in the* nm*-plane.*

Summing these terms at patch size $r = 1$ gives the approximate evolution of

$$\frac{dU}{dt} \approx -2.4663\,KU, \quad \text{to compare with} \quad \frac{dU}{dt} = -\frac{\pi^2}{4}KU = -2.4674\,KU$$

of the gravest (slowest decaying) mode $u = \cos(\pi x/2)\exp(-\pi^2 Kt/4)$ of the diffusion PDE on the entire domain $-1 < x < 1$. This truncation of the slow manifold model is correct to nearly three decimal places even when evaluated at the extreme patch size of $r = 1$.[68]

5.5 ▪ Flexible truncations empower adaptable modeling

Most modeling problems have fixed parameters: the strength of gravity, the Reynolds number, the coefficients of chemical interactions, heating rates, coupling constants, and so on. In specific applications some parameters are more important than others, and thus we would want to model the effects of some parameters more accurately than others. Similarly with dynamical variables: we want to model some dynamical variables more accurately than others, and more accurately than some parameters, and vice versa. Such modeling requires flexible truncation of the asymptotic approximation to center manifold models. Theorem 4.10 empowers us to model flexibly.

Example 5.1 (introduction). Consider the specific pitchfork bifurcation problem (4.19): $\dot{x} = ax - xy$ and $\dot{y} = -y + x^2$. The general form of approximations to the shape of the slow manifold is that of a multinomial series in a and x, namely

$$\begin{aligned} y = h(a,x) &= (1 - 2a + 4a^2 + \cdots)x^2 \\ &\quad + (2 - 16a + 88a^2 + \cdots)x^4 \\ &\quad + (12 - 192a + 1920a^2 + \cdots)x^6 + \cdots. \end{aligned}$$

This is the double sum $h = \sum_{p,q=0}^{\infty} h_{pq} a^p x^q$. Graphically, place these coefficients h_{pq} on a grid in the nm-plane as shown in Figure 5.7, called a Newton diagram by Murdock (2003). The diagram has zero elements in the bottom left because the slow manifold $\mathcal{M}$ is tangent to the slow subspace $y = 0$. Statements about the error in $h^{(1)}$ are then determined by the two terms that occur in

[68] A Domb and Sykes (1957) plot of the ratio of the coefficients indicates that the domain of convergence in patch size is at least $r < \sqrt{2}$.

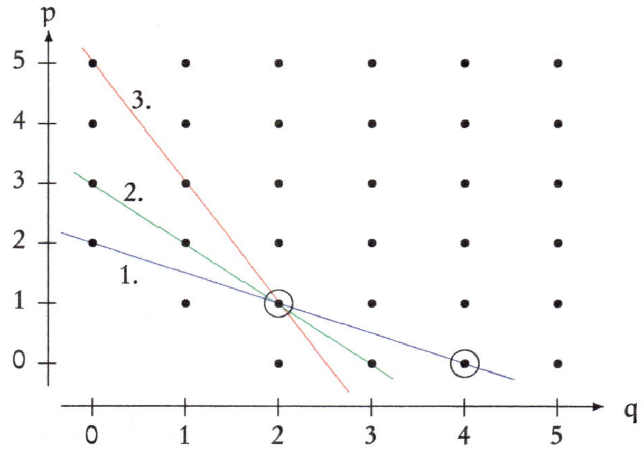

Figure 5.8. *Identifies the nonzero coefficients in* $h^{(2)} - h^{(1)}$, *circles, which indicate the error in the approximation* $h^{(1)}$ *to* $\mathcal{M}$; *also plotted are three lines each indicating a different allowable statement of the error in* $h^{(1)}$: *blue* 1, $p/2 + q/4 = 1$; *green* 2, $p/3 + q/3 = 1$; *and red* 3, $p/5 + 2q/5 = 1$.

$h^{(2)} - h^{(1)} = -2ax^2 + 2x^4$, which are shown circled in Figure 5.8. The Flexible Approximation Theorem 4.10 indicates that *any* straight line in the qp-plane which is to the left of and below both of these two terms is the basis of a valid statement of the error in $h^{(1)}$. For example, Figure 5.8 illustrates the following valid cases:

1. the (blue) line $p/2 + q/4 = 1$ shows $h = h^{(1)} + \mathcal{O}(a^2 + x^4)$;
2. the (green) line $p/3 + q/3 = 1$ shows $h = h^{(1)} + \mathcal{O}(a^3 + x^3)$;
3. the (red) line $p/5 + 2q/5 = 1$ shows $h = h^{(1)} + \mathcal{O}(a^5 + x^{5/2})$.

The Flexible Approximation Theorem 4.10 asserts that one may approximate the center manifold $\mathcal{M}$, and the long-term dynamics on $\mathcal{M}$, to an order of accuracy that is different for the parameters and the dynamical variables of the model. Such flexibility is useful because although sometimes we are only interested in the leading-order effect of a parameter, sometimes we are interested in high-order expansions in an artificial parameter and low order with respect to dynamic variables.

Section 5.5.1 reminds us of the extreme case when a slow manifold model is global in some variables and local in others. In such cases there is "no order of error" in the global variables, only in the local variables.

5.5.1 ▪ Base global models upon a subspace of equilibria

Global models often occur in chemical reactions, for example. Typically some chemical rate constants are large, corresponding to rapid reactions, and some are small, corresponding to slow reactions. Adopting the time scale of the fast reactions, write the chemical system as[69]

$$\dot{x} = \epsilon f(x, y) \quad \text{and} \quad \dot{y} = g(x, y),$$

[69]Many call such systems "singularly perturbed systems" (see section 4.6.3) and apply simpler methods of analysis and approximation to obtain leading-order approximate models.

where ϵ is the small ratio of the time scales, and where, for fixed x, the y equation would have solutions rapidly decaying to some manifold of quasi-equilibria. The x equation then governs the long-term evolution along this manifold.

We met some global models in Example 4.7 on nonlinear mixing; in equivalent modeling, (4.24) and (4.25), of the prototype pitchfork bifurcation; in section 4.7 on a chemical kinetics problem; and in section 4.8 on the Stefan problem of melting. Carr (1981) describes the following realistic chemical example.

Example 5.2. Consider the toy system

$$\dot{x} = \epsilon(x - xy) \quad \text{and} \quad \dot{y} = -y + x^2 . \tag{5.25}$$

At $\epsilon = 0$ there is a whole manifold of equilibria $\mathcal{M}_0$, $y = x^2$. Center manifold theory supports the model around each and every point on $\mathcal{M}_0$; just adjoin $\dot{\epsilon} = 0$. On $\mathcal{M}_0$, and recognizing it is approximate, then the evolution $\dot{x} = \epsilon x(1 - x^2) + \mathcal{O}(\epsilon^2)$.

A further iteration refines the slow manifold: from (5.25) and using the chain rule,

$$\begin{aligned} y &= x^2 - \dot{y} \\ &= x^2 - 2x\dot{x} \\ &= x^2 - 2\epsilon x^2(1 - x^2) + \mathcal{O}(\epsilon^2). \end{aligned}$$

Substituting this new approximation into (5.25), the corresponding evolution

$$\begin{aligned} \dot{x} &= \epsilon\left\{x - x\left[x^2 - 2\epsilon x^2(1 - x^2) + \mathcal{O}(\epsilon^2)\right]\right\} \\ &= \epsilon x(1 - x^2)(1 + 2\epsilon x^2) + \mathcal{O}(\epsilon^3). \end{aligned}$$

There is no component of $\mathcal{O}(x^p)$ in the errors as the model is global in x: the *order of error is uniform in* $\mathcal{M}_0$.[70]

However, now compare the two statements

$$\dot{x} = \epsilon x(1 - x^2) + \mathcal{O}(\epsilon^2) \quad \text{and} \quad \dot{x} = \epsilon x(1 - x^2)(1 + 2\epsilon x^2) + \mathcal{O}(\epsilon^3):$$

in the evolution the $\mathcal{O}(\epsilon^2)$ error is approximately $-2\epsilon^2 x^3(1 - x^2)$ and so the multiplicative "constant" of the error varies with x, and in particular, it becomes very large for large x. The neighborhood of theoretical support will correspondingly vary in size across $\mathcal{M}_0$: generally the neighborhood is small when the error "constant" is large.

Example 5.3. Consider the system $\dot{x} = -xy$ and $\dot{y} = -y - xy + \epsilon$ for small parameter ϵ. When $\epsilon = 0$ the variable y decays exponentially quickly to $y = 0$ (provided $x > -1$); once on $y = 0$ then the variable x no longer evolves. Thus $y = 0$ for every $x > -1$ is a subspace of attractive equilibria. One could construct the slow manifold based around $\epsilon = x = y = 0$. However, more useful is the global slow manifold

$$y = \frac{\epsilon}{1 + x} - \frac{\epsilon^2 x}{(1 + x)^4} + \mathcal{O}(\epsilon^3). \tag{5.26}$$

[70] At least, the error is of uniform order on any compact subset of $\mathcal{M}_0$.

Find this global slow manifold from the subspace of equilibria $\mathbb{E}_0 = \{(x, y, \epsilon) \mid \epsilon = y = 0\}$—a subspace parametrized by x. Adjoining the trivial $\dot\epsilon = 0$ and linearizing about the equilibria $(x, y, \epsilon) = (s, 0, 0)$ leads to

$$\text{Jacobian} = \begin{bmatrix} 0 & -s & 0 \\ 0 & -1-s & 1 \\ 0 & 0 & 0 \end{bmatrix}.$$

As the Jacobian is upper triangular, the eigenvalues about each equilibria are along the diagonal, namely 0 (twice) and $-1 - s$. Thus there exists a slow manifold, $y = h(x, \epsilon)$, about each equilibrium for $s > -1$. Consequently seek the global approximation $y = \epsilon h_1(x) + \epsilon^2 h_2(x) + \cdots$ to find the above global slow manifold (5.26). There is no approximation in x, only in ϵ. Generally expect such global models to be more accurate.

Corresponding to the global slow manifold (5.26), the model is the evolution of variable x on the slow manifold, namely

$$\dot x = -\frac{\epsilon x}{1+x} + \mathcal{O}(\epsilon^2),$$

local in ϵ, but global in x as *the order of error is uniform in* x.

The domain of global validity has to be bounded away from the singular point $x = -1$; that is, a notional "constant" multiplying the order of error ϵ^2 varies with x. The neighborhood of the manifold of equilibria in which the center manifold theory supports the model will vary correspondingly in extent.

After this brief excursion into local errors of global approximations, we now turn to the flexibility arising in asymptotic truncations when multiple small parameters appear in a slow manifold.

5.5.2 ▪ Newton diagrams guide errors

Example 5.4 introduces how to view information in a Newton diagram. The flow of information in a Newton diagram determines valid truncations of the multivariate asymptotic approximations in complicated modeling.

Example 5.4. Reconsider the global approximation of Example 5.2. Now make the parameter ϵ vary slowly in time, at rate δ say, so that the system (5.25) is here

$$\dot\epsilon = \delta, \quad \dot\delta = 0, \quad \dot x = \epsilon(x - xy) \quad \text{and} \quad \dot y = -y + x^2. \tag{5.27}$$

From Example 5.2, *an* approximation to the slow manifold is

$$y = h^{(1)}(x, \epsilon, \delta) := x^2 - 2\epsilon x^2(1 - x^2).$$

Let us explore the next step in the iterative construction of the slow manifold. Given $\dot x \approx \epsilon(x - xh^{(1)})$, the residual of the y equation is

$$\text{Res}_{(5.27)} = 2x^2(1-x^2)\left[\delta + \epsilon^2(2 - 6x^2)\right] + \mathcal{O}(\epsilon^3).$$

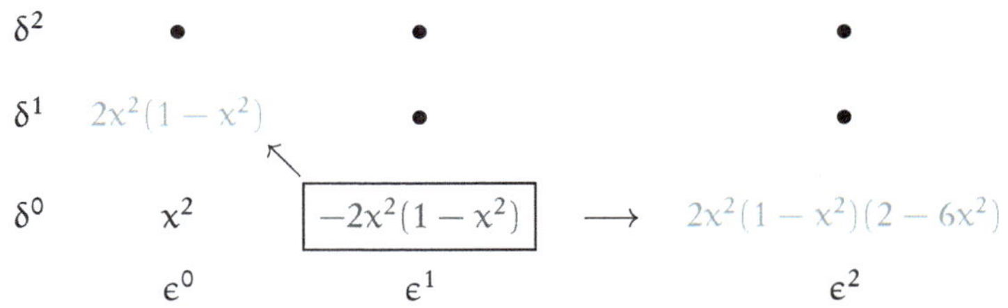

Figure 5.9. *A type of partial Newton diagram (Murdock, 2003) with the arrows denoting how two terms in the multivariate asymptotic approximation are derived from the "earlier" term (framed) at order* $\epsilon^1\delta^0$.

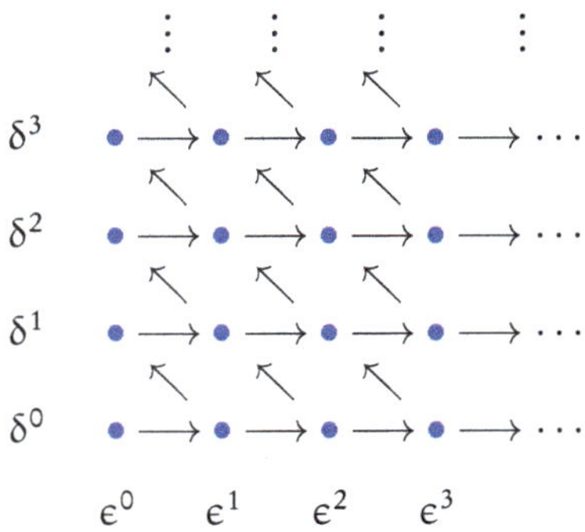

Figure 5.10. *Information flow across the Newton diagram with arrows showing how terms of various multivariate orders depend upon other terms; the bullets,* •, *denote some algebraic expressions in the global variable* x.

In this problem, simply add this residual to $h^{(1)}$ to improve the approximation of the slow manifold: $h^{(2)} := h^{(1)} + \text{Res}_{(5.27)}$. Figure 5.9 graphically shows how the term $-2\epsilon x^2(1-x^2)$ in $h^{(1)}$ generates two new terms in the approximation:

- the nonlinearity in the equations generates terms to the right, as shown by the right-pointing arrow,[71]
- whereas the slowly varying ϵ replaces ϵ factors with δ factors and hence generates terms to the above left, as shown by another arrow.

That is, these two arrows suggest how information and influences flow across the Newton diagram.

The arrows of Figure 5.10 determine valid truncations of the multivariate asymptotic expansions.[72] A finite truncation of the multivariate asymptotic expansions means we keep a finite number of terms in the expansion. These finite number of terms correspond to some subset of the terms represented in a Newton diagram. Thus to compute terms correctly in the expansion, we must ensure that the arrows of influence *only point out* of the set of terms retained in the expansion. If this were not so, that is, if we tried to compute terms that depended upon neglected

[71]The nonlinearity also generates terms $\mathcal{O}(\epsilon^3)$, which would appear farther to the right on the diagram. However, for our purposes only the leading-order effect in the various directions is significant.

[72]At the very least, the arrows determine valid truncations when computing the multivariate expansions.

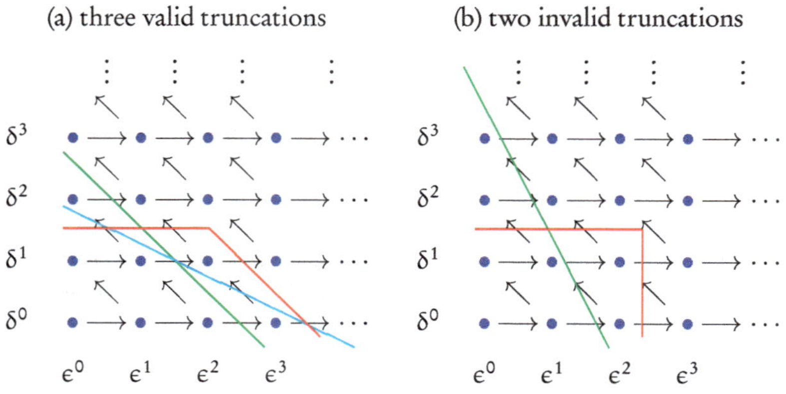

Figure 5.11. *Valid and invalid truncations of a multivariate asymptotic expansion drawn on the Newton diagram: the green, cyan, and red lines give the boundaries of computed terms—terms below and to the left are those retained. In the valid truncations of* (a)*, the errors may be algebraically expressed as green,* $\mathcal{O}(\epsilon^3 + \delta^3)$*; cyan,* $\mathcal{O}(\epsilon^4 + \delta^2)$*; and red,* $\mathcal{O}(\epsilon^4 + \delta^4, \delta^2)$.

terms, then almost surely we must incur errors. Influence arrows must not point from unknown terms to supposedly found terms.

Figure 5.11 schematically show some valid and invalid truncations of a particular multivariate asymptotic expansion. The colored lines divide the computed terms, those to the lower left, and the discarded terms, those to the upper right, in the computation of the slow manifold. The invalid truncations have influence arrows entering the computed terms from the region of discarded terms. Thus the computed terms will have errors from the discards. When the influence arrows only leave the computed term region, then the computed terms will be correct.[73]

Let's proceed to explore the general dynamical system

$$\dot{\mathbf{u}} = \mathcal{L}\mathbf{u} + \mathsf{f}(\mathbf{u}), \tag{5.28}$$

where $\mathbf{f}$ is strictly nonlinear so that the origin is an equilibrium. Suppose the system's center manifold model is

$$\mathbf{u} = \mathbf{v}(\mathbf{s}) = \mathcal{E}\mathbf{s} + \mathcal{O}(\mathbf{s}^2) \quad \text{such that} \quad \dot{\mathbf{s}} = \mathbf{g}(\mathbf{s}) = \mathcal{G}\mathbf{s} + \mathcal{O}(\mathbf{s}^2)\,. \tag{5.29}$$

Although it is not the only aspect we may need to consider, the operator (matrix) $\mathcal{G}$ of the model (5.29) often dictates feasible truncations of the asymptotic approximations.

In this discussion we only consider "local" amplitudes in $\mathbf{s}$, such as ϵ and δ in the preceding example; "global" amplitudes have no error terms associated with them and so do not enter this discussion of truncations of asymptotic approximations.

[73]This discussion on "valid" asymptotic truncations is predicated on those obtained via computation to any specified order of error. In principle, one could compute to a high order of error and then truncate to some lower order that I labeled "invalid." For example, in the problem illustrated in Figure 5.11 one might compute an approximation to $\mathcal{O}(\epsilon^4 + \delta^4)$, valid, and then choose to truncate the approximation to error $\mathcal{O}(\epsilon^4 + \delta^2)$, labeled "invalid" in the figure. Such a model would correctly have error $\mathcal{O}(\epsilon^4 + \delta^2)$. The issue here is that the Flexible Approximation Theorem 4.10 could not *directly* certify the error $\mathcal{O}(\epsilon^4 + \delta^2)$. Our discussion of "valid" and "invalid" is only in the context of direct certification by the Flexible Approximation Theorem 4.10.

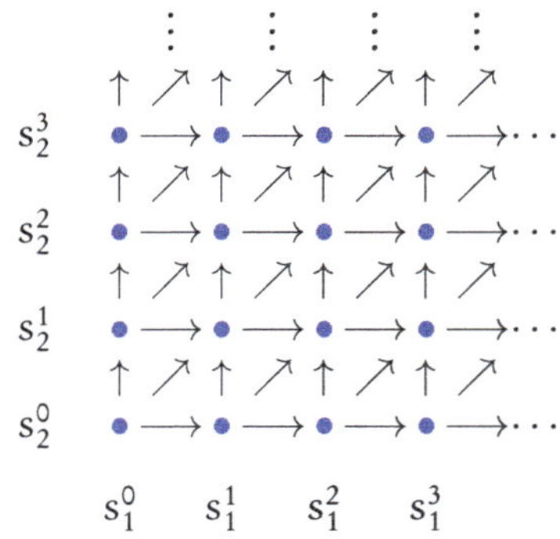

Figure 5.12. *Information flow across the Newton diagram with arrows indicating how, through the nonlinearity* $f(u)$ *in the dynamical system* (5.28), *terms of various multivariate orders generally influence other terms.*

Nonlinearity places little restriction The question is: if we modify a term in the asymptotic approximation to the center manifold model, what terms will be affected in the residual? Since the residual drives corrections to the approximation of the center manifold, via the homological equation

$$-\mathcal{L}\hat{v} + \frac{\partial \hat{v}}{\partial s}\mathcal{G}s + \mathcal{E}\hat{g} = \text{Res}\,, \tag{5.30}$$

the answer to this question determines the dependencies in the Newton diagram.

Now, each term in the asymptotic expansions for the center manifold model is a product of powers of m (local) amplitudes in s. Thus each term corresponds to a lattice point in some general m-dimensional Newton diagram: a term in $s_1^{p_1} \cdots s_m^{p_m}$ corresponds to the lattice point $(p_1, \ldots, p_m)$. Suppose we modify the term in $s_1^{p_1} \cdots s_m^{p_m}$; what terms are affected by such a modification?

First, through the linear operator $\mathcal{L}$ in the dynamical system (5.28), any change at order $s_1^{p_1} \cdots s_m^{p_m}$ has a direct effect at that order. Indeed, this is how the iterative construction works: when a term of a given order occurs in the residual, the homological equation (5.30) guides us to choose modifications that cancel out the terms in the residual. The linear operators $\mathcal{L}$ and $\mathcal{E}$ on the left-hand side of the homological equation (5.30) reflect this direct dependence.

Second, the strictly nonlinear function $f(u)$ in the dynamical system (5.28) influences only higher-order terms. Any modification in the approximation $u = v(s)$ at a term $s_1^{p_1} \cdots s_m^{p_m}$ will, because of the necessarily multiplicative nonlinearities in f, cause the function f to generate terms of higher order in the residual. Here, higher order means a term $s_1^{q_1} \cdots s_m^{q_m}$ where the exponents $q_j \geqslant p_j$ and at least one of these inequalities is a strict inequality. Thus on a Newton diagram, two-dimensional for simplicity, the nonlinearity represented by the nonlinear function f only influences upward and/or to the right—Figure 5.12 illustrates these typical influences.

Figure 5.12 shows that direct nonlinearity places relatively mild restrictions on the truncation of the asymptotic approximations: we just need to consistently neglect terms that are of higher order in at least one variable of the center manifold model.

Third, the time derivative $\dot{u}$ has subtle effects that deserve careful consideration.

Linear time dependence dominates possible truncations The immediately previous discussion explored effects of modifying some term in the asymptotic approximation to the center manifold model (5.29). In particular, it explored the dependencies generated through the linear term $\mathcal{L}\mathbf{u}$ and nonlinear terms $\mathbf{f}(\mathbf{u})$ in the dynamical system (5.28). Now we explore the intriguing dependencies generated through the time derivative $\dot{\mathbf{u}}$ in the dynamical system (5.28).

Suppose the residual has a term in $s_1^{p_1}\cdots s_m^{p_m}$: given such a term in the residual we modify, through choosing $\hat{\mathbf{v}}$ or $\hat{\mathbf{g}}$, the asymptotic approximation to the center manifold in order to cancel the term in the residual. However, depending upon operator $\mathcal{G}$, time derivatives, represented by the term $\partial\hat{\mathbf{v}}/\partial\mathbf{s}\,\mathcal{G}\mathbf{s}$ of the homological equation (5.30), couple modifications.

Consider some low-dimensional examples.

- The operator $\mathcal{G}$ being diagonal is the simplest case. In particular, for the common case of approximating slow manifolds the operator $\mathcal{G} = 0$ is even simpler. In this case the influence of any change is always to strictly higher-order terms, up and to the right in a 2D Newton diagram as in Figure 5.12, and so we may be extremely flexible about truncation.

- If $\mathcal{G}$ is block diagonal, then the variables in each block are fully coupled together, determined simultaneously, and the error must be expressed in their overall equally weighted amplitude.

- Alternatively, the operator $\mathcal{G}$ is in Jordan form.

Example 5.5. For example, consider the system of Example 5.4, namely $\dot\epsilon = 0$, $\dot x = \epsilon - xy$, and $\dot y = -y + x^2$, for which, with slow variables $\mathbf{s} = (\epsilon, x)$,

$$\mathcal{G} = \begin{bmatrix} 0 & 0 \\ 1 & 0 \end{bmatrix}.$$

Hence a modification in a term $x^p\epsilon^q$ generates, via the time derivative $\dot x$, a term in $x^{p-1}\epsilon^{q+1}$. An influence flows upward-left in a Newton diagram on the pq-plane, such as for those plotted in Figures 5.10 and 5.11.

Such a toy example is analogous to crucial parameter dependencies occurring in the gravitational forcing of thin films, or when there is a source/sink of contaminant in its dispersion along a channel.

Example 5.6. Similar to Example 5.4, we could have a slowly varying bifurcation parameter as in the system

$$\dot\delta = 0, \quad \dot\epsilon = \delta, \quad \dot x = \epsilon x - xy, \quad \dot y = -y + x^2.$$

Then, with the slow variable vector $\mathbf{s} = (\delta, \epsilon, x)$, the operator

$$\mathcal{G} = \begin{bmatrix} 0 & 0 & 0 \\ 1 & 0 & 0 \\ 0 & 0 & 0 \end{bmatrix}.$$

Consequently, as in Example 5.4, we can truncate asymptotic approximations so that the order in δ is equal to or lower than the order in ϵ. But there is no dependency between slow variables x and (δ, ϵ), so we are free to truncate the asymptotic expansion in variables x independent of how we truncate it in ϵ and δ.

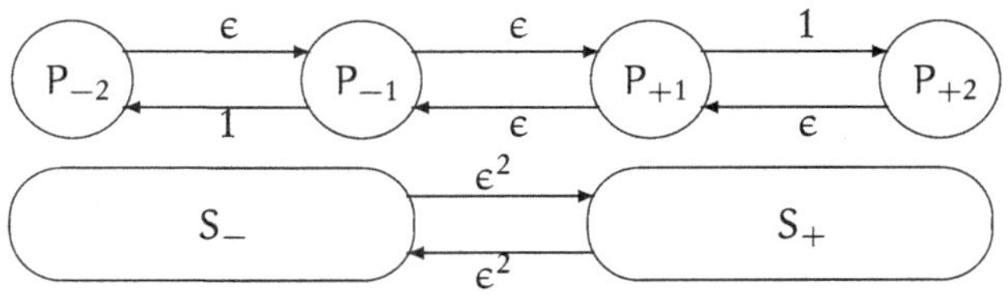

Figure 5.13. *Top is the four-state, continuous-time Markov chain of Exercise* 5.1*: the rates of change of state are shown on the arrows. Bottom is the two-state, quasi-stationary, slow manifold model of the long-term metastable evolution.*

Example 5.7 (a mistake). I made this unnecessarily complicated application of theory in the gravitational forcing of a thin fluid layer. Consider the toy system $\dot\epsilon = 0$, $\dot x = \epsilon - xy$, and $\dot y = -y + x^2$, where the ϵ forcing of x is analogous to gravitational acceleration of the fluid. In an effort to avoid having the ϵ forcing in the linear dynamics, I set $\epsilon = \delta^2$ and adjoined $\dot\delta = 0$. Consequently, the "linear" forcing became "nonlinear" and could surely be treated differently. Alas no. The $\partial\hat{v}/\partial s\,\tilde{g}$ term in the equation for corrections then involves $\delta^2\partial\hat{v}/\partial x$ instead of $\epsilon\partial\hat{v}/\partial x$. But we gain nothing, as the information flow across the Newton diagram is still fundamentally the same.

This recognition of flexible truncation empowers extraordinarily useful extensions of much current practice. With currently extant methods, such as that of multiple scales, people have to try very hard to get the relative scaling of one variable "correct" compared to another. For example, to analyze the pitchfork bifurcation one is normally forced to assume properties like "$\epsilon \sim x^2$" at the very start of the analysis. With center manifold theory we need assert no such thing. In pitchfork bifurcations, the fact that "$\epsilon \sim x^2$" is an emergent property of the dynamics in the eventual model; it *need not* affect the center manifold modeling, which is primarily concerned with the separation of transients from long-term evolution. Flexible Approximation Theorem 4.10, in essence, empowers one to scale different effects in a very general way. That is, the center manifold model is valid over a wide range of scaling regimes.

5.6 ▪ Exercises

Exercise 5.1 (metastable states). Adapt the quasi-stationary probability arguments to the modeling of metastable states. The top of Figure 5.13 shows a four-state, continuous-time Markov chain: we aim to derive the two-state, quasi-stationary, slow manifold model of the long-term metastable evolution. The interest here is that most stochastic averaging theorems just predict the ultimate stasis of equal likelihood in both the strongly attracting states $P_{\pm 2}$, whereas we want to know about the dynamics of the long-time switching between metastable states $P_{\pm 2}$. The two-state model in Figure 5.13 tells us about such switching.

The differential equation for the four-state Markov chain is

$$\dot p = \mathcal{L}_\epsilon p \quad \text{for} \quad \mathcal{L}_\epsilon = \begin{bmatrix} -\epsilon & 1 & 0 & 0 \\ \epsilon & -1-\epsilon & \epsilon & 0 \\ 0 & \epsilon & -1-\epsilon & \epsilon \\ 0 & 0 & 1 & -\epsilon \end{bmatrix}. \tag{5.31}$$

1. Show that $\mathcal{L}_0$ has eigenvalues $\lambda = 0$ (twice) and $\lambda = -1$ (twice). Hence deduce that there exists a 2D emergent slow manifold for the system (5.31) (3D if one counts the parameter ϵ) for small parameter ϵ.
2. Confirm that slow eigenvectors are $e_+ = (0, 0, 0, 1)$ and $e_- = (1, 0, 0, 0)$ and corresponding left eigenvectors are $z_+ = (0, 0, 1, 1)$ and $z_- = (1, 1, 0, 0)$. Hence define appropriate "probability" amplitudes $s_\pm$ and write down the initial approximation to the slow manifold.
3. Iteratively construct the slow manifold model to find that the two metastable states evolve according to $\dot{s}_\pm \approx \epsilon^2(s_\mp - s_\pm)$.
4. Interpret your modeling in terms of the original four-state system.
5. Verify the rate in the exchange, $\dot{s}_\pm \approx \epsilon^2(s_\mp - s_\pm)$, by finding the characteristic polynomial for matrix $\mathcal{L}_\epsilon$, finding that its factors are λ, $\lambda + 1 + \epsilon$, and a quadratic factor that predicts $\lambda \approx 2\epsilon^2$ corresponding to the rate of switching. What are higher-order corrections to this rate of switching?

Exercise 5.2 (negative probabilities OK). The exact analysis of the quasi-stationary distribution problem in the example in section 5.3.6 is not only linear, but straightforwardly linear. Let's explore and end up with physical meaning for negative probabilities!

> When you have eliminated the impossible, whatever remains, however improbable, must be the truth.
> *A. C. Doyle*

Recall that the continuous-time, Markov chain differential equation for this example is

$$\dot{\mathbf{p}} = \mathcal{L}_\epsilon \mathbf{p} \quad \text{for} \quad \mathcal{L}_\epsilon = \begin{bmatrix} 0 & 2\epsilon & 0 \\ 0 & -\frac{1}{2} - \epsilon & \frac{1}{2} \\ 0 & \frac{1}{2} - \epsilon & -\frac{1}{2} \end{bmatrix}. \tag{5.32}$$

1. Deduce that the matrix $\mathcal{L}_\epsilon$ has eigenvalues $0, -\epsilon$, and -1 (exactly). Find that the corresponding eigenvectors are $\mathbf{v}_1 = (1, 0, 0)$, $\mathbf{v}_2 = (-2, 1, 1)$, and $\mathbf{v}_3 = (2\epsilon, -1, 1 - 2\epsilon)$ and the corresponding left eigenvectors are $\mathbf{z}_1 = (1, 1, 1)$, $\mathbf{z}_2 = (0, 1 - 2\epsilon, 1)$, and $\mathbf{z}_3 = (0, 1, -1)$.
2. Hence deduce that a general exact solution of (5.32) is that the probabilities
$$\mathbf{p}(t) = c_1\mathbf{v}_1 + c_2\mathbf{v}_2 e^{-\epsilon t} + c_3\mathbf{v}_3 e^{-t}.$$
3. In modeling dynamics our fundamental principle is to neglect rapidly decaying transients, such as e^{-t}, and retain slowly varying components, such as $e^{-\epsilon t}$. Neglect the e^{-t} term in the above general solution and deduce, from the definition of $\mathbf{s}$, that in modeling we would have $s_1 = c_1 - 2c_2 e^{-\epsilon t}$ and $s_2 = 2c_2 e^{-\epsilon t}$. Differentiate to confirm that the slow manifold, quasi-stationary, model $\dot{s}_1 = \epsilon s_2$ and $\dot{s}_2 = -\epsilon s_2$ is "exact" (that is, exact upon neglecting rapid transients).
4. Suppose the Markov chain is initially in state 3 so that the initial probability vector is $\mathbf{p}(0) = (0, 0, 1)$. Find that the particular solution arises from $c_1 = 1$, $c_2 = c_3 = 1/[2(1-\epsilon)]$. Use the exact model of the previous part to argue that

the corresponding initial condition for the model is thus $s_1(0) = -\epsilon/(1-\epsilon)$ and $s_2(0) = 1/(1-\epsilon)$.

Amazingly, in order to make the correct long-time predictions, we *must* start the model with negative probability in s_1 and probability in $s_2 > 1$.[74] This appears a more startling example of initial slip than Example 4.10 noted.

Exercise 5.3 (elementary slow manifolds). For each of the following dynamical systems, where possible,

- argue for the existence of a slow manifold based upon the linear dynamics near the origin,
- find a linear approximation for the shape of the slow manifold,
- apply an iteration scheme to derive a low-dimensional model, and
- identify any bifurcations.

1.
$$\begin{aligned}\dot{x} &= -\tfrac{1}{2}x - y - x^2y\,,\\ \dot{y} &= -x - 2y + \epsilon y - y^2.\end{aligned}$$

2. We sometimes need to expand transcendental functions in a Taylor series:
$$\begin{aligned}\dot{\theta} &= -\theta + \epsilon v + v^2,\\ \dot{v} &= \sin\theta\,.\end{aligned}$$

3.
$$\begin{aligned}\dot{x} &= x - 2y + \epsilon x\,,\\ \dot{y} &= 3x - 6y + x^3.\end{aligned}$$

4.
$$\begin{aligned}\dot{x} &= 2x + y + 2z + yz\,,\\ \dot{y} &= x - y + z - xz\,,\\ \dot{z} &= -3x - y - 3z - xy\,.\end{aligned}$$

5. In this system "forced" by ϵ (Chicone, 2006, p. 347), invoke a generalized eigenvector (Table 4.1):
$$\begin{aligned}\dot{\epsilon} &= 0\,,\\ \dot{z} &= \epsilon - z + w + \tfrac{1}{4}\left[(1+\epsilon)z^2 - 2\epsilon wz - (1-\epsilon)w^2\right],\\ \dot{w} &= \epsilon + z - w - \tfrac{1}{4}\left[(1+\epsilon)z^2 - (2\epsilon-4)wz + (3+\epsilon)w^2\right].\end{aligned}$$

6. Invoke a generalized eigenvector (Table 4.1):
$$\begin{aligned}\dot{x} &= 2x + y + 2z + \epsilon yz\,,\\ \dot{y} &= x + z - \epsilon xz\,,\\ \dot{z} &= -3x - y - 3z - \epsilon xy\,.\end{aligned}$$

[74] There is a type of "quantum mechanical uncertainty" in these negative probabilities. The apparent violation of probability is small, of $\mathcal{O}(\epsilon)$, and only lasts over a fast time scale. On the long time scale of interest to our modeling, times $\sim 1/\epsilon$, the negative probabilities disappear rapidly.

Answer:

1 Possibly $x \approx (1-\frac{2}{25}\epsilon)s - (\frac{1}{25}+\frac{18}{625}\epsilon)s^2$ and $y \approx -(\frac{1}{2}+\frac{4}{25}\epsilon)s - (\frac{2}{25}+\frac{36}{625}\epsilon)s^2$, on which $\dot s \approx \frac{1}{5}\epsilon s + (\frac{1}{10}+\frac{12}{125}\epsilon)s^2$.

2 Possibly $\theta \approx \epsilon s + (1-3\epsilon)s^2$ and $v = s$, where $\dot s \approx \epsilon s + (1-3\epsilon)s^2$.

3 Perhaps $x \approx (2-\frac{2}{25}\epsilon)s + (\frac{16}{25}-\frac{368}{625}\epsilon)s^3$ and $y \approx (1-\frac{6}{25}\epsilon)s + (\frac{48}{25}-\frac{1104}{625}\epsilon)s^3$, where $\dot s \approx \frac{6}{5}\epsilon s - (\frac{8}{5}-\frac{72}{125}\epsilon)s^3$.

4 Perhaps $x = s$, $y \approx s^2 - 4s^3$, and $z \approx -s - 2s^3$, where $\dot s \approx s^2 - 9s^3$.

5 Perhaps express a slow manifold as $z \approx s + \frac{1}{2}s^2$, $w \approx s - \frac{1}{2}s^2$ such that $\dot s \approx \epsilon - s^2 + \frac{1}{2}s^3$.

6 Perhaps parametrize the slow manifold as $x = s_1$, $y = s_2$, $z = -s_1 + \epsilon(-2s_1s_2 + 2s_2^2) + \mathcal{O}(\epsilon^2)$, on which the evolution is $\dot s_1 = s_2 + \epsilon(-5s_1s_2 + 4s_2^2) + \mathcal{O}(\epsilon^2)$ and $\dot s_2 = \epsilon(s_1^2 - 2s_1s_2 + 2s_2^2) + \mathcal{O}(\epsilon^2)$.

Exercise 5.4 (competing nonlinearities I). For what values of parameter α does the nonlinear advection and diffusion stabilize the nonlinear growth in the PDE

$$\frac{\partial u}{\partial t} + u\frac{\partial u}{\partial x} = u + \alpha u\left(\frac{\partial u}{\partial x}\right)^2 + \frac{\partial^2 u}{\partial x^2}\ ?$$

Use boundary conditions $u(0,t) = u(\pi,t) = 0$. Modify the computer algebra from section 5.2 to answer. What does the slow manifold model predict if parameter α is exactly at the critical value dividing stability from instability?

Answer: $u = a\sin x - \frac{1}{6}a^2\sin 2x + \frac{1}{32}(1+\alpha)a^3\sin 3x + \mathcal{O}(a^4,\alpha^2)$ such that $\dot a = (-\frac{1}{12}+\frac{1}{4}\alpha)a^3 + \mathcal{O}(a^4,\alpha^2)$. Nonlinear instability.

Exercise 5.5 (competing nonlinearities II). For what values of parameter α does the nonlinear advection and diffusion stabilize the nonlinear reaction in the PDE

$$\frac{\partial u}{\partial t} + u\frac{\partial u}{\partial x} = u + \alpha u^3 + \frac{\partial^2 u}{\partial x^2}\ ?$$

Use boundary conditions $u(0,t) = u(\pi,t) = 0$. Modify the computer algebra from section 5.2 to answer. What does the slow manifold model predict if parameter α is exactly at the critical value dividing stability from instability?

Answer: $u = a\sin x - \frac{1}{6}a^2\sin 2x + \frac{1}{32}(1-\alpha)a^3\sin 3x + \mathcal{O}(a^4,\alpha^2)$ such that $\dot a = (-\frac{1}{12}+\frac{3}{4}\alpha)a^3 + \mathcal{O}(a^4,\alpha^2)$. Remarkably, there is an exact manifold of equilibria, namely $u = 36a\sin x/(36 + a^2 + 12a\cos x)$.

Exercise 5.6 (ecological competition). Constable et al. (2013) considered interactions of two related species, X and Y. We aim to find a slow manifold model of the long-term interaction.

The dynamics of the species are the following, in nondimensional terms:

- individuals of both species reproduce at rate $1-\mu$, that is, $X \xrightarrow{1-\mu} 2X$ and $Y \xrightarrow{1-\mu} 2Y$;
- but the offspring mutate with small probability μ into an individual of the other species, that is, $X \xrightarrow{\mu} X+Y$ and $Y \xrightarrow{\mu} Y+X$;

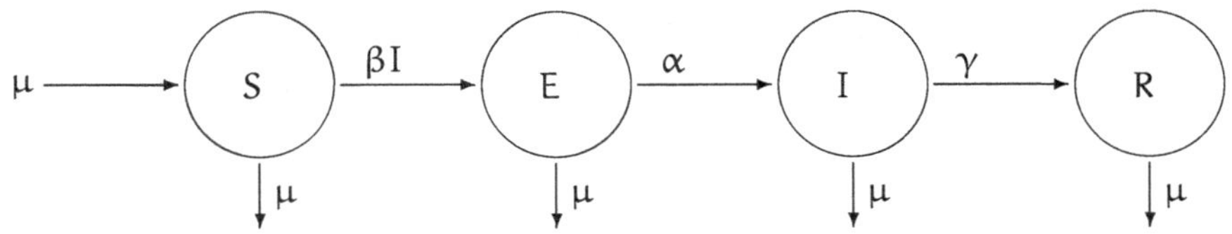

Figure 5.14. *Exercise 5.7 explores this network of interactions in an epidemiology model of susceptible (S), exposed (E), infected (I), and recovered (R) individuals.*

- the species prey on each other with coefficient 5/6, that is, $X + Y \xrightarrow{5/6} X$ and $X + Y \xrightarrow{5/6} Y$; and
- the species are also a little cannibalistic, but prefer the other species, eating each other with coefficient 1/6, that is, $2X \xrightarrow{1/6} X$ and $2Y \xrightarrow{1/6} Y$.

1. Let $x(t)$ and $y(t)$ be proportional to the number of individuals of each of the two species, X and Y, respectively. Argue that a mathematical model of the interaction between the species is coded by the ODEs

$$\frac{dx}{dt} = x - \mu(x - y) - \tfrac{1}{6}x^2 - \tfrac{5}{6}xy\,,$$
$$\frac{dy}{dt} = y + \mu(x - y) - \tfrac{5}{6}xy - \tfrac{1}{6}y^2.$$

2. Show that extinction $(0, 0)$ and coexistence $(1, 1)$ are equilibria, extinction is always unstable, and coexistence is stable for mutation $\mu > \frac{1}{3}$ and unstable for $\mu < \frac{1}{3}$.
3. Deduce the existence and emergence of a slow manifold through $x = y = 1$ for mutation μ near the critical 1/3.
4. Approximate the slow manifold and the evolution thereon to find that, for mutation $\mu = \frac{1}{3} - \mu'$, they are

$$x = 1 + s + (\tfrac{2}{3} - \tfrac{8}{3}\mu')s^2 - \tfrac{4}{27}s^4 + \mathcal{O}(s^6 + \mu'^3),$$
$$y = 1 - s + (\tfrac{2}{3} - \tfrac{8}{3}\mu')s^2 - \tfrac{4}{27}s^4 + \mathcal{O}(s^6 + \mu'^3),$$
$$\dot{s} = 2\mu' s - (\tfrac{2}{9} - \tfrac{8}{9}\mu')s^3 + \tfrac{4}{81}s^5 + \mathcal{O}(s^6 + \mu'^3).$$

5. Interpret the predicted equilibria in this slow manifold model.

Exercise 5.7 (SEIR epidemiology model). Schwartz and Smith (1983) proposed a mathematical model of the dynamics of an infectious disease. Figure 5.14 summarizes the model. A population is divided into four classes:

- a fraction $S(t)$ of the population are susceptible to the disease;
- a fraction $E(t)$ have been exposed to the infection but are not yet infectious themselves;
- a fraction $I(t)$ have become infectious after being exposed; and

- a fraction $R(t)$ of the population have recovered and are now immune to the disease.

Individuals move from one class to the next at varying rates, as indicated in Figure 5.14:

- each susceptible is likely to be exposed to the disease at a rate proportional to the fraction of infected, βI;
- each exposed becomes infectious at a characteristic rate α;
- each infected recovers at a characteristic rate γ;
- a supply of fresh individuals arrive at a flat rate μ; and conversely
- each individual leaves the system (dies) at a rate μ, independent of state, so that the total population is steady.

Consequently, the ordinary differential equations for the fractions are

$$\begin{aligned}\dot{S} &= \mu - \beta I(t)S(t) - \mu S(t),\\ \dot{E} &= \beta I(t)S(t) - (\alpha+\mu)E(t),\\ \dot{I} &= \alpha E(t) - (\gamma+\mu)I(t),\\ \dot{R} &= \gamma I(t) - \mu R(t).\end{aligned} \tag{5.33}$$

Notice that $S + E + I + R = 1$ is an invariant of the dynamics; consequently we ignore the ODE for $R(t)$ as it may be deduced from the others, and because it does not feed back into any of the other variables.

1. For simplicity suppose parameters $\alpha = \gamma = 1$ and $\beta = 2$, but that the supply rate μ may vary. Write out the three ODEs for (S, E, I) parametrized by μ.
2. Our modeling needs a convenient equilibrium. Show that when the supply $\mu = 0$, then $(S, 0, 0)$ forms a subspace of equilibria.
3. Indeed, argue that $(S, 0, 0)$ is an invariant subspace for nonzero supply μ, with evolution $\dot{S} = \mu(1 - S)$. Interpret the predictions of the dynamics on this subspace.

   However, this subspace is only partly emergent.
4. Linearize about the equilibria, for negligible supply μ, and deduce that the subspace $(S, 0, 0)$ is only attractive for $S < 1/2$. Further deduce that there exists an emergent 2D slow manifold near $S = 1/2$ for small μ.
5. Invoke a generalized eigenvector to derive that the 2D slow manifold is tangent to $(S, E, I) = (1/2 + s, i/2, i/2)$ for variables s and i.
6. Defining amplitudes $s = S - 1/2$ and $i = E + I$, find the slow manifold and that the evolution on the slow manifold is

$$\begin{aligned}\frac{di}{dt} &= i(s-\mu) + \mathcal{O}(s^3 + i^3 + \mu^3),\\ \frac{ds}{dt} &= -\tfrac{3}{2}\mu - \tfrac{1}{2}i + \mu(-s + \tfrac{3}{16}i) + i(-\tfrac{3}{4}s + \tfrac{1}{16}i) + \mathcal{O}(s^3 + i^3 + \mu^3).\end{aligned}$$

7. What are the meaningful equilibria predicted in this model as supply parameter μ varies? Which are stable? Unstable? What do these predictions mean for the original model?

Exercise 5.8 (a moving boundary). Consider a Stefan problem, similar to that of section 4.8, where there is a reaction front separating reacting material from unreacting (Kim et al. (2010) introduced a stochastic version of this problem): suppose the reacting material occupies the domain $0 < x < s(t)$ and the unreacting material occupies the domain $x > s(t)$; that is, the moving interface $x = s(t)$ separates the two.

Let the field $u(x,t)$ characterize the reaction. Suppose the unreacting material is $u(x,t) = 0$ for $x > s(t)$ and the reacting material is characterized by the field $u(x,t) > 0$. In the reacting region we suppose that the field satisfies the (nondimensional) linear PDE $\partial u/\partial t = \partial^2 u/\partial x^2 + \alpha u$ for some parameter α. Set the physical boundary condition on $x = 0$ to be that of fixed flux $u_x(0,t) = -1$. Similarly pose the interface conditions that not only is $u(s(t),t) = 0$ but also the gradient $u_x(s(t),t) = -1$.

1. Show that for reaction parameter $\alpha = 0$ there is an equilibrium field for all interface locations s.
2. Hence deduce that about each equilibrium there must be a zero eigenvalue corresponding to a simple shift in interface location s. Argue that all other eigenvalues are negative (heuristically or rigorously), and hence deduce that a slow manifold exists.
3. Construct the slow manifold model to a low order in parameter α and parametrized by the interface location s.
4. Write a computer algebra program to construct a model to high order in parameter α. Comment on the model.

Answer: Eigenmodes are $u_n = \cos(n\pi x/s)$ and $s_n = -(-1)^n$ with eigenvalue $\lambda_n = -n^2\pi^2/s^2$. The slow manifold is $u = s - x + \alpha\frac{1}{12}(s-x)^2(2x+s) + \mathcal{O}(\alpha^2)$ such that $\dot{s} = \frac{1}{2}\alpha s + \mathcal{O}(\alpha^2)$.

Exercise 5.9 (modify the moving boundary). The Stefan moving boundary in Exercise 5.8 is unrealistic in that it supposes that the influx of stuff, $-u_x$, at $x = 0$ exactly balances the outflux, u_x, required at the interface $x = s$, namely $u_x = -1$ at both. Redo the modeling of Exercise 5.8 with the influx $-u_x(0,t)$ fixed but not exactly 1. For simplicity you may set the internal parameter $\alpha = 0$.[75]

Answer: Let $u_x(0,t) = -1-\beta$; then $u = s-x+\frac{1}{2}\beta(s-x)^2/s - \frac{1}{24}(s^2-x^2)^2/s^3 + \mathcal{O}(\beta^3)$ such that $s\dot{s} = \beta - \frac{1}{3}\beta^2 + \mathcal{O}(\beta^3)$.

Exercise 5.10 (Kuramoto–Sivashinsky). The Kuramoto–Sivashinsky PDE is

$$\frac{\partial u}{\partial t} + u\frac{\partial u}{\partial x} + \frac{\partial^2 u}{\partial x^2} + \Lambda\frac{\partial^4 u}{\partial x^4} = 0, \tag{5.34}$$

with, for our purposes, boundary conditions of $u(0) = u_{xx}(0) = u(\pi) = u_{xx}(\pi) = 0$. Use center manifold theory to elucidate the dynamics in the bifurcation that occurs as Λ decreases through 1. Specifically, write and run a Reduce program (based on our analysis of the Burgers-like PDE) to obtain approximations

[75] In this case of reaction $\alpha = 0$, the simplicity of the slow manifold suggests that a self-similar solution exists. Find it.

to the slow manifold and the evolution thereon. Comment on the application of the theorems.

Answer: Letting $\Lambda = 1-\epsilon$, the slow manifold field $u = a\sin x - (\frac{1}{24}+\frac{7}{144}\epsilon)a^2\sin 2x + (\frac{1}{1152}+\frac{1}{512}\epsilon)a^3\sin 3x + \mathcal{O}(a^4,\epsilon^2)$ such that $\dot a = \epsilon a - (\frac{1}{48}+\frac{7}{288}\epsilon)a^3 + \mathcal{O}(a^4,\epsilon^2)$.

Exercise 5.11 (Kuramoto–Sivashinsky extension). What is a slow manifold model of the Kuramoto–Sivashinsky PDE (5.34) if the boundary conditions are changed to $u_x(0) = u_{xxx}(0) = u_x(\pi) = u_{xxx}(\pi) = 0$? Hint: The solution of $\hat u_{xxxx} + \hat u_{xx} =$ residual has to cope with the residual being in the form of a sum of polynomials in x times sine and cosine functions. This may be done, albeit slowly, using the Green's function

$$G = \frac{1}{\pi}\left[1+\cos\xi - (\pi-\xi)^2/2 + (2+2\cos\xi - (\pi-\xi)\sin\xi)\cos x\right] + \begin{cases} 0, & 0<x<\xi, \\ x-\xi-\sin(x-\xi), & \xi<x<\pi, \end{cases}$$

whence the solution is $\hat u = \int_0^\pi G(x,\xi)\,\mathrm{Res}(\xi)\,d\xi$. If you have done everything else right, the boundary conditions and amplitude conditions will hold.

Exercise 5.12 (Lorenz bifurcation). Consider the famous Lorenz equations

$$\begin{aligned} \dot x &= \rho(-x+y), \\ \dot y &= \sigma x - y - xz, \\ \dot z &= -\beta z + xy. \end{aligned}$$

For some values of the parameters these equations exhibit chaos. But before the onset of chaos, the zero solution loses stability as the parameter σ crosses 1.

1. Argue for the existence of a slow manifold for σ near 1.
2. Find a linear approximation $\mathbf{u} = \mathbf{v}(s)$ for the shape of the slow manifold given that a parameter of the slow manifold is to be $s = x$.
3. Apply the iteration scheme to derive a low-dimensional model for the Lorenz equations for σ near 1, for the particular values $\beta = \rho = 1$, and where a parameter of the description is $s = x$. Hence describe qualitatively the structure of the bifurcation that occurs as σ crosses 1.

Exercise 5.13 (a sine-diffusion equation). Consider the dynamics of

$$\frac{\partial\phi}{\partial t} = \sin\phi + \nu\frac{\partial^2\phi}{\partial x^2}, \quad \phi(0) = \phi(\pi) = 0.$$

Find the critical value of the dissipation ν for which the spectrum indicates that center manifold theory applies. Verify that the Sturm–Liouville properties in Table 5.1 hold. Construct the Landau equation model that displays the pitchfork bifurcation as the dissipation is varied through the critical value.

Exercise 5.14 (Swift–Hohenberg bifurcation). The Swift–Hohenberg PDE,

$$\frac{\partial u}{\partial t} = \epsilon u - (1 + \partial_x^2)^2 u - u^3 ,$$

is a toy model of Rayleigh–Benard convection of a fluid in a domain of large extent in x. It is often used as an example system for pattern formation and evolution (Greenside and Coughran, 1984; Cross and Hohenberg, 1993, e.g.).

1. Analyze the linear dynamics to determine the critical value of the parameter ϵ at which a nontrivial solution may bifurcate from $u = 0$.
2. Use center manifold theory to establish the basis for finding the Landau equation modeling the dynamics of the bifurcation. Assume that the field $u(x,t)$ has boundary conditions $u = u_{xx} = 0$ at $x = 0, 2\pi$.
3. Write, debug, and execute a Reduce program to construct the Landau equation.

Exercise 5.15 (Swift–Hohenberg bifurcation extension). Modify your analysis of Exercise 5.14 to model the dynamics of the version with a nonlocal nonlinearity:

$$\frac{\partial u}{\partial t} = \epsilon u - (1 + \partial_x^2)^2 u - u\, G \star u^2 ,$$

where $G\star$ denotes convolution with the kernel $G(x) = \exp(-|x|)$:

$$G \star f = \int_{-\infty}^{\infty} G(\xi) f(x - \xi)\, d\xi .$$

But here use boundary conditions that the field $u(x,t)$ is 2π-periodic in x.

Exercise 5.16 (2D Navier–Stokes like dynamics). Laminie et al. (1993) used a system of two PDEs to explore a "multiscale" numerical method on fluid-like dynamics. Modifying the PDEs slightly, consider two fields $u(x,y,t)$ and $v(x,y,t)$ evolving according to

$$\frac{\partial u}{\partial t} + u\frac{\partial u}{\partial x} + v\frac{\partial u}{\partial y} + \tfrac{1}{2}u\left(\frac{\partial u}{\partial x} + \frac{\partial v}{\partial y}\right) = \nabla^2 u + Ru ,$$

$$\frac{\partial v}{\partial t} + u\frac{\partial v}{\partial x} + v\frac{\partial v}{\partial y} + \tfrac{1}{2}v\left(\frac{\partial u}{\partial x} + \frac{\partial v}{\partial y}\right) = \nabla^2 v + Rv ,$$

with boundary conditions for the domain $[0,\pi]^2$ of

$$u = \frac{\partial v}{\partial x} = 0 \text{ on } x = 0, \pi , \quad v = \frac{\partial u}{\partial y} = 0 \text{ on } y = 0, \pi .$$

Based upon the equilibrium $u = v = 0$, argue that there exists a slow manifold, in two dynamic variables, for parameter R near 1. Construct the slow manifold and describe briefly the long-term dynamics of the system for parameter $R \approx 1$.

Answer: $u \approx a \sin x$ and $v \approx b \sin y$ such that $\dot a \approx (R-1)a - \frac{3}{16}a^3 - \frac{1}{8}ab^2$ and $\dot b \approx (R-1)b - \frac{1}{8}a^2 b - \frac{3}{16}b^3$.

Exercise 5.17 (advection on small patches). Recall the small-patch modeling of macroscale dynamics introduced in section 5.4. The slow manifold modeling supported applying the gap-tooth scheme to reaction-diffusion PDE (5.16). Here extend the modeling to include advection. Generalize the PDE to

$$\frac{\partial u}{\partial t} = K\frac{\partial^2 u}{\partial x^2} + f(u) + h(u)\frac{\partial u}{\partial x}$$

for some general advection function $h(u)$. Rescale this PDE to microscale variables $\xi = x/r$ and $\tau = t/r^2$. Modify the computer algebra of Algorithm 5.5 to include the nonlinear advection, and execute to find the slow manifold model that

$$u = U(1 - r^2\xi^2) - r^3\tfrac{1}{3K}\xi(1-\xi^2)Uh(U) + \mathcal{O}(r^4),$$
$$\frac{dU}{dt} = -2KU + f(U) + \tfrac{1}{6}r^2\left[-2KU + f(U) - Uf'(U) - \tfrac{1}{K}Uh(U)^2\right] + \mathcal{O}(r^3).$$

Observe that the advection enhances dissipation.

Exercise 5.18 (nonhomogeneous boundaries around patches). Exercise 5.17 only discerns small effects of the microscale advection term $h(u)\partial u/\partial x$. The reason is the high degree of symmetry in the problem. Here break symmetry by applying nonhomogeneous physical boundaries at $x = \pm 1$. Suppose the field is required to be $u = u_l$ at $x = -1$ and $u = u_r$ at $x = +1$ for some prescribed values u_l and u_r. Revise the parabolic interpolation of section 5.4 to pass through $(x, u) = (-1, u_l)$, $(0, U)$ and $(+1, u_r)$. Use this new interpolation to revise the boundary conditions (5.18) for the microscale patch. Modify the computer algebra of Algorithm 5.5 to include the nonlinear advection of Exercise 5.17 and include the new patch boundary conditions. Execute to find the slow manifold model that

$$u = U + r\tfrac{1}{2}(u_r - u_l)\xi + r^2\tfrac{1}{2}(u_r - 2U + u_l)\xi^2 + \mathcal{O}(r^3),$$
$$\frac{dU}{dt} = K(u_r - 2U + u_l) + f(U) + \tfrac{1}{2}h(U)(u_r - u_l) + \mathcal{O}(r).$$

Observe classic first- and second-order differences that appropriately govern the small-patch dynamics in the gap-tooth scheme.

Exercise 5.19 (small Neumann patches). In the gap-tooth scheme it is important that the nature of the microscale patch boundary conditions not be tied to the nature of the macroscale boundary conditions (Roberts and Kevrekidis, 2007). Section 5.4 assumed that the microscale simulator required Dirichlet boundary conditions on the patch. In this exercise assume that the microscale simulator requires Neumann boundary conditions of prescribed gradient $u_x(\pm r, t)$.

- Use the amplitude definition (5.17), and parabolic interpolation with the Dirichlet boundary conditions on the macroscale of $u(\pm 1, t) = 0$, to derive patch Neumann boundary conditions $u_x(\pm r, t) = \mp 2ru(0, t)$.[76]
- Rescale space and time to microscale variables in the reaction-diffusion PDE (5.16) and the patch boundary conditions.

[76] Be aware that these patch boundary conditions are not pure Neumann boundary conditions as they nonlocally involve $u(0, t)$ as well as the gradients on the patch boundary.

- Identify a subspace of equilibria at patch size $r = 0$.
- Linearize and find eigenvalues about each equilibrium (confirm that this eigenproblem is self-adjoint so we know that the eigenvalues are real).
- Deduce that there exists an emergent slow manifold model; in what qualitative domain does it exist?
- Derive the leading approximation to the slow manifold and its dynamics to confirm the efficacy of the gap-tooth scheme in this scenario.
- Modify Algorithm 5.5 to compute higher-order approximations to the macroscale dynamics of the slow manifold.
- Incorporate into your computer algebra the nonlinear advection of Exercise 5.17 and the nonhomogeneous boundary conditions of Exercise 5.18. Interpret the resulting slow manifold model of the macroscale dynamics.

Exercise 5.20 (craftily fudge boundary conditions). In many physical problems we need to approximate nontrivial "microscale" internal structures: using polynomials is desirable as they are simple, but how? Let's explore an example.

Consider the dynamics of suspended sediment in a flood—equivalent to the dynamics of large molecules in so-called field flow fractionation (Suslov and Roberts, 2000, 1999, e.g.). We wish to primarily model the spread of sediment in the large horizontal direction. However, the sediment has significant vertical structure, roughly an exponential. It is very difficult to model the horizontal evolution in a complex fluid flow when complicated by such a nontrivial vertical distribution. Theoretically there is no problem, but the direct practical construction is difficult: the difficulty being that complicated interactions between other fields/physics and the natural exponential sediment distribution leads to a slow manifold typically expressed as double sums in exponentials and polynomials, schematically $\sum_{m,n} z^m e^{-nz}$. Such double sums quickly explode combinatorially and thus bog down the practical modeling. A practical route is instead to modify the problem so that the vertical exponentials are approximated by polynomials.

The basic challenge is to invent a base that, using polynomials, describes well enough the slow subspace of exponentials. Only then can we hope to reasonably model complicated and nonlinear multiphysics interactions. The canonical problem for the concentration field $c(z,t)$, on the layer $0 \leqslant z \leqslant 1$, is the linear drift-diffusion equation

$$\frac{\partial c}{\partial t} = w\frac{\partial c}{\partial z} + \frac{\partial^2 c}{\partial z^2} \quad \text{such that} \quad wc + \frac{\partial c}{\partial z} = 0 \text{ on } z = 0, 1, \tag{5.35}$$

where w is the nondimensional downward drift speed of the sediment, and the diffusive mixing has nondimensional coefficient 1. The boundary conditions are conservative in that the exponential $c \propto e^{-wz}$ is an equilibrium distribution. Our challenge is to modify (5.35) so that polynomials approximate this exponential equilibrium in a way that can be embedded into a much more complicated nonlinear physical problem.

1. The key property of sediment is the decreasing concentration from the bed to the surface. We could crudely approximate the exponential decay e^{-wz} by the linear $c \propto (1-z)$; that is, $c = 2(1-z)C$, where we define the cross-layer average $C(t) := \int_0^1 c(z,t)\,dz$.

(a) Noting that for this distribution $c = 0$, $\partial c/\partial z = -2C$ at $z = 1$, and $c = 2C$, $\partial c/\partial z = -2C$ at $z = 0$, argue that this distribution is an equilibrium of the PDE and boundary conditions

$$\frac{\partial c}{\partial t} = \frac{\partial^2 c}{\partial z^2}, \quad c + \frac{\partial c}{\partial z} = 0 \text{ on } z = 0, \quad c = 0 \text{ on } z = 1.$$

(b) Show that the following problem embeds the above together with the physical (5.35) in this family parametrized by γ and w:

$$\begin{aligned} &\frac{\partial c}{\partial t} = wc + \frac{\partial^2 c}{\partial z^2}, \\ &[w + (1-\gamma)]c + \frac{\partial c}{\partial z} = 0 \text{ on } z = 0, \\ &[w + (1-\gamma)]c + \gamma\frac{\partial c}{\partial z} = 0 \text{ on } z = 1. \end{aligned}$$

(c) Write Reduce code to derive, or verify by substitution, that the slow subspace of this system is approximately

$$\begin{aligned} c &= 2(1-z)C + \gamma(-2+4z)C \\ &\quad + w\left(\tfrac{7}{12} - \tfrac{31}{12}z + \tfrac{5}{2}z^2 - \tfrac{1}{2}z^3\right)C \\ &\quad + w\gamma\left(\tfrac{3}{2} - \tfrac{13}{12}z - 8z^2 + \tfrac{5}{2}z^3\right)C + \mathcal{O}(w^2,\gamma^2), \\ \dot C &= \left[\tfrac{3}{2}w - \tfrac{9}{2}w\gamma\right]C + \mathcal{O}(w^2,\gamma^2). \end{aligned}$$

(d) Although the structure in z is nicely polynomial, argue that the small corrections are woeful even for drift velocity $w = 0$. Further, comment on the failure to construct nontrivial equilibria.

2. The woeful performance even for drift $w = 0$ is discouraging. But let's impose more requirements: in addition to requiring $c \propto (1-z)$ as the base, let's require a good solution for drift $w = 0$ for all parameters γ. Let's require $c = [\gamma + (1-\gamma)2(1-z)]C$ to be exact equilibria for drift $w = 0$.

   Repeat analogous steps to those of the previous part to argue that a suitable embedding is the family of problems

$$\begin{aligned} &\frac{\partial c}{\partial t} = wc + \frac{\partial^2 c}{\partial z^2}, \\ &[w + 2(1-\gamma)]c + (2-\gamma)\frac{\partial c}{\partial z} = 0 \text{ on } z = 0, \\ &[w + 2(1-\gamma)]c + \gamma\frac{\partial c}{\partial z} = 0 \text{ on } z = 1. \end{aligned}$$

   As in the previous part, write Reduce code to derive, or verify by substitution, that the slow subspace of this system is

$$\begin{aligned} c &= 2(1-z)C + \gamma(-1+2z)C + w\left(\tfrac{1}{3} - \tfrac{4}{3}z + z^2\right)C \\ &\quad + w\gamma\left(\tfrac{1}{2} - \tfrac{1}{3}z - z^2\right)C + w\gamma^2\left(-\tfrac{1}{3} + \tfrac{2}{3}z\right)C + \mathcal{O}(w^2), \\ \dot C &= \mathcal{O}(w^2). \end{aligned}$$

   Argue that this approximation looks reasonable for drift up to $w \approx 1$, or 2.

3. Using Reduce, or otherwise, find that the problem posed in the previous part has undesirable nontrivial evolution $\dot{C} \approx -\frac{1}{4}w^2C$. Through modifying the boundary conditions by replacing wc with $[w + A(1-\gamma)w^2]c$, for some unknown A, show that one can find slow subspace approximations with evolution $\dot{C} = \mathcal{O}(w^4)$ by choosing $A = \frac{1}{6}$. Hence argue that this may provide a good base for introducing sediment dynamics within a more complicated multiphysics problem.

Exercise 5.21. Revisit Example 5.3: $\dot{x} = -xy$, $\dot{y} = -y - xy + \epsilon$, for small parameter ϵ. Construct slow manifolds and models for small (ϵ, x) and also just for small ϵ. Compare and contrast the two different approximations for the dynamics and their theoretical support.

Exercise 5.22 (revisit melting). Revisit the example Stefan problem of melting introduced in section 4.8. Recall that the bar and melt stretch from $x = 0$ to $x \to \infty$ with temperature field $u(x,t)$. Heat the end at $x = 0$ to maintain the fixed temperature boundary condition $u(0,t) = 1$ (nondimensionally). At any time the bar will have melted to a position $x = s(t)$ so that temperature $u(x,t) > 0$ for the melt $0 \leqslant x < s$, whereas temperature $u = 0$ for the solid $s < x < \infty$. Assume continuity of temperature at the melt-solid interface so that $u(s,t) = 0$. Heat diffuses through the melt according to the diffusion equation

$$\frac{\partial u}{\partial t} = \frac{\partial^2 u}{\partial x^2}\,.$$

The interesting interface, $x = s(t)$, between the melt and the solid moves in proportion to the flux of heat diffusing from the melt into the interface:

$$\frac{ds}{dt} = \dot{s} = -k\frac{\partial u}{\partial x} \quad \text{at } x = s\,.$$

Recall that section 4.8 finds a slow manifold model by assuming that the interface constant k is asymptotically small.

This exercise pursues a different approach. Recall that the quasi-equilibrium temperature distribution is the linear $u = 1 - x/s$. Is there another way to make this an exact equilibrium? Yes. We could artificially introduce a counterflux into the interface condition that "pins" the interface. We could change the interface condition to $ds/dt = -k\left(\partial u/\partial x + 1/s\right)$ at $x = s$, as then $u = 1 - x/s$ becomes a set of equilibria for all interface locations s. But this is nonphysical and so surely irrelevant. Perhaps, but we access the physically relevant case by introducing a homotopy parameter γ into the interface condition. Consider

$$\frac{ds}{dt} = -k\left(\frac{\partial u}{\partial x} + \frac{1-\gamma}{s}\right) \quad \text{at } x = s:$$

when parameter $\gamma = 1$ the artifice disappears and we recover the physical problem, but when $\gamma = 0$ we find that the system has a set of equilibria on which to base a slow manifold.

1. Argue, based upon the set of equilibria at $\gamma = 0$, that there exists a slow manifold[77] for finite parameter γ. Why could evaluating at $\gamma = 1$ give an

[77] The emergence of the slow manifold is problematical as here the spectrum of the linear operator is hard to establish.

approximate model for the physics? Argue that the resulting model may be global in parameter k (but see section 4.6).

2. Derive that the first nontrivial approximation to the slow manifold model for small γ is

$$u \approx 1 - \frac{x}{s} + \frac{\gamma k}{6(1+k/3)}\left(\frac{x^3}{s^3} - \frac{x}{s}\right) \quad \text{such that} \quad \dot{s} \approx \frac{\gamma k}{s(1+k/3)}.$$

3. Modify the computer algebra of Algorithm 4.3 to construct higher-order corrections. Curiously, in the update rules one only needs to modify `linv(x,x)`. Also, invoking the switch `on gcd;` helps cancel common factors.

Answer: Letting $\xi = x/s$,

$$\begin{aligned} u &= 1 - \xi + \frac{\gamma k}{2(3+k)}\left(\xi^3 - \xi\right) \\ &\quad - \frac{\gamma^2 3k^2}{40(3+k)^3}\xi(1-\xi^2)[1+3k-(9+3k)\xi^2] + \mathcal{O}(\gamma^3), \\ s\dot{s} &= \frac{\gamma 3k}{3+k} + \frac{\gamma^2 6k^3}{5(3+k)^3} + \mathcal{O}(\gamma^3). \end{aligned}$$

Part II Summary

- Center manifold theorems support the low-dimensional modeling of dynamical systems (even for infinite-dimensional, nonautonomous systems; see section 4.9).
  - The existence of a model (Theorem 4.3) is based upon the eigenvalues and eigenvectors at a base equilibrium or set of equilibria.
  - The model emerges over a finite domain in state space because exponentially quick attraction is maintained upon finite perturbations in smooth dynamical systems (Theorem 4.5).
  - We can construct a model to any specified degree of asymptotic approximation (Theorem 4.8) by solving the governing equations to the same specified degree, with a great deal of flexibility (Theorem 4.10).

  Importantly, these theorems support modeling in a finite-sized domain: a pure mathematician phrases it as "there exists a neighborhood," whereas the applied mathematics optimism principle conjectures that the neighborhood is big enough to include the regime of interest. Indeed it is straightforward to construct examples where the "neighborhood" is the entirety of state space.

- The center manifold theorems are based upon an equilibrium. But often in modeling we find a family of equilibria, such as the chemical kinetics of section 4.7 or a film of water that can be of any constant depth. In such a case the models that are immediately local around each equilibrium unite to form a model that is global across the whole family to be valid in a neighborhood of the entire set of equilibria.

- Some people argue that alternative approaches give a better model. But remember that all methods that adhere to the principle of fidelity to the original dynamics must necessarily be describing the same set of solutions and so share the same domain of validity. Section 4.6 showed how the same results are obtained from different scalings, and from a singular perturbation approach.

- The coordinate-free approach of section 5.3 empowers users to find center/slow manifold models in any convenient mathematical expression of the physical problem.

- Approximations are found so straightforwardly that in many applications we easily program computers to handle most of the messy details. The iteration relies on solving the homological equation (5.10), $-\mathcal{L}\hat{\mathbf{v}} + \mathcal{E}\hat{\mathbf{g}} = \text{Res}$, for corrections driven by the residual of the equations evaluated at the current approximation.

- Many people are misled into regarding the amplitude (sometimes called an order parameter) as almost preordained. Instead the center manifold approach shows that one can be extremely flexible in choosing an amplitude parameter; see Lemma 5.1 for example. Out of the almost limitless possibilities for a measure of amplitude in a model (see section 5.3.7 for example), the best choice is typically one with good physical relevance, such as the two quasi-stationary probabilities of section 5.3.6. You should explicitly define what you want as an amplitude parametrization for your model, reflecting physical niceties, rather than letting the parametrization be implicit and uncontrolled.

- Global models are very powerful in approximating dynamics over a wide domain. Global modeling is based upon a subspace of manifold of equilibria and often arises via a regularizing rescaling of singular perturbation problems. As seen in the examples of section 5.5.1, and elsewhere, the error in global models varies across its domain.

- The Newton diagrams of section 5.5.2 help determine reliable truncations of the asymptotic approximations to center or slow manifold models. In particular, the linear dynamics of a model, $\dot{\mathbf{s}} \approx \mathcal{G}\mathbf{s}$ as in (5.29), often dictates valid truncations of a model via the appearance of $\mathcal{G}$ in the general homological equation (5.30), $-\mathcal{L}\hat{\mathbf{v}} + \partial\hat{\mathbf{v}}/\partial\mathbf{s}\mathcal{G}\mathbf{s} + \mathcal{E}\hat{\mathbf{g}} = \text{Res}$.

Part III

Macroscale spatial variations emerge from microscale dynamics

Part III Contents

We now turn to the modeling of dynamics on large spatial scales. The term "large" is meaningless unless it is compared to something. The physical situations addressed in this chapter are those with two disparate length scales. Examples include dispersion in a tube where the length of the tube is much larger than the relatively small cross-section, and the flow of water over some surface when the water is relatively thin compared to the large lateral extent of the surface. The challenge is to derive a model for the dynamics over the large scale that accurately captures the macroscale effects of the microscale processes.

Two disparate length scales occur in many interesting examples of continuum fluid and solid mechanics. Chapter 6 introduces briefly the basic PDEs of fluid mechanics that subsequent sections use. The principles of conservation of mass (section 6.2) and conservation of momentum (section 6.3) lead to PDEs that generally hold for many important fluid flows.

To greatly simplify the analysis, Chapter 7 investigates a basic situation, that of the dispersion of material carried in the flow of a viscous fluid along a pipe. Physically, pipes are long and thin: we model the large-scale variations in concentration of the material along the pipe, seeking to ignore small-scale dynamics across the small cross-section of the pipe. The challenge is to provide a sound rationale for modeling the 3D microscale advection-diffusion as a 1D advection-dispersion along the pipe. Center manifold theory provides the route: first via the Fourier transform, in section 7.2, and second, directly, in section 7.3.

Such models of dispersion in pipes are important in chemical engineering, where many reactions takes place in pipes. Similar models are important in the modeling of dispersion in rivers and estuaries. For example, in 1948 Henry Stommel and Lewis Fry Richardson dropped parsnips into Loch Long from the pier at Blairmore to experimentally test models of dispersion in the ocean (Summers, 2005, e.g.): parsnips were chosen because they barely float and so will be carried by the ocean without influence by the wind, are easily visible, and are cheap and biodegradable.

Most problems have nonlinear interactions, even problems as basic as predicting the spread of coffee spilled onto a table. The challenge here is to derive large-scale models of such nonlinear dynamics. The convenient example developed in Chapter 8 is that of the modeling of a relatively thin film of fluid flowing over a solid substrate, such as the spilled cup of coffee. Here the microscale is the small thickness of the fluid, while the macroscale is the relatively large lateral extent of the variations in thickness.

Chapter 8 argues that the key to modeling such situations is simply to treat a lateral space derivative as a small "parameter" and then use the center manifold approach of Part II to construct and support a model of the nonlinear dynamics. For a thin fluid film, the analysis should be carried out to fourth order to obtain a good model (section 8.4). The approach applies to many other physical problems.

But, in practice, many people want a model of the fluid flow that includes the lateral velocity as a dynamic variable, as well as the fluid thickness. Such a model is needed, for example, in simulating drop formation, wave transitions, three-dimensional instabilities, Faraday waves, viscous hydraulic jumps, flow vortices in a compound channel, and flow down and up a step. Chapter 9 shows how to embed physical equations in a family of cognate problems to empower the application of center manifold techniques.

In the particular case of thin fluid layer flow, section 9.2 uses center manifold techniques to derive a model that conserves mass and accurately resolves the effects of inertia, bed drag, hydrostatic gradients, and enhanced effective viscosity. The methodology applies to many physical situations where we want models over large length scales when the small scales have crucial features of the modeling.

Chapter 6

Conservation underlies mathematical modeling of fluids

Contents

This chapter briefly introduces how to describe mathematically the movement of material such as air, water, or heat. We derive differential equations that they must satisfy as they move. The equations are based on the physical principle that material is neither created nor destroyed except by identified processes. In any particular problem it is these differential equations that are solved to make predictions. Alternatively, as in many examples invoked so far, we may model these differential equations with the aid of center manifold theory. Those familiar with basic continuum mechanics may skip this chapter.

For those that want more discussion and illustrative examples than the all too brief introduction of this chapter, I suggest my earlier book on a one-dimensional introduction to continuum mechanics (Roberts, 1994). Most of the principal concepts of continuum dynamics are more easily understood in one-dimensional dynamics rather than in the three-dimensional dynamics introduced here. Among many, another useful introductory book is that by Dean and Dalrymple (1991), especially Chapter 2.

6.1 ▪ Eulerian flow description

Without flows, biological systems are dead. *Eisenberg* (2012)

We describe fluid flow primarily in terms of velocity, density, and pressure fields. Of these, the velocity field, the actual motion of the fluid, is the one of prime importance. Other physical properties arise in certain applications.

Let's adopt a so-called Eulerian description of field variables. The Eulerian approach describes how physical quantities vary in space and time by considering them at specific points in space-time.

- For example, the pressure p is considered as a function of space and time, which in Cartesian coordinates is written $p(x, y, z, t)$. You may see isobars on weather maps; these are contours of the pressure field on the Earth's surface. They show how the pressure field varies on the surface. Over successive days the isobars move, indicating that the pressure field varies in time. Our observations of the pressure field are largely derived from fixed observation stations, that is, we record the pressure at a specific point as time varies—this is the essence of the Eulerian description.

  By contrast, a Lagrangian description of the flow describes conditions experienced by particles of the fluid as they move through space. But we do not adopt the Lagrangian approach.

- Similarly, the density is described by its dependence upon space and time as the function $\rho(x, y, z, t)$.

- The velocity field of a fluid flow is a vector field $\mathbf{q}(x, y, z, t)$. At every point in space the fluid moves with some speed in some direction and is consequently described as a vector. As above, we write this dependence as a function of position and time: $\mathbf{q}(x, y, z, t) = u(x, y, z, t)\mathbf{i} + v(x, y, z, t)\mathbf{j} + w(x, y, z, t)\mathbf{k}$ in Cartesian coordinates.

- Some problems we deal with have circular symmetry and then it is more useful to use cylindrical coordinates (r, θ, z) to parametrize spatial positions. For example, then we write the density as a function $\rho(r, \theta, z, t)$. For vector fields we then must write the field in terms of the unit vectors of the cylindrical coordinate systems: for example, $\mathbf{q} = q_r \mathbf{e}_r + q_\theta \mathbf{e}_\theta + q_z \mathbf{k}$, where the components are functions of (r, θ, z, t), whereas the unit vectors are a function of angle θ.

Example 6.1 (four velocity fields). Some simple velocity fields illustrate possibilities. We describe the velocity field because this is the quantity that most directly reflects the motion of the fluid and thus is often the quantity of prime interest.

- Uniform flow in the x-direction is described simply by $\mathbf{u} = U\mathbf{i}$. The velocity is the same at every point in the fluid, as shown in Figure 6.1(a).
- Solid body rotation, as seen on a turntable or a wheel, is $\mathbf{u} = r\mathbf{e}_\theta = -y\mathbf{i} + x\mathbf{j}$, as seen in Figure 6.1(b). Both the formula and the figure illustrate that the speed of the motion increases linearly with distance from the axis of rotation.

  Although more characteristic of the motion of solids rather than fluids, this flow is precisely that needed in mercury mirrors for astronomic telescopes. Vibration-induced distortions of the flow, and hence of the mirror surface, from bearings have recently been reduced by suspending the backing of the mirror on an air cushion.

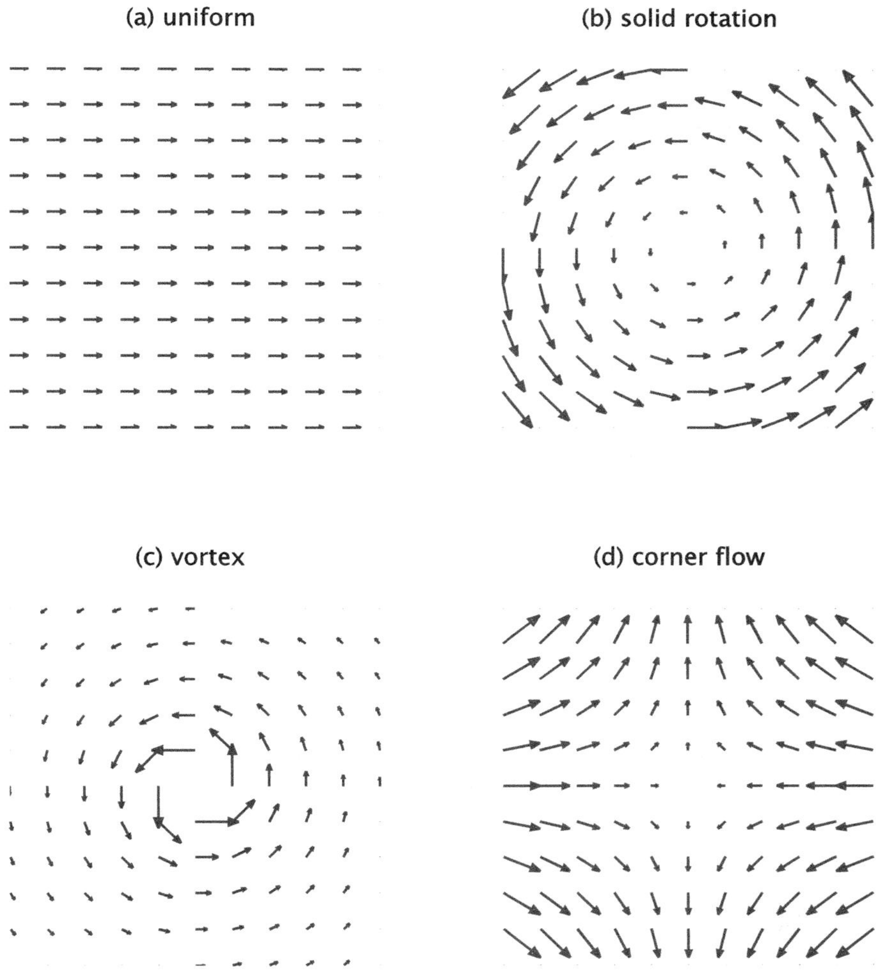

Figure 6.1. *Four example flow fields:* (a) *velocity field of uniform flow in the* x*-direction;* (b) *velocity field of solid body rotation;* (c) *velocity field of the "bathtub" vortex;* (d) *velocity field of a corner flow.*

- A more interesting flow is that of a vortex, as seen in Figure 6.1(c) and in Plate 101 of the album by van Dyke (1982). In this circularly symmetric flow, the velocity field is $\mathbf{q} = \frac{1}{r}\mathbf{e}_\theta = \frac{-y}{x^2+y^2}\mathbf{i} + \frac{x}{x^2+y^2}\mathbf{j}$.

  Although sometimes called the "bathtub" vortex, this flow is more easily seen in a laundry trough, where the plug-hole is more centrally placed. Put some water in the trough, give it a swirl with your hand, pull out the plug, and watch the flow. Floating plastic building blocks on the surface will show the particle motion. By and large, such a building block does not rotate as it revolves around the center of the vortex; it typically maintains the same orientation. Such preservation of orientation indicates an irrotational flow—an important class of flows in fluid dynamics. In contrast, a building block placed on a turntable rotates as it revolves, showing that its motion is rotational.
- As a last example the velocity field $\mathbf{q} = -x\mathbf{i} + y\mathbf{j}$ may be seen as flow turns a corner due to some obstruction. A quadrant of the velocity field shown in

Figure 6.1(d) may be seen in a corner of the flow over a rectangular obstruction shown in Plate 5 by van Dyke (1982).

Underlying molecular characteristics These ideas of velocity and density fields overlook that real fluids are made of molecular particles. In principle the physical fields we use are defined by averages over volumes, with each volume centered on some particular point in space-time.[78] Thus

$$\rho(x,y,z,t) = \frac{1}{\|V\|}\sum(\text{mass of particles in } V),$$

where the volume V is centered on (x,y,z,t). For this averaging to be acceptable we need

- the volume V to be small compared with the phenomenon of interest, where typically our interest lies on a scale between say 1 mm and the 1000 km of the atmosphere, and
- large compared with the molecular scale so that the averages over V are not sensitive to molecules popping in and out of V.

Since there are roughly 10^{23} molecules in each liter of air, it follows that the typical separation between air molecules is on the length scale of 10^{-5} mm. This is very much smaller than the length scales of our interest, so the averaging process appears well defined. The concepts and equations of fluid mechanics are generally accurate.

It is in quite esoteric situations, such as space vehicle reentry into the thin high-altitude atmosphere, that we have to worry about the detailed molecular nature of the flow. The molecular nature of fluids may also need reconsidering in the exploding field of nanomechanics.

6.1.1 ▪ The state space connects with dynamical systems

This book investigates some continuum dynamics using tools of analysis from the modern theory of dynamical systems. Consequently we here put the equations of continuum mechanics into the context of a dynamical system.

The key concept, introduced by Poincaré, is that of a state space. The state space of a system is the set of all possible states of the system. For example, as explained by Abraham and Shaw (1983, §2.1), the state of an ideal pendulum oscillating in two physical dimensions is its position and velocity. When we measure these by the angle from the vertical θ and the angular velocity ω, then the state space is the 2D set of all possible values of (θ, ω).

However, if the pendulum, instead of being swung on the end of a thin rod, is hung by an elastic spring, then two more variables are needed to describe its state: namely, the extension of the spring, say ℓ, and the rate of change of the extension, say w. Thus the state space is enlarged to the set of all possible values of $(\theta, \omega, \ell, w)$, a 4D space.

At any given instant of time, a dynamical system will be in precisely one state. In other words, the state of the dynamical system at any time is represented by one point in state space; call such a state $\mathbf{u}$ in general. Evolution of the system in time, its state

[78] Statistical mechanics provides a more principled route to deriving the partial differential equations of continuum mechanics. But the subtle beautiful details of statistical mechanics are too involved for our level of treatment.

changing, is then represented by movement of a point through state space (Abraham and Shaw, 1983, §1.2), that is, the state $u = u(t)$. This is the basis of the abstract view of an evolving system: no matter how complex the original physics, we represent its evolution by the movement of a point through the state space.

If the mathematical model of the system is consistent, then the system can only evolve in one way from any given state. After all, if the system could evolve in more than one way, then we could not use the model to make predictions because any one of the possible futures could occur. Indeed this is a modeling issue—we only address those models that are predictive; that is, we only address models with a unique future from any given state.[79]

Consequently, at any time and at any state the dynamical system must be moving through state space in some definite direction and at some definite speed. Thus the rate of change of the state is describable by a vector function of position:

$$\dot{u} = \frac{du}{dt} = f(u),$$

where f is some definite function characteristic of the particular dynamical system. We analyze such evolution equations throughout this book.

The equations of incompressible fluid dynamics, the continuity equations together with either the Euler or the Navier–Stokes equations, are placed in this framework. First, the state space is the set of all possible fields, that is, of all possible combinations of velocity $q(x, y, z)$, pressure $p(x, y, z)$, and density $\rho(x, y, z)$. Each configuration of these fields throughout all space is one state in state space. In other words, each point in state space represents the entire velocity and pressure fields throughout all space. Section 1.3 of the fourth volume by Abraham and Shaw (1988) graphically illustrates this, where you can see that each configuration of the fluid flow corresponds in state space to just the position of one point. Then a fluid flow that is unsteady, namely evolves in time, is represented by the movement of a point through state space. Such a state space representation of fluid flow is an enormous abstraction; for example, all the complex interaction and development of a turbulent flow field is represented by the movement of one point in a state space.

The complication is that because states of the fluid flow are described in terms of functions of space, usually velocity $q(x, y, z)$ and pressure $p(x, y, z)$ (as typically we assume the density is constant), the dimensionality of the state space is infinite. Crudely speaking, at each point in space there is the freedom for the fluid to have any pressure and velocity, but there is an infinite number of spatial points and hence there is an infinite number of "variables" needed to describe the state of the fluid flow.

Second, the equations of fluid mechanics govern the evolution through state space. Although, strictly speaking, the continuity equation and Navier–Stokes equations are not in the dynamical system form $\dot{u} = f(u)$, they are effectively equivalent. You can imagine that rather than letting the state space be the entire set of all possible q and p, we may let the state space be the set of all such fields that satisfy the extra proviso that the velocity field is divergence free. That is, the state space is effectively a subset of the $q(x, y, z), p(x, y, z)$-space. Then the Navier–Stokes equation (or Euler equation) effectively describes the vector field, f, in this effective state space; the pressure field may be viewed as being used to keep the velocity field divergence free.

[79]Except for the chapters on stochastic differential equations and their modeling.

6.2 ▪ Conservation of mass conserves material

We need to find equations that reflect how fields such as the fluid velocity $\mathbf{q}$ and density ρ evolve in time. The evolution comes from physical laws of conservation. First, the material making up the fluid can neither appear nor disappear. This conservation of mass leads to the so-called continuity equation.

Quick and dirty An extremely quick and dirty method is to adapt the one-dimensional continuity equation (Roberts, 1994, eq. (2.1)),

$$\frac{\partial \rho}{\partial t} + \frac{\partial (u\rho)}{\partial x} = 0,$$

to three dimensions. We simply recognize that time derivatives and density remain unchanged in the adaptation, whereas the one-dimensional scalar velocity u becomes a vector quantity $\mathbf{q}$ in three dimensions, and one-dimensional spatial gradients $\partial/\partial x$ become one of the vector differential operators grad ∇, or div $\nabla\cdot$, as appropriate. Thus the above one-dimensional continuity equation translates to

$$\frac{\partial \rho}{\partial t} + \nabla \cdot (\mathbf{q}\rho) = 0.$$

Choosing the divergence is indicated because the time derivative shows that the equation is a scalar equation, and the div operator is an appropriate way to produce a scalar quantity from the vector flux $\mathbf{q}\rho$.

6.2.1 ▪ Conserving fluid gives the continuity equation

We now turn to a more acceptable argument for deriving the above continuity equation.[80] As in one dimension (Roberts, 1994, §2.1), I prefer the clarity of an integral argument.

Consider any fixed volume V within the fluid. The mass of fluid inside the volume is $m = \int_V \rho\, dV$. Over time this mass varies as fluid is carried into or out of the volume by the flow of the fluid. The rate of change of mass is, by definition,

$$\frac{dm}{dt} = \int_V \frac{\partial \rho}{\partial t}\, dV.$$

But this rate of change must be equal to the net rate of fluid entering the volume across the surface, S, of the volume. Across any small area dS of the surface, fluid mass crosses into V at a rate $\rho\mathbf{q} \cdot (-\hat{\mathbf{n}})$ (where $\hat{\mathbf{n}}$ is the unit vector normal to dS but pointing out of V), namely

- proportional to the local density on that surface, and
- proportional to the velocity component normal to the surface.

Consequently, integrating over the entire surface, it must also be true that

$$\begin{aligned}\frac{dm}{dt} &= -\int_S \rho\mathbf{q} \cdot \hat{\mathbf{n}}\, dS \\ &= -\int_V \nabla \cdot (\rho\mathbf{q}) dV \quad \text{by the divergence theorem.}\end{aligned}$$

[80] Dean and Dalrymple (1991, §2.2.1) give a similar development but in differential terms.

Equating these two expressions for dm/dt and combining them into the one integral leads to the penultimate result that

$$\int_V \left[\frac{\partial \rho}{\partial t} + \nabla \cdot (\rho \mathbf{q})\right] dV = 0 .$$

This identity holds for all volumes V in the fluid.

Now, the only way that the integral of a particular continuous integrand can be zero for *all* volumes V is if the integrand is *identically zero.* Consider, if the integrand is ever nonzero at a point, then by continuity it must be the same sign in some small volume V' around that point, and so the integral over V' would be nonzero. But such nonzero-ness contradicts that the integral is always zero. Thus the integrand must be zero everywhere. Consequently the previous equation assures us that everywhere in the fluid

$$\frac{\partial \rho}{\partial t} + \nabla \cdot (\rho \mathbf{q}) = 0 . \tag{6.1}$$

This is the general continuity equation.

In most familiar circumstances the continuity equation (6.1) simplifies. For example, if the density of the fluid is everywhere the same for all time, ρ is constant, then all the derivatives of ρ are zero and the continuity equation reduces to the requirement that the velocity field must be divergence free:

$$\nabla \cdot \mathbf{q} = 0 . \tag{6.2}$$

This version of the continuity equation (6.2) is much more general than suggested; as well as applying to constant-density fluids, it also applies to so-called incompressible flows. For example, the water in the ocean is effectively incompressible and its flow satisfies (6.2); yet oceanic water has significant density variations. Indeed such density variations are vitally important in the structure of the world's oceans:

- cold Arctic and Antarctic water is dense and sinks below the warmer surface layers of midlatitude oceans to form the cold abyssal ocean water;

- warmish but very saline and consequently dense Mediterranean water flows out of the Gibraltar Straits and so it also sinks below the surface layers of the Atlantic ocean;

- conversely, fresh and thus relatively light water from the Amazon flows many hundreds of kilometers out from the South American coast as surface water.

The important characteristic here is that a small volume of water maintains its identity for very long periods of time. Indeed oceanographers deduce the source of a sample of the ocean simply by analyzing its salinity and trace elements.

So consider the general continuity equation (6.1) from the point of view of a fixed parcel of fluid rather than from a fixed point in space. Expanding the divergence of the product gives

$$\frac{\partial \rho}{\partial t} + \mathbf{q} \cdot \nabla \rho + \rho \nabla \cdot \mathbf{q} = 0 .$$

Now the rate of change of density of the parcel of fluid is, by the chain rule,

$$\begin{aligned}\frac{d\rho}{dt} &= \frac{\partial\rho}{\partial t} + \frac{\partial\rho}{\partial x}\frac{dx}{dt} + \frac{\partial\rho}{\partial y}\frac{dy}{dt} + \frac{\partial\rho}{\partial z}\frac{dz}{dt} \\ &= \frac{\partial\rho}{\partial t} + \frac{\partial\rho}{\partial x}u + \frac{\partial\rho}{\partial y}v + \frac{\partial\rho}{\partial z}w \\ &= \frac{\partial\rho}{\partial t} + \mathbf{q}\cdot\nabla\rho,\end{aligned}$$

as appears in the first two terms of the expanded continuity equation above. Thus if *each and every* parcel of fluid is itself incompressible, its density does not change, $d\rho/dt = 0$ throughout the fluid, and so the general continuity equation (6.1) reduces to the simpler divergence-free requirement (6.2). This simplification occurs even though the density may vary in space and time, as it does in the ocean, so long as each parcel of fluid is itself incompressible.

Each of the four velocity fields of Example 6.1 is divergence free. For example, the vortex flow field satisfies

$$\nabla\cdot\mathbf{q} = \frac{\partial}{\partial x}\left(\frac{-y}{x^2+y^2}\right) + \frac{\partial}{\partial y}\left(\frac{x}{x^2+y^2}\right) = \frac{y2x}{(x^2+y^2)^2} + \frac{-x2y}{(x^2+y^2)^2} = 0,$$

by cancellation. Hence each of the four could occur in the flow of an incompressible fluid.

The material derivative Appearing within the above derivation is the very important concept of the *material derivative*, namely the derivative with respect to time of some quantity as seen by a parcel of the material; the above paragraph discussed the density as seen by a fluid parcel. Where needed the material derivative of some quantity, c say, will be denoted by

$$\frac{dc}{dt} = \frac{\partial c}{\partial t} + \mathbf{q}\cdot\nabla c. \tag{6.3}$$

It has two components:

- $\partial c/\partial t$ from the rate of change of c at a fixed location in space, and
- $\mathbf{q}\cdot\nabla c$ from the movement of the fluid parcel through space with velocity $\mathbf{q}$, and so it experiences change due to spatial gradients of c.

At first acquaintance the difference between the material derivative and the derivative at a fixed point often causes confusion. Part of this confusion arises from deficient mathematical notation. In particular, the notation for differentiation carries implicit assumptions with it: the problem being that these assumptions change when we change from a material derivative to a derivative at a fixed point. Chemistry has an answer, albeit too cumbersome for common use. Chemists routinely denote which variables are fixed as the differentiation is performed. For example, $(\partial\rho/\partial t)_{xyz}$ denotes the derivative with respect to time of the density, keeping position, (x, y, z), fixed. Conversely, if the locations of individual fluid particles are described by functions $(X(\mathbf{a}, t), Y(\mathbf{a}, t), Z(\mathbf{a}, t))$ and the density of fluid at that particle is $\rho(\mathbf{a}, t)$, then $(\partial\rho/\partial t)_{\mathbf{a}}$ denotes the derivative with respect to time of the density, keeping the fluid particle under consideration fixed. This last is the same as the material derivative. That is,

$$\left(\frac{\partial c}{\partial t}\right)_{\mathbf{a}} = \left(\frac{\partial c}{\partial t}\right)_{xyz} + \mathbf{q}\cdot(\nabla c)_t\,,$$

as usually expressed more concisely by the material derivative (6.3).

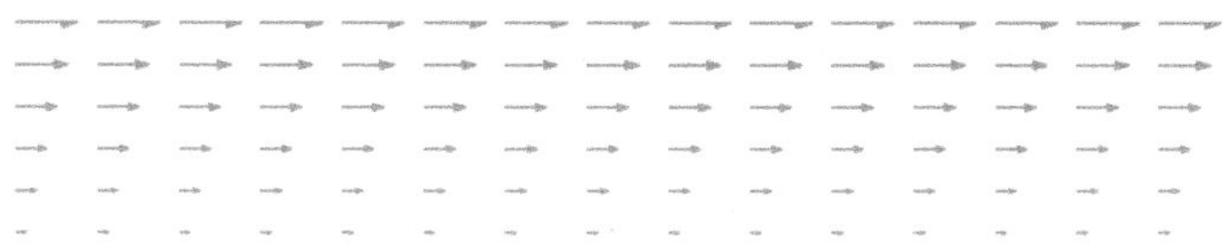

Figure 6.2. *Example velocity field of a shear flow of fluid to the right with a velocity proportional to the vertical location.*

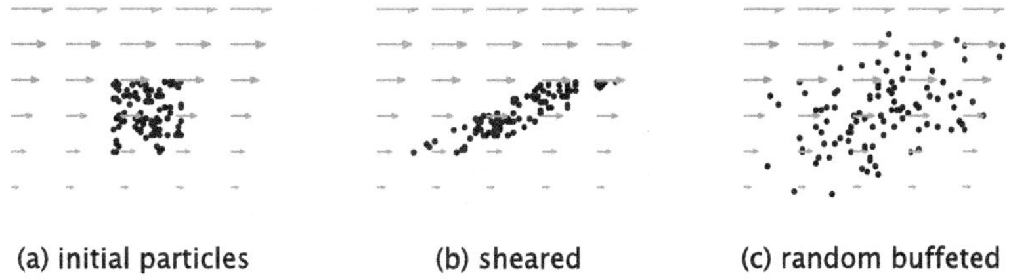

Figure 6.3. *In the shear flow of Figure* 6.2*:* (a) *shows an initial clump of* 100 *particles;* (b) *being carried by the different velocities of the shear flow, this clump spreads out in the horizontal;* (c) *but also random molecular buffeting causes the particles to spread (diffuse) in all directions.*

6.2.2 ▪ Molecules, particles, and heat move and diffuse

As a fluid moves it carries with it material and properties such as salts and heat energy. Such material also diffuses through the fluid. Such transport of material is crucial: for example, blood carries nutrients to our cells and carries away waste. Let's derive the basic PDEs describing such transport.

As in section 6.2.1, conservation of material is the inspiring principle. This section follows the same approach but generalizes section 6.2.1.

First, define the concentration field c of the material under consideration. Be careful to clarify whether the concentration is the number of particles per unit volume, or the mass per unit volume (as in the density), or the amount of heat energy per unit volume (otherwise known as the temperature). However, apart from systems expressing chemical reactions, the differential equations are often linear in the concentration so in exploratory studies one is often casual about the precise units of the concentration field. For definiteness, suppose we measure concentration c by the mass per unit volume.

Second, consider the material in any fixed volume V within the fluid. Given a concentration field c, the mass of material inside the volume is $m = \int_V c\,dV$. Over time this mass varies as the material moves into or out of the volume. The rate of change with time of the mass of material is, by definition,

$$\frac{dm}{dt} = \int_V \frac{\partial c}{\partial t}\,dV\,.$$

But this rate of change must be equal to the net rate of material entering the volume across the surface, S, of the volume. Figure 6.3 illustrates the two principal mechanisms for movement of material: the fluid flow (as for the fluid density) and additionally the diffusion of material.

- Figure 6.3(b) illustrates how particles are carried by a flow. Across any small area dS of the surface, material is carried into V at a rate $c\mathbf{q}\cdot(-\hat{\mathbf{n}})$ (where $\hat{\mathbf{n}}$ is the unit vector normal to dS but pointing out of V), namely

- proportional to the local concentration on that surface, and
- proportional to the velocity component normal to the surface.

Consequently, integrating over the entire surface of the volume V, the flow contribution to the material entering the volume (leaving if negative) is $-\int_S c\mathbf{q}\cdot\hat{\mathbf{n}}\,dS = -\int_V \boldsymbol{\nabla}\cdot(c\mathbf{q})dV$ by the divergence theorem.

- Small particles of material, such as fine sediment or pollen, are unceasingly buffeted by random molecular impacts. Consequently, such small particles undergo small random motions just like molecules themselves. Such small random motions accumulate over time into *Fickian diffusion*, as illustrated in Figure 6.3(c). For example, imagine two small neighboring regions in space, one of which has high concentration and one with low concentration; then small random motion of all the particles results in more particles moving from the high-concentration region to the low-concentration region than vice versa simply because there are more particles that are moving; consequently there is a net movement of material from high to low that is proportional to the difference in concentration. The analogue for continuous variations is that random movements generate a flux of material proportional, but opposite, to the gradient of concentration, leading to the Fickian model that the flux of material is $-\kappa\boldsymbol{\nabla}c$ for some characteristic diffusion constant κ.

  Now recall our fixed volume V. Across any small area dS of the surface, material diffuses into V at a rate $(-\kappa\boldsymbol{\nabla}c)\cdot(-\hat{\mathbf{n}})$ (where $\hat{\mathbf{n}}$ is the unit vector normal to dS but pointing out of V). Consequently, integrating over the entire surface of the volume V, the diffusive contribution to the material entering the volume (leaving if negative) is $\int_S \kappa(\boldsymbol{\nabla}c)\cdot\hat{\mathbf{n}}\,dS$. By the divergence theorem, this contribution is $\int_S \kappa(\boldsymbol{\nabla}c)\cdot\hat{\mathbf{n}}\,dS = \int_V \boldsymbol{\nabla}\cdot(\kappa\boldsymbol{\nabla}c)dV$.

Putting together these two contributions to the rate of change in the mass of material in volume V, we deduce

$$\frac{dm}{dt} = \int_V \frac{\partial c}{\partial t}\,dV = -\int_V \boldsymbol{\nabla}\cdot(c\mathbf{q})dV + \int_V \boldsymbol{\nabla}\cdot(\kappa\boldsymbol{\nabla}c)dV.$$

Rearrange this equation to have just one integral: $\int_V \partial c/\partial t + \boldsymbol{\nabla}\cdot(c\mathbf{q}) - \boldsymbol{\nabla}\cdot(\kappa\boldsymbol{\nabla}c)\,dV = 0$. As before, the only way that the integral of a continuous integrand can be zero for *all* volumes V is if the integrand is *identically zero.* Consequently the previous equation assures us that everywhere in the fluid

$$\frac{\partial c}{\partial t} + \boldsymbol{\nabla}\cdot(c\mathbf{q}) - \boldsymbol{\nabla}\cdot(\kappa\boldsymbol{\nabla}c) = 0. \tag{6.4}$$

This is the general PDE for the conservation of material as particles are carried by the fluid and buffeted by random molecular collisions.

In many applications the diffusivity κ is constant in space-time and is taken outside the divergence. Further, in many applications the fluid is incompressible so the velocity field is divergence free, PDE (6.2). When both hold, the conservation PDE (6.4) reduces to the commonly used form

$$\frac{\partial c}{\partial t} + \mathbf{q}\cdot\boldsymbol{\nabla}c - \kappa\nabla^2 c = 0. \tag{6.5}$$

When heat energy is conserved, the same PDEs apply to the temperature field. Sometimes, as in chemical reactions or the temperature field, there is a source of material (chemicals or heat) expressed as extra terms on the right-hand sides, often nonlinear terms.

6.3 ▪ Navier–Stokes conserves fluid momentum

The second physical principle, which leads to a governing differential equation, is that of conservation of momentum. This principle leads us to either the Navier–Stokes equation or the Euler equation depending upon the importance of internal friction within the fluid.

Quick and dirty An extremely quick and dirty method is to adapt the one-dimensional momentum equation (Roberts, 1994, eq. (3.2))

$$\rho\left(\frac{\partial u}{\partial t}+u\frac{\partial u}{\partial x}\right)=F+\frac{\partial \sigma}{\partial x}$$

to three dimensions. On the left-hand side we observe the material derivative of the velocity. On the right-hand side, F denotes body forces, such as gravity $\rho \mathbf{g}$, and so becomes a vector, whereas the end forces σ could remain as the scalar pressure field, $-p$, which exerts a normal force across any surface in the fluid. Consequently the above one-dimensional momentum equation translates to

$$\rho\frac{d\mathbf{q}}{dt}=\rho\left(\frac{\partial \mathbf{q}}{\partial t}+\mathbf{q}\cdot\nabla\mathbf{q}\right)=\mathbf{F}-\nabla p\,.$$

However, this equation is only valid for fluids that are not "sticky," that is, not viscous. In a viscous fluid, the fluid particles exert tangential forces across surfaces; for example, they experience a tangential drag when slipping over a solid substrate. For this common case the stress becomes a rank-two tensor, denoted $\boldsymbol{\sigma}$.[81] Then for a viscous fluid the above momentum equation generalizes to

$$\rho\frac{d\mathbf{q}}{dt}=\rho\left(\frac{\partial \mathbf{q}}{\partial t}+\mathbf{q}\cdot\nabla\mathbf{q}\right)=\mathbf{F}+\nabla\cdot\boldsymbol{\sigma}\,.$$

Interestingly, this momentum equation is directly analogous to Newton's second law, which asserts that (mass) × (acceleration) = (net force). In the correspondence between the momentum equation and Newton's second law,

- mass is represented by the density ρ;
- acceleration of the mass represents the acceleration of fluid particles, which is given by the material derivative of the velocity, $d\mathbf{q}/dt$; and
- net force is represented by the body force $\mathbf{F}$ and internal stresses $\boldsymbol{\sigma}$.

We turn to a more conventional derivation of the momentum equation, one in which we see more readily the meaning of new concepts such as the stress (Batchelor, 1979, §1.3, e.g.).

[81] A rank-n tensor is some physical quantity with n directional properties; for example, a scalar is a rank-zero tensor; a vector is a rank-one tensor; and the array of second derivatives of a scalar, $\partial^2\phi/\partial x_i \partial x_j$ say, is a rank-two tensor, as each derivative may be in two different directions.

6.3.1 ▪ Stress tensor

Consider a small tetrahedral parcel of fluid with three faces normal to the coordinate axes and the fourth face with normal $\hat{\mathbf{n}}$. Across each face, the fluid outside the tetrahedron exerts a force, or traction, on the fluid inside; call the forces $A_x \mathbf{T}_x$, $A_y \mathbf{T}_y$, $A_z \mathbf{T}_z$, and $A_n \mathbf{T}_n$, respectively, where A_i is the area of the ith face and $\mathbf{T}_i$ are the stresses.

The stress is linear in the normal First, if there is any imbalance in the net force applied by these stresses, then the fluid tetrahedron must accelerate. Newton's second law then indicates that

$$\rho V \frac{d\mathbf{q}}{dt} = A_x \mathbf{T}_x + A_y \mathbf{T}_y + A_z \mathbf{T}_z + A_n \mathbf{T}_n ,$$

where V is the volume of the tetrahedron. Hence, crudely, the acceleration must be proportional to (surface area)/(volume), which becomes infinite for arbitrarily small volumes. Such infinite acceleration cannot be permitted, and hence there cannot be any net imbalance of the forces on the tetrahedron. Thus

$$\begin{aligned} \mathbf{T}_n &= -\frac{A_x}{A_n}\mathbf{T}_x - \frac{A_y}{A_n}\mathbf{T}_y - \frac{A_y}{A_n}\mathbf{T}_z \\ &\qquad \text{(by projection of areas)} \\ &= (\hat{\mathbf{n}} \cdot \mathbf{i})\mathbf{T}_x + (\hat{\mathbf{n}} \cdot \mathbf{j})\mathbf{T}_y + (\hat{\mathbf{n}} \cdot \mathbf{k})\mathbf{T}_z . \end{aligned}$$

This equation determines that the stress across a plane with any normal $\hat{\mathbf{n}}$ is linear in $\hat{\mathbf{n}}$: for some stress tensor $\boldsymbol{\sigma}$,

$$\mathbf{T}_n = \hat{\mathbf{n}} \cdot \boldsymbol{\sigma} . \tag{6.6}$$

Indeed, in a matrix representation, $\mathbf{T}_x$, $\mathbf{T}_y$, and $\mathbf{T}_z$ form the rows of the stress tensor:

$$\boldsymbol{\sigma} = \begin{bmatrix} \sigma_{xx} & \sigma_{xy} & \sigma_{xz} \\ \sigma_{yx} & \sigma_{yy} & \sigma_{yz} \\ \sigma_{zx} & \sigma_{zy} & \sigma_{zz} \end{bmatrix} ,$$

where σ_{ij} is the component of the stress in the jth coordinate direction upon a plane with normal in the direction of the ith coordinate.

The stress tensor is symmetric By similar arguments, one reasons that to keep the angular acceleration of a fluid parcel finite as one considers smaller and smaller sizes, the moments of the forces on a fluid particle must vanish. Consider the moments about the vertical axis passing through the centroid of the face A_n of the tetrahedron:

- there are no contributions to the moment from the face A_n or A_z;
- the face A_x contributes $\frac{\Delta x}{3} A_x \mathbf{T}_x \cdot \mathbf{j} = V\sigma_{xy}$;
- the face A_y contributes $\frac{\Delta y}{3} A_y \mathbf{T}_y \cdot (-\mathbf{i}) = -V\sigma_{yx}$.

The net moment is thus $V(\sigma_{xy} - \sigma_{yx})$. However, the moment of inertia is proportional to $V^{5/3}$, and so to avoid infinite angular accelerations of very small parcels of fluid we must require that $\sigma_{xy} = \sigma_{yx}$, and similarly for other axes so that in general $\sigma_{ij} = \sigma_{ji}$. Hence the stress tensor is symmetric.

The pressure By convention, the pressure of the fluid is defined such that the trace of the stress tensor is −3 times the pressure p. That is,

$$p = -\frac{1}{3}\left(\sigma_{xx} + \sigma_{yy} + \sigma_{zz}\right).$$

Then the stress tensor may be written

$$\sigma = \begin{bmatrix} \sigma_{xx} & \sigma_{xy} & \sigma_{xz} \\ \sigma_{yx} & \sigma_{yy} & \sigma_{yz} \\ \sigma_{zx} & \sigma_{zy} & \sigma_{zz} \end{bmatrix} = \begin{bmatrix} -p + \tau_{xx} & \tau_{xy} & \tau_{xz} \\ \tau_{yx} & -p + \tau_{yy} & \tau_{yz} \\ \tau_{zx} & \tau_{zy} & -p + \tau_{zz} \end{bmatrix}.$$

The deviatoric stress tensor τ_{ij} measures the departure from isotropy of the stress tensor. If τ_{ij} are all zero, then the stress tensor $\boldsymbol{\sigma} = -pI$ and hence the stress across *any* surface, normal $\hat{\mathbf{n}}$, is $\boldsymbol{\sigma} = \hat{\mathbf{n}} \cdot (-pI) = -p\hat{\mathbf{n}}$. That is, the force across any surface is directly inward and of magnitude p. This is precisely the accepted understanding of pressure.

Conversely, if not all τ_{ij} are zero, then a local change of coordinates[82] will put the stress tensor in diagonal form (always possible for a symmetric matrix):

$$\begin{bmatrix} -p + \tau'_{xx} & 0 & 0 \\ 0 & -p + \tau'_{yy} & 0 \\ 0 & 0 & -p + \tau'_{zz} \end{bmatrix}.$$

Hence a nonzero deviatoric stress tensor implies that in some directions the stresses will be stronger than in others, causing a deformation of parcels of fluid.

6.3.2 ▪ Momentum equation

Consider any fixed volume V within the fluid. The net momentum of the fluid inside the volume is $\mathbf{p} = \int_V \rho \mathbf{q}\, dV$. Over time this momentum varies as it is carried into or out of the volume by the flow of the fluid. The rate of change of net momentum is, by definition,

$$\frac{d\mathbf{p}}{dt} = \int_V \frac{\partial}{\partial t}(\rho \mathbf{q})\, dV.$$

But this must be equal to the net rate of momentum

- entering the volume across the surface, S, of the volume;
- generated internally by external body forces such as gravity; and
- induced by stresses imposed on the surface of the volume by the fluid immediately outside.

Consider each of these in turn. Across any small area dS of the surface, fluid momentum crosses into V at a rate $\rho \mathbf{q}(-\hat{\mathbf{n}}) \cdot \mathbf{q}$ (where $\hat{\mathbf{n}}$ is the unit vector normal to dS but pointing out of V), namely

- proportional to the local momentum density, $\rho \mathbf{p}$, on that surface, and
- proportional to the velocity component normal to the surface.

[82]The pressure, being proportional to the trace of the stress tensor, is invariant to such changes in coordinates.

External body forces[83] generate momentum inside V at a rate $\int_V \mathbf{F}\,dV$, whereas the stresses applied on each small part of the surface, dS, are $\mathbf{T}\,dS = \hat{\mathbf{n}}\cdot\boldsymbol{\sigma}\,dS$. Consequently, summing all these contributions, it must also be that

$$\begin{aligned}\frac{d\mathbf{p}}{dt} &= -\int_S \rho\mathbf{q}\hat{\mathbf{n}}\cdot\mathbf{q}\,dS + \int_V \mathbf{F}\,dV + \int_S \hat{\mathbf{n}}\cdot\boldsymbol{\sigma}\,dS \\ &\quad\text{(by the divergence theorem)} \\ &= \int_V -\nabla\cdot(\rho\mathbf{q}\mathbf{q}) + \mathbf{F} + \nabla\cdot\boldsymbol{\sigma}\,dV.\end{aligned}$$

Equating these two expressions for $d\mathbf{p}/dt$ and combining them into the one integral leads to the penultimate result that

$$\int_V \left[\frac{\partial}{\partial t}(\rho\mathbf{q}) + \nabla\cdot(\rho\mathbf{q}\mathbf{q}) - \mathbf{F} - \nabla\cdot\boldsymbol{\sigma}\right] dV = 0$$

for all volumes V in the fluid. As in section 6.2.1, such an integral can only be zero *for all* volumes V if the integrand is everywhere zero. Thus we deduce the basic form of the *momentum equation*

$$\frac{\partial}{\partial t}(\rho\mathbf{q}) + \nabla\cdot(\rho\mathbf{q}\mathbf{q}) = \mathbf{F} + \nabla\cdot\boldsymbol{\sigma}. \tag{6.7}$$

In conjunction with continuity However, in virtually all cases we only use this equation in conjunction with the continuity equation (6.1). In this case, after expanding the derivatives of the products on the left-hand side as

$$\frac{\partial}{\partial t}(\rho\mathbf{q}) + \nabla\cdot(\rho\mathbf{q}\mathbf{q}) = \rho\frac{\partial\mathbf{q}}{\partial t} + \mathbf{q}\frac{\partial\rho}{\partial t} + \mathbf{q}\nabla\cdot(\rho\mathbf{q}) + \rho(\mathbf{q}\cdot\nabla)\mathbf{q},$$

we recognize that the middle two terms are simply $\mathbf{q}$ times the continuity equation and so vanish identically. This gives us the more usual form of conservation of momentum, directly analogous to Newton's second law:

$$\rho\frac{d\mathbf{q}}{dt} = \rho\left(\frac{\partial\mathbf{q}}{\partial t} + (\mathbf{q}\cdot\nabla)\mathbf{q}\right) = \mathbf{F} + \nabla\cdot\boldsymbol{\sigma}. \tag{6.8}$$

6.3.3 ▪ The Euler equations for an ideal fluid

In a "frictionless" fluid, one of low viscosity or one that is *inviscid*, parcels of fluid cannot resist uneven stress distributions: if there is any imbalance of stress, indicated by a non-isotropic stress tensor, then the fluid rapidly deforms in response to the uneven forces. Consequently the only stress supportable by an inviscid fluid is pressure, that is, the stress $\boldsymbol{\sigma} = -p\mathbf{I}$. In this case the momentum equation reduces to

$$\frac{d\mathbf{q}}{dt} = \frac{\partial\mathbf{q}}{\partial t} + (\mathbf{q}\cdot\nabla)\mathbf{q} = \frac{1}{\rho}\mathbf{F} - \frac{1}{\rho}\nabla p. \tag{6.9}$$

This PDE is called *Euler's equation*.

For example, in a fluid at rest, $\mathbf{q} = \mathbf{0}$, and acted upon by gravity $\mathbf{F} = -\rho g\mathbf{k}$, we deduce that the pressure must be hydrostatic: $p = p_0 - \rho g z$, where p_0 is the pressure at $z = 0$.

[83] Such as the body force $\mathbf{F} = -\rho g\mathbf{k}$ for downward gravity.

Often, as for this gravity, the body forces are *apparently* removed from Euler's equation by incorporation into the pressure. Any body force that can be written as the gradient of a scalar potential, $F = -\nabla\Phi$,[84] may be absorbed by a pseudopressure $p' = p + \Phi$, and Euler's equation then appears apparently without the body force.

All four flows of Example 6.1 are solutions of Euler's equation for an inviscid fluid.

Irrotational flows are gradient flows We obtain an extraordinarily important class of flows[85] by writing $q = \nabla\phi$ for some *velocity potential* $\phi(x, y, z, t)$. By the continuity equation, the potential ϕ must satisfy Laplace's equation as

$$\nabla^2\phi = \nabla \cdot q = 0 .$$

Euler's equation is automatically satisfied for this class of flow; it reduces to Bernoulli's equation for the pressure since Euler's equation becomes (by a remarkable integration of the nonlinear Euler equation)

$$\begin{aligned} 0 &= \frac{\partial}{\partial t}\nabla\phi + \nabla\left(\frac{1}{2}|q|^2\right) + \frac{1}{\rho}\nabla\Phi + \frac{1}{\rho}\nabla p , \\ &= \nabla\left[\frac{\partial\phi}{\partial t} + \frac{1}{2}|q|^2 + \frac{\Phi}{\rho} + \frac{p}{\rho}\right] . \end{aligned}$$

The only way that the gradient of a quantity can be zero in some spatial domain is if it is constant in space; that is, we obtain Bernoulli's equation,

$$\frac{p}{\rho} + \frac{\partial\phi}{\partial t} + \frac{1}{2}|q|^2 + \frac{\Phi}{\rho} = B(t) , \qquad (6.10)$$

for some function $B(t)$. This equation gives the pressure field for any given flow field of an ideal fluid.

Example 6.2 (two potential flows). Three of the flows in Example 6.1 are potential flows. Consider two.

(a) $q = Ui$ has velocity potential $\phi = Ux$ and the pressure is reduced from hydrostatic: $p = -\Phi - \frac{1}{2}\rho U^2$.

(c) The vortex flow $q = (-yi + xj)/(x^2 + y^2)$ has velocity potential $\phi = \arctan(y/x) = \theta$ in cylindrical coordinates and a pressure $p = -\Phi - \frac{1}{2}\rho/r^2$.[86]

6.3.4 ▪ Newtonian fluids obey the Navier–Stokes equations

Real fluids, such as treacle, paint, and toothpaste, exhibit a startling range of interesting behavior. In such flows the relation between the pattern of flow and the stress tensor can be very complicated. However, the most common fluids found around us, namely air

[84] The body force $F = -\nabla(\rho g z)$ for downward gravity.

[85] Curiously, one can only justify the importance of such inviscid flows by invoking viscous dissipation of rotational energy.

[86] This flow is seen in the swirling water of the bathtub vortex (van Dyke, 1982, Plate 101, e.g.). The funnel shape of the central core of air is determined once you know that in effect air exerts a constant pressure, p_0 say, everywhere upon the free surface of the water. If the body force potential is due to gravity, $\Phi = \rho g z$, and hence the free surface of the water must be such that $p_0 = -\rho g z - \frac{1}{2}\rho/r^2$. Rearranging shows that the height of the free surface in the funnel must be of the form $z = A - B/r^2$.

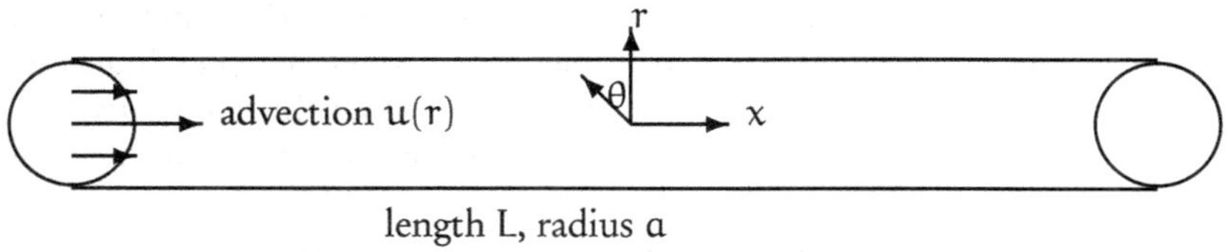

Figure 6.4. *Schematic diagram of pipe flow and the coordinate system.*

and water, are relatively simple. To a good approximation the deviatoric stress tensor is simply a linear function of the local velocity gradients:

$$\tau_{ij} = A_{ijk\ell}\frac{\partial q_k}{\partial x_\ell},$$

where repeated subscripts indicate summation over that subscript. Now, the tensor $A_{ijk\ell}$ has $3^4 = 81$ elements, each of which could be potentially different and would need to be determined for each particular fluid by experiments. However, the triple requirements of isotropy, symmetry, and zero trace of the deviatoric stress tensor (Batchelor, 1979, §3.3) result in the great simplification that

$$A_{ijk\ell} = \mu\left(\delta_{ik}\delta_{j\ell} + \delta_{i\ell}\delta_{jk} - \frac{2}{3}\delta_{ij}\,\delta_{k\ell}\right),$$

where μ is a constant characteristic of the fluid and is called the *viscosity*. Hence, in terms of the flow field, the stress at any point is

$$\sigma_{ij} = -p\delta_{ij} + \mu\left(\frac{\partial q_i}{\partial x_j} + \frac{\partial q_j}{\partial x_i} - \frac{2}{3}\nabla\cdot q\delta_{ij}\right). \tag{6.11}$$

A fluid obeying this relationship between stress and velocity field is called a *Newtonian fluid*.

For an incompressible fluid, a little algebra shows that the momentum equation then reduces to the *Navier–Stokes equation*,

$$\frac{dq}{dt} = \frac{\partial q}{\partial t} + (q\cdot\nabla)q = \frac{1}{\rho}F - \frac{1}{\rho}\nabla p + \nu\nabla^2 q, \tag{6.12}$$

where $\nu := \mu/\rho$ is called the *kinematic viscosity*. For water and air under common conditions the kinematic viscosity is 0.0114 cm^2 / sec and 0.145 cm^2 / sec, respectively. Observe that the term due to viscosity, $\nabla^2 q$, describes an essentially Fickian diffusion of momentum. This diffusion is reflected in many of the flow properties that we later explore.

Example 6.3 (Poiseuille pipe flow). The Navier–Stokes and continuity equations determine the velocity field of the flow of a viscous fluid along a pipe, as shown schematically in Figure 6.4.

Cylindrical coordinates (x, r, θ) are convenient, where x is the distance along the pipe, r is a radius, and θ is an angle around the pipe (see Figure 6.4). Let the fluid velocity have components (u, v, w); that is, $\mathbf{q} = u\mathbf{i} + v\mathbf{e}_r + w\mathbf{e}_\theta$. Then the Navier–Stokes equation and continuity equation are

$$\begin{aligned}
\frac{\partial u}{\partial t} + \mathbf{q}\cdot\nabla u &= -\frac{1}{\rho}\frac{\partial p}{\partial x} + \nu\nabla^2 u\,, \\
\frac{\partial v}{\partial t} + \mathbf{q}\cdot\nabla v - \frac{w^2}{r} &= -\frac{1}{\rho}\frac{\partial p}{\partial r} + \nu\left(\nabla^2 v - \frac{v}{r^2} - \frac{2}{r^2}\frac{\partial w}{\partial \theta}\right), \\
\frac{\partial w}{\partial t} + \mathbf{q}\cdot\nabla w + \frac{vw}{r} &= -\frac{1}{r\rho}\frac{\partial p}{\partial \theta} + \nu\left(\nabla^2 w + \frac{2}{r^2}\frac{\partial v}{\partial \theta} - \frac{w}{r^2}\right), \\
\frac{\partial u}{\partial x} + \frac{1}{r}\frac{\partial (rv)}{\partial r} + \frac{1}{r}\frac{\partial w}{\partial \theta} &= 0\,,
\end{aligned}$$

where

$$\mathbf{q}\cdot\nabla = u\frac{\partial}{\partial x} + v\frac{\partial}{\partial r} + \frac{w}{r}\frac{\partial}{\partial \theta}\,, \quad \nabla^2 = \frac{\partial^2}{\partial x^2} + \frac{1}{r}\frac{\partial}{\partial r}\left(r\frac{\partial}{\partial r}\right) + \frac{1}{r^2}\frac{\partial^2}{\partial \theta^2}\,.$$

Consider the ideal configuration of a perfectly circular pipe of infinite length.

- Rotational and translation symmetry show that there is no angular dependence, $\partial/\partial\theta = 0$, and no longitudinal dependence, $\partial/\partial x = 0$, in the flow, except that the pressure may vary linearly in x.
- Reflection symmetry, about any plane through the pipe's axis, shows that $w = 0$—there is no reason to favor helical flow in any direction, so zero angular velocity is a possibility. Choosing $w = 0$ ensures that the angular component of the Navier–Stokes equation is satisfied.
- The continuity equation then shows that $rv =$ constant. But since the fluid cannot cross the pipe's surface, the radial velocity v must be zero at $r = a$. Hence the constant must be zero in $rv =$ constant, and thus $v = 0$ everywhere.
- The radial component of the Navier–Stokes equation is automatically satisfied provided the pressure is independent of r, that is, if the pressure is constant across any given cross-section.
- The axial/longitudinal component of the Navier–Stokes equation then shows that
$$\frac{\partial u}{\partial t} = -\frac{1}{\rho}\frac{\partial p}{\partial x} + \frac{\nu}{r}\frac{\partial}{\partial r}\left(r\frac{\partial u}{\partial r}\right).$$
Every term appearing here is independent of x; hence the pressure p must be linear in x—there is a constant pressure gradient along the pipe. If the imposed pressure gradient is independent of time, then u will also not depend upon t and hence $\partial u/\partial t = 0$. If a pressure change of Δp is imposed across the ends of a pipe of length L, then
$$\frac{1}{r}\frac{\partial}{\partial r}\left(r\frac{\partial u}{\partial r}\right) = \frac{\Delta p}{\nu\rho L} = \frac{\Delta p}{\mu L}\,.$$
Integrating twice leads to a velocity profile of
$$u(r) = \frac{\Delta p}{4\mu L}r^2 + A\log r + B\,.$$

There cannot be infinite velocity in the center of the pipe, so $A = 0$, and $u = 0$ on the pipe wall, as it is a viscous fluid, so $B = -\frac{\Delta p}{4\mu L}a^2$.

Thus Poiseuille pipe flow has a parabolic velocity distribution[87]

$$q = -\frac{\Delta p}{4\mu L}\left(a^2 - r^2\right)\mathbf{i}. \tag{6.13}$$

It is this background fluid flow that we use to explore dispersion of material in a pipe.[88]

The flux of fluid through the pipe due to a given pressure drop is of interest. The fluid flux is

$$Q = \int_{\text{cross-section}} u\, dA = \int_0^a \int_0^{2\pi} u(r)\, r d\theta\, dr = -\frac{a^4 \Delta p\, \pi}{8\mu L}.$$

The interesting feature here is that for a given pressure gradient the flux is proportional to the fourth power of the radius. For example, the flux in 19-mm (3/4″) water pipe is some five times as much as the flux in 13-mm (1/2″) water pipe for the same pressure gradient. Alternatively, a 19-mm pipe can be five times the length of 13-mm pipe and still deliver the same flow of water for a given pressure drop.

6.4 ▪ Exercises

Exercise 6.1 (vorticity and circulation). Do some research on and then write a one-page summary of vorticity, irrotational flow, and the Kelvin circulation theorem.

Exercise 6.2 (coordinate systems). Find out and write down the continuity, Euler, and Navier–Stokes equations in component form in both Cartesian and spherical coordinates.

Exercise 6.3 (advection diffusion). Figures 6.5–6.7 show three different velocity fields. The figures also show an initial clump of 100 particles. These particles are advected by the fluid flow and diffuse by random molecular buffeting.

1. In the case of Figure 6.5, sketch the region occupied by the clump of particles after the time it takes the center of the clump to go a quarter of the way around the center (roughly), and sketch the region occupied when three quarters of the way around the center.
2. In the case of Figure 6.6, sketch the region occupied by the clump of particles after the time it takes the center of the clump to go a quarter of the way around the center (roughly), and sketch the region occupied when three quarters of the way around the center.
3. In the case of Figure 6.7, sketch the region occupied by the clump of particles after the time it takes the first particle to reach the right-hand boundary of the plotted region.

[87] The minus sign in front of the pressure change Δp shows that fluid flows from high pressure to low.

[88] This Poiseuille pipe flow is significantly modified by twists and turns in curving pipes. Computer algebra helped elaborate details of the flow in an arbitrarily curving pipe, provided that the curvature and torsion are not too rapid (Roberts, 2004).

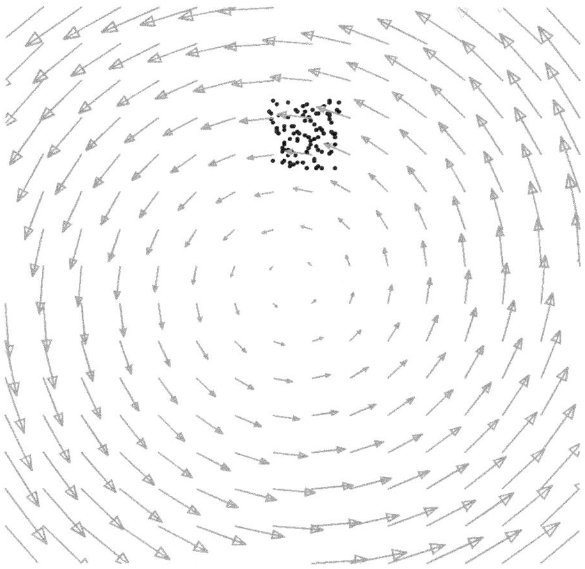

Figure 6.5. *A clump of particles is to diffuse as it is advected by the shown fluid flow. Exercise* 6.3 *asks where the clump goes.*

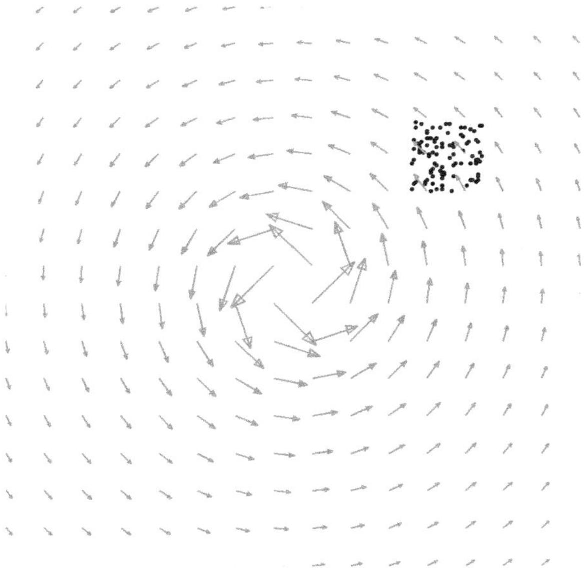

Figure 6.6. *A clump of particles is to diffuse as it is advected by the shown fluid flow. Exercise* 6.3 *asks where the clump goes.*

Exercise 6.4 (a swirling flow). In a viscous fluid of constant density with no body forces, consider a swirling flow that in cylindrical coordinates has velocity components $u = w = 0$ and $v = v(r,t)$ and pressure $p = p(r,t)$. What do the Navier–Stokes and continuity equations reduce to? Seeking a solution of the form

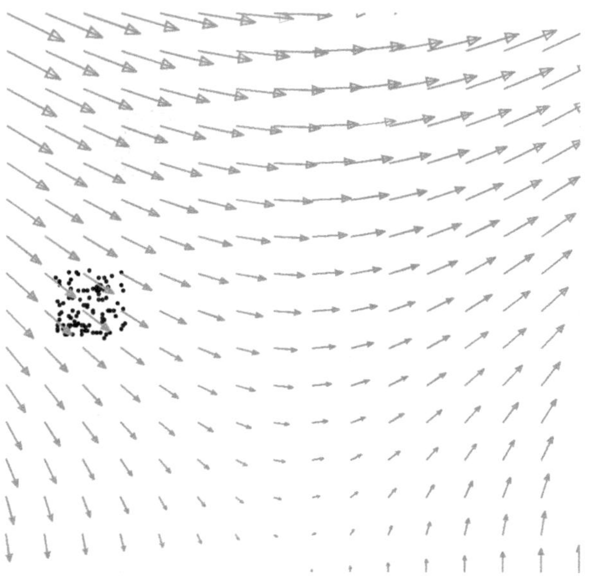

Figure 6.7. *A clump of particles is to diffuse as it is advected by the shown fluid flow. Exercise* 6.3 *asks where the clump goes.*

$v(r,t) = A(t)\exp\left[-r^2/D(t)\right]$, determine $A(t)$ and $D(t)$. Explain what this solution may physically describe.

Exercise 6.5 (plane Poiseuille flow). Analogous to the flow in a pipe of Example 6.3, consider the flow, generated by a pressure gradient, between two parallel and rigid planes.

- Write down the Navier–Stokes and continuity equations in Cartesian coordinates.
- Let coordinate y measure location normal to the two plane walls with the fluid flow within $|y| < a$. Let coordinate x be aligned with the pressure gradient along the plane walls, and let coordinate z be normal to the pressure gradient.
- Use symmetry to argue that gradients $\partial/\partial x = 0$, except for the nonzero pressure gradient, and $\partial/\partial z = 0$.
- Use continuity to deduce $v = 0$, and use a component of the Navier–Stokes equation to deduce $\partial p/\partial y = 0$.
- Hence deduce that the fluid flow has a parabolic form proportional to the pressure gradient.

Chapter 7
Cross-stream mixing causes longitudinal dispersion along pipes

Contents

This chapter justifies and constructs models of the long-term spread of a material along a long pipe that is carried by Poiseuille flow down the pipe (Mercer and Roberts, 1994). The modeling surprise is that the *effective* rate of dispersion along the pipe is vastly faster than is predicted by simple cross-sectional averaging. Instead, the effective dispersion along the pipe is a combination of microscale mixing and the fluid flow along the pipe as originally recognized in the modeling by Taylor (1953,4).

The approach here may be readily extended to cater for reactive (Wright, 2012) or sedimenting (Suslov and Roberts, 2000) material in the flow along a pipe of complex and varying geometry (Mercer and Roberts, 1994; Roberts, 2004; Rosencrans, 1997). The approach also extends to other applications, such as homogenization and beam and shell modeling.

Consider some material being carried by a fluid flow. We assume that the material is neutrally buoyant; that is, the material has no affect on the density of the fluid and so concentration differences do not cause density differences that could generate secondary flows. Here the along-pipe fluid flow is the Poiseuille flow of Example 6.3:

$$u(r) := 2U(1 - r^2/a^2),$$

where $U := -\Delta p\, a^2/8\mu L$ is the cross-sectionally averaged downstream velocity when driven by a constant pressure drop Δp.

As introduced by section 6.2.2, the governing partial differential equation (6.4), which represents conservation of some material, is $\partial c/\partial t + \nabla \cdot (\mathbf{q}c) = \nabla \cdot (\kappa \nabla c)$, where c is the concentration field of the material and κ is the coefficient of diffusivity. For example, for common salt diffusing in water (Batchelor, 1979, p. 595), $\kappa_{NaCl} = 1.1 \times 10^{-5}\ cm^2/sec$, whereas $\kappa_{KMnO_4} = 1.4 \times 10^{-5}\ cm^2/sec$. These diffusivities are one thousandth the size of the kinematic viscosity of water, $\nu = 1.14 \times 10^{-2}\ cm^2/sec$, and so the viscous dynamics that establish Poiseuille flow occur a thousand times faster than the dynamics of dispersion that we examine here. This disparity in rates of diffusivity justifies the assumption of an established Poiseuille flow. For Poiseuille flow the conservation PDE (6.4) reduces to

$$\frac{\partial c}{\partial t} + u(r)\frac{\partial c}{\partial x} = \kappa\nabla^2 c = \kappa\left[\frac{\partial^2 c}{\partial x^2} + \frac{1}{r}\frac{\partial}{\partial r}\left(r\frac{\partial c}{\partial r}\right) + \frac{1}{r^2}\frac{\partial^2 c}{\partial \theta^2}\right]. \tag{7.1}$$

Typically, the material cannot cross the wall of the pipe (although there are physical situations where the material is allowed to diffuse through the wall or to react with chemicals on the wall). Consequently, the boundary condition on the concentration is that of zero diffusive flux on the wall:

$$\frac{\partial c}{\partial r} = 0 \quad \text{on } r = a. \tag{7.2}$$

Cross-sectional averaging is inadequate Obtain a simple but misleading model of the long-term spread of material along the pipe by integrating the PDE (7.1) over a cross-section. Let $C(x,t)$ be the cross-sectional average of the concentration c, namely $C(x,t) := \iint c\, r\, d\theta\, dr/(\pi a^2)$. In a long and thin pipe, the cross-pipe diffusion has a long time to act and so we may expect that $c(x,r,\theta,t) \approx C(x,t)$ throughout. Now averaging the advection-diffusion PDE (7.1) over a cross-section and using the boundary condition (7.2) leads to

$$\frac{\partial C}{\partial t} + \frac{1}{\pi a^2}\iint u(r)\frac{\partial c}{\partial x} r\, d\theta\, dr = \kappa\frac{\partial^2 C}{\partial x^2}.$$

Under the assumption that $c \approx C$, then $\partial c/\partial x \approx \partial C/\partial x$, which is independent of r and θ and so may be taken outside the remaining integral to give

$$\frac{\partial C}{\partial t} + U\frac{\partial C}{\partial x} \approx \kappa\frac{\partial^2 C}{\partial x^2},$$

where U is the average velocity down the pipe. This is a simple model for the transport of material along the pipe by the flow. It asserts that the material is carried at the average velocity U and spreads out along the pipe by molecular diffusion κ. Whereas the bulk transport at the average velocity is a reasonable first approximation, the spread or dispersion along the pipe is typically immensely faster than that due to molecular diffusion, sometimes by a factor of a million.[89] To discover the correct dispersion, and lesser effects, we need to use a better modeling approach—that provided by the center manifold theory of Part II.

Nondimensionalize Before undertaking further analysis, we find it convenient to nondimensionalize this problem. As usual, I recommend the regular perturbation approach of nondimensionalizing with respect to microscale processes and then seeking

[89] This is one example illustrating that cross-sectional averaging is incomplete as a modeling paradigm.

the relatively slow variations in space-time via center manifold theory. Here the microscale length is the pipe width, and the microscale time is the cross-pipe diffusion time. Thus[90]

- scale r and x with respect to the pipe radius, that is, $r^* = r/a$ and $x^* = x/a$ give nondimensional locations, and
- scale time with respect to a cross-pipe diffusion time, $\tau = a^2/\kappa$, that is, the nondimensional time $t^* = t\kappa/a^2$.

Then the nondimensional mean velocity is $U^* := Ua/\kappa$; this U^* is typically called a *Péclet number* and denoted Pe, but we continue to use U^*. This Péclet number Ua/κ is often large, thousands or millions. For example,

$$\mathrm{Pe} = \frac{\nu}{\kappa}\frac{Ua}{\nu} \approx 1000\,\mathrm{Re}\,.$$

Upon this scaling, and upon henceforth omitting the asterisks that indicate nondimensional quantities, the material advection of the PDE (7.1) is described by the nondimensional PDE

$$\frac{\partial c}{\partial t} + 2U(1-r^2)\frac{\partial c}{\partial x} = \frac{\partial^2 c}{\partial x^2} + \frac{1}{r}\frac{\partial}{\partial r}\left(r\frac{\partial c}{\partial r}\right) + \frac{1}{r^2}\frac{\partial^2 c}{\partial \theta^2} \tag{7.3}$$

in a pipe of radius 1 with boundary condition

$$\frac{\partial c}{\partial r} = 0 \quad \text{on } r = 1\,. \tag{7.4}$$

We seek a model, supported by center manifold theory, for the long-term dispersion of material along the pipe governed by these equations.

Applies to particle probability distributions Consider a particle executing a random walk in the cross-section of the pipe and carried along by a radial, r, dependent drift down the pipe. Then PDE (7.3) may also be interpreted as the Fokker–Planck equation for the probability density function of the particle. In this case our modeling here justifies and constructs a model of the long-term evolution of the probability density function. The analysis we undergo here also generalizes to apply to the evolution of other stochastic systems. Section 21.2 develops further the modeling of Fokker–Planck equations in the context of modeling stochastic dynamics.

7.1 ▪ There is no spectral gap!

Of course this problem of material dispersion along a pipe is linear in the concentration c and could be solved as such. We could, for example, seek solutions in the separated form

$$c = c_{nk}(r)e^{ikx+in\theta+\lambda t}$$

[90] Many people prefer to scale the along-pipe variable x with respect to an advection distance in a cross-pipe diffusion time, that is, $x^* = x\kappa/(a^2U)$. It eventuates that this is indeed the correct scaling. However, assuming a longitudinal scaling, even if correct, imposes our preconceptions on the modeling. Such preconceptions reduce flexibility. Instead, let the problem tell us what the relevant length scales are in a model. That is, we impose as little as possible on the problem and let the complex physical interactions, expressed in the mathematics, lead to the appropriate modeling.

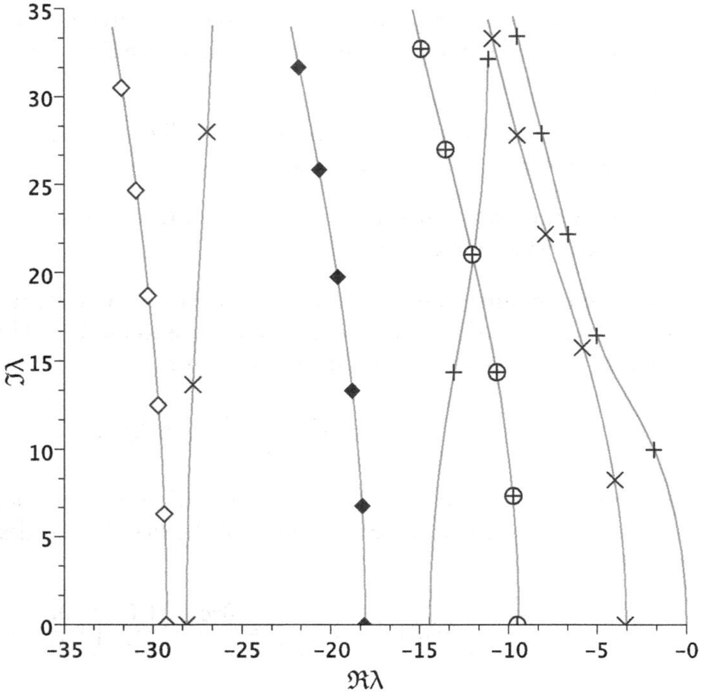

Figure 7.1. *The spectrum of complex eigenvalues for dispersion in a pipe. Each branch is plotted for longitudinal wavenumbers* $0 \leqslant kU < 60$*; negative* k *is just the complex conjugate. The angular mode is indicated by a symbol plotted at intervals* $U\Delta k = 10$*:* $+$ *for* $n = 0$*,* $\times$ *for* $n = 1$*,* $\oplus$ *for* $n = 2$*,* ♦ *for* $n = 3$*, and* ◇ *for* $n = 4$*.*

for real longitudinal wavenumber k and for integer azimuthal wavenumber n. The nondimensional PDE (7.3) then becomes the eigenproblem

$$\lambda c_{nk} = -2U(1-r^2)ikc_{nk} - k^2c_{nk} + \frac{1}{r}\frac{\partial}{\partial r}\left(r\frac{\partial c_{nk}}{\partial r}\right) - \frac{n^2}{r^2}\frac{\partial^2 c_{nk}}{\partial \theta^2}. \tag{7.5}$$

Solving this eigenproblem numerically (omitting the k^2c_{nk} term for simplicity), Figure 7.1 plots the complex eigenvalues λ as integer n and continuous k are varied. Observe that all modes decay since the eigenvalue's real part $\Re(\lambda) < 0$, except for one mode. However, the very right-hand branch in the figure shows that there is a continuum of modes with decay rate near zero. It is these modes that dominate the long-term dynamics of solutions to the advection-diffusion PDE (7.3). These modes form the basis of a low-dimensional model of the dispersion along the pipe.

Unfortunately there is no clear spectral gap in this problem: there is a continuum of eigenvalues ranging from 0 to large negative values. Thus we cannot simply assert that a certain number of modes will form the basis for the slow manifold model, and the rest of the modes are exponentially decaying, as required by Center Manifold Theorems 4.3, 4.5, 4.8, and 4.10: there is not a clear dividing line between the two. However, a crucial characteristic of the modes of long-term importance, small decay rate, is that they are of low wavenumber k, say $Uk < 5$ or so. Thus by focusing attention on low wavenumbers k, large longitudinal length scales, we make progress.

7.2 ▪ A Fourier transform justifies

The application of center manifold theory appears rigorous when we take the Fourier transform along the pipe (Mercer and Roberts, 1990). Denote Fourier-transformed quantities by a tilde: for example, the Fourier transform of the concentration field is

$$\tilde{c}(r,\theta,k,t) := \frac{1}{2\pi}\int_{-\infty}^{\infty} c(r,\theta,x,t)e^{-ikx}\,dx\,.$$

Under this Fourier transform, quantities $c \mapsto \tilde{c}$ and $\partial/\partial x \mapsto ik$. Consequently the advection-diffusion PDE (7.3) transforms to

$$\frac{\partial\tilde{c}}{\partial t} = \underbrace{\frac{1}{r}\frac{\partial}{\partial r}\left(r\frac{\partial\tilde{c}}{\partial r}\right)+\frac{1}{r^2}\frac{\partial^2\tilde{c}}{\partial\theta^2}}_{\text{linear}} \underbrace{-2Ui(1-r^2)k\tilde{c}-k^2\tilde{c}}_{\text{nonlinear}}\,. \tag{7.6}$$

Recall that in analyses of bifurcations, such as sections 4.5.1 and 5.1, we examine the structure of the solutions for some small parameter ϵ: as the parameter ϵ crosses through zero, the bifurcation occurs. To place the bifurcation problems within the province of center manifold theory we adjoin the trivial dynamic equation $\dot{\epsilon} = 0$. Then we could construct a slow manifold model that is rigorously valid for (ϵ, u) sufficiently small. Thus the model makes predictions about the dynamics for small but finite ϵ. We do the same here, but here the small parameter is the longitudinal wavenumber k. Thus adjoin to (7.6) the trivial dynamic equation

$$\frac{\partial k}{\partial t} = 0\,. \tag{7.7}$$

Then (7.6)–(7.7) is a dynamical system for evolution in $(k, \tilde{c}(r,\theta))$ state space.

7.2.1 ▪ There exists a slow manifold

Apply the established approach (Table 4.2).

Equilibria In $(k, \tilde{c}(r,\theta))$ state space, the subspace $(k,\tilde{c}) = (0,\tilde{C})$ comprises equilibria for any $\tilde{C}$ constant with respect to r and θ (the pipe cross-section variables). Because the equilibria form a subspace parametrized by $\tilde{C}$, the slow manifold we find is local in wavenumber k but global in $\tilde{C}$.

Because the problem is linear in the concentration, we just analyze dynamics about the origin, the equilibrium $\tilde{c} = 0$ and $k = 0$, as the analysis is identical for all equilibria with field $\tilde{c}$ constant across the pipe.

Linearization Near the origin, the linearized dynamics are governed by the terms labeled as "linear" in (7.6); the terms involving $k\tilde{c}$ and $k^2\tilde{c}$ are nonlinear. Not surprisingly, the linearized dynamics may be written in terms of eigenmodes involving Bessel functions. Substituting $\tilde{c} = e^{in\theta+\lambda t}J_n(\mu r)$, one may deduce that eigenvalue $\lambda_{mn} = -{j'_{n,m}}^2$ and radial wavenumber $\mu_{mn} = j'_{n,m}$, where $j'_{n,m}$ is the mth root of $J'_n(\mu) = 0$. Table 7.1 lists the linearized spectrum from data listed by Abramowitz and Stegun (1965, p. 411)—these eigenvalues match the wavenumber $k = 0$ points that Figure 7.1 plots along the negative real axis. Table 7.1 displays the one 0 eigenvalue (and there is another 0 eigenvalue from $\dot{k} = 0$), and all the remaining eigenvalues are negative, being bounded away from 0 by $-\beta = -3.39$. The critical mode with eigenvalue 0 is constant across the

Table 7.1. *Spectrum of the linearized dynamics at zero wavenumber, that is, the spectrum of cross-pipe diffusion when there is no longitudinal variation in concentration.*

λ_{mn}	$n=0$	$n=1$	$n=2$	$n=3$
$m=0$	0	−3.39	−9.33	−17.65
$m=1$	−14.68	−28.42	−44.97	−64.24
$m=2$	−49.22	−72.87	−99.39	−128.3

pipe, $\mu_{00}=0$. Let's choose to measure the amplitude of the critical mode by $\tilde C :=$ the cross-pipe average of $\tilde c$. Thus there exists an emergent slow manifold parametrized by wavenumber k and cross-pipe average $\tilde C$.

Such a slow manifold is useful to us because it is certainly valid and emergent for the range of small wavenumbers identified earlier as making up the long-term evolution.

7.2.2 ▪ The slow manifold cut-off

Section 7.3 details how to construct the slow manifold: it finds the constructed slow manifold model to be

$$\begin{aligned}\tilde c &= \tilde C + ikU\tilde C\frac{-2+6r^2-3r^4}{24}\\ &\quad - k^2U^2\tilde C\frac{31-180r^2+300r^4-200r^6+45r^8}{11520} + \mathcal{O}(k^3).\end{aligned} \tag{7.8a}$$

On this slow manifold the evolution is

$$\frac{\partial\tilde C}{\partial t} = -ikU\tilde C - k^2\tilde C\left(1+\frac{U^2}{48}\right) + \mathcal{O}(k^3). \tag{7.8b}$$

To check that these form a slow manifold model of the specified order of error, simply substitute these expressions into the Fourier-transformed PDE (7.6) and boundary condition (7.4).

Synthesize a global model We now synthesize a global physical model by integrating together these Fourier space models. Center manifold theory guarantees that the model (7.8) applies for finite domain of small wavenumbers k. Thus there exists a cut-off wavenumber K such that the model (7.8) is valid for all wavenumbers $|k|<K$. Let us combine the information about these wavenumbers. Define the inverse transform $C_K(x,t) := \int_{-K}^{K}\tilde C(k,t)e^{ikx}\,dk$; then taking the time derivative and invoking integration by parts leads to

$$\begin{aligned}\frac{\partial C_K}{\partial t} &= \int_{-K}^{K}\frac{\partial\tilde C}{\partial t}e^{ikx}\,dk\\ &= \int_{-K}^{K}\left[-ikU\tilde C - k^2\tilde C(1+U^2/48)+\mathcal{O}(k^3)\right]e^{ikx}\,dk\\ &= \int_{-K}^{K}\left[-U\frac{\partial\tilde Ce^{ikx}}{\partial x} + (1+U^2/48)\frac{\partial^2\tilde Ce^{ikx}}{\partial x^2} + \mathcal{O}(k^3)\right]dk,\\ &= -U\frac{\partial C_K}{\partial x} + (1+U^2/48)\frac{\partial^2 C_K}{\partial x^2} + \mathcal{O}(K^3C_K).\end{aligned}$$

A little care with the order of error in wavenumber k in the integrand leads to the given order of error in cut-off wavenumber K.[91] Provided there is negligible material in the high-wavenumber structures, $|k| > K$, then the cross-pipe average $C(x,t) = \overline{c(r,\theta,x,t)} \approx C_K(x,t)$. Thus this rigorous model for C_K becomes the classic advection-diffusion model

$$\frac{\partial C}{\partial t} \approx -U\frac{\partial C}{\partial x} + (1 + U^2/48)\frac{\partial^2 C}{\partial x^2}$$

for the shear dispersion along the pipe: the enhancement of lateral dispersion by $U^2/48$ is often enormous. Our Fourier transform approach is rigorous.

Furthermore, the Emergence Theorem 4.5 guarantees that there is a finite domain of initial conditions from which solutions described by the above models will emerge exponentially quickly. A challenge is to map a specified initial distribution of material, $c_0(r,\theta,x)$, to a specific initial condition for the models, either $\tilde{C}(k,0)$ or $C_K(x,0)$. Recall that Example 4.10 demonstrated that in general there is some initial slip between initial conditions of the full system and the models: here, despite defining $\tilde{C}(k,t)$ to be the cross-pipe average of $\tilde{c}(r,\theta,k,t)$, generally $\tilde{C}(k,0)$ is *not* the cross-pipe average of $\tilde{c}(r,\theta,k,0)$. The apparent paradox is resolved by the normal-form transformations of section 12.2.

7.2.3 ▪ Summary

In essence, the Fourier transform splits the physical problem into a number, albeit an infinite number, of subproblems, each with a given value of wavenumber k. Then for small wavenumber k we identify the existence of a low-dimensional slow manifold. Constructing a slow manifold model for each k gives a model of the long-term evolution for each wavenumber k. Then the inverse Fourier transform puts all these models together, albeit only accurate for small wavenumber k, to create a model in physical space for the long-term evolution of dispersion in a pipe (Mercer and Roberts, 1990, e.g.).

The high-wavenumber modes are not relevant to the long-term evolution because the spectrum shows they decay exponentially. In this way we justify the construction of a slow manifold model for the long-term evolution of dispersion that, because it is based on small wavenumbers k, is a model for solutions that vary slowly along the pipe.

7.3 ▪ Direct construction follows the same pattern

Although the Fourier approach puts the slow manifold analysis on a rigorous foundation, the Fourier transforms and their inverse are unwieldy—and impractical in nonlinear systems. Precisely the same results are obtained by a direct and cleaner approach—an approach that also applies to nonlinear systems. The key is to recall that in Fourier space multiplication by the wavenumber k makes a term small, as it is nonlinear, because we notionally adjoined $\dot{k} = 0$. But recall that multiplication by wavenumber k arises from the longitudinal derivative $\partial/\partial x$; thus treating multiplication by wavenumber k as small

[91] Consider the order of error in the integrand, error $\varepsilon = \mathcal{O}(k^3)$. By Definition 2.2 this means that ε/k^3 is bounded as $k \to 0$. Thus there exists a constant c_3 such that $|\varepsilon| \leqslant c_3|k|^3$ for all $|k| \leqslant K$. Hence the magnitude of the integral of the error $|\int_{-K}^{K} \varepsilon\, dk| \leqslant \int_{-K}^{K} |\varepsilon|\, dk \leqslant \int_{-K}^{K} c_3|k|^3\, dk \leqslant c_3K^4/2$. But also, $C_K = \mathcal{O}(K)$, so the integrated error is $\mathcal{O}(K^3 C_K)$ as claimed.

in Fourier space corresponds in physical space to treating longitudinal derivatives $\partial/\partial x$ as small nonlinear operators.[92]

Recent research (Roberts, 2013) proposes how to put onto a rigorous foundation the direct formal approach developed in this section.

Equilibria Analogous to section 7.2.1, setting wavenumber $k = 0$ corresponds to setting the longitudinal derivatives $\partial/\partial x = 0$. Then $c(r, \theta, x, t) =$ constant are equilibria of the PDE (7.6). We find a slow manifold about each equilibrium and hence deduce a model global in concentration and local to small $\partial/\partial x$.

Linearization In the linearization, it is the presence of $\partial/\partial x$ that flags whether a term is to be treated as a nonlinear perturbation or not. Thus rewrite the physical PDE (7.3) as

$$\frac{\partial c}{\partial t} = \underbrace{\frac{1}{r}\frac{\partial}{\partial r}\left(r\frac{\partial c}{\partial r}\right) + \frac{1}{r^2}\frac{\partial^2 c}{\partial \theta^2}}_{\text{linear } \mathcal{L}c} \underbrace{-2U(1-r^2)\frac{\partial c}{\partial x} + \frac{\partial^2 c}{\partial x^2}}_{\text{nonlinear}} \quad \text{and} \quad \frac{\partial c}{\partial r} = 0 \text{ at } r = 1\,. \tag{7.9}$$

Now analyze the dynamics based on this split and seek solutions that are assumed *slowly varying* in x so that $\partial/\partial x$ is small; this corresponds to low wavenumber k.

Linearly, then, we ignore all terms with an x-derivative and examine

$$\frac{\partial c}{\partial t} = \mathcal{L}c = \frac{1}{r}\frac{\partial}{\partial r}\left(r\frac{\partial c}{\partial r}\right) + \frac{1}{r^2}\frac{\partial^2 c}{\partial \theta^2} \quad \text{and} \quad \frac{\partial c}{\partial r} = 0 \text{ at } r = 1\,.$$

This PDE is just the zero-wavenumber version of a problem examined earlier, and we know that its spectrum is that eigenvalues $\lambda_{mn} = -{j'_{n,m}}^2$ (Table 7.1). All of these eigenvalues are strictly negative, representing exponentially decaying modes, except for $m = n = 0$, for which $\lambda_{00} = 0$, representing the critical mode of long-term importance.[93] Physically, this problem is simply that of cross-pipe diffusion. Hence concentration fluctuations across the pipe decay by diffusion, in the long term leaving a uniform concentration. Thus the cross-pipe structure of the critical mode is simply that of a constant with respect to r and θ. Choosing to write the model in terms of the cross-pipe average, to an initial approximation, the slow manifold model is the trivial

$$c \approx C(x,t) \quad \text{such that} \quad \frac{\partial C}{\partial t} \approx 0\,. \tag{7.10}$$

We only know that concentration c is approximately constant with respect to the cross-pipe variables r and θ; thus we have to cater for the slow manifold amplitude C to depend upon longitudinal position x as well as time t. Because the amplitude C depends upon position x, and there is an infinite number of positions x, the slow manifold is formally of infinite dimensions.[94]

[92]Theoretically, a derivative such as $\partial/\partial x$ cannot be small as it is an unbounded operator. One way to avoid this theoretical difficulty is to assert that the operator $\partial/\partial x$ actually denotes a low-band pass filtered derivative. Then the phrase "$\partial/\partial x$ is small" means that the low-band pass filter only passes sufficiently low wavenumbers. Such filtering is equivalent to the wavenumber cut-off invoked by section 7.2.2.

[93]The existence of the critical mode is a direct consequence of the overall conservation of material.

[94]This infinite dimensionality of the slow manifold is an issue; however, existence, emergence, and approximation theorems have been established by Gallay (1993), Aulbach and Wanner (1996, 2000), and Potzsche and Rasmussen (2006) for systems with infinite-dimensional center manifolds in Banach spaces.

One also needs to define precisely the amplitude C. Typically amplitudes are defined via linear projections. Thus here we first have to define the inner product, $\langle v, w\rangle := \int_0^1\int_0^{2\pi} vw\, r\, d\theta\, dr$, of integration across the pipe. Then we choose to define the amplitude $C(x,t) := \frac{1}{\pi}\langle 1, c(r,\theta,x,t)\rangle$, which is the cross-pipe average. Exercise 7.1 shows that defining other choices is equally acceptable.

Iteratively refine Now seek to iteratively refine the model (7.10). Suppose that we have an approximate description

$$c(r,\theta,x,t) \approx c(C,r,\theta) \quad \text{such that} \quad \frac{\partial C}{\partial t} \approx g(C)\,,$$

where the context differentiates between the space-time field $c(r,\theta,x,t)$ and its slow manifold description $c(C,r,\theta)$, in which the "amplitude" C depends upon x and t. Seek corrections $\hat{c}(C,r,\theta)$ and $\hat{g}(C)$, where

$$c \approx c(C,r,\theta) + \hat{c}(C,r,\theta) \quad \text{such that} \quad \frac{\partial C}{\partial t} \approx g(C) + \hat{g}(C)$$

is a more refined model. Substituting into the advection-diffusion PDE (7.9) and neglecting products of the small corrections gives

$$-\frac{\partial c}{\partial C}\hat{g} - \frac{\partial \hat{c}}{\partial C}g - u\frac{\partial \hat{c}}{\partial x} + \mathcal{L}\hat{c} + \frac{\partial^2 \hat{c}}{\partial x^2} = \frac{\partial c}{\partial C}g + u\frac{\partial c}{\partial x} - \mathcal{L}c - \frac{\partial^2 c}{\partial x^2}\,.$$

As noted in similar applications of this procedure, the right-hand side is just the residual of the governing equations for the current approximation. Before using this residual to update the approximation we need to simplify the left-hand side operator. Approximate the left-hand side by its dominant terms:

- because we only seek slowly varying solutions, anything involving the longitudinal derivative $\partial/\partial x$ is small and so we neglect $-u\partial\hat{c}/\partial x + \partial^2\hat{c}/\partial x^2$;
- neglect the term $\partial\hat{c}/\partial C\, g$ because the initial approximation for g is 0, from (7.10);
- last, simplify the term $\partial c/\partial C\,\hat{g}$ to just $\hat{g}$ because the initial approximation to the field c is just the cross-sectional average C.

Thus we solve the homological equation

$$\mathcal{L}\hat{c} - \hat{g} = \text{Res}_{(7.9)} = \frac{\partial c}{\partial C}g + u\frac{\partial c}{\partial x} - \mathcal{L}c - \frac{\partial^2 c}{\partial x^2} \tag{7.11}$$

for corrections $\hat{c}$ and $\hat{g}$ to the slow manifold and the evolution thereon.

First iteration To improve $c \approx C$ and $g \approx 0$, the homological equation (7.11) requires us to solve

$$\mathcal{L}\hat{c} - \hat{g} = \text{Res}_{(7.9)} = u(r)\frac{\partial C}{\partial x} - \frac{\partial^2 C}{\partial x^2}\,.$$

As always, $\mathcal{L}$ is singular (from the critical mode of zero eigenvalue), and so we cannot find $\hat{c}$ without first determining $\hat{g}$ to put the right-hand side in the range of $\mathcal{L}$. There are two equivalent ways of phrasing the same solvability condition:

- one could observe that by cross-sectionally integrating this equation, $\iint \cdots r\,d\theta\,dr$, and using the conservative boundary conditions, $\partial\hat{c}/\partial r = 0$, the $\mathcal{L}\hat{c}$ term disappears to give
$$-\pi\hat{g} = \iint u\,r\,d\theta\,dr\frac{\partial C}{\partial x} - \pi\frac{\partial^2 C}{\partial x^2},$$
which simplifies to
$$\hat{g} = -U\frac{\partial C}{\partial x} + \frac{\partial^2 C}{\partial x^2},$$
as $u(r) = 2U(1-r^2)$ has cross-sectional average of U by the nondimensionalization;

- or one could note that $\mathcal{L}$ is self-adjoint, upon defining the inner product $\langle v, w\rangle := \iint vw\,r\,d\theta\,dr$, and so an adjoint eigenvector corresponding to the critical mode is just $z(r,\theta) = \text{constant}$, so that $\langle z, \text{equation}\rangle$ is the same solvability condition as just used above.

Using this update $\hat{g}$, the homological equation for the first correction $\hat{c}$ becomes
$$\mathcal{L}\hat{c} = U(1-2r^2)\frac{\partial C}{\partial x}.$$

In the absence of any θ variations in the right-hand side, the solution to this homological equation is found by integration in r:
$$\frac{1}{r}\frac{\partial}{\partial r}\left(r\frac{\partial\hat{c}}{\partial r}\right) = (1-2r^2)U\frac{\partial C}{\partial x} \quad\Longrightarrow\quad \hat{c} = \left(\tfrac{1}{4}r^2 - \tfrac{1}{8}r^4\right)U\frac{\partial C}{\partial x} + A,$$
where A is an arbitrary integration constant (with respect to r and θ). Determine such arbitrary constants by the definition chosen for the amplitude of the model. Here we write the model in terms of C, the cross-sectional average of c. Since the initial approximation (7.10) satisfies this cross-sectional average, we therefore require that the cross-sectional average of every correction $\hat{c}$ be zero. Here this requires that $A = -U\partial C/\partial x/12$. Thus
$$\hat{c} = \tfrac{1}{24}\left(-2+6r^2-3r^4\right)U\frac{\partial C}{\partial x}.$$

Correcting the initial model (7.10) with these $\hat{g}$ and $\hat{c}$ we deduce the more refined model[95]
$$c \approx C + \tfrac{1}{24}\left(-2+6r^2-3r^4\right)U\frac{\partial C}{\partial x} \quad \text{such that} \quad \frac{\partial C}{\partial t} \approx -U\frac{\partial C}{\partial x} + \frac{\partial^2 C}{\partial x^2}. \tag{7.12}$$

The evolution predicted by this model, $\partial C/\partial t \approx -U\partial C/\partial x + \partial^2 C/\partial x^2$, is the same as that obtained in the introduction: advection at the mean velocity of the fluid, and along-pipe molecular diffusion. However, there is one important difference: here we recognize that there is structure in the cross-pipe concentration, proportional to $w(r) = \frac{1}{24}\left(-2+6r^2-3r^4\right)$ and proportional to mean concentration gradients $\partial C/\partial x$, and it is this structure that leads in the next iteration to realistic estimates for the longitudinal pipe dispersion at large times.

[95]The Approximation Theorem suggests that the $\partial^2 C/\partial x^2$ term needs correction because the residuals in the equation resulting from this approximate model (see the next iteration) are of this form.

Second iteration Using the updated approximation, equation (7.11) then requires us to solve the new homological equation[96]

$$\begin{aligned}\mathcal{L}\hat{c} - \hat{g} &= \left[1 + wU\frac{\partial}{\partial x}\right]\left(-U\frac{\partial C}{\partial x} + \frac{\partial^2 C}{\partial x^2}\right) + u\left[\frac{\partial C}{\partial x} + w\frac{\partial^2 C}{\partial x^2}\right] \\ &\quad - (1-2r^2)U\frac{\partial C}{\partial x} - \frac{\partial^2 C}{\partial x^2} - wU\frac{\partial^3 C}{\partial x^3} \\ &= (u-1)w\frac{\partial^2 C}{\partial x^2}.\end{aligned}$$

Importantly, the derivative $\partial c/\partial C$ has to be done in the generalized sense so that here, for example, it becomes the operator $1 + wU\partial/\partial x$; this operator then acts on g so that $(\partial c/\partial C)g$ just realizes the chain rule applied to $\partial c/\partial t$. The solvability condition, $\iint \ldots r\,d\theta\,dr$, then determines

$$\hat{g} = \frac{U^2}{48}\frac{\partial^2 C}{\partial x^2}.$$

We do not bother to determine the corresponding correction $\hat{c}$ by hand.

This slow manifold analysis shows that the long-term model of dispersion in a pipe is[97]

$$\frac{\partial C}{\partial t} \approx -U\frac{\partial C}{\partial x} + \left(1 + \frac{U^2}{48}\right)\frac{\partial^2 C}{\partial x^2}. \tag{7.13}$$

This simple advection-diffusion equation has an effective diffusion coefficient of $1 + U^2/48$. Typical solutions of (7.13) are Gaussian distributions with mean location traveling at the mean velocity of the flow and spreading according to the effective diffusion coefficient. In most applications the along-pipe molecular diffusion, here 1, is negligible by many orders of magnitude when compared with the $U^2/48$ contribution. Where does this extra dispersion come from? It is called shear dispersion because it arises from the shear in the basic Poiseuille pipe flow. Imagine for a moment that there is no molecular diffusion at all. Then a small slug of material released into the flow across some section of the pipe will be differentially carried along according to the velocities at the different radii inside the pipe: the part of the slug released next to the pipe wall will barely move, while the part of the slug in the center of the pipe will be carried a long way down the pipe. Thus the spread along the pipe, as measured by the standard deviation in the x-direction, for example, must increase with the mean advection velocity U. Such shear-induced spread is ameliorated by the cross-pipe diffusion, which causes individual contaminant particles to eventually sample the entire velocity distribution. The dimensional form of the model (7.13) shows these effects:[98]

$$\frac{\partial C}{\partial t} \approx -U\frac{\partial C}{\partial x} + \left(\kappa + \frac{U^2 a^2}{48\kappa}\right)\frac{\partial^2 C}{\partial x^2}, \tag{7.14}$$

where the shear dispersion coefficient $U^2a^2/(48\kappa)$ is proportional to the square of the velocity and inversely proportional to the cross-pipe diffusivity κ.

[96] That the residual on the right-hand side here is proportional to $\partial^2 C/\partial x^2$ indicates that the slow manifold model (7.12) is accurate to errors $\mathcal{O}(\partial_x^2)$.

[97] The higher-order analysis of section 7.4 demonstrates that the error in this approximation is $\mathcal{O}(\partial_x^3)$.

[98] Observe, through the division by κ, the apparent paradox that the smaller the molecular diffusion, the larger the shear dispersion! This apparent paradox is well established.

Algorithm 7.1 *Iteration finds the slow manifold of the nondimensional advection-diffusion PDE (7.3) for the lateral dispersion of a material in pipe flow.*

```
% update operators: zero mean is inbuilt in linv
operator mean; linear mean;
let { mean(1,r) => 1
    , mean(r,r) => 2/3
    , mean(r^~p,r) => 2/(p+2)};
operator linv; linear linv;
let { linv(1,r) => (r^2-1/2)/4
    , linv(r,r) => (r^3-2/5)/9
    , linv(r^~p,r) => (r^(p+2)-2/(p+4))/(p+2)^2};
% create cross-sectional average concentration
depend cc,x,t; let df(cc,t)=>g;
factor df;
u:=2*uu*(1-r^2);
% linear approximation
c:=cc;  g:=0;
% iteration, only retain terms up to order 4 in d/dx
let df(cc,x,~p) => 0 when numberp(p) and p>4;
for iter:=1:99 do begin
   write
   res:=df(c,t)+u*df(c,x)-df(r*df(c,r),r)/r-df(c,x,2);
   g:=g+(gd:=-mean(res,r));
   c:=c+linv(res+gd,r);
   if res=0 then iter:=1000000;
end;
write resamp:=mean(c,r)-cc;
write resbc:=sub(r=1,df(c,r));
```

7.4 ▪ Higher orders converge

Computer algebra easily verifies the above results and computes higher-order corrections. For example, executing Algorithm 7.1 informs us that a higher-order model for the long-term evolution is

$$\frac{\partial C}{\partial t} = -U\frac{\partial C}{\partial x} + \left(1+\frac{U^2}{48}\right)\frac{\partial^2 C}{\partial x^2} - \frac{1}{2880}U^3\frac{\partial^3 C}{\partial x^3} - \frac{41}{2580480}U^4\frac{\partial^4 C}{\partial x^4} + \mathcal{O}(\partial_x^5). \quad (7.15)$$

Physically, these third- and fourth-derivative corrections describe how the skewness and the kurtosis of the along-pipe distribution evolve in time. They show, for example, that a contaminant pulse is generally not Gaussian but will be skewed somewhat. The skewness does decay, but only algebraically slowly, and so is captured in this slow manifold model. Chatwin (1970) extensively discussed this issue of algebraic approach to Gaussian.

Because of the linearity in concentration c, the only small parameter in this problem is $\partial/\partial x$, that is, the wavenumber k. Seeking solutions of the model (7.15) proportional to $e^{ikx+\lambda t}$, we find

$$\lambda \approx -ikU - \left(\frac{1}{U^2}+\frac{1}{48}\right)(kU)^2 + \frac{i}{2880}(kU)^3 - \frac{41}{2580480}(kU)^4. \quad (7.16)$$

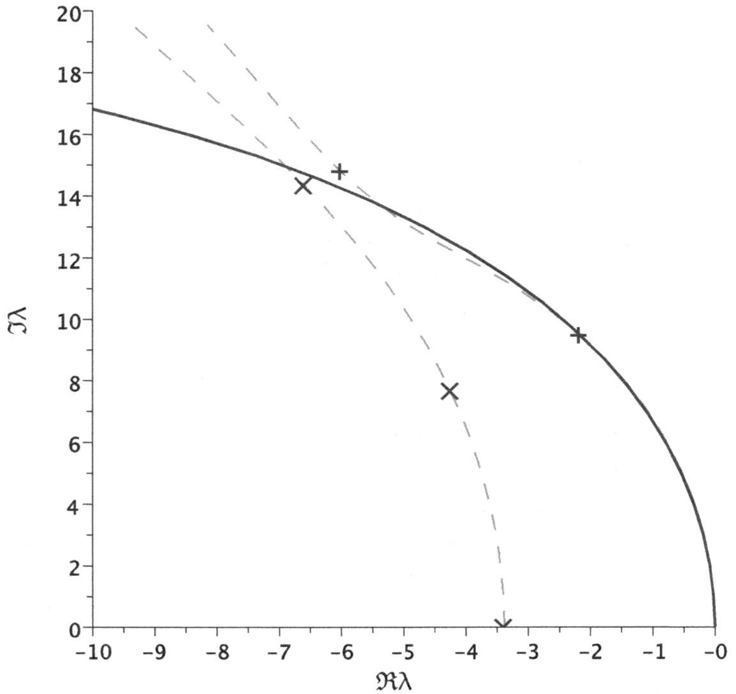

Figure 7.2. *The spectrum of complex eigenvalues for dispersion in a pipe compared with that of the slowly varying slow manifold model (solid line). Each branch is plotted for longitudinal wavenumbers* $0 \leqslant kU < 30$*; the angular mode is indicated by a symbol plotted at intervals* $U\Delta k = 10$*:* $+$ *for* $n = 0$ *and* $\times$ *for* $n = 1$. *Observe the good fit between the model and the leading branch up to wavenumber* 10 *or so.*

This is simply the Taylor series of the $m = n = 0$ branch of the spectrum of the full problem shown in Figure 7.1: the first two terms are a parabolic approximation leading to an advection-diffusion model; higher-order approximations lead to higher-order models. The slow manifold model neglects all branches of the spectrum except the dominant $m = n = 0$ branch, and then only accurately approximates that part of the branch (see Figure 7.2) corresponding to small eigenvalue and small wavenumber.

Thus the slow manifold model is limited by its dependence upon

- the neglected transients of the neglected branches in the spectrum—the most significant of these decays approximately like $e^{-3.39t}$—showing that the model is valid on dimensional time scales $t \gg a^2/\kappa/3.39$,
- and the necessarily slowly varying dependence along the pipe from the effective Taylor expansion in wavenumber k.

In this particular problem we used the computer algebra to discover *quantitative* bounds on the spatial resolution.[99] Computing to 24th order in wavenumber k, Mercer and Roberts (1994) showed that the above expansion for λ in wavenumber k appears to converge for nondimensional wavenumber $|kU| < 13.8$. This sharp limit on the convergence

[99]This sharp result for the spatial resolution is some 10 times as good as that heuristically estimated by Taylor (1954).

strongly suggests that we only attempt to use such a model to resolve along-pipe structures on a dimensional length scale greater than $\frac{2\pi}{13.8}U^*a = 0.45Ua^2/\kappa$. Because of the smallness of molecular diffusion κ, this length is typically quite large in applications.

7.5 ▪ Discussion

> The infinite! No other question has ever moved so profoundly the spirit of man; no other idea has so fruitfully stimulated his intellect; yet no other concept stands in greater need of clarification than that of the infinite.
>
> *David Hilbert*

There are many facets of interest in this modeling.

- The nontrivial models (7.13) and (7.15) are straightforward to understand and to use to predict dispersion along a pipe.
- The models are putatively low dimensional, and yet, being phrased in terms of the time evolution of a function of x, are actually infinite dimensional! Nonetheless, they are considerably simpler than the original mathematical description of the advection and diffusion in a pipe because we have systematically eliminated all the cross-pipe dynamics to result in a description of the along-pipe dispersion.
- The slowly varying, low-wavenumber, long-wavelength approximation, as used here for pipe dispersion, occurs in many applications. The essence of the approximation is to get the spectrum of the low-wavenumber, long-wavelength modes correct and to ignore the dynamics on all other branches of exponentially decaying modes.
- The convergence that we see in the slow manifold model here is rare and primarily due to the linearity in concentration field c. Most slow manifold models of nonlinear dynamics are only asymptotic. Here, the convergence engenders sharp estimates on the applicability of the model.
- Then the question arises of how can one extend the domain of applicability of the model. In systems with slowly varying structures in space there are two similar but different answers.
  - If the challenge is to improve the temporal resolution to more rapid dynamics in time, then one has to somehow retain more of the leading rightmost (asymmetric) modes seen in the spectrum of eigenvalues of Figure 7.1 (Watt and Roberts, 1995,6, e.g.).
  - Conversely, if the challenge is to improve the spatial resolution, then one just has to somehow retain modes that interact with the leading mode. Figure 7.2 shows how the slowly varying model tracks the leading branch of the exact spectrum well, until the exact spectrum "kinks" at wavenumber $k \approx 13$. This kink is due to interaction (at complex wavenumber!) with the next axisymmetric mode seen in Figure 7.1 (emanating from $\Re\lambda \approx 15$ and also marked with $+$). To improve spatial resolution, we only need to develop a model with just these two modes (Smith, 1987, e.g.).

These comments are for linear dynamics. In nonlinear dynamics the symmetric and asymmetric modes are more likely to all interact with each other anyway, in which case improving space or time resolution is likely to be the same challenge.

- Although, via the Fourier transform, this modeling is rigorous in linear problems, such as pipe dispersion and many exercises of this section, it has been much harder to be rigorous in nonlinear problems that invoke the slowly varying approximation (Roberts, 2013).

- Remember that modeling dynamics is a nonlinear process. It is nonlinear even when the underlying dynamics are linear, as in this modeling of shear dispersion in a pipe. The underlying PDE (7.9) is linear in the concentration field. But the shear dispersion model (7.13) of the dynamics contains the square of the mean velocity: the modeling process from (7.9) to (7.13) is nonlinear. This nonlinearity in the modeling process occurs because we model out-of-equilibrium dynamics: the chain rule for $\partial c/\partial t$ leads to the product $(\partial c/\partial C)(\partial C/\partial t)$ of two unknowns to find. Equilibrium dynamics may be modeled with linear arguments, but out-of-equilibrium modeling is inherently nonlinear.

- Another view of the advection-diffusion PDE (7.3) is that it is a Fokker–Planck equation for the probability density function of a particle carried along the pipe at velocity $u(r)$ and undergoing a stochastic random walk across the pipe (coefficient κ). Then the modeling we have done in this section derives an effective Fokker–Planck equation, namely (7.14), for an effective stochastic advection-dispersion along the pipe. The effective Fokker–Planck equation (7.14) then corresponds to a stochastic differential equation that, in effect, models the long-term random advection of a particle along the pipe. Section 21.2 discusses how to model the Fokker–Planck equation of a stochastic system.

  Camassa et al. (2010) discussed quite extensively such a stochastic view of shear dispersion.[100]

7.6 ▪ Exercises

Exercise 7.1 (different amplitudes). Algorithm 7.1 implements the definition of the amplitude being the cross-pipe average concentration, $C(x,t) := \operatorname{mean} c(r,\theta,x,t)$. Modify only the `linv` operator to implement different amplitude definitions, execute the algorithm, and compare the resulting models:

1. the center pipe value, $C(x,t) := c(0,\theta,x,t)$;
2. the average around the pipe wall, $C(x,t) := \frac{1}{2\pi}\int_0^{2\pi} c(1,\theta,x,t)\,d\theta\,$;
3. the "cup concentration," proportional to the flux of material, past any station x, namely $C(x,t) := \frac{1}{\pi}\int_0^1 2(1-r^2)c(r,\theta,x,t)\,r d\theta\,dr\,$.

Observe in the resulting models that only the expressions for the slow manifold field $c(r,\theta,x,t)$ differ; the evolution equations for all of the amplitudes are identical. *In linear dynamics the evolution is independent of any linear redefinition of the amplitude (Lemma* 5.1*).*

[100]One might inquire what the stochastic counterpart role of the higher-order terms in higher-order models such as (7.15) is. Such higher-order effects correspond to non-Gaussian characteristics in the effective longitudinal random walk.

Exercise 7.2 (dispersion in a shear thinning flow). Suppose the flow down the pipe is of a shear thinning fluid. Then the flow is more like that of a plug flow: there are rapid shear layers near the pipe wall, and the flow in the center of the pipe is more like a plug. Assume the shear thinning fluid flow has nondimensional velocity $u = \frac{3}{2}U(1-r^4)$. What changes need to be made to the analysis and computer algebra to cater for this along-stream advection velocity? Modify Algorithm 7.1 and execute to construct the slowly varying model of dispersion.

Answer: The slow manifold is $c = C - \frac{1}{96}(5-12r^2+4r^6)U\partial C/\partial x + \mathcal{O}(\partial_x^2)$. On this slow manifold the evolution is $\partial C/\partial t = -U\partial C/\partial x + (1+\frac{U^2}{96})\partial^2 C/\partial x^2 - \frac{13}{483840}U^3\partial^3 C/\partial x^3 + \mathcal{O}(\partial_x^4)$.

Exercise 7.3 (leaky pipe dispersion). Suppose the material being carried along the pipe is not perfectly conserved but leaks through the pipe wall; for example, the material might be heat energy that radiates out of the pipe, or the material might be extremely fine suspended particles that are weakly attracted to the pipe walls by van der Waals forces but then remain stuck once attracted. Then the boundary condition on the pipe wall becomes

$$\frac{\partial c}{\partial r} + \epsilon c = 0 \text{ at } r = 1$$

for some nondimensional parameter ϵ measuring the strength of the material loss. Comment on the theoretical support for a model when the parameter ϵ is small. Modify Algorithm 7.1 and execute to construct the slowly varying model of dispersion in the presence of such small loss of material. Hint: In addition to working to some error $\mathcal{O}(\epsilon^p)$, you only need to code a change to the update of the evolution $\hat{g}$ in proportion to the residual of the new boundary condition—the update $\hat{v}$ of the concentration field then looks after itself.

Answer: The slow manifold is $c = [1+\epsilon(\frac{1}{4}-\frac{1}{2}r^2)]C - \frac{1}{24}[2-6r^2+3r^6+\frac{1}{24}\epsilon(5-36r^2+54r^4-20r^6)]\partial C/\partial x + \mathcal{O}(\partial_x^2, \epsilon^2)$. On this slow manifold the evolution is $\partial C/\partial t = -2\epsilon C - (1+\frac{1}{6}\epsilon)\partial C/\partial x + (\delta + \frac{1}{48} + \frac{1}{1440}\epsilon)\partial^2 C/\partial x^2 + \mathcal{O}(\partial_x^3, \epsilon^2)$.

Exercise 7.4 (heat exchanger). Consider two thin pipes in which a fluid carries heat at different velocities, u_a and u_b (Roberts, 1988, 2013). Suppose the pipes are next to each other so that heat is exchanged from the hotter fluid to the cooler. Letting $a(x,t)$ and $b(x,t)$ denote the temperature in each fluid at position x at time t, suppose that the heat transport and exchange are governed by the coupled PDEs

$$\begin{aligned}\frac{\partial a}{\partial t} + u_a\frac{\partial a}{\partial x} &= b - a,\\ \frac{\partial b}{\partial t} + u_b\frac{\partial b}{\partial x} &= a - b.\end{aligned}$$

1. Take the Fourier transform in space and find the spectrum in terms of the wavenumber k and the mean and difference velocities: $v = (u_a + u_b)/2$ and $w = (u_a - u_b)/2$.
2. Hence justify a slow manifold model of the heat exchanger for small wavenumbers (slowly varying in space).
3. Construct in Fourier space the slowly varying model of the heat exchanger.

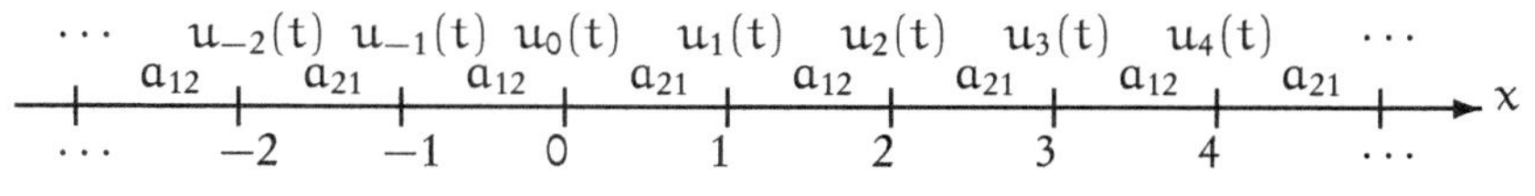

Figure 7.3. *Schematic diagram of inhomogeneous diffusion on a lattice of material $u_n(t)$, and two-periodic diffusivity a_{21} and a_{12} between even and odd lattice points.*

4. Given the square-root singularity in the spectrum you found in the first part, what do you expect to be the radius of convergence of the power series in wavenumber k?
5. By taking the sum and difference of the original governing PDEs, writing in terms of mean and difference variables, $c = (a+b)/2$ and $d = (a-b)/2$, directly construct in physical variables a slow manifold model. Compare with the previous part.

Answer: Spectrum $\lambda = -1 - i\nu k \pm \sqrt{1 - w^2k^2}$. Slow manifold evolution is $\partial c/\partial t \approx -\nu\partial c/\partial x + \frac{1}{2}w^2\partial^2 c/\partial x^2 - \frac{1}{8}w^4\partial^4 c/\partial x^4$.

Exercise 7.5 (homogenize lattice dynamics). The Fourier transform of section 7.2 also rigorously justifies homogenizing diffusion. Here consider dynamics on the lattice of Figure 7.3. At each point of the lattice the field $u_n(t)$ evolves by diffusive exchange with its neighbors

$$\frac{du_n}{dt} = a_n^-(u_{n-1} - u_n) + a_n^+(u_{n+1} - u_n), \tag{7.17}$$

where the coefficients of interaction are periodic in n with period 2 ($a_n^\pm$ are alternatively a_{12} and a_{21}, as shown in Figure 7.3). These equations model the diffusion of quantity u through a microscale laminate of alternating material and with alternating diffusivities. Our challenge is to deduce the effective macroscale diffusion equation for the field u over large distances x.

Proceed by taking the Fourier transform, but of every second lattice point. Figure 7.4 shows how the field at every second point is smooth, whereas the field at every point has detailed sawtooth structure. That is, we seek lattice fields for even n and odd n separately: for subscript 1 denoting odd points, and subscript 2 denoting even points, and for spatial Fourier wavenumber k in x-variations seek

$$u_n(t) = \int_{-\pi/2}^{\pi/2} e^{ikn}\tilde{u}_1(k,t)\,dk, \quad \text{odd } n;$$

$$u_n(t) = \int_{-\pi/2}^{\pi/2} e^{ikn}\tilde{u}_2(k,t)\,dk, \quad \text{even } n.$$

- Substitute into the microscale lattice dynamics (7.17), and since on the lattice the modes e^{ikn} are linearly independent over the range of wavenumbers $-\pi/2 < k < \pi/2$, then equate coefficients of the integrand to deduce that the Fourier-transformed version of equation (7.17) is the coupled system of odd/even dynamics

$$\frac{d\tilde{u}_1}{dt} = a_{21}(e^{-ik}\tilde{u}_2 - \tilde{u}_1) + a_{12}(e^{ik}\tilde{u}_2 - \tilde{u}_1),$$

$$\frac{d\tilde{u}_2}{dt} = a_{12}(e^{-ik}\tilde{u}_1 - \tilde{u}_2) + a_{21}(e^{ik}\tilde{u}_1 - \tilde{u}_2).$$

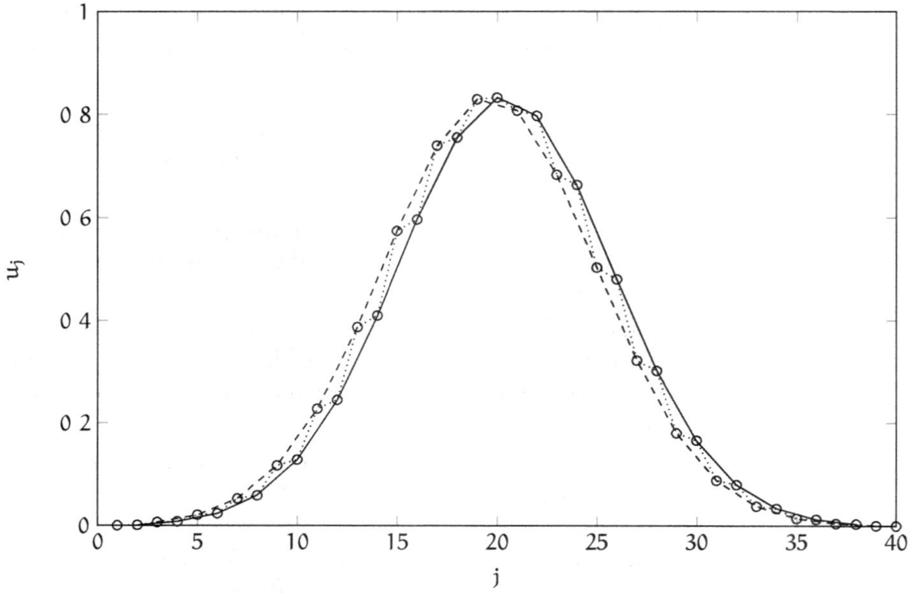

Figure 7.4. *An example solution of diffusion (7.17) with alternating diffusivity. The dotted line shows that the complete field has a "complicated" sawtooth structure. However, sampling at every second point (solid and dashed lines) shows the smooth macroscale structure.*

- When wavenumber $k = 0$ (extremely long wavelengths in x), show that this system has eigenvalues 0 and $-2(a_{21} + a_{21})$. Hence deduce that there exists a slow manifold model for at least some small but finite range of wavenumbers k.
- Define the mean Fourier component $\tilde{U}(k,t) = [\tilde{u}_1(k,t) + \tilde{u}_2(k,t)]/2$ and then argue that the slow manifold is $\tilde{u}_1 \approx \tilde{u}_2 \approx \tilde{U}$.
- Seek the slow manifold in a power series in small wavenumber k, expanding $e^{\pm ik} = 1 \pm ik - \frac{1}{2}k^2 + \cdots$, to deduce that the slow manifold evolution of the mean Fourier component is $d\tilde{U}/dt \approx -k^2\bar{a}\tilde{U}$ for the harmonic mean coefficient of diffusivity $\bar{a} = 2a_{12}a_{21}/(a_{12} + a_{21})$.[101]
- Linearly superpose mean Fourier components over a range of small wavenumbers k to deduce that an emergent slow manifold model of the diffusion on the lattice is the classic "homogenized equation"

$$\frac{\partial U}{\partial t} \approx \bar{a}h^2\frac{\partial^2 U}{\partial x^2} \quad \text{for lattice spacing } h.$$

  What microscale field is predicted in terms of $U(x,t)$?
- Extend the above analysis to at least quartic terms in wavenumber k to find higher-order corrections to the homogenized diffusion model.

[101]The 2×2 matrix equations that need to be solved are the analogues of the "cell problem" that one reads about in traditional approaches to homogenization.

Exercise 7.6 (dispersion in a channel). Model the slowly varying, long-time dispersion of material in a two-dimensional channel $-1 < y < 1$ when governed by the nondimensional advection-diffusion PDE

$$\frac{\partial c}{\partial t} + \tfrac{3}{2}U(1-y^2)\frac{\partial c}{\partial x} = \frac{\partial^2 c}{\partial y^2} + \frac{\partial^2 c}{\partial x^2}, \quad \frac{\partial c}{\partial y} = 0 \text{ on } y = \pm 1 .$$

1. Assuming slow variations in x, find the spectrum of the cross-channel diffusion operator to justify a slow manifold model.
2. Modify Algorithm 7.1 to construct models of the dispersion.
3. Derive the above nondimensional advection-diffusion PDE from a dimensional problem. Then transform the nondimensional model into dimensional terms. Interpret.

Answer: Spectrum is $\lambda_n = -n^2\pi^2/4$ for $n = 0, 1, 2, \ldots$. The slow manifold is $c = C - \frac{1}{120}(7-30y^2+15y^4)U\partial C/\partial x + \mathcal{O}(\partial_x^2)$. On this slow manifold the evolution is $\partial C/\partial t = -U\partial C/\partial x + (1+\frac{2U^2}{105})\partial^2 C/\partial x^2 + \frac{4}{17325}U^3\partial^3 C/\partial x^3 + \mathcal{O}(\partial_x^4)$.

Exercise 7.7 (turbulent dispersion in a channel). Model the slowly varying, long-time dispersion of material in a two-dimensional channel $-1 < y < 1$ when governed by the nondimensional advection-diffusion PDE

$$\frac{\partial c}{\partial t} + \tfrac{3}{2}(1-y^2)\frac{\partial c}{\partial x} = \frac{\partial}{\partial y}\left[(1-y^2)\frac{\partial c}{\partial y}\right], \quad c \text{ bounded on } y = \pm 1 .$$

For example, the cross-channel diffusion coefficient $(1-y^2)$ may represent turbulent mixing in a channel where the mixing by small eddies is smaller near the sides than the mixing by bigger eddies in the middle of the channel.[102]

1. Assuming slow variations in x, use Sturm–Liouville theory (Table 5.1 or standard texts such as Kreyszig (2011, §5.7, e.g.)) on an ODE in y to justify a slow manifold model. In particular, since the second eigenvector $v_2(y)$ must have one zero, Table 5.1, item 3, use the rough approximation $v_2(y) \approx y$ and the Rayleigh quotient, Table 5.1, item 6, to roughly estimate the leading negative eigenvalue λ_2.[103]
2. Modify Algorithm 7.1 to construct models of the dispersion. Hint: The operator `linv` does not have to be exactly correct; solving $(-y^2c_y)_y = \text{Res}$ (and ignoring any constant terms in the Res) is sufficient to ensure convergence, albeit in more iterations.[104]
3. Derive the above nondimensional advection-diffusion PDE from a dimensional problem. Then transform the nondimensional model into dimensional terms. Interpret.

Answer: The slow manifold is $c = C - \frac{1}{12}(1-3y^2)\partial C/\partial x + \mathcal{O}(\partial_x^2)$. On this slow manifold the evolution is $\partial C/\partial t = -\partial C/\partial x + \frac{1}{30}\partial^2 C/\partial x^2 + \frac{1}{630}\partial^3 C/\partial x^3 + \mathcal{O}(\partial_x^4)$.

---

[102] Setting the mixing proportional to the advection velocity is characteristic of eddy viscosity models of turbulence. Strictly, the model asserts that the mixing is proportional to the velocity gradient, but in a relatively wide channel or river the dominant gradient is the one over the depth of the fluid that we have not resolved.

[103] This approximation of $v_2(y)$ is the exact eigenvector, but we do not need to know that in order to estimate λ_2, especially as theoretical support only needs the rough value of the eigenvalue.

[104] The reason this hack of the update works is that it incurs an error, but the error is of lower powers in y. Thus successive iterates drive the errors to lower and lower powers in y until the error disappears at y^0 (provided the update to the evolution is correct).

Exercise 7.8 (turbulent dispersion in a pipe). Model the slowly varying, long-time dispersion of material along a pipe. Assume the pipe has nondimensional radius 1 and the material concentration is governed by the nondimensional advection-diffusion PDE (7.3) adapted to when the "eddy diffusivity"[105] is $\kappa \propto 1 - r^2$. The boundary condition on the pipe wall is here that c is bounded.

1. Assuming slow variations in x, use Sturm–Liouville theory (higher-dimensional analogues of Table 5.1 or Kreyszig (2011, §5.7, e.g.)) on a PDE in r and θ to justify a slow manifold model. In particular, use the rough approximation that the second eigenvector $v_2(r, \theta) \approx r\cos\theta$ and the Rayleigh quotient, a generalization of Table 5.1, item 6, to roughly estimate the leading negative eigenvalue λ_2.
2. Modify Algorithm 7.1 to construct models of the dispersion. Hint: The operator `linv` does not have to be exactly correct; solving $r^{-1}(-r^3 c_r)_r = \text{Res}$ (and ignoring any constant terms in the Res) is sufficient to ensure convergence, albeit in more iterations.[106]
3. Interpret the model.

Answer: Eigenvalue $\lambda_2 \approx -4$. The slow manifold is $c = C - \frac{1}{8}(1 - 2r^2)\,\text{Pe}\,\partial C/\partial x + \mathcal{O}(\partial_x^2)$. On this slow manifold the evolution is $\partial C/\partial t = -\text{Pe}\,\partial C/\partial x + (\frac{1}{2} + \frac{1}{24}\text{Pe}^2)\partial^2 C/\partial x^2 - \frac{1}{24}\text{Pe}\,\partial^3 C/\partial x^3 + \mathcal{O}(\partial_x^4)$.

Exercise 7.9 (field-flow fractionation). Consider the transport of some molecules in a symmetric field-flow fractionation channel as analyzed by Giddings (1986) and Schure et al. (1989), for example. This exercise models the transport and dispersion of the molecules using center manifold theory following the approach of Suslov and Roberts (1999). Consider the advection-diffusion of the molecules with concentration field $c(x, y, t)$ described by the conservation PDE (6.5) in the shear velocity field $\mathbf{q} = \dot{\gamma} y \mathbf{i} - v_0 \mathbf{j}$:

- a small downward vertical velocity v_0 is applied in field-flow fractionation to gather the molecules near the $y = 0$ bottom boundary—the fluid is forced through a porous bed at $y = 0$, but the molecules are too big to penetrate the bed;
- because the molecules are gathered near $y = 0$, only the shear $\dot{\gamma} y \mathbf{i}$ is important in the advection; the "distant" upper wall is negligible.

1. Nondimensionalize the system with respect to the length κ/v_0, where diffusion balances downward advection, and the time v_0^2/κ for diffusion to act across such a length. Derive the PDE

$$\frac{\partial c}{\partial t} + Uy\frac{\partial c}{\partial x} - \frac{\partial c}{\partial y} = \frac{\partial^2 c}{\partial x^2} + \frac{\partial^2 c}{\partial y^2}$$

[105] As in Exercise 7.7, setting the mixing proportional to the advection velocity is characteristic of eddy viscosity models of turbulence.

[106] The reason this hack of the update works is that it incurs an error, but the error is of lower powers in r. Thus successive iterates drive the errors to lower and lower powers in r until the error disappears at r^0 (provided the update to the evolution is correct).

Table 7.2. *Typical set of physical parameters for field-flow fractionation and the consequent parameters (in the second part) appearing in the analysis. The data are for the Cow Pea Mosaic Virus (Litzén,* 1993, *p.* 464*) in the field-flow fractionation channel of Wahlund and Giddings (*1987*).*

Parameter		Value
Channel height		0.05 cm
Kinematic viscosity		$0.01\ \mathrm{cm}^2/\mathrm{s}$
Mean longitudinal velocity		0.1 cm / s
Cross-flow velocity	ν_0	5×10^{-4} cm / s
Molecular diffusivity	κ	$2 \times 10^{-7}\ \mathrm{cm}^2/\mathrm{s}$

for nondimensional parameter $U = \dot\gamma\kappa/\nu_0^2$. Argue that parameter U may be viewed as a Péclet number.[107]

Estimate the characteristic parameters for the field-flow fractionation data of Table 7.2.

2. Boundary conditions are that the concentration $c \to 0$ as $y \to \infty$, and that there is no flux of molecules through the bed. Derive that this no-flux condition nondimensionalizes to

$$c + \frac{\partial c}{\partial y} = 0 \quad \text{on } y = 0 .$$

3. Find the slow subspace of equilibria $c = Ce^{-y}$.
4. Verify that there are homogeneous solutions of the PDE and boundary conditions for all (real) $\lambda < 0$ of the form

$$c = e^{\lambda t}\begin{cases} e^{-y/2}\big[\cos \ell y - (2\ell)^{-1}\sin \ell y\big] & \text{for } \ell > 0, \\ e^{-y/2}(1-y/2), & \\ e^{-y/2}\big[\cosh \ell y - (2\ell)^{-1}\sinh \ell y\big] & \text{for } \tfrac12 > \ell > 0 . \end{cases}$$

This apparently continuous spectrum affects our results, but derive that any distant top wall, say with boundary conditions $c + \partial c/\partial y = 0$ at $y = B$, discretizes the spectrum so that all eigenvalues $\lambda < -1/4$ except for the zero eigenvalue of the mode e^{-y}. Deduce that there exists an emergent slow manifold model for solutions varying slowly enough in the lateral x-direction.

5. Derive that the slow manifold, parametrized by $C(x,t) := c(x,0,t)$ say, is

$$c = e^{-y}C - \tfrac12 U y^2 e^{-y}\frac{\partial C}{\partial x} + U^2(y^2 + \tfrac13 y^3 + \tfrac18 y^4)e^{-y}\frac{\partial^2 C}{\partial x^2} + \mathcal{O}(\partial_x^3),$$

such that

$$\frac{\partial C}{\partial t} = -U\frac{\partial C}{\partial x} + (1+2U^2)\frac{\partial^2 C}{\partial x^2} + \mathcal{O}(\partial_x^3).$$

6. Discuss how molecules with different diffusivity are advected at different speeds (relative to the reference speed ν_0) along the channel, hence reaching the end of the channel at different times, and are thus separated. Discuss how the effective lateral dispersion affects the separation of different molecules.

[107] Some prefer to also scale the lateral x-direction differently from the transverse y-direction. But let's not assume so a priori.

7. Write, debug, and execute a Reduce program to compute high-order approximations to the slow manifold model. Observe that the evolution equation is roughly $\partial C/\partial t \sim \sum_n (-a)^n n! \partial^n C/\partial x^n$ and hence is at best asymptotic for all wavenumbers in the lateral structure (reflecting the potential for a continuous spectrum noted above).

Exercise 7.10 (dispersion in porous media). In flow in porous media such as soil or underground aquifers, the fluid flows through the highly tortuous "pipes" of the pores between granules of material. In such flow there is no regular physical structure on which to base analysis. Instead one posits a microscopic modeling parametrizing the cross-stream direction by the advection velocity v rather than the physical cross-stream direction. One argues the following.

- The concentration of interest $c(x, v, t)$ is the concentration of particles at position x at time t that are traveling at velocity v through nanoscale pores.
- The nanopores occupy a fraction ϕ of the physical space, and the distribution of nanopore sizes is such that the probability of a particle traveling at velocity v through its pore is $\psi(v)$ (where $\int \psi(v)\, dv = 1$). For example, take $\psi = e^{-v}$ for $v \geqslant 0$.
- The particles may stick to the pore's surface or may be stuck in a constriction, and the probability of this happening is proportional to how far the particles travel; that is, absorption of particles occurs at a rate $\mu v c$ for $v \geqslant 0$.
- Because of such nanopore physics, particles may be injected or removed from traveling at the equilibrium velocity distribution ψ. But mixing at nanochannel junctions tends to restore the equilibrium velocity distribution ψ. Particles tend to meet such pore junctions at a rate $\bar{v}/\ell$, where ℓ is a typical pore-channel length and $\bar{v}$ is the mean advection velocity. Thus a perturbed velocity distribution will relax to the equilibrium distribution ψ at this rate $\bar{v}/\ell$.
- Putting this all together, only allowing positive velocities $v \geqslant 0$, the microscale continuum model for movement of particles in the porous media is

$$\phi \frac{\partial c}{\partial t} + v \frac{\partial c}{\partial x} = -\mu v c + \frac{\bar{v}}{\ell} \left[\psi(v) \int_0^\infty c\, dv - c \right]. \tag{7.18}$$

1. Argue that for slow variations in longitudinal direction x and for weak absorption, small μ, the spectrum of the dynamics of (7.18) has one zero eigenvalue and the rest are identical, namely $-\bar{v}/(\ell\phi)$. Hint: Decompose $c(v, t)$ into a part that has zero mean and a part proportional to $\psi(v)$; that is, substitute $c(v, t) = C(t)\psi(v) + \tilde{c}(v, t)$, where $\int_0^\infty \tilde{c}\, dv = 0$.
2. Choose the particular equilibrium distribution that $\psi = e^{-v}$ for $v \geqslant 0$ (and zero otherwise). Set $\ell = 1$ for simplicity. Then construct a low-order approximation of the slow manifold of slowly varying advection-dispersion (separately consider the components with zero mean and the component proportional to $\psi(v)$). Express the slow manifold model in terms of the total $C(x, t) = \int_0^\infty c\, dv$. Interpret.

3. The slow manifold model is

$$\phi\frac{\partial C}{\partial t} = -\mu C - (1-2\mu)\frac{\partial C}{\partial x} + (1-6\mu)\frac{\partial^2 C}{\partial x^2} + \mathcal{O}(\mu^2, \partial_x^3).$$

Now physical advection velocities are $v \geqslant 0$, so the coefficient in parentheses in front of $\partial C/\partial x$ surely has to be positive. Further, physically the coefficient of the dispersion term $\partial^2 C/\partial x^2$ needs to be positive for a sound model. However, for parameter $\mu > 1/2$ both coefficients are unphysically negative. Modify the slow manifold model so that it is still consistent with the asymptotic analysis *and* both coefficients are physically positive.

4. Extension: Explore the slow manifold slowly varying models for other equilibrium velocity distributions $\psi(v)$.

Answer: $c = e^{-v}[1+\mu(1-v)]C + e^{-v}[1-v+\mu(-4v+2v^2)]\partial C/\partial x + e^{-v}[-2v+v^2+\mu(3-3v+9v^2-3v^3)]\partial^2 C/\partial x^2 + \mathcal{O}(\mu^2, \partial_x^3)$ such that $\phi\partial C/\partial t = -\mu C - (1-2\mu)\partial C/\partial x + (1-6\mu)\partial^2 C/\partial x^2 + \mathcal{O}(\mu^2, \partial_x^3)$. There are indefinitely many possible physical modifications that maintain asymptotic correctness: two are $\phi\partial C/\partial t = -\mu C - (1+2\mu)^{-1}\partial C/\partial x + (1+6\mu)^{-1}\partial^2 C/\partial x^2 + \mathcal{O}(\mu^2, \partial_x^3)$ and $\phi\partial C/\partial t = -\mu C - e^{-2\mu}\partial C/\partial x + e^{-6\mu}\partial^2 C/\partial x^2 + \mathcal{O}(\mu^2, \partial_x^3)$.

Exercise 7.11 (slowly varying melt-solid interface). Reconsider the Stefan problem of the melting of a solid such as ice. Section 4.8 showed how the dynamics of melting was modeled by a slow manifold. But there the melting was into a one-dimensional solid. What if the melt and solid had two or three dimensions?

Define the mathematical problem. Let coordinate z measure distance into the melting solid (ice at 0°C for example), let coordinate x measure lateral distance across the melt-solid, and define $u(x, z, t)$ to be the temperature field in the melt-solid. From time $t = 0$ we start and maintain heating of the surface $z = 0$ for all x at nondimensional temperature 1: $u(x, 0, t) = 1$ is a boundary condition. The solid, initially at temperature $u(x, z, 0) = 0$, heats by diffusion governed by the classic (nondimensional) diffusion equation

$$\frac{\partial u}{\partial t} = \nabla^2 u = \frac{\partial^2 u}{\partial x^2} + \frac{\partial^2 u}{\partial z^2} \tag{7.19}$$

and melts wherever the temperature $u > 0$. Let the interface between the solid and the melt be at $z = s(x, t)$; then modeling indicates that the boundary conditions at the interface are that the melt and solid temperature $u(x, s, t) = 0$ and that the interface moves according to[108]

$$\frac{\partial s}{\partial t} = k\left(\frac{\partial u}{\partial z} - \frac{\partial s}{\partial x}\frac{\partial u}{\partial x}\right) \quad \text{at } z = s^-. \tag{7.20}$$

These define the physical fields and the governing equations.

[108] When the interface depends upon lateral coordinate x, then melting normal to the interface is at an angle to the z-direction. Since a normal to the interface $z = s(x, t)$ is $\mathbf{k} - \mathbf{i}s_x$, the distance in z corresponding to the interface moving a distance in the normal is $\Delta s = \Delta z = \sqrt{1+s_x^2}\Delta n$. The solid melts at a rate proportional to the heat flux across the interface from the melt, that is, $\Delta n/\Delta t$ is proportional to the normal gradient of temperature $(\nabla u)\cdot(\mathbf{k}-\mathbf{i}s_x)/\sqrt{1+s_x^2} = (u_z - s_x u_x)/\sqrt{1+s_x^2}$. Putting these together gives boundary condition (7.20).

Now seek a slow manifold model given slow variations laterally in space.

1. As always, first find interesting equilibria. As in section 4.8, argue that there is a family of equilibria when melt-solid coupling parameter $k = 0$, but now when there are also no lateral variations, $\partial/\partial x = 0$.
2. Linearize the dynamics about these equilibria to find that the spectrum supports the existence of a slow manifold—one that is global in $s(x,t)$ provided parameter k is small and the lateral derivatives $\partial/\partial x$ are small.
3. Analogous to section 4.8, construct a slow manifold model, assuming small k and slow variations in x, using a couple of iterations.

Answer: The leading approximation of temperature $u = 1 - z/s$ leads to the PDE model $\partial s/\partial t \approx (-k/s)(1 + s_x^2)$, but a second iteration suggests instead the PDE $\partial s/\partial t \approx (-k/s)\left[1 + \frac{1}{3}k + \frac{1}{3}s_x^2 + \frac{2}{3}ss_{xx}\right]$.

Exercise 7.12 (basic pattern diffusion). The Swift–Hohenberg PDE

$$\frac{\partial u}{\partial t} + (1-\epsilon)u + 2\nabla^2 u + \nabla^4 u + u^3 = 0 \tag{7.21}$$

is an interesting toy problem for exploring pattern formation and evolution (Swift and Hohenberg, 1977; Cross et al., 1986; Hutt, 2008, e.g.). As a first suggestive indication of its pattern formation, let's consider its marginal ($\epsilon = 0$), linear, one space dimension dynamics by modeling the field $u(x,t)$ governed by the PDE

$$\frac{\partial u}{\partial t} + u + 2\frac{\partial^2 u}{\partial x^2} + \frac{\partial^4 u}{\partial x^4} = 0. \tag{7.22}$$

Just as for shear dispersion, consider the x spatial dimension to be very large; it is the evolution of solutions to the PDE (7.22) over large space scales that is of interest.[109]

1. Analogous to shear dispersion (section 7.2), take the Fourier transform in space by writing the field $u(x,t) = \int_{-\infty}^{\infty} e^{ikx}\tilde u(k,t)\,dk$, for wavenumber k, and deriving the transformed system $d\tilde u/dt = -(1-k^2)^2\tilde u$.
2. Notionally adjoin the ODE $dk/dt = 0$ and explore the dynamics in $(k,\tilde u)$-space. Deduce that there are two subspaces of equilibria for wavenumbers ± 1.
3. Deduce that near each subspace of equilibria, there are corresponding slow subspaces where $d\tilde u/dt \approx -4\kappa^2\tilde u$ for wavenumbers $k = \pm(1+\kappa)$ and small κ. Deduce that for other wavenumbers, $\tilde u$ decays exponentially quickly; find a useful bound on the rate of decay for $|\kappa| > K$ for any given K.
4. Analogous to shear dispersion (section 7.2.2), combine these slow subspaces.
   (a) For any given small K, define the complex-amplitude field $U(x,t) = \int_{-K}^{K} e^{i\kappa x}\tilde u(1+\kappa,t)\,d\kappa$ and its complement $\bar U(x,t) = \int_{-K}^{K} e^{-i\kappa x}\tilde u(-1-\kappa,t)\,d\kappa$. Deduce that these fields satisfy the diffusion PDEs $\partial U/\partial t \approx 4\partial^2 U/\partial x^2$ and $\partial \bar U/\partial t \approx 4\partial^2 \bar U/\partial x^2$.

[109]This exercise suggests how one might model complex pattern-forming systems without the ad hoc mathematical introduction of multiple spatial scales to represent the one physical space.

(b) Further, deduce that the original field is approximately

$$u(x,t) \approx U(x,t)e^{ix} + \bar{U}(x,t)e^{-ix} + \mathcal{O}(e^{-\beta t})$$

for some rate β.

5. Discuss how this expression for the field $u(x,t)$ predicts that from any initial condition a pattern quickly forms with characteristic length 2π, and then the pattern diffuses out in space. Perhaps discuss specifically the case where the initial field $u(x,0)$ is nonzero only in some finite bounded part of the spatial domain. You might like to reread section 3.6 on complex-amplitude descriptions of oscillations.

Exercise 7.13 (pattern diffusion emerges). I introduced (Roberts, 1992b, eqs. (34)–(35)) a pattern-forming system with more microscale structure than the Swift–Hohenberg equation (7.21). In this exercise explore the large-scale spatial dynamics of the two fields $u(x,t)$ and $v(x,t)$ coupled by the one spatial dimension linear system

$$\begin{aligned} \frac{\partial u}{\partial t} &= -u - \frac{\partial^4 u}{\partial x^4} - 2\frac{\partial^2 v}{\partial x^2}, \\ \frac{\partial v}{\partial t} &= -2\frac{\partial^2 u}{\partial x^2} - v - \frac{\partial^4 v}{\partial x^4}. \end{aligned} \tag{7.23}$$

Repeat Exercise 7.12 but with the two component fields u and v instead of just one field.

Exercise 7.14 (embed to homogenize heterogeneous diffusion). Consider materials with complicated microstructure: we want to model their large-scale dynamics by equations with effective "average" coefficients. Heterogeneous diffusion in 1D is the simplest such example: suppose the material has structure so that "heat" $u(x,t)$ diffuses according to

$$\frac{\partial u}{\partial t} = \frac{\partial}{\partial x}\left\{\kappa(x)\frac{\partial u}{\partial x}\right\}, \tag{7.24}$$

where diffusion coefficient $\kappa(x)$ is ℓ-periodic in space: $\kappa(x)$ characterizes the heterogeneity. Our challenge is to derive the effective PDE

$$\frac{\partial U}{\partial t} = \bar{\kappa}\frac{\partial^2 U}{\partial x^2} + \cdots \tag{7.25}$$

for some effective mean field $U(x,t)$, some effective macroscale diffusivity $\bar{\kappa}$, and possibly with some higher-order corrections.

Techniques such as homogenization or multiple scales provide a mechanism for deriving effective models such as (7.25), but only in the limit of infinite scale separation between the microscale "cell" size ℓ and the macroscale domain size, say L. Instead, this exercise uses embedding and our approach to slowly varying solutions to provide rigorous support to the modeling at finite scale separation; that is, L is larger than ℓ, but not infinitely so. This exercise illustrates the general principle that models of microscale structures are often best phrased as ensemble averages (Young et al., 2001; Birch and Young, 2006, e.g.).

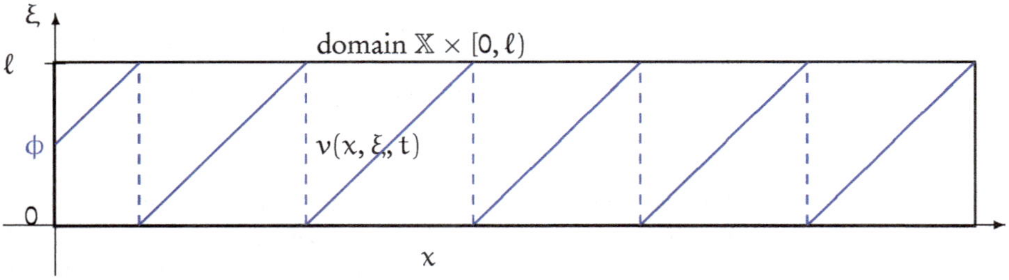

Figure 7.5. *Cylindrical domain of the embedding PDE (7.25) for field* $v(x,\xi,t)$*. Obtain solutions of the heterogeneous diffusion PDE (7.24) on the solid diagonal line as* $u(x,t) = v(x, x+\phi, t)$ *for any constant phase* ϕ*.*

1. Invent an embedding by considering a field $v(x,\xi,t)$ satisfying the PDE

$$\frac{\partial v}{\partial t} = \left(\frac{\partial}{\partial x} + \frac{\partial}{\partial \xi}\right)\left\{\kappa(\xi)\left(\frac{\partial v}{\partial x} + \frac{\partial v}{\partial \xi}\right)\right\} \tag{7.26}$$

   in the domain $\xi \in [0,\ell]$ and with ℓ-periodic boundary conditions in ξ, as indicated by Figure 7.5. Show that the field $u(x,t) := v(x, x+\phi, t)$ (that is, the field v evaluated on the solid diagonal lines in Figure 7.5) satisfies the heterogeneous diffusion PDE

$$\frac{\partial u}{\partial t} = \frac{\partial}{\partial x}\left\{\kappa(x+\phi)\frac{\partial u}{\partial x}\right\}. \tag{7.27}$$

   Hence deduce that a solution of (7.26) gives a solution of (7.24).

2. Now suppose you are given a solution $u(x,t)$ of (7.24). Embed this given solution in any set of fields $u_\phi(x,t)$ that solve (7.27) with $u_0(x,t) = u(x,t)$ and with the field u_ϕ being smooth in ϕ, x, and t. Show that $v(x,\xi,t) := u_{\xi-x}(x,t)$ satisfies (7.26). Hence deduce that (7.26) and (7.27) are equivalent and that they give us a set of solutions for an ensemble of materials all with the same diffusivity structure, but with the structural phase of the material shifted through all possibilities.

3. Let's analyze the embedding PDE (7.26) for a useful slow manifold. Show that for negligible wavenumber in x (via the Fourier transform) or negligible variations in x (directly), the PDE (7.26) possess a 1D subspace of equilibria, $v \propto$ constant in ξ.

   To check the spectrum about all of these equilibria, argue that the eigenvalue ODE from (7.26) is Sturm–Liouville and hence has only real eigenvalues. Take $\int_0^\ell \cdots v\,d\xi$ of the eigenvalue problem and deduce eigenvalues

$$\lambda \leqslant -\kappa_{\min}\frac{\int_0^\ell v_\xi^2\,d\xi}{\int_0^\ell v^2\,d\xi}.$$

   Use the calculus of variations to minimize the numerator subject to the denominator being constant, and subject to the necessary condition that $v(\xi)$ has zero mean (being orthogonal to the eigenvector corresponding to $\lambda = 0$). Deduce that all nonzero eigenvalues satisfy $\lambda \leqslant -4\pi^2\kappa_{\min}/\ell^2$.

Hence argue that a slow manifold model of the embedding PDE (7.26) exists and generally emerges on a cross-cell diffusion time.[110]

4. For definiteness, let the diffusivity be say $\kappa = 1/(1 + a\cos x)$ for cell size $\ell = 2\pi$ and some constant a. Construct a low-order approximation to the slow manifold model of the embedding PDE (7.26), and, in terms of a "mean" field $V(x, t)$, find that $v = V + a\sin\xi\,\partial V/\partial x + a\cos\xi\,\partial^2 V/\partial x^2 + \cdots$ such that V evolves according to $\partial V/\partial t = \partial^2 V/\partial x^2 + \cdots$. Relate this slow manifold model to the proposed effective diffusion PDE (7.25).

[110] As an extension to the exercise, discuss the apparent paradox of emergent order from an ensemble of physically disconnected solutions of different phases ϕ.

Chapter 8

Thin fluid films evolve slowly over space and time

Contents

The flow of rainwater on a road or a windscreen, paint, coating flows, and the flow of many protective biological fluids are all examples of the flow of thin layers of fluid. This chapter aims to create accurate yet relatively simple models for the nonlinear dynamics of such thin layers of fluid moving on a solid bed. The techniques developed herein apply much more generally, as indicated by some of the exercises.

When the fluid layer is thin, then, as in dispersion in a pipe, the only important dynamics occurs laterally along the fluid. Across the thin fluid film, viscosity acts quickly to damp almost all cross-film structure. This distinction between the longitudinal and cross-fluid dynamics is analogous to that of dispersion in a pipe, and we proceed with a similar analysis. However, the equations are very different, as indicated in Figure 8.1 and introduced in section 8.1. In particular, this problem has many nonlinearities: not only is the advection in the Navier–Stokes equation described by a nonlinear term, but also the unknown thickness of the fluid is another source of nonlinearity.

Assuming no longitudinal variations, the linear analysis of section 8.2 shows that there is one critical mode in the cross-fluid dynamics; all others decay due to viscosity. This critical mode is associated with conservation of the fluid. Consequently it is natural to express the low-dimensional model in terms of the fluid film thickness $h(x,t)$. Seeking solutions that vary slowly along the film, the center manifold analysis of section 8.4 creates an effective model of the dynamics.

Interestingly, our modeling is global in the thickness of the fluid layer. The analysis, when properly done, is valid for arbitrarily large variations in the thickness of the fluid film! The caveat is that the variations must be sufficiently slow. Of course, when the fluid film is too thick, then the attraction to the slow manifold will be too slow to form an effective model.

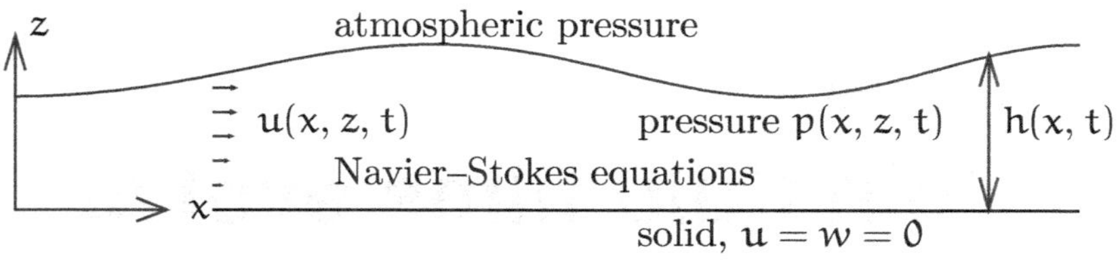

Figure 8.1. *Schematic diagram of a thin fluid film flowing along a solid bed.*

8.1 ▪ Governing fluid equations

Figure 8.1 illustrates that we seek to model the fluid flow with velocity field $\mathbf{q}(x,z,t)$ with a pressure field $p(x,z,t)$ of a fluid layer of thickness $h(x,t)$ lying on a solid substrate. This section derives the fluid equations and boundary conditions and then nondimensionalizes the equations for convenient analysis. If you wish to proceed directly to the analysis of the mathematical modeling of the thin fluid flow, then skip this section and proceed directly to section 8.2.

For a Newtonian fluid, base the modeling upon the incompressible Navier–Stokes (6.12) and continuity (6.2) equations: namely $\partial\mathbf{q}/\partial t+\mathbf{q}\cdot\boldsymbol{\nabla}\mathbf{q}=-\frac{1}{\rho}\boldsymbol{\nabla}p+\nu\boldsymbol{\nabla}^2\mathbf{q}$ and $\boldsymbol{\nabla}\cdot\mathbf{q}=0$. We restrict attention to two-dimensional flow taking place in the xz-plane; that is, assume there is no dependence upon the other lateral coordinate y. The x-axis is aligned along the solid bed of the flow; the z-axis is perpendicular. The viscous flow must stick to the solid bed; this physical condition gives the bed boundary condition that velocity

$$\mathbf{q}=0 \quad \text{on } z=0. \tag{8.1}$$

The fluid surface is complicated The free surface of the fluid, $z=h(x,t)$, evolves with the flow. Because the free surface is unknown we not only need two boundary conditions for the Navier–Stokes equations, we need a third boundary equation in order to be able to determine h. One condition is that the fluid flow as given by the velocity, $\mathbf{q}$, must follow the free surface $z=h(x,t)$. A straightforward way to derive such an equation[111] starts by noting that $z-h$ must be constant, specifically 0, for the fluid particles on the free surface. Thus the material derivative (6.3), $d(z-h)/dt$, must be zero on the free surface. Hence

$$0=\frac{\partial}{\partial t}(z-h)+\mathbf{q}\cdot\boldsymbol{\nabla}(z-h)=-\frac{\partial h}{\partial t}+w-u\frac{\partial h}{\partial x}$$
$$\Longrightarrow \frac{\partial h}{\partial t}=w-u\frac{\partial h}{\partial x} \quad \text{on } z=h. \tag{8.2}$$

This free surface condition is called the *kinematic condition.*

Other boundary conditions come from forces (stresses) acting across the free surface. Above the fluid film we suppose there is a very light and almost inviscid fluid such as air (one thousandth the density of water).

- Since air is inviscid, it cannot sustain any tangential stress across the surface; the tangential stress $T_t=0$.

[111] This derivation applies to other types of moving boundaries also.

Recall that the fluid stress tensor (6.6) is

$$\sigma = \begin{bmatrix} -p + \tau_{xx} & \tau_{xz} \\ \tau_{zx} & -p + \tau_{zz} \end{bmatrix},$$

where the components of the deviatoric stress tensor are

$$\tau_{xz} = \tau_{zx} = \mu \left(\frac{\partial u}{\partial z} + \frac{\partial w}{\partial x} \right), \quad \tau_{xx} = 2\mu \frac{\partial u}{\partial x}, \quad \tau_{zz} = 2\mu \frac{\partial w}{\partial z}.$$

Now a unit normal vector across the free surface $z = h$ is the unit vector $\hat{\mathbf{n}} = (-h'\mathbf{i} + \mathbf{k}) / \sqrt{1 + h'^2}$, where for brevity let's introduce $h' := \partial h / \partial x$. Thus the fluid stress across the free surface is

$$\mathbf{T} = \sigma \cdot \hat{\mathbf{n}} = \frac{1}{\sqrt{1 + h'^2}} \begin{bmatrix} ph' - \tau_{xx} h' + \tau_{xz} \\ -h'\tau_{zx} - p + \tau_{zz} \end{bmatrix}.$$

For the tangential stress to be zero, and since a unit tangent vector is $\hat{\mathbf{t}} = (\mathbf{i} + h'\mathbf{k}) / \sqrt{1 + h'^2}$,

$$T_t = \mathbf{T} \cdot \hat{\mathbf{t}} = \frac{1}{1 + h'^2} \left[h'(\tau_{zz} - \tau_{xx}) + \left(1 - h'^2\right) \tau_{xz} \right] = 0.$$

Thus the physical condition of zero tangential stress on the free surface requires the boundary condition

$$2h' \left(\frac{\partial w}{\partial z} - \frac{\partial u}{\partial x} \right) + \left(1 - h'^2\right) \left(\frac{\partial u}{\partial z} + \frac{\partial w}{\partial x} \right) = 0 \quad \text{on } z = h. \tag{8.3}$$

- Bernoulli's equation (6.10), $p_a + \rho_a \left(\partial \phi / \partial t + \frac{1}{2} |q|^2 \right) = B(t)$, also applies in the air flow above the fluid. Since the air density ρ_a is very low, then Bernoulli's equation (6.10) asserts that $B(t) \approx p_a$, the air pressure; that is, any air movement has little effect. Hence the fluid stress exerted normally across the surface has to be constant, say $T_n = -p_a$, equal and opposite to air pressure.

  However, surface tension is often an important force on the thin films we address. The effect of surface tension is like that of an elastic membrane; it causes a pressure jump if the surface is curved: positive if the fluid surface is convex; negative if it is concave.[112] The jump in pressure is proportional to the curvature $1/R = -h'' / \left(1 + h'^2\right)^{3/2}$. With σ denoting the coefficient of surface tension, then $T_n + \sigma / R = -p_a$. For a clean water/air interface the coefficient $\sigma = 72.8$ dyn / cm (Batchelor, 1979, p. 597), although this coefficient varies markedly with temperature and surface contamination.

  For the normal stress and surface tension to oppose atmospheric air pressure[113]

$$\left(1 + h'^2\right) (p - p_a) = \tau_{zz} - 2h'\tau_{xz} + h'^2 \tau_{xx} - \frac{\sigma h''}{\sqrt{1 + h'^2}}.$$

[112]Experimentally, surface tension is difficult to deal with as it is sensitive to surface contamination and temperature. Even small gradients of contaminants supply tangential stresses, driving fluid flow. The temperature dependence causes the patterns you see when you heat a thin layer of oil in a pan on the stove—it is called Marangoni convection.

[113]If the fluid is inviscid, as would be the case for large-scale flows, then this condition reduces to $p = p_a$ on the free surface.

Without loss of generality we take $p_a = 0$ (since it is only gradients of pressure that are important); thus

$$\left(1+h'^2\right) p = 2\rho\nu \left[\frac{\partial w}{\partial z} + h'^2\frac{\partial u}{\partial x} - h'\left(\frac{\partial u}{\partial z} + \frac{\partial w}{\partial x}\right)\right] - \frac{\sigma h''}{\sqrt{1+h'^2}} \quad \text{on } z = h\,. \tag{8.4}$$

Nondimensionalize these equations For a reference length, suppose that H is a characteristic thickness of the thin film, as shown schematically in Figure 8.1. Then nondimensionalize by writing the equations with respect to the reference length H, the reference time H^2/ν being a cross-film diffusion time, the reference velocity ν/H, and the reference pressure $\rho\nu^2/H^2$. Nondimensionalizing with respect to small-scale variables empowers us to later make regular perturbations supported by center manifold theory and thus avoid singular perturbations.

With these choices, and in nondimensional quantities, we solve the Navier–Stokes and continuity equations

$$\frac{\partial q}{\partial t} + q\cdot\nabla q = -\nabla p + \nabla^2 q\,, \tag{8.5}$$

$$\nabla\cdot q = 0\,, \tag{8.6}$$

subject to the bed boundary condition (8.1), and on the free surface the kinematic condition (8.2), and two dynamic conditions (8.3) and (the nondimensional version of (8.4))

$$\left(1+h'^2\right) p = 2\left[\frac{\partial w}{\partial z} + h'^2\frac{\partial u}{\partial x} - h'\left(\frac{\partial u}{\partial z} + \frac{\partial w}{\partial x}\right)\right] - \frac{\mathrm{We}\, h''}{\sqrt{1+h'^2}} \quad \text{on } z = h\,, \tag{8.7}$$

where $\mathrm{We} := \sigma H/(\rho\nu^2)$ is a form of *Weber number* characterizing the importance of surface tension.

In the Navier–Stokes equation we neglect body forces. If the presence of gravity were to be modeled, then it would appear in the nondimensional combination gH^3/ν^2. This may be neglected when gH^3/ν^2 is genuinely small (that is, when $H \ll 0.01$ cm for water). Including gravity is not a great complication, and its inclusion is left as an exercise for you.

8.2 ▪ Linear dynamics suggests a slow manifold model

This section takes as given the set of nondimensional PDEs and boundary conditions: the Navier–Stokes (8.5), the continuity (8.6), and boundary condition (8.1) on the bottom, and the fluid surface conditions of (8.2), (8.3), and (8.7).

8.2.1 ▪ Equilibria

The construction of a slow manifold rests upon an understanding of the dynamics linearized about some equilibrium. Here the equilibrium is a film of constant thickness, dimensionally H, and of zero velocity and pressure. Without loss of generality we take the equilibrium film thickness to be just 1; in this problem other thicknesses are similar.

8.2.2 ▪ Linearize the dynamics

The dynamics linearized about this equilibrium is based on velocities, pressure and $h-1$ being small. Thus, as well as neglecting products of small terms, evaluation of boundary conditions on the free surface is approximated by their evaluation on the approximate surface $z=1$. The linearized equations are

$$\begin{aligned}
&\frac{\partial q}{\partial t} = -\nabla p + \nabla^2 q\,, \quad \nabla\cdot q = 0 \quad \text{in the fluid},\\
&q = 0 \quad \text{on } z=0\,,\\
&\frac{\partial h}{\partial t} = w\,, \quad \frac{\partial u}{\partial z}+\frac{\partial w}{\partial x} = 0\,, \quad p = 2\frac{\partial w}{\partial z} - \text{We}\, h'' \quad \text{on } z=1\,.
\end{aligned} \tag{8.8}$$

The appearance of this system of equations is too complicated to make immediate progress.

Assume slowly varying solutions in space To investigate the dynamics of "long waves" on this thin film we treat $\partial/\partial x$ as small.[114] Thus "linearly" means we base the construction of a slow manifold model on the limit when there are *no* longitudinal variations: $\partial/\partial x = 0$. Neglecting all x-derivatives, and reordering, the linear system (8.8) becomes

$$\frac{\partial w}{\partial z} = 0\,, \qquad w = 0 \text{ on } z=0\,, \tag{8.9a}$$

$$\frac{\partial w}{\partial t} = -\frac{\partial p}{\partial z} + \frac{\partial^2 w}{\partial z^2}\,, \qquad p = 2\frac{\partial w}{\partial z} \text{ on } z=1\,, \tag{8.9b}$$

$$\frac{\partial u}{\partial t} = \frac{\partial^2 u}{\partial z^2}\,, \qquad u = 0 \text{ on } z=0\,, \quad \frac{\partial u}{\partial z} = 0 \text{ on } z=1\,, \tag{8.9c}$$

$$\frac{\partial h}{\partial t} = w \text{ on } z=1\,. \tag{8.9d}$$

In this linearization the vertical velocity $w=0$, as the continuity PDE (8.9a) asserts $w=$ constant in z, and the bottom boundary condition then insists that the constant is zero. Consequently the pressure $p=0$ also, as the vertical momentum PDE (8.9b) then reduces to $p=$ constant in z, and the surface boundary condition on pressure insists that the constant is zero. The only nonzero components can be the lateral velocity u and the fluid surface h.

An emergent slow manifold exists Seeking solutions of (8.9c)–(8.9d) for u and h proportional to $e^{\lambda t}$ leads to the following. First, from (8.9d), $\lambda_0 = 0$ is an eigenvalue associated with the mode $h=$ constant and $u=w=p=0$. It is the presence of this eigenvalue that indicates the existence of a useful slow manifold model.

The other eigenmodes have negative eigenvalues from viscous decay. With no horizontal pressure gradient, the horizontal component (8.9c) of the Navier–Stokes equation, $\lambda u = \partial^2 u/\partial z^2$, is to be solved with boundary conditions $u=0$ on the bed $z=0$ and $\partial u/\partial z = 0$ on the fluid surface $z=1$ (from the tangential stress condition). This has solutions $u_n = \sin\pi(n-\frac12)z$ for negative eigenvalues $\lambda_n = -\pi^2(n-\frac12)^2$ for $n=1,2,\ldots$.

[114] Remember that in strict analysis it is nonsense to say "$\partial/\partial x$ is small," as a derivative is an unbounded operator. As commented in the analysis of dispersion, one approach is to define the operator $\partial/\partial x$ to mean a low-band pass filter of the spatial derivative. Then "$\partial/\partial x$ is small" has some meaning (Roberts, 2013).

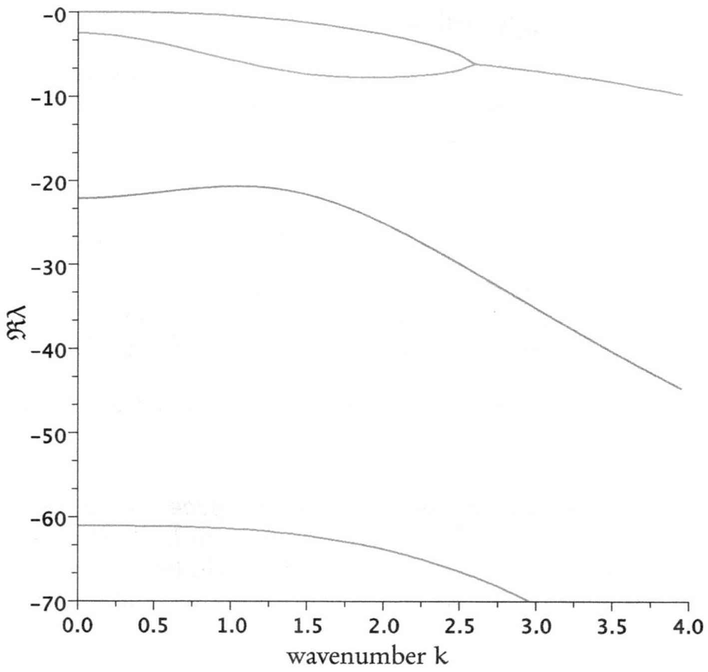

Figure 8.2. *The numerically determined spectrum of eigenvalues, $\Re(\lambda)$, for small disturbances to a thin film with Weber number* $\mathrm{We} = 3$, *plotted against horizontal wavenumber* k. *When the two upper branches merge, the eigenvalues become a complex conjugate pair.*

From the other fluid surface boundary conditions, the depth field h_n corresponding to these shear modes is zero. This analysis shows that for slowly varying solutions in space, long waves, there exists one zero eigenvalue of the dynamics, whereas the rest of the eigenvalues are strictly negative, bounded above by $-\beta = -\pi^2/4$ (divide this nondimensional rate by the time scale to get the dimensional rate of $-\pi^2\nu/(4H^2)$). Thus expect the dynamics of thin films to exponentially quickly approach a low-dimensional slow manifold parametrized by the fluid depth corresponding to the zero eigenvalue and characterized by slow variations in x.

Numerical solutions cater for x variations Numerical solutions, for arbitrary finite lateral wavenumber k, confirm the above analysis of the linear eigenproblem: simply seek solutions of the system (8.8) proportional to $e^{\lambda t + ikx}$ and use finite differences to approximately solve for the cross-film structure. Figure 8.2 plots the resultant eigenvalues for the case where the nondimensional surface tension $\mathrm{We} = 3$. Observe that for zero wavenumber the numerical results agree with the previous analytical arguments: $\lambda_0 = 0$, $\lambda_1 = -2.4674$, $\lambda_2 = -22.2066$, and $\lambda_3 = -61.6850$. For finite wavenumber k, all the eigenvalues are negative and so decay exponentially rapidly. However, the low wavenumber k, long-wavelength modes on the leading branch have small negative eigenvalues and so only decay slowly. Thus the long-term dynamics is dominated by solutions that are associated with the zero mode *and* that vary slowly in space; all other modes decay rapidly.

As in shear dispersion, we are justified in constructing a low-dimensional model of the dynamics via center manifold techniques.

8.3 ▪ Temporarily simplify to linearized surface conditions

A major complication in detail is the presence of the moving surface of the fluid in this problem. It seems useful, as a first step, to analyze a simpler version. This section simplifies by using only the linearized versions of the boundary conditions on the free surface, given in (8.8), while retaining the nonlinearity in the Navier–Stokes equation. Section 8.4, below, develops a model appropriate to the exact boundary conditions: if confident, you may skip directly to Section 8.4.

Parametrize the slow dynamics The first task is to decide how to parametrize the slow manifold model. Since the critical mode is $h = \text{constant}$, $p = u = w = 0$, it is appropriate to use the film thickness h as the parameter (although it is actually $h - 1$, which is *small* in this section). This parametrization is especially useful since h has the simple physical meaning of the thickness/depth of the fluid. Thus a linear description of the slow manifold is the fluid fields

$$u = w = p = 0\,, \tag{8.10}$$

and the fluid depth h is free to vary according to $\partial h/\partial t \approx 0$.

Iteration improves Now we organize an iteration scheme to refine the description of the slow manifold. Suppose that $q(x, z, h)$ and $p(x, z, h)$ are an approximation to the slow manifold fluid fields, with the associated approximate evolution $\partial h/\partial t \approx g(h)$. Seek equations for corrections to this description by

- substituting into the governing equations $q = q(x, z, h) + \hat{q}(x, z, h)$, $p = p(x, z, h) + \hat{p}(x, z, h)$, such that $\partial h/\partial t = g(h) + \hat{g}(h)$;
- omitting products of corrections;
- omitting x-derivatives of corrections as both corrections and $\partial/\partial x$ are small;
- approximating other terms involving corrections by replacing the current approximation with the initial linear approximation;
- rearranging the equations.

The upshot is to solve the following PDEs for corrections:

$$\frac{\partial^2 \hat{u}}{\partial z^2} = \frac{\partial u}{\partial t} + q \cdot \nabla u + \frac{\partial p}{\partial x} - \nabla^2 u\,, \tag{8.11a}$$

$$\frac{\partial^2 \hat{w}}{\partial z^2} - \frac{\partial \hat{p}}{\partial z} = \frac{\partial w}{\partial t} + q \cdot \nabla w + \frac{\partial p}{\partial z} - \nabla^2 w\,, \tag{8.11b}$$

$$\frac{\partial \hat{w}}{\partial z} = -\nabla \cdot q\,, \tag{8.11c}$$

subject to the boundary conditions

$$\frac{\partial \hat{u}}{\partial z} = -\left(\frac{\partial u}{\partial z} + \frac{\partial w}{\partial x}\right) \quad \text{on } z = 1\,, \tag{8.11d}$$

$$\hat{p} - 2\frac{\partial \hat{w}}{\partial z} = -p + 2\frac{\partial w}{\partial z} - \mathrm{We}\, h'' \quad \text{on } z = 1\,, \tag{8.11e}$$

$$\hat{q} = 0 \quad \text{on } z = 0\,, \tag{8.11f}$$

$$g + \hat{g} = w + \hat{w} \quad \text{on } z = 1\,. \tag{8.11g}$$

The first three iterations are as follows.

1. Substitute (8.10) into the right-hand sides of (8.11a)–(8.11g).

   - The continuity PDE (8.11c) becomes $\partial\hat{w}/\partial z = 0$, which with the boundary condition (8.11f) that $\hat{w} = 0$ on the bed leads to $\hat{w} = 0$, and hence the new $w = 0$.
   - The vertical momentum PDE (8.11b) then becomes $-\partial\hat{p}/\partial z = 0$ (since $\hat{w} = 0$) such that $\hat{p} = -\mathrm{We}\, h''$ on the approximate surface $z = 1$,[115] from the normal stress boundary condition (8.11e). The solution is $p = \hat{p} = -\mathrm{We}\, h''$ constant in the vertical.
   - The horizontal momentum PDE (8.11a) becomes $\partial^2\hat{u}/\partial z^2 = 0$ with boundary conditions (8.11f), that $\hat{u} = 0$ on the bed $z = 0$, and (8.11d), that $\partial\hat{u}/\partial z = 0$ on the surface $z = 1$. The solution is thus $u = \hat{u} = 0$.

   Consequently, the free surface evolution from the kinematic boundary condition (8.11g) is still $\partial h/\partial t \approx g + \hat{g} = 0$. All that has changed in this first iteration is that the pressure field in the fluid is corrected to account for the pressure induced by curvature of the surface.

2. Now substitute the above approximation into the right-hand sides of (8.11a)–(8.11g).

   - The continuity PDE (8.11c) is again $\partial\hat{w}/\partial z = 0$, which with the boundary condition (8.11f), that $\hat{w} = 0$ on the bed, leads again to $\hat{w} = 0$, and hence the new $w = 0$.
   - The vertical momentum PDE (8.11b) then becomes $-\partial\hat{p}/\partial z = 0$ (again) such that, from the normal stress boundary condition (8.11e), $\hat{p} = 0$ (zero now because we have accounted for the necessary pressure jump caused by surface curvature) on the approximate surface $z = 1$. The solution is $\hat{p} = 0$, leaving the pressure field unaltered at $p = -\mathrm{We}\, h''$.
   - The horizontal momentum PDE (8.11a) becomes $\partial^2\hat{u}/\partial z^2 = -\mathrm{We}\, h'''$ (from the horizontal pressure gradients induced by surface curvature variations) with boundary conditions (8.11f), that $\hat{u} = 0$ on the bed $z = 0$, and (8.11d), that $\partial\hat{u}/\partial z = 0$ on the surface $z = 1$.[116] The solution is $u = \hat{u} = (z - \frac{1}{2}z^2)\,\mathrm{We}\, h'''$.

   Consequently, the free surface evolution from the kinematic boundary condition (8.11g) is yet again $\partial h/\partial t \approx g + \hat{g} = 0$. It is only at the next iteration that we find nontrivial evolution.

3. On the third iteration, substituting the latest approximation into the right-hand sides of (8.11a)–(8.11g) leads to the following.

   - The continuity PDE (8.11c) is now $\partial\hat{w}/\partial z = (-z + \frac{1}{2}z^2)\,\mathrm{We}\, h^{iv}$, which with the boundary condition (8.11f), that $\hat{w} = 0$ on the bed, leads to $w = \hat{w} = (-\frac{1}{2}z^2 + \frac{1}{6}z^3)\,\mathrm{We}\, h^{iv}$.

[115] This is just hydrostatic pressure (given the absence of gravity here).

[116] Pressure gradients induce a parabolic shear flow, just as in Poiseuille flow in a pipe.

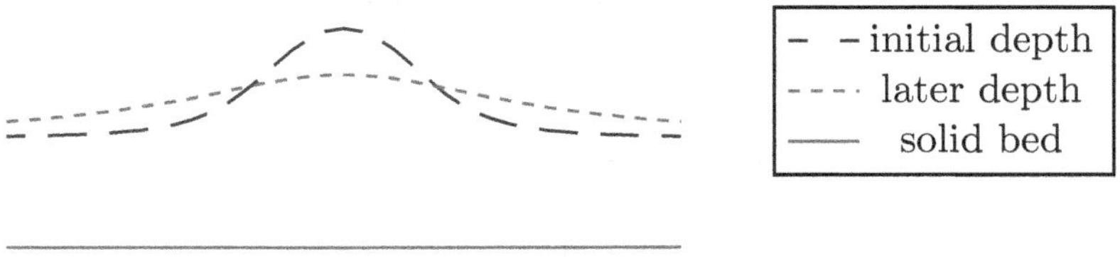

Figure 8.3. *Schematic diagram of a hump in a fluid film spreading out longitudinally under surface tension.*

- The vertical and horizontal momentum PDEs (8.11a) and (8.11b) do not need to be considered as this is the last iteration done here by hand. (In this iteration nonlinear terms arise through the advection $u\partial u/\partial x$ term in the horizontal momentum equation.)

  Using the above vertical velocity, we deduce the model

$$\frac{\partial h}{\partial t} \approx [w]_{z=1} = -\frac{We}{3} h^{iv} = -\frac{We}{3} \frac{\partial^4 h}{\partial x^4} . \tag{8.12}$$

Interpret the model The PDE (8.12) is a model for the long-term evolution of thin films under the action of surface tension. At this level of approximation the model is linear; however, further iterations would generate nonlinear terms. But for now if we seek solutions proportional to $e^{\lambda t + ikx}$, then observe that $\lambda = -\frac{We}{3} k^4$, in agreement with the small-k behavior of the leading eigenvalue branch shown in Figure 8.2. This PDE predicts that all variations in thickness of a fluid film will eventually decay, to result in a film of uniform thickness, although the decay, driven by surface tension, is very slow for long-wavelength disturbances.

Another way of saying the same thing is that the model (8.12) is rather like a diffusion equation, but with a much slower diffusion (fourth order) than the normal Fickian diffusion (second order). A hump of fluid in the film slowly spreads out due to surface tension, as Figure 8.3 shows schematically, but the spread becomes *very* slow over large distances.

The center manifold theorems reasonably assure us that this model is indeed relevant to the long-term dynamics of thin films. Recent research proposes one way to make the assurance rigorous (Roberts, 2013).

With a little more craft the above iteration may be restructured so that new information about the structure of the solution fields is used as soon as it is discovered. If we had first used the vertical momentum PDE (8.11b) to find the hydrostatic pressure, second used the horizontal momentum PDE (8.11a) to find the lateral velocity field driven by the new pressure, and third used the continuity PDE (8.11c) to find the vertical fluid velocity required to compensate for variations in that lateral velocity, then fourth we would have deduced the leading-order evolution (8.12) in just one iteration. Such restructuring is akin to Gauss–Seidel iteration and results in fewer iterations being needed.

Computer algebra confirms and extends We instruct a computer to carry out the algebra using Algorithm 8.1. The structure of Algorithm 8.1 is exactly the same as always, but it uses the Gauss–Seidel like iteration just described. One innovation is that we need to count the number of spatial derivatives so that terms with many x-derivatives, here

Algorithm 8.1 *Iteration finds the slow manifold of the lubrication flow for the lateral spread of a thin film of fluid in the case of linearized boundary conditions.*

```
% Use d to count the number of derivatives in x
let d^9=>0;  factor d,df,w;
% solves df(u,z,2)=rhs s.t. sub(z=0,u)=0=sub(z=1,df(u,z))
operator usol; linear usol;
let { usol(z^~~n,z) => (z^(n+2)/(n+2)-z)/(n+1)
    , usol(1,z) => (z^2/2-z) };
% linear solution
depend h,x,t; let df(h,t)=>g;
u:=w:=p:=g:=0;
% iteration (Gauss-Seidel like)
for iter:=1:9 do begin
   % vertical mom determines the pressure
   write resw:=df(w,t)+u*df(w,x)*d+w*df(w,z)+df(p,z)
       -df(w,x,2)*d^2-df(w,z,2);
   write restn:= sub(z=1,-p+2*df(w,z)-we*df(h,x,2)*d^2);
   p:=p+int(resw,z,z,1)+restn;
   % horiz mom determines horiz velocity
   write resu:=df(u,t)+u*df(u,x)*d+w*df(u,z)+df(p,x)*d
       -df(u,x,2)*d^2-df(u,z,2);
   write restt:=-sub(z=1,df(u,z)+df(w,x)*d);
   u:=u+usol(resu,z)+restt*z;
   % continuity finds vertical velocity
   write resc:=-df(u,x)*d-df(w,z);
   w:=w+int(resc,z,0,z);
   % kinematic bc determines evolution
   write g:=sub(z=1,w);
   if {resw,restn,resu,restt,resc}={0,0,0,0,0}
   then write iter:=1000000+iter;
end;
```

nine or more, are removed from the computation to restrict the algebra to the significant terms.

Upon running this code observe that a high-order model is

$$\frac{\partial h}{\partial t} \approx -\frac{\text{We}}{3}h^{iv} - \frac{3\,\text{We}}{5}h^{vi} - \left(\frac{23\,\text{We}}{21} + \frac{2\,\text{We}^2}{45}\right)h^{viii} + \frac{37\,\text{We}^2}{840}\left(h'''h^{iv}\right)' . \quad (8.13)$$

The linear terms correct the linear spectrum; the nonlinear term comes from the horizontal advection term, $u\partial u/\partial x$, as this is the main nonlinearity in a film with linearized boundary conditions.[117]

[117] If we truncate the linearized version of the model (8.13) to the first two terms, then for modes $e^{\lambda t+ikx}$ we observe $\lambda = -\text{We}\,k^4(\frac{1}{3}-\frac{3}{5}k^2)$. In *using* the model we would observe modes with wavenumbers $k > \sqrt{5}/3$ growing exponentially! Such growth is definitely unphysical. This unphysical growth is always a risk with a long-wavelength, slowly varying model: at finite wavenumber the approximation may be so inaccurate that high-wavenumber modes grow instead of decay. Two things may be done: either include higher-order terms in the model to see if they fix the problem or recast the linear dynamics. Consider an example of the latter here.

8.4 ▪ Thin fluid films vary in thickness

This section constructs the model for thin fluid film dynamics that caters for large variations in the film thickness. The modeling is global in fluid thickness. The global validity is due to all the equilibria when the fluid is of arbitrary constant depth: these equilibria form a slow subspace about which the analysis is valid. But the variations in fluid depth must be slow in x; that is, the lateral derivatives, $\partial^n/\partial x^n$, are small.

The unknown location of the free surface of the film is the major technical difficulty. To proceed, scale the vertical coordinate, $Z = z/h$, so that the free surface corresponds to $Z = 1$ precisely. Unfortunately, because the fluid thickness h varies with x and t, this scaling of z affects space and time derivatives and so plays havoc with details of the governing equations. Fortunately we relegate such intricate details to the computer.

Under the change of coordinates from (x, z, t) to

$$X = x, \quad Z = z/h(x,t), \quad T = t,$$

the chain rule shows that derivatives transform according to

$$\begin{aligned}
\frac{\partial}{\partial z} &= \frac{\partial T}{\partial z}\frac{\partial}{\partial T} + \frac{\partial X}{\partial z}\frac{\partial}{\partial X} + \frac{\partial Z}{\partial z}\frac{\partial}{\partial Z} = \frac{1}{h}\frac{\partial}{\partial Z},\\
\frac{\partial}{\partial x} &= \frac{\partial T}{\partial x}\frac{\partial}{\partial T} + \frac{\partial X}{\partial x}\frac{\partial}{\partial X} + \frac{\partial Z}{\partial x}\frac{\partial}{\partial Z} = \frac{\partial}{\partial X} - Z\frac{\partial h/\partial x}{h}\frac{\partial}{\partial Z},\\
\frac{\partial}{\partial t} &= \frac{\partial T}{\partial t}\frac{\partial}{\partial T} + \frac{\partial X}{\partial t}\frac{\partial}{\partial X} + \frac{\partial Z}{\partial t}\frac{\partial}{\partial Z} = \frac{\partial}{\partial T} - Z\frac{\partial h/\partial t}{h}\frac{\partial}{\partial Z}.
\end{aligned}$$

Algorithm 8.2 implements these rules in lines 7–13, where `xx` denotes X, `zz` denotes Z, and `tt` denotes T. Then wherever we need these rules they are automatically invoked by Reduce: within the iteration they are invoked in the evaluation of the residuals of the nonlinear PDEs and the residuals of the boundary conditions.

The only places where we now need to *explicitly* consider these derivative rules are in the terms in the PDEs for the corrections $\hat{u}$, $\hat{w}$, and $\hat{p}$. However, these only involve z-derivatives (see (8.11a)–(8.11e)), which transform to $\frac{1}{h}\partial/\partial Z$. Thus we multiply the residuals on the right-hand sides of these equations (8.11a)–(8.11e) by the appropriate power of h, as seen in lines 11, 17, and 20 of Algorithm 8.3.

For consistency in counting the derivatives in any term, and for brevity of the printed results, I use `h(m)` to denote $\partial^m h/\partial x^m$. This is implemented in lines 4–6, along with the interchange of time and space derivatives in line 16.

The iteration loop uses new information as it becomes available: first, the vertical momentum PDE determines the pressure correction; second, the horizontal momentum PDE determines the u correction; third, continuity gives w; and last, the kinematic free surface condition determines the latest version of the model evolution $\partial h/\partial t = g(h)$.

Instead of approximating the linear dynamics by a Taylor series in wavenumber k, as done up to now, we may approximate by a rational function, viz. $\lambda = -\mathrm{We}\,k^4/3(1+\frac{9}{5}k^2)$, which has physically acceptable behavior for *all* wavenumbers k *and* is of the same order of accuracy for small wavenumbers. By rewriting this spectrum as $\lambda(1+\frac{9}{5}k^2) = -\frac{\mathrm{We}}{3}k^4$ we appreciate that it arises from the differential equation model

$$\frac{\partial h}{\partial t} - \frac{9}{5}\frac{\partial^3 h}{\partial x^2 \partial t} = -\frac{\mathrm{We}}{3}\frac{\partial^4 h}{\partial x^4}.$$

Including nonlinear terms in this equation, we then would obtain a model with the correct low-wavenumber and nonlinear behavior, *and* with stable high-wavenumber dynamics so that useful simulations are attained.

Algorithm 8.2 *Initialization and preamble of the iteration, Algorithm 8.3, to find the slow manifold of the lubrication flow for the lateral spread of a thin film of fluid.*

```
factor d,we;
% Use d to count the number of derivatives of x
let d^7=>0;
% use operator h(m) to denote df(h,x,m)
operator h; hx:=d*h(1); hxx:=d^2*h(2);
let { df(h,xx)=>h(1), df(h(~m),xx)=>h(m+1) };
% use stretched coordinates: zz=z/h(x,t), xx=x, tt=t
depend xx,x,z,t;
depend zz,x,z,t;
depend tt,x,z,t;
let { df(~a,x) => d*df(a,xx)-zz*hx/h*df(a,zz)
    , df(~a,t) => df(a,tt)-zz*g/h*df(a,zz)
    , df(~a,z) => df(a,zz)/h };
% linear solution
depend h,xx,tt;
let { df(h,tt)=>g, df(h(~m),tt)=>df(g,xx,m) };
u:=w:=p:=g:=0;
rds:=1; % approximates 1/sqrt(1+h_x^2)
% solves df(u,zz,2)=rhs s.t. u(zz=0)=0=df(u,z)(zz=1)
operator usol; linear usol;
let { usol(zz^~~n,zz) => (zz^(n+2)/(n+2)-zz)/(n+1)
    , usol(1,zz) => (zz^2/2-zz) };
```

Algorithms 8.2 and 8.3 find that the long-term evolution of long-wavelength modes is approximately described by[118]

$$\begin{aligned}\frac{\partial h}{\partial t} &= -\tfrac{1}{3}\,\mathrm{We}\,\frac{\partial}{\partial x}\left(h^3h'''\right)\\ &\quad - \mathrm{We}\,\frac{\partial}{\partial x}\left(\tfrac{3}{5}h^5h^{v} + 3h^4h'h^{iv} + h^4h''h''' + \tfrac{11}{6}h^3h'^2h''' - h^3h'h''^2\right)\\ &\quad + \mathcal{O}(\partial_x^8).\end{aligned} \tag{8.14}$$

As is well established, the leading-order model, $\partial h/\partial t = -\frac{1}{3}\,\mathrm{We}\,\partial/\partial x(h^3h''')$, is a fourth-order, nonlinear, diffusion-like equation. It describes the flattening of the fluid surface under surface tension. The sixth-order terms on the second line in the model (8.14) may be most useful as an error estimator.

The slow manifold itself, in terms of the scaled vertical coordinate $Z = z/h$, comprises the corresponding fields of nearly constant pressure $p \approx -\mathrm{We}\,h''$, the approximately parabolic lateral velocity $u \approx \mathrm{We}\,h^2(Z - \frac{1}{2}Z^2)$, and the very small vertical velocity $w \approx -\frac{1}{2}\,\mathrm{We}\,h^2h'h'''Z^2 - \frac{1}{2}\,\mathrm{We}\,h^3h^{iv}(Z^2 - \frac{1}{3}Z^3)$. Higher-order terms describe finer details of the fluid fields.

Gravity Another important effect of thin fluid films is gravity. One simply incorporates gravity into our modeling by adding a downward (normal) gravitational force, say `+gz`, into the computation of the w PDE on lines 7–8 of Algorithm 8.3, and subtracting a lateral gravitational force, say `-gx`, in the computation of the u PDE on

[118] The right-hand side can always be written as a gradient due to conservation of fluid.

Algorithm 8.3 *After the initialisation of Algorithm 8.2, this iteration finds the slow manifold of the lubrication flow for the lateral spread of a thin film of fluid.*

```
1      % iteration
2      for iter:=1:9 do begin
3         % update rds:=1/sqrt(1+h_x^2) expansion
4         write resds:=(1+hx^2)*rds^2-1;
5         rds:=rds-resds/2;
6         % vertical mom & normal stress
7         write resw:=df(w,t)+u*df(w,x)+w*df(w,z)+df(p,z)
8             -df(w,x,2)-df(w,z,2);
9         write restn:= sub(zz=1,-p*(1+hx^2) +2*(df(w,z)
10            +hx^2*df(u,x)-hx*(df(u,z)+df(w,x))) -we*hxx*rds);
11        p:=p+h*int(resw,zz,zz,1) +restn;
12        % horizontal mom & bed & tang stress
13        write resu:=df(u,t)+u*df(u,x)+w*df(u,z)+df(p,x)
14            -df(u,x,2)-df(u,z,2);
15        write restt:=-sub(zz=1, (1-hx^2)*(df(u,z)+df(w,x))
16            +2*hx*(df(w,z)-df(u,x)) );
17        u:=u+h^2*usol(resu,zz)+h*restt*zz;
18        % continuity & bed
19        write resc:=df(u,x)+df(w,z);
20        w:=w-h*int(resc,zz,0,zz);
21        g:=sub(zz=1,w-u*hx);
22        showtime;
23        if {resds,resw,restn,resu,restt,resc}={0,0,0,0,0,0}
24        then write iter:=iter+1000000;
25     end;
```

lines 13–14.[119] That is all that is necessary (although to avoid being swamped by detail you might like to truncate to lower order in ∂_x; change line 3 of Algorithm 8.2). Execute the new code to derive models with gravitational forcing.

Limitation Although this model describes the long-term dynamics of thin films, it is limited in its usefulness (even with gravitational effects included). For example, in the linearized problem at finite wavenumber, (8.8), the leading branch of the spectrum, shown in Figure 8.2, merges with the next branch, that of the gravest shear mode $u \approx \sin \pi Z/2$, at a wavenumber $k \approx 2.5$ to become oscillatory decaying modes. Such oscillations are the remnants of the waves that surface tension can support, but viscosity damps. Such decaying waves are important in applications, as reviewed by Chang (1994), for example. However, the model (8.14) *cannot* describe the necessary oscillations because it has only one component and is only first order in time.

But with some imagination we could modify the governing equations so that a slow manifold model is formed based on the *two* leading branches of the spectrum (Roberts, 1996, e.g.). Such a model has much wider applicability because it is a much improved description at finite wavenumber. Chapter 9 develops this modeling.

[119] In this nondimensionalization, these coefficients of gravity are proportional to the gravity number. Since the reference length is h and the reference time is h^2/ν, the nondimensional gravity terms here are proportional to $g(h^2/\nu)^2/h = gh^3/\nu^2$.

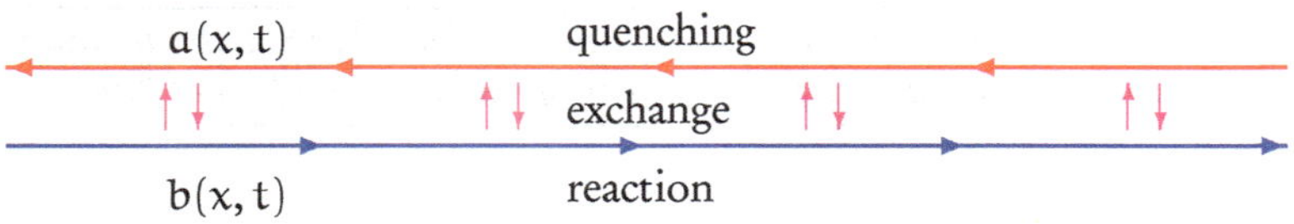

Figure 8.4. *Schematic diagram of two pipes (red and blue) carrying heat to the left and the right, with temperature fields* a *and* b*, and exchanging heat.*

8.5 ▪ Exercises

Exercise 8.1 (intermolecular forces). On very thin fluid films, many people contend that intermolecular van der Waals forces are significant (Oron et al., 1997; Kitavtsev et al., 2011, e.g.). Modify Algorithms 8.2 and 8.3 to include such an effect and deduce the change to the slow manifold of lubrication flow.

Introduce into the vertical momentum equation a long-range attraction and short-range repulsion between the fluid and the substrate; for the sake of this example say the force is $\epsilon(-\delta^4/z^4 + \delta^5/z^5)$. This force is attractive, negative, for $z > \delta$, and repulsive, positive, for small $z < \delta$; thus δ is the equilibrium distance for fluid molecules from the substrate molecules. The size of the force is proportional to parameter ϵ; assume ϵ is small by working to errors $\mathcal{O}(\epsilon^2)$. Execute the modified Algorithms 8.2 and 8.3 to deduce, for example, the pressure field

$$p \approx -\mathrm{We}\, h'' + \epsilon\delta\left[\tfrac{1}{3}\delta^3(z^{-3} - h^{-3}) - \tfrac{1}{4}\delta^4(z^{-4} - h^{-4})\right].$$

(Try not to worry about the infinite pressures predicted in this continuum model of what is physically discrete molecular dynamics.) Hence deduce that the fluid thickness evolves according to

$$\frac{\partial h}{\partial t} \approx -\tfrac{1}{3}\frac{\partial}{\partial x}\left[h^3\frac{\partial}{\partial x}\left(\mathrm{We}\,\frac{\partial^2 h}{\partial x^2} - \Pi(h)\right)\right] \quad \text{for } \Pi = \epsilon\delta\left(-\frac{\delta^3}{3h^3} + \frac{\delta^4}{4h^4}\right).$$

The extra term $\Pi(h)$ appears as a small extra "pressure" in the evolution equation, but in this model it physically represents the varying effects of the intermolecular forces across the whole fluid layer.

Exercise 8.2 (nonlinear heat exchanger). Consider the heat exchanger schematically depicted in Figure 8.4 with temperature fields $a(x,t)$ in the top pipe and $b(x,t)$ in the bottom pipe (Roberts, 2013, §4). Suppose there is a generative reaction in the bottom pipe b and a quenching reaction in the top pipe a so that the nondimensional governing PDEs are

$$\frac{\partial a}{\partial t} = +\frac{\partial a}{\partial x} + \tfrac{1}{2}(b - a) - \tfrac{1}{2}a^2, \quad \frac{\partial b}{\partial t} = -\frac{\partial b}{\partial x} + \tfrac{1}{2}(a - b) + \tfrac{1}{2}b^2.$$

Over long times will the generating reaction overcome the quenching and spreading? Or not?

1. Change to mean and difference fields, $c(x,t) = \tfrac{1}{2}(a + b)$ and $d(x,t) = \tfrac{1}{2}(a - b)$, respectively.
2. Linearize about the origin, seek solutions proportional to $e^{ikx+\lambda t}$, and observe that there is no strict spectral gap.

3. As well as linearizing in amplitude, treat lateral derivatives $\partial/\partial x$ as small and then argue for the existence of a slow manifold model $d = d(c)$.
4. Construct the slow manifold model to find
$$\frac{\partial c}{\partial t} \approx -2c\frac{\partial c}{\partial x} + \frac{\partial^2 c}{\partial x^2} + \tfrac{1}{2}c^3.$$
5. Is the nonlinear reaction stabilizing? Or not?

Exercise 8.3 (a forest fire model). We wish to make a model of the advance along one spatial dimension of a forest fire through unburnt forest. Suppose the *nondimensional* governing equation for the forest fire is the reaction-diffusion equation
$$\frac{\partial u}{\partial t} = \frac{\partial^2 u}{\partial x^2} + \frac{\partial^2 u}{\partial y^2} + rf(u),$$
where

- x and y are horizontal and vertical coordinates, respectively;
- the nondimensional temperature of the forest is proportional to $u(x, y, t)$ relative to some reference temperature;
- the forest and the fire take place above the flat ground $y = 0$ in a layer of vegetation $0 < y < 1$;
- the diffusion terms model the radiant spread of the fire's energy;
- the reaction term $f(u)$ of strength r represents the energy released by the forest burning (or otherwise) at a temperature u;
- the ground acts as a good insulator so use the boundary condition that $\partial u/\partial y = 0$ on $y = 0$; and
- at the top of the forest heat is radiated out into the atmosphere, so use the boundary condition $\partial u/\partial y = -bu$ on $y = 1$.

Assume the forest fire is very large in the horizontal direction, much larger than the height of the trees, so that $\partial/\partial x$ is small. Also assume that the strength r of the reaction f is small (relative to the transport of heat) and assume the radiation coefficient b is small.

1. In these circumstances use the eigenvalues of the vertical heat diffusion to argue for the existence of a slow manifold, say
$$u = v(U, y, b, r) \quad \text{such that } \frac{\partial U}{\partial t} = g(U, b, r),$$
where $U(x, t)$ measures the intensity of the fire at any location x.
2. Using the initial approximation that $u \approx U$ such that $\partial U/\partial t \approx 0$, deduce equations for the first correction to this initial approximation. Solve the equations to construct a first model of the forest fire dynamics.

Exercise 8.4 (a project on fixed heat flux convection). The sun shining on the Earth's surface heats the air near the surface. The warmer air then expands and becomes lighter than the cooler air above. Consequently the warm light air moves upward to be replaced by cool denser air from above. Such overturning motions, called convection, occur in many fluid flows, including cooking pots when heated from below and in the Earth's mantle. When the heating is done with fixed energy flux, then the onset of fluid motion involves large length scales that we analyze in this exercise (Chapman and Proctor, 1980; Chapman et al., 1980, e.g.).

Because the fluid expands and contracts we need to start with the compressible continuity equation (6.1), the Navier–Stokes equation (6.12) with gravitational body force $\mathbf{F} := -\rho g \mathbf{k}$, and the conservation equation (6.5) applying to the heat energy as represented by the temperature field T. Let's assume that the density varies linearly with temperature; that is, $\rho := \rho_0[1 - \alpha(T - T_0)]$ for temperatures near some reference T_0.

1. Multiply the Navier–Stokes equation (6.12) by ρ, assume that density variations have no effect except when multiplied by gravity (as gravity is so strong), and derive that the convection flow satisfies the PDEs

$$\begin{aligned}
&\boldsymbol{\nabla} \cdot \mathbf{q} = 0, \\
&\frac{\partial \mathbf{q}}{\partial t} + (\mathbf{q} \cdot \boldsymbol{\nabla})\mathbf{q} = \alpha g(T - T_0)\mathbf{k} - \frac{1}{\rho_0}\boldsymbol{\nabla}(p + \rho_0 g z) + \nu \boldsymbol{\nabla}^2 \mathbf{q}, \\
&\frac{\partial T}{\partial t} + \mathbf{q} \cdot \boldsymbol{\nabla} T = \kappa \nabla^2 T.
\end{aligned}$$

2. Solve these PDEs in a layer of fluid, $-d < z < d$, between two solid plates at $z = \pm d$. On each plate invoke the no-slip boundary condition that $\mathbf{q} = \mathbf{0}$ on $z = \pm d$. Also assume a fixed heating flux at the bottom plate, $\partial T/\partial z = -\beta$, and correspondingly fixed cooling at the upper plate; thus we also apply the boundary condition $\partial T/\partial z = -\beta$ on $z = \pm d$. Substitute that the temperature field $T := T_0 - \beta z + \theta(x, z, t)$ and that the pressure is relative to hydrostatic $p \mapsto p - \rho_0 g z - \frac{1}{2}\rho_0 g \alpha \beta z^2$; nondimensionalize with respect to reference length d, reference time d^2/κ, reference temperature change βd, and reference pressure $\rho_0 \nu \kappa / d^2$; and derive that the convection satisfies the equations (Chapman and Proctor, 1980, (2.1)–(2.3))

$$\begin{aligned}
&\boldsymbol{\nabla} \cdot \mathbf{q} = 0, \\
&\frac{1}{\mathrm{Pr}}\left[\frac{\partial \mathbf{q}}{\partial t} + (\mathbf{q} \cdot \boldsymbol{\nabla})\mathbf{q}\right] = \mathrm{Ra}\,\theta \mathbf{k} - \boldsymbol{\nabla} p + \boldsymbol{\nabla}^2 \mathbf{q}, \\
&\frac{\partial \theta}{\partial t} + \mathbf{q} \cdot \boldsymbol{\nabla} \theta = w + \nabla^2 \theta, \\
&\frac{\partial \theta}{\partial z} = 0 \quad \text{and} \quad \mathbf{q} = \mathbf{0} \quad \text{on } z = \pm 1
\end{aligned}$$

for Prandtl number $\mathrm{Pr} := \nu/\kappa$ characteristic of the fluid, and Rayleigh number $\mathrm{Ra} := \alpha \beta g d^4/(\nu\kappa)$ characterizing the strength of the heating through the background temperature gradient β.

3. Now we are ready to start using a center manifold approach to model the dynamics of convection. For simplicity restrict attention to two space dimensions: x horizontal and z vertical.

As always start with any equilibrium: argue that there is a subspace of equilibria $\theta = \text{constant}$, $p = \text{Ra}\,\theta z$, and $\mathbf{q} = 0$. But this freedom corresponds to just changing the background reference temperature T_0 so we may as well consider only the equilibrium with $\theta = 0$.

4. Linearize by omitting the advection terms in $\mathbf{q} \cdot \boldsymbol{\nabla}$. From the previous part we know that $\theta = \text{constant}$, $p = \text{Ra}\,\theta z$, and $\mathbf{q} = 0$ is an eigenvector corresponding to eigenvalue $\lambda = 0$. Show that the shear flow $\theta = cx$, $p = c\,\text{Ra}\,xz$, $u = c\,\text{Ra}\,\frac{1}{6}(z^3 - z)$, and $w = 0$ is also an eigenvector corresponding to the eigenvalue $\lambda = 0$.

5. So far we have no indication of the required level of heating via the Rayleigh number Ra, but now show that at the critical Rayleigh $\text{Ra} = 45$ there is another eigenvector corresponding to eigenvalue $\lambda = 0$. Seek a temperature change $\theta = c\frac{1}{2}x^2 + cQ(z)$:[120] first, the vertical momentum PDE determines the pressure $p = c\,\text{Ra}\,\frac{1}{2}x^2 z + c\,\text{Ra}\int Q\,dz$; second, horizontal momentum determines lateral velocity $u = c\,\text{Ra}\,x\frac{1}{6}(z^3 - z)$; third, continuity determines vertical velocity $w = -c\,\text{Ra}\,\frac{1}{24}(z^2 - 1)^2$; and last, the heat equation determines that a solution for Q exists only for critical Rayleigh number $\text{Ra}_c := 45$.

   For heating near this critical Rayleigh number this extra eigenvector is an extra slow mode that leads to an interesting model of convection motion with large lateral extent.

6. We would normally try to determine the entire spectrum, or at least to bound the eigenvalues away from zero. But the analysis is already complicated so let's assume all other eigenvalues are negative.

   Argue that the eigenvectors found in previous parts suggest that we seek a slow manifold parametrized by the vertical mean heat change, $\Theta(x, t) := \frac{1}{2}\int_{-1}^{1} \theta(x, z, t)\,dz$. Further, the eigenvectors suggest we should seek the slow manifold in a parameter regime of Rayleigh number near 45 and for relatively slow variations in x.

7. In the further nonlinear analysis the pressure field is a painful complication.[121] Let's eliminate the pressure from consideration by writing the PDEs in terms of a stream function $\psi(x, z, t)$ such that the velocity $\mathbf{q} = (\partial\psi/\partial z, 0, -\partial\psi/\partial x)$. Verify that the continuity PDE is satisfied by any such velocity field.

   Also introduce the scalar vorticity $\omega = \partial w/\partial x - \partial u/\partial z\,(= -\nabla^2\psi)$. Combine the horizontal and vertical components of the Navier–Stokes PDEs to deduce the vorticity PDE

$$\frac{1}{\text{Pr}}\left[\frac{\partial\omega}{\partial t} + \mathbf{q} \cdot \boldsymbol{\nabla}\omega\right] = \text{Ra}\,\frac{\partial\theta}{\partial x} + \nabla^2\omega\,,$$

   in which the pressure has been eliminated. We solve this vorticity PDE in conjunction with the heat PDE

$$\frac{\partial\theta}{\partial t} + \mathbf{q} \cdot \boldsymbol{\nabla}\theta = w + \nabla^2\theta\,.$$

[120] The use of which PDE determines which field in the solution is the same as that for thin film fluid flow.

[121] Unlike thin fluid layers, where the pressure is "tied down" by the constant atmospheric pressure on the free surface.

Algorithm 8.4 *Given the definition of the operators in Algorithm* 8.5, *this algorithm constructs the slow manifold model of slowly varying, large lateral extent, fixed heat flux convection of Exercise* 8.4. *Variable* `q` *denotes* θ, *variable* `qq(n)` *denotes* $\partial^n\Theta/\partial x^n$, *and constant* `rp` *denotes* $1/\mathrm{Pr}$.

```
factor d,rad,rp;
depend xx,x;
let df(~a,x)=>d*df(a,xx);
operator qq; depend qq,xx,t;
let { df(qq,t)=>gq , df(qq(~n),t)=>df(gq,xx,n)
    , df(qq,xx)=>qq(1) , df(qq(~n),xx)=>qq(n+1) };
q:=qq;
gq:=psi:=omega:=u:=w:=0;
ra:=45*(1+rad);
let { d^5=>0, rad^2=>0 };
for iter:=1:9 do begin
    resy:=ra*df(q,x)+df(omega,x,x)+df(omega,z,z)
        -rp*(df(omega,t)+u*df(omega,x)+w*df(omega,z));
    psi:=psi+solv(resy,z);
    u:=df(psi,z); w:=-df(psi,x); omega:=df(w,x)-df(u,z);
    resq:=+w+df(q,x,x)+df(q,z,z)
        -df(q,t)-u*df(q,x)-w*df(q,z);
    resqt:=sub(z=+1,df(q,z)); resqb:=sub(z=-1,df(q,z));
    gq:=gq+(gdq:=mean(resq,z)+(resqt-resqb)/2);
    q:=q-solq(resq-gdq,z);
    showtime;
    if {resy,resq,resqt,resqb}={0,0,0,0}
    then write iter:=100000+iter;
end;
resamp:=mean(q,z)-qq;
resut:=sub(z=+1,u); resub:=sub(z=-1,u);
resyt:=sub(z=+1,w); resyb:=sub(z=-1,w);
```

Argue that the boundary conditions for these PDEs are

$$\frac{\partial\theta}{\partial z}=\frac{\partial\psi}{\partial z}=\psi=0 \quad \text{on } z=\pm1 .$$

8. Explain how Algorithms 8.4 and 8.5 construct a slow manifold model of slowly varying, large lateral extent, fixed heat flux convection in 2D space.
9. Execute Algorithms 8.4 and 8.5 to find the evolution on the slow manifold

$$\frac{\partial\Theta}{\partial t}=-\epsilon\frac{\partial^2\Theta}{\partial x^2}-\tfrac{34}{231}\frac{\partial^4\Theta}{\partial x^4}+\tfrac{30}{7}\left(\frac{\partial\Theta}{\partial x}\right)^2\frac{\partial^2\Theta}{\partial x^2}+\mathcal{O}\big(\epsilon^2+\partial_x^5\big) \tag{8.15}$$

for Rayleigh number $\mathrm{Ra}=45(1+\epsilon)$.

(a) Relate the leading-order structure in the horizontal velocity u to the eigenvectors we found earlier.

(b) Similarly relate the leading order of the vertical velocity w.

Algorithm 8.5 *This algorithm defines operators for Algorithm* 8.4 *to construct the slow manifold model of slowly varying, large lateral extent, fixed heat flux convection of Exercise* 8.4.

```
operator solv; linear solv;
let { solv(z^~~p,z)=>(z^(p+4)
        -(1+(-1)^p)/2*((p+4)*(z^2-1)/2+1)
        -(1-(-1)^p)/2*((p+4)*(z^3-z)/2+(3*z-z^3)/2)
        )/(p+1)/(p+2)/(p+3)/(p+4)
    , solv(1,z)=>(z^4-2*z^2+1)/24 };
operator solq; linear solq;
let { solq(z^~~p,z)=>(z^(p+2)
          -(1+(-1)^p)/2/(p+3)
          -(1-(-1)^p)/2*(p+2)*z )/(p+1)/(p+2)
    , solq(1,z)=>(z^2-1/3)/2 };
operator mean; linear mean;
let { mean(z^~~p,z)=>(1+(-1)^p)/2/(p+1)
    , mean(1,z)=>1 };
```

(c) Show that the PDE (8.15) predicts a linearly unstable band of long-wavelength modes, with wavelengths $\mathcal{O}(1/\sqrt{\epsilon})$.

(d) Show that the PDE (8.15) conserves Θ by writing the known part of the RHS as a gradient ∂_x of some Θ-flux.

Exercise 8.5 (stress-free fixed heat flux convection). Repeat Exercise 8.4, with modified algorithms, in the case of so-called stress-free top and bottom boundaries (Chapman and Proctor, 1980, e.g.); that is, replace the boundary conditions that the velocity $\mathbf{q} = 0$ at $z = \pm 1$ with the boundary conditions that $w = \partial u/\partial z = 0$ at $z = \pm 1$. Compare and contrast the analysis and results for the two exercises.

Exercise 8.6 (asymmetric fixed heat flux convection). Repeat Exercise 8.4, with modified algorithms, in the asymmetric case of a stress-free top boundary and a no-slip bottom boundary (Chapman and Proctor, 1980, e.g.); that is, replace the boundary conditions that the velocity $\mathbf{q} = 0$ at $z = 1$ with the boundary condition that $w = \partial u/\partial z = 0$ at $z = 1$, and retain the boundary condition that $\mathbf{q} = 0$ at $z = -1$. Compare and contrast the analysis and results for the two exercises (and possibly Exercise 8.5). Comment on the extra term in the macroscale Θ-PDE.

Exercise 8.7 (canonical pattern formation). The Swift–Hohenberg PDE is a well-known prototypical PDE for studying issues in pattern formation and evolution (Swift and Hohenberg, 1977; Greenside and Coughran, 1984; Cross et al., 1986; Cross and Hohenberg, 1993; Hutt, 2008, e.g.): for a field $u(x,t)$ in one spatial dimension, and nondimensionally, the Swift–Hohenberg PDE is

$$\frac{\partial u}{\partial t} = ru - \left(1 + \frac{\partial^2}{\partial x^2}\right)^2 u - u^3. \tag{8.16}$$

Exercise 7.12 explored its marginal ($\epsilon = 0$), linear, one space dimension dynamics, and the exercise found that patterns diffuse. Let's adapt a proposal (Roberts, 2013, §3.4) to show how a little magic derives the Ginzburg–Landau equation,

$$\frac{\partial a}{\partial t} \approx ra - 3|a|^2 a + 4\frac{\partial^2 a}{\partial x^2}, \tag{8.17}$$

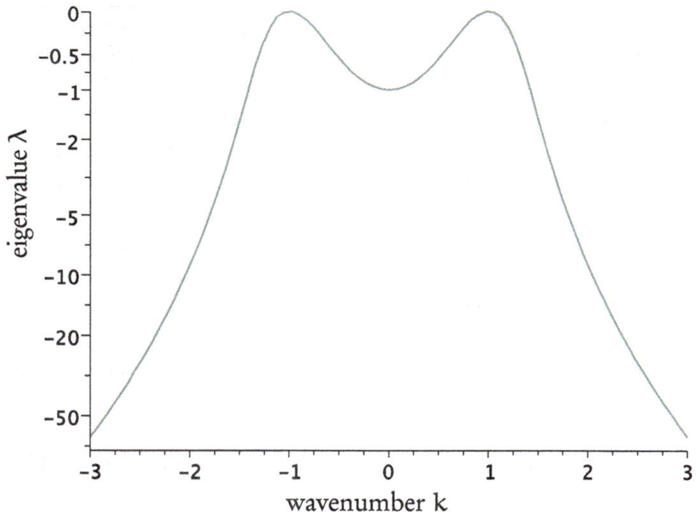

Figure 8.5. *Spectrum of the linearized Swift–Hohenberg PDE* (8.16) *for parameter* $r = 0$ *(with nonuniform vertical axis). There are two neutral modes at wavenumbers* $k = \pm 1$.

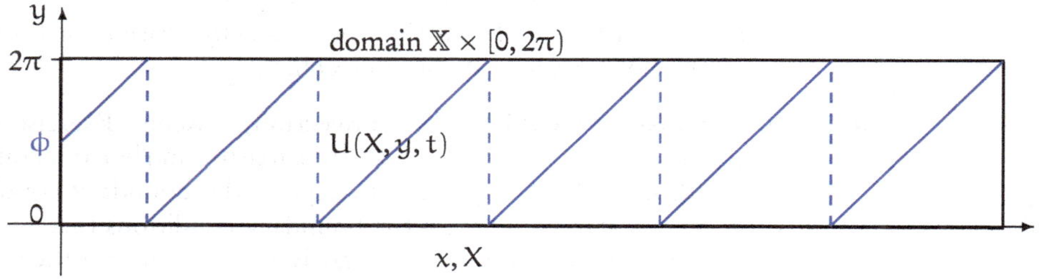

Figure 8.6. *Cylindrical domain of the embedding PDE* (8.18) *for field* $U(X, y, t)$. *Obtain solutions of the Swift–Hohenberg PDE* (8.16) *on the solid diagonal line as* $u(x, t) = U(x, x + \phi, t)$ *for any constant phase* ϕ.

as a long-term description of the pattern formation and evolution of the nonlinear Swift–Hohenberg PDE.

1. Linearize the Swift–Hohenberg PDE (8.16), seek solutions $u \propto e^{\lambda t + ikx}$ for wavenumber k, and find the spectrum $\lambda = r - (1 - k^2)^2$. As illustrated in Figure 8.5, this spectrum has two zero eigenvalues when parameter $r = 0$. Argue that there should be a long-term solution dominated by a pattern made from linear combinations of $e^{\pm ix}$. But, as for shear dispersion and thin film flow, there is no spectral gap!
2. Following the general approach of Table 4.2, let's embed the Swift–Hohenberg PDE (8.16) in a family of cognate problems. But here the cognate problems are an ensemble of identical copies of the Swift–Hohenberg PDE (8.16)! Figure 8.6 illustrates that we "helically wrap" solutions of the Swift–Hohenberg PDE (8.16) around a cylinder. Consider a field $U(X, y, t)$, 2π-periodic in y, and satisfying the PDE

$$\frac{\partial U}{\partial t} = rU - \left(1 + \frac{\partial^2}{\partial y^2} + 2\frac{\partial^2}{\partial y \partial X} + \frac{\partial^2}{\partial X^2}\right)^2 U - U^3 \tag{8.18}$$

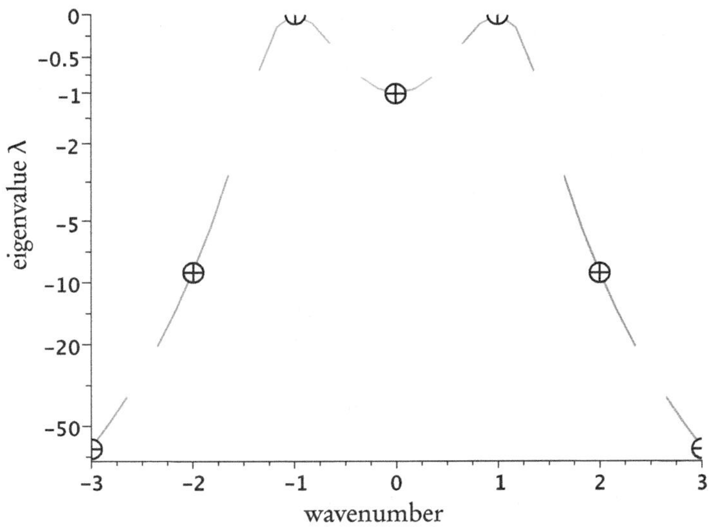

Figure 8.7. *Spectrum of the linearized Swift–Hohenberg ensemble* (8.18) *for parameter* $r = 0$. *Assuming slow variations in* X, *the spectrum "breaks apart." For no variations in* X, *circles, the spectrum is discrete, with two neutral modes at wavenumbers* ± 1.

for $(X, y) \in \mathbb{X} \times [0, 2\pi)$. Show that $u(x,t) := U(x, x + \phi, t)$ satisfies the Swift–Hohenberg PDE (8.16) for all constant phases ϕ. That is, solving the embedding PDE (8.18) for U is equivalent to finding an ensemble of solutions to the Swift–Hohenberg PDE (8.16), one for each ϕ.

3. The Swift–Hohenberg ensemble (8.18) has equilibrium $U(X, y, t) = 0$, and we consider those at parameter $r = 0$ for simplicity. Now the magic happens: assume slow variations in longitudinal variable X, and base analysis about the case when there are no variations in X, that is, $\partial/\partial X = 0$. By seeking solutions $U \propto e^{\lambda t + i\ell y}$ show that the spectrum of the linearized Swift–Hohenberg ensemble (8.18) discretizes as indicated in Figure 8.7.

   Argue that the linearized dynamics of slow variations, $U \propto e^{iKX}$ for small K, fills out the spectrum of the Swift–Hohenberg PDE (8.16), as also indicated by Figure 8.7.

4. Hence deduce that there exists an emergent slow manifold of the Swift–Hohenberg ensemble (8.18) for solutions that vary slowly in X.

5. Deduce that the slow subspace is $U \approx a e^{iy} + \bar{a} e^{-iy}$ for independent complex amplitudes a and $\bar{a}$. (To describe real solutions we find that a and $\bar{a}$ are complex conjugates of each other, as in Chapter 3.)

6. Starting from the slow subspace approximation, construct the slow manifold of slowly varying solutions to be $U \approx a e^{iy} + \bar{a} e^{-iy} - \frac{1}{64} a^3 e^{i3y} - \frac{1}{64} \bar{a}^3 e^{-i3y}$ such that $\partial a/\partial t \approx ra - 3\bar{a}a^2 + 4\partial^2 a/\partial X^2$ and $\partial \bar{a}/\partial t \approx r\bar{a} - 3a\bar{a}^2 + 4\partial^2 \bar{a}/\partial X^2$. When a and $\bar{a}$ are complex conjugates, as needed for real solutions U, these become the Ginzburg–Landau equation (8.17).

7. Summarize how the Ginzburg–Landau equation (8.17) models the pattern evolution of the Swift–Hohenberg PDE (8.16).

Algorithm 8.6 *Constructs the slow manifold of pattern evolution in the one-dimensional Swift–Hohenberg* PDE (8.16). *Variable* `aa(n)` *denotes the* n*th spatial derivative* $\partial^n a/\partial x^n$*, and similarly for* `bb(n)`. *Variable* `d` *counts the number of spatial derivatives of amplitudes* a *and* b *in every term, while variable* `small` *counts the number of factors of amplitudes* a *and* b *in every term, and of the parameter* $\sqrt{r}$.

```
depend xx,x;
operator aa;  depend aa,xx,t;
operator bb;  depend bb,xx,t;
let { df(aa,t)=>ga,                 df(bb,t)=>gb
    , df(aa(~n),t)=>df(ga,xx,n),  df(bb(~n),t)=>df(gb,xx,n)
    , df(aa,x)=>d*aa(1),          df(bb,x)=>d*bb(1)
    , df(aa(~n),x)=>d*aa(n+1),    df(bb(~n),x)=>d*bb(n+1)
    , df(aa,xx)=>aa(1),           df(bb,xx)=>bb(1)
    , df(aa(~n),xx)=>aa(n+1),     df(bb(~n),xx)=>bb(n+1)   };
operator cis;
let { df(cis(~v),~u)=>i*cis(v)*df(v,u)
    , cis(~u)*cis(~v)=>cis(u+v)
    , cis(~u)^~p=>cis(p*u)
    , cis(0)=>1  };
operator linv; linear linv;
let { linv(cis(~k*x),cis)=>-cis(k*x)/(1-k^2)^2
    , linv(1,cis)=>-1  };
u:=small*aa*cis(x)+small*bb*cis(-x);
ga:=gb:=0;
let { d^3=>0, small^4=>0 };
for iter:=1:99 do begin
  resu:=df(u,t)+(1-small^2*r)*u+2*df(u,x,2)+df(u,x,4)+u^3;
  ga:=ga+(gad:=-coeffn(resu,cis(+x),1))/small;
  gb:=gb+(gbd:=-coeffn(resu,cis(-x),1))/small;
  u:=u+linv(resu+gad*cis(+x)+gbd*cis(-x),cis);
  if resu=0 then write iter:=iter+10000;
  showtime;
end;
```

Exercise 8.8 (higher-order patterns). Recall that the Swift–Hohenberg PDE (8.16) is an interesting toy problem for exploring pattern formation and evolution (Swift and Hohenberg, 1977; Greenside and Coughran, 1984; Cross et al., 1986; Cross and Hohenberg, 1993; Hutt, 2008, e.g.). Let's use computer algebra to neatly account for the growth induced for parameter $r > 0$ and limited by the nonlinearity, to any order we please. The neat trick here is to automatically and naturally account for the embedding of Exercise 8.7.

1. Execute Algorithm 8.6 to find, first, that the slow manifold field is more accurately $u \approx ae^{ix} + \bar{a}e^{-ix} - \frac{1}{64}(a^3e^{i3x} + \bar{a}^3e^{-i3x}) - \frac{9}{128}i(a^2\partial a/\partial x\, e^{i3x} - \bar{a}^2\partial\bar{a}/\partial x\, e^{-i3x})$ for complex amplitude $a(x,t)$ and its complex conjugate $\bar{a}(x,t)$, which both vary slowly in space, and, second, that the evolution on the slow manifold predicts that patterns evolve according to the Ginzburg–Landau equation (8.17).

2. Explain how Algorithm 8.6 constructs an asymptotic expansion for the slow manifold of the Swift–Hohenberg PDE (8.16). In particular, discuss how the algorithm automatically caters for the embedding of Exercise 8.7 by symbolically using two x-coordinate variables.
3. Modify the algorithm to derive a higher-order accurate slow manifold model.

Exercise 8.9 (other nonlinear patterns). Modify Algorithm 8.6 to model reactions in the Swift–Hohenberg equation with nonlinearities different from the usual cubic nonlinearity, such as $u\partial u/\partial x$. Execute the code and discuss the resulting slow manifold model(s) for the complex amplitude $a(x, t)$.

Chapter 9

Resolve inertia in thicker faster fluid films

Contents

This chapter reconsiders the flow of a thin layer of fluid. Recall that the 2D fluid has depth $h(x,t)$ and moves with depth-averaged velocity $\bar{u}(x,t)$. In many applications the modeling in Chapter 8 of the slow flow of a thin fluid film is too limited. The aim here, realized in section 9.2, is to systematically derive the model that, in a nondimensional form, these macroscale variables evolve according to (Roberts, 1996, p. 69)

$$\frac{\partial h}{\partial t} = -\frac{\partial(h\bar{u})}{\partial x}, \tag{9.1a}$$

$$\frac{\partial \bar{u}}{\partial t} + 1.504\bar{u}\frac{\partial \bar{u}}{\partial x} \approx -\frac{\pi^2}{4h^2\,\mathrm{Re}}\bar{u} - \frac{\pi^2}{12}\frac{\partial h}{\partial x} + \frac{4.093}{\mathrm{Re}}\frac{\partial^2 \bar{u}}{\partial x^2} \tag{9.1b}$$

for Reynolds number Re. This model resolves waves through the interaction of depth and inertia in the terms $\partial h/\partial t \approx -h\partial\bar{u}/\partial x$ and $\partial\bar{u}/\partial t \approx -\frac{\pi^2}{12}\,\partial h/\partial x$. Such a model is needed to resolve faster wave-like dynamics in many situations, such as falling films (Nguyen and Balakotaiah, 2000; Chang, 1994, p. 110), wave transitions (Chang et al., 2002) to solitary waves (Ruyer-Quil and Manneville, 2000), higher Reynolds number flows (Prokopiou et al., 1991, eq. (19)), rising film flow and a slot coater (Kheshgi, 1989, eq. (37)), and rivulets under a sloping cylinder (Alekseenko et al., 1996).

Importantly, although written in terms of the depth-averaged velocity $\bar{u}$, section 9.1.4 argues that one should *not* derive the model (9.1) by depth averaging the fluid equations. Instead one should use center manifold theory. For example, depth averaging predicts that the coefficient of $-\partial h/\partial x$ in (9.1b) is just one, but $\pi^2/12$ ($= 0.8225$) is correct, as section 9.1.4 verifies in several predictions.

But the lateral velocity mode, represented by $\bar{u}$, is linearly decaying: the PDE (9.1b) shows that the decay rate is $\pi^2/(4h^2\,\text{Re})$. How then can the lateral velocity $\bar{u}$ be in the slow manifold? Section 9.1 introduces how we can be justifiably flexible in practice.

9.1 ▪ Many approaches create a slow manifold

There are more ways to kill a cat than choking it with butter.

Anonymous allegory

This section introduces a general methodology to extend the slow manifold model of thin fluid films to faster, more interesting flows. Section 8.2 found the spectrum of thin fluid films to be $\{0, -\pi^2/(4h^2), -9\pi^2/(4h^2), -25\pi^2/(4h^2), \ldots\}$, due to conservation of fluid and viscous dissipation. The zero eigenvalue implies the existence of the slow manifold model discussed in that section. But the leading negative eigenvalue, $-\pi^2/(4h^2)$, is much smaller, one ninth the size, than the next negative eigenvalue, $-9\pi^2/(4h^2)$. Perhaps this leading negative eigenvalue could be effectively zero when compared to the remaining ones, which are at least nine times as big. Upon this observation we base a more interesting model of thin fluid films.

The modeling is based upon treating the eigenvalue $-\pi^2/(4h^2)$ as small compared to the eigenvalue $-9\pi^2/(4h^2)$. Let us see this treatment work in a toy linear problem.

In this section consider, for a field $u(z,t)$, the linear diffusion dynamics

$$\frac{\partial u}{\partial t} = \frac{\partial^2 u}{\partial z^2}, \quad 0 < z < 1, \tag{9.2}$$

analogous to the viscous diffusion (8.9c) across a fluid layer of thickness one, with boundary conditions, akin to (8.9c) at the bed and the free surface, of

$$u = 0 \text{ on } z = 0 \quad\text{and}\quad \frac{\partial u}{\partial z} = 0 \text{ on } z = 1\,. \tag{9.3}$$

The exact solution to this problem serves as a reference for our approximate techniques. Separation of variables leads to the exact solution

$$u(z,t) = \sum_{n=1}^{\infty} c_n \sin\left[(n - \tfrac{1}{2})\pi z\right] \exp\left[-(n - \tfrac{1}{2})^2\pi^2 t\right]$$

for some coefficients c_n depending upon initial conditions. For large time t the longest-lasting mode is $u \to c_1 \sin(\pi z/2)\exp(-\pi^2 t/4)$. Emerging from transients decaying like $\exp(-9\pi^2 t/4)$, it is this mode that we aim to model with center manifold theory.

Center manifold theory by Aulbach and Wanner (1996, 2000) assures us that there exists an emergent slow manifold that is C^8 (since the ratio of eigenvalues $\frac{-9\pi^2/4}{-\pi^2/4} = 9$), and Potzsche and Rasmussen (2006) assure us that we can approximate the slow manifold.

The next three subsections introduce three different practical ways to generate such a slow manifold model, represented schematically by Figure 9.1. These three ways are

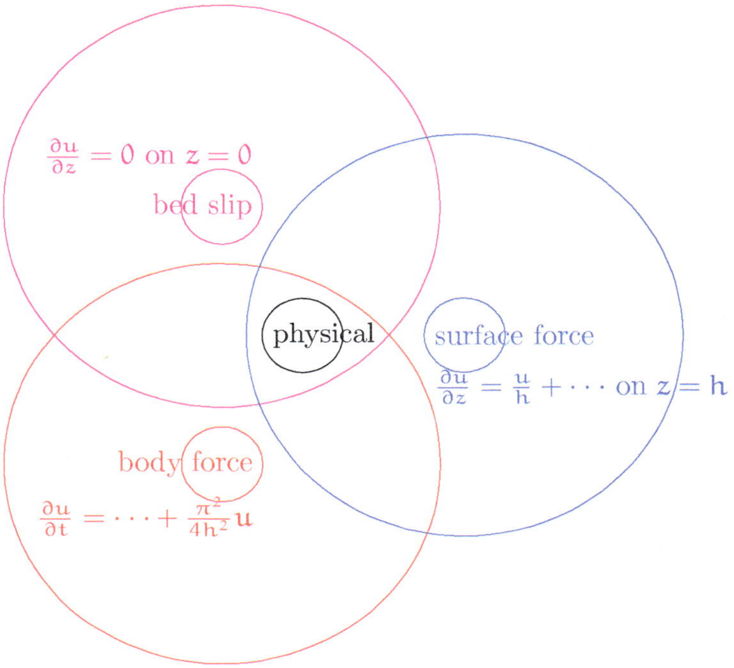

Figure 9.1. *Schematic diagram of three approaches to modeling the physical fluid flow: the black circle and text represent the physical problem; the different-colored discs represent models based upon different artificial problems that are connected to the physical by a parameter. The three larger circles schematically denote the domain of convergence of the three models: these domains are shown to all include the physical problem.*

examples of a vast range of possibilities available to us. Here all three ultimately predict the same model, the physical model at the center of Figure 9.1. The difference lies in the varying amount of work required to achieve a specified accuracy.

9.1.1 ▪ Direct forcing creates a slow manifold

The simplest fudge to generate a slow manifold model of the diffusion dynamics is to simply artificially force the mode across the domain: here modify the diffusion PDE (9.2) to

$$\frac{\partial u}{\partial t} = \frac{\partial^2 u}{\partial z^2} + (1-\gamma)\tfrac{1}{4}\pi^2 u;$$

setting the artificial parameter $\gamma = 1$ recovers the original diffusion dynamics; setting the parameter $\gamma = 0$ gives the basis of a slow manifold. This approach was the initial choice for constructing slow manifold models of fluid layers (Roberts, 1996).

Now find the slow manifold. When parameter $\gamma = 0$, separation of variables shows that the spectrum of the diffusion operator shifts by an amount $+\pi^2/4$ to $\{0, -7\pi^2/2, -6\pi^2, \ldots\}$. Thus this PDE with boundary conditions (9.3) has a slow manifold about $\gamma = 0$. Here, substitute to see that the exact slow manifold model is

$$u = \bar{u}\tfrac{\pi}{2}\sin(\pi z/2) \quad \text{such that} \quad \dot{\bar{u}} = -\gamma\tfrac{\pi^2}{4}\bar{u},$$

where $\bar{u}(t)$ denotes the mean $u(z,t)$ over the layer. Set the artificial parameter $\gamma = 1$ to recover immediately the precisely correct model of the long-time decay of the original PDE (9.2).

This artifice gives the exact long-time model. But introducing trigonometric structure across the layer, while exact, can be algebraically inconvenient in complex situations, and sometimes impossible in practice. Thus we look at two alternatives that only involve polynomials across the layer. These alternatives are less accurate but are often computationally more convenient.[122]

9.1.2 ▪ Forcing the surface also forces a slow manifold

Another simple fudge to generate a slow manifold model of the diffusion dynamics is to artificially force the surface. That is, solve the PDE (9.2) but modify the top boundary condition (9.3) to

$$\frac{\partial u}{\partial z} = (1-\gamma)u \quad \text{on } z = 1; \tag{9.4}$$

when parameter $\gamma = 0$, the mode $u \propto z$ is neutral, corresponds to eigenvalue zero, and hence implies the existence of a slow manifold for nonzero parameter γ; when $\gamma = 1$ we recover the original problem.

Substitute the following into the governing equations to check that the slow manifold for this variation is

$$\begin{aligned} u &= \bar{u}\left[2z + \gamma\tfrac{1}{2}(z - 2z^3) + \gamma^2\tfrac{1}{40}(-z - 2z^3 + 6z^5) + \mathcal{O}(\gamma^3)\right], \\ \text{where} \quad \dot{\bar{u}} &= \left[-3\gamma + \tfrac{3}{5}\gamma^2 + \mathcal{O}(\gamma^3)\right]\bar{u}. \end{aligned}$$

Evaluated at parameter $\gamma = 1$, the quintic polynomial shape of $u(z,t)$ approximates that of the longest-lasting mode $\sin(\pi z/2)$; the evolution $\dot{\bar{u}} = -2.4\bar{u}$ approximates the correct rate of decay of $\pi^2/4 = 2.4674$. Algorithm 9.1 generates higher-order expansions in parameter γ. Such higher orders are more accurate.

Such a polynomial slow manifold is significantly easier to use in complex problems that involve many other physical interactions. The cost is that higher orders have to be computed.

9.1.3 ▪ Eliminating bed dissipation slips in a slow manifold

A third and final simple fudge to generate a slow manifold model of the diffusion dynamics is to artificially let the bed boundary condition slip or insulate. That is, solve the PDE (9.2) with modified bed boundary condition (9.3) of

$$(1-\gamma)\frac{\partial u}{\partial z} = 2\gamma u \quad \text{on } z = 0;$$

when parameter $\gamma = 0$, the mode $u = \text{constant}$ is neutral, corresponds to eigenvalue zero, and hence implies the existence of a slow manifold for nonzero parameter γ, and when $\gamma = 1$ we recover the original problem.

[122] I conjecture the following heuristic (by analogy with the Rayleigh quotient of Table 5.1): the most accurate artifices are those that best preserve the shape of the relevant eigenfunction. Research may establish or refute this conjecture.

Algorithm 9.1 *Modify the boundary condition at $z=1$ to generate a slow manifold model for the diffusion* PDE (9.2) *with boundary conditions* (9.3).

```
on div; off allfac; on revpri; factor gamma,uu;
% solves df(u,z,2)=rhs s.t. sub(z=0,u)=0 & mean(u,z)=0
operator usol; linear usol;
let {usol(z^~~n,z)=>(z^(n+2)-2*z/(n+3))/(n+2)/(n+1)
    , usol(1,z)=>(z^2-2*z/3)/2 };
operator mean; linear mean;
let {mean(z^~~n,z)=>1/(n+1), mean(1,z)=>1 };

depend uu,t; let df(uu,t)=>g;
u:=2*uu*z; g:=0;
let gamma^10=>0;
for iter:=1:99 do begin
   write pde:=-df(u,t)+df(u,z,2);
   write bed:=sub(z=0,u);
   write bcs:=sub(z=1, df(u,z)-(1-gamma)*u );
   write amp:=mean(u,z)-uu;
   g:=g+(gd:=mean(3/2*z*pde,z)-3/2*bcs);
   u:=u-usol(pde-2*z*gd,z);
   if {pde,bed,bcs,amp}={0,0,0,0}
   then write iter:=1000000+iter;
end;
```

Substitute the following into the governing equations to check that the slow manifold for this variation is

$$\begin{aligned} u &= \bar{u}\left[1+\gamma\tfrac{1}{3}(-2+6z-3z^3)+\gamma^2\tfrac{1}{90}(-28+60z+30z^2-60z^3+15z^4)\right. \\ &\quad \left.+\mathcal{O}(\gamma^3)\right], \\ \text{where}\quad \dot{\bar{u}} &= \left[-2\gamma+\tfrac{2}{3}\gamma^2+\mathcal{O}(\gamma^3)\right]\bar{u}\,. \end{aligned}$$

Evaluated at parameter $\gamma=1$, the quartic polynomial shape of $u(z,t)$ approximates that of the longest-lasting mode $\sin(\pi z/2)$; the evolution $\dot{\bar{u}}=-2.6666\,\bar{u}$ approximates the correct rate of decay of $\pi^2/4=2.4674$. Algorithm 9.2 generates more-accurate higher-order expansions in parameter γ.

9.1.4 ▪ What is the fuss? Just average

The following apparently exact derivation using traditional averaging gives incorrect results. But to highlight the most glaring error, force the diffusion (9.2) by a spatially distributed constant "body force" g:

$$\frac{\partial u}{\partial t}=\frac{\partial^2 u}{\partial z^2}+g \quad \text{in } 0<z<1\,.$$

Now a traditional, supposedly exact, derivation averages by integrating over the layer, $\int_0^1\cdot\,dz$:

$$\dot{\bar{u}}=\frac{\partial}{\partial t}\int_0^1 u\,dz=\int_0^1\frac{\partial^2 u}{\partial z^2}+g\,dz=\left[\frac{\partial u}{\partial z}\right]_0^1+g=-\frac{\partial u}{\partial z}(0,t)+g\,.$$

Algorithm 9.2 *Modify the boundary condition at $z = 0$ to generate a slow manifold model for the diffusion PDE (9.2) with boundary conditions (9.3).*

```
on div; off allfac; on revpri; factor gamma,uu;
% solves df(u,z,2)=rhs s.t. sub(z=1,u_z)=0 & mean(u,z)=0
operator usolv; linear usolv;
let {usolv(z^~~n,z)=>((z^(n+2)-1/(n+3))/(n+2)-z+1/2)/(n+1)
    , usolv(1,z)=>(z^2-1/3)/2-z+1/2 };
operator mean; linear mean;
let {mean(z^~~n,z)=>1/(n+1), mean(1,z)=>1 };

depend uu,t; let df(uu,t)=>g;
u:=uu;  g:=0;
let gamma^10=>0;
for iter:=1:99 do begin
   write pde:=-df(u,t)+df(u,z,2);
   write bed:=sub(z=0,(1-gamma)*df(u,z)-2*gamma*u);
   write bcs:=sub(z=1, df(u,z) );
   write amp:=mean(u,z)-uu;
   g:=g+(gd:=mean(pde,z)+bed);
   u:=u-usolv(pde-gd,z);
   if {pde,bed,bcs,amp}={0,0,0,0}
   then write iter:=1000000+iter;
end;
```

We know that the shape of the mode is $u = \bar{u}\frac{\pi}{2}\sin(\pi z/2)$, so the bed boundary contribution $u_z(0,t) = \bar{u}\pi^2/4$. The averaged differential equation then gives the "exact" averaged model $\dot{\bar{u}} = -\frac{\pi^2}{4}\bar{u} + g$. True, the decay rate $\pi/4$ is exact. But look at the predicted equilibrium: the averaged model predicts an equilibrium mean $\bar{u}_* = g/(\pi^2/4)$, which is some 22% too high. The true equilibrium field from the PDE is $u_* = g(z - z^2/2)$, which has mean $\bar{u}_* = g/3$.

We could get the equilibrium correct. Assuming that the field is the parabolic $u = 3(z - z^2/2)\bar{u}$, then the bed boundary contribution is $u_z(0,t) = 3\bar{u}$. Consequently the model is $\dot{\bar{u}} = -3\bar{u} + g$, which predicts the correct equilibrium. Unfortunately for averaging, the predicted decay rate of 3 is now wrong by 22%.

Although sometimes viewed as an exact process, averaging generally has subtle and uncontrollable errors.[123] Nonlinearity generally compounds such errors. In contrast, any of the previous three slow manifold approaches systematically approximates the correct model, $\dot{\bar{u}} = -\frac{\pi^2}{4}\bar{u} + \frac{\pi^2}{12}g$, with its correct decay rate and its correct equilibrium.

9.2 ▪ Fluid films with inertia have waves

Let's proceed to form a slow manifold model of fluid film dynamics. The model resolves the effect of inertia to model faster or thicker flows than the lubrication approximation

[123] As mentioned elsewhere, the problem with averaging is that in *modeling* evolving out-of-equilibrium dynamics, the operator $\partial/\partial t$ is nonlinear! In the modeling here we seek a field u as a function $u(z,\bar{u})$ of the macroscale coarse variable $\bar{u}$ such that $\dot{\bar{u}} = g(\bar{u})$. Both $u(z,\bar{u})$ and $g(\bar{u})$ are unknowns to be found. By the chain rule, the time derivative $\partial u/\partial t = (\partial u/\partial \bar{u})g$ is a nonlinear product of the two modeling unknowns.

and hence resolves waves on the fluid film. This section chooses the last of the three alternative approaches introduced in the previous section; as in section 9.1.3 we choose to base the analysis on allowing slip on the bed. Exercises invoke different choices.

9.2.1 ▪ Embed the fluid equations in a useful family

Consider the equations for a 2D fluid layer flowing over a 1D flat bed. We want to resolve gravity waves albeit damped by viscosity. Thus here nondimensionalize with respect to a typical fluid layer thickness H, the gravity wave velocity $\sqrt{gH}$, the gravity wave time $\sqrt{H/g}$, and the reference pressure $\rho g H$. For a Newtonian fluid, base the modeling upon the incompressible Navier–Stokes (6.12) and continuity (6.2) PDEs, which are nondimensionally

$$\frac{\partial \mathbf{q}}{\partial t} + \mathbf{q}\cdot\nabla\mathbf{q} = -\mathbf{k} - \nabla p + \frac{1}{\mathrm{Re}}\nabla^2\mathbf{q}, \tag{9.5a}$$

$$\nabla\cdot\mathbf{q} = 0 \tag{9.5b}$$

for Reynolds number $\mathrm{Re} = \nu/\sqrt{gH^3}$ and gravitational forcing downward normal $-\mathbf{k}$ to the bed. On the flat bed the boundary condition is of no-slip,

$$u = 0 \quad \text{on } z = 0,$$

and no fluid penetration

$$w = 0 \quad \text{on } z = 0. \tag{9.5c}$$

There are three boundary conditions on the free surface, located at $z = h(x,t)$, as this section is limited to 2D flow. First, the kinematic condition (8.2) is nondimensionally

$$\frac{\partial h}{\partial t} = w - u\frac{\partial h}{\partial x} \quad \text{on } z = h. \tag{9.5d}$$

Second, zero tangential stress (in the absence of wind effects) requires (8.3), repeated here, and recall we use $h' := \partial h/\partial x$ for conciseness,

$$2h'\left(\frac{\partial w}{\partial z} - \frac{\partial u}{\partial x}\right) + (1 - h'^2)\left(\frac{\partial u}{\partial z} + \frac{\partial w}{\partial x}\right) = 0 \quad \text{on } z = h. \tag{9.5e}$$

Third, neglecting surface tension, the constant atmospheric normal stress requires (8.4), which here becomes, nondimensionally,

$$(1 + h'^2)\,p = \frac{2}{\mathrm{Re}}\left[\frac{\partial w}{\partial z} + h'^2\frac{\partial u}{\partial x} - h'\left(\frac{\partial u}{\partial z} + \frac{\partial w}{\partial x}\right)\right] \quad \text{on } z = h. \tag{9.5f}$$

We need to embed these physical equations in a family of problems to access the physics from a useful base.

Embed in an artificial family Section 9.1 considered three ways to artificially modify cross-sectional dissipation to construct a model of the longest-lasting mode. Any of the three ways would serve here.[124] Let's invoke the approach of section 9.1.3 by simply

[124] However, Exercise 9.4 shows that the free surface modification of section 9.1.2 is remarkably more effective, and just as straightforward.

modifying the viscous drag on the bed. Replace the no-slip condition, $u = 0$ on $z = 0$, by the somewhat slippery condition that

$$(1-\gamma)\frac{\partial u}{\partial z} = \frac{2\gamma}{h} u \text{ on } z = 0. \tag{9.5g}$$

As in section 9.1, the introduced γ parametrizes the family; when $\gamma = 1$ this boundary condition becomes the physical no-slip condition that $u = 0$, whereas when $\gamma = 0$ we find a useful tractable base for modeling.

9.2.2 ▪ Linearization about equilibria implies a model

Equilibria anchor the analysis As in many problems, there is a variety of equilibria, but only some are useful. For example, the system (9.5) has equilibria of no motion, $q = 0$, $p = h - z$, $h =$ constant, and γ arbitrary. However, these equilibria are not useful here because there is only *one* neutral mode, namely the fluid thickness h, and hence the modeling would be the same as in Chapter 8. The equilibria of the system (9.5) that are useful are those for parameter

$$\gamma = 0 \quad \text{when} \quad u = \text{constant}, \quad h = \text{constant}, \quad p = h - z, \quad \text{and} \quad w = 0. \tag{9.6}$$

This subspace of equilibria, denoting the subspace by $\mathbb{E}_0$, is that of arbitrary lateral mean flow of a fluid layer of arbitrary thickness (and hydrostatic pressure).

Because the subspace $\mathbb{E}_0$ of equilibria has two parameters, a mean lateral velocity u and fluid depth h, we know there are going to be (at least) two eigenvalues of 0 in the linearization at each equilibrium. These two eigenvalues indicate that there are two slow modes in the model, namely the mean lateral velocity and fluid depth, leading to a model that will represent dynamic interactions between the lateral velocity and the fluid depth. Let's look at these modes more analytically.

Linearization of slowly varying solutions Consider the dynamics linearized about any one of the equilibria (9.6) in $\mathbb{E}_0$. We also focus on solutions slowly varying in space so we assume the lateral derivative $\partial/\partial x$ is a small operator that is negligible in the linearization (this assumption suits the equilibria, which are all constant laterally). The linearization is thus to substitute into the modified fluid equations (9.5) the fields $h \mapsto h + \hat{h}(t)$, $u \mapsto \bar{u} + \hat{u}(z,t)$, $w \mapsto \hat{w}(z,t)$, and $p \mapsto h - z + \hat{p}(z,t)$ and parameter $\gamma \mapsto 0$ for hatted quantities being small. The system (9.5) becomes

$$\frac{\partial \hat{u}}{\partial t} = \frac{1}{\text{Re}}\frac{\partial^2 \hat{u}}{\partial z^2}, \quad \frac{\partial \hat{w}}{\partial t} = -\frac{\partial \hat{p}}{\partial z} + \frac{1}{\text{Re}}\frac{\partial^2 \hat{w}}{\partial z^2}, \tag{9.7a}$$

$$\frac{\partial \hat{w}}{\partial z} = 0, \tag{9.7b}$$

$$\hat{w} = 0 \quad \text{on } z = 0, \tag{9.7c}$$

$$\frac{\partial \hat{h}}{\partial t} = \hat{w} \quad \text{on } z = h, \tag{9.7d}$$

$$\frac{\partial \hat{u}}{\partial z} = 0 \quad \text{on } z = h, \tag{9.7e}$$

$$\hat{p} = \frac{2}{\text{Re}}\frac{\partial \hat{w}}{\partial z} \quad \text{on } z = h, \tag{9.7f}$$

$$\frac{\partial \hat{u}}{\partial z} = 0 \text{ on } z = 0. \tag{9.7g}$$

Now, equations (9.7b) and (9.7c) require that the perturbation normal velocity $\hat{w} = 0$. Then the second equation of (9.7a) together with the boundary condition (9.7f) requires the perturbation pressure $\hat{p} = 0$. Then the kinematic equation (9.7d) reduces to simply $\partial\hat{h}/\partial t = 0$, which gives us one eigenvalue of 0 corresponding to changes in depth. This leaves the lateral momentum equation for $\hat{u}$, the first equation of (9.7a), together with slippery boundary conditions (9.7e) and (9.7g). Straightforward separation of variables leads to solutions $\hat{u} = \cos(\ell\pi z/h)e^{\lambda_\ell t}$ for eigenvalues $\lambda_\ell = -\ell^2\pi^2/(h^2\,\mathrm{Re})$ for $\ell = 0, 1, 2, \ldots$. That is, the spectrum about each equilibrium is $\lambda_0 = 0$ (multiplicity two), and $\lambda_\ell \leqslant -\pi^2/(h^2\,\mathrm{Re})$ for $\ell = 1, 2, \ldots$.

Establish existence and emergence Assuming the Navier–Stokes system (9.5) is sufficiently well behaved, this pattern of eigenvalues indicates that center manifold theory assures us that there exists a slow manifold model. The model will be global in depth h and mean lateral velocity $\bar{u}$ and emerges on a nondimensional time scale $h^2\,\mathrm{Re}/\pi^2$. Crudely, this existence and emergence holds at all x to support a model across all x (Roberts, 1988, 2013). However, the modeling only applies to solutions varying slowly enough in space.

Another caveat is that the modeling is local to artificial parameter $\gamma = 0$. That is, we need to find evidence that evaluating the model at the physically relevant parameter $\gamma = 1$ is reasonable.

9.2.3 ▪ Start with the first iteration

To start constructing the slow manifold model we seek corrections to the equilibrium fields forced by residuals of the governing equations (9.5) when we account for the small derivatives $\partial/\partial x$ and parameter γ. The equations for corrections are the linear (9.7) but forced by the small residuals. Let's consider the corrections in the following useful sequence, and starting with the equilibrium approximations that $u = \bar{u}$, $w = 0$, and $p = h - z$.

Continuity determines w The continuity equation (9.5b) has residual $\mathrm{Res}_{(9.5b)} = \bar{u}'$ upon using primes for derivatives $\partial/\partial x$. Adding this as forcing to the LHS of (9.7b) means we solve $\partial\hat{w}/\partial z + \bar{u}' = 0$ subject to the bed boundary condition (9.7c) of $\hat{w} = 0$ on $z = 0$. The solution is simply $\hat{w} = -z\bar{u}'$. That is, the leading-order change to the vertical velocity is due to the lateral gradient of the mean flow:

$$w \approx 0 + \hat{w} = -z\frac{\partial\bar{u}}{\partial x}. \tag{9.8}$$

Vertical momentum determines p The vertical component of the Navier–Stokes equation (9.5a) has residual $\mathrm{Res}_{(9.5a)} = (\bar{u}'^2 - u\bar{u}'')z$ from the nonlinear advection. Adding this as forcing to the LHS of the second equation of (9.7a) means we solve $-\partial\hat{p}/\partial z = (\bar{u}'^2 - u\bar{u}'')z$. The necessary boundary condition is nontrivial: the normal stress equation (9.5f) has residual $\mathrm{Res}_{(9.5f)} = 2\bar{u}'/\mathrm{Re}$, so we need to also add this forcing to the LHS of the linear boundary condition (9.7f) and seek $\hat{p} + 2\bar{u}'/\mathrm{Re} = 0$ on $z = h$. The solution is straightforwardly $\hat{p} = -2\bar{u}'/\mathrm{Re} + \frac{1}{2}(\bar{u}'^2 - u\bar{u}'')(h^2 - z^2)$. That is, the leading-order change to the pressure field is due to viscous stresses at the free surface and the need to accelerate the flow vertically:

$$p \approx h - z + \hat{p} = h - z - \frac{2}{\mathrm{Re}}\frac{\partial\bar{u}}{\partial x} + \frac{1}{2}\left[\left(\frac{\partial\bar{u}}{\partial x}\right)^2 - u\frac{\partial^2\bar{u}}{\partial x^2}\right](h^2 - z^2). \tag{9.9}$$

Kinematic condition evolves h Just evaluate the kinematic condition (9.5d) to determine the current approximation for the evolution of the fluid depth. With the improved vertical velocity (9.8) we determine

$$\frac{\partial h}{\partial t} \approx -h\frac{\partial \bar{u}}{\partial x} - \bar{u}\frac{\partial h}{\partial x} = -\frac{\partial}{\partial x}(h\bar{u}). \tag{9.10}$$

This PDE for h is the classic form to conserve fluid and is maintained in all consistent approximations.

Lateral momentum equation The lateral component of the Navier–Stokes equation (9.5a) has residual $\text{Res}_{(9.5a)} = h' + \bar{u}\bar{u}' - 3\bar{u}''/\text{Re}$ from hydrostatic pressure gradients, nonlinear advection, and viscous stresses. Adding this as forcing to the LHS of the first equation of (9.7a) means we solve $\partial\bar{u}/\partial t + h' + \bar{u}\bar{u}' - 3\bar{u}''/\text{Re} = \partial^2\hat{u}/\partial z^2$. The necessary boundary conditions are both nontrivial:

- the tangential stress equation (9.5e) has residual $\text{Res}_{(9.5e)} = h\bar{u}'' + 4h'\bar{u}'$, so we need to also add this forcing to the LHS of the corresponding linear boundary condition (9.7e) and seek $\partial\hat{u}/\partial z + h\bar{u}'' + 4h'\bar{u}' = 0$ on $z = h$;
- the bed slippery condition (9.5g) has residual $\text{Res}_{(9.5g)} = -\gamma 2\bar{u}/h$, so add this to the LHS of the corresponding linear boundary condition (9.7g) and seek $\partial\hat{u}/\partial z - \gamma 2\bar{u}/h = 0$ on $z = 0$; and
- we further must be definite about what we mean by the amplitude, namely define $\bar{u} := \frac{1}{h}\int_0^h u\,dz$, so that all corrections $\hat{u}$ have to have zero mean.

Once we approximate the evolution

$$\frac{\partial \bar{u}}{\partial t} \approx -\frac{2\gamma}{\text{Re}\,h^2}\bar{u} - \frac{\partial h}{\partial x} - \bar{u}\frac{\partial \bar{u}}{\partial x} + \frac{4}{\text{Re}\,h}\frac{\partial}{\partial x}\left(h\frac{\partial \bar{u}}{\partial x}\right) \tag{9.11}$$

by integrating over the water depth, the solution of the PDE is straightforward. Then $\hat{u} = \gamma(-\frac{2}{3} + 2z/h - z^2/h^2)\bar{u} - (\frac{1}{6}\bar{u}'' + \frac{2}{3}\bar{u}'h'/h)(h^2 - 3z^2)$. That is, the leading-order change to the pressure field is due to viscous stresses at the free surface and the need to accelerate the flow vertically:

$$u \approx \bar{u} + \gamma\left(-\frac{2}{3} + 2\frac{z}{h} - \frac{z^2}{h^2}\right)\bar{u} - \left(\frac{1}{6}\frac{\partial^2\bar{u}}{\partial x^2} + \frac{2}{3h}\frac{\partial\bar{u}}{\partial x}\frac{\partial h}{\partial x}\right)(h^2 - 3z^2). \tag{9.12}$$

The leading model (9.11) for the mean lateral velocity has interesting terms:

- $-\frac{2\gamma}{\text{Re}\,h^2}\bar{u}$ represents the viscous drag on the bed;
- $-\partial h/\partial x$ arises from hydrostatic pressure gradients and upon interacting with the mass conservation equation (9.10) gives rise to water waves;
- $-\bar{u}\partial\bar{u}/\partial x$ represents horizontal advection of momentum; and
- $\frac{4}{\text{Re}\,h}\partial/\partial x\,(h\partial\bar{u}/\partial x)$ represents internal viscous dissipation of the flow, but enhanced by a factor of four in analogy with the Trouton viscosity (Ribe, 2001, p. 143, e.g.).

But we do not yet know that the results derived above are asymptotically correct because we have not verified that the residuals of the governing equations are as yet small enough. The details become tedious and are best left to the computer algebra of the next section.

9.2.4 ▪ Computer algebra handles higher-order details

This section codes into Reduce the computer algebra necessary to continue the construction of the slow manifold model of the fluid layer: Algorithms 9.3 and 9.4 list the code.

Preliminaries As in section 8.4, the unknown location of the free surface of the film is the major technical difficulty. As before, scale the vertical coordinate, $Z = z/h$, so that the free surface corresponds to $Z = 1$ precisely. That is, transform coordinates from (x, z, t) to

$$X = x, \quad Z = z/h(x,t), \quad T = t.$$

The chain rule shows that derivatives in the fluid equations (9.5) then transform according to

$$\begin{aligned}\frac{\partial}{\partial z} &= \frac{1}{h}\frac{\partial}{\partial Z},\\ \frac{\partial}{\partial x} &= \frac{\partial}{\partial X} - Z\frac{\partial h/\partial x}{h}\frac{\partial}{\partial Z},\\ \frac{\partial}{\partial t} &= \frac{\partial}{\partial T} - Z\frac{\partial h/\partial t}{h}\frac{\partial}{\partial Z}.\end{aligned}$$

These transformations would be painful to do by hand, but the computer handles them for us with ease: the transformation is invoked in Reduce when necessary from the definitions at the start of Algorithm 9.3. Also observe that this initial code inserts a count of the number of lateral derivatives in every term, d, so that we can truncate the asymptotic approximations consistently in the slowly varying approximation.

Second, Algorithm 9.3 reflects that the slow manifold is parametrized by the fluid depth h = `h` and the mean lateral velocity $\bar{u}$ = `uu`, which both vary in X and T. For conciseness, represent spatial derivatives of these by `h(m)` and `uu(m)`. Code that as well as evolving time, $\partial h/\partial t$ = `gh` and $\partial uu/\partial t$ = `gu`; the time derivative commutes with X-derivatives so that, for example, ∂`h(m)`$/\partial t = \partial^m gh/\partial X^m$.

Third, Algorithm 9.3 defines four operators to efficiently solve the various depth PDEs and integrals necessary to find corrections from the PDE residuals: `wsolv` solves $\partial\hat{w}/\partial Z = \text{Res}$ such that $\hat{w} = 0$ at the bed $Z = 0$, `psolv` solves $-\partial\hat{p}/\partial Z = \text{Res}$ such that $\hat{p} = 0$ at the free surface $Z = 1$, `mean` computes the depth average over the fluid layer, and `usolve` solves $\partial^2\hat{u}/\partial Z^2 = \text{Res}$ such that $\hat{u} = 0$ at the bed $Z = 0$ and the depth average of $\hat{u}$ is zero.

Iterative construction After the initialization of Algorithm 9.3, then Algorithm 9.4 iteratively constructs the slow manifold model starting from the linear approximation coded in the first two lines. The iterative loop terminates when all fluid equations (9.5) have residuals that are zero to the order of error specified in terms of the number of lateral derivatives as counted by d, and the order of the artificial parameter γ.

Within the iteration, Algorithm 9.4 considers the fluid equations in the same order as in the first iteration explicitly performed by section 9.2.3. In all cases, the residuals of the fluid equations are written in a familiar physical form to give confidence in the

Algorithm 9.3 *Preliminary definitions for the construction of the basic inertia-resolving model of a thin layer of fluid. Algorithm 9.4 completes this code.*

```
factor d,re,gamma;
depend xx,x,z,t;
depend zz,x,z,t;
depend tt,x,z,t;
let { df(~a,x) => df(a,xx)*d-zz*hx/h*df(a,zz)
    , df(~a,t) => df(a,tt)-zz*gh/h*df(a,zz)
    , df(~a,z) => df(a,zz)/h };

operator h; operator uu;
depend h,xx,tt;
depend uu,xx,tt;
hx:=h(1)*d; hxx:=h(2)*d^2;
let { df(h,xx)=>h(1),     df(h(~m),xx)=>h(m+1)
    , df(h,tt)=>gh,       df(h(~m),tt)=>df(gh,xx,m)
    , df(uu,xx)=>uu(1),   df(uu(~m),xx)=>uu(m+1)
    , df(uu,tt)=>gu,      df(uu(~m),tt)=>df(gu,xx,m) };

operator wsolv; linear wsolv;
let {wsolv(zz^~~n,zz)=>zz^(n+1)/(n+1)
    ,wsolv(1,zz)=>zz };
operator psolv; linear psolv;
let {psolv(zz^~~n,zz)=>(1-zz^(n+1))/(n+1)
    ,psolv(1,zz)=>(1-zz) };
operator mean; linear mean;
let {mean(zz^~~n,zz)=>1/(n+1)
    ,mean(1,zz)=>1 };
operator usolv; linear usolv;
let {usolv(zz^~~n,zz)=>(zz^(n+2)-1/(n+3))/(n+2)/(n+1)
    ,usolv(1,zz)=>(zz^2-1/3)/2 };
```

results. The residual of the continuity PDE (9.5b) corrects the vertical velocity field w, remembering that $\partial/\partial z = \frac{1}{h}\partial/\partial Z$. The residuals of the vertical momentum PDE (9.5a) and the normal stress condition (9.5f) correct the pressure field p. The kinematic condition (9.5d) corrects the evolution of the fluid depth. Last, the residuals of the lateral momentum PDE (9.5a) together with the tangential stress condition (9.5e) and the artificially slippery bed condition (9.5g) determine corrections for both the lateral velocity field u and, via a solvability condition, the evolution of the mean velocity $\bar{u}$.

Resultant model Algorithms 9.3 and 9.4 take 12 iterations to determine a slow manifold to residuals and hence errors $\mathcal{O}(\gamma^{10}, \partial_x^3)$. Looking at the coefficients of $\partial\bar{u}/\partial t = gu$ of terms in γ^9, one sees that they are mostly up to about 0.02 in magnitude. This suggests that the model is determined to coefficients correct to nearly two decimal places. Thus evaluating at the physical parameter $\gamma = 1$, one finds that the slow manifold model is the conservation of mass, $\partial h/\partial t = -\partial(h\bar{u})/\partial x$, together with a somewhat complicated PDE

Algorithm 9.4 *Following the preliminary definitions of Algorithm 9.3, this code iteratively constructs the slow manifold model of a thin layer of fluid. The iteration is driven by the residuals of the differential equations and boundary conditions.*

```
u:=uu;  p:=h*(1-zz);
w:=gh:=gu:=0;
let { d^3=>0, gamma^10=>0 };
for it:=1:99 do begin

resc:=df(u,x)+df(w,z);
w:=w-h*wsolv(resc,zz);
resw:=df(w,t)+u*df(w,x)+w*df(w,z)
    +df(p,z) +1 -1/re*(df(w,x,2)+df(w,z,2));
restn:= sub(zz=1,-p*(1+hx^2) +2/re*(df(w,z)
    +hx^2*df(u,x)-hx*(df(u,z)+df(w,x))) );
p:=p+h*psolv(resw,zz)+restn;
gh:=sub(zz=1,w-u*hx);
resu:=df(u,t)+u*df(u,x)+w*df(u,z)
    +df(p,x) -1/re*(df(u,x,2)+df(u,z,2));
restt:=-sub(zz=1,  (1-hx^2)*(df(u,z)+df(w,x))
    +2*hx*(df(w,z)-df(u,x)) );
resbed:=sub(zz=0,(1-gamma)*df(u,z)-2*gamma*u/h);
gu:=gu+(gud:=-mean(resu,zz)+(restt+resbed)/h/re);
u:=u+usolv(resu+gud,zz)*h^2*re-h*resbed*(zz-1/2);

showtime;
if {resw,restn,resu,restt,resc,resbed}={0,0,0,0,0,0}
then write it:=1000000+it;
end;
```

for the mean lateral velocity, namely

$$\begin{aligned}\mathrm{Re}\,\frac{\partial\bar{u}}{\partial t} &\approx -2.47\,\bar{u}/h^2 - 0.83\,\mathrm{Re}\,h' - \mathrm{Re}(1.44\bar{u}\bar{u}' + 0.22h'\bar{u}^2/h)\\ &\quad + (4.09\bar{u}'' + 4.86h'\bar{u}'/h + 0.02h'^2\bar{u}/h^2 - 0.54h''\bar{u}/h)\\ &\quad + \mathrm{Re}^2(0.05h^2\bar{u}\bar{u}'^2 + 0.01h^2\bar{u}^2\bar{u}'' - 0.05hh'\bar{u}^2\bar{u}'\\ &\qquad + 0.06h'^2\bar{u}^3 - 0.04hh'^2\bar{u} + 0.01hh''\bar{u}^3 - 0.01h^2h''\bar{u})\\ &\quad + \mathcal{O}(\partial_x^3). \end{aligned} \tag{9.13}$$

This model contains the important well-recognized effects of bed drag, $-2.47\,\bar{u}/h^2$; interaction with depth variations, $-0.83\,\mathrm{Re}\,h'$; advection, $-\mathrm{Re}\,1.44\bar{u}\bar{u}'$; and Trouton viscosity, $4.09\bar{u}''$. The model also includes a lot of smaller effects from more subtle interactions, especially in faster flows of higher Reynolds numbers when the Re^2 terms become more significant.

9.3 ▪ Why the mysterious factor of two?

Recall that section 9.2.1 embedded the physical fluid equations in the one-parameter family (9.5). The embedding introduced the variously and artificially slippery bed boundary

Algorithm 9.5 *Following modified Algorithm* 9.2, *plot two graphs that indicate the best Euler parameter for the modified diffusion problem of section* 9.1.3.

```
    load_package gnuplot;
    gu:=sub(gamma=1,g/uu);
    plot(gu,eu=(0 .. 4));
    cs:=coeff(g/uu,gamma);
    k:=length(cs)-1;
    bb:=(part(cs,k+1)*part(cs,k-1)-part(cs,k   )^2)
       /(part(cs,k   )*part(cs,k-2)-part(cs,k-1)^2)$
    plot(min(max(0,bb),1),eu=(0 .. 4));
```

condition (9.5g), namely $(1-\gamma)\partial u/\partial z = \frac{2\gamma}{h}u$ on $z=0$. The question is: why is the RHS of the bed boundary condition $\frac{2\gamma}{h}u$ instead of say γu, or $2\gamma u$, or $\frac{\gamma}{h}u$, or $\frac{4\gamma}{h^2}u$?

Half the answer is that we divide by the fluid depth h due to dimensional reasons. Suppose we had not yet nondimensionalized. Then the LHS $(1-\gamma)\partial u/\partial z$ would have dimensions of velocity divided by distance, namely an inverse time. To make sense, the RHS also needs to have the same dimensions; hence the velocity u on the RHS needs to be divided by a length. The only convenient length is the fluid depth h. Such dimensional considerations always need to be respected.

The outstanding question is then, why the factor of 2 on the RHS of the slippery (9.5g), instead of say 1 or 4 or -2? After all, when artificial parameter $\gamma = 1$, the LHS of the bed condition (9.5g) vanishes, so an arbitrary multiplicative constant would make no difference to the nature of the RHS. True. The issue is not with what happens to the equations at parameter $\gamma = 1$; the issue is how well approximations developed about $\gamma = 0$ approximate the solutions at $\gamma = 1$.

To explore the possibilities let's consider the bed boundary condition with a general constant on the RHS, say

$$(1-\gamma)\frac{\partial z}{\partial u} = \frac{\mathrm{Eu}\,\gamma}{h}u \quad \text{on } z=0, \tag{9.14}$$

for some constant Eu, which I call an Euler parameter.[125] Let's simplify enormously our exploration by considering the linear diffusion problem of section 9.1.3 rather than the full nonlinear fluid system (9.5).

To model the diffusion problem with the Euler-parametrized bed condition (9.14), simply modify Algorithm 9.2 in just the line

```
bed:=sub(z=0,(1-gamma)*df(u,z)-eu*gamma*u)
```

Execute the algorithm to higher order, say to errors $\mathcal{O}(\gamma^{20})$. The resulting series solution in γ as a function of Eu shows how well the series approximates the solution at the physically relevant $\gamma = 1$ for different Eu.

[125]The name comes from the equivalence between introducing such a parameter Eu and performing an Euler transform upon the series solution after construction; see Exercise 9.3. Van Dyke (1964, p. 148) attributes the Euler transform to Bellman (1955), but the name suggests a centuries-old history.

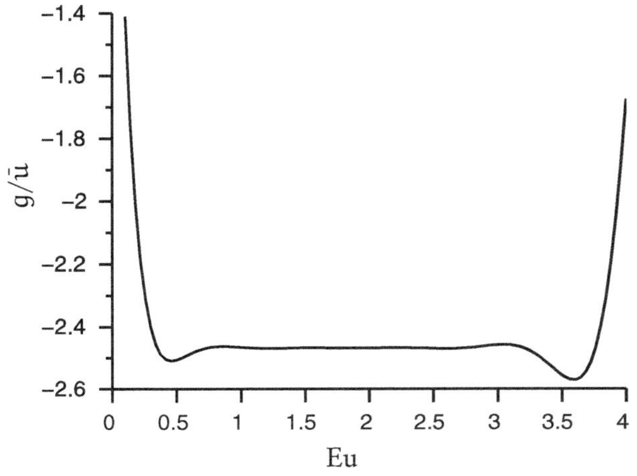

Figure 9.2. *Plot of* $g/\bar{u}$ *as a function of Euler parameter* Eu *for the construction of Algorithm* 9.2. *This plot finds that a good Euler parameter is* $\text{Eu} \approx 2$ *to reproduce the correct constant* $-\pi^2/4$.

Simply evaluate the series To assess the performance, choose any one characteristic of the solution as a diagnostic: almost all choices are equivalent. For the diffusion problem let's consider $g/\bar{u}$, from the evolution $\partial\bar{u}/\partial t = g$. We know from the $\gamma = 1$ analytic solution that $g = -(\pi^2/4)\bar{u} = -2.4674\,\bar{u}$ (section 9.1). One simple assessment is to simply sum the constructed series at $\gamma = 1$ and plot the coefficient as a function of Eu; see the first few lines of Algorithm 9.5. Figure 9.2 plots the coefficient. Over the domain $\text{Eu} \in (0, 4)$ observe that there are some rapid variations in the predicted coefficient, but that there is a region of Eu for which the predicted value is not only nearly constant, but also nearly correct, namely $-\pi^2/4 = -2.4674$. The simple, generic rule of thumb is to choose an Euler parameter Eu in the middle of such a constant region, namely here choose $\text{Eu} \approx 2$.

The appearance of Figure 9.2 is typical for a wide range of problems and diagnostics. There is some region of Euler parameter Eu in which the diagnostic variable is reasonably well predicted, but outside that region the predictions are hopeless.

Radius of convergence Let's explore a little more deeply the issue of choosing an appropriate Euler parameter. For a series solution in γ to converge well at $\gamma = 1$, we generally want the series to have as large a radius of convergence as possible.[126] Recall that the radius of convergence in γ is the distance to the closest singularity in the complex γ-plane. The Euler parameter moves the singularities around in the complex γ-plane and hence affects the radius of convergence.

Van Dyke (1984) describes many examples in fluid mechanics of using the Euler transform to move singularities around and hence usefully improve the radius of convergence of series approximations.[127]

[126] Although we expect asymptotic series approximations to center manifold models to be divergent in dynamic variables, I generally find the coefficients in such asymptotic series to be convergent in parameters for at least some domain.

[127] Van Dyke (1984) also discusses other methods of convergence acceleration. But these methods do not appear to be useful in our context of deriving models, although the methods could be used to improve the evaluation of coefficients at artificial parameter $\gamma = 1$.

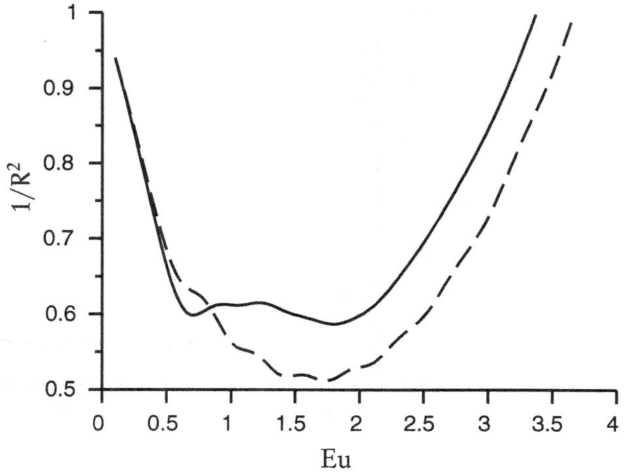

Figure 9.3. *Solid line: estimate of the radius of convergence* R *as a function of* Eu *via a generalized Domb–Sykes plot. Dashed line: crude estimate using the ratio* (9.15).

To delve more deeply into the best Euler parameter we might choose Eu to maximize the radius of convergence. A Domb–Sykes plot may estimate the radius of convergence for any given Eu (Domb and Sykes, 1957; van Dyke, 1984, p. 293). But a Domb–Sykes plot only works if the pattern of signs of the coefficients in the series is either all the same, or alternating. Here the pattern of signs is not so simple. So instead we might invoke the generalized Domb–Sykes plot of Mercer and Roberts (1990, App.). To do so for each parameter Eu is a little tedious and error prone: Figure 9.3 plots some computed estimates, but the data on the left are dodgy for technical reasons. Nonetheless the curve indicates optimal convergence for $\text{Eu} \approx 2$, or maybe a little smaller. Experience suggests that there is no point being too precise about the optimum, so $\text{Eu} \approx 2$ suffices.

A simple crude alternative, based upon the Mercer and Roberts (1990) plot, is to take the last quadruple of coefficients in a series, say the coefficients are c_{k-2}, c_{k-1}, c_k, and c_{k+1} for some k, and plot

$$B_k^2 := \frac{c_{k+1}c_{k-1} - c_k^2}{c_k c_{k-2} - c_{k-1}^2} \tag{9.15}$$

as a function of the Euler parameter. The last few lines of Algorithm 9.5 do this. Figure 9.3 plots the result here: observe a minimum in B_k^2, suggesting a maximum in the radius of convergence (as $1/B_k$ crudely estimates the radius of convergence), at Euler parameter around 1.7. But B_k^2 is only a crude indicator. That is, choosing an Euler parameter $\text{Eu} \approx 1.7 \approx 2$ suffices.

9.4 ▪ Exercises

Exercise 9.1 (restrict to smaller velocities). The model (9.13) for the mean lateral velocity $\bar{u}$ is quite complicated because it is global in $\bar{u}$. Here simplify the modeling by restricting analysis to small velocities.

Suppose we think we want the model to apply with notionally $\bar{u} \sim \sqrt{\partial_x}$. In the computer algebra of Algorithm 9.4, introduce a parameter to measure the size of $\bar{u}$, say called ε here, and so set the initial approximation to the lateral velocity field $u = \varepsilon\bar{u}$: to obtain corrections correctly simply divide updates of the evolution by ε, namely $gu := gu + \hat{g}_u/\varepsilon$. Then count derivatives twice in terms of ε to reflect the chosen regime of $\bar{u} \sim \sqrt{\partial_x}$, so set $d = \varepsilon^2$. Truncate your analysis to residuals of $\mathcal{O}(\varepsilon^6)$. Execute the code to find models correct to errors $\mathcal{O}(\bar{u}^6 + \partial_x^3)$, models that systematically omit some of the complications in the model (9.13).

Exercise 9.2 (include more physical effects). In a fluid flowing over a substrate, the substrate is often sloped. Also, we may want to incorporate surface tension into the modeling. Let's do so. Derive nondimensional physical fluid equations for a fluid on a bed with angle θ to the horizontal and with surface tension acting on the fluid surface (see section 8.1): the new equations just add a couple of terms to the system (9.5). Correspondingly modify the code of Algorithm 9.4 for the surface tension and for *small* slope $\tan\theta$. Execute the modified code to derive the new model. Physically interpret the main new terms.

Exercise 9.3 (Euler transform). Imagine that as part of an extensive system of equations, the equation $(1-\gamma)\partial u/\partial z = \gamma u$ occurs. But this equation is the only place that parameter γ occurs. The solution u of the entire system will depend upon parameter γ. Imagine we construct a series solution in γ. Substitute the Euler transform (Van Dyke, 1964, p. 148) $\gamma = \mathrm{Eu}\,\gamma'/[1 + (\mathrm{Eu} - 1)\gamma']$ and show that the equation becomes the generalized version $(1-\gamma')\partial u/\partial z = \mathrm{Eu}\,\gamma' u$. That is, solving with this generalized equation is equivalent to Euler transforming the solution of the original.

Exercise 9.4 (Euler parameter). Recall that section 9.1.2 modeled diffusion in a layer by fudging the top boundary condition to $\partial u/\partial z = (1-\gamma)u$ on $z = 1$. Let's see that it is hard to find a better variation on this condition. Introduce an Euler parameter into this approach by modifying the top boundary condition to $(1+\mathrm{Eu}\,\gamma)\partial u/\partial z = (1-\gamma)u$ on $z = 1$.

- Argue that this modified boundary condition still empowers us to model the diffusion across the layer at $\gamma = 1$ from a tractable base at $\gamma = 0$.
- Modify and execute the code of Algorithm 9.1 to construct a model for the diffusion across the layer, but now also parametrized by Eu. Constructing to errors $\mathcal{O}(\gamma^{10})$ is sufficient.
- Plot the predicted rate of decay, g/uu, at the physical $\gamma = 1$ for Euler parameter $-0.7 \leqslant \mathrm{Eu} \leqslant 0.5$ and see a nice flat region.
- As a crude indicator of the radius of convergence, plot B_k^2, equation (9.15), for Euler parameter $-0.8 \leqslant \mathrm{Eu} \leqslant 0.6$; see Algorithm 9.5. Deduce that $\mathrm{Eu} = 0$ is near optimal, although $\mathrm{Eu} \approx -1/8$ may be marginally better.
- Discuss why the embedding of section 9.1.2 appears significantly better than the embedding of section 9.1.3.

Exercise 9.5 (better convergence to thin layer flow). Exercise 9.4 indicates that the embedding of section 9.1.2 is better than that used for the fluid layer modeling of section 9.2, which used the embedding of section 9.1.3. Modify Algorithms 9.3 and 9.4 to implement the embedding of section 9.1.2, and adapt code from Algorithm 9.1 to approximate the lateral flow from the base of a linear shear $u \approx 2Z\bar{u}$. Cautiously determine the correct updating rules by neglecting all lateral derivatives, $\partial_x = 0$, and checking your modifications to errors $\mathcal{O}(\gamma^2)$ and $\mathcal{O}(\gamma^3)$. Then compute the model to errors $\mathcal{O}(\partial_x^3, \gamma^6)$ and observe that most coefficients in the fluid model are determined to three decimal places (see (9.1b)).

Exercise 9.6 (more physical effects). Do Exercise 9.2 including the effects of slope and surface tension for the approach of Exercise 9.5.

Exercise 9.7 (3D fluid layer). Model the dynamics of a layer of a 3D fluid flowing over a 2D flat bed (Roberts and Li, 2006).

- Derive the 3D version of the nondimensional Navier–Stokes equations (9.5) and boundary conditions for a fluid with velocity $\mathbf{q} = (u, v, w)$ varying in (x, y, z, t). Two tangential stress conditions are needed at the free surface.
- By an artifice such as (9.5g) (or, following Exercise 9.5, the better artifice (9.4)) for both lateral velocities, embed the system in an artificial family such that $\gamma = 0$ is a tractable base and $\gamma = 1$ is physical.
- Modify Algorithms 9.3 and 9.4 (or that from Exercise 9.5) to derive a model for the dynamics of the fluid depth h and mean lateral velocities $\bar{u}$ and $\bar{v}$. Evaluated at $\gamma = 1$, the model should include the following terms (Roberts and Li, 2006, eqs. (64)–(66)):

$$
\begin{aligned}
\frac{\partial h}{\partial t} &= -\boldsymbol{\nabla} \cdot [(\bar{u}, \bar{v})h], \\
\mathrm{Re}\frac{\partial \bar{u}}{\partial t} &\approx -\frac{\pi^2}{4h^2}\bar{u} + \frac{\pi^2}{12}\mathrm{Re}\frac{\partial h}{\partial x} - \mathrm{Re}\left(1.504\bar{u}\frac{\partial \bar{u}}{\partial x} + 1.346\bar{v}\frac{\partial \bar{u}}{\partial y}\right) \\
&\quad + 4.093\frac{\partial^2 \bar{u}}{\partial x^2} + \frac{\partial^2 \bar{u}}{\partial y^2} + 3.093\frac{\partial^2 \bar{v}}{\partial x \partial y} + 4.833\frac{h_x}{h}\frac{\partial \bar{u}}{\partial x} \\
&\quad + \frac{h_y}{h}\frac{\partial \bar{u}}{\partial y} + 1.917\frac{h_x}{h}\frac{\partial \bar{v}}{\partial y} + 1.917\frac{h_y}{h}\frac{\partial \bar{v}}{\partial x} + \cdots,
\end{aligned}
$$

  and a corresponding PDE for $\partial\bar{v}/\partial t$.

Exercise 9.8 (patterns from multiple PDEs). Instead of the single field of the Swift–Hohenberg equation (8.16), consider the following coupled pair of PDEs for the two fields $u(x,t)$ and $v(x,t)$:

$$
\begin{aligned}
\frac{\partial u}{\partial t} &= \epsilon u - \left(1 + \frac{\partial^2}{\partial x^2}\right)^2 u - uv, \\
\frac{\partial v}{\partial t} &= -\left(1 - \frac{\partial^2}{\partial x^2}\right)^2 v + u^2.
\end{aligned}
$$

1. Argue that there should be an emergent pattern forming a slow manifold based upon modes $(u, v) \propto (e^{\pm ix}, 0)$.
2. Modify Algorithm 8.6 and execute the code to construct a slow manifold model of the spatial pattern with slowly varying amplitudes.
3. Discuss the resulting slow manifold model for the complex amplitude $a(x,t)$.

Exercise 9.9 (pattern mean-flow interaction). Modify the coupled pair of PDEs of Exercise 9.8 for the two fields $u(x,t)$ and $v(x,t)$ to

$$\frac{\partial u}{\partial t} = \epsilon u - \left(1 + \frac{\partial^2}{\partial x^2}\right)^2 u - uv,$$
$$\frac{\partial v}{\partial t} = v - \left(1 - \frac{\partial^2}{\partial x^2}\right)^2 v + u^2.$$

1. Argue that there should be an emergent pattern forming a slow manifold based upon three modes: $(u,v) \propto (e^{\pm ix}, 0)$ and $(u,v) \propto (0,1)$. The mode proportional to $(0,1)$ is analogous to a mean flow in fluids.
2. Modify Algorithm 8.6, introducing an extra amplitude for the mean-flow mode, and execute the code to construct a slow manifold model of the spatial pattern and mean-flow interaction with slowly varying amplitudes.

   Find that to low order the slow manifold evolution is

$$\dot{a}_0 \approx 2a_1\bar{a}_1 + 2\frac{\partial^2 a_0}{\partial x^2},$$
$$\dot{a}_1 \approx (\epsilon - a_0)a_1 - \tfrac{1}{24}\bar{a}_1 a_1^2 + 4\frac{\partial^2 a_1}{\partial x^2}.$$

3. Discuss the resulting emergent slow manifold model: emergence, fields, predicted evolution, and so on.

Exercise 9.10 (patterns and mean flow in the Swift–Hohenberg PDE). Recall the Swift–Hohenberg PDE (8.16) for a field $u(x,t)$ in one spatial dimension. It is a system for the evolution of patterns in space (Swift and Hohenberg, 1977; Greenside and Coughran, 1984; Cross et al., 1986; Cross and Hohenberg, 1993; Hutt, 2008, e.g.). Let's model possible pattern and mean-flow interactions inherent in the Swift–Hohenberg PDE.

Following the general approach of Table 4.2, we seek to embed the Swift–Hohenberg PDE (8.16) in a family of cognate problems. Recall that after the ensemble embedding of Exercise 8.7, the spectrum magically discretizes to that plotted in Figure 8.7: the discrete eigenvalues are $\{0, -1, -9, -64, \ldots\}$. Exercises 8.7–8.9 exploited the spectral gap between eigenvalues 0 and -1. Here, let's exploit the spectral gap between the trio of eigenvalues $\{0, -1\}$ and the rest of the eigenvalues headed by -9.

1. Bring the mean-flow mode into the slow manifold by inventively fudging the Swift–Hohenberg PDE (8.16), or equivalently the Swift–Hohenberg ensemble (8.18). For example, perhaps modify the Swift–Hohenberg PDE (8.16) by adding the term $(1-\gamma)(u + \partial^2 u/\partial x^2)$ for homotopy parameter $0 \leqslant \gamma \leqslant 1$; feel free to try other embeddings. Argue that the ensemble then has an emergent slow manifold in the neighborhood of the equilibrium $u = r = \gamma = 0$ based upon the slow subspace $a_0 + a_1 e^{ix} + \bar{a}_1 e^{-ix}$.
2. Find a low-order model by hand. For example, for the modification suggested above, find $\dot{a}_1 \approx ra_1 - 3a_0^2 a_1 - 3\bar{a}_1 a_1^2 + 2i(1-\gamma)\partial a_1/\partial x$ and $\dot{a}_0 \approx (r - \gamma)a_0 - a_0^3 - 6\bar{a}_1 a_1 a_0$.

3. Modify Algorithm 8.6 to construct models to relatively high order in the homotopy parameter γ and relatively low order in amplitudes and bifurcation parameter r.
4. Discuss the convergence of the modeling in artificial γ, and the predictions of the modeling.
5. Extension: Modify the nonlinearity in the Swift–Hohenberg PDE and compare and contrast the resultant models.

Part III Summary

- This part demonstrated how to put approximations based upon slow variation in space within the center manifold framework. The resulting models are still infinite dimensional in that the models are PDEs in xt-space. However, the model PDEs are considerably simpler than the original dynamical equations because the original, being expressed as PDEs in xyt- and $xyzt$-space, are of infinitely higher dimension again.

- Generally we invoke the Navier–Stokes (6.12) and continuity (6.1) equations to together describe the small-scale motion of a viscous incompressible fluid. Our primary fields of interest are the velocity field $\mathbf{q}(x, y, z, t)$ and a concentration field $c(x, y, z, t)$, and to some extent the pressure field $p(x, y, z, t)$ (section 6.1). Typically we treat incompressible fluids except in special circumstances.

- Another flow property of great interest in fluid dynamics is the vorticity, $\nabla \times \mathbf{q}$, but for simplicity I choose not to expound on it and its effects in this introduction.

- In dispersion of material in a pipe, Chapter 7 showed that the advection-diffusion within the 3D space of the pipe, $\partial c/\partial t + \mathbf{q} \cdot \nabla c = \kappa \nabla^2 c$, could be modeled accurately in the long term by a very much simpler 1D advection-dispersion, $\partial C/\partial t + U \partial C/\partial x = D \partial^2 C/\partial x^2$, along the pipe. Hence, following an initial localized release, the long-term spread is approximately Gaussian. Interestingly, the effective dispersion coefficient, (7.13), is typically very much larger than the underlying diffusion coefficient. Such nonlinear dependence of the coefficients is typical in modeling, even for linear systems; see section 7.5.

  Section 7.4 straightforwardly derived higher-order corrections to the advection-dispersion model. Indeed, we put a quantitative limit on the spatial resolution of such a model via Fourier space.

- We treated the dynamics of thin fluid films similarly (Chapter 8). The two-dimensional fluid dynamics of a thin film, through the action of cross-film viscosity, is systematically modeled by the evolution of just the film thickness: $\partial h/\partial t \approx -\frac{1}{3} \mathrm{We}\, \partial/\partial x(h^3 \partial^3 h/\partial x^3)$ in the absence of gravity. The key is to treat a lateral derivative as a small parameter (operator) and then apply standard center manifold ideas.

- The same slow manifold modeling techniques apply to many interesting physical problems: for example, the exercises of section 8.5 indicate that applications include heat exchangers, forest fires, fluid convection, and pattern formation.

- Cross-sectional averaging is unsound as a *quantitative* method of modeling: section 9.1.4 establishes this statement for nonlinear fluid modeling, and Chapter 7 establishes it even in linear systems.

- There are many ways to tweak a physical problem to apply center manifold techniques to derive and justify a model. Usually one just needs to artificially modify the system to change the spectrum to suit center manifold theory (section 9.1). In any physical problem, some ways are better than others, as section 9.3 found by introducing the Euler parameter to systematically explore possibilities.
- In the particular case of modeling the dynamics of a thin layer of fluid, this approach lets the problem's PDEs tell us the velocity profile (9.12) across the fluid (instead of being imposed as a guess). Then the approach derives, via the computer algebra of section 9.2, the conservation of mass and the mean momentum equations (9.1) for the fluid flow. Such PDEs accurately describe the emergent fluid flow over large lateral scales, scales much larger than the relatively small thickness of the fluid layer.
- These models, being infinite-dimensional PDEs, are rich enough to describe a wide range of interesting dynamics. In contrast, models of the pitchfork and Hopf bifurcation are just pale shadows of the full range of dynamics inherent in the spatiotemporal dynamics of many fluid problems.
- We explored a rigorous justification in Fourier space (for linear problems such as shear dispersion) for this center manifold approach to modeling slow variations in space. The techniques generalize immediately to nonlinear problems even though the theory does not—at least not yet (Roberts, 2013).

An interesting aspect of this use of center manifold theory is a comparison with the method of multiple scales. In contrast to the method of multiple scales, our center manifold approach puts very few a priori assumptions on the size of various physical effects. The main requirement in a center manifold analysis is to *qualitatively* identify which are dominant terms, included in the linear operator $\mathcal{L}$, and which are perturbation terms, such as nonlinear terms or slowly varying effects. In contrast, the method of multiple scales puts strong constraints on the size of various effects: for example, in order to derive the Taylor model of dispersion, one has to scale space with ϵ and time with ϵ^2 and then change to a moving coordinate system; having done all that, there is no good rationale for generating higher-order corrections to the obtained leading-order model, whereas the center manifold approach naturally generates higher-order approximations from iterative refinements. I contend that the method of multiple scales is a poorer paradigm for modeling dynamics.

Part IV

Normal forms illuminate many modeling issues

Part IV Contents

It may give some practical results, but that's not why we do it.
R. P. Feynman

This part discusses how coordinate transforms underpin mathematical modeling. In any given example, one constructs a coordinate transform to put a system into a normal form in which key modeling concepts are clearly seen. The definition of a normal form for a system of equations is subjective. It all depends upon the purpose we have in mind. Here our aim is to simplify the dynamics in a way that clearly and beautifully extracts any useful low-dimensional model.

Chapter 10 begins to explore how to disentangle effects in nonlinear dynamical systems to empower ready qualitative interpretation. We develop the nonlinear analogue of matrix diagonalization. The analogue is a nonlinear coordinate transformation of the variables to what is called a normal form. Only an introduction is given, but, for example, we see how linearization is indeed valid in two-variable systems.

Chapter 11 then generalizes the coordinate transform from two variables to any finite number of variables. Further, the analysis caters for dynamics with stable, center and unstable variables, and caters for systems that are slightly different from the ideal to be more practically useful. The construction of such coordinate transforms (section 11.2) immediately establishes both the existence and emergence of center manifold dynamics as a long-term model of the system dynamics.

Further, the existence of such coordinate transforms leads to a new approach to confirming that these crucial modeling properties of existence and emergence hold in a finite domain (section 11.3).

Center manifold theory, underpinned by nonlinear coordinate transforms, assures us that long-time behavior of many dynamical systems may be effectively predicted by a low-dimensional model. Chapter 12 considers how to equip such a low-dimensional model with correct initial conditions. The natural assumption, that many people make automatically, is only a coarse leading approximation. Instead, the so-called isochrons of the coordinate transform (Chapter 12) provide initial conditions to make forecasts accurate over long times.

In the example of shear dispersion, section 12.2 shows how this approach distinguishes between a release of material near the side of a channel and a release in the middle of the channel.

Two practical issues arise: why transform the many fast variables when all we want are initial conditions for the slow variables, and how can we cater for systems in general form $\dot{\mathsf{u}} = \mathcal{L}\mathsf{u} + \mathsf{f}(\mathsf{u})$? Section 12.3 answers both challenges by developing an algorithmically efficient approach to determining initial conditions via the adjoint of the right-hand side operators.

Importantly, section 12.4 argues that the same projection that determines initial conditions also empowers us to determine the effects on the model of any uncertainty in the original system. Furthermore, the projection also provides a leading-order estimate of the effects of time-dependent forcing.

Normal-form coordinate transforms empower us to explore other delicate issues in modeling. Many physical situations have large-scale, slow dynamics of prime interest on which there are potentially fast waves or oscillations that we wish to neglect. Examples are incompressible fluid flow that neglects sound, beam models that neglect elastic waves, the quasi-geostrophic atmosphere that neglects "gravity" waves, and Newtonian dynamics that neglects the gravity waves of general relativity (section 13.1). But the existence and emergence of such slow dynamics from fast oscillations/waves (on a so-called subcenter slow manifold) are questionable.

Sections 13.3, 13.4, and 13.6 introduce three examples that appear to show the existence of useful subcenter slow manifold models. But section 13.5 uses a full coordinate transform to demonstrate three general results:

- subcenter slow manifolds may exist but usually only for systems exponentially close to any specified system (section 13.5.1);
- within the ideal mathematical framework, such subcenter slow manifolds are not emergent (section 13.5.2); and
- the evolutions on and off a subcenter slow manifold are different (sections 13.5.3 and 13.5.4).

This part establishes that coordinate transforms to nonlinearly separate modeled modes from modes of unwanted detail empower our understanding of the strengths and limitations of the modeling process.

Chapter 10

Normal-form transformations simplify evolution

Contents

Consider dynamics described by a set of ordinary differential equations. A coordinate transformation of variables to a normal form tries to simplify the algebraic description of the dynamics so that we easily deduce relevant features such as stability and classification.

Example 10.1 (bend coordinates to simplify). Consider the pair of ODEs

$$\dot{x} = -2x \quad \text{and} \quad \dot{y} = -y + x^2\,; \tag{10.1}$$

Figure 10.1 shows some trajectories.

Linearization We might explore the ODEs as follows: the only equilibrium (fixed point) of the dynamical system is $x = y = 0$; the linearization about this equilibrium is

$$\dot{x} \approx -2x \quad \text{and} \quad \dot{y} \approx -y\,; \tag{10.2}$$

both of these ODEs describe decaying dynamics and so the equilibrium $(x, y) = (0, 0)$ is a stable node.

But how do we really know that we can neglect the nonlinear term x^2 in the y equation? True, the argument that the term x^2 is negligible near $(0, 0)$ is plausible; but is it correct? Normal-form coordinate transformations provide a sound route to answer this question.

Alternative coordinates For the system (10.1) try changing to new coordinate variables (X, Y) such that

$$x = X \quad \text{and} \quad y = Y - \tfrac{1}{3}X^2. \tag{10.3}$$

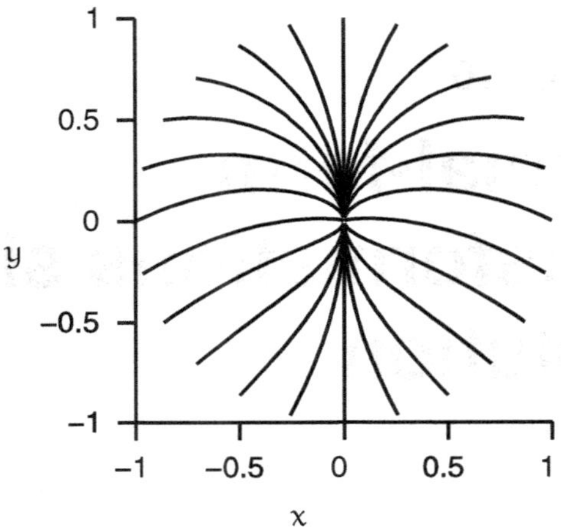

Figure 10.1. *Some trajectories of the pair of coupled ODEs* (10.1) *evolve toward the stable equilibrium at the origin.*

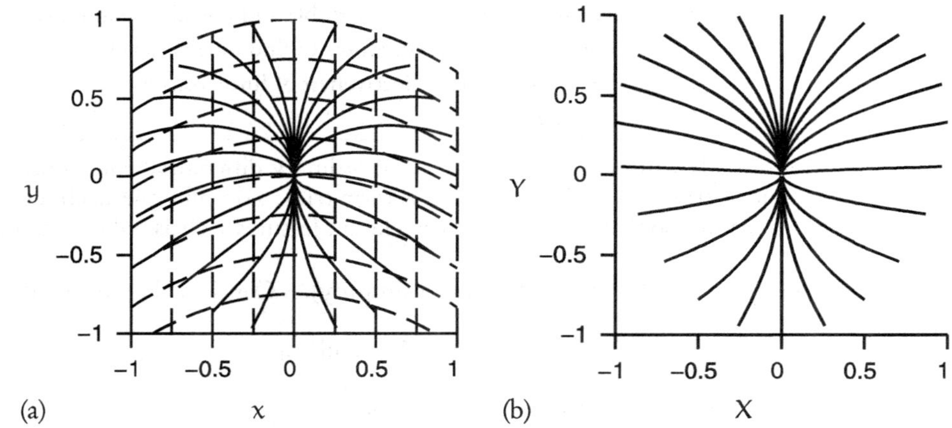

Figure 10.2. (a) *trajectories of the ODEs* (10.1) *with* XY *coordinate curves superimposed;* (b) *trajectories replotted in the transformed* XY*-plane.*

This coordinate transform slightly relabels the points in the xy-plane: the origin is still the origin, but, for example, the point with xy-label $(1, \frac{2}{3})$ has XY-label $(1, 1)$. The coordinate transform is a slight relabeling because near the equilibrium of interest, the origin, the coordinate transform is a near identity, $x \approx X$ and $y \approx Y$. But the nonlinear part of the transform "bends" the xy-plane upward to "straighten out" the dynamics in the XY-plane, as shown in Figure 10.2. Now algebraically find the dynamics of (10.1) in the new coordinates.

- First, $X = x$ so $\dot{X} = \dot{x} = -2x = -2X$. The variable X evolves according to the linearization (10.2).
- Second, from the ODE (10.1), $\dot{y} = -y + x^2 = -Y + \frac{1}{3}X^2 + X^2 = -Y + \frac{4}{3}X^2$, whereas from the transformation, $\dot{y} = \dot{Y} - \frac{2}{3}X\dot{X} = \dot{Y} - \frac{2}{3}X(-2X) = \dot{Y} + \frac{4}{3}X^2$.

Equating these two expressions for $\dot{y}$, we deduce that the variable Y evolves according to simply $\dot{Y} = -Y$. This is also the linearization (10.2).

The near identity coordinate transform (10.3) brings out that $\dot{X} = -2X$ and $\dot{Y} = -Y$. Hence the dynamics (solutions) of the original system (10.1) corresponds exactly to the dynamics (solutions) of this simple pair of linear and uncoupled ODEs. That is, the nonlinearities in (10.3) simply bend the dynamics of $\dot{X} = -2X$ and $\dot{Y} = -Y$; the nonlinearities do not affect its qualitative nature—see the bending by comparing Figures 10.1 and 10.2(b). The original linearization (10.2) is indeed valid.

The important point of this example is that *our normal-form coordinate transformation seeks new variables in which the dynamical properties are plain for all to see.*

Recall that the diagonalization of matrices does analogous simplification for linear problems. To use the ideas of normal-form transformations we usually, but not always, start with a system that already has its linear terms diagonalized. Then *the task of the normal-form transformation is to simplify the nonlinear terms, eliminating them if possible, so we readily see the qualitative nature of the dynamics.*

Example 10.2 (bend x). How do the nonlinearities in the system

$$\dot{x} = y^2 \quad \text{and} \quad \dot{y} = -y \tag{10.4}$$

affect the linearization $\dot{x} \approx 0$ and $\dot{y} \approx -y$? The answer here is, hardly at all. However, it is not obvious because, as we see in subsequent chapters, the slow evolution of x, namely $\dot{x} \approx 0$, is exquisitely sensitive to nonlinearities. It just so happens that here the nonlinear term y^2 is not significant.

To see this, transform the dynamics to new variables X and Y such that $x = X - \frac{1}{2}Y^2$ and $y = Y$. These new variables are nearly the same as x and y near the equilibrium at the origin, but not quite. Now consider the dynamics.

- First, since $y = Y$, $\dot{Y} = \dot{y} = -y = -Y$. The Y dynamics is identical to that for y.
- Second, $\dot{x} = y^2 = Y^2$ and also from the transform $\dot{x} = \dot{X} - Y\dot{Y} = \dot{X} + Y^2$. Equating these two expressions for $\dot{x}$ we deduce that $\dot{X} = 0$. The X dynamics is exactly trivial.

The near identity coordinate transform demonstrates that the dynamics of (10.4) is just a bent version of the dynamics of its linearization—the linearization $\dot{x} \approx 0$ and $\dot{y} \approx -y$ is identical to the transformed system $\dot{X} = 0$ and $\dot{Y} = -Y$. Figure 10.3 shows that trajectories of the ODEs (10.4) appear straightened in the XY-plane. Thus we may indeed use the linearization to report not only the qualitative dynamics of (10.4), but also the quantitative.

These applications are straightforward. In particular, I mysteriously knew the necessary change of variables. Section 10.1 explores how to find the change of variables, the coordinate transform, for a given dynamical system. Usually we find the transform as a multivariate power series. In this introduction to the topic, except for the next example, we only explore two-dimensional dynamical systems.

Historically, normal forms were primarily used to classify the low-dimensional dynamics of a system that had already been reduced to the center manifold. The next example is one such case.

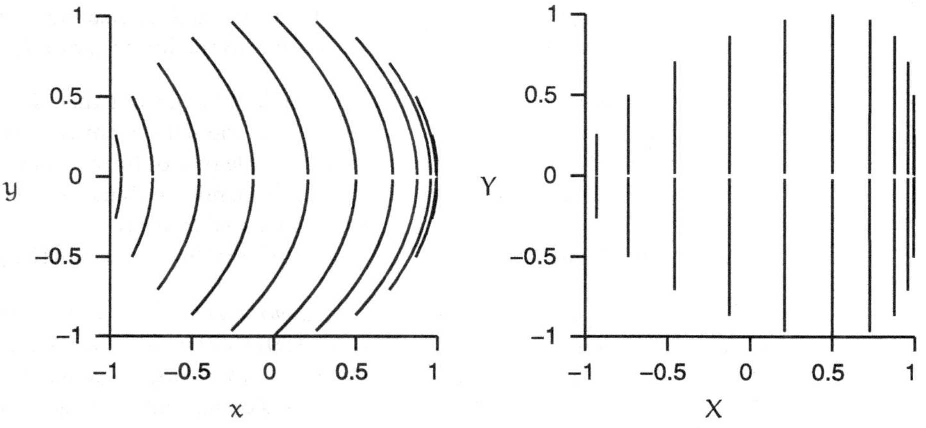

Figure 10.3. *Trajectories of the pair of coupled ODEs* (10.4) *approach the line* $y = 0$ *(left), but seen in the transformed* XY*-plane (right) they appear to do so in a "straight fashion."*

Example 10.3 (transform on a center manifold). Chow et al. (1990) considered the system

$$\dot{x}_1 = -x_2 - (x_3 - x_1)^2, \quad \dot{x}_2 = x_1, \quad \dot{x}_3 = -(x_3 - x_1)^2. \tag{10.5}$$

Its linearization is $\dot{x}_1 = -x_2$, $\dot{x}_2 = x_1$, and $\dot{x}_3 = 0$ with eigenvalues $\pm i$ and 0 that indicate that the whole state space is the center manifold. The computer algebra (Macsyma) code of Chow et al. (1990) then constructs a normal form that is easier to interpret. Correspondingly, my web service for center manifolds (Roberts, 2012) constructs the change of variables $x_1 = se^{it} + \bar{s}e^{-it} + \cdots$, $x_2 = -ise^{it} + i\bar{s}e^{-it} + \cdots$, $x_3 = s_3 + \cdots$. The new variable s_3 is approximately x_3, but less clear is that the new variable s is the complex amplitude of the oscillations in x_1 and x_2 (see section 3.6). In these new variables, the system (10.5) becomes

$$\begin{aligned}
\dot{s} &\approx s_3 s + i\big(\tfrac{3}{2}s_3^2 s - \tfrac{1}{6}s^2\bar{s}\big) + \big(-\tfrac{17}{4}s_3^3 s + \tfrac{41}{9}s_3 s^2\bar{s}\big),\\
\dot{s}_3 &\approx \big(-s_3^2 - 2s\bar{s}\big) + \big(4s_3^4 + \tfrac{9}{2}s_3^2 s\bar{s} - \tfrac{271}{18}s^2\bar{s}^2\big).
\end{aligned}$$

The leading terms are enough to predict that s_3 will generally evolve to become negative and then grow rapidly negative, since $\dot{s}_3 \approx -s_3^2 - 2|s|^2$. But such negative s_3 stabilizes the oscillations by causing their complex amplitude s to decay, since $\dot{s} \approx s_3 s$.

But our aim is different from such classification on the center manifold. Here our aim is to use coordinate transforms to create, support, and interpret the formation of center manifold models of dynamical systems.

10.1 ▪ Unstable and stable manifolds of saddles

There is a variety of methods proposed to construct stable and unstable manifolds: Krauskopf et al. (2006) surveyed some methods. However, it appears hard to use most such methods in complex systems where high dimensionality is typical. Thus herein we restrict our attention to algebraic methods that empower us to cope with high dimensions, albeit only to some level of asymptotic approximation.

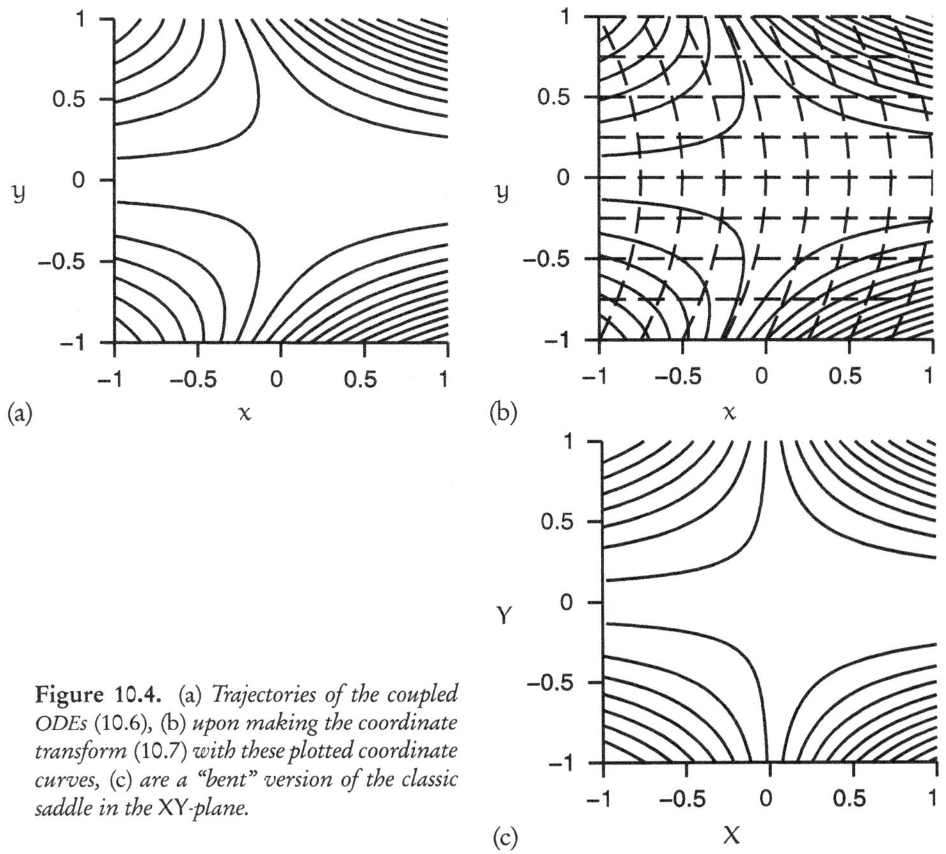

Figure 10.4. (a) *Trajectories of the coupled ODEs* (10.6), (b) *upon making the coordinate transform* (10.7) *with these plotted coordinate curves,* (c) *are a "bent" version of the classic saddle in the* XY*-plane.*

Example 10.4. Find a near identity change of variables, near the origin, to place the system

$$\dot{x} = 2x + y^2 \quad \text{and} \quad \dot{y} = -y \tag{10.6}$$

into its normal form $\dot{X} = 2X$ and $\dot{Y} = -Y$. That is, find the change of variable that straightens the dynamics of Figure 10.4(a) into the classic linear saddle shown in Figure 10.4(c).

Solution: The y equation is already in its normal form, so let us just try to transform the x variable (as illustrated in Figure 10.4(b)): pose $x = X + \hat{x}(X, Y)$ and $y = Y$ for some as yet unknown $\hat{x}(X, Y)$. Since this change of variables is to be a near identity we want $\hat{x}(X, Y)$ to be small (compared with the dominant linear term $x \approx X$). Substitute into the right-hand side of the x equation (10.6):

$$2x + y^2 = 2X + 2\hat{x} + Y^2 .$$

But also $x = X + \hat{x}(X, Y)$ so, using the chain rule, the time derivative on the left-hand side of the x equation (10.6) is

$$\dot{x} = \dot{X} + \frac{\partial \hat{x}}{\partial X}\dot{X} + \frac{\partial \hat{x}}{\partial Y}\dot{Y} = 2X + 2X\frac{\partial \hat{x}}{\partial X} - Y\frac{\partial \hat{x}}{\partial Y} .$$

Equate these two expressions for the time derivative $\dot{x}$ to deduce

$$2X\frac{\partial\hat{x}}{\partial X} - Y\frac{\partial\hat{x}}{\partial Y} - 2\hat{x} = Y^2.$$

This is a first example of a homological equation to seek a normal form. Such homological equations may be quite tricky to solve. Let's use the following method of undetermined coefficients.

- If the modification $\hat{x}$ had any X dependence, then $\partial\hat{x}/\partial X$ would be nonzero; hence the term on the left, $2X\partial\hat{x}/\partial X$, would generate some X-dependent terms on the left-hand side. But there are no X-dependent terms on the right-hand side so we discard any thought of X dependence in the modification $\hat{x}$.
- Thus try $\hat{x} = cY^q$ for some coefficient c and exponent q; then the left-hand side becomes $-(q+2)cY^q$. This left-hand side must match the right-hand side of simply Y^2 and hence the exponent $q = 2$ and the coefficient $c = -1/4$.

This analysis implies that the near identity change of variables

$$x = X - \tfrac{1}{4}Y^2 \quad\text{and}\quad y = Y \tag{10.7}$$

transforms the ODE (10.6) into the simple normal form $\dot{X} = 2X$ and $\dot{Y} = -Y$. Consequently, *the dynamics of the* ODE (10.6) *is simply a curved or bent version of that of the simple linear system* $\dot{X} = 2X$ *and* $\dot{Y} = -Y$.

This linear system is a saddle, since X grows exponentially as Y decays exponentially. Thus the origin in the ODE (10.6) is also a saddle, albeit bent by the nonlinearity.

In coupled pairs of nonlinear ODEs, there are special curves in state space that guide the shape of the overall evolution. In Figure 10.4(a) of Example 10.4,

- the horizontal line $y = 0$ attracts all solutions of the ODEs (10.6)—we call the line $y = 0$ the unstable manifold (of the equilibrium at the origin)—whereas
- the parabola $x = -y^2/4$ separates the solutions that grow to $x \to \infty$ and those that grow to $x \to -\infty$—we call the curve $x = -y^2/4$ the stable manifold (of the equilibrium at the origin).

Interestingly, converse statements hold when we run time backward: the stable manifold is attractive in reversed time, and the unstable manifold separates solutions between $y \to \pm\infty$ in reversed time. Also, the stable and unstable manifolds of an equilibrium have other roles in other situations: for example, in chaotic maps, they tangle together in a way that characterizes crucial aspects of the chaos. But their complementary nature always holds and underlies their definition (Kuznetsov, 1995, §2.2).

Definition 10.1 (hyperbolic equilibrium). *An equilibrium of a dynamical system* (10.8) *is termed* **hyperbolic** *if the linearization of the dynamics has no pure imaginary eigenvalues (more generally, if the eigenvalues are bounded away from the imaginary axis).*

For example, any saddle point is a hyperbolic equilibrium: the positive and negative eigenvalues characteristic of a saddle are bounded away from zero.

Definition 10.2 (un/stable manifold). *Consider a dynamical system of* ODEs

$$\dot{u} = f(u)\,, \quad u \in \mathbb{R}^n\,, \tag{10.8}$$

with a hyperbolic equilibrium at u_*:[128]

[128] Often the origin is placed at the equilibrium.

- *the set of initial conditions whose subsequent evolution approaches the equilibrium is called its* **stable manifold**,

$$\mathcal{M}_s = \{\mathbf{u}(0) : \mathbf{u}(t) \to \mathbf{u}_*\,,\ t \to +\infty\}; \tag{10.9}$$

- *the set of initial conditions whose* backward *evolution approaches the equilibrium is called its* **unstable manifold**,

$$\mathcal{M}_u = \{\mathbf{u}(0) : \mathbf{u}(t) \to \mathbf{u}_*\,,\ t \to -\infty\}. \tag{10.10}$$

At this stage we have no right to call these sets manifolds because we do not know that they are smooth or that they have a definite dimension; Theorem 10.3, below, provides the assurance.

Example 10.5 (stable and unstable manifolds).

- In the system (10.6), the unstable manifold $\mathcal{M}_u$ of the origin is the line $y = 0$, whereas the stable manifold $\mathcal{M}_s$ is the curve $x = -y^2/4$.
- In the system (10.1), the entire xy-plane is the stable manifold, as all solutions are attracted to the origin; the unstable manifold is empty.
- For the system (10.4), the equilibrium at the origin is not hyperbolic, as one eigenvalue is zero. Later we generalize the definition of a stable manifold $\mathcal{M}_s$ so that here it is the curve $x = -y^2/2$.

Observe in these examples that the stable and unstable manifolds are indeed smooth curves in the plane and hence are justifiably called manifolds. Also the nontrivial manifolds are tangent to the eigenvectors of the linearized evolution near the equilibrium. The next theorem formalizes these general properties.

Theorem 10.3 (local manifolds). *Let* $\mathbf{u}_*$ *be a hyperbolic equilibrium of a dynamical system* (10.8). *Then within some neighborhood of the equilibrium* $\mathbf{u}_*$ *the stable and unstable manifolds* $\mathcal{M}_s$ *and* $\mathcal{M}_u$ *are smooth manifolds. Moreover, these manifolds are tangent, respectively, to the eigenspaces of the linearized dynamics corresponding to eigenvalues with negative real part and positive real part.*

Kuznetsov (1995, §2.2) gave more discussion and an outline of a proof.

10.2 ▪ Normal forms display the stable and unstable manifolds

Let's return to coordinate transforms. Reconsider the stable and unstable manifolds of Examples 10.2 and 10.4, but now let's clearly see these manifolds in the normal form of the dynamics.

- The normal form of the system (10.6) is $\dot{X} = 2X$ and $\dot{Y} = -Y$. The general solution of this pair of linear equations is $(X, Y) = (X_0 e^{2t}, Y_0 e^{-t})$. To identify the stable manifold of solutions observe that if X_0 is nonzero then the solution grows. Hence for a solution to decay to zero as $t \to +\infty$ we must have $X_0 = 0$. That is, the Y-axis is the stable manifold. Conversely, for the unstable manifold, for the solutions $\to 0$ as $t \to -\infty$ we must have $Y_0 = 0$; namely the X-axis is the unstable manifold.

  Then, since we know the coordinate transform (10.7) that $x = X - \frac{1}{4}Y^2$ and $y = Y$, we know the stable manifold $X_0 = 0$ becomes $x_0 = -\frac{1}{4}y_0^2$, as commented upon earlier; similarly, the unstable manifold $Y_0 = 0$ is simply $y_0 = 0$.

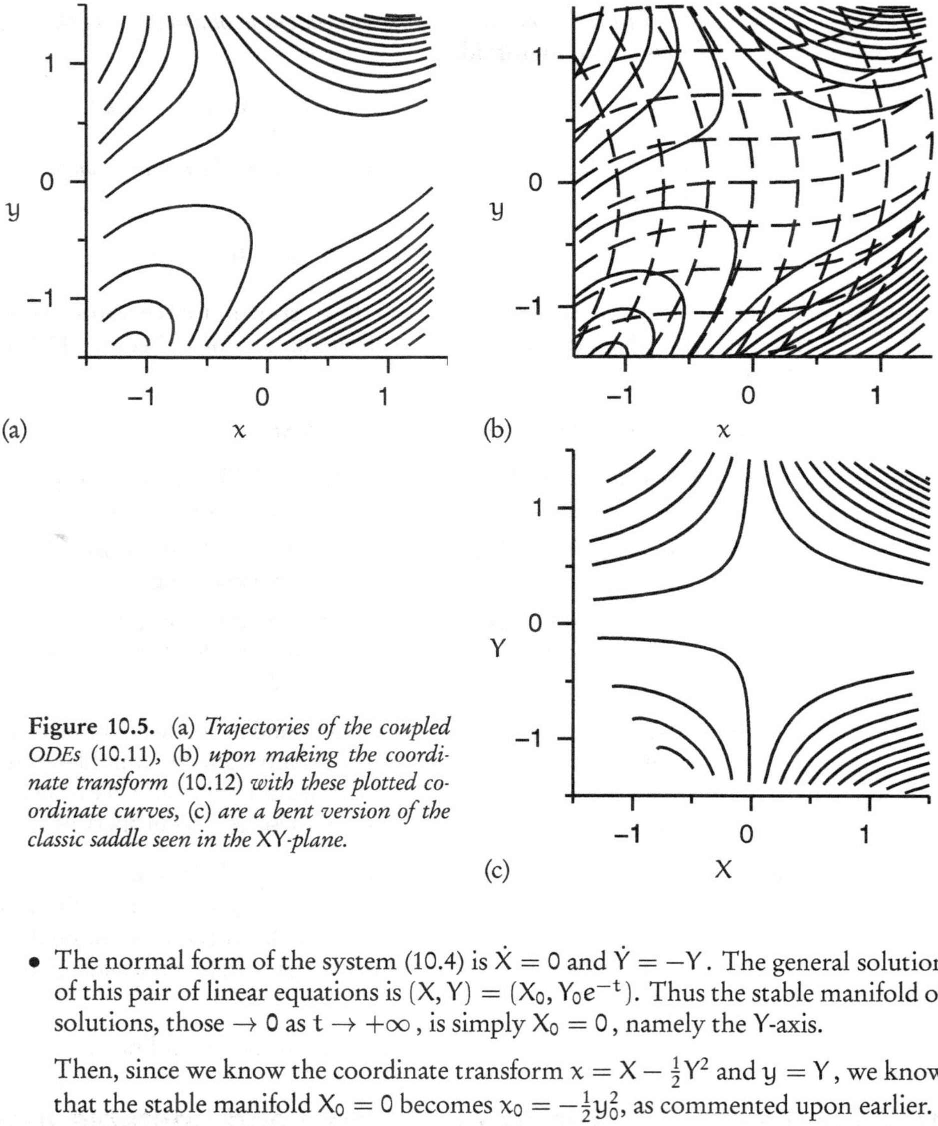

Figure 10.5. (a) *Trajectories of the coupled ODEs* (10.11), (b) *upon making the coordinate transform* (10.12) *with these plotted coordinate curves,* (c) *are a bent version of the classic saddle seen in the* XY*-plane.*

- The normal form of the system (10.4) is $\dot{X} = 0$ and $\dot{Y} = -Y$. The general solution of this pair of linear equations is $(X, Y) = (X_0, Y_0 e^{-t})$. Thus the stable manifold of solutions, those $\to 0$ as $t \to +\infty$, is simply $X_0 = 0$, namely the Y-axis.

  Then, since we know the coordinate transform $x = X - \frac{1}{2}Y^2$ and $y = Y$, we know that the stable manifold $X_0 = 0$ becomes $x_0 = -\frac{1}{2}y_0^2$, as commented upon earlier.

In the normal form, the unstable and stable manifolds are the corresponding linear eigenspaces corresponding to eigenvalues with positive and negative real parts, respectively.

10.3 ▪ Solve homological equations to find normal forms

The previous section uses coupled pairs of ODEs that are sufficiently simple that we could straightforwardly construct the normal-form transformation. In general, a normal form is harder to construct. Moreover, in general, a normal form has to be constructed as an asymptotic series. Here we explore how iteration empowers us to asymptotically approximate normal forms.

For simplicity we restrict attention to pairs of coupled ODEs. Let us start with a modified version of Example 10.4.

Example 10.6 (first approximation). Use iteration to find a near identity change of variables, in a neighborhood of the origin, to place the system

$$\dot{x} = 2x + y^2 \quad \text{and} \quad \dot{y} = -y + x^3 \tag{10.11}$$

into its normal form $\dot{X} = 2X$ and $\dot{Y} = -Y$. That is, find a change of variables that straightens the dynamics into the classic linear saddle shown in Figure 10.5. First we appear to succeed, but second we show we actually fail, and last we show how to always succeed.

Solution: Let us do the first step in a systematic construction of the normal-form transformation. Since the change of variables is to be a near identity we first seek $x = X + \hat{x}(X, Y)$ and $y = Y + \hat{y}(X, Y)$, where the corrections $\hat{x}$ and $\hat{y}$ to the identity transformation are small. Substitute into the x ODE of (10.11), using the chain rule for time derivatives:

$$\begin{aligned}
&\dot{x} = 2x + y^2 \\
\implies\ & \dot{X} + \frac{\partial \hat{x}}{\partial X}\dot{X} + \frac{\partial \hat{x}}{\partial Y}\dot{Y} = 2X + 2\hat{x} + Y^2 + 2Y\hat{y} + \hat{y}^2 \\
\implies\ & 2X + 2X\frac{\partial \hat{x}}{\partial X} - Y\frac{\partial \hat{x}}{\partial Y} = 2X + 2\hat{x} + Y^2 + 2Y\hat{y} + \hat{y}^2 \\
\implies\ & +2X\frac{\partial \hat{x}}{\partial X} - Y\frac{\partial \hat{x}}{\partial Y} - 2\hat{x} = Y^2 + 2Y\hat{y} + \hat{y}^2 \\
\implies\ & +2X\frac{\partial \hat{x}}{\partial X} - Y\frac{\partial \hat{x}}{\partial Y} - 2\hat{x} \approx Y^2
\end{aligned}$$

upon neglecting the very small $\hat{y}^2$ and significantly smaller $Y\hat{y}$ when compared to the corrections retained on the left-hand side. This is the same homological equation we solved in Example 10.4. Its solution here is the same, namely $\hat{x} = -Y^2/4$.

Now do the analogous analysis for the y ODE of (10.11). Using the chain rule for time derivatives deduce

$$\begin{aligned}
&\dot{y} = -y + x^3 \\
\implies\ & \dot{Y} + \frac{\partial \hat{y}}{\partial X}\dot{X} + \frac{\partial \hat{y}}{\partial Y}\dot{Y} = -Y - \hat{y} + X^3 + 3X^2\hat{x} + 3X\hat{x}^2 + \hat{x}^3 \\
\implies\ & -Y + 2X\frac{\partial \hat{y}}{\partial X} - Y\frac{\partial \hat{y}}{\partial Y} = -Y - \hat{y} + X^3 + 3X^2\hat{x} + 3X\hat{x}^2 + \hat{x}^3 \\
\implies\ & +2X\frac{\partial \hat{y}}{\partial X} - Y\frac{\partial \hat{y}}{\partial Y} + \hat{y} = X^3 + 3X^2\hat{x} + 3X\hat{x}^2 + \hat{x}^3 \\
\implies\ & +2X\frac{\partial \hat{y}}{\partial X} - Y\frac{\partial \hat{y}}{\partial Y} + \hat{y} \approx X^3
\end{aligned}$$

upon neglecting the very small $\hat{x}^2$ and significantly smaller $X\hat{x}$ when compared to the corrections retained on the left-hand side. This homological equation has solution $\hat{y} = X^3/7$.

This analysis implies that the near identity change of variables

$$x \approx X - \tfrac{1}{4}Y^2 \quad \text{and} \quad y \approx Y + \tfrac{1}{7}X^3 \tag{10.12}$$

transforms the ODE (10.11) into the simple normal form $\dot{X} = 2X$ and $\dot{Y} = -Y$. Consequently, *the dynamics of the* ODE (10.11) *is simply a curved or bent version of that of the simple saddle system* $\dot{X} = 2X$ *and* $\dot{Y} = -Y$, as shown in the transformation of Figure 10.5.

But the transformation is not quite so simple, as seen in this example. Note that at one stage for each ODE in the pair (10.11) we neglected small terms in the corrections. This neglect implies the transformation (10.12) is only approximate, as indicated. We seek higher-order approximations by iteration with the aid of computer algebra.

Example 10.7 (computer algebra iterates effortlessly). The iteration, whether by computer algebra or not, depends upon solving homological equations of the form

$$2X\frac{\partial\hat{x}}{\partial X} - Y\frac{\partial\hat{x}}{\partial Y} - 2\hat{x} = -\operatorname{Res}_{(10.11),x}$$
$$\text{and } 2X\frac{\partial\hat{y}}{\partial X} - Y\frac{\partial\hat{y}}{\partial Y} + \hat{y} = -\operatorname{Res}_{(10.11),y}\,.$$

These homological equations appeared in Example 10.6 with the residuals on the right-hand side of Y^2 and X^3, respectively. At higher orders in the asymptotic solution the residuals on the right-hand side involve a sum of terms in X^pY^q.

- Consider the x homological equation with X^pY^q in the right-hand side residual. We guess a component in the correction $\hat{x}$ of cX^pY^q for some constant c to be determined. Substitute into the homological equation to seek to solve $(2p-q-2)cX^pY^q = -X^pY^q$. Clearly we choose $c = -1/(2p-q-2)$ for each component X^pY^q in the residual. For example, in the first iterate, the residual was $-Y^2$ ($p = 0$ and $q = 2$), so $c = \frac{1}{4}$, and hence the correction $\hat{x} = -\frac{1}{4}Y^2$.
- Similarly consider the y homological equation with X^pY^q in the right-hand side residual. Guess a component in the correction $\hat{y}$ of cX^pY^q for some constant c to be determined. Substitute into the homological equation to seek to solve $(2p-q+1)cX^pY^q = -X^pY^q$. Clearly we choose $c = -1/(2p-q+1)$ for each component X^pY^q in the residual. For example, in the first iterate, the residual was $-X^3$ ($p = 3$ and $q = 0$), so $c = -\frac{1}{7}$, and hence the correction $\hat{y} = +\frac{1}{7}X^3$.

Algorithm 10.1 derives such corrections from the residuals, iteratively, until the residual is of high order. It finds that the transform from the ODEs (10.11) to the normal form $\dot{X} = 2X$ and $\dot{Y} = -Y$ is

$$x = X - \tfrac{1}{4}Y^2 + \tfrac{2}{21}X^3Y + \tfrac{1}{2}X^2Y^3 + \mathcal{O}(X^6+Y^6)\,, \tag{10.13a}$$
$$y = Y + \tfrac{1}{7}X^3 - \tfrac{1}{4}X^2Y^2 - \tfrac{3}{16}XY^4 + \mathcal{O}(X^6+Y^6)\,. \tag{10.13b}$$

Algorithm 10.1 computes the residual for each equation and then updates the transformation by

- placing a term $-X^pY^q/(2p-q-2)$ into the x transform for each term X^pY^q in the x residual;
- placing a term $-X^pY^q/(2p-q+1)$ into the y transform for each term X^pY^q in the y residual.

Algorithm 10.1 *Reduce code to find the normal-form transformation to put* ODEs (10.11) *into the normal form* $\dot{X} = 2X$ *and* $\dot{Y} = -Y$. *Here* xx *denotes* X *and* yy *denotes* Y. *Resonance occurs at* $\mathcal{O}(X^6Y^{10} + X^5Y^{11})$.

```
depend xx,t; depend yy,t;
let { df(xx,t)=>2*xx, df(yy,t)=>-yy };
% inverse homological operator
operator linv; linear linv;
% linear approximation, count order with small
factor small;
x:=xx*small; y:=yy*small;
% iterate to solve
let small^6=>0;
for it:=1:99 do begin
  resx:=df(x,t)-2*x-y^2;
  x:=x-(linv(resx,t) where linv(~r,t)=>
    r^2/(2*xx*df(r,xx)-yy*df(r,yy)-2*r));
  resy:=df(y,t)+y-x^3;
  y:=y-(linv(resy,t) where linv(~r,t)=>
    r^2/(2*xx*df(r,xx)-yy*df(r,yy)+r));
  if {resx,resy}={0,0} then it:=1000+it;
end;
```

For each individual term X^pY^q, dividing by the homological operator acting upon X^pY^q, then dividing by X^pY^q, gives the correct coefficient. The variable `small` counts the number of X and Y factors in each term; hence we truncate to a consistent order in X and Y by discarding terms higher than some specified power in the counting variable `small`.

This beautiful little program provides a simple and flexible base for analyzing almost any pairs of ODEs. Unfortunately, the program is also defective!

Try executing Algorithm 10.1 to construct an asymptotic approximation to 16th order by using `let small^17=>0`: 15th order is fine, but when you attempt to find the 16th order you encounter a fatal error through an attempt to divide by zero!

Example 10.8 (resonance affects the normal form). Explore this critical issue of zero division in the normal form of the ODEs (10.11). Recall that the iteration depends upon solving homological equations of the form

$$2X\frac{\partial\hat{x}}{\partial X} - Y\frac{\partial\hat{x}}{\partial Y} - 2\hat{x} = -\operatorname{Res}_{(10.11),x}$$

$$\text{and } 2X\frac{\partial\hat{y}}{\partial X} - Y\frac{\partial\hat{y}}{\partial Y} + \hat{y} = -\operatorname{Res}_{(10.11),y}\,.$$

The residuals on the right-hand side involve a sum of terms in X^pY^q:

- in the x homological equation these lead to division by $2p - q - 2$, which is zero for terms, such as X^2Y^2 and X^3Y^4, for which $2p - q - 2 = 0$;
- similarly in the y homological equation these lead to division by $2p - q + 1$, which is zero for terms, such as XY^3 and X^2Y^5, for which $2p - q + 1 = 0$.

Such problematic zero divisors occur frequently in asymptotic approximation of dynamics.[129] Generically they are called resonances because historically these zero divisors were first encountered in analyzing oscillations and the zero divisors physically signaled a resonant interaction between component oscillators.

The resolution of the zero divisor problem is either that we must abandon high-order smoothness, or that some of the terms in the ODEs for x *and* y *must remain in the ODEs for the evolution of the new variables* X *and* Y. That is, if we insist on smoothness,[130] then we can no longer insist that the normal form be precisely that of a *linear* saddle. Instead there are inescapable nonlinear modifications to the evolution.

Solution: Transform the x and y variables *and* the X- and Y-evolution: pose

$$\begin{gathered} x = x(X,Y) \approx X \quad \text{and} \quad y = y(X,Y) \approx Y \\ \text{such that } \dot X = F(X,Y) \approx 2X \quad \text{and} \quad \dot Y = G(X,Y) \approx -Y \end{gathered}$$

for some as yet unknown near identity transform $x(X,Y)$ and $y(X,Y)$, and for some minimal set of terms remaining in the evolution $\dot X = F$ and $\dot Y = G$. Now adopt the approach of supposing we know an approximation to the transformation and then seeking corrections.

Suppose that at some iteration we have a current approximation to the transform and also to the normal-form ODEs. Seek small corrections $\hat x(X,Y)$ and $\hat y(X,Y)$ to the transform and corrections $\hat F(X,Y)$ and $\hat G(X,Y)$ to the evolution such that after the corrected transformation $x = x(X,Y) + \hat x(X,Y)$ and $y = y(X,Y) + \hat y(X,Y)$ the resultant evolution $\dot X = F(X,Y) + \hat F(X,Y)$ and $\dot Y = G(X,Y) + \hat G(X,Y)$ *better* describes the original dynamics of the ODEs (10.11). The major complication with substituting the coordinate transformation into (10.11) is the time derivatives: consider the x-derivative

$$\begin{aligned} [\dot x]_{\text{new}} &= \left(\frac{\partial x}{\partial X} + \frac{\partial \hat x}{\partial X}\right)\dot X + \left(\frac{\partial x}{\partial Y} + \frac{\partial \hat x}{\partial Y}\right)\dot Y \\ &= \left(\frac{\partial x}{\partial X} + \frac{\partial \hat x}{\partial X}\right)(F+\hat F) + \left(\frac{\partial x}{\partial Y} + \frac{\partial \hat x}{\partial Y}\right)(G+\hat G) \\ &= \underbrace{\frac{\partial x}{\partial X}F + \frac{\partial x}{\partial Y}G}_{=\dot x} + \frac{\partial x}{\partial X}\hat F + \frac{\partial \hat x}{\partial X}F + \frac{\partial x}{\partial Y}\hat G + \frac{\partial \hat x}{\partial Y}G \\ &\quad + \underbrace{\frac{\partial \hat x}{\partial X}\hat F + \frac{\partial \hat x}{\partial Y}\hat G}_{\text{negligible}} \\ &\approx [\dot x]_{\text{current}} + \hat F + 2X\frac{\partial \hat x}{\partial X} - Y\frac{\partial \hat x}{\partial Y} \end{aligned}$$

upon recognizing

- that the time derivative is $\dot x$ for the current approximation,
- that the products of small corrections are negligible, and

[129]Zero divisors have plagued applications of calculus for nearly 300 years: the early application of calculus to the problem of three bodies interacting under gravity introduced such zero divisors.

[130]Recall that the term *smooth* means we can differentiate as many times as is necessary for the order of accuracy to which we work.

- that we approximate the coefficients of corrections by their first order approximations ($\partial x/\partial X \approx 1$, $F \approx 2X$, $\partial x/\partial Y \approx 0$ and $G \approx -Y$).

The right-hand side of the x equation is straightforward:

$$\begin{aligned}[2x + y^2]_{\text{new}} &= 2x + 2\hat{x} + y^2 + 2y\hat{y} + \hat{y}^2 \\ &\approx [2x + y^2]_{\text{current}} + 2\hat{x}\end{aligned}$$

upon neglecting products of corrections and approximating coefficients of corrections by their first-order approximation ($2y \approx 0$). Equate the two sides and rearrange to

$$\hat{F} + 2X\frac{\partial\hat{x}}{\partial X} - Y\frac{\partial\hat{x}}{\partial Y} - 2\hat{x} = -\operatorname{Res}_{(10.11),x},$$

which is exactly as before *except we now have the option of avoiding dividing by zero through placing any troublesome terms in the evolution via* $\hat{F}$ *rather than trying to absorb them into the coordinate transform via* $\hat{x}$.

The retained terms on the left-hand side, $2X\partial\hat{x}/\partial X - Y\partial\hat{x}/\partial Y - 2\hat{x}$, are all of "size" $\hat{x}$ as the multiplications by small X and Y are countered by the "divisions" by X and Y in the derivatives.

Similarly for the y ODE: the complicated y time derivative

$$[\dot{y}]_{\text{new}} \approx [\dot{y}]_{\text{current}} + \hat{G} + 2X\frac{\partial\hat{y}}{\partial X} - Y\frac{\partial\hat{y}}{\partial Y},$$

and the right-hand side is straightforwardly

$$[-y + x^3]_{\text{new}} \approx [-y + x^3]_{\text{current}} - \hat{y}\,.$$

Equate the two sides and rearrange to

$$\hat{G} + 2X\frac{\partial\hat{y}}{\partial X} - Y\frac{\partial\hat{y}}{\partial Y} + \hat{y} = -\operatorname{Res}_{(10.11),y},$$

which is exactly as before *except we now have the option of avoiding dividing by zero through placing any troublesome terms in the evolution via* $\hat{G}$ *rather than trying to absorb them into the coordinate transform via* $\hat{y}$.

Execute Algorithm 10.2 to find that the normal form of the ODEs (10.11) is

$$\dot{X} = 2X + \tfrac{235829}{677376}X^6Y^{10} + \mathcal{O}(X^{17} + Y^{17}) \tag{10.14a}$$

$$\text{and}\quad \dot{Y} = -Y - \tfrac{21439}{112896}X^5Y^{11} + \mathcal{O}(X^{17} + Y^{17})\,. \tag{10.14b}$$

Procedures `homx` and `homy` compute the homological operators that are used in several places to flag whether a particular term will give rise to a zero divisor or not. This algorithm works to arbitrarily high order; it is limited only by computer memory and speed.

The conclusion is: we cannot *smoothly* transform the ODEs (10.11) into the linear saddle $\dot{X} = 2X$ and $\dot{Y} = -Y$; terms such as those in (10.14) may be inescapable.[131]

[131] Allowing nonsmooth logarithms into the coordinate transform typically empowers us to remove all nonlinear terms, such as X^6Y^{10} and X^5Y^{11} (just as logarithms enable us to form power series solutions of ODEs when the indicial equation has repeated roots). However, we generally do not pursue this possibility here.

Algorithm 10.2 *Correct Reduce code to find the normal-form transformation to put* ODEs (10.11) *into the normal form* (10.14). *Resonance occurs at* $\mathcal{O}(X^6Y^{10}+X^5Y^{11})$.

```
depend xx,t; depend yy,t;
let { df(xx,t)=>ff, df(yy,t)=>gg };
% inverse homological operator
operator linv; linear linv;
procedure homx(r); 2*xx*df(r,xx)-yy*df(r,yy)-2*r;
procedure homy(r); 2*xx*df(r,xx)-yy*df(r,yy)+r;
% linear approximation, count order with small
factor small;
x:=xx*small; y:=yy*small;
ff:=2*xx; gg:=-yy;
% iterate to solve
let small^17=>0;
for it:=1:99 do begin
  resx:=df(x,t)-2*x-y^2;
  ff:=ff+(ffd:=-(linv(resx,t) where {
    linv(~r,t)=>r when homx(r)=0,
    linv(~r,t)=>0 when homx(r) neq 0 }))/small;
  x:=x-(linv(resx+ffd,t) where linv(~r,t)=>r^2/homx(r));
  resy:=df(y,t)+y-x^3;
  gg:=gg+(ggd:=-(linv(resy,t) where {
    linv(~r,t)=>r when homy(r)=0,
    linv(~r,t)=>0 when homy(r) neq 0 }))/small;
  y:=y-(linv(resy+ggd,t) where linv(~r,t)=>r^2/homy(r));
  if {resx,resy}={0,0} then write it:=1000+it;
end;
```

However, recall that the terms that cannot be smoothly removed from the X-evolution are those of the form X^pY^q for $2p = q+2$, and since exponent $q \geqslant 0$, thus $p \geqslant 1$. Consequently, $X = 0$ is invariant in the normal-form evolution (10.14), as the variable X will always be a factor in $\dot{X}$. Hence the stable manifold of the normal form is precisely the linear space $X = 0$. The stable manifold in the xy-plane then comes from (10.13a), namely $x = -\frac{1}{4}y^2 + \mathcal{O}(y^6)$, as (10.13b) reduces to $y = Y + \mathcal{O}(Y^6)$ when $X = 0$.

Conversely, recall that the terms that cannot be removed from the Y-evolution are those of the form X^pY^q for $q = 2p+1$, and since exponent $p \geqslant 0$, thus $q \geqslant 1$. Consequently, $Y = 0$ is invariant in the normal-form evolution (10.14), as the variable Y will always be a factor in $\dot{Y}$. Hence the unstable manifold of the normal form is precisely the linear space $Y = 0$. The unstable manifold in the xy-plane then comes from (10.13b), namely $y = \frac{1}{7}x^2 + \mathcal{O}(x^6)$, as (10.13a) reduces to $x = X + \mathcal{O}(X^6)$ when $Y = 0$.

10.4 ▪ The unreasonable effectiveness of linearization

Although we have not demonstrated it in general, similar characteristics hold for general systems of coupled nonlinear ODEs: a specially crafted nonlinear coordinate transform straightens the stable and unstable manifolds.

The following theorem invokes a homeomorphic coordinate transformation, not a diffeomorphism. That is, the coordinate transform need not be smooth, where smooth is in the sense of being differentiable as often as we would like.

Theorem 10.4 (Hartman–Grossman). *If* u_* *is a* hyperbolic equilibrium *for the autonomous differential equation* $\dot{u} = f(u)$, *then there is both an open set* $\mathcal{D}$ *containing the equilibrium and also a homeomorphism* H *with domain* $\mathcal{D}$ *such that the trajectories of the differential equation are mapped by* H *to trajectories of the linearized system* $\dot{u} = \mathcal{L}(u - u_*)$ *in the set* $\mathcal{D}$, *where the Jacobian* $\mathcal{L} = \partial f/\partial u$ *at* $u = u_*$ *(Chicone, 2006, Thm. 1.47).*

Importance Linearization is an incredibly important tool in applications of mathematics. This theorem asserts that linearization is not just valid close to the equilibrium. The theorem asserts that the linearized dynamics hold for the nonlinear dynamics in a finite domain; it is just that the nonlinear dynamics are bent by the homeomorphic coordinate transformation. That is, linearization is valid qualitatively over a much larger domain than its notional quantitative validity.

The caveat in the Hartman–Grossman Theorem 10.4 is that it applies only to hyperbolic equilibria, namely those with eigenvalues that have real part nonzero.[132] Recall that Part II argued that the case of an equilibrium where one or more eigenvalues has real part zero is of vital interest in modeling. The rest of Part IV focuses on this case of some eigenvalues of zero real part.

10.5 ▪ Exercises

Exercise 10.1 (slow and stable manifolds). Adapt the arguments of this section to untangle the dynamics near the origin of the system

$$\dot{x} = ay - x^2y \quad\text{and}\quad \dot{y} = -y - ay + xy^2,$$

where the parameter a is small.

1. Deduce that the x-axis is a slow manifold for all parameters a, and that the y-axis is a stable manifold for parameter $a = 0$.
2. Use one iteration to approximate the coordinate transform $x \approx X$ and $y \approx Y$ such that $\dot{X} = 0$ and $\dot{Y} = -(1 + a)Y$. Treat parameter $a = \mathcal{O}(X^2 + Y^2)$.
3. Hence approximate the shape of the slow manifold and stable manifold from the coordinate transform.
4. Repeat for a second iteration.

Exercise 10.2 (hyperbolic decoupling). Consider the dynamical system $\dot{x} = x - xy + y^3$ and $\dot{y} = -2y + x^2$ near the equilibrium at the origin. Figure 10.6 shows trajectories of the dynamical system.

- Seek a coordinate transform $x = X + \hat{x}(X, Y)$ that, along with $y = Y + \hat{y}(X, Y)$, simplifies the x ODE to its linearization $\dot{X} = X$, but only find a first approximation to $\hat{x}$.

[132]Remarkably, Aulbach and Wanner (2000) generalized the Hartman–Grossman theorem to infinite-dimensional systems in Banach spaces, to nonautonomous systems, and to systems that may or may not have any center modes corresponding to eigenvalues of zero real part.

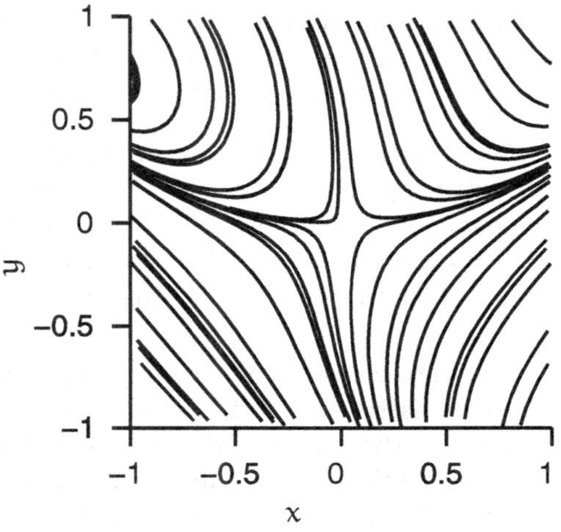

Figure 10.6. *Trajectories for Exercise* 10.2.

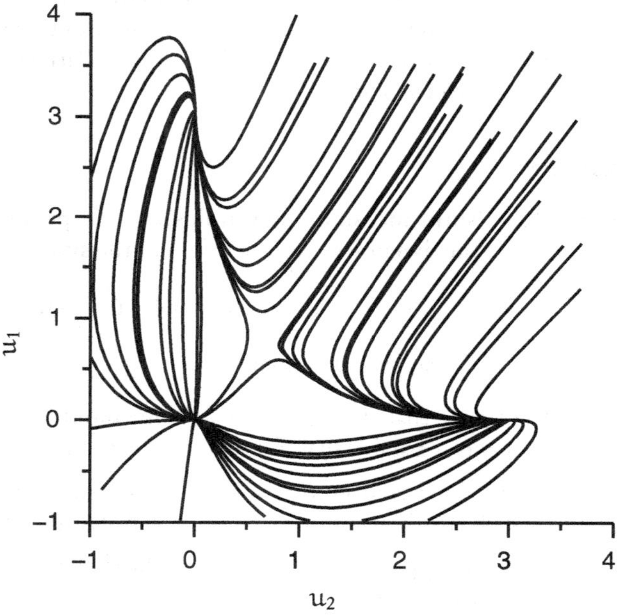

Figure 10.7. *Trajectories for Exercise* 10.3.

- Then seek a coordinate transform $y = Y + \hat{y}(X, Y)$ that simplifies the y ODE to its linearization $\dot{Y} = -2Y$, but only find a first approximation to $\hat{y}$.
- Based upon your local coordinate transform, sketch on Figure 10.6 some representative (X, Y) coordinate curves.

Answer: $x \approx X + \frac{1}{2}XY - \frac{1}{7}Y^3$, $y \approx Y + \frac{1}{4}X^2 + \frac{1}{2}X^2Y$.

Exercise 10.3 (graphical coordinates). Figure 10.7 shows trajectories of a dynamical system with four equilibria. Based upon these trajectories, at the equilibria near $(1,1)$, sketch on this figure some representative coordinate curves of a local coordinate system. The local coordinate system is to be one that you think would simplify the algebraic representation of the local dynamics to its local linearization.

Exercise 10.4 (resonance). Modify Algorithm 10.2 to find the normal form of the pair of ODEs

$$\dot{x} = 2x + a_x x^2 + b_x xy + c_x y^2 ,$$
$$\dot{y} = -y + a_y x^2 + b_y xy + c_y y^2$$

for some general coefficients a_x through to c_y. Find the coefficients in their normal form

$$\dot{X} \approx 2X + AX^2Y^2 \quad \text{and} \quad \dot{Y} \approx -Y + BXY^3 .$$

Exercise 10.5 (stable and unstable manifolds). Modify Algorithm 10.2 to find the normal-form transformation of the pair of ODEs

$$\dot{x} = 2x + x^2 + y^2 ,$$
$$\dot{y} = -3y + x^2 - y^2 .$$

Hence deduce asymptotic approximations to the shape of their stable and unstable manifolds.

Answer: $\dot{X} \approx 2X + \frac{1}{49}X^4Y^2$, $\dot{Y} \approx -3Y - \frac{4}{147}X^3Y^3$, $x = X + \frac{1}{2}X^2 - \frac{1}{8}Y^2 + \mathcal{O}(X^3 + Y^3)$, and $y = Y + \frac{1}{7}X^2 + \frac{1}{3}Y^2 + \mathcal{O}(X^3 + Y^3)$: the stable manifold is parametrically $(x, y) \approx (-\frac{1}{8}Y^2, Y + \frac{1}{3}Y^2)$, and the unstable manifold is $(x, y) \approx (X + \frac{1}{2}X^2, \frac{1}{7}X^2)$.

Chapter 11

Separating fast and slow dynamics proves modeling

Contents

We must never cease from exploration. And the end of our exploration will be to arrive where we began and to know the place for the first time.

T. S. Eliot

Chapter 10 introduced the idea that nonlinear coordinate transforms enhance our understanding of dynamics by simplifying the differential equations. This chapter uses such nonlinear coordinate transforms to illuminate vital modeling issues. We not only find the existence and emergence of center manifold models, we also find accurate initial conditions for such models and deduce that the modeling holds over finite domains.

This chapter focuses on autonomous dynamics. Nonautonomous and stochastic dynamics are considerably more complicated; Part VII explores their even more subtle modeling issues.

Example 11.1 (multinomial approximation). For example, suppose we aim to model the emergent dynamics of

$$\dot{x} = 0.1\,x - xy \quad \text{and} \quad \dot{y} = -y + x^2, \tag{11.1}$$

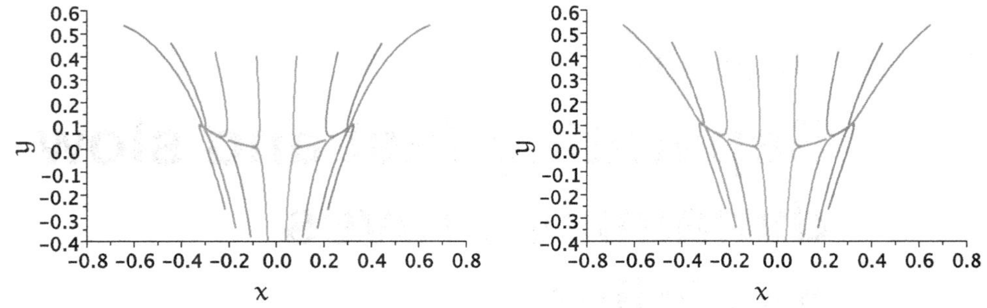

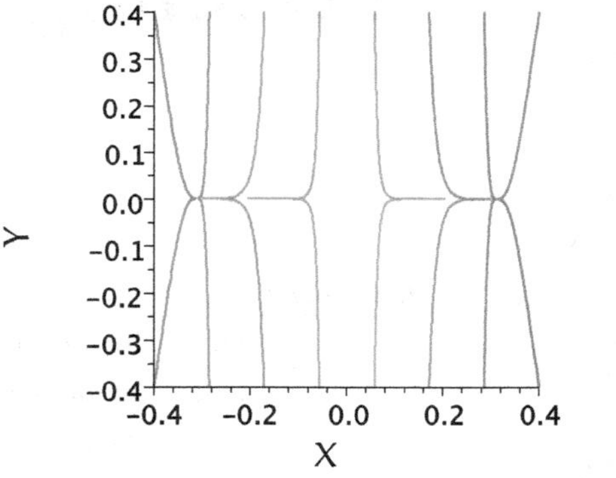

Figure 11.1. *To model the dynamics of* (11.1) *(trajectories top left), we might use the normal-form system* (11.3) *(trajectories bottom right), mapped into the original state space by the coordinate transform* (11.2) *(mapped trajectories top right). The original and the approximate systems agree well (top row) at least in this finite domain.*

whose trajectories are plotted in the top-left graph of Figure 11.1. We might use the multinomial coordinate transform

$$\begin{aligned} x &= X + XY + \tfrac{1}{2}XY^2 + \tfrac{1}{6}XY^3 + 2X^3Y, \\ y &= Y + 0.84\,X^2 - 2X^2Y^2 + 2X^4 \end{aligned} \tag{11.2}$$

to see how well the multinomial normal-form system

$$\dot{X} = 0.1\,X - 0.84\,X^3 - 2X^5 \quad\text{and}\quad \dot{Y} = (-1 + 2X^2 + 4X^4)Y \tag{11.3}$$

predicts the long-term evolution of the system (11.1). Figure 11.1 plots trajectories of (11.3) in the bottom right. Then the coordinate transform (11.2) maps these into the original xy-plane: see the top right of Figure 11.1. The two sets of trajectories in the xy-plane, the original and the approximate, agree well throughout the plotted domain.

Outside this plotted finite domain the agreement breaks down because not only is the normal-form system (11.3) approximate, but so is the mapping (11.2) between the systems. One of the aims of this chapter is to estimate the size of the domain of such agreement.

11.1 ▪ Introduce separation of slow, stable, and unstable dynamics

This section uses the simpler case of just a three-variable dynamical system to introduce the general transformational analysis of section 11.2. One new feature is that whereas previous chapters associated slow variables with eigenvalues precisely zero, here the slow variables are associated with modes of small eigenvalues (as in the example system (11.1)).

Catering for such small nonzero eigenvalues suits many applications, as in Chapter 9's modeling of the lateral velocity of a thin fluid film.

Readers happy to grapple immediately with the issues in the dynamics of an arbitrary number of variables could skip to section 11.2.

Consider a physical dynamical system, in dependent variables $\mathbf{u}(t) \in \mathbb{R}^3$, expressed as the autonomous differential equations $\dot{\mathbf{u}} = \mathcal{L}\mathbf{u} + \mathbf{f}(\mathbf{u})$, with an equilibrium at the origin and where the 3×3 matrix $\mathcal{L}$ has eigenvalues α (small), $-\beta$ (finite), and $+\gamma$ (finite). Suppose we have already used a linear change of basis to separate the physical system into variables $x(t)$, $y(t)$, and $z(t)$, where

$$\dot{x} = \alpha x + f(x,y,z), \quad \dot{y} = -\beta y + g(x,y,z), \quad \text{and } \dot{z} = +\gamma z + h(x,y,z). \tag{11.4}$$

We now seek to analyze the dynamics where x is the slow variable of long-term interest, y is the fast stable variable, and z is the fast unstable variable. Eventually we concentrate upon the case where the unstable z mode is not present, in which case we expect the slow x mode to dominate the emergent long-time dynamics. But considering unstable modes empowers analysis of boundary conditions to spatial modeling (Roberts, 1992a), so we retain unstable modes in this analysis.

Let's seek a near identity nonlinear coordinate change to new coordinates (X, Y, Z),[133]

$$x = x(X,Y,Z), \quad y = y(X,Y,Z), \quad \text{and } z = z(X,Y,Z). \tag{11.5}$$

In these new variables, the separated dynamical system (11.4) is to be approximated by the normal-form system

$$\begin{aligned}
\dot{X} &= \alpha X + F(X,Y,Z) = \alpha X + F_c(X) + F_h(X,Y,Z)YZ\,, \\
\dot{Y} &= [-\beta + G(X,Y,Z)]Y \quad \text{and } \dot{Z} = [+\gamma + H(X,Y,Z)]Z\,.
\end{aligned} \tag{11.6}$$

As foreshadowed in Chapter 10, in such a form we readily identify important qualitative features of nonlinear dynamics.

Definition 11.1. *Let* $\mathbb{O}(p)$ *denote terms* $\mathcal{O}(X^p + Y^p + Z^p)$ *as* $(X,Y,Z) \to 0$.

For example, $X^\ell Y^m Z^n = \mathbb{O}(p)$ if and only if $\ell + m + n \geqslant p$.

Assumption 11.2 (smooth, slow-fast dynamics). *In the dynamical system* (11.4), *for some order* $\mathfrak{p} \geqslant 2$ *assume the two rates* $\beta, \gamma > \mathfrak{p}|\alpha|$; *the three functions* f, g, *and* h *are* $C^{\mathfrak{p}}(\mathbb{d})$[134] *for some finite, connected, domain* $\mathbb{d}$ *containing the origin; and* f, g *and* h *are* $\mathbb{O}(2)$.

Lemma 11.3, below, establishes the existence of a useful coordinate transformation up to order $\mathfrak{p}$. Since the order $\mathfrak{p}$ is primarily limited by $\min(\beta,\gamma)/|\alpha|$, the order increases for smaller rate α. That is, the greater the relative spectral gap between slow and fast modes, the higher order the potential separation between slow and fast variables.

Lemma 11.3 (normal-form transforms exist). *For all orders* p, $2 \leqslant p \leqslant \mathfrak{p}$, *there exists a near identity multinomial coordinate transform* (11.5) *such that a multinomial normal-form system* (11.6), *together with* (11.5), *is equivalent to the separated system* (11.4) *to residuals* $\mathbb{O}(p)$.

---

[133]Throughout this discourse, the symbols x, y, and z have many meanings that I aim to be apparent from the context: unadorned they generally give coordinates in the xyz-space; with a single argument, such as $x(t)$, or an overdot they denote solutions of the system (11.4); with several arguments, such as $x(X,Y,Z)$, or with corresponding partial derivative operators, such as x_Y, they denote the coordinate transform (11.5).

[134]Recall that $C^{\mathfrak{p}}(\mathbb{d})$ denotes the set of smooth functions that are continuously differentiable at least $\mathfrak{p}$ times throughout the domain $\mathbb{d}$.

Proof. Prove by induction. First, the lemma is trivially true for $p = 2$ under the identity transform $x = X$, $y = Y$, and $z = Z$ with normal-form system $\dot{X} = \alpha X$, $\dot{Y} = -\beta Y$, and $\dot{Z} = +\gamma Z$. We only change this identity transform by higher-order, multinomial corrections and so in this sense they are all near identity.

Second, assume the lemma is true for some p, $2 \leqslant p < \mathfrak{p}$. Seek a correction $\mathbb{O}(p)$, indicated by hats, that results in a higher-order residual of the system (11.4), namely of $\mathbb{O}(p+1)$: let the corrected coordinate transform be denoted

$$\begin{aligned} x &= x(X,Y,Z) + \hat{x}(X,Y,Z), \\ y &= y(X,Y,Z) + \hat{y}(X,Y,Z), \\ z &= z(X,Y,Z) + \hat{z}(X,Y,Z). \end{aligned}$$

Correspondingly, let the corrected evolution in these coordinates be

$$\begin{aligned} \dot{X} &= \quad \alpha X + F(X,Y,Z) + \hat{F}(X,Y,Z), \\ \dot{Y} &= -\beta Y + G(X,Y,Z)Y + \hat{G}(X,Y,Z)Y, \\ \dot{Z} &= +\gamma Z + H(X,Y,Z)Z + \hat{H}(X,Y,Z)Z. \end{aligned}$$

Substitute these into the physical system (11.4) and first consider the x equation. In the right-hand side $f(x+\hat{x}, y+\hat{y}, z+\hat{z}) = f(x,y,z) + \mathbb{O}(p+1)$ by a multivariable Taylor's remainder theorem since $\hat{x}, \hat{y}, \hat{z} = \mathbb{O}(p)$ and derivatives $f_x, f_y, f_z = \mathbb{O}(1)$ by Assumption 11.2. But the time derivative on the left-hand side is more complicated, as by the chain rule the corrected time derivative

$$\begin{aligned} \dot{x} &= (x_X + \hat{x}_X)\dot{X} + (x_Y + \hat{x}_Y)\dot{Y} + (x_Z + \hat{x}_Z)\dot{Z} \\ &= (x_X + \hat{x}_X)(\alpha X + F + \hat{F}) + (x_Y + \hat{x}_Y)(-\beta + G + \hat{G})Y \\ &\quad + (x_Z + \hat{x}_Z)(\gamma + H + \hat{H})Y \\ &= x_X(\alpha X + F) + x_Y(-\beta + G)Y + x_Z(\gamma + H)Z + \underbrace{\hat{x}_X(\alpha X + F)}_{=\alpha X \hat{x}_X + \mathbb{O}(p+1)} \\ &\quad + \underbrace{x_X \hat{F}}_{=\hat{F} + \mathbb{O}(p+1)} + \underbrace{\hat{x}_X \hat{F}}_{\mathbb{O}(2p-1)} + \underbrace{\hat{x}_Y(-\beta + G)Y}_{=-\beta Y \hat{x}_Y + \mathbb{O}(p+1)} + \underbrace{x_Y \hat{G} Y}_{\mathbb{O}(p+2)} + \underbrace{\hat{x}_Y \hat{G} Y}_{\mathbb{O}(2p)} \\ &\quad + \underbrace{\hat{x}_Z(\gamma + H)Z}_{=\gamma Z \hat{x}_Z + \mathbb{O}(p+1)} + \underbrace{x_Z \hat{H} Z}_{\mathbb{O}(p+2)} + \underbrace{\hat{x}_Z \hat{H} Z}_{\mathbb{O}(2p)}. \end{aligned}$$

Equating the two sides with a little rearrangement, the x equation becomes

$$\begin{aligned} &- \alpha\hat{x} + \alpha X \hat{x}_X - \beta Y \hat{x}_Y + \gamma Z \hat{x}_Z + \hat{F} \\ &= \underbrace{\alpha x + f - x_X(\alpha X + F) - x_Y(-\beta + G)Y - x_Z(\gamma + H)Z}_{\text{Res}_{x,p}} + \mathbb{O}(p+1). \end{aligned} \tag{11.7}$$

Similarly, the y equation becomes

$$\begin{aligned} &+ \beta\hat{y} + \alpha X \hat{y}_X - \beta Y \hat{y}_Y + \gamma Z \hat{y}_Z + Y\hat{G} \\ &= \underbrace{-\beta y + g - y_X(\alpha X + F) - y_Y(-\beta + G)Y - y_Z(\gamma + H)Z}_{\text{Res}_{y,p}} + \mathbb{O}(p+1), \end{aligned} \tag{11.8}$$

and the z equation becomes

$$-\gamma\hat{z}+\alpha X\hat{z}_X-\beta Y\hat{Z}_Y+\gamma Z\hat{z}_Z+Z\hat{H}$$
$$=\underbrace{+\gamma z+h-z_X(\alpha X+F)-z_Y(-\beta+G)Y-z_Z(\gamma+H)Z}_{\text{Res}_{z,p}}+\mathbb{O}(p+1). \tag{11.9}$$

By Assumption 11.2 and the induction assumption, all three residuals, $\text{Res}_{x,p}$, $\text{Res}_{y,p}$, and $\text{Res}_{z,p}$, are $\mathbb{O}(p)$; hence there are coefficients $a_{\ell mn}$, $b_{\ell mn}$, and $c_{\ell mn}$ such that

$$(\text{Res}_{x,p},\text{Res}_{y,p},\text{Res}_{z,p})=\sum_{\ell+m+n=p}(a_{\ell mn},b_{\ell mn},c_{\ell mn})X^\ell Y^m Z^n+\mathbb{O}(p+1).$$

There are many possible ways to choose the coordinate transform updates $\hat{x}$, $\hat{y}$, and $\hat{z}$, and complementary evolution updates $\hat{F}$, $Y\hat{G}$, and $Z\hat{H}$, as equations (11.7)–(11.9) form an underdetermined system for the updates. It is up to our qualitative aims to decide what updates are desirable to implement among all the possibilities. For the arguments that follow let's choose the following updates: these updates form in some sense the minimal coordinate transform necessary to achieve our modeling aims. For any constant C_p, satisfy the $\mathbb{O}(p)$ terms in the x equation (11.7) by choosing

$$\hat{F}=\left[a_{p00}-(p-1)\alpha C_p\right]X^p+\sum_{\ell+m+n=p,\ m,n>0}a_{\ell mn}X^\ell Y^m Z^n,$$
$$\hat{x}=C_pX^p+\sum_{\ell+m=p,\ m>0}\frac{a_{\ell m0}X^\ell Y^m}{(\ell-1)\alpha-m\beta}+\sum_{\ell+n=p,\ n>0}\frac{a_{\ell 0n}X^\ell Z^n}{(\ell-1)\alpha+n\gamma}.$$

By Assumption 11.2 the divisors in these sums are never zero. For example,

$$\begin{aligned}(\ell-1)\alpha-m\beta &\leqslant(\ell-1)\alpha-\beta && \text{as } m\geqslant 1\\ &<(\ell-1)\alpha-p|\alpha| && \text{by Assumption 11.2}\\ &=-(p\pm(\ell-1))|\alpha| && \text{where } \pm=\text{sign}\,\alpha\\ &\leqslant 0 && \text{as } \ell\leqslant p.\end{aligned}$$

Similarly, the divisor $(\ell-1)\alpha+n\gamma>0$. Notice that the form of the update $\hat{F}$ is a sum of a term purely in X and terms with a common factor of YZ, as required by the normal form (11.6).

Moving on to the y-dynamics, for any constant C'_p, satisfy the $\mathbb{O}(p)$ terms in equation (11.8) by choosing

$$Y\hat{G}=-(p-1)\alpha C'_pX^{p-1}Y+\sum_{\ell+m+n=p,\ m>0}b_{\ell mn}X^\ell Y^m Z^n,$$
$$\hat{y}=C'_pX^{p-1}Y+\sum_{\ell+n=p}\frac{b_{\ell 0n}X^\ell Z^n}{\ell\alpha+\beta+n\gamma}.$$

The divisor appearing here is similarly never zero:

$$\begin{aligned}\ell\alpha+\beta+n\gamma &\geqslant\ell\alpha+\beta && \text{as } n\geqslant 0\\ &>\ell\alpha+p|\alpha| && \text{by Assumption 11.2}\\ &=(p\pm\ell)|\alpha| && \text{where } \pm=\text{sign}\,\alpha\\ &\geqslant 0 && \text{as } \ell\leqslant p.\end{aligned}$$

Last, in the z-dynamics, for any constant C''_p, satisfy the $\mathcal{O}(p)$ terms in equation (11.9) by choosing

$$Z\hat{H} = -(p-1)\alpha C''_p X^{p-1}Z + \sum_{\ell+m+n=p,\ n>0} c_{\ell mn}X^\ell Y^m Z^n,$$

$$\hat{z} = C''_p X^{p-1}Z + \sum_{\ell+m=p} \frac{c_{\ell m0}X^\ell Y^m}{\ell\alpha - m\beta - \gamma},$$

where, similarly to the previous case, these divisors are never zero. The above choices ensure all the pth-order terms in the residuals are canceled, and also ensures we maintain the normal form (11.6).

With these $\mathcal{O}(p)$ updates, we obtain a multinomial coordinate transform (indeed a family of transforms parametrized by C_p, C'_p, and C''_p) and corresponding multinomial normal form with residual $\mathcal{O}(p+1)$. Induction then proves the lemma. □

Example 11.2 (continue Example 11.1). The system (11.1) has an equilibrium at the origin. Linearized about the origin, $\dot{x} \approx 0.1x$ and $\dot{y} = -y$, the system has one positive eigenvalue and one negative. Two possibilities follow.

- We could let x in the system be a fast unstable variable and separate the stable and unstable dynamics, with corresponding rates $\beta = 1$ and $\gamma = 0.1$. However, in the constructed multinomials this would entail divisions by powers of the small eigenvalue 0.1, which generate large coefficients and a small domain of applicability of the multinomial approximations.
- Instead, we treat x as a slow variable, by setting rate $\alpha = 0.1$, and then y is a fast stable variable with rate $\beta = 1$. Then the order of construction of a *smooth* multinomial approximation and map is limited to ninth order as $p < 10 = \beta/\alpha$. By avoiding small divisors, such multinomial transforms should be valid over a useful domain, such as that plotted in Figure 11.1.

Example 11.3. Potzsche and Rasmussen (2006, Ex. 5.4) sought traveling wave solutions, $u(x-ct)$ with wave speed c, of the Korteweg–de Vries equation, $u_t + u_{xxx} - 6uu_x = 0$ (although they explored varying coefficients, here we choose simpler constant coefficients). Our normal form captures in the slow modes the spatial structure of such traveling waves. Seeking solutions $u(x-ct)$, letting $u_1 = u$, $u_2 = u'$, and $u_3 = u''$, where primes denote derivatives of the spatial structure, leads to the system

$$u_1' = u_2\,, \quad u_2' = u_3\,, \quad u_3' = cu_2 + 6u_1u_2\,.$$

For simplicity suppose the wave speed $c = 1$ (which it is not usually). A transform to diagonalize the linear part into slow variable x, stable y, and unstable z is that $u_1 = x+y+z$, $u_2 = z-y$, and $u_3 = z+y$. This coordinate transform leads to the system

$$\begin{aligned} x' &= \quad -6(x+y+z)(z-y),\\ y' &= -y + 3(x+y+z)(z-y),\\ z' &= +z + 3(x+y+z)(z-y). \end{aligned}$$

This system is in the form of (11.4) with slow rate $\alpha = 0$, fast rates $\beta = \gamma = 1$, and infinitely differentiable nonlinearity. Lemma 11.3 then assures us that approximate coordinate transforms exist to separate the slow, stable, and unstable dynamics. Either by hand, or via my current web service (Roberts, 2009d), we find that the coordinate transform

$$x = X - 3(2XY + 2XZ + Y^2 + Z^2),$$
$$y = Y + \tfrac{3}{2}XZ + 3Y^2 + Z^2,$$
$$z = Z + \tfrac{3}{2}XY + Y^2 + 3Z^2,$$

together with the new variable "evolution" of the spatial structure

$$X' = 0, \quad Y' = -(1+3X)Y, \quad Z' = +(1+3X)Z,$$

reproduces the original system to errors $\mathcal{O}(3)$. In spatial structures that exist for all x, there cannot be any Z component as these are unbounded as $x \to +\infty$, and there cannot be any Y component as these are unbounded as $x \to -\infty$. The only bounded solutions for all space lie in the slow manifold and so must here obey $X' = 0$: although here boringly constant, in other situations such as $c \neq 1$, the slow manifold is nontrivial and captures interesting solitary waves.

11.2 ▪ Iteratively construct a useful coordinate transform

This section generalizes the analysis of section 11.1 to the coupled dynamics of many variables. This section shows we can generally separate the nonlinear dynamics of the center, stable, and unstable variables. The multivariable coordinate transforms constructed in this section apply to the dynamics of any sufficiently smooth system of ODEs in some domain around each equilibrium. Section 11.3 and Chapter 12 show one way to use these transformations in establishing the existence and emergence properties of reduced-dimensional, relatively simple, long-term models.

Start by distinguishing between various potential variables depending upon the corresponding linearized dynamics (consistent with previous usage).

Definition 11.4 (variable names). *Consider the autonomous dynamical system* $\dot{u} = f(u)$ *with equilibrium* u^*. *Suppose a linear change of basis,* $u = u^* + P(v, w)$, *transforms the linearized dynamics about* u^* *to the* separated *form* $\dot{v} = Av$ *and* $\dot{w} = Bw$ *for some matrices* A *and* B. *If* all *the eigenvalues of matrix* A *and* none *of matrix* B

- *have (significant) positive real part, then* v *are called the* **unstable variables;**
- *have (significant) negative real part, then* v *are called the* **stable variables;**
- *have (near) zero real part, then* v *are called the* **center variables;**
- *have no (near) zero real part, then* v *are called the* **hyperbolic variables;**
- *are (near) zero, then* v *are called the* **slow variables;**
- *are not (near) zero, then* v *are called the* **fast variables.**

In this definition the concepts of "significant" and "near" are a subjective decision based upon the modeling context: purists could ignore the "significant" and "near" qualification. Which numbers are "near" zero or not "near" zero primarily depends upon two aspects: upon the time scales on which we choose to use a model and upon whether the dynamical system has a spectral gap to suit. Assumption 11.6, below, quantifies nearness in terms of potential spectral gaps in the sets of eigenvalues (see Figure 11.2 below).

Definition 11.5. *Consistent with Definition* 11.1, *let* $\mathbb{O}(p)$ *denote terms* $\mathcal{O}\big(|X|^p + |Y|^p + |Z|^p\big)$ *as* $(X, Y, Z) \to 0$.

Generalizing section 11.1 to arbitrary finite dimensions, consider a physical dynamical system expressed in dependent variables $u(t) \in \mathbb{R}^{m+n+n'}$ as the autonomous differential equations $\dot{u} = \mathcal{L}u + f(u)$, with an equilibrium at the origin ($u^* = 0$) and where the matrix $\mathcal{L}$ has eigenvalues λ_j, $j = 1, \ldots, m + n + n'$, repeated according to multiplicity. Suppose we have used a linear change of basis to linearly separate this physical system into m center variables $x(t) \in \mathbb{R}^m$, n stable variables $y(t) \in \mathbb{R}^n$, and n' unstable variables $z(t) \in \mathbb{R}^{n'}$. Then the physical system is described by the linearly separated but nonlinearly coupled ODEs

$$\dot{x} = Ax + f(x, y, z), \quad \dot{y} = By + g(x, y, z), \quad \text{and } \dot{z} = Cz + h(x, y, z). \tag{11.10}$$

Assumption 11.6 (spectral trichotomy). *The matrices in the separated system* (11.10) *are assumed to satisfy the following:*

- $A \in \mathbb{R}^{m \times m}$ *with eigenvalues* $\alpha_1, \ldots, \alpha_m$, *possibly complex, with real part of magnitude* $|\Re\alpha_j| \leqslant \alpha$*;*
- $B \in \mathbb{R}^{n \times n}$ *with eigenvalues* $\beta_1, \ldots, \beta_n$, *possibly complex, with real part* $\Re\beta_j \leqslant -\beta$*;*
- $C \in \mathbb{R}^{n' \times n'}$ *with eigenvalues* $\gamma_1, \ldots, \gamma_{n'}$, *possibly complex, with real part* $\Re\gamma_j \geqslant \gamma$*;*
- *for simplicity, section* 11.2 *assumes the matrices* A, B, *and* C *have been diagonalized (an assumption not imposed in subsequent sections);*
- *an extension of subsequent arguments also covers the important cases when matrices* A, B, *and* C *are triangular, such as in a Jordan form due to repeated eigenvalues, rather than just diagonalized.*

Figure 11.2 illustrates a suitable spectrum of such eigenvalues, sometimes called an *exponential trichotomy*. As usual, at the equilibrium the strictly nonlinear functions f, g, and h are $\mathbb{O}(2)$. We aim to construct a smooth coordinate transform to separate the nonlinear dynamics into the same center, stable, and unstable dynamics—at least as far as possible.

A significant change in symbology from that in section 11.1 is that here the unsubscripted symbols α, β, and γ no longer denote eigenvalues. Instead, as shown in Figure 11.2, the unsubscripted symbols α, β, and γ denote bounds on the real parts of the eigenvalues associated with the center variables, stable variables, and unstable variables, respectively. In section 11.2, for some parameter $p \geqslant 2$, we separate the spectrum of eigenvalues into three disjoint subsets, each subset separated by a "gap":

$$\begin{aligned}
&|\Re\lambda_j| \leqslant \alpha && \text{for the center variables,}\\
&\Re\beta_j \leqslant -\beta < -p\alpha && \text{for the stable variables,}\\
&\Re\gamma_j \geqslant \gamma > p\alpha && \text{for the unstable variables.}
\end{aligned}$$

The maximum possible value of parameter p characterizes the size of the spectral gap separating the center variables from the hyperbolic stable and unstable variables; indeed $p < \min(\beta, \gamma)/\alpha$. Assumption 11.7 may also constrain parameter p by the smoothness of the nonlinear functions in the ODEs (11.10).

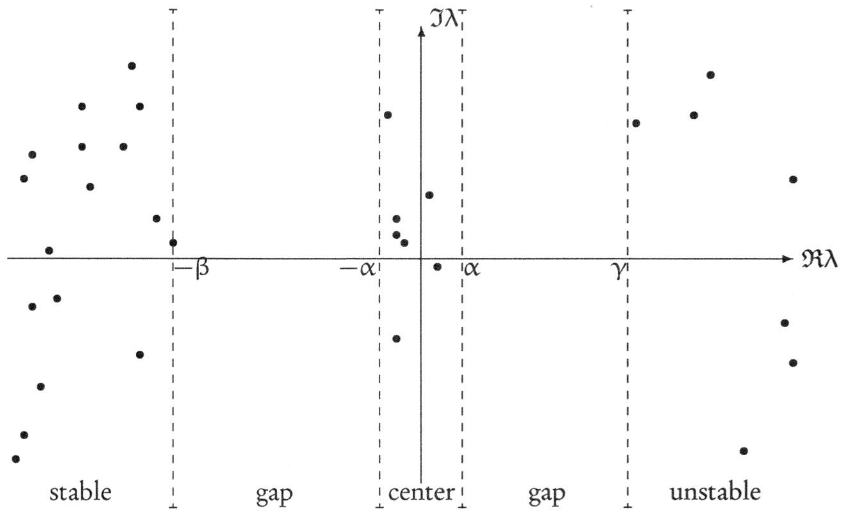

Figure 11.2. *Schematic picture of the complex plane of eigenvalues λ_j (discs) forming three separate sets characterized by bounding parameters α, β, and γ of the center, stable, and unstable modes, respectively.*

Assumption 11.7 (smooth nonlinearity). *The nonlinear functions appearing in the separated system* (11.10) *satisfy that*

- $\mathsf{f}, \mathsf{g}, \mathsf{h} \in C^{\mathsf{p}}(\mathsf{d})$ *(having continuous* p*-derivatives) for some open connected domain* d *about the equilibrium at the origin, and*
- $\mathsf{f}, \mathsf{g}, \mathsf{h}$ *are* $\mathcal{O}(2)$.

Let's proceed by seeking a near identity nonlinear coordinate change to new coordinates $(\mathsf{X}, \mathsf{Y}, \mathsf{Z})$,[135]

$$\mathsf{x} = \mathsf{x}(\mathsf{X}, \mathsf{Y}, \mathsf{Z}), \quad \mathsf{y} = \mathsf{y}(\mathsf{X}, \mathsf{Y}, \mathsf{Z}), \quad \text{and } \mathsf{z} = \mathsf{z}(\mathsf{X}, \mathsf{Y}, \mathsf{Z}). \tag{11.11a}$$

In these new variables, the separated dynamical system (11.10) is to be approximated by the normal-form system

$$\dot{\mathsf{X}} = A\mathsf{X} + \mathsf{F}(\mathsf{X}, \mathsf{Y}, \mathsf{Z}), \quad \dot{\mathsf{Y}} = B\mathsf{Y} + \mathsf{G}(\mathsf{X}, \mathsf{Y}, \mathsf{Z}), \quad \dot{\mathsf{Z}} = C\mathsf{Z} + \mathsf{H}(\mathsf{X}, \mathsf{Y}, \mathsf{Z}), \tag{11.11b}$$

such that, and crucially,

$$\mathsf{F}(\mathsf{X}, 0, \mathsf{Z}) = \mathsf{F}(\mathsf{X}, \mathsf{Y}, 0) = \mathsf{F}_c(\mathsf{X}), \quad \mathsf{G}(\mathsf{X}, 0, \mathsf{Z}) = 0, \quad \mathsf{H}(\mathsf{X}, \mathsf{Y}, 0) = 0, \tag{11.11c}$$

and hence there exist matrix functions G and H such that $\mathsf{G} = G(\mathsf{X}, \mathsf{Y}, \mathsf{Z})\mathsf{Y}$ and $\mathsf{H} = H(\mathsf{X}, \mathsf{Y}, \mathsf{Z})\mathsf{Z}$. As foreshadowed in Chapter 10, section 11.3 and Chapter 12 use the normal form (11.11) to readily identify important qualitative features of the nonlinear dynamics.

The next lemma constructively establishes the existence of a useful coordinate transformation up to order p.

[135] Similar to the previous section, the symbols x, y, and z have many meanings dependent upon the context: unadorned they generally give coordinates in the u state space; with a single argument, such as $\mathsf{x}(t)$, or an overdot they denote solutions of the system (11.10); with several arguments, such as $\mathsf{x}(\mathsf{X}, \mathsf{Y}, \mathsf{Z})$, or with corresponding partial derivative operators, such as $\partial\mathsf{x}/\partial\mathsf{Y}$, they denote the coordinate transform (11.11a).

Lemma 11.8 (normal-form transforms exist). *For all orders* p, $2 \leqslant p \leqslant \mathfrak{p}$, *there exists a near identity multinomial coordinate transform* (11.11a) *such that a multinomial normal-form system* (11.11b), *together with* (11.11a) *and* (11.11c), *is equivalent to the linearly separated system* (11.10) *to residuals* $\mathcal{O}(p)$.

Proof. Prove by a constructive induction. First, the lemma is trivially true for $p = 2$ under the identity transform $x = X$, $y = Y$, and $z = Z$ with normal-form system $\dot{X} = AX$, $\dot{Y} = BY$, and $\dot{Z} = CZ$. We only change this identity transform by higher-order, multinomial corrections and so in this sense they are all near identity.

Second, assume the lemma is true for some p, $2 \leqslant p \leqslant \mathfrak{p}-1$. Seek a correction $\mathcal{O}(p)$, indicated by hats, that results in a higher-order residual, namely of $\mathcal{O}(p+1)$; let the corrected coordinate transform be

$$\begin{aligned} x &= x(X,Y,Z) + \hat{x}(X,Y,Z), \\ y &= y(X,Y,Z) + \hat{y}(X,Y,Z), \\ z &= z(X,Y,Z) + \hat{z}(X,Y,Z), \end{aligned} \tag{11.12}$$

where $\hat{x}$, $\hat{y}$, and $\hat{z}$ are multinomial in X, Y, and Z. Correspondingly, also find corrections to the updates to the evolution, say the multinomial $\hat{F}$, $\hat{G}$, and $\hat{H}$:

$$\dot{X} = AX + F(X,Y,Z) + \hat{F}(X,Y,Z), \tag{11.13a}$$
$$\dot{Y} = BY + G(X,Y,Z) + \hat{G}(X,Y,Z), \tag{11.13b}$$
$$\dot{Z} = CZ + H(X,Y,Z) + \hat{H}(X,Y,Z). \tag{11.13c}$$

The homological equation guides stable corrections Substitute the corrected transform (11.12) into the governing ODE (11.10) for the stable variables; by the chain rule

$$\begin{aligned} &\frac{\partial y}{\partial X}\dot{X} + \frac{\partial \hat{y}}{\partial X}\dot{X} + \frac{\partial y}{\partial Y}\dot{Y} + \frac{\partial \hat{y}}{\partial Y}\dot{Y} + \frac{\partial y}{\partial Z}\dot{Z} + \frac{\partial \hat{y}}{\partial Z}\dot{Z} \\ &= By + B\hat{y} + g(x+\hat{x}, y+\hat{y}, z+\hat{z}). \end{aligned}$$

Then into this equation substitute the corrected evolution (11.13)—as always the time derivative of out-of-equilibrium dynamics is voluminous:

$$\begin{aligned} &\frac{\partial y}{\partial X}(AX + F + \hat{F}) + \frac{\partial y}{\partial Y}(BY + G + \hat{G}) + \frac{\partial y}{\partial Z}(CZ + H + \hat{H}) \\ &+ \frac{\partial \hat{y}}{\partial X}(AX + F + \hat{F}) + \frac{\partial \hat{y}}{\partial Y}(BY + G + \hat{G}) + \frac{\partial \hat{y}}{\partial Z}(CZ + H + \hat{H}) \\ &= By + B\hat{y} + g(x+\hat{x}, y+\hat{y}, z+\hat{z}). \end{aligned}$$

Now omit products of small corrections as being negligible, and approximate coefficients of small corrections by their leading-order term (e.g., $\partial y/\partial X \approx 0$ and $\partial y/\partial Y \approx I$):

$$\begin{aligned} &\hat{G} - B\hat{y} + \frac{\partial \hat{y}}{\partial X}AX + \frac{\partial \hat{y}}{\partial Y}BY + \frac{\partial \hat{y}}{\partial Z}CZ \\ &= \underbrace{-\frac{\partial y}{\partial X}(AX + F) - \frac{\partial y}{\partial Y}(BY + G) - \frac{\partial y}{\partial Z}(CZ + H)}_{-dy/dt} + By + g(x,y,z). \end{aligned}$$

Then the equation for the ℓth component of the correction to the transform of the stable variable and the new stable dynamics is the homological equation

$$\hat{G}_\ell - \beta_\ell \hat{y}_\ell + \sum_{i=1}^{m} \alpha_i X_i \frac{\partial \hat{y}_\ell}{\partial X_i} + \sum_{j=1}^{n} \beta_j Y_j \frac{\partial \hat{y}_\ell}{\partial Y_j} + \sum_{k=1}^{n'} \gamma_k Z_k \frac{\partial \hat{y}_\ell}{\partial Z_k} = \text{Res}_{(11.10),\ell}\,, \tag{11.14}$$

where $\text{Res}_{(11.10),\ell}$ denotes the current residual of the ℓth component of the stable ODE in the system (11.10) (the right-hand side of the previous equation).

We discover what sort of terms may be assigned to the transformation $\mathbf{y}(\mathbf{X},\mathbf{Y},\mathbf{Z})$ and what terms remain in the $\mathbf{Y}$-evolution by considering the possibilities for the right-hand side residual. The transform is constructed as a multivariate asymptotic expansion about the origin in $(\mathbf{X},\mathbf{Y},\mathbf{Z})$-space so we put the right-hand side residual in the form of a multinomial in $\mathbf{X}$, $\mathbf{Y}$, and $\mathbf{Z}$. Since the nonlinearity function $\mathbf{g} \in C^p(d)$ we expand it in a multivariable Taylor polynomial with remainder $\mathcal{O}(p)$ in $\mathbf{X}$, $\mathbf{Y}$, and $\mathbf{Z}$. Suppose the right-hand side, the residual $\text{Res}_{(11.10),\ell}$, has, potentially among many others, a term in the multinomial of form

$$c\mathbf{X}^{\mathbf{p}}\mathbf{Y}^{\mathbf{q}}\mathbf{Z}^{\mathbf{r}} = c\prod_{i=1}^{m} X_i^{p_i} \prod_{j=1}^{n} Y_j^{q_j} \prod_{k=1}^{n'} Z_k^{r_k}$$

for some vectors of (real) exponents $\mathbf{p}$, $\mathbf{q}$, and $\mathbf{r}$, almost always integers, and some coefficient c. The exponents sum to the order p of interest ($p = \sum_{i=1}^{m} p_i + \sum_{j=1}^{n} q_j + \sum_{k=1}^{n'} r_k$). Because of the special form of the homological operator on the left-hand side of (11.14), seek contributions to the corrections of $\hat{G}_\ell = \mathfrak{g}\mathbf{X}^{\mathbf{p}}\mathbf{Y}^{\mathbf{q}}\mathbf{Z}^{\mathbf{r}}$ and $\hat{y}_\ell = \mathfrak{y}\mathbf{X}^{\mathbf{p}}\mathbf{Y}^{\mathbf{q}}\mathbf{Z}^{\mathbf{r}}$. Then this component of (11.14) becomes

$$\mathfrak{g} + \mu\mathfrak{y} = c, \quad \text{where rate } \mu = -\beta_\ell + \sum_{i=1}^{m} p_i\alpha_i + \sum_{j=1}^{n} q_j\beta_j + \sum_{k=1}^{n'} r_k\gamma_k\,. \tag{11.15}$$

Two cases arise depending upon the real part of the rate μ.

1. In the resonant case $|\Re\mu| \leqslant \alpha$, we set $\mathfrak{g} = c$ and $\mathfrak{y} = 0$; that is, assign $c\mathbf{X}^{\mathbf{p}}\mathbf{Y}^{\mathbf{q}}\mathbf{Z}^{\mathbf{r}}$ to the $\mathbf{Y}$-evolution and nothing to the coordinate transform $\mathbf{y}$.

   Since $\Re\beta_\ell$ is negative ($\leqslant -\beta$), this case of $|\Re\mu| \leqslant \alpha$ could potentially occur with exponents $\mathbf{q} = \mathbf{0}$, but let's see that this potential case only occurs at a high enough order to be excluded by Lemma 11.8. The lowest-order possibility is when exponents $\mathbf{r} = \mathbf{0}$ for the unstable variables. Then

$$\begin{aligned}\Re\mu &= -\Re\beta_\ell + \sum_{i=1}^{m} p_i\Re\alpha_i \\ &\geqslant \beta - \alpha\sum_{i=1}^{m} p_i = \beta - \alpha p \\ &> \beta - \alpha(\beta/\alpha - 1) = \alpha\,.\end{aligned}$$

   Instead, the case $|\Re\mu| \leqslant \alpha$ can only arise when at least one of the exponents $\mathbf{q}$ of $\mathbf{Y}$ is positive. Hence, there will be at least one Y_j factor in updates $\hat{\mathbf{G}}$ to the $\mathbf{Y}$-evolution, and so we maintain that $\mathbf{G} = \mathbf{0}$ when $\mathbf{Y} = \mathbf{0}$.

2. When $|\Re\mu| > \alpha$, an acceptable solution of (11.15) is to place all the forcing into the coordinate transform, $\mathfrak{y} = c/\mu$ (we never divide by zero as $\Re\mu \neq 0$), and not change the $\mathbf{Y}$-evolution, $\mathfrak{g} = 0$.

When the residual on the right-hand side of the homological equation (11.14) is a sum of such multinomial terms, then we sum the corrections, found as above, for all those terms $\mathbb{O}(p)$. Consequently, we always find a coordinate transform, to any order $p < \mathfrak{p}$, that maintains that $\mathbf{G} = 0$ when $\mathbf{Y} = 0$, as required by condition (11.11c).

Off-diagonal neglect delays the induction So far the argument requires the matrices in system (11.10) to be diagonalized. But recall that the fourth part of Assumption 11.6 commented that triangular matrices were sufficient. Let's see that the argument still applies for triangular matrices. For example, if matrix A is upper triangular, $A_{ij} \neq 0$ only for $j \geqslant i$, then the homological equation (11.14) has extra terms. When these terms are put on the right-hand side, (11.14) becomes

$$\begin{aligned}\hat{G}_\ell - \beta_\ell \hat{y}_\ell + \sum_{i=1}^{m} \alpha_i X_i \frac{\partial \hat{y}_\ell}{\partial X_i} + \sum_{j=1}^{n} \beta_j Y_j \frac{\partial \hat{y}_\ell}{\partial Y_j} + \sum_{k=1}^{n'} \gamma_k Z_k \frac{\partial \hat{y}_\ell}{\partial Z_k} \\ = \operatorname{Res}_{(11.10),\ell} - \sum_{i=1}^{m} \sum_{j=i+1}^{m} A_{i,j} X_j \frac{\partial \hat{y}_\ell}{\partial X_i} .\end{aligned} \tag{11.16}$$

Then corrections, as described above, for stable variable $\hat{y}_\ell$ involving center variable X_i generate extra terms in the residual on the right-hand side that are still of order p and hence appear to confound the induction and algorithm. However, for upper triangular matrix A, the off-diagonal terms only delay, not confound. Such extra terms on the right-hand side are of one lower order in variable X_i and one higher order in variable X_j, crucially always for $j > i$. Thus successive iterations of analysis at order p eventually correct all expressions involving variable X_i while simultaneously "sweeping" extra generated terms to higher indexed center variables, which are in turn corrected eventually. "Eventually" is achieved in a finite number of iterations and similarly for the hyperbolic variables $\mathbf{Y}$ and $\mathbf{Z}$, provided the matrices B and C are triangular, such as a Jordan form due to repeated eigenvalues. Consequently, discussing the homological equation (11.14) for corrections is sufficient.[136] Analogous comments apply to the center and the unstable dynamics, to which we now turn.

Transform the unstable dynamics By symmetry, the analysis of the coordinate transform and evolution for the unstable variables is the same as that for the stable variables. One just swaps the role of the symbols relating to the stable and unstable variables.

The homological equation guides center corrections For the center dynamics, each iteration toward constructing a coordinate transform substitutes corrections to the transform (11.12) and the evolution (11.13) into the governing ODE (11.10) for the center variables. Then, as for the stable equation, omit products of small corrections, and approximate coefficients of the remaining small corrections by their leading-order term.[137]

[136] Such a simple approach should also work in principle for many infinite-dimensional systems provided that each Jordan block is of finite dimension.

[137] Also neglect off-diagonal components of the matrices at the expense of possibly increasing the number of required induction steps and iterations.

Analogous to the stable dynamics, the equation for the ℓth component of the correction to the transform of the center variable is then the homological equation

$$\hat{F}_\ell - \alpha_\ell X_\ell + \sum_{i=1}^{m} \alpha_i X_i \frac{\partial \hat{x}_\ell}{\partial X_i} + \sum_{j=1}^{n} \beta_j Y_j \frac{\partial \hat{x}_\ell}{\partial Y_j} + \sum_{k=1}^{n'} \gamma_k Z_k \frac{\partial \hat{x}_\ell}{\partial Z_k} = \text{Res}_{(11.10),\ell}\,, \qquad (11.17)$$

where $\text{Res}_{(11.10),\ell}$ denotes the residual of the ℓth component of the center ODE of the system (11.10) evaluated at the current approximation. The crucial difference with the previous discussion of the stable variables is that the left-hand side of (11.17) has the term $-\alpha_\ell X_\ell$ instead of the $-\beta_\ell \hat{y}_\ell$ term.

Consider the range of possibilities for the right-hand side of the homological equation (11.17). In general, the multinomial right-hand side residual $\text{Res}_{11.10,\ell}$ is a sum of multinomial terms of the form $c\mathbf{X}^{\mathbf{p}}\mathbf{Y}^{\mathbf{q}}\mathbf{Z}^{\mathbf{r}}$ for some vectors of integer exponents $\mathbf{p}$, $\mathbf{q}$, and $\mathbf{r}$, and some coefficient c. Because of the special form of the homological operator on the left-hand side of (11.17), seek corresponding corrections $\hat{F}_\ell = \mathfrak{f}\mathbf{X}^{\mathbf{p}}\mathbf{Y}^{\mathbf{q}}\mathbf{Z}^{\mathbf{r}}$ and $\hat{x}_\ell = \mathfrak{x}\mathbf{X}^{\mathbf{p}}\mathbf{Y}^{\mathbf{q}}\mathbf{Z}^{\mathbf{r}}$. Then (11.17) becomes

$$\mathfrak{f} + \mu\mathfrak{x} = c, \quad \text{where rate} \quad \mu = -\alpha_\ell + \sum_{i=1}^{m} p_i\alpha_i + \sum_{j=1}^{n} q_j\beta_j + \sum_{k=1}^{n'} r_k\gamma_k\,. \qquad (11.18)$$

Two cases typically arise depending upon the real part of the rate μ.

1. In the resonant case $|\Re\mu| \leqslant \alpha$ we set $\mathfrak{f} = c$ and $\mathfrak{x} = 0$; that is, the multinomial term corrects the evolution in the new variables. This case only arises when either there are purely center variables in the multinomial term or there is at least one each of stable and unstable variables. For example, this case cannot arise with exponents $\mathbf{q} \neq 0$ and $\mathbf{r} = 0$. Suppose the exponents $\mathbf{q} \neq 0$ and $\mathbf{r} = 0$; then

$$\begin{aligned} \Re\mu &= -\alpha_\ell + \sum_{i=1}^{m} p_i\alpha_i + \sum_{j=1}^{n} q_j\beta_j \\ &\leqslant \alpha\left(1 + \sum_{i=1}^{m} p_i\right) - \beta\sum_{j=1}^{n} q_j \\ &\leqslant \alpha \times p - \beta \times 1 \\ &\leqslant \alpha(p-1) - \beta \\ &< \alpha(\beta/\alpha - 1) - \beta = -\alpha\,. \end{aligned}$$

   Thus for a stable variable to be forced into the evolution of the new center variables $\mathbf{X}$, it must occur in a term that also has at least one unstable variable. Consequently, we are assured that $\mathbf{F}$ is purely a function of $\mathbf{X}$ when either $\mathbf{Y} = 0$ or $\mathbf{Z} = 0$, as required by condition (11.11c).

2. In the remaining case when $|\Re\mu| > \alpha$, we assign $\mathfrak{x} = c/\mu$ and do not change the $\mathbf{X}$-evolution, $\mathfrak{f} = 0$.

   However, in these cases when a multinomial term is purely a function of $\mathbf{X}$, that is, exponents $\mathbf{q} = 0$ and $\mathbf{r} = 0$, then we may still prefer to assign the term to the evolution. It is a subjective decision: you have to weigh the complexity cost of the algebraic expression of the center variables' evolution against modeling considerations like the following. In applications, the eigenvalues α_i of the center variables typically vary with interesting parameters, such as a bifurcation parameter. In such applications, terms that are resonant for one parameter value may often be nonresonant for another parameter value. To avoid the "shifting sands" of differing coordinate transforms and differing models for closely related but slightly different parameter values, we often choose to include more pure $\mathbf{X}$ terms in the evolution of the center variables than are necessary. Such inclusion does no harm to the induction proof.

The same considerations arise for each multinomial term that is of $\mathbb{O}(p)$, but not $\mathbb{O}(p+1)$, in the residual. Consequently, we are always able to find a coordinate transform with residual $\mathbb{O}(p+1)$ that maintains a center $\mathbf{X}$-evolution satisfying condition (11.11c).

Induction With such $\mathbb{O}(p)$ updates, we obtain a multinomial coordinate transform (indeed, typically with some flexibility) and corresponding multinomial normal form that agree to a residual $\mathbb{O}(p+1)$ with the original system (11.10). Induction then proves the lemma. □

Before section 11.3 explores the usefulness of Lemma 11.8 and its constructive proof, the following example illustrates some of the properties of the induction. Section 11.3 subsequently develops this example further.

Example 11.4 (spatial structures). Consider the spatial structures of boundary value problems involving the spatial ODE $u_{xx} - u_{xxxx} + u^2 = 0$ on some spatial domain. To avoid confusion with center variables, rewrite the independent spatial variable as time t, and rewrite the fourth-order ODE as the system of four coupled ODEs for $\mathbf{u} = (u, \dot{u}, \ddot{u}, \dddot{u})$ as

$$\dot{\mathbf{u}} = \mathcal{L}\mathbf{u} + \mathbf{f}(\mathbf{u}) = \begin{bmatrix} 0 & 1 & 0 & 0 \\ 0 & 0 & 1 & 0 \\ 0 & 0 & 0 & 1 \\ 0 & 0 & 1 & 0 \end{bmatrix} \mathbf{u} + \begin{bmatrix} 0 \\ 0 \\ 0 \\ u^2 \end{bmatrix}. \tag{11.19}$$

- There is an equilibrium at the origin upon which we base analysis.
- The eigenvalues of the linearization are $\alpha_1 = \alpha_2 = 0$, $\beta_1 = -1$, and $\gamma_1 = 1$. These eigenvalues indicate that the spatial evolution has two slow modes, one stable mode and one unstable mode: within the context of a boundary value problem expect that the right-hand boundary conditions will control the unstable mode via a boundary layer on the right, whereas the left-hand boundary conditions will control the stable mode via a boundary layer on the left.
- Attempting to diagonalize the matrix $\mathcal{L}$ we find there is a generalized eigenvector and so find the Jordan form using the transform

$$\mathbf{u} = \underbrace{\begin{bmatrix} 1 & 0 & -1 & 1 \\ 0 & 1 & 1 & 1 \\ 0 & 0 & -1 & 1 \\ 0 & 0 & 1 & 1 \end{bmatrix}}_{=P} \begin{bmatrix} x_1 \\ x_2 \\ y \\ z \end{bmatrix}, \quad \text{whence } P^{-1}\mathcal{L}P = \begin{bmatrix} 0 & 1 & & \\ 0 & 0 & & \\ & & -1 & \\ & & & 1 \end{bmatrix}.$$

Hence the system (11.19) is the same as

$$\begin{aligned}\dot{x} &= \begin{bmatrix}0 & 1\\0 & 0\end{bmatrix}x + \begin{bmatrix}0\\-(x_1-y+z)^2\end{bmatrix},\\ \dot{y} &= -y + \tfrac{1}{2}(x_1-y+z)^2,\\ \dot{z} &= +z + \tfrac{1}{2}(x_1-y+z)^2.\end{aligned} \tag{11.20}$$

- Analysis to residuals $\mathcal{O}(3)$, perhaps via my current web service (Roberts, 2009d), recommends the coordinate transform

$$\begin{aligned}x_1 &= X_1 + \left(4X_2Y + 4X_2Z + 2X_1Y - 2X_1Z - \tfrac{1}{4}Y^2 - \tfrac{1}{4}Z^2\right),\\ x_2 &= X_2 + \left(-2X_2Y + 2X_2Z - 2X_1Y - 2X_1Z + \tfrac{1}{2}Y^2 - \tfrac{1}{2}Z^2\right),\\ y &= Y + \left(X_2^2 - X_2X_1 - \tfrac{1}{4}X_2Z + \tfrac{1}{2}X_1^2 + \tfrac{1}{2}X_1Z - \tfrac{1}{2}Y^2 - YZ + \tfrac{1}{6}Z^2\right),\\ z &= Z + \left(-X_2^2 - X_2X_1 + \tfrac{1}{4}X_2Y - \tfrac{1}{2}X_1^2 + \tfrac{1}{2}X_1Y - \tfrac{1}{6}Y^2 + YZ + \tfrac{1}{2}Z^2\right),\end{aligned} \tag{11.21a}$$

together with the system

$$\begin{aligned}\dot{X}_1 &= X_2,\\ \dot{X}_2 &= \left(-X_1^2 + 2YZ\right),\\ \dot{Y} &= -Y - X_1Y,\\ \dot{Z} &= Z + X_1Z.\end{aligned} \tag{11.21b}$$

As required by the normal form (11.11b), $Y = 0$ is invariant, $Z = 0$ is invariant, and the X-evolution only involves Y and Z in terms where they are multiplied together.

System (11.21)—the ODEs (11.21a) together with the coordinate transform (11.21b)—reproduces the original system (11.19) to errors $\mathcal{O}(3)$ (cubic).

- Exercise 11.3 asks you to derive the above quadratic transformation and evolution.

11.3 ▪ Existence and emergence proved nearby

All differential equations are imperfect models. *Richard E. Meyer,* 1992

Our theory is based upon the dynamics near an equilibrium, conventionally assumed at the origin. In that sense the approach is local. Nonetheless, by continuity in various bounds the locale of theoretical support is finite in size (as section 4.6 introduced). Thus the theoretical support is not "just asymptotic," our theory supports evaluation at the finite-parameter values appropriate to specific applications.

Sometimes the domain of theoretical support is infinite in size, that is, global. One important class of global support is when there is a whole subspace/manifold of equilibria, each of which satisfies the criteria for the theory (as in the examples of sections 4.7 and 4.8).

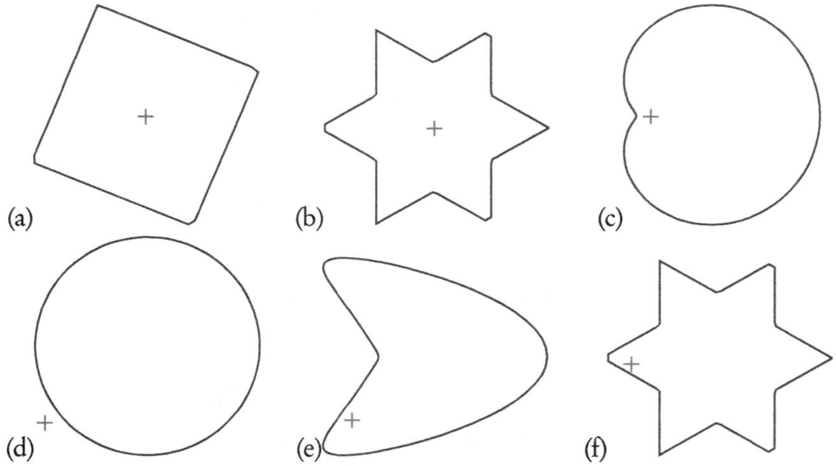

Figure 11.3. (a,b,c), *the boundary of three example star-shaped domains in 2D; and* (d,e,f), *the boundary of three example domains that are not star shaped. In each a cross marks the reference equilibrium at the origin.*

11.3.1 ▪ Time scales remain separated in a domain

For potential straightforward use of the multivariate Taylor remainder theorem, this section restricts domains to be star shaped relative to the basic equilibrium. That is, each point in a domain must be connected to the equilibrium (assumed to be at the origin) by a straight line segment that stays within the domain. If you want to relax the star-shaped constraint, then you will need to adapt applications of the multivariate Taylor remainder theorem.

Definition 11.9 (star shaped). *An open subset* $S \subseteq \mathbb{R}^N$ *is* **star shaped** *if for each* $u \in S$, $\gamma u \in S$ *for all* $0 \leqslant \gamma \leqslant 1$.

Figure 11.3 plots three example star-shaped domains in the plane and three examples that are not star shaped.

The major results of this section are that key invariant manifolds exist and emerge from the dynamics over a domain that is at least as large as that identified by the following definition. The size of the domain is primarily limited by the typical increase in $|G(u)|$ and $|H(u)|$ as $|u|$ increases.

Definition 11.10 (emergence-preserving domain). *Consider a normal-form system* (11.11b). *Choose a positive parameter* $\mu < \min(\beta, \gamma)$.[138] *Let matrices* P *and* Q *diagonalize* B *and* C, *respectively, in the sense that* $P^{-1}BP$ *and* $Q^{-1}CQ$ *are in (real) Jordan form.*[139] *Then define domain* $\mathbb{D}_\mu \subseteq \mathbb{R}^{m+n+n'}$ *to be a star-shaped domain for which parameter* $\mu < \min\left(\beta - \operatorname{cond} P \, \|G\|, \gamma - \operatorname{cond} Q \, \|H\|\right)$.[140]

---

[138] We usually choose μ in the spectral gap of Figure 11.2, that is, $\alpha < \mu < \min(\beta, \gamma)$.

[139] Typically we wish to avoid complex matrices when the system is real. Hence we seek matrices P and Q that reduce matrices B and C to block-diagonal form, where the diagonal includes 2×2 blocks of the real form $\begin{bmatrix} \mathfrak{R} & \mathfrak{I} \\ -\mathfrak{I} & \mathfrak{R} \end{bmatrix}$ for each pair of complex conjugate eigenvalues $\mathfrak{R} \pm i\mathfrak{I}$, as well as Jordan blocks for degenerate eigenvalues.

[140] In this definition $\|\cdot\|$ denotes the 2-norm of the matrix, $\|G\| := \max_{|v|=1} |Gv|$, and as usual the condition number $\operatorname{cond} U := \|U^{-1}\| \cdot \|U\|$. Exercise 11.6 refines this definition to show that crucial modeling properties often hold over usefully larger domains $\mathbb{D}_\mu$. Exercise 11.7 shows that in degenerate cases we need μ to be bounded away from the minimum; for example, there exists an $\epsilon > 0$ such that $\mu + 2\epsilon \leqslant \beta - \operatorname{cond} P \, \|G\|$.

Assumption 11.6 assures us that the domain $\mathbb{D}_\mu$ contains the origin. Continuity of the multinomial $\mathbf{G}$ and $\mathbf{H}$ assures us that $\mathbb{D}_\mu$ is a finite neighborhood about the origin. This definition only addresses the definite specific class of normal-form systems (11.11b): true, the machinations of section 11.2 derive the system (11.11b) as part of an approximation, but its role as an approximation is not relevant in this definition.

Example 11.5 (continue Example 11.4). Recall the derived quadratic normal-form system (11.21b) in Example 11.4. The spectrum is $\{-1, 0, 1\}$, so $\beta = \gamma = 1$; say we choose separation parameter $\mu = 3/4$. Recall that the normal form (11.21b) has $\dot{Y} = -Y - X_1 Y$ and $\dot{Z} = Z + X_1 Z$, so here $-B = C = 1$ and $-G = H = X_1$. Being scalar, B and C are trivially diagonalizable by $P = Q = 1$, and hence trivially $\operatorname{cond} P = \operatorname{cond} Q = 1$. The condition on the domain $\mathbb{D}_{3/4}$ is then that $\mu = 3/4 < 1 - 1 \cdot |X_1|$; that is, $|X_1| < 1/4$. This condition preserves the star-shaped nature, so define $\mathbb{D}_{3/4} = \{(X_1, X_2, Y, Z) \mid |X_1| < 1/4\}$.

For arbitrary separation parameter μ the Y and Z ODEs restrict the domain $\mathbb{D}_\mu$ to be $|X_1| < 1 - \mu$.

A useful refinement of the definition of the domain $\mathbb{D}_\mu$ (see Exercise 11.6) finds that a suitable star-shaped domain is the entire half-space $X_1 > -(1 - \mu)$. I use this domain $\mathbb{D}_\mu$ in further use of this example system.

The following important manifolds are invariant across the whole $\mathbf{X}, \mathbf{Y}, \mathbf{Z}$-space—hence they may be subspaces—but their characterization with regard to the dynamics about the equilibrium is only certain within $\mathbb{D}_\mu$ (Murdock, 2003, Def. 5.1.2, cf.).

Lemma 11.11 (invariant manifolds). *The equilibrium at the origin of the multinomial normal-form system* (11.11b) *has the following invariant manifolds (subspaces) in* $(\mathbf{X}, \mathbf{Y}, \mathbf{Z}) \in \mathbb{D}_\mu$:[141]

- *a* center manifold *(subspace)* $\mathbb{M}_c$ *is* $(\mathbf{X}, \mathbf{0}, \mathbf{0}) \in \mathbb{D}_\mu$*;*
- *a* stable manifold *(subspace)* $\mathbb{M}_s$ *is* $(\mathbf{0}, \mathbf{Y}, \mathbf{0}) \in \mathbb{D}_\mu$*;*
- *an* unstable manifold *(subspace)* $\mathbb{M}_u$ *is* $(\mathbf{0}, \mathbf{0}, \mathbf{Z}) \in \mathbb{D}_\mu$*;*
- *a* center-stable manifold *(subspace)* $\mathbb{M}_{cs}$ *is* $(\mathbf{X}, \mathbf{Y}, \mathbf{0}) \in \mathbb{D}_\mu$*;*
- *a* center-unstable manifold *(subspace)* $\mathbb{M}_{cu}$ *is* $(\mathbf{X}, \mathbf{0}, \mathbf{Z}) \in \mathbb{D}_\mu$.

Proof. The proof is straightforward from the conditions (11.11c) and domain Definition 11.10. For example, the stable manifold is the set of solutions $(\mathbf{X}, \mathbf{Y}, \mathbf{Z}) = \mathcal{O}(e^{-\mu t})$ as $t \to \infty$. Now $\mathbf{X} = \mathbf{0}$ and $\mathbf{Z} = \mathbf{0}$ are invariant by the conditions (11.11c) for all $\mathbf{Y}$. But only while solutions stay in $\mathbb{D}_\mu$ are we assured that additionally $|\mathbf{Y}| = \mathcal{O}(e^{-\mu t})$ through Lemma 11.16 (see below). Conversely, if either $\mathbf{X} \neq \mathbf{0}$ or $\mathbf{Z} \neq \mathbf{0}$, then by comparison theorems either $|\mathbf{X}| e^{\mu t} \to \infty$ or $|\mathbf{Z}| e^{\mu t} \to \infty$ or the system leaves $\mathbb{D}_\mu$, and so neither would be in the stable manifold.

One can proceed similarly for the other invariant manifolds. Further, all the manifolds are connected (to the equilibrium at the origin) because the domain $\mathbb{D}_\mu$ is star shaped (Definition 11.9). □

[141] Re notation: usually one uses calligraphic $\mathcal{M}$ for manifolds, and we do so for the typically curved manifolds in the original state space. But to distinguish the manifolds in the original space from the straight subspace manifolds of our normal form, we use the straight blackboard $\mathbb{M}$ in this latter case.

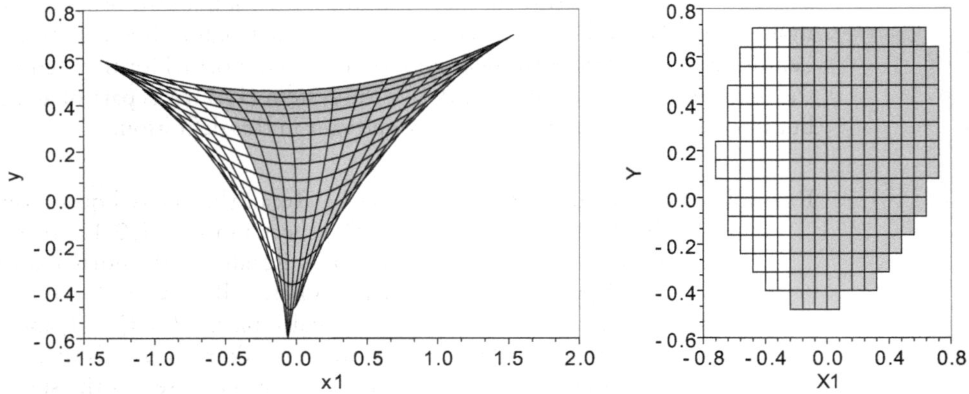

Figure 11.4. *The left-hand picture is the* x_1y*-plane of the example system* (11.20)*, and the right-hand picture is the* X_1Y*-plane. The right-hand* X_1Y*-plane shows a cross-section of the domain in which the specific coordinate transform* (11.21a) *is a* C^∞*-diffeomorphism and (shaded gray) the cross-section of the subset* $D_{3/4}$ *in which the separation trichotomy is preserved. The left-hand picture shows the corresponding domain cross-section in the* x_1y*-plane,* $d_{3/4}$ *(shaded gray).*

The next step is to map these invariant manifolds of the normal-form system (11.11b) into the original variables $\mathbf{x}$, $\mathbf{y}$, and $\mathbf{z}$. Then we relate the invariant manifolds of the normal form with those of the original system (11.10).[142] To understand the mapping, we characterize the near identity, multinomial coordinate transform (11.11a) constructed by the induction of section 11.2.

We continue to use capital letters for most quantities in the **XYZ**-space, such as the domains D_μ, and lowercase letters for most quantities in the $\mathbf{xyz}$-space, such as corresponding domains d_μ.

Definition 11.12 (diffeomorphic domain). *For some given order* $p \geqslant 2$*, define the domain* $D_\mu \subseteq \mathbb{D}_\mu$ *such that the coordinate transform* (11.11a) *is a* C^p*-diffeomorphism*[143] *onto a domain* $d_\mu \subseteq d$.

Since the coordinate transform (11.11a) is near identity, then by continuity of derivatives, this domain D_μ is a (finite) neighborhood of the equilibrium at the origin. In our usual case where the coordinate transform (11.11a) is multinomial, then we may as well take the order $p = \infty$.

Example 11.6 (continue Example 11.4). Two-dimensional plots perhaps best illustrate the domain of a coordinate transform. The coordinate transform (11.21a) is 4D so we view it in the 2D X_1Y-plane ($X_2 = Z = 0$). In the X_1Y-plane the coordinate

[142]Our approach is a little different from that of Murdock (2003, Chap. 5). In essence Murdock uses the coordinate transform (11.11a) to map the original dynamics (11.10) into the **XYZ**-space, calls that the *full system*, and typically denotes it by $\dot{x} = a(x)$, with x denoting the transformed variables (Murdock, 2003, p. 296). He then calls the normal-form system (11.11b) the *truncated system*, denoted $\dot{y} = \hat{a}(y)$. Murdock then compares invariant manifolds of these two systems; that is, he compares the invariant manifolds in what we call the **XYZ**-space. In contrast, here we use the coordinate transformation (11.11a) to map the invariant manifolds of the normal form (11.11b) into the $\mathbf{xyz}$ variables of the original physical system (11.10) and make comparisons in that original space.

[143]That is, the transform and its inverse exist and are differentiable up to derivatives of order p.

transform (11.21a) reduces to

$$x_1 = X_1 + \left(2X_1Y - \tfrac{1}{4}Y^2\right), \quad y = Y + \left(\tfrac{1}{2}X_1^2 - \tfrac{1}{2}Y^2\right).$$

Via the determinant of the Jacobian, the grid in Figure 11.4 plots the domain in which this coordinate transform is invertible and hence is a diffeomorphism: the right-hand plot is the domain in the X_1Y-plane, and the left-hand plot is the domain in the x_1y-plane.

But we also want to maintain the separation between center, stable, and unstable variables in the corresponding dynamical system (11.21b). Using the results of Example 11.5, the shaded region in Figure 11.4 indicates the part of the transform's domains that forms $D_{3/4}$, the domain where the trichotomy is preserved with parameter $\mu = 3/4$.

11.3.2 ▪ Invariant manifolds exist

Everything we know is only some kind of approximation, because *we know that we do not know all the laws* as yet. *Feynman* (1998, *p.* 2)

Instead of proving there exists a center manifold for a specified system (Chicone, 2006, Thm. 4.1, e.g.), which is then approximately constructed, Proposition 11.13, below, establishes that there is a system "close" to the specified system that has a center manifold, and other invariant manifolds, which we know exactly. This is an example of a backward error theory (Grcar, 2011, e.g.), whereas all previous center manifold theory addresses forward errors.

The above quote from Feynman, one of the great mid-20th-century physicists, relates to our backward theorems. In applications we *never* know the exact mathematical models: *all* our mathematical models are approximate. Thus a small enough perturbation to a prescribed mathematical model is as accurate a description of reality as the prescribed model. The backward approach here establishes properties about such slightly perturbed models. Thus the backward theorem proved here is equally as valid in applications as forward theorems.

Proposition 11.13 (invariant manifolds exist). *Consider the dynamical system* (11.10) *satisfying the specified conditions. For all orders* $2 \leqslant p \leqslant \mathfrak{p}$ *and a given rate* μ, *there exists a dynamical system that both is* $\mathcal{O}(p)$ *close to the system* (11.10) *and in the domain* d_μ *possesses center, stable, unstable, center-stable, and center-unstable manifolds, denoted, respectively, by* $\mathcal{M}_c$, $\mathcal{M}_s$, $\mathcal{M}_u$, $\mathcal{M}_{cs}$, *and* $\mathcal{M}_{cu}$.[144]

Proof. The hard work has already been done. For any order p, Lemma 11.8 establishes that there exists such a dynamical system that is $\mathcal{O}(p)$ close, namely the combination of the normal form (11.11b) with the coordinate transform (11.11a) defines sufficiently close dynamics in the state space of (11.10).

Lemma 11.11 establishes the existence of the requisite invariant manifolds for the normal form (11.11b) in D_μ. The coordinate transform (11.11a) maps these into corresponding invariant manifolds in d_μ for the corresponding system (11.11) in the original state space. □

[144]Although Proposition 11.13 only asserts that there is a dynamical system, there are vastly many such dynamical systems. First, there is the freedom identified at lower orders in the inductive proof of Lemma 11.8. Second, there is the freedom to choose higher-order terms in the multinomials to be anything provided they satisfy conditions (11.11c).

Some people are attracted by the possibility of exponentially small errors (Jones et al., 1996; Cotter and Reich, 2006; Iooss and Lombardi, 2010, e.g.). This possibility follows immediately from Proposition 11.13 under two further restrictions. In the case when the eigenvalues of the center modes have precisely zero real part, $\alpha = 0$, and when the specified dynamical system (11.10) is infinitely differentiable, $\mathfrak{p} = \infty$, then the iterative construction of the normal form that proves Lemma 11.8 may be continued to arbitrarily high order p, in principle: the reason being that the two constraints on the order p are that first $p < \min(\beta, \gamma)/\alpha = \infty$ and second $p \leqslant \mathfrak{p} = \infty$. To achieve the exponential closeness, for brevity, let $\epsilon = |(X, Y, Z)|$, and for any given ϵ choose order $p = -c/(\epsilon \log \epsilon)$ for some constant c. Then in Proposition 11.13 the order of closeness

$$\mathbb{O}(p) = \mathcal{O}\big(|X|^p + |Y|^p + |Z|^p\big) = \mathcal{O}\big(\epsilon^p\big) = \mathcal{O}\big(e^{p \log \epsilon}\big) = \mathcal{O}\big(e^{-c/\epsilon}\big) \quad \text{as } \epsilon \to 0.$$

This exponentially small closeness is a bit slippery as it requires higher and higher order construction as $\epsilon \to 0$. Nonetheless the argument of this paragraph establishes the following corollary of Proposition 11.13.

Corollary 11.14 (exponentially small closeness). *Suppose the dynamical system* (11.10) *satisfies Assumption* 11.6 *but with* $\alpha = 0$ *and Assumption* 11.7 *but with infinite differentiability,* $\mathfrak{p} = \infty$. *Then there exists a dynamical system exponentially close to the system* (11.10)*; the difference is* $\mathcal{O}\big(\exp\big[-c/|(X, Y, Z)|\big]\big)$ *as* $(X, Y, Z) \to 0$ *for some* c*, with center, stable, unstable, center-stable, and center-unstable manifolds in the domain* $\mathbb{d}_\mu$.

An immediate partnering consequence of the existence Proposition 11.13 addresses the evolution of the approximating system (11.11) on its invariant manifolds in the original state space.

Corollary 11.15 (evolution on manifolds). *The evolution of the approximate system* (11.11) *in* $\mathbb{d}_\mu$ *on any of the invariant manifolds* $\mathcal{M}_i$, $i \in \{c, s, u, cs, cu\}$*, is described by the system* (11.11b) *restricted to* $\mathbb{M}_i$ *and transformed by* (11.11a).

Center manifolds are very interesting in applications. Because of the properties (11.11c), the evolution on the center manifold $\mathbb{M}_c$ ($Y = 0$ and $Z = 0$) is $\dot{X} = F_c(X)$. The coordinate transform (11.11a) maps this evolution into the original state space to give the parametric description

$$\begin{bmatrix} x \\ y \\ z \end{bmatrix} = \begin{bmatrix} x(X, 0, 0) \\ y(X, 0, 0) \\ z(X, 0, 0) \end{bmatrix} \quad \text{such that} \quad \dot{X} = F_c(X). \tag{11.22}$$

Most people further simplify the representation of the evolution in the state space by choosing the new coordinate X to be equal to the original x on the center manifold. Example 11.4 makes this choice. Many people appear to make this choice implicitly, as in the specific constructive induction that proves Lemma 11.8. But it is a choice, and one may use the flexibility about how to parametrize the center manifold if desired and as also discussed by section 5.3.3. Nonetheless, in the case when the coordinate transform is chosen so that $x(X, 0, 0) = X$, then the evolution on the center manifold $\mathcal{M}_c$ is that on the graph

$$y = y(x, 0, 0), \quad z = z(x, 0, 0) \quad \text{such that} \quad \dot{x} = F_c(x) \tag{11.23}$$

for $(x, 0, 0) \in \mathbb{d}_\mu$ (at least).

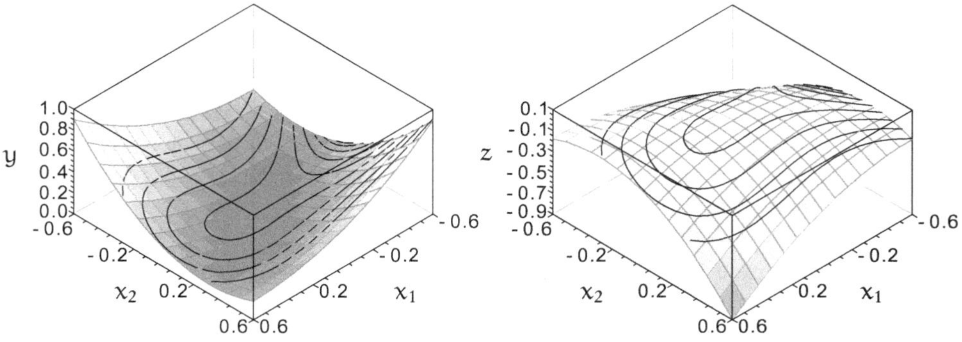

Figure 11.5. *Center manifold $\mathcal{M}_c$ of Example* 11.7*: left,* y*; right,* z*. Some trajectories in the center manifold are plotted on the center manifold.*

Example 11.7 (continue Example 11.4). Start with the quadratic approximating system (11.21b). Its center manifold $Y = Z = 0$ is mapped by the quadratic coordinate transform (11.21a) to $\mathcal{M}_c$ described parametrically by

$$x_j = X_j\,, \quad y = X_2^2 - X_2X_1 + \tfrac{1}{2}X_1^2, \quad z = -X_2^2 - X_2X_1 - \tfrac{1}{2}X_1^2.$$

Figure 11.5 plots this center manifold. On this center manifold $\mathcal{M}_c$ the evolution is governed by (11.21b), namely $\dot{X}_1 = X_2$ and $\dot{X}_2 = -X_1^2$. Figure 11.5 also shows some of the trajectories of these ODEs. However, as mentioned above, here $x_j = X_j$ on $\mathcal{M}_c$ and so the description of the center manifold $\mathcal{M}_c$ and the evolution thereon simplifies to $\dot{x}_1 = x_2$ and $\dot{x}_2 = -x_1^2$ on the graph

$$y = x_2^2 - x_2x_1 + \tfrac{1}{2}x_1^2, \quad z = -x_2^2 - x_2x_1 - \tfrac{1}{2}x_1^2.$$

11.3.3 ▪ The center manifold dynamics emerges

This book's theme is that center manifolds best model emergent dynamics in complex systems. This section establishes the crucial theorem that for a range of initial conditions, all solutions of the detailed full system approach a solution on the center manifold exponentially quickly. That is, as appealed to throughout the book, the dynamics on the center manifold exactly models the dynamics of the full system apart from exponentially decaying transients.

Again we establish this emergence property in a backward theorem. But another departure from other extant theorems (Carr, 1981, e.g.), is that we relax the "straitjacket" that solutions are required to remain in the neighborhood of the reference equilibrium for all time (as also relaxed by Kobayasi and Takagi (2003)). The following lemma establishes that solutions are exponentially quickly attracted to the center manifold over a finite time—although all time is preferred. Thus even if some solutions eventually exit the domain of validity of the center manifold model, we are empowered to use the center manifold model until they do exit.[145]

[145]Such a theorem should be especially useful in generalizations to stochastic dynamics. There the problem is that an inevitable rare event will eventually occur to push the stochastic system out of the domain of validity. For example, Berglund and Gentz (2003) do prove theory up until the first exit time (see their Theorem 2.4). Nonetheless, the almost certain eventual occurrence of such rare events plagues most established forward theorems on stochastic modeling.

Lemma 11.16 (emergent dynamics). *Consider the class of normal-form systems* (11.11b), *and define a domain* $\mathbb{D}_\mu$ *(see Definition* 11.10*). For any initial condition* $(\mathbf{X}_0, \mathbf{Y}_0, 0) \in \mathbb{M}_{cs} \subseteq \mathbb{D}_\mu$,[146] *solutions* $(\mathbf{X}(t), \mathbf{Y}(t), 0)$ *of the normal form* (11.11b) *are exponentially quickly attracted to the solution* $(\mathbf{X}, 0, 0)$ *on the center manifold* $\mathbb{M}_c$ *in the sense that* $|(\mathbf{X},\mathbf{Y}) - (\mathbf{X}, 0)| \leqslant c|\mathbf{Y}_0|e^{-\mu t}$ *for some constant* c *and for all* $0 \leqslant t < T_\mu$*, where the "first exit" time* T_μ *is such that both* $(\mathbf{X}(t), \mathbf{Y}(t)) \in \mathbb{M}_{cs}$ *and* $(\mathbf{X}(t), 0) \in \mathbb{M}_{cs}$ *for all* $0 \leqslant t < T_\mu$ *(in many cases* $T_\mu = \infty$*).*

Example 11.8 (a cylinder of attraction). Consider the normal-form system in center variables $\mathbf{X} = (X_1, X_2)$ and stable variables $\mathbf{Y} = (Y_1, Y_2)$:

$$\frac{d\mathbf{X}}{dt} = \mathbf{F}(\mathbf{X}), \quad \frac{d\mathbf{Y}}{dt} = \begin{bmatrix} -1 & 1 \\ -1 & -1 \end{bmatrix} \mathbf{Y} + \begin{bmatrix} X_1^2 Y_1 - X_1 X_2 Y_2 \\ -X_1 X_2 Y_1 + X_2^2 Y_2 \end{bmatrix}.$$

There are no unstable variables $\mathbf{Z}$. Here matrices determining the stable dynamics are

$$B = \begin{bmatrix} -1 & 1 \\ -1 & -1 \end{bmatrix} \quad \text{and} \quad G = \begin{bmatrix} X_1^2 & -X_1 X_2 \\ -X_1 X_2 & X_2^2 \end{bmatrix}.$$

The linear matrix B of the stable variables is already in real Jordan form, corresponding to eigenvalues $-1 \pm i$, so choose diagonalizing matrix $P = I$ for which the condition number $\operatorname{cond} P = 1$ (see Definition 11.10). As the nonlinearity matrix G is symmetric we find its 2-norm from the largest eigenvalue: its two eigenvalues are zero and $X_1^2 + X_2^2$; the larger gives the norm $\|G\| = X_1^2 + X_2^2$. The eigenvalues of B are $\beta_j = -1 \pm i$, so an upper bound on their real part is $-\beta = -1$. For any decay rate $0 < \mu < 1 = \beta$, the domain $\mathbb{D}_\mu$ is then constrained by $\mu < \beta - \operatorname{cond} P \|G\| = 1 - (X_1^2 + X_2^2)$. That is, domain $\mathbb{D}_\mu$ is the cylinder $X_1^2 + X_2^2 < 1 - \mu$ for all $\mathbf{Y}$. Lemma 11.16 asserts that while solutions stay within the cylindrical $\mathbb{D}_\mu$, solutions decay to the center manifold $\mathbf{Y} = 0$ through being bounded by $|\mathbf{Y}| \leqslant |\mathbf{Y}_0|e^{-\mu t}$.

Exercise 11.5 explores an interesting example of $\mathbf{F}$ for which this bound is reasonably sharp.

Conversely, Exercise 11.6 modifies the following proof of Lemma 11.16 in order to find a criterion that often empowers the domain $\mathbb{D}_\mu$ of emergence to be much larger than that guaranteed by Definition 11.10.

Sketch proof. The case of initial condition $\mathbf{Y}_0 = 0$ is trivial, so consider $\mathbf{Y}_0 \neq 0$. For simplicity consider only the case when matrix B is diagonalizable; Exercise 11.7 extends the proof to degenerate matrices B.

Let's use a Lyapunov function to show that $\mathbf{Y} = 0$ is exponentially quickly attractive.[147] Code the Lyapunov function in a weighted norm: define the vector norm $|\mathbf{Y}|_P := |P^{-1}\mathbf{Y}|$ in terms of the usual 2-norm $|z| = \sqrt{z^T z}$, and in terms of a matrix P that diagonalizes matrix B (as introduced in Definition 11.10). Noting that $BP = PD$ and recalling from condition (11.11c) that $\mathbf{G} = G\mathbf{Y}$, straightforward algebra derives that

$$\frac{d}{dt}|\mathbf{Y}|_P^2 = (P^{-1}\mathbf{Y})^T \left[D + D^T + (P^{-1}GP) + (P^{-1}GP)^T \right] (P^{-1}\mathbf{Y}). \tag{11.24}$$

Consider two parts of the right-hand side in turn.

[146] Recall that the center stable manifold $\mathbb{M}_{cs} = \mathbb{D}_\mu$ in the important case when there are no unstable modes; that is, when $\mathbf{Z}(t)$ is absent.

[147] C. Chicone (private communication, 2013) suggested a Lyapunov function approach.

- Since matrix P transforms B to real Jordan form (Definition 11.10), then $D^T + D = 2\,\mathrm{diag}(\Re\beta_j)$ for eigenvalues β_j of B. Consequently,

$$\begin{aligned}(P^{-1}\mathbf{Y})^T \left[D + D^T\right] (P^{-1}\mathbf{Y}) &= 2\sum_j \Re\beta_j (P^{-1}\mathbf{Y})_j^2 \\ &< -2\beta|P^{-1}\mathbf{Y}|^2 = -2\beta|\mathbf{Y}|_P^2.\end{aligned}$$

- The remaining part of the right-hand side is usefully bounded since

$$\begin{aligned}&\left|(P^{-1}\mathbf{Y})^T \left[(P^{-1}GP) + (P^{-1}GP)^T\right] (P^{-1}\mathbf{Y})\right| \\ &\leqslant 2|P^{-1}\mathbf{Y}| \cdot \|P^{-1}GP\| \cdot |P^{-1}\mathbf{Y}| \\ &\leqslant 2\|P^{-1}\| \cdot \|G\| \cdot \|P\| \cdot |P^{-1}\mathbf{Y}|^2 = 2\,\mathrm{cond}\,P\,\|G\| \cdot |\mathbf{Y}|_P^2 .\end{aligned}$$

Consequently, in the domain $\mathbb{D}_\mu$ equation (11.24) ensures

$$\frac{d}{dt}|\mathbf{Y}|_P^2 < 2(-\beta + \mathrm{cond}\,P\,\|G\|)|\mathbf{Y}|_P^2 < -2\mu|\mathbf{Y}|_P^2 .$$

A comparison theorem then ensures $|\mathbf{Y}|_P^2 \leqslant |\mathbf{Y}_0|_P^2 e^{-2\mu t}$; that is, $|\mathbf{Y}|_P \leqslant |\mathbf{Y}_0|_P e^{-\mu t}$. Thus the stable variables $\mathbf{Y}$ decay to zero exponentially quickly in the P-norm.

In the usual 2-norm there may be some transient growth constrained by the condition number of P. Returning to the 2-norm,

$$\begin{aligned}|\mathbf{Y}| &= |PP^{-1}\mathbf{Y}| \\ &\leqslant \|P\| \cdot |P^{-1}\mathbf{Y}| = \|P\| \cdot |\mathbf{Y}|_P \\ &\leqslant \|P\| \cdot |\mathbf{Y}_0|_P e^{-\mu t} = \|P\| \cdot |P^{-1}\mathbf{Y}_0| e^{-\mu t} \\ &\leqslant \|P\| \cdot \|P^{-1}\| \cdot |\mathbf{Y}_0| e^{-\mu t} = \mathrm{cond}\,P\,|\mathbf{Y}_0| e^{-\mu t}\end{aligned}$$

for as long as solutions stay in the domain $\mathbb{D}_\mu$ (Murdock, 2003, Lem. 5.3.1, cf.).

Last, we need to prove not just attraction of the slow manifold, but that *solutions off* $\mathbf{Y} = \mathbf{0}$ approach a *solution on* $\mathbf{Y} = \mathbf{0}$. Given any solution $(\mathbf{X}(t), \mathbf{Y}(t))$ of the normal form (11.11b), because the $\mathbf{X}$ equation is independent of $\mathbf{Y}$, we have that $(\mathbf{X}(t), \mathbf{0})$ is also a solution of the normal form (11.11b). Further, the distance between them is $|(\mathbf{X}, \mathbf{Y}) - (\mathbf{X}, \mathbf{0})| = |\mathbf{Y}| \leqslant c|\mathbf{Y}_0| e^{-\mu t}$ for $0 \leqslant t \leqslant T_\mu$ and constant $c = \mathrm{cond}\,P$. □

Lemma 11.16 establishes a finite-size domain in which a center manifold model emerges in time. However, it only applies to systems in the special normal form (11.11b). Proposition 11.17, below, uses the multinomial diffeomorphism of Lemma 11.8 to prove similar emergence in a wide class of dynamical systems. Lemma 11.8 establishes that there is a member of this class asymptotically close to any reasonably general specified dynamical system.

Proposition 11.17 (emergent dynamics). *Consider the system* (11.11) *restricted to its center-stable manifold* $\mathcal{M}_{cs}$ *in the* xyz*-space. For all initial conditions* $(x(0), y(0), z(0)) \in \mathcal{M}_{cs}$, *there exists a constant* c *and a center manifold solution* $\dot{\mathbf{X}} = A\mathbf{X} + \mathbf{F}_c(\mathbf{X})$ *with some initial condition* $\mathbf{X}(0) = \mathbf{X}_0$ *such that the distance*

$$\left|(x(t), y(t), z(t)) - (x(\mathbf{X}(t), 0, 0), y(\mathbf{X}(t), 0, 0), z(\mathbf{X}(t), 0, 0))\right| \leqslant ce^{-\mu t}$$

for as long as the solutions stay in the domain d_μ.

Sketch proof. To realize the exponential decay of distance between general solutions of the system (11.11) and the slow manifold solution, consider the trajectory starting from initial condition $(\mathbf{x}_0, \mathbf{y}_0, \mathbf{z}_0) \in \mathcal{M}_{cs} \subseteq d_\mu$:

- it maps to the trajectory of (11.11b) starting from some point $(\mathbf{X}_0, \mathbf{Y}_0, 0) \in D_\mu \subseteq \mathbb{D}_\mu$;
- by Lemma 11.16, this trajectory approaches exponentially quickly the solution starting from $(\mathbf{X}_0, 0, 0)$;
- hence starting the model $\dot{\mathbf{s}} = A\mathbf{s} + \mathbf{F}_c(\mathbf{s})$ with initial condition $\mathbf{s}(0) = \mathbf{X}_0$ gives the requisite solution on the center manifold $\mathcal{M}_c$ approached by the trajectory from the specified initial condition.

The constant $c := \operatorname{cond} P\,|\mathbf{Y}_0|\,\text{Lip}$, where P is the similarity matrix introduced in the proof of Lemma 11.16 and Lip is a Lipschitz constant of the coordinate transform (11.11a): $|(\mathbf{x}(\mathbf{X},\mathbf{Y},0), \mathbf{y}(\mathbf{X},\mathbf{Y},0)) - (\mathbf{x}(\mathbf{X},0,0), \mathbf{y}(\mathbf{X},0,0))| \leqslant \text{Lip}\,|\mathbf{Y}|$ for $(\mathbf{X},\mathbf{Y},0) \in \mathbb{M}_{cs} \subseteq \mathbb{D}_\mu$. □

Corollary 11.18 (center-unstable dynamics). *Lemma* 11.16 *and Proposition* 11.17 *immediately also apply to the center-unstable manifold of system* (11.10) *when considered backward in time.*

Further research is needed. This section establishes quantitative information about the finite size of domains of model validity and attraction in terms of a given constructed coordinate transform (11.11a) and normal-form system (11.11b). A currently outstanding challenge is to establish such quantitative information directly from the algebraic form of the original physical system (11.10). Further, this section only establishes that there are normal-form systems *asymptotically close* to a specified smooth system. Another research challenge is to quantify a bound on the error, over some domain, between a constructed normal-form system and a specified system.

11.3.4 ▪ Case study: Classic 2D pitchfork bifurcation

Consider the system

$$\dot{x} = \alpha x - xy\,, \quad \dot{y} = -y + x^2 \tag{11.25}$$

for small parameter α, as there exists a pitchfork bifurcation as parameter α increases through zero. In a domain around the equilibrium at the origin, we seek to find a new simpler description of the bifurcation dynamics: say $\dot{X} = \cdots$ and $\dot{Y} = \cdots$ in terms of new variables X and Y. Such a simplification exists because the rates of the linearized system $\dot{x} \approx \alpha x$ and $\dot{y} \approx -y$, namely rates α and -1, are well separated when parameter α is small (in the sense illustrated by Figure 11.2).

First find a coordinate transform For this simple system relate the original and the new variables by a *near identity transform*

$$x = x(X, Y) \approx X \quad \text{and} \quad y = y(X, Y) \approx Y.$$

The evolution in these new variables will be of the form

$$\dot{X} = F(X) \quad \text{and} \quad \dot{Y} = -Y + G(X, Y),$$

where again F and G are nonlinear functions of their arguments.

Maintain fidelity with the original dynamics by substituting these assumptions into the original dynamical equations and solving to some order of error. The two dynamical equations (11.25) then become the homological equations

$$\begin{aligned} F - \alpha\hat{x} + \alpha X\frac{\partial\hat{x}}{\partial X} - Y\frac{\partial\hat{x}}{\partial Y} &= \alpha X - (X+\hat{x})(Y+\hat{y}) - F\frac{\partial\hat{x}}{\partial X} - G\frac{\partial\hat{x}}{\partial Y}, \\ G + \hat{y} + \alpha X\frac{\partial\hat{y}}{\partial X} - Y\frac{\partial\hat{y}}{\partial Y} &= (X+\hat{x})^2 - F\frac{\partial\hat{y}}{\partial X} - G\frac{\partial\hat{y}}{\partial Y}. \end{aligned}$$

To solve these homological equations, iterate by substituting the current approximation in the right-hand side and finding the appropriate left-hand side.

- The initial trivial approximation is that $F = G = \hat{x} = \hat{y} = 0$.
- Substituting this initial approximation in the right-hand side, the above homological equations become

$$\begin{aligned} F - \alpha\hat{x} + \alpha X\frac{\partial\hat{x}}{\partial X} - Y\frac{\partial\hat{x}}{\partial Y} &= \alpha X - XY, \\ G + \hat{y} + \alpha X\frac{\partial\hat{y}}{\partial X} - Y\frac{\partial\hat{y}}{\partial Y} &= X^2. \end{aligned}$$

  Simplify the evolution by putting as much as we can in the coordinate transform $\hat{x}$ *and* $\hat{y}$*, and as little as possible in the evolution* F *and* G.

  - Observe that $\hat{x} = XY$ and $F = \alpha X$ satisfies the first x equation. Attempting to set the desirable $F = 0$ here fails because $\hat{x} \propto X$ is a homogeneous solution of the left-hand side.
  - For the second y equation, the right-hand side is quadratic in X, so let us see what a quadratic can do for us: try $\hat{y} = aX^2$, and then the equation becomes

$$G + a(1+2\alpha)X^2 = X^2,$$

    so we choose $a = 1/(1+2\alpha)$ and the evolution $G = 0$. The first nontrivial approximation is then

$$x = X + XY, \quad y = Y + \frac{X^2}{1+2\alpha}, \quad \dot{X} = \alpha X, \quad \dot{Y} = -Y.$$

- Then in the second iteration the dynamical equations become

$$\begin{aligned} F - \alpha\hat{x} + \alpha X\frac{\partial\hat{x}}{\partial X} - Y\frac{\partial\hat{x}}{\partial Y} &= -XY^2 - \frac{X^3}{1+2\alpha} + \text{h.o.t.}, \\ G + \hat{y} + \alpha X\frac{\partial\hat{y}}{\partial X} - Y\frac{\partial\hat{y}}{\partial Y} &= 2X^2Y + \text{h.o.t.} \end{aligned}$$

Look at the cubic terms in the x equation:

- first, the $-XY^2$ term in the right-hand side is matched by including the term $\frac{1}{2}XY^2$ in $\hat{x}$;
- second, the $-X^3/(1+2\alpha)$ in the right-hand side cannot be generated by $Y\partial\hat{x}/\partial Y$ because of the multiplication by Y and so if put into $\hat{x}$ would generate a small divisor α (that is, X^3 is nearly a homogeneous solution); thus we must include it in the evolution by setting $F = -X^3/(1+2\alpha)$.

Now turn to the cubic term in the y equation:

- the $2X^2Y$ in the right-hand side cannot be generated by $\hat{y} - Y\partial\hat{y}/\partial Y$ because *terms linear in* Y *are nearly homogeneous solutions* of the left-hand side operator; thus we must include it in the evolution by setting $G = +2X^2Y$.

The second approximation is then that

$$x = X + XY + \tfrac{1}{2}XY^2, \quad y = Y + \frac{X^2}{1+2\alpha}, \tag{11.26a}$$

$$\text{such that} \quad \dot{X} = \alpha X - \frac{X^3}{1+2\alpha}, \quad \dot{Y} = -Y + 2X^2Y. \tag{11.26b}$$

The normal form illuminates the original Now explore crucial features of this normal-form system (11.26b) and the coordinate transformation (11.26a) that relates it to the original system (11.25). Because of the construction, we know that the original system (11.25) is asymptotically close to the combined (11.26) to quartic $\mathbb{O}(4)$ errors.

1. Observe that $Y = 0$ is an invariant manifold of (11.26b): specifically, it is a center manifold of the system; more specifically, it is a slow manifold.

   Setting $Y = 0$ in the transformation (11.26a) shows that the shape of this slow manifold in the original variables is $x = X$ and $y = X^2/(1+2\alpha)$. This slow manifold is smooth enough to be twice differentiable provided $|\alpha| < 1/2$, which corresponds to the spectral gap condition required by Lemma 11.8. Higher-order models with higher-order smoothness require stronger constraints of the parameter α. (Roberts, 1989b, identified these slow manifold approximations.)

2. We obtain indications of the extent of emergence of the slow manifold model by looking for where the separation of time scales and the coordinate transform break down.

   Emergence in the new variables Separation of variables in the Y-evolution (11.26b) indicates that $Y \to 0$ provided $|X| < 1/\sqrt{2}$. More specifically, we are assured by Lemma 11.16 of the decay $Y = \mathcal{O}(e^{-\mu t})$ for as long as solutions satisfy $1 - 2X^2 > \mu$; that is, they stay in the star-shaped domain $\mathbb{D}_\mu = \{(X, Y) \mid |X| < \sqrt{(1-\mu)/2}\}$.

   Degenerate coordinate transform But the modeling must break down for the original variables when the coordinate transform is degenerate. This certainly occurs when the determinant is zero of the Jacobian

   $$J = \begin{bmatrix} 1 + \partial\hat{x}/\partial X & \partial\hat{x}/\partial Y \\ \partial\hat{y}/\partial X & 1 + \partial\hat{y}/\partial Y \end{bmatrix} = \begin{bmatrix} 1 + Y + \frac{1}{2}Y^2 & X(1+Y) \\ 2X/(1+2\alpha) & 1 \end{bmatrix}.$$

   That is, the coordinate transform degenerates at

   $$2X^2 = (1+2\alpha)\left[1 + \frac{Y^2}{2(1+Y)}\right].$$

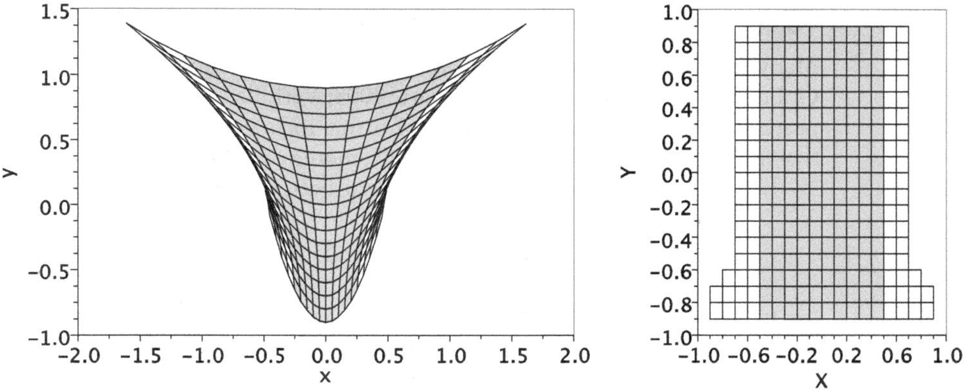

Figure 11.6. *The left-hand picture is the* xy*-plane of the example system* (11.25), *and the right-hand picture is the* XY*-plane, both for parameter* $\alpha = 0$*. The right-hand* XY*-plane shows the domain in which the specific coordinate transform* (11.26a) *is a* C^∞*-diffeomorphism, and (shaded) the subset* $D_{1/2}$ *in which emergence is preserved. The left-hand picture shows the corresponding domain* $d_{1/2}$ *in the* xy*-plane (shaded).*

Exclude such degeneracies by restricting attention to the domain $D_\mu = \{(X, Y) \mid |X| < \sqrt{[1 - \max(\mu, 2\alpha)]/2},\ Y > -1\}$.

Thus the modeling is valid in the domain d_μ that is the image of D_μ under the coordinate transform (11.26a), as shown in Figure 11.6.

This guarantees a minimum extent of the domain to which the modeling applies, albeit as yet only guaranteed to be accurate near the origin.

Higher-order models may increase or decrease the indicated domain of validity.

3. Setting $Y = 0$ in the X-evolution (11.26b) gives that the evolution on the slow manifold is $\dot{X} = \alpha X - X^3/(1 + 2\alpha)$. Since $x = X$ on the slow manifold, this evolution is the pitchfork bifurcation model $\dot{x} = \alpha x - x^3/(1 + 2\alpha)$.

   The relevance of the model follows because the X-evolution (11.26b) is identical for all Y. Thus the evolution of X is independent of whether a state is on or off the slow manifold $Y = 0$. Consequently, all states in the domain D_μ starting with the same X_0 have exactly the same long-term evolution.

   In the original variables, under the transformation (11.26a), this says that an initial condition (x_0, y_0) must be projected along curves of constant Y (the near vertical lines in Figure 11.6) to the slow manifold in order for the slow manifold model to make correct long-term predictions. Chapter 12 discusses further such sound initial conditions.

Summary This basic example illustrates how normal forms

1. capture a center manifold model;
2. indicate the size of the domain in which the center manifold model applies;
3. illustrate the emergence theorem and how one projects initial conditions onto a model (Chapter 12 develops this projection to practical generality).

11.3.5 ▪ Singular perturbation problems emerge instantaneously

Many people are attracted to phrasing problems as singular perturbations (Bykov and Gol'dshtein, 2013; Pavliotis and Stuart, 2008; Verhulst, 2005, e.g.). Let's look at singular perturbation model reduction via our normal-form coordinate transformations. We find that slow manifolds exist and "instantly" emerge over finite domains (sometimes global).

Let's consider the class of singular perturbation dynamics governed by

$$\dot{u} = \mathcal{F}(u,v) \quad \text{and} \quad \dot{v} = \frac{1}{\epsilon}\mathcal{G}(u,v) \tag{11.27}$$

for $u(t) \in \mathbb{R}^m$, $v(t) \in \mathbb{R}^n$, and the regime where parameter ϵ is small. The singular perturbation argument is that as parameter $\epsilon \to 0$ the $v(t)$ dynamics is very fast and will rapidly settle onto an equilibrium of the v ODE.[148] Hence solving $\mathcal{G}(u,v) = 0$ gives quasi-equilibria $v = \mathcal{V}(u)$ parametrized by the "frozen" slow variable u. Then the argument is that the slow variables are not truly frozen but instead evolve according to the u ODE, namely $\dot{u} \approx \mathcal{F}(u, \mathcal{V}(u))$. Indeed some beautiful theorems (Pavliotis and Stuart, 2008; Verhulst, 2005, e.g.) establish the slow manifold model that

$$v = \mathcal{V}(u) + \mathcal{O}(\epsilon) \quad \text{such that} \quad \dot{u} = \mathcal{F}(u, \mathcal{V}(u)) + \mathcal{O}(\epsilon). \tag{11.28}$$

Let's view this scenario using our normal-form coordinate transformations. In particular, we do not use the limit $\epsilon \to 0$ but treat parameter ϵ as a fixed finite value, albeit notionally small.

To apply our coordinate transform approach most systematically, we embed the original singular (11.27) as the $\theta = 1$ member of the family of systems

$$\dot{u} = \theta\mathcal{F}(u,v) \quad \text{and} \quad \dot{v} = \frac{1}{\epsilon}\mathcal{G}(u,v) \tag{11.29}$$

for homotopy parameter θ for at least $0 \leqslant \theta \leqslant 1$. Then we analyze this family and set parameter $\theta = 1$ to recover results about the original (11.27). One might imagine parameter θ is a sort of "temperature" in that when $\theta = 0$ the slow variables u are "frozen," but when $\theta = 1$ the system has "warmed" to become the original.

Now proceed along familiar lines. First, find equilibria: system (11.29) has a manifold of equilibria for parameter $\theta = 0$ and $v^* = \mathcal{V}(u^*)$. Second, change to coordinates local to each equilibrium. We introduce new slow and fast variables, familiarly called x and y, according to the linear transformation

$$\begin{bmatrix} u \\ v \end{bmatrix} = \begin{bmatrix} u^* \\ \mathcal{V}(u^*) \end{bmatrix} + \begin{bmatrix} I & 0 \\ L & I \end{bmatrix} \begin{bmatrix} x \\ y \end{bmatrix}$$

for $x(t) \in \mathbb{R}^m$ and $y(t) \in \mathbb{R}^n$. The submatrix $L(u^*) := -\left[\partial\mathcal{G}/\partial v\right]^{-1} \partial\mathcal{G}/\partial u$ evaluated at $(u^*, \mathcal{V}(u^*))$. This choice for the submatrix $L(u^*)$ ensures the system (11.29) becomes linearly separated:

$$\dot{x} = \theta f(x,y) \quad \text{and} \quad \dot{y} = \frac{1}{\epsilon}By + \frac{1}{\epsilon}g(x,y), \tag{11.30}$$

where matrix $B(u^*) := \partial\mathcal{G}/\partial v$ evaluated at $(u^*, \mathcal{V}(u^*))$; the function $f(x,y) := \mathcal{F}(u^* + x, \mathcal{V}(u^*) + Lx + y)$ is also implicitly a function of u^*, as is the function

[148] We restrict attention to this scenario of rapid attraction of fast variables v to an equilibrium. In other singular perturbation scenarios the fast variable v is rapidly attracted to an invariant distribution reflecting rapid oscillations, chaos, or stochasticity in the v ODE (Berglund and Gentz, 2003, e.g.).

$g(x,y) := \mathcal{G}(u^* + x, \mathcal{V}(u^*) + Lx + y) - By - \epsilon\theta f(x,y)$, which is also implicitly a function of the small product $\epsilon\theta$ (only the leading dependence in parameters θ and small ϵ is explicit). The function $g = \mathbb{O}(2)$ by the choice of L and B. Further, functions $\mathbf{f}$ and $\mathbf{g}$ are as smooth as $\mathcal{F}$ and $\mathcal{G}$ in the corresponding domains except that we require $\det \partial\mathcal{G}/\partial v$ to be bounded away from zero as is consistent with the singular perturbation assumption that the fast variables v evolve rapidly to a quasi-equilibrium. In the linearly separated form (11.30) we readily apply the normal-form results.

To quantify the separation of time scales, suppose all eigenvalues of B have negative real part bounded away from zero: $\Re\beta_j \leqslant -\beta^* < 0$ (β^* depends upon u^*); thus the β invoked in previous sections is here β^*/ϵ (large since ϵ is small). Based about the equilibria $\theta = 0$, the linear matrix for the slow variables x is zero, with eigenvalues that are zero, so $\alpha = 0$.[149] Section 11.2, via Lemma 11.8, shows the existence of coordinate transforms that together with the normal form

$$\dot{X} = \theta F(X) \quad \text{and} \quad \dot{Y} = \frac{1}{\epsilon}BY + \frac{1}{\epsilon}G(X,Y)Y \tag{11.31}$$

are asymptotically close to the original (11.30) to any specified order (the order limited only as noted in the previous paragraph). By rescaling time with ϵ, one can see that the coordinate transform and the normal form (11.31), apart from the explicit factors shown above, depend only upon parameters ϵ and θ via the small product $\epsilon\theta$.

Now determine the domain of emergence from the results of section 11.3 to this point. We need to choose a rate parameter $\mu^* < \beta^*/\epsilon$, say choose $\mu^* := \beta^*/\sqrt{\epsilon}$. Then for the normal forms (11.31) the star-shaped domain $\mathbb{D}^*_\mu$ must satisfy $\mu^* < \beta^*/\epsilon - \operatorname{cond} P^* \|G\|/\epsilon$; that is, $(X,Y) \in \mathbb{D}^*_\mu$ must satisfy

$$\|G(X,Y)\| < \frac{\beta^*(1-\sqrt{\epsilon})}{\operatorname{cond} P^*} \to \frac{\beta^*}{\operatorname{cond} P^*} \quad \text{as } \epsilon \to 0\,.$$

I use the superscript $*$ on the domain to indicate that it too depends upon the location u^* of the base equilibria of the analysis; that is, $\mathbb{D}^*_\mu := \mathbb{D}_\mu(u^*)$. As $G = \mathbb{O}(1)$, these local domains $\mathbb{D}^*_\mu$ exist for all values of the singular perturbation parameter $\epsilon < 1$. Further, the domain exists for homotopy parameter $\theta = 1$ for sufficiently small ϵ, as the homotopy parameter only occurs in the combination $\epsilon\theta$. Define the global domain $\mathbb{D}_\mu := \bigcup_{u^*} \mathbb{D}^*_\mu$, which also contains $\theta = 1$ for sufficiently small ϵ; since the attractiveness of the slow manifold is ensured inside each $\mathbb{D}^*_\mu$, the slow manifold is attractive in the union $\mathbb{D}_\mu$. Letting $\beta := \min_{u^*} \beta^*$ and $\mu := \beta/\sqrt{\epsilon}$, Lemma 11.16 asserts that all solutions of the normal form (11.31) in the union $\mathbb{D}_\mu$ are attracted to a slow manifold solution at least as fast as $\mathcal{O}(e^{-\mu t}) = \mathcal{O}(e^{-\beta t/\sqrt{\epsilon}})$. That is, as the singular perturbation parameter $\epsilon \to 0$, this attraction happens almost instantaneously in time.

But how do the existence and attraction translate to dynamics (11.27) in the original slow/fast variables (u,v)? We proceed via the linearly transformed dynamics (11.30) of the local variables (x,y). The coordinate transforms to/from variables (X,Y) are near identity (multinomial) and so will be C^p-diffeomorphic in some domain $D_\mu \subseteq \mathbb{D}_\mu$. As in Definition 11.12, let the domain $d_\mu \subseteq \mathbb{R}^{m+n}$ be the image of D_μ under the coordinate transform. Since the coordinate transform is near identity and depends upon ϵ and θ only in the combination $\epsilon\theta$, the physically relevant case of parameter $\theta = 1$ lies in domain d_μ for sufficiently small ϵ. The linear transformation then maps the domain d_μ into the original variables (u,v). Thus we are assured that there is a global domain about the set of equilibria (u^*, v^*) found at $\epsilon = 0$ (provided $\partial\mathcal{G}/\partial v$ has eigenvalues with real part

[149] Strictly, one lets $\mathbb{O}(p)$ denote terms $\mathcal{O}(\theta^{p/2} + |X|^p + |Y|^p + |Z|^p)$ as $(\theta, X, Y, Z) \to 0$.

bounded away from zero), in which a slow manifold exists and in which all solutions are attracted exponentially quickly, at least as fast as $\mathcal{O}(e^{-\mu t/\sqrt{\epsilon}})$, to solutions on the slow manifold.

This coordinate transform view reaffirms the existence and rapid emergence of slow manifolds in singular perturbation problems.

11.4 ▪ Exercises

Exercise 11.1 (competing species). Consider the dynamical system for two species with populations $u(t)$ and $v(t)$. These two species compete for resources according to the system

$$\dot{u} = u(1 - u - v) \quad \text{and} \quad \dot{v} = v(1 - v - au)$$

for some fixed coefficient a, $0 < a < 1$.

1. Find all three equilibria in this system.
2. Linearize to find that two of the equilibria are linearly unstable and the other is marginal.
3. Linearly change variables to $u = x$ and $v = 1 + y - ax$, and find that the system is equivalent to
$$\dot{x} = -(1 - a)x^2 - xy \quad \text{and} \quad \dot{y} = -y - a(1 - a)x^2 - y^2.$$
4. Find a near identity, quadratic coordinate transform (Cox and Roberts, 1995) to put this system in the normal form
$$\dot{X} \approx -(1 - a)X^2 \quad \text{and} \quad \dot{Y} \approx -Y.$$
5. Hence describe the stability of this equilibrium.

Exercise 11.2 (small constant forcing). Construct the normal form of the slow-fast system $\dot{x} = -xy$ and $\dot{y} = -y + x^2 + \epsilon$ for small parameter ϵ. For a strict application of the approach of this section, consider parameter ϵ as a new dependent variable (x_2 say) that satisfies $\dot{\epsilon} = 0$. Interpret the normal-form. Discuss the support the normal-form transformation provides to the slow manifold model of this system.

Exercise 11.3 (quadratic terms). Reconsider Example 11.4 and derive the given quadratic transformation and evolution. Given the initial approximation that $x_i = X_i$, $y = Y$, and $z = Z$, apply the approach of section 11.2. Start with the x_2 ODE, and find that its transformation gives the information we need to subsequently find the x_1-transformation.

Exercise 11.4 (irregular intersecting slow manifold). Exercise 4.14 considered a model of heterogeneous catalysis (attributed to Yablonsky et al. (2003)). In a relevant parameter regime this example possesses two intersecting global slow subspaces. We use coordinate transforms to discover under what nonlinearities such intersecting slow subspaces persist as intersecting slow manifolds.

For variables $x(t)$ and $y(t)$, restricted to $x, y \geqslant 0$ for physical relevance, consider the ODEs

$$\dot{x} = -xy + \epsilon f(x, y) \quad \text{and} \quad \dot{y} = -xy + \epsilon g(x, y), \tag{11.32}$$

where nonlinear functions f and g are small in some sense so we control their effects by the small parameter ϵ.

1. For negligible f and g, $\epsilon = 0$, deduce that the x-axis and the y-axis, $x, y > 0$, are subspaces of stable equilibria.
2. For small ϵ, seek a near identity coordinate transform

   $$x = X + \hat{x}(X, Y) \quad \text{and} \quad y = Y + \hat{y}(X, Y)$$

   to new simpler ODEs

   $$\dot{X} = -XY + \hat{F}(X, Y) \quad \text{and} \quad \dot{Y} = -XY + \hat{G}(X, Y),$$

   where we aim for $\hat{F}(0, Y) = \hat{G}(X, 0) = 0$ to maintain that $X = 0$ and $Y = 0$ as invariant subspaces. Deduce that the corrections, with hats, need to satisfy the homological equations

   $$\begin{aligned} \hat{F} + Y\hat{x} - XY\hat{x}_X - XY\hat{x}_Y + X\hat{y} &= \text{Res}_x(X, Y), \\ \hat{G} + Y\hat{x} - XY\hat{y}_X - XY\hat{y}_Y + X\hat{y} &= \text{Res}_y(X, Y), \end{aligned}$$

   where Res denotes the residual of the respective ODEs (in general in multinomial form and from the current approximation).
3. Deduce that any constant term in the residuals must be assigned to the evolution corrections, $\hat{F}$ and $\hat{G}$, which would destroy the aim of maintaining $X = 0$ and $Y = 0$ as invariant subspaces.
4. Deduce that linear terms in the residual, say $\text{Res}_x(X, Y) = f_{10}X + f_{01}Y$ and $\text{Res}_y(X, Y) = g_{10}X + g_{01}Y$, generate corrections with $\hat{x}$ and $\hat{y}$ constant, and with $\hat{F}$ linear in X and $\hat{G}$ linear in Y, consistent with maintaining $X = 0$ and $Y = 0$ as invariant subspaces.
5. Argue that a subsequent iterative refinement to the coordinate transform would then generically destroy the aim of maintaining $X = 0$ and $Y = 0$ as invariant subspaces (as shown in Figure 4.22 for Exercise 4.14).
6. More generally, argue that a term of Y^p in Res_x forces a term in $\hat{x} \propto Y^{p-1}$, and correspondingly for $\text{Res}_y \propto X^p$. Hence argue that such residuals generically destroy maintaining $X = 0$ and $Y = 0$ as invariant subspaces.

Exercise 11.5 (growth despite apparent decay). In nonautonomous systems, such as $\dot{y} = G(t)y$, just because eigenvalues of $G(t)$ have real part negative for all t does not mean that all solutions $y(t)$ decay.

1. Consider the Markus–Yamabe system (Chicone, 2006, p. 197)
$$\frac{dy}{dt} = G(t)y \quad \text{for } G = \begin{bmatrix} -1+\frac{3}{2}\cos^2 t & 1-\frac{3}{2}\sin t\cos t \\ -1-\frac{3}{2}\sin t\cos t & -1+\frac{3}{2}\sin^2 t \end{bmatrix}.$$
Find that the eigenvalues of $G(t)$ are $\frac{1}{4}(-1 \pm \sqrt{7}i)$ (independent of time). Despite the eigenvalues having negative real part, there are growing solutions: show that $y = (-\cos t, \sin t)e^{t/2}$ is a solution of the ODE.

2. Consider the normal-form system in center variables $x = (x_1, x_2)$ and stable variables $y = (y_1, y_2)$:
$$\frac{dx}{dt} = \begin{bmatrix} 0 & -1 \\ 1 & 0 \end{bmatrix} x, \quad \frac{dy}{dt} = \begin{bmatrix} -1 & 1 \\ -1 & -1 \end{bmatrix} y + \begin{bmatrix} x_1^2 y_1 - x_1 x_2 y_2 \\ -x_1 x_2 y_1 + x_2^2 y_2 \end{bmatrix}.$$
Observe that $y = 0$ is a center subspace. Show that $x = a(\cos t, \sin t)$ are solutions for the center variables x. Substitute this solution into the y ODE and find that the resulting linear equation for y has eigenvalues with negative real part for $a^2 < 2$. Nonetheless, show that there are solutions that grow exponentially, like $\exp[(a^2-1)t]$, for $a^2 > 1$.
Hence deduce the general result that the center subspace may be attractive only for a range of x more limited than indicated by the eigenvalues of the y ODE.

3. Discuss the relationship between Example 11.8 and the results of the previous part.

Exercise 11.6 (larger domain of emergence). Recall that Lemma 11.16 establishes that solutions in a finite domain $\mathbb{D}_\mu$ are exponentially quickly attracted to the center manifold over at least a finite time. The aim of this exercise is to enlarge the domain $\mathbb{D}_\mu$ and still prove an exponential attraction. Assume there are no unstable variables $\mathbf{Z}$.

In the class of normal-form systems (11.11b) where real matrix P diagonalizes real B in the sense that $P^{-1}BP$ is in (real) Jordan form, suppose matrix P also approximately diagonalizes G in a rough sense, namely that the product $P^{-1}GP = \Delta + E + \mathcal{E}$, where diagonal $\Delta = \operatorname{diag}[\delta_j(\mathbf{X},\mathbf{Y})]$, matrix $E(\mathbf{X},\mathbf{Y})$ is symmetric, and matrix $\mathcal{E}(\mathbf{X},\mathbf{Y})$ is antisymmetric. Modify the proof of Lemma 11.16 to deduce that $|\mathbf{Y}|_P \leqslant |\mathbf{Y}_0|_P e^{-\mu t}$ for as long as $(\mathbf{X},\mathbf{Y})$ stays in the domain $\mathbb{D}_\mu$ such that
$$\max_{j,(\mathbf{X},\mathbf{Y})\in\mathbb{D}_\mu} \left[\Re\beta_j + \delta_j(\mathbf{X},\mathbf{Y})\right] + \operatorname{cond} P\,\|E(\mathbf{X},\mathbf{Y})\| < -\mu\,.$$

Reconsider Example 11.5 and use the new definition of the domain $\mathbb{D}_\mu$ to derive that $\mathbb{D}_\mu$ may be the entire half-space $X_1 > -(1-\mu)$.

Exercise 11.7 (degenerate matrix emergence). Prove emergence (that is, Lemma 11.16) in the degenerate case where matrix B is not diagonalizable but is instead similar to a (real) Jordan-form matrix.

1. Consider the Jordan block $J = \begin{bmatrix} \lambda & 1 \\ 0 & \lambda \end{bmatrix}$: our aim is to show that J is similar to another Jordan-form matrix with bounded quadratic form. Choose parameter $0 < \epsilon < 1$ and let matrix $Q = \operatorname{diag}(1, \epsilon)$; recalling that the norm of a symmetric matrix is its largest eigenvalue, deduce that the condition number $\operatorname{cond} Q = 1/\epsilon$. Use Q to show that matrix J is similar to $J_\epsilon = \begin{bmatrix} \lambda & \epsilon \\ 0 & \lambda \end{bmatrix}$. Derive the bound $\mathbf{Z}^T(J_\epsilon^T + J_\epsilon)\mathbf{Z} \leqslant 2(\Re\lambda + \epsilon)|\mathbf{Z}|^2$. That is, this quadratic form of the Jordan form is bounded by no worse than a potentially small perturbation of the eigenvalues, at the cost of invoking a similarity matrix with large condition number.

2. Repeat part 1 for the $k \times k$ Jordan block

$$J = \begin{bmatrix} \lambda & 1 & & 0 \\ & \lambda & \ddots & \\ 0 & & \ddots & 1 \\ 0 & 0 & & \lambda \end{bmatrix}.$$

Find a diagonal matrix Q, with condition number $\operatorname{cond} Q = 1/\epsilon^k$, that shows J is similar to

$$J_\epsilon = \begin{bmatrix} \lambda & \epsilon & & 0 \\ & \lambda & \ddots & \\ 0 & & \ddots & \epsilon \\ 0 & 0 & & \lambda \end{bmatrix}.$$

Derive the bound $\mathbf{Z}^T(J_\epsilon^T + J_\epsilon)\mathbf{Z} \leqslant 2(\Re\lambda + 2\epsilon)|\mathbf{Z}|^2$ (note the extra factor of two).

3. Extend the proof of Lemma 11.16 to (real) nondiagonalizable B. Hint: Since there exists a similarity matrix P such that $D = P^{-1}BP$ is in real Jordan form J, then there exists a similarity matrix P_ϵ such that $D_\epsilon = P_\epsilon^{-1}BP_\epsilon$ is in real Jordan form J_ϵ. Argue that the "cost" is that $\operatorname{cond} P_\epsilon = \epsilon^{-k} \operatorname{cond} P$, where k is the size of the largest Jordan block in D.

Exercise 11.8 (robust sparse modeling). Typically, constructed models rapidly become complicated at higher orders or with increased complexity of the physical system, and rapidly become more susceptible to artificial instabilities. One avenue to restore robustness is to seek sparse models, models with relatively few and well-understood coefficients. That is, we could seek a sparse model system and a coordinate transform (not necessarily sparse) that minimizes differences with the original system over some desired domain. This principle differs from that employed so far, which is that the asymptotic expansions must agree. Furthermore, the principle appears closer to models derived by variational methods (Ferguson, 2004; Lebedev and Cloud, 2003, e.g.). I am not aware of any such development. Research beckons.

As a speculative toy example of some of the ideas mentioned, let's explore a simple example of a sparse model. Consider the system in variables $(x(t), y(t))$

$$\dot{x} = -\tfrac{1}{8}xy\,, \quad \dot{y} = -y + x^2. \tag{11.33}$$

Normal-form modeling For reference, the coordinate transform $x \approx X + \frac{1}{8}XY + \frac{1}{128}XY^2$ and $y \approx Y + X^2 + \frac{1}{4}X^4 - \frac{1}{32}X^2Y^2$ transforms the system (11.33) into the normal form

$$\dot{X} \approx -\tfrac{1}{8}X^3, \quad \dot{Y} \approx (-1 + \tfrac{1}{4}X^2)Y.$$

Further analysis generates higher-order nonlinear terms in the slow X ODE, but the increase in asymptotic accuracy is achieved at a problematic increase in nonlinear complexity.

Sparse modeling Instead of the transform and normal form in the previous paragraph, find an approximation with a "sparser" slow X ODE: what could be sparser than the above $\dot{X} = -X^3$? Here let's seek a linear model $\dot{X} = F_1X$ for some constant F_1. Further, let's require the approximation to be of known accuracy in a specified domain, say in the domain $-1 \leqslant X, Y \leqslant 1$. Consequently, and because of the symmetry in the system (11.33), seek to approximate the system (11.33) by the dynamics of

$$\begin{aligned} \dot{X} &= F_1X, & x &= (1 + a_{11}Y)X + (a_{30} + a_{31}Y)X^3, \\ \dot{Y} &= (G_0 + G_2X^2)Y, & y &= Y + (b_{20} + b_{21}Y)X^2. \end{aligned} \tag{11.34}$$

Use the ODEs (11.33) to find reasonable coefficients in the sparse model (11.34).

1. Substituting the assumed form (11.34), the residuals of the ODEs (11.33), find the (eighth-order) multinomial residuals.
2. Use Chebyshev polynomials, $T_n(z) := \cos(n \arccos z)$, to minimize these residuals over the domain $-1 \leqslant X, Y \leqslant 1$. With some computer algebra find that the residuals are

$$\begin{aligned} \text{Res}_x &= (-F_1 - \cdots - \tfrac{3}{64}a_{11}b_{21})T_1(X)T_0(Y) + \cdots - \tfrac{1}{256}a_{31}b_{21}T_5(X)T_2(Y), \\ \text{Res}_y &= (\tfrac{1}{2} - \tfrac{1}{2}b_{20} + \cdots + \tfrac{1}{4}a_{11}^2)T_0(X)T_0(Y) + \cdots + \tfrac{1}{64}a_{31}^2T_6(X)T_2(Y). \end{aligned} \tag{11.35}$$

3. The beauty of Chebyshev polynomials is that they are bounded, $|T_n(z)| \leqslant 1$ for $|z| \leqslant 1$, and achieve this bound $(n + 1)$ times in $|z| \leqslant 1$. Numerically minimize the sum of squares of the 20 coefficients of $T_m(X)T_n(Y)$ to obtain the (near) optimal approximation

$$\begin{aligned} \dot{X} &= -0.12X, & x &= (1 + 0.11Y)X + (-0.01 - 0.03Y)X^3, \\ \dot{Y} &= (-0.99 + 0.09X^2)Y, & y &= Y + (1.28 - 0.50Y)X^2. \end{aligned} \tag{11.36}$$

4. From the sum of the absolute value of the resultant coefficients in the residual (11.35), deduce that this model (11.36) is guaranteed to have error less than 0.07 throughout the domain $-1 \leqslant X, Y \leqslant 1$.

5. From (11.36) deduce the existence and emergence of the slow manifold $x = X - 0.01X^3$ and $y = 1.28X^2$ throughout the domain $-1 \leqslant X, Y \leqslant 1$. Compare with the normal-form slow manifold of $Y \approx X^2 + \frac{1}{4}X^4$.

6. Compare the evolution on this slow manifold, $\dot{X} = -0.12X$, with that of the normal-form slow manifold.

This example illustrates how one might construct sparse models that are guaranteed to fit the original dynamics to some known error over some known domain. Further, the approach confirms the emergence of the model dynamics and provides a rationale for determining good initial conditions for the model.

Chapter 12

Appropriate initial conditions empower accurate forecasts

Contents

For models of high dimension, initial conditions often have long-lasting effects and need to be estimated correctly. This is seen in examples such as the approach to limit cycles (Winfree, 1974; Guckenheimer, 1975), the quasi-geostrophic approximation (Lorenz, 1986, e.g.), long-wave models of fluid films (Roy et al., 2002, e.g.), dispersion in channels (Mercer and Roberts, 1994, e.g.), and the concept of initial slip in some disciplines (Grad, 1963; Haake and Lewenstein, 1983; Geigenmüller et al., 1983, e.g.). Such dynamics can only be *forecast* accurately with the correct initial conditions.

Let's explore the surprisingly nontrivial issue of providing appropriate initial conditions for a low-dimensional model, given some initial condition for the full dynamics. Our aim is to choose initial conditions that ensure that long-time predictions are accurate.

Example 12.1 (pitchfork bifurcation). As a preliminary example, extend the example system (11.25) of section 11.3.4 by asking the question: what is the appropriate forecast if the system has initial condition $(x(0), y(0)) = (\frac{1}{4}, \frac{1}{2})$, say? The analysis of section 11.3.4 found that the system

$$\dot{x} = \alpha x - xy\,, \quad \dot{y} = -y + x^2$$

could be modeled by the combined coordinate transform and dynamics

$$x = X + XY + \tfrac{1}{2}XY^2, \quad y = Y + \frac{X^2}{1+2\alpha},$$
$$\text{such that} \quad \dot{X} = \alpha X - \frac{X^3}{1+2\alpha}, \quad \dot{Y} = -Y + 2X^2Y.$$

In this normal form, we easily see that the slow manifold $Y = 0$ emerges exponentially quickly from all initial conditions with $|X| < 1/\sqrt{2}$. Since $x = X$ on the slow manifold we are tempted to answer the question by solving the slow manifold evolution with the initial condition that $X(0) = x(0) = \frac{1}{4}$. However, this initial condition is incorrect.

Consider the following argument. The X-dynamics are independent of the decaying fast variable Y and hence the solutions from initial value (X_0, Y_0) approach the solution on the slow manifold starting from $(X_0, 0)$. Thus, given the initial value (X_0, Y_0), the correct initial condition for the slow manifold model is simply that $X(0) = X_0$. This initial condition for the model guarantees long-time fidelity with the solution of the full normal-form system.

How does this straightforward result appear in original variables? After all, in a practical problem we would be given an initial value (x_0, y_0), not the value in transformed variables. The procedure is to revert the coordinate transform to find (X_0, Y_0) corresponding to any given (x_0, y_0); then the model is to be started with initial condition $X(0) = X_0$. An iterative approximate reversion usually suffices. Here rewrite the coordinate transform as the following and iterate:

$$\begin{aligned}
& X = x_0 - XY - \tfrac{1}{2}XY^2, && Y = y_0 - \frac{X^2}{1+2\alpha} \\
\Longrightarrow\quad & X \approx x_0, && Y \approx y_0 \\
\Longrightarrow\quad & X \approx x_0 - x_0y_0 - \tfrac{1}{2}x_0y_0^2, && Y \approx y_0 - \frac{x_0^2}{1+2\alpha} \\
\Longrightarrow\quad & X \approx x_0 - x_0y_0 + \tfrac{1}{2}x_0y_0^2 + \frac{x_0^3}{1+2\alpha}, && Y \approx y_0 - \frac{x_0^2}{1+2\alpha} + \frac{2x_0^2y_0}{1+2\alpha}.
\end{aligned}$$

Discarding the cubic and higher-order terms in the above, the correct initial value for the slow variable is $X(0) = X_0 \approx x_0 - x_0y_0$, not just x_0.

Returning to the original initial condition $(x(0), y(0)) = (\frac{1}{4}, \frac{1}{2})$, we now predict that the slow manifold model should be solved with initial condition $X(0) \approx \frac{1}{4} - \frac{1}{4}\frac{1}{2} = \frac{1}{8}$ (higher-order terms in the approximation improve the asymptotic accuracy).

In essence, in this example the mapping of an initial condition projects any given initial value along the near vertical curves in Figure 11.6 to the slow manifold.[150] These near vertical curves are called *isochrons* (Guckenheimer, 1975, e.g.).[151] This graphic of such

[150]Surprisingly, Exercise 5.2 argues that such projections imply that there is a role in modeling for negative probabilities!

[151]Winfree (1974) considered a nonlinear oscillator with a stable limit cycle Γ as a model for a biological clock. Each point on Γ is assigned a phase ϕ that increases uniformly in time, $\dot{\phi} = \omega$. If a solution on Γ is perturbed away from Γ, then it will relax back to Γ—the biological clock will reset itself—but with an altered phase. Winfree proposed the concept of isochrons to describe how different disturbances induce different changes to

Figure 12.1. *Photograph of a small branch and attached leaves inspires an analogy to a slow manifold and the isochrons attached to each point of the slow manifold.*

vertical curves being based upon the slow manifold brings to mind the botanical picture of leaves emanating from the branch of a tree, as shown in Figure 12.1, and suggests some terminology.

As in Example 12.1, to fully appreciate the provision of initial conditions one should explore the issue using coordinate transforms, as discussed in sections 12.1 and 12.2 below. However, for just an efficient algorithm to determine the initial conditions one could skip directly to section 12.3.

12.1 ▪ The normal form provides good initial conditions

Following Example 12.1, this section establishes that our normal form of a system provides the required initial conditions for a model using the isochrons (Cox and Roberts, 1995; Elphick et al., 1987). Section 12.3, below, develops a computationally practical version of the principle established here.

Inspired by Example 12.1, the near vertical curves of Figure 11.6 filling the state space, and the picture of leaves emanating from a branch shown in Figure 12.1, we define isochrons as follows.

Definition 12.1 (foliation). *Let a dynamical system for* $\mathsf{u}(t) \in \mathbb{R}^n$ *have general solution/flow* $\mathsf{u} = \mathsf{U}(t, t_0, \mathsf{u}_0)$ *when provided with initial condition* $\mathsf{u}(t_0) = \mathsf{u}_0$. *Suppose the*

---

the phase of the solution. To each point Q on Γ, with phase ϕ say, he associated a surface $\Sigma(\phi)$, transverse to Γ, passing through Q. This isochron $\Sigma(\phi)$ consisted of all points P that, after their transient approach to Γ, have the same phase as Q at the same time. Guckenheimer (1975) subsequently proved the existence of isochrons for a nonlinear oscillator using invariant manifolds. Fenichel (1974) established the same idea more generally.

domain $d \subset \mathbb{R}^n$ *of interest is decomposed into a family of disjoint,* $(n-m)$*-dimensional manifolds*[152] $\mathcal{I}(t, s)$*; that is, the domain* $d = \bigcup_s \mathcal{I}(t, s)$ *for relevant* $s \in \mathbb{R}^m$*. If the flow always maps, at all later times, an* $\mathcal{I}$ *into another* $\mathcal{I}$ *(for as long as solutions stay in the domain)—that is, for all* t, t'*, and* s'*, there exists* s *such that* $d \cap \mathbf{U}(t, t', \mathcal{I}(t', s')) \subset \mathcal{I}(t, s)$*—then the decomposition into the family* $\mathcal{I}(t, s)$ *is called a* **foliation** *of the system and the members of the family* $\mathcal{I}(t, s)$ *are called* **leaves** *or* **isochrons.**[153]

In autonomous systems, one normally chooses the leaves/isochrons to be constant in time, $\mathcal{I}(s)$, as in Figure 11.6.

The foliations of prime interest to us are those where variable $s \in \mathbb{R}^m$ parametrizes an emergent center manifold. Due to the attraction of the center manifold, and because isochrons are mapped into isochrons, it follows that all points in an isochron are attracted to the same solution in the center manifold. Hence they have the same long-term evolution.

Corollary 12.2 (isochrons). *In the center-stable manifold,* $Z = 0$*, isochrons of the normal-form system* (11.11) *are the manifolds for fixed* X *and parametrized by* Y *in the transform* (11.11a)*, and similarly for isochrons in the center-unstable manifold.*

Proof. We restrict attention to the center-stable manifold $Z = 0$; for notational simplicity in this proof and without loss of generality we proceed as if there are no unstable modes Z. Notionally let $s = X$ and define $\mathbb{I}(s) := \{(s, Y) \in D_\mu\}$: these are the isochrons of the normal-form system (11.11b) (Murdock, 2003, Thm. 5.1.5). The reason is that all initial conditions in $\mathbb{I}(X_0)$ give the same $X(t)$ in the solution and so map into $\mathbb{I}(X(t))$, as long as the solution stays in D_μ. For any given parameter s, define $\mathcal{I}(s) := \{(x, y) \in d_\mu \mid x = x(s, Y),\ y = y(s, Y),\ (s, Y) \in D_\mu\}$ to be the image of the isochron $\mathbb{I}(s)$ under the coordinate transform (11.11a). Clearly $\mathcal{I}(s)$ are then isochrons of (11.11).

One can proceed similarly for isochrons in the center-unstable manifold. □

Consider only the center-stable manifold For simplicity we henceforth consider systems on the center-stable manifold. That is, there are no strongly unstable modes. The absence of such unstable modes is either because there are none in the physical system, or because we are only concerned with the dynamics on the center-stable manifold. To recap, instead of the (x, y, z) system (11.10), we consider nonlinear systems in the separated form

$$\begin{aligned} \dot{x} &= Ax + f(x, y), \quad x \in \mathbb{R}^m, \\ \dot{y} &= By + g(x, y), \quad y \in \mathbb{R}^n, \end{aligned} \tag{12.1}$$

where the eigenvalues satisfy the exponential trichotomy Assumption 11.6, which here is just a dichotomy, namely that eigenvalues α_k of the $m \times m$ matrix A satisfy $|\Re\alpha_k| \leqslant \alpha$, and the eigenvalues β_k of the $n \times n$ matrix B satisfy $\Re\beta_k \leqslant -\beta < -p\alpha$. The smooth functions f and g are strictly nonlinear, of $\mathcal{O}(2)$, and are $C^{\mathfrak{p}}$.

Under these conditions, recall that Lemma 11.8 ensures that for any chosen order p $(2 \leqslant p < \mathfrak{p})$, there exist both a near identity multinomial coordinate transform, say

$$x = x(X, Y) \quad \text{and} \quad y = y(X, Y), \tag{12.2a}$$

[152]Generally isochrons are manifolds (hence symbolized by calligraphic I: I for isochron, and calligraphic as a manifold). However, being a manifold is not strictly necessary, and there may be unusual cases when one invokes isochrons that are not strictly manifolds.

[153]This concept of foliation by leaves/isochrons is connected to the concepts of fibers, fibrations, fiber maps, and fiber bundles (Murdock, 2003, pp. 300–302, e.g.). Isochrons have also been termed **"isochronic manifolds"** (Roberts, 1989a, e.g.).

and a companion multinomial dynamical system, say

$$\dot{X} = AX + F(X) \quad \text{and} \quad \dot{Y} = BY + G(X, Y), \tag{12.2b}$$

where $G(X, 0) = 0$ and such that the dynamics of (12.2) is $\mathbb{O}(p)$ close to that of the specified system (12.1).

- Observing that $Y = 0$ is invariant under (12.2b), Proposition 11.13 then asserts the existence of an m-dimensional center manifold $\mathcal{M}_c$ for the system (12.2). Corollary 11.15 then asserts that the $\dot{X}$ ODE of (12.2b) gives the evolution on $\mathcal{M}_c$:

$$\dot{X} = AX + F(X) \quad \text{on } (x, y) = (x(X, 0), y(X, 0)). \tag{12.3}$$

  Let's call this the *model*.

- Choose a positive rate $\mu < \beta$. Then Proposition 11.17 assures us that there exists a finite domain d_μ, about the origin, such that for each solution $P(t) := (x(t), y(t))$ of the system (12.2) there is a solution $X(t)$ of the model (12.3), such that upon defining $Q(t) := (x(X, 0), y(X, 0))$,

$$|P(t) - Q(t)| = \mathcal{O}(e^{-\mu t}) \quad \text{for as long as } P, Q \in d_\mu\,. \tag{12.4}$$

Therefore solutions of the system (12.2) are exponentially quickly attracted to $\mathcal{M}_c$ (at least for solutions in d_μ), and at large times the essential dynamical behavior of the system (12.2) is captured by the model (12.3). But it is only with correct initial conditions that predictions made by the low-dimensional model (12.3) will be accurate over long times.

Corollary 12.3 (good initial conditions). *Suppose that* $(x(0), y(0)) = (x_0, y_0) \in d_\mu$ *is the specified initial condition for the original system* (12.1). *Use the coordinate transform* (12.2a) *to map* (x_0, y_0) *to* $(X_0, Y_0) \in D_\mu$. *The initial condition for the model* (12.3) *that* $x(0) = X_0$ *then ensures that the model and the original agree with errors* $\mathbb{O}(p)$ *and transients* $\mathcal{O}(e^{-\mu t})$ *for as long as they stay in* d_μ.

Proof. First, if the initial value X_0 is fixed, then all solutions of the transformed system (12.2b) from initial conditions (X_0, Y_0) have the same long-term dynamics, regardless of the value of Y_0, provided $(X(t), Y(t)) \in D_\mu$. This is because the evolution equation (12.2b) for X is independent of Y. Therefore $(X_0, 0)$ is the initial condition on $\mathbb{M}_c$ for a solution that evolves identically to the solution from the initial condition (X_0, Y_0) off $\mathbb{M}_c$ apart from transients $\mathcal{O}(e^{-\mu t})$.

Second, the coordinate transformation (12.2a) maps this property to the system (12.2), provided only that we determine the initial condition $(X_0, Y_0) \in D_\mu$ corresponding to any specified $(x_0, y_0) \in d_\mu$. Since the coordinate transform is determined to reproduce the dynamics of the original (12.1) to errors $\mathbb{O}(p)$ for some chosen order p, there will be an additional error of $\mathbb{O}(p)$. □

The $\mathbb{O}(p)$ errors may evolve to become significant if the initial point is close to a boundary between basins of attraction of different attractors on $\mathcal{M}_c$ or if the solution on $\mathcal{M}_c$ has a positive Lyapunov exponent. An analysis of this limitation is beyond the scope of this chapter.

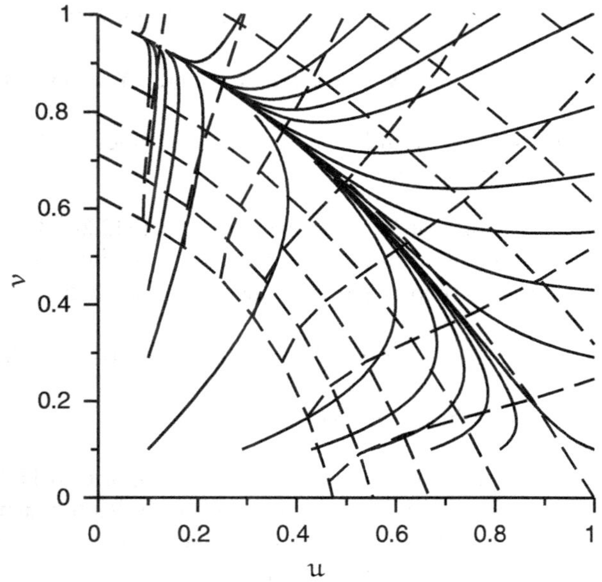

Figure 12.2. *(Solid) trajectories of the competing species system* (12.5) *for parameters* $\alpha = 1/2$ *and* $\beta = 1$*; solutions* $(u(t), v(t))$ *rapidly converge on a slow manifold curving diagonally across the plane. Dashed curves approximate the coordinate system* (12.7) *that gives the normal form for this system.*

Example 12.2 (competing species). Consider the competing species system for populations $(u(t), v(t))$ specified by

$$\dot{u} = u(1 - u - \beta v) \quad \text{and} \quad \dot{v} = v(1 - v - \alpha u), \tag{12.5}$$

which models a range of phenomena from population dynamics to competing modes near a multiple bifurcation point (Cox and Roberts, 1995). Figure 12.2 plots trajectories showing a rapid approach to a 1D slow manifold in uv-space. Our challenge is to determine the correct initial condition on this slow manifold for any given initial condition off the slow manifold.

To illustrate our approach we restrict analysis to the degenerate case of $\beta = 1$ and make the linear change of variables $u = x$ and $v = 1 + y - \alpha x$ in order to bring (12.5) into the form (12.1), with

$$\dot{x} = \quad -(1-\alpha)x^2 - xy, \tag{12.6a}$$

$$\dot{y} = -y - \alpha(1-\alpha)x^2 - y^2. \tag{12.6b}$$

Apply the first nonlinear stage of our normal-form construction, to remove unnecessary quadratic terms from (12.6), by seeking the near identity coordinate transformation

$$x = X + aX^2 + bXY + cY^2 \quad \text{and} \quad y = Y + dX^2 + eXY + fY^2, \tag{12.7}$$

with constants a–f to be determined. Under this change of variables the system (12.6) becomes

$$\dot{X} - bXY - 2cY^2 = -(1-\alpha)X^2 - XY + \mathbb{O}(3),$$
$$\dot{Y} - eXY - 2fY^2 = -Y - [d + \alpha(1-\alpha)]\,X^2 - eXY - [f+1]\,Y^2 + \mathbb{O}(3).$$

To simplify the equations that govern $\dot{X}$ and $\dot{Y}$ we choose $b = 1$, $c = 0$, $d = -\alpha(1-\alpha)$, and $f = 1$. Coefficients a and e remain at our disposal and we choose to set them to zero. The evolution equations for X and Y are then

$$\dot{X} = -(1-\alpha)X^2 \quad \text{and} \quad \dot{Y} = -Y. \tag{12.8}$$

By construction we know that system (12.7)–(12.8) reproduces the dynamics of the original system (12.6) to errors $\mathbb{O}(3)$.

Suppose now that we are given the initial condition $(x(0), y(0)) = (x_0, y_0)$ (small) for (12.6). Then by asymptotically reverting the coordinate transform (12.7), the corresponding values for X and Y are

$$X_0 = x_0 - x_0 y_0 + \mathbb{O}(3),$$
$$Y_0 = y_0 + \alpha(1-\alpha)x_0^2 - y_0^2 + \mathbb{O}(3).$$

According to (12.8), $Y \to 0$ exponentially as $t \to \infty$, and X evolves independently of Y, so the long-term evolution in (12.8) is the same as from the initial condition $(X_0, 0)$. Thus the appropriate initial condition for the model (12.8) to make accurate forecasts is that $X(0) = X_0 = x_0 - x_0 y_0 + \mathbb{O}(3)$.

In the original state space, for (12.6), this initial condition on $\mathcal{M}_c$ is (x_0^*, y_0^*), where from the transform (12.7)

$$x_0^* = X_0 = x_0 - x_0 y_0 + \mathbb{O}(3) \quad \text{and} \quad y_0^* = -\alpha(1-\alpha)x_0^2 + \mathbb{O}(3).$$

Cox and Roberts (1995) compared this normal-form approach to determining initial conditions to two other approaches. The two alternative approaches are first to solve the system exactly and second to use explicit perturbation expansions for *solutions* (instead of the governing *equations*, as in this normal-form method). The conclusion was that, in general, for nonlinear problems the first approach is not available to us, and the algebraic details of the second quickly become overwhelming. Geigenmüller et al. (1983) derived a third method for determining correct initial conditions through a systematic perturbation procedure, but only for linear systems. Such adjustment of the initial conditions has been called the initial slip (Grad, 1963, e.g.). Our nonlinear coordinate transformation empowers us to compute the initial slip by solving a succession of algebraic problems, instead of the differential problems associated with other methods. Nonetheless, for practical purposes section 12.3, below, greatly simplifies our approach by focusing on tangent plane approximations to the isochrons.

12.2 ▪ An infinite-dimensional example: Shear dispersion

We now determine initial conditions for the Taylor model of shear dispersion in a pipe or channel (Taylor, 1953, e.g.). Chapter 7 introduced and discussed this modeling of longitudinal dispersion along pipes, as an example of dispersion in other long thin fluid flows such as in rivers or channels. Here we explore the dispersion along a 2D channel, with the water flow being fastest in the center of the channel, as shown schematically by Figure 12.3. Chapter 7 shows how the variation in water speed with distance from the bank increases material gradients, while cross-channel diffusion tends to smooth gradients out. A balance between these competing physical mechanisms occurs at large enough times after the release of material; then the material concentration varies slowly in

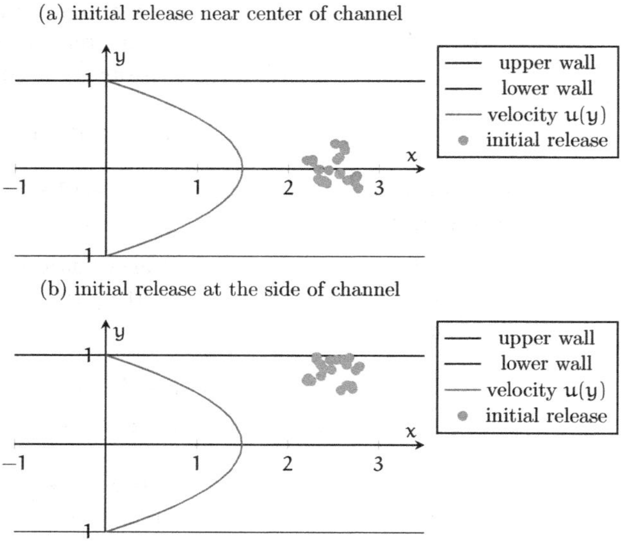

Figure 12.3. *Schematic picture of a 2D channel with fluid carrying material at velocity* $u(y) = \frac{3}{2}(1-y^2)$ *to the right:* (a) *particles (gray) are released in the center of the channel;* (b) *particles (gray) are released on a side.*

the downstream direction. The slow evolution of the concentration obeys an advection-diffusion PDE that was derived in Chapter 7 through the techniques of center manifold theory.

Our challenge here is to understand the difference ensuing from different initial conditions. What is the difference between releasing particles in the middle of the channel, Figure 12.3(a), and releasing them near a side, Figure 12.3(b)? Physically, we know a release in the middle will carry the particles farther downstream than a release at a side, but how does that show up in the mathematics? The answer is by choosing initial conditions correctly (section 12.2.3), as shown in a normal-form transform (section 12.2.2) of the governing equations (section 12.2.1).

The following analysis can be made rigorous in Fourier space, as outlined by section 7.2, but such rigor is as yet generally unavailable in nonlinear, infinite-dimensional problems.[154] In the current absence of rigorous theory, we apply the formal techniques of Cox and Roberts (1995) (who also discussed other methods, such as the adiabatic elimination of Haake and Lewenstein (1983)).

12.2.1 ▪ The linear advection-diffusion of material

Consider the dispersion of a material, with concentration $c(x,y,t)$, in a channel $-1 < y < 1$ for a large domain in x. As in Chapter 7, we nondimensionalize the system with respect to the width of the channel and the cross-channel diffusion time. Thus we take the detailed concentration c to satisfy the nondimensional PDE

$$\frac{\partial c}{\partial t} = -u(y)\frac{\partial c}{\partial x} + \frac{\partial^2 c}{\partial y^2}, \quad -1 < y < 1, \tag{12.9a}$$

[154]Although I recently proposed a new and rigorous theoretical approach to this wide class of modeling problems (Roberts, 2013).

subject to the boundary conditions that no material escapes through the sides of the channel,

$$\frac{\partial c}{\partial y} = 0 \quad \text{at } y = \pm 1 . \tag{12.9b}$$

The PDE (12.9) has two space dimensions; we perform a normal-form transformation to derive a slow manifold involving just one space dimension. Here we focus on the *structure* of the normal-form transformation.

The problem (12.9) is linear in the concentration. The concentration evolves so that after some time the spatial variations in x are small, and so we treat all x-derivatives as nonlinear perturbation terms. With this interpretation, the linearized dynamics is

$$\frac{\partial c}{\partial t} = \frac{\partial^2 c}{\partial y^2} ,$$

which physically describes cross-channel diffusion. The eigenmodes of this PDE are

$$c_n = e^{\lambda_n t} \cos[n\pi(y+1)/2] , \quad n = 0, 1, 2, \ldots ,$$

with corresponding eigenvalues $\lambda_n = -n^2\pi^2/4$. According to this linear picture, all modes decay except c_0. The system is therefore analogous to (12.1), with one critical slow mode, c_0, but infinitely many exponentially damped modes. Since all other modes have zero average in y, for a given concentration field c we determine the component in the critical slow space by taking a y-average to give $\bar{c}(x,t)$; then the component in the stable space is $c'(x,y,t) := c - \bar{c}$.

12.2.2 ▪ A normal form separates mean from decay

Here we derive a normal form for the PDE (12.9) in the channel. In the normal-form transformation we choose new field variables $C(x,t)$ and $D(x,y,t)$ in terms of which the governing PDE (12.9) is simplified; we seek a near identity transformation between $(\bar{c}, c')$ and (C, D) variables. The coordinate transform becomes well defined by requiring a zero cross-sectional average of D over y; that is, $\overline{D} = 0$. Without going into details, it eventuates that the coordinate transform is of the form (using subscripts x and t to denote partial derivatives)

$$\bar{c} = C - \overline{h_1(y) D_x} + \mathcal{O}(\partial_x^2) \quad \text{and} \quad c' = D + h_1(y) C_x + \mathcal{O}(\partial_x^2) \tag{12.10}$$

for some cross-sectional structure function $h_1(y)$, of zero average $\overline{h_1} = 0$, that depends upon the advection velocity profile $u(y)$. In summary, we decompose the concentration field as

$$c(x,y,t) = C + D - \overline{h_1(y) D_x} + h_1(y) C_x + \mathcal{O}(\partial_x^2), \tag{12.11}$$

in terms of a slow field $C(x,t)$ and a fast stable field $D(x,y,t)$.

Let's confirm the decomposition (12.11): substitute it into the governing advection-diffusion PDE (12.9a) to find we need

$$C_t + D_t - \overline{h_1 D_{tx}} + h_1 C_{tx} = -uC_x - uD_x - uh_1 C_{xx} + D_{yy} + h_1'' C_x + \mathcal{O}(\partial_x^2), \tag{12.12}$$

where a D_{xx} second-order derivative is absorbed into the error term. Look at equation (12.12) in two ways: taking the cross-channel average extracts how the slow C evolves, whereas subtracting the average determines how the decaying fast D evolves.

- First average (12.12), remembering that $\overline{D} = \overline{h_1} = 0$ at the boundary $y = \pm 1$ gradients $D_y = h_1' = 0$ and that C is independent of y:

$$C_t - \overline{h_1 D_{tx}} \approx -\overline{u} C_x - \overline{u D_x} - \overline{u h_1} C_{xx}\,. \tag{12.13}$$

The aim of the coordinate transform is to cause this PDE for the slow C to be independent of fast D, so we want $\overline{h_1 D_{tx} - u D_x} = 0$. Now, to leading order the fast variables mix across the channel, $D_t \approx D_{yy}$, so we replace the time derivative D_t by D_{yy} to require $\overline{h_1 D_{xyy} - u D_x} = 0$. Integrate by parts twice to find $\overline{(h_1'' - u) D_x} = 0$. To be zero for all D, we require $h_1'' - u =$ constant. Integrating this ODE across the channel determines that the constant is the average $-\bar{u}$. That is, to eliminate fast D from the slow C we must choose the cross-channel structure function to satisfy the ODE $h_1'' = u - \bar{u}$, with conditions $h_1'(\pm 1) = 0$ and $\overline{h_1} = 0$. Then the slow C equation becomes the classic advection-dispersion PDE

$$C_t \approx -\overline{u} C_x - \overline{u h_1} C_{xx}\,. \tag{12.14}$$

For example, if the velocity is the parabolic $u := \frac{3}{2}(1 - y^2)$ on $-1 \leqslant y \leqslant 1$, then $h_1 := (-7 + 30y^2 - 15y^4)/120$. It follows that $\overline{u h_1} = -2/105$ and so the slow C equation becomes the PDE $C_t \approx -C_x + \frac{2}{105} C_{xx}$.

- Second, determine the evolution of the fast D. Subtract the average (12.13) from the PDE (12.12) to find

$$D_t = -u C_x + \bar{u} C_x - u D_x + \overline{u D_x} + D_{yy} + h_1'' C_x + \mathcal{O}(\partial_x^2), \tag{12.15}$$

where second derivatives C_{xx} and $C_{xt} \approx -\bar{u} C_{xx}$ are absorbed into the error term. To make this fast PDE independent of the slow C we require $(-u + \bar{u} + h_1'') C_x = 0$. This requirement is satisfied as the first step sets $h_1'' = u - \bar{u}$. Hence the fast D field satisfies the PDE

$$D_t \approx D_{yy} - u D_x + \overline{u D_x}\,. \tag{12.16}$$

In summary, the coordinate transform (12.11) separates the advection-diffusion PDE (12.9) into the normal form of the fast-decaying dynamics (12.16) and the advection-dispersion model (12.14) for the slow field $C(x,t)$.

12.2.3 ▪ Project to an appropriate initial condition

The aim is to use the advection-dispersion PDE (12.14) to make predictions for the mean concentration $C(x,t)$. But to make predictions we need initial conditions. These initial conditions must come from the initial concentration field that $c(x,y,t) = c_0(x,y)$ at time $t = 0$. The initial conditions must be able to account for the differences ensuing from release of material in the center of the channel and the side of the channel, as illustrated in Figure 12.3.

The normal-form decomposition (12.10) answers the challenge. Any initial field $c_0(x,y)$ corresponds to an initial field for the normal-form variables $C_0(x)$ and $D_0(x,y)$. From the decomposition (12.10), and in terms of the initial mean field $\bar{c}_0(x) := \overline{c_0(x,y)}$ and initial difference field $c_0'(x,y) := c_0(x,y) - \bar{c}_0(x)$,

$$\bar{c}_0 = C_0 - \overline{h_1(y) D_{0x}} + \mathcal{O}(\partial_x^2) \quad \text{and} \quad c_0' = D_0 + h_1(y) C_{0x} + \mathcal{O}(\partial_x^2).$$

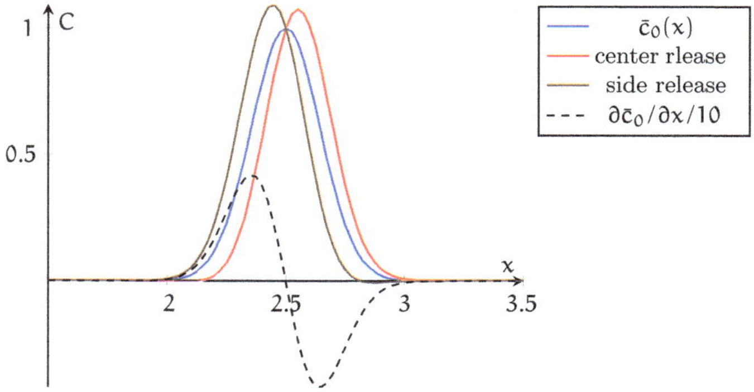

Figure 12.4. *Schematic diagram of the two releases illustrated in Figure 12.3: (blue) the initial cross-channel average concentration $\bar{c}_0(x)$; (dashed) its scaled derivative; (red) the effective initial condition when released in the center of the channel; and (brown) the effective initial condition when released at the side of the channel.*

Since these expressions are formally asymptotic in ∂_x, revert the series to determine

$$C_0 = \bar{c}_0 + \overline{h_1(y)D_{0x}} + \mathcal{O}(\partial_x^2) = \bar{c}_0 + \overline{h_1(y)c'_{0x}} + \mathcal{O}(\partial_x^2),$$
$$D_0 = c'_0 - h_1(y)C_{0x} + \mathcal{O}(\partial_x^2) = c'_0 - h_1(y)\bar{c}_{0x} + \mathcal{O}(\partial_x^2).$$

Since $h_1(y)$ has zero mean, the first of these two is the same as

$$C_0 = \overline{c_0} + \overline{h_1(y)c_{0x}} + \mathcal{O}(\partial_x^2). \tag{12.17}$$

Crucially, the normal-form PDEs (12.14) and (12.16) for the fields C and D are decoupled. So the evolution of C from the initial condition C_0 given by (12.17) holds for all time. Moreover, the predictions made by solving the PDE (12.14) with initial condition (12.17) apply exponentially quickly from any release because the difference field D decays to zero exponentially quickly.

For example, consider a channel with parabolic velocity profile, as shown in Figure 12.3. The two different releases shown in Figure 12.3 have the same cross-channel average $\bar{c}_0(x)$, such as that shown by Figure 12.4 (blue). But because one release is in the center and the other release at the side, we need two different initial conditions for the two releases: equation (12.17) shows that the initial condition depends upon the longitudinal gradient c_{0x}, as shown in Figure 12.4 (dashed). Recall that here the cross-channel structure function $h_1 = (-7 + 30y^2 - 15y^4)/120$, so at the center of the channel $h_1(0) = -7/120 < 0$ and at the sides $h_1(\pm 1) = 8/120 > 0$. Thus from equation (12.17) the correct initial condition for the mean C is shifted downstream when released in the center of the channel, as shown by Figure 12.4 (red). Conversely, the initial condition is shifted upstream when released at a side, Figure 12.4 (brown). The initial conditions (12.17) are essential to ensure correct long-time predictions.

12.3 ▪ Efficiently derive correct initial conditions

The previous sections show that a normal-form coordinate transform provides a wonderful view of the correct initial conditions for a low-dimensional model of detailed dynamics. However, there are two major problems with the normal form in practice:

- we are most interested in applications when there are huge numbers of negligible modes, so the algebra of a complete normal-form transformation is huge as it grapples with all the modes—is there a way to derive initial conditions with vastly less algebra?
- so far we have only derived normal-form transformations where the center modes and the fast modes are explicitly linearly separated, which is inconvenient in most applications—how can we avoid requiring a linearly separated system?

This section answers both challenges (Roberts, 1997, 2000).

Example 12.3 (shear dispersion revisited). Recall that section 12.2 explored predicting dispersion in a channel with the advection-diffusion PDE (12.14) for the cross-channel average concentration $C(x,t)$. Section 12.2 uses a full coordinate transform to derive that a release of material with initial concentration $c_0(x,y)$ implies that the initial condition for the cross-channel average should be (12.17), namely

$$C(x,0) = C_0(x) = \overline{c_0} + \overline{h_1(y)c_{0x}} + \mathcal{O}(\partial_x^2).$$

Defining the inner product $\langle z, u\rangle := \frac{1}{2}\int_{-1}^{1} zu\,dy = \overline{zu}$, this initial condition is of the form

$$C(x,0) = \langle z, c_0\rangle\,, \quad \text{where } z := 1 + h_1(y)\partial_x + \mathcal{O}(\partial_x^2).$$

The significance of this form is that despite the infinite-dimensionality of the cross-channel dynamics, we eventually express the initial condition as a simple projection by a "vector" z that has a simple polynomial structure in y. The aim of this section is to derive such straightforward projection formulas in general without having to invoke the extensive machinations of a full normal-form transformation.

There is a close relationship between this provision of initial conditions and the correct projection of forcing onto a model. In a task taken up more fully in Part VII, this section also shows how the effects of forcing should be properly included in a dynamical model. Computer algebra can handle many of the details.

Recall from section 4.2 that we generally address dynamics for physical state variables $\mathbf{u}(t) \in \mathbb{R}^n$ and some parameters represented by ϵ. The dynamics are governed by ordinary (partial) differential equations written in the form (4.6), repeated here for convenience,

$$\dot{\mathbf{u}} = \mathcal{L}\mathbf{u} + \mathbf{f}(\mathbf{u},\epsilon), \tag{12.18}$$

where the overdot denotes d/dt, $\mathcal{L}\mathbf{u}$ denotes linear terms and $\mathbf{f}$ denotes the strictly nonlinear terms. Based on the equilibrium at the origin, and under the assumptions of a spectral gap in the eigenvalues, Theorem 4.3, there exists an m-dimensional center manifold $\mathcal{M}$ model

$$\mathbf{u} = \mathbf{v}(\mathbf{s},\epsilon) \quad \text{such that} \quad \dot{\mathbf{s}} = \mathbf{g}(\mathbf{s},\epsilon), \quad \mathbf{s} \in \mathbb{R}^m. \tag{12.19}$$

Crucially, given the spectrum at the equilibrium, the Emergence Theorem 4.5 asserts that all solutions "near" the center manifold $\mathcal{M}$ are attracted exponentially quickly to a solution on $\mathcal{M}$. This emergence theorem on the exponential collapse to a smooth invariant manifold lies at the heart of the usefulness of center manifold theory in applications. But to make a forecast, an initial condition has to be found for the model (12.19), say

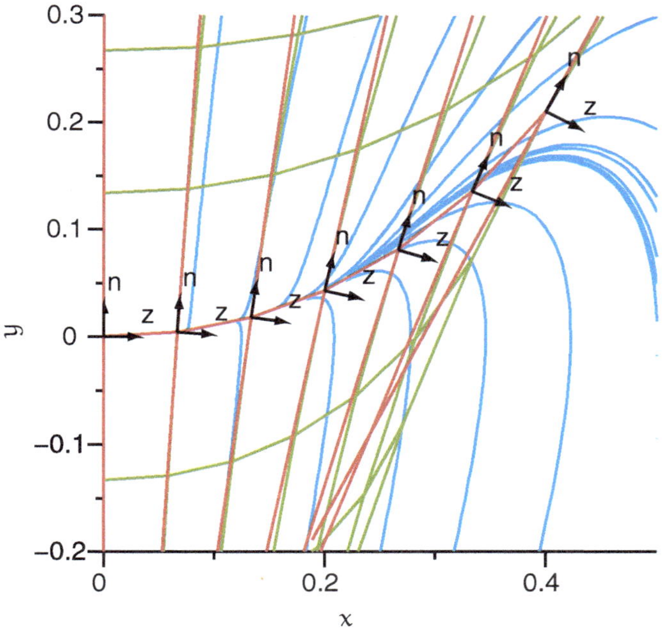

Figure 12.5. *(Blue) trajectories of the dynamical system of Example* 12.1, *namely* $\dot{x} = -xy$ *and* $\dot{y} = -y + x^2$, *exponentially quickly approach a slow manifold; (green) coordinate curves of the normal-form coordinate system; (red lines) tangent line approximations to the isochrons (Y-coordinate curves) as they cross the slow manifold* $\mathcal{M}$ *(red curve); the plotted vectors* $\mathbf{z}$ *are the normal to these tangents and define a projection of initial conditions onto the slow manifold.*

$s(0) = s_0$ for some s_0. The geometric picture of a center manifold and the evolution nearby (Roberts, 1989a), illuminated by normal forms as plotted for two examples in Figures 12.5 and 12.6, is the key to being able to analyze the initial condition problem for low-dimensional models.

The analysis is based upon how trajectories near the center manifold evolve. We implicitly identify which trajectories approach each other exponentially quickly, thus lying on the same isochron (green lines in Figures 12.5 and 12.6), and hence have the same long-term evolution. The tangent space of the isochrons (red lines) is characterized by a linearly independent set of normal vectors, $\mathbf{z}_j$; see Figures 12.5 and 12.6 for two examples.

Theorem 12.4 (initial condition). *The correct initial state on a center manifold* $\mathcal{M}$, *parametrized by* $\mathbf{s}_0$, *is determined by the projection from* $\mathbf{u}_0$ *to* $\mathbf{v}(\mathbf{s}_0, \epsilon)$ *(a displacement* $\mathbf{u}_0 - \mathbf{v}(\mathbf{s}_0, \epsilon)$*) orthogonal to* $\mathbf{z}_j(\mathbf{s}, \epsilon)$,

$$\big\langle \mathbf{z}_j(\mathbf{s}_0, \epsilon), \mathbf{u}_0 - \mathbf{v}(\mathbf{s}_0, \epsilon)\big\rangle = 0 \quad \textit{for } j = 1, \dots, m, \tag{12.20}$$

where $\langle\cdot,\cdot\rangle$ *is a suitable inner product on the state space. The error in* $\mathbf{s}_0$ *is* $\mathcal{O}\big(|\mathbf{u}_0 - \mathbf{v}(\mathbf{s}_0, \epsilon)|^2\big)$ *as* $\mathbf{u}_0 \to \mathcal{M}$. *Section* 12.3.1 *shows that we find the normal vectors* $\mathbf{z}_j(\mathbf{s}, \epsilon)$ *through solving*

$$\mathcal{D}\mathbf{z}_j = \sum_{k=1}^{m} \big\langle \mathcal{D}\mathbf{z}_j, \mathbf{e}_k \big\rangle \mathbf{z}_k\,, \tag{12.21}$$

where, as defined by (12.25), $\mathcal{D}$ *is a dual operator of the dynamical system* (12.18) *linearized about* $\mathcal{M}$, *and where* $\mathbf{e}_k(\mathbf{s}, \epsilon)$ *are tangent vectors to* $\mathcal{M}$ *defined by* (12.26).

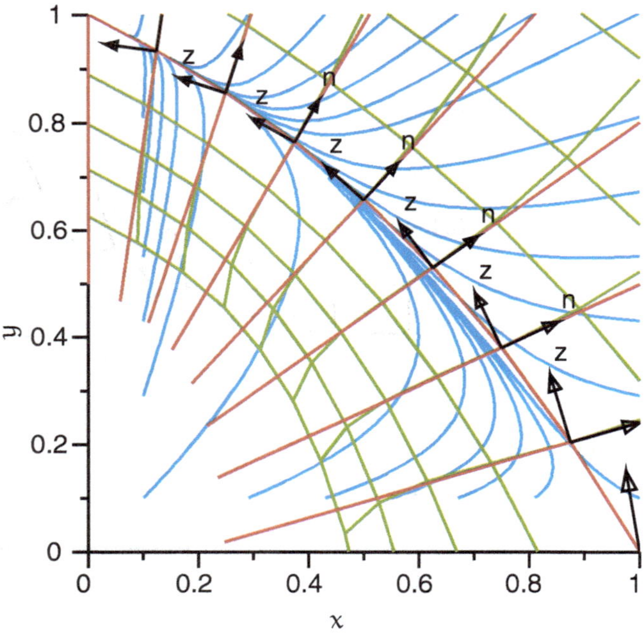

Figure 12.6. *Reconsider the competing species of Example* 12.2. *Adding to the (blue) trajectories and (green) coordinate curves of Figure* 12.2, *(red lines) are the tangent line approximations to the isochrons (Y-coordinate curves) as they cross the slow manifold $\mathcal{M}$ (red curve); the vectors $\mathbf{z}$ are the normal to these tangents and define a projection of initial conditions onto the slow manifold.*

In a small change of notation, $\mathbf{e}_j$ here denotes tangent vectors to $\mathcal{M}$ that vary along $\mathcal{M}$, whereas the basis vectors for the center subspace $\mathbb{E}$ are now denoted $\mathbf{e}_j^0$. Section 12.3.3 then shows how to solve this equation using an iterative scheme based on that developed in Chapter 5 to construct the center manifold model. The iteration seeks corrections $\hat{\mathbf{z}}_j$ to $\mathbf{z}_j$ by solving

$$-\mathcal{L}^\dagger \hat{\mathbf{z}}_j = \operatorname{Res}_{(12.21)}\,, \tag{12.22}$$

where $\mathcal{L}^\dagger$ is the adjoint of the linear operator in the original system (12.18).[155] (Recall that Table 5.2 summarizes essential information about inner products, adjoints, and left eigenvectors.) The iteration terminates when the residual of (12.21) is zero to some order of error.

This approach is efficient because we

- approximate the details of the curving isochrons by just their tangent, and
- completely characterize the projection along the tangents by just m normal vectors $\mathbf{z}_j$.

For simplicity, we assume here that the critical eigenvalues of linear operator $\mathcal{L}$ are all zero and so explicitly only address slow manifolds, denoted $\mathcal{M}_0$. The case of nonzero, pure imaginary eigenvalues has extra complicating details that are addressed by Exercise 12.6.

[155] Such an adjoint also arises in the deliberations of meteorologists on providing suitable initial conditions for weather forecasting (Rabier et al., 1996; Lorenz and Emanuel, 1998, e.g.).

12.3.1 ▪ Evolution near the center manifold

Consider the evolution of a point on the slow manifold $\mathcal{M}_0$, and the evolution of any neighboring point (Roberts, 1989a, §5.1). Let n denote the small vector joining these points; some special examples are illustrated by Figures 12.5 and 12.6. Then, under the flow of the dynamics of the system (12.18), the vector $n(t)$ satisfies the linear equation

$$\frac{dn}{dt} = \mathcal{J}n, \quad \text{where } \mathcal{J} = \mathcal{L} + \mathcal{N}, \tag{12.23}$$

which is the Jacobian of (12.18) evaluated on $\mathcal{M}_0$ and where $\mathcal{N} = \partial f/\partial u$ is the Jacobian of the nonlinear and parameter-dependent terms. The Jacobians $\mathcal{N}$ and hence $\mathcal{J}$ are functions of position s on the slow manifold. Now consider any given n to be a function of position on $\mathcal{M}_0$ rather than time, that is, consider as $n(s)$ where s evolves in time. In this case the chain rule deduces that for equation (12.23), and using hereafter the summation convention that repeated subscripts in a term are summed over 1 to m,

$$\frac{dn}{dt} = \frac{\partial n}{\partial s}(\mathcal{G}s + g) = \left(\mathcal{G}_{jk}s_k + g_j\right)\frac{\partial n}{\partial s_j}\,.$$

Under the evolution in the neighborhood of $\mathcal{M}_0$ the small displacement vector n will do either of two things:

- the tip of n off $\mathcal{M}_0$ will approach $\mathcal{M}_0$ exponentially quickly but in general it will slip from the base point on $\mathcal{M}_0$ and thus n will exponentially quickly become tangent to $\mathcal{M}_0$;

- alternatively, if n is aligned just right, then the tip will approach the base on $\mathcal{M}_0$ while remaining transverse to $\mathcal{M}_0$—these are the special n shown schematically in Figures 12.5 and 12.6.

This latter alternative is of interest, as an initial condition of the full dynamics at the tip of n will have a long-term evolution that is indistinguishable from that of the evolution of the base point on $\mathcal{M}_0$. Thus the base point forms the appropriate initial condition for the low-dimensional dynamics on $\mathcal{M}_0$.

12.3.2 ▪ Project onto the slow manifold

To characterize the projection of given initial conditions u_0 onto $\mathcal{M}_0$, we use the normal vectors z to the isochron tangents. More specifically, as Figures 12.5 and 12.6 show, we use the normal vectors of $\mathcal{I}$ where it crosses $\mathcal{M}_0$. Let $n_\alpha = u_0 - v(s_0, \epsilon)$ be a family of small vectors that span the tangent space of $\mathcal{I}$ at $\mathcal{M}_0$. Then in terms of an inner product $\langle\cdot,\cdot\rangle$, we seek m linearly independent vectors, say r_i as a function of position s on $\mathcal{M}_0$ (in Figures 12.5 and 12.6 $r \propto z$), such that

$$\langle r_i, n_\alpha\rangle = 0\,. \tag{12.24}$$

It appears much easier to find the m vectors r_i than the n vectors n_α (possibly infinite in number). The orthogonality (12.24) for n that satisfy the dynamical (12.23) is assured by requiring

$$\mathcal{D}r_i = 0, \quad \text{where} \quad \mathcal{D} := \frac{d}{dt} + \mathcal{J}^\dagger \tag{12.25}$$

is called the *dual* and $\mathcal{J}^\dagger$ denotes the adjoint of operator $\mathcal{J}$ in the specified inner product. The orthogonality is because the dual (12.25)

$$\begin{aligned}
&\implies \left\langle \frac{d\mathbf{r}_i}{dt} + \mathcal{J}^\dagger \mathbf{r}_i, \mathbf{n}_\alpha \right\rangle = 0 \\
&\implies \left\langle \frac{d\mathbf{r}_i}{dt}, \mathbf{n}_\alpha \right\rangle + \langle \mathbf{r}_i, \mathcal{J}\mathbf{n}_\alpha \rangle = 0 \\
&\implies \left\langle \frac{d\mathbf{r}_i}{dt}, \mathbf{n}_\alpha \right\rangle + \left\langle \mathbf{r}_i, \frac{d\mathbf{n}_\alpha}{dt} \right\rangle = 0 \quad \text{by (12.23)} \\
&\implies \frac{d}{dt}\langle \mathbf{r}_i, \mathbf{n}_\alpha \rangle = 0,
\end{aligned}$$

which ensures (12.24) provided it holds somewhere on each trajectory; typically, holding at the origin will be sufficient. Equation (12.25) describes how the normal vectors $\mathbf{r}_i$ vary over $\mathcal{M}_0$.

We need to solve (12.25). But in general the vectors $\mathbf{r}_i$ inconveniently vary quickly in magnitude and direction over $\mathcal{M}_0$, whereas all we are interested in is the space spanned by $\mathbf{r}_i$. That is, we just need the space of normals to isochrons $\mathcal{I}$ at $\mathcal{M}_0$. Consequently, we seek a basis for the space, one that varies relatively slowly over $\mathcal{M}_0$, by the invertible transformation $\mathbf{r}_i = Q_{ij}\mathbf{z}_j$ for some basis vectors $\mathbf{z}_j$ that are also a function of position on $\mathcal{M}_0$. For a reason discussed in section 12.4, below, we seek the particular basis such that

$$\langle \mathbf{z}_i, \mathbf{e}_j \rangle = \delta_{ij}\,, \quad \text{where} \quad \mathbf{e}_j(\mathbf{s}, \epsilon) := \frac{\partial \mathbf{v}}{\partial s_j} \tag{12.26}$$

are local tangent vectors to $\mathcal{M}_0$ depending upon the chosen parametrization of $\mathcal{M}_0$; typically $\mathbf{e}_j \to \mathbf{e}_j^0$ as $(\mathbf{s}, \epsilon) \to \mathbf{0}$. Substituting $\mathbf{r}_i = Q_{ij}\mathbf{z}_j$ into (12.25) then leads to

$$\begin{aligned}
&\frac{dQ_{ij}}{dt}\mathbf{z}_j + Q_{ij}\mathcal{D}\mathbf{z}_j = 0 && (\sharp) \\
\implies &\frac{dQ_{ik}}{dt} + Q_{ij}\langle \mathcal{D}\mathbf{z}_j, \mathbf{e}_k \rangle = 0 && \text{upon taking } \langle\ ,\mathbf{e}_k\rangle \\
\implies &-Q_{ij}\langle \mathcal{D}\mathbf{z}_j, \mathbf{e}_k \rangle \mathbf{z}_k + Q_{ij}\mathcal{D}\mathbf{z}_j = 0 && \text{putting } dQ_{ij}/dt \text{ in } (\sharp).
\end{aligned}$$

But Q_{ij} is invertible, and so we must solve

$$\mathcal{D}\mathbf{z}_j - \langle \mathcal{D}\mathbf{z}_j, \mathbf{e}_k \rangle \mathbf{z}_k = 0 \tag{12.27}$$

in conjunction with the orthonormality condition (12.26) in order to find the basis vectors $\mathbf{z}_j$ for the space of normals to the isochron tangents.

Equation (12.27) has a reasonable interpretation (Roberts, 2000). The left-hand side projects the residual of the dual (12.25) onto the tangent space (take $\langle \text{LHS}, \mathbf{e}_\ell \rangle$), and hence (12.27) requires these components of the residual to be zero. The null space of the projection is normal to the isochron (substitute $\sum_j a_j \mathbf{z}_j$ for $\mathcal{D}\mathbf{z}_j$ in the LHS) as residuals in this space just contribute to twisting the basis without changing the actual space. The orthonormality condition (12.26) then closes the problem to give a unique solution for the basis vectors $\mathbf{z}_j$.

To determine the projection of a given initial state $\mathbf{u}_0$ onto an initial state $\mathbf{s}_0$ for the model (12.19), once the basis vectors $\mathbf{z}_j$ are found we solve

$$\langle \mathbf{z}_j(\mathbf{s}_0, \epsilon), \mathbf{u}_0 - \mathbf{v}(\mathbf{s}_0, \epsilon) \rangle = 0 \quad \text{for all } j \tag{12.28}$$

for s_0. This projection is linear in distance away from the center manifold $\mathcal{M}_0$. There will be errors quadratic in the distance. However, in many applications the stable manifold $\mathcal{M}_s$ is precisely the linear stable subspace $\mathbb{E}_s$, as seen in Figure 12.5 for example; hence at least near the origin we may expect that a linear projection onto $\mathcal{M}_0$ will be quite good.

This concludes the derivation of Theorem 12.4.

Example 12.4 (basic projection). Consider the system $\dot{x} = -xy$ and $\dot{y} = -y + x^2 - 2y^2$ (which is close to that depicted in Figure 12.5). This system has the exact slow manifold parametrized by say $x = s$ and $y = s^2$ and with evolution $\dot{s} = -s^3$ (knowing the slow manifold $\mathcal{M}_0$ exactly avoids one source of complication).

- Choose the natural inner product $\langle \mathbf{z}, \mathbf{e} \rangle := \mathbf{z} \cdot \mathbf{e}$.
- The tangent (12.26) to the slow manifold in this parametrization is $\mathbf{e}(s) = d/ds(x, y) = d/ds(s, s^2) = (1, 2s)$.
- The Jacobian (12.23) of the system on $\mathcal{M}_0$ is

$$\mathcal{J} := \begin{bmatrix} -y & -x \\ 2x & -1 - 4y \end{bmatrix}_{(x,y)=(s,s^2)} = \begin{bmatrix} -s^2 & -s \\ 2s & -1 - 4s^2 \end{bmatrix}.$$

- In this inner product, the dual operator (12.25) just involves the transpose of the Jacobian to become

$$\mathcal{D} := \frac{d}{dt} + \begin{bmatrix} -s^2 & 2s \\ -s & -1 - 4s^2 \end{bmatrix}.$$

- The orthogonality condition (12.26) for the basis of projection vectors $\mathbf{z}_j$ here becomes the normalization condition that $\mathbf{z} \cdot \mathbf{e} = \mathbf{z} \cdot (1, 2s) = 1$. Let's satisfy this normalization by seeking vector $\mathbf{z} = (1 - 2sz(s), z(s))$ for some as yet unknown scalar function $z(s)$.
- We determine scalar $z(s)$ so as to satisfy the dual equation (12.27). For the immediately above vector $\mathbf{z}$, straightforward algebra gives, with $z' = dz/ds$,

$$\mathcal{D}\mathbf{z} = \begin{bmatrix} 2sz - s^2 + 4s^3 z + 2s^4 z' \\ -z - s - 2s^2 z - s^3 z' \end{bmatrix}.$$

From this $\mathcal{D}\mathbf{z}$ we find $\langle \mathcal{D}\mathbf{z}, \mathbf{e} \rangle = -3s^2$, and thence $\langle \mathcal{D}\mathbf{z}, \mathbf{e} \rangle \mathbf{z} = \big(-3s^2 + 6s^3 z, -3s^2 z\big)$. Thus the residual

$$\text{Res}_{(12.27)} = \begin{bmatrix} 2s \\ -1 \end{bmatrix} \left((1 - s^2)z + s + s^3 z'\right).$$

To zero this residual, we solve the ODE $s^3 z' + (1 - s^2)z + s = 0$ for $z(s)$. This ODE is first order and linear so obtain the general solution by multiplying by an integrating factor to find $z = -s + Cse^{+1/(2s^2)}$. The exponential is viciously unbounded as $s \to 0$, so we must set integration constant $C = 0$; that is, the solution is $z = -s$. Hence the projection vector $\mathbf{z} = (1 + 2s^2, -s)$, in qualitative agreement with Figure 12.5 (which is a slightly different system: one notable difference here is that this projection vector is globally valid in s).

- The projection (12.20) then determines initial conditions. For any given initial value (x_0, y_0), we seek initial value s_0 such that

$$\langle z(s_0), (x_0, y_0) - (s_0, s_0^2) \rangle = (1 + 2s_0^2)(x_0 - s_0) - s_0(y_0 - s_0^2) = 0 .$$

This condition is most easily solved iteratively after rearranging to

$$s_0 = x_0 - \frac{s_0(y_0 - s_0^2)}{1 + 2s_0^2} .$$

For example, if $(x_0, y_0) = (\frac{1}{2}, \frac{1}{2})$, then an initial approximation and one iteration give

$$s_0 \approx \tfrac{1}{2} \implies s_0 \approx \tfrac{1}{2} - \frac{\frac{1}{2}(\frac{1}{2} - \frac{1}{4})}{1 + 2 \cdot \frac{1}{4}} = \tfrac{5}{12} .$$

For another example, if $(x_0, y_0) = (\frac{1}{2}, 0)$, then an initial approximation and one iteration give

$$s_0 \approx \tfrac{1}{2} \implies s_0 \approx \tfrac{1}{2} - \frac{\frac{1}{2}(0 - \frac{1}{4})}{1 + 2 \cdot \frac{1}{4}} = \tfrac{7}{12} .$$

The same x_0 corresponds to different s_0 because of the differing initial transients.

12.3.3 ▪ Iteration constructs the projection

Recall that Chapter 5 develops how to construct a center manifold model iteratively, typically empowered by computer algebra. This section develops a corresponding iteration scheme to determine the vectors z_j that govern the projection of initial conditions (Roberts, 2000, §3). Section 12.3.4 then applies the scheme to supplement the pitchfork bifurcation in the Burgers-like problem of section 5.1.

We aim to solve the dual equation (12.27) subject to the orthonormality condition (12.26). The method of solution is to iteratively improve an approximation based upon the residuals of these two equations. We start the iteration with the linear approximation of z_j by eigenvectors z_j^0 (possibly generalized eigenvectors) of $\mathcal{L}^\dagger$, the adjoint linear operator.

Suppose that at any stage in the iteration we know an approximation z_j. We then seek a correction $\hat{z}_j$ to form an improved approximation $z_j + \hat{z}_j$. First, substituting $z_j + \hat{z}_j$ into the orthonormality condition (12.26) gives $\langle \hat{z}_i, e_j \rangle = \delta_{ij} - \langle z_i, e_j \rangle$. Approximating the coefficient of the correction quantity then shows that we impose on corrections $\hat{z}_i$ the requirement that

$$\left\langle \hat{z}_i, e_j^0 \right\rangle = \operatorname{Res}_{(12.26)} := \delta_{ij} - \langle z_i, e_j \rangle . \tag{12.29}$$

Second, substituting $z_j + \hat{z}_j$ into the dual equation (12.27) and omitting products of correction terms gives

$$-\mathcal{D}\hat{z}_j + \langle \mathcal{D}z_j, e_k \rangle \hat{z}_k + \langle \mathcal{D}\hat{z}_j, e_k \rangle z_k = \mathcal{D}z_j - \langle \mathcal{D}z_j, e_k \rangle z_k .$$

Approximating all coefficients of correction quantities, all appearing on the left-hand side, by their leading approximation (superscript zero), we seek a correction such that

$$-\mathcal{L}^\dagger\hat{z}_j + \left\langle \mathcal{L}^\dagger z_j^0, e_k^0 \right\rangle \hat{z}_k + \left\langle \mathcal{L}^\dagger \hat{z}_j, e_k^0 \right\rangle z_k^0 = \mathcal{D}z_j - \langle \mathcal{D}z_j, e_k \rangle z_k .$$

Algorithm 12.1 *Outline of the general iteration to construct the projection of initial conditions onto a slow manifold. The iteration for the initial condition projection could be intertwined with the iteration for the center manifold model; however, we keep them separate for clarity.*

find the center manifold and the evolution thereon, Algorithm 5.1;
initialization and linear approximation;
repeat
    compute residuals of normality (12.26) and adjoint (12.27);
    compute projected adjoint residual;
    solve (12.29) and (12.30) for the correction;
    update approximation;
until residuals are small enough.

The two inner products on the left-hand side vanish, as they both may be transformed to a form $\langle\cdot, \mathcal{L}\mathbf{e}_k^0\rangle$, which is zero as $\mathbf{e}_k^0$ is a critical eigenvector of $\mathcal{L}$. Thus we solve the linear equation

$$-\mathcal{L}^\dagger \hat{\mathbf{z}}_j = \operatorname{Res}_{(12.27)} := \mathcal{D}\mathbf{z}_j - \langle \mathcal{D}\mathbf{z}_j, \mathbf{e}_k\rangle \mathbf{z}_k \tag{12.30}$$

for the corrections $\hat{\mathbf{z}}_j$. These corrections are simply driven by the residual of the projected dual (12.27) evaluated at the current approximation. Based upon this derivation, Algorithm 12.1 outlines the general scheme.

One uncomfortable feature of (12.30) is that typically during the course of the iteration the right-hand side is not in the range of $\mathcal{L}^\dagger$. This occurs because we approximate $\mathcal{J}^\dagger$, the true adjoint at an arbitrary point on $\mathcal{M}$, by $\mathcal{L}^\dagger$, the tractable adjoint at the origin. To proceed we simply project the residual appearing on the right-hand side into the range of $\mathcal{L}^\dagger$. This projection is readily done by subtracting $\langle \operatorname{Res}_{(12.27)}, \mathbf{e}_k^0\rangle \mathbf{z}_k^0$ from $\operatorname{Res}_{(12.27)}$ on the right-hand side. This projection introduces the same order of error as the other approximations made in the iterative scheme, and so the residuals still systematically decrease. After a finite number of iterations, the dual equation (12.27) will be solved to some asymptotic order in amplitudes $\mathbf{s}$ and parameters $\boldsymbol{\epsilon}$.

12.3.4 ▪ Example: Pitchfork bifurcation in Burgers' PDE

A principal reason for adopting iteration is that the iteration is simply implemented in computer algebra. This section implements the iterative Algorithm 12.1 applied to determining initial conditions for the model (5.5),

$$\dot{a} = \epsilon a - \frac{1}{12}a^3 + \mathcal{O}(\epsilon^2 + a^4),$$

of the Burgers-like PDE (5.1),

$$\frac{\partial u}{\partial t} = -u\frac{\partial u}{\partial x} + (1+\epsilon)u + \frac{\partial^2 u}{\partial x^2}, \quad u(0) = u(\pi) = 0.$$

A key ingredient to code for the dual is the adjoint of the Jacobian of the PDE. Here the (Fréchet) derivative with respect to u of the RHS of the PDE (5.1) is the Jacobian $\mathcal{J} = -u\partial_x - u_x + (1+\epsilon) + \partial_x^2$. Choosing the inner product $\langle z, e\rangle := \frac{2}{\pi}\int_0^\pi z(x)e(x)\,dx$, we

Algorithm 12.2 *Following after Algorithm* 5.2, *this iteration finds the projection of initial conditions onto the slow manifold of the forced Burgers* PDE (5.1).

```
1   operator mean; linear mean;
2   let { mean(1,x)=>2
3       , mean(cos(~~k*x),x)=>0 };
4   operator lainv; linear lainv;
5   let { lainv(sin(~k*x),x)  => sin(k*x)/(1-k^2)
6       , lainv(sin(x),x)=>0 };
7   es:=df(u,a);
8   z:=sin(x);
9   for it:=1:99 do begin
10      write norm:=trigsimp( mean(z*es,x)-1 ,combine);
11      dz:= df(z,t)+(1+eps)*z+u*df(z,x)+df(z,x,2);
12      write resz:=trigsimp( dz-mean(dz*es,x)*z ,combine);
13      z:=z-lainv(resz,x)-norm*sin(x);
14      if {resz,norm}={0,0} then it:=1000000;
15  end;
```

find the adjoint from

$$\begin{aligned}
\langle z, \mathcal{J}v\rangle &= \tfrac{2}{\pi}\int_0^\pi z\left[-uv_x - u_x v + (1+\epsilon)v + v_{xx}\right]\,dx \\
&= \tfrac{2}{\pi}\int_0^\pi \underbrace{-zuv_x - zu_x v}_{=-(uv)_x z} + (1+\epsilon)zv + zv_{xx}\,dx \\
&\qquad \text{(integrate by parts, with } z = v = 0 \text{ at } x = 0, \pi) \\
&= \tfrac{2}{\pi}\int_0^\pi +uvz_x + (1+\epsilon)zv + z_{xx}v\,dx \\
&= \tfrac{2}{\pi}\int_0^\pi \left[uz_x + (1+\epsilon)z + z_{xx}\right]v\,dx \\
&= \langle \mathcal{J}^\dagger, z, v\rangle
\end{aligned}$$

upon defining the adjoint $\mathcal{J}^\dagger z := uz_x + (1+\epsilon)z + z_{xx}$. This adjoint is coded in the heart of Algorithm 12.2.

Implemented in Reduce for the PDE (5.1), Algorithm 12.1 may be realized as Algorithm 12.2.

- **Lines 1–3** For the inner product we need the operator `mean` to compute the integral average over x.
- **Lines 4–6** Corrections arise from solving (12.30), which is of the form $-\mathcal{L}^\dagger\hat{z}_j =$ RHS. Define the operator `lainv` to act as the inverse of $\mathcal{L}^\dagger$. Since operator $\mathcal{L}$ is self-adjoint, $\mathcal{L}^\dagger$ has the same inverse as $\mathcal{L}$, except that here we explicitly set to zero any component in the RHS that is outside the range of $\mathcal{L}^\dagger$, namely any $\sin x$ component.
- **Line 7** From the slow manifold (5.4), the local tangent vector to $\mathcal{M}_0$ depends upon position on $\mathcal{M}_0$ as

$$e(x) = \sin x - \frac{1}{3}a\sin(2x) + \frac{3}{32}a^2\sin(3x) + \mathcal{O}\big(a^3 + \epsilon^{3/2}\big).$$

- **Line 8** This gives the initial linear approximation to the projection (which is the same as $e^0(x)$ as $\mathcal{L}$ is self-adjoint):

$$z \approx z^0 = \sin x .$$

- **Lines 9–15** These perform the iteration until the residuals are negligible according to the order of error set in Algorithm 5.2.
  - **Lines 10–11** These compute residuals of normality (12.26) and dual (12.25) equations for the current approximation.
  - **Line 12** This computes the projected dual residual (12.27).
  - **Line 13** This solves for the correction and updates the approximation to the projection vector $z(x)$.

Executing Algorithm 12.2 derives that the projection vector

$$z(a,x) := \left(1+\tfrac{1}{18}a^2\right)\sin x + \tfrac{1}{6}a\sin(2x) + \tfrac{1}{96}a^2\sin(3x) + \mathcal{O}\left(a^3+\epsilon^{3/2}\right). \qquad (12.31)$$

To find a correct initial condition for the model (5.5) we then do the following. Given an initial condition for the Burgers' PDE (5.1), namely that $u = u_0(x)$ at $t = 0$, we project onto the slow manifold by solving for the amplitude a_0 in the nonlinear equation

$$\langle z(a_0,x), u_0(x) - v(a_0,x)\rangle = 0 . \qquad (12.32)$$

An iterative approach usually suffices. Starting with the first approximation $a_0 \approx \langle z^0, u_0\rangle$, the next approximation is

$$a_0 \approx \langle z^0, u_0\rangle + \langle z(a_0,x), u_0(x) - v(a_0,x)\rangle ,$$

where the first approximation is used on the RHS. For example, if $u_0 = \alpha\sin x$ for some particular α, then

$$a_0 = \alpha + \frac{1}{36}\alpha^3 + \mathcal{O}(\alpha^4, \epsilon^2) ,$$

and *not* simply $a_0 = \alpha$, as would be implied by a direct application of the definition of amplitude a.

In this particular application the issue of the precisely correct initial condition is not of great interest because the ultimate fate of the dynamics is absorption by a stable equilibrium, and an incorrect initial condition just causes a small error in the timing of the absorption. However, in more complicated dynamical models, such as in the shear dispersion of material in a pipe or channel discussed in section 12.2.1, errors in the initial condition can cause significant long-term errors in the predictions of a model.

12.4 ▪ Model forcing or uncertainty by the same projection

This section considers the general dynamical system (12.18) with a small, of $\mathcal{O}(\epsilon)$, forcing superimposed. Alternatively, the small forcing may be due to uncertainty in the system (12.18). For another alternative, the forcing may be due to a small refinement of the detailed modeling, but one does not want to redo all the gory details of the modeling. We argue here that any forcing or uncertainty should be projected into the model using the same projection derived for initial conditions. The error in such a projection is quadratic in amplitude of the forcing or uncertainty but is global over the center manifold.

Example 12.5 (basic projection). Recall that Example 12.4 considered the system $\dot{x} = -xy$ and $\dot{y} = -y + x^2 - 2y^2$ (which is close to that depicted in Figure 12.5). This system has the exact slow manifold parametrized by say $x = s$ and $y = s^2$ and with evolution thereon of $\dot{s} = -s^3$. Example 12.4 illustrates that we find initial conditions on the slow manifold using the projection $\langle \mathbf{z}, \cdot \rangle$, where $\mathbf{z} = (1 + 2s^2, -s)$.

Now suppose that there is a small forcing of $\mathcal{O}(\epsilon)$ added to the fast stable y ODE. Alternatively, suppose that the y ODE is uncertain by some constant ϵ. That is, we ask what is the influence of ϵ if the y ODE becomes $\dot{y} = -y + x^2 - 2y^2 - \epsilon$ for some small ϵ. The usual heuristic people employ is that the ϵ term is a small effect in the least significant variable, namely y, so surely we can neglect this small term. But such small effects can have significant consequences for the slow variables.

This section argues that under such a forcing the evolution on the slow manifold becomes

$$\dot{s} = -s^3 + \langle \mathbf{z}, (0, -\epsilon) \rangle + \mathcal{O}(\epsilon^2) = -s^3 + \epsilon s + \mathcal{O}(\epsilon^2).$$

That is, this small modification (or uncertainty) of the least significant ODE causes the stable equilibrium at $s = 0$ to bifurcate into two equilibria of the remarkably large amplitude $s = \pm\sqrt{\epsilon}$.

Let's turn to a reasonably general case. Modify the general dynamical system (12.18) and now consider

$$\dot{\mathbf{u}} = \mathcal{L}\mathbf{u} + \mathbf{f}(\mathbf{u}, \epsilon) + \varepsilon \mathbf{p}(\mathbf{u}, t, \epsilon) \tag{12.33}$$

for small forcing/uncertainty/refinement $\varepsilon\mathbf{p}$ (be careful to distinguish the parameters ϵ from the amplitude ε of the modification). The new modifying term could be nonautonomous, deterministic (Roberts, 1989a; Cox and Roberts, 1991), or stochastic (Chao and Roberts, 1996). Our aim is to transform the modification of the detailed system (12.33) into a corresponding modification of the slow manifold model (12.19). That is, we seek the forced center manifold and the evolution thereon in the form

$$\begin{aligned} \mathbf{u} &= \mathbf{v}(\mathbf{s}, \epsilon) + \varepsilon \mathbf{w}(\mathbf{s}, t, \epsilon) + \mathcal{O}(\varepsilon^2), \\ \text{such that} \quad \dot{\mathbf{s}} &= \mathcal{G}\mathbf{s} + \mathbf{g}(\mathbf{s}, \epsilon) + \varepsilon \mathbf{q}(\mathbf{s}, t, \epsilon) + \mathcal{O}(\varepsilon^2), \end{aligned} \tag{12.34}$$

where $\mathbf{w}$ describes the displacement of $\mathcal{M}_0$ and $\mathbf{q}$ is our main interest, as it describes the correct modification to be used in the model.

To find the effects of the modification $\mathbf{q}$, simply substitute (12.34) into the original system (12.33) and group all terms linear in ε to deduce

$$\frac{d\mathbf{w}}{dt} + \mathcal{E}\mathbf{q} = \mathcal{J}\mathbf{w} + \mathbf{p}, \tag{12.35}$$

where $\mathcal{E} = [\mathbf{e}_j]$ is the matrix of tangent vectors $\mathbf{e}_j$ (see equation (5.7)), and here

$$\frac{d\cdot}{dt} = \frac{\partial \cdot}{\partial t} + \frac{\partial \cdot}{\partial \mathbf{s}} (\mathcal{G}\mathbf{s} + \mathbf{g}),$$

due to the direct dependency upon time introduced by any nonautonomous component in the modification $\mathbf{p}(\mathbf{u}, t, \epsilon)$, as well as due to the indirect dependence through the evolution of the field $\mathbf{u}$. Taking $\langle \mathbf{z}_i, \cdot \rangle$ of equation (12.35) and using the adjoint properties,

$$\frac{d}{dt} \langle \mathbf{z}_i, \mathbf{w} \rangle - \left\langle \frac{d\mathbf{z}_i}{dt}, \mathbf{w} \right\rangle + \langle \mathbf{z}_i, \mathbf{e}_j \rangle q_j = \left\langle \mathcal{J}^\dagger \mathbf{z}_i, \mathbf{w} \right\rangle + \langle \mathbf{z}_i, \mathbf{p} \rangle. \tag{12.36}$$

Using the orthonormality (12.26), the term $\langle z_i, e_j \rangle q_j = q_i$. Without loss of generality, we choose the parametrization of positions *near* the slow manifold $\mathcal{M}_0$ so that

$$\langle z_i, w \rangle = 0 . \tag{12.37}$$

Indeed, Cox and Roberts (1991) showed that this is the *only* choice for w that removes clumsy history-dependent integrals from the modification q of the model.[156] Then (12.36) becomes

$$q_i = \langle z_i, p \rangle + \left\langle \frac{dz_i}{dt} + \mathcal{J}^\dagger z_i, w \right\rangle .$$

The last term involves $\mathcal{D}z_i$, which, by the projected dual (12.27), must lie in the space normal to the isochron tangents spanned by the z_k's; $\mathcal{D}z_i$ is thus orthogonal to w, and so the last term vanishes. Hence the appropriate linear approximation due to the modification of the model is simply the projection

$$q_i = \langle z_i, p \rangle ; \quad \text{that is,} \quad \dot{s} = \mathcal{G}s + g + \varepsilon \left[\langle z_i, p \rangle \right] + \mathcal{O}(\varepsilon^2) \tag{12.38}$$

for any given modification p in terms of the vectors z_i determined for the projection of initial conditions.[157]

Example 12.6 (forced Burgers PDE). Recall that section 12.3.4 found the projection of initial conditions for a modified Burgers PDE. Here we add some further forcing or uncertainty or other small modification $\varepsilon p(u, x, t)$ to the PDE (5.1) and consider

$$\frac{\partial u}{\partial t} = (1 + \epsilon)u - u\frac{\partial u}{\partial x} + \frac{\partial^2 u}{\partial x^2} + \varepsilon p(u, x, t). \tag{12.39}$$

Then by the projection vector (12.31) the forcing of the model (5.5) turns it into

$$\dot{a} = \epsilon a - \tfrac{1}{12}a^3 + \varepsilon \langle z, p(v, x, t) \rangle + \mathcal{O}(\epsilon^2 + a^4, \varepsilon^2). \tag{12.40}$$

Some example modifications are the following.

- A spatially uniform, additive forcing $\varepsilon p(t)$ induces a nonlinear forcing in the model: from the projection vector (12.31),

$$\begin{aligned}\langle z, 1 \rangle &= \frac{2}{\pi}\int_0^\pi z\,dx \\ &\approx \frac{2}{\pi}\int_0^\pi (1 + \tfrac{1}{18}a^2)\sin x + \tfrac{1}{6}a\sin(2x) + \tfrac{1}{96}a^2\sin(3x)\,dx \\ &\approx \frac{1}{\pi}(4 + \tfrac{17}{72}a^2);\end{aligned}$$

  hence the slow manifold evolution (12.40) becomes

$$\dot{a} = \epsilon a - \tfrac{1}{12}a^3 + \varepsilon\left(\tfrac{4}{\pi} + \tfrac{17}{72\pi}a^2\right)p + \mathcal{O}(\epsilon^2 + a^4 + \varepsilon^2).$$

[156] Part VII explores in detail how this issue arises and why we require the orthogonality (12.37).

[157] There is an interesting inverse problem associated with forcing. It is straightforward to argue that a given forcing of the model corresponds to some forcing of the full system. Thus, any reasonable forcing of the model used in a simulation does actually correspond to some physical forcing of the original system.

- As in Example 12.4, in Burgers' PDE an uncertainty proportional to $p = \sin(2x)$ may change the stability of the origin as
$$\langle z, \sin 2x\rangle = \frac{1}{\pi}\int_0^\pi z\sin 2x\,dx \approx \tfrac{1}{6}a\,,$$
leading to the slow manifold evolution (12.40) becoming
$$\dot a = (\epsilon + \tfrac{1}{6}\varepsilon)a - \tfrac{1}{12}a^3 + \mathcal{O}(\epsilon^2 + a^4 + \varepsilon^2).$$
- If we had omitted to include variations in the bifurcation parameter r in Burgers' PDE (5.1) (that is, we had set $r = 1$ precisely, not $r = 1 + \epsilon$), then we could include the effects without rederiving the slow manifold model. Simply choose modification $p = u$, which on the slow manifold (5.4) becomes $p \approx a\sin x - \frac{1}{6}a^2\sin 2x$. Then, after adding $\langle z, p\rangle$, the slow manifold evolution (12.40) is the same with parameter ϵ replaced by the strength ε of the modification.

12.5 ▪ Exercises

Exercise 12.1 (initial conditions and uncertainty). Recall Exercise 4.2, which found the slow manifold $x = s$, $y \approx 2s^2$, and $z \approx 3s^2$ for the system
$$\dot x = -xy\,,$$
$$\dot y = -2y + z + x^2\,,$$
$$\dot z = y - z + x^2\,.$$

1. In the usual inner product, derive the dual and the orthogonality conditions for a projection vector $\mathbf{z}$.
2. By starting from the initial approximation that $\mathbf{z} \approx (1,0,0)$, iterate to derive that $\mathbf{z} \approx (1 + 10s^2, -s, -s)$.
3. Given the initial condition that $x(0) = y(0) = z(0) = \frac{1}{4}$, determine that the corresponding initial condition for the slow manifold model $\dot s \approx -2s^3$ is $s(0) \approx \frac{7}{52}$.
4. Suppose the coefficient -2 of the term $-2y$ in the y ODE is uncertain by an amount ε. Use the projection to derive that the model is uncertain by a term $-2\varepsilon s^3$.
5. Suppose all four coefficients of all four linear terms in the ODEs are uncertain by independent small errors of the same standard deviation σ. Use the projection to deduce that the coefficient of the model $\dot s \approx -2s^3$ has an error with standard deviation approximately $\sqrt{26}\sigma$.

Exercise 12.2 (damped Lorenz). Recall that Example 4.16 modelled a damped version of the Lorenz system (Lorenz, 1986), namely
$$\dot u = -vw + bvz\,,$$
$$\dot v = uw - buz\,,$$
$$\dot w = -uv\,,$$
$$\dot x = -x - z\,,$$
$$\dot z = +x - z + buv\,.$$

With its damping, this system has an emergent 3D slow manifold $x \approx -\frac{1}{2}buv$ and $z \approx \frac{1}{2}buv$, on which $\dot{u} \approx -vw$, $\dot{v} \approx uw$, and $\dot{w} \approx -uv$.

1. In the usual inner product, derive the dual and the orthogonality conditions for the three projection vectors $\mathbf{z}_j$.
2. By starting from the initial approximation that $\mathbf{z}_1 \approx (1,0,0,0,0)$, $\mathbf{z}_2 \approx (0,1,0,0,0)$, and $\mathbf{z}_3 \approx (0,0,1,0,0)$, iterate to derive that $\mathbf{z}_1 \approx (1,0,0,\frac{1}{2}bv,\frac{1}{2}bv)$, $\mathbf{z}_2 \approx (0,1,0,-\frac{1}{2}bu,-\frac{1}{2}bu)$, and $\mathbf{z}_3 \approx (0,0,1,\frac{1}{2}b(v^2-u^2),0)$, all to error $\mathcal{O}(s^3)$, where $s = |(u,v,w)|$.
3. When dealing with initial conditions it is confusing to parametrize the slow manifold with the original variables: reparametrize the slow manifold in terms of (s_1,s_2,s_3). Given the initial state that $u(0)=\frac{1}{2}$, $v(0)=1$, $w(0)=x(0)=z(0)=0$, determine that the corresponding initial condition for the slow manifold model is $s_1(0) \approx \frac{1}{2}$ and $s_2(0) \approx 1$, but $s_3(0) \approx \frac{3}{32}b^2$.
4. The damping in this system is artificial, so we are interested in the effects of the damping: let's indicate that the effect is weak. Weaken the damping by an amount ε through modifying the system by adding $\varepsilon\mathbf{p} = \varepsilon(0,0,0,x,z)$. Evaluate this modification on the slow manifold, project, and deduce that the slow manifold evolution is only affected at $\mathcal{O}(\varepsilon b^2 s^4)$.

Exercise 12.3 (nonseparated equations). Recall that Exercise 5.3, part 4, considered the following system:

$$\begin{aligned}\dot{x} &= 2x+y+2z+yz\,,\\ \dot{y} &= x-y+z-xz\,,\\ \dot{z} &= -3x-y-3z-xy\,.\end{aligned}$$

It has a 1D slow manifold $x \approx s-s^2$, $y \approx s^2$, and $z \approx -s+s^2$, where $\dot{s} \approx s^2$.

1. In the usual inner product, derive the dual and the orthogonality conditions for the projection vector $\mathbf{z}$.
2. Starting from the initial approximation that $\mathbf{z} \approx (4,1,3)$, use an iteration to derive that $\mathbf{z} \approx (4-16s, 1-9s, 3-16s)$.
3. For the given initial condition that $x(0)=z(0)=0$ and $y(0)=y_0$, use the projection to determine first that $s(0) \approx y_0$ and second that $s(0) \approx y_0 - 9y_0^2$.
4. Suppose we add a small time-dependent forcing to the original system of $\varepsilon\mathbf{p}(t) = \varepsilon(1,-1,-1)p(t)$. Use the projection to deduce that the slow manifold model should be modified to $\dot{s} \approx 9\varepsilon p(t)s + s^2$.
5. Write computer algebra code for Algorithm 12.1 to derive higher-order corrections to these projection vectors; for example, $\mathbf{z} \approx (4-16s+258s^2, 1-9s+93s^2, 3-16s+240s^2)$.

Exercise 12.4 (generalized eigenvectors). Generalized eigenvectors appear to require more iterations in constructing the projection vectors. Recall that Exercise 5.3, part 6, explored the system

$$\begin{aligned}\dot{x} &= 2x+y+2z+yz\,,\\ \dot{y} &= x+z-xz\,,\\ \dot{z} &= -3x-y-3z-xy\,.\end{aligned}$$

A different parametrization of the 2D slow manifold model is $x = s_1+2s_1s_2-2s_2^2$, $y = s_2+2s_1s_2-2s_2^2$, $z = -s_1-4s_1s_2+4s_2^2$, on which the evolution is $\dot s_1 = s_2-3s_1s_2$ and $\dot s_2 = s_1^2 - 2s_1s_2$, all to errors $\mathcal{O}(|s|^3)$.

1. In the usual inner product, derive the dual and the orthogonality conditions for the two projection vectors $\mathbf{z}$.
2. Starting from the initial approximation that $\mathbf{z}_1 \approx (2,0,1)$ and $\mathbf{z}_2 \approx (1,1,1)$, use one or more iterations to derive that $\mathbf{z}_1 \approx (2+4s_1+7s_2, 0, 1+4s_1+7s_2)$ and $\mathbf{z}_2 \approx (1-s_1+2s_2, 1, 1-s_1+2s_2)$.
3. Write computer algebra code for Algorithm 12.1 to derive the quadratic corrections to these projection vectors. It may take six iterations to derive.

Exercise 12.5 (boundary condition forcing). Reconsider the projection of modifications to Burgers' PDE developed in Example 12.6. Let's now consider applying nonzero boundary conditions: instead of requiring the homogeneous $u(0,t) = u(\pi,t) = 0$, we require $u(0,t) = \varepsilon p_0(t)$ and $u(\pi,t) = \varepsilon p_\pi(t)$ for some given functions p_0, p_π (Roberts, 2000, §3.3).

1. Argue that these nonzero boundary conditions may be met by solving the PDE with homogeneous boundary conditions but with "interior forcing" in vanishingly thin layers next to each boundary of
$$\varepsilon p(x,t) = -\varepsilon p_0(t)\delta'(x-0^+) + \varepsilon p_\pi(t)\delta'(x-\pi^-),$$
where δ' denotes the first derivative of the Dirac delta function. Hint: Integrate the forced PDE across these thin layers.
2. Hence deduce that the inhomogeneous boundary condition leads to the model
$$\dot a = \epsilon a - \frac{1}{12}a^3 + \frac{2}{\pi}\varepsilon(p_\pi + p_0) + \frac{1}{3\pi}a\varepsilon(p_\pi - p_0) + \mathcal{O}(\epsilon^2 + a^4 + \varepsilon^2).$$

Exercise 12.6 (general center projection). Chapter 16 develops how to derive center manifold models when there are some pure imaginary eigenvalues, and when the model is to be parametrized by Cartesian variables. This exercise derives the iteration needed to correctly project initial conditions.

Consider a general dynamical system of the form (12.18), with center manifold model (12.19), and suppose there is nontrivial linear dynamics on the center manifold (Roberts, 2000, §4). That is, there is some nonzero $m \times m$ matrix $\mathcal{G}$ such that $\dot{\mathbf{s}} \approx \mathcal{G}\mathbf{s}$ near the equilibrium at the origin; the matrix is the Jacobian, $\mathcal{G} = \partial g/\partial s$, evaluated at $\mathbf{s} = \mathbf{0}$. Given that we seek to satisfy the dual equation (12.21) and the orthogonality condition (12.26), derive that a suitable iteration would be of the form
$$-\mathcal{L}^\dagger\hat{\mathbf{z}}_j + \mathcal{G}_{jk}\hat{\mathbf{z}}_k - \mathcal{G}_{k\ell}s_\ell\frac{\partial\hat{\mathbf{z}}_j}{\partial s_k} = \mathcal{D}\mathbf{z}_j - \langle\mathcal{D}\mathbf{z}_j, \mathbf{e}_k\rangle\mathbf{z}_k \tag{12.41}$$
for corrections $\hat{\mathbf{z}}_j$ to approximate projection vectors $\mathbf{z}_j$ (Roberts, 2000, §4.2).

Exercise 12.7 (projecting onto oscillations). Consider the system (Roberts, 2000, §4)

$$\dot{u} = \begin{bmatrix} -1 & -1 & 0 \\ 2 & 1 & 0 \\ 1 & 2 & -1 \end{bmatrix} u + \begin{bmatrix} -2u_1u_3 \\ 2u_1u_3 \\ u_2^2 \end{bmatrix}.$$

You are given that the center manifold of this system may be parametrized by

$$u = \begin{bmatrix} 2x \\ -2x + 2y \\ -3x + y \quad + \frac{42}{5}y^2 + \frac{22}{5}xy + \frac{88}{5}x^2 \end{bmatrix} + \mathcal{O}(x^3 + y^3),$$

with consequent evolution on the center manifold of

$$\dot{x} = -y - 2xy + 6x^2 + \mathcal{O}(x^3 + y^3), \quad \dot{y} = x + \mathcal{O}(x^3 + y^3).$$

Use equation (12.41) to derive that projection vectors for initial conditions are

$$z_x \approx \begin{bmatrix} \frac{1}{2} + \frac{2}{5}y - \frac{6}{5}x \\ 0 - \frac{1}{5}y + \frac{3}{5}x \\ 0 + \frac{2}{5}y - \frac{6}{5}x \end{bmatrix}, \quad z_y \approx \begin{bmatrix} \frac{1}{2} - \frac{4}{5}y + \frac{2}{5}x \\ \frac{1}{2} + \frac{2}{5}y - \frac{1}{5}x \\ 0 - \frac{4}{5}y + \frac{2}{5}x \end{bmatrix}.$$

Chapter 13

Subcenter slow manifolds are useful but do not emerge

Contents

The case of slow dynamics among fast oscillations/waves is very common (section 13.1), and people often concentrate on the slow dynamics on what is called a subcenter slow manifold. There is little theory that helps justify the construction and utility of subcenter slow manifolds. Unfortunately, the theory I am aware of is extremely limited and is mentioned more for its conceptual support than in any hope of rigor in an interesting application (section 13.2). Perhaps the best support comes from the normal-form coordinate transformations of Chapters 11 and 12. This chapter explores analogous coordinate transformations (section 13.5) to understand three example basic models: the quasi-geostrophic approximation (section 13.3), a drifting electron in a magnetic field (section 13.4), and a spinning top (section 13.6).

One crucial facet of modeling is the following, which is in contrast to the center modes being able to be completely decoupled from stable modes; here, although slow manifolds may exist (Proposition 13.5), slow modes generally cannot be completely decoupled from any fast oscillation/wave modes.

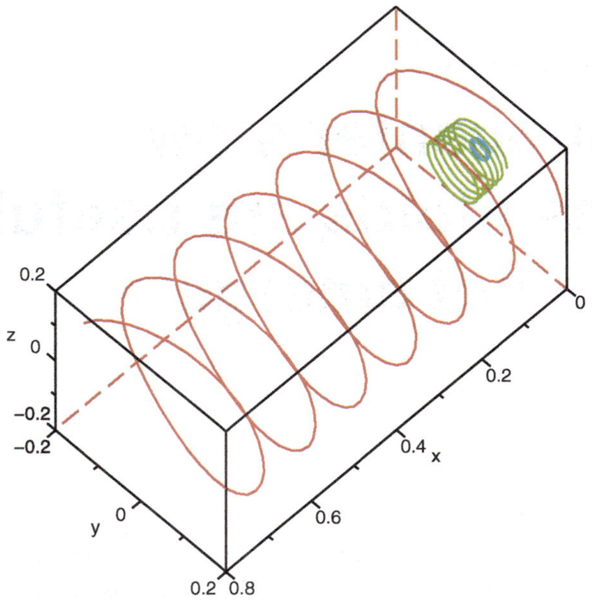

Figure 13.1. *Trajectories of the system* (13.1) *for three different oscillation amplitudes: blue, small amplitude; green, medium; red, larger. The slow* x *variable drifts differently for each.*

Example 13.1 (drift is endemic). Consider the 3D toy system introduced by Cox and Roberts (1995):

$$\dot{x} = z^2, \quad \dot{y} = -z, \quad \dot{z} = y\,. \tag{13.1}$$

Observe in the system (13.1) and in the trajectories of Figure 13.1 that y and z oscillate (linearly), and that these oscillations feed into the slow x-evolution. Simulating such systems over long times can be costly because one would have to resolve the rapid oscillations. But the slow manifold easily empowers long-term predictions: the slow manifold is $y = z = 0$ and then $\dot{x} = 0$, so it predicts that slow x is constant over the long term. Eliminating the fast oscillations results in no evolution on the 1D slow manifold model. The problem is that such a prediction is not an accurate reflection of general solutions of the system, as is readily seen in Figure 13.1!

Wave resonance forces a slow drift The mean effect of the oscillations is a drift in x. Transform to new variables denoted by capital letters:

$$X = x + yz/2\,, \quad Y = y\,, \quad Z = z\,. \tag{13.2}$$

In the new variables we find Y and Z oscillating identically to y and z, as the variables are identical. Consider the slow X-evolution: $\dot{X} = \dot{x} + \dot{y}z/2 + y\dot{z}/2 = z^2 - z^2/2 + y^2/2 = (Y^2 + Z^2)/2$. This evolution identifies that X drifts proportional to the square of the oscillation amplitude:

$$\dot{X} = \tfrac{1}{2}(Y^2 + Z^2), \quad \dot{Y} = -Z\,, \quad \dot{Z} = Y\,. \tag{13.3}$$

The significant generic feature of this example is that the presence of fast oscillations (waves) causes a drift along $x \approx X$ that is not described by the slow manifold model. The drift depends upon the amplitude of the oscillations, $(Y^2 + Z^2)$, as in Figure 13.1. Consequently, there cannot be a strong emergence theorem for slow manifold models when fast oscillatory modes are eliminated. This is an example of the general feature that the evolutions on and off a slow manifold are fundamentally incompatible.

13.1 ▪ Incompressible flow and other slow approximations

> In the study of atmospheric wave motion, the problem of integration is greatly complicated by the simultaneous existence of a discrete set of wave motions.... This extreme generality whereby the equations of motion apply to the entire spectrum of possible motions—to sound waves as well as to cyclone waves—constitutes a serious defect of the equations from a meteorological point of view.... It means that the investigator must take into account modifications to the large-scale motions of the atmosphere which are of little meteorological importance and which only serve to make the integration of the equations a virtual impossibility. *Jule Charney,* 1948

In most common fluid flows we assume that the fluid is incompressible. How can this be realistic when we know that the atmosphere and the ocean are bathed in sound? For example, we use sound to communicate and yet the wind sweeps by apparently unaffected. Many physical systems potentially possess fast oscillations/waves superimposed on the slow large-scale motions that are of main interest. This section introduces many important similar physical situations.

Sound First investigate sound. Consider the continuity and inviscid Euler equations

$$\frac{\partial \rho}{\partial t} + \nabla \cdot (\rho \mathbf{q}) = 0, \tag{13.4}$$

$$\frac{\partial \mathbf{q}}{\partial t} + (\mathbf{q} \cdot \nabla)\mathbf{q} = -\frac{1}{\rho}\nabla p. \tag{13.5}$$

In order to examine the properties of sound we need an equation of state relating pressure to density. For example, we know that the density of a parcel of air will increase if we raise the pressure acting on the parcel. Perhaps the simplest realistic quantitative relation embodying this experience is that of a *polytropic fluid*, one for which[158]

$$p = A\rho^{\gamma}, \tag{13.6}$$

where γ and A are characteristics of the fluid and its temperature. For example, in air $\gamma \approx 7/5$, as is appropriate for a gas of diatomic molecules. The polytropic equation of state (13.6) implies that

$$\nabla p = \frac{\partial p}{\partial \rho}\nabla \rho = \gamma A \rho^{\gamma-1}\nabla \rho = c^2 \nabla \rho,$$

where $c^2 = \partial p/\partial \rho$, in which c turns out to be the speed of sound. The equation of state (13.6) eliminates the pressure from the Euler equation (13.5).

[158]The constant γ characterizes the number of vibrational degrees of freedom of a molecule of the fluid.

A stationary fluid, $\mathbf{q} = 0$, of constant density, $\rho = \rho_0$, is an equilibrium or fixed point of the inviscid Euler equations above.

Consider small-amplitude fluctuations, otherwise known as sound, to such a density and velocity field. That is, seek perturbations

$$\mathbf{q} = 0 + \mathbf{q}', \quad \rho = \rho_0 + \rho',$$

where primed quantities are small. The continuity equation (13.4) gives

$$\begin{aligned} 0 &= \frac{\partial \rho}{\partial t} + \nabla \cdot (\rho \mathbf{q}) \\ &= \frac{\partial \rho'}{\partial t} + \nabla \cdot [(\rho_0 + \rho')\mathbf{q}'] \\ &\qquad \text{(neglecting products of perturbations)} \\ &\approx \frac{\partial \rho'}{\partial t} + \rho_0 \nabla \cdot \mathbf{q}' \\ \implies \quad \frac{\partial \rho'}{\partial t} &= -\rho_0 \nabla \cdot \mathbf{q}' . \end{aligned}$$

Similarly, the Euler equation (13.5) gives

$$\begin{aligned} \frac{\partial \mathbf{q}'}{\partial t} + \mathbf{q}' \cdot \nabla \mathbf{q}' &= -\gamma A \left(\rho_0 + \rho'\right)^{\gamma - 2} \nabla \rho' \\ &\qquad \text{(upon neglecting products of perturbations)} \\ \implies \quad \frac{\partial \mathbf{q}'}{\partial t} &\approx -\gamma A \rho_0^{\gamma-2} \nabla \rho' = -\frac{c_0^2}{\rho_0} \nabla \rho' . \end{aligned}$$

Then consider

$$\begin{aligned} \frac{\partial^2 \rho'}{\partial t^2} &= -\rho_0 \nabla \cdot \left(\frac{\partial \mathbf{q}'}{\partial t} \right) \quad \text{by continuity} \\ &= -\rho_0 \nabla \cdot \left(-\frac{c_0^2}{\rho_0} \nabla \rho' \right) \quad \text{by Euler} \\ \implies \quad \frac{\partial^2 \rho'}{\partial t^2} &= c_0^2 \nabla^2 \rho' \quad \text{(a wave equation).} \end{aligned}$$

Sound arises because density fluctuations in a quiescent atmosphere satisfy a wave equation such as

$$\frac{\partial^2 \rho'}{\partial t^2} = c_0^2 \nabla^2 \rho' .$$

That is, the fluctuations of density we call sound travel through the fluid according to the above wave equation with wave speed c_0.

As seen in university courses, solutions of the wave equation in one spatial dimension may take the form $\rho' = f(x - c_0 t) + g(x + c_0 t)$—a superposition of right- and left-traveling waves. This solution describes waves traveling without change in form or amplitude—exactly our experience with sound and exactly what makes it practical to communicate with sound over long distances.

Slow incompressible fluid flow Now consider the flow of a fluid in the presence of fast sound waves. That is, the speed of sound, c_0, is much faster than the relatively slow speed of the fluid flow. Because sound oscillates very fast, we expect that it has negligible effect on the slow bulk flow. Further, the slow flow evolves according to the incompressible equations of fluid dynamics.

Now a large parameter such as c_0, the speed of sound, is a little inconvenient to deal with in the following derivation. Instead let's use the *Mach number* $M = U/c_0$, where U is a typical speed of the fluid flow. Here imagine we have chosen units, or nondimensionalized, so that the fluid flows with velocities of size 1. Then the Mach number, $M = 1/c_0$, is a *small* parameter.

To investigate flow in the presence of superposed sound, decompose the velocity and density fields as

$$\rho = \rho_0 + M^2\tilde{\rho} + M^3\rho' + \cdots, \tag{13.7}$$

$$q = \tilde{q} + M^2 q' + \cdots, \tag{13.8}$$

where it eventuates that ρ_0 is constant, quantities with tildes are those of the slow bulk flow, and primed quantities describe the sound field.

One oddity of the approximation process is due to sound: namely, time derivatives of any sound quantity are large. Specifically, the time derivatives of a primed quantity are taken to be of size c_0, or $1/M$, times the quantity. Thus write

$$\frac{\partial \rho}{\partial t} = \frac{\partial \rho_0}{\partial t} + M^2\frac{\partial \tilde{\rho}}{\partial t} + M^2 \cdot M\frac{\partial \rho'}{\partial t} + \cdots,$$

where the $M^3\partial\rho'/\partial t$ term is of order M^2 because $M\partial/\partial t$ of a primed quantity is of size 1, order M^0, in the limit of large sound speed.

- The continuity equation (13.4) then becomes

$$\frac{\partial \rho_0}{\partial t} + M^2\frac{\partial \tilde{\rho}}{\partial t} + M^2.M\frac{\partial \rho'}{\partial t} + \nabla \cdot \left[\rho_0\tilde{q} + M^2\left(\rho_0 q' + \tilde{\rho}\tilde{q}\right)\right] = \mathcal{O}(M^3).$$

  All terms of order M^0 lead to

$$\frac{\partial \rho_0}{\partial t} + \nabla \cdot (\rho_0\tilde{q}) = 0,$$

  which if ρ_0 is constant, to be shown soon, then reduces to the familiar incompressible equation that the divergence of the velocity field is zero: $\nabla \cdot \tilde{q} = 0$.

  However, all terms of order M^2 lead to (if ρ_0 is constant)

$$M\frac{\partial \rho'}{\partial t} + \rho_0\nabla \cdot q' = -\frac{\partial \tilde{\rho}}{\partial t} - \nabla \cdot (\tilde{\rho}\tilde{q}).$$

  The left-hand side is precisely the terms derived previously for the propagation of sound. The right-hand side represents the generation of sound by the slow flow, wind noise for example.

- Euler's equation for a polytropic fluid,

$$\frac{\partial q}{\partial t} + q \cdot \nabla q = -\frac{c^2}{\rho}\nabla\rho,$$

  becomes

$$\frac{\partial \tilde{q}}{\partial t} + M \cdot M\frac{\partial q'}{\partial t} + \tilde{q} \cdot \nabla\tilde{q} = -\frac{1}{M^2}\frac{c^2}{c_0^2\rho}\nabla\left(\rho_0 + M^2\tilde{\rho} + M^3\rho'\right) + \mathcal{O}(M^2).$$

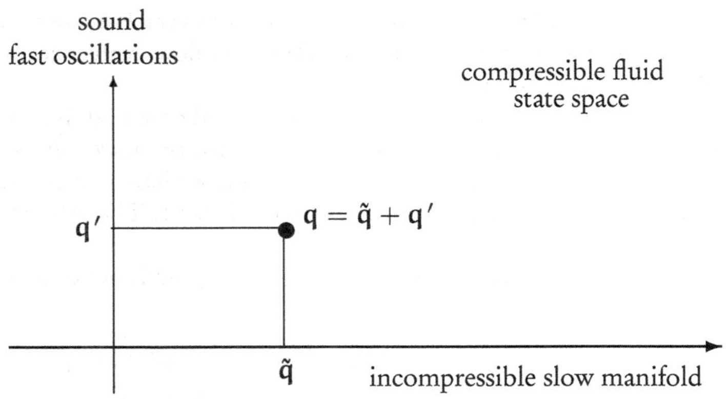

Figure 13.2. *Schematic diagram of compressible fluid dynamics separated into the slow manifold of incompressible flow and the fast manifold of sound waves. Each axis actually represents an infinity of dimensions corresponding to complex fluid flows. Fast oscillations of sound take place in the "vertical" subspace; these oscillations are centered on the horizontal axis representing the slow manifold of incompressible dynamics.*

The largest term in this equation is of order $1/M^2$ and asserts ($c \approx c_0$ to leading order) that

$$0 = \frac{1}{\rho}\nabla \rho_0 ,$$

which as promised earlier ensures that ρ_0 = constant; that is, to leading order the density of the fluid is constant. The next-largest terms in the equation are of order M^0 and ensure

$$\frac{\partial \tilde{q}}{\partial t} + \tilde{q} \cdot \nabla \tilde{q} = -\frac{1}{\rho}\nabla \tilde{p} .$$

This is essentially in the form of the Euler equation, with an effective bulk pressure of $\tilde{p} = p_0 + \tilde{\rho} \approx p_0 + \frac{\partial p}{\partial \rho}(\rho - \rho_0)$, where p_0 is a constant reference pressure. Thus, as claimed by the incompressible approximation, Euler's equation applies to slow flow in the presence of fast sound waves.

Finally, terms of order M (noting that variations of c with density are still insignificant) lead to

$$M\frac{\partial q'}{\partial t} = -\frac{1}{\rho}\nabla \rho' ,$$

which is the companion equation for sound waves.

The derivation shows how we may filter out fast sound waves from a slow fluid flow of interest. The separation, shown schematically in Figure 13.2, is best achieved for a relatively large speed of sound, that is, small Mach number flows.

The argument asserts that one can treat the sound as being superimposed upon the slow flow. However, it turns out that this is only strictly true when the waves are of small amplitude or low frequency. Ultrasonic waves of large enough amplitude interact with themselves through the nonlinearities in the system to generate large-scale motion. Such resonant interaction is a generic feature of slow manifold approximations and is examined in the next sections in more depth with the aid of normal-form transformations.

Subcenter slow manifolds act as a center Center manifolds are relevant to emergent low-dimensional models of dynamical systems because they are exponentially attractive. That is, all nearby dynamics are captured by the center manifold model exponentially quickly.

In contrast, slow manifolds, such as the incompressible approximation, are important, not because they are attractive, but because they act as a center for all the nearby dynamics. For example, sound waves are just oscillations about incompressible fluid flow without significant interaction. In a review of dynamical modeling, van Kampen (1985) asserts that such slow manifold approximations are based upon the principle of a "guiding center."

Consider some other examples.

Rigid body dynamics Think of a ball thrown, hit, or kicked through the air. For most practical purposes we treat the ball as a rigid body, just using equations for its translational and rotational degrees of freedom. At its most complicated we would only consider it as a dynamical system with a twelve dimensional state space: three for spatial position, three for velocity, three for angular orientation, and three for angular velocity. However, a real ball is an elastic or viscoelastic body—it has infinitely many internal dynamical modes of vibration. In practice we ignore the infinitely many fast vibrational modes and just consider the slow translational and rotational dynamics of a rigid body. Such a model is justified because the internal vibrational modes appear to have no effect on the flight of a ball (Muncaster, 1983, e.g.).

Beams Beams are used all around us in construction. To predict the load a beam may safely carry, and how a beam deforms under load, engineers do not solve the 3D elasticity equations for the beam as an elastic solid. Instead they simply solve a beam equation, of the form

$$\frac{\partial^2 \eta}{\partial t^2} = \frac{AE}{\rho} \frac{\partial^4 \eta}{\partial x^4},$$

for the sideways deflection, η, of the centerline of a beam, with x measuring distance along the beam. Similar to the rigid body approximation, such a beam model neglects the fast vibrational modes in the cross-section of the beam; instead it just resolves the much slower dynamics of bending (Roberts, 1993). Compare this with shear dispersion in a pipe where cross-pipe diffusional modes are neglected because they are exponentially damped; here we neglect cross-beam vibration because it appears as rapid oscillations about the shape of the bent beam.

Tides Tidal flow around the world occurs on a 12-hour time scale. Computational models of tides need to resolve the dynamics on this time scale and on a height scale of the order of meters. Such models are used to predict not only tidal heights but also the flushing of effluent, or the lack thereof, and the migration of sea organisms. Also on the ocean there exist short water waves with a period of the order of tens of seconds—waves of interest in their own right for swimmers, sailors, and surfers. However, how can tidal models neglect the short-period waves? Short-period waves are neither exponentially decaying nor particularly small. The only reasonable rationale is to consider that models of tidal dynamics act as a center for actual realizations in that short-period waves are just oscillations about tidal flow.

But visit the beach and see that the situation is more complicated. Ocean waves approach the shore, break, and convert their momentum into a rush up the beach. Look closer, stand in the shallows, and you will generally see and feel an along-shore current. Such a current is enhanced/generated by some of the momentum in the waves, the along-shore momentum. That is, the presence of waves generates a large-scale current contribution that would not be present if it were not for the waves. Such a macroscale mean effect of fast oscillations/waves is typical, as also seen as drift in Example 13.1, and is one effect we illuminate with normal-form coordinate transforms.

Quasi-geostrophy In atmospheric dynamics, weather forecasters need to predict the evolution of the high- and low-pressure patterns that we see on the daily weather maps. These patterns, called Rossby waves, have a characteristic time scale of one week as they propagate from west to east. However, in the dynamics of a rotating and vertically stratified atmosphere there also exist so-called gravity waves with a much shorter time scale of the order of tens of minutes. Mostly, these gravity waves do not appear significantly in the atmosphere;[159] instead the atmosphere is in a familiar balance characterized by winds traveling parallel to the isobars seen on weather maps, called geostrophic balance. However, numerical models of the atmosphere are plagued by the fast gravity waves when all meteorologists want to resolve is the slow Rossby wave dynamics. The slow Rossby waves are called quasi-geostrophic because nonlinearities have a perceptible and important influence on the balance. Thus meteorologists want to construct and use models of the slow manifold of quasi-geostrophic dynamics in an atmosphere that also supports the much faster dynamics of gravity waves.

Irrotational fluid flow Consider 2D fluid flow. The Navier–Stokes equations with zero viscosity may be rearranged to show that $d\omega/dt = 0$, where $\omega = \nabla \times \mathbf{q}$ is the vorticity of the flow; that is, the fluid carries vorticity without change. Consequently, if $\omega = 0$ initially, then $\omega = 0$ for all time; this is a special case of Kelvin's circulation theorem (Batchelor, 1979, p. 276, e.g.). People often invoke this result to justify the extensive study of irrotational flows, that is, fluid flow on a manifold characterized by $\omega = 0$, where one may write $\mathbf{q} = \nabla\phi$ for some velocity potential ϕ. However, in an inviscid fluid there is no mechanism for the decay of vorticity—zero vorticity is not an attractive manifold. Conversely, if we consider viscosity significant, so that zero vorticity is attractive, then Kelvin's theorem no longer applies. Is there any way to properly rationalize the study of irrotational fluid dynamics? After all, planes still fly through turbulent, vorticity-"saturated" air despite a lot of the design work being done with irrotational theory.

Let's sketch an argument for 2D flow. Any velocity field of a fluid flow may be written as the sum of irrotational and rotational parts:

$$\mathbf{q} = \underbrace{\nabla\phi}_{\text{irrotational}} + \underbrace{\frac{1}{2\pi}\iint \frac{-(y-y')\mathbf{i} + (x-x')\mathbf{j}}{(x-x')^2 + (y-y')^2}\,\omega(x',y')\,dx'\,dy'}_{\text{rotational}},$$

where ϕ is a potential, satisfying $\nabla^2\phi = 0$, and ω is the vorticity field; time dependence is implicit in ϕ and ω. Now, the vorticity is carried by the fluid; thus if we imagine it is made up of a large number of discrete lumps called vortices, $\omega = \sum_i \omega_i \delta(\mathbf{x} - \mathbf{x}_i)$,[160] then the location of these vortices evolves according to the following adaptation of the

[159] Gravity waves may be important in local-scale phenomena such as the patchiness of rainfall. If gravity waves occur, then they propagate upward at an angle and eventually break and dissipate in the low density of the stratosphere.

[160] Such a vortex decomposition is the basis for some high Reynolds number numerical simulations.

earlier integral:

$$\frac{dx_i}{dt} = \nabla\phi + \frac{1}{2\pi}\sum_{j\neq i}\frac{-(y_i - y_j)\mathbf{i} + (x_i - x_j)\mathbf{j}}{(x_i - x_j)^2 + (y_i - y_j)^2}\omega_j \,.$$

This is a Hamiltonian system. Its dynamics may be incredibly complex, chaotic for example. However, it does indicate that there is not much interaction between the rotational and irrotational parts of the flow. The irrotational part carries the vortices around, but the vortices do not affect the irrotational part (except through interaction with boundaries). Thus it seems we may ignore the vorticity by assuming that its contribution to the dynamics averages to zero over the time scale of interest. In this sense irrotational flow acts as a center for fluid dynamics.[161]

Einstein versus Newton in cosmology Recall that Einstein's theory of relativity is a more accurate description of solar and cosmic dynamics than is Newtonian dynamics. However, for most practical purposes one approximates the theory of relativity by considering that the speed of light, c, is very fast. Done with care, the theory of relativity reduces to Newtonian dynamics in the limit as $c \to \infty$ (Oliynyk, 2010, e.g.). However, the situation is not that simple. Einstein's theory of relativity contains gravity waves, which are nowhere evident in Newtonian dynamics. Somehow the gravity waves are lost in the Newtonian limit: the reason is that they become infinitely fast. Thus Newtonian dynamics forms some sort of slow manifold within the slow evolution and fast gravity waves of Einstein's theory of relativity.

Summary These examples convince us of the need to study the modeling of slow dynamics in the presence of fast oscillating dynamics.

13.2 ▪ A little subcenter manifold theory

Building upon earlier work by Lyapunov, Sijbrand (1985, §7) looked briefly at dynamical systems with purely oscillatory dynamics. As usual, consider a dynamical system in the form

$$\dot{u} = \mathcal{L}u + f(u), \tag{13.9}$$

where here the linear spectrum is purely imaginary, eigenvalues $\pm i\omega_j$, as shown schematically in Figure 13.3. Provided the eigenvalues are simple, then linearly the dynamics is purely oscillatory. A complete linear solution is

$$u = \sum_j \left[a_j e^{i\omega_j t} + \bar{a}_j e^{-i\omega_j t}\right].$$

This solution is viewed as being composed of vibrational modes $a_j e^{i\omega_j t} + \bar{a}_j e^{-i\omega_j t}$, sometimes called normal modes. The oscillations of such a normal mode all take place in an invariant subspace $\mathbb{E}_j = \operatorname{span}\{\Re(a_j), \Im(a_j)\}$, and all take the form of cycles. That is, associated with each complex conjugate eigenvalue pair is an invariant subspace.

[161] This last part of the argument, that the contribution of vorticity averages to zero, may be better made in 3D fluid flow where vorticity naturally gets tangled up in that incredible mess we call turbulence. In 2D fluid flow, conservation principles place strong constraints on the dynamics and cause the natural generation of large-scale coherent vortex structures (Rhines, 1979, e.g.). It is not appropriate to rely on the average over a coherent structure to be zero.

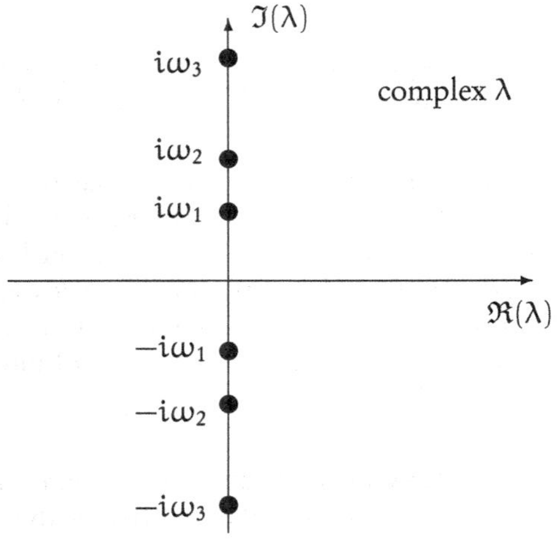

Figure 13.3. *Schematic complex eigenvalue spectrum for a dynamical system that linearly has purely oscillatory dynamics.*

The question is: how much of this manifold and cycle structure is maintained in the actual nonlinear dynamics (13.9)?

The essence of the existence theorem is as follows. Provided that (Sijbrand, 1985, Thm. 7.1):

- the spectrum is not degenerate (no multiple eigenvalues);
- the linear frequencies are nonresonant, that is $m\omega_j \neq n\omega_k$ for any j, k, m, n (roughly); and
- there exists a nontrivial integral of the motion, such as a conserved energy integral,

then for each j there exists an invariant manifold, $\mathcal{M}_j$,[162] of (13.9) tangent to $\mathbb{E}_j$ at the origin and composed of cycles. Such an invariant manifold is termed a *subcenter manifold* because it is the nonlinear analogue of a subspace of the center eigenspace. Such a theorem more or less justifies the construction of nonlinear normal modes for vibrational systems, as done by Shaw and Pierre (1993, 1994) and Shaw (1994), for example.

However, does the theory give any support for using subcenter manifolds as the basis for low-dimensional models of high-dimensional dynamical systems? The answer is: certainly not directly. For a sobering example, consider the following system (Sijbrand, 1985, eq. (7.23)), which in bipolar coordinates, $u_1 = r\cos\phi$, $u_2 = r\sin\phi$, $u_3 = \rho\cos\psi$, and $u_4 = \rho\sin\psi$, is

$$\begin{aligned} \frac{dr}{dt} &= -r\left(r^4 - 10r^2\rho^2 + 5\rho^4\right), \\ \frac{d\rho}{dt} &= -\rho\left(5r^4 - 10r^2\rho^2 + \rho^4\right), \\ \frac{d\phi}{dt} &= \omega_1, \\ \frac{d\psi}{dt} &= \omega_2. \end{aligned} \tag{13.10}$$

[162]Incidentally, if the nonlinear terms f are analytic, then $\mathcal{M}_j$ is analytic!

Linearly, the u_1u_2-plane ($\rho = 0$) is the invariant subspace $\mathbb{E}_1$, whereas the u_3u_4-plane ($r = 0$) is $\mathbb{E}_2$. Nonlinearly, the following happens.

- $\rho \sim Ar^5$ is a small r approximation to the corresponding $\mathcal{M}_1$, but it is a valid approximation for any A—subcenter manifolds may differ by some power. However, the case $A = 0$ is the only such subcenter manifold that is analytic in u_1 and u_2.
- Further, *in* $\mathcal{M}_1$ all solutions starting sufficiently close to the origin evolve according to $\dot{r} \sim -r^5$; thus a model restricted to $\mathcal{M}_1$ would predict that the origin is stable; however, the manifold $\rho = r$ is also invariant, and on this manifold solutions evolve according to $\dot{r} = 4r^5$, which proves that the origin is unstable.

Of course, this system does not have a conserved quantity and so a precondition of the theorem is not satisfied. Nonetheless, the example is rather sobering in that basic properties, such as emergence, needed for subcenter manifolds to form low-dimensional models, are not assured.

Nonetheless, this theory shows that at least in some circumstances one can extract a subset of oscillatory dynamics from a high-dimensional dynamical system. The above definition of a subcenter manifold applies to the case when one extracts the dynamics associated with some particular nonzero frequency. One of the very limiting preconditions to do this is the condition of nonresonance. However, there is one class of manifolds for which the nonresonant precondition is trivially satisfied and that is when the associated eigenvalues are zero. Section 13.1 outlined the importance of the case where linearly the modes of interest have zero eigenvalue, and the modes we wish to ignore are associated with the pure imaginary eigenvalues of fast oscillations. This chapter focuses on this case of systems with a *slow manifold* among fast oscillations.

There is great scope for developing more theory. Nonetheless the practical utility of assuming the existence and importance of slow manifolds is indisputable.

13.3 ▪ A quasi-geostrophic slow manifold

The present state of supporting theory being unsatisfactory, let's justify slow manifold models of dynamical systems via a normal-form coordinate transformation. This section, section 13.4, and section 13.6 introduce three interesting example systems with slow manifold models among fast oscillations. Section 13.5 proceeds to illuminate our view of such slow manifold models by employing normal-form coordinate transforms.

Lorenz (1986) proposed the following system of five coupled equations to model some important characteristics of atmospheric dynamics:

$$\dot{u} = -vw + bvz, \tag{13.11a}$$
$$\dot{v} = \quad uw - buz, \tag{13.11b}$$
$$\dot{w} = -uv, \tag{13.11c}$$
$$\dot{x} = -z, \tag{13.11d}$$
$$\dot{z} = \quad x + buv. \tag{13.11e}$$

Figure 13.4 shows two trajectories of the system: one evolves very smoothly around an elliptical shape; the other follows the ellipse but has fast oscillations superimposed. In atmospheric dynamics, weather forecasters need to predict the evolution of the high- and low-pressure patterns that we see on the daily weather maps. These patterns, the so-called Rossby waves, have a characteristic time scale of one week as they drift from west to east. The atmosphere, rotating and stratified, also supports so-called gravity waves, which

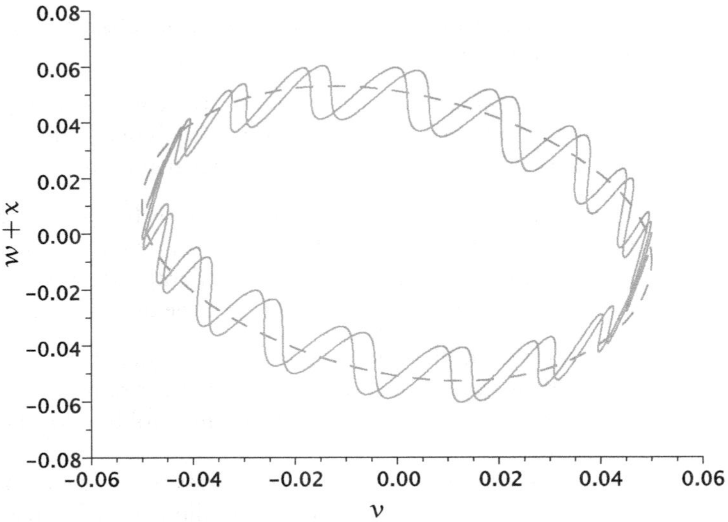

Figure 13.4. *Two simulations of the Lorenz (1986) system* (13.11) *showing (dashed) a guiding center evolution and (solid) a nearby trajectory with fast oscillations.*

have a much faster characteristic time of the order of tens of minutes. Lorenz's idea was to model the nonlinear interaction between the slow Rossby waves and the fast gravity waves by the system (13.11). In this system there is a very clear-cut distinction between the slow modes, u, v, and w, and the fast modes, x and z.

- The fast waves are represented by variables x and z, which, linearly, oscillate with frequency 1, that is, like $e^{\pm it}$.
- The slow waves are represented by u, v, and w, which, linearly, are fixed. However, nonlinearly and in the absence of the fast waves, that is, $\dot u = -vw$, $\dot v = uw$, and $\dot w = -uv$, these variables have an exact solution in terms of elliptic functions that show that u, v, and w possess slow oscillations with an amplitude-dependent period that becomes arbitrarily long for small enough amplitudes.
- The fast and the slow waves are coupled by the nonlinear interaction terms with coefficient b. If coefficient b is zero, then there is no interaction, and the separation between fast and slow dynamics is clear. The interesting case is nonzero b when there is interaction. Then we aim to extract a slow manifold approximation from the dynamical system.
- A further feature of the model is that it possesses the two invariants

  $$H = u^2 + v^2 \quad \text{and} \quad I = v^2 + w^2 + x^2 + z^2 .$$

  These invariants arise from analogies of the conservation of energy and potential enstrophy. The analysis here does not use these invariants.

Numerical simulations, such as the two plotted in Figure 13.4, easily show the fast oscillations, predominantly in x and z, superimposed upon a slow oscillation, which is predominantly in u, v, and w. However, the nonlinear coupling causes solutions without any noticeable fast oscillations to lie in a curved manifold, seen in Figure 13.5, rather than

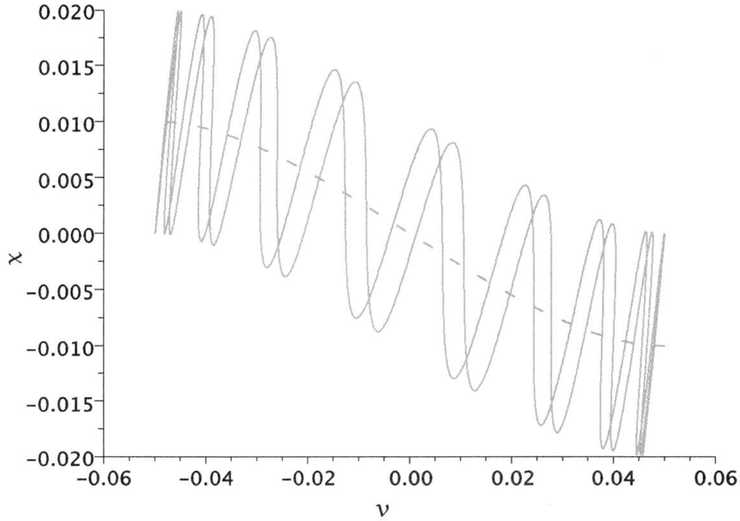

Figure 13.5. *Two simulations of the Lorenz (1986) system (13.11) showing (dashed line) a cross-section of the slow manifold, and (solid line) a nearby trajectory with fast oscillations.*

in the uvw-subspace. We seek an asymptotic description of this slow manifold because, as seen in simulations, it appears to act as a center for all the nearby dynamics.

Let's straightforwardly construct a slow manifold, $\mathcal{M}_0$. The construction proceeds via the same algorithm as used for constructing center manifolds, but with none of the complications associated with a Hopf bifurcation because the eigenvalues associated with a slow manifold are zero. We simply seek a slow manifold parametrized in terms of the slow variables

$$x = x(u, v, w) \quad \text{and} \quad z = z(u, v, w)\,, \tag{13.12}$$

on which the evolution is given by the first three equations of the Lorenz system (13.11a)–(13.11c). That is, instead of considering fast variables x and z as direct functions of time, we consider them as functions of the slow variables u, v, and w, which in turn are slow functions of time. The dashed-line trajectory in Figure 13.5 shows a cross-section through the slow manifold (13.12). Then

$$\begin{aligned}
\dot{x} &= -z \quad \text{from (13.11d)},\\
\dot{x} &= \frac{\partial x}{\partial u}\dot{u} + \frac{\partial x}{\partial v}\dot{v} + \frac{\partial x}{\partial w}\dot{w} \quad \text{by (13.12) and the chain rule}\\
&= -\frac{\partial x}{\partial u}(vw - bvz) + \frac{\partial x}{\partial v}(uw - buz) - \frac{\partial x}{\partial w}uv\,,
\end{aligned}$$

and similarly

$$\begin{aligned}
\dot{z} &= x + buv \quad \text{from (13.11e)},\\
\dot{z} &= \frac{\partial z}{\partial u}\dot{u} + \frac{\partial z}{\partial v}\dot{v} + \frac{\partial z}{\partial w}\dot{w} \quad \text{by (13.12) and the chain rule}\\
&= -\frac{\partial z}{\partial u}(vw - bvz) + \frac{\partial z}{\partial v}(uw - buz) - \frac{\partial z}{\partial w}uv\,.
\end{aligned}$$

Algorithm 13.1 *Simple Reduce code to compute the quasi-geostrophic slow manifold of the Lorenz (1986) five-equation toy of the atmosphere* (13.11).

```
% Let s count the order of approximation.
let s^6=0;   factor s;
depend u,t; depend v,t; depend w,t;
let { df(u,t) => -s*v*w+b*v*z
    , df(v,t) =>  s*u*w-b*u*z
    , df(w,t) => -s*u*v};
x:=z:=0;
for it:=1:99 do begin
    x:=x-(resz:=-df(z,t)+x+s^2*b*u*v);
    z:=z+(resx:=-df(x,t)-z);
    if {resx,resz}={0,0} then it:=1000000;
end;
```

Rearranging the first and last lines of each of these two equations gives

$$x = -buv - \frac{\partial z}{\partial u}(vw - bvz) + \frac{\partial z}{\partial v}(uw - buz) - \frac{\partial z}{\partial w}uv\,, \tag{13.13a}$$

$$z = \frac{\partial x}{\partial u}(vw - bvz) - \frac{\partial x}{\partial v}(uw - buz) + \frac{\partial x}{\partial w}uv\,. \tag{13.13b}$$

Solve these by iteration as in section 4.4. Substitute the initial linear approximation, that $x^{(0)} = z^{(0)} = 0$, in the right-hand side to derive the first nontrivial approximation that

$$x^{(1)} = -buv \quad \text{and} \quad z^{(1)} = 0\,.$$

A second iteration leads to

$$x^{(2)} = -buv \quad \text{and} \quad z^{(2)} = b\left(u^2 - v^2\right)w\,,$$

and so forth: Algorithm 13.1 easily computes arbitrarily high orders. Thus our algebra concludes that the slow manifold is

$$x = -buv + \mathcal{O}(s^4) \quad \text{and} \quad z = b(u^2 - v^2)w + \mathcal{O}(s^4), \tag{13.14}$$

where $s = |(u, v, w)|$. This shape is seen in numerical simulations where, since $H = u^2 + v^2$ is constant, the slow evolution takes place on a curve $x = -bu\sqrt{H - u^2}$ for some H. On the slow manifold $\mathcal{M}_0$ the evolution takes place according to (13.14) substituted into (13.11a)–(13.11c), namely

$$\begin{aligned} \dot{u} &= -vw\left[1 - b^2(u^2 - v^2)\right] + \mathcal{O}(s^5)\,, \\ \dot{v} &= uw\left[1 - b^2(u^2 - v^2)\right] + \mathcal{O}(s^5)\,, \\ \dot{w} &= -uv\,. \end{aligned} \tag{13.15}$$

This model of the dynamics is simpler because it is low dimensional: it is three-dimensional instead of the original five. In many applications this is where an enormous simplification occurs. However, in some applications, such as atmospheric dynamics, the gain in the dimensional reduction is rather small: as here, one reduces the dimensionality by simply a factor of two or three rather than an infinite amount, as in many other applications.

However, a further and important gain is attained on the slow manifold. In atmospheric dynamics, the evolution on the slow manifold is of the order of a week, whereas that of the fast dynamics is of the order of tens of minutes; thus in the absence of fast waves, explicit numerical simulations can take a time step some one hundred times as large as would otherwise be the case. An important characteristic of a slow manifold is the relatively long time scale of all its dynamics.

13.4 ▪ The slow manifold of electron drift in a magnetic field

This is the second of three physical examples of the construction of a slow manifold model among fast oscillations.

Consider an electron with position $x(t)$ and $y(t)$ in the two-dimensional plane and moving with velocity $u(t)$ and $v(t)$. Apply a large magnetic field normal to the plane (in the z-direction): any movement in the plane by the electron is deflected by the magnetic field but stays within the plane. Why should the electron move? Because we also apply an inhomogeneous electric field in the plane with components $E(x, y)$ and $F(x, y)$. The dynamics of the electron has a slow manifold (van Kampen, 1985, p. 110).

Based upon the above description, and with a large magnetic field, physicists might write the dynamical equations as

$$\dot{x} = u, \quad \dot{y} = v, \quad \dot{u} = -\frac{1}{\varepsilon}v + E(x, y), \quad \dot{v} = +\frac{1}{\varepsilon}u + F(x, y),$$

where $1/\varepsilon$ measures the strength of the large magnetic field.[163]

Rescaling time regularizes Typically, as here, physicists tend to write the equations with a large parameter so that the fast dynamics is asymptotically unresolvable as it occurs at an infinite rate in a singular perturbation. We prefer to be able to discover the more subtle effects of the fast dynamics by rescaling time to cater for the fast dynamics on an $\mathcal{O}(1)$ time scale in a regular perturbation. As usual we rescale time and variables. Here let new time $t^* = t/\varepsilon$ and scale velocities $u = u^*/\varepsilon$ and $v = v^*/\varepsilon$, and leave positions x and y unscaled. Substitute and drop stars to derive the equivalent ODEs

$$\dot{x} = u, \quad \dot{y} = v, \quad \dot{u} = -v + \epsilon E(x, y), \quad \dot{v} = +u + \epsilon F(x, y), \tag{13.16}$$

where small parameter $\epsilon = \varepsilon^2$. Now we seek a slow manifold.

Seek equilibria Formally adjoin $\dot{\epsilon} = 0$, as the magnetic field is large, so parameter ϵ is small. Then immediately equilibria are $u = v = \epsilon = 0$ for any positions x and y. There is a two-dimensional subspace of equilibria $\mathbb{E}_0$, namely the xy-plane of electron position.

Linearization determines the nature of the model Ignore the trivial, formally adjoined $\dot{\epsilon} = 0$. Then linearized about any position x and y, the electron dynamics (13.16) has linear operator

$$\mathcal{L} = \begin{bmatrix} 0 & 0 & 1 & 0 \\ 0 & 0 & 0 & 1 \\ 0 & 0 & 0 & -1 \\ 0 & 0 & 1 & 0 \end{bmatrix}.$$

[163] Alternatively, some physicists would write down the ODEs for small mass ε, namely $\varepsilon\dot{u} = -v + E(x, y)$ and $\varepsilon\dot{v} = +u + F(x, y)$. It makes no difference to us; the consequent rescaling and analysis are equivalent.

Straightforward algebra derives the characteristic equation $\det(\lambda I-\mathcal{L}) = \lambda^2(\lambda^2+1) = 0$. Consequently, the spectrum is $\lambda = 0$ (twice) and $\lambda = \pm i$. There exists a two-dimensional slow manifold among fast oscillations of frequency one.

Since there is a two-dimensional subspace of equilibria, and a two-dimensional slow manifold, the subspace of equilibria—here the xy-plane of fixed position—globally approximates the slow manifold. The only changes to the slow manifold arise through dependence upon the small finite parameter ϵ parametrizing the influence of the electric field.

Construct the slow manifold The slow variables are positions x and y. The ODEs (13.16) indicate that any fast dynamics are primarily in velocity oscillations through $\dot{u} = -v$ and $\dot{v} = +u$. Thus we seek a slow manifold

$$u = u(x,y) \quad \text{and} \quad v = v(x,y);$$

instead of considering velocities u and v directly as functions of time, we consider them as functions of position x and y, where x and y in turn vary slowly in time. Rearrange the last two ODEs of (13.16) to give the basis for an iteration scheme to construct the slow manifold:

$$v = \epsilon E(x,y) - \dot{u} \quad \text{and} \quad u = -\epsilon F(x,y) + \dot{v}.$$

- The leading approximation to the slow electron dynamics is then drift with velocity orthogonal to the applied electric and magnetic fields $u \approx -\epsilon F(x,y)$ and $v \approx +\epsilon E(x,y)$. The resultant slow position drift is $\dot{x} \approx -\epsilon F(x,y)$ and $\dot{y} \approx +\epsilon E(x,y)$. This is a well-known adiabatic approximation.

  For example, when the electric fields E and F arise as the gradient of a potential $\mathcal{V}$, namely $(E,F) = \nabla\mathcal{V}$, then drift of this adiabatic approximation takes the electron along equipotentials of the potential $\mathcal{V}$.

- The next-order approximation comes from substituting for nontrivial time dependence in the right-hand side of the above iteration. For example,

$$\begin{aligned}\dot{u} &= \frac{\partial u}{\partial x}\dot{x} + \frac{\partial u}{\partial y}\dot{y} \\ &\approx -\epsilon\frac{\partial F}{\partial x}(-\epsilon F) - \epsilon\frac{\partial F}{\partial y}(+\epsilon E) \\ &= \epsilon^2(FF_x - EF_y).\end{aligned}$$

  Hence the corrected velocities of the electron on the slow manifold are

$$u \approx -\epsilon F + \epsilon^2(-FE_x + EE_y) \quad \text{and} \quad v \approx +\epsilon E + \epsilon^2(-FF_x + EF_y).$$

  The evolution $\dot{x} = u$ and $\dot{y} = v$ then forms a quasi-equilibrium model of the electron's slow motion.

It is straightforward to continue to higher order. Van Kampen (1985) calls this sort of slow manifold model a guiding center.

The unresolved question is: what, if any, is the influence of any undamped fast oscillations of the electron? The normal-form coordinate transform of section 13.5.4 provides the answer.

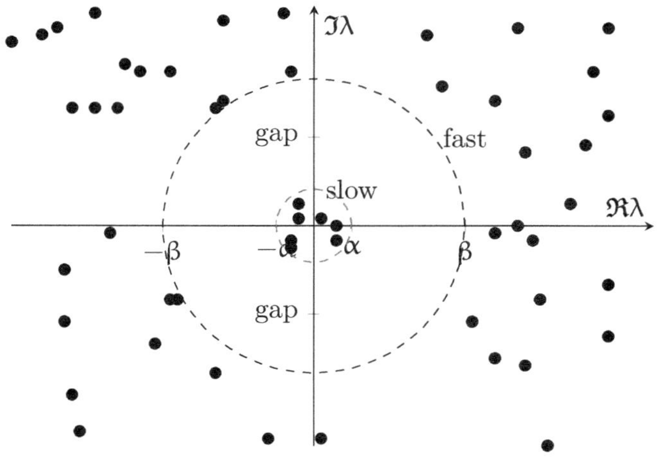

Figure 13.6. *Schematic picture of the complex plane of eigenvalues λ_j (discs) forming two separate sets characterized by bounding parameters α and β of the slow and fast modes, respectively.*

13.5 ▪ Normal-form coordinate transforms highlight resonant drift

The construction of section 13.3 suggests that there is indeed a slow manifold, $\mathcal{M}_0$, of the dynamics of the Lorenz system (13.11). Analogously, section 13.4 constructed a slow manifold $\mathcal{M}_0$ of electron drift in a transverse magnetic field, and section 13.6 finds a slow manifold of a spinning top. This section addresses the following question: how can we be sure, in the absence of exponential contraction to $\mathcal{M}_0$, that the evolution on $\mathcal{M}_0$ serves as a good low-dimensional model for the dynamics of the system? A normal-form transformation shows that the evolution on $\mathcal{M}_0$ is a model for all nearby trajectories, but that generically there are irreducible discrepancies, in the form of a long-term drift, when any fast oscillations are present. Example 13.1, at the start of this chapter, is a 3D dynamical system exhibiting this forced drift generated by fast oscillations.

This section focuses on the dynamics of interesting slow variables when among many uninteresting fast variables. Recall that Definition 11.4 introduced the terminology of slow and fast variables: in a linearization about some convenient equilibrium, the slow/fast variables denote those variables associated with eigenvalues near zero, and not near zero, respectively. We require there to be a gap between the slow and fast eigenvalues, as shown in Figures 13.6 and 13.7. Suppose we used a linear change of basis to linearly separate the physical system into m slow variables $x(t) \in \mathbb{R}^m$ and n fast variables $y(t) \in \mathbb{R}^n$. Then the physical system is described by the linearly separated but nonlinearly coupled ODEs

$$\dot{x} = Ax + f(x, y) \quad \text{and} \quad \dot{y} = By + g(x, y). \tag{13.17}$$

This system is analogous to the system (11.10) explored in section 11.2 but only has two classes of variables (slow x and fast y) instead of three classes (stable, center and unstable).

13.5.1 ▪ Slow manifolds exist

Much of the algebra of section 11.2 still applies and is still useful, but we have to modify assumptions, interpretations, and results. Instead of Assumption 11.6, to study slow manifolds among fast oscillations we here require the following.

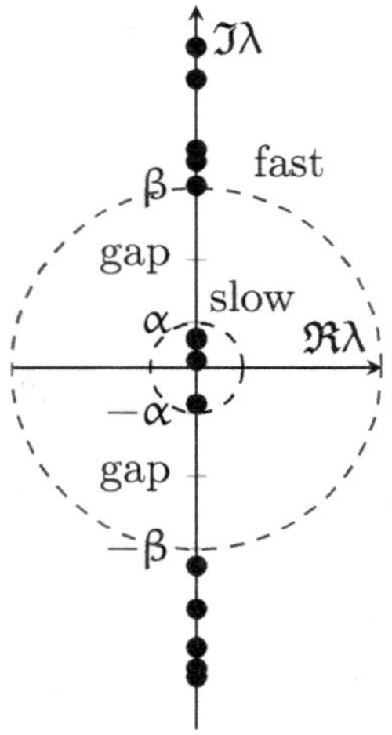

Figure 13.7. *Schematic picture of the complex plane of eigenvalues λ_j (discs) forming two separate sets characterized by bounding parameters α and β of the slow and fast modes, respectively. This picture is of the case of main interest when there are slow modes among fast oscillations.*

Assumption 13.1 (fast-slow dichotomy). *The matrices in the separated system* (13.17) *are assumed to satisfy the following:*

- $A \in \mathbb{R}^{m\times m}$ *with eigenvalues* $\alpha_1, \dots, \alpha_m$, *possibly complex, with modulus* $|\alpha_j| \leqslant \alpha$;
- $B \in \mathbb{R}^{n\times n}$ *with eigenvalues* $\beta_1, \dots, \beta_n$, *possibly complex, with modulus* $|\beta_j| \geqslant \beta$;
- *for simplicity, section* 13.5 *assumes that the matrices* A *and* B *have been diagonalized, but extensions also cover the important cases when matrices* A *and* B *are triangular, such as in a Jordan form due to repeated eigenvalues, rather than just diagonalized.*

Figures 13.6 and 13.7 illustrate suitable spectra of such eigenvalues, sometimes called a *dichotomy*. As usual and invoking Assumption 11.7, at the origin the strictly nonlinear functions $\mathbf{f}$ and $\mathbf{g}$ are $\mathbb{O}(2)$, and they are sufficiently smooth, $C^p(d)$ for $\beta/\alpha > p \geqslant 2$. We aim to construct a coordinate transform to separate the nonlinear dynamics of the system (13.17) into the same slow and fast dynamics as the linearized system—at least as far as possible.

Proceed by seeking a near identity nonlinear coordinate change to new coordinates $(\mathbf{X}, \mathbf{Y})$,

$$\mathbf{x} = \mathbf{x}(\mathbf{X}, \mathbf{Y}) \quad \text{and} \quad \mathbf{y} = \mathbf{y}(\mathbf{X}, \mathbf{Y}), \tag{13.18a}$$

where in these new variables, the separated dynamical system (13.17) is approximated by the normal-form system

$$\dot{\mathbf{X}} = A\mathbf{X} + \mathbf{F}(\mathbf{X}, \mathbf{Y}) \quad \text{and} \quad \dot{\mathbf{Y}} = B\mathbf{Y} + \mathbf{G}(\mathbf{X}, \mathbf{Y}) \tag{13.18b}$$

such that

$$\mathbf{F}(\mathbf{X}, \mathbf{Y}) = \mathbf{F}_0(\mathbf{X}) + \mathcal{O}(|\mathbf{Y}|^2), \quad \mathbf{G}(\mathbf{X}, \mathbf{0}) = \mathbf{0}, \tag{13.18c}$$

and hence there exists a matrix function G such that $\mathbf{G} = G(\mathbf{X}, \mathbf{Y})\mathbf{Y}$. Lemma 13.2 constructively establishes the existence of a useful coordinate transformation up to order p.

Lemma 13.2 (slow/fast normal-form transforms exist). *For all orders* p, $2 \leqslant p \leqslant \mathfrak{p}$, *there exists a near identity multinomial coordinate transform* (13.18a) *such that a multinomial normal-form system* (13.18b), *together with* (13.18a) *and* (13.18c), *is equivalent to the system* (13.17) *to errors* $\mathbb{O}(p)$.

Proof. A proof is by a constructive induction, which is directly analogous to that used to prove Lemma 11.8, and is left as Exercise 13.7. □

Lemma 13.3 (invariant slow manifold). *The equilibrium at the origin of the multinomial normal-form system* (13.18b) *has an invariant slow manifold (subspace)* $\mathbb{M}_0$ *of* $(\mathbf{X}, 0) \in \mathbb{R}^{m+n}$.[164]

Proof. The proof is straightforward from the conditions (13.18c): since $\mathbf{G}(\mathbf{X}, 0) = 0$, then $\mathbf{Y} = 0$ satisfies the $\mathbf{Y}$ ODE of the system (13.18b). □

Generally there is no "fast manifold," as $\mathbf{X} = 0$ does not in general satisfy the $\mathbf{X}$ ODE of the system (13.18b), because of $\mathcal{O}(|\mathbf{Y}|^2)$ terms.

The next step is to map this slow manifold of the normal-form system (13.18b) into the original variables x and y. Since the coordinate transform (13.18a) is near identity, then by continuity of derivatives, the following domain D is a (finite) neighborhood of the equilibrium at the origin. In the common case where the coordinate transform (13.18a) is multinomial, then we may take the order $p = \infty$ in this definition.

Definition 13.4 (diffeomorphic domain). *For some given order* $p \geqslant 2$, *define the domain* $D \subseteq \mathbb{R}^{m+n}$ *such that the coordinate transform* (13.18a) *is a* C^p*-diffeomorphism*[165] *onto a domain* $d \subseteq \mathbb{R}^{m+n}$.

A nearby slow manifold exists Proposition 13.5 establishes that there is a system "close" to the specified and with a slow manifold that we know exactly. As in section 11.3.2, this is an example of a backward error theory (Grcar, 2011, e.g.). Such backward error theory is particularly useful for slow manifolds because slow manifolds among fast oscillations are quite delicate; for example, Lorenz (1986) discussed the slow manifold (13.14) for system (13.11), but Lorenz and Krishnamurty (1987) subsequently proved that such a slow manifold could not exist for the system. Such a proof of nonexistence is disturbing given that the concept of a slow manifold is immensely useful in understanding geophysical fluid dynamics.

Proposition 13.5 (slow manifolds exist). *Consider the dynamical system* (13.17) *satisfying the specified conditions. For all orders* $2 \leqslant p \leqslant \mathfrak{p}$, *there exists a dynamical system that both is* $\mathbb{O}(p)$ *close to the system* (13.17)[166] *and in the domain* d *possesses a slow manifold, denoted by* $\mathcal{M}_0$.

Proof. For any order p, Lemma 13.2 establishes that there exists such a dynamical system that is $\mathbb{O}(p)$ close; namely the combination of the normal form (13.18b) with the coordinate transform (13.18a) defines sufficiently close dynamics in the state space of (13.17). Lemma 13.3 establishes the existence of the slow manifold for the normal form (13.18b).

[164] Recall that Proposition 11.13 constrained the center manifold $\mathbb{M}_c \subset \mathbb{D}_\mu$. Lemma 13.3 does not place any such constraints on $\mathbf{X}$ because the nature of slowness in out-of-equilibrium dynamics is very slippery. The structurally stable hyperbolicity within $\mathbb{D}_\mu$ for center manifolds has no generic counterpart for slow manifolds among fast oscillations, that I am aware of.

[165] That is, the transform and its inverse exist and are differentiable up to derivatives of order p.

[166] As in Corollary 11.14, under the common circumstances of both infinite differentiability of the system and the slow eigenvalues being precisely zero ($\alpha = 0$), one may claim that there is a system *exponentially* close.

The coordinate transform (13.18a) maps these into corresponding invariant manifolds in d for the corresponding system (13.18) in the original state space. □

13.5.2 ▪ Subcenter slow manifolds do not knowably emerge

Now we focus more specifically on the dynamics of systems where the linearization has pure imaginary eigenvalues and there is a gap between slow modes and fast oscillating modes. Figure 13.7 plots such a spectrum. The examples discussed in sections 13.1, 13.3, and 13.4 all fall into this category. Indeed, most Hamiltonian physical systems fall into this category. But there are two interesting modeling issues in this scenario.

1. Here there is no analogue of Lemma 11.16 asserting that the slow manifold is attractive. Instead, in the general normal-form system (13.18) the fast modes $\mathbf{Y}$ are only known to oscillate rapidly because of their linearization $\dot{\mathbf{Y}} \approx B\mathbf{Y}$. True, the nonlinear terms $\mathbf{G}$ may make the slow manifold $\mathbf{Y} = \mathsf{0}$ attractive, but if so the attraction would only be relatively weak. Further, the nonlinear terms $\mathbf{G}$ may cause the fast variables to evolve away from the slow manifold $\mathbf{Y} = \mathsf{0}$, perhaps to a limit cycle or strange attractor. The best we can hope for is the common view that the slow manifold acts as a sort of center for all neighboring dynamics, as suggested by Figures 13.4 and 13.5.

2. But even acting as a center is generally not assured. Recall that Figure 13.1 shows fast oscillations about a slow mode, but the slow mode evolves *differently* depending upon the amplitude of the fast oscillations. The evolution on a slow manifold is generally not representative of all nearby solutions. The normal form shows this through the slow variable ODE (13.18b) having the generic form $\dot{\mathbf{X}} = A\mathbf{X} + \mathbf{F}_0(\mathbf{X}) + \mathcal{O}(|\mathbf{Y}|^2)$: the presence of fast oscillations $\mathbf{Y}$ generally forces the slow modes. The amplitude of such forcing is generally quadratic. In many physical applications this forcing by the oscillations is acknowledged as a wave/mean-flow effect, such as Stokes drift in water waves.

Outstanding question If subcenter slow manifolds do not generally emerge, then why is the concept of a subcenter slow manifold so useful in practice? I suggest some reasons.

- The fast oscillations generally have a quadratic effect on the mean flow (sometimes quartic), which may be usefully negligible in applications.
- In spatially extended systems, the fast oscillations appear as waves, which may propagate out of the domain of interest, leaving slow dynamics behind: water waves propagate to a beach, break, and dissipate energy there, leaving a largely "flat" ocean; gravity waves in the atmosphere propagate to the upper atmosphere and break there to dissipate their energy, leaving a geostrophic slow manifold in the lower atmosphere.
- Also, in spatially extended systems, propagating fast waves may scatter from microscale inhomogeneities (such as ice floes in water waves), become decoherent, and potentially have their mean effects then average to zero.
- Physical systems often have weak dissipation, neglected in the mathematical modeling, and if the fast oscillations are not strongly generated, then the weak dissipation may be enough to maintain dynamics on the slow manifold. For example, ideal

elastic beams and shells are usually slightly viscoelastic and so the "ringing" of any sharp impact is eventually dissipated (if not radiated, as in the bell on a bicycle).

Forecast from initial conditions Integrating the fast/slow system (13.17) numerically is costly when the rapid oscillations (potentially of large amplitude) are present. Practically, we want to integrate numerically just the slow modes. Unfortunately, since the evolution (13.18b) of the slow modes cannot in general be made independent of $\mathbf{Y}$, solutions on and off the slow manifold (where $\mathbf{Y}=0$ and $\mathbf{Y}\neq 0$, respectively) evolve in essentially different ways. It is therefore impossible in general to project an initial condition $(\mathbf{x}_0,\mathbf{y}_0)$ to $\mathcal{M}_0$ so as to maintain the same essential dynamical behavior for all time.

However, it is possible to match the behaviors on and off the slow manifold for a time of $o(|\mathbf{Y}|^{-2})$ in the presence of fast oscillations of small enough amplitude $|\mathbf{Y}|$. As discussed in section 12.1, for any given initial condition $(\mathbf{x}(0),\mathbf{y}(0))=(\mathbf{x}_0,\mathbf{y}_0)$, one transforms the initial value $(\mathbf{x}_0,\mathbf{y}_0)\mapsto(\mathbf{X}_0,\mathbf{Y}_0)$ using the coordinate transform (13.18a) and projects this value to $(\mathbf{X}_0,0)$ onto the slow manifold $\mathbb{M}_0$, and then $\mathbf{X}(0)=\mathbf{X}_0$ is perhaps the best appropriate initial condition for the slow manifold model (13.18b) with $\mathbf{Y}=0$.

This projection of initial conditions accounts for the linear dynamics of oscillations about the slow manifold. As such, it is equivalent to the linear projection of initial conditions discussed in section 12.3 (Roberts, 1989a). However, for practical purposes, instead of using normal-form coordinate transforms, the algorithms of section 12.3 more efficiently construct the projection of initial conditions.

Oscillation amplitudes are slow variables There is a fundamental problem with the concept of a slow manifold among fast oscillations/waves. The problem is that there are more slow variables than just those associated with a zero eigenvalue. The amplitudes of the oscillations are also slow variables. Recall that Chapter 3 exploited the slow evolution of oscillation amplitudes to derive useful models of their long-term behavior. The same considerations hold more generally. The next three examples illustrate how and why, in principle, oscillation/wave amplitude should be included in a model of the slow modes.

In practice, one may not want to include such fast oscillation/wave amplitudes in a model, but at least we now recognize the cost in loss of accuracy.

Example 13.2 (slow oscillation amplitude). Recall that the 3D toy system (13.1) of Example 13.1,

$$\dot{x}=z^2,\quad \dot{y}=-z,\quad \dot{z}=y,$$

under the coordinate transform (13.2),

$$x=X-\tfrac{1}{2}YZ,\quad y=Y,\quad z=Z,$$

becomes the normal-form system (13.3):

$$\dot{X}=\tfrac{1}{2}(Y^2+Z^2),\quad \dot{Y}=-Z,\quad \dot{Z}=Y.$$

Alternatively, we bring to view that the amplitude of the oscillations is a slow variable by the cylindrical coordinate transform (analogous to that of section 3.1)

$$x=X-\tfrac{1}{4}R^2\sin 2\Theta,\quad y=R\cos\Theta,\quad z=R\sin\Theta. \tag{13.19}$$

Such cylindrical coordinates seem particularly appropriate given the typical trajectories of Figure 13.1. Because y and z are just linear oscillators, then the evolutions of radius R and phase θ are just $\dot R = 0$ and $\dot\theta = 1$. Consider the x-evolution $\dot x = z^2$: given $\dot R = 0$ and $\dot\theta = 1$, the time derivative $\dot x = \dot X - \frac{1}{2}R^2\cos 2\theta$, whereas the right-hand side $z^2 = R^2\sin^2\theta = \frac{1}{2}R^2(1-\cos 2\theta)$. Canceling and rearranging gives

$$\dot X = \tfrac{1}{2}R^2, \quad \dot R = 0, \quad \dot\Theta = 1. \tag{13.20}$$

The phase Θ evolves quickly, whereas both X and R evolve slowly (for small wave amplitude R). That is, strictly, the oscillation amplitude R is a slow variable as is X.

13.5.3 ▪ Near quasi-geostrophic normal form

Consider again the quasi-geostrophic slow manifold of section 13.3. Here seek a normal-form coordinate transformation with the aim of simplifying the dynamics (13.11):

$$\begin{aligned}&\dot u = -vw + bvz, \quad \dot v = uw - buz, \quad \dot w = -uv,\\ &\dot x = -z, \quad \dot z = x + buv.\end{aligned}$$

Here "simplifying" means decoupling as far as possible the fast oscillations from the slow dynamics in the rewritten form of the equations. Proposition 13.5 promises that such a separation exists, to any order, because the eigenvalues of this system are 0 and $\pm i$ to fit the schematic distribution of Figure 13.7. Let's rewrite the evolution in terms of five new independent variables $\mathbf{s} = (U, V, W, X, Z)$. These are related to the original variables by the near identity transformation

$$\begin{aligned}&u = u(\mathbf{s}) \approx U, \quad v = v(\mathbf{s}) \approx V, \quad w = w(\mathbf{s}) \approx W,\\ &x = x(\mathbf{s}) \approx X, \quad z = z(\mathbf{s}) \approx Z.\end{aligned} \tag{13.21}$$

Thus X and Z encapsulate the fast dynamics in the transformed variables, whereas U, V, and W encapsulate the slow. With such a transformation we seek to warp our view of the state space to find a simpler view of the dynamics.

Since u, v, w, x, and z evolve in time, then so must U, V, W, X, and Z. They do so according to some prescription such as $\dot{\mathbf{s}} = \mathbf{g}(\mathbf{s})$, which, written in components is, say,

$$\begin{aligned}&\dot U = A(\mathbf{s}), \quad \dot V = B(\mathbf{s}), \quad \dot W = C(\mathbf{s}),\\ &\dot X = F(\mathbf{s}) \approx -Z, \quad \dot Z = G(\mathbf{s}) \approx X,\end{aligned} \tag{13.22}$$

where A, B, and C are to be nonlinear functions of $\mathbf{s}$. The last two equations incorporate the dominant behavior of X and Z, that of fast oscillation, through the given linear approximation. Let's choose the coordinate transformation (13.21) so that A, B, C, F, and G are as simple as possible.

As in the construction of a low-dimensional invariant manifold, whether it be center or slow, we approximate the normal-form transformation by iteration. Let's detail just the first iteration. Given the linear approximation, we simplify any quadratic terms arising in the evolution equation (13.22) through the transform (13.21).

- First consider the fast modes. Substitute the linear (or identity) approximation together with small corrections (indicated by hats as for earlier procedures) into the x-evolution equation (13.11d):

$$
\begin{aligned}
\dot{x} &= -z = -Z - \hat{z} \quad \text{by (13.21), and also} \\
\dot{x} &= \frac{\partial x}{\partial U}\dot{U} + \frac{\partial x}{\partial V}\dot{V} + \frac{\partial x}{\partial W}\dot{W} + \frac{\partial x}{\partial X}\dot{X} + \frac{\partial x}{\partial Z}\dot{Z} \quad \text{by the chain rule} \\
&\quad \text{(substituting for time derivatives)} \\
&= \frac{\partial \hat{x}}{\partial U}\hat{A} + \frac{\partial \hat{x}}{\partial V}\hat{B} + \frac{\partial \hat{x}}{\partial W}\hat{C} \\
&\quad + \left(1 + \frac{\partial \hat{x}}{\partial X}\right)(-Z + \hat{F}) + \frac{\partial \hat{x}}{\partial Z}(X + \hat{G}) \\
&\quad \text{(ignoring products of small corrections)} \\
&\approx -Z + \hat{F} - Z\frac{\partial \hat{x}}{\partial X} + X\frac{\partial \hat{x}}{\partial Z}.
\end{aligned}
$$

Rearranging the equality between the first and last lines leads to

$$
\hat{F} + \hat{z} - Z\frac{\partial \hat{x}}{\partial X} + X\frac{\partial \hat{x}}{\partial Z} = 0. \tag{13.23}
$$

Similarly, substituting the linear (or identity) approximation together with corrections into the z-evolution equation (13.11e),

$$
\begin{aligned}
\dot{z} &= x + buv \\
&= X + \hat{x} + b(U + \hat{u})(V + \hat{v}) \quad \text{by (13.21)} \\
&= X + \hat{x} + bUV \quad \text{to leading order, and also} \\
\dot{z} &= \frac{\partial z}{\partial U}\dot{U} + \frac{\partial z}{\partial V}\dot{V} + \frac{\partial z}{\partial W}\dot{W} + \frac{\partial z}{\partial X}\dot{X} + \frac{\partial z}{\partial Z}\dot{Z} \quad \text{by the chain rule} \\
&\quad \text{(substituting for time derivatives)} \\
&= \frac{\partial \hat{z}}{\partial U}\hat{A} + \frac{\partial \hat{z}}{\partial V}\hat{B} + \frac{\partial \hat{z}}{\partial W}\hat{C} \\
&\quad + \frac{\partial \hat{z}}{\partial X}(-Z + \hat{F}) + \left(1 + \frac{\partial \hat{z}}{\partial Z}\right)(X + \hat{G}) \\
&\quad \text{(ignoring products of small corrections)} \\
&\approx X + \hat{G} - Z\frac{\partial \hat{z}}{\partial X} + X\frac{\partial \hat{z}}{\partial Z}.
\end{aligned}
$$

Rearranging the equality between the third and last lines leads to

$$
\hat{G} - \hat{x} - Z\frac{\partial \hat{z}}{\partial X} + X\frac{\partial \hat{z}}{\partial Z} = bUV. \tag{13.24}
$$

Now solve the two coupled equations (13.23) and (13.24).[167] A solution keeping $\hat{F}$ and $\hat{G}$ as simple as possible is $\hat{F} = \hat{G} = \hat{z} = 0$ and $\hat{x} = -bUV$. Thus, we remove the quadratic coupling term in the z equation, coupling the fast to the slow dynamics, by choosing the transformation $x \approx X + \hat{x} = X - bUV$.

Of course there are induced cubic or higher-order couplings between the dynamics to be found by further iterations.

[167] Akin to equations for the center manifold of a Hopf bifurcation; see Chapter 16.

- Second, consider the slow modes. Similar arguments to the above lead to the following equations for the corrections, upon substituting, using the chain rule, and neglecting products of hatted correction quantities or products of such corrections and small variables:

$$\begin{aligned}\hat{A} - Z\frac{\partial \hat{u}}{\partial X} + X\frac{\partial \hat{u}}{\partial Z} &= -VW + bVZ,\\ \hat{B} - Z\frac{\partial \hat{v}}{\partial X} + X\frac{\partial \hat{v}}{\partial Z} &= UW - bUZ,\\ \hat{C} - Z\frac{\partial \hat{w}}{\partial X} + X\frac{\partial \hat{w}}{\partial Z} &= -UV.\end{aligned}$$

  These equations are independent of each other, so solve them each in turn. In the first equation, we want to keep $\hat{A}$ as simple as possible. We may choose $\hat{u}$ to contain the term $-bVX$ to balance the bVZ term on the right-hand side. But $\hat{u}$ cannot be used to balance the $-VW$ term on the right-hand side; it can only balance terms with Z or X in them due to the multiplication by these variables in the terms involving $\hat{u}$ on the left-hand side. Thus we are forced to set $\hat{A} = -VW$ and consequently $\hat{u} = -bVX$. Similarly for the other equations: we are forced to put $\hat{B} = UW$ and $\hat{C} = -UV$ and consequently $\hat{v} = bUX$ and $\hat{w} = 0$.

- Finally, to this level of approximation, the coordinate transformation

$$\begin{aligned}&u = U - bVX + \mathcal{O}(s^3), \quad v = V + bUX + \mathcal{O}(s^3), \quad w = W + \mathcal{O}(s^3),\\ &x = X - bUV + \mathcal{O}(s^3), \quad z = Z + \mathcal{O}(s^3)\end{aligned} \tag{13.25a}$$

  leads to evolution equations

$$\begin{aligned}&\dot{U} = -VW + \mathcal{O}(s^3), \quad \dot{V} = UW + \mathcal{O}(s^3), \quad \dot{W} = -UV + \mathcal{O}(s^3),\\ &\dot{X} = -Z + \mathcal{O}(s^3), \quad \dot{Z} = X + \mathcal{O}(s^3).\end{aligned} \tag{13.25b}$$

  In this transformed version of the dynamics, the quadratic interaction terms have been removed. With possibly cubic errors, the fast wave terms have been removed from the slow dynamics, and the slow wave terms have been removed from the fast dynamics. Thus they evolve independently—*the slow dynamics of* U, V, *and* W *are independent of the fast waves in* X *and* Z. In this sense the three-dimensional evolution of U, V, and W forms a low-dimensional model for the five-dimensional Lorenz system.

The slow manifold In the normal-form equations, the slow manifold is easily apparent. To this order of accuracy, the subspace $X = Z = 0$ is invariant under the dynamics, has no fast waves, and hence forms the slow manifold $\mathbb{M}_0$. From the transformation given above, as $X = Z = 0$ on $\mathbb{M}_0$, then $u \approx U$, $v \approx V$, and $x \approx -bUV$; thus with possibly cubic errors, $\mathcal{M}_0$ is $x \approx -buv$ and $z \approx 0$, exactly as determined in section 13.3.

It would appear that the normal-form transformation supports the existence of a slow manifold and its relevance as a model of the dynamics. But all we really know, by Proposition 13.5, is that the combined system (13.25), of the evolution (13.25a) and coordinate transform (13.25b), possesses a slow manifold and is close to the original system.

Wave-wave interactions cause drift At higher orders in the analysis we encounter difficulties in trying to separate *completely* the slow dynamics from the fast. Consider what the x equation may look like at some iteration:

$$\hat{A} - Z\frac{\partial \hat{u}}{\partial X} + X\frac{\partial \hat{u}}{\partial Z} = \sum_{m,n} f_{mn}(U, V, W) X^m Z^n \,,$$

where f_{mn} is some multinomial expression in U, V, and W. Typically, nonlinear terms will generate a wide variety of such terms. Specifically, suppose that terms quadratic[168] in X and Z are generated by wave-wave interactions so that we need to solve

$$\hat{A} - Z\frac{\partial \hat{u}}{\partial X} + X\frac{\partial \hat{u}}{\partial Z} = f_{20}X^2 + f_{11}XZ + f_{02}Z^2 \,. \tag{13.26}$$

The homological operator on the left-hand side, $-Z\partial\hat{u}/\partial X + X\partial\hat{u}/\partial Z$, maps quadratics in X and Z into quadratics, cubics into cubics, and so on. Thus it is sufficient to consider a quadratic form for $\hat{u}$ in an attempt to simplify $\hat{A}$. Trying the general form $\hat{u} = aX^2 + bXZ + cZ^2$, the equation becomes

$$\hat{A} + bX^2 + 2(c - a)XZ - bZ^2 = f_{20}X^2 + f_{11}XZ + f_{02}Z^2 \,.$$

Observe that the homological operator is singular because it can only balance a right-hand side for which $f_{20} + f_{02} = 0$. If $f_{20} + f_{02} \neq 0$, then we are forced into making $\hat{A}$ have a component of $\frac{1}{2}(f_{20} + f_{02})(X^2 + Z^2)$. Hence the evolution equation for the slow variable U is forced to be, in general, of the form[169]

$$\dot{U} = A_0(U, V, W) + A_2(U, V, W)(X^2 + Z^2) + \cdots \,, \tag{13.27}$$

where A_0 and A_2 are functions known to some order in their asymptotic expansion. Such a form implies that the slow evolution is, in general, necessarily coupled to the fast oscillations. The coupling only involves terms of the form $X^2 + Z^2$, which is approximately invariant in the fast oscillations—it is the amplitude squared, say R^2—and so U still evolves slowly. However, in the presence of fast oscillations, U necessarily drifts away from solutions without fast oscillations. Consequently, the evolution off the slow manifold $\mathcal{M}_0$ is necessarily different from that on $\mathcal{M}_0$—it drifts away at a rate proportional to R^2. In general, the slow manifold cannot strictly serve as a low-dimensional model—the long-term trends on and off $\mathcal{M}_0$ are necessarily different.

The consoling feature of this analysis for slow manifold approximations is that the discrepancy between the model and the actual dynamics is proportional to the square of the amplitude of the fast oscillations and so may be quite small.

On the nonexistence of a slow manifold Lorenz and Krishnamurty (1987) followed up the work of Lorenz (1986) with an article in which they argued that the slow manifold cannot exist at all! Exact numerical solutions of (13.11) showed that resonances between the fast and slow oscillations inevitably generated fast waves that break up slow solutions on the supposedly slow manifold. It appears that an infinite sequence of discontinuities occurs in the slow manifold. Cox and Roberts (1994) argued that these discontinuities are exponentially small and so do not affect asymptotic expressions for $\mathcal{M}_0$, such as (13.15).

[168] Quartic terms in X and Z are also typically intractable.

[169] Here A_2 is at least linear because no such quadratic term exists.

Analogously expect that in general the hypothesis that a slow manifold exists is incorrect due to exponentially small effects.

Recall the more positive view promised by Proposition 13.5. Coordinate transforms generally lead to evolution equations analogous to (13.27). In such a normal form a simple slow manifold clearly exists, namely the fast oscillations being zero (here $X = Z = 0$). Because, in principle, such a normal form is obtainable to arbitrarily high order, it follows that the difference between its dynamics and the dynamics of the original system must be exponentially small. The positive deduction is that even if the original system does not strictly have a slow manifold, *there exists a system exponentially close to the original that does possess a slow manifold.*

13.5.4 ▪ Near the slow manifold of electron drift

The previous example investigated a subcenter slow manifold using nonlinear modifications of the original variables. But using oscillation amplitude and phase variables is more appealing, as these explicitly recognize the oscillation amplitude as a slow variable that interacts with other slow processes in the system, as in the analysis of oscillations of Chapter 3. Here we reconsider the electron in a magnetic field, section 13.4, and invoke computer algebra to derive the interaction between any electron fast oscillations and the electron slow drift.

Recall the dynamic equations (13.16) for the position $x(t)$ and $y(t)$ in the plane of the electron that moves with velocity $u(t)$ and $v(t)$; write the equations more specifically as

$$\dot{x} = u, \quad \dot{y} = v, \quad \dot{u} = -v - \epsilon\mathcal{V}_x, \quad \dot{v} = +u - \epsilon\mathcal{V}_y, \quad \mathcal{V} = \tfrac{1}{4}(x^4 + y^4) \qquad (13.28)$$

for the nondimensional quartic electric potential $\mathcal{V}(x, y)$ giving forces $-\epsilon\boldsymbol{\nabla}\mathcal{V}$. We transform these dynamics to new variables (X, Y, R, Θ) representing the mean position (X, Y) of the electron, the amplitude R of any fast oscillations of the electron, and the phase Θ of such oscillations:

$$x \approx X + R\sin\Theta, \quad y \approx Y + R\cos\Theta, \quad u \approx R\cos\Theta, \quad v \approx -R\sin\Theta. \qquad (13.29)$$

The aim of the transform (13.29) is to find the following frequency shift $\dot{\Theta}$ and the slow evolution of position and amplitude:

$$\begin{aligned}
\dot{\Theta} &= 1 + \epsilon\tfrac{3}{2}(X^2 + Y^2 + \tfrac{1}{2}R^2) + \mathcal{O}(\epsilon^2),\\
\dot{R} &= -\epsilon^2\tfrac{3}{2}RXY(X^2 + Y^2 + 3R^2) + \mathcal{O}(\epsilon^3),\\
\dot{X} &= -\epsilon Y(Y^2 + \tfrac{3}{2}R^2) + \mathcal{O}(\epsilon^2),\\
\dot{Y} &= +\epsilon X(X^2 + \tfrac{3}{2}R^2) + \mathcal{O}(\epsilon^2).
\end{aligned} \qquad (13.30)$$

This transformed version of the ODEs looks more complicated, but crucially the ODE dynamics is slow (apart from $\dot{\Theta} \approx 1$) so we can easily simulate over long times. This transformed version is not limited by the fast time scale of the electron oscillations. Repeating the main message: due to R^2 drift terms in the mean position evolution, $\dot{X}$ and $\dot{Y}$, the mean evolution off a slow manifold in the presence of fast oscillations, $R \neq 0$, is different from the mean evolution on a slow manifold, $R = 0$. A slow manifold might well be useful, but it generally cannot be a complete, closed, and accurate model.

Computer algebra iterates to a transform Algorithm 13.2 lists the overall structure of computer algebra to iteratively find the details of the coordinate transform (13.29).

Algorithm 13.2 *Reduce code to deduce the normal-form coordinate transform of electron dynamics* (13.28) *to position, amplitude, and phase variables* X, Y, R, *and* Θ, *respectively. Invokes the code fragments of Algorithms* 13.3, 13.4, *and* 13.5.

```
on div; off allfac; on revpri; factor eps,rr;
in "emagvars.red"$
in "emagoperators.red"$
let eps^3=>0;
for iter:=1:99 do begin
   in "emagupdates.red"$
   if {resu,resv,resx,resy}={0,0,0,0}
   then iter:=1000000;
end;
in "emagpost.red"$
end;
```

Algorithm 13.3 *Definitions and initialization for Algorithm* 13.2 *to invoke.*

```
% electric potential
vv:=(x^4+y^4)/4;
vvx:=df(vv,x); vvy:=df(vv,y);
% new amplitude and phase angle variables
depend rr,t; depend q,t;
depend xx,t; depend yy,t;
let { df(rr,t)=>drrdt, df(q,t)=>dqdt
    , df(xx,t)=>dxxdt, df(yy,t)=>dyydt
    };
% initial linear approximation
u:=rr*cos(q);      v:=-rr*sin(q);
x:=xx+rr*sin(q); y:=yy+rr*cos(q);
drrdt:=dxxdt:=dyydt:=0;
dqdt:=1;
end;
```

First, the algorithm initializes variables and makes definitions as detailed in Algorithm 13.3. Second, the algorithm iteratively updates approximations, as detailed in Algorithm 13.4, until the residuals are zero for all four equations of the electron's dynamical system (13.28). The iteration continues until the residuals are $\mathcal{O}(\epsilon^3)$ to give the next higher order terms to (13.30). Postprocessing follows the iteration.

Initialization Algorithm 13.3 first initializes the electric potential $\mathcal{V}(x, y)$, called `vv`, and the resultant electric field. The potential could be any multinomial form in x and y without having to make any adaptations elsewhere. Second, the dynamics are to be written in terms of four new evolving variables—X, Y, R, and Θ, which are called `xx`, `yy`, `rr`, and `q`—so inform Reduce that these variables depend upon time. Then define the time derivatives to be expressions that we construct. Third, encode the initial linear approximation (13.29) to the coordinate transform. Last, set the initial time derivatives $\dot{X} = \dot{Y} = \dot{R} = 0$ and the frequency $\dot{\Theta} = 1$.

Algorithm 13.4 *To update the coordinate transformation, Algorithm* 13.2 *invokes this code fragment. Algorithm* 13.5 *defines the necessary operators used here.*

```
        resu:=trigsimp( df(u,t)-v+eps*vvx ,combine);
        resv:=trigsimp( df(v,t)+u+eps*vvy ,combine);
        res:= resv+df(resu,t);
        drrdt:=drrdt+(gd:=coeffn(res,sin(q),1)/2);
        dqdt:=dqdt+(hd:=coeffn(res,cos(q),1)/2/rr);
        u:=u+(pd:=linv(res-2*gd*sin(q)-2*hd*rr*cos(q),q));
        v:=v+resu+df(pd,t);

        resx:=-df(x,t)+u;
        dxxdt:=dxxdt+(gd:=mean(resx,q));
        x:=x+ints(resx-gd,q);
        resy:=-df(y,t)+v;
        dyydt:=dyydt+(gd:=mean(resy,q));
        y:=y+ints(resy-gd,q);
    end;
```

Iteratively update approximations First consider the fast oscillations. They arise from the u and v components of the electron system (13.28). We need to put these oscillations in terms of amplitude/phase variables. Chapter 3 introduced this sort of transformation and it is utilized for Duffing's ODE (3.4) in Algorithm 3.1. We adapt code from Algorithm 3.1. Algorithm 13.4 computes the residuals of the u and v equations (using `trigsimp` to linearize products of trigonometric functions): we need to seek updates such that $\dot{\hat{u}} - \hat{v} + \text{Res}_u = 0$ and $\dot{\hat{v}} + \hat{u} + \text{Res}_v = 0$. Eliminate update $\hat{v}$ by adding the time derivative of the first to the second to seek to solve

$$\ddot{\hat{u}} + \hat{u} + \text{Res} = 0 \quad \text{where} \quad \text{Res} = \text{Res}_v + \frac{\partial}{\partial t}\text{Res}_u\,.$$

This equation for updates is identical to that we used for Duffing's ODE (3.4), so the next three lines in Algorithm 13.4 are copied with appropriate name changes from the Duffing Algorithm 3.1. All that remains, as coded, is to then compute the update for v from $\dot{\hat{u}} - \hat{v} + \text{Res}_u = 0$. This first block of code corrects the coordinate transform to account for the fast oscillations.

The second block of code in Algorithm 13.4 updates the transform based upon the dynamics of position. Code for each coordinate is equivalent, so just describe the x coordinate. It satisfies $\dot{x} = u$, so denoting $\dot{X}$ by g, the equation for updates is $\hat{g} + \partial\hat{x}/\partial\Theta = \text{Res}_x$ for computed residual $\text{Res}_x = u - \dot{x}$. Components in the residual Res_x that oscillate in phase Θ are integrated into the coordinate transform $\hat{x}$; this is done by the operator `ints`. It is only the mean component of the residual Res_x that has to go into the update of $\dot{X}$;[170] Algorithm 13.5 details the operator `mean` that extracts the mean. The same process updates the y coordinate.

Operators integrate or extract Algorithm 13.4 invokes three operators to perform its magic: `linv`, `mean`, and `ints`. Algorithm 13.5 defines these operators. First, `linv` is

[170] If we integrated the mean component into the coordinate transform, then there would be a component in the transform that grew linearly in time. Such growth is unallowable, as it makes the coordinate transform nonuniformly valid in time.

Algorithm 13.5 *Some operators for Algorithm* 13.2, *which invokes this code fragment.*

```
% to solve x_qq+x+res=0
operator linv; linear linv;
let { linv(1,q)=>-1
    , linv(cos(~n*q),q)=>cos(n*q)/(n^2-1)
    , linv(sin(~n*q),q)=>sin(n*q)/(n^2-1) };
% to solve x_t=res
operator mean; linear mean;
let { mean(1,q)=>1
    , mean(cos(~~n*q),q)=>0
    , mean(sin(~~n*q),q)=>0 };
operator ints; linear ints;
let { ints(cos(~~n*q),q)=>sin(n*q)/n
    , ints(sin(~~n*q),q)=>-cos(n*q)/n };
end;
```

a direct copy from Algorithm 3.1, and later exercises in Chapter 3, of the same operator for the oscillations of Duffing's ODE (3.4). To compute the mean of an expression linear in trigonometric functions, the operator `mean` just keeps anything constant in Θ and discards anything trigonometric in Θ. The operator `ints` just needs to compute an integral of its argument. Algorithm 13.2 defines these operators before the iterative updates begin to find the coordinate transform.

The normal-form coordinate transform For this particular potential, the result is that the new variables evolve according to (13.30), where the physical variables are, in gory detail,

$$\begin{aligned}
x &= X + R\sin\Theta + \epsilon\Big[-\tfrac{3}{2}R(X^2+Y^2)\sin\Theta + \tfrac{1}{4}R^2(Y\sin 2\Theta - 2X\cos 2\Theta) \\
&\quad - \tfrac{1}{48}R^3(36\sin\Theta + \sin 3\Theta)\Big] + \mathcal{O}(\epsilon^2), \\
y &= X + R\cos\Theta + \epsilon\Big[-3RX^2\cos\Theta + \tfrac{1}{4}R^2(X\sin 2\Theta + 2Y\cos 2\Theta) \\
&\quad - \tfrac{1}{48}R^3(36\sin\Theta - \sin 3\Theta)\Big] + \mathcal{O}(\epsilon^2), \\
u &= R\cos\Theta + \epsilon\Big[-Y^3 + \tfrac{1}{2}R^2(-3Y + 2X\sin 2\Theta + Y\cos 2\Theta) \\
&\quad - \tfrac{1}{16}R^3\cos 3\Theta\Big] + \mathcal{O}(\epsilon^2), \\
v &= -R\sin\Theta + \epsilon\Big[X^3 + \tfrac{1}{2}R(X^2 - Y^2)\sin\Theta \\
&\quad + \tfrac{1}{2}R^2(3X - 2Y\sin 2\Theta + X\cos 2\Theta) - \tfrac{1}{16}R^3\sin 3\Theta\Big] + \mathcal{O}(\epsilon^2).
\end{aligned}$$

Higher-order results are easily computed by changing the order of truncation of the analysis in Algorithm 13.2. By simply changing the potential in Algorithm 13.3 one obtains coordinate transforms and normal-form electron evolution for different applied electric fields.

The existence of such coordinate transforms assures us that a system exists that is close to the specified system and possesses a slow manifold.

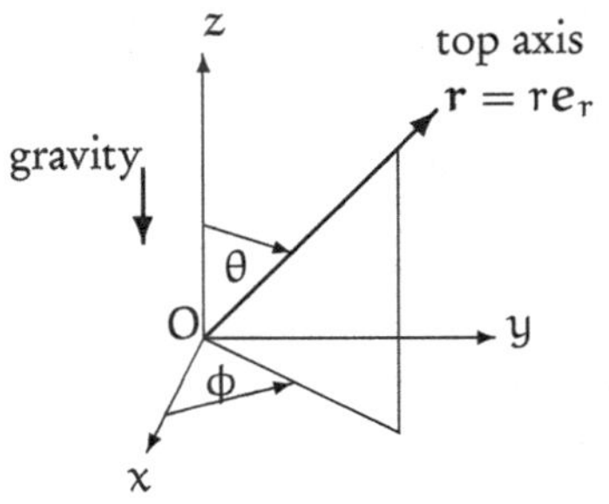

Figure 13.8. *A top spins about an axis* $\mathbf{r}$ *with its tip at the origin O. Downward gravity causes the axis to precess around the vertical z-axis: the azimuthal angle* ϕ *varies linearly at a rate that depends upon the polar inclination* θ.

No matter what potential you try, just like the evolution (13.30), almost every nonlinear interaction insists on the existence of oscillation-induced drift effects in the evolution off the slow manifold. These induced drift effects imply that the evolution on a slow manifold is necessarily different from that off the slow manifold.

13.6 ▪ Case study: Precession is a slow manifold of a spinning top

Consider a spinning top on a table. Depending upon the angle of its inclination to the vertical, the axis of the top will precess. Berry and Shukla (2011) interpreted this precession as a slow manifold. Let's detail and extend the analysis of a spinning top that is mentioned in Exercise 4.17.

Define Cartesian (x, y, z) and spherical (r, θ, ϕ) coordinates as in Figure 13.8, with the axis of the spinning top being aligned along the radial unit vector $\mathbf{e}_r$. By equating the rate of change of angular momentum to the applied torque, Berry and Shukla (2011, eq. (3.6)) derive that the axis $\mathbf{r}(t)$ of the top satisfies the nondimensional second-order vector differential equation

$$\mathbf{r} \times \ddot{\mathbf{r}} + G\dot{\mathbf{r}} + \mathbf{r} \times \mathbf{e}_z = 0, \tag{13.31}$$

where parameter $G := A\Omega/\sqrt{Bmg\ell}$, constant A is the moment of inertia about the rotation axis, Ω is the angular velocity of the spinning top, B is the moment of inertia perpendicular to the rotation axis, m is the top mass, g is the strength of gravity in the direction $-\mathbf{e}_z$, and ℓ is the distance of the center of mass of the top along the axis.

Rapid rotation fixes the top Initially, one assumes that rotation parameter G is large enough so the first and third terms of the differential equation (13.31) are negligible. The differential equation then reduces to $G\dot{\mathbf{r}} \approx 0$. Hence we deduce that the axis is approximately fixed in space. However, this initial approximation inevitably envelops us in a singular perturbation as it neglects the second-order time derivative.

Rescaling time regularizes Instead we change the nondimensionalization by rescaling time: $t = t^*/G$ so that $d/dt = G\,d/dt^*$; then dropping $*$'s leads to the equivalent differential equation

$$\mathbf{r} \times \ddot{\mathbf{r}} + \dot{\mathbf{r}} + \frac{1}{G^2}\mathbf{r} \times \mathbf{e}_z = 0 .$$

Now large parameter G corresponds to small $\epsilon := 1/G^2$—highlighting that the small parameter is $1/G^2$, not $1/G$. Last, a dynamical systems approach prefers a system of first-order differential equations, so let's introduce a "velocity" $\mathbf{u}(t)$ so that the system becomes

$$\dot{\mathbf{r}} = \mathbf{u} \quad \text{and} \quad \mathbf{r} \times \dot{\mathbf{u}} + \mathbf{u} + \epsilon \mathbf{r} \times \mathbf{e}_z = \mathbf{0}. \tag{13.32}$$

Now the small parameter ϵ, corresponding to rapid rotation, appears as a regular perturbation.

13.6.1 ▪ A slow manifold exists

To discover whether slow dynamics exists, perform the usual routine (Table 4.2): find equilibria, linearize, and characterize the spectrum.

Seek equilibria Set parameter $\epsilon = 0$ corresponding to extremely fast rotation (G large) or negligible gravity. There is a family of equilibria of the top pointing in any direction: $\mathbf{r} = \mathbf{R}$ and $\mathbf{u} = \mathbf{0}$ for every constant vector $\mathbf{R}$.

Find the slow subspaces Linearize the dynamics about each of these equilibria: seek $\mathbf{r}(t) = \mathbf{R} + \hat{\mathbf{r}}(t)$ and $\mathbf{u}(t) = \mathbf{0} + \hat{\mathbf{u}}(t)$ for hatted quantities small. Neglecting ϵ and products of small quantities, the top system (13.32) becomes

$$\dot{\hat{\mathbf{r}}} = \hat{\mathbf{u}} \quad \text{and} \quad \mathbf{R} \times \dot{\hat{\mathbf{u}}} + \hat{\mathbf{u}} = \mathbf{0}.$$

To find eigenvalues and eigenvectors, seek solutions proportional to $e^{\lambda t}$ and write in Cartesian coordinates as

$$\begin{bmatrix} \lambda & 0 & 0 & 1 & 0 & 0 \\ 0 & \lambda & 0 & 0 & 1 & 0 \\ 0 & 0 & \lambda & 0 & 0 & 1 \\ 0 & 0 & 0 & 1 & -R_z\lambda & R_y\lambda \\ 0 & 0 & 0 & R_z\lambda & 1 & -R_x\lambda \\ 0 & 0 & 0 & -R_y\lambda & R_x\lambda & 1 \end{bmatrix} \begin{bmatrix} \hat{\mathbf{r}} \\ \hat{\mathbf{u}} \end{bmatrix} = \mathbf{0}.$$

Nontrivial solutions exist only when the determinant of this matrix is zero: expanding in the first three columns of the matrix the determinant is

$$\lambda^3 \begin{vmatrix} 1 & -R_z\lambda & R_y\lambda \\ R_z\lambda & 1 & -R_x\lambda \\ -R_y\lambda & R_x\lambda & 1 \end{vmatrix} = \lambda^3 \left[1 + \lambda^2(R_x^2 + R_y^2 + R_z^2)\right].$$

Thus the eigenvalues about each equilibrium are $\lambda = 0$ (three times) and $\lambda = \pm i/|\mathbf{R}|$. As all eigenvalues of the system have zero real part, the entire state space is the center manifold. As in much of classical physics, the primary interest lies in the slow dynamics associated with the eigenvalue zero, here of multiplicity three, among the fast oscillations, here of frequency $1/|\mathbf{R}|$. From both the eigenvectors corresponding to the eigenvalue zero and the equilibria for all $\mathbf{R}$, the slow subspace is composed of all spin orientations $\mathbf{r}$ with zero velocity $\mathbf{u} = \mathbf{0}$, as parameter $\epsilon = 0$.

Curiously, there are only five eigenvalues despite the system (13.32) appearing to have six components, $\mathbf{r}(t)$ and $\mathbf{u}(t)$. The reason is that one component of the equations is *implicitly* an algebraic equation, not differential. The algebraic equation arises because the tip of the top is fixed on the table top: take the inner product $\mathbf{r}\cdot$ of the second equation of (13.32); since $\mathbf{r}\cdot\mathbf{r}\times\mathbf{a}=0$ for all $\mathbf{a}$ the equation reduces to the algebraic $\mathbf{r}\cdot\mathbf{u}=0$. That is, the velocity $\mathbf{u}$ must always be orthogonal to the axis direction $\mathbf{r}$. This constraint is not a conservation law: it does not follow from a time derivative being zero. Instead, the algebraic $\mathbf{r}\cdot\mathbf{u}=0$ is a constraint that must be satisfied at all times, including the initial conditions. Thus there are only five dynamic modes.

One of the five dynamic modes is a distinct but associated conservative mode. Since we have just shown $\mathbf{r}\cdot\mathbf{u}=0$, then

$$\frac{d(r^2)}{dt}=\frac{d}{dt}(\mathbf{r}\cdot\mathbf{r})=2\mathbf{r}\cdot\dot{\mathbf{r}}=2\mathbf{r}\cdot\mathbf{u}=0.$$

That is, the required orthogonality of the velocity also ensures that r is conserved in the time evolution. We could set $r=1$ without loss of generality, but let's carry it with us for the moment.

13.6.2 ▪ A beautiful slow manifold gives precession

Unusually, the spinning top system (13.32) has an algebraically exact slow manifold. Look at the spinning top system (13.32): on the slow manifold the time derivatives are to be small; hence the second of system (13.32) indicates that the slow manifold is velocity $\mathbf{u}\approx-\epsilon\mathbf{r}\times\mathbf{e}_z$ for all orientations $\mathbf{r}$. Since $\mathbf{r}=r\mathbf{e}_r$ and $\mathbf{e}_z=\cos\theta\,\mathbf{e}_r-\sin\theta\,\mathbf{e}_\theta$, then $\mathbf{r}\times\mathbf{e}_z=-r\sin\theta\,\mathbf{e}_\phi$ and hence the slow manifold velocity indicates $\dot{\mathbf{r}}\approx\epsilon r\sin\theta\,\mathbf{e}_\phi$. That is, the spin axis revolves about the vertical z-axis, a distance $r\sin\theta$ away, at a frequency of approximately ϵ. This describes the precession.

Now make into exact equalities all the approximations of the preceding paragraph. Seek an exact slow manifold of the form

$$\mathbf{u}=-\beta(\mathbf{r}\cdot\mathbf{e}_z)\,\mathbf{r}\times\mathbf{e}_z$$

for some as yet unknown function $\beta(\mathbf{r}\cdot\mathbf{e}_z)=\beta(r\cos\theta)$. We expect $\beta\approx\epsilon$ for small parameter ϵ. Differentiating with respect to time gives the acceleration (omitting the argument to function β for symbolic simplicity)

$$\begin{aligned}\dot{\mathbf{u}}&=-\beta'[\dot{\mathbf{r}}\cdot\mathbf{e}_z]\mathbf{r}\times\mathbf{e}_z-\beta\dot{\mathbf{r}}\times\mathbf{e}_z\\&=-\beta'[\mathbf{u}\cdot\mathbf{e}_z]\mathbf{r}\times\mathbf{e}_z-\beta\mathbf{u}\times\mathbf{e}_z\\&=-\beta'\underbrace{[(-\beta\mathbf{r}\times\mathbf{e}_z)\cdot\mathbf{e}_z]}_{=0}\mathbf{r}\times\mathbf{e}_z-\beta(-\beta\mathbf{r}\times\mathbf{e}_z)\times\mathbf{e}_z\\&=\beta^2(\mathbf{r}\times\mathbf{e}_z)\times\mathbf{e}_z\\&=\beta^2\left[(\mathbf{r}\cdot\mathbf{e}_z)\mathbf{e}_z-\mathbf{r}\right]\end{aligned}$$

using the cross-product rule that $\mathbf{a}\times(\mathbf{b}\times\mathbf{c})=\mathbf{b}(\mathbf{a}\cdot\mathbf{c})-\mathbf{c}(\mathbf{a}\cdot\mathbf{b})$. Substitute this acceleration $\dot{\mathbf{u}}$ into the second equation of (13.32) and substitute the slow manifold $\mathbf{u}$; then factor to obtain

$$\left[(\mathbf{r}\cdot\mathbf{e}_z)\beta^2-\beta+\epsilon\right]\mathbf{r}\times\mathbf{e}_z=0.$$

Hence the quadratic in brackets must be zero, which gives

$$\beta=\frac{1-\sqrt{1-4\epsilon\mathbf{r}\cdot\mathbf{e}_z}}{2\mathbf{r}\cdot\mathbf{e}_z}=\frac{1-\sqrt{1-4\epsilon r\cos\theta}}{2r\cos\theta}. \tag{13.33}$$

This minus square root[171] correctly gives $\beta \approx \epsilon$ for small parameter ϵ. The evolution on the slow manifold is then

$$\dot{\mathbf{r}} = \mathbf{u} = -\beta\mathbf{r} \times \mathbf{e}_z = \beta r \sin\theta \, \mathbf{e}_\phi \,.$$

This exact solution beautifully predicts unceasing precession at frequency precisely β.[172]

13.6.3 ▪ Nutation is the fast oscillations

The linear analysis about equilibria shows that there are potential fast oscillations, frequency $1/r$, about the slow manifold of precession. In the classic spinning top these fast waves cause a wobble called nutation. Here we recast some of the analysis to see where nutation occurs.

To most clearly distinguish the components of the dynamics, let's turn to the spherical coordinate system (r, θ, ϕ) of Figure 13.8. First, denote the position vector and velocity of the top's center of mass by

$$\mathbf{r} = r\mathbf{e}_r \quad \text{and} \quad \mathbf{u} = u\mathbf{e}_r + v\mathbf{e}_\theta + w\mathbf{e}_\phi \tag{13.34}$$

for velocity components (u, v, w). Then the first equation of system (13.32), namely $\dot{\mathbf{r}} = \mathbf{u}$, becomes dynamical equations for the position/orientation of the top in terms of these velocity components:

$$\dot{r} = u\,, \quad r\dot{\theta} = v\,, \quad \text{and} \quad r\sin\theta\,\dot{\phi} = w\,. \tag{13.35}$$

The second equation of system (13.32) involves the acceleration $\dot{\mathbf{u}}$. This acceleration is complicated by the dependence upon time of the unit vectors $\mathbf{e}_r$, $\mathbf{e}_\theta$, and $\mathbf{e}_\phi$. Without going into all the details one may derive that the acceleration

$$\begin{aligned}
\dot{\mathbf{u}} &= \left(\dot{u} + r\dot{\theta}^2 - r\dot{\phi}^2\sin^2\theta\right)\mathbf{e}_r \\
&\quad + \left(\dot{v} + 2\dot{r}\dot{\theta} - r\dot{\phi}^2\sin\theta\,\cos\theta\right)\mathbf{e}_\theta \\
&\quad + \left(\dot{w} + 2\dot{r}\dot{\phi}\sin\theta + 2r\dot{\theta}\dot{\phi}\cos\theta\right)\mathbf{e}_\phi \\
&= \left(\dot{u} + \frac{1}{r}v^2 - \frac{1}{r}w^2\right)\mathbf{e}_r \\
&\quad + \left(\dot{v} + \frac{1}{r}uv - \frac{1}{r}\cot\theta\,w^2\right)\mathbf{e}_\theta \\
&\quad + \left(\dot{w} + \frac{1}{r}uw + \frac{1}{r}\cot\theta\,vw\right)\mathbf{e}_\phi\,.
\end{aligned}$$

[171]The plus alternative is another exact solution that is not yet of interest.

[172]We could include small dissipative effects simply by letting parameter ϵ grow slowly in time. Such growth corresponds to slow decay of the rate Ω of rotation of the top. Slow exponential decay of Ω corresponds to slow exponential growth of ϵ. Does the exact slow manifold "break up" under such dissipation? Certainly the system eventually "falls off the edge" of the slow manifold as parameter ϵ grows past the critical $1/(4r\cos\theta)$. This question really asks whether some fast waves develop due to small dissipative effects on this exact slow manifold. Another dissipation could be air drag in proportion to the velocity. This dissipation could be modeled through replacing $\dot{\mathbf{u}}$ by $\dot{\mathbf{u}} + \varepsilon\mathbf{u}$ for small parameter ε. Linearly, this only affects the fast waves by changing their eigenvalues to the dissipative $\lambda = -\varepsilon \pm i/|\mathbf{R}|$. The effect on the nonlinear dynamics is yet to be determined.

Then in system (13.32) the cross-product $\mathbf{r}\times\dot{\mathbf{u}} = r\mathbf{e}_r\times\dot{\mathbf{u}}$ nullifies the $\mathbf{e}_r$ component of the above to transform system (13.32) into

$$\begin{aligned}&+\left(r\dot{v}+uv-\cot\theta\,w^2\right)\mathbf{e}_\phi\\ &-\left(r\dot{w}+uw+\cot\theta\,vw\right)\mathbf{e}_\theta\\ &+u\mathbf{e}_r+v\mathbf{e}_\theta+w\mathbf{e}_\phi-\epsilon r\sin\theta\,\mathbf{e}_\phi=\mathbf{0}.\end{aligned}$$

Look at the three components in turn.

$\mathbf{e}_r$ This component asserts the algebraic equation $u=0$. That is, the top cannot move off the surface it spins upon. Then the radial component of (13.35) gives $\dot{r}=u=0$. That is, the radial position of the center of mass of the top is conserved. Without loss of generality, let's now set $r=1$ by the nondimensionalization. We could have set $r=1$ earlier, but in this approach we see constant r arise explicitly as a conservation result consequent to the algebraic $u=0$.

$\mathbf{e}_\theta$ **and** $\mathbf{e}_\phi$ Setting $u=0$ and $r=1$, these components become

$$\dot{w}=+v-vw\cot\theta \quad\text{and}\quad \dot{v}=-w+w^2\cot\theta+\epsilon\sin\theta\,. \tag{13.36}$$

Observe that there is no dependence upon the azimuth ϕ. Hence the azimuth ϕ does not interact with the other three variables; the azimuth ϕ may be determined after all the other three variables by straightforwardly integrating the last equation of (13.35): $\dot{\phi}=w/\sin\theta$.

Consequently, the orientation of a spinning top has only three interacting variables: the velocities v and w and the polar angle θ. The three dynamical equations are (13.36) and the second equation of (13.35), namely $\dot{\theta}=v$.

The slow manifold of precession The three dynamical equations, (13.36) and $\dot{\theta}=v$, under the gravitational forcing, here represented by $\epsilon\sin\theta$, possess the equilibrium

$$v=0,\quad \theta=\Theta\ \text{(constant)},\quad\text{and}\quad w=\beta(\cos\Theta)\sin\Theta\ \text{(constant)},$$

where, as before, the variable β satisfies the quadratic $-\beta+\beta^2\cos\Theta+\epsilon=0$ and hence as before β is given by (13.33).

Nutation is dominantly v and w oscillation About these equilibria of precession are oscillations in velocities v and w because equation (13.36) has the classic oscillation components $\dot{w}=+v$ and $\dot{v}=-w$. Such pairs of differential equations predict oscillations of frequency one. In conclusion, spherical coordinates largely but not completely separate the fast oscillations in the top's dynamics from the slow precession.

At finite parameter ϵ (equivalently, finite G) the nutation frequency will be slightly different due to the linearization at finite ϵ.

13.7 ▪ Exercises

Exercise 13.1 (basic complex amplitudes). Consider the dynamical system of Example 4.13, namely

$$\dot{x}=-x^3+xy^2+xz^2,\quad \dot{y}=-z,\quad \dot{z}=y\,.$$

Transform this system to normal-form coordinates with complex amplitudes for the fast oscillations.

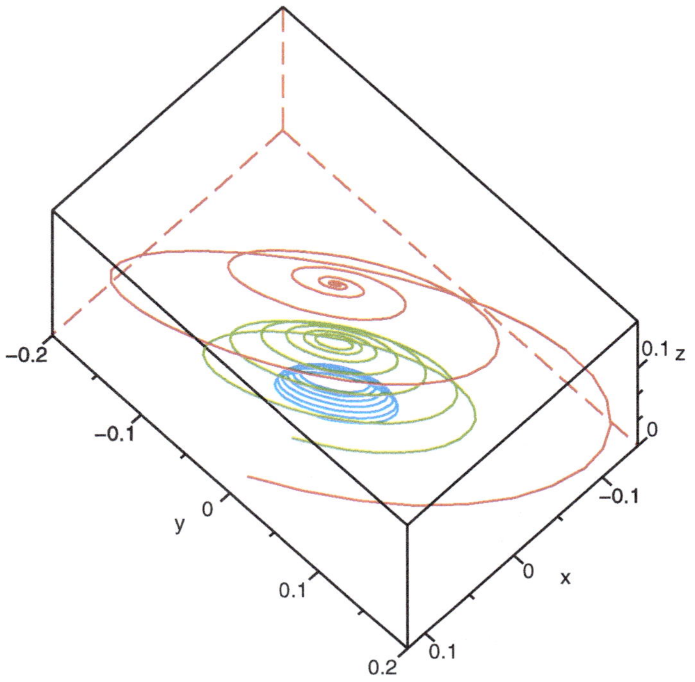

Figure 13.9. *Three trajectories of the system in Exercise* 13.3 *showing quantitatively different drift in* z *depending upon the amplitude of oscillations in* x *and* y.

Exercise 13.2 (quartic resonance). In a coordinate transformation of the toy quasi-geostrophic problem of section 13.5.3, we find we need to solve homological equations such as (13.26). There we saw that the homological equation generally requires components in $\dot{U}$ proportional to $(X^2 + Z^2)$. Consider the homological operator and determine what is generally forced into $\dot{U}$ from the fourth-order quartic terms.

Answer: The null space of the homological operator contains $X^4 + 2X^2Z^2 + Z^4 = (X^2 + Z^2)^2$ and so no component of $\hat{u}$ can eliminate this component from the right-hand side. Hence, terms in $(X^2 + Z^2)^2$ have to go into the slow evolution $\dot{U}$.

Exercise 13.3 (resonance on the slow manifold). Consider the 3D system

$$\dot{x} = -x - y - 2xz, \quad \dot{y} = 2x + y + 2xz, \quad \dot{z} = y^2.$$

Figure 13.9 shows trajectories of fast oscillations and slow drift in z, but as usual the slow drift depends upon the amplitude of the oscillations.

1. Show that the system, linearized about the origin, has eigenvalues $\pm i$ and 0. We aim to find a coordinate transform to separate the fast oscillations and the slow mode.
2. Show that the fast oscillations are elliptical in x and y by showing that the linearized dynamics conserve "energy" $2x^2 + 2xy + y^2$.
3. Now seek a nonlinear separated normal form by the near identity transform

$$x = x(X, Y, Z) \approx X, \quad y = y(X, Y, Z) \approx Y, \quad z = z(X, Y, Z) \approx Z$$

such that the new variables evolve according to

$$\begin{aligned}\dot{X} &= F(X,Y,Z) \approx -X-Y,\\ \dot{Y} &= G(X,Y,Z) \approx 2X+Y,\\ \dot{Z} &= H(X,Y,Z) \approx 0.\end{aligned}$$

Suppose we know some approximation to the transform; seek additive improvements, denoted by hats. Substitute into the governing system to deduce that the corrections should satisfy

$$\begin{aligned}\hat{x}+\hat{y}+(-X-Y)\hat{x}_X+(2X+Y)\hat{x}_Y+\hat{F} &= \text{Res}_x \approx -2XZ,\\ -2\hat{x}-\hat{y}+(-X-Y)\hat{y}_X+(2X+Y)\hat{y}_Y+\hat{G} &= \text{Res}_y \approx +2XZ,\\ (-X-Y)\hat{z}_X+(2X+Y)\hat{z}_Y+\hat{H} &= \text{Res}_z \approx Y^2.\end{aligned}$$

For example, deduce the residual based upon the linear approximation as is listed on the RHS of the above equations.

4. Seek $\hat{x}$ and $\hat{y}$ together. Since Z appears nowhere in the LHS operator, seek quadratic terms that are linear in Z, namely

$$\hat{x} = c_1XZ + c_2YZ, \quad \hat{y} = c_3XZ + c_4YZ.$$

Substitute, equate coefficients, and try to solve equations, but the four equations have rank two and so some terms must be put into the evolution $\hat{F}$ and/or $\hat{G}$. Argue that perhaps the simplest solution is

$$\hat{F} = YZ, \quad \hat{x} = XZ, \quad \hat{G} = \hat{y} = 0.$$

5. Seek $\hat{z} = c_1'X^2 + c_2'XY + c_3'Y^2$ and find that some oscillation energy must be put into the slow mode. Hence seek $\hat{H} \propto 2X^2 + 2XY + Y^2$ to make a near constant forcing of the slow mode, and justify that the general form of the z-correction is $\hat{z} = X^2 + c_3'(2X^2 + 2XY + Y^2)$.

Exercise 13.4 (quasi-geostrophy amplitude). Consider the normal-form coordinate transform for the quasi-geostrophic toy of section 13.5.3. Adapt the computer algebra of section 13.5.4 to express the dynamical system in terms of slow variables involving amplitude and phase variables for the fast oscillations. Interpret your model.

Exercise 13.5 (electron complex). Consider the normal-form coordinate transform for the electron drift of section 13.5.4. Merge the computer algebra with computer algebra of section 3.6 to express the dynamical system in terms of slow variables involving complex-amplitude variables for the fast electron oscillations. Interpret your model.

Exercise 13.6 (quasi-geostrophy complex). Consider the normal-form coordinate transform for the quasi-geostrophic toy of section 13.5.3. Merge and adapt the computer algebra of section 13.5.4 and the computer algebra of section 3.6 to express the dynamical system in terms of slow variables involving complex-amplitude variables for the fast oscillations. Interpret your model.

Exercise 13.7 (prove slow/fast separation). Prove Lemma 13.2 by adapting the induction of the proof of Lemma 11.8.

Part IV Summary

- The near identity change of variables introduced in Chapter 10 simplifies the algebraic description of a nonlinear dynamical system to a normal form that empowers us to easily interpret the dynamics. The approach is to use the homological equations to iteratively construct the coordinate transform (section 10.3).

- These coordinate transforms highlight the concept of hyperbolic equilibria and their stable and unstable manifolds (section 10.1). The Hartman–Grossman theorem of section 10.4 then assures us that linearization is unreasonably effective: the linearization is qualitatively correct in a finite domain.

- Normal-form coordinate transforms clearly separate variables into stable, unstable, and center variables. Thus such coordinate transforms not only show the center manifold model, they also show how to treat initial conditions and indicate the domain of relevance of the model. The key requirement is the spectral trichotomy of Assumption 11.6 that the eigenvalues at a useful equilibrium separate into three well-separated sets depending upon their real part. Then, limited only by the eigenvalues' separation and the smoothness of the original dynamical system (Assumption 11.7), Lemma 11.8 establishes that such normal-form coordinate transforms exist.

- Section 11.3 then establishes criteria for the domain in which the nonlinear separation of dynamics is guaranteed to exist (Proposition 11.13). The limiting factors are first, the maintenance of the spectral trichotomy to nearby out-of-equilibrium dynamics (Definition 11.10), and second, the domain in which the coordinate transform is smoothly invertible (Definition 11.12). Lemma 11.16 then establishes a domain in which the center manifold model is known to exist and emerge.

  One novel aspect of section 11.3 is the development of backward theorems (Grcar, 2011, e.g.), whereas all previous center manifold theory is forward. Further, although some results may only hold for a finite time (e.g., Proposition 11.13), such finite time may be all that is needed. Both of these aspects may make theory more useful in application.

- Section 11.3.5 demonstrates that singular perturbations are often a special case of the general normal-form coordinate transforms developed herein.

- The normal-form coordinate transforms of Chapter 11 provide a useful foliation of state space (Definition 12.1). The $\mathbf{X} =$ constant surfaces are the isochrons (Corollary 12.2) of a given system. Hence, given some initial condition, one finds its isochron, and then the correct initial condition for a center manifold model is the intersection of the isochron with the center manifold (Corollary 12.3).

For example, in the shear dispersion, given an initial concentration field $c_0(x, y)$ in the channel, section 12.2 derives that the initial cross-channel average concentration should be (12.17), namely $C_0(x) \approx \overline{c_0} + \overline{h_1(y)c_{0x}}$ for $h_1 := (-7+30y^2-15y^4)/120$.

- An algorithmically efficient scheme is to find a projection, determined by normal vectors z_j, that determines the model's initial condition s_0 from a given u_0 by solving $\langle z_j, u_0 - v(s_0)\rangle = 0$ for all j (Theorem 12.4). This projection approximates the isochrons by their tangent, but such an approximation typically appears good over reasonably large regions of state space and is global on the center manifold. The dual equation (12.21) determines the projection vectors z_j in terms of the adjoint of the Jacobian of the dynamical systems.

  Algorithm 12.1 outlines general computer algebra code to be used in practice to determine the projection.

- The same projection empowers us to assess effects of uncertainty and of time-dependent forcing. Section 12.4 considers general systems in the form (12.33), $\dot{u} = \mathcal{L}u + f(u)$, with a center manifold model $u = v(s)$ such that $\dot{s} = g(s)$. The section shows that given some small change to the system, to say $\dot{u} = \mathcal{L}u + f(u) + \varepsilon p(u, t)$, the center manifold model is modified to $u \approx v(s)$ such that $\dot{s} = g(s) + \varepsilon q(s, t)$, where $q_j = \langle z_j, p\rangle$; see (12.38). Since the change p may represent either uncertain or known additional terms, this projection quantifies both uncertainty and forcing.

- Chapter 13 considered the broad class of subcenter manifolds. Section 13.2 discussed the theory of Sijbrand (1985), which asserts that subcenter manifolds do exist when based upon oscillations at some natural frequency of the linearized dynamics. However, the existence is subject to there being no resonant interaction with other oscillations. In the interesting case of subcenter slow manifolds, resonance generally occurs, as shown by the mean drift forced by oscillations in Example 13.1.

- Such subcenter slow manifolds, based upon zero eigenvalues among many fast oscillations, are commonly invoked in applications (section 13.1). Such slow manifold dynamics are routinely approximately constructed. Proposition 13.5 assures us that generally such subcenter slow manifolds at least exist in a system close to any system we may specify.

  However, there is no emergence theorem supporting the relevance of slow dynamics to the full dynamics when the full dynamics also includes fast oscillations (section 13.5.2).

- Instead, normal-form coordinate transforms establish that generally the evolution off a slow manifold, in the presence of fast oscillations, is different from the slow dynamics on the subcenter slow manifold and cannot be encompassed by the slow dynamics (sections 13.5.3 and 13.5.4).

Part V

High-fidelity discrete models use slow manifolds

Part V Contents

Suppose you need to solve a PDE, such as the Navier–Stokes equations for fluid flow, or a reaction-diffusion PDE, such as the Ginzburg–Landau equation $\partial u/\partial t = \nabla^2 u - u^3$. In general, an analytic solution is impossible. So you need to resort to numerical solutions with a computer.

Numerical solutions typically involve discretizing in space to predict the solution field $u(\mathbf{x}, t)$ at discrete points $\mathbf{X}_j$ spread out over the domain of interest. That is, numerics replaces the task of predicting the infinite number of components of a field $u(\mathbf{x}, t)$ by the task of predicting just the finite number of values $u(\mathbf{X}_j, t)$. This replacement of the infinite by the finite is a process of dimensional reduction, of modeling, that is the core theme of this book. Surely we must be able to adapt our center manifold techniques to construct, illuminate, and improve numerical discretizations.

For example, consider the nondimensional linear diffusion PDE for a field $u(x, t)$ on a finite domain:

$$\frac{\partial u}{\partial t} = \frac{\partial^2 u}{\partial x^2} \quad \text{such that} \quad u(0, t) = u(\pi, t) = 0\,. \tag{V.1}$$

Classic separation of variables tells us that there is an infinite number of spatial modes in the general solution of the full diffusion PDE (V.1):

$$u(x,t) = \sum_{k=1}^{\infty} c_k e^{-k^2 t} \sin kx\,.$$

To form a spatial discretization with $m+1$ intervals, define grid points $X_j = jH$, where the grid size $H := \pi/(m+1)$, and define grid values $U_j(t) = u(X_j, t)$. Then a simple finite difference approximation is the spatially discrete model

$$\dot{U}_j = \frac{U_{j+1} - 2U_j + U_{j-1}}{H^2} \quad \text{for} \quad j = 1, 2, \ldots, m\,. \tag{V.2}$$

This discrete model has just m modes, one for each grid value U_j, whereas the original continuum diffusion (V.1) has an infinity of modes. The spatial discretization (V.2) is somehow a low-dimensional model of the diffusion dynamics (V.1). This part proposes that center manifold theory provides a marvelous alternative method for deriving accurate spatial discretizations with wonderful support.

To introduce the basic ideas, Chapter 14 divides the solution domain into just two finite elements. The key is to couple the solution fields within each element with a coupling parameter controlling the flow of information between the two elements. The result is a model expressed in the evolution of two grid values, one for each element, but that resolves the subgrid scale field in each element by systematically approximating solutions of the governing PDE. This contrasts with other discretization methods, which *impose assumed* subgrid fields, whereas here the PDE tells us appropriate subgrid fields.

Chapter 15 divides arbitrarily sized domains into m elements. Using nonlocal coupling conditions inspired by Chapter 14, section 15.1 starts by discretizing the basic diffusion PDE (V.1) with support from center manifold theory for typical physical boundary conditions. The center manifold resolves subgrid scale structures (section 15.1.2). Section 15.1.3 then proves that the particular coupling conditions we use also assure us of high-order consistency between the discrete model and the original PDE, for a wide range of linear PDEs. The same approach discretizes nonlinear PDEs (section 15.2), with similar center manifold theory support (section 15.2.2), and resolves nonlinear subgrid scale structures, apparently with similar high-order consistency (although only weaker results have as yet been proved). Thus two different aspects support our approach: center manifold theory and classic consistency.

The construction of a slow manifold spatial discretization involves significant algebra, generally best left to a computer. In the common case of spatially homogeneous PDEs, the analysis is essentially identical in each element and the construction algorithms are straightforward (sections 15.1.4 and 15.2.3). However, in the important case of the linear advection-diffusion PDE, section 15.3, the details are of interest, so we construct the slow manifold model by hand algebra. The resulting discrete model is amazingly robust for all advection speeds.

The development presented in this part is restricted to one spatial dimension for simplicity. The techniques and theory can be generalized to higher spatial dimensions (Roberts et al., 2013; MacKenzie and Roberts, 2003; MacKenzie, 2005); however, more research and application development is needed.

Chapter 14

Introduce holistic discretization on just two elements

Contents

This chapter illuminates the process of analyzing a PDE to develop a spatial discretization for numerical simulation. For example, for a field $u(x,t)$, this chapter analyzes the nonlinear diffusion equation (on some finite domain)

$$\frac{\partial u}{\partial t} = u\frac{\partial^2 u}{\partial x^2} \quad \text{and discretizes it to} \quad \dot{U}_j = U_j\frac{U_{j-1} - 2U_j + U_{j+1}}{H^2}.$$

Discretization approximates the dynamics of an infinite-dimensional system, $u(x)$, evolving in time, by a finite-dimensional system, U_j, evolving in time for $j = 1, \ldots, m$. Thus discretization is another example of the process of dimensional reduction of a dynamical system. Center manifold theory provides a powerful method that first supports such numerical discretizations and second extends other approaches.

Section 14.1 introduces some key concepts in an example with just one numerical grid point, one finite element, in the domain.

Section 14.2 explores an example of relatively naive coupling between adjacent elements when the domain is split into many finite elements.

Section 14.3 shows an asymptotic expansion in the coupling strength converges.

Section 14.4 introduces a nonlocal interelement coupling that ensures classic consistency as well as rigorous support from center manifold theory.

14.1 ▪ Nonlinear diffusion in one element

Let us analyze an introductory example to introduce some key concepts. The aim of this section is to model the long-term dynamics of the field $u(x,t)$ governed by the nonlinear diffusion PDE

$$\frac{\partial u}{\partial t} = u\frac{\partial^2 u}{\partial x^2} \quad \text{such that} \quad u(\pm 1, t) = 0. \tag{14.1}$$

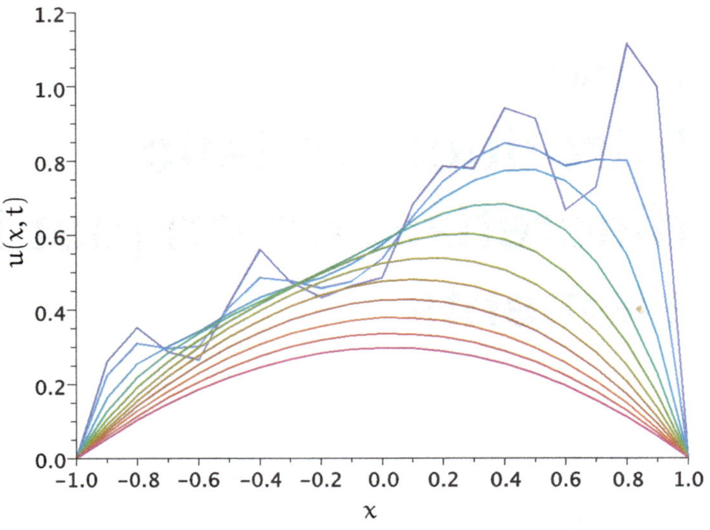

Figure 14.1. *Numerical simulation of the PDE* (14.1) *shows that the nonlinear diffusion rapidly smooths the fluctuating initial conditions (blue) to an emergent slowly decaying parabola (red). The field* $u(x, t)$ *is plotted at times* $t = 0, 0.01, 0.04, \ldots, 0.81, 1$.

Figure 14.1 illustrates how the diffusion $\partial^2 u/\partial x^2$ on the spatial interval $[-1, 1]$ smooths out any spatial variations in the field $u(x, t)$: the small-scale spatial variations will smooth the fastest. However, the smoothing is nonlinear: at locations where the field u is largest the diffusion smooths the quickest.

A special exact solution But the field u also decays over long times; see Figure 14.1. For example, the parabolic initial condition $u(x, 0) = a_0(1 - x^2)$ leads to the exact solution $u(x, t) = a_0(1 - x^2)/(1 + 2a_0 t)$ for all time (to check, just substitute into the PDE (14.1)). That is, the algebraically decaying parabola is an exact solution of the nonlinear diffusion equation (14.1). Now rewrite this algebraically decaying parabola in the equivalent form

$$u = a(t)(1 - x^2), \quad \text{where} \quad \dot{a} = -2a^2 . \tag{14.2}$$

Let's interpret this form as a low-dimensional model of the full system (14.1): the model (14.2) replaces the spatiotemporal dynamics of the field $u(x, t)$ by simply the dynamics of the scalar $a(t)$. The long-term evolution of the field $u(x, t)$ should tend to this parabolic form because any wiggles in the initial shape of the field u should smooth by diffusion to this parabolic shape, as shown by Figure 14.1. This section constructs the model (14.2) as a slow manifold. But very importantly, center manifold theory *guarantees* that starting from a wide range of initial conditions $u(x, 0)$ the evolution of the field $u(x, t)$ will be described, exponentially quickly in time, by the emergent model (14.2).

Artificially create a slow manifold The key to applying center manifold theory is to identify simple equilibria at a critical parameter value. Physically, equilibria usually correspond to some conserved mode such as total amount of material or momentum. Here for the nonlinear diffusion problem (14.1) there is no exactly conserved mode and thus no exact equilibrium.[173] Thus *adopt the artifice of inventing* a basic equilibrium and embedding the original problem (14.1) as one member of a useful family of problems (Table 4.2). Such invention is a glorious fudge: this section aims to convince you that it is effective and valuable.

We base the analysis upon a nearby system that does form a tractable basis for center manifold theory and perturb it to obtain the original system (14.1). Schematically,

$$\begin{array}{ccc} \text{invented base systems} & \to & \text{original physical system (14.1)}, \\ \gamma = 0 & \to & \gamma = 1\,. \end{array}$$

The parameter γ varies from the artificial invented base system to the original physical system: $\gamma = 0$ corresponds to the artificial invented base system, and $\gamma = 1$ corresponds to the original. Here create a nearby base system with a conserved mode by inventively modifying the boundary conditions at $x = \pm 1$. Instead of $u = 0$, require that the insulating and conserving $\partial u/\partial x = 0$ for which $u =$ constant become an exact solution of the nonlinear diffusion PDE (14.1). Now connect these different boundary conditions with a parameter γ by the almost convex combination in the so-called Robin boundary conditions

$$(1-\gamma)\frac{\partial u}{\partial x} = \mp 2\gamma u \quad \text{on } x = \pm 1\,. \tag{14.3}$$

These Robin boundary conditions reproduce the physical boundary conditions $u(\pm 1, t) = 0$ when the artificial parameter $\gamma = 1$ and reduce to a conservative boundary condition $u_x(\pm 1, t) = 0$ when the artificial parameter $\gamma = 0$.

The choice of signs, "$\mp$", on the right-hand side of (14.3) is determined by requiring the physical meaning to be the same for the boundary condition at each end of the domain: at $x = +1$ the flux of u out of the domain is proportional to $-\partial u/\partial x$, whereas at $x = -1$ the flux of u out of the domain is proportional to $+\partial u/\partial x$. Exercise 14.1 explores the reason for the coefficient 2 on the right-hand side of (14.3): other coefficient values are poorer. There is an enormous scope in inventing other bases for the analysis. We just explore the dynamics of the PDE (14.1) with the boundary conditions (14.3).

Proceed with slow manifold analysis (Table 4.2): find equilibria, check for decay in the linear dynamics, construct the model to some order in the parameters, and finally interpret the resulting model.

Find equilibria With $\gamma = 0$, the boundary conditions (14.3) are insulating and the nonlinear diffusion of the PDE (14.1) smooths solutions in the domain to $u =$ constant, as seen in Figure 14.2. Thus for every constant a, $u = a$, and $\gamma = 0$ is a one-parameter family of equilibria. We base the slow manifold analysis around these equilibria by notionally adjoining to the PDE the trivial $\dot\gamma = 0$. Upon perturbing the system to account for effects at nonzero γ, the variable a will no longer be constant but will evolve slowly to parametrize the slow evolution on the slow manifold. Because γ is the only small parameter, although the model will be local in γ, the model will be global in a, which is a good thing, as the model is then justified for all sizes of the field u.

Linear perturbations decay Seek the field $u = a + \hat u(x)e^{\lambda t}$ and $\gamma = \hat\gamma$, where $\hat u$ and $\hat\gamma$ are small. Substitute into the PDE (14.1) and boundary conditions (14.3), and neglect

[173]True, $u(x,t) = 0$ is an equilibrium, but then the linearization of the PDE is the trivial $\partial u/\partial t = 0$ (as the right-hand side is strictly nonlinear at $u = 0$). Because $\partial u/\partial t = 0$, then all modes do not evolve and have zero eigenvalue, and thus every $u(x,t)$ is the slow manifold. Consequently, there is no simplification, no dimensional reduction, about the equilibrium $u = 0$. It is useless to us.

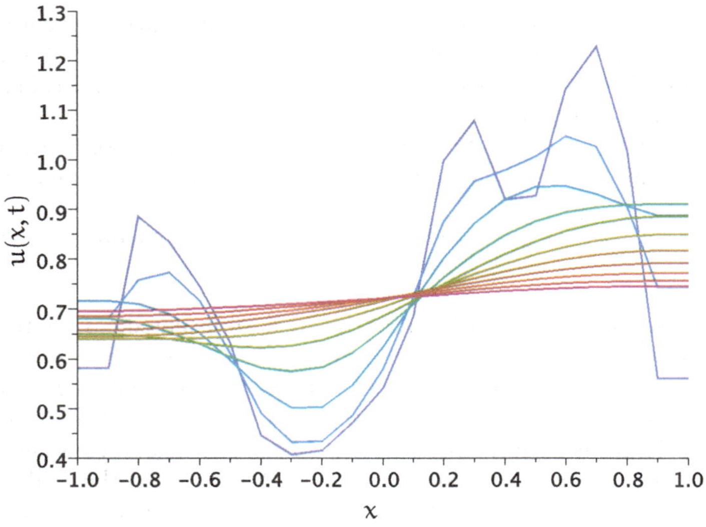

Figure 14.2. *Numerical simulation of the PDE* (14.1) *but with insulating boundary conditions* $\partial u/\partial x = 0$ *at* $x = \pm 1$ *shows that the nonlinear diffusion rapidly smooths the fluctuating initial conditions (blue) to an emergent constant (red). The field* $u(x,t)$ *is plotted at times* $t = 0, 0.01, 0.04, \ldots, 0.81, 1$.

products of small quantities to find that any perturbation $\hat{u}(x)$ must satisfy

$$\lambda\hat{u} = a\frac{\partial^2\hat{u}}{\partial x^2} \quad \text{such that} \quad \frac{\partial\hat{u}}{\partial x} = \mp 2\hat{\gamma}a \text{ on } x = \pm 1\,. \tag{14.4}$$

Nonzero eigenvalues must be

$$\lambda_k = -\frac{\pi^2k^2}{4}a\,, \quad \text{corresponding to modes}$$
$$\hat{u}_k \propto \cos\left[\pi k(x+1)/2\right]\,, \quad \hat{\gamma} = 0\,.$$

The diffusion causes all wiggles in the field u to decay exponentially quickly. As shown by Figure 14.2, where $a \approx 0.7$, the slowest nonconstant mode is $\hat{u}_1 \propto \sin(\pi x/2)$, with its decay rate determined by the eigenvalue $\lambda_1 = -a\pi^2/4 \approx -0.7\pi^2/4 \approx -1.7$. Due to the particular nonlinear diffusion, the decay rate to the equilibria differs: the decay is more rapid for larger amplitudes a, and is small for small a (vanishing as $a \to 0$). An interestingly unusual feature is that $a = 0$ is a singular equilibrium in the analysis in that all the decay rates become zero. It is only for finite $a > 0$ and positive u that the slow manifold is valid.

But the neutral modes corresponding to zero eigenvalue are not simple. We must remember to look for generalized eigenmodes. Solving (14.4) with eigenvalue $\lambda = 0$, we find that $\hat{u}$ is an arbitrary constant only when $\hat{\gamma} = 0$; we apparently cannot find a *two*-dimensional slow subspace. But we require a two-dimensional slow subspace: one dimension for the freedom to specify the parameter γ and the other dimension to form the space in which the amplitude a will slowly evolve. Consequently, the slow subspace must contain generalized eigenmodes. Now $(\hat{u}, \hat{\gamma}) \propto (1, 0)$ satisfies (14.4) with zero eigenvalue. To seek a generalized eigenmode, try to solve (14.4) with $\lambda\hat{u}$ replaced by a constant b corresponding to the constant first component of the ordinary eigenmode; then find that

$\hat{u} = bx^2/(2a)$ and $\hat{\gamma} = -b/(2a^2)$. Set the arbitrary $b = -2a^2B$. Then the slow subspace is the linear combination of the eigenmode and generalized eigenmode

$$(\hat{u}, \hat{\gamma}) = A(1, 0) + B(-ax^2, 1) = (A - Bax^2, B).$$

Substitute into the linearized differential equations to deduce that in the slow subspace the system evolves according to the Jordan form

$$\begin{bmatrix} \dot{A} \\ \dot{B} \end{bmatrix} = \begin{bmatrix} 0 & -2a^2 \\ 0 & 0 \end{bmatrix} \begin{bmatrix} A \\ B \end{bmatrix}.$$

This Jordan form identifies in the B term that a small nonzero parameter γ causes the field $\hat{u}$ to leak out of the domain across the boundaries. The critical aspect is that there is nonetheless a two-dimensional slow subspace parametrized by an overall measure of the field u, namely a, and by the trivial artificial parameter γ. This two-dimensional slow subspace forms the basis of the two-dimensional slow manifold.

Obtain a first approximation Seek a slow manifold model

$$u = a + \hat{u}(a, x, \gamma) \quad \text{such that} \quad \dot{a} = \hat{g}(a, \gamma)$$

for small corrections $\hat{u}$ and $\hat{g}$. Substitute into the PDE (14.1) and the boundary conditions (14.3), and then omit products of small quantities to require the corrections to satisfy

$$\hat{g} = a\frac{\partial^2 \hat{u}}{\partial x^2} \quad \text{and} \quad \frac{\partial \hat{u}}{\partial x} = \mp 2\gamma a \quad \text{on } x = \pm 1.$$

Straightforward algebra gives the solution

$$\hat{u} = A - \gamma a x^2 \quad \text{and} \quad \hat{g} = -2\gamma a^2.$$

Determine the arbitrary integration constant A by precise definition of amplitude a. To match amplitude a of the earlier quoted exact solution (14.2), define amplitude

$$a(t) := u(0, t). \tag{14.5}$$

Consequently, for updates, $\hat{u}(0, t) = 0$. Thus the integration constant $A = 0$. The first nontrivial model is consequently

$$u = a - \gamma a x^2 \quad \text{such that} \quad \dot{a} = -2\gamma a^2. \tag{14.6}$$

This model is so good that it is an exact slow manifold model: check the PDE (14.1) and Robin boundary conditions (14.3) to verify.

Interpret Evaluate at $\gamma = 1$ to get a model of the original system with the original boundary conditions. Observe that the slow manifold model (14.6) then becomes precisely the correct model (14.2). This is wonderful! By inventing a suitable base for analysis at parameter $\gamma = 0$, at the finite $\gamma = 1$ we recover an exact slow manifold model of the long-term evolution.

Remarkably, center manifold theory asserts that *there is a finite neighborhood in which all solutions of the nonlinear diffusion PDE approach the model;* we assume that the neighborhood is sufficiently big to include the physical case of $\gamma = 1$. Pause to consider the strength of this result: the theory guarantees that the decaying parabola (14.2) will emerge exponentially quickly as the long-term solution from a wide range of initial conditions. The theory not only provides a framework for constructing the model but also assures us that the model is relevant as it describes the emergent dynamics.

Convinced? This example shows one case where we can manipulate boundary conditions to our advantage by empowering center manifold theory to support modeling at some finite boundary condition parameter. If you are convinced that this sort of manipulation is valid, then skip to Chapter 15, where a similar trick couples finite elements to empower center manifold theory to support discretizations in space of PDEs. If you are not yet convinced by this boundary condition manipulation, then keep reading this chapter.

14.2 ▪ Couple two elements together

Section 14.1 introduced modeling the dynamics on a spatial domain by creatively fudging the boundary conditions. But spatial discretizations of PDEs generally divide the spatial domain into many finite elements or employ a finite-sized grid. Thus our next step in using center manifold theory to support spatial discretizations is to move from one element on a spatial domain to two elements on the domain. Chapter 15 subsequently develops the approach for many elements.

In this section, let us explore two elements that are coupled together. Focus on the coupling by exploring the simple linear diffusion PDE: seek to create a model for the dynamics of field $u(x,t)$ governed by

$$\frac{\partial u}{\partial t} = \nu\frac{\partial^2 u}{\partial x^2} \quad \text{on } -H < x < H \quad \text{such that } \frac{\partial u}{\partial x} = 0 \text{ at } x = \pm H\,. \tag{14.7}$$

Here the spatial domain is of size $2H$, and constant ν is the coefficient of diffusion.

Create two elements Introduce an artificial internal boundary at the mid-domain point $x = 0$:

- the left-hand element is then $-H < x < 0$ with midpoint $X_1 = -H/2$, and
- the right-hand element is $0 < x < H$ with midpoint $X_2 = +H/2$.

Define U_j to be the two grid values of the field u: $U_j(t) = u(X_j, t)$. Our aim is to deduce the evolution of the grid values U_j as a model for the coupled dynamics on these elements.

The resultant model must be of this form As the diffusion equation is linear, the model will be linear, and we preview the desired model. Due to the left-right symmetry, the model must be of the form

$$\dot U_1 = aU_1 + bU_2\,, \quad \dot U_2 = bU_1 + aU_2 \tag{14.8}$$

for some constants a and b to be determined (by symmetry, swapping the subscripts 1 and 2 must give the same ODEs). Use separation of variables to deduce more details of the dynamics and hence determine the result of our slow manifold modeling. The exact solution of the diffusion (14.7) over the whole domain involves an expansion in orthogonal Fourier modes (Kreyszig, 2011, §12.6, e.g.):

$$u(x,t) = \sum_{k=0}^{\infty} A_k \cos\left[\frac{\pi k}{2H}(x+H)\right] \exp\left[-\frac{\nu\pi^2 k^2}{4H^2} t\right]$$

for some constants A_k depending upon initial conditions. A long-term model must resolve the slowest dynamics in this solution. The more rapidly decaying modes, $k \geqslant 2$, correspond to the ignored exponential transients in the rapid evolution to the slow manifold. That is, we want the slow manifold modeling to resolve

$$u = A_0 + A_1 \cos\left[\frac{\pi}{2H}(x+H)\right] \exp\left[-\frac{\nu\pi^2}{4H^2}t\right] + \text{ignored transients}, \tag{14.9}$$

where the ignored transients are the terms for $k = 2, 3, \ldots$ in the above full infinite sum. Consequently, since at the midpoints $\cos[\pi(X_1 + H)/(2H)] = 1/\sqrt{2}$ and $\cos[\pi(X_2 + H)/(2H)] = -1/\sqrt{2}$, then *after* the transients have decayed,

$$U_1(t) = u(X_1, t) = A_0 + \frac{A_1}{\sqrt{2}} \exp\left[-\frac{\nu\pi^2}{4H^2}t\right],$$
$$U_2(t) = u(X_2, t) = A_0 - \frac{A_1}{\sqrt{2}} \exp\left[-\frac{\nu\pi^2}{4H^2}t\right].$$

Observe that $U_1 - U_2 = \sqrt{2}A_1 \exp\left[-\nu\pi^2 t/(4H^2)\right]$, and so a little algebra gives the time derivatives

$$\dot{U}_1 = \frac{\nu\pi^2}{8H^2}(-U_1 + U_2) \quad \text{and} \quad \dot{U}_2 = \frac{\nu\pi^2}{8H^2}(+U_1 - U_2). \tag{14.10}$$

This is the exact long-term model we wish to obtain, but our challenge is to obtain it without knowing the full solution. Center manifold theory together with innovative use of boundary conditions to couple the two elements provides the systematic methodology to derive this model without knowing the full details of all the infinite number of spatial modes.

This modeling is analogous to the modeling of quasi-stationary probability distributions undertaken in section 5.3.6. There the system is ultimately absorbed into the state P_1, whereas here the system ultimately ends up with field u being boringly constant. Section 5.3.6 introduced the quasi-stationary probability distribution and its dynamics to ascertain the longest-lasting dynamics before attaining the ultimate state P_1. Analogously, here we seek the slowest-decaying structure in the field u in order to see the longest-lasting dynamics of equation (14.10). Our challenge now is the same as for quasi-stationary probability: how can we embed the problem of interest into a family of problems that has a slow manifold that will most simply tell us of these long-time dynamics?

Create an internal boundary Place an internal boundary at $x = 0$ and then across this internal boundary couple the field u in the two elements on either side.

- First, recover the original fully coupled system by seeking that the field u and its first derivative u_x are continuous across the internal boundary $x = 0$. That is, we seek

  $$u^- = u^+ \quad \text{and} \quad u_x^- = u_x^+,$$

  where superscripts $\pm$ denote evaluation at $x = \pm 0$ (that is, at $x = 0$ from the respective element).

- Second, create an artificial slow manifold base for the model by requiring that the two elements be insulated from each other with conservative boundary conditions:

  $$u_x^- = 0 \quad \text{and} \quad u_x^+ = 0.$$

Then in each element the solution will exponentially quickly tend to a constant field, but the constant will in general be different in each element.

- Coupling the elements perturbatively then leads to internal boundary conditions that imply a model for the long-term evolution. Here interpolate between the above two internal boundary conditions by embedding them in the one-parameter family[174]

$$u_x^- = u_x^+ \quad \text{and} \quad (1-\gamma)(u_x^- + u_x^+) = 2\gamma(u^+ - u^-)/H \tag{14.11}$$

  for some coupling parameter γ:[175]

  - when $\gamma = 0$ the two elements are uncoupled by the effectively insulating internal boundary condition between them;
  - when $\gamma = 1$ these conditions request that the field and its first derivative should be continuous and hence fully couples the two elements together; and
  - $0 < \gamma < 1$ interpolates between these extremes.

We base a slow manifold model on the equilibria that exist when $\gamma = 0$—the model is expressed as a power series in γ—and then evaluate the constructed model at $\gamma = 1$ to obtain a model of the original system.

Find equilibria Put $\gamma = 0$ and the coupling condition (CC) becomes insulating so the diffusion equation (14.7) has the equilibria of the field u being constant in each element:

$$\gamma = 0 \quad \text{and} \quad u(x,t) = \begin{cases} U_1, & -H < x < 0, \\ U_2, & 0 < x < H, \end{cases} \tag{14.12}$$

where U_1 and U_2 are independent constants. These two parameters, constant when $\gamma = 0$, become the two evolving variables in the model when we couple the elements together with coupling parameter $\gamma \neq 0$.

Other equilibria are $\gamma \neq 0$ and $u =$ constant, but we do not use these equilibria as the base of the modeling: the reason is that then there is only *one* variable associated with the field u and hence we would not derive a two-variable model.

Exponential decay assures relevance Consider the dynamics when the coupling parameter $\gamma = 0$. The two identical elements are totally isolated from each other and have the same diffusion equation within each and the same boundary conditions (reflection symmetry); thus the linearized dynamics are the same in each element. Consider just the right-hand element for simplicity; the left-hand element is the same. Separation of variables derives that the modes in the right-hand element are the decaying Fourier modes (Kreyszig, 2011, §12.6, e.g.)

$$u \propto \cos\left[\frac{k\pi}{H}x\right] \exp\left[-\frac{\nu\pi^2k^2}{H^2}t\right]$$

[174]Sometimes called a homotopy because there is a continuous change from one to another.

[175]Why choose $2/H$ in the RHS of the coupling condition (CC) (14.11)? Divide the CC by two; then the CC compares the mean gradient of the field u at the internal boundary with the gradient estimated from the two surrounding grid values when the solution is approximately constant in each element.

for integer wavenumbers $k = 0, 1, 2, \ldots$. Thus the eigenvalues of the diffusion dynamics are $\lambda = 0, -\nu\pi^2/H^2, -4\nu\pi^2/H^2$, and so on. These are all negative except for the zero eigenvalue of the neutral mode that $u =$ constant in the element. There is also a neutral mode corresponding to the coupling parameter γ being constant. Consequently, center manifold theory assures us that a model exists for nonzero coupling parameter γ and that solutions of the diffusion equation approach the model on a cross-element diffusion time of approximately $H^2/(\nu\pi^2)$.

Derive the first model Now seek a first improvement to the boring insulated dynamics of $\gamma = 0$. This first step toward the exact model (14.9)–(14.10) ends up deriving the terms in γ^1 that

$$\dot{U}_1 = \gamma\frac{\nu}{H^2}(-U_1 + U_2) \quad \text{and} \quad \dot{U}_2 = \gamma\frac{\nu}{H^2}(U_1 - U_2)\,. \tag{14.13}$$

For parameter $\gamma = 1$, these ODEs approximate the exact model (14.10) as the exact cross-diffusion coefficient $\pi^2/8 = 1.2337 \approx 1$, which is the coefficient appearing in this first approximation.

The usual iterative improvement Suppose that at some iteration the approximate field in the left element is $u = u_1(x, \mathbf{U}, \gamma)$ such that $\dot{\mathbf{U}} = \mathbf{g}(\mathbf{U}, \gamma)$; then seek small improvements to u_1 and $\mathbf{g}$. Substitute $u = u_1 + \hat{u}_1$ such that $\dot{\mathbf{U}} = \mathbf{g} + \hat{\mathbf{g}}$ into the PDE (14.7):

$$\frac{\partial}{\partial U_1}(u_1 + \hat{u}_1)(g_1 + \hat{g}_1) + \frac{\partial}{\partial U_2}(u_1 + \hat{u}_1)(g_2 + \hat{g}_2) = \nu\frac{\partial^2}{\partial x^2}(u_1 + \hat{u}_1)\,.$$

Drop products of small quantities and rearrange to

$$-\frac{\partial u_1}{\partial U_1}\hat{g}_1 - \frac{\partial u_1}{\partial U_2}\hat{g}_2 - \frac{\partial \hat{u}_1}{\partial U_1}g_1 - \frac{\partial \hat{u}_1}{\partial U_2}g_2 + \nu\frac{\partial^2 \hat{u}_1}{\partial x^2} = \frac{\partial u_1}{\partial U_1}g_1 + \frac{\partial u_1}{\partial U_2}g_2 - \nu\frac{\partial^2 u_1}{\partial x^2}\,.$$

Observe that the right-hand side is just the residual of the PDE (14.7) evaluated at the current approximation; in this first iterate $\text{Res}_{(14.7)} = 0$. Simplify this equation for the small improvements by approximating the coefficients of improvements by their initial approximation, that is, $\partial u_1/\partial U_1 \approx 1$, $\partial u_1/\partial U_2 \approx 0$, and $g_j \approx 0$. Then the PDE for the improvements is

$$-\hat{g}_1 + \nu\frac{\partial^2 \hat{u}_1}{\partial x^2} = \text{Res}_{(14.7)}\,. \tag{14.14}$$

We need boundary conditions for the improvement $\hat{u}_1$: the coupling condition (14.11) is nontrivial; similar analysis leads to requiring that at the internal boundary $x = 0$

$$\frac{\partial \hat{u}_1}{\partial x} = \frac{\partial \hat{u}_2}{\partial x} \quad \text{and} \quad -\frac{\partial \hat{u}_1}{\partial x} - \frac{\partial \hat{u}_2}{\partial x} = \text{Res}_{(14.11)}\,, \tag{14.15}$$

where in this first iterate $\text{Res}_{(14.11)} = 2\gamma(U_1 - U_2)/H$. Eliminate $\partial\hat{u}_2/\partial x$ from (14.15); then the nontrivial boundary condition for the field in the left-hand element is $\partial\hat{u}_1/\partial x = -\frac{1}{2}\text{Res}_{(14.11)}$ at $x = 0$. Now *invoke the classic solvability condition*: integrate (14.14) over the element to deduce

$$-H\hat{g}_1 + \nu\left[\frac{\partial \hat{u}_1}{\partial x}\right]_{-H}^{0} = \int_{-H}^{0} \text{Res}_{(14.7)}\,dx\,.$$

With the boundary conditions this implies the improvement to the evolution

$$\hat{g}_1 = \frac{1}{H}\left[-\int_{-H}^{0} \mathrm{Res}_{(14.7)}\,dx - \nu\tfrac{1}{2}\,\mathrm{Res}_{(14.11)}\right]. \tag{14.16}$$

In the first iterate, this correction to the evolution gives the first approximation (14.13) to the exact model (14.10) for the left-hand element, and similarly for the right-hand element.

Now determine the solution field within each element. Integrate (14.14) twice to determine the improvement to the field: first,

$$\frac{\partial \hat{u}_1}{\partial x} = \frac{1}{\nu}\int_{-H}^{x} \mathrm{Res}_{(14.7)} + \hat{g}_1\,dx$$

ensures that the derivative $\partial\hat{u}_1/\partial x = 0$ on the left-hand extreme of the domain, and second,

$$\hat{u}_1 = \int_{-H/2}^{x} \frac{\partial \hat{u}_1}{\partial x}\,dx = \frac{1}{\nu}\int_{-H/2}^{x}\int_{-H}^{x} \mathrm{Res}_{(14.7)} + \hat{g}_1\,dx\,dx \tag{14.17}$$

ensures that $\hat{u}_1 = 0$ at the midelement grid point $x = -H/2$ so that the iteration maintains $u_1(-H/2, t) = U_1(t)$. In the first iterate, this formula updates the left-hand field to

$$u_1 \approx U_1 + \gamma\tfrac{1}{8}(1 + 2x/H)(3 + 2x/H)(U_2 - U_1),$$

and similarly for the field in the right-hand element. This procedure constructs the model (14.13), which evaluated at $\gamma = 1$ models the original diffusion PDE (14.7).

We completed the first step. Further iterations would derive more accurate approximations. But such algebra is very tedious and so best assigned to a computer by the algorithm of the next section.

14.3 ▪ The two-element asymptotic model converges

Obtaining the first approximate model was straightforward. However, more accurate approximations involve much more algebra. Let's resort to computer algebra. For example, Algorithm 14.1 derives that

$$\dot{U}_j = \left(\gamma + \tfrac{1}{3}\gamma^2 + \tfrac{1}{45}\gamma^3 - \tfrac{13}{189}\gamma^4 - \tfrac{809}{14175}\gamma^5\right)\frac{\nu}{H^2}(-U_j + U_{3-j}) + \mathcal{O}(\gamma^6), \tag{14.18}$$

which when evaluated at the physically relevant coupling $\gamma = 1$ estimates the correct cross-diffusion coefficient to two decimal place accuracy, namely 1.2297. Let us explore how Algorithm 14.1 works.

Iterate to achieve some order of error The core of Algorithm 14.1 is the loop in the middle that iterates until the three nontrivial governing equations are satisfied to some order of error. The three nontrivial equations are the diffusion PDE (14.7) in each of the two elements and the interelement coupling condition (14.11). Specify the order of error by the statement `let gamma^6=>0;`; that tells the computer algebra to

Algorithm 14.1 *Derive the coupled model for diffusion between two finite elements, the PDE* (14.7).

```
factor hh,nu,uu;
% parametrize the model in evolving uu(1) and uu(2)
operator uu; depend uu,t;
let { df(uu(1),t)=>g1, df(uu(2),t)=>g2 };
% initial slow subspace
u1:=uu(1); u2:=uu(2);
g1:=g2:=0;
% iterate to this order of error
let gamma^6=>0;
for iter:=1:99 do begin
  % compute residuals
  res1:=df(u1,t)-nu*df(u1,x,2);
  res2:=df(u2,t)-nu*df(u2,x,2);
  cc:=sub(x=0,(1-gamma)*(df(u2,x)+df(u1,x))
              -2*gamma*(u2-u1)/hh);
  % update left element from residuals
  g1:=g1+(gd:=-int(res1,x,-hh,0)/hh-nu*cc/2/hh);
  u1:=u1+int(int(res1+gd,x,-hh,x),x,-hh/2,x)/nu;
  % update right element from residuals
  g2:=g2+(gd:=-int(res2,x,0,+hh)/hh+nu*cc/2/hh);
  u2:=u2+int(int(res2+gd,x,+hh,x),x,+hh/2,x)/nu;
  if {res1,res2,cc}={0,0,0} then iter:=1000000;
end;
% check other requirements
ccc:=sub(x=0,df(u1,x)-df(u2,x));
bc1:=sub(x=-hh,df(u1,x));
bc2:=sub(x=+hh,df(u2,x));
amp1:=sub(x=-hh/2,u1)-uu(1);
amp2:=sub(x=+hh/2,u2)-uu(2);
```

discard all terms $\mathcal{O}(\gamma^6)$. Thus the algorithm derives the approximate model to an error $\mathcal{O}(\gamma^6)$.

Parametrize the model Use the variable `u1` to store the field $u_1(x,t)$ of the left-hand element and the variable `u2` to store the field of the right-hand element. Then `u1` and `u2` depend upon position x and upon the evolving grid values U_j denoted by `uu(j)` for $j = 1, 2$. Tell Reduce that `uu(j)` evolves in time with the `depend` statement. Then use a `let` statement to tell Reduce to replace time derivatives of U_j by an expression $g_j(\mathbf{U})$ stored in `g1` and `g2`.

Initial solution Initialize the fields u_j in each element to be simply a constant in space x, namely the grid value U_j. The initial approximation to the evolution in time is that there is none; hence initially $g_j = 0$.

Compute residuals Inside the iterative loop, the first task is to compute the residuals of the nontrivial equations. Here there are three nontrivial equations: the PDE (14.7) in

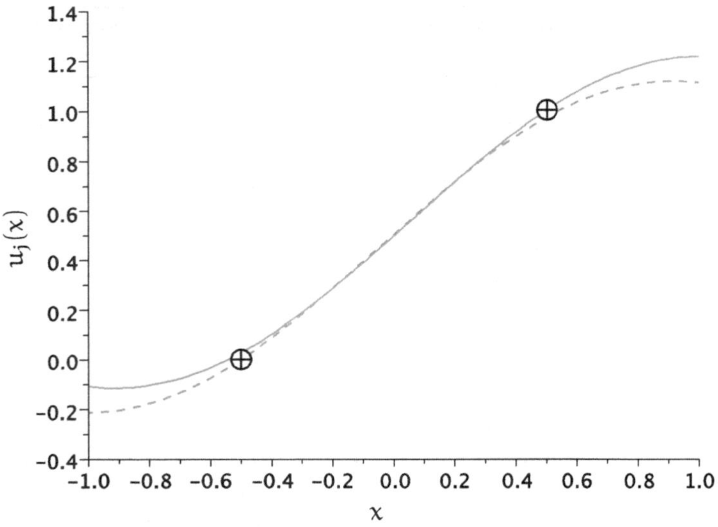

Figure 14.3. *The left-hand field* $u_1(x)$ *(dashed) and the right-hand field* $u_2(x)$ *(solid) for the physical* $\gamma = 1$ *when grid values* $U_1 = 0$ *and* $U_2 = 1$. *The extrapolation from the left-hand field does a reasonable job of extrapolating into the right-hand element, and vice versa.*

the left-hand element, the PDE (14.7) in the right-hand element, and the coupling condition (14.11).

Update left and right elements from the residuals After computing the residuals, improve the current approximation using the formulas (14.16) and (14.17). See these coded in the Reduce code of Algorithm 14.1.

Check other requirements The last five lines in Algorithm 14.1 confirm that the straightforward coupling, boundary, and amplitude conditions are indeed met by the solution derived in the iteration loop.

Outcome Executing Algorithm 14.1 gives not only the model (14.18), but also the fields $u_j(x, \mathbf{U})$ internal to each element. These are the subgrid fields that approximate those in the exact model (14.9), for example. Figure 14.3 plots these subgrid fields. See how beautifully smoothly the subgrid fields match in the neighborhood of the internal boundary at $x = 0$. Despite the crude chopping up of the domain into finite elements, high-order analysis pieces the solution field smoothly back together.

Interestingly, Figure 14.3 shows that the extrapolation of the left-hand field into the right-hand element very nearly goes through the grid value $U_2 = 1$ at $x = H/2$. Similarly, the extrapolation of the right-hand field into the left-hand element very nearly goes through the grid value $U_1 = 0$ at $x = -H/2$. Indeed, truncating to errors $\mathcal{O}(\gamma^6)$ and evaluating at both $\gamma = 1$ and the grid point $x = H/2$, the left-hand field at the right-hand grid point $u_1(H/2, \mathbf{U}) = 0.028\, U_1 + 0.922\, U_2$. This is within 8% of the correct grid value. Higher-order analysis gets closer and closer. Perhaps we can use the requirement that the left-hand field should extrapolate through the right-hand grid value as an alternative to the coupling conditions (14.11). The next section shows we can, and subsequent proofs establish why it is wonderful to do so.

14.4 ▪ Prefer nonlocal internal boundary conditions

Inspired by Figure 14.3, let us replace the coupling conditions (14.11) between the two elements with

$$\begin{aligned} u_1(+H/2,\mathbf{U}) &= \gamma u_2(+H/2,\mathbf{U}) + (1-\gamma)u_1(-H/2,\mathbf{U}), \\ u_2(-H/2,\mathbf{U}) &= \gamma u_1(-H/2,\mathbf{U}) + (1-\gamma)u_2(+H/2,\mathbf{U}). \end{aligned} \tag{14.19}$$

Why? First, look at the fully coupled case $\gamma = 1$: these nonlocal coupling conditions reduce to

$$u_1(H/2,\mathbf{U}) = u_2(H/2,\mathbf{U}) = U_2 \quad \text{and} \quad u_2(-H/2,\mathbf{U}) = u_1(-H/2,\mathbf{U}) = U_1.$$

That is, the left-hand field should extrapolate through the right-hand grid value, and the right-hand field should extrapolate through the left-hand grid value. If the two subgrid fields are to merge very smoothly, then surely this condition is a suitable requirement. Second, look at what happens when the coupling parameter $\gamma = 0$: these nonlocal coupling conditions reduce to

$$u_1(H/2,\mathbf{U}) = u_1(-H/2,\mathbf{U}) \quad \text{and} \quad u_2(-H/2,\mathbf{U}) = u_2(+H/2,\mathbf{U}). \tag{14.20}$$

That is, the extrapolation of the left-hand field is the left-hand grid value, and vice versa. Thus at $\gamma = 0$, not only are the two elements decoupled from each other, as in the previous section, but the piecewise constant *equilibrium* solution (14.12) still holds. Thus this family of equilibria still acts as the basis for a slow manifold model of the dynamics.

The outcome Before looking at the theoretical support and the minor changes to the algorithm, let us see some of the results. Solving the diffusion PDE (14.7) with the nonlocal coupling conditions (14.19) leads to the model

$$\dot{U}_j = \left(\gamma + \tfrac{1}{6}\gamma^2 + \tfrac{2}{45}\gamma^3 + \tfrac{1}{70}\gamma^4 + \tfrac{8}{1575}\gamma^5\right)\frac{\nu}{H^2}(-U_j + U_{3-j}) + \mathcal{O}(\gamma^6). \tag{14.21}$$

This model converges a little quicker in the coupling parameter γ: summing these terms at $\gamma = 1$ gives the interelement interaction constant as 1.2305, which agrees with the exact model (14.10) to an error of 0.003.

The subgrid scale fields $u_1(x,\mathbf{U})$ and $u_2(x,\mathbf{U})$ are more impressive. Figure 14.4 plots the fields of both the left-hand and the right-hand elements for the model with errors $\mathcal{O}(\gamma^6)$ evaluated at $\gamma = 1$: the extrapolations of the fields agree with each other much better than those of the previous coupling condition, shown in Figure 14.3. Evidently the nonlocal coupling conditions (14.19) cause neighboring subgrid fields to merge more smoothly.

Derive the model with computer algebra Only small changes are required in the Reduce code to use the new coupling conditions (14.19). Algorithm 14.2 lists the iterative loop with the required changes: change the computation of the coupling condition, modify the update of the correction $\hat{g}_j$, and only terminate the loop when both residuals of the coupling condition are zero. Easy!

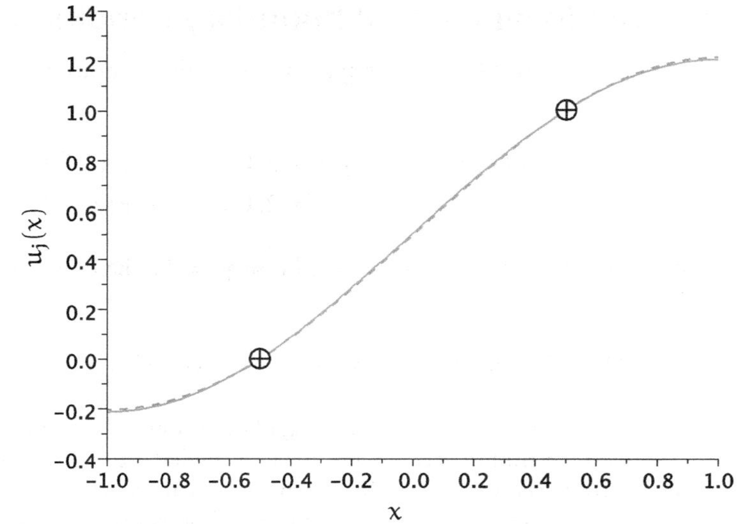

Figure 14.4. *The left-hand field* $u_1(x)$ *(dashed) and the right-hand field* $u_2(x)$ *(solid) for the physical* $\gamma = 1$ *and the nonlocal coupling conditions* (14.19) *when grid values* $U_1 = 0$ *and* $U_2 = 1$. *The extrapolation from the left-hand field does a beautiful job of extrapolating into the right-hand element, and vice versa.*

Algorithm 14.2 *Derive the coupled model for diffusion between two finite elements, the PDE* (14.7), *utilizing the nonlocal coupling conditions* (14.19). *The only differences with Algorithm* 14.1 *lie within the iteration loop, so the loop is the only part listed here.*

```
for iter:=1:99 do begin
  % compute residuals
  res1:=df(u1,t)-nu*df(u1,x,2);
  res2:=df(u2,t)-nu*df(u2,x,2);
  cc1:=sub(x=+hh/2,u1-uu(1)-gamma*(uu(2)-uu(1)));
  cc2:=sub(x=-hh/2,u2-uu(2)-gamma*(uu(1)-uu(2)));
  % update left element from residuals
  g1:=g1+(gd:=-int(res1,x,-hh,0)/hh-nu*cc1/hh^2);
  u1:=u1+int(int(res1+gd,x,-hh,x),x,-hh/2,x)/nu;
  % update right element from residuals
  g2:=g2+(gd:=-int(res2,x,0,+hh)/hh-nu*cc2/hh^2);
  u2:=u2+int(int(res2+gd,x,+hh,x),x,+hh/2,x)/nu;
  if {res1,res2,cc1,cc2}={0,0,0,0} then iter:=1000000;
end;
```

Theory supports the model The dynamics of the linearized problem is the base of the application of center manifold theory. Here the nonlinear terms are those involved in the interelement coupling. Thus the linear dynamics are those with coupling parameter $\gamma = 0$. When $\gamma = 0$ the two elements are completely decoupled. Hence just investigate the linear dynamics in the left-hand element. The linear dynamics in the right-hand element are the same by symmetry.

The linear problem in the left-hand element is

$$\frac{\partial u_1}{\partial t} = \nu \frac{\partial^2 u_1}{\partial x^2}$$
$$\text{such that } \frac{\partial u_1}{\partial x} = 0 \text{ at } x = -H$$
$$\text{and } u_1(H/2, t) = u_1(-H/2, t). \tag{14.22}$$

The PDE and the first boundary condition require all solutions to be of the form $u_1 = \exp(\lambda t)\cos[k(x+H)]$ for some wavenumbers k and consequent growth rate $\lambda = -\nu k^2$. The nonlocal condition then implies $\cos(3kH/2) = \cos(kH/2)$. Expanding and rearranging leads to the requirement

$$2\sin(kH/2)\sin(kH) = 0. \tag{14.23}$$

This requirement is satisfied whenever $kH = n\pi$; hence the family of wavenumbers comprises $k = n\pi/H$, with corresponding eigenvalues $\lambda = -\nu\pi^2 n^2/H^2$, as in (14.2) of the previous coupling condition. As before, for both elements combined, but uncoupled, the spectrum has two zero eigenvalues, and all other eigenvalues are negative and less than $-\nu\pi^2/H^2$. Consequently, center manifold theory assures us that a slow manifold model exists in some domain, is emergent in some domain, and may be constructed asymptotically such as by Algorithm 14.2.

14.5 ▪ Exercises

Exercise 14.1 (other artifices). The magical model (14.6) is no accident: why implement the factor of two in the right-hand side of the artificial boundary conditions (14.3)? The answer is that it is equivalent to an optimal Euler parameter (see section 9.3). Let us see again that such invention can range from the wonderful to the disastrous. Change the artificial boundary conditions (14.3) to

$$(1-\gamma)\frac{\partial u}{\partial x} = \mp \mathrm{Eu}\,\gamma u \quad \text{on } x = \pm 1 \tag{14.24}$$

for some parameter Eu. By substitution show that $u = a(t)(1 - Bx^2)$ such that $\dot{a} = -2Ba^2$, for some B depending upon the parameters Eu and γ, forms an *exact* solution of the nonlinear diffusion PDE (14.1) with Robin boundary conditions (14.24). Deduce $B(\gamma, \mathrm{Eu})$ as an *exact* analytic function of Eu and γ. Confirm that setting $\gamma = 1$ gives the correct $B(1, \mathrm{Eu}) = 1$, independent of the artificial parameter Eu.

But in practice we construct such models as power series in the artificial parameter γ, so write out some terms of the expansion of $B(\gamma, \mathrm{Eu})$ in powers of small parameter γ (perhaps obtain the Maclaurin series of $B(\gamma, \mathrm{Eu})$ via a geometric series) and observe that

1. the power series truncates for the magical $\mathrm{Eu} = 2$ used previously, in boundary conditions (14.3);

2. the power series converges at $\gamma = 1$ for some domain of Eu (containing Eu = 2) and so could be used to approximate the model to some accuracy;[176] and
3. for other values of the parameter Eu, the power series diverges at $\gamma = 1$ and so is practically useless.

Exercise 14.2 (Galerkin method). Show that the Galerkin method with linear interpolation structure function $\phi(x) = 1 - |x|$ generates the quantitatively wrong model $\dot{a} = -3a^2$ when applied to the PDE (14.1).

Exercise 14.3 (linear diffusion). Create a one-variable model for the long-term dynamics of linear diffusion of the field $u(x,t)$ in a small element.[177] Aim to model

$$\frac{\partial u}{\partial t} = \nu \frac{\partial^2 u}{\partial x^2} \quad \text{such that} \quad u_x(0,t) = 0 \quad \text{and} \quad u(H,t) = 0$$

for some diffusion coefficient ν. Verify that the exact long-term dynamics are $u = a(t)\cos[\pi x/(2H)]$ such that $\dot{a} = -\nu\pi^2/(4H^2)\,a$. Create a slow manifold–based model by replacing the right-hand boundary condition by the artificial

$$u(H,t) = (1-\gamma)u(0,t)\,. \tag{14.25}$$

Analyze, solve algebraically for the exact slow manifold and model, interpret, and discuss the dependence upon the artificial parameter γ.

Answer: The field $u = a\cos(kx/H)$ such that $\dot{a} = -\nu k^2 a/H^2$ where $k = \cos^{-1}(1-\gamma)$.

Exercise 14.4 (Burgers' PDE on two elements). Derive some two-element models of various accuracy for the nonlinear Burgers PDE on the domain $[-H, H]$

$$\frac{\partial u}{\partial t} + u\frac{\partial u}{\partial x} = \frac{\partial^2 u}{\partial x^2} \quad \text{such that} \quad \frac{\partial u}{\partial x} = 0 \text{ at } x = \pm H\,. \tag{14.26}$$

Modify the Reduce program of Algorithm 14.1. The easiest way to control the combinatorial explosion of terms generated by the nonlinearity is to simply multiply the nonlinearity by some power of the coupling parameter γ; that is, solve

$$\frac{\partial u}{\partial t} + \gamma^p u\frac{\partial u}{\partial x} = \frac{\partial^2 u}{\partial x^2}$$

to some order in the coupling/nonlinearity parameter γ. Then evaluating at $\gamma = 1$ is not only a fully coupled model but also a model for the original Burgers PDE (14.26). Investigate the nonlinear terms in the model for various orders of truncation.

---

[176] Recall that the radius of convergence of a power series is the distance to the nearest singularity of the function—when considered as an analytic function in the complex plane. For example, $f(x) = \sqrt{x^2/2 + x + 1}$ has power series $f = 1 + \frac{1}{2}x + \frac{1}{8}x^2 - \frac{1}{16}x^3 + \cdots$. Its radius of convergence is $\sqrt{2}$. The reason is that the argument of the square root is zero at $x = -1 \pm i$ and hence the two square-root singularities at $x = -1 \pm i$ limit the radius of convergence.

[177] Again, this problem is a little like finding a quasi-stationary probability distribution: here the ultimate state is the "absorbing" field $u = 0$; we aim to find the longest-lasting dynamics that occurs before this ultimate absorption.

Answer: For an example, set the exponent $p = 1$. Then as well as similar expressions for $\dot{U}_1$ and $u_1(x,t)$, the evolution and the field in the right-hand element are

$$\begin{aligned}\dot{U}_2 &= (\gamma + \tfrac{1}{3}\gamma^2)\frac{1}{H^2}(U_1 - U_2) + \gamma^2\frac{1}{2H}U_2(U_1 - U_2) + \cdots,\\ u_2(x,t) &= U_2 + \gamma(U_1 - U_2)\left[\tfrac{3}{8} - (\tfrac{x}{H}) + \tfrac{1}{2}(\tfrac{x}{H})^2\right]\\ &\quad + \gamma^2(U_1 - U_2)\left[\tfrac{9}{64} - \tfrac{1}{4}(\tfrac{x}{H}) - \tfrac{5}{24}(\tfrac{x}{H})^2 + \tfrac{1}{3}(\tfrac{x}{H})^3 - \tfrac{1}{12}(\tfrac{x}{H})^4\right]\\ &\quad + \gamma^2 H U_2(U_1 - U_2)\left[\tfrac{1}{24} - \tfrac{1}{4}(\tfrac{x}{H})^2 + \tfrac{1}{6}(\tfrac{x}{H})^3\right] + \cdots.\end{aligned}$$

Exercise 14.5 (Burgers' variation). Repeat Exercise 14.4 but now control the nonlinearity *independently* of the coupling by instead solving

$$\frac{\partial u}{\partial t} + \epsilon u\frac{\partial u}{\partial x} = \frac{\partial^2 u}{\partial x^2}.$$

Truncate to various powers of ϵ and γ and compare the terms in the models.

Answer: As well as similar expressions for $\dot{U}_1$ and $u_1(x,t)$, the evolution and the field in the right-hand element are

$$\begin{aligned}\dot{U}_2 &= \gamma\frac{1}{H^2}(U_1 - U_2) + \gamma\epsilon\frac{1}{2H}U_2(U_1 - U_2) + \cdots,\\ u_2(x,t) &= U_2 + \gamma(U_1 - U_2)\left[\tfrac{3}{8} - (\tfrac{x}{H}) + \tfrac{1}{2}(\tfrac{x}{H})^2\right]\\ &\quad + \gamma\epsilon H U_2(U_1 - U_2)\left[\tfrac{1}{24} - \tfrac{1}{4}(\tfrac{x}{H})^2 + \tfrac{1}{6}(\tfrac{x}{H})^3\right] + \cdots.\end{aligned}$$

Exercise 14.6 (nonlinear diffusion on two elements). Make other modifications of Algorithm 14.1 to model the nonlinear diffusion PDE (14.1) by the dynamics in two elements on the domain $[-1, 1]$. First, model the dynamics with insulating boundary conditions at $x = \pm 1$. Second, make further significant modifications of the computer algebra to model the dynamics with the original fixed-value boundary conditions of (14.1). Investigate the predictions of your model.

Exercise 14.7 (nonlocal coupling). Repeat Exercises 14.4 and 14.6 but with the nonlocal coupling (14.19).

Exercise 14.8 (three-zone diffusion). Consider the linear diffusion dynamics of a field $u(x,t)$ on the interval $-1 < x < 1$ satisfying the diffusion PDE

$$\frac{\partial u}{\partial t} = \frac{\partial^2 u}{\partial x^2} \quad \text{such that} \quad \frac{\partial u}{\partial x} = 0 \text{ at } x = \pm 1.$$

This exercise seeks to create a three-element, three-zone model for the diffusion to compare with the general spectral solution

$$u(x,t) = \sum_{k=0}^{\infty} c_k \cos[k\pi(x+1)/2]\exp(-k^2\pi^2 t/4).$$

In particular, the decay rates of the first four modes are $\lambda = 0, -2.4674, -9.8696,$ and -22.2066, and this exercise aims to model dynamics associated with the first three of these modes. Define three notional elements: $-1 \leqslant x \leqslant -\frac{1}{3}$, $-\frac{1}{3} \leqslant x \leqslant \frac{1}{3}$, and $\frac{1}{3} \leqslant x \leqslant 1$. Let $u_j(x,t)$ be the field in each element, $j = -1, 0, +1$, respectively. Couple the elements by the nonlocal coupling (14.19) with grid points $X_j = \frac{2}{3}j$; that is,

$$u_0(\pm\tfrac{2}{3}, t) = \gamma u_{\pm1}(\pm\tfrac{2}{3}, t) + (1-\gamma)u_0(0,t),$$
$$u_{\pm1}(0,t) = \gamma u_0(0,t) + (1-\gamma)u_{\pm1}(\pm\tfrac{2}{3}, t).$$

1. Argue that there exists an emergent 3D slow manifold in a neighborhood of coupling parameter $\gamma = 0$ (4D if you count the parameter γ). In the argument, use that the eigenmodes are purely trigonometric, such as $\cos[k(x \pm 1)]$ and $A\cos kx + B\sin kx$, and deduce that the eigenvalues (of various multiplicities) correspond to wavenumbers $k = 3n\pi/2$ for integer n.
2. Construct the first nontrivial approximation to the slow manifold and the evolution thereon of

$$u_{\pm1} \approx U_{\pm1} + \gamma(U_0 - U_{\pm1})\tfrac{9}{8}(1 \mp x)^2,$$
$$u_0 \approx U_0 + \gamma\left[(U_{+1} - U_{-1})\tfrac{3}{4}x + (U_{+1} - 2U_0 + U_{-1})\tfrac{9}{8}x^2\right],$$
$$\dot U_{\pm1} \approx \gamma\tfrac{9}{4}(-U_{\pm1} + U_0) \quad\text{and}\quad \dot U_0 \approx \gamma\tfrac{9}{4}(U_{-1} - 2U_0 + U_{+1}),$$

   upon defining the amplitudes $U_j(t) = u_j(X_j, t)$, that is, the midvalue and the two edge values. Compare the eigenvalues predicted here for $\gamma = 1$ with the exact eigenvalues.
3. Write, debug, and run a computer algebra program to iteratively construct the slow manifold and evolution thereon to high order in coupling parameter γ. Comment on evidence indicating that the asymptotic series converges at the physically relevant coupling $\gamma = 1$.

Exercise 14.9 (Euler accelerates). An Euler transformation may improve convergence of a series solution (Van Dyke, 1964; Bender and Orszag, 1981, e.g.).

1. For an introductory example, consider some quantity $f = (1-\gamma/2)^{-1} = 1 + \frac{1}{2}\gamma + \frac{1}{4}\gamma^2 + \frac{1}{8}\gamma^3 + \cdots$ that we wish to estimate at $\gamma = 1$ from the series. The series converges for $|\gamma| < 2$. Change the series parameter to a new parameter γ' by the specific Euler transform $\gamma = 3\gamma'/(2+\gamma')$, noting that $\gamma = 1$ corresponds to $\gamma' = 1$. Expand to see that

$$\gamma = \tfrac{3}{2}\gamma' - \tfrac{3}{4}\gamma'^2 + \tfrac{3}{8}\gamma'^3 + \cdots.$$

   Substitute this expansion into the series for f to determine that as a function of γ',

$$f = 1 + \tfrac{3}{4}\gamma' + \tfrac{3}{16}\gamma'^2 + \tfrac{3}{64}\gamma'^3 + \cdots.$$

   This series for f now appears to converge for $|\gamma'| < 4$ and hence converges quicker at the interesting $\gamma' = 1$: here the four-term series in γ' for $f \approx 1.9844$ at $\gamma' = 1$, whereas the four-term series in γ for $f \approx 1.8750$ at $\gamma = 1$. The reason the Euler transform improves convergence here is that it moves the singularity in f, at $\gamma = 2$, farther away to $\gamma' = 4$.

   However, if there were a (hidden) singularity in f at say $\gamma = -3$, then this particular Euler transform would have moved it to $\gamma' = -1$, which would ruin convergence for $|\gamma'| > 1$. General Euler transforms, $\gamma = (1+\mathrm{Eu})\gamma'/(1+\mathrm{Eu}\,\gamma')$ for some constant Eu, may or may not improve convergence depending upon both Eu and the location of singularities in the complex γ-plane.

2. Return to Exercise 14.8. Show that under the Euler transform $\gamma = (1+\mathrm{Eu})\gamma'/(1+\mathrm{Eu}\,\gamma')$ the coupling conditions become

$$(1+\mathrm{Eu}\,\gamma')u_0(\pm\tfrac{2}{3},t) = (1+\mathrm{Eu})\gamma' u_{\pm 1}(\pm\tfrac{2}{3},t) + (1-\gamma')u_0(0,t),$$
$$(1+\mathrm{Eu}\,\gamma')u_{\pm 1}(0,t) = (1+\mathrm{Eu})\gamma' u_0(0,t) + (1-\gamma')u_{\pm 1}(\pm\tfrac{2}{3},t).$$

Argue that the slow manifold existence and emergence are as before.

3. In the computer algebra construction of Exercise 14.8, change only the computation of the residuals of the coupling conditions to construct the slow manifold. Show that the Euler parameter $\mathrm{Eu} = 0$ generates the slow manifold as before, that $\mathrm{Eu} = 1/2$ appears to improve convergence in γ', but that $\mathrm{Eu} = 1$ is hopeless.

Exercise 14.10 (slow/fast zone diffusion). Some problems have symmetries to maintain. For example, in shear dispersion along a pipe, in Chapter 7, the long-term evolution and model have only radial dependence. So the resolution of the long-term model depends only upon the resolution of radially dependent dynamics. Thus Chikwendu and Ojiakor (1985) suggested modeling the dynamics in two modes based on fast and slow zones in the flow.

Analogously, in the three-zone diffusion of Exercise 14.8 the asymmetric modes are completely decoupled from the symmetric modes. So the resolution of the long-term model depends mainly on the symmetric modes (although transients from general initial conditions need asymmetric modes as well). This exercise introduces a "two-zone," i.e., a "fast/slow zone," model of the diffusion dynamics of Exercise 14.8. It can be adapted to a two-zone model of shear dispersion in a pipe or channel.

1. Here the two zones are notionally the fast central zone $-\frac{1}{3} < x < \frac{1}{3}$ and the slow edge zone $\frac{1}{3} < |x| < 1$ together. It may seem weird having two distinct physical domains lumped together into the one zonal element, but let's see how it works. In the slow edge zone write the dynamics in terms of the symmetric and asymmetric modes, respectively:

$$v(x,t) = \tfrac{1}{2}[u_{+1}(x,t) + u_{-1}(-x,t)],$$
$$w(x,t) = \tfrac{1}{2}[u_{+1}(x,t) - u_{-1}(-x,t)].$$

Then we only analyze for positive x, as the negative x field $u_{-1}(-x,t) = v(x,t) - w(x,t)$. For a little simplicity, instead of $u_0(x,t)$ let $u(x,t)$ denote the field in the fast central zone.

By various sum and difference combinations of the coupling conditions in Exercise 14.8, deduce that the equivalent coupling and boundary conditions for these fields are

$$\begin{aligned}
&\tfrac{1}{2}\left[u(\tfrac{2}{3},t)+u(-\tfrac{2}{3},t)\right]=\gamma v(\tfrac{2}{3},t)+(1-\gamma)u(0,t),\\
&\tfrac{1}{2}\left[u(\tfrac{2}{3},t)-u(-\tfrac{2}{3},t)\right]=\gamma w(\tfrac{2}{3},t),\\
&\frac{\partial v}{\partial x}=\frac{\partial w}{\partial x}=0 \text{ at } x=1\,,\\
&v(0,t)=\gamma u(0,t)+(1-\gamma)v(\tfrac{2}{3},t),\\
&w(0,t)=0,
\end{aligned}$$

except for this last condition where here we omit a $(1-\gamma)w(\tfrac{2}{3},t)$ term. This omission generates the two-zone model.

2. Find that equilibria of the system of diffusion equations in each field with these coupling conditions are $\gamma = 0$, $u(x,t) = U$, $v(x,t) = V$, and $w(x,t) = 0$. Consider the linearized dynamics about these equilibria and deduce that there exists a slow manifold parametrized globally in U and V and locally in coupling γ.

3. Derive by hand the $\mathcal{O}(\gamma^2)$ model. If you define the amplitudes $U(t)=u(0,t)$ and $V(t)=v(1,t)$, then the model is $w=0$ and

$$\begin{aligned}
&u=U+\gamma\tfrac{9}{4}x^2(V-U)+\mathcal{O}(\gamma^2),\\
&v=V-\gamma\tfrac{9}{8}(1-x)^2(V-U)+\mathcal{O}(\gamma^2),\\
&\dot U=\gamma\tfrac{9}{2}(V-U)+\mathcal{O}(\gamma^2)\quad\text{and}\quad \dot V=\gamma\tfrac{9}{4}(U-V)+\mathcal{O}(\gamma^2).
\end{aligned}$$

4. Modify the computer algebra of Exercise 14.8 or 14.9, or write anew, to construct the two-zone model of diffusion to arbitrary specified order in γ.

Exercise 14.11 (finite advection). Model the linear advection-diffusion PDE

$$\frac{\partial u}{\partial t}+2\frac{\partial u}{\partial x}=\frac{\partial^2 u}{\partial x^2}\quad\text{such that}\quad\frac{\partial u}{\partial x}=0 \text{ at } x=\pm1\,,$$

with two elements on the interval $[-1,1]$. Use the nonlocal coupling (14.19) to connect the dynamics in each interval. Modify the computer algebra in Algorithm 14.2 to be able to construct models to high orders in the coupling parameter γ. Interpret your model and the fields in each element.

Answer: This is a straightforward task if you multiply the advection term by a small parameter and seek an expansion in that small parameter. However, the question does not encourage this easy route. The hard task is to cater for the linear operator $\partial^2u/\partial x^2-2\partial u/\partial x$ without perturbation. The result for the right-hand element is

$$\begin{aligned}
\dot U_2&=\frac{4\gamma}{2-e^{-1}+e^{-3}}(U_1-U_2)+\mathcal{O}(\gamma^2)\,,\\
u_2(x,t)&=U_2+\frac{\gamma}{2-e^{-1}+e^{-3}}\Big[(1-e^{-1}-2x+e^{2(x-1)})U_1\\
&\quad-(1-2x-e^{-1}+e^{2(x-1)})U_2\Big]+\mathcal{O}(\gamma^2)\,.
\end{aligned}$$

Exercise 14.12 (adjoint). Consider the linear diffusion equations (14.22) for the field in the left of two coupled elements. Although the solvability condition encoded in Algorithm 14.2 does work, notice that the iterations converge slower than necessary. This exercise seeks the best solvability condition for (14.22).

Recall that solvability conditions are that the right-hand side r of a system of equations $\mathcal{L}u = r$ must be in the range of the operator $\mathcal{L}$. Consequently the right-hand side r must be orthogonal to the null space of the adjoint $\mathcal{L}^\dagger$ of the operator. We seek a basis for the null space of the adjoint of the linear diffusion operator in (14.22).

Define the spatial diffusion operator

$$\begin{aligned} \mathcal{L}u &= \nu \frac{\partial^2 u}{\partial x^2} \\ \text{such that } \frac{\partial u}{\partial x} &= 0 \text{ at } x = -H \quad \text{and} \quad u(H/2, t) = u(-H/2, t)\,. \end{aligned}$$

Using the inner product $\langle u, v\rangle := \int_{-H}^{H/2} uv\,dx$, argue that the adjoint

$$\begin{aligned} \mathcal{L}^\dagger v &= \nu \frac{\partial^2 v}{\partial x^2} \\ \text{such that } \frac{\partial v}{\partial x} &= 0 \text{ at } x = -H \quad \text{and} \quad v(H/2, t) = 0\,, \\ \text{with jumps } \left[\frac{\partial v}{\partial x}\right]_{-H/2^-}^{-H/2^+} &= \left.\frac{\partial v}{\partial x}\right|_{H/2} \quad \text{and} \quad [v]_{-H/2^-}^{-H/2^+} = 0\,. \end{aligned}$$

Hence deduce that a basis for the null space of the adjoint $\mathcal{L}^\dagger$ is

$$v = \begin{cases} 1, & x < -H/2, \\ \frac{1}{2} - \frac{x}{H}, & -H/2 < x < H/2\,. \end{cases}$$

Similarly find the adjoint and a basis for the null space of the linear diffusion dynamics in the right of the two elements. Finally, improve the coding of the solvability condition for g_1 and g_2 in Algorithm 14.2. Compare the old and the new versions.

Exercise 14.13 (double roots). Discuss the theoretical implications of the double roots at wavenumbers $k = 2m\pi/H$ of the requirement (14.23), and the adjoints in Exercise 14.12.

Exercise 14.14 (ensure continuity). The approach described in this chapter creates models based upon piecewise constant functions. That is, in each element, the starting approximation is that the field u is constant. Surely we can create models from a *continuous piecewise linear basis.*

Let's see one way to form a piecewise linear basis. Model the linear diffusion PDE, $\partial u/\partial t = \partial^2 u/\partial x^2$, on the domain $[-1, 1]$ by notionally dividing the domain into two artificial elements and creating a piecewise linear basis. Set the diffusion coefficient and each element size to 1 for simplicity. Let the physical boundary conditions be that $u(\pm 1, t) = 0$.

1. Use separation of variables to solve this PDE in terms of an infinite sum of Fourier modes. What are the longest-lasting mode and its dynamics? What is the rate of decay of the next longest lasting mode? In later parts of the exercise, use this solution as a background reference.
2. Now divide the domain into two elements by placing artificial internal coupling conditions at the midpoint $x = 0$. Ensure continuity of field u between the elements by requiring $u(-0,t) = u(+0,t)$. Additionally couple the two elements with the *nonlocal* coupling condition

$$u_x(-0,t) - (1-\gamma)u_x(-1,t) = u_x(+0,t) - (1-\gamma)u_x(+1,t)\,. \quad (14.27)$$

(You may find arguments easier when expressed in terms of two subfields: $u_1(x,t) = u(x,t)$ when $x < 0$ and $u_2(x,t) = u(x,t)$ when $x > 0$; whether you do so or not is up to you.) Argue briefly that when the coupling parameter $\gamma = 1$, these conditions ensure continuity of u and its derivative. Argue that when the coupling parameter $\gamma = 0$, the diffusion PDE has a neutral mode that has field u continuous (but discontinuous derivative u_x) and is piecewise linear across the domain.
3. Find the complete spectrum of the diffusion dynamics on the domain with the basic coupling, $\gamma = 0$, and hence deduce that a slow manifold exists and will be relevant for nonzero coupling.
4. Using $U(t) = u(0,t)$ as the amplitude parameter, construct (by hand) the slow manifold model to errors $\mathcal{O}(\gamma^2)$. Compare with the original analytic solution obtained by separation of variables.

Solution: The longest-lasting mode is $\cos(\pi x/2)\exp(-\pi^2 t/4)$; the next longest lasting mode is $\sin(\pi x)\exp(-\pi^2 t)$. The spectrum of the basically coupled diffusion is $\lambda_k = -\pi^2 k^2$ for integer k, with the eigenvalues of even $k \geqslant 2$ having multiplicity 2. The leading approximation of the slow manifold is $u = (1-|x|)U + \gamma(\frac{2}{3}|x| - x^2 + \frac{1}{3}|x|^3)U + \mathcal{O}(\gamma^2)$, on which the evolution is $\dot{U} = -2\gamma U + \mathcal{O}(\gamma^2)$.

Exercise 14.15 (ensure continuity II). Repeat Exercise 14.14 with the one change that the physical boundary condition $u = 0$ at $x = +1$ is replaced by the insulating condition $u_x = 0$ at $x = +1$. All other aspects of the question, including the coupling condition (14.27), remain the same.

Exercise 14.16 (continuity with computer algebra). Write, test, and execute a computer algebra program to construct the slow manifold model of Exercise 14.14 to arbitrary order in coupling parameter γ. Interpret and discuss.

1. Inventively introduce an Euler parameter into the coupling condition (14.27) to improve convergence of the slow manifold model in γ. Justify recommending the choice of your Euler parameter.
2. Modify the program to cater for nontrivial physical boundary conditions of $u(-1,t) = a$ and $u(+1,t) = b$. The variables a and b do not need to be assumed small. Observe the stable equilibrium predicted by the model of $U = (a+b)/2$.
3. Now allow the boundary values a and b to vary slowly in time. For simplicity, assume that their nth time derivatives scale as γ^n. Interpret and discuss.

4. Reset $a = b = 0$ and modify the program to model Burgers' PDE $u_t = u_{xx} - \alpha u u_x$ on this domain. Manage the nonlinearity by treating α as an independent small parameter. Argue that the effects of $\mathcal{O}(\alpha^2)$ nonlinearly enhance the decay of solutions within the element.

Exercise 14.17 (discrete lattice dynamics). Consider *discrete* diffusive dynamics on a small "chunk" of lattice: $\frac{du_j}{dt} = u_{j-1} - 2u_j + u_{j+1}$ for $j = 1 : 5$ and $u_0 = u_1$, $u_6 = u_5$. What are the eigenvalues and eigenvectors of this five-dimensional linear system? *Now model this "fine"-scale system by a two-dimensional "coarse"-scale discrete system.* Embed its five-dimensional dynamics by dividing the lattice into two overlapping elements, $v_j = u_{j+2}$ and $w_j = u_{j+4}$ for $j = -2 : 2$, but consider the "overlapping" variables, v_0 and w_{-2}, v_1 and w_{-1}, and v_2 and w_0, as independent variables. Couple the two elements with $v_2 = \gamma w_0 + (1-\gamma)v_0$ and $w_{-2} = \gamma v_0 + (1-\gamma)w_0$. Argue that based around the coupling parameter $\gamma = 0$ there is a two-dimensional slow manifold of the enlarged system (3D if you count γ) $\frac{dv_j}{dt} = v_{j-1} - 2v_j + v_{j+1}$ and $\frac{dw_j}{dt} = w_{j-1} - 2w_j + w_{j+1}$ for $j = -1 : 1$ and $v_{-2} = v_{-1}$, $w_2 = w_1$. Construct and interpret the slow manifold model. Relate your results to the original eigenvalues and eigenvectors.

Chapter 15

Holistic discretization in one space dimension

Contents

Chapter 14 showed that we can divide a domain into two finite elements, couple them again, and rigorously support a model of the dynamics using center manifold theory. This chapter divides a spatial domain into any number of elements and uses specially crafted coupling conditions to empower center manifold theory to support the resulting discretization of the dynamics. Interspersed throughout this chapter we prove that nonlocal coupling conditions inspired by (14.19) ensure that the discrete model is also *consistent* with the original PDE as the element size $H \to 0$. That is, the discretization and the PDE are the same to some order of error in H. Analysis to higher orders in the coupling parameter γ results in a better order of error in H. As shown schematically in Figure 15.1, this dual support from *both* consistency and center manifold theory for the discrete models we derive is remarkable.

The approach developed here is based purely upon the *local* dynamics on finite-sized elements while maintaining fidelity with the solutions of the original PDE. Being local, the approach flexibly caters for complicated geometries, varying resolutions, and physical boundaries.

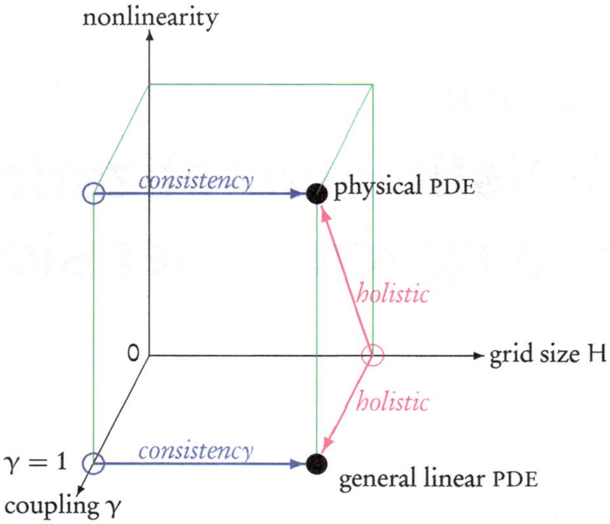

Figure 15.1. *This conceptual diagram shows that traditional finite difference modeling (right arrows) approaches physical PDEs (discs) via consistency as the grid size $H \to 0$ (left circles), whereas the holistic method (forward arrows) approaches physical problems via asymptotics in nonlinearity and the interelement coupling γ (from right circle). Our specially crafted holistic method supports the model from* both *approaches.*

15.1 ▪ Model diffusion-like dynamics consistently

Let's start by exploring perhaps the simplest PDE for a field $u(x,t)$ in one space dimension, the (nondimensional) diffusion PDE

$$\frac{\partial u}{\partial t} = \frac{\partial^2 u}{\partial x^2}. \tag{15.1}$$

We aim to develop classic spatially discrete models for this PDE, such as the simple centered difference approximation to the right-hand side,

$$\frac{\partial U_j}{\partial t} \approx \frac{U_{j-1} - 2U_j + U_{j+1}}{H^2},$$

where the grid values $U_j(t) := u(X_j, t)$ for an equispaced grid, say $X_j := jH$. Most important, we aim to underpin these models with center manifold theory so that the approach also supports the modeling of complex systems of PDEs.

Introduce a regular grid As shown in Figure 15.2, place grid points a distance H apart, $X_j := jH$, for example. Express the field in the jth element by $u = u_j(x,t)$ so that the fields u_j in all elements also satisfy the diffusion PDE (15.1). We do not restrict the fields u_j to disjoint domains but extend each element analytically out to the two adjacent grid points, as shown in Figures 15.2 and 15.3. That is, the jth element is $X_{j-1} \leqslant x \leqslant X_{j+1}$.

Suppose that we form m elements in the physical domain. For definite simplicity, assume that the field $u(x,t)$ is spatially periodic with the period thus being $L := mH$, or equivalently the number of elements $m := L/H$. Corollary 15.3, below, additionally includes the cases of Dirichlet and Neumann boundary conditions on the physical domain.

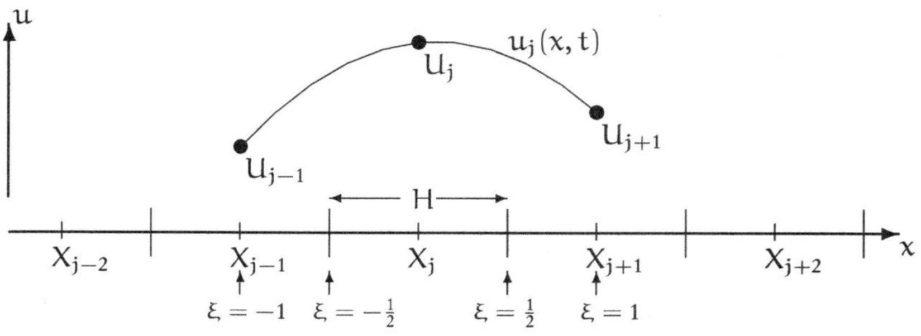

Figure 15.2. *Schematic picture of the equispaced grid, X_j, of spacing H; the unknowns U_j; notional internal boundaries between elements (vertical lines); and in the neighborhood of X_j the field $u_j(x,t)$ that extends outside the element and, when coupling $\gamma = 1$, is aimed to pass through the neighboring grid values $U_{j\pm1}$.*

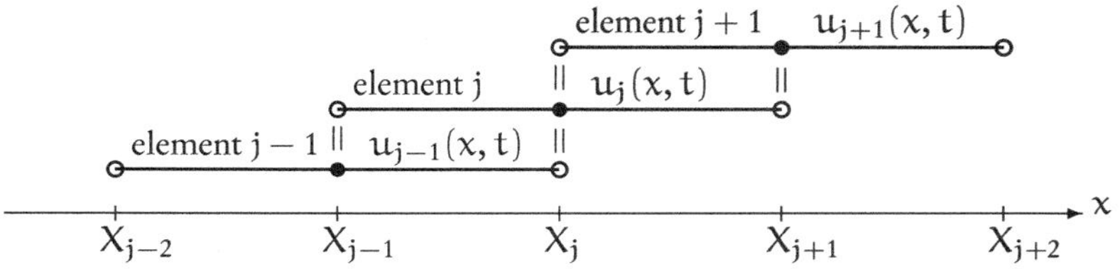

Figure 15.3. *Schematic diagram of three consecutive overlapping elements coupled by the conditions* (15.2). *The coupling conditions* (15.2) *link values shown by the circles and discs. In particular, when the coupling parameter $\gamma = 1$, the values of adjacent elements must be equal, as indicated.*

Use nonlocal coupling conditions Here we explore how an amazingly good discrete approximation follows from using nonlocal coupling conditions inspired by (14.19), namely the edge values on each element

$$u_j(X_{j\pm1},t) = \gamma u_{j\pm1}(X_{j\pm1},t) + (1-\gamma)u_j(X_j,t). \tag{15.2}$$

That is, the field of the jth element when extrapolated to the surrounding gridpoints, $u_j(X_{j\pm1},t)$, is a weighted combination of two vital extremes:

- when $\gamma = 1$, the jth element field $u_j(X_{j\pm1},t)$ is to equal the field in the neighboring elements at their grid points, $u_{j\pm1}(X_{j\pm1},t)$, to in effect recover the physical continuity, as shown schematically in Figure 15.3;
- but when $\gamma = 0$, the edge value $u_j(X_{j\pm1},t)$ of each element is to be just identical to the midelement value $u_j(X_j,t)$ so that each element becomes isolated from all other elements.

The parameter γ controls the coupling between neighboring elements.

15.1.1 ▪ The dynamics collapses onto a slow manifold

Equilibria? Center manifold theory is based on equilibria. Where are the equilibria in the coupled system of the diffusion PDE (15.1) with coupling condition (15.2)? A useful subspace of equilibria occurs when coupling parameter $\gamma = 0$; then the diffusion PDE (15.1) has equilibria of $u_j =$ constant. This constant is generally different in each of the m elements, as each element is isolated from each other. Thus there is an m-dimensional subspace of equilibria, $\mathbb{E}_0$, of piecewise constant fields, and coupling parameter $\gamma = 0$.

There may well be other equilibria, but we do not need them; we only need the subspace $\mathbb{E}_0$.

Linearize to find the spectrum The spectrum of the linearized dynamics implies the existence of a slow manifold. All element are identical, because of the spatial periodicity, and isolated from all other elements, as the coupling parameter $\gamma = 0$. Hence, *linearly*, the dynamics in each element determines that of the set of m elements. In any one element we find it is convenient to introduce the subgrid spatial variable $\xi := (x - X_j)/H$, as indicated in Figure 15.2: subgrid coordinate $\xi = 0$ corresponds to the element midpoint $x = X_j$, whereas subgrid coordinates $\xi = \pm 1$ correspond to the element edges $x = X_{j\pm1}$. Further, the diffusion PDE (15.1) becomes $\partial u/\partial t = \partial^2 U/\partial \xi^2 /H^2$. Because this PDE is constant coefficient, separation of variables in the diffusion PDE (15.1) finds solutions $u_j = e^{\lambda t}[A\sin(k\xi) + B\cos(k\xi)]$ for eigenvalues $\lambda = -k^2/H^2$. Determine the allowed wavenumbers k from the isolating condition (15.2) that holds when coupling $\gamma = 0$, namely $u_j(X_{j\pm1}, t) = u_j(X_j, t)$. Substitution into these isolation conditions gives equations $\pm A \sin k + B\cos k = B$:

- subtracting the two equations gives $A \sin k = 0$ and hence $k = n\pi$ for $n = 1, 2, 3, \ldots$, whereas
- adding the two equations gives $B\cos k = B$ and hence $k = 2n\pi$ for $n = 0, 1, 2, 3, \ldots$ (with multiplicity 2 for $n > 0$).

Thus the spectrum of eigenvalues is $\{0, -\pi^2/H^2, -4\pi^2/H^2, -9\pi^2/H^2, \ldots\}$. Since the analysis is identical for all elements, each of these eigenvalues has multiplicity m or $3m$. This spectrum of m eigenvalues 0 and all the rest with real part negative, $\leqslant -\pi^2/H^2$, implies that there exists a slow manifold (Theorem 4.3) that is exponentially quickly attractive (Theorem 4.5). Since the linear transients decay like $e^{-\pi^2 t/H^2}$, the slow manifold emerges on the cross-element diffusion time scale of $\mathcal{O}(H^2)$.

Importantly, in principle, there exists an exact slow manifold. That is, there exists a discretization that predicts exact solutions of the continuum diffusion PDE. Consequently, apart from initial transients there need not be any error in the discretization—in principle.

Define amplitudes Since there are m zero eigenvalues, one for each element, the slow manifold will be m-dimensional and the evolution thereon described by m variables (or "amplitudes").[178] We could define any of a wide variety of variables to parametrize the slow manifold (see section 5.3.3). Let's make the convenient choice to use the evolving grid values

$$U_j(t) := u_j(X_j, t). \tag{15.3}$$

With m elements, this defines m variables with which to parametrize the m-dimensional slow manifold.

[178] Actually, the slow manifold is $(m + 1)$-dimensional, as there is also a zero eigenvalue corresponding to the implicitly adjoined equation $\dot\gamma = 0$. However, this extra dimension is not that of a truly evolving variable, as the coupling parameter γ is constant. For the purposes of discussing dynamical variables we treat the slow manifold as being parametrized by m dynamical variables and hence being m-dimensional.

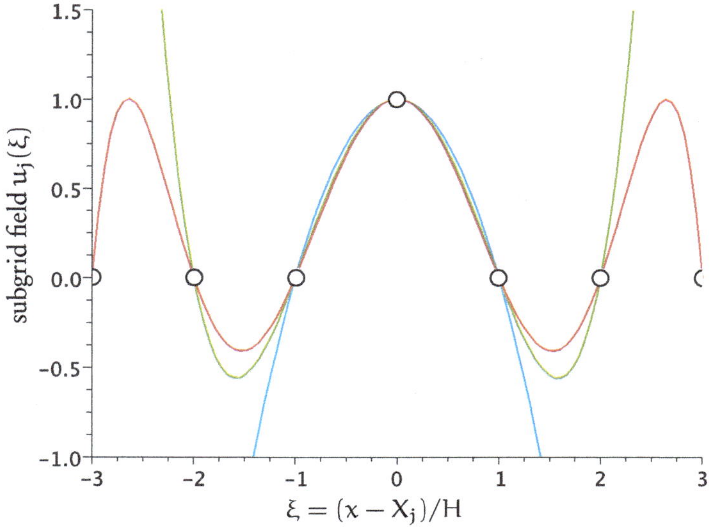

Figure 15.4. *The slow manifold* (15.4) *of the diffusion PDE* (15.1) *is a classic polynomial interpolation. These are the subgrid fields* $u_j(\xi)$ *for the case when grid value* $U_j = 1$ *and all other grid values are zero (but only* $-1 < \xi < 1$ *is relevant):* $\mathcal{O}(\gamma^2)$ *error, blue parabola;* $\mathcal{O}(\gamma^3)$ *error, green quartic; and* $\mathcal{O}(\gamma^4)$ *error, red sixth-order polynomial.*

15.1.2 ▪ Classic formulas form the slow manifold

Before discussing how computer algebra constructs the slow manifold discretization (section 15.1.4), let's discuss the eventual slow manifold model. Algorithm 15.1 (below) constructs the slow manifold as a power series in the coupling parameter γ:

$$u_j = U_j + \gamma\left[\xi\mu\delta + \tfrac{1}{2}\xi^2\delta^2\right]U_j + \gamma^2(\xi^2 - 1)\left[\tfrac{1}{6}\xi\mu\delta^3 + \tfrac{1}{24}\xi^2\delta^4\right]U_j + \mathcal{O}(\gamma^3), \quad (15.4)$$

where $\xi := (x - X_j)/H$ is the convenient subgrid spatial variable in place of x, and where Algorithm 15.2 (below) converts the expressions to use the centered mean operator $\mu U_j := \frac{1}{2}(U_{j+1/2} + U_{j-1/2})$ and the centered difference operator $\delta U_j := U_{j+1/2} - U_{j-1/2}$ (Table 15.1, below). The slow manifold (15.4) describes the subgrid field in the jth element as a constant, to crude leading order, corrected by interactions with its nearest neighbors through the γ terms, and corrected by interactions with its next to nearest neighbors through the γ^2 terms, and so on. When evaluated at full coupling, $\gamma = 1$, these fields reduce to classic polynomial interpolation. For example, truncating to errors $\mathcal{O}(\gamma^2)$ and evaluating at full coupling $\gamma = 1$ predicts the subgrid field $u_j(x,t) = \left[1 + \xi\mu\delta + \frac{1}{2}\xi^2\delta^2\right]U_j$, which is parabolic, as it involves up to ξ^2 (that is x^2), and evaluates correctly to U_j and $U_{j\pm1}$ at $\xi = 0, \pm1$ (that is, $x = X_j, X_{j\pm1}$). Figure 15.4 plots this parabola for the case when grid value $U_j = 1$ and all other $U_{j\pm p} = 0$. Figure 15.4 also plots the slow manifold interpolation for truncations with errors $\mathcal{O}(\gamma^3)$ and $\mathcal{O}(\gamma^4)$ to show their classic, higher-order polynomial shape. Such a close relation with classic polynomial interpolation in such simple problems is marvelous. In the more difficult class of nonisotropic and/or nonlinear problems, our slow manifold subgrid fields instead adapt specifically to the subgrid dynamics of the problem.

The evolution on the slow manifold (15.4) forms the discrete model. We use the grid values U_j to parametrize the slow manifold. Thus evolution on the slow manifold corresponds to evolution of the grid values and hence gives the discrete model. The construction of Algorithm 15.1 finds the corresponding evolution of the grid values on the slow manifold, in terms of centered differences $\delta^2 U_j$, $\delta^4 U_j$, and so on, to be

$$\dot{U}_j = \frac{1}{H^2}\left[\gamma\delta^2 - \tfrac{1}{12}\gamma^2\delta^4 + \tfrac{1}{90}\gamma^3\delta^6\right] U_j + \mathcal{O}(\gamma^4), \quad j = 1, \ldots, m. \tag{15.5}$$

When truncated at any order in coupling γ and then evaluated at full coupling, $\gamma = 1$, this slow manifold evolution (15.5) forms a classic finite difference approximation to the diffusion PDE (15.1). For two examples:

$$\mathcal{O}(\gamma^2), \quad \dot{U}_j = \frac{1}{H^2}\delta^2 U_j\,;$$
$$\mathcal{O}(\gamma^3), \quad \dot{U}_j = \frac{1}{H^2}\left[\delta^2 U_j - \tfrac{1}{12}\delta^4 U_j\right].$$

These right-hand sides are the classic second- and fourth-order accurate finite difference approximations to the second derivative in the diffusion PDE (15.1). The higher the order in coupling γ of the truncation, the more accurate the model in terms of consistency.

We also see this increase of accuracy in the equivalent differential equation for the discrete model (15.5). Replace centered differences according to (see Table 15.1)

$$\delta = 2\sinh \tfrac{1}{2}H\partial_x = H\partial_x + \tfrac{1}{24}H^3\partial_x^3 + \tfrac{1}{1920}H^5\partial_x^5 + \mathcal{O}(H^7);$$

then the discrete slow manifold model (15.5) transforms into the equivalent differential equation

$$\frac{\partial u}{\partial t} = \gamma\frac{\partial^2 u}{\partial x^2} + \tfrac{1}{12}H^2\gamma(1-\gamma)\frac{\partial^4 u}{\partial x^4} + \tfrac{1}{360}H^4\gamma(1-\gamma)(1-4\gamma)\frac{\partial^6 u}{\partial x^6} + \mathcal{O}(H^6, \gamma^4). \tag{15.6}$$

For full coupling $\gamma = 1$ this equivalent differential equation reduces to the diffusion PDE (15.1) with error $\mathcal{O}(H^6)$. Truncating to lower orders in coupling parameter γ results in an error of lower order in grid spacing H because the $(1-\gamma)$ factors in the equivalent differential equation will not have been fully formed in the discrete model (15.5). Computing to higher orders in coupling parameter γ results in an error of higher order in H. This consistency between the slow manifold discrete model (15.5) and the original differential diffusion PDE (15.1) follows from the specific form of the coupling conditions (15.2). As Figure 15.1 illustrates, our modeling has dual support: consistency and center manifold theory.

15.1.3 ▪ Operator algebra proves consistency

This section proves that high-order consistency follows from the specific nonlocal interelement coupling conditions (15.2). Consistency is established between a discretization and the PDE as the grid spacing, the element size, $H \to 0$. The theorems contribute support to our conjecture that the particular nonlocal coupling conditions (15.2) are great for a wide range of PDEs.

Prelude Let's start with a simplified version of the subsequent lemma and theorem. The key to all three is the coupling conditions (15.2), which with the definition of amplitudes (15.3) reduce to

$$u_j(X_{j\pm1}, t) = \gamma U_{j\pm1} + (1-\gamma)U_j\,. \tag{15.7}$$

Using operator δ_x to denote differences in space x and reserving operator δ for differences in index j, the coupling conditions (15.7) imply that

$$\begin{aligned}
\delta_x^2 u_j(X_j,t) &= u_j(X_j+H,t) - 2u_j(X_j,t) + u_j(X_j-H,t) && \text{(definition of } \delta_x\text{)}\\
&= u_j(X_{j+1},t) - 2u_j(X_j,t) + u_j(X_{j-1},t) && \text{(grid spacing)}\\
&= \gamma U_{j+1} + (1-\gamma)U_j - 2U_j + \gamma U_{j-1} + (1-\gamma)U_j && \text{from (15.7)}\\
&= \gamma(U_{j+1} - 2U_j + U_{j-1})\\
&= \gamma\delta^2 U_j\,. && (15.8)
\end{aligned}$$

That is, the coupling conditions transform spatial differences of the subgrid fields u_j into differences of the grid values U_j, ameliorated by the coupling parameter γ.

Now consider what the transformation of difference (15.8) means for the coupled fields. Recall that classic finite difference arguments[179] approximate the second derivative $\partial_x^2 u_j \approx (\delta_x^2 - \frac{1}{12}\delta_x^4)u_j/H^2$. As a preliminary example, instead of the diffusion PDE, suppose we were solving the PDE $\partial_t u_j = (\delta_x^2 - \frac{1}{12}\delta_x^4)u_j/H^2$ with the coupling conditions (15.7). Multiply by H^2 to write $H^2\partial_t u_j = (\delta_x^2 - \frac{1}{12}\delta_x^4)u_j$. In this equation the dominant balance is $H^2\partial_t u_j \approx \delta_x^2 u_j$, so we can replace $\delta_x^4 u_j \approx (H^2\partial_t)^2 u_j = H^4\partial_t^2 u_j$. Move this term to the left-hand side and consider the equivalent PDE

$$\left(H^2\partial_t + \tfrac{1}{12}H^4\partial_t^2\right)u_j \approx \delta_x^2 u_j.$$

Now evaluate this PDE at the grid point $x = X_j$: on the left-hand side the time derivatives commute with evaluation at X_j; on the right-hand side the transformation of differences (15.8) applies; thus the PDE and coupling conditions ensure the grid value equation

$$\left(H^2\partial_t + \tfrac{1}{12}H^4\partial_t^2\right)U_j \approx \gamma\delta^2 U_j\,.$$

Now reverse some of the above: here the dominant balance is $H^2\partial_t U_j \approx \gamma\delta^2 U_j$ and so the second time derivative on the left-hand side is $(H^4\partial_t^2)U_j \approx (\gamma\delta^2)^2 U_j = \gamma^2\delta^4 U_j$. Moving this term to the right-hand side, and dividing by H^2, we derive that the grid values must satisfy the evolution equation

$$\partial_t U_j \approx \frac{1}{H^2}\left[\gamma\delta^2 U_j - \tfrac{1}{12}\gamma^2\delta^4 U_j\right].$$

This matches the slow manifold model (15.5) and classic formulas. It shows that this desirable connection is due to the particular form of the magical nonlocal coupling conditions (15.2), via their form (15.8). We now proceed to prove that this connection applies to every order of analysis.

Lemma 15.1. *The slow manifold discretization of the diffusion* PDE (15.1) *upon elements with interelement coupling condition* (15.2)*, when truncated to errors* $\mathcal{O}(\gamma^p)$ *and evaluated at* $\gamma = 1$*, is consistent with the diffusion* PDE (15.1) *to* $\mathcal{O}(H^{2p-2})$ *as the grid size* $H \to 0$.

Proof. Consider the diffusion PDE (15.1). A classic operator relationship is the spatial derivative $\partial_x = \frac{2}{H}\sinh^{-1}\frac{1}{2}\delta_x$ (see Table 15.1). That is, in some sense $\partial_x \approx \frac{1}{H}[\delta_x -$

[179]For example, the second difference $\delta_x^2 u(x) = u(x+H) - 2u(x) + u(x-H)$. Expand the right-hand side by the Taylor series $u(x \pm H) = u(x) \pm hu'(x) + \frac{1}{2}H^2u''(x) \pm \frac{1}{6}H^3u'''(x) + \frac{1}{24}H^4u^{iv}(x) + \mathcal{O}(H^5)$. The second difference then is expressed as $\delta_x^2 u = H^2u'' + \frac{1}{12}H^4u^{iv} + \mathcal{O}(H^5)$. In this last equation we recognize the dominant term $\delta_x^2 u \approx H^2u''$, so we replace the fourth derivative, $H^4u^{iv} \approx \delta_x^4$, and move it to the left-hand side to derive $\delta_x^2 u - \frac{1}{12}\delta_x^4 u \approx H^2u''$, the second derivative.

$\frac{1}{24}\delta_x^3 + \frac{3}{640}\delta_x^5 - \cdots]$. Hence the diffusion PDE (15.1) on the jth element is equivalent to

$$\partial_t u_j = \partial_x^2 u_j = \left[\frac{2}{H}\sinh^{-1}\tfrac{1}{2}\delta_x\right]^2 u_j \qquad \left(\partial_t u_j \approx \frac{1}{H^2}\left[\delta_x^2 - \tfrac{1}{12}\delta_x^4 + \cdots\right]u_j\right).$$

Now invoke a beautiful trick: inverting the operator function on the right-hand side derives that the diffusion PDE is equivalent to

$$\left[2\sinh\tfrac{H}{2}\sqrt{\partial_t}\right]^2 u_j = \delta_x^2 u_j \qquad \left(\left[H^2\partial_t + \tfrac{1}{12}H^4\partial_t^2 + \cdots\right]u_j \approx \delta_x^2 u_j\right).$$

Evaluate this relationship at the grid point $x = X_j$: the left-hand side involves time derivatives, which commute with the evaluation at the grid point and hence reduce to those of the grid value $U_j(t)$; the coupling conditions, in the form (15.8), replace the right-hand side. Thus,

$$\left[2\sinh\tfrac{H}{2}\sqrt{d/dt}\right]^2 U_j = \gamma\delta^2 U_j \qquad \left(\left[H^2\frac{d}{dt} + \tfrac{1}{12}H^4\frac{d^2}{dt^2} + \cdots\right]U_j \approx \gamma\delta_x^2 U_j\right).$$

Now revert the operator relationship to deduce that the slow manifold evolution of the grid values is

$$\dot{U}_j = \frac{dU_j}{dt} = \left[\frac{2}{H}\sinh^{-1}\tfrac{1}{2}\sqrt{\gamma}\delta\right]^2 U_j \qquad \left(\frac{dU_j}{dt} \approx \frac{1}{H^2}\left[\gamma\delta^2 - \tfrac{1}{12}\gamma^2\delta^4 + \cdots\right]U_j\right).$$

Evaluating at coupling parameter $\gamma = 1$, this right-hand side is precisely the correct centered difference operator formula for the diffusion PDE (15.1). However, generally we truncate at some power in coupling γ, as in (15.5): when truncating to errors $\mathcal{O}(\gamma^p)$ the right-hand side contains the correct terms up to, but not including, centered differences of $\mathcal{O}(\delta^{2p})$. Applied to a smooth solution, the difference $\delta^{2p} = \mathcal{O}(H^{2p})$ for grid spacing H. Then noting the division by H^2 in the above-displayed equation, the resulting error is $\mathcal{O}(H^{2p-2})$. That is, the discrete, slow manifold model is consistent with the diffusion PDE (15.1) to this order of error. □

This is a slick proof. Its power is that it is not constructive; nowhere in the proof did we need to construct any of the subgrid fields. Thus we readily generalize the proof to the following more general theorem. The following theorem (Roberts et al., 2013, Thm. 4) is independent of whether one seeks a slow manifold model or not: it addresses useful general consistency that follows from the specific coupling conditions (15.2).

Theorem 15.2 (isotropic consistency). *Let $\mathcal{L} = \ell(\delta_x^2) = \sum_{k=0}^{\infty}\ell_k\delta_x^{2k}$ be some linear, constant-coefficient isotropic operator with coefficient $\ell_1 \neq 0$. Then the discretization of $u_t = \mathcal{L}u$ upon elements with interelement coupling condition* (15.2), *when truncated to errors $\mathcal{O}(\gamma^p)$ and evaluated at $\gamma = 1$, is consistent with $u_t = \mathcal{L}u$ to errors $\mathcal{O}(\ell_p H^{2p})$.*

For example, Lemma 15.1 is the special case when $\mathcal{L} = \partial_x^2 = \left[\frac{2}{H}\sinh^{-1}\frac{1}{2}\delta_x\right]^2$.

Proof. Recall that the coupling condition (15.2) empowers us to replace differences of the subgrid field by differences of the element amplitudes, albeit moderated by the coupling parameter γ, as in equation (15.8). Consider the dynamic equation $u_t = \mathcal{L}u$. On the jth element it is formally equivalent to

$$\partial_t u_j = \mathcal{L}u_j = \ell_0 u_j + \ell'(\delta_x^2)u_j, \quad \text{where} \quad \ell'(\delta_x^2) = \sum_{k=1}^{\infty}\ell_k\delta_x^{2k}.$$

Put $\ell_0 u_j$ on the left-hand side and invert the operator ℓ' on the right-hand side (the inverse formally exists as $\ell_1 \neq 0$) to derive that the equation is equivalent to

$$\ell'^{-1}(\partial_t - \ell_0)u_j = \delta_x^2 u_j\,.$$

Evaluate this relationship at the grid point $x = X_j$: the left-hand side involves time derivatives, which commute with evaluation at $x = X_j$ and so reduce to those of grid value $U_j(t)$; the coupling conditions (15.8) replace the right-hand side. Thus

$$\ell'^{-1}(\partial_t - \ell_0)U_j = \gamma\delta^2 U_j\,.$$

Now revert the operator relationship to deduce that the evolution of the grid values is

$$\dot{U}_j = \frac{dU_j}{dt} = \ell_0 U_j + \ell'(\gamma\delta^2)U_j = \ell(\gamma\delta^2)U_j.$$

Evaluating at coupling parameter $\gamma = 1$, this right-hand side is precisely the correct centered difference operator formula for the equation $u_t = \mathcal{L}u$. So when truncating to errors $\mathcal{O}(\gamma^p)$, the right-hand side contains the correct terms up to errors $\mathcal{O}(\ell_p\delta^{2p})$. For smooth fields u this error is $\mathcal{O}(\ell_p H^{2p})$. That is, the discrete model is consistent with the equation $u_t = \mathcal{L}u$ to this order of error. □

Consequently, when we use the particular coupling conditions (15.8) to discretize PDEs we obtain dual support: consistency and center manifold theory, as illustrated in Figure 15.1. Example 15.2, below, shows how this theorem reaffirms the usefulness of the coupling conditions (15.2) in the coarse-scale discrete modeling of fine-scale lattice dynamics, not just partial differential equations.

15.1.4 ▪ Computer algebra constructs the slow manifold discretization

Algorithm 15.1 constructs the slow manifold discretisation for the diffusion PDE (15.1) on overlapping elements coupled by the conditions (15.2). The iterative corrections are driven by the residuals of the governing equations. Upon termination, the output of the algorithm, `uj` and `gj`, gives the slow manifold discretization (15.4)–(15.5).

- The first line of Algorithm 15.1 just improves the format of printed output.
- Algorithm 15.1 defines the subgrid scale variable `xi` $= \xi := (x - X_j)/H$ and its relevant property that $d\xi/dx = 1/H$.
- To find updates to the subgrid field on the slow manifold, Algorithm 15.1 needs to solve $d^2\hat{u}_j/d\xi^2 =$ RHS for corrections $\hat{u}_j$. Operator `linv` solves this when the RHS is a polynomial in ξ. It invokes the two conditions $\hat{u}_j(0,t) = 0$, from the amplitude definition, and $\hat{u}_j$ at $\xi = \pm 1$ are equal to partially account for the coupling conditions.
- To find updates to the evolution on the slow manifold, Algorithm 15.1 invokes the operator `mean`, which computes the mean over the jth element. Using the mean to determine the update to the evolution is not correct, but it is correct enough.[180] Section 15.2 justifies a better procedure.

[180] Yet again, one benefit of this iterative computer algebra approach is that one can adopt the quick and dirty approach of just try it and see. It is only the end that is important: if the residuals are zero to the specified error, then the Flexible Approximation Theorem 4.10 guarantees that the slow manifold model is correct to the specified error. The details of how the algorithm reaches this end point of zero residuals only reflects efficiency: the end justifies the means.

Algorithm 15.1 *Construct the slow manifold discretization of diffusion* PDE (15.1).

```
on div; off allfac; on revpri; factor gamma,hh;
% function of subgrid variable xi=(x-x_j)/hh
depend xi,x;  let df(xi,x)=>1/hh;
% solves u''=RHS such that u(0)=0 and u(+1)=u(-1)
operator linv; linear linv;
let { linv(xi^~~p,xi)=>(xi^(p+2)-(1-(-1)^p)*xi/2)/(p+1)/(p+
    , linv(1,xi)=>(xi^2)/2 };
% compute mean over an element
operator mean; linear mean;
let { mean(xi^~~p,xi)=>(1+(-1)^p)/2/(p+1)
    , mean(1,xi)=>1 };
% parametrize with evolving grid values uu(j)
operator uu;  depend uu,t;
let df(uu(~k),t)=>sub(j=k,gj);
% initial approximation in jth element
uj:=uu(j);  gj:=0;
% iterative refinement to specified error
let gamma^4=>0;
for iter:=1:99 do begin
   pde:=-df(uj,t)+df(uj,x,2);
   rcc:=-sub(xi=+1,uj)+gamma*sub({xi=0,j=j+1},uj)
       +(1-gamma)*sub(xi=0,uj);
   lcc:=-sub(xi=-1,uj)+gamma*sub({xi=0,j=j-1},uj)
       +(1-gamma)*sub(xi=0,uj);
   gj:=gj+(gd:=(rcc+lcc)/hh^2-mean(pde,xi));
   uj:=uj-hh^2*linv(pde-gd,xi)+xi*(rcc-lcc)/2;
   if {pde,rcc,lcc}={0,0,0} then iter:=1000000;
end;
amp:=sub(xi=0,uj)-uu(j);
```

- The slow manifold is parametrized by grid values `uu(j)` $= U_j(t)$ and knows that their time derivative is stored in variable `gj`: `duu(j)/dt` $= dU_j/dt =$ `gj`.
- The initial approximation to the subgrid field is that it is piecewise constant, `uj` $= u_j(x,t) = U_j$, and there is no evolution, $dU_j/dt =$ `gj` $=$ `0`.
- In the iterative refinement, iterate until the residuals of the diffusion PDE (15.1) and the coupling conditions (15.2) are zero to the specified order of error, here $\mathcal{O}(\gamma^4)$:
  - for the current approximation to the slow manifold `uj` and the evolution thereon `gj`, compute the residuals of the diffusion PDE and the coupling conditions;
  - then update the evolution via a magic recipe involving the residuals of the diffusion PDE and coupling conditions—the simplest way to find the coefficients in these formulas for the updates is to use the method of undetermined coefficients (section 1.3);
  - update the subgrid field from the residuals.

Table 15.1. *Interrelations between operators (National Physical Lab,* 1961, *p.* 65, *e.g.): shift operator* $\mathcal{E}u(x) = u(x+H)$ *or* $\mathcal{E}U_j = U_{j+1}$, *centered difference* δ, *spatial derivative* $\partial u(x) = \partial u/\partial x$, *forward difference* Δ, *backward difference* ∇, *and centered mean* μ.

	$\mathcal{E}$	δ	$H\partial$	Δ	∇
$\mathcal{E} =$	$\mathcal{E}$	$1+\frac{1}{2}\delta^2+\delta\sqrt{1+\frac{1}{4}\delta^2}$	$e^{H\partial}$	$1+\Delta$	$(1-\nabla)^{-1}$
$\delta =$	$\mathcal{E}^{\frac{1}{2}}-\mathcal{E}^{-\frac{1}{2}}$	δ	$2\sinh\frac{1}{2}H\partial$	$\Delta/\sqrt{1+\Delta}$	$\nabla/\sqrt{1-\nabla}$
$H\partial =$	$\log\mathcal{E}$	$2\sinh^{-1}\frac{1}{2}\delta$	$H\partial$	$\log(1+\Delta)$	$-\log(1-\Delta)$
$\Delta =$	$\mathcal{E}-1$	$\delta\sqrt{1+\frac{1}{4}\delta^2}+\frac{1}{2}\delta^2$	$e^{H\partial}-1$	Δ	$\nabla/(1-\nabla)$
$\nabla =$	$1-\mathcal{E}^{-1}$	$\delta\sqrt{1+\frac{1}{4}\delta^2}-\frac{1}{2}\delta^2$	$1-e^{-H\partial}$	$\Delta/(1+\Delta)$	∇
$\mu = \frac{1}{2}(\mathcal{E}^{\frac{1}{2}}+\mathcal{E}^{-\frac{1}{2}})$		$\sqrt{1+\frac{1}{4}\delta^2}$	$\cosh\frac{1}{2}H\partial$	$\frac{1+\frac{1}{2}\Delta}{\sqrt{1+\Delta}}$	$\frac{1-\frac{1}{2}\nabla}{\sqrt{1-\nabla}}$

Algorithm 15.2 *Convert the slow manifold of* linear *problems to centered operator form, where* μ *is the centered mean operator and* δ *is the centered difference operator: invokes the result, implicit in Table* 15.1, *that the shift operator* $\mathcal{E}^{\pm 1} = 1 \pm \mu\delta + \frac{1}{2}\delta^2$. *Second, also convert the evolution to an equivalent differential equation (also applies to nonlinear discretizations) using, again from Table* 15.1, *that the shift operator* $\mathcal{E} = \exp(H\partial_x)$, *expanded in its Taylor series in* $H\partial_x$.

```
% convert to central difference operator form
rules:={ mu^2=>1+delta^2/4, uu(j)=>1
    , uu(j+~p)=>(1+sign(p)*mu*delta+delta^2/2)^abs(p)}$
gop:=(gj where rules);
uop:=(uj where rules);

% convert to equivalent DE using Taylor expansion
let hh^9=>0;
depend uu,x;
rules:={uu(j)=>uu, uu(j+~p)=>uu+(for n:=1:9 sum
                df(uu,x,n)*(hh*p)^n/factorial(n)) }$
gde:=(gj where rules);
```

- Last, Algorithm 15.1 just confirms the required parametrization that `uu(j)` is indeed the grid value: `uu(j)` $= U_j = u_j(X_j, t)$.

The output of Algorithm 15.1 is not in the form recorded in the slow manifold model (15.4)–(15.5): these equations express the model using centered discrete operators. The postprocessing Algorithm 15.2 transforms the output into the form reported. It replaces all the shift operations by the equivalent discrete operator form

$$U_{j+p} = \mathcal{E}^p U_j = \left[1 + \operatorname{sign}(p)\mu\delta + \tfrac{1}{2}\delta^2\right]^{|p|} U_j$$

and invokes the difference operator identity $\mu^2 = 1 + \frac{1}{4}\delta^2$. For reference, Table 15.1 summarizes the operator identities used here, and other useful identities besides.

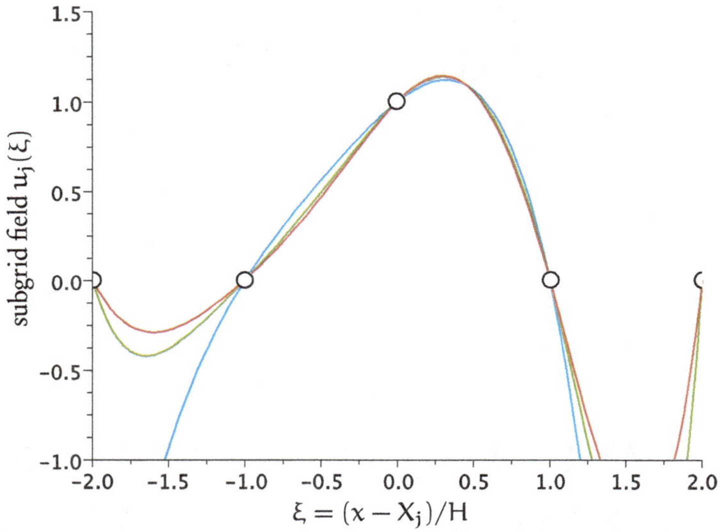

Figure 15.5. *The slow manifold* (15.4) *of the advection-diffusion PDE of Example* 15.1 *is* not *classic polynomial interpolation. These are the subgrid fields* $u_j(\xi)$ *for the case of significant advection* $cH = 2$ *and when grid value* $U_j = 1$ *and all other grid values are zero (but only* $-1 < \xi < 1$ *is relevant):* $\mathcal{O}(\gamma^2, c^3)$ *error, blue quartic;* $\mathcal{O}(\gamma^3, c^3)$ *error, green sixth order; and* $\mathcal{O}(\gamma^4, c^3)$ *error, red eighth order.*

The postprocessing Algorithm 15.2 also generates the equivalent differential equation (15.6) for sufficiently smooth "grid value fields."[181] Table 15.1 identifies $\mathcal{E} = e^{H\partial}$, that is $\mathcal{E}^p = e^{pH\partial}$, so using the Taylor series for the exponential, we have

$$U_{j+p} = U(X_j + pH, t) = \sum_{n=0}^{\infty} \frac{1}{n!}(pH)^n \frac{\partial^n U}{\partial x^n} .$$

Example 15.1. (small but significant advection). Instead of the diffusion PDE (15.1), let's discretize the linear advection-diffusion PDE

$$\frac{\partial u}{\partial t} = -c\frac{\partial u}{\partial x} + \frac{\partial^2 u}{\partial x^2} .$$

Section 15.3 explores the fascinating case of finite advection speed c. Here we suppose that the advection speed c is small enough so that $c\partial u/\partial x$ may be treated as a perturbation. Even here a surprise awaits.

We only need to make two modifications to Algorithm 15.1. First, treat advection speed c as small by truncating in powers of c: include `let c^3=>0;` before the for-loop in order to obtain a model correct to errors $\mathcal{O}(c^3)$. Second, modify the computation of the residual of the PDE to the case of the advection-diffusion

[181] It may seem ludicrous to regard U as a differentiable function of x when it is only defined at discrete grid points X_j. There are two ways to proceed. First, define field $U(x)$ to be a smooth reasonable interpolate of the known grid values U_j; then we can differentiate. Second, and perhaps better from a theoretical view, construct a smooth ensemble of grid values U_j on an ensemble of grids X_j that are all spatial shifts of each other for all real shifts; then define the field $U(x)$ to be U_j for the member of the ensemble that has $x = X_j$.

PDE, namely to `pde:=-df(uj,t)-c*df(uj,x)+df(uj,x,x);`. Execute the code to find the slow manifold discretization. Figure 15.5 shows some examples of the slow manifold subgrid field demonstrating that in this approach the subgrid interpolation adapts to the dynamics of the PDE rather than being imposed on the system (as in the assumptions of other methodologies). The algorithm also finds that the evolution on the slow manifold, namely the discrete model, is

$$\dot{U}_j = \gamma\frac{1}{H^2}\delta^2 U_j - \gamma\frac{c}{H}\mu\delta U_j + \gamma\frac{c^2}{12}\delta^2 U_j + \mathcal{O}(\gamma^2, c^3).$$

Evaluated at physical coupling $\gamma = 1$, the first two terms on the right-hand side are standard finite difference approximations to diffusion, δ^2/H^2, and advection, $c\mu\delta/H$. The third term on the right-hand side is surprising: it increases dissipation for finite advection c; such increased dissipation helps maintain stability for finite advection speed and helps avoid unphysical zig-zags in solutions. Section 15.3 explores more.

Such a desirable term arises because even though the PDE is linear, the process of accurate modeling out-of-equilibrium dynamics is nonlinear (through the chain rule that $\partial u/\partial t = \sum_k \partial u_j/\partial U_k \cdot dU_k/dt$) and generates such c^2 terms that are nonlinear in the PDE's coefficients.

Example 15.2. (coarse discrete model of discrete dynamics). Instead of the diffusion PDE (15.1), suppose that the subgrid scale dynamics is itself on a fine-scale grid (a microscale lattice):

$$\dot{u}_i = \nu(u_{i-1} - 2u_i + u_{i+1}), \tag{15.9}$$

where, duplicating notation slightly, u_i is the value of the field at the ith fine-scale grid point x_i. Suppose there are some integer n fine-scale grid points in each coarse-scale element; that is, the fine-scale grid spacing is $h = H/n$. We might then write the linear operator $\mathcal{L}u = \nu(u|_{x-H/n} - 2u|_x + u|_{x+H/n})$. Divide this fine grid into overlapping elements, as in section 15.1. One may check that the spectrum of the discrete dynamics with coupling $\gamma = 0$ ensures an emergent slow manifold exists for nonzero coupling. The Isotropic Consistency Theorem 15.2 also assures us that the coarse-grid model will be, in some sense, consistent with the fine-grid dynamics.

Almost the identical computer algebra to Algorithm 15.1 generates the coarse-grid model for the discrete dynamics of (15.9), and with similar theoretical support. To apply Algorithm 15.1 here, imagine the subgrid variable $\xi \propto i/n$, so changing fine-grid index i by 1 is the same as changing subgrid variable ξ by $1/n$. For simplicity choose the coefficient $\nu = n^2/H^2$ so the fine-scale equation (15.9) is approximately the continuum diffusion PDE (15.1). Then just change the computation of the residual of the governing subgrid scale equation from the continuum diffusion PDE (15.1) to the fine-scale discrete equation (15.9):

```
nu:=n^2/hh^2;
pde:=-df(uj,t)+nu*(sub(xi=xi+1/n,uj)
          -2*uj+sub(xi=xi-1/n,uj));
```

The modified algorithm successfully terminates. Because the original updates are not crafted for this particular subgrid scale operator, the number of iterations is larger. But because the two subgrid scale operators are the same to leading effects, the algorithm does terminate.

In this simple fine-grid problem, the subgrid scale structure, the slow manifold, remains as the classic interpolation (15.4) through the coarse-scale grid points. However, the coarse-scale evolution on the slow manifold appropriately changes to

$$\begin{aligned}\dot{U}_j = \frac{1}{H^2}\Big[&\gamma\delta^2 - \frac{\gamma^2}{12}\left(1-\frac{1}{n^2}\right)\delta^4 \\ &+ \frac{\gamma^3}{360}\left(1-\frac{1}{n^2}\right)\left(4-\frac{1}{n^2}\right)\delta^6\Big]U_j + \mathcal{O}(\gamma^4). \qquad (15.10)\end{aligned}$$

Evaluate at coupling $\gamma = 1$ to recover the physical coarse-grid model. This example demonstrates novel and powerful support for mapping fine-scale lattice dynamics to coarser-scale lattice dynamics.

Potentially one could map to even coarser-scale dynamics, and so on, to generate a hierarchy of models across a wide range of scales (Roberts, 2009c, e.g.)!

Recall Example 0.1 in the preface. It is a combination of the previous two examples because it combines the advection of Example 15.1 with the fine-scale lattice of Example 15.2. The preface discussed how this combination mapped advection-diffusion on a fine lattice to an equivalent advection-diffusion on a coarser lattice. Such mapping can be continued indefinitely to create a multiscale hierarchy of models (Roberts, 2009c, e.g.). Importantly, the nonlinear nature of modeling from one scale to another, as seen in c^2 terms, empowers the whole to be more than the sum of the parts.

Example 15.3 (gap-tooth dynamics are consistent). You might regard the overlapping elements as a wanton extravagance. Algebraically the overlap is immaterial. However, when the microscale dynamics are only known numerically, one can more efficiently resolve just small patches of space, sometimes called the "teeth" (Gear et al., 2003; Samaey et al., 2005, e.g.). Instead of overlapping elements, consider elements I_j that only cover part of space: $|x - X_j| < rH$ for some fraction r that we aim to be small for computational efficiency. Correspondingly, there is a gap between adjacent elements of size $(1-2r)H$. Suppose we solve the system equation $\partial u/\partial t = \mathcal{L}u$ for some isotropic linear operator $\mathcal{L}$ but only solve it on the teeth I_j.

Following research that established center manifold support for the perturbed diffusion equation, Roberts and Kevrekidis (2007) recommend applying Dirichlet-like interelement coupling conditions

$$\begin{aligned}u_j(X_j \pm rH, t) = \big[&1 \pm r\mu\delta + \tfrac{1}{2}r^2\delta^2 \\ &+ \tfrac{1}{6}r(r^2-1)\mu\delta^3 + \tfrac{1}{24}r^2(r^2-1)\delta^4 + \cdots\big]U_j.\end{aligned}$$

Introducing a coupling strength γ, in terms of operators this coupling condition is equivalent to the $\gamma = 1$ case of[182]

$$\mathcal{E}_x^r u_j = \left[1 + \gamma\left(\pm\mu\delta + \tfrac{1}{2}\delta^2\right)\right]^r u_j = \left[1 + \gamma(\mathcal{E}^{\pm 1} - 1)\right]^r u_j \quad \text{at } x = X_j. \quad (15.11)$$

[182]At times one needs to be careful: the operators on the right-hand side apply to only the index j if applied to $u_j(X_j, t)$, but apply to both index j and space x if applied to $u_j(x, t)$, which is then subsequently evaluated at $x = X_j$.

Previous research proved center manifold support for the coarse-grid modeling of the dynamics on such teeth coupled across the gaps with (15.11) (Roberts and Kevrekidis, 2007).

Now let us prove that the resultant models have classic consistency. The isotropic system equation is

$$\partial_t u_j = \mathcal{L} u_j = \ell_0 u_j + \ell'(\eth^2) u_j \quad \text{on } I_j$$

for some operator function ℓ' of the tooth-sized centered difference operator $\eth := \mathcal{E}_x^{r/2} - \mathcal{E}_x^{-r/2}$, so that $\eth^2 = \mathcal{E}_x^r - 2 + \mathcal{E}_x^{-r}$. Inverting the operators in the above equation leads to

$$\ell'^{-1}(\partial_t - \ell_0) u_j = \eth^2 u_j = (\mathcal{E}_x^r - 2 + \mathcal{E}_x^{-r}) u_j\,.$$

Evaluate at the grid point $x = X_j$ and use the coupling condition (15.11) to deduce

$$\ell'^{-1}(\partial_t - \ell_0) U_j = \left\{ [1 + \gamma(\mathcal{E} - 1)]^r - 2 + \left[1 + \gamma(\mathcal{E}^{-1} - 1)\right]^r \right\} U_j\,.$$

In constructing a slow manifold discrete model we truncate the analysis with errors $\mathcal{O}(\gamma^p)$. Thus our model is the above to a difference $\mathcal{O}(\gamma^p)$. By symmetry such error terms are $\mathcal{O}(\gamma^p \delta^{2p})$. Thus the slow manifold discrete model is

$$\begin{aligned}\ell'^{-1}(\partial_t - \ell_0) U_j = {} & \left\{ [1 + \gamma(\mathcal{E} - 1)]^r - 2 + \left[1 + \gamma(\mathcal{E}^{-1} - 1)\right]^r \right\} U_j \\ & + \mathcal{O}(\gamma^p \delta^{2p} U_j).\end{aligned}$$

Now restore full coupling by setting $\gamma = 1$ to obtain

$$\ell'^{-1}(\partial_t - \ell_0) U_j = \{\mathcal{E}^r - 2 + \mathcal{E}^r\} U_j + \mathcal{O}(\delta^{2p}) = \eth^2 U_j + \mathcal{O}(\delta^{2p} U_j).$$

Inverting the operators, we expect the error to maintain its order:

$$\partial_t U_j = [\ell_0 + \ell'(\eth^2)] U_j + \mathcal{O}(\delta^{2p}) = \mathcal{L} U_j + \mathcal{O}(\delta^{2p} U_j).$$

Hence the slow manifold model on the teeth is consistent with the original dynamics when coupled across the gaps by (15.11). Potentially there is a great saving in computation when the system is only known numerically.

15.2 ▪ Discretize nonlinear PDEs

Burgers' equation introduces the modifications Burgers' PDE for a field $u(x, t)$,

$$\frac{\partial u}{\partial t} = -\alpha u \frac{\partial u}{\partial x} + \frac{\partial^2 u}{\partial x^2}, \tag{15.12}$$

contains the physically important mechanisms of nonlinear advection, $u u_x$, and diffusive dissipation, u_{xx}. It is thus widely used to introduce and test many numerical and analytical methods (Roberts, 2001, e.g.). The nonlinearity parameter α controls the importance of the nonlinear advection term (Exercises 14.4 and 14.5 show how this parameter helps control approximations). This section shows how to derive holistic discretizations of

Algorithm 15.3 *Derive the slow manifold "holistic" discretization for Burgers' PDE* (15.12) *on finite elements. Simply make the following modifications to Algorithm* 15.1*: improve printing, change the order of error to also truncate in small nonlinearity parameter* α*, and change the computation of the residual to Burgers' PDE* (15.12).

```
factor alpha;
...
let {gamma^4=>0, alpha^3=>0};;
...
pde:=-df(uj,t)-alpha*uj*df(uj,x)+df(uj,x,2);
```

Burgers' PDE. The methods are easily adaptable to discretize other reaction-advection-diffusion PDEs in one space dimension. Straightforward extensions apply to PDEs in multiple space dimensions (MacKenzie, 2005; Roberts et al., 2013).

Algorithm 15.3 gives the two required changes to holistically discretize Burgers' PDE (15.12). We have the choice of either recognizing that the advection is nonlinear or treating as small the coefficient of the advection term. The latter is the easier to implement: in the Reduce code we just add a command to work to errors $\mathcal{O}(\alpha^3)$, for example. To derive discretizations for Burgers' PDE we then just code the nonlinear advection term into the residual. Execute the code to find the slow manifold model.

Our modeling is consistent to high order Let's look briefly at the discrete model, say use Algorithm 15.3 to generate the slow manifold model to errors $\mathcal{O}(\gamma^4, \alpha^3)$. The resultant expression `gj` for dU_j/dt contains 99 terms.[183] With so many terms the model is hard to interpret, although it would be routine to output the code to a file for subsequent input to a numerical simulation. But let's look at the equivalent PDE, stored in `gde` by the postprocessing Algorithm 15.2, considering increasing powers of the grid spacing H.

- Execute `coeffn(gde,hh,0);` to find that the equivalent PDE to the discrete model contains
$$\frac{\partial U}{\partial t} = \gamma\left[-\alpha U\frac{\partial U}{\partial x} + \frac{\partial^2 U}{\partial x^2}\right] + \cdots .$$
These form Burgers' PDE (15.12) in the limit of grid spacing $H \to 0$ when evaluated at the physical coupling $\gamma = 1$. The slow manifold modeling is consistent.

- Odd powers of H do not appear in the equivalent PDE; for example, verify by executing `coeffn(gde,hh,1);`.

- Execute `coeffn(gde,hh,2);` to find that the leading order of error in the equivalent PDE is the $\mathcal{O}(H^2)$ terms
$$\tfrac{1}{12}H^2\gamma(1-\gamma)\left[\frac{\partial^4 U}{\partial x^4} - 2\alpha U\frac{\partial^3 U}{\partial x^3} + \alpha^2 U^2\frac{\partial^2 U}{\partial x^2}\right].$$
If we constructed the model to errors $\mathcal{O}(\gamma^2)$, then the factor $\gamma(1-\gamma) = \gamma + \mathcal{O}(\gamma^2)$ would appear only as γ, so this displayed term would appear as the leading error when parameter $\gamma = 1$. These indicate that the errors in the slow manifold model would be $\mathcal{O}(H^2)$. However, if we construct the model to errors $\mathcal{O}(\gamma^3)$, then

[183] Incidentally, the corresponding expression for the subgrid field u_j contains 580 terms. Execute `length(uj);` to count the terms.

upon evaluation at the physical coupling of $\gamma = 1$ these errors vanish, due to the factor $(1-\gamma)$, indicating a higher-order consistency as grid spacing $H \to 0$.

- Execute `coeffn(gde,hh,4);` to find that the $\mathcal{O}(H^4)$ errors in the equivalent PDE are

$$\tfrac{1}{720}H^4\gamma(1-\gamma)\left[(2-8\gamma)\frac{\partial^6 u}{\partial x^6} - \cdots - 10\alpha^2\gamma u^2\left(\frac{\partial u}{\partial x}\right)^2\frac{\partial^2 u}{\partial x^2}\right].$$

  If we construct the model to errors $\mathcal{O}(\gamma^3)$, then these indicate that the errors in the slow manifold model are $\mathcal{O}(H^4)$. However, if we construct the model to errors $\mathcal{O}(\gamma^4)$, then upon evaluation at the physical coupling of $\gamma = 1$ these errors all vanish, again due to the factor $(1-\gamma)$, indicating a higher-order consistency as grid spacing $H \to 0$.

- And so on, although as yet we have no proof of indefinitely high-order consistency for nonlinear PDEs—only consistency to errors $\mathcal{O}(H^2)$ has yet been proved for nonlinear PDEs (Roberts et al., 2013, Lemma 5).

Our modeling is invariant to algebraic rewriting Burgers' PDE (15.12) contains the nonlinear advection term $u\partial u/\partial x$. In traditional numerical discretizations, a controversy over many decades was whether one should discretize this term directly, as $U_j(U_{j+1} - U_{j-1})/2H$, or whether one should discretize in conservative form, namely, recognizing $u\partial u/\partial x = \partial(\frac{1}{2}u^2)/\partial x$, one would use $(U_{j+1}^2 - U_{j-1}^2)/4H$. Other people recommended a weighted mix of the two (Fornberg, 1973; Foias et al., 1991, e.g.). Which is correct? Or is something else better? Such controversy bedevils traditional discretizations. Let's use our code to confirm that in our methodology, any valid rewriting of such nonlinear terms generates the same model. For example, change the computation of the residual to invoke the general linear combination of the two nonlinear representations, parametrized by β,

$$\frac{\partial u}{\partial t} = -\alpha\left[\beta u\frac{\partial u}{\partial x} + (1-\beta)\frac{\partial}{\partial x}(\tfrac{1}{2}u^2)\right] + \frac{\partial^2 u}{\partial x^2}.$$

Execute the corresponding version of Algorithm 15.1. Observe in the results that there is no dependence whatsoever upon the convex combination parameter β; for example, compute $\partial u_j/\partial\beta$ to see that it is zero. The reason for this independence is simple. Center manifold methodology creates a model from the solutions of the governing equations, not their algebraic form. Thus the constructed center manifold models are invariant to valid algebraic rewriting of the equations. Such invariance is desirable.

15.2.1 ▪ Improve the solvability integral

Recall that Algorithm 15.1 uses the mean operator over the jth element to update the evolution $\hat{g}_j$. The mean operator is not correct, but nonetheless it is correct enough to work. This section derives the best correct update rule in that it reduces the number of iterations in the computer algebra. Nonetheless, the eventual slow manifold model is identical irrespective of which update rule is used.

Interestingly, this fully correct solvability condition appears to link to the simplest basis functions used in the finite element method.

Consider the equations we need to solve for the updates at any iteration in Algorithm 15.1. The linearized PDE and coupling conditions are

$$\frac{\partial^2 \hat{u}_j}{\partial x^2} = \hat{g}_j + \text{Res}_j \quad \text{such that} \quad \hat{u}_j(X_{j\pm1}, t) - \hat{u}_j(X_j, t) = \text{Res}_\pm \,. \tag{15.13}$$

In terms of the subgrid variable $\xi = (x - X_j)/H$ for the jth element, these equations become

$$\frac{1}{H^2}\frac{\partial^2 \hat{u}_j}{\partial \xi^2} = \hat{g}_j + \text{Res}_j \quad \text{such that} \quad \hat{u}_j(\pm1, t) - \hat{u}_j(0, t) = \text{Res}_\pm \,. \tag{15.14}$$

Integrate this PDE for $\hat{u}_j$ over an "extended" jth element with weight function $w(\xi) = 1 - |\xi|$:

$$\int_{-1}^{1} (1 - |\xi|)\frac{1}{H^2}\frac{\partial^2 \hat{u}_j}{\partial \xi^2}\, d\xi = \int_{-1}^{1} (1 - |\xi|)\left[\hat{g}_j + \text{Res}_j\right] d\xi \,.$$

Integrating the left-hand side by parts twice, over each half of the element, and also simplifying the right-hand side slightly, gives

$$\frac{1}{H^2}\left\{[(1+\xi)u_\xi - u]_{-1}^{0} + [(1-\xi)u_\xi + u]_0^1\right\} = \hat{g}_j + \int_{-1}^{1} (1 - |\xi|)\,\text{Res}_j\; d\xi \,.$$

Since u_ξ, equivalently Hu_x, is assumed continuous (implicitly) at the center of the element $\xi = 0$, the left-hand side of this equation becomes

$$\begin{aligned} \text{LHS} &= \frac{1}{H^2}\left\{\hat{u}_j|_{\xi=-1} - 2\hat{u}_j|_{\xi=0} + \hat{u}_j|_{\xi=+1}\right\} \\ &= \frac{1}{H^2}\{\text{Res}_+ + \text{Res}_-\}, \end{aligned}$$

from the nonlocal coupling conditions. Consequently, rearranging the whole equation gives

$$\hat{g}_j = \frac{1}{H^2}(\text{Res}_+ + \text{Res}_-) - \int_{-1}^{1} (1 - |\xi|)\,\text{Res}_j\; d\xi \,. \tag{15.15}$$

This is the correct solvability condition to update the evolution of the jth grid value (via modifying the operator `mean`). Our earlier crude formula for updating g_j does work, as we discovered, but the iterations converge slower than is necessary because the crude formula only approximates this correct update.

Interestingly, the weight function $w(\xi) = 1 - |\xi|$ is the same weight function used for decades to generate finite element models via Galerkin method projection. Exercise 15.7 asks you to deduce this weight function from the adjoint of the subgrid problem (15.13).

15.2.2 ▪ Theory supports discretizing general PDEs

Center Manifold Theorems 4.3, 4.5, and 4.10 lead to the following general corollary for the discretization of reaction-advection-diffusion PDEs. However, we do need to invoke the more widely applicable, infinite-dimensional theorems of Vanderbauwhede and Iooss (1988) or Haragus and Iooss (2011).[184] The first development from previous sections is

[184] Extension to nonautonomous PDEs with time-dependent boundary conditions would invoke the support of the existence and emergence theorems of Aulbach and Wanner (1996, 2000) and the approximation theorems of Potzsche and Rasmussen (2006).

that here we address general nonlinearity, which, as long as it is smooth enough, makes little change to the application of center manifold theory. A second novel feature is to address several different macroscale domain boundary conditions, not only periodic domains invoked previously, but also Dirichlet and Neumann. The approach is a straightforward synthesis of Chapter 14 with the previous parts of this section.

Corollary 15.3. *There exists an emergent slow manifold discrete model of the PDE*

$$\frac{\partial u}{\partial t} = f_2(u, u_x)\frac{\partial^2 u}{\partial x^2} + \epsilon f_0(u, u_x) \tag{15.16}$$

with the nonlocal coupling conditions (15.2)*, for smooth enough* f_i*, with* $f_2 > 0$*, and with the domain divided into* m *overlapping elements by* (15.2)*, and domain boundary conditions of any of the following:*

- *periodic boundary conditions on* u*,* $u(x + mH, t) = u(x, t)$*; or*
- *Dirichlet boundary conditions of fixed value* $u = a$ *at* $x = X_j$ *for* $j = 0$ *and/or* $j = m + 1$*; or*
- *Neumann boundary conditions of fixed derivative* $\partial u/\partial x = b$ *at* $x = X_{j+1/2}$ *for* $j = 0$ *and/or* $j = m$*; or*
- *potentially other alternatives.*

The m*-dimensional slow manifold discrete model may be written as*

$$u = u_j(x, \mathbf{U}, \gamma, \epsilon) \quad \textit{such that} \quad \dot{U}_j = g_j(\mathbf{U}, \gamma, \epsilon), \tag{15.17}$$

where U_j *measures the amplitude of the field* u *in the* j*th element. The discrete model* (15.17) *of the* PDE (15.16) *may be constructed as an asymptotic expansion in two small parameters: the coupling parameter* γ *and the nonlinearity parameter* ϵ*.*

Proof. The different domain boundary conditions may be incorporated into the center manifold framework in a variety of ways (Roberts, 2003a). Here let's choose the simplest one consistent with the established theoretical base provided by coupling (15.2). Without loss of generality, let's just consider the left end of the domain. Consider the following two cases:

- Dirichlet boundary conditions specify $u = a$ at $x = X_0$, so apply

$$u_1(X_0, t) = \gamma a + (1 - \gamma)u_1(X_1, t).$$

- Neumann boundary conditions specify $\partial u/\partial x = b$ at $x = X_{1/2}$, so apply

$$u_1(X_0, t) = \gamma \left[-Hb + H\frac{\partial u_1}{\partial x}(X_{1/2}, t) + u_1(X_0, t) \right] + (1 - \gamma)u_1(X_1, t).$$

When the coupling parameter $\gamma = 1$, these domain boundary conditions reduce to the desired Dirichlet or Neumann conditions. Conversely, when the coupling parameter $\gamma = 0$, both of these cases reduce to the same isolating case, $u_1(X_0, t) = u_1(X_1, t)$, and similarly at the right end of the domain. So the same equilibria and analogous linearized analysis apply as in section 15.1.

When parameters $\epsilon = \gamma = 0$, there exists an m-dimensional subspace of equilibria of piecewise constant subgrid fields $u_j(x,t) = U_j$ (constant). Linearized about each of these equilibria, and assuming constant grid spacing H for simplicity, the linearized dynamics on each element is $u_t = f_{2j}u_{jxx}$, where constant $f_{2j} = f_2(U_j, 0)$. The spectrum of these linear dynamics is proportional to that of the spectrum that section 15.1 established for the diffusion equation, namely, $\{0, -\pi^2/(f_{2j}H^2), -4\pi^2/(f_{2j}H^2), -9\pi^2/(f_{2j}H^2), \ldots\}$.

Over all of the m elements, the zero eigenvalue has multiplicity m. The center manifold Existence Theorem 4.3 then assures us that for smooth enough f there exists an m-dimensional slow manifold (here together with a couple of extra passive dimensions for the two parameters ϵ and γ), say parametrized by grid values U_j. Further, the Emergence Theorem 4.5 assures us that the slow manifold model is emergent and exponentially attractive, roughly like $\exp[-\pi^2 t/(f_{2,\min}H^2)]$, where $f_{2,\min} = \min_j f_2(U_j, 0) > 0$. Finally, the Flexible Approximation Theorem 4.10 asserts that the slow manifold model may be constructed globally in grid values U_j (provided $f_{2,\min} > 0$) but locally in the small parameters γ and ϵ. □

Iterative construction of the slow manifold discretization for nonlinear PDEs (15.16) proceeds via straightforward adaptations of Algorithm 15.1. As seen in the case of Burgers' PDE (15.12), for periodic domains the main change is simply modifying the computation of the residual. Indeed, for these straightforward cases, my web service currently provides you with the service of deriving such discretizations (Roberts, 2009b). The cases of Dirichlet boundary conditions and/or Neumann boundary conditions require more changes (Roberts, 2003a, e.g.). In particular, a boundary forcing, U_0 or $U'_{1/2}$, that depends upon time may also be naturally incorporated.

Cater for finite nonlinearity For some problems, such as Burgers' PDE (15.12), the nonlinearity vanishes for the entire subspace of piecewise constant fields. Then the nonlinearity parameter ϵ need not be assumed small. As the nonlinearity vanishes in the m-dimensional subspace, then the m eigenvalues of zero remain zero for finite parameter ϵ. By continuity, the negative eigenvalues in the spectrum will remain negative for some finite domain in parameters. Hence the slow manifold exists and is attractive globally in this finite domain of nonlinearity parameter ϵ. For such problems the slow manifold need be local only in the coupling parameter γ. To date there is very little established research on the applications of such modeling. Section 15.3, below, shows some of the beauty of the approach in discrete modeling of the linear advection-diffusion.

15.2.3 ▪ Write function code for computational integration

Ultimately we need to use this modeling to execute computer simulations or other systems analysis tasks. The center manifold approach typically generates complicated algebraic expressions as it resolves subgrid processes. The last thing you want is to have to postprocess such expressions by hand. You need methods to systematically program the transformation of the model into other useful forms. For example, this subsection shows how to transform the modeling of Burgers' PDE (15.12) into a simulation to draw solutions such as that of Figure 15.6. This subsection describes the systematic programmable steps you need to turn the Reduce derivation into function code for later simulation or other analysis.

For definiteness, let's suppose we have done the discrete modeling of Algorithm 15.1 or 15.3 for the heat or Burgers' PDE. Further suppose we want to transform the evolution equations into an Octave/MATLAB/Scilab function script for subsequent simulation.

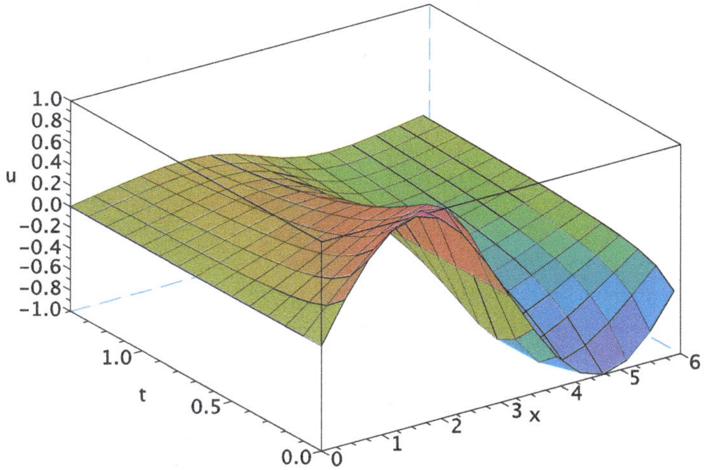

Figure 15.6. *Space-time simulation of Burgers' PDE* (15.12) *with nonlinearity parameter* $\alpha = 6$*: the discrete field* $u(x,t)$ *is plotted vertically, space* x *increases to the upper right, and time increases to the upper left. The initial condition is* $u(x,0) = \sin x$*, seen at the bottom right extreme of the surface. Observe the nonlinear steepening of the sine wave and then the dissipation quickly damping the profile.*

Algorithm 15.4 *Write Reduce expressions to a file for later processing: here write the expression for the time derivatives* $\dot{\mathbf{u}}$*.*

```
gamma:=1;
off nat; out "dudt.red";
write dudt:=gj;
shut "dudt.red"; on nat;
```

For an example, we describe the steps needed to draw solutions of Burgers' PDE (15.12), as shown in Figure 15.6.

Output model to a file The first task is to execute the discrete modeling such as in Algorithm 15.1 or 15.3. We are only interested in the physical model obtained at full coupling $\gamma = 1$, so write the model to a file as in Algorithm 15.4.

For example, suppose we executed Algorithm 15.3 to model Burgers' PDE (15.12) to errors $\mathcal{O}(\gamma^3, \alpha^2)$. Then the code of Algorithm 15.4 generates the file `dudt.red` shown in Dialogue 15.1.

Edit the file to a function Copy the written code of Dialogue 15.1 to a file, say `dburgdt.m`, and do the following edits:

- `:=` to `=` for assignment;
- optional, remove all the blank characters (␣);
- `**` to `.^` to ensure exponentiation is done elementwise;
- `*` to `␣.*` to ensure multiplication is done elementwise—some nonlinear problems also require division to be explicitly elementwise;

Dialogue 15.1 *Our discretization of Burgers' PDE (15.12) generates this computer-readable code when output to a file by Algorithm 15.4. It is not our purpose here, but note that this file, as is, may be later read by Reduce to further analyze the model in computer algebra.*

```
dudt := hh**(-2)*( - 5/2*uu(j) + 4/3*uu(1 + j) - 1/12*uu(2 + j)
 + 4/3*uu( - 1 + j) - 1/12*uu( - 2 + j)) + hh**(-1)*alpha*( - 11
/24*uu(1 + j)*uu(j) - 1/8*uu(1 + j)**2 + 1/24*uu(2 + j)*uu(j) +
1/24*uu(2 + j)*uu(1 + j) + 11/24*uu( - 1 + j)*uu(j) + 1/8*uu( -
1 + j)**2 - 1/24*uu( - 2 + j)*uu(j) - 1/24*uu( - 2 + j)*uu( - 1
+ j))$
```

Dialogue 15.2 *The expression given in Dialogue 15.1 for the time derivatives of the discretization of Burgers' PDE (15.12) needs to be edited to this form for use in Octave/MATLAB/Scilab.*

```
dudt=hh.^(-2) .*(-5/2 .*uu(j)+4/3 .*uu(1+j)-1/12 .*uu(2+j) ...
+4/3 .*uu(-1+j)-1/12 .*uu(-2+j))+hh.^(-1) .*alpha .*(-11 ...
/24 .*uu(1+j) .*uu(j)-1/8 .*uu(1+j).^2+1/24 .*uu(2+j) .*uu(j)+ ...
1/24 .*uu(2+j) .*uu(1+j)+11/24 .*uu(-1+j) .*uu(j)+1/8 .*uu(- ...
1+j).^2-1/24 .*uu(-2+j) .*uu(j)-1/24 .*uu(-2+j) .*uu(-1 ...
+j));
```

Algorithm 15.5 *Unix script to do all the edits necessary to change Reduce output such as that shown in Dialogue 15.1 to an Octave/MATLAB/Scilab statement such as that of Dialogue 15.2 to be included in the function definition of Algorithm 15.6. This script uses the Unix stream editor* `sed`.

```
#!/bin/bash
sed 's/:=/=/' dudt.red |\
sed 's/ //g' |\
sed 's/\*\*/.^/g' |\
sed 's/\*/ .*/g' |\
sed 's/\// .\//g' |\
sed 's/$/ \.\.\./' |\
sed 's/^ \.\.\.//' |\
sed 's/\$ \.\.\./;/' >dburgdt.m
```

- each end of line to the continuation `␣...`;
- `$␣...` to `;`.

For example, the results of these edits on the example of Dialogue 15.1 are the lines of Dialogue 15.2. Algorithm 15.5 shows a script that will do all these edits for you on computers with a Unix operating system. In the cycle of exploration and refinement one may need to do these edits repeatedly; such an automatic script is then very useful.

Then wrap around these lines of Dialogue 15.2, the assignment statement, the statements of Algorithm 15.6. The general pattern in the wrapper of Algorithm 15.6 is the following:

- the function statement defines the input and output variables; here time t and the m grid values u are to be input as a column vector, and their time derivatives $\dot{\mathsf{u}}$ output;

Algorithm 15.6 *Wrap the following code around the assignment statement of Dialogue* 15.2 *to form a function in Octave/MATLAB/Scilab. This function computes time derivatives of the* m *grid values assuming spatially periodic conditions.*

```
function dudt=dburgdt(uu,t)
m=length(uu);
alpha=6;  hh=2*pi/m;
uu=[uu(m-1:m);uu;uu(1:2)];
j=3:m+2;
...
end
```

Algorithm 15.7 *Octave code to integrate and draw in Figure* 15.6 *a simulation of Burgers' PDE* (15.12) *for nonlinear parameter* $\alpha = 6$ *on the domain* $[0, 2\pi]$. *It uses the function defined by Algorithm* 15.6. *Minor changes would be needed if using others in the Octave/MATLAB/Scilab family.*

```
m=16;
xs=linspace(0,2*pi,m+1)'; xs=xs(1:m);
ts=linspace(0,1.4,15);
u0=sin(xs);
us=lsode(@dburgdt,u0,ts);
surf(xs,ts,us);
xlabel('x'),ylabel('t'),zlabel('u')
print -depsc2 dudtsim.eps
```

- the length of uu is the number of grid values;
- set any parameters such as the grid spacing and the strength of the nonlinearity—often you will prefer to use a global statement to achieve the same end;
- to most easily implement periodic boundary conditions, copy the end of the grid value vector to the front, and copy the front to the end;
- define the pointer index vector j to just point to the relevant m grid values in the interior;
- set the edited statement to be output from Reduce;
- and, last, add some statement to end the function.

Utilize the function to make numerical predictions The usual purpose of forming a discrete model is to make numerical predictions. For example, we may simulate our discretization of Burgers' PDE (15.12) to see examples of the space-time evolution such as that shown in Figure 15.6. Such simulations are most easily made using time integration algorithms inbuilt into Octave/MATLAB/Scilab. As an example, Algorithm 15.7 gives the code that used our discrete model to draw Figure 15.6. Similar code may be used to explore any of the discrete models we develop in this chapter.

15.3 ▪ Model advection-diffusion robustly

This section explores perhaps the simplest significant example PDE: the linear advection-diffusion equation

$$\frac{\partial u}{\partial t} = -c\frac{\partial u}{\partial x} + \frac{\partial^2 u}{\partial x^2} \tag{15.18}$$

for a field $u(x,t)$, where constant c is the finite advection speed. We discover consistency for small element size H and find interesting and stable upwind approximations for large advection cH (Roberts, 2003b, §3). The associated sophisticated dependence upon advection speed c indicates that odd, anisotropic operators act different from even isotropic operators. For variety, and because details of low orders are of interest, we do not resort to computer algebra.

15.3.1 ▪ The slow manifold discrete model emerges

As in section 15.1, embed the dynamics of the advection-diffusion PDE (15.18) on the overlapping elements of Figure 15.2. Let there be m elements in a periodic spatial domain. Using $u_j(x,t)$ to denote the concentration field in the jth element, we consider

$$\frac{\partial u_j}{\partial t} = -c\frac{\partial u_j}{\partial x} + \frac{\partial^2 u_j}{\partial x^2}, \tag{15.19}$$

subject to the beautiful coupling conditions (15.2). We base the existence of a slow manifold discrete model on the decoupled case by considering the coupling as a small perturbation via notionally adjoining $\dot\gamma = 0$.

Seek equilibria As before for interelement coupling $\gamma = 0$, piecewise constant fields $u_j = \text{constant}$ are equilibria of the advection-diffusion PDE (15.19) with coupling conditions (15.2).

Spectrum implies a discrete model Analogous to before, as the PDE is linear, the PDE for significant perturbations about each of these equilibria is just the advection-diffusion PDE (15.19) subject to the isolated element conditions $u_j(X_{j\pm1},t) = u_j(X_j,t)$. Seeking perturbation solutions to (15.19) of the form

$$\exp[\lambda t + c(x-X_j)/2][A\cos k(x-X_j) + B\sin k(x-X_j)] \quad \text{for } k>0$$

leads to the eigenvalue $\lambda = -k^2 - \frac{1}{4}c^2$. Substituting these solutions into the isolated element conditions and seeking nontrivial solutions leads to the requirement that

$$\sin(kH)\left[\cosh(\tfrac{1}{2}cH) - \cos(kH)\right] = 0.$$

This determinant is zero in two cases:

- first, when $\sin(kH) = 0$, which occurs when $kH = n\pi$ for some integer n. This leads to eigenvalues $\lambda_n = -n^2\pi^2/H^2 - \frac{1}{4}c^2$, $n = 1,2,3,\ldots$.
- second, when $\cos kH = \cosh(\frac{1}{2}cH)$, which occurs for complex wavenumbers k. Substituting wavenumber $kH = \mathfrak{R} + i\mathfrak{I}$, the condition becomes $\cos kH = \cos(\mathfrak{R} + i\mathfrak{I}) = \cos\mathfrak{R}\cosh\mathfrak{I} + i\sin\mathfrak{R}\sinh\mathfrak{I} = \cosh(\frac{1}{2}cH)$: equating the imaginary parts leads to $\sin\mathfrak{R} = 0$; that is, $\mathfrak{R} = n\pi$ for integer n; equating the real parts leads to $(-1)^n\cosh\mathfrak{I} = \cosh(\frac{1}{2}cH)$, which implies integer n must be even, and $\mathfrak{I} = \pm\frac{1}{2}cH$. That is, wavenumber $k = n\pi/H \pm i\frac{1}{2}c$ for even n. The corresponding eigenvalues are $\lambda_n = -n^2\pi^2/H^2 \pm icn\pi/H$ for even n.

Putting these two cases together, the spectrum of the advection-diffusion PDE on an isolated element is $\{0, -\pi^2/H^2 - \frac{1}{4}c^2, -4\pi^2/H^2 \pm ic2\pi/H, -4\pi^2/H^2 - \frac{1}{4}c^2, \dots\}$. Hence an emergent mD slow manifold exists for the advection-diffusion PDE (15.19) on m overlapping elements.

Interestingly, the rate of attraction to the slow subspace, and hence also to the slow manifold, is enhanced due the advection by an amount $c^2/4$. Although this enhancement may become a major contribution to the rate of attraction, it is limited in effect for $cH > 2\sqrt{3}\pi$ by the next gravest eigenvalues, which have real part $-4\pi^2/H^2$.

15.3.2 ▪ Construct the advection-diffusion model

As previously, seek a slow manifold and the evolution thereon in a power series in the interelement coupling parameter γ. Usually we construct models using iteration in computer algebra. The distinction here is that to do the calculations by hand we explicitly identify each term in the power series solution by writing

$$u(x,t) = u_j(x, \mathbf{U}) = \sum_{k=0}^{\infty} \gamma^k u_j^k(x, \mathbf{U})$$

$$\text{such that } \dot{U}_j = g_j(\mathbf{U}) = \sum_{k=1}^{\infty} \gamma^k g_j^k(\mathbf{U}), \tag{15.20}$$

where the superscripts on u_j and g_j are an index and do *not* denote exponentiation.

Now look at each equation. Substitute the expansion (15.20) into the advection-diffusion PDE (15.18) and equate like powers of coupling γ to derive

$$\sum_i \sum_{l=1}^{k} g_i^l \frac{\partial u_j^{k-l}}{\partial U_i} = -c\frac{\partial u_j^k}{\partial x} + \frac{\partial^2 u_j^k}{\partial x^2}, \quad k = 0, 1, 2, \dots. \tag{15.21a}$$

Similarly, substituting the expansion (15.20) into the amplitude condition (15.3) and equating powers of γ requires

$$u_j^0(X_j, \mathbf{U}) = U_j \quad \text{and} \quad u_j^k(X_j, \mathbf{U}) = 0 \quad \text{for } k \geqslant 1, \tag{15.21b}$$

whereas substituting the expansion (15.20) into the coupling conditions (15.2) requires

$$u_j^k(X_{j\pm1}, \mathbf{U}) - u_j^k(X_j, \mathbf{U}) = \begin{cases} \pm\delta U_{j\pm1/2}, & k = 1, \\ 0, & k \neq 1; \end{cases} \tag{15.21c}$$

recall that the centered difference $\delta U_j := U_{j+1/2} - U_{j-1/2}$ (Table 15.1). This section proceeds to solve the first two steps in this sequence of equations and interprets the resultant sequence of discrete models.

First-order coupling First explore the holistic model with only first-order interactions between adjacent elements; that is, errors are $\mathcal{O}(\gamma^2)$. To start, the index $k = 0$ equations of (15.21) are

$$\begin{aligned} -c\frac{\partial u_j^0}{\partial x} + \frac{\partial^2 u_j^0}{\partial x^2} &= 0, \\ u_j^0(X_j, \mathbf{U}) &= U_j, \\ u_j^0(X_{j\pm1}, \mathbf{U}) - u_j^0(X_j, \mathbf{U}) &= 0. \end{aligned}$$

These have solutions that are piecewise constant: in the jth element

$$u_j^0(x, \mathbf{U}) = U_j \,. \tag{15.22}$$

As before, these piecewise constant solutions form the base of the subgrid fields in the slow manifold model.

Find the $\mathcal{O}(\gamma)$ corrections by solving the index $k = 1$ version of equations (15.21):

$$\begin{gathered} -c\frac{\partial u_j^1}{\partial x} + \frac{\partial^2 u_j^1}{\partial x^2} = g_j^1 \,, \\ u_j^1(X_j, \mathbf{U}) = 0 \,, \\ u_j^1(X_{j\pm1}, \mathbf{U}) - u_j^1(X_j, \mathbf{U}) = \pm\delta U_{j\pm1/2} \,. \end{gathered}$$

For convenience and as before, we write the solution of these equations in terms of a subgrid scale spatial variable $\xi := (x - X_j)/H$. Variable ξ ranges over $[-1, 1]$ in the jth element, as shown in Figure 15.2. The solution of these equations of first order in coupling γ are then

$$u_j^1 = \left[\frac{\exp(cH\xi) - 1}{4\sinh^2(cH/2)} - \frac{\cosh(cH/2)}{2\sinh(cH/2)}\xi\right]\delta^2 U_j + \xi\mu\delta U_j \tag{15.23a}$$

$$\text{and} \quad g_j^1 = -c\frac{\mu\delta}{H}U_j + \nu_1\frac{\delta^2}{H^2}U_j \,, \tag{15.23b}$$

$$\text{where} \quad \nu_1 = \frac{cH\cosh(cH/2)}{2\sinh(cH/2)} \,. \tag{15.23c}$$

The subgrid field $u_j \approx U_j + u_j^1$ approximates the blue curve shown in Figure 15.5. Figure 15.7 plots the coefficient ν_1 as a function of advection speed cH relative to grid spacing H. Recall that the centered mean operator μ appearing above is defined as $\mu U_j := \frac{1}{2}(U_{j+1/2} + U_{j-1/2})$ (Table 15.1).

Substitute full coupling $\gamma = 1$ into $\dot{U}_j = \gamma g_j^1$ to obtain our first finite difference model for the linear advection-diffusion PDE (15.18):

$$\dot{U}_j = -c\frac{\mu\delta}{H}U_j + \nu_1\frac{\delta^2}{H^2}U_j \,. \tag{15.24}$$

As the grid spacing $H \to 0$, expansion of the difference operators in terms of differentiation (Table 15.1) shows that this spatial discretization is equivalent to

$$u_t = -cu_x + u_{xx} + \frac{H^2}{12}(c - \partial_x)^2 u_{xx} + \mathcal{O}(H^4) \tag{15.25}$$

and so is indeed consistent to $\mathcal{O}(H^2)$, independent of advection speed c, with the linear advection-diffusion PDE (15.18). Such consistency excellently reaffirms the dual support of this approach to discrete modeling.

This analysis naturally introduces the novel enhancement of the dissipation $\delta^2 U_j$ by the factor $\nu_1(cH)$, which grows from 1 as cH increases from 0. Alternatively, for small cH, from (15.23c) $\nu_1 - 1 \approx \frac{c^2}{12}\delta^2 U_j$ involves the square of the advection speed; thus one may also view this term as an upwind correction to the finite difference approximation of the first derivative (Table 15.1):

$$-\frac{c}{H}\mu\delta + \frac{c^2}{12}\delta^2 = -\frac{c}{H}\left[\left(\frac{1}{2} - \frac{cH}{12}\right)\mathcal{E}^{1/2} + \left(\frac{1}{2} + \frac{cH}{12}\right)\mathcal{E}^{-1/2}\right]\delta \,,$$

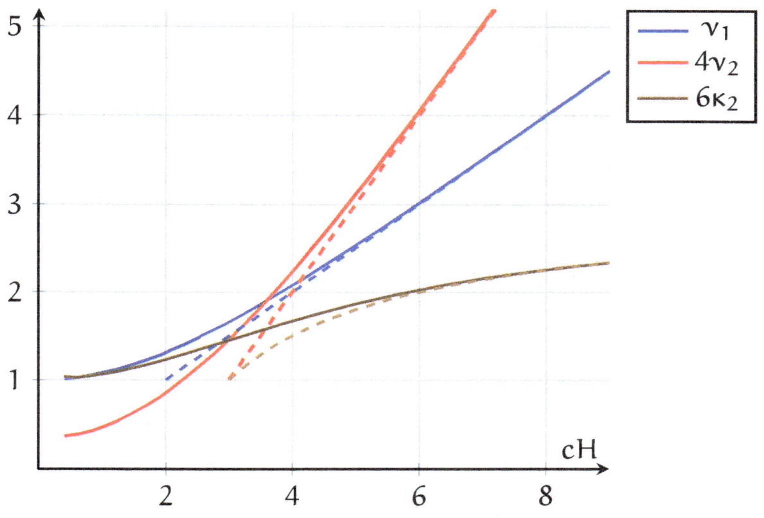

Figure 15.7. *Coefficients of the slow manifold models* (15.24) *and* (15.29) *as functions of advection speed and grid spacing,* cH. *The dashed lines are the large* cH *asymptotes:* $\nu_1 \approx \frac{1}{2}cH$, $\nu_2 \approx \frac{1}{4}cH - \frac{1}{2}$, *and* $\kappa_2 \approx \frac{1}{2} - \frac{1}{cH}$.

which increases the weight of the upwind grid point ($\mathcal{E}^{-1/2}$ is upwind if c is positive). Either interpretation, as enhanced dissipation or upwind correction, is well known to be stabilizing for finite advection speeds c.

Let's explore the large-cH limit when there is strong advection across each element. From (15.23c), $\nu_1 \sim cH/2$ as $cH \to \infty$ and is indeed within a few percent of this value for $cH > 4$; see Figure 15.7. Thus for large advection speed c on a finite-width grid, the slow manifold analysis promotes the upwind model[185] (written in terms of the backward difference operator $\nabla := 1 - \mathcal{E}^{-1}$; see Table 15.1)

$$\dot{U}_j \approx -\frac{c}{H}\nabla U_j = -\frac{c}{H}\left(U_j - U_{j-1}\right). \tag{15.26}$$

This evolution of the grid values is not, and need not be, consistent with the advection-diffusion PDE (15.18) as $H \to 0$ because it applies for finite cH. That the upwind model (15.26) should be relevant to the advection-diffusion PDE (15.18) comes from the slow manifold expansion in the coupling parameter γ, albeit evaluated at $\gamma = 1$ (via the holistic arrows in Figure 15.1).

Exact solutions of the model (15.26) are readily obtained. For example, consider a point release from $j = 0$ at $t = 0$: $U_j(0) = 1$ if $j = 0$ and is 0 otherwise. Defining moment-generating function $G(z,t) := \sum_{j=0}^{\infty} z^j U_j(t)$, some straightforward algebra shows that here $G(z,t) = \exp[(z-1)ct/H]$. This is the same moment-generating function as that for a Poisson probability distribution with parameter ct/H. Hence the mean location and variance of U_j are

$$\mu_j = \sigma_j^2 = \frac{ct}{H} \quad \Longrightarrow \quad \mu_x = ct \quad \text{and} \quad \sigma_x^2 = cHt. \tag{15.27}$$

[185]If the advection velocity is negative, $c < 0$, then various signs change and the large-cH model remains an *upwind* model but is consequently written in terms of forward differences. This also occurs for the later model (15.31) accurate to errors $\mathcal{O}(\gamma^3)$.

Thus for cH *not small*, the extreme upwind model (15.26) has precisely the correct advection speed c, and although the variance is quantitatively wrong, cHt instead of $2t$, at least it is qualitatively correct for finite cH. The slow manifold model (15.24) is consistent for small H and has the virtue of always being stable and always maintaining nonnegativity of solutions no matter how large the advection speed c.

This slow manifold approach to discretizing dynamical PDEs *ameliorates the tyrannical restrictions imposed by the Courant–Friedrichs–Lewy condition.*[186]

Second-order coupling Second, explore the holistic finite difference model with second-order interactions between adjacent elements; that is, the errors are $\mathcal{O}(\gamma^3)$. The details of u_j^2 are of little direct interest here but contribute to, for example, the green curve in Figure 15.5. The discrete model depends directly upon

$$\begin{aligned} g_j^2 &= +\frac{c}{H}\kappa_2\mu\delta^3 U_j + \frac{1}{H^2}\left(\delta^2 - \nu_1\delta^2 - \nu_2\delta^4\right) U_j\,, \qquad (15.28)\\ \nu_2 &= \frac{cH\cosh(cH/2)}{4\sinh(cH/2)} - \frac{1}{2} + \frac{cH\cosh(cH/2)}{8\sinh^3(cH/2)} - \frac{1}{4\sinh^2(cH/2)}\,,\\ \kappa_2 &= \frac{1}{2} + \frac{1}{2\sinh^2(cH/2)} - \frac{\cosh(cH/2)}{cH\sinh(cH/2)}\,. \end{aligned}$$

Observe the marvelous feature that when we form $\dot U_j = \gamma g_j^1 + \gamma^2 g_j^2$ evaluated at $\gamma = 1$, the terms in $\nu_1(cH)\delta^2$ neatly cancel. The discrete model for the linear advection-diffusion thus reduces to

$$\dot U_j = -\frac{c}{H}\left(\mu\delta - \kappa_2\mu\delta^3\right) U_j + \frac{1}{H^2}\left(\delta^2 - \nu_2\delta^4\right) U_j\,. \qquad (15.29)$$

As it involves third and fourth differences, this model is a five-point-wide spatial discretization.

As the element size $H \to 0$, the hyperdiffusion coefficient $\nu_2 \sim 1/12$ and the dispersion coefficient $\kappa_2 \sim 1/6$ give the classic second order in H corrections to the central difference approximations of the first two derivatives. Furthermore, as $H \to 0$, expansion of the difference operators in terms of differentiation (Table 15.1) shows that the discretization (15.29) is equivalent to

$$u_t = -cu_x + u_{xx} + \frac{H^4}{90}(c - \partial_x)^3 u_{xxx} + \mathcal{O}(H^6) \qquad (15.30)$$

and so is consistent to $\mathcal{O}(H^4)$, independent of advection speed c, with the advection-diffusion PDE (15.18). With the specific coupling conditions (15.2), as we compute to higher order in coupling parameter γ it appears that we may marvelously get correspondingly higher order consistency with the original PDE (15.18).

For large advection speed or grid size, large cH, the discretization (15.29) behaves astonishingly well. Using the large-cH asymptotes plotted in Figure 15.7 for ν_2 and κ_2, the model (15.29) reduces to simply

$$\begin{aligned} \dot U_j &= -\frac{c}{H}\left(\nabla + \frac{1}{2}\nabla^2\right) U_j + \frac{1}{H^2}\nabla^2 U_j\\ &= -\frac{c}{2H}\left(U_{j-2} - 4U_{j-1} + 3U_j\right) + \frac{1}{H^2}\left(U_{j-2} - 2U_{j-1} + U_j\right). \qquad (15.31) \end{aligned}$$

[186] `http://en.wikipedia.org/wiki/Cfl_condition` [25 Sept. 2014].

This large-cH model uses only backward differences to incorporate second-order upwind estimates of the derivative, $H\partial_x \approx \nabla + \frac{1}{2}\nabla^2$, and the second derivative, $H^2\partial_x^2 \approx \nabla^2$. To show its good properties,[187] consider again a point release from $j = 0$ at time $t = 0$. The moment-generating function $G(z,t) := \sum_{j=0}^{\infty} z^j U_j(t)$ for the evolution governed by (15.31) is readily shown to be

$$G(z,t) = \exp\left[-\frac{ct}{2H}(z-1)(z-3) + \frac{t}{H^2}(z-1)^2\right]. \qquad (15.32)$$

Then since

$$\left.\frac{\partial G}{\partial z}\right|_{z=1} = \frac{ct}{H} \quad\text{and}\quad \left.\frac{\partial^2 G}{\partial z^2}\right|_{z=1} = \left(\frac{ct}{H}\right)^2 - \frac{ct}{H} + \frac{2t}{H^2},$$

we determine the mean position and variance of the spread in U_j to be

$$\mu_j = \frac{ct}{H} \text{ and } \sigma_j^2 = \frac{2t}{H^2} \quad\Longrightarrow\quad \mu_x = ct \text{ and } \sigma_x^2 = 2t. \qquad (15.33)$$

These predicted mean and variance following a point release are *exactly* correct *for all* time for the advection-diffusion PDE (15.18). This specific result applies to all finite advection speeds c and all finite grid spacings H whenever cH is large enough.

Summary Creating spatial discretizations that, as shown in Figure 15.1, are both consistent for small element size H and also holistically derived via center manifold theory leads to slow manifold models that are remarkably accurate and stable over a wide range of parameters.

15.4 ▪ Exercises

Exercise 15.1 (consistency). Adapt the proof of Theorem 15.2 on isotropic consistency to prove the following consistency of general operators. For some coefficients ℓ_k, let the operator $\mathcal{L} = \ell(\mu_x\delta_x) = \sum_{k=0}^{\infty} \ell_k(\mu_x\delta_x)^k$ be some linear, constant-coefficient operator with coefficient $\ell_1 \neq 0$. Then the discretization of $u_t = \mathcal{L}u$ upon elements with interelement coupling condition (15.2), when truncated to errors $\mathcal{O}(\gamma^p)$ and evaluated at $\gamma = 1$, is consistent with $u_t = \mathcal{L}u$ to errors $\mathcal{O}(\ell_p H^p)$.[188]

Exercise 15.2 (advection on a lattice). Modify Algorithm 15.1 to generate the slow manifold discrete model of the *lattice* advection-diffusion differential-difference equation

$$\dot{u}_i = \frac{n^2}{H^2}(u_{i-1} - 2u_i + u_{i+1}) - c\frac{n}{2H}(u_{i+1} - u_{i-1})$$

for *small* advection speeds c, that is, construct to errors $\mathcal{O}(c^3)$, say. Recalling Example 15.1, describe and interpret your results.

---

[187] The upwind difference model (15.31) is only stable for $cH > 2/3$. However, from Figure 15.7 the approximation (15.31) is only applicable to (15.29) for cH greater than about 4: its instability for very much smaller cH is irrelevant.

[188] There are two limitations in this result: first, in practice we find the order of error is $\mathcal{O}(\ell_p H^{2p})$, which is considerably better; second, it relies upon division by ℓ_1 (in forming the inverse ℓ'^{-1}) and so is nonuniform in the $\mu_x\delta_x$ term when the major term for the existence and attraction of a slow manifold model is the dissipation in the δ_x^2 term. However, at least the proof is a straightforward adaptation.

Exercise 15.3 (nonlinear reaction-diffusion). Modify Algorithm 15.1 to generate the slow manifold discrete model of a reaction-diffusion PDE for *small* nonlinearity α, say the Ginzburg–Landau equation

$$\frac{\partial u}{\partial t} = \frac{\partial^2 u}{\partial x^2} + \epsilon(u - u^3).$$

Perhaps construct to errors $\mathcal{O}(\epsilon^2)$. Explore the form of the discrete model as a prelude of the discussion of section 15.2 on the nonlinear Burgers PDE. Compare with the models produced by my web service (Roberts, 2009b): your models should be the same.

Answer: Resolving subgrid scale dynamics generates new discretizations of such nonlinear reaction processes:

$$\dot{U}_j = H^{-2}\gamma\delta^2 U_j + \epsilon(U_j - U_j^3) + \gamma\epsilon\tfrac{1}{12}\left[-3U_j^2\delta^2 U_j + \delta^2(U_j^3)\right] + \mathcal{O}(\gamma^2, \epsilon^2).$$

Exercise 15.4 (correct solvability). Revise Algorithm 15.3 to implement the correct solvability condition (15.15). Compare the resultant discrete models to see that they are identical, and compare the number of iterations each version takes to obtain the discretizations.

Exercise 15.5 (nonlinear advection). Break the spatial domain into m elements of size H with nonlocal coupling conditions (15.2). Write computer algebra to construct the discrete models of the PDE

$$\frac{\partial u}{\partial t} = \frac{\partial^2 u}{\partial x^2} - \epsilon u^3 \frac{\partial u}{\partial x}.$$

Write to truncate to specified powers in small parameters ϵ and γ. Compare with the models produced by my web service (Roberts, 2009b): your models should be the same.

Exercise 15.6 (nonlinear diffusion). Modify Algorithm 15.1 to generate the slow manifold discrete model of the nonlinear diffusion PDE

$$\frac{\partial u}{\partial t} = u\frac{\partial^2 u}{\partial x^2}.$$

The challenge here is to appropriately modify the update rules, usually by appropriate divisions by U_j. Explore the form of the discrete model. Compare with the models produced by my web service (Roberts, 2009b): your models should be the same.

Exercise 15.7 (adjoint). To holistically discretize advection-reaction-diffusion equations we often build the discrete model upon diffusive dissipation in each finite element. That is, in the jth element we analyze the dynamics of PDEs in the form

$$\frac{\partial u}{\partial t} = \frac{\partial^2 u}{\partial x^2} + \cdots$$

coupled to their neighbors by the nonlocal coupling conditions (15.2), where the dots denote some specific perturbations of the advection and reaction processes.

An iterative construction of the slow manifold of holistic discretization solves problems of the form

$$\mathcal{L}\hat{u}_j = \frac{\partial^2 \hat{u}_j}{\partial x^2} = \mathrm{Res}_j, \quad \text{where} \quad \hat{u}_j(X_{j\pm1},t) - \hat{u}_j(X_j,t) = \mathrm{Res}_\pm .$$

Recalling Exercise 14.12, find the null space of the adjoint of this problem. Hence determine the solvability condition (15.15) on the above residuals for this problem.

Exercise 15.8 (periodic spatial structures). In atom-scale physics or in modern metamaterials the underlying space has periodic structures: periodic potential wells in an atomic lattice or layers of fibers in composite materials. In such a scenario it is tempting to create a spatial discretization of the continuum dynamics with one variable to characterize the physics in each potential well or physical layer. Let's do so here.

Consider the diffusion PDE in one spatial dimension with an additive "potential" $V(x)$ governing a field $u(x,t)$,[189]

$$\frac{\partial u}{\partial t} = \frac{\partial^2 u}{\partial x^2} - V(x)u,$$

where the potential has strength ν and we nondimensionalize space to be 1-periodic. For definiteness, say $V(x) = \nu \sin^2 \pi x = \nu(1-\cos 2\pi x)/2$. When strength $\nu = 0$ the PDE reduces to the simple diffusion equation, so we use this as a base for constructing discrete models for nonzero potentials.

1. Define a grid of spacing $H = 1$ with grid points X_j at the bottom of each well, the middle of each layer. Divide space into elements centered upon each grid point and couple the elements with coupling conditions (15.2).
2. Verify that the theoretical support established in section 15.1.1 still applies to the equilibria at parameters $\gamma = \nu = 0$.
3. Modify Algorithm 15.1 to construct the slow manifold of a discrete model for nonzero parameters γ and ν.

   The modification has to account for subgrid structures that involve trigonometric functions $\cos m\pi x$ and $\sin m\pi x$, with polynomial coefficients. A quick approach for humans to code, albeit dirty and slower for the computer, caters for such terms by treating the polynomial coefficient as constant! Then code for updates that the trigonometric factors would generate.
4. Execute your code to derive that the grid values $U_j(t) := u_j(X_j,t)$ satisfy the discrete equation

   $$\dot{U}_j = (-\tfrac{1}{2}\nu + \tfrac{1}{32\pi^2}\nu^2)U_j + \gamma(1 - \tfrac{1}{32\pi^2}\nu^2)\delta^2 U_j + \mathcal{O}(\gamma^2, \nu^3).$$

   Why does this predict that the potential/layering $V(x)$ reduces interaction between the fields in each "well" or layer? Plot the subgrid fields for say parameter $\nu = 30$ and some choice for $\mathbf{U}$.

[189] In atomic physics one would solve the nonlinear Schrödinger equation (Alfimov et al., 2002, e.g.), but for simplicity we neglect nonlinear effects for the moment and the multiplication of time derivatives by $-i$. Adjustments for both effects are near trivial after establishing the approach.

5. Change the definition of amplitude to $U_j := \int_{X_j-1/2}^{X_j+1/2} u_j(x,t)\,dx$, so U_j directly measures the total amount of "stuff" in the jth interval. Derive exactly the same discrete equation as above.
6. Modify your code to derive discrete models of the nonlinear reaction-diffusion PDE

$$\frac{\partial u}{\partial t} = \frac{\partial^2 u}{\partial x^2} - V(x)u + \epsilon(u - u^3).$$

Exercise 15.9 (Kuramoto–Sivashinsky PDE). Write a page on the theoretical support for discrete models of the Kuramoto–Sivashinsky PDE

$$\frac{\partial u}{\partial t} + 4\frac{\partial^4 u}{\partial x^4} + \alpha\left(\frac{\partial^2 u}{\partial x^2} + u\frac{\partial u}{\partial x}\right) = 0$$

with periodic boundary conditions that $u(x,t) = u(x+mH,t)$. Break the domain into m elements of size H with nonlocal coupling conditions

$$u(X_{j\pm1},t) - u(X_j,t) = \gamma\left[u_{j\pm1}(X_{j\pm1},t) - u_j(X_j,t)\right], \tag{15.34}$$

$$u_{xx}(X_{j\pm1},t) - u_{xx}(X_j,t) = \frac{\gamma}{H^2}\left[\delta^2 u_{j\pm1}(X_{j\pm1},t) - \delta^2 u_j(X_j,t)\right]$$
$$= \frac{\gamma}{H^2}\left[u_{j\pm2}(X_{j\pm2},t) - 3u_{j\pm1}(X_{j\pm1},t) + 3u_j(X_j,t) - u_{j\mp1}(X_{j\mp1},t)\right]. \tag{15.35}$$

Write computer algebra code to construct the consequent discrete models of this Kuramoto–Sivashinsky PDE. Hint: The solvability condition for updates to the evolution will be of the form

$$\hat{g}_j \propto \int_{-1}^{1} (1 - |\xi|)\,\text{Residual}\;d\xi + \cdots.$$

Write to truncate to specified powers in small parameters α and γ. Compare with the models produced by my web service (Roberts, 2009b): differences arise because the coupling conditions are different.

With $\alpha = 0$, corresponding to pure linear decay, you should see that the leading-order (in coupling γ) model is correct. Explore the consistency of higher-order corrections.

Exercise 15.10 (adjoint for fourth order PDEs). To holistically discretize PDEs with fourth-order dissipation, such as the nonlinear Kuramoto–Sivashinsky PDE, we build the discrete model upon dissipation in each finite element. That is, in the jth element we analyze the dynamics of PDEs in the form

$$\frac{\partial u}{\partial t} = -\frac{\partial^4 u}{\partial x^4} + \cdots$$

coupled to their neighbours by nonlocal coupling conditions (15.34)–(15.35).

Thus to construct the slow manifold of holistic discretization we iteratively solve problems of the form

$$\begin{aligned} \mathcal{L}\hat{u}_j &= \frac{\partial^4 \hat{u}_j}{\partial x^4} = \mathrm{Res}_j, \\ \text{where} \quad \hat{u}_j(X_{j\pm1}, t) - \hat{u}_j(X_j, t) &= \mathrm{Res}_0^{\pm}, \\ \hat{u}_{jxx}(X_{j\pm1}, t) - \hat{u}_{jxx}(X_j, t) &= \mathrm{Res}_2^{\pm}. \end{aligned}$$

Recalling Exercise 15.7, use integration by parts to find that the adjoint of this operator $\mathcal{L}$ is

$$\begin{aligned} \mathcal{L}^\dagger v &= \frac{\partial^4 v}{\partial x^4}, \\ \text{with } v = v_{xx} &= 0 \text{ at } x = X_{j\pm1}, \\ [v] = [v_{xx}] &= 0 \text{ at } x = X_j, \\ v_x|_{X_{j+1}} + [v_x]_{X_j} - v_x|_{X_{j-1}} &= 0, \\ \text{and } v_{xxx}|_{X_{j+1}} + [v_{xxx}]_{X_j} - v_{xxx}|_{X_{j-1}} &= 0. \end{aligned}$$

Determine that the null space of the adjoint is spanned by $v = H - |x - X_j|$, and thus determine the solvability condition, a variant of (15.15), on the above residuals for this problem.

Exercise 15.11 (another adjoint). Repeat Exercise 15.10 but replace the coupling conditions there with the coupling conditions

$$u(X_{j\pm1}, t) - u(X_j, t) = \gamma\left[u_{j\pm1}(X_{j\pm1}, t) - u_j(X_j, t)\right], \tag{15.36a}$$

$$u(X_{j\pm2}, t) - u(X_j, t) = \gamma\left[u_{j\pm2}(X_{j\pm2}, t) - u_j(X_j, t)\right]. \tag{15.36b}$$

The integration by parts is more horrible, but with $\xi := (x - X_j)/H$ you should end up with the null space of the adjoint being spanned by the classic cubic spline kernel

$$v = \begin{cases} 4 - 6\xi^2 + 3|\xi|^3, & |\xi| \leqslant 1, \\ (2 - |\xi|)^3, & 1 \leqslant |\xi| \leqslant 2. \end{cases}$$

Part V Summary

- Artificially modifying boundary conditions empowers us to rigorously model PDEs on small domains (section 14.1).

- Chopping up a larger domain into finite elements using coupling conditions between the elements similarly empowers us to form discrete models of PDEs (section 14.2). Wonderfully, the governing PDE itself determines the physical field within each finite element. In contrast, traditional finite differences and finite elements impose an assumed shape for the subgrid field.

  Section 14.4 introduces the nonlocal coupling condition that the extrapolation of a subgrid field to the neighboring grid point must equal the corresponding grid value. This coupling condition performs very well.

- Chapter 15 establishes that the nonlocal interelement coupling conditions (15.2), namely

  $$u_j(X_{j\pm1}, t) = \gamma u_{j\pm1}(X_{j\pm1}, t) + (1-\gamma)u_j(X_j, t),$$

  appear to be great for spatially discretizing a wide range of PDEs and microscale lattice equations (and analogously for multiple space dimensions (Roberts et al., 2013; MacKenzie and Roberts, 2003; MacKenzie, 2005)). The predicted field $u(x, t)$ is the collection of fields $u_j(x, t)$ on each of the overlapping elements.

- Theoretical support for this approach has two facets. First, center manifold theory assures us (Corollary 15.3) that for a wide range of nonlinear reaction-advection-diffusion PDEs, there exists an emergent slow manifold discretization. That is, in principle there exists a spatially discrete model that has *no closure error* apart from the neglect of exponentially decaying transients. The main caveat is that it is hard to be sure whether the assurance extends enough from $\gamma = 0$ to include the crucial case of full coupling $\gamma = 1$. However, Theorem 15.2 provides the second facet and establishes that the discrete models of linear isotropic PDEs are consistent to high order with the PDEs as the element size $\to 0$; further, this high-order consistency is also seen in nonisotropic advection-diffusion (section 15.3), and in nonlinear problems (section 15.2), although it is as yet only partially proved.

- In application, section 15.2 discussed some marvelous properties of the spatial discretization of Burgers' PDE. In particular, it showed that this center manifold approach is invariant to algebraic rewriting of the PDE, thus resolving the longstanding controversies over whether one should discretize nonlinear terms in conservative form, or not, and what form in particular. In application to the important advection-diffusion PDE (section 15.3) the approach generates discrete models that overcome the Courant–Friedrichs–Lewy condition by automatically morphing from classically consistent approximations for small advection speeds to robust and stable upwind approximations for large advection speeds.

Part VI

Hopf bifurcation: Oscillations within the center manifold

Part VI Contents

In all the examples of previous parts, the dynamics on the center manifold is based on the eigenvalue of zero, not the more general case where just the real part of the eigenvalue is zero but the imaginary part is nonzero. But a huge number of applications need analysis of oscillations on the center manifold, including Hopf bifurcations of neuronal models, delayed control, jitter in machining, disease cycles, electronic oscillators, and nanodevices. The need extends to multiple-mode oscillatory bifurcations and wave propagation. This part develops practical techniques to construct models of such oscillations in complex systems.

To be practically useful, the techniques apply to systems where the center modes are not linearly separated from the stable modes.

Constructing center manifold models with oscillations on the center manifold is more difficult technically, as introduced in Chapter 16. The difficulty primarily lies in the form of the homological equation that needs to be solved for corrections to the center manifold. A first issue that is often overlooked is the need to decide how to parametrize the center manifold, based upon the center subspace (section 16.1). Chapter 16 develops techniques when the chosen parametrization is in Cartesian-like variables. Section 16.2 shows how the homological equation generally requires many correction terms to be determined simultaneously through a set of coupled linear equations. Computer algebra readily solves the coupled components of such homological equations (sections 16.3–16.4).

When the critical eigenvalues are pure imaginary, the basic dynamics on a center manifold is that of oscillation. These oscillations are usually relatively fast at some finite frequency determined by the imaginary eigenvalues. Rather than resolving the details of these oscillations, we are usually much more interested in the relatively slower evolution of the amplitude of the oscillations. The same is true for spatially propagating waves. Recall from Chapter 3 that a coordinate transformation to complex amplitudes puts oscillations into a readily interpreted normal form. Correspondingly, Chapter 17 explores how to construct the center manifold model *directly* in complex-amplitude variables. Computer algebra empowers reliable construction (section 17.3). The result is a model that is in a readily interpreted normal form. Hopf bifurcations—general, multiple, or degenerate—and nonlinear normal modes can be predicted depending upon the problem.

Delay differential equations are popular in control and in biochemical modeling. Using complex-amplitude variables also empowers us to model straightforwardly the onset of instabilities in delay differential equations (DDEs; section 17.4).

Chapter 16

Directly model oscillations in Cartesian-like variables

Contents

In all the examples of Part II, the dynamics on the center manifold is based on the eigenvalue of zero; hence they are more specifically examples of a slow manifold. The more general case is where just the real part of the eigenvalue is zero but the imaginary part is nonzero. This chapter introduces this latter case.

Although it is not discussed further here, the same techniques algebraically construct center manifolds when the real parts of the center eigenvalues are nonzero, that is, near zero but not assumed asymptotically small (Roberts, 1989b, e.g.). This case is mostly associated with what people call an inertial manifold (Temam, 1990; Foias et al., 1988; Chueshov and Scheutzow, 2001, e.g.) or an (un)stable manifolds.

When the critical eigenvalues are nonzero, it is more difficult to construct the center manifold. However, because an eigenvalue with a nonzero imaginary part is always associated with oscillations, such cases are important, as they arise in the common transition from steady to oscillatory dynamics. Many of the same issues also arise in wave propagation.

Example 16.1 (simple polar coordinates). Before moving on to discuss generic issues, let us construct a center manifold and the evolution thereon for an example. We twist this special example to analyze as before. Consider the three-variable system

$$\dot{u} = -wu - v\,, \quad \dot{v} = +u - wv\,, \quad \dot{w} = -w + u^2 + v^2\,. \tag{16.1}$$

Very near the equilibrium at the origin, as $\dot{w} \approx -w$, the w variable decays exponentially quickly to zero. Then the u and v variables oscillate according to $\dot{u} = -v$ and $\dot{v} = +u$. As shown by the trajectories of the system in Figure 16.1, we derive the more precise statement that, near the origin, the system is exponentially quickly attracted to the parabolic-like center manifold

$$w \approx u^2 + v^2 - 2(u^2 + v^2)^2, \tag{16.2a}$$

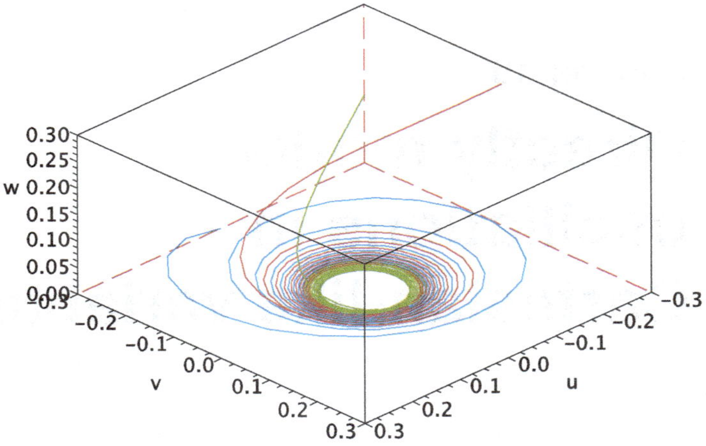

Figure 16.1. *Trajectories of the simple oscillatory system* (16.1) *showing the quick attraction of the bowl-like center manifold on which the solutions undergo long-lasting, near circular oscillations.*

on which the long-term evolution is

$$\dot{u} \approx -(u^2+v^2)u - v\,, \quad \dot{v} \approx -(u^2+v^2)v + u\,. \tag{16.2b}$$

How? The system (16.1) is very special; transforming to polar coordinates empowers us to apply previous methods.

To find the center manifold, convert the uv-plane to polar coordinates (r,θ): $u = r\cos\theta$ and $v = r\sin\theta$. Differentiating these and setting equal to the right-hand sides of the system (16.1) leads to

$$\dot{u} = \dot{r}\cos\theta - r\sin\theta\,\dot{\theta} = -wr\cos\theta - r\sin\theta\,,$$
$$\dot{v} = \dot{r}\sin\theta + r\cos\theta\,\dot{\theta} = -wr\sin\theta + r\cos\theta\,.$$

Two polar equations arise. Subtracting $\sin\theta$ times the first equation above from $\cos\theta$ times the second leads to $\dot{\theta} = 1$. That is, the phase angle increases linearly with time—the oscillations have constant frequency.

The radial equation arises from adding $\cos\theta$ times the first equation above to $\sin\theta$ times the second and leads to $\dot{r} = -wr$. Put this together with the w equation from system (16.1) to form the system

$$\dot{r} = -wr \quad \text{and} \quad \dot{w} = -w + r^2.$$

Recognize that these two coupled equations have an exponentially attractive slow manifold in a neighborhood of the origin. Using earlier methods, one constructs the slow manifold to be $w = r^2 - 2r^4 + \mathcal{O}(r^6)$, on which the system evolves according to $\dot{r} = -r^3 + \mathcal{O}(r^5)$.

Using $u = r\cos\theta$ and $v = r\sin\theta$, this converts back to u and v variables, giving the center manifold (16.2a) and the evolution thereon of (16.2b).

The previous example fell to our analysis because its strong angular symmetry transforms it into polar coordinates. Such huge simplification in polar coordinates is rare. Generally we must undertake more complicated analysis, as suggested by the following example.

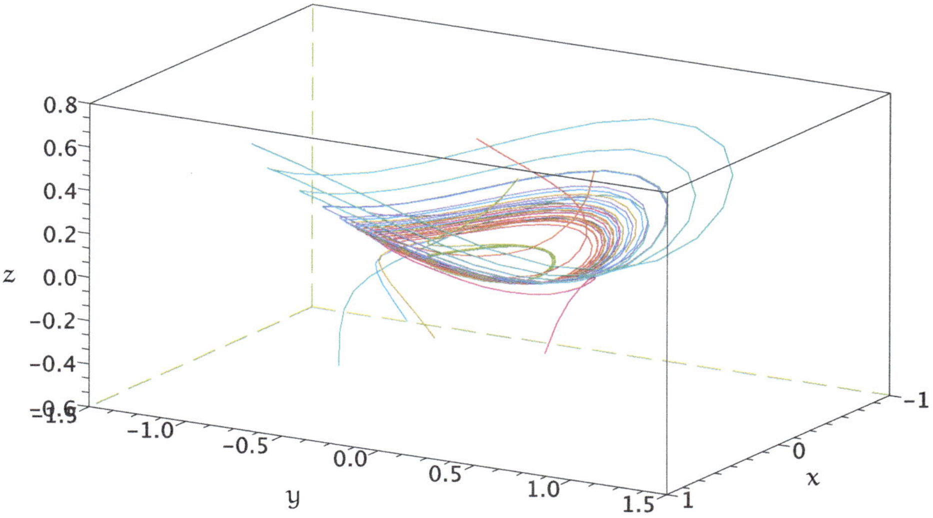

Figure 16.2. *Eight example trajectories of the system* (16.3) *with parameter* $\alpha = 0.2$ *showing the rapid attraction to a deformed bowl-shaped center manifold on which the oscillations evolve in deformed near circles.*

Example 16.2 (center subspace is a coordinate plane). Deduce the center manifold dynamics of the system

$$\dot{x} = y\,, \quad \dot{y} = -x - xz\,, \quad \dot{z} = -z + 5\alpha x^2\,. \tag{16.3}$$

Trajectories plotted in Figure 16.2 show that the dynamics settles onto a curved surface. The surface is evidently a deformed bowl. Because of the lack of angular symmetry, transforming to polar coordinates is no help here. Instead we analyze directly.

Solution: The origin is an equilibrium. Linearize the dynamics about the origin to see the eigenvalue -1 associated with z and the complex conjugate pair of eigenvalues $\pm i$ associated with variables x and y. Hence seek a center manifold

$$z = h(x,y) \quad \text{such that} \quad \dot{x} = y\,, \quad \dot{y} = -x - xh\,. \tag{16.4}$$

Substitute the center manifold (16.4) into the governing z equation in (16.3) and rearrange to

$$h + y\frac{\partial h}{\partial x} - x\frac{\partial h}{\partial y} = 5\alpha x^2 + xh\frac{\partial h}{\partial y}\,. \tag{16.5}$$

Solve this homological equation iteratively. Initially the right-hand side is approximately $5\alpha x^2$ via the center subspace approximation $h \approx 0$. Then seek solutions for a better h on the left-hand side in the form $h = ax^2 + bxy + cy^2$ to deduce

$$(a-b)x^2 + (2a+b-2c)xy + (b+c)y^2 = 5\alpha x^2.$$

Equating coefficients in variables x and y gives $a = 3\alpha$, $b = -2\alpha$, and $c = 2\alpha$. Hence the first nontrivial approximation to the shape of the center manifold is

$$h \approx \alpha(3x^2 - 2xy + 2y^2)\,. \tag{16.6a}$$

Substituting this into (16.4), the evolution on the center manifold is then approximately

$$\dot{x} \approx y, \quad \dot{y} \approx -x - \alpha(3x^3 - 2x^2y + 2xy^2). \tag{16.6b}$$

In the second iteration, substitute the approximation (16.6a) into the right-hand side of (16.5) and rearrange to

$$h + y\frac{\partial h}{\partial x} - x\frac{\partial h}{\partial y} = 5\alpha x^2 + \alpha^2(-6x^4 + 16x^3y - 12x^2y^2 + 8xy^3). \tag{16.7}$$

Solve this homological equation for a better h. We already know that the component (16.6a) matches the $5\alpha x^2$ term on the right-hand side. To match the new quartic terms on the right-hand side, seek a quartic $h = ax^4+bx^3y+cx^2y^2+dxy^3+ey^4$. Substitute this quartic into the left-hand side; then equate coefficients of variables x and y in

$$\begin{aligned}&(a-b)x^4 + (4a+b-2c)x^3y + (3b+c-3d)x^2y^2\\ &+ (2c+d-4e)xy^3 + (d+e)y^4\\ &= \alpha^2(-6x^4 + 16x^3y - 12x^2y^2 + 8xy^3)\end{aligned}$$

to find that a second approximation to the shape of the center manifold is

$$\begin{aligned}h \approx{}& \alpha(3x^2 - 2xy + 2y^2)\\ &+ \alpha^2\left(-\tfrac{142}{85}x^4 + \tfrac{368}{85}x^3y - \tfrac{156}{17}x^2y^2 + \tfrac{448}{85}xy^3 - \tfrac{448}{85}y^4\right).\end{aligned} \tag{16.8}$$

Algorithm 16.1 shows how to use computer algebra to construct, to any specified order, the center manifold of this basic oscillation problem.

The technical difficulty Observe that in constructing the center manifold approximations (16.6a) and (16.8) we had to find all the quadratic terms simultaneously and all the quartic terms simultaneously. This simultaneous deduction is not hard in such a simple example, but it can be confoundedly difficult in problems of real complexity and interest.

Having to solve all terms of a given order simultaneously arises from the coupling in the form of the homological operator $h + yh_x - xh_y$: the term yh_x couples with terms of one order higher in x and one order lower in y, whereas the term xh_y couples terms of one order lower in x and one order higher in y. The homological operator always has this coupling for imaginary eigenvalues in the center manifold. In simple examples the coupling is plain to see. In real problems it is usually hidden, but the coupling is always there, causing significant technical difficulties in constructing a center manifold model.

Chapter 17 explores how using complex-amplitude variables in the center manifold avoids the difficult coupling in the homological equation. Consequently most people use complex-amplitude variables for modeling oscillatory dynamics. However, dynamics may be modeled over a wider range of parameters using Cartesian parameters such as the x and y above. Why can Cartesian parametrization hold over a wider range? The reason is that in complex-amplitude variables the dynamics on the center manifold *must* be close to simple harmonic oscillations, whereas in Cartesian variables the dynamics need not. In Cartesian variables the description of the center manifold is uniform as the

Algorithm 16.1 *Iteratively construct the center manifold of the system* (16.3). *The method of undetermined coefficients works well in such basic problems. Use the parameter* α *to control the truncation of the asymptotic approximations. This algorithm uses the useful utility procedure defined in Algorithm* 16.2.

```
on div; off allfac; on revpri; factor alpha;
in "coeffs.red"$
depend x,t; depend y,t;
let { df(x,t)=>y, df(y,t)=>-x-x*z };
z:=0; % initial approximation
% define generic multinomial and homological operator
maxo:=3;
for n:=maxo:maxo do let alpha^(n+1)=>0;
operator c;
h:=for i:=2:2*maxo sum for j:=0:i sum c(i-j,j)*x^(i-j)*y^j$
cs:=for i:=2:2*maxo join for j:=0:i collect c(i-j,j)$
homo:=h+y*df(h,x)-x*df(h,y)$
% iterate to improve
for iter:=1:9 do begin
  write res:=-df(z,t)-z+5*alpha*x^2;
  eqns:=coeffs({homo-res},{x,y});
  z:=z+sub(solve(eqns,cs),h);
  if res=0 then iter:=1000000;
end; end;
```

Algorithm 16.2 *Useful procedure to find the coefficients of a multivariate multinomial. Let's store it in the file* `coeffs.red` *for other algorithms.*

```
procedure coeffs(exps,vars);
    if vars={} then exps else coeffs(
        foreach z in exps join coeff(z,first vars),
        rest vars);                                        end;
```

frequency becomes small, whereas in complex-amplitude variables the description usually involves divisions by frequency, indicating nonuniformity of applicability. Hence we explore modeling in Cartesian variables a little further, despite its extra complexity.[190]

Develop a general approach Most physical problems do *not* have oscillating modes linearly decoupled from the other modes, as used in Example 16.2. Thus we develop an approach for the general case when the modes are implicit in the mathematical description, not explicit. Let us develop the method intertwined with a specific example. We explore the dynamics of the ODEs (Wiggins, 1990, p. 239, Ex. 2.1j)

$$\dot{u} = \underbrace{\begin{bmatrix} -1 & -1 & 0 \\ 2 & 1 & 0 \\ 1 & 2 & -1 \end{bmatrix}}_{\mathcal{L}} u + \underbrace{\begin{bmatrix} u_3^2 \\ \epsilon u_2 - u_3^2 \\ 0 \end{bmatrix}}_{f(\epsilon,u)}, \tag{16.9}$$

[190] I conjecture that such Cartesian variable parametrizations of center manifold oscillations may be related to the so-called mild-slope approximation of water waves (Toledo and Agnon, 2010, e.g.).

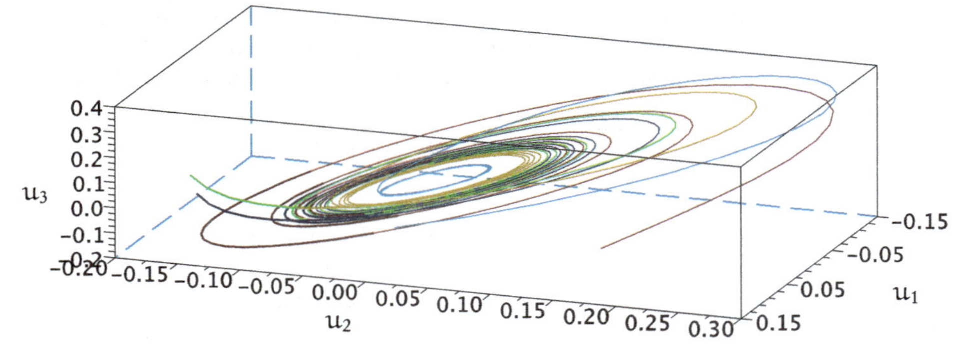

Figure 16.3. *Trajectories of the system* (16.9) *for parameter* $\epsilon = 0$ *showing a quick approach to a center manifold of oscillations.*

where ϵ is a control parameter. Figure 16.3 shows some trajectories. Our first task is to determine eigenvalues of the coefficient matrix $\mathcal{L}$ of the linear terms.[191] It is straightforward to discover that matrix $\mathcal{L}$ has eigenvalues $\pm i$ and -1 (when $\epsilon = 0$, the critical value). Thus there exists a center manifold corresponding to the two eigenvalues $\pm i$. The mode with eigenvalue -1 represents the many exponentially decaying modes we find in real applications.

To unfold the dynamics near the critical value of $\epsilon = 0$, formally adjoin the differential equation $\dot\epsilon = 0$ as discussed previously (section 4.5.1, e.g.) in unfolding pitchfork bifurcations. This empowers us to treat ϵ as a small factor in the analysis. In particular we now group any ϵ terms into the nonlinear parts of the dynamics, as already shown in (16.9).

16.1 ▪ The basis of the center manifold

First, find the center subspace The eigenvectors corresponding to the critical eigenvalues form a basis for the center subspace, $\mathbb{E}_c$, the linear approximation to the center manifold. The eigenvectors of $\lambda = \pm i$ are $\mathbf{e} = (1, -1-i, (-3-i)/2)$ and its complex conjugate $\bar{\mathbf{e}}$ (and incidentally, that for $\lambda = -1$ is $(0,0,1)$). These two complex conjugate eigenvectors span a complex-valued center subspace $\mathbb{E}_c$.[192] But we are only interested in real values. Thus we use[193] the real and imaginary parts of these eigenvectors to span $\mathbb{E}_c$:

$$\mathbf{e}_1 = \begin{bmatrix} 1 \\ -1 \\ -3/2 \end{bmatrix}, \quad \mathbf{e}_2 = \begin{bmatrix} 0 \\ 1 \\ 1/2 \end{bmatrix}.$$

Then linearly approximate the center manifold by the center subspace

$$\mathbb{E}_c = \operatorname{span}\{\mathbf{e}_1, \mathbf{e}_2\} = \left\{ x\mathbf{e}_1 + y\mathbf{e}_2 \mid \text{for all } x, y \right\}.$$

Check that the linear subspace $\mathbb{E}_c$ is correct by applying $\mathcal{L}$ to vectors in $\mathbb{E}_c$ and seeing that the result is still in $\mathbb{E}_c$:

$$\mathcal{L}(x\mathbf{e}_1 + y\mathbf{e}_2) = \cdots = x\mathbf{e}_2 - y\mathbf{e}_1.$$

[191]For $\epsilon \neq 0$ the eigenvalues are $\lambda = \epsilon/2 \pm i\sqrt{1 - \epsilon - \epsilon^2/4}$ and $\lambda = -1$.

[192]Later we find it more convenient to actually use this complex-valued description.

[193]Equivalently, we use the linear combinations $(\mathbf{e} + \bar{\mathbf{e}})/2$ and $i(\mathbf{e} - \bar{\mathbf{e}})/2$.

This linear subspace is then invariant to the linear evolution because if $\mathbf{u} = x\mathbf{e}_1 + y\mathbf{e}_2$, then

$$\dot{x}\mathbf{e}_1 + \dot{y}\mathbf{e}_2 = \dot{\mathbf{u}} = \mathcal{L}\mathbf{u} = x\mathbf{e}_2 - y\mathbf{e}_1 .$$

But $\mathbf{e}_1$ and $\mathbf{e}_2$ are linearly independent vectors and so we justifiably equate coefficients on the left-hand and right-hand sides to obtain

$$\dot{x} = -y \quad \text{and} \quad \dot{y} = x .$$

These two coupled equations describe the dynamics of the classic linear oscillator with frequency 1. These oscillations must occur because these particular dynamics correspond precisely to the two complex conjugate eigenvalues, $\lambda = \pm i$, of the original system exactly at the critical parameter.

Second, decide a parametrization We need to seek the nonlinear center manifold and the long-term evolution thereon. To parametrize the center manifold $\mathcal{M}_c$ we clearly need to use $\mathbf{s} = (x, y)$. But what do x and y mean precisely? *One* possibility is to choose, as we do here, that $x := u_1/2$ and $y := (u_1 + u_2)/2$. This agrees with what we have used for the linear dynamics. We now insist that this choice of definition of x and y also hold for the nonlinear analysis. Then the nonlinear shape of $\mathcal{M}_c$ follows from the third component of $\mathbf{u}$,

$$u_3 = -\frac{3}{2}x + \frac{1}{2}y + h(\epsilon, \mathbf{s}) , \tag{16.10}$$

for some as yet unknown nonlinear function h.[194] The nonlinear evolution on $\mathcal{M}_c$ must be a perturbation of the linear dynamics found above and so will be of the form

$$\dot{\mathbf{s}} = \underbrace{\begin{bmatrix} 0 & -1 \\ 1 & 0 \end{bmatrix} \mathbf{s}}_{\mathcal{G}\mathbf{s}} + \mathbf{g}(\epsilon, \mathbf{s}) , \tag{16.11}$$

where we need to find $\mathbf{g}$. These unknown nonlinear functions h and $\mathbf{g}$ determine the low-dimensional model of the dynamics near the Hopf bifurcation that occurs as ϵ crosses 0.

16.2 ▪ The homological equation

The next task is to find the nonlinear corrections to the shape of the center manifold and the evolution thereon. This then forms a more useful and accurate model of the long-term dynamics.

Now substitute the assumptions (16.10) and (16.11) into the original governing equations (16.9) to give

$$\begin{aligned} &\left\{ \begin{bmatrix} 2 & 0 \\ -2 & 2 \\ -3 & 1 \end{bmatrix} + \begin{bmatrix} 0 & 0 \\ 0 & 0 \\ \partial h/\partial x & \partial h/\partial y \end{bmatrix} \right\} (\mathcal{G}\mathbf{s} + \mathbf{g}) = \begin{bmatrix} 0 & -2 \\ 2 & 2 \\ 1 & -3 \end{bmatrix} \mathbf{s} \\ &+ \begin{bmatrix} 0 \\ 0 \\ -h \end{bmatrix} + \begin{bmatrix} 1 \\ -1 \\ 0 \end{bmatrix} (-3x + y + h)^2 + \begin{bmatrix} 0 \\ 1 \\ 0 \end{bmatrix} \epsilon(-x + y) . \end{aligned} \tag{16.12}$$

[194] In systems with a few dimensions, this approach of defining the amplitudes to be a couple of the coordinates is feasible. For high-dimensional systems it is unlikely that one can practically do so, and slightly more general techniques are needed.

When we solve this, albeit just to some order of accuracy, then center manifold theory assures us that the long-term evolution will be described by (16.11).

First, rearrange (16.12) into a form suitable for iteration. Do this rearrangement by putting all the *dominant* terms on the left-hand side:

$$\begin{bmatrix} 1 & 0 \\ -1 & 1 \\ -3/2 & 1/2 \end{bmatrix} \mathbf{g} + \begin{bmatrix} 0 \\ 0 \\ h - y\partial h/\partial x + x\partial h/\partial y \end{bmatrix} = -\begin{bmatrix} 0 & 0 \\ 0 & 0 \\ \partial h/\partial x & \partial h/\partial y \end{bmatrix} \mathbf{g} + \begin{bmatrix} 1 \\ -1 \\ 0 \end{bmatrix} (-3x + y + h)^2 + \begin{bmatrix} 0 \\ 1 \\ 0 \end{bmatrix} \epsilon(-x + y). \tag{16.13}$$

Recognize which terms in $\mathbf{g}$ and h to put on the left-hand side as all those terms that involve $\mathbf{g}$ and h *and* that are *lowest order* in $\mathbf{s}$ and ϵ. This includes anything linear in $\mathbf{g}$ and h and anything of the form $s_i\partial h/\partial s_j$ because the derivative lowers the order by 1 but the multiplication raises it by 1. However, other terms in $\mathbf{g}$ and h are put on the right-hand side because they are either quadratic in h or multiplied by an s_i and so are of order higher than the lowest.

Second, start the iteration by putting $h = h^{(0)} = 0$ and $\mathbf{g} = \mathbf{g}^{(0)} = \mathbf{0}$ into the right-hand side of (16.13) to give

$$\begin{bmatrix} 1 & 0 \\ -1 & 1 \\ -3/2 & 1/2 \end{bmatrix} \mathbf{g} + \begin{bmatrix} 0 \\ 0 \\ h - y\partial h/\partial x + x\partial h/\partial y \end{bmatrix} = \begin{bmatrix} 1 \\ -1 \\ 0 \end{bmatrix} (-3x + y)^2 + \begin{bmatrix} 0 \\ 1 \\ 0 \end{bmatrix} \epsilon(-x + y).$$

The first two components of this equation dictate that

$$\mathbf{g}^{(1)} = \begin{bmatrix} \frac{1}{4}(-3x + y)^2 \\ \epsilon(y - x) \end{bmatrix}.$$

The last component then leads to a so-called *homological equation*:[195]

$$h - y\frac{\partial h}{\partial x} + x\frac{\partial h}{\partial y} = \frac{\epsilon}{2}x - \frac{\epsilon}{2}y + \frac{27}{8}x^2 - \frac{18}{8}xy + \frac{3}{8}y^2.$$

In general, such equations as these for the shape of a center manifold are solved in the space of multinomials in $\mathbf{s}$. This is generic because we seek a function $h(\mathbf{s})$ whose shape is a multinomial in $\mathbf{s}$, here $\sum_{m,n} h_{mn}x^m y^n$, and so it is the coefficients of this multinomial, h_{mn}, that must be found. It is only in simple circumstances, such as that of the direct instability of a pitchfork bifurcation, that we can overlook this feature of the solution process; here we *must* cater for it. Because the right-hand side is a quadratic in x and y we therefore try a quadratic form for h, namely[196]

$$h = Ax + By + Cx^2 + Dxy + Ey^2.$$

[195]The homological equation is a hyperbolic PDE with characteristic curves being trajectories of the basic oscillation. This matches with the property that a center manifold is a union of suitable trajectories.

[196]The functions x, y, x^2, xy, and y^2 are linearly independent multinomials that we use as a basis.

The equation then becomes

$$(A+B)x+(B-A)y+(C+D)x^2+(2C+D-2E)xy+(D+E)y^2$$
$$=\frac{\epsilon}{2}x-\frac{\epsilon}{2}y+\frac{27}{8}x^2-\frac{18}{8}xy+\frac{3}{8}y^2.$$

Equating coefficients of terms in the linearly independent $x^m y^n$ we obtain the set of linear equations[197]

$$\begin{bmatrix} 1 & 1 & 0 & 0 & 0 \\ -1 & 1 & 0 & 0 & 0 \\ 0 & 0 & 1 & 1 & 0 \\ 0 & 0 & -2 & 1 & 2 \\ 0 & 0 & 0 & -1 & 1 \end{bmatrix} \begin{bmatrix} A \\ B \\ C \\ D \\ E \end{bmatrix} = \begin{bmatrix} \epsilon/2 \\ -\epsilon/2 \\ 27/8 \\ -18/8 \\ 3/8 \end{bmatrix}$$

with solution $A=\epsilon$, $B=0$, $C=21/8$, $D=6/8$, and $E=9/8$. Thus an approximate description of the displacement of the center manifold from the center subspace is

$$h^{(1)}=\frac{\epsilon}{2}x+\frac{21}{8}x^2+\frac{3}{4}xy+\frac{9}{8}y^2. \tag{16.14}$$

The homological equation, although a challenge to organize clearly, is not too complicated. For a pitchfork bifurcation it is diagonal. For a simple Hopf bifurcation it is always effectively tridiagonal. However, for higher-order bifurcations the homological equation may become much more complicated.

To conclude this section, observe that a first approximation to the nonlinear shape of the center manifold is

$$u=\begin{bmatrix} x \\ -x+y \\ -\frac{3}{2}x+\frac{1}{2}y+\frac{\epsilon}{2}x+\frac{21}{8}x^2+\frac{3}{4}xy+\frac{9}{8}y^2 \end{bmatrix}.$$

On this manifold the evolution is approximately described by

$$\begin{aligned} \dot{x} &= -y+\tfrac{1}{4}(-3x+y)^2, \\ \dot{y} &= x+\epsilon(y-x). \end{aligned} \tag{16.15}$$

Figure 16.4 plots segments of trajectories of the model (16.15) showing the noncircular nature of the nonlinear oscillations on the center manifold. However, *as is generally the case for Hopf bifurcations*, such a quadratic approximation is almost always unreliable. For positive ϵ the origin of the above model is linearly unstable. The quadratic nonlinear terms may or may not stabilize the dynamics. Here the dynamics is stabilized, but it is just as easy for quadratic models to have all solutions eventually evolve to infinity. The quadratic model is deficient. I emphasize that this is generally true for Hopf bifurcations. To obtain a reliable model, one must perform one more iteration to obtain the correct cubic nonlinear terms in the model.

[197]Observe that the set of linear equations decouples into a pair of equations for the linear coefficients and a triple of equations for the quadratic coefficients. Such partitioning is generic for all orders in most systems.

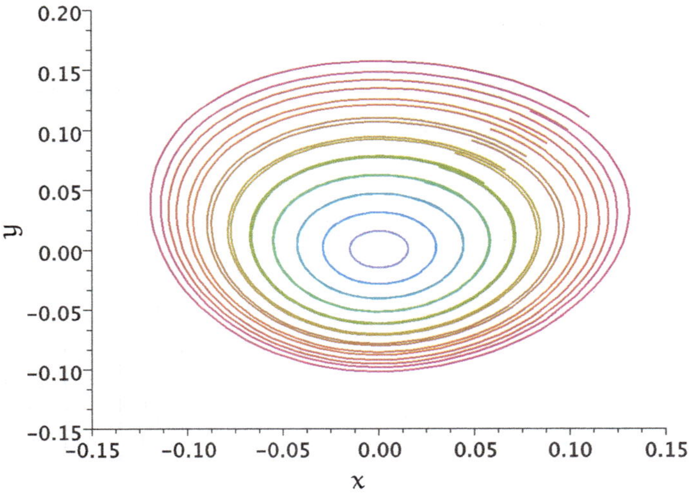

Figure 16.4. *Some trajectories of the model system* (16.15) *for parameter* $\epsilon = 0$. *The counterclockwise evolution very gradually decays.*

16.3 ▪ Further iteration

Computer algebra may routinely perform higher-order computations; we employ Reduce.

It is appealing to recast the iteration scheme. Start with an approximation to the center manifold and the evolution thereon: $\mathbf{u} \approx \tilde{\mathbf{u}}(\mathbf{s})$ such that $\dot{\mathbf{s}} \approx \tilde{\mathbf{g}}(\mathbf{s})$. For example, initially $\tilde{\mathbf{u}} = \mathcal{E}\mathbf{s}$ and $\tilde{\mathbf{g}} = \mathcal{G}\mathbf{s}$, where the columns of $\mathcal{E}$ span the center eigenspace $\mathbb{E}_c$. Then seek a better approximation

$$\mathbf{u} = \tilde{\mathbf{u}}(\mathbf{s}) + \hat{\mathbf{u}}(\mathbf{s}) \quad \text{such that } \dot{\mathbf{s}} = \tilde{\mathbf{g}}(\mathbf{s}) + \hat{\mathbf{g}}(\mathbf{s}).$$

Substituting into the governing differential equation $\dot{\mathbf{u}} = \mathcal{L}\mathbf{u} + \mathbf{f}(\mathbf{u})$ results in

$$\frac{\partial \tilde{\mathbf{u}}}{\partial \mathbf{s}}\left(\tilde{\mathbf{g}} + \hat{\mathbf{g}}\right) + \frac{\partial \hat{\mathbf{u}}}{\partial \mathbf{s}}\left(\tilde{\mathbf{g}} + \hat{\mathbf{g}}\right) = \mathcal{L}\tilde{\mathbf{u}} + \mathcal{L}\hat{\mathbf{u}} + \mathbf{f}\left(\tilde{\mathbf{u}} + \hat{\mathbf{u}}\right).$$

Ignoring nonlinear terms in hatted quantities, and approximating $\tilde{\mathbf{g}}$ and $\tilde{\mathbf{u}}$ wherever they multiply a hatted correction quantity by $\mathcal{G}\mathbf{s}$ and $\mathcal{E}\mathbf{s}$, respectively, this equation becomes the *homological equation*

$$-\mathcal{L}\hat{\mathbf{u}} + \frac{\partial \hat{\mathbf{u}}}{\partial \mathbf{s}}\mathcal{G}\mathbf{s} + \mathcal{E}\hat{\mathbf{g}} = -\frac{\partial \tilde{\mathbf{u}}}{\partial \mathbf{s}}\tilde{\mathbf{g}} + \mathcal{L}\tilde{\mathbf{u}} + \mathbf{f}(\tilde{\mathbf{u}}). \tag{16.16}$$

Appearing in this equation are some old friends.

- The right-hand side is precisely the residual of the dynamical system, $-\dot{\mathbf{u}} + \mathcal{L}\mathbf{u} + \mathbf{f}(\mathbf{u})$, evaluated for the current approximation. It may be calculated very easily and directly in the computer algebra.
- On the left-hand side, $-\mathcal{L}\hat{\mathbf{u}} + \partial\hat{\mathbf{u}}/\partial\mathbf{s}\,\mathcal{G}\mathbf{s}$ is the general *homological operator*. Given that the residual on the right-hand side is usually expressed as a multinomial in $\mathbf{s}$, we solve the homological equation in the space of such multinomials.

Algorithm 16.3 *Constructs the center manifold oscillations of the system* (16.9). *Uses* `coeffs` *function of Algorithm* 16.2.

```
1  on div; off allfac; on revpri; factor eps,small;
2  in "coeffs.red"$
3  % form of the center manifold
4  u1:=x;
5  u2:=-x+y;
6  u3:=-3*x/2+y/2+h;
7  depend x,t; let df(x,t) => gx;
8  depend y,t; let df(y,t) => gy;
9  % seek corrections dh as multinomial in x and y
10 o:=3;
11 operator c;
12 dh:=for m:=0:o sum for n:=0:o-m sum c(m,n)*x^m*y^n$
13 clist:=for m:=0:o join for n:=0:o-m collect c(m,n)$
14 % count order of terms with small
15 for n:=o:o do let small^n=>0;
16 % linear approximation, iteratively corrected
17 h:=0; gx:=-y; gy:=x;
18 for iter:=1:9 do begin
19    res1:=-df(u1,t)-u1-u2+small*u3^2;
20    res2:=-df(u2,t)+2*u1+u2+small*(eps*u2-u3^2);
21    % solve first two components for evolution g
22    gx:=gx+res1;
23    gy:=gy+(res1+res2);
24    % form and solve homological equation for h
25    res3:=-df(u3,t)+u1+2*u2-u3;
26    eqns:=coeffs({dh-y*df(dh,x)+x*df(dh,y)-res3},{x,y});
27    h:=h+sub(solve(eqns,clist),dh);
28    if {res1,res2,res3}={0,0,0} then iter:=1000000;
29 end;                                                   end;
```

- However, the homological equation in full is generally singular. Also appearing on the left-hand side is $\mathcal{E}\hat{\mathfrak{g}}$, which enables us to choose $\hat{\mathfrak{g}}$, the correction to the evolution on the center manifold, so that the homological equation can be solved.

In short, the left-hand side is precisely the same operator as appeared before, in equation (16.13), and the right-hand side is just the governing differential equation.

Algorithm 16.3 performs the iteration as follows.

Lines 4–6 Specify the shape of the center manifold to be the linear center subspace bent by function $h(x, y)$.

Lines 7–8 Tell Reduce that x and y depend upon t and their time derivatives are components of $\mathfrak{g}$.

Lines 10–13 Set up some quantities for solving the homological equation in the space of multinomials in x and y up to the specified order.

Line 15 Tell Reduce to discard all terms higher than cubic in ϵ, x, and y.

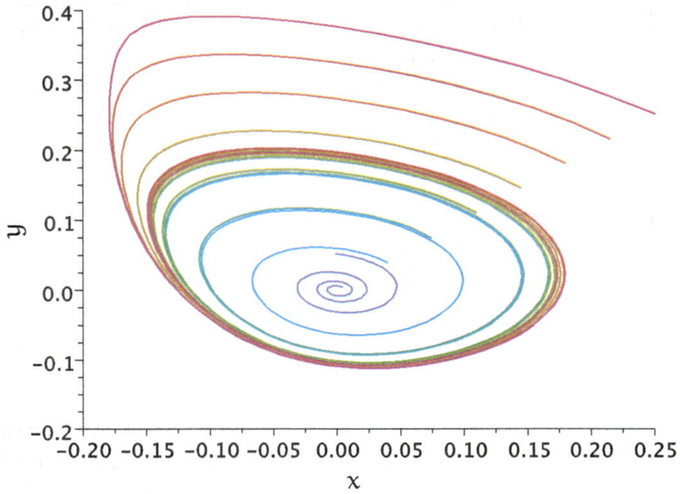

Figure 16.5. *Trajectories of the center manifold model system* (16.17) *at parameter* $\epsilon = 0.2$ *show the stable limit cycle.*

Line 17 Start iterative construction with the linear approximation.

Lines 18–29 Perform the iteration.

- Compute the residual of the dynamical equations into `res`.
- Solve for $\hat{g}$ and update g.
- Form the homological equation in terms of the unknown coefficients of h, `c(m,n)`.
- Equate all the coefficients, here implicitly to zero, and substitute the solution into the correction for u.

The output of this program shows that the center manifold has nonlinear shape

$$\begin{aligned} h = {} & \frac{\epsilon}{2}x + \frac{21}{8}x^2 + \frac{3}{2}xy + \frac{9}{8}y^2 \\ & - \frac{3}{80}\left(311x^3 + 319x^2y + 217xy^2 + 177y^3\right) + \mathcal{O}\left(\epsilon^2 + s^4\right). \end{aligned}$$

The corresponding model on this center manifold for the evolution is

$$\begin{aligned} \dot{x} = {} & -y + \tfrac{9}{4}x^2 - \tfrac{3}{2}xy + \tfrac{1}{4}y^2 \\ & - \tfrac{1}{8}\left(63x^3 - 3x^2y + 21xy^2 - 9y^3\right) + \mathcal{O}\left(\epsilon^2 + s^4\right), \\ \dot{y} = {} & x + \epsilon(y - x) + \mathcal{O}\left(\epsilon^2 + s^4\right). \end{aligned} \tag{16.17}$$

As is generally the case in Hopf bifurcations, with these cubic nonlinearities this model is usefully predictive. Numerical simulations, such as those in Figure 16.5, show the existence of a finite-amplitude limit cycle for positive ϵ and the birth of the limit cycle as ϵ crosses through zero.

Center manifold theory assures us that the above model is relevant to the original system. Why is this cubic truncation relevant and not the quadratic (16.15)? Although I do not prove it, the reason is that the cubic model (16.17) is *structurally stable*; that is, small perturbations only have small effects on the evolution. The quadratic model (16.15) was not so structurally stable. For example, the cubic corrections found above may change the evolution from one of "escape to infinity" to one of a limit cycle, and vice versa. Structural stability is important because we only ever obtain low-order approximations to the exact (infinite order) center manifold model.

Given that the cubic model (16.17) is structurally stable, we deduce from the stability of orbits about the origin in (s, ϵ)-space that the origin of the center manifold model is stable. Hence, the emergence theorem applies for all time, and all trajectories close enough to the center manifold approach a solution of the center manifold model.

16.4 ▪ Synchronization manifolds may attract oscillators

Networks of near identical oscillators model a wide variety of pattern-forming systems, such as neural networks, fluid convection, interacting lasers, and coupled biochemical systems. These networks exhibit rich collective behavior, including synchrony, traveling waves, spatiotemporal chaos, and incoherence (Abrams and Strogatz, 2006; Moon et al., 2005). In particular, much interest continues in the synchronization of nonlinear oscillators (Sun et al., 2009, e.g.).

Example 16.3 (coupled van der Pol oscillators). This example is a simple case: the system is just a pair of oscillators with one forcing the other, but not vice versa.

A lone van der Pol oscillator has governing equations

$$\dot{x}_1 = y_1\,,$$
$$\dot{y}_1 = -x_1 + \mu(1 - x_1^2)y_1 \qquad (16.18)$$

for some nonlinearity parameter μ. For example, for small μ this system oscillates with an amplitude of $\sqrt{x_1^2 + y_1^2} \approx 2$. Now let this oscillator drive a similar van der Pol oscillator

$$\dot{x}_2 = y_2 + 2(x_1 - x_2)\,,$$
$$\dot{y}_2 = -x_2 + (\mu + \epsilon)(1 - x_2^2)y_2\,, \qquad (16.19)$$

where ϵ characterizes the natural difference between the two oscillators, and the constant 2 parametrizes the strength of the influence of the first oscillator upon the second. In this simple system introduced by Sun et al. (2009), the second oscillator is not coupled back to the first.

Find the center subspace The linearized system, upon also neglecting small μ terms, has eigenvalues $\pm i$ and -1 (twice). The center space spanned by the eigenvectors of $\pm i$ is the span of $(1, 0, 1, 0)$ and $(0, 1, 0, 1)$. That is, the center subspace is one of synchronized in-phase oscillations of equal amplitude.

The synchronised manifold emerges This center subspace of linear synchronization is linearly attractive like $\exp(-t)$. Consequently, in the full nonlinear

Algorithm 16.4 *Constructs the center manifold of two coupled van der Pol oscillators as a simple example of a synchronous manifold. Here for small* μ *and* ϵ.

```
on div; off allfac; on revpri;
factor small,eps,mu;
% Use small to control truncation in order (mu,eps).
let small^2=>0;
% Form homological equation to high enough order.
n:=3*deg((1+small)^9,small);
operator c;
cs:=for each k in {x,y} join
    for i:=1:n join for j:=0:i collect c(i-j,j,k)$
hx:=for i:=1:n sum for j:=0:i sum c(i-j,j,x)*x1^(i-j)*y1^j$
hy:=for i:=1:n sum for j:=0:i sum c(i-j,j,y)*x1^(i-j)*y1^j$
homx:=-hy +2*hx+y1*df(hx,x1)-x1*df(hx,y1)$
homy:=+hx      +y1*df(hy,x1)-x1*df(hy,y1)$

% dependencies and initial center subspace
depend x1,t; depend y1,t;
let { df(x1,t)=>y1 , df(y1,t)=>-x1+small*mu*(1-x1^2)*y1 };
x2:=x1; y2:=y1;

% iterate to center manifold
for it:=1:99 do begin
    write resx:=-df(x2,t)+y2+2*(x1-x2);
    write resy:=-df(y2,t)-x2+small*(mu+eps)*(1-x2^2)*y2;
    eqns:=for i:=0:n join append(
        coeff(coeffn(homx-resx,x1,i),y1)
       ,coeff(coeffn(homy-resy,x1,i),y1));
    soln:=solve(eqns,cs);
    x2:=x2+sub(soln,hx);
    y2:=y2+sub(soln,hy);
    showtime;
    if {resx,resy}={0,0} then write it:=it+10000;
end;
```

dynamics we expect that nonlinear synchronized oscillations emerge from among transients with similar such decay.

Construct the synchronized manifold Algorithm 16.4 constructs the center manifold of synchronized oscillations. Since there is no coupling from the second oscillator to the first in this simple example, the dynamics of the synchronized oscillation is simply that of the first oscillator (16.18). The computer algebra finds that the synchronized response of the other oscillator is then

$$\begin{aligned} x_2 &= x_1 + \epsilon\left(\tfrac{1}{2}x_1 - \tfrac{1}{50}y_1^3 - \tfrac{4}{50}x_1y_1^2 + \tfrac{3}{50}x_1^2y_1 - \tfrac{7}{50}x_1^3\right) + \mathcal{O}\left(\epsilon^2+\mu^2\right), \\ y_2 &= y_1 + \epsilon\left(\tfrac{1}{2}y_1 + x_1 - \tfrac{6}{50}y_1^3 + \tfrac{2}{50}x_1y_1^2 - \tfrac{7}{50}x_1^2y_1 - \tfrac{17}{50}x_1^3\right) \\ &\quad + \mathcal{O}\left(\epsilon^2+\mu^2\right). \end{aligned}$$

16.5 ▪ Exercises

Exercise 16.1 (exact center manifold I). For 3D dynamics, in variables $(u(t), v(t), w(t))$, consider the system, modified from (16.1),

$$\dot{u} = -wu - v\,, \quad \dot{v} = +u - wv\,, \quad \dot{w} = -w + u^2 + v^2 - 2w^2\,.$$

Use polar coordinates to describe the uv-plane and hence deduce that there exists an exact center manifold, $w = u^2 + v^2$. What is the evolution on the center manifold?

Exercise 16.2 (exact center manifold II). For 3D dynamics, in variables $(u(t), v(t), w(t))$, consider the system (Romanovski et al., 2013, §3.4)

$$\begin{aligned}\dot{u} &= -v + a(u^2 + v^2) + dvw,\\ \dot{v} &= +u + b(u^2 + v^2) - duw,\\ \dot{w} &= -w + u^2 + v^2 + 2auw + 2bvw\end{aligned}$$

for any given constants a, b, and d. Deduce that there is a 2D center manifold through the origin tangent to the uv-plane at the origin. Verify that the parabolic surface $w = u^2 + v^2$ is invariant to the dynamics of the system, and hence deduce that this surface is an exact center manifold for the system. What are the ODEs for the dynamics on this center manifold?

Exercise 16.3 (exact center manifold III). For 3D dynamics, in variables $(u(t), v(t), w(t))$, consider the system (Romanovski et al., 2013, §3.2)

$$\begin{aligned}\dot{u} &= -v - \tfrac{1}{2}abuw - \tfrac{1}{2}b^2vw,\\ \dot{v} &= +u + \tfrac{1}{2}a^2uw + \tfrac{1}{2}abvw,\\ \dot{w} &= -w + u^2 + v^2 + (a + b)uw + (b - a)vw\end{aligned}$$

for any given constants a and b. Deduce that there is a 2D center manifold through the origin tangent to the uv-plane at the origin. Verify that the surface

$$w = \frac{u^2 + v^2}{1 - au - bv}$$

is invariant to the dynamics of the system, and hence deduce that this surface is an exact center manifold for the system. What are the ODEs for the dynamics on this center manifold?

Exercise 16.4 (Lorenz 86 model). Lorenz (1986) proposed the following system of ODEs to explore the existence of a slow manifold among fast atmospheric waves:

$$\begin{aligned}\dot{u} &= -vw\,,\\ \dot{v} &= uw\,,\\ \dot{w} &= -uv - bx\,,\\ \dot{x} &= -y + bw\,,\\ \dot{y} &= x\,.\end{aligned}$$

Assuming parameter b is small, argue that the oscillations in x and y do not form the basis of a center manifold model because the whole state space is the center subspace.

Exercise 16.5 (Lorenz 96 model). Lorenz (1996) proposed a system of ODEs to explore predictability in weather and climate forecasting. Many others adopted this system to further explore predictability (Boffetta et al., 2002, e.g.). Although the real interest in this model lies in its chaotic regime, we explore the onset of its oscillatory dynamics via a Hopf bifurcation. Conceptually, Lorenz divided the atmosphere on the Earth into K cells of large-scale high- and low-pressure systems. These systems would decay at rate 1, nondimensionally, but are forced by the sun's heating, represented by F, and interact nonlinearly with their neighbors. Lorenz defined $x_k(t)$ to be the strength of cell k, for $k = 1, 2, \dots, K$. As a system to explore he proposed

$$\frac{dx_k}{dt} = -x_k + x_{k-1}(x_{k+1} - x_{k-2}) + F, \quad k = 1, 2, \dots, K, \tag{16.20}$$

where the x_k are to be K-periodic, that is, $x_k = x_{k\pm K}$. In this exercise explore the simple case of $K = 4$ cells.

1. Show that there is an equilibrium at $x_k = F$.
2. Justify the existence and emergence of a Hopf bifurcation at forcing $F = 1$.
3. Find the center subspace at $F = 1$.
4. Find by hand the terms quadratic in amplitudes s_i of the shape of the center manifold.
5. There is no easy way: explain how Algorithms 16.5 and 16.6 construct approximations to the center manifold and the evolution thereon.
6. Approximate the center manifold of this Hopf bifurcation. Interpret the center manifold model.

Answer: With forcing $F = 1 + f$ for small f, the center manifold may be parametrized as

$$\begin{aligned}
x_1 &= 1 + f - s_1 + \tfrac{1}{13}(-6s_1^2 + 5s_1s_2 - 7s_2^2) + \mathcal{O}(s^3),\\
x_2 &= 1 + f + s_2 + \tfrac{1}{13}(-7s_1^2 - 5s_1s_2 - 6s_2^2) + \mathcal{O}(s^3),\\
x_3 &= 1 + f + s_1 + \tfrac{1}{13}(-6s_1^2 + 5s_1s_2 - 7s_2^2) + \mathcal{O}(s^3),\\
x_4 &= 1 + f - s_2 + \tfrac{1}{13}(-7s_1^2 - 5s_1s_2 - 6s_2^2) + \mathcal{O}(s^3),\\
\dot s_1 &= fs_1 - (1+f)s_2 + \tfrac{1}{13}(-7s_1^3 + s_1^2s_2 - 11s_1s_2^2 + 7s_2^3) + \mathcal{O}(s^4),\\
\dot s_2 &= (1+f)s_1 + fs_2 + \tfrac{1}{13}(-7s_1^3 - 11s_1^2s_2 - s_1s_2^2 - 7s_2^3) + \mathcal{O}(s^4).
\end{aligned}$$

Exercise 16.6 (Lorenz 96 model extended I). Do Exercise 16.5, modifying Algorithms 16.5 and 16.6, for the case of $K = 6$ cells.

Answer: The center subspace may be, for example, $x = 1 + (2, 1, -1, -2, -1, 1)s_2 + (0, -1, -1, 0, 1, 1)s_2$, and so on.

Exercise 16.7 (Lorenz 96 model extended II). Do Exercise 16.5, modifying Algorithms 16.5 and 16.6, for the case of $K = 8$ cells.

Answer: The center subspace may be, for example, $x = 1 + (1, 0, -1, 0, 1, 0, -1, 0)s_2 + (0, 1, 0, -1, 0, 1, 0, -1)s_2$, and so on.

Algorithm 16.5 *Reduce code to begin constructing the center manifold of the onset of oscillations in the model* (16.20) *of Lorenz (*1996*): Algorithm* 16.6 *codes the subsequent iteration. Uses Algorithm* 16.2.

```
factor small;
kk:=4;  maxo:=3; ff:=1+small*f;
% Set storage, index arrays, and eigenvectors
array x(kk),res(kk),p(kk),m(kk),e1(kk),e2(kk),dx(kk),homo(kk);
for k:=1:kk do p(k):=(if k=kk then 1 else k+1);
for k:=1:kk do m(k):=(if k=1 then kk else k-1);
e1(1):=-1;e1(3):=1;e2(2):=1;e2(4):=-1;
% set initial linear approximation
depend s1,t; let df(s1,t)=>g1;
depend s2,t; let df(s2,t)=>g2;
for k:=1:kk do x(k):=ff+small*(e1(k)*s1+e2(k)*s2);
g1:=-s2; g2:=+s1;
% collect expressions and coefficients of
% homological equation and amplitude condition
operator c;
for k:=1:kk do dx(k):=for m:=0:maxo sum
    for n:=0:maxo-m sum c(k,m,n)*s1^m*s2^n$
clist:=for k:=1:kk join for m:=0:maxo join
    for n:=0:maxo-m collect c(k,m,n)$
for k:=1:kk do homo(k):=s2*df(dx(k),s1)-s1*df(dx(k),s2)
    -dx(k)+dx(p(k))-dx(m(m(k)))$
in "coeffs.red"$
damp:=coeffs({for k:=1:kk sum e1(k)*dx(k)
             ,for k:=1:kk sum e2(k)*dx(k)},{s1,s2})$
zeros:=for k:=1:kk collect 0;
```

Algorithm 16.6 *Reduce code to construct the center manifold of the onset of oscillations in the model* (16.20) *of Lorenz (*1996*). After initializing with Algorithm* 16.5*, this code performs the iteration.*

```
for iter:=2:9 do begin
  if iter<=maxo then let small^(iter+1)=>0;
  for k:=1:kk do write res(k):=
    -df(x(k),t)-x(k)+x(m(k))*(x(p(k))-x(m(m(k))))+ff;
  ress:=for k:=1:kk collect res(k);
  g1:=g1+(dg1:=(for k:=1:kk sum e1(k)*res(k))/2)/small;
  g2:=g2+(dg2:=(for k:=1:kk sum e2(k)*res(k))/2)/small;
  eqns:=for k:=1:kk join coeffs(
    {homo(k)+res(k)-e1(k)*dg1-e2(k)*dg2},{s1,s2});
  sol:=solve(append(eqns,damp),clist);
  for k:=1:kk do x(k):=x(k)+sub(sol,dx(k));
  if ress=zeros then write iter:=1000000+iter;
end;
```

Exercise 16.8 (Lorenz 96 model double Hopf). Extend Exercise 16.5 to the case of $K = 12$ cells, but note that the bifurcation at $F = 1$ is a double Hopf bifurcation. More extensively modify Algorithms 16.5 and 16.6.

Answer: The center subspace may be, for example,

$$\begin{aligned} x = 1 &+ (1, 0, -1, 0, 1, 0, -1, 0, 1, 0, -1, 0)s_2 \\ &+ (0, 1, 0, -1, 0, 1, 0, -1, 0, 1, 0, -1)s_2 \\ &+ (-1, -2, -1, 1, 2, 1, -1, -2, -1, 1, 2, 1)s_3 \\ &+ (1, 0, -1, -1, 0, 1, 1, 0, -1, -1, 0, 1)s_4,, \end{aligned}$$

and so on.

Chapter 17

Model the modulation of oscillations

Contents

The low-dimensional center manifold model (16.17) is, in an important sense, a failure. True it is low dimensional and of guaranteed accuracy and relevance. However,

- it is very hard for us to discern any qualitative features of the evolution from the form of (16.17); and
- any numerical simulations must resolve the fast oscillations on the center manifold and so we must specify short time steps, ones on the scale of the fast oscillation rather than ones on the scale of the long-term evolution.

This chapter builds on the coordinate transform analysis of oscillations in Chapter 3 to show how to derive effective models for the modulation of the fast oscillations. The result remedies both the above problems. Further, the method greatly simplifies the homological equation and its solution process!

Let's start by revisiting Example 16.2. But for generality, sections 17.2 and 17.3 then establish constructive algorithms for the case of a general spectrum of center eigenvalues to cater for when the center manifold is composed of an arbitrary number of oscillatory and slow modes.

Delay differential equations (DDEs) are important in many application areas. Section 17.4 uses our techniques to model the oscillatory instabilities generated by delays.

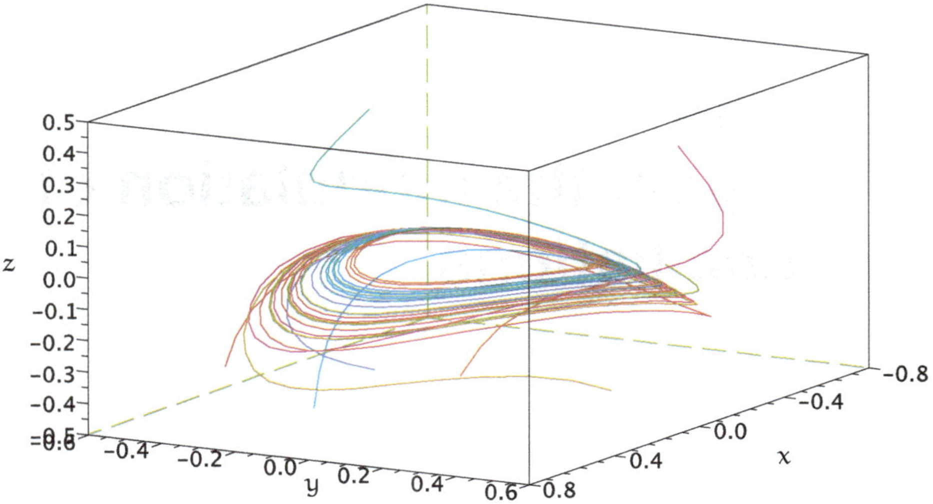

Figure 17.1. *Eight example trajectories of system* (17.1) *for parameter* $\alpha = -0.2$ *showing the trajectories being attracted exponentially quickly to an inverted bowl-like center manifold, and thereafter showing the slowly decaying amplitude of their oscillations on the center manifold.*

Example 17.1 (Example 16.2 revisited). Reconsider the center manifold dynamics of the system (16.3), namely

$$\dot{x} = y\,, \quad \dot{y} = -x - xz\,, \quad \dot{z} = -z + 5\alpha x^2\,. \tag{17.1}$$

Figures 16.2 and 17.1 plot some sample trajectories showing the quick approach to a bowl-like center manifold that depends upon the parameter α. But rather than describe the center manifold in the Cartesian $z = h(x, y)$, now we seek to express the center manifold model in terms of complex amplitudes of the oscillations on the center manifold that are seen in Figure 17.1. (Exercise 17.1 correspondingly asks you to find the center manifold in terms of amplitude-angle variables.)

Recall from Example 16.2 that the origin is an equilibrium, and that linearizing the dynamics about the origin gave eigenvalue -1 associated with z, and the complex conjugate pair of eigenvalues $\pm i$ associated with variables x and y. Hence this section seeks a center manifold in terms of the amplitude of the complex oscillations of the imaginary eigenvalues:

$$\begin{aligned} &x \approx a e^{it} + \bar{a} e^{-it}, \quad y \approx i a e^{it} - i\bar{a} e^{-it}, \quad z \approx 0 \\ &\text{such that} \quad \dot{a} = g(a, \bar{a}). \end{aligned} \tag{17.2}$$

Via the magical procedure developed in this section, Algorithm 17.1, below, finds that the complex amplitude a evolves according to

$$\dot{a} = \alpha(1 + \tfrac{11}{2}i)a^2\bar{a} + \mathcal{O}(\alpha^2) = \alpha(1 + \tfrac{11}{2}i)|a|^2 a + \mathcal{O}(\alpha^2). \tag{17.3}$$

In contrast to the xy-form of the center manifold evolution, in this complex-amplitude form we readily interpret the evolution on the center manifold: the sign of the real part of the coefficient on the right-hand side is the same as the sign of the parameter α; hence for $\alpha < 0$ the origin is stable (with the oscillations decaying algebraically), whereas for $\alpha > 0$ the origin is unstable.

Importantly, in the complex-amplitude form (17.3) we readily recognize that a model cubic in the amplitude a is not adequate to form a structurally stable model for all small α. We have to proceed to higher order to find out how the system stabilizes the as yet unstable prediction for parameter $\alpha > 0$. The Cartesian modeling of Chapter 16 in Example 16.2 does not make this important attribute apparent in the algebraic model. Complex-amplitude variables are typically more useful.

Recall that section 3.6 discussed how expressing oscillations in terms of complex exponentials forms a powerful and flexible view of the dynamics. Also recall how this view is closely connected to the amplitude-phase description of oscillations, as the complex amplitude is typically $a = \frac{1}{2}re^{i(\theta-t)}$ for amplitude r and phase θ. This chapter incorporates the methods of section 3.6 into the case when the oscillations occur embedded in a center manifold of a more complex dynamical system.[198] Sections 17.2 and 17.3 generalize the approach to multiple oscillation modes together with multiple slow modes all interacting within some complex system.

17.1 ▪ Linearly separated systems

Recall that we use the term "linearly separated" to mean dynamical systems in the form

$$\dot{\mathbf{x}} = A\mathbf{x} + \mathbf{f}(\mathbf{x}, \mathbf{y}), \quad \dot{\mathbf{y}} = B\mathbf{y} + \mathbf{g}(\mathbf{x}, \mathbf{y}), \tag{17.4}$$

where the eigenvalues of matrix A all have zero real part, whereas the eigenvalues of matrix B all have negative real part, and where functions $\mathbf{f}$ and $\mathbf{g}$ are strictly nonlinear.[199] In this form, the dynamics linearized about the origin possesses decoupled center variables $\mathbf{x}$ and stable (hyperbolic) variables $\mathbf{y}$. In the separated form (17.4) the construction of a center manifold in terms of slowly varying complex amplitudes is a straightforward combination of the complex modulation coordinate transformation of section 3.6 and the techniques from section 4.4 to construct a slow manifold.

The reason the slow manifold techniques from section 4.4 apply is simply that in a complex-amplitude parametrization of oscillations the complex amplitudes are slowly varying. An example suffices to show how to merge the transformation of section 3.6 and the techniques from section 4.4.

Example 17.2. (continue Example 17.1). The specific system (17.1) is in the separated form (17.4) when we group the variables into center variables (x, y) and stable variable z. The reason is that then the system (17.1) becomes

$$\begin{bmatrix}\dot{x}\\ \dot{y}\end{bmatrix} = \underbrace{\begin{bmatrix}0 & 1\\ -1 & 0\end{bmatrix}}_{A}\begin{bmatrix}x\\ y\end{bmatrix} + \begin{bmatrix}0\\ -xz\end{bmatrix}, \quad \dot{z} = \underbrace{-1}_{B}\,xz + 5\alpha x^2$$

for matrices A with eigenvalues $\pm i$ and matrix B with eigenvalue -1. Our aim is to find the bowl-like shape of Figures 16.2 and 17.1, and the evolution thereon. Algorithm 17.1 does the algebra. Algorithm 17.1 predicts that the center manifold has the shape plotted in Figure 17.2, but crucially the description is time dependent, as seen in the rotating mesh of Figure 17.2. The rotating mesh visually represents the approximate tracking of trajectories by the complex-amplitude parametrization.

[198] A similar approach empowers us to analyze the evolution of the spatial modulation of periodic patterns in space as seen in fluid convection and Taylor vortices.

[199] More generally, the eigenvalues λ_A of A have small real part, $|\Re\lambda_A| \leqslant \alpha$, and the eigenvalues of B are hyperbolic, $|\Re\lambda_B| \geqslant \beta$, with a significant spectral gap, $\beta > p\alpha$ for large enough $p \geqslant 2$.

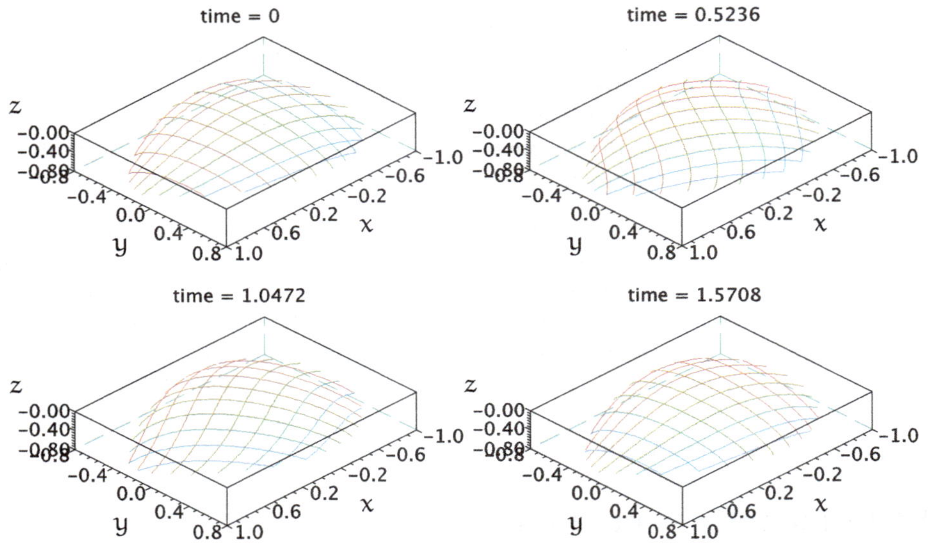

Figure 17.2. *Equation* (17.5) *predicts the center manifold for the system* (17.1) *for parameter* $\alpha = -0.2$. *Four times are shown, covering a quarter period, reflecting that the parametrization* (17.5) *rotates within the center manifold approximately in synchronization with the trajectories so that a slowly varying amplitude emerges. Plotted are the grid of lines of fixed* ξ *and* η, *where* $\xi + i\eta = ae^{it}$ *for complex amplitude* a.

Start by copying Algorithm 3.3, which section 3.6 uses to construct the complex-amplitude description of oscillations—using `b` to denote amplitude $\bar{a}$. Algorithm 17.1 is mostly Algorithm 3.3 but with two differences. First, the computation of the residual is split into the two parts, lines 20 and 26, to suit the x and y parts of our system (17.1). Also we omit the amplitude-counting variable `small` as we have no need of it in this example: parameter α serves to conveniently truncate the asymptotics.

Second, Algorithm 17.1 includes slow manifold code to determine the stable variable z. Line 13 initializes the center manifold to the center subspace $z = 0$. Lines 16–19 update $z(a, \bar{a}, t)$ from the residual of the z ODE in the system (17.1). But there is one new feature: the update for $z(a, \bar{a}, t)$ has to account for the time dependence in the oscillating modes. Let's see how: we seek an update to $z(a, \bar{a}, t)$ by substituting $z = z(a, \bar{a}, t) + \hat{z}(a, \bar{a}, t)$ for small $\hat{z}$ into the z equation (17.1). By neglecting small quantities we deduce the homological equation

$$\frac{\partial \hat{z}}{\partial t} + \frac{\partial \hat{z}}{\partial a}\dot{a} + \frac{\partial \hat{z}}{\partial \bar{a}}\dot{\bar{a}} + \hat{z} = \text{Res}_{(17.1)z} \,.$$

Since the complex amplitudes are slowly varying, $\dot{a} \approx 0 \approx \dot{\bar{a}}$, the complicated terms on the LHS are neglected. Thus we solve $\partial\hat{z}/\partial t + \hat{z} = \text{Res}_{(17.1)z}$ for updates. Hence for each term in the residual involving the factor $e^{i\omega t}$ we straightforwardly include the same term in the update but divided by the factor $(i\omega + 1)$, as coded in line 19.

The iteration in Algorithm 17.1 terminates when the residuals of the ODEs are zero to the order in parameter α specified in line 14. For example, the algorithm

Algorithm 17.1 *Iteratively construct the center manifold of the elementary Hopf bifurcation system* (17.1). *Here use complex amplitudes, and obtain code fragments from the normal-form Algorithm* 3.3.

```
factor alpha;
operator cis;
let { df(cis(~u),~t)  => i*df(u,t)*cis(u)
    , cis(~u)*cis(~v)  => cis(u+v)
    , cis(~u)^~p => cis(p*u)
    , cis(0)=>1 };
operator linv; linear linv;
depend a,t; depend b,t;
let { df(a,t)=>ga, df(b,t)=>gb };
ga:=gb:=0;
x:=a*cis(t)+b*cis(-t);
y:=df(x,t);
z:=0;
let alpha^3=>0;
for iter:=1:99 do begin
  write resz:=-df(z,t)-z+5*alpha*x^2;
  z:=z+(linv(resz,cis)
    where { linv(1,cis)=>1,
    linv(cis(~w*t),cis)=>cis(w*t)/(i*w+1) });
  write resy:=df(y,t)+x+x*z;
  ga:=ga+i/2*(ca:=coeffn(resy,cis(+t),1));
  gb:=gb-i/2*(cb:=coeffn(resy,cis(-t),1));
  x:=x+(linv(resy-ca*cis(t)-cb*cis(-t),cis)
    where { linv(1,cis)=>-1,
    linv(cis(~w*t),cis)=>cis(w*t)/(w^2-1) });
  y:=df(x,t);
  if {resz,resy}={0,0} then iter:=1000000;
end;
```

deduces the center manifold

$$\begin{aligned} x &= a e^{it} + \bar a e^{-it} + \alpha\big[(\tfrac18 - \tfrac14 i)a^3 e^{i3t} + (\tfrac18 + \tfrac14 i)\bar a^3 e^{-i3t}\big] + \mathcal{O}(\alpha^2), \\ y &= i a e^{it} - i\bar a e^{-it} + \alpha\big[(\tfrac34 + \tfrac38 i)a^3 e^{i3t} + (1 + \tfrac{11}{2} i)a^2 \bar a e^{it} \\ &\qquad + (1 - \tfrac{11}{2} i)a\bar a^2 e^{-it} + (\tfrac34 - \tfrac38 i)\bar a^3 e^{-i3t}\big] + \mathcal{O}(\alpha^2), \\ z &= \alpha\big[(1-2i)a^2 e^{i2t} + 10 a\bar a + (1+2i)\bar a^2 e^{-i2t}\big] + \mathcal{O}(\alpha^2). \end{aligned} \tag{17.5}$$

Figure 17.2 plots the center manifold surface predicted by (17.5). The figure plots the curves of fixed ξ and η, where $\xi + i\eta = a e^{it}$, at four different times to demonstrate how the complex-amplitude coordinate system rotates around the center manifold. The construction chooses this rotation around the center manifold to be approximately in synchrony with the trajectories so that the complex amplitude evolves relatively slowly.

The iteration of Algorithm 17.1 correspondingly deduces that the slowly varying amplitude evolves according to

$$\dot a = \alpha(1 + \tfrac{11}{2} i)a^2 \bar a + \alpha^2(-\tfrac{73}{5} - \tfrac{2199}{80} i)a^3 \bar a^2 + \mathcal{O}(\alpha^3), \tag{17.6}$$

and correspondingly for the complex conjugate $\bar{a}$. In this slowly varying form we readily predict long-term dynamics. Recall that the imaginary part of coefficients determines any nonlinear frequency shifts. But more significant is the real part of the coefficients: here the positive coefficient in the cubic term $a^2\bar{a}$ is stabilizing for parameter $\alpha < 0$ and destabilizing for parameter $\alpha > 0$. However, the negative real part of the coefficient of the quintic term $a^3\bar{a}^2$ indicates that the dynamics restabilizes for small $\alpha > 0$.

Example 17.2 has two special attributes not found in general separated systems. First, the center subspace $z = 0$ is in its entirety exactly invariant when parameter $\alpha = 0$, so that we could construct the center manifold in an asymptotic expansion only in α. In a more general example, the center manifold would be asymptotic in the amplitudes as well. Second, in the system (17.1) the ODE $\dot{x} = y$ is simple enough to empower us to directly use code from section 3.6. If this component ODE were significantly more complicated, then we would have to derive corrections based upon its residual in a nontrivial way. Be aware of these extra needs in more general applications.

Example 17.3 (not linearly separated). Consider the $\epsilon = 0$ version of the system of ODEs (16.9) for variable $\mathbf{u} = (u_1(t), u_2(t), u_3(t))$:

$$\dot{u}_1 = -u_1 - u_2 + u_3^2, \quad \dot{u}_2 = 2u_1 + u_2 - u_3^2, \quad \dot{u}_3 = u_1 + 2u_2 - u_3\,.$$

This system has an equilibrium at the origin. The linearization about the origin is $\dot{\mathbf{u}} = \mathcal{L}\mathbf{u}$ for matrix

$$\mathcal{L} = \begin{bmatrix} -1 & -1 & 0 \\ 2 & 1 & 0 \\ 1 & 2 & -1 \end{bmatrix}.$$

The system is not linearly separated because this matrix couples all three components. Thus modeling the dynamics of the system requires a more general approach than that of Example 17.2.

17.2 ▪ Linear terms are generally not separated

In applications we generally prefer to model systems in the original variables. Although a linear transform to separate the eigenmodes is trivial mathematics, such a transformation often confounds application-specific interpretation. Thus we continue to develop methods to analyze systems in the variables as given (apart from perhaps shifting the origin).

Consider dynamical systems for variable $\mathbf{u}(t) \in \mathbb{R}^n$ in the general physical form

$$\dot{\mathbf{u}} = \mathcal{L}\mathbf{u} + \mathbf{f}(\epsilon, \mathbf{u}) \tag{17.7}$$

for linear operator $\mathcal{L}$ and nonlinear/perturbation effects in the sufficiently smooth function $\mathbf{f}$. Assume $\mathbf{f}(0, \mathbf{0}) = \mathbf{0}$ so that the origin is our base equilibrium. Suppose there exist m eigenvalues of the linear operator $\mathcal{L}$ with zero real part, counted according to their multiplicity. More specifically, let the m eigenvalues be $\lambda = i\omega_j$, for index $j = 1, \ldots, m$, for some frequencies ω_j. In real problems these frequencies occur in complex conjugate pairs ($\omega_j = -\omega_{j'}$ for some j, j').

We also suppose that all the other eigenvalues of operator $\mathcal{L}$ have real part bounded away from zero, that is, hyperbolic. But recall that we are most interested in the case where all the other eigenvalues of operator $\mathcal{L}$ have real part negative because in this case the center manifold dynamics emerges exponentially quickly as a model of the long-term dynamics. Nonetheless, the procedure described herein not only applies to the case of hyperbolic y, it also applies to the construction of subcenter manifolds (Kelley, 1967a, e.g.), such as slow manifolds or nonlinear normal modes in Hamiltonian-like systems (Lamarque et al., 2012; Boivin et al., 1996; Leith, 1980, e.g.).

Recall that a center manifold is tangent to the center subspace $\mathbb{E}_c$. For the general system (17.7) suppose we have found a complete set of linearly independent eigenvectors of $\mathcal{L}$ corresponding to the m center eigenvalues. Denote these eigenvectors by e_j, $j = 1, \ldots, m$. Then the linear approximation to the center manifold is that u lies in the center subspace:

$$u = \sum_{j=1}^{m} a_j e_j e^{i\omega_j t} \tag{17.8}$$

for some, typically complex valued, amplitudes a_j.[200] The difference between the analysis of Chapter 16 and that of this chapter is primarily that there we set $u = \sum_{j=1}^{m} s_j e_j$ so that the amplitudes s_j oscillate rapidly in the linear approximation. Here, the amplitudes a_j are constant in the linear approximation (17.8) because the $e^{i\omega_j t}$ factor explicitly accounts for the oscillations. In essence, (17.8) is a time-varying basis for the center subspace. Because the amplitudes a_j are constant in the linear approximation, the amplitudes are slowly varying in the nonlinear and perturbed modeling, and consequently crucial aspects of the construction become simpler.

To model the nonlinear and perturbed dynamics of the system (17.7) we modify the center subspace (17.8). Generalizing the center subspace (17.8), seek a center manifold in the form

$$u(t) = v(\epsilon, a, t) \quad \text{such that } \dot{a} = g(\epsilon, a) \tag{17.9}$$

for some parameters ϵ measuring departure from some critical values of physical parameters. The new feature here is the *explicit* time dependence in the shape of the center manifold, $v(\epsilon, a, t)$. This explicit time dependence resolves the basic oscillations whose amplitude and frequency shifts we aim to describe: in essence we describe the center manifold in a time-dependent basis, just as section 3.6 introduced for describing oscillations in a 2D state space. However, the time dependence introduces the extra complication that by the chain rule[201]

$$\frac{du}{dt} = \frac{\partial v}{\partial t} + \sum_{j=1}^{m} \frac{\partial v}{\partial a_j}\frac{da_j}{dt} = \frac{\partial v}{\partial t} + \sum_{j=1}^{m} \frac{\partial v}{\partial a_j} g_j \,,$$

where the partial derivative $\partial/\partial t$ is done keeping a_j constant, whereas $\partial/\partial a_j$ is done keeping t constant. As usual, the total derivative d/dt accounts for time variations both indirectly through the complex amplitude a and directly in time t. This meaning of the partial derivatives is an important feature in ensuing analysis. Thus to find the center manifold we solve equations of the form

$$\frac{\partial v}{\partial t} + \sum_{j=1}^{m} g_j \frac{\partial v}{\partial a_j} = \mathcal{L}v + f(\epsilon, v) \tag{17.10}$$

for the center manifold shape $v(\epsilon, a, t)$ and the slow evolution thereon $\dot{a} = g(\epsilon, a)$.

[200] As seen in Example 17.1, in real applications any complex conjugate frequencies will generally have corresponding amplitudes that are also the complex conjugate of each other. The algebraic analysis and construction generally should and do respect complex conjugate symmetry.

[201] This complication is akin to the material derivative in fluid dynamics and represents the flow of information along the nontrivial trajectories of the underlying oscillations.

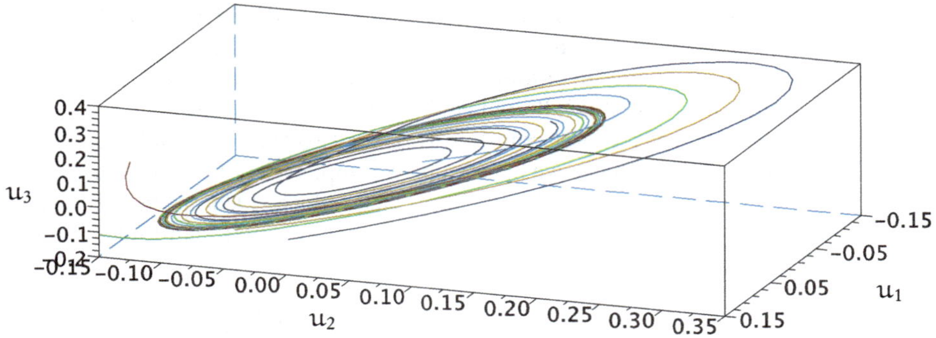

Figure 17.3. *Trajectories of the system* (16.9) *for parameter* $\epsilon = 0.1$ *illustrating a quick approach to a center manifold of oscillations. Then the system slowly evolves to a finite-amplitude limit cycle.*

Example 17.4 (a Hopf bifurcation). Reconsider the system of ODEs (16.9) for variable $\mathbf{u} = (u_1(t), u_2(t), u_3(t))$, with parameter ϵ:

$$\dot{u}_1 = -u_1 - u_2 + u_3^2, \quad \dot{u}_2 = 2u_1 + (1+\epsilon)u_2 - u_3^2, \quad \dot{u}_3 = u_1 + 2u_2 - u_3\,. \tag{17.11}$$

Figure 17.3 illustrates that trajectories quickly approach a center manifold of oscillations. This system has an equilibrium at the origin. We base the analysis and construction about this equilibrium. Further, when parameter $\epsilon = 0$, the linear operator

$$\mathcal{L} = \begin{bmatrix} -1 & -1 & 0 \\ 2 & 1 & 0 \\ 1 & 2 & -1 \end{bmatrix}$$

has eigenvalues $\lambda = -1, \pm i$. Thus the system has a center manifold, at least near the origin, that emerges from transients decaying like $\mathcal{O}(e^{-t})$. Recall from section 16.1 that the complex eigenvectors corresponding to the imaginary eigenvalues are $\mathbf{e} = (1, -1-i, (-3-i)/2)$ and its complex conjugate $\bar{\mathbf{e}} = (1, -1+i, (-3+i)/2)$. These combine to form the center subspace

$$\mathbf{v} \approx a \begin{bmatrix} 1 \\ -1-i \\ \frac{-3-i}{2} \end{bmatrix} e^{it} + \bar{a} \begin{bmatrix} 1 \\ -1+i \\ \frac{-3+i}{2} \end{bmatrix} e^{-it} \tag{17.12}$$

as the initial linear approximation to the center manifold—in a time-varying basis.

17.3 ▪ Iterative refinement forms the center manifold

Now we improve the linear center subspace approximation (17.8): first by hand, then via computer algebra.

Given any approximation to the center manifold and the evolution thereon, $\mathbf{u} \approx \mathbf{v}(\epsilon, a, t)$ such that $\dot{a} \approx g(\epsilon, a)$, seek a better approximation

$$\mathbf{u} = \mathbf{v}(\epsilon, a, t) + \hat{\mathbf{v}}(\epsilon, a, t) \quad \text{such that } \dot{a} = g(\epsilon, a) + \hat{g}(\epsilon, a)$$

for small corrections $\hat{v}$ and $\hat{g}$. Substitute these expressions into the governing differential equation (17.7):

- ignore nonlinear terms in small (hatted) corrections as being very small;
- recognize that g is here small because to leading approximation the amplitudes a_j are constant; and
- approximate v wherever it multiplies a (hatted) correction by the linear center subspace approximation (17.8).

Then the governing differential equation (17.7) becomes the center manifold homological equation for corrections[202]

$$-\mathcal{L}\hat{v}+\frac{\partial\hat{v}}{\partial t}+\sum_{j=1}^{m}\hat{g}_j e_j e^{i\omega_j t}=\underbrace{-\frac{\partial v}{\partial t}-\sum_{j=1}^{m}g_j\frac{\partial v}{\partial a_j}}_{=-dv/dt}+\mathcal{L}v+f(\epsilon,v)=\text{Res}_{(17.7)}\,. \quad (17.13)$$

As typical, the right-hand side is just the residual of the dynamical system, $-\dot{u}+\mathcal{L}u+f(\epsilon,u)$, evaluated for the current approximation. In contrast to the approach of Chapter 16, the left-hand side no longer has the term $\partial\hat{v}/\partial s\,\mathcal{G}s$ that greatly complicates the solution of the homological equation for oscillations. Consequently here the solution process is simpler. However, one increase in complication occurs in other detail due to the presence of the partial derivative $\partial\hat{v}/\partial t$ term arising from the time dependence in the basis vectors $e_j e^{i\omega_j t}$.

17.3.1 ▪ A first iteration illustrates the iterative construction

This section develops algebraic techniques to perform the iterative improvement of a center manifold model.

The driver for corrections is the homological equation (17.13). As usual suppose that, if not already in multinomial form, the nonlinearity f of the governing system (17.7) has been expanded in a Taylor series in small (ϵ, u). Then at every iteration of the construction, on the RHS of the homological equation (17.13) the residual $\text{Res}_{(17.7)}$ will be in multinomial form. In each and every term in such a multinomial, the product of exponentials will collapse into one exponential factor. Thus in each iteration the homological equation (17.13) takes the form

$$-\mathcal{L}\hat{v}+\frac{\partial\hat{v}}{\partial t}+\sum_{j=1}^{m}\hat{g}_j e_j e^{i\omega_j t}=\text{Res}_{(17.7)}=\sum_{\omega}r_\omega e^{i\omega t} \quad (17.14)$$

for some set of frequencies $\{\omega\}$ and for some coefficient vectors $r_\omega(\epsilon, a)$ that are multinomial in parameters ϵ and amplitudes a (for example, see equation (17.20) in the Hopf bifurcation Example 17.5, below).

Solving the homological equation (17.14) splits into two cases depending upon whether a particular term has a frequency that is resonant or nonresonant. The set of resonant frequencies, $\Omega := \{\omega_j \mid j = 1, \dots, m\}$ are those corresponding to the eigenvalues of the center subspace.

---

[202] A further significant simplification in deriving the homological equation (17.13) is that of neglecting aspects of generalized eigenvectors on the left-hand side. As discussed elsewhere, albeit not strictly correct, such neglect of generalized eigenvectors just slows the convergence of the iteration to the center manifold model. The neglect does not generally ruin the iterative construction.

Nonresonant, $\omega \notin \Omega$: Consider just those terms on the RHS of the homological equation (17.14) that are nonresonant, $\omega \notin \Omega$. For each such term, the linear operator $(-\mathcal{L} + \partial/\partial t)$ is invertible (the inverse is unique). Thus the sum of such terms on the RHS of the homological equation (17.14) generates components in the shape of the center manifold of the sum

$$\hat{v}' = \sum_{\omega \notin \Omega} \left(-\mathcal{L} + \frac{\partial}{\partial t}\right)^{-1} \left[r_\omega e^{i\omega t}\right]. \tag{17.15}$$

Resonant, $\omega \in \Omega$: Consider the remaining resonant terms on the RHS of the homological equation (17.14). Such resonant terms force corrections to the evolution via $\hat{g}_j$. They force evolution corrections because the linear operator $(-\mathcal{L} + \partial/\partial t)$ is singular for such resonant frequencies, and so the RHS must be projected in the range of $(-\mathcal{L} + \partial/\partial t)$ using $\hat{g}_j$.

To do such projection we generally need to define an appropriate inner product such as

$$\langle v_1, v_2 \rangle := \frac{1}{T} \int_0^T \bar{v}_1^T v_2 \, dt \tag{17.16}$$

for sufficiently large T, which is the average over many periods of the normal vector inner product.[203] In such an inner product, the operator $(-\mathcal{L} + \partial/\partial t)$ has adjoint $(-\mathcal{L}^T - \partial/\partial t)$. Determine the adjoint eigenvector (generalized when necessary) corresponding to the center eigenvalue $\lambda = i\omega_j$, and denote it by $z_j(t)$. For convenience in formulas we put the time periodic dependence into these adjoint eigenvectors, $z_j \propto e^{i\omega_j t}$, while continuing to write the direct eigenvectors as the explicit product $e_j e^{i\omega_j t}$. Also for convenience we normalize the adjoint eigenvectors so that

$$\left\langle z_k(t), e_j e^{i\omega_j t} \right\rangle = \begin{cases} 1 & \text{if } j = k, \\ 0 & \text{if } j \neq k. \end{cases} \tag{17.17}$$

Now a projection empowers us to solve the resonant components of the homological equation (17.14). For each mode $k = 1, \ldots, m$ consider the projection $\langle z_k(t), \cdot \rangle$ of the homological equation (17.14):

- the projection of the LHS is

$$\begin{aligned} &\left\langle z_k(t), -\mathcal{L}\hat{v} + \frac{\partial \hat{v}}{\partial t} \right\rangle + \sum_{j=1}^m \hat{g}_j \left\langle z_k(t), e_j e^{i\omega_j t} \right\rangle \\ &= \left\langle -\mathcal{L}^T z_k - \frac{\partial z_k}{\partial t}, \hat{v} \right\rangle + \hat{g}_k = 0 + \hat{g}_k, \end{aligned}$$

- whereas the projection of the RHS is

$$\left\langle z_k(t), \sum_\omega r_\omega e^{i\omega t} \right\rangle = \sum_\omega \left\langle z_k, r_\omega e^{i\omega t} \right\rangle = \left\langle z_k, r_{\omega_k} e^{i\omega_k t} \right\rangle$$

[203] One subtlety is that the time integral in this inner product is done keeping amplitudes a and $\bar{a}$ constant. The reason for this "freezing" of the otherwise time-varying amplitudes is that the time integral arises as the inverse of the time differentiation in the left-hand side operator $-\mathcal{L} + \partial/\partial t$, and in this operator the time derivative is done keeping the amplitudes constant.

because the averaging over time in the inner product eliminates all frequencies except the frequency ω_k in the adjoint eigenvector $\mathbf{z}_k(t)$.

Thus, by the adjoint and orthogonality, the projection of the homological equation (17.14) simplifies to just

$$\hat{g}_k = \langle \mathbf{z}_k(t), \mathbf{r}_{\omega_k} e^{i\omega_k t} \rangle, \quad k = 1, \ldots, m, \tag{17.18}$$

which uniquely determines all the corrections to the evolution. Recall that this is the solvability condition.

Substituting these into the homological equation (17.14), putting all known terms into the RHS, we find an equation for resonant corrections to be

$$-\mathcal{L}\hat{\mathbf{v}}'' + \frac{\partial \hat{\mathbf{v}}''}{\partial t} = \sum_{j=1}^{m} \left[\mathbf{r}_{\omega_j} - \langle \mathbf{z}_j, \mathbf{r}_{\omega_j} e^{i\omega_j t} \rangle \mathbf{e}_j \right] e^{i\omega_j t}.$$

The RHS is now in the range of the operator $(-\mathcal{L} + \partial/\partial t)$. Consequently, the resonant updates are formally

$$\hat{\mathbf{v}}'' = \sum_{j=1}^{m} \left(-\mathcal{L} + \frac{\partial}{\partial t} \right)^{-1} \left[\mathbf{r}_{\omega_j} - \langle \mathbf{z}_j, \mathbf{r}_{\omega_j} e^{i\omega_j t} \rangle \mathbf{e}_j \right] e^{i\omega_j t} \tag{17.19}$$

for any generalized inverse $(-\mathcal{L} + \partial/\partial t)^{-1}$. It is best to choose the specific generalized inverse to respect your definition of the amplitudes a_j.

After finding the resonant and nonresonant contributions, the correction to the center manifold shape is just $\hat{\mathbf{v}} = \hat{\mathbf{v}}' + \hat{\mathbf{v}}''$. Equation (17.18) determines the correction to the evolution.

As usual, each iteration determines corrections to higher and higher order in small parameters and amplitudes. We terminate the iteration when the residuals are zero to any specified order. Then the Approximation Theorems 4.8 and 4.10 assure us that the center manifold is constructed to the same order of error.

Example 17.5 (continue Example 17.4). First, substitute the linear center subspace approximation (17.12) into the iterative homological equation (17.13) to find here that corrections satisfy

$$\begin{aligned} &-\mathcal{L}\hat{\mathbf{v}} + \frac{\partial \hat{\mathbf{v}}}{\partial t} + \left(\mathbf{e}\hat{g} e^{it} + \bar{\mathbf{e}}\hat{\bar{g}} e^{-it} \right) \\ &= \begin{bmatrix} 1 \\ -1 \\ 0 \end{bmatrix} \left[\frac{4+3i}{2} a^2 e^{i2t} + 5a\bar{a} + \frac{4-3i}{2} \bar{a}^2 e^{-i2t} \right] \\ &\quad + \epsilon \begin{bmatrix} 0 \\ 1 \\ 0 \end{bmatrix} \left[(-1-i) a e^{it} + (-1+i) \bar{a} e^{-it} \right]. \end{aligned} \tag{17.20}$$

The right-hand side residual is of the form of a sum of terms each involving a complex exponential $e^{i\omega t}$ for frequencies $\omega = -2, \ldots, 2$, including both resonant frequencies, $\omega \in \Omega = \{\pm 1\}$, and nonresonant frequencies, $\omega \in \{-2, 0, 2\}$.

Nonresonant corrections Let's use the method of undetermined coefficients. For each such term in the residual, say $\mathbf{r}e^{i\omega t}$, pose that the component gives rise to a component in $\hat{\mathbf{v}}$ of the form $\mathbf{c}e^{i\omega t}$. Substitute into the left-hand side of (17.20) to obtain a linear equation for the vector coefficient $\mathbf{v}$ of the form

$$[-\mathcal{L} + i\omega I]\,\mathbf{c} = \mathbf{r}\,.$$

For the matrix $\mathcal{L}$ in system (17.11), this linear equation has solution

$$\mathbf{c} = \frac{1}{(1+i\omega)(\omega^2-1)}\begin{bmatrix} 1+\omega^2 & 1+i\omega & 0 \\ -2(1+i\omega) & (\omega-i)^2 & 0 \\ -(3+i\omega) & -(1+2i\omega) & \omega^2-1 \end{bmatrix}\mathbf{r}\,. \tag{17.21}$$

Solving for the $\omega = \pm 2$ and $\omega = 0$ components in the homological equation (17.20) we deduce that the update $\hat{\mathbf{v}}$ contains the terms

$$\begin{bmatrix} (1-\frac{4}{3}i)a^2e^{i2t} + (1+\frac{4}{3}i)\bar{a}^2e^{-i2t} \\ (-\frac{5}{3}+\frac{5}{6}i)a^2e^{i2t} + 5a\bar{a} + (-\frac{5}{3}-\frac{5}{6}i)\bar{a}^2e^{-i2t} \\ (-\frac{1}{3}+i)a^2e^{i2t} + 10a\bar{a} + (-\frac{1}{3}-i)\bar{a}^2e^{-i2t} \end{bmatrix}.$$

Resonant corrections However, the $\omega = \pm 1$ components are not so straightforward. The matrix inverse is singular for frequencies $\omega = \pm 1$, as seen in the division by $\omega^2 - 1$, which is zero for $\omega = \pm 1$. As is *always* the case for the construction of a center manifold, the operator on the left-hand side, here $-\mathcal{L} + \partial/\partial t$, is singular at those frequencies that characterize the center manifold. It has to be singular because that is the only way a nontrivial set of eigenvectors can span a center eigenspace. As is always the case, we choose terms in the evolution, $\hat{g}$, so that the right-hand side is put in the range of the operator $-\mathcal{L} + \partial/\partial t$ (the solvability condition).

A projection involving an inner product with adjoint eigenvectors makes the homological equation solvable. We obtain the adjoint eigenvector $\mathbf{z}$ corresponding to the critical oscillation by solving $\partial\mathbf{z}/\partial t = -\mathcal{L}^T\mathbf{z}$. Here let's choose $\mathbf{z} = (\frac{1-i}{2}, -\frac{i}{2}, 0)e^{it}$ and its complex conjugate $\bar{\mathbf{z}}$. These adjoint eigenvectors are normalized so that $\langle \mathbf{z}, \mathbf{e}e^{it}\rangle = 1$ and $\langle \mathbf{z}, \bar{\mathbf{e}}e^{-it}\rangle = 0$, and similarly for the complex conjugate. Then the solvability projection (17.18) applied to (17.20) finds the update to the evolution to be

$$\hat{g} = \langle \mathbf{z}, \text{RHS}\rangle = \frac{1-i}{2}\epsilon a\,,$$

and correspondingly for the complex conjugate.

Consequently, the first nontrivial approximation to the nonlinear/perturbed evolution is $\dot{a} = \frac{1-i}{2}\epsilon a$. This ODE has solution $a(t) = a_0e^{\epsilon(1-i)t/2} = a_0e^{\epsilon t/2}e^{-i\epsilon t/2}$, which through $e^{\epsilon t/2}$ describes exponentially growing or decaying solutions depending upon whether the parameter ϵ is positive or negative, respectively. The factor $e^{-i\epsilon t/2}$ just shifts the frequency of the basic oscillations. Contrast this model with the equivalent evolution equation (16.15): there it is complicated by the quadratic terms, which here we see have no significant effect on the long-term evolution of the oscillation amplitudes—they just affect the shape of the oscillations, which here is accounted for in the update $\hat{\mathbf{v}}$.

Having chosen $\hat{g}$, the right-hand side is now in the range of the singular linear operator $-\mathcal{L} + \partial/\partial t$. However, the right-hand side still contains components in $e^{\pm it}$; it is just that an update $\hat{v}$ can be found to suit. There will be two degrees of freedom in the solution: the freedom corresponds to an arbitrary linear combination of the critical eigenvectors. We must impose an extra condition upon the updates $\hat{v}$ in order to make the solution unique. For example, we might decide to seek solutions containing the form

$$\hat{v} = ce^{\pm it} = \begin{bmatrix} 0 \\ c_2 \\ c_3 \end{bmatrix} e^{\pm it},$$

that is, with $\hat{u}_1 = 0$ in any $e^{\pm it}$ components. In essence this decision[204] defines the oscillation amplitude a to be the component of e^{it} in $u_1(t)$ of the original variables. Similarly, $\bar{a}$ is the component of e^{-it} in $u_1(t)$. Seeking solutions in this form, and ignoring the first component of the equation, which is allowed because we know that the right-hand side is in the range, components in the right-hand side of the form $re^{\pm it}$ give

$$c = \begin{bmatrix} 0 & 0 & 0 \\ 0 & \frac{1}{-1\pm i} & 0 \\ 0 & -1 & \frac{1}{1\pm i} \end{bmatrix} r. \tag{17.22}$$

Here the components in $e^{\pm it}$ give rise to components in $\hat{v}$ of

$$\begin{bmatrix} 0 \\ \frac{-1+i}{2} \\ \frac{1+i}{4} \end{bmatrix} \epsilon a e^{it} + \begin{bmatrix} 0 \\ \frac{-1-i}{2} \\ \frac{1-i}{4} \end{bmatrix} \epsilon \bar{a} e^{-it}.$$

Gathering all components gives that the first nonlinear approximation to the shape of the center manifold and the rapid oscillations thereon are approximately described by

$$u \approx \begin{bmatrix} 1 \\ -1 - i + \epsilon\frac{-1+i}{2} \\ \frac{-3-i}{2} + \epsilon\frac{1+i}{4} \end{bmatrix} ae^{it} + \begin{bmatrix} (1 - \frac{4}{3}i) \\ (-\frac{5}{3} + \frac{5}{6}i) \\ (-\frac{1}{3} + i) \end{bmatrix} a^2 e^{i2t} + \begin{bmatrix} 0 \\ 5a\bar{a} \\ 10a\bar{a} \end{bmatrix} + \text{c.c.},$$

where c.c. denotes the complex conjugate of the first two terms. Further iterations, using analogous solutions, generate higher-order corrections that discern the emergent Hopf bifurcation in this problem. But the algebra is both routine and tedious, so it is ideal for computer algebra.

17.3.2 ▪ Computer algebra speeds iteration

Computer algebra routinely performs high-order computations. Indeed, what follows underlies my web service (Roberts, 2012), which constructs center manifold models for any (multinomial) system input by a user.[205] We express the iteration discussed in section 17.3.1 in the algorithm applied to Examples 17.4 and 17.5.

[204] One must always define the amplitudes, whether implicitly (as before) or explicitly.

[205] `http://www.maths.adelaide.edu.au/anthony.roberts/gencm.php` provides the service at the time of writing.

Algorithm 17.2 *Iteration forms the center manifold model of the elementary Hopf bifurcation Examples 17.4 and 17.5. This preamble defines operators for the problem. Algorithm 17.3 lists the code for the iteration.*

```
operator cis;
let { cis(~m)*cis(~n) => cis(m+n)
    , cis(~m)^~p => cis(p*m)
    , df(cis(~th),~x) => i*cis(th)*df(th,x)
    , cis(0)=>1    };
operator ee;  depend ee,cis;
ea:=cis(+t)*(ee(1)+(-1-i)*ee(2)+(-3-i)/2*ee(3));
eb:=cis(-t)*(ee(1)+(-1+i)*ee(2)+(-3+i)/2*ee(3));
infix #; precedence #,^;
let ~u#~j => coeffn(u,ee(j),1);

operator ll; linear ll;
let {ll(ee(1),ee)=>-ee(1)+2*ee(2)+ee(3)
    ,ll(ee(2),ee)=>-ee(1)+ee(2)+2*ee(3)
    ,ll(ee(3),ee)=>-ee(3)    };

operator linv; linear linv;
let {linv(ee(1)*cis(~w*t),cis)=>
        (w^2+1)*cis(w*t)/(w^2-1)/(i*w+1)*ee(1)
        -2*cis(w*t)/(w^2-1)*ee(2)
        -(i*w+3)*cis(w*t)/(w^2-1)/(i*w+1)*ee(3)
    ,linv(ee(2)*cis(~w*t),cis)=>cis(w*t)/(w^2-1)*ee(1)
        +(w-i)^2*cis(w*t)/(w^2-1)/(i*w+1)*ee(2)
        -(2*i*w+1)*cis(w*t)/(w^2-1)/(i*w+1)*ee(3)
    ,linv(ee(3)*cis(~w*t),cis)=>cis(w*t)/(i*w+1)*ee(3)
    ,linv(ee(1),cis)=>-ee(1)+2*ee(2)+3*ee(3)
    ,linv(ee(2),cis)=>-ee(1)+ee(2)+ee(3)
    ,linv(ee(3),cis)=>+ee(3)
    ,linv(ee(1)*cis(~w),cis)=>0 when w^2=t^2
    ,linv(ee(2)*cis(~w),cis)=>-cis(w)*ee(3)
        +cis(w)/(i*df(w,t)-1)*ee(2) when w^2=t^2
    ,linv(ee(3)*cis(~w),cis)=>
        cis(w)/(i*df(w,t)+1)*ee(3) when w^2=t^2    };
```

Algorithms 17.2 and 17.3 perform the iterative construction of a center manifold model. Before exploring how the algorithms implement the iteration of section 17.3, let's briefly discuss the results. The algorithms deduce that the modulation of the oscillations is described by

$$\dot{a} = \frac{1-i}{2}\epsilon a - \left(15 + \frac{20}{3}i\right)\bar{a}a^2 + \mathcal{O}\big(\epsilon^2 + a^4\big)\,. \tag{17.23}$$

This model is considerably simpler in structure and in detail than the analogous evolution equation (16.17). Earlier we saw that linearly in amplitude a the origin is exponentially attractive for $\epsilon < 0$ and exponentially unstable for $\epsilon > 0$. The nonlinear model (17.23) straightforwardly determines the stable limit cycle that exists for parameter $\epsilon > 0$—something not discernible previously except via numerical simulations—that character-

Algorithm 17.3 *Iteration forms the center manifold model of the elementary Hopf bifurcation Examples* 17.4 *and* 17.5. *Following the preamble definitions of Algorithm* 17.2, *this code uses the residual of the governing* ODEs *to derive corrections to the center manifold model.*

```
factor ee,small,eps;
depend a,t; depend b,t;
let { df(a,t)=>ga, df(b,t)=>gb };
u:=small*( a*ea+b*eb );
ga:=gb:=0;
let {eps^2=>0, small^4=>0};
for iter:=1:99 do begin
    res:=-df(u,t)+ll(u,ee)+eps*ee(2)*u#2+(ee(1)-ee(2))*u#3^2;
    ga:=ga+(gad:=coeffn((1+i)/2*res#1+i/2*res#2,cis(+t),1))/small;
    gb:=gb+(gbd:=coeffn((1-i)/2*res#1-i/2*res#2,cis(-t),1))/small;
    u:=u+linv(res-gad*ea-gbd*eb ,cis);
    if res=0 then write iter:=1000000+iter;
end;
```

izes the Hopf bifurcation. To see this limit cycle, just seek a solution $a = re^{i\phi}$, where r and the phase ϕ are real, slowly varying functions of time: r measures the amplitude of the oscillations and $\dot\phi$ is the frequency shift of the oscillations. Substituting into (17.23) and dividing by $e^{i\phi}$ leads to

$$\dot r + ir\dot\phi = \frac{1-i}{2}\epsilon r - \left(15 + \frac{20}{3}i\right) r^3 .$$

- The real part of this equation gives

$$\dot r = \frac{1}{2}\epsilon r - 15r^3 = \frac{1}{2}r\left(\epsilon - 30r^2\right),$$

and consequently if $\epsilon < 0$, then r evolves to zero, which represents the fixed point at the origin; if $\epsilon > 0$, then r evolves to $r_\circ = \sqrt{\epsilon/30}$, which is the amplitude of the stable limit cycle. Similar to the pitchfork bifurcation, the amplitude of this nonlinear state grows like $\sqrt{\epsilon}$ near critical.

- Upon dividing by r, the imaginary part gives that

$$\dot\phi = -\frac{\epsilon}{2} - \frac{20}{3}r^2 .$$

This frequency shift $-\epsilon/2$ at amplitude $r = 0$ is also determined by linear analysis, whereas on the limit cycle, $r = r_\circ$, the frequency shift is

$$\dot\phi = -\frac{\epsilon}{2} - \frac{20}{3}r_\circ^2 = -\frac{13}{18}\epsilon .$$

This frequency shift is different from that for the origin because the nonlinear limit cycle is distinctly different.

Now explore the details of Algorithms 17.2 and 17.3. First consider the preamble Algorithm 17.2.

Lines 1–5 Set up `cis(w*t)` to denote the complex exponential $e^{i\omega t}$. It is easier to do this than use the inbuilt exponential.

Lines 6–8 The matrix facility in Reduce is primitive so instead we invoke the operator `ee` to algebraically represent basis vectors for the vector $\mathbf{u}$. For example, store the basic linear oscillation on the center eigenspace in `ea` and `eb` using these basis vectors.

Lines 9–10 The infix operator `#` usefully extracts a specified component from a vector using the `ee` basis.

Lines 12–15 Define the action of the matrix $\mathcal{L}$ as the linear operator `ll`. Since $\mathcal{L}\mathbf{e}_j$ is the jth column of matrix $\mathcal{L}$, we code `ll(ee(j),ee)` to transform to the `j`th column of $\mathcal{L}$.

Lines 17–33 Last, in this preamble, define the operator `linv` to be the inverse of $-\mathcal{L}+\partial/\partial t$ when applied to an RHS with a sum of terms in $e^{i\omega t}$. The definition is split into three cases: $|\omega| \neq 0, 1$, $\omega = 0$, and $|\omega| = 1$. In the last of the three cases we cater for the singular nature of the operator $-\mathcal{L}+\partial/\partial t$.

Lines 18–25 Code the inverse $(-\mathcal{L}+\partial/\partial t)^{-1}$, as given by equation (17.21), column by column for the case of nonzero and nonresonant frequency.

Lines 26–28 Code the inverse for the specific case of zero frequency, obtained by setting $\omega = 0$ in the previous.

Lines 29–33 Code a generalized inverse $(-\mathcal{L}+\partial/\partial t)^{-1}$, as given by equation (17.21), column by column for the singular case of resonant frequency. It assumes the RHS is properly in the range of the singular operator.

Second, consider the constructive iteration coded in Algorithm 17.3.

Line 1 Structure the printing of results.

Lines 2–3 Parametrize the center manifold by two complex amplitudes, a and b, that evolve slowly in time according to expressions stored in `ga` and `gb`.

Lines 4–5 Start from the linear center subspace approximation of the oscillations `ea` and `eb` with amplitudes a and b being constant.

Line 6 Truncate the construction to residuals and errors $\mathcal{O}(\epsilon^2, |a|^4)$. Use the variable `small` to count the number of amplitudes a and b in each term (as already invoked in the initial approximation).

Lines 7,12,13 Iterate until the residual of the ODEs is zero to the specified error.

Line 8 Compute the residual of the ODEs.

Lines 9–10 Extract the resonant components from the residual using the adjoint eigenvectors, and correct the evolution on the center manifold.

Line 11 Such corrections put the RHS of the homological equation in the range of $-\mathcal{L}+\partial/\partial t$, so use the coded (generalized) inverse of $-\mathcal{L}+\partial/\partial t$ to correct the shape of the center manifold.

Algorithms 17.2 and 17.3 may be adapted to form the center manifold model of a wide range of systems. For example, simply modify the computation of the residual to find a center manifold model for systems with the same linear operator $\mathcal{L}$ but different nonlinear/perturbation functions $\mathbf{f}(\epsilon, \mathbf{u})$.

17.4 ▪ Case study: Delay differential equations lose stability to oscillations

As very nicely introduced by Erneux (2009), delay differential equations (DDEs) are often of interest in applications, especially in control or biological systems (Verdugo and Rand, 2008; Yamaguchi et al., 2011, e.g.). This section illuminates how DDEs often exhibit oscillatory instabilities. The first example is the canonical system for delays. Subsequent applications invoke the first case as a key part of the analysis.

17.4.1 ▪ The canonical delay differential equation

When first introduced to differential equations, you are sure to have been introduced to the solutions $u(t)$ of the ODE $du/dt = \alpha u$. Recall that for all (constant) parameters α, solutions to this ODE are $u = ce^{\alpha t}$ for some integration constant c. This exponential solution is the key to developing solutions for higher-order ODEs (involving trigonometric functions as well as exponentials). These in turn form a foundation for our knowledge of ordinary differential equations.

In DDEs, a similarly simple system is the key to further development. For some time delay τ, the fundamental DDE is

$$\frac{du}{dt} = \alpha u(t-\tau), \tag{17.24}$$

where the left-hand side is the time derivative at time t, and the right-hand side is the function u evaluated at a time delay τ before the current time. However, let's make our analysis easier by nondimensionalizing the problem so that the time delay is set to 1: set $t^* = t/\tau$ and $\alpha^* = \alpha\tau$; then the DDE (17.24) becomes $du/dt^* = \alpha^* u(t-1)$. Omitting stars for simplicity, the nondimensional canonical DDE is

$$\frac{du}{dt} = \alpha u(t-1). \tag{17.25}$$

Let's seek to solve this canonical DDE (17.25). Because it is constant coefficient we seek solutions $u = ce^{\lambda t}$. Substituting this exponential into the DDE (17.25) gives $c\lambda e^{\lambda t} = \alpha c e^{\lambda(t-1)} = \alpha c e^{\lambda t} e^{-\lambda}$. Dividing by $ce^{\lambda t}$ leads to the characteristic equation for the eigenvalues that

$$\lambda = \alpha e^{-\lambda}. \tag{17.26}$$

This transcendental equation for the eigenvalues λ is hard to solve (except for the trivial case when parameter $\alpha = 0$). We find parametric solutions.

Real eigenvalues Rearrange the characteristic equation (17.26) to $\alpha = \lambda e^{\lambda}$; this equation expresses the coefficient α as a function of the eigenvalue. Figure 17.4 plots these real eigenvalues as the black curve. For coefficient $\alpha > 0$ there are growing solutions, $\lambda > 0$, as befits the traditional exponential character of the DDE (17.25).

But the transcendental nature of the characteristic equation (17.26), via the exponential term, gives many more modes. Indeed, it gives an infinity of modes.

Complex eigenvalues Separately identify the real and imaginary parts of the eigenvalue $\lambda = \mu + i\omega$; then the characteristic equation (17.26) becomes

$$\mu + i\omega = \alpha e^{-\mu - i\omega} = \alpha e^{-\mu}\cos\omega - i\alpha e^{-\mu}\sin\omega.$$

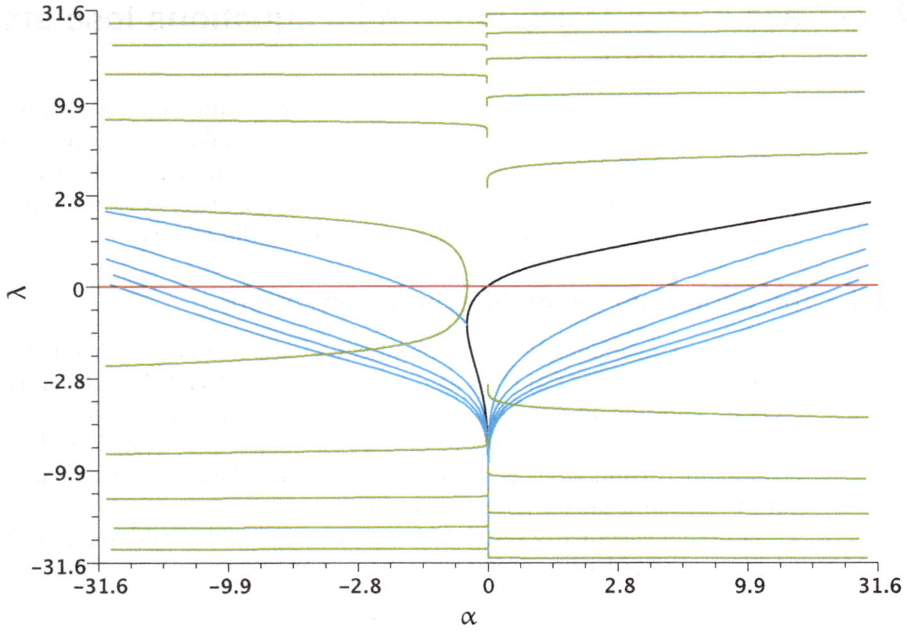

Figure 17.4. *(Black) the real eigenvalues, and (blue) real parts and (green) imaginary parts of the eigenvalues of the canonical DDE* (17.25) *as a function of coefficient* α. *There are an infinite number of branches of eigenvalues; this plot shows the first five complex conjugate pairs. The axes are stretched to form a quasi-loglog plot.*

Equate real and imaginary parts to find we need to simultaneously solve the pair of real equations $\mu = \alpha e^{-\mu}\cos\omega$ and $\omega = -\alpha e^{-\mu}\sin\omega$. From the second equation, $\alpha e^{-\mu} = -\omega/\sin\omega$, which when substituted into the first equation gives the real part

$$\mu = -\omega\cot\omega\,, \quad \text{and then coefficient} \quad \alpha = -\omega e^{\mu}/\sin\omega. \tag{17.27}$$

Equation (17.27) gives real part μ and coefficient α as a function of the imaginary part ω of the eigenvalue. Figure 17.4 plots the first five branches of complex conjugate pairs of eigenvalues.[206]

Oscillatory bifurcations appear Figure 17.4 shows that for small coefficient α the real parts of all the eigenvalues are all negative (blue curves), and hence the solutions of the DDE (17.25) are stable. However, for large enough negative coefficient α, the real parts become positive to predict unstable growing solutions.[207] Now, a mode becomes unstable when the real part of its eigenvalue crosses zero. From equation (17.27) this zero real part occurs when $\omega\cot\omega = 0$, that is, when $\omega = -\pi/2 + k\pi$ for any integer k. The case $\omega = -\pi/2$ gives the first instability for negative coefficients, $\alpha < 0$; then the

[206] Where do the infinite number of eigenvalues come from? The answer is that a DDE is an infinite-dimensional dynamical system, even a DDE as innocuously simple looking as the canonical (17.24). The infinite dimensionality arises because to compute the right-hand side of the DDE (17.24) one has to remember the values of the field u for all times $[t-\tau, t]$. The DDE (17.24) is actually a prescription for updating u over this interval. Such an interval has an infinite number of components.

[207] There are also growing modes for large enough positive coefficient a, but their growth rate is always less than that of the real eigenvalue (black).

second equation in (17.27) gives $\alpha = -(-\frac{\pi}{2})e^0/\sin(-\pi/2) = -\pi/2$. That is, it is only for parameter $-\pi/2 < \alpha < 0$ that the canonical DDE (17.25) has only decaying solutions.

As the coefficient α crosses the critical value $-\pi/2$, to more negative rates, an oscillatory growth occurs in solutions to the DDE (17.25); at onset the oscillations have frequency $\pi/2$. Correspondingly, for the dimensional DDE (17.24), an oscillatory instability occurs for $\alpha\tau < -\pi/2$ with frequency at onset of $\pi/(2\tau)$.

17.4.2 ▪ Oscillations appear in the delayed logistic model

In population dynamics the logistic differential equation is one of the basic models for the number of animals in a population. Let $n(t)$ denote the number at time t. The logistic ODE is $dn/dt = rn(1 - n/c)$, which arises from postulating that animals breed at rate r and die due to predation or disease so that the natural carrying capacity is c. This logistic model predicts smooth growth in population $n(t)$ to the carrying capacity c.

But often animals breed seasonally. Instead of the continuous time dependence of the logistic differential equation, the logistic map is a discrete time model $n_{k+1} = (1+r)n_k - rn_k^2/c$ postulated to reflect seasonal breeding. For small to moderate breeding growth rates r it also predicts smooth population growth to the carrying capacity c. However, at larger breeding rates, the population becomes unstable to oscillations, whereas chaos occurs at still larger breeding rates.

This section shows that population oscillations are also predicted by a DDE population model. The logistic DDE we explore is

$$\frac{dn}{dt} = rn(t)[1 - n(t-\tau)/c]. \tag{17.28}$$

There are two equilibria of this logistic DDE: $n = 0$ and $n = c$. Linearize about each in turn. First, for n near zero, the logistic DDE becomes the classic $\dot{n} = rn$, which is unstable for the positive breeding rates r under consideration. Second, for population n near the carrying capacity c, change variables to $n(t) = c + u(t)$, and the logistic DDE becomes

$$\frac{du}{dt} = -r[1 + u(t)/c]u(t-\tau). \tag{17.29}$$

For small departures u from the carrying capacity, this DDE is $\dot{u} \approx -ru(t-\tau)$, that is, the canonical DDE (17.24) with parameter $a = -r$ and delay τ. Section 17.4.1 then assures us that oscillations of frequency $\pi/(2\tau)$ will be established in the population model, via some form of Hopf bifurcation, for parameters such that $r\tau > \pi/2$.

17.4.3 ▪ The center manifold of the Hopf bifurcation

We construct the center manifold model of the Hopf bifurcation that occurs in the logistic DDE (17.29) for parameters $r\tau \approx \pi/2$. It is most convenient to nondimensionalize so that the oscillations at the onset of the bifurcation have frequency 1. Scaling time to $t^* = t\pi/(2\tau)$ and scaling the population count with the carrying capacity, $u^*(t^*) = u(t)/c$, the DDE (17.29) becomes $du^*/dt^* = -r^*(1+u^*)u^*(t^*-\pi/2)$ for parameter $r^* = r\tau 2/\pi$. Omitting the stars, henceforth we construct the center manifold of

$$\frac{du}{dt} = -r[1 + u(t)]u(t - \pi/2) \tag{17.30}$$

for parameter near 1, $r = 1 + \epsilon$, to find this DDE's Hopf bifurcation.

Hereafter we work toward explaining how Algorithm 17.4, below, approximates a center manifold of the DDE (17.30) as

$$u = ae^{it} + \tfrac{2-i}{5}a^2e^{i2t} + \tfrac{1-3i}{20}a^3e^{i3t} + \text{c.c.} + \mathcal{O}(a^4+\epsilon^2). \tag{17.31a}$$

On this center manifold the complex amplitude a evolves according to the complex Landau equation

$$\dot a = \frac{1}{1+i\pi/2}\left[i\epsilon a - \tfrac{1-3i}{5}|a|^2a\right] + \mathcal{O}(a^4+\epsilon^2). \tag{17.31b}$$

Multiplying the numerator and denominator by $1-i\pi/2$, the real part of the growth rate on the right-hand side becomes $\frac{\pi}{2}\epsilon - \frac{1}{5}(\frac{3\pi}{2}-1)|a|^2$ so that this model of the evolution on the center manifold (17.31a) predicts oscillations in u of amplitude $|a| = \sqrt{5\epsilon/(3-2/\pi)}$.

Existence and emergence To construct the center manifold model (17.31a) we first identify the linearized dynamics about some equilibrium. Recall that the dynamics we explore is based about the equilibrium at the carrying capacity: the nondimensional number of animals is $n(t) = 1 + u(t)$. About the equilibrium, $n = 1$ or equivalently $u = 0$, the linearized dynamics are that of the canonical DDE (17.24). Figure 17.4 shows the spectrum. From the spectrum, the existence theorem asserts that a center manifold model may be built from the center subspace of oscillations with frequency 1 that exists when nondimensional parameter $r = 1$ in (17.30).[208] The emergence theorem assures us that all initial conditions in some finite neighborhood of the equilibrium $(u, r) = (0, 1)$ are attracted exponentially quickly to the oscillating dynamics on the center manifold (17.31a): Figure 17.4 suggests that the rate of decay of the leading fast mode is $\beta \approx 1.60$, which in our nondimensionalization is rate $2\beta/\pi \approx 1$, and hence the decay to the center manifold (17.31a) should be roughly $\mathcal{O}(e^{-t})$.

Construction Algorithm 17.4 constructs the center manifold as follows.

- Define the operator `cis` $\theta = e^{i\theta}$. The first four properties defined for `cis`, in the `let` statement, are as invoked before. The last two properties are new and due to the delay in the DDE: we want `cis`$(\theta + k\pi/2) = i^k e^{i\theta}$ but have to cater for whether there is an overall denominator of 2 in the argument or not.

  We address operator `linv` later.

- The center manifold (17.31a) is best described with two time-varying basis "vectors" $e^{\pm it}$:[209] linearly, $u \approx ae^{it} + \bar a e^{-it}$. As before, it is most convenient, in computer algebra, to represent the complex conjugate of the amplitude a as an independent amplitude `b`. A real center manifold is obtained by ultimately setting $b = \bar a$; that the expressions we develop preserve complex conjugate symmetry is a useful error check.

  As before, specify that amplitudes `a` and `b` depend upon time and that their time derivatives are $\dot a =$ `ga` and $\dot b =$ `gb` for some expressions we need to find.

[208] The frequent caveat is that this DDE is infinite dimensional whereas the theorems as quoted are finite dimensional. We should invoke the more widely applicable infinite-dimensional theorems of Vanderbauwhede and Iooss (1988) or Haragus and Iooss (2011).

[209] In a DDE these vectors are a little unusual. At any time t the symbol "$e^{\pm it}$" denotes the entire immediate past dependence of $e^{\pm i\theta}$ for times θ over $t - \pi/2 \leqslant \theta \leqslant t$. It is this dependence upon the immediate past that the symbol represents.

Algorithm 17.4 *Iteratively construct the center manifold of the logistic* DDE (17.30) *using complex-amplitude modulation.*

```
factor cis,a,b,small;
operator cis;
let { cis(0)=>1
    , cis(~u)*cis(~v)=>cis(u+v)
    , cis(~u)^~p=>cis(p*u)
    , df(cis(~u),~v)=>i*cis(u)*df(u,v)
    , cis(~u+~~k*pi)=>(-1)^k*cis(u)
    , cis((~u+~~k*pi)/2)=>i^(k)*cis(u/2) };
operator linv; linear linv;
let { linv(cis(~k*t),cis)=>-cis(k*t)/(i*k+(-i)^k)
    , linv(1,cis)=>-1 };
depend a,t; depend b,t;
let { df(a,t)=>ga, df(b,t)=>gb };
u:=a*cis(t)+b*cis(-t);
ga:=gb:=0;
let small^5=>0;   % use small to count order of a term
r:=1+small^2*eps; % scale r-1 with amplitudes
% Express solution in terms of two coefficients
% ca:=1/(1+i*pi/2); cb:=1/(1-i*pi/2);
let { pi*ca=>-2*i+2*i*ca
    , pi*cb=>+2*i-2*i*cb
    , ca*cb=>(ca+cb)/2 };
for iter:=1:9 do begin
    as:=a+(for n:=1:5 sum (-pi/2)^n*df(a,t,n)/factorial(n));
    bs:=b+(for n:=1:5 sum (-pi/2)^n*df(b,t,n)/factorial(n));
    resdde:=df(u,t)+r*(1+small*u)*sub({t=t-pi/2,a=as,b=bs},u);
    ga:=ga-ca*(cad:=coeffn(resdde,cis(+t),1));
    gb:=gb-cb*(cbd:=coeffn(resdde,cis(-t),1));
    u:=u+linv(resdde-cad*cis(t)-cbd*cis(-t),cis);
    if resdde=0 then write iter:=1000000+iter;
end;
```

- The initial approximation to the center manifold (17.31a) is that $u = ae^{it} + be^{-it}$ such that $\dot{a} = \dot{b} = 0$.

- The parameter r has to be close to critical, so we set $r = 1 + \epsilon$. Computer algebra must truncate expressions to some order of error: let's use the artificial variable `small` to count the number (less 1) of amplitudes in a term and work, for example, to residuals and errors $\mathcal{O}(\texttt{small}^5)$ via its `let` statement. We also need to truncate in the parameter ϵ: one way is to count it with `small` also, as done here by setting `r:=1+small^2*eps` so that each `eps` is "worth" two amplitudes.

- In the construction of the center manifold (17.31a) we have to divide by $1 \pm i\pi/2$. Such divisions generate awkward rational function forms—rational functions slow every computer algebra system—so instead use the two variables `ca` and `cb` to denote these quotients and use `let` to ensure that products of these variables are linearized to a canonical form.

- Then the for-loop iterates until the residual of the DDE (17.30) is zero to the specified order of amplitude (as controlled by truncation in `small`).

- To compute the residual `resdde` of the DDE (17.30) we need to evaluate parts of the DDE at the delayed time $t-\pi/2$. Recognize that there are three sources of time dependence where the delay has to be accounted for: first, clearly in $u(t-\pi/2)$, but second and third also in the amplitudes a and b. Fortunately, these amplitudes are to be slowly varying in time—that is the aim of writing the center manifold in terms of a complex pair of rotating basis vectors $e^{\pm it}$—so we expand them in a Taylor series about the current time:

$$a(t-\pi/2)=\sum_{n=0}^{\infty}\frac{1}{n!}\left(-\frac{\pi}{2}\right)^n\frac{d^n a(t)}{dt^n},$$

  and similarly for $b(t-\pi/2)$. Algorithm 17.4 codes a finite truncation of the sum into `as` and `bs`; the order of truncation needs to be increased for higher-order analysis.

  Then the statement to compute `resdde` is a near direct transcription of the DDE (17.30) with `small` inserted to count nonlinear combinations, and with the `sub()` function used to compute the delayed factor.

- Now update the shape of the center manifold (17.31a). Recognize that u is not a function of just time t, but also a function of amplitudes a and b, which in turn depend upon time. Thus suppose $u(t)=v(t,a,b)$ is an approximation to the center manifold (17.31a) and seek a small correction so that $u(t)=v(t,a,b)+\hat v(t,a,b)$ is a better approximation. Correspondingly, suppose corrections $\dot a=g_a+\hat g_a$ and $\dot b=g_b+\hat g_b$ will improve the description (17.31b) of the oscillations on the center manifold. Substitute into the DDE (17.30), and then neglect some small terms to find

$$\begin{aligned}&\frac{du}{dt}+r(1+u)u(t-\tfrac{\pi}{2})\\&=\text{Res}_{(17.30)}+\frac{\partial\hat v}{\partial t}+\frac{\partial\hat v}{\partial a}g_a+\frac{\partial v}{\partial a}\hat g_a+\frac{\partial\hat v}{\partial b}g_b+\frac{\partial v}{\partial b}\hat g_a\\&\quad+\hat v(t-\tfrac{\pi}{2},a(t-\tfrac{\pi}{2}),b(t-\tfrac{\pi}{2}))\\&\quad+\left.\frac{\partial v}{\partial a}\right|_{t-\pi/2}(-\tfrac{\pi}{2})\hat g_a+\left.\frac{\partial v}{\partial b}\right|_{t-\pi/2}(-\tfrac{\pi}{2})\hat g_b\\&=\text{Res}_{(17.30)}+\frac{\partial\hat v}{\partial t}+\hat v(t-\tfrac{\pi}{2},a,b)+e^{it}(1+i\tfrac{\pi}{2})\hat g_a+e^{-it}(1-i\tfrac{\pi}{2})\hat g_b\,.\end{aligned}$$

  Subtly, terms like $\partial v/\partial a|_{t-\pi/2}(-\frac{\pi}{2})\hat g_a$ come from the recognition that $u(t-\frac{\pi}{2})=v(t-\frac{\pi}{2},a(t-\frac{\pi}{2}),b(t-\frac{\pi}{2}))\approx v(t-\frac{\pi}{2},a-\frac{\pi}{2}\dot a,b-\frac{\pi}{2}\dot b)$ is affected by the corrections to the evolution $\dot a$ and $\dot b$.[210] In the last step we recognize first that

[210]One easily misses such subtleties. However, the method of undetermined coefficients in section 1.3 rescues us from such oversights. If unaware of how changes in evolution also feed through the time delay, then we would approximate the equation for the corrections as $\text{Res}_{(17.30)}+\partial\hat v/\partial t+\hat v(t-\frac{\pi}{2},a,b)+e^{it}\hat g_a+e^{-it}\hat g_b=0$. Consequently we would choose the corrections $\hat g_a$ and $\hat g_b$ as exactly the coefficients of $e^{\pm it}$ in the residual. Trying such a correction in the iteration does not work, so we resort to checking the coefficient of the corrections, thinking perhaps we have the sign wrong. Surprisingly, the method of undetermined coefficients tells us that the necessary coefficients are $-1/(1\pm i\pi/2)$. It works, so we use it. Eventually one may find the subtle correct explanation.

$\partial v/\partial a \approx e^{it}$, and second that amplitudes a and b are slowly varying so to a good enough approximation $a(t-\frac{\pi}{2}) \approx a(t)$ when part of a correction, and similarly for b. Further, g_a and g_b are quadratic in the small variables (including ϵ), so $\partial\hat{v}/\partial a\, g_a$ may be neglected.

To find corrections, there are two parts to solving the equation

$$\text{Res}_{(17.30)} + \frac{\partial\hat{v}}{\partial t} + \hat{v}(t-\tfrac{\pi}{2}, a, b) + e^{it}(1+i\tfrac{\pi}{2})\hat{g}_a + e^{-it}(1-i\tfrac{\pi}{2})\hat{g}_b = 0\,.$$

- First, for each component in e^{ikt} in $\text{Res}_{(17.30)}$ we seek a component in $\hat{v}$ of ce^{ikt}. Then elementary algebra gives the constant $c = -1/[ki + (-i)^k]$. Thus we code the operator `linv` and use it to update the center manifold u.
- Second, components $e^{\pm it}$ are a homogeneous solution of the linear operator for $\hat{v}$ (as expected, and as seen in the zero divisor for the constant c above). Thus terms $e^{\pm it}$ in $\text{Res}_{(17.30)}$ must generate changes to the evolution on the center manifold through $\hat{g}_a$ and $\hat{g}_b$. These considerations lead to the code for updates in Algorithm 17.4.

- Executing Algorithm 17.4 generates a higher-order version of the emergent center manifold (17.31a) and the evolution (17.31b) thereon.

17.4.4 ▪ Other approaches model instability in delay differential equations

The amplitude modulation approach developed here, based upon time-dependent basis vectors, is the simplest. However, one may wonder whether the instabilities could be modeled as in Chapter 16 using an xy-model; Campbell (2009) took this latter approach in modeling drill vibrations. Recall that the Taylor series expansion of the modulation complex amplitude $a(t-\pi/2)$ works in Chapter 16 because the amplitude is slowly varying. However, a difficulty is that x and y would not be slowly varying, so the corresponding delay terms are problematic.

In another approach one could rephrase DDE problems properly with a second coordinate to represent the immediate past. For example, the logistic DDE (17.28) is equivalent to the system for field $n(t,\theta)$ satisfying PDE and nonlocal boundary condition

$$\frac{\partial n}{\partial t} = \frac{\partial n}{\partial \theta} \quad\text{and}\quad \frac{\partial n}{\partial t}(t,0) = rn(t,0)[1-n(t,\tau)/c];$$

the nonlocal boundary condition is effectively the DDE; the PDE has the general solution $n(t,\theta) = f(t-\theta)$ to store the immediate past of $n(t)$ for $0 \leqslant \theta \leqslant \tau$. In this form, the center manifold analysis of the DDE could be either through modulations or through xy variables (Campbell, 2009).

There is also interest in DDEs for large delay τ, large compared to the period of the natural oscillations (Erneux, 2009, §6.3.2, e.g.). However, unless the timing of the delay is very precise, then the delay would "sample" the past oscillations as an effective weighted average over the oscillations (which are fast compared to the delay time). I suggest that the idealization of a precise delay differential is rarely useful in such a context. If the delay is subject to uncertainties that "blur" the delay over oscillations, then surely one should use a delay model that represents such blurring. We can invoke such a model simply by writing a few intermediaries that feed back into the system (Mocek et al., 2005, e.g.). Exercise 17.14 shows that just two intermediaries generate similar oscillatory instabilities as are generated by a DDE.

17.5 ▪ Outlook

As promised, the modulation model is more straightforward to derive and has significant advantages over the Cartesian-like xy-models derived in Chapter 16: it is generally more compact, easier to analyze, and far quicker to numerically simulate.

Because of these advantages, similar analysis is frequently performed in much more general settings. For example, one may derive evolution equations for the modulation of spatial patterns or for the modulation of waves. Such derivation is traditionally done using the method of multiple scales, a method that appears distinctly inferior to the center manifold approach.

However, the modulation approach has one very interesting limitation: it is not valid for small frequency ω (the basic oscillation)—there are divisions by frequency ω in the model, which compound at high order and cause the asymptotic expansion to perform poorly. This small frequency deficiency is *not* so for the xy-model of Chapter 16; there the description is uniformly valid should conditions include $\omega \to 0$.[211]

After studying these two sorts of bifurcations, the pitchfork and the Hopf, we now have the analytical and computer algebra tools to rationally analyze almost any autonomous dynamical system to extract a finite-dimensional model. No matter how many modes make up an emergent center manifold, they all have eigenvalues that are either zero or pair up with a complex conjugate. These are the two cases we have examined. The general analysis and construction are then just as this chapter has investigated: for autonomous systems there are no new features, just greater complexity of detail.

17.6 ▪ Exercises

For many of these exercises you could construct and/or check the center manifold model using my web service to do the computer algebra (Roberts, 2012).[212]

Exercise 17.1 (amplitude phase version). Deduce the center manifold dynamics of the system (17.1), but now express the center manifold model in terms of the real amplitude and phase angle of the oscillations on the center manifold. Hint: Recall section 3.3.

Answer: $x = r\cos\theta + \frac{1}{32}\alpha r^3(\cos 3\theta + 2\sin 3\theta) + \mathcal{O}(\alpha^2)$, $y = -r\sin\theta + \frac{1}{32}\alpha r^3(8\cos\theta - 3\sin\theta + 6\cos 3\theta - 44\sin 3\theta) + \mathcal{O}(\alpha^2)$, $z = \frac{1}{2}\alpha r^2(5 + \cos 2\theta + 2\sin 2\theta) + \mathcal{O}(\alpha^2)$ such that $\dot r = \frac{1}{4}\alpha r^3 - \frac{73}{80}\alpha^2 r^5 + \mathcal{O}(\alpha^3)$ and $\dot\theta = 1 + \frac{11}{8}\alpha r^2 - \frac{2199}{1280}\alpha^2 r^4 + \mathcal{O}(\alpha^3)$.

Exercise 17.2 (separated oscillations). Modify Algorithm 17.1 slightly to model the dynamics of the following system for small parameter α:

$$\dot x = -\alpha x + y\,,$$
$$\dot y = -x + yz\,,$$
$$\dot z = -z + 5\alpha x^2\,.$$

[211] In the theory of water waves, a nonlinear Schrödinger equation describes the evolution of deep-water waves via a modulation model akin to the analysis of Hopf bifurcations. However, models of waves in shallow water are instead based on analysis analogous to that employed in the pitchfork bifurcation. These two models of water waves are completely dissimilar. What is missing is a model that can describe both deep-water and shallow-water dynamics uniformly; such a model would be analogous to the above xy-model of the Hopf bifurcation. There is a so-called mild-slope equation (Toledo and Agnon, 2010, e.g.), but it only describes static wave configurations; it is not a dynamical equation. An interesting research project is to find a model that includes both deep- and shallow-water wave dynamics in analogy to the xy-model above.

[212] `http://www.maths.adelaide.edu.au/anthony.roberts/gencm.php` provided the service at the time of writing.

For what values of the parameter α is the origin stable in the dynamical system? When is there a nearby stable limit cycle?

Answer: On the center manifold $x = ae^{it} + \bar{a}e^{-it} + \mathcal{O}(\alpha)$, $y = iae^{it} - i\bar{a}e^{-it} + \mathcal{O}(\alpha)$, and $z = \mathcal{O}(\alpha)$, the evolution is $\dot{a} = \alpha[\frac{1}{2}a - (\frac{9}{2} + i)a^2\bar{a}] + \mathcal{O}(\alpha^2)$. The origin ($a = 0$) is stable for $\alpha < 0$ and unstable for $\alpha > 0$. When $\alpha > 0$ the limit cycle at $|a| \approx 1/3$ is predicted to be stable.

Exercise 17.3 (separated system I). Reconsider the 3D system of Exercise 16.2. Noting that the system is of separated form, modify Algorithm 17.1 to construct the center manifold and evolution thereon in terms of complex amplitude $a(t)$. Because the ODEs for both u and v are nontrivial, the updates for them and the evolution involve combining the residuals of both the u and v ODEs. Execute the modified algorithm to find that all the determined coefficients are pure imaginary in the asymptotic approximations for $\dot{a}$. Writing amplitude $a = r(t)e^{i\theta(t)}$ for real radius r and real phase angle θ, then deduce $\dot{r} = 0$ and $\dot{\theta} = 1 + \cdots$. Consequently argue that the long-term evolution on the center manifold is that of a family of concentric limit cycles—to as high an order as can be determined.[213]

Exercise 17.4 (separated system II). Do Exercise 17.3 but for the 3D system of Exercise 16.3.

Exercise 17.5 (frequency two). Consider the dynamics of the 3D system

$$\begin{aligned}
\dot{u}_1 &= u_2u_1 + u_1\epsilon - 2u_3 - 2u_2\,,\\
\dot{u}_2 &= -2u_3 - 3u_2 + u_1\,,\\
\dot{u}_3 &= 2u_3 + 3u_2 + u_1\,.
\end{aligned}$$

1. Argue that there exists a center manifold based upon on the equilibrium at the origin and based upon parameter $\epsilon = 0$.
2. Determine that some center eigenvectors and adjoint eigenvectors are $(1, 1, -1 \mp i)$ and $(1, \pm i, \pm i)$, respectively.
3. By hand or computer algebra determine that the complex amplitude of the oscillations evolves according to

$$\dot{a} \approx +\tfrac{1}{2}a\epsilon - \tfrac{11}{51}\bar{a}a^2 i - \tfrac{35}{34}\bar{a}a^2$$

   on a center manifold

$$\begin{aligned}
u_1 &\approx e^{-i2t}\bar{a} + e^{i2t}a + \tfrac{1}{3}e^{-i4t}\bar{a}^2 i - \tfrac{1}{3}e^{i4t}a^2 i\,,\\
u_2 &\approx e^{-i2t}\bar{a} + e^{i2t}a - 2\bar{a}a + \tfrac{5}{51}e^{-i4t}\bar{a}^2 i - \tfrac{1}{17}e^{-i4t}\bar{a}^2\\
&\quad - \tfrac{5}{51}e^{i4t}a^2 i - \tfrac{1}{17}e^{i4t}a^2\,,\\
u_3 &\approx e^{-i2t}\bar{a}i - e^{-i2t}\bar{a} - e^{i2t}ai - e^{i2t}a + 3\bar{a}a\\
&\quad - \tfrac{5}{51}e^{-i4t}\bar{a}^2 i - \tfrac{11}{102}e^{-i4t}\bar{a}^2 + \tfrac{5}{51}e^{i4t}a^2 i - \tfrac{11}{102}e^{i4t}a^2\,.
\end{aligned}$$

[213] Indeed the appearance of a center manifold made up of only limit cycles is precisely the reason Romanovski et al. (2013) were interested in this and other systems.

Exercise 17.6 (toward Rossler chaos). The Rossler dynamical system was invented to exhibit chaos. A modified version of the Rossler system is

$$\begin{aligned}\dot{x} &= -y - z\,,\\ \dot{y} &= x + cy\,,\\ \dot{z} &= -xy/5 - 5z + xz\,.\end{aligned}$$

For parameter $c \approx 0.3$ this system displays classic chaos, but let's explore a bifurcation that occurs before chaos is reached. Justify and construct a two-dimensional model for the Hopf bifurcation that occurs before the onset of chaos as parameter c is increased. What are the approximate size and frequency of the predicted limit cycles as a function of parameter c?

Exercise 17.7 (wave mean-flow interaction). In many important applications, especially geophysical fluid dynamics, wave and mean-flow interactions are vital (Mcphaden and Ripa, 1990, e.g.). The simplest prototypical example of such a system is the interaction between an oscillator and a slow mode. Thus explore the long-term modeling of the following system that arbitrarily couples the oscillator of Exercise 17.5 with the slow mode of Example 2.1:

$$\begin{aligned}\dot{u}_1 &= u_4u_1 + u_2u_1 - 2u_3 - 2u_2\,,\\ \dot{u}_2 &= -2u_3 - 3u_2 + u_1\,,\\ \dot{u}_3 &= 2u_3 + 3u_2 + u_1\,,\\ \dot{u}_4 &= -u_4^2 - u_2u_1 + u_5 - u_4\,,\\ \dot{u}_5 &= u_5^2 - u_5 + u_4\,.\end{aligned}$$

1. Comment on the equilibrium at the origin.
2. Linearizing, show that there are two pure imaginary complex conjugate eigenvalues, one zero eigenvalue, and two negative eigenvalues.
3. Find eigenvectors and deduce that there exists a three-dimensional center manifold tangent to some definite center subspace, and emergent at some rate.
4. By hand or via computer algebra construct the following approximate center manifold, parametrized in terms of a complex amplitude a and a slow mode amplitude b,

$$\begin{aligned}u_1 &\approx +e^{-i2t}\bar{a} + e^{i2t}a + \tfrac{1}{3}e^{-i4t}\bar{a}^2i + \tfrac{1}{8}e^{-i2t}b\bar{a}i - \tfrac{1}{3}e^{i4t}a^2i\\ &\quad - \tfrac{1}{8}e^{i2t}bai\,,\\ u_2 &\approx +e^{-i2t}\bar{a} + e^{i2t}a + \tfrac{5}{51}e^{-i4t}\bar{a}^2i - \tfrac{1}{17}e^{-i4t}\bar{a}^2 - \tfrac{11}{40}e^{-i2t}b\bar{a}i\\ &\quad - \tfrac{1}{5}e^{-i2t}b\bar{a} - \tfrac{5}{51}e^{i4t}a^2i - \tfrac{1}{17}e^{i4t}a^2 + \tfrac{11}{40}e^{i2t}bai\\ &\quad - \tfrac{1}{5}e^{i2t}ba - 2\bar{a}a\,,\\ u_3 &\approx +e^{-i2t}\bar{a}i - e^{-i2t}\bar{a} - e^{i2t}ai - e^{i2t}a - \tfrac{5}{51}e^{-i4t}\bar{a}^2i\\ &\quad - \tfrac{11}{102}e^{-i4t}\bar{a}^2 + \tfrac{11}{40}e^{-i2t}b\bar{a}i + \tfrac{13}{40}e^{-i2t}b\bar{a} + \tfrac{5}{51}e^{i4t}a^2i\\ &\quad - \tfrac{11}{102}e^{i4t}a^2 - \tfrac{11}{40}e^{i2t}bai + \tfrac{13}{40}e^{i2t}ba + 3\bar{a}a\,,\\ u_4 &\approx +b - \tfrac{1}{2}b^2 - \tfrac{1}{2}\bar{a}a - \tfrac{9}{40}e^{-i4t}\bar{a}^2i - \tfrac{1}{20}e^{-i4t}\bar{a}^2 + \tfrac{9}{40}e^{i4t}a^2i\\ &\quad - \tfrac{1}{20}e^{i4t}a^2\,,\\ u_5 &\approx +b + \tfrac{1}{2}b^2 + \tfrac{1}{2}\bar{a}a - \tfrac{1}{40}e^{-i4t}\bar{a}^2i + \tfrac{1}{20}e^{-i4t}\bar{a}^2 + \tfrac{1}{40}e^{i4t}a^2i\\ &\quad + \tfrac{1}{20}e^{i4t}a^2\,,\end{aligned}$$

and determine that the evolution on this center manifold, or its equivalent, is

$$\dot{a} \approx +\tfrac{1}{2}ba - \tfrac{1}{16}b^2ai - \tfrac{1}{4}b^2a - \tfrac{421}{4080}\bar{a}a^2i - \tfrac{887}{680}\bar{a}a^2\,,$$
$$\dot{b} \approx -\bar{a}a + b^3 + \tfrac{6}{5}b\bar{a}a\,.$$

5. From this nonlinear center manifold evolution, deduce that the oscillations reinforce the instability of the mean mode and that the effect of the mean mode (before it explodes) on the oscillations depends upon its initial condition.

Exercise 17.8 (modulate the Lorenz 96 model). Recall that Exercise 16.5 introduced a system of ODEs proposed by Lorenz (1996) to explore predictability in weather and climate forecasting. In the case of $K = 4$ cells, Algorithms 16.5 and 16.6 construct approximations to the center manifold of the equilibrium $x_k = F = 1$, and the evolution thereon. Explain how Algorithms 17.5 and 17.6 form the center manifold model in terms of the slow modulation of complex amplitudes. Base your explanation on Algorithms 17.2 and 17.3 of section 17.3.2, noting that the Toeplitz structure of the equations empowers us to use a shift operator to simplify the representation of some operations. Execute Algorithms 17.5 and 17.6 and interpret the model's predictions.

Answer: $\dot{a} = (F-1)(1+i)a - \frac{32+22i}{13}a^2\bar{a} + \mathcal{O}(|a|^5)$, which predicts a Hopf bifurcation as parameter F crosses 1.

Exercise 17.9 (modulate the Lorenz 96 model again). Modify Algorithms 17.5 and 17.6 to extend Exercise 17.8 to the case of $K = 6$ cells. Compare with the Cartesian-like model of Exercise 16.6.

Exercise 17.10 (modulate the Lorenz 96 model double Hopf). Extend Exercise 17.9 to the case of $K = 12$ cells, but note that the bifurcation at $F = 1$ is a double Hopf bifurcation. Compare with the Cartesian-like model of Exercise 16.8.

Exercise 17.11 (Stoleriu's center manifold). Stoleriu (2012) considered the following pair of coupled, nonlinear, second-order ODEs for $x(t)$ and $y(t)$ with parameter σ:

$$\ddot{x} = 1 - \cos x + \sigma y \quad \text{and} \quad \ddot{y} = (y + 1/\sigma)\sin x\,. \tag{17.32}$$

Analyze the center manifold of this system, keeping parameter σ as some finite fixed quantity.

1. Rewrite as a system of four coupled nonlinear first-order ODEs.
2. Show that there is an equilibrium at the origin.
3. Linearizing about the origin, find that the eigenvalues of the linear problem are ± 1 and $\pm i$. Hence deduce that there is a center manifold of quasi-oscillations within a decaying and a growing mode.
4. Find the eigenvectors of the imaginary eigenvalues.
5. Using complex-amplitude variables to parametrize the center manifold, approximate the center manifold. You will need to expand the trigonometric functions in (17.32) in an appropriate low-order Taylor series.
6. Interpret a low-order model to predict that the oscillations on the center manifold persist with a nonlinear frequency shift.

Algorithm 17.5 *Preamble to model the Hopf bifurcation in the system of Lorenz (1996) for Exercise 17.8. Continued by the iterative construction in Algorithm 17.6.*

```
kk:=4; ff:=1+small^2*f; factor ee,small;
operator cis;
let { cis(~m)*cis(~n) => cis(m+n)
    , cis(~m)^~p => cis(p*m)
    , df(cis(~th),~x) => i*cis(th)*df(th,x)
    , cis(0)=>1    };
operator ee;  depend ee,cis;
ones:=for k:=1:kk sum ee(k);
ea:=cis(+t)*(ee(1)+i*ee(2)-ee(3)-i*ee(4));
eb:=cis(-t)*(ee(1)-i*ee(2)-ee(3)+i*ee(4));
infix #; precedence #,^;
let ~u#~j => coeffn(u,ee(j),1);
% vector multiplication is componentwise
let {ee(~j)^2=>ee(j), ee(~j)*ee(~k)=>0 when j neq k};
% define shift operator for x_{j-p} in x_j equation et al.
operator ss; linear ss;
let { ss(ee(~j),ee,~p)=> ee(j+p) when j+p>0 and j+p<=kk
    , ss(ee(~j),ee,~p)=> ee(j+p-kk) when j+p>kk
    , ss(ee(~j),ee,~p)=> ee(j+p+kk) when j+p<=0    };
operator linv; linear linv;
let {linv(ee(~j)*cis(~w*t),cis)=>(
        (1+2*i*w-3*w^2-i*w^3)*ee(j)
        +(-1-2*i*w)*ss(ee(j),ee,1)
        +(1-i*w+w^2)*ss(ee(j),ee,2)
        +(2+2*i*w-w^2)*ss(ee(j),ee,3)
        )*cis(w*t)/(1-w^2)/(3+4*i*w-w^2)
    ,linv(ee(~j),cis)=>( ee(j)-ss(ee(j),ee,1)
        +ss(ee(j),ee,2)+2*ss(ee(j),ee,3) )/3
    ,linv(ee(~j)*cis(+t),cis)=>( (8-11*i)*ee(j)
        +(-3-4*i)*ss(ee(j),ee,1) +(8-i)*ss(ee(j),ee,2)
        +(7-4*i)*ss(ee(j),ee,3) )*cis(+t)/40
    ,linv(ee(~j)*cis(-t),cis)=>( (8+11*i)*ee(j)
        +(-3+4*i)*ss(ee(j),ee,1) +(8+i)*ss(ee(j),ee,2)
        +(7+4*i)*ss(ee(j),ee,3) )*cis(-t)/40    };
```

Exercise 17.12 (Rikitake system oscillates unusually). The Rikitake system

$$\dot{x} = -\mu x + yz, \quad \dot{y} = -y + (z - c)x, \quad \dot{z} = 1 - xy$$

was proposed for the study of the reversal of the Earth's magnetic field (de Carvalho Braga et al., 2010, e.g.). Show that it has oscillatory dynamics for parameters $\mu = 2$ and $c = 1$ that do not harbor a Hopf bifurcation.

1. Set the parameters and find an equilibrium at $(1, 1, 2)$.
2. Linearize about this equilibrium to find eigenvalues $\lambda = -3, \pm\sqrt{2}i$. Hence deduce the existence and emergence of a center manifold in the neighborhood of the equilibrium.

Algorithm 17.6 *Following the preamble of Algorithm 17.5, this is the iterative construction to model the Hopf bifurcation in the system of Lorenz (1996) for Exercise 17.8.*

```
depend a,t; depend b,t;
let { df(a,t)=>ga, df(b,t)=>gb };
x:=ones+small*( a*ea+b*eb );
ga:=gb:=0;
let small^4=>0;
for iter:=1:9 do begin
    res:=-df(x,t)-x+ss(x,ee,1)*(ss(x,ee,-1)-ss(x,ee,2))
        +ff*ones;
    ga:=ga+(gad:=coeffn(
        res#1-i*res#2-res#3+i*res#4,cis(+t),1)/4)/small;
    gb:=gb+(gbd:=coeffn(
        res#1+i*res#2-res#3-i*res#4,cis(-t),1)/4)/small;
    x:=x+linv(res-0*gad*ea-0*gbd*eb ,cis);
    if res=0 then write iter:=1000000+iter;
end;
```

3. Confirm that $(1, 1, \pm\sqrt{2}i)$ are eigenvectors spanning the center subspace. Hence construct a center manifold model in complex amplitudes to cubic terms in the amplitudes.
4. Interpret the model to deduce that whatever bifurcation happens near these parameter values, it is not a Hopf bifurcation.
5. Extend the center manifold construction by setting parameter $c = 1 + c'$ for small c'. Deduce that the equilibrium $(1, 1, 2)$ is stable for $c > 1$ and unstable for $c < 1$.

Answer: Approximate the center manifold as $x \approx 1 + ae^{i\sqrt{2}t} + be^{-i\sqrt{2}t}$, $y \approx 1 + ae^{i\sqrt{2}t} + be^{-i\sqrt{2}t}$, and $z \approx 2 + a\sqrt{2}e^{i\sqrt{2}t} - b\sqrt{2}e^{-i\sqrt{2}t}$. Then the evolution is $da/dt \approx -i\frac{2\sqrt{2}}{3}a^2b$ and $db/dt \approx +i\frac{2\sqrt{2}}{3}ab^2$.

Exercise 17.13 (traveling waves in a Hasegawa–Wakatani PDE). With a parameter κ, the nondimensional Hasegawa–Wakatani PDEs we analyze are

$$\xi = \nabla^2\phi\,, \tag{17.33a}$$
$$\frac{\partial\xi}{\partial t} + \{\phi, \xi\} = 2(\phi - \rho) - \nabla^4\xi\,, \tag{17.33b}$$
$$\frac{\partial\rho}{\partial t} + \{\phi, \rho\} = 2(\phi - \rho) - \kappa\frac{\partial\phi}{\partial y} - \nabla^4\rho\,, \tag{17.33c}$$

where the Poisson brace

$$\{\phi, \cdot\} := \frac{\partial\phi}{\partial x}\frac{\partial\cdot}{\partial y} - \frac{\partial\phi}{\partial y}\frac{\partial\cdot}{\partial x}\,.$$

Physically, these PDEs model the dynamics of a plasma in a tokamak (Farrell and Ioannou, 2009, e.g.): ϕ is proportional to an electrostatic potential, ξ is an ion vorticity, ρ denotes density fluctuations, and parameter κ represents a background gradient of density.

Spatial domain? Solve the PDEs in a 2D bounded spatial domain (x, y), say the nondimensional domain is $0 < x < \pi$ and 2π-periodic in y. For simplicity, apply Dirichlet-like boundary conditions of $\phi = \xi = \rho = \xi_{xx} = \rho_{xx} = 0$ on $x = 0, \pi$.

1. Check that $\phi = \xi = \rho = 0$ is an equilibrium of the Hasegawa–Wakatani PDEs (17.33).
2. For what critical parameter κ do the modes $\phi = \sin x \exp[\pm i(y - 2t)]$ form a center subspace at the equilibrium? Use that for any given ϕ, first, equation (17.33a) gives ξ; second, equation (17.33b) gives ρ; and last, equation (17.33c) determines under what conditions a solution exists.
3. For this critical parameter κ, confirm that modes $\phi = \sin(px)\exp(iqy + \lambda t)$ have eigenvalue λ with negative real part for wavenumbers say $p = 1, 2, 3$ and $q = 0, 1, 2, 3$. Hence deduce that there exists an emergent center manifold for small (ϕ, ξ, ρ). Estimate the rate of decay of transients.
4. Construct, using computer algebra or otherwise, that the center manifold at the critical κ is
$$\phi \approx A \sin x\, e^{i(-y+2t)} + B \sin x\, e^{i(y-2t)} + 0.07692\, AB \sin 2x,$$
where real solutions follow for amplitude B being the complex conjugate of A. Justify why this center manifold describes traveling waves. A corresponding center manifold model is
$$\frac{dA}{dt} \approx +(0.07505 + i0.09381)A^2 B\,,$$
and correspondingly for the evolution of B.
5. Hence deduce that the nonlinearity in the Hasegawa–Wakatani PDEs (17.33) engenders finite-amplitude instabilities of the traveling waves near the critical parameter.

Answer: The critical parameter $\kappa = 10$. From the eigenvalue of the mode $\phi = \sin x$, a center manifold emerges with transients roughly $\mathcal{O}(e^{-t/4})$.

Exercise 17.14 (intermediaries approximate delay). Recall that DDEs like the canonical (17.24) are an idealization: usually there is a feedback loop, but the feedback need not be precisely at the time lag τ; instead the feedback is often through some intermediate steps (intermediate chemical or biomolecular reactions, for example, as discussed by Mocek et al. (2005)). This exercise demonstrates that an oscillatory instability arises for as little as two intermediate steps.

Consider the following system of ODEs for $\mathbf{u}(t) = (u_0(t), u_1(t), u_2(t))$:

$$\dot{u}_0 = cu_2\,, \quad \tfrac{1}{2}\dot{u}_1 = -u_1 + u_0\,, \quad \tfrac{1}{2}\dot{u}_2 = -u_2 + u_1\,. \tag{17.34}$$

Observe that mode u_0 feeds into u_1, which then feeds into u_2, which only then feeds back into u_0 to promote growth or decay. Thus the variable u_2 models a version of the delay $u_0(t-1)$:[214] the feedback is not crisp and sharp like the delay term, but "smudged" over time by the intermediate processes.

1. For this linear system, seek solutions proportional to $e^{\lambda t}$ and find the characteristic equation $\lambda(1+\lambda/2)^2 - c = 0$.
2. Seek parameter values for the (likely) onset of instability by seeking imaginary eigenvalues $\lambda = i\omega$ and deducing that either $\omega = c = 0$ or $\omega = \pm 2$ and $c = -4$.
3. Let eigenvalues $\lambda = \mu + i\omega$ and express frequency ω and coefficient c as functions of μ. Plot μ and ω as a function of c.
4. When is the system (17.34) stable? Unstable? Oscillatory unstable? Compare with the canonical DDE (17.25).

Exercise 17.15 (nonlinear quasi-DDE). Extend Exercise 17.14 to model the Hopf bifurcation in the nonlinear dynamics of

$$\dot u_0 = cu_2 - 2u_0^2 - u_0^3\,, \quad \tfrac{1}{2}\dot u_1 = -u_1 + u_0\,, \quad \tfrac{1}{2}\dot u_2 = -u_2 + u_1 \tag{17.35}$$

for parameter $c \approx -4$. As in Exercise 17.14, system (17.35) approximates, through the two intermediaries (Mocek et al., 2005, e.g.), the nonlinear DDE $\dot u = cu(t-1) - 2u(t) - u(t)^3$.

Answer: Setting parameter $c = -4 - \alpha$, the evolution on the center manifold is $\dot a \approx +\frac{1}{5}a\alpha i + \frac{1}{10}a\alpha - \frac{16}{15}\bar a a^2 i - \frac{1}{5}\bar a a^2$, when we parametrize the center manifold as $u_0 \approx +e^{-i2t}\bar a + e^{i2t}a - \frac{7}{12}e^{-i4t}\bar a^2 i + \frac{1}{12}e^{-i4t}\bar a^2 + \frac{7}{12}e^{i4t}a^2 i + \frac{1}{12}e^{i4t}a^2 - \bar a a$, $u_1 \approx +\frac{1}{2}e^{-i2t}\bar a i + \frac{1}{2}e^{-i2t}\bar a - \frac{1}{2}e^{i2t}a i + \frac{1}{2}e^{i2t}a - \frac{1}{12}e^{-i4t}\bar a^2 i + \frac{1}{4}e^{-i4t}\bar a^2 + \frac{1}{12}e^{i4t}a^2 i + \frac{1}{4}e^{i4t}a^2 - \bar a a$, $u_2 \approx +\frac{1}{2}e^{-i2t}\bar a i - \frac{1}{2}e^{i2t}a i + \frac{1}{12}e^{-i4t}\bar a^2 i + \frac{1}{12}e^{-i4t}\bar a^2 - \frac{1}{12}e^{i4t}a^2 i + \frac{1}{12}e^{i4t}a^2 - \bar a a$.

Exercise 17.16 (periodic chronic myelogenous leukemia). Ion and Georgescu (2013) explored Hopf bifurcations in a DDE modeling leukemia:[215]

$$\dot x = -\frac{x(t)}{1+x(t)^n} - \delta x(t) + \frac{kx(t-r)}{1+x(t-r)^n}\,.$$

For simplicity we fix upon parameters $n = 2$, $\delta \approx 1/8$, $k = 3/2$, and time delay $r = 64/3$; that is,

$$\dot x = -\frac{x(t)}{1+x(t)^2} - (\tfrac{1}{8} + \delta')x(t) + \frac{\frac{3}{2}x(t-r)}{1+x(t-r)^2}\,.$$

Near these parameters the equilibrium $x = x_* := \sqrt{3}$ perhaps undergoes a Hopf bifurcation. "Perhaps" because instead of a precise time delay, we model $x(t-r)$

[214] Why do we write $u_0(t-1)$ instead of $u_0(t-\tau)$ for some other delay τ? The answer is that the two time constants of $\frac{1}{2}$ in the intermediaries add up to one. It eventuates that the *mean time* of the "smudged delay" through the intermediaries is then 1 to correspond on average to the delay term $u_0(t-1)$.

[215] Their parameter β_0 is absorbed in a time scaling.

by introducing two intermediaries in the system as in Exercise 17.14 (Mocek et al., 2005, e.g.). After defining $x(t) = x_* + u_1(t)$, analyze[216]

$$\begin{aligned}
\dot{u}_1 &= -\frac{(x_* + u_1)}{1 + (x_* + u_1)^2} - \left(\tfrac{1}{8} + \delta'\right)(x_* + u_1) + \frac{\frac{3}{2}(x_* + u_3)}{1 + (x_* + u_3)^2}, \\
\dot{u}_2 &= \frac{3}{32}(u_1 - u_2), \\
\dot{u}_3 &= \frac{3}{32}(u_2 - u_3)
\end{aligned}$$

to find that this system does undergo a Hopf bifurcation as δ' decreases through zero.

1. Confirm that $\mathbf{u} = 0$ is an equilibrium when we set $x_* = \sqrt{3}$.
2. Analyze the Jacobian of the system at $\mathbf{u} = 0$ to show that there exists an emergent center manifold for $\delta' \approx 0$ based upon oscillations of frequency $\omega = 3/32$.
3. Expand the rational function $x/(1 + x^2)$ in a Taylor series in $\mathbf{u}$, expanding to errors $\mathcal{O}(|\mathbf{u}|^4)$ as we typically must account for cubic terms in a Hopf bifurcation.
4. Construct a center manifold model in terms of complex amplitudes, $u_1 \approx ae^{i\omega t} + be^{-i\omega t}$, to predict the existence of oscillations for negative δ' of amplitude $|a| \approx \sqrt{-5\delta'}$, roughly.

Exercise 17.17 (nonlinearly stabilized DDE). Consider the DDE system

$$\dot{x} = -[1 + \alpha - y(t)]x(t - \pi/2) \quad \text{and} \quad \dot{y} = -y(t) + x(t)^2.$$

The origin $x = y = 0$ is an equilibrium. From the analysis of the canonical DDE of section 17.4.1, we know that without the coupling to the fast variable y the x ODE would be at marginal criticality when parameter $\alpha = 0$. With the coupling, any oscillations in x should drive a positive y, which then helps stabilize the oscillations. Let's see this in the center manifold modeling.

1. Modify Algorithm 17.4 to construct that the center manifold is
$$\begin{aligned}
x &= ae^{it} + be^{-it}, \\
y &= b^2e^{-i2t}\left(\tfrac{2}{5}i + \tfrac{1}{5}\right) + a^2e^{i2t}\left(-\tfrac{2}{5}i + \tfrac{1}{5}\right) + 2ab + \mathcal{O}(\epsilon^3)
\end{aligned}$$
for ϵ measuring the size of amplitudes a and b. Correspondingly, the evolution on this center manifold is
$$\dot{a} = ba^2\frac{-\frac{4}{5}i\pi - \frac{36}{5}i - \frac{18}{5}\pi + \frac{8}{5}}{\pi^2 + 4} + a\alpha\frac{4i + 2\pi}{\pi^2 + 4} + \mathcal{O}(\epsilon^4),$$
and similar for the amplitude b.
2. Deduce the amplitude and frequency of the oscillations that emerge for parameter $\alpha > 0$.

[216] As in Exercise 17.14 the two rates $3/32$ in the intermediary ODEs are chosen so their reciprocals sum to the time delay $r = 64/3$ in the original system.

Exercise 17.18 (destabilized center DDE). Explore a DDE system where the center modes involve all variables. Consider the system

$$\dot u_1 = u_2(t-\pi/2) - u_1(t)^2 \quad\text{and}\quad \dot u_2 = u_1(t-\pi/2) + u_2(t)^2.$$

Our aim is to deduce that the quadratic reaction destabilizes the delay-induced oscillations about the origin $\mathbf{u} = 0$.

1. Deduce that the characteristic equation for dynamics linearized about the origin is $\lambda^2 - e^{-\pi\lambda} = 0$. The following quick and dirty MATLAB/Octave code finds some eigenvalues:

```
ce=@(z) z.^2-exp(-pi*z)
lams=fsolve(ce,randn(100,2)*[2;10*i])
plot(real(lams),imag(lams),'o')
```

   Execute this code to find that there are one unstable mode and two center modes, $\lambda = \pm i$, and all the rest are stable modes. What are the eigenvalues of the unstable mode and leading stable modes?

2. Develop from Algorithm 17.4 a computer algebra algorithm to derive that the center manifold may be written as

$$u_1 \approx ae^{it} + be^{-it} + b^2\big(-\tfrac{2}{5}i + \tfrac{1}{5}\big)e^{-i2t} - 2ba + a^2\big(\tfrac{2}{5}i + \tfrac{1}{5}\big)e^{i2t},$$
$$u_2 \approx -ae^{it} - be^{-it} + b^2\big(\tfrac{2}{5}i - \tfrac{1}{5}\big)e^{-i2t} + 2ba + a^2\big(-\tfrac{2}{5}i - \tfrac{1}{5}\big)e^{i2t}.$$

   Correspondingly find that the evolution is

$$\dot a \approx a^2b\frac{-\frac{36}{5}i\pi - \frac{16}{5}i - \frac{8}{5}\pi + \frac{72}{5}}{\pi^2+4},$$
$$\dot b \approx ab^2\frac{+\frac{36}{5}i\pi + \frac{16}{5}i - \frac{8}{5}\pi + \frac{72}{5}}{\pi^2+4}.$$

   Hence deduce that this model predicts nonlinear growth of the center modes, in addition to the growth of the unstable mode.

Answer: $\lambda = 0.4745$ and $\lambda = -0.6846 \pm i2.8499$. The center modes are nonlinearly unstable due to the positive real part of the coefficient in $\dot a \approx (0.6758 \pm i1.8616)|a|^2a$.

Exercise 17.19 (two limit cycles in a DDE). Ion (2007) identified parameter values where two limit cycles branch off an equilibrium in a degenerate Hopf bifurcation. For a delay $\tau = \pi/2$, consider the DDE

$$\dot u + \alpha u(t-\tau) = \epsilon\left[u(t)^2 + cu(t)u(t-\tau)\right]$$

for parameter $\alpha = 1 + \epsilon^4\alpha'$ and some constant c.

1. Modify Algorithm 17.4 to construct that the center manifold is

$$u = ae^{it} + be^{-it} + 2\epsilon ab - \tfrac{1}{5}(1+2i+(2-i)c)\epsilon a^2e^{i2t}$$
$$- \tfrac{1}{5}(1-2i+(2+i)c)\epsilon b^2e^{-i2t} + \mathcal{O}(\epsilon^2).$$

   Correspondingly, the evolution on this center manifold is

$$\dot a = \epsilon^2a^2b\frac{18-4i+(9-7i)c+(1-3i)c^2}{5(1+i\pi/2)} + \mathcal{O}(\epsilon^3).$$

2. Deduce that, from the cubic term in the above model, the real part of the coefficient vanishes for the two values of the parameter (Ion, 2007)

$$c = c_{\pm} = \frac{18 - 7\pi \pm \sqrt{36 + 212\pi + \pi^2}}{2(3\pi - 2)}.$$

Argue that near these two parameter values we need to compute higher-order terms on the center manifold.

3. Construct the center manifold to errors $\mathcal{O}(\epsilon^5)$, set $c = c_+ + \epsilon^2 c'$, and deduce that the higher-order center manifold evolution is

$$\begin{aligned} \dot{a} &= (0.453 + 0.2884i)\epsilon^4\alpha' a \\ &\quad - \left[4.34i + (0.7695 + 2.025i)\epsilon^2 c'\right]\epsilon^2 a^2 b \\ &\quad + (2.103 - 70.89i)\epsilon^4 a^3 b^2 + \mathcal{O}(\epsilon^5). \end{aligned}$$

For what parameter regime, in α' and c', does this model predict two limit cycles? What is their predicted stability?

Exercise 17.20 (nonlinear normal modes). Renson et al. (2012) explored the nonlinear normal modes of a pair of coupled oscillators: in particular, they employed a finite element method to construct the invariant manifolds of independent nonlinear oscillations. One of their example systems is

$$\begin{aligned} \dot{x}_1 &= x_3\,, \\ \dot{x}_2 &= x_4\,, \\ \dot{x}_3 &= -2x_1 + x_2 - \tfrac{1}{2}x_1^3 + \tfrac{3}{10}(-x_3 + x_4)\,, \\ \dot{x}_4 &= x_1 - 2x_2 + \tfrac{3}{10}(x_3 - 2x_4)\,. \end{aligned} \tag{17.36}$$

In the following analysis, assume the two factors of 3/10 in the above ODEs, the linear damping, are effectively small. Linearly, we discover that this system has two independent oscillators. The challenge is to discover two nonlinearly independent oscillators—the so-called nonlinear normal modes. Although the entire state space is the center manifold in this system, the techniques for constructing modulation models of oscillations empower us to find nonlinear normal modes.

1. Linearize system (17.36) about the origin, neglecting the terms with factors of 3/10 as being small, and deduce that the system has two oscillatory modes of two different frequencies.

2. For the linearized system, find that eigenvectors and adjoint eigenvectors may be chosen to be $(1, 1, \pm i, \pm i)$ and $(1, -1, \pm i\sqrt{3}, \mp i\sqrt{3})$.

3. Hence deduce that a time-dependent coordinate transform to complex amplitudes of

$$\begin{aligned} x_1 &\approx e^{-i\sqrt{3}t}\bar{b} + e^{-it}\bar{a} + e^{i\sqrt{3}t}b + e^{it}a\,, \\ x_2 &\approx -e^{-i\sqrt{3}t}\bar{b} + e^{-it}\bar{a} - e^{i\sqrt{3}t}b + e^{it}a\,, \\ x_3 &\approx -\sqrt{3}e^{-i\sqrt{3}t}\bar{b}i - e^{-it}\bar{a}i + \sqrt{3}e^{i\sqrt{3}t}bi + e^{it}ai\,, \\ x_4 &\approx \sqrt{3}e^{-i\sqrt{3}t}\bar{b}i - e^{-it}\bar{a}i - \sqrt{3}e^{i\sqrt{3}t}bi + e^{it}ai \end{aligned}$$

transforms the system (17.36) to

$$\dot{a} \approx \tfrac{3}{4}\bar{b}bai + \tfrac{3}{8}\bar{a}a^2i - \tfrac{3}{40}a\,,$$
$$\dot{b} \approx \tfrac{1}{8}\sqrt{3}\bar{b}b^2i + \tfrac{1}{4}\sqrt{3}b\bar{a}ai - \tfrac{3}{8}b\,.$$

4. In this form discuss how $a = 0$ and $b = 0$, separately, are invariant subspaces, and hence predict nonlinear invariant manifolds for the original variables—the nonlinear normal modes.
5. Extend the analysis to higher order in complex amplitudes a and b, and confirm the continuing existence of nonlinear normal modes.

Exercise 17.21 (double Hopf in a DDE). Erneux (2009, §7.2) explored an example of a laser subject to optoelectronic feedback. For certain parameter values it has a two-frequency Hopf bifurcation.

Erneux (2009, eq. (7.42)) transforms the laser system to the nondimensional[217]

$$\frac{d^2 \log[1+y]}{dt^2} = -\theta^2\left[(1-\eta)y(t) + \eta y(t-\pi)\right]$$

for parameters η and θ. Erneux (2009) identifies double Hopf bifurcations from the origin at parameters (η, θ) of $(3/8, 2)$, $(7/32, 4)$, $(-5/8, 2)$, and $(-9/32, 4)$, among others.

1. Here we work with a system of first-order DDEs, so transform the DDE to the coupled system

$$\dot{x} = -\theta^2\left[(1-\eta)y(t) + \eta y(t-\pi)\right],$$
$$\dot{y} = [1+y(t)]x(t).$$

   Observe that this system has an equilibrium at the origin.
2. The following quick and dirty Octave/MATLAB code plots the spectrum for the equilibrium at the origin.

```
eta=3/5, theta=2
ce=@(z) z.^2+theta^2*(1-eta+eta*exp(-pi*z))
lams=fsolve(ce,randn(100,2)*[2;10*i])
plot(real(lams),imag(lams),'o')
```

   Justify the characteristic equation. Then argue that the results indicate that in all four cases mentioned the center manifold is attractive.
3. For parameters $(\eta, \theta) = (3/8, 2)$ find that the center subspace is spanned by eigenvectors $(1, \mp i/\omega)e^{\pm i\omega t}$ for frequencies $\omega = 1, 2$. Hence, develop from Algorithm 17.4 a computer algebra algorithm to construct a model of the center manifold dynamics in terms of the four complex amplitudes of $e^{\pm it}$ and $e^{\pm i2t}$.

   Make small perturbations to the parameters (η, θ) to find a model that unfolds this double Hopf bifurcation.

[217]This exercise invokes a slightly different definition of the parameter η: here $\eta := \eta_{\text{Erneux}}/(1+\eta_{\text{Erneux}})$.

4. For parameters $(\eta, \theta) = (7/32, 4)$ find that the center subspace is spanned by eigenvectors $(1, \mp i/\omega)e^{\pm i\omega t}$ for frequencies $\omega = 3, 4$. Hence modify the computer algebra algorithm for the previous part, including small perturbations to the parameters (η, θ), to construct a model of the center manifold dynamics in terms of the four complex amplitudes of $e^{\pm i3t}$ and $e^{\pm i4t}$. Discuss why the model here is somewhat simpler.

Answer: The gravest eigenvalue are, respectively, $-0.69 \pm i3.87$, $-0.38 \pm i1.02$, -0.31, and $-0.41 \pm i2.03$.

Exercise 17.22 (crane control seemed intractable). Nayfeh (2008) used three examples to compare two approaches for determining Hopf bifurcations in DDEs: the method of multiple scales and a center manifold approach. In the third example of controlling a crane, Nayfeh (2008) could not apply the center manifold approach "because we do not know how to convert the governing equations into an operator system." However, there is no need to do such a conversion—it is only a theoretical nicety, not a practical requirement. This exercise leads you through a direct, practical center manifold analysis.

Nayfeh (2008) expressed controlling the payload "pendulum" in a container crane using delayed position feedback by the DDE[218]

$$\begin{aligned}&\ddot\phi(t) + \phi(t) + 2\mu\dot\phi(t) + k\ddot\phi(t-\tau)\\ &+ \epsilon\alpha_3\phi^3(t) - \epsilon k\left[\phi(t-\tau)\dot\phi^2(t-\tau) + \tfrac{1}{2}\phi^2(t-\tau)\ddot\phi(t-\tau)\right]\\ &= 0\,. \end{aligned} \tag{17.37}$$

Here μ, k, α_3, and delay τ are parameters of the crane and control. Nayfeh (2008) used the parameter ϵ to conveniently label the nonlinear terms; let's continue to do so. Nayfeh's difficulty arose via the linear delayed derivative term $\ddot\phi(t-\tau)$, but it is not a problem for our center manifold construction.

For simplicity, hereafter set the delay to $\tau = 3\pi/2$.

1. As always, first find equilibria: the system (17.37) has an equilibrium for $\phi = 0$.
2. Show that neutral linear solutions $\phi \propto e^{i\omega t}$ occur for frequency $\omega = 1$ when $2\mu = k$. That is, the critical delay strength could be $k_c = 2\mu$.
3. By seeking linear solutions $\phi \propto e^{\lambda t}$, verify that other eigenvalues have real part negative for $k_c < 0.9$, perhaps using quick and dirty numerical code adapted from Exercise 17.21. Argue that an emergent Hopf bifurcation occurs as parameter k increases through $k_c = 2\mu$, provided $k_c = 2\mu < 0.9$.
4. For simplicity set parameter $k_c = 2\mu = 2/3$. Using the linear subspace approximation that $\phi \approx ae^{it} + be^{-it}$, where complex amplitudes a and b evolve slowly in time, $\dot a \approx \dot b \approx 0$, find that the residual of the system (17.37) is then

$$\begin{aligned}\text{Res} = \epsilon\big\{&\big[-i\delta a + (\tfrac{1}{3}i + 3\alpha_3)a^2b\big]e^{it} + (\alpha_3 - i)a^3e^{i3t}\\ &+\big[i\delta b + (-\tfrac{1}{3}i + 3\alpha_3)ab^2\big]e^{-it} + (\alpha_3 + i)a^3e^{-i3t}\big\}.\end{aligned}$$

[218]Simplify the DDE somewhat by scaling time so that Nayfeh's parameter $\alpha_1 = 1$ and by omitting some nonlinear terms by setting parameters $\alpha_4 = \alpha_5 = 0$.

Hence deduce that corrections $\hat{\phi}$ to the center manifold, and the evolution thereon, $\dot{a}$ and $\dot{b}$, will be $\mathcal{O}(\epsilon)$.

5. By analyzing the linear part of the system, the first line of (17.37), deduce update rules for corrections to the center manifold and evolution thereon. Consequently deduce that here the center manifold is

$$\begin{aligned}\phi &= ae^{it} + \epsilon\tfrac{1}{16}(1 - i + (1+i)\alpha_3)a^3e^{i3t} \\ &\quad + be^{-it} + \epsilon\tfrac{1}{16}(1 + i + (1-i)\alpha_3)b^3e^{-i3t} + o(\epsilon)\end{aligned}$$

such that the complex amplitudes satisfy

$$\begin{aligned}\dot{a} &= \epsilon\frac{18 + 9\pi - 6i}{40 + 36\pi + 9\pi^2}\delta a \\ &\quad + \epsilon\frac{-6 - 3\pi + 2i + (18 + 54i + 27\pi i)\alpha_3}{40 + 36\pi + 9\pi^2}a^2b + o(\epsilon)\end{aligned} \tag{17.38}$$

and its complex partner for $\dot{b}$.

The Landau equation (17.38) is the same model as derived by Nayfeh (2008) but for the particular parameters chosen here.

6. Modify the above steps for general $k_c = 2\mu$.

7. Modify Algorithm 17.4 to write, debug, and execute a computer algebra program to iteratively construct a center manifold model of the crane system (17.37) (there is no need to invoke `small` as here variable ϵ naturally characterizes the perturbations). Perhaps also calculate the next $\mathcal{O}(\epsilon^2)$ corrections.

8. Modify the governing DDE in your computer algebra to show that the algorithm constructs center manifold models of the system (17.37) with other $\mathcal{O}(\epsilon)$ cubic nonlinearities, such as those other terms by Nayfeh (2008, eq. (168)).

Further, set small parameter $\epsilon = \varepsilon^2$ and introduce some $\mathcal{O}(\varepsilon)$ quadratic nonlinearities: show that your algorithm constructs the appropriate Landau equation for these systems.

Part VI Summary

- Hopf bifurcations may be modeled using center manifold theory. The same techniques apply to more complex systems with multiple interacting oscillators. A direct application of the theory generates a model in an abstract phase space (the xy-model). But one has to solve nontrivial forms of the homological equation (section 16.2); canonically

$$h + y\frac{\partial h}{\partial x} - x\frac{\partial h}{\partial y} = \text{Res} \,.$$

  One must also compute the cubic terms in the center manifold to obtain a structurally stable model (section 16.3).

- However, a simpler and usually more useful model of oscillations uses the complex amplitude. For example, the model then describes the modulation of the basic oscillations involved in a Hopf bifurcation. The modulation equations straightforwardly predict the amplitude and frequency of the stable limit cycles that occur for parameters near critical (e.g., section 17.3.2).

- Many people require that the center variables be separated from the stable (hyperbolic) variables (as in section 17.1). However, it is usually more convenient to work in original physical variables, for which the system is not usually separated. The analysis then works from the center eigenvectors to construct a model (sections 17.2–17.4).

- Using the complex amplitude to parametrize the center manifold, section 17.3 found the general homological equation (17.13) to be of the form

$$-\mathcal{L}\hat{v} + \frac{\partial \hat{v}}{\partial t} = \text{Res}_{(17.7)} - \sum_{j=1}^{m} \hat{g}_j e_j e^{i\omega_j t}.$$

  Using $\hat{g}_j$ to put the right-hand side into the range of the left-hand side—the solvability condition—one then inverts $-\mathcal{L} + \partial/\partial t$ to find corrections to the center manifold (remembering here that $\partial/\partial t$ is done while keeping amplitudes constant, that is, frozen). The simplification is that each term on the right-hand side is considered separately as there is no coupling between the terms in the left-hand side operator.

- Delay differential equations (DDEs) typically harbor oscillatory instabilities, which we model with the approach developed in section 17.4. There is no need to explicitly embed the history of the delay system, as done by some people (Nayfeh, 2008, e.g.); instead it is more straightforward to account for the history directly (section 17.4.3).

Part VII

Avoid memory in modeling nonautonomous systems, including stochastic

Part VII Contents

Noise is all around us and often should be incorporated into modeling. In finance, the jagged fluctuations in the price of shares, exchange rates, and aggregated indices are essentially unpredictably random. In mechanical systems, small-scale freedom in moving parts leads to vibrations that may combine chaotically to generate a noisy environment. In the cell, there may be so few of the relatively large proteins that key protein reactions occur at random times depending upon random molecular motion. These examples, and many more, are just some cases where noisy, random fluctuations in time are important aspects of dynamics.

Many books (Baxter and Rennie, 1996; Stampfli and Goodman, 2001; Roberts, 2009a, e.g.) introduce the calculus of noisy systems in the context of financial applications. This part develops concepts and techniques for noisy systems in applications in the classic dynamics of science and engineering. The part starts by exploring dynamical modeling issues in deterministic nonautonomous systems. But the aim of modeling noisy stochastic systems underlies the development and choices made in that development.

Averaging is an immensely popular methodology for macroscale modeling of situations with complex microstructures. Averaging is frequently invoked in analyzing dynamical systems subject to forces that fluctuate on microscale times. Chapter 18 briefly discusses how to alternatively view averaging as the leading term in a nonlinear time-dependent coordinate transform. Such a coordinate transform is chosen to separate the microscale fluctuations in time from the macroscale mean effects: it clarifies what it means to "freeze" a variable, identifies emergence to the macroscale model, and potentially derives corrections to the averaging. The methodology here provides an alternative view to powerfully complement the results obtainable by averaging.

Part IV established that coordinate transforms are a powerful way to derive models of autonomous deterministic systems. Chapter 19 develops the techniques to disentangle slow modes from uninteresting fast modes in nonautonomous systems while accounting for the mean effect of the microscale interactions among the fast modes. With time-dependent fluctuations in the system, there are many subjective decisions to be made: five principles guide those decisions. Section 19.1 explores a 2D system to introduce many general and fascinating subtleties in modeling nonautonomous systems: the coordinate transform necessarily involves integrals of the fluctuations, with some being integrals of future fluctuations; to remove microscale fluctuations from a model the slow mode is not quite what we might define; and generically quadratic interactions among fluctuations generate important mean effects. For an example in applications, Hasselmann (1976) explained how slow changes of climate are an integral response to excitation by short-period weather disturbances. Section 19.2 then establishes that for reasonably general systems, coordinate transforms generally separate slow from fast dynamics.

But as for autonomous systems, we are primarily interested in constructing just the slow manifold: constructing a full coordinate transform is too onerous in high-dimensional systems. Section 19.3.3 details the most general efficient method to algebraically construct slow manifold models of nonautonomous dynamics. Computer algebra empowers us to handle the many details of microscale fast-time interactions that accumulate to generate mean and fluctuating effects on the interesting slow modes.

Among nonautonomous systems, noisy stochastic systems both are interesting and have many unique features. For example, a "white noise" forcing formally contains fluctuations on all time scales: how then can we separate fast from slow in a modeling transformation? To address such issues we need to understand some basics about stochastic systems: Chapter 20 is a basic introduction to stochastic processes and their calculus.

Stochastic processes in general are built from the Wiener process, often called a random walk (section 20.1). But stochastic calculus is so delicate that there are different interpretations of what the algebraic symbols mean (section 20.2): although the Itô interpretation is most convenient for establishing theoretical properties, for many applications the Stratonovich interpretation is used, as then familiar classic calculus identities hold. With basic calculus established, we are empowered to solve some stochastic differential equations algebraically (section 20.3) and numerically (section 20.4). Last, section 20.5 differentiates between strong solutions, which describe each random realization, and weak solutions, which describe the probability density distribution. Weak solutions and approximations appear to be the only way to eventually remove all fast processes from a stochastic slow model.

Chapter 21 then explores issues in modeling stochastic systems. Using time-dependent coordinate transforms, section 21.1 invokes a backward approach to find that stochastic slow manifold models exist and are emergent (section 21.3 summarizes alternative extant forward theory). The clarity of the coordinate transform approach illuminates (section 21.1.1) that noises are transformed by nonlinearities, that when modeling one must anticipate the noise, that the slow variables are not quite what you might want to define, and that the average stochastic slow manifold is not the deterministic slow manifold. The coordinate transform also illuminates how to treat initial conditions for a stochastic model (section 21.1.2), whether the initial conditions are deterministic or themselves stochastic.

Appearing in the coordinate transform and consequent slow manifold model are irreducible noise-noise interactions involving fast-time convolutions (section 21.1.3). To model these irreducible effects without fast-time convolutions we necessarily must resort to "weak" models, that is, to modeling just statistics of the irreducible components. Analysis of the corresponding Fokker–Planck equation empowers us to do this weak modeling (section 21.2.1) by using techniques from the slowly varying modeling of Chapter 7. The weak modeling analysis of Fokker–Planck equations generalizes to a variety of stochastic differential equations (sections 21.2.2 and 21.2.3). Given computer algebra algorithms may be readily adapted to similarly model any in a wide range of stochastic systems.

Chapter 18

Averaging is often a good first modeling approximation

Contents

Averaging is often the first recourse for modeling problems with difficult rapid variations in time. There is a huge body of theory, applications, and variations on the theme (Verhulst, 2005, Chaps. 11–13, e.g.). This chapter aims to show how to *view averaging as the leading approximation in a systematic normal-form coordinate transform approach.*

Although Part VII's ultimate destination is to model stochastic dynamics, we first develop the theory, methods, and techniques of deterministic nonautonomous dynamics. Especially relevant is when the fast-time dynamics is chaotic: the reason is that deterministic chaos looks stochastic on long time scales. So we start with averaging in a deterministic setting.

Example 18.1 (simple averaging). Consider the slow variable $x(t)$ and fast variable $y(t)$ in the system[219]

$$\dot{x} = \epsilon(x^3 - xy^2) \quad \text{and} \quad \dot{y} = -y + x + a\sin t\,.$$

Figure 18.1, for parameter $\epsilon = 0.2$, indicates that the variable y undergoes fast finite oscillations. Nonetheless, over long times, from all nearby initial conditions, the variable x collapses to zero. Our challenge is to describe, to model, this slow x-evolution despite the fast fluctuating y.

We assume the parameter ϵ is small. Then the x-evolution is slow. In contrast, the y variable evolves quickly: not only does it tend to decay at rate 1, but its sinusoidal forcing has frequency 1. The averaging argument is that the rapid fluctuations in the fast variable y can be replaced in the x equation by the mean, namely $y^2 \mapsto \overline{y^2}$.

[219]Throughout this section on averaging, I use trigonometric functions in the forcing. In principle, one may apply the techniques to forcing with almost arbitrarily complicated time dependence, such as the stochastic forcing of later sections, or a chaotic subsystem. However, trigonometric forcing has the advantage that we relatively easily perform algebra while still displaying generic features of the averaging methodology.

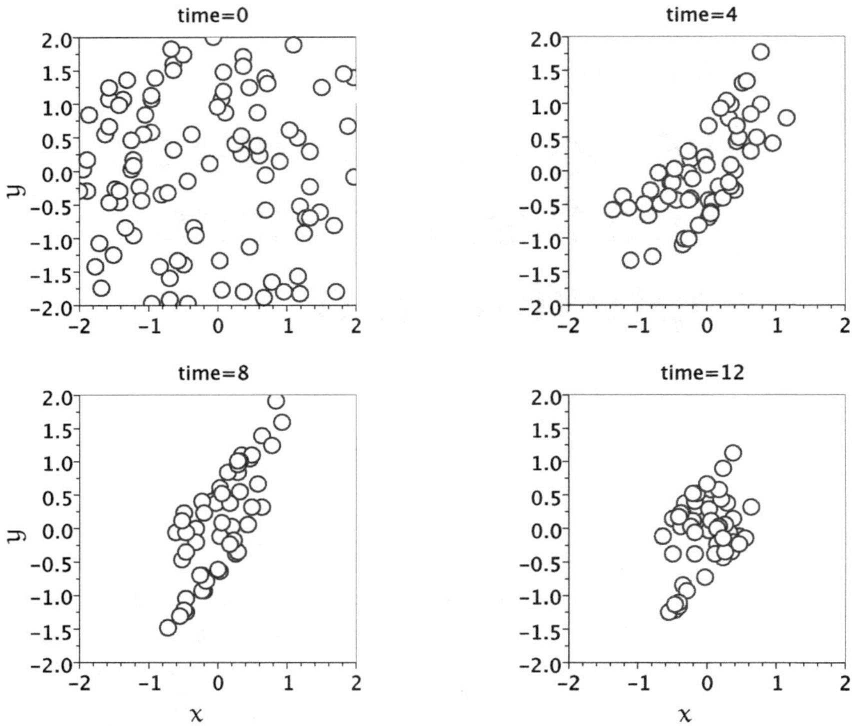

Figure 18.1. *Locations at four times in the* xy*-plane of solutions to the system of Example* 18.1, *with nonautonomous forcing* $a \sin t$, *starting from* 100 *random initial conditions and parameters* $a = 1$ *and* $\epsilon = 0.2$*: the system slowly decays to* $x = 0$ *among fast* y *fluctuations.*

Consider the y-dynamics. Since x is slow then the variable x may as well be constant in the fast y-dynamics. Upon invoking this heuristic of fixing, or freezing, the slow variable x, basic calculus readily solves the y ODE to give

$$y \approx \tilde{y}(x,t) := x + \tfrac{1}{2}a(\sin t - \cos t).$$

Figure 18.1 shows the overall emergence of this linear dependence upon x, with the superposed oscillations appearing as scatter around $y = x$. From this equation the mean effect of the fluctuations in the fast y on the slow x depends upon

$$\begin{aligned}\overline{y^2} &\approx \overline{\tilde{y}^2} \\ &\approx \overline{x^2} + \overline{ax(\sin t - \cos t)} + \tfrac{1}{4}a^2\left[\overline{\sin^2 t - 2\sin t\,\cos t + \cos^2 t}\right] \\ &\approx x^2 + \tfrac{1}{4}a^2.\end{aligned}$$

Substitute into the slow x ODE to obtain the model

$$\dot{x} \approx \epsilon(x^3 - x\overline{y^2}) \approx -\epsilon\tfrac{1}{4}a^2 x\,.$$

The finite fluctuations in the approximation $y \approx \tilde{y}$ make a major difference to the stability of the slow x. If we had ignored the fluctuations, then $y \approx x^2$, the model would be that $\dot{x} \approx 0$, and hence x would be constant. Instead, recognizing the fluctuations in y predicts that the slow x decays (exponentially), $\dot{x} \approx -\frac{1}{4}\epsilon a^2 x$, as seen in the simulations of Figure 18.1.

This example indicates that we have to grapple with time-dependent fluctuations of finite size. Finite, time-dependent, *out-of-equilibrium* fluctuations are common in modeling—turbulence is just one example.

A vast amount has been developed and written on averaging and its theory (Verhulst, 1989, 2005; Sanders et al., 2007; Pavliotis and Stuart, 2008, e.g.), including a huge number of successful applications of averaging. There are also a huge number of applications where averaging fails to produce a useful model, and then modelers adjoin corrections, often called "homogenization" or "deviations," to obtain a useful model. This chapter does not aim to replicate any of this well-established approach. Instead the aim is to illuminate and enhance the method of averaging through normal-form coordinate transforms. This exposition builds upon Part IV, which explores deterministic autonomous normal forms.

Recall that this chapter justifies averaging as a leading approximation to an illuminating coordinate transform. The big complication, as in Chapter 3 on modeling oscillations and implicitly in Example 18.1, is that the coordinate transformation must be time dependent.

When seeking to apply averaging, most people express their ODEs as a singular perturbation problem as some parameter, conventionally ϵ, tends to zero. For example, in the system

$$\frac{dx}{dt} = f(x, y, t/\epsilon) \quad \text{and} \quad \frac{dy}{dt} = \frac{1}{\epsilon} g(x, y, t/\epsilon),$$

the fast y-derivative gets infinitely large as $\epsilon \to 0$. As done almost everywhere throughout this book we avoid singular perturbations by converting to a regular perturbation problem: the advantage is that then we straightforwardly resolve subtle fast-time effects that accumulate to impact the slow long-term evolution. Here simply scale time: introduce the faster time $t' = t/\epsilon$ so that $d/dt' = \epsilon\, d/dt$. Then, upon omitting primes, the above singular perturbation system becomes the regular

$$\frac{dx}{dt} = \epsilon f(x, y, t) \quad \text{and} \quad \frac{dy}{dt} = g(x, y, t). \tag{18.1}$$

Example 18.1 is in this form. Let's use this general system of ODEs (18.1) to establish the connection between the method of averaging and coordinate transforms. We aim to model the long-time dynamics of the system (18.1) by transforming to a new coordinate system in new variables (X, Y).

18.1 ▪ Find fast y for a frozen x

Consider the ODE system (18.1): for small parameter ϵ the x variable will only evolve slowly. Thus a key reference solution is obtained by freezing the slow x, say set $x(t) = X$ constant, and solving the second ODE in (18.1) for the fast y. Let $y = \tilde{y}(X, t)$ be a solution of $d\tilde{y}/dt = g(X, \tilde{y}, t)$ for some initial condition to be considered later.

For Example 18.1, recall that $\tilde{y}(X, t) = X + \frac{1}{2}a(\sin t - \cos t)$.[220]

[220]Perhaps one might consider including a general homogeneous solution, namely $(y_0 - X + \frac{1}{2}a)e^{-t}$ for an initial condition $\tilde{y}(0) = y_0$. But as such a homogeneous solution decays exponentially quickly, it has no effect on the long-time averages and so is negligible.

18.2 ▪ Establish emergence in a y neighborhood

Let's coordinate transform the fast y variable. Seek to reparametrize the y-space in the form

$$y = \tilde{y}(X,t) + \hat{y}_1(X,t)Y + \mathcal{O}(\epsilon + Y^2)$$

in terms of a new fast variable Y. The linear term in Y, namely $\hat{y}_1(X,t)Y$, is as yet unknown but empowers us to describe solutions in the neighborhood of the reference $\tilde{y}$ for frozen x. Now seek the dynamics of the system (18.1) in this new fast variable Y in the form

$$\dot{Y} = -\mu Y + \mathcal{O}(\epsilon + Y^2).$$

The order of error terms captures effects that are nonlinear in Y, and hence are small near the reference $\tilde{y}$, or are of the size of the small parameter ϵ. The constant $-\mu$ may be viewed as a so-called Lyapunov exponent of the system. We need to determine the dynamics expressed in this new fast variable Y instead of y.

The y ODE (18.1) tells us how to determine the Lyapunov exponent $-\mu$ and the function $\hat{y}_1$. We also acknowledge that the slow x variable X is not actually frozen, but instead evolves slowly; that is, we invoke that $\dot{X} = \mathcal{O}(\epsilon)$. By the chain rule, upon substituting the y-transformation, the time derivative

$$\begin{aligned}\dot{y} &= \tilde{y}_t + Y\hat{y}_{1t} + (\tilde{y}_X + Y\hat{y}_{1X})\dot{X} + \hat{y}_1(-\mu Y) + \mathcal{O}(\epsilon + Y^2)\\ &= \tilde{y}_t + Y(\hat{y}_{1t} - \mu\hat{y}_1) + \mathcal{O}(\epsilon + Y^2).\end{aligned}$$

As in Chapters 3 and 17 on modeling oscillations, one has to be very careful with partial derivatives: as before, here subscripts t, X, and y denote partial derivatives, keeping the other two of the threesome constant.[221] By the ODE (18.1), the time derivative $\dot{y}$ has to equal the right-hand side

$$\begin{aligned}&g\left(X + \mathcal{O}(\epsilon), \tilde{y} + Y\hat{y}_1 + \mathcal{O}(\epsilon + Y^2), t\right)\\ &= g(X,\tilde{y},t) + Y\hat{y}_1 g_y(X,\tilde{y},t) + \mathcal{O}(\epsilon + Y^2).\end{aligned}$$

Now, by definition of the reference solution, $\tilde{y}_t = g(X,\tilde{y},t)$, so these two terms cancel. Equating the coefficients of terms linear in the new coordinate Y leads to requiring

$$\hat{y}_{1t} - \mu\hat{y}_1 = \hat{y}_1 g_y(X,\tilde{y},t). \tag{18.2}$$

Example 18.2 (simple averaging continued). For Example 18.1, the right-hand side of the fast y equation is $g = -y + x + a\sin t$ and hence $g_y = -1$, a constant. In this case, equation (18.2) for part of the coordinate transform reduces to $\hat{y}_{1t} = (\mu - 1)\hat{y}_1$—the classic ODE. Its solution is $y_1 \propto e^{(\mu-1)t}$. Thus in Example 18.1, the coordinate transform would be $y = \tilde{y} + Y\hat{y}_1 + \mathcal{O}(\epsilon + Y^2) = \left[X + \frac{1}{2}a(\sin t - \cos t)\right] + Ye^{(\mu-1)t} + \mathcal{O}(\epsilon + Y^2)$ (where variable Y absorbs the constant of integration in $\hat{y}_1$). But the correction term $Ye^{(\mu-1)t}$ is generally not uniformly valid:

- if $\mu - 1 > 0$, then the correction grows without bound as time $t \to \infty$;
- if $\mu - 1 < 0$, then the correction grows without bound as time $t \to -\infty$;

[221] Remember that the partial derivatives in t, X, and y are performed keeping the other two variables constant. It is in this rigorous sense that we freeze X in many of the computations. For example, later time integrals come from solving ODEs in time where the time derivative is partial, done keeping X constant, and so the integrals have to be calculated with X constant (frozen).

- but fortunately, when we choose $\mu = 1$ then the correction $Ye^{(\mu-1)t} = Y$ for all time, is uniformly small, and is a valid term in the asymptotics.

Thus for the problem of Example 18.1, the time-dependent coordinate transform of the fast variables is

$$y = \underbrace{\left[X + \tfrac{1}{2}a(\sin t - \cos t)\right]}_{\tilde{y}(X,t)} + Y + \mathcal{O}(\epsilon + Y^2) \quad \text{such that } \dot{Y} = -Y + \mathcal{O}(\epsilon + Y^2).$$

That is, trajectories near the reference solution $\tilde{y}(X,t)$ are attracted to it roughly like e^{-t} through the decay of Y. We deduce that in Example 18.1 $y = \tilde{y}(X,t)$ is emergent.

Now let's use the analogous argument for the more general system (18.2). For equation (18.2), move the term $\mu\hat{y}_1$ to the right-hand side and divide by $\hat{y}_1$ to get

$$\frac{\hat{y}_{1t}}{\hat{y}_1} = \mu + g_y(X, \tilde{y}, t).$$

Then integrate with respect to time t, keeping X frozen, as the time derivative on the left-hand side was defined as keeping X constant, to get

$$\hat{y}_1 \propto \exp\left[\int \mu + g_y(X, \tilde{y}, t)\, dt\right].$$

The integral appearing here will generally grow either for large positive time t or for large negative time t, depending upon whether the parameter μ is chosen less than or greater than the mean $-\overline{g_y}$, respectively. Such growth is not allowed. Thus we must define the parameter μ to be the mean decay rate: we choose $\mu(X) = -\overline{g_y(X, \tilde{y}, t)}$. Consequently, the function $\hat{y}_1 \propto \exp\left[\int g_y - \overline{g_y}\, dt\right]$ (and choose equality as Y absorbs any integration constant). Then the general time-dependent coordinate transform of the fast variable is

$$y = \tilde{y}(X,t) + Y\hat{y}_1(X,t) + \mathcal{O}(\epsilon + Y^2) \quad \text{such that } \dot{Y} = -\mu(X)Y + \mathcal{O}(\epsilon + Y^2).$$

That is, $-\mu = \overline{g_y(X, \tilde{y}, t)}$ is a Lyapunov exponent: it is the mean rate of decay to the reference solution $\tilde{y}$.[222] If $-\mu \geqslant 0$, then $\tilde{y}$ is not attractive and therefore averaging over it is generally irrelevant, as the system would not spend time near $\tilde{y}$. However, when $-\mu < 0$, the reference solution $\tilde{y}$ is attractive to nearby trajectories, and hence emergent.

Example 18.3 (no unique emergence). Consider the system $\dot{x} = \epsilon(-x^3 + y\sin t)$ and $\dot{y} = g(x, y, t) = xy\cos t$. The y ODE for frozen $x = X$ is separable, leading to the general solution $\tilde{y} = y_0 e^{X\sin t}$ as the reference to be averaged over. However, the local rate of attraction to $\tilde{y}$ is the derivative $g_y(X, \tilde{y}, t) = X\cos t$. Averaging this over all time gives the Lyapunov exponent $-\mu = 0$. As $\mu = 0$ there is no attraction to a unique long-term reference solution, and hence there can be no unique averaged equation for the slow x.

This lack of uniqueness is apparent in the reference solution $\tilde{y} = y_0 e^{X\sin t}$ because $\tilde{y}$ does not "forget" the initial condition y_0, the integration constant. In contrast, in Example 18.1 the effect of the integration constant decays exponentially in time, and hence a unique averaged model exists.

[222]There are several caveats here: the functions have to be smooth enough, the mean has to exist, and so on.

18.3 ▪ A transform of x leads to averaging

Now consider the slow x-evolution of the system (18.1), namely $\dot{x} = \epsilon f(x, y, t)$. We seek to simplify this by transforming to a new slow coordinate, namely X, to partner the new fast coordinate Y. Seek the new coordinate X such that

$$x = X + \epsilon\hat{x}(X, Y, t) + \mathcal{O}(\epsilon^2) \quad \text{and} \quad \dot{X} = \epsilon\hat{F}(X) + \mathcal{O}(\epsilon^2).$$

Substitute this ansatz into the slow x ODE and use the chain rule to derive the left-hand side time derivative

$$\begin{aligned} \dot{x} &= \dot{X} + \epsilon\hat{x}_X\dot{X} + \epsilon\hat{x}_Y\dot{Y} + \epsilon\hat{x}_t + \mathcal{O}(\epsilon^2) \\ &= \epsilon\hat{F} + \mathcal{O}(\epsilon^2) - \epsilon\mu Y\hat{x}_Y + \epsilon\hat{x}_t + \mathcal{O}(\epsilon^2). \end{aligned}$$

Complementing this, the right-hand side becomes

$$\begin{aligned} \dot{x} &= \epsilon f(x, y, t) \\ &= \epsilon f\left(X + \mathcal{O}(\epsilon), \tilde{y} + \hat{y}_1(X, t)Y + \mathcal{O}(\epsilon + Y^2), t\right) \\ &= \epsilon f_0(X, t) + \epsilon f_1(X, t)\hat{y}_1(X, t)Y + \mathcal{O}(\epsilon^2 + \epsilon Y^2), \end{aligned}$$

where $f_0(X, t) = f(X, \tilde{y}(X, t), t)$ and $f_1(X, t) = f_y(X, \tilde{y}(X, t), t)$. Equating the coefficients of ϵ on the two sides leads to requiring

$$\hat{F} + \hat{x}_t - \mu Y\hat{x}_Y = f_0 + f_1\hat{y}_1 Y + \mathcal{O}(Y^2)\,.$$

Seek a solution for small Y in the form $\hat{x} = \hat{x}_0(X, t) + Y\hat{x}_1(X, t) + \mathcal{O}(Y^2)$, and equate the coefficients of the constant terms in Y and the linear terms in Y to deduce

$$\hat{F} + \hat{x}_{0t} = f_0 \quad \text{and} \quad \hat{x}_{1t} - \mu\hat{x}_1 = f_1\hat{y}_1\,. \tag{18.3}$$

Consider the first of these equations. To have the evolution for X as simple as possible, we wish to incorporate as much as possible of f_0 into the coordinate transform $\hat{x}_0$. Hence we want $\hat{x}_0 = \int f_0(X, t)\,dt$ (recall that such integrals are done keeping X frozen, as the partial derivative $\hat{x}_{0t}$ is defined with X constant). But in general this integrand has nonzero mean; hence it will generate an arbitrarily large integral $\hat{x}_0$ either forward or backward in time, or both. An arbitrarily large $\hat{x}_0$ is not allowed, as it destroys the uniform validity of the approximation in time. Thus we must use $\hat{F}$ to remove the mean: set $\hat{F}(X) = \overline{f_0} = \overline{f(X, \tilde{y}(X, t), t)}$. This argument deduces the classic averaged model as the leading approximation

$$\dot{X} = \epsilon\hat{F}(X) + \mathcal{O}(\epsilon^2) = \epsilon\overline{f(X, \tilde{y}(X, t), t)} + \mathcal{O}(\epsilon^2).$$

But we now know more. We know that the slow variable X in the model is not quite the original slow variable x: it is different by the amount $\epsilon\hat{x}_0$, as $x \approx X + \epsilon\hat{x}_0$. With the above choice of $\hat{F}$, we solve the first equation in (18.3) to deduce, formally, that $\hat{x}_0 = \int f_0(X, t) - \overline{f_0(X, t)}\,dt$.

Example 18.4 (simple averaging continued again). Revisit Example 18.1. There, the reference $\tilde{y} = X + \frac{1}{2}a(\sin t - \cos t)$, the right-hand side of the slow x ODE is $\epsilon f(x, y, t) = \epsilon(x^3 - xy^2)$, and $\hat{y}_1 = 1$. To find the coordinate transform of the slow x, from the above $\mathcal{O}(\epsilon)$ analysis we need to solve

$$\hat{F} + \hat{x}_t - \mu Y\hat{x}_Y = X^3 - X\tilde{y}^2 - 2X\tilde{y}Y + \mathcal{O}(Y^2).$$

Equating the coefficients of terms constant in Y, we seek to solve

$$\begin{aligned}\hat{F} + \hat{x}_{0t} &= X^3 - X\tilde{y}^2 \\ &= X^3 - X\left[X + \tfrac{1}{2}a(\sin t - \cos t)\right]^2 \\ &= -aX^2(\sin t - \cos t) - \tfrac{1}{4}a^2X(1 - \sin 2t).\end{aligned}$$

As before, the average of this right-hand side determines the average model $\dot{X} = \epsilon\hat{F} = -\frac{1}{4}\epsilon a^2 X$.

After subtracting the average, the right-hand side determines the coordinate transform component $\hat{x}_0 = aX^2(\cos t + \sin t) - \frac{1}{8}a^2X\cos 2t$. That is, the real x cannot be precisely the slow averaged X but instead

$$x = X + \epsilon\left[aX^2(\cos t + \sin t) - \tfrac{1}{8}a^2X\cos 2t\right] + \mathcal{O}(\epsilon^2 + \epsilon Y).$$

While continuing this example, let's look at the terms linear in Y. Then the ODE (18.3) for the slow coordinate x in the vicinity of the reference solution, $\hat{x}_{1t} - \mu\hat{x}_{1t} = f_1\hat{y}_1$, becomes

$$\hat{x}_{1t} - \hat{x}_1 = -2X\tilde{y} = -2X^2 - aX(\sin t - \cos t).$$

The solution bounded for all time is

$$\hat{x}_1 = 2X^2 + aX\sin t.$$

That is, the time-dependent coordinate system (X, Y) near the long-term process is, more accurately, that the real

$$\begin{aligned}x = X &+ \epsilon\left[aX^2(\cos t + \sin t) - \tfrac{1}{8}a^2X\cos 2t\right] \\ &+ \epsilon Y\left[2X^2 + aX\sin t\right] + \mathcal{O}(\epsilon^2 + \epsilon Y^2).\end{aligned}$$

Recall that to accompany this we have the coordinate transform

$$y = \underbrace{\left[X + \tfrac{1}{2}a(\sin t - \cos t)\right]}_{\tilde{y}(X,t)} + Y + \mathcal{O}(\epsilon + Y^2).$$

Because such relations between (x, y) and (X, Y) are just coordinate transforms, they provide a one-to-one correspondence between the original dynamics and the dynamics of the model.

Here we know that in the new (X, Y) coordinate system,

$$\dot{X} = -\tfrac{1}{4}\epsilon a^2 X + \mathcal{O}(\epsilon^2) \quad \text{and} \quad \dot{Y} = -Y + \mathcal{O}(\epsilon + Y^2).$$

Hence for all initial conditions (in some neighborhood), $Y \to 0$ exponentially quickly so that the averaged model emerges. Further, the slow variable X decaying over long times is the averaged dynamics.

To finish our short discourse on averaging, return to the general case and the coordinate transform of x in the neighborhood of the long-term process. In general we found the transform $x \approx X + \epsilon\hat{x}_0$; now let's find the term linear in Y in the more accurate $x \approx X + \epsilon\hat{x}_0 + \epsilon Y\hat{x}_1$. From equation (18.3), the function $\hat{x}_1$ must satisfy $\hat{x}_{1t} - \mu\hat{x}_1 = f_1\hat{y}_1$, where we must have $\mu > 0$ for an emergent averaged model. Using the integrating factor $e^{-\mu t}$, the solution of this ODE is

$$x_1 = \int^t e^{\mu(t-\tau)} f_1(X,\tau)\hat{y}_1(X,\tau)\,d\tau\,.$$

What about the end point of the integration? Since x_1 must be bounded for all time, the end point has to be the disturbing $+\infty$: that is,

$$x_1 = -\int_t^\infty e^{\mu(t-\tau)} f_1(X,\tau)\hat{y}_1(X,\tau)\,d\tau\,.$$

Why disturbing? Because the coordinate transform then has to "look into the future": it depends upon future values of f_1 and $\hat{y}_1$, although through the exponential decay of $e^{\mu(t-\tau)}$ the integral only looks over a time scale $1/\mu$ into the future. In working out Example 18.4 we were completely oblivious to the future dependence (as it was just trigonometric functions in the algebra). To maintain a precise correspondence between model and original we must accept such short-time future dependence.

If you reject such future dependence, then you must accept the unpalatable consequence that the averaged X has an unknowable relation to the real x, albeit $\mathcal{O}(\epsilon)$ close.

I phrased the arguments and analysis of this averaging in terms of a system (18.1) composed of a pair of scalar ODEs. However, there is nothing in the argument and analysis that cannot be straightforwardly generalized to the case of a system of many coupled ODEs. All of the above generic features are seen in modeling nonautonomous and stochastic systems.

We conclude that averaging can be put within the framework of modeling by coordinate transforms. The benefits are that we obtain a clear view of the modeling and that we obtain a systematic methodology for proceeding to higher-order effects when necessary.

18.4 ▪ Exercises

Exercise 18.1 (forcing frequency). Redo the averaging modelling of Example 18.1 but with the sinusoidal forcing of a general frequency ω. Discuss limitations appearing for small frequency.

Answer: $\hat{y} = x + a(\sin\omega t - \omega\cos\omega t)/(1+\omega^2)$ and then $\dot{x} \approx -\epsilon[x^3 + a^2x/(1+\omega^2)]$.

Exercise 18.2 (simple pair). A classically simple pair of fast/slow ODEs is

$$\dot{x} = -\epsilon xy \quad\text{and}\quad \dot{y} = -y + x^2 + a\sin t\,.$$

Use averaging to deduce the model that $\dot{x} \approx -\epsilon x^3$ independent of the forcing.

Exercise 18.3 (future dependence). An interesting pair of fast/slow ODEs (derived from advection/dispersion problems) is

$$\dot{x} = -\epsilon y\sin t \quad\text{and}\quad \dot{y} = -y + x\sin t\,. \tag{18.4}$$

Use averaging to deduce the model that $\dot{X} \approx -\frac{1}{4}\epsilon X$ for some $X \approx x$: determine more details of this approximate equality. Replace $\sin t$ by the general function $w(t)$ and, using integrating factors, find integrals that formally solve equations to deduce that the approximate coordinate transform between (x,y) and (X,Y) involves an integral involving future dependence.

Exercise 18.4 (rescale variables). In explaining an approach to stochastic modeling of atmospheric dynamics, Monahan and Culina (2011) introduced a system similar to

$$\frac{dx}{dt} = -x + \Sigma(x)y^2 \quad \text{and} \quad \frac{dy}{dt} = -\frac{1}{\epsilon}y + a\sin\omega t$$

for arbitrary smooth functions $\Sigma(x)$. Rescale time, forcing amplitude a and frequency ω in terms of ϵ to change this singular perturbation system, for small ϵ, into a regular perturbation form. Then apply averaging and deduce the model that $x \approx X$ where, in the original time, $dX/dt \approx -X + c\Sigma(X)$ for some constant c quadratic in the amplitude a and dependent upon frequency ω. Determine the relevant details.

Exercise 18.5 (not ensemble averages). Many people automatically presume ensemble averaging is the same as time averaging—the ergodic hypothesis. This exercise shows that we do need to be at least a little discerning. Consider the simple system

$$\dot{x} = -\epsilon x y^2 \quad \text{and} \quad \dot{y} = -y + \cos\omega\,,$$

where parameter ω parametrizes the realizations of the ensemble (say $0 \leqslant \omega \leqslant 2\pi$ or equivalent).

- Show that averaging predicts that the slow model is
  - either $\dot{X} \approx -\frac{1}{2}\epsilon X$ for ensemble averaging
  - or $\dot{X} \approx -\epsilon X\cos^2\omega$ when the averaging is over time.

  Which is correct? The answer is that both are correct statements, but the second is the more useful one for accurate long-term modeling.
- Argue that the fast y coordinate transform is $y \approx \cos\omega + Y$ such that $\dot{Y} \approx -Y$.
- Then deduce that the leading equation for the slow x coordinate transform is $\hat{F} + \hat{x}_{0t} = -\epsilon X\cos^2\omega$.
  - If we choose the ensemble average $\hat{F} = -\frac{1}{2}\epsilon X$, then show that necessarily $x_0 = \epsilon X(\frac{1}{2} - \cos^2\omega)t$, which grows in time without bound and thus destroys any uniform validity of the coordinate transform.
  - Alternatively, to uniformly solve $\hat{F} + \hat{x}_{0t} = -\epsilon X\cos^2\omega$, argue that we must choose $\hat{x}_0 = 0$ (or constant anyway) and $\hat{F} = -\epsilon X\cos^2\omega$.

Conclusion: Using ensemble averaging requires the ergodic hypothesis to be correct—that time and ensemble averages are the same—in order for the long-term modeling to be accurate.

Chapter 19

Coordinate transforms separate slow from fast in nonautonomous dynamics

Contents

One great advantage of basing modeling upon coordinate transforms is that exactly transformed dynamics fully reproduces the original dynamics for all time and all state space. It is only when we approximate the transformed dynamics that errors occur. Consequently, modeling errors are much better controlled.

As developed in Part IV, normal-form coordinate transformations provide a sound basis for simplifying multiscale nonlinear dynamics (Elphick et al., 1987; Cox and Roberts, 1995; Aulbach and Wanner, 2000, e.g.). In systems with fast and slow dynamics, one seeks a coordinate transform that decouples the slow from the fast. The decoupled slow modes then provide accurate predictions for the long-term dynamics. Such normal-form coordinate transformations provide new insight into simplifying dynamics over that provided by other techniques. Such insight is especially needed for the tricky case of nonautonomous systems.

Principles we require in a coordinate transform

1. Avoid unbounded (secular) terms in the transformation and the evolution (to ensure uniform asymptotic approximations).
2. Decouple all the slow processes from the fast processes (to ensure a valid long-term model).
3. Insist that the slow manifold is precisely the transformed fast modes being zero.
4. Ruthlessly eliminate as many as possible of the terms in the evolution (to simplify at least the algebraic form of the ODEs or SDEs).
5. Avoid as far as possible fast-time memory integrals in the evolution (to endeavor to remove all fast-time processes from the slow modes).

In general we can meet all these principles, although the last two are only phrased as "as far/many as possible": section 19.1 explores the issues in a particular example nonautonomous forced system, whereas section 19.2 presents general theory for finite-dimensional, nonlinear, nonautonomous dynamical systems. The broad applicability of coordinate transforms empowers a web service to construct such time-dependent normal forms for us (Roberts, 2009d): enter any suitable system of ODEs or SDEs into the web page and it provides you with an appropriate normal form constructed according to the above principles to separate slow and fast nonautonomous dynamics.

19.1 ▪ Model an example nonlinear nonautonomous system

This section uses a basic example system to introduce the modeling of the dynamics of nonlinear, multi-time-scale, nonautonomous/noisy systems. For some specified forcing $w(t)$ with overall strength σ, we explore the dynamics of $(x(t), y(t))$ according to the coupled ODEs

$$\dot{x} = -xy \quad \text{and} \quad \dot{y} = -y + x^2 - 2y^2 + \sigma w(t). \tag{19.1}$$

Figure 19.1 plots some representative trajectories of the ODE system (19.1). In the plotted finite domain near the origin, the y variable decays exponentially quickly to fluctuate about $y \approx x^2$, whereas the x variable evolves relatively slowly over long times, albeit affected by the rapidly fluctuating y. Our challenge is to separate, amid the confusing fluctuations, the slow $x(t)$ from the fast $y(t)$.

The modeling issues raised, and their resolution, in this relatively simple nonautonomous system are generic, as discussed in section 19.2. *The detailed analysis of this example demonstrates the inevitability of the proposed methodology when you set out, as we do here, to systematically and accurately separate slow and fast processes in nonlinear nonautonomous or noisy systems.*

The forcing $w(t)$ could be as regular as a periodic oscillator, as used in Figure 19.1, or it could be the output of a deterministic chaotic system (Just et al., 2001, e.g.), or it could be random, such as the white noise formal derivative of a random walk Wiener process. The last case of a noisy (stochastic) forcing is both particularly interesting and particularly delicate. Indeed, the delicacies of the stochastic case determine the methodology we use for all cases.[223]

[223] Any stochastic interpretation of the algebra is to be the Stratonovich interpretation, not Itô (Chapter 20), because we use ordinary calculus for all analysis (as did Arnold and Imkeller (1998) and Arnold (2003)). Chapter 21 develops modeling specifically for stochastic systems.

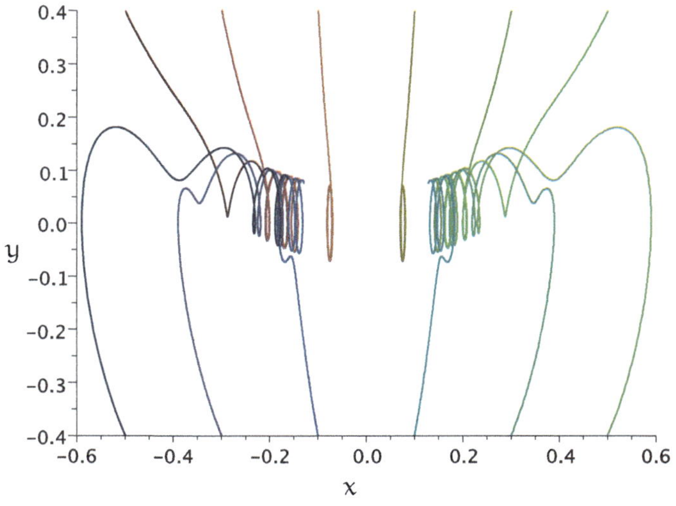

Figure 19.1. *Trajectories of the example system* (19.1) *from different initial conditions for the forcing* $w(t) = \cos t$ *of strength* $\sigma = 0.1$. *The trajectories are attracted to* $y \approx x^2 + \sigma(\cos t + \sin t)/2$.

The challenge is to adapt the autonomous, deterministic, normal-form coordinate transformation from Part IV to the nonautonomous system (19.1) in order to not only decouple the interesting slow mode x but simplify its evolution as far as possible. The analysis and argument are detailed in order to demonstrate in a relatively simple setting how Principles 1–5 above may be realized. Those familiar with the concept of normal forms could skip to section 19.2 for generic arguments.

Let's construct a coordinate transform that decouples the fast and slow dynamics of the example ODE (19.1) in the presence of its time-dependent, nonautonomous forcing $\sigma w(t)$. Sections 19.1.1–19.1.4 detail how to find a coordinate transform $(x, y) \mapsto (X, Y)$ such that in the new (X, Y) variables the system (19.1) becomes

$$\dot{X} = -X^3 - \sigma w X + \cdots \quad \text{and} \quad \dot{Y} = -[1 + 4\sigma w + \cdots]Y. \tag{19.2}$$

The Y component of this system clearly shows that provided the forcing $w(t)$ is not too extreme, then $Y \to 0$ exponentially quickly in time. Thus the long-term dynamics is given simply by the slow X component of the ODEs (19.2). The coordinate transform then maps this modeling picture into the original xy-plane. But, in order to cater for the fluctuations in time due to the forcing $w(t)$, the coordinate transform must be time dependent, as shown schematically in the coordinate system of Figure 19.2.

This section is detailed in order to argue that no alternatives go unrecognized. As usual (Roberts, 1997), our method is to iteratively refine a coordinate transform based upon the residuals of the governing ODEs (19.1). In principle such a coordinate transform captures all the dynamics of the original system; hence deductions from the transformed system apply to the original system.

19.1.1 ▪ Simplifying fast dynamics involves memory

We start from the linearized dynamics in a near identity transform: $x \approx X$, $y \approx Y$, $\dot{X} \approx 0$, and $\dot{Y} \approx Y$.

With $x \approx X$ and $\dot{X} \approx 0$, seek a change to the y coordinate of the form

$$y = Y + \hat{y}(t, X, Y) + \cdots \quad \text{and} \quad \dot{Y} = -Y + \hat{G}(t, X, Y) + \cdots, \tag{19.3}$$

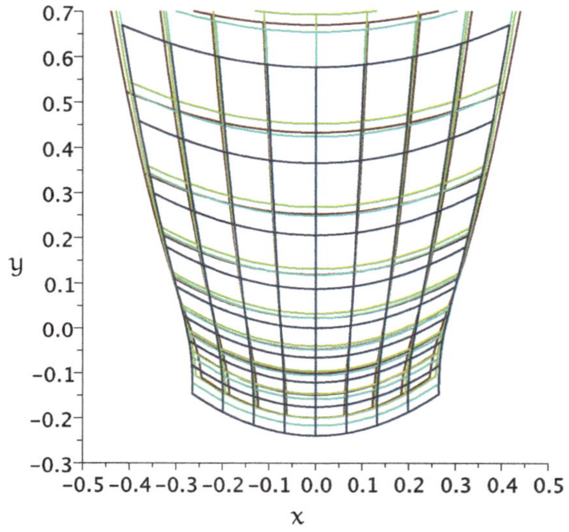

Figure 19.2. *Four different (colored) meshes represent four instants of the coordinate curves in the* xy*-plane of the time-dependent* (X, Y) *coordinate system that simplifies the algebraic description of the dynamics of the system* (19.1) *to* (19.14b) *and* (19.11b) *(below) in the case of forcing* $w(t) = \cos t$ *of intensity* $\sigma = 0.3$.

where $\hat{y}$ and $\hat{G}$ are small, namely $\mathbb{O}(2)$, corrections to the transform and the corresponding evolution. The notation $\mathbb{O}(p)$ compactly denotes terms of $\mathcal{O}(X^p + Y^p + \sigma^p)$. Consequently, in this section all asymptotic expressions are *local* to $(X, Y, \sigma) = (0, 0, 0)$, that is, valid in some neighborhood of the origin; later sections give examples *globally* valid in one or more variables. Substitute (19.3) into the y ODE (19.1) and drop products of small corrections to recognize that we need to solve the homological equation

$$\hat{G} + \frac{\partial \hat{y}}{\partial t} + \hat{y} - Y\frac{\partial \hat{y}}{\partial Y} = \sigma w(t) + X^2 - 2Y^2 + \cdots ; \tag{19.4}$$

the partial derivatives are here done keeping constant the other two variables of X, Y, and t. Since equation (19.4) is linear in the corrections—the hatted quantities on the left—we are empowered to consider the effects of the right-hand side forcing terms one by one, and then combine them.

- First, consider the autonomous part of (19.4): $\hat{G} + \hat{y}_t + \hat{y} - Y\hat{y}_Y = X^2 - 2Y^2$. Keep the evolution as simple as possible (Principle 4) by not changing the evolution at all, $\hat{G} = 0$, and instead using the method of undetermined coefficients to modify the coordinate transform by setting $\hat{y} = X^2 + 2Y^2$.

- Second, consider the remaining nonautonomous forcing of (19.4): $\hat{G} + \hat{y}_t + \hat{y} - Y\hat{y}_Y = \sigma w(t)$. The term $Y\hat{y}_Y$ on the left-hand side cannot help us as Y does not appear on the right-hand side. Keep the Y-dynamics as simple as possible (Principle 4) by choosing $\hat{G} = 0$ and adapting the coordinate transform by solving $\hat{y}_t + \hat{y} = \sigma w(t)$. Its solution is the convolution $\hat{y} = \sigma e^{-t}\star w$ (also used in Example 20.1), where the convolution is defined as

$$e^{-t}\star w = \int_0^t e^{\tau - t} w(\tau)\, d\tau \quad \text{so that} \quad \frac{d(e^{-t}\star w)}{dt} = w - e^{-t}\star w. \tag{19.5}$$

Finally, combining the two effects, the new approximation of the coordinate transform and the dynamics is

$$y = Y + X^2 + 2Y^2 + \sigma e^{-t}\star w + \cdots \quad \text{and} \quad \dot{Y} = -Y + \cdots\,. \tag{19.6}$$

In these leading-order terms of the coordinate transform, (19.6) predicts that the nonautonomous slow manifold $Y = 0$ corresponds to the vertically fluctuating parabola $y \approx X^2 + \sigma e^{-t}\star w$, as seen in the overall vertical displacements of the coordinate mesh in Figure 19.2.

19.1.2 ▪ Split forcing to eliminate memory from slow dynamics

Now seek a correction to the slow x component of the time-dependent coordinate transform in the form

$$x = X + \hat{x}(t, X, Y) + \mathcal{O}(3) \quad \text{and} \quad \dot{X} = \hat{F}(t, X, Y) + \mathcal{O}(3), \tag{19.7}$$

where $\hat{x}$ and $\hat{F}$ are $\mathcal{O}(2)$ corrections to the transform and the corresponding evolution. Substitute into the x equation of ODE (19.1) and omit small products to recognize that we need to solve the homological equation

$$\hat{F} + \frac{\partial \hat{x}}{\partial t} - Y\frac{\partial \hat{x}}{\partial Y} = -XY - \sigma X e^{-t}\star w + \mathcal{O}(3). \tag{19.8}$$

Again, since (19.8) is linear in the corrections—the hatted quantities on the left—we are empowered to consider the effects of the right-hand side forcing terms one by one, and then combine them.

- First consider the autonomous part of the homological equation, $\hat{F} + \partial\hat{x}/\partial t - Y\partial\hat{x}/\partial Y = -XY$; keep the evolution unchanged, $\hat{F} = 0$, by choosing $\hat{x} = XY$ in the coordinate transform.

- Second, consider the fluctuating part of the homological equation, $\hat{F} + \hat{x}_t - Y\hat{x}_Y = -\sigma X e^{-t}\star w$. The term $\hat{x}_Y$ cannot help us solve this fluctuating part as there is no Y factor in the right-hand side term. We do not necessarily integrate the forcing w into the coordinate transform $\hat{x}$ as then terms in $\hat{x}$ may grow unboundedly like $\int w\,dt$ (Principle 1):

  - for some forcing, the integral $\int w\,dt$ would not grow, for example $w = \cos t$, and so could be integrated into the coordinate transform;
  - but for other forcing it would grow; for example, $w = \cos\sqrt{t}$ has zero mean (and finite variance, namely 1/2), but the integral $\int w\,dt = 2\cos\sqrt{t} + 2\sqrt{t}\sin\sqrt{t}$ grows in time like $\sqrt{t}$;
  - and for the stochastic case of white noise $w(t)$, the integral $\int w\,dt$ is a random walk Wiener process that also grows like $\sqrt{t}$.

  Thus we *choose* to not integrate all the forcing w into the coordinate transform $\hat{x}$. To avoid a fast-time convolution in the slow evolution $\hat{F}$ (Principle 5), use identity (19.5) to split $e^{-t}\star w = w - d(e^{-t}\star w)/dt$ in the right-hand side of (19.4), and hence choose components $\hat{F} = -\sigma X w$ and $\hat{x} = \sigma X e^{-t}\star w$.

Finally, combining the components of the corrections gives

$$x = X + XY + \sigma X e^{-t}\star w + \mathcal{O}(3) \quad\text{and}\quad \dot X = -\sigma X w + \mathcal{O}(3)\,. \tag{19.9}$$

Observe that the slow dynamics being $\dot X \approx -\sigma X w$ is one example of additive forcing of the original system (19.1) appearing as a multiplicative forcing in a slow variable.

The splitting to arrive at (19.9) applies to all reasonable forcing $w(t)$, even stochastic forcing. However, if you are analyzing a system where the forcing w has some a priori known special structure, then you may instead choose to allocate the mean of w to the evolution $\dot X$ and integrate all the fluctuations into the coordinate transform $\hat x$. Choosing to remove all fluctuations from $\dot X \approx -\sigma X w$ will result in a model for X that is easily integrated. But on the other hand, for systems with fluctuating terms, recall that there are well-established good methods, Filon quadrature for example, for efficient integration of highly oscillatory integrals (Iserles, 2004, e.g.); recent methods for integrating differential equations with highly oscillatory parts (Condon et al., 2010; Bunder and Roberts, 2013, e.g.); and well-known good methods for efficient integration of SDEs (Kloeden and Platen, 1992, e.g.).

Earlier approaches to stochastic bifurcation did not split the stochastic forcing; consequently they derived more complicated normal forms that additionally have fast-time memory processes. For example, Arnold (2003) analyzed a two-variable fast/slow stochastic system and presents in his equation (8.5.47) a normal form for the slow mode. Whereas Arnold's normal form is perfectly good for exploring bifurcations, it is inadequate for long-time, macroscale modeling as it contains several fast-time memory integrals of the sort we choose to avoid (Principle 5).

This discussion identifies that the splitting between coordinate transform and resultant model dynamics is a highly subjective decision by the modeler. The decision depends upon the particular characteristics of the time-dependent forcing in the original system and upon how the resultant dynamical system is to be used. We proceed from the choice leading to (19.9) by continuing to invoke a splitting that is suitable for a wide range of forcing, including the case of stochastic forcing.

19.1.3 ▪ Split forcing to avoid memory in fast dynamics

Now revisit the fast y ODE to find second-order effects of the forcing. Seek corrections, $\hat y$ and $\hat G$, to the y transform and Y-evolution (19.6). The corrections are driven by the updated residual of the y equation in the ODE (19.1):

$$\hat G + \frac{\partial \hat y}{\partial t} + \hat y - Y\frac{\partial \hat y}{\partial Y} = -4\sigma Y e^{-t}\star w - 2\sigma^2(e^{-t}\star w)^2 + \mathcal{O}(3)\,. \tag{19.10}$$

Separately consider the two fluctuating forcing terms on the right-hand side.

- First, to solve $\hat G + \hat y_t + \hat y - Y\hat y_Y = -4\sigma Y e^{-t}\star w$ we must seek $\hat G$ and $\hat y$ proportional to Y, whence the last two terms on the left-hand side cancel, as $\hat y - Y\hat y_y = 0$, leaving just $\hat G + \hat y_t = -4\sigma Y e^{-t}\star w$. As before, the convolution identity (19.5) empowers us to split the convolution on the right-hand side as $e^{-t}\star w = w - d(e^{-t}\star w)/dt$ and thus to choose $\hat G = -4\sigma Y w$ and $\hat y = 4\sigma Y e^{-t}\star w$ to avoid growing secular terms in $\hat y$ (Principle 1) and to avoid fast-time convolution in the Y-evolution (Principle 5).

- Second, the quadratic forcing term on the right-hand side is no problem as it is constant with respect to Y: keep the evolution unchanged, $\hat G = 0$ (Principle 4), and then the convolution $\hat y = -2\sigma^2 e^{-t}\star(e^{-t}\star w)^2$ corrects the coordinate transform.

Consequently, the fast-time transform and dynamics are more accurately

$$y = Y + X^2 + 2Y^2 + \sigma\left[e^{-t}\star w + 4Ye^{-t}\star w\right] - 2\sigma^2 e^{-t}\star(e^{-t}\star w)^2 + \mathcal{O}(3), \quad (19.11a)$$

$$\dot{Y} = -Y - 4\sigma Y w + \mathcal{O}(3). \quad (19.11b)$$

19.1.4 ▪ Quadratic forcing normally appears in the slow dynamics

Interesting third-order effects now show up in the slow dynamics. Seek corrections, $\hat{x}$ and $\hat{F}$, to the transform and evolution (19.9). Drive the corrections by the updated residual of the x equation of ODE (19.1):

$$\hat{F} + \frac{\partial \hat{x}}{\partial t} - Y\frac{\partial \hat{x}}{\partial Y} = -X^3 - 3XY^2 + \sigma XY(5w - 6e^{-t}\star w)$$
$$+ \sigma^2 X\left[we^{-t}\star w - (e^{-t}\star w)^2 + 2e^{-t}\star(e^{-t}\star w)^2\right] + \mathcal{O}(4). \quad (19.12)$$

The solution of this one equation is rather involved. Consider the right-hand side in three separate groups of terms.

- Driven by the terms $-X^3 - 3XY^2$, let's choose $\hat{F} = -X^3$ and $\hat{x} = \frac{3}{2}XY^2$ as in the approach for autonomous dynamics (Chapter 11).
- To match the term linear in forcing, $\hat{F} + \hat{x}_t - Y\hat{x}_Y = \sigma XY(5w - 6e^{-t}\star w)$, we must seek $\hat{F}$ and $\hat{x}$ proportional to XY, whence $\hat{x}_t - Y\hat{x}_y \mapsto \hat{x}_t - \hat{x}$. Consequently solve $\hat{F} + \hat{x}_t - \hat{x} = \sigma XY(5w - 6e^{-t}\star w)$. We now argue that foreknowledge, anticipation, of the forcing's future necessarily appears. Consider the two cases.
  - Implementing Principle 4, we set $\hat{F} = 0$ and split, using (19.5):

$$\begin{aligned} 5w - 6e^{-t}\star w &= 5w - 3e^{-t}\star w - 3e^{-t}\star w \\ &= 5w - 3w + 3\frac{d(e^{-t}\star w)}{dt} - 3e^{-t}\star w \\ &= 2w + 3\frac{d(e^{-t}\star w)}{dt} - 3e^{-t}\star w\,. \end{aligned}$$

    Hence $\hat{x}$ has a component $3e^{-t}\star w$ from the last two terms and a component $2e^{+t}\star w$ from the first, where this convolution is defined as

$$e^{+t}\star w = \int_t^{\infty} e^{t-\tau} w(\tau)\, d\tau\,, \quad \text{so that} \quad \frac{d(e^{+t}\star w)}{dt} = -w + e^{+t}\star w\,. \quad (19.13)$$

    Consequently $\hat{x} = \sigma XY(3e^{-t}\star w - 2e^{+t}\star w)$. The convolution (19.13) invokes anticipation of the forcing in accord with the arguments of Arnold and Imkeller (1998) and Arnold (2003) for stochastic systems.
  - Alternatively, if we disallow anticipation, then we must assign all of the right-hand side term $\sigma XY(5w - 6e^{-t}\star w)$ into the X-evolution by assigning $\hat{F} = \sigma XY(5w - 6e^{-t}\star w)$ and $\hat{x} = 0$. The difficulty here is that the evolution to the slow manifold then depends upon Y, contradicting Principle 2. To be faithful to Principle 2 we must invoke anticipation in the coordinate transformation.

Many more anticipatory convolutions appear in higher-order terms, namely those terms with Y factors. However, anticipatory convolutions need never occur in the evolution on the slow manifold itself as $Y = 0$ there (Theorem 19.2).

- Now consider the quadratic forcing term in (19.12), namely

$$\hat{F} + \hat{x}_t - Y\hat{x}_Y = \sigma^2 X\left[we^{-t}\star w - (e^{-t}\star w)^2 + 2e^{-t}\star(e^{-t}\star w)^2\right].$$

Due to the factor X, we must seek contributions to the solution that are proportional to X; consequently, on the left-hand side $-Y\hat{x}_Y = 0$. In order to avoid growing (secular) components (Principle 1), at least part of the fluctuations on the right-hand side generally must be assigned into the evolution $\hat{F}$.

The term $(e^{-t}\star w)^2 = (e^{-t}\star w)(e^{-t}\star w)$ and the identity (19.5) enable us to split the second factor of these two convolutions as $e^{-t}\star w = w - d(e^{-t}\star w)/dt$. Thus the term

$$\begin{aligned}(e^{-t}\star w)^2 &= (e^{-t}\star w)\left(w - \frac{d(e^{-t}\star w)}{dt}\right)\\ &= we^{-t}\star w - \tfrac{1}{2}\frac{d}{dt}\left[(e^{-t}\star w)^2\right].\end{aligned}$$

Invoking identity (19.5) again for the outer convolution, and the above expression for $(e^{-t}\star w)^2$, the term

$$\begin{aligned}e^{-t}\star(e^{-t}\star w)^2 &= (e^{-t}\star w)^2 - \frac{d}{dt}[e^{-t}\star(e^{-t}\star w)^2]\\ &= we^{-t}\star w - \frac{d}{dt}\left[\tfrac{1}{2}(e^{-t}\star w)^2 + e^{-t}\star(e^{-t}\star w)^2\right].\end{aligned}$$

The quadratic forcing term in (19.12) is then rewritten as

$$\hat{F} + \hat{x}_t - Y\hat{x}_Y = \sigma^2 X\left[2we^{-t}\star w - \frac{d}{dt}\left(\tfrac{1}{2}(e^{-t}\star w)^2 + 2e^{-t}\star(e^{-t}\star w)^2\right)\right].$$

This expression implies updates to the evolution of $\hat{F} = 2\sigma^2 Xwe^{-t}\star w$ and to the coordinate transform of $\hat{x} = -\sigma^2 X(\tfrac{1}{2} + 2e^{-t}\star)(e^{-t}\star w)^2$.

The upshot is that the x-transformation and X-evolution are more accurately

$$\begin{aligned}x &= X + XY + \tfrac{3}{2}XY^2 + \sigma\left[Xe^{-t}\star w + XY(3e^{-t}\star w - 2e^{+t}\star w)\right]\\ &\quad - \sigma^2 X(\tfrac{1}{2} + 2e^{-t}\star)(e^{-t}\star w)^2 + \mathcal{O}(4), \qquad (19.14a)\\ \dot{X} &= -X^3 - \sigma Xw + 2\sigma^2 Xwe^{-t}\star w + \mathcal{O}(4). \qquad (19.14b)\end{aligned}$$

Why does a convolution remain in $\dot{X}$? Unfortunately, there appears no way to *generally* remove the fast-time convolution appearing in the $\dot{X}$ ODE (19.14b) *while* maintaining fidelity with the original system. The convolution multiplying the forcing w seems a generally irreducible quadratic effect. Of course, if we know enough about the forcing $w(t)$, then we may be able to remove the convolution. For example, if w is constant, then $we^{-t}\star w = w^2$ is also constant and hence not an issue. For another example, if $w = \sin 2t$, then $we^{-t}\star w = \sin 2t\,(-\tfrac{2}{5}\cos 2t + \tfrac{1}{5}\sin 2t) = -\tfrac{1}{5}\sin 4t + \tfrac{1}{10} - \tfrac{1}{10}\cos 4t$ with no convolution apparent. For the example of a stochastic noise forcing w

we may plausibly replace $we^{-t}\star w$ by its long-term statistics to derive a stochastic weak model (Chao and Roberts, 1996, e.g.) (section 21.2). But in general it seems there is no way to remove the convolution while maintaining fidelity to the original dynamics.

19.1.5 ▪ Higher-order modeling

Further such algebra constructs higher-order, time-dependent coordinate transforms from the original (x, y) variables to the new (X, Y) variables—for example, via my web service (Roberts, 2009d). Expect that the series solutions we generate for a coordinate transform and normal form are divergent—it is rare to find convergence of these sorts of asymptotic expansions of nonlinear dynamics. Nonetheless, there exist a coordinate transform and normal form in a finite neighborhood to which they are the asymptotic approximations; see section 19.2 and also Theorem 4.1 of Aulbach and Wanner (2000). The arguments there show we can continue the construction of a time-dependent coordinate transform to arbitrarily high order (provided the system of equations is smooth enough). Throughout we would maintain the normal form

$$\dot{X} = F(t, X) \quad \text{and} \quad \dot{Y} = YG(t, X, Y),$$

where the coefficient G is negative for (X, Y) near $(0, 0)$, at least. Such a coordinate transform shows that there is a system arbitrarily close to the original that does possess an emergent slow manifold, the time-dependent image of $Y = 0$, on and near which the evolution is governed by the decoupled $\dot{X} = F(t, X)$. Section 19.2 establishes such modeling for reasonably general nonautonomous dynamical systems.

19.1.6 ▪ Singular perturbations ignore convolutions

Recall that section 11.3.5 considered the singular perturbation form of slow/fast systems. There we found that slow manifold dynamics emerged instantaneously because transients decayed not only exponentially quickly in time, but at a rate that is asymptotically large as the perturbation parameter $\epsilon \to 0$. The implication here for nonautonomous systems is that in singular perturbation formulations the convolution integrals found would occur over a time scale that goes to zero as parameter $\epsilon \to 0$. That is, the convolutions would not appear in a singular perturbation approach.

Consequently, for nonautonomous systems the singular perturbation approach does not see that at finite ϵ the slow manifold involves convolutions over the immediate history of the fluctuations, or that the true slow variable X is slightly different from the original slow x, and it does not see the mean drift effects of fluctuations through the nonlinearity. At least, such effects cannot be seen without invoking additional subtle arguments.

19.2 ▪ Coordinate transforms support long-term modeling

This section uses formal arguments to establish a couple of key generic properties of nonautonomous, time-dependent normal forms seen in the example system (19.1) of section 19.1. Following recent research (Roberts, 2008, e.g.), this section establishes first that a time-dependent coordinate transform can decouple slow modes from fast modes, to make the time-dependent slow manifold easy to see, and second that although anticipation of the time dependence may be necessary in the full transform, no anticipation need appear on the slow manifold itself.[224] Aulbach and Wanner (2000) proved that such separating coordinate transforms exist for general nonautonomous systems, even in infinite-dimensional Banach spaces.

Consider a reasonably general system of ODEs in $m+n$ dimensions for slow variables $x(t) \in \mathbb{R}^m$ and fast stable variables $y(t) \in \mathbb{R}^n$:

$$\dot{x} = Ax + f(t, x, y), \tag{19.15a}$$

$$\dot{y} = By + g(t, x, y), \tag{19.15b}$$

where[225]

- the spectrum of matrix A is zero and for simplicity we assume A is upper triangular with all elements zero except possibly $A_{i,j}$ for $j > i$ (such as in the Jordan form appropriate for position and velocity variables of a mechanical system);
- for simplicity assume matrix B has been diagonalized with diagonal elements $\beta_1, \ldots, \beta_n$, possibly complex, with $\Re\beta_j < 0$;[226]
- f and g are time-dependent functions that are nonlinear in the sense that, for all time, f and g and their gradients with respect to x and y are all zero at the origin;
- assume the time dependence is implicitly a linear combination of some number of independent forcing processes $w_k(t)$ (which in the stochastic case may be Stratonovich white noises, namely formal derivatives of independent Wiener processes).

For example, the example nonautonomous system (19.1) takes the form (19.15) with slow variables $x = (\sqrt{\sigma}, x)$, fast variable $y = y$, and

$$A = \begin{bmatrix} 0 & 0 \\ 0 & 0 \end{bmatrix}, \quad f = \begin{bmatrix} 0 \\ -xy \end{bmatrix}, \quad B = -1, \quad g = x^2 - 2y^2 + \sqrt{\sigma}^2 w(t).$$

A time-dependent coordinate transform We transform the ODE system (19.15) in (x, y) to a new (X, Y) coordinate system by a time-dependent near identity coordinate transform

$$x = x(t, X, Y) \quad \text{and} \quad y = y(t, X, Y). \tag{19.16}$$

This time-dependent coordinate transform is chosen such that the ODE (19.15) transforms to a simpler form for multiscale modeling. Aulbach and Wanner (2000, Thm. 4.1) assure us that such a coordinate transform generally exists (including in infinite-dimensional

[224] In nonautonomous dynamical systems, many researchers (Aulbach and Wanner, 2000; Potzsche and Rasmussen, 2006, e.g.) prefer the term "integral manifold" in place of the term "invariant manifold." Since we generally discuss definite manifolds, mainly the slow manifold, let's choose not to invoke the distinction.

[225] In principle we could analyze a system with locally unstable variables as well as local slow and local stable variables (Aulbach and Wanner, 2000, e.g.). However, let's concentrate on the most relevant case for modeling applications. Here we exclude the general case of center modes associated with nonzero pure imaginary eigenvalues because then the details begin to get horrendous (Roberts, 2008, §5).

[226] If matrix B is in Jordan form, rather than diagonalizable, then extensions of the arguments lead to the same results. In principle, the matrices A and B could also depend upon time (Chicone and Latushkin, 1997; Aulbach and Wanner, 2000; Potzsche and Rasmussen, 2006, e.g.). When the Lyapunov exponents of the corresponding linear dynamics are zero and negative, respectively, then a slow manifold should still exist and have nice properties (even in the stochastic case (Boxler, 1989; Arnold, 2003, e.g.)). However, for simplicity let's focus on the algebraically tractable case when the basic linear operators A and B are constant. Indeed, in constructing a nonautonomous slow manifold, the *only* definite example I recall that is not based upon constant operators A and B is by Potzsche and Rasmussen (2008).

Banach spaces). Based upon the experience of the example of section 19.1, we seek to simplify the ODEs according to Principles 1–5: in particular we allow anticipation.

Require convolutions The example of section 19.1 invokes the convolutions (19.5) and (19.13), as does Example 20.1. In general we need to invoke convolutions that depend upon the eigenvalues β_j of the matrix B in order to encapsulate memory or anticipation over the various fast time scales in the fast variable y. For any parameter μ (possibly complex), and for sufficiently well-behaved time-dependent functions $u(t)$, define the *convolution*

$$e^{\mu t}\star u := \begin{cases} \int_{-\infty}^{t} \exp[\mu(t-\tau)]u(\tau)\,d\tau\,, & \Re\mu < 0\,, \\ \int_{t}^{+\infty} \exp[\mu(t-\tau)]u(\tau)\,d\tau\,, & \Re\mu > 0\,, \\ \text{undefined}, & \Re\mu = 0\,. \end{cases} \tag{19.17}$$

Henceforth we assume $\Re\mu \neq 0$ to avoid the tricky borderline case ($\Re\mu$ is to be bounded away from zero). Then the convolution is always bounded for at least a bounded $u(t)$ (Principle 1). Such convolutions are used throughout Part VII. Three useful properties of this convolution are

$$\frac{d}{dt}e^{\mu t}\star u = \mu e^{\mu t}\star u - (\operatorname{sgn}\Re\mu)u\,, \tag{19.18a}$$

$$e^{\mu t}\star 1 = \frac{\operatorname{sgn}\Re\mu}{\mu}\,, \tag{19.18b}$$

$$e^{\mu t}\star e^{\nu t}\star = \frac{(\operatorname{sgn}\Re\mu)e^{\nu t}\star - (\operatorname{sgn}\Re\nu)e^{\mu t}\star}{\mu-\nu}\,, \quad \mu \neq \nu\,. \tag{19.18c}$$

Property (19.18a) is the general case of the second parts of equations (19.5) and (19.13), which proved crucial for the example of section 19.1. Subsequent chapters additionally invoke the following two stochastic convolution properties useful for modeling noisy dynamics:

$$E[e^{\mu t}\star u] = e^{\mu t}\star E[u] \quad \text{for a stochastic process } u, \tag{19.18d}$$

$$E\left[|e^{\mu t}\star w|^2\right] = \frac{1}{2|\Re\mu|} \quad \text{for white noise } w. \tag{19.18e}$$

Whether stochastic or deterministic, a crucial property is that

- with $\Re\mu < 0$ the convolution $e^{\mu t}\star$ integrates over the past,
- with $\Re\mu > 0$ the convolution $e^{\mu t}\star$ integrates into the future, and
- both integrate over a time scale of order $1/|\Re\mu|$.

19.2.1 ▪ Transform the fast dynamics

As in the analysis of the example system of section 19.1, let's generally seek a simplifying coordinate transformation by iterative improvement. This section addresses the general issues raised by transforming the fast dynamics of y. Section 19.2.2 addresses the slow dynamics of x.

Suppose (19.16) is some approximation to the desired coordinate transform. Iteratively we seek *small* corrections $\hat{x}$ and $\hat{y}$ to the transform, namely

$$x = x(t,X,Y) + \hat{x}(t,X,Y) \quad \text{and} \quad y = y(t,X,Y) + \hat{y}(t,X,Y), \tag{19.19}$$

where $\hat{x}$ and $\hat{y}$ are multinomial in X and Y. Correspondingly, also find corrections to the updates to the evolution, say the *small* multinomial $\hat{F}$ and $\hat{G}$:

$$\dot{X} = AX + F(t, X, Y) + \hat{F}(t, X, Y), \tag{19.20a}$$

$$\dot{Y} = BY + G(t, X, Y) + \hat{G}(t, X, Y). \tag{19.20b}$$

We aim for these evolution corrections $\hat{F}$ and $\hat{G}$ to be as simple as possible (Principle 4), albeit via subjective decisions of what in the model is simple and what is not.

The homological equation guides corrections For the fast dynamics, the iteration rests on the following derivation. Substitute the posed corrected transform (19.19) into the governing ODE (19.15b) for the fast variables: by the chain rule

$$\frac{\partial y}{\partial t} + \frac{\partial \hat{y}}{\partial t} + \frac{\partial y}{\partial X}\dot{X} + \frac{\partial \hat{y}}{\partial X}\dot{X} + \frac{\partial y}{\partial Y}\dot{Y} + \frac{\partial \hat{y}}{\partial Y}\dot{Y} = By + B\hat{y} + g(t, x + \hat{x}, y + \hat{y}).$$

Then into this equation substitute the corrected evolution (19.20a)–(19.20b)—as always the time derivative of out-of-equilibrium dynamics is voluminous:

$$\begin{aligned}
&\frac{\partial y}{\partial t} + \frac{\partial y}{\partial X}(AX + F + \hat{F}) + \frac{\partial y}{\partial Y}(BY + G + \hat{G}) \\
&+ \frac{\partial \hat{y}}{\partial t} + \frac{\partial \hat{y}}{\partial X}(AX + F + \hat{F}) + \frac{\partial \hat{y}}{\partial Y}(BY + G + \hat{G}) \\
&= By + B\hat{y} + g(t, x + \hat{x}, y + \hat{y}).
\end{aligned}$$

Now omit products of small corrections as being negligible, and approximate coefficients of small corrections by their leading-order term (for example, $\partial y/\partial X \approx 0$ and $\partial y/\partial Y \approx I$):

$$\begin{aligned}
&\hat{G} + \frac{\partial \hat{y}}{\partial t} - B\hat{y} + \frac{\partial \hat{y}}{\partial Y}BY + \frac{\partial \hat{y}}{\partial X}AX \\
&= \underbrace{-\frac{\partial y}{\partial t} - \frac{\partial y}{\partial X}(AX + F) - \frac{\partial y}{\partial Y}(BY + G) + By}_{-dy/dt} + g(t, x, y).
\end{aligned}$$

As explained later, we also neglect the last term on the left-hand side, $(\partial \hat{y}/\partial X)AX$; it eventuates that extra iterations compensate for such apparently wanton neglect. Then the equation for the jth component of the correction to the transform of the fast variable and the new fast dynamics is the time-dependent version of the usual homological equation

$$\hat{G}_j + \frac{\partial \hat{y}_j}{\partial t} - \beta_j \hat{y}_j + \sum_{\ell=1}^{n} \beta_\ell Y_\ell \frac{\partial \hat{y}_j}{\partial Y_\ell} = \text{Res}_{(19.15b),j}\,, \tag{19.21}$$

where $\text{Res}_{(19.15b),j}$ denotes the current residual of the jth component of the ODE (19.15b) (the right-hand side of the previous equation). In constructing a coordinate transform we repeatedly reevaluate the residual on the right-hand side and then solve this homological equation to find iterative corrections.

Correct from each multinomial term What sort of terms may be put into the transformation $\mathfrak{y}(t, \mathbf{X}, \mathbf{Y})$, and what terms have to remain in the $\mathbf{Y}$-evolution? The answer comes from considering the range of possibilities for the right-hand side residual. The coordinate transform is constructed as a multivariate asymptotic expansion about the origin in $(\mathbf{X}, \mathbf{Y})$-space, so we put the right-hand side residual in the form of a multinomial in $\mathbf{X}$ and $\mathbf{Y}$; we cater for time dependence by allowing the coefficients in the multinomial to be functions of time t. Suppose the jth component of the right-hand side, the residual $\text{Res}_{(19.15b),j}$, has, potentially among many others, a term of the multinomial form

$$c(t)\mathbf{X}^{\mathbf{p}}\mathbf{Y}^{\mathbf{q}} := c(t)\prod_{i=1}^{m} X_i^{p_i}\prod_{j=1}^{n} Y_j^{q_j}$$

for some vectors of exponents $\mathbf{p}$ and $\mathbf{q}$, usually integers, and some coefficient $c(t)$ that is typically time dependent. Because of the special form of the homological operator on the left-hand side of (19.21), seek contributions to the corrections of $\hat{G}_j = \mathfrak{g}(t)\mathbf{X}^{\mathbf{p}}\mathbf{Y}^{\mathbf{q}}$ and $\hat{y}_j = \mathfrak{y}(t)\mathbf{X}^{\mathbf{p}}\mathbf{Y}^{\mathbf{q}}$. Then this component of (19.21) becomes

$$\mathfrak{g} + \dot{\mathfrak{y}} - \mu\mathfrak{y} = c, \quad \text{where the rate } \mu := \beta_j - \sum_{\ell=1}^{n} q_\ell\beta_\ell\,. \tag{19.22}$$

Three cases arise depending upon the real part of the rate μ.

1. In the resonant case $\Re\mu = 0$, we need to satisfy $\mathfrak{g} + \dot{\mathfrak{y}} = c$, where we want to put as much into $\mathfrak{y}$ as possible (Principle 4). The mean and some types of fluctuating components of the forcing $c(t)$ must be generally assigned to $\mathfrak{g}$, as generally they would give rise to unbounded secular terms in $\mathfrak{y}$ (Principle 1):

   - when the fluctuating part of $c(t)$ is noisy (stochastic), then integrating it into the coordinate transform $\mathfrak{y}$ will almost surely generate unallowable square-root growth;[227]
   - when the forcing $c(t)$ is periodic, instead of noisy, then one may choose the mean zero fluctuations in the forcing $c(t)$ to be integrated into the coordinate transform $\mathfrak{y}$ instead of being assigned to the evolution $\mathfrak{g}$, but to maintain consistency with the noisy case we choose not to do so here.

   Thus the generic solution is $\mathfrak{g} = c$ and $\mathfrak{y} = 0$; that is, assign $c(t)\mathbf{X}^{\mathbf{p}}\mathbf{Y}^{\mathbf{q}}$ to the $\mathbf{Y}$-evolution and nothing to the coordinate transform $\mathbf{y}$.

   Since $\Re\beta_\ell$ are *all* negative, this case of $\Re\mu = 0$ can only arise when at least one of the exponents $\mathbf{q}$ of $\mathbf{Y}$ is positive in order for the sum in (19.22) to have real part zero. Hence, there will be at least one Y_ℓ factor in updates $\hat{\mathbf{G}}$ to the $\mathbf{Y}$-evolution, and so we maintain that $\mathbf{Y} = \mathbf{0}$ is the slow manifold.

2. When $\Re\mu < 0$, a solution of (19.22) is to place all the forcing into the coordinate transform, $\mathfrak{y} = e^{\mu t}\star c$, and not to introduce a component into the $\mathbf{Y}$-evolution, $\mathfrak{g} = 0$. As $\Re\mu < 0$, the convolution is over the history of the forcing $c(t)$; the convolution represents a memory of the forcing over a time scale of $1/|\Re\mu|$.

[227] Recall that the sum of independent random numbers has a standard deviation that grows as the square root of the number of terms in the sum.

But the ultimate intended use of a macroscopic model defines a slow time scale and consequently affects which terms appear in the model. For large enough exponents $\mathbf{q}$, that is, for high enough order in Y, the rate $\Re\mu$ must become positive. In the transition from negative to positive, the rate $\Re\mu$ may become close to zero. Then the time scale $1/|\Re\mu|$ becomes large and may be as large as the macroscopic time scale of the slow dynamics of interest. In that case set the transform $\mathfrak{y} = 0$ and assign this term in the forcing to the $\mathbf{Y}$-evolution with $\mathfrak{g} = c$. Again this case can only occur with at least one fast variable $\mathbf{Y}$, and so we maintain $\mathbf{Y} = 0$ as the slow manifold.

3. When $\Re\mu > 0$, and accepting anticipation in the transform, modify the coordinate transform by setting $\mathfrak{y} = e^{\mu t}\star c$, and do not change the $\mathbf{Y}$-evolution, $\mathfrak{g} = 0$.

Consequently, *we are always able to find a coordinate transform, to any specified order, that maintains that* $\mathbf{Y} = 0$ *is the invariant slow manifold.*

Off-diagonal neglect Recall the earlier recommendation to omit a term in (19.21): the term $(\partial\hat{\mathfrak{y}}/\partial\mathbf{X})A\mathbf{X}$, equivalently $(\partial\hat{\mathfrak{y}}_j/\partial X_\ell)A_{\ell,i}X_i$, should appear in the left-hand side. However, its omission is acceptable when the matrix A is upper triangular, as specified earlier, as then any term introduced that involves X_ℓ only generates extra terms that are lower order in X_ℓ. Although such extra terms increase the order of X_i for $i > \ell$, through the off-diagonal terms in A, successive iterations generate new terms involving fewer factors of X_ℓ and so iteration steadily accounts for the introduced terms. Similarly for the Y variables when the linear operator B is in Jordan form due to repeated eigenvalues. Discussing equation (19.21) for corrections is sufficient.[228] Analogous comments apply to the slow dynamics, to which we now turn.

19.2.2 ▪ Transform the slow dynamics

The homological equation guides corrections For the slow dynamics, each iteration toward constructing a time-dependent coordinate transform substitutes corrections to the transform (19.19) and the evolution (19.20) into the governing ODE (19.15a) for the slow variables. Then, as for the fast equations, omit products of small corrections, approximate coefficients of the remaining small corrections by their leading-order term, and neglect matrix A on the left-hand side (at the expense of probably increasing the number of required iterations). Analogous to the fast dynamics, the equation for the jth component of the correction to the transform of the slow variable is then the homological equation

$$\hat{F}_j + \frac{\partial\hat{x}_j}{\partial t} + \sum_{\ell=1}^{n} \beta_\ell Y_\ell \frac{\partial\hat{x}_j}{\partial Y_\ell} = \text{Res}_{(19.15a),j}\,, \qquad (19.23)$$

where $\text{Res}_{(19.15a),j}$ denotes the residual of the jth component of the ODE (19.15a) evaluated at the current approximation. The small but crucial difference from the previous discussion of the fast variables is that the left-hand side of (19.23) does not have an analogue of the $-\beta_j\hat{\mathfrak{y}}_j$ term.

Correct from each multinomial term Consider the range of possibilities for the right-hand side. In general, due to the asymptotic approach, the multinomial right-hand side

[228]Such a simple approach should also work in principle for infinite-dimensional systems provided that each Jordan block is of finite dimension.

residual $\operatorname{Res}_{(19.15a),j}$ is a sum of terms of the form $c(t)\mathbf{X}^{\mathbf{p}}\mathbf{Y}^{\mathbf{q}}$ for some vectors of integer exponents $\mathbf{p}$ and $\mathbf{q}$, and typically time-dependent coefficient $c(t)$. Because of the special form of the homological operator on the left-hand side of (19.23), seek corresponding corrections $\hat{F}_j = \mathfrak{f}(t)\mathbf{X}^{\mathbf{p}}\mathbf{Y}^{\mathbf{q}}$ and $\hat{x}_j = \mathfrak{x}(t)\mathbf{X}^{\mathbf{p}}\mathbf{Y}^{\mathbf{q}}$. Then (19.23) becomes

$$\mathfrak{f} + \dot{\mathfrak{x}} - \mu\mathfrak{x} = c, \quad \text{where} \quad \mu := -\sum_{\ell=1}^{n} q_\ell \beta_\ell \,. \tag{19.24}$$

Two cases typically arise depending upon the real part of the rate μ.[229]

1. The resonant case $\Re\mu = 0$ only arises when the $\mathbf{Y}$ exponents $\mathbf{q} = \mathbf{0}$, as the exponents have to be nonnegative and $\Re\beta_\ell < 0$. Consequently, the case $\Re\mu = 0$ implies $\mu = 0$. Thus we need to solve $\mathfrak{f} + \dot{\mathfrak{x}} = c$, where we want to put as much into $\mathfrak{x}$ as possible (Principle 4). Since the forcing $c(t)$ generally has nonzero mean and possibly noisy fluctuating components, at first sight the generic solution is $\mathfrak{f} = c$ and $\mathfrak{x} = 0$; that is, assign $c(t)\prod_{i=1}^{m} X_i^{p_i}$ to the $\mathbf{X}$-evolution and nothing to the coordinate transform.

   But recall Principle 5: we want to avoid fast-time integrals in the slow evolution. Consider the case when the forcing $c(t)$ has the form of a fast-time convolution $c = e^{\nu t}\star C(t)$ for some $C(t)$ and some rate ν. From (19.18a) deduce the split

   $$c = e^{\nu t}\star C = \frac{\operatorname{sgn}\Re\nu}{\nu} C + \frac{1}{\nu}\frac{d}{dt}(e^{\nu t}\star C).$$

   Hence to avoid fast-time memory integrals in the slow $\mathbf{X}$-evolution (Principle 5), set $\mathfrak{f} = C(t)(\operatorname{sgn}\Re\nu)/\nu$ and $\mathfrak{x} = (e^{\nu t}\star C)/\nu$. If $C(t)$ in turn is a fast-time convolution, then continue the above splitting. This splitting corresponds to that used by section 19.1 to avoid fast-time memory convolutions in the slow evolution.

   When the forcing c is a quadratic product of convolutions, then similar splittings eliminate all memory from the slow variables except for terms having coefficients of the form $c_1(t)e^{\nu t}\star c_2(t)$, where c_1 has no convolutions. Algebraic transformations cannot eliminate such terms; for now accept the violation of Principle 5 in such quadratic forcing terms.[230] Such quadratic terms encode the mechanisms that cause fast-time fluctuations to generate important mean drift effects on the macroscale model—we return to these later.

   Since the case $\Re\mu = 0$ only arises for terms in the residual with no $\mathbf{Y}$ dependence, we maintain that the slow evolution of the $\mathbf{X}$ variables is independent of $\mathbf{Y}$, and this holds both on and off the slow manifold.[231]

2. The remaining case in the solution of (19.24), when the real part $\Re\mu > 0$, occurs when at least one of the exponents $\mathbf{q}$ of $\mathbf{Y}$ is positive. Accepting anticipation in the coordinate transform, we assign $\mathfrak{x} = e^{\mu t}\star c$ and do not change the $\mathbf{X}$-evolution, $\mathfrak{f} = 0$.

[229]The case $\Re\mu < 0$ cannot arise, as all the decay rates $-\Re\beta_j > 0$, since we assume that there are no fast unstable modes.

[230]Similar considerations apply to higher-order terms in the time dependence, but for simplicity this book neglects to pursue such higher orders. You may know from physical applications that often quadratic fluctuations or oscillations drive the main mean drift effects, so we need to know quadratic effects in fluctuations or oscillations, but not necessarily higher order.

[231]However, when the fast dynamics contains rapidly oscillating, nondecaying modes, the corresponding eigenvalues occur as complex conjugate pairs, which typically interact to cause $\mu = 0$; the slow modes cannot be completely decoupled from fast oscillations (Cox and Roberts, 1995). Physically, waves do interact with mean flow.

By anticipating the time dependence in the system we are *always able to find a coordinate transform, to any specified order, that maintains a slow* $\mathbf{X}$*-evolution that is independent of whether the system is on or off the slow manifold.* Thus the projection of initial conditions (Chapter 12), and the exponential approach to a solution of the slow variables (section 11.3), is assured only by invoking such anticipation.

The preceding arguments are phrased in the context of an iteration scheme to construct the time-dependent coordinate transform and the corresponding evolution. Each step in the iterative process satisfies the governing ODEs to a higher order in the asymptotic expansions. By induction, we immediately deduce the following theorem.

Theorem 19.1 (Roberts, 2008, Prop. 1). *With anticipation of the forcing allowed, a near identity time-dependent coordinate transformation exists to convert the nonautonomous system* (19.15) *into the normal form*

$$\dot{\mathbf{X}} \simeq A\mathbf{X} + \mathbf{F}(t,\mathbf{X}), \tag{19.25a}$$

$$\dot{\mathbf{Y}} \simeq \left[B + G(t,\mathbf{X},\mathbf{Y})\right]\mathbf{Y}, \tag{19.25b}$$

where $\simeq$ *denotes that these are equalities to some specified order of* $(\mathbf{X},\mathbf{Y})$ *in an asymptotic expansion about the origin.*

Significantly, $\mathbf{F}$ *and* G *need only contain fast-time memory integrals in terms that are quadratic, or higher power, in the forcing.*

This theorem forms a basis for justifying models of nonautonomous dynamics (Aulbach and Wanner (2000) give an analogous but much more wide ranging theorem).

19.2.3 ▪ Slow dynamics do not need to anticipate the forcing

Despite the presence of anticipatory convolutions appearing in the coordinate transform (19.16), this section argues that none of them appear in the slow dynamics because the anticipatory convolutions always involve fast variables. Bensoussan and Flandoli (1995) correspondingly show that it is not necessary to anticipate noise on a stochastic inertial manifold.

In the previous sections, the anticipatory convolutions only occur when the rate $\Re\mu > 0$. But for both the slow and the fast components, this occurrence is only generated when at least one fast Y_j variable appears in the term under consideration. Moreover, there is no ordinary algebraic operation that reduces the number of $\mathbf{Y}$ factors in any term; potentially the time derivative operator might,

$$\frac{d}{dt} = \frac{\partial}{\partial t} + \sum_{\ell,k} Y_k B_{\ell,k}\frac{\partial}{\partial Y_\ell} + \sum_{\ell,k} X_k A_{\ell,k}\frac{\partial}{\partial X_\ell},$$

but, for nondiagonal matrices A and B, although in the algebra X_ℓ variables may be replaced by X_k, and Y_ℓ variables may be replaced by Y_k, nonetheless the same number of variables is retained in each term and a $\mathbf{Y}$ variable is never replaced by an $\mathbf{X}$ variable. The reason is that the slow and fast dynamics are *linearly* decoupled in the original system (19.15). Consequently all anticipatory convolutions appear in terms with at least one component of the fast variables $\mathbf{Y}$.

Since the evolution (19.25a) of the slow modes $\mathbf{X}$ is free of fast $\mathbf{Y}$ variables, the evolution is also free of anticipatory convolutions. However, as seen in examples, there generally are anticipatory convolutions in the evolution (19.25a) of the fast modes $\mathbf{Y}$. Further, although the stochastic coordinate transform (19.16) has anticipatory convolutions, on the slow manifold $\mathbf{Y} \simeq 0$ there are none. Consequently the preceding formal analysis leads to the following theorem.

Theorem 19.2 (Roberts, 2008, Prop. 2). *Although* anticipation *may be invoked, there* need not *be any* anticipation *in the dynamics* (19.25a) *of the slow modes in the nonautonomous normal form of the system* (19.15). *Moreover, on the slow manifold,* $\mathbf{Y} \simeq 0$, *the time-dependent* coordinate transform (19.16) *need not have anticipation.*

19.2.4 ▪ Example: Avoid homogenization

With the iterative procedure established, this section shows how it is straightforwardly applied to an example that is otherwise difficult. Pavliotis and Stuart (2008, §11.6.7) developed a combination of averaging and homogenization for the modeling of stochastic dynamics. One of their examples requiring their homogenization is the following five-variable system (after scaling time by a factor of ϵ^2 to remove the singular perturbation):

$$\begin{aligned} \dot{x}_1 &= \epsilon y_1, \quad \dot{x}_2 = \epsilon y_2, \quad \dot{x}_3 = \epsilon(x_1 y_2 - x_2 y_1), \\ \dot{y}_1 &= -y_1 - \alpha y_2 + w_1(t), \quad \dot{y}_2 = -y_2 + \alpha y_1 + w_2(t). \end{aligned} \tag{19.26}$$

This system has two independent forcing sources $w_i(t)$ and, for small parameter ϵ, has three slow variables $x_i(t)$ and two fast variables $y_i(t)$. The spiraling decay of the fast variables, combined with the form of the slow dynamics of x_3, generates subtle effects that averaging does not resolve but homogenization does. Our normal-form coordinate transform proceeds without any difficulty for either deterministic or noisy forcing $w_i(t)$.

This system is particularly simple, as the only nonlinearity is in the x_3 equation. Only nonlinearity? Not quite: remember that the chain rule hidden in time derivatives is also a source of nonlinear complication. Such nonlinear complication reflects the out-of-equilibrium dynamics.

The system (19.26) is relatively simple because it may be diagonalized when written in terms of complex variables $y, \bar{y} = y_1 \pm i y_2$, and $x, \bar{x} = x_1 \pm i x_2$, and when we also write the forcing as $w, \bar{w} = w_1 \pm i w_2$. Let's do so. Then the system (19.26) becomes the complex-variable system

$$\dot{x} = \epsilon y, \quad \dot{\bar{x}} = \epsilon \bar{y}, \quad \dot{x}_3 = i\epsilon(x\bar{y} - \bar{x}y)/2, \tag{19.27a}$$

$$\dot{y} = (-1 + i\alpha)y + w(t), \quad \dot{\bar{y}} = (-1 - i\alpha)\bar{y} + \bar{w}(t). \tag{19.27b}$$

From the ODEs (19.27b) for the two fast variables, the two eigenvalues of the matrix B for this dynamical system are $\beta = -1 + i\alpha$ and $\bar{\beta} = -1 - i\alpha$.

The fast-variable coordinate transform The fast variable y satisfies the forced linear ODEs with straightforward solution $y = e^{\beta t}\star w$; similarly for $\bar{y}$. Recall that this means we change fast coordinates from $(y, \bar{y})$ to $(Y, \bar{Y})$ via the transform

$$y = Y + e^{\beta t}\star w \quad \text{such that } \dot{Y} = -Y, \tag{19.28}$$

and similarly for the complex conjugate variable $\bar{Y}$. Due to the simplicity of the ODEs (19.27b) we now know that in the new coordinate system $\dot{Y} = \beta Y = (-1 + i\alpha)Y$. Thus $Y(t) \to 0$ exponentially quickly as $t \to \infty$ from *all* initial conditions. Hence there is an emergent model in terms of the slow variables $\mathbf{X}$ for the time-dependent dynamics.

The only proviso is that the forcing $w(t)$ has to be sufficiently integrable.

The two simple slow variables Substitute this fast-variable transform into the first two ODEs of the slow x equation (19.27a), and substitute the corrected coordinate transform $x = X + \hat{x}(t, \mathbf{X}, \mathbf{Y})$. This leads to the homological equation (19.23) here being

$$\begin{aligned}\dot{X} + \frac{\partial \hat{x}}{\partial t} + \beta Y \frac{\partial \hat{x}}{\partial Y} + \bar{\beta}\bar{Y}\frac{\partial \hat{x}}{\partial \bar{Y}} &= \epsilon Y + \epsilon e^{\beta t}\star w \\ &= \epsilon Y - \frac{\epsilon}{\beta} w + \frac{\epsilon}{\beta}\frac{d(e^{\beta t}\star w)}{dt},\end{aligned}$$

where the splitting of the convolution on the right-hand side uses the identity (19.18a). Choosing to absorb into the coordinate transform $\hat{x} = (\epsilon/\beta)(Y + e^{\beta t}\star w)$ results in

$$x \approx X + \frac{\epsilon}{\beta}(Y + e^{\beta t}\star w) \quad \text{such that} \quad \dot{X} \approx -\frac{\epsilon}{\beta} w, \tag{19.29}$$

and similarly for the complex conjugate variables $\bar{x}$ and $\bar{X}$. This solution of the homological equation shows that in order to disentangle the slow X dynamics from the fast Y dynamics we need to use a coordinate system that not only fluctuates a little via the $\epsilon e^{\beta t}\star w/\beta$ term, but also slopes a little in Y via the $\epsilon Y/\beta$ term. If we were using averaging, then we would find only that $\dot{X} \approx 0$ (assuming that the average of w is zero). As Pavliotis and Stuart (2008) discuss, such averaging is generally inadequate because the fluctuations in X and $\bar{X}$ are needed to interact with those in Y and $\bar{Y}$ in order to deduce the nontrivial dynamics of x_3.

The third interesting slow variable To derive the model of x_3, seek $x_3 = X_3 + \hat{x}_3(t, \mathbf{X}, \mathbf{Y})$ in the left-hand side of the third equation in (19.27a), and substitute the coordinate transforms (19.28)–(19.29) into the right-hand side. This leads to the homological equation (19.23) here being

$$\begin{aligned}&\dot{X}_3 + \frac{\partial \hat{x}_3}{\partial t} + \beta Y \frac{\partial \hat{x}_3}{\partial Y} + \bar{\beta}\bar{Y}\frac{\partial \hat{x}_3}{\partial \bar{Y}} \\ &= \epsilon(x_1 y_2 - x_2 y_1) \\ &= \frac{i\epsilon}{2}\left(X\bar{Y} - \bar{X}Y + Xe^{\bar{\beta} t}\star\bar{w} - \bar{X}e^{\beta t}\star w\right) \\ &\quad + \frac{\epsilon^2\alpha}{|\beta|^2}\left(Y\bar{Y} + Ye^{\bar{\beta} t}\star\bar{w} + \bar{Y}e^{\beta t}\star w + e^{\bar{\beta} t}\star\bar{w}\, e^{\beta t}\star w\right).\end{aligned}$$

Let's look in turn at each of these contributions on the right-hand side.

- Absorb the terms $X\bar{Y} - \bar{X}Y$ into the coordinate transform $\hat{x}_3$ by including $\frac{1}{2}i\epsilon(X\bar{Y}/\bar{\beta} - \bar{X}Y/\beta)$.
- Split the term $\bar{X}e^{\beta t}\star w = \bar{X}(-\frac{1}{\beta}w + \frac{1}{\beta}d(e^{\beta t}\star w)/dt)$ using the convolution identity (19.18a). Along with a corresponding splitting of the term $Xe^{\bar{\beta} t}\star\bar{w}$, this leads to an integrated component in $\hat{x}_3$ of $\frac{1}{2}i\epsilon(Xe^{\bar{\beta} t}\star\bar{w}/\bar{\beta} - \bar{X}e^{\beta t}\star w/\beta)$. The nonintegrable component contributes to the evolution $\dot{X}_3$ by including $\frac{1}{2}i\epsilon(\bar{X}w/\beta - X\bar{w}/\bar{\beta})$.
- Absorb the three terms involving Y and $\bar{Y}$ into the coordinate transform $\hat{x}_3$ by including $\epsilon^2\alpha/|\beta|^2(-\frac{1}{2}Y\bar{Y} + Ye^{-\beta t}\star e^{\bar{\beta} t}\star\bar{w} + \bar{Y}e^{-\bar{\beta} t}\star e^{\beta t}\star w)$.

- The last term involving $e^{\bar\beta t}\star\bar w\, e^{\beta t}\star w$ could contribute directly to the evolution $\dot X_3$ or could be split into some more canonical form. For our purposes, just allocate $\epsilon^2\alpha/|\beta|^2 e^{\bar\beta t}\star\bar w\, e^{\beta t}\star w$ into $\dot X_3$.

The upshot is that the x_3 coordinate transform

$$\begin{aligned} x_3 &\approx X_3 + \tfrac{1}{2}i\epsilon\left(X\bar Y/\bar\beta - \bar X Y/\beta + Xe^{\bar\beta t}\star\bar w/\bar\beta - \bar X e^{\beta t}\star w/\beta\right) \\ &\quad + \frac{\epsilon^2\alpha}{|\beta|^2}\left(-\tfrac{1}{2}Y\bar Y + Ye^{-\beta t}\star e^{\bar\beta t}\star\bar w + \bar Y e^{-\bar\beta t}\star e^{\beta t}\star w\right) \end{aligned} \tag{19.30}$$

leads to the slow variable X_3 evolving according to (recall $\beta = -1 + i\alpha$)

$$\dot X_3 \approx \tfrac{1}{2}i\epsilon(\bar X w/\beta - X\bar w/\bar\beta) + \frac{\epsilon^2\alpha}{|\beta|^2} e^{\bar\beta t}\star\bar w\, e^{\beta t}\star w. \tag{19.31}$$

This nonautonomous ODE (19.31) governs the interesting long-term dynamics of the system. The term quadratic in the forcing, the last term above, typically generates a nonzero mean driving force on the $x_3 \approx X_3$ mode.

- For example, if the forcing just happens to be the fast oscillations $w(t) = e^{i\alpha t}$, then $e^{\beta t}\star w = \cdots = e^{i\alpha t}$. Consequently, $e^{\bar\beta t}\star\bar w\, e^{\beta t}\star w = 1$ and the mean driving of X_3 is $\epsilon^2\alpha/|\beta|^2$ generated by the fast-time oscillations interacting nonlinearly with each other.

- In the example of stochastic forcing where $w(t)$ is white noise, then as $|\beta|^2 = 1+\alpha^2$ and by the identity (19.18e), this X_3 forcing term has mean, or expectation,

$$\operatorname{E}\left[\frac{\epsilon^2\alpha}{|\beta|^2} e^{\bar\beta t}\star\bar w\, e^{\beta t}\star w\right] = \frac{\epsilon^2\alpha}{1+\alpha^2}\operatorname{E}\left[|e^{\beta t}\star w|^2\right] = \frac{\epsilon^2\alpha}{2(1+\alpha^2)}\,.$$

  This term again generates a mean drift of X_3.

Summary A time-dependent coordinate transform straightforwardly disentangles the slow x-dynamics from the fast y-dynamics in the example system (19.26). The dynamics in the transformed X and Y variables immediately establishes the existence and emergence, through the decay $Y \to 0$, of a time-dependent slow manifold. The nonautonomous dynamics on this slow manifold then models the original dynamics for all time, and in particular shows that nonlinear interactions of fluctuations generate mean forcing effects.

Extend to other invariant manifolds The nonautonomous coordinate transforms developed here also extend straightforwardly to the separation of other invariant manifolds (often called integral manifolds) in nonautonomous systems. One can generally construct coordinate transforms to separate center, stable, unstable, center-stable, or center-unstable modes for nonautonomous systems, just as Chapter 11 established for autonomous systems. My web service (Roberts, 2009d) does such construction for any in a wide range of systems that you might enter.

For example, such transforms of a system could aid in the control of stable and unstable manifolds (Balasuriya and Padberg-Gehle, 2013, e.g.) and aid in the management of mixing in environmental or microscale flows (Froyland and Padberg, 2009, e.g.).

19.3 ▪ Construct slow manifold models efficiently

It is not practical to resolve all the fast dynamics only to subsequently set them to zero. Sections 19.1–19.2 sought a coordinate transform of the full state space. This full transform has the advantage that we resolve all the dynamics of the original system, at least in principle. It readily leads to the existence and emergence properties of section 19.3.1.

However, in modeling we are primarily interested in describing the long-term dynamics of the system on the slow manifold as characterized by the slow variables. That is, we are primarily interested in just the slow manifold in the state space and the evolution thereon. Thus a more practical route to modeling is to construct *only* the coordinate transform on the slow manifold, as introduced in the example of section 19.3.2. Such a restricted construction is immensely useful in most real applications where typically there is an enormous number of fast modes of little interest whose dynamics is modeled only as needed in this approach.

Sections 19.3.3–19.3.4 simplify the arguments and analysis of the preceding sections to construct only the nonautonomous slow manifold. The vast simplification is to neglect analyzing the dynamics of the fast variables when off the slow manifold.

We require that the system be in linearly separated form (19.15). More restrictively, we also require the conditions following the ODEs (19.15). These requirements lead to tractable algebra.

19.3.1 ▪ Backward existence and emergence

Theorem 19.1 immediately suggests the following two backward propositions. Recall that in applications of mathematics one never knows the precise differential equations that govern the physical system. Consequently, a property of a differential equation that is close to the supposed mathematical system is as valid a statement about the physical system as a property of the precise mathematical system. Such backward propositions are useful in applications (Grcar, 2011, e.g.).

Proposition 19.3 (existence). *For any in the class of nonautonomous systems* (19.15) *there exists an asymptotically close system, to any specified order limited by the smoothness of* f *and* g, *that possesses in a finite domain a nonautonomous slow manifold of the multinomial form*

$$\mathsf{x} = \mathsf{x}(t, \mathsf{X}) = \mathsf{X} + \mathcal{O}(|\mathsf{X}|^2) \quad \textit{and} \quad \mathsf{y} = \mathsf{y}(t, \mathsf{X}). \tag{19.32}$$

On this slow manifold the evolution is of the multinomial form

$$\dot{\mathsf{X}} = A\mathsf{X} + \mathsf{F}(t, \mathsf{X}). \tag{19.33}$$

At the time of writing, this and the subsequent sketch proof await further validation before being established as a full proof.

Sketch proof. For any given nonautonomous system (19.15) the construction of a multinomial coordinate transform (19.16) in section 19.2, to the specified order, leads to a multinomial normal form (19.25) of Theorem 19.1. The normal-form dynamics (19.25) together with the coordinate transform (19.16) describe some system of nonautonomous dynamics in (x, y). By construction, these nonautonomous dynamics in (x, y) are close to the original system (19.15), to the specified asymptotic order. By the form of (19.25b), the subspace $\mathsf{Y} = 0$ is a global invariant manifold; because of the properties of the operators A and B, $\mathsf{Y} = 0$ is a slow manifold. Since the multinomial coordinate transform is near identity, the transform of this global slow manifold into (x, y)-space, namely (19.32), will exist for at least some finite domain about the origin.

By (19.25a) the evolution on the slow manifold is immediately (19.33). □

Proposition 19.4 (emergent dynamics). *For any nonautonomous system in the class* (19.15) *and for any chosen rate* β *in the spectral gap* $\max_j \Re\beta_j < -\beta < 0$, *the nonautonomous slow manifold of the close system of Proposition* 19.3 *attracts all its solutions in a finite domain* d_β *at the rate* β *in the sense that for all* $(x(t), y(t)) \in d_\beta$ *for* $0 \leqslant t \leqslant T$ *there exists an* $X(t)$ *solution to* (19.33) *and a constant* c *such that*

$$\left|(x(t), y(t)) - (x(t, X(t)), y(t, X(t)))\right| \leqslant ce^{-\beta t} \quad \text{for } 0 \leqslant t \leqslant T. \tag{19.34}$$

Sketch proof. Recall that the close system of Proposition 19.3 is the multinomial normal-form system (19.25) mapped into (x, y)-space by the constructed multinomial coordinate transform (19.16). Choose a rate β in the spectral gap $\max_j \Re\beta_j < -\beta < 0$. Let the domain D'_β be the domain in (X, Y) where the state transition matrix $\Psi(t, 0)$ of $B + G(t, X, Y)$ for fixed (X, Y) (Chicone, 2006, p. 151, e.g.) has norm bounded by $Ce^{-\beta t}$ for some C for $0 \leqslant t \leqslant T$. The domain D'_β is finite sized about the origin, since $G(t, 0, 0) = 0$ and by continuity of the multinomial G. Let domain $D_\beta \subseteq D'_\beta$ such that the coordinate transform (19.16) is a diffeomorphism with Jacobian of norm $\leqslant J$ for $0 \leqslant t \leqslant T$, and let d_β be the corresponding domain in (x, y)-space. While solutions of the close system stay in d_β,

$$\begin{aligned}
&\left|(x(t), y(t)) - (x(t, X(t)), y(t, X(t)))\right| \\
&= \left|(x(t, X(t), Y(t)), y(t, X(t), Y(t))) - (x(t, X(t), 0), y(t, X(t), 0))\right| \\
&= \left|(x(t, X(t), Y(t)) - x(t, X(t), 0), y(t, X(t), Y(t)) - y(t, X(t), 0))\right| \\
&\leqslant J\left|(X(t), Y(t)) - (X(t), 0)\right| = J|Y(t)| \leqslant JY_0Ce^{-\beta t}
\end{aligned}$$

for $Y_0 := |Y(0)|$. This establishes the exponential attractiveness (19.34) (with constant $c = JY_0C$) of the slow manifold solutions. □

These two propositions are important. All our constructed nonautonomous slow manifolds are the attractive exact slow manifolds, in a finite domain, for a nonautonomous system close to the one specified. Currently "close" means in a high-order asymptotic sense: further research may establish bounds on the closeness in terms of the spectral and nonlinear properties of the original system.

19.3.2 ▪ Example of a nonlinear nonautonomous system

Let's reconsider the example system of section 19.1. Recall that the example two-variable system (19.1) is

$$\dot{x} = -xy \quad \text{and} \quad \dot{y} = -y + x^2 - 2y^2 + \sigma w(t). \tag{19.35}$$

Figure 19.1 plots some representative trajectories for an indicative periodic forcing $w(t)$. Evidently, the finite domain d_β of existence and emergence of a slow manifold for (19.35) includes the region shown in the figure. Section 19.1's construction of a full coordinate transform resolved transients in the fast variable $y(t)$. Our challenge here is to separate, amid the fluctuations, the slow $x(t)$ without resolving the fast $y(t)$ except for its components driven by slow $x(t)$ and forcing $w(t)$.

But do not seek the nonautonomous slow manifold in the form $y = y(t, x)$. One of the lessons from the full coordinate transform is that we should seek both x and y as functions of a slow parameter X. For example, the slow-variables transform of equation (19.9) shows that we want to parametrize $x \approx X + \sigma X e^{-t} \star w$. The reason is to avoid fast-time memory convolutions in the model: it is only by parametrizing the slow variable x like this that equation (19.9) writes the model evolution without convolutions as $\dot{X} \approx -\sigma X w$. Thus we must seek the nonautonomous slow manifold as $x = x(t, X)$ and $y = y(t, X)$, where $x = X + \mathcal{O}(X^2)$ and $y = \mathcal{O}(X^2)$.

The analysis here parallels that of section 19.1 but with the difference that we focus only on the time-dependent slow manifold.

Simplifying fast dynamics involves memory This approach is much simpler than the full coordinate transform analysis of section 19.1.1. With $x \approx X$ and $\dot{X} \approx 0$, seek a time-dependent slow manifold in the form

$$y = \hat{y}(t, X) + \mathbb{O}(3), \tag{19.36}$$

where now $\mathbb{O}(p)$ compactly denotes terms of $\mathcal{O}(X^p + \sigma^p)$. Substitute (19.36) into the y ODE (19.35) and drop products of small corrections to recognize that we need to solve

$$\frac{\partial \hat{y}}{\partial t} + \hat{y} = \sigma w(t) + X^2 + \cdots, \tag{19.37}$$

where the partial derivative in t is done keeping constant X.

First, consider the autonomous part of (19.37): $\hat{y}_t + \hat{y} = X^2$. Since the partial derivative in t is done keeping constant X, the solution is $\hat{y} = X^2$. Second, consider the remaining nonautonomous forcing of (19.37): $\hat{y}_t + \hat{y} = \sigma w(t)$. Its solution is the convolution $\hat{y} = \sigma e^{-t} \star w$. Consequently the first nontrivial approximation of the nonautonomous slow manifold is

$$y = X^2 + \sigma e^{-t} \star w + \cdots. \tag{19.38}$$

This nonautonomous slow manifold is a vertically fluctuating parabola, as also derived in section 19.1.1.

Eliminate memory from the slow dynamics Now consider the slow x component of the dynamics. Let's seek a small difference between the original slow variable x and the new X in the form

$$x = X + \hat{x}(t, X) + \mathbb{O}(3) \quad \text{such that} \quad \dot{X} = \hat{F}(t, X) + \mathbb{O}(3), \tag{19.39}$$

where $\hat{x}$ and $\hat{F}$ are $\mathbb{O}(2)$. As in section 19.1.2, but now given the time-varying parabola (19.38), substitute into the x equation of ODE (19.1) and omit small products to find we need a solution of

$$\hat{F} + \frac{\partial \hat{x}}{\partial t} = -\sigma X e^{-t} \star w + \mathbb{O}(3). \tag{19.40}$$

Consider the fluctuating part. We do not integrate the forcing w into the coordinate transform $\hat{x}$, as then terms in $\hat{x}$ may grow unboundedly like $\int w \, dt$ (Principle 1); recall the following:

- for some forcing, the integral $\int w \, dt$ will not grow, for example, $w = \cos t$, and so can be integrated into the coordinate transform;

- but for other forcing it will grow; for example, $w = \cos\sqrt{t}$ has zero mean (and finite variance, namely 1/2), but the integral $\int w\,dt = 2\cos\sqrt{t}+2\sqrt{t}\sin\sqrt{t}$ grows in time like $\sqrt{t}$;

- and for the stochastic case of white noise $w(t)$, the integral $\int w\,dt$ is a random walk Wiener process that also grows like $\sqrt{t}$.

Thus we choose not to integrate all the forcing w into the relation between x and X.

But to avoid a fast-time convolution in the slow evolution $\hat{F}$ (Principle 5), use identity (19.5) to split $e^{-t}\star w = w - d(e^{-t}\star w)/dt$ and hence choose components $\hat{F} = -\sigma X w$ and $\hat{x} = \sigma X e^{-t}\star w$. Consequently,

$$x = X + \sigma X e^{-t}\star w + \mathbb{O}(3) \quad \text{and} \quad \dot{X} = -\sigma X w + \mathbb{O}(3). \tag{19.41}$$

As in section 19.1.2, the slow dynamics being $\dot{X} \approx -\sigma X w$ is one example of additive forcing of the original system (19.1) appearing as a multiplicative forcing in a slow variable.

As discussed in section 19.1.2, the splitting to arrive at (19.41) applies to all reasonable forcing $w(t)$, even stochastic forcing. Earlier approaches to stochastic bifurcation did not split the stochastic forcing and consequently they derived more complicated normal forms that additionally have fast-time memory processes. Such splitting between coordinate transform and resultant model dynamics is a subjective decision by the modeler and may be varied to suit particular applications.

Higher-order modeling Further such algebra constructs higher-order direct approximations of the nonautonomous slow manifold. For example, via my current web service (Roberts, 2009e) one finds the following slow manifold for the nonautonomous system (19.1):

$$\begin{aligned}
y &\approx X^2 + \sigma e^{-t}\star w - 2\sigma X^2 e^{-t}\star e^{-t}\star w + 2\sigma X^2 e^{-t}\star w,\\
x &\approx X + \sigma X e^{-t}\star w,\\
\dot{X} &\approx -X^3 - \sigma X w + 2\sigma^2 X w e^{-t}\star w.
\end{aligned}$$

In this higher-order modeling we discern the deterministic stabilization of $X = 0$ via the $-X^3$ term, and the likely enhanced stability of $X = 0$ via the mean effect of the quadratic forcing interaction factor $we^{-t}\star w$.

Although finding just the slow manifold itself is a major simplification, by omitting the Y-dynamics (19.25b) we do miss out on estimating the domain of attraction of the slow manifold model.

19.3.3 ▪ Slow manifold modeling for nonautonomous dynamics

This section adapts and streamlines the normal-form arguments of section 19.2 to establish a general algorithm for constructing slow manifold models without having to resolve all the unnecessary fast dynamics. The importance of this section is to establish a practical methodology to create accurate, finite-dimensional, slow manifold models of the long-term dynamics of high-dimensional, nonautonomous, deterministic or stochastic systems.

This nonautonomous modeling is the most complicated in this book and involves fearsome algebraic detail, which we enlist a computer to do. This section develops a general computer algebra technique together with a definite specific example (Roberts 2006, 2005).

The example is the forced system for a field $u(x,t)$ on the spatial domain $0 \leqslant x \leqslant \pi$, namely

$$\frac{\partial u}{\partial t} = \frac{\partial^2 u}{\partial x^2} + u - u\frac{\partial u}{\partial x} + \sigma w(x,t) \\ \text{such that} \quad u = 0 \text{ at } x = 0, \pi, \tag{19.42}$$

which involves the important physical processes of linear diffusion u_{xx}, nonlinear advection uu_x, and some general additive distributed forcing $w(x,t)$ of strength σ. Our aim is to model the interaction of the linear dissipation, the nonlinear combination, and the space-time dependent forcing. We resolve effects of the forcing w up to quadratic terms, as such quadratic terms often generate important long-term mean driving of the field u.

The eigenproblem must be accessible In general, the approach developed here applies straightforwardly to systems, for a field $u(t)$ in some Banach space, expressible in the form

$$\frac{\partial u}{\partial t} = \mathcal{L}u + f(u) + \sigma F(u)w(t) \tag{19.43}$$

for some linear operator $\mathcal{L}$, some smooth nonlinear function $f(u)$, and some smooth, potentially nonlinear, time-dependent distributed forcing $F(u)w(t)$ of strength σ. Functions f and F have to be smooth so we can expand in a multinomial in the field u. Crucially, we need to know and be able to manipulate all eigenvalues and eigenvectors of the linear operator $\mathcal{L}$ (for simplicity we only consider time-invariant linear operators $\mathcal{L}$). In the system (19.42) the eigenvalues and eigenvectors of $\mathcal{L} = 1 + \partial^2/\partial x^2$ with its Dirichlet boundary conditions are

$$\lambda_k = -(k^2 - 1), \quad e_k = \sin kx, \quad k = 1, 2, \ldots. \tag{19.44}$$

The rules for linearization of products of trigonometric functions empower us to manipulate the eigenvectors in their nonlinear combinations. In general, for the operator $\mathcal{L}$ in system (19.43) we need to know the complete spectrum of eigenvalues λ_k, repeated according to multiplicity, and corresponding (generalized) eigenvectors e_k, for $k = 1, 2, \ldots$. Lastly, to cater for nonlinear interactions, we need to be able to write products of eigenvectors as a sum of eigenvectors just as we know how to do for trigonometric functions.

Assume there are m zero eigenvalues, $\lambda_1, \ldots, \lambda_m$, and the rest with negative real parts, $\Re\lambda_k < -\beta < 0$ for $k = m+1, m+2, \ldots$. In the system (19.42) there is one zero eigenvalue, $\lambda_1 = 0$, and the rest are negative, $\lambda_k \leqslant -3$ for $k = 2, 3, \ldots$. Center manifold theory (Chicone and Latushkin, 1997; Aulbach and Wanner, 2000, e.g.) and the normal-form approach of section 19.2 (Roberts, 2008) then assure us of the existence and emergence of a time-dependent slow manifold model of the dynamics.[232]

An equilibrium anchors the analysis In the absence of forcing, $\sigma = 0$, both the specific system (19.42) and the general system (19.43) have an equilibrium at field $u = 0$. We base the slow manifold modeling about this equilibrium. Consequently the results are a local asymptotic expansion in the amplitude of u and the strength of the forcing σ.

[232]Herein we restrict attention to the case of a slow manifold where all the critical eigenvalues are precisely zero. For example, we do not address the case of Hopf bifurcations, where two eigenvalues have zero real part but nonzero imaginary part (Roberts, 2008, §5). Hopf bifurcations, as with other oscillatory modes in the center manifold, typically have intricate interactions with the forcing modes. Let's not get involved yet.

Algorithm 19.1 *Iterative loop in computer algebra to construct a time-dependent slow manifold model of the system* (19.42), *starting from a linear approximation. The iteration continues until the* PDE *is satisfied to a residual of the specified order in amplitude* a *and forcing magnitude* σ. *The two ellipses denote code detailed subsequently.*

```
factor a,sigma;
let {a^4=>0, sigma^3=>0};
...
depend a,t;
let df(a,t)=>g;
u:=a*sin(x);
g:=0;
operator w; depend w,tt; depend tt,t;
wx:=(for k:=1:3 sum w(k)*sin(k*x));
...
for it:=1:20 do begin
    res:=-df(u,t)+u+df(u,x,2)-u*df(u,x)+sigma*wx;
    res:=trigsimp(res,combine);
    write length_res:=length(res);
    g:=g+(gd:=secular(res,xt));
    u:=u+linv(res-gd*sin(x),xt);
    if res=0 then it:=1000+it;
end;
```

Using the parameter a to measure the amplitude of the field u, Algorithm 19.1 truncates our asymptotic analysis of the system (19.42) to residuals and errors of $\mathcal{O}(a^4, \sigma^3)$.

Develop the model from a linear approximation As usual, the linear slow subspace forms the leading approximation to a slow manifold model. In the system (19.42) the slow mode is $e_1 = \sin x$. Measure its amplitude by a variable a so the linear subspace is $u = a \sin x$, as coded in Algorithm 19.1. But under the nonlinearity and forcing w the amplitude a varies in time, so Algorithm 19.1 codes the evolution of a by $da/dt = g(a, t)$, where the initial approximation of the rate is $da/dt = g \approx 0$, as coded. In general, one would write the slow subspace as the linear combination $u = \sum_{k=1}^{m} a_k(t) e_k$ for amplitudes a_k of the various eigenvectors e_k.

The forcing $w(t)$ may have an infinite number of modes. For example, in principle the forcing in the system (19.42) could be the infinite sum $w = \sum_{k=1}^{\infty} w_k(t) \sin kx$. However, such infinite sums are (usually) intractable for computer algebra, so we truncate. Algorithm 19.1 codes a finite sum for the system (19.42) for general time-dependent coefficients $w_k(t)$. If you have a specific forcing to analyze, then use a more specific expression.

Iterate to form the slow manifold model As usual, iteration constructs successive approximations to the forced slow manifold model. Algorithm 19.1 shows the general scheme for the system (19.42). Within a for-loop,

1. compute the residual `res` of the PDE (19.43)—here the system (19.42)—rewriting any nonlinear products of basis eigenvectors as linear sums of the eigenvectors—here by invoking the well-known linearization of trigonometry;

Algorithm 19.2 *Update time-independent effects in a slow manifold model of the system (19.42). Temporarily add code to cater for* slowly varying forcing.

```
depend x,xt;
procedure lamb(k); (k^2-1);
operator linv; linear linv;
let linv(sin(~k*x),xt)=>-sin(k*x)/lamb(k);

operator secular; linear secular;
let { secular(sin(~k*x),xt)=>0
    , secular(sin(x),xt)=>1 };

% invoke slowly varying forcing
let df(w(~k),t,~p)=>0 when fixp(p) and p>1;
```

2. update the evolution $da/dt = g$ by extracting any resonant `secular` terms from the residual, whatever they may be in the specific application;

3. update the slow manifold field u using the remaining terms in the residual and a coded inverse, `linv`, of the operator $\mathcal{L}$;

4. and finally exit the loop when the residual is zero.

The difficult part, as we soon find, is discovering the nature, form, and coding of the operators `secular` and `linv` that determine the updates. Here let's code these operators in three stages: first for autonomous dynamics (which also works for slowly varying forcing), second for linear forcing effects, and finally for quadratic forcing effects.

Update autonomous dynamics: $\sigma = 0$ For autonomous dynamics we just invoke the details from section 5.1, and in particular from the computer algebra of section 5.2. Recall that the equation for updates $\hat{u}$ and $\hat{g}$ to the field and evolution is

$$\mathcal{L}\hat{u} - \hat{g}\sin x + \text{Res}_{(19.42)} = 0\,. \tag{19.45}$$

Thus for each component of $e_k = \sin kx$ in the residual, obtain an update $\hat{u} = -e_k/\lambda_k$ as coded in the linear operator `linv` of Algorithm 19.2. The extra feature here over that of section 5.1 is the use of variable `xt`, which we will soon need for time-dependent forcing.

Recall that this updates fails, via a horrible division by zero, for zero eigenvalues. Thus the code must first remove any component in the residual in the slow modes $e_1, \ldots, e_m$, which for the system (19.42) is just $e_1 = \sin x$. We remove such components using the update $\hat{g}$ to the slow evolution via the operator `secular`, which for the system (19.42) extracts from an expression the coefficient of $e_1 = \sin x$. The iteration in Algorithm 19.1 uses this to update the evolution via $\hat{g}$. Add the code in Algorithm 19.2 to that in Algorithm 19.1. For no time-dependent forcing, replace `sigma^3=>0` with `sigma=>0` and then execute. The computer algebra finds, as in section 5.2, that the autonomous slow manifold model is

$$u = a\sin x - \tfrac{1}{6}a^2\sin 2x + \tfrac{1}{32}a^3\sin 3x + \mathcal{O}(a^4, \sigma)$$
$$\text{such that}\quad \dot{a} = -\tfrac{1}{12}a^3 + \mathcal{O}(a^4, \sigma).$$

Slowly varying forcing But the point of this chapter is to address imposed time fluctuations, whether forcing or parametric. The case where the forcing is slowly varying is easily handled here—in contrast to forcing with rapid time variations. As Algorithm 19.2 shows in its last line, to cater for slowly varying fluctuations just add code to neglect sufficiently high order time derivatives of the forcing w. Slow variations justify such neglect as slow variations mean that the high-order derivatives are very small. For example, in Algorithms 19.1 and 19.2, replace `sigma^3=>0` with `sigma^2=>0` so we retain effects linear in the slowly varying forcing. Then execute to find the time-dependent slow manifold. Even with just three forcing components ($\sin x$, $\sin 2x$, $\sin 3x$), the slow manifold field u has 40 terms and so is not recorded here. However, the corresponding evolution on the time-dependent slow manifold of the system (19.42) is

$$\begin{aligned}\dot a &= -\tfrac{1}{12}a^3 + \sigma w_1 + a\sigma(\tfrac{1}{6}w_2 - \tfrac{1}{18}\dot w_2)\\ &\quad + a^2\sigma(\tfrac{1}{18}w_1 - \tfrac{1}{54}\dot w_1 + \tfrac{1}{96}w_3 - \tfrac{19}{2304}\dot w_3)\\ &\quad + a^3\sigma(\tfrac{1}{108}w_2 - \tfrac{85}{41472}\dot w_2) + \mathcal{O}(a^4, \sigma^2, \ddot w).\end{aligned} \tag{19.46}$$

This modeling shows the direct forcing of the slow mode via the σw_1 term, but the modeling also shows that forcing interacts nonlinearly with the slow mode to generate multiplicative forcing effects in the $a^p\sigma$ terms. Algebraic complications grow enormously quickly when time-varying forcing is introduced, especially when accounting for quadratic effects of the forcing. Nonetheless, computer algebra empowers the modeling.

Account for linear forcing effects But stochastic noise forcing is never going to be slowly varying. Neither is any forcing that is on a comparable time scale to that of the fast-decaying modes. In general we have to abandon the restriction of *slow* time fluctuations and address how to construct a slow manifold model with rapidly varying forcing. The first change is to recognize that the homological equation (19.45) changes to

$$-\frac{\partial \hat u}{\partial t} + \mathcal{L}\hat u - \hat g \sin x + \text{Res}_{(19.42)} = 0 \tag{19.47}$$

to account for rapid time fluctuations in $\hat u$. Here let's start by exploring linear effects of such fluctuations.

There are two tasks: first we need to handle time convolutions and their properties (19.17)–(19.18c), and second we need to adapt the solution of the homological equations from that done in sections 19.2.1–19.2.2. These two tasks are reflected in the two blocks of additional code in Algorithm 19.3.

First, let's now define the representation and its properties applicable to PDEs with real eigenvalues. To represent the convolutions (19.17) in the computer algebra, define the linear operator `z(u,tt,mu)` $:= e^{\mu t}\star u$. Then the time derivative (19.18a) of such a convolution is coded in the second line of Algorithm 19.3. Subsequent lines code the transforms (19.18b) and (19.18c). The symbol `tt` empowers us to distinguish (in the algebra) the two different sources of time variations: `tt` represents the fast time scale of the quickly decaying variables, as notionally distinct from the slow time scale of the slow modes in $a(t)$. Nonetheless, `tt` is just a time and so depends upon time `t` as coded in Algorithm 19.1 (previously an unexplained dependence). In essence, `tt` is the time t used in partial derivatives that are done keeping slow variables constant.

Second, we need to extend the solution of the homological equation (19.47) to the case when the residual contains fluctuating $w_k(t)$ and their convolutions. Since the fluctuating $w_k(t)$ need to be accounted for in the operators `secular` and `linv`, the first

Algorithm 19.3 *Code to adjoin to Algorithms* 19.1 *and* 19.2 *in order to model linear effects in time-varying forcing in the system* (19.42) *(but omit the last line of Algorithm* 19.2*).*

```
operator z; linear z;
let { df(z(~f,tt,~mu),t)=>-sign(mu)*f+mu*z(f,tt,mu)
    , z(1,tt,~mu)=>1/abs(mu)
    , z(z(~r,tt,~nu),tt,~mu)
      =>(sign(mu)*z(r,tt,nu)-sign(nu)*z(r,tt,mu))/(mu-nu)
        when mu neq nu };

depend w,xt;
let { secular(sin(~k*x)*~r,xt)=>0
    , linv(sin(~k*x)*~r,xt)=>z(r,tt,lamb(k))*sin(k*x)
    , secular(sin(x)*w(~k),xt)=>w(k)
    , secular(sin(x)*z(~r,tt,~mu),xt)
      =>secular(sin(x)*r,xt)/abs(mu)
    , linv(sin(x)*z(~r,tt,~mu),xt)
      =>sin(x)*z(~r,tt,~mu)/mu+linv(sin(x)*r,xt)/abs(mu) };
```

line in the second block of Algorithm 19.3 asserts that the fluctuations depend upon the fast variable `xt`. The next two lines cater for forcing fluctuations in the stable modes ($e_k = \sin kx$ for $k \geqslant m+1 = 2$): the first asserts that such forcing has no effect on the evolution because the second asserts that we integrate their effects into the location of the slow manifold. The location is determined via a convolution over the past at a rate depending upon the eigenvalue of the mode.

Terms involving the slow mode, $e_1 = \sin x$, are more complicated. Sections 19.1.2 and 19.3.2 identified that it is a subjective decision about whether an effect is to be assigned to modify the evolution or to modify the slow manifold shape. In any particular application you may choose to make different decisions from those presented here, in which case you will correspondingly modify the methods.

- A direct forcing of the slow mode, as in $w_k \sin x$, goes to directly forcing the slow dynamics by choosing $\hat{g} = w_k$. Algorithm 19.3 codes this by the operator `secular` and by ignoring it in `linv` as it should not appear.[233]

- However, a forcing of the slow mode with a convolution gives us more flexibility: say a term $(\sin x)e^{\mu t}\star r$ occurs in the residual for some r. From the time derivative (19.18a) we know that

$$(\sin x)e^{\mu t}\star r = (\sin x)\frac{1}{\mu}\frac{d}{dt}(e^{\mu t}\star r) + (\sin x)\frac{1}{|\mu|}r\,.$$

  Operator `linv` puts the first term on the right-hand side into the shape of the slow manifold via $\hat{u}$. The second term on the right-hand side may generate either shape changes or slow evolution changes depending upon r, so both `secular` and `linv` consider it again, recursively as specified in Algorithm 19.3.

[233]If the integral $\int w_k\,dt$ is bounded, then it could all be assigned into the slow manifold update $\hat{u}$. But here we assume that the integral is not necessarily bounded.

Applying Algorithms 19.1–19.3 to (19.42) predicts that the time-dependent nonautonomous slow manifold is

$$\begin{aligned} u &= a\sin x - \tfrac{1}{6}a^2\sin 2x + \sigma\big[e^{-3t}\star w_2\sin 2x + e^{-8t}\star w_3\sin 3x\big] \\ &\quad + a\sigma\big[-\tfrac{1}{6}e^{-3t}\star w_2\sin x + (\tfrac{1}{3}e^{-3t}\star w_1 - \tfrac{1}{5}e^{-8t}\star w_3 + \tfrac{1}{5}e^{-3t}\star w_3)\sin 2x \\ &\quad + \tfrac{3}{10}(e^{-8t}\star - e^{-3t}\star)w_2\sin 3x + \tfrac{2}{7}(e^{-15t}\star - e^{-8t}\star)w_3\sin 4x\big] \\ &\quad + \mathcal{O}(a^3+\sigma^2). \end{aligned}$$

The first line contains the direct effects of the forcing. The second and third lines contain information about the microscale interaction between forcing, nonlinearity, and dissipation. Algorithms 19.1–19.3 find that the corresponding evolution on this nonautonomous slow manifold is[234]

$$\dot a = -\tfrac{1}{12}a^3 + \sigma w_1 + \tfrac{1}{6}a\sigma w_2 + \mathcal{O}(\sigma^2 + a^4).$$

Being able to keep the evolution $\dot a$ relatively simple like this, despite the fearsome nonlinear interactions of the forcing, is wonderful.

Example 19.1 (modified Burgers PDE). To see the sort of changes necessary, let's apply the same modeling to the PDE

$$\frac{\partial u}{\partial t} = \frac{\partial^2 u}{\partial x^2} - u\left(\frac{\partial u}{\partial x}\right)^2 + \sigma w(x,t), \quad \frac{\partial u}{\partial x} = 0 \text{ at } x = 0, \pi. \tag{19.48}$$

- Anchor the analysis on the equilibrium at $u = 0$ (although more fun could be had by anchoring the analysis at arbitrarily constant u).
- With the insulating boundary conditions, the eigenvalues and eigenfunctions of the linear diffusion PDE, $\partial u/\partial t = \partial^2 u/\partial x^2$, are $\lambda_k = -k^2$ and $e_k = \cos kx$ for $k = 0, 1, 2, \ldots$ (the index k starts from zero).
- Hence the slow subspace is that of constant fields, spanned by $e_0 = 1$, and since $\lambda_k \leqslant \lambda_1 = -1$ the slow manifold emerges from transients roughly like e^{-t}.
- To construct the slow manifold model, modify Algorithms 19.1–19.3.
  - Change throughout all `sin(x)` to 1 to reflect the change in the basis of the slow subspace.
  - Change the eigenvalue function `lamb` from $1 - k^2$ to just $-k^2$.
  - In the pattern matching, change all `sin(~k*x)` to `cos(~~k*x)` to reflect the change in eigenfunctions to the cosines, and include the extra tilde to also match the identity case of $k = 1$, as here $\cos x$ is one of the fast stable modes.
  - Correspondingly change all `sin(k*x)` on the right-hand sides into `cos(k*x)`. Including the forcing $w(x,t)$ so that, for example, we force with $w = \sum_{k=1}^{3} w_k \cos kx$, say.
  - Omit the line for slowly varying forcing, as before.

[234]The extra time derivative terms that appear in the slowly varying model (19.46) are here subsumed into the slow manifold itself (for example, the $a\sigma e^{-3t}\star w_2 \sin x$ term), leaving a simpler evolution. The two views are rationalized by noting that the amplitude a here is slightly different from the amplitude a used in (19.46).

Algorithm 19.4 *Code to adjoin to Algorithms* 19.1–19.3 *to model quadratic effects in time-varying forcing in the system* (19.42).

```
1  let { secular(sin(x)*w(~k)*z(~r,tt,~mu),xt)=>w(k)*z(r,tt,mu)
2      , secular(sin(x)*z(~r,tt,~mu)*z(~s,tt,~nu),xt)
3        =>-secular(sin(x)*(r*z(s,tt,nu)+s*z(r,tt,mu)),xt)
4          /(mu+nu)
5      , linv(sin(x)*z(~r,tt,~mu)*z(~s,tt,~nu),xt)
6        =>sin(x)*z(r,tt,mu)*z(s,tt,nu)/(mu+nu)
7          -linv(sin(x)*(r*z(s,tt,nu)+s*z(r,tt,mu)),xt)
8          /(mu+nu)
9      , secular(sin(x)*z(~r,tt,~mu)^2,xt)
10       =>-secular(sin(x)*r*z(r,tt,mu),xt)/mu
11     , linv(sin(x)*z(~r,tt,~mu)^2,xt)
12       =>sin(x)*z(~r,tt,~mu)^2/(2*mu)
13         -linv(sin(x)*r*z(r,tt,mu),xt)/mu };
```

  - Last, change the computation of the residual to that appropriate for the PDE (19.48).

- Execute your revised code to find that the iteration takes just two steps to find the nonautonomous slow manifold

$$u = a + \sigma(e^{-t}\star w_1 \cos x + e^{-4t}\star w_2 \cos 2x + e^{-9t}\star w_3 \cos 3x) + \mathcal{O}(a^4, \sigma^2).$$

  On this slow manifold the evolution is $\dot{a} = 0 + \mathcal{O}(a^4, \sigma^2)$.

This appears to be a dull slow manifold: just a constant field with some imposed fluctuations, but no evolution at all. The reason is that the fluctuations $w(x,t)$ do not have an interesting linear effect on the evolution; they only have an effect at second order in the strength σ. We have to resolve how fluctuations interact with each other through the nonlinearity to drive slow evolution. We need to discover the quadratic effects of forcing fluctuations.

Account for quadratic forcing effects The good news is that for the fast modes we do not have to do anything extra: the code of Algorithm 19.4 is sufficient. But to resolve quadratic forcing effects we do have to provide extra code for the slow modes.

Again, what to do with the slow modes is a subjective decision. The simplest decision is to simply pass any component of the slow modes, $e_1 = \sin x$ for system (19.42), to the evolution on the slow manifold. The result would be a most complicated model. Instead we pursue trying to keep the evolution as simple as possible: in linear forcing effects we are able to eliminate all memory convolutions from the evolution; in quadratic forcing effects we must accept some memory convolutions, but only those in the form $w_k e^{\mu t}\star w_\ell$.

Let's consider each term in the residual of the system (19.42) as an example of the general PDE (19.43). Each term in the residual generates corrections via the homological equation (19.47).

- Our first extra code is to pass on to the evolution, via the correction $\hat{g}$ and the operator `secular`, any component of $w_k e^{\mu t}\star r(t)\sin x$. The first line of Algorithm 19.4 does this.

- The second and third pattern matchings in Algorithm 19.4 cater for products of convolutions. From the time derivative (19.18a) (for negative μ and ν) and the product rule,
$$\frac{d}{dt}\left[(e^{\mu t}\star r)(e^{\nu t}\star s)\right] = (\mu+\nu)(e^{\mu t}\star r)(e^{\nu t}\star s) + re^{\nu t}\star s + se^{\mu t}\star r\,.$$
Rearranging,
$$(e^{\mu t}\star r)(e^{\nu t}\star s) = \frac{1}{\mu+\nu}\frac{d}{dt}\left[(e^{\mu t}\star r)(e^{\nu t}\star s)\right] - \frac{re^{\nu t}\star s + se^{\mu t}\star r}{\mu+\nu}.$$
The first term on the right-hand side integrates into the shape of the slow manifold via $\hat{u}$ and the operator `linv`. The second term on the right-hand side needs further consideration as to whether it generates shape changes or evolution changes or both, depending upon r and s; thus we pass it on to further consideration by both `secular` and `linv`.

- The fourth and fifth pattern matchings in Algorithm 19.4 are just the previous product patterns in the specific case where $\mu = \nu$.

Merge the pattern-matching code of Algorithm 19.4 into Algorithms 19.1–19.3, change `sigma^2=>0` to `sigma^3=>0` to retain quadratic interactions, and execute. Seven iterations of copious algebra converge to an answer, via residuals with up to 695 terms. The time-dependent slow manifold is a very large algebraic expression not worth recording here. The corresponding evolution on the nonautonomous slow manifold is

$$\begin{aligned}\dot{a} &= -\tfrac{1}{12}a^3 + \sigma w_1 + \tfrac{1}{6}a\sigma w_2 + a^2\sigma(\tfrac{1}{18}w_1 + \tfrac{1}{96}w_3)\\
&\quad + \sigma^2\big(\tfrac{1}{6}w_1e^{-3t}\star w_2 + \tfrac{1}{22}w_2e^{-8t}\star w_3 + \tfrac{1}{22}w_3e^{-3t}\star w_2\big)\\
&\quad + a\sigma^2\big(\tfrac{1}{18}w_1e^{-3t}\star w_1 - \tfrac{2}{55}w_2e^{-3t}\star w_2 + \tfrac{3}{220}w_2e^{-8t}\star w_2\\
&\quad + \tfrac{1}{110}w_3e^{-3t}\star w_3 - \tfrac{2133}{141680}w_3e^{-8t}\star w_3 + \tfrac{1}{161}w_3e^{-15t}\star w_3\\
&\quad + \tfrac{7}{2640}w_1e^{-8t}\star w_3 + \tfrac{1}{30}w_1e^{-3t}\star w_3 + \tfrac{1}{66}w_3e^{-3t}\star w_1\big)\\
&\quad + \mathcal{O}(\sigma^3 + a^4).\end{aligned} \tag{19.49}$$

With forcing spread across many modes, the effects of the forcing-forcing interactions become very complicated. Computer algebra appears essential.

Fortunately, in many applications it is only the mean effect of these forcing-forcing interactions that is significant. When only the mean effect is desired, then these complicated sums typically collapse to a vastly simpler form.[235] For example, if the forcing is $w_k = \cos(t)/k^2$, then the previous evolution on the slow manifold collapses to the much simpler

$$\begin{aligned}\dot{a} &= -\tfrac{1}{12}a^3 + \sigma\big(1 + \tfrac{1}{24}a + \tfrac{49}{864}a^2\big)\cos t\\
&\quad + \sigma^2\big(\tfrac{25}{3456} + \tfrac{1363}{138240}a\big)(1+\cos 2t) + \mathcal{O}(\sigma^3 + a^4).\end{aligned}$$

This model predicts that the mean quadratic forcing displaces the equilibrium from a mean $\bar{a} = 0$ to $\bar{a} \approx \frac{1}{2}(5\sigma/6)^{2/3}$, which, being $\mathcal{O}(\sigma^{2/3})$, is a somewhat larger effect than that of the $\mathcal{O}(\sigma)$ cosine fluctuations. Mean quadratic effects often accumulate to be significant.

[235] At the time of writing, an unresolved research issue is to find a *general* way to determine the mean effect of forcing-forcing interactions without having to compute and manipulate the combinatorially exploding details of the full residual.

Example 19.2 (modified Burgers PDE continued). Reconsider the modeling of the PDE (19.48) of Example 19.1. Adapt Algorithm 19.4 to the cosines of this case and merge with the code of Example 19.1. Execute the code to find that the slow manifold is the same as given in Example 19.1 but with extra complicated terms with factor $a\sigma^2$. The corresponding evolution on the nonautonomous slow manifold is now the nontrivial

$$\dot a = -\tfrac{1}{2}a\sigma^2\left(w_1 e^{-t}{\star}w_1 + w_2 e^{-4t}{\star}w_2 + w_3 e^{-9t}{\star}w_3\right) + \mathcal{O}\left(a^4,\sigma^2\right). \quad (19.50)$$

But what does this modeling predict? For example, when the forcing $w_k = \cos(t)/k^2$, then the slow manifold model (19.50) becomes

$$\dot a = -a\sigma^2 c(1+\cos 2t) + \mathcal{O}\left(a^4,\sigma^2\right) \quad \text{for constant } c = \tfrac{47449}{186624}\,.$$

The mean effect on the right-hand side is stabilizing as seen from the exact solution: neglecting unknowns $\mathcal{O}\left(a^4,\sigma^2\right)$, separating variables, and integrating gives

$$a(t) = a_0 \exp\left[-\sigma^2 c\left(t + \tfrac{1}{2}\sin 2t\right)\right].$$

For this particular forcing the model predicts $a \to 0$ as $t \to \infty$.

Typically, $w_k e^{\mu t}{\star}w_k > 0$ on average for most forcing; for example, it is positive for constant forcing and positive on average for white noise forcing. Thus, typically and on average, the model (19.50) is $\dot a \approx -(\text{const})\sigma^2 a$, which predicts that the amplitude a decays to zero on a time scale of $1/\sigma^2$. That is, typically the forcing interacts via the nonlinearity to stabilize $u = 0$ in this example.

Beware of nonlinear combinations of basis eigenvectors The two examples worked through in this subsection had an enormously useful symmetry. First, the basis eigenfunctions are the well-known trigonometric functions. Second, the nonlinearity had the required odd/even symmetry to ensure that nonlinear combinations of the trigonometric basis eigenfunctions are expressible in terms of the addition of just a couple of other trigonometric basis eigenfunctions. Without this symmetry the nonlinear analysis would involve sums over all basis eigenfunctions, sums that would be infinite in the case of PDEs. For example, if the system (19.42) were modified to have a quadratic reaction, say $-u^2$, then this reaction would act on the $\sin kx$ basis to generate, via trigonometric rules, terms in $\cos kx$; such cosine terms would have to be expressed as the infinite sum of a sine series half-range Fourier expansion; such infinite sums are very awkward. Similarly, for example, if the other example PDE (19.48) were modified to include the usual advective nonlinearity of uu_x, then this advection would act on the $\cos kx$ basis to generate terms in $\sin kx$ by the trigonometric rules; such sine terms would have to be expressed as the infinite sum of a cosine series half-range Fourier expansion, and again such infinite sums are very awkward. In examples and exercises we avoid such complication. In practice one would truncate such infinite sums to a finite sum just as we have for the forcing $w(x,t)$.

19.3.4 ▪ Truncate asymptotics flexibly

The combinatorial explosion of terms is a major challenge in modeling nonautonomous dynamics. Recall that section 4.5.3, albeit for autonomous dynamics, assures us that we can truncate the asymptotic expressions quite flexibly. We often need to do so to contain the combinatorial explosion. Most of the explosion arises through the multitude of interactions of the range of forcing terms through the nonlinearity. Thus we are most likely to want to discard terms of higher order in both forcing σ and nonlinearity. Either one of the two methods described in this section helps contain the combinatorial explosion of terms.

The first method is to introduce an arbitrary parameter, ε say; multiply nonlinear terms in the equation and the forcing terms by the parameter; and then truncate to some order in ε. That is, instead of modeling PDE (19.43) to some order in $\|u\|$ and σ, one models

$$\frac{\partial u}{\partial t} = \mathcal{L}u + \varepsilon f(u) + \varepsilon\sigma F(u)w(t)$$

to some order in parameter ε. Or one could make finer adjustments by choosing exponents p and q and modeling

$$\frac{\partial u}{\partial t} = \mathcal{L}u + \varepsilon^p f(u) + \varepsilon^q \sigma F(u)w(t)$$

to some order in parameter ε. This approach is very simple and flexible, but a downside is that the parameter ε has little physical meaning, if any.

A second alternative is to introduce a variable that simply counts the number of amplitudes in each term, analogous to how section 3.6 introduced the counting of oscillation amplitudes. Then we truncate the asymptotic expansion to an order determined by the counting. Suppose parameter ε is our counting variable. The procedure is to set the slow subspace to, for example, $u = \varepsilon a \sin x$ or in general to $u = \varepsilon \sum_{k=1}^{m} a_k e_k$. Then in the governing equations every multiplication by a u generates an extra power of the counting variable ε. To control the forcing one then also replaces parameter σ by say $\varepsilon^p\sigma$ for some exponent p. Consequently, truncating the construction to residuals $\mathcal{O}(\varepsilon^q)$ results in errors $\mathcal{O}(|a|^q + \sigma^{q/p})$. But there is one extra implementation detail to remember: since $\dot{a} = g(a,t)$, whenever g is invoked in a term there is an extra ε due to the amplitude in the term that was differentiated; thus g has to be stored with one less factor of the counting variable ε; that is, when computing updates, $\hat{g} = \cdots$, now compute them as $\hat{g} = (\cdots)/\varepsilon$. Being able to assert the errors $\mathcal{O}(|a|^q + \sigma^{q/p})$ is a nice feature of this second approach.

19.4 ▪ Exercises

Exercise 19.1 (basic nonautonomous slow manifold). Consider the system

$$\dot{x} = ax \quad \text{and} \quad \dot{y} = -y + w(t)x^2$$

for some small parameter a and some given forcing $w(t)$ (deterministic or stochastic). Solve the x ODE, and then derive that the emergent fast y-dynamics is $y = x^2 e^{-(1+2a)t} \star w(t)$, upon neglecting exponential decaying transients in y, and where $\star$ denotes convolution over the past. Hence deduce that this is the nonautonomous slow manifold for the system.[236]

[236] Even this basic system leads people into error. For example, Boxler (1991) appears to have overlooked some time dependence and would deduce incorrectly that the slow manifold is $y = x^2 w(t)/(2a+1) + \mathcal{O}(x^4)$—in this particular case for a stochastic w (Boxler, 1991, §5.3).

Exercise 19.2 (linear modeling). Monahan and Culina (2011) introduced this linear forced system for small parameter τ:

$$\frac{dx}{dt} = -x + \frac{a}{\sqrt{\tau}}y \quad \text{and} \quad \frac{dy}{dt} = \frac{1}{\sqrt{\tau}}x - \frac{1}{\tau}y + \frac{b}{\tau}w(t/\tau).$$

To avoid singular perturbations, let $\tau = \epsilon^2$ and rescale time, $t = \tau t' = \epsilon^2 t'$, so that $d/dt = \frac{1}{\tau}d/dt'$. Derive that the system becomes, upon dropping primes, the regular system

$$\frac{dx}{dt} = -\epsilon^2 x + a\epsilon y \quad \text{and} \quad \frac{dy}{dt} = \epsilon x - y + bw(t).$$

Use the arguments of section 19.1, for small ϵ and b, to derive that the time-dependent coordinate transform

$$y \approx Y + \epsilon X + be^{-t}\star w - \epsilon^2 ab(e^{-t}\star w + e^{-t}\star e^{-t}\star w),$$
$$x \approx X - \epsilon aY - \epsilon abe^{-t}\star w$$

leads to the normal-form system

$$\dot{Y} \approx -[1 + \epsilon^2 a]Y \quad \text{and} \quad \dot{X} = \epsilon^2(a-1)X + \epsilon abw.$$

Exercise 19.3 (fluctuation interaction). An interesting pair of fast/slow differential equations inspired by advection-dispersion problems is

$$\dot{x} = -\sigma yw(t) \quad \text{and} \quad \dot{y} = -y + \sigma xw(t),$$

where $w(t)$ denotes some specified fluctuations in time. Use the arguments of section 19.1 to derive that the time-dependent coordinate transform[237]

$$x \approx X + \sigma Ye^{t}\star w \quad \text{and} \quad y \approx Y + \sigma Xe^{-t}\star w$$

leads to the normal-form system

$$\dot{X} \approx -\sigma^2 Xwe^{-t}\star w \quad \text{and} \quad \dot{Y} \approx -Y + \sigma^2 Ywe^{t}\star w.$$

For the specific case of forcing $w = \sin\omega t$, determine from this normal-form system the range of parameters σ and ω for which a simple model emerges ($Y \to 0$), and in this parameter regime solve for the long-term dynamics of X.

Answer: Since $e^{\pm t}\star\sin\omega t = (\sin\omega t \pm \omega\cos\omega t)/(1+\omega^2)$, the $\dot{Y}$ equation asserts that we need $\sigma^2 < 2(1+\omega^2)$ for $Y \to 0$. Then solving the $\dot{X}$ equation predicts $X = c\exp\left[-\sigma^2(\frac{1}{2}t + \frac{1}{4}\cos\omega t - \frac{1}{4\omega}\sin\omega t)/(1+\omega^2)\right]$.

Exercise 19.4 (parametric fluctuations). Bonckaert et al. (2014) considered the following example of a strongly damped Duffing–van der Pol ODE for $u(t)$ under the influence of a nonautonomous parametric perturbation:

$$\ddot{u} = (\alpha + \sigma w(t))u - u - u^3 - u^2\dot{u}.$$

[237] One of the interesting aspects of this example is the quickness with which we could go to even higher-order interactions in the fluctuations, higher orders in the strength σ.

Find a normal-form coordinate transform to separate fast and slow dynamics when parameters α and σ are small.

1. Write the ODE as a linearly separated system by changing to slow $x := u + \dot{u}$ and fast $y := \dot{u}$ coordinates: namely (Bonckaert et al., 2014, eq. (50))

$$\begin{bmatrix} \dot{x} \\ \dot{y} \end{bmatrix} = \begin{bmatrix} 0 \\ -y \end{bmatrix} + (\alpha + \sigma w(t)) \begin{bmatrix} 1 & -1 \\ 1 & -1 \end{bmatrix} \begin{bmatrix} x \\ y \end{bmatrix} - \begin{bmatrix} 1 \\ 1 \end{bmatrix} (x^3 - 2x^2 y + xy^2).$$

2. Use the arguments of section 19.1 to derive that the time-dependent coordinate transform from (X, Y) to (x, y),

$$x \approx X + \alpha Y + \tfrac{1}{2} XY^2 - 2X^2 Y + \sigma e^{t} \star w\, Y,$$
$$y \approx Y + \alpha X + XY^2 - X^3 + \sigma e^{-t} \star w\, X,$$

leads to the normal-form system

$$\dot{X} \approx (\alpha + \sigma w - 2\alpha\sigma w) X - (1 - 3\sigma w) X^3 - \sigma^2 w e^{-t} \star w\, X^3,$$
$$\dot{Y} \approx (-1 - \alpha + 2X^2) Y - \sigma w (1 - 2\alpha + 7X^2) Y + \sigma^2 w e^{t} \star w\, Y.$$

3. Discuss the emergence of a slow manifold and the influence of the parametric fluctuations on the long-term dynamics.

Exercise 19.5 (nongeneric bifurcation). Consider for small parameter ϵ

$$\dot{x} = \epsilon x + x^3 - (1 - \sigma w) xy,$$
$$\dot{y} = -y + x^2 + y^2 + \sigma y w.$$

Autonomously ($\sigma = 0$), there is a bifurcation to two equilibria for small $\epsilon > 0$. The nonautonomous forcing w affects this bifurcation somehow.

Use the arguments of section 19.1 to derive a time-dependent coordinate transform from (X, Y) to (x, y), an algebraically complicated transform, that leads to the normal-form system

$$\dot{X} \approx \epsilon X - X^5 - 2\sigma X^5 w - 3\sigma^2 X^5 w e^{-t} \star w,$$
$$\dot{Y} \approx (-1 + 4X^2 + 6X^4) Y + \sigma (1 + 2X^2 + 22X^4) Y w.$$

From this system, argue that small forcing w makes little difference to the deterministic bifurcation.

Answer:

$$\begin{aligned} x \approx{}& X + XY + 2X^3 Y \\ &+ \sigma \big[(-XY^2 + 3X^3 Y) e^{+t} \star + (XY^2 + XY^3) e^{2t} \star \\ &\quad + X^3 e^{-t} \star - XY^3 e^{3t} \star \big] w, \\ y \approx{}& Y - Y^2 + Y^3 - Y^4 + X^2 - 7X^2 Y^2 + X^4 \\ &+ \sigma \big[(-Y + 2Y^3 - 3Y^4 - 10X^2 Y^2 - 4X^2 Y^2 e^{+t} \star) e^{+t} \star \\ &\quad + (X^2 - 2X^2 Y + 3X^2 Y^2 + X^4 + 4X^4 e^{-t} \star) e^{-t} \star + 2X^2 Y^2 e^{2t} \star \big] w. \end{aligned}$$

Exercise 19.6 (control in enzyme kinetics). Lebiedz and Rehberg (2013) considered the control of a simple differential equation model of enzyme kinetics. They phrased the Michaelis–Menten enzyme kinetics as the singularly perturbed problem

$$\dot{x} = -x + (x + \tfrac{1}{2})y + u(t) \quad \text{and} \quad \epsilon\dot{y} = x - (x+1)y\,.$$

They implemented a numerical construction of the slow manifold for parameter $\epsilon = 1/100$; we simply divide the y equation by $\epsilon = 1/100$ to get a fast/slow system with fast rate of the order of 100 and slow rate of order 1.

Treat the resultant term $-100y$ on the right-hand side as the dominant term in the two equations, and for simplicity treat every other term as a small perturbing influence. Treat the unknown control $u(t)$ as some forcing term. Use our procedure to nonlinearly separate the fast/slow dynamics to find that the coordinate transform

$$\begin{aligned} x &\approx X - \tfrac{1}{100}(Y - XY) - \tfrac{1}{100}Ye^{100t}\star u\,, \\ y &\approx Y + X - e^{-100t}\star u \end{aligned}$$

transforms the Michaelis–Menten enzyme kinetics system to the equivalent

$$\begin{aligned} \dot{X} &\approx -\tfrac{1}{2}X + X^2 + u(t) - \tfrac{1}{100}(X + \tfrac{1}{2})u(t), \\ \dot{Y} &\approx -(100.5)Y - 101XY - \tfrac{1}{100}Yu(t). \end{aligned}$$

Comment on the appearance of a slow manifold for general imposed control $u(t)$.

Exercise 19.7 (convolutions). By differentiating (19.17), prove the identities (19.18a) and (19.18b). Recall that convolutions are commutative. Considering in two ways a composition of two convolutions, use (19.18a) to prove the identity (19.18c).

Exercise 19.8 (forced delay differential equation). Hutt et al. (2012) discussed stochastic bifurcation in the delay differential equation (DDE)

$$\dot{u}_0 = -u_0(t) + u_0(t-\tau) - \eta u_0(t-\tau)^3 + \sigma w(t)$$

for a delay τ, nonlinearity coefficient η, and some additive noise process $w(t)$ of strength σ. Let's simplify by fixing upon the delay $\tau = 32/3$ and considering the forcing $w(t)$ to be deterministic.

1. But a DDE is infinite dimensional. For many practical purposes we can model the delay term by two intermediaries. Let's do so. Introduce $u_1(t)$ to approximately represent $u_0(t-\tau/2)$, and introduce $u_2(t)$ to approximately represent $u_0(t-\tau)$. Then argue that an appropriate system to consider for $\mathbf{u}(t) = (u_0, u_1, u_2)$ is

$$\dot{\mathbf{u}} = \begin{bmatrix} -1 & 0 & 1 \\ 2/\tau & -2/\tau & 0 \\ 0 & 2/\tau & -2/\tau \end{bmatrix}\mathbf{u} + \begin{bmatrix} 1 \\ 0 \\ 0 \end{bmatrix}(-\eta u_2^3 + \sigma w).$$

2. Choosing the delay $\tau = 32/3$, diagonalize the linear part via the transform
$$u = \frac{1}{16}\begin{bmatrix} 16 & 16 & 16 \\ 16 & -12 & -4 \\ 16 & 9 & 1 \end{bmatrix} x,$$
where $x = (x_1, y_1, y_2)$, to find that the system becomes
$$\dot{x} = \begin{bmatrix} 0 & 0 & 0 \\ 0 & -\frac{7}{16} & 0 \\ 0 & 0 & -\frac{15}{16} \end{bmatrix} x + \frac{1}{35}\begin{bmatrix} 3 \\ -2 \\ 42 \end{bmatrix} \left[-\eta(x_1 + \tfrac{9}{16}y_1 + \tfrac{1}{16}y_2)^3 + \sigma w\right].$$

3. Use the techniques of section 19.3 to construct a slow manifold model for this system for general forcing $w(t)$ of small strength σ. Deduce that the forced slow manifold evolution is
$$\begin{aligned}\dot{X} &\approx +\tfrac{3}{35}\sigma w - \tfrac{3}{35}\eta X^3 - \tfrac{72}{42875}\eta\sigma X^2 w \\ &\quad + \eta\sigma^2 X\Big(-\tfrac{1404}{336875}we^{-15t/16}\star w + \tfrac{11664}{3301375}we^{-7t/16}\star w\Big)\end{aligned}$$
for slow variable $x_1 \approx X$.

Exercise 19.9 (forced Dirichlet PDEs). Revise section 19.3.3's construction of a time-dependent slow manifold to model the dynamics of each of the following PDEs for a field $u(x,t)$ with Dirichlet boundary conditions of $u = 0$ at $x = 0, \pi$:
$$\frac{\partial u}{\partial t} = ru + \frac{\partial^2 u}{\partial x^2} - u^3 + \sigma w(x,t); \tag{19.51}$$
$$\frac{\partial u}{\partial t} = \sin u + r\frac{\partial^2 u}{\partial x^2} + \sigma u^2 w(x,t); \tag{19.52}$$
$$\frac{\partial u}{\partial t} = u + \frac{\partial^2 u}{\partial x^2} - u\frac{\partial u}{\partial x} + \sigma u w(x,t). \tag{19.53}$$
For PDEs (19.51) and (19.52) analyze for parameter r close to one: in (19.51) there is a bifurcation for $r > 1$ when forcing $\sigma = 0$, whereas in (19.52) there is a bifurcation for $r < 1$ when forcing $\sigma = 0$. In PDE (19.52) expand $\sin u$ in a Taylor series in u. For PDE (19.53) cater for the multiplicative forcing by writing the forcing as a sum of cosines, say $w = \sum_{k=0}^{2} w_k(t)\cos kx$.

Exercise 19.10 (forced Neumann PDEs). Revise Example 19.1's construction of a time-dependent slow manifold to model the dynamics of the PDEs (19.51)–(19.53) for a field $u(x,t)$ with Neumann boundary conditions of $\partial u/\partial x = 0$ at $x = 0, \pi$.

Why does the construction fail, using only straightforward modifications, for the third PDE, (19.53)? Discuss how you might circumvent this failure.

Exercise 19.11 (forcing Kuramoto–Sivashinsky). Explore a modified version of the well known Kuramoto–Sivashinsky PDE
$$\frac{\partial u}{\partial t} + \sigma w(x,t)u\frac{\partial u}{\partial x} + \frac{\partial^2 u}{\partial x^2} + \frac{\partial^4 u}{\partial x^4} = 0 \tag{19.54}$$

for a field $u(x,t)$ with Neumann boundary conditions of $\partial u/\partial x = 0$ at $x = 0, \pi$. The modification is that instead of a fixed nonlinear advection uu_x, the PDE has nonlinear advection moderated by a fluctuating factor $w(x,t)$.

- With no fluctuating advection, $\sigma = 0$, argue that there exists a 2D slow subspace spanned by $\cos x$ and 1.
- Expand the factor $w(x,t)$ in a truncated sine series, say just two modes.
- Modify Example 19.1's construction to model the nonlinear dynamics of PDE (19.54) with its fluctuations:
  - invoke two amplitudes, say a and b;
  - since the nonlinearity is ameliorated by parameter σ, only truncate the asymptotic expansion in σ, not a and b;
  - perhaps use both basis functions, $\cos x$ and 1, to track potentially secular effects, in both `secular` and `linv`, into `gd`; then separate using the `coeffn` function into two updates for the evolution;
  - be careful with signs of the updates.
- The model is complicated, but pick out effects that suggest some w-w interactions stabilize $a = b = 0$; argue a case for such stabilization effects.
- Modify your code, execution, and interpretation for the case when the fourth-order dissipation is moderated by a coefficient to $r\partial^4 u/\partial x^4$ for parameter r near 1.

Exercise 19.12 (forcing Swift–Hohenberg in 1D). The Swift–Hohenberg PDE is of much interest in pattern formation (Greenside and Coughran, 1984; Cross and Hohenberg, 1993, e.g.). Model its forced dynamics for a field $u(x,t)$ in 1D space with boundary conditions of being 2π-periodic:

$$\frac{\partial u}{\partial t} + (1-\epsilon)u + 2\frac{\partial^2 u}{\partial x^2} + \frac{\partial^4 u}{\partial x^4} + \epsilon u^3 = \sigma w(x,t). \tag{19.55}$$

- With no forcing and at critical, $\sigma = \epsilon = 0$, argue that there exists a 2D slow subspace of PDE (19.55) spanned by $\cos x$ and $\sin x$, or equivalently by $e^{\pm ix}$.
- Modify section 19.3.3's construction of a time-dependent slow manifold to model the nonlinear dynamics of PDE (19.55) with its fluctuations (see also exercise 5.14):
  - decide whether to use basis eigenfunctions $\cos x$ and $\sin x$, or equivalently $e^{\pm ix}$ (using `cis` analogously to section 3.6);
  - follow analogous hints to those given for Exercise 19.11.
- Interpret the resulting model.

Exercise 19.13 (forcing Swift–Hohenberg in 2D). The Swift–Hohenberg PDE is of more interest in multiple space dimensions. Here model its forced dynamics for a field $u(x,y,t)$ in 2D space with boundary conditions of being 2π-periodic in each of x and y, but here scale the PDE a little differently:

$$\frac{\partial u}{\partial t} + 4(1-\epsilon)u + 4\nabla^2 u + \nabla^4 u + \epsilon u^3 = \sigma w(x,y,t). \tag{19.56}$$

Follow the outline of Exercise 19.12 but here with slow manifold spanned by four basis functions, such as $\cos(x)\cos(y)$ and partners, or $\cos(x \pm y)$ and $\sin(x \pm y)$, or analogous complex exponentials; decide which. Develop and execute Reduce code to construct a forced slow manifold for some spectrum of forcing w, and interpret the resulting model.

Exercise 19.14 (existence and emergence proofs). Challenge: Establish good proofs of Propositions 19.3 and 19.4 on the backward existence and emergence of slow manifolds for nonautonomous systems.

Chapter 20

Introducing basic stochastic calculus

Contents

Nothing in Nature is random. . . . A thing appears random only through the incompleteness of our knowledge. *Spinoza*

Chapters 18 and 19 explored the foundation and construction of reduced models of nonautonomous dynamical systems. This chapter establishes the basics of stochastic processes that we need to similarly construct and interpret reduced models of noisy stochastic systems. Those comfortable with basic stochastic calculus of noisy systems could proceed directly to Chapter 21, which addresses the issues of modeling emergent stochastic dynamics.

We model noise in a dynamical system with stochastic differential equations (SDEs).[238] Using SDEs implicitly admits ignorance of the nature of the fluctuations in the dynamics. Implicitly one assumes that there are a multitude of unknown processes influencing whatever is of interest. Under the central limit theorem these unknown influences accumulate to be normally distributed (Gaussian). But there are many traps for the unwary modeler.

Just as for deterministic systems, stochastic coordinate transforms tease apart the dynamics of stochastic systems. The following example introduces the key ideas. As usual, the key is to use a *time-dependent* coordinate transform, as introduced in Chapter 3 for the modeling of oscillations and in Chapter 17 to model Hopf bifurcations, and as developed by Chapters 18 and 19 for deterministic nonautonomous dynamics.

Example 20.1 (coordinate transforms clarify modeling). Consider briefly the coupled differential equations for variables $x(t)$ and $y(t)$, where ϵ is a small number like 0.1 or 0.01,

$$\dot{x} = \epsilon y \qquad \text{so } x(t) \text{ jitters about slowly,} \tag{20.1a}$$

$$\dot{y} = -y + w(t) \qquad \text{compared to the fast jitter of } y(t). \tag{20.1b}$$

We imagine $w(t)$ to be some noise process that has rapid, jittery fluctuations with zero mean. Nonetheless, for this example suppose that the noise w is smooth enough so that at least some classic calculus applies to expressions involving w.

Figure 20.1 shows some snapshots of the evolution of the system (20.1), all from $x(0) = y(0) = 0$, for many different realizations of the noisy forcing w. See that initially, up to time one, the system spreads in y to some equilibrium distribution. Thereafter, individual fluctuations of trajectories within this distribution feed into x, which then spreads sideways.

Challenge Separate the long-time dynamics of the spread of x from

- the noisy fluctuations of $w(t)$ on many time scales, and
- the $\mathcal{O}(1)$ time scale of the Ornstein–Uhlenbeck process y.

Because the system (20.1) is linear, some linear arguments clearly make this separation.

However, averaging fails us: averaging only predicts $d\bar{x}/dt = 0$ as, because $w(t)$ has zero mean, the mean $\bar{y} = 0$, and hence $d\bar{x}/dt = \epsilon\bar{y} = 0$.

Instead of averaging, decouple the two differential equations in the system (20.1) with a time-dependent, noisy coordinate transform to (X, Y) variables, where

$$x = X - \epsilon Y - \epsilon \int_{-\infty}^{t} e^{\tau - t} w(\tau)\, d\tau \quad \text{and} \quad y = Y + \int_{-\infty}^{t} e^{\tau - t} w(\tau)\, d\tau\,.$$

[238]An alternative is to treat so-called random dynamical systems (Arnold and Crauel, 1991; Duan et al., 2003; Lu and Schmalfuss, to appear, e.g.) instead of SDEs. However, the theoretical advantages do not appear to warrant the extra complications in practical applications.

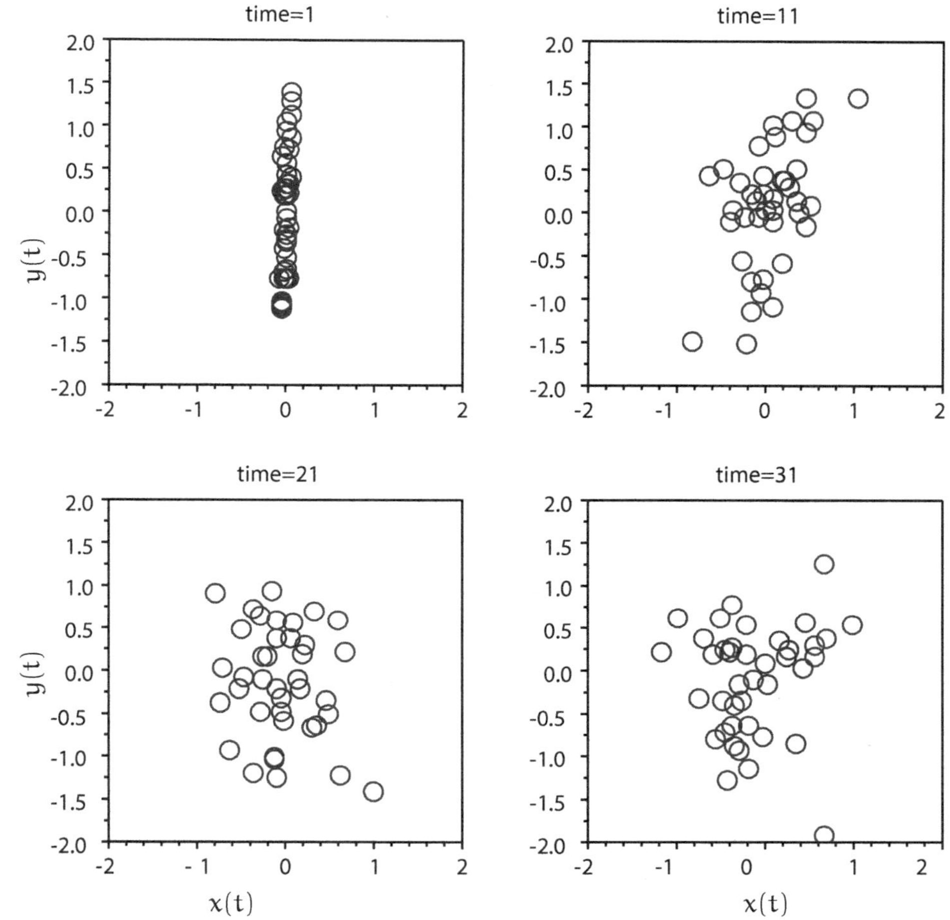

Figure 20.1. *Evolution of the system* (20.1), *all from* $x(0) = y(0) = 0$, *for many different realizations of the noisy forcing* $w(t)$.

These integrals appear to be the same convolutions as those introduced in section 19.1.1 for deterministic forcing. Straightforward calculus then shows that the system (20.1) becomes, in these new (X, Y) coordinates, simply

$$\dot{X} = \epsilon w \quad \text{and} \quad \dot{Y} = -Y. \tag{20.2}$$

This time-dependent coordinate transform does some marvelous things for us because most of the fluctuations caused by the noisy $w(t)$ are absorbed by the coordinate transform.

- First, $Y \to 0$ exponentially fast. Thus the long-term dynamics only involves X.
- Second, solving the X equation by direct integration, the solution $X(t) = X_0 + \epsilon \int_0^t w(\tau)\, d\tau$ describes the long-term dynamics. If $w(t)$ is genuinely noise-like, then $X(t)$ undergoes a slow random walk—slow because of the multiplication by the small ϵ.

- Third, in constructing the coordinate transform to reduce the dynamics, we only need to *integrate* $w(t)$. Thus the analysis applies to functions $w(t)$ that behave "badly," such as nondifferentiable noise.[239]

Even though $X(t)$ has many time scales from the noise $w(t)$, the SDE $\dot{X} = \epsilon w$ is the long-time model, as there exist good methods for integrating SDEs despite all the time scales in the noise (although in this example direct integration suffices to solve the SDE model).[240]

In general, time-dependent coordinate transforms decouple fast and slow dynamics, even when the time dependence arises as stochastic noise that has all time scales. Such time-dependent coordinate transforms then empower accurate modeling of the nonautonomous or stochastic slow dynamics over long times. They are called normal-form transformations because this decoupling of stochastic dynamics is analogous to corresponding simplifications in autonomous deterministic systems (Murdock, 2003; Arnold, 2003, e.g.). We explore the useful and intriguing properties of such stochastic normal-form coordinate transformations in modeling multiscale nonlinear stochastic dynamical systems.

But to make progress specific to stochastic systems, we need this chapter to establish some basic properties of stochastic dynamics.

20.1 ▪ White noise sums to a classic random walk

> It had been a long night. Twelve empty beer bottles stood or lay on the bar. I couldn't focus or understand what the guy was saying. Something about being thrown out. I lost consciousness. When I came to, I was out in the middle of nowhere. What a lamppost was doing out in the middle of nowhere, I'll never know, but I was leaning against it. As good a place to start as any [$W(t = 0) = 0$]. I took a step to the right [$\Delta W_1 = +1$, $W(t_1) = 1$]. Then another [$\Delta W_2 = +1$, $W(t_2) = 2$]. Then I stepped to the left [$\Delta W_3 = -1$, $W(t_3) = 1$]. I kept walking back and forth, never knowing which direction I would go next [$\langle \Delta W_i, \Delta W_j \rangle = 0$ for $i \neq j$]. It seemed like I spent as much time on one side of the lamppost as the other [$\langle W(t) \rangle = 0$], but the longer I staggered around the farther away from the lamppost I got until I started drifting back to it ($\langle W^2 \rangle^{1/2}$ increasing with time). If I'd been able to remember all this when I sobered up, I'd have had a pretty good understanding of the Wiener process (the distance from the lamppost) and white noise (the individual steps). There's nothing like an experiment.
>
> *Penland* (2003)

A starting point to describe noise, that is, to describe a stochastic process, is Brownian motion or, more technically, a Wiener process. Figure 20.2 shows five realizations from an example Brownian motion/Wiener process (generated by Algorithm 20.1). Such a

[239] The analysis potentially applies to discontinuous "jump" processes $w(t)$. But we do not develop this potential herein.

[240] Interestingly, the system (20.1) could represent the pollen grains observed by Brown in 1826: a pollen grain with position $x(t)$ and momentum $y(t)$. Newton's second law would change a pollen grain's momentum through buffeting by molecular collisions, the noise $w(t)$, and the momentum reduced by drag $-y$. The slow manifold model (20.2) shows that the emergent dynamics are that the pollen grain's position follows the white noise buffeting.

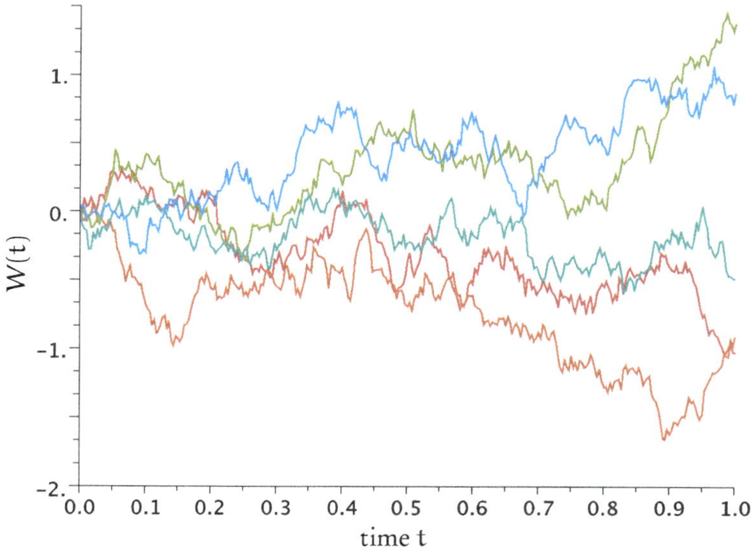

Figure 20.2. *Five realizations of Brownian motion (Wiener process) $W(t)$ generated by Algorithm 20.1. Herein, most figures of stochastic systems plot five realizations of a random process to hint that stochastic processes, such as this Wiener process, are a composite of an uncountable infinity of individual realizations.*

Algorithm 20.1 *Octave/MATLAB/Scilab plots* m *realizations of a Brownian motion/Wiener process as shown in Figure 20.2. In Scilab use* `rand(.,.,"n")` *instead of* `randn(.,.)` *for* $N(0,1)$ *distributed random numbers.*

```
m=5; % realizations
n=300; % time steps
t=linspace(0,1,n+1)';
h=diff(t(1:2));
dw=sqrt(h)*randn(n,m);
w=cumsum([zeros(1,m);dw]);
plot(t,w)
```

Wiener process is notionally the integral of a white noise: a Wiener process $W(t) = \int w(t)\,dt$ for some white noise $w(t)$. It is such a Wiener process that is mathematically tractable.

20.1.1 ▪ The Wiener process

The stochastic process illustrated in Figure 20.2 is named after the British botanist Robert Brown, who reported such motion in 1826 when observing in his microscope the movement of tiny pollen grains: their fluctuating jagged paths are due to small but incessant and random impacts of water molecules. Subsequently, Wiener formalized the properties of such a stochastic process in the 20th century. Algorithm 20.1 generates the realizations in Figure 20.2 by dividing time into small steps of length $\Delta t = h$ so the jth time step reaches time $t_j = jh$ (assuming $t = 0$ is the start of the period of interest). Then it determines W_j, the value of the process at time t_j, by adding up many independent and normally distributed increments:[241]

$$W_{j+1} = W_j + \sqrt{h} Z_j, \quad \text{where } Z_j \sim N(0,1) \quad \text{and with } W_0 = 0 .$$

We generate a Brownian motion or Wiener process, $W(t)$, in the limit as the time step size $h \to 0$, so that the number of steps becomes large like $1/h$.

The random increments to W, namely $\Delta W_j = \sqrt{h} Z_j$, are chosen to scale with $\sqrt{h}$, because this scaling is precisely what is needed to generate a reasonable limit as the time step $h \to 0$, as we now see. Consider the process W at some fixed time $t = nh$; that is, we take n steps of size h to reach t and hence $W(t)$ is approximated by n random increments of variance h:

$$\begin{aligned} W_n &= W_{n-1} + \sqrt{h} Z_{n-1} \\ &= W_{n-2} + \sqrt{h} Z_{n-2} + \sqrt{h} Z_{n-1} \\ &= W_{n-3} + \sqrt{h} Z_{n-3} + \sqrt{h} Z_{n-2} + \sqrt{h} Z_{n-1} \\ &\vdots \\ &= W_0 + \sum_{j=0}^{n-1} \sqrt{h} Z_j . \end{aligned}$$

First, $W_0 = 0$. Second, we know that a sum of normally distributed random variables is a normally distributed random variable with mean given by the sum of means and variance given by the sum of the variances. Since all the increments $Z_j \sim N(0,1)$, then $\sqrt{h} Z_j \sim N(0,h)$, and the sum of n such increments is

$$W(t) = \sum_{j=0}^{n-1} \sqrt{h} Z_j \sim N(0, nh) = N(0,t) .$$

Thus we deduce that $W(t) \sim N(0,t)$. By taking random increments $\propto \sqrt{h}$, the distribution of W at any time is fixed—that is, independent of the number of discrete steps to get to that time. Thus considering the step size $h \to 0$ appears a reasonable limit.

Consider the alternatives to the $\sqrt{h}$ scaling. If we were to take random increments $\propto h^p$ for exponent $p > \frac{1}{2}$, then as the time step $h \to 0$ the resultant realizations would have infinitely large fluctuations. Conversely, if we were to take random increments $\propto h^p$ for exponent $p < \frac{1}{2}$, then as the time step $h \to 0$ all the resultant realizations would approach the boring zero function. Modify Algorithm 20.1 and see. Only random increments $\propto \sqrt{h}$ give a nontrivial, and very interesting, limit.

Similar arguments show that the change in W over any time interval of length t, say the change $W(t+s) - W(s)$, is also normally distributed, $N(0,t)$, and, *further*, is independent of any details of the process that occurred for times before time s. This property of independence is vitally important in various crucial places in the development of the theory of SDEs.

Other microtime increments are equivalent The Wiener process, Brownian motion, has a prime role in stochastic calculus because of the central limit theorem. If the increments in W follow some other distribution on the smallest micro times, then provided the variance of the micro-increments is finite the process always looks Brownian on the

[241]Recall that asserting that a random variable $Z \sim N(a,b)$ means that Z is normally distributed with mean a and variance b (standard deviation $\sqrt{b}$).

macroscale. This Brownian macroscale is assured because the sum of many random variables with finite variance tends to a normal distribution. For example, often we take recourse to approximating the increments as the binary choice of either a step up or a step down: choosing $Z_j = \pm 1$ each with probability $\frac{1}{2}$—as then the mean and variance of such Z_j are zero and one respectively—the cumulative sum $\sum_{j=0}^{n-1} \sqrt{h} Z_j$ has the appearance on the macroscale of an $N(0, nh)$ random variable, as required by the Brownian motion of a Wiener process.

Wiener proved that the following process exists and is unique in a stochastic sense.[242]

Definition 20.1 (Wiener process). *Define a* **Brownian motion** *or* **Wiener process** *or* **random walk**, *usually denoted* $W(t)$, *by the following properties:*

- $W(t)$ *is continuous;*
- $W(0) = 0$;
- *over any time step* t, *the change* $W(t+s) - W(s) \sim N(0, t)$ *for* $t, s \geqslant 0$; *and*
- $W(t+s) - W(s)$ *is independent of any details of the process for times earlier than* s.

An immediate consequence of the second and third properties in the definition, set $s = 0$, is that the Wiener process $W(t) \sim N(0, t)$. That is, the graph of $W(t)$ spreads wider and wider as time t increases, roughly like $\sqrt{t}$.

Continuous but not differentiable To show continuity, start by recalling that $W(t + s) - W(s) \sim N(0, t)$, so we now imagine t as small and write this as

$$W(t+s) - W(s) = \sqrt{t} Z_t \,,$$

where the random variables $Z_t \sim N(0, 1)$. Now although Z_t will vary with t, it comes from a normal distribution with mean 0 and variance 1, so as $t \to 0$ then almost surely the right-hand side $\sqrt{t} Z_t \to 0$. Thus, almost surely $W(t+s) \to W(s)$ as $t \to 0$, and hence W is continuous (almost surely).

Although continuous, as Figure 20.3 (left) illustrates, a Wiener process is too jagged to be differentiable. Recall that for a smooth function such as $f(t) = e^t$, as Figure 20.3 (right) shows, we generally see a linear variation near any point: $f(t+s) - f(s) \approx t f'(s)$. We are familiar with the change $f(t+s) - f(s)$ decreasing linearly with time step t, and upon this is based all the familiar rules of differential and integral calculus. In contrast, in the Wiener process, and generally for solutions of SDEs, Figure 20.3 (left) shows $W(t+s) - W(s)$ decreasing much slower, like $\sqrt{t}$. These changes in W are much steeper and vastly more jagged than smooth differentiable functions. Notionally the Wiener process has infinite slope and is thus nowhere differentiable. This feature generates lots of marvelous new effects that make the theory of stochastic calculus enormously intriguing.

Summary Brownian motion, the Wiener process $W(t)$, is *the* basic fluctuating stochastic process. The Wiener process is continuous but not differentiable; its independent random fluctuations, ΔW, scale with the square root of a time step, $\sqrt{\Delta t}$. Last, and notionally, a Wiener process is the integral of white noise.

[242] Probabilists will miss herein the comfort of a proper definition of a probability space and the associated concepts, such as σ-algebras, filtrations, and adapted processes. Such a missing rigorous background would enable greater insight into deeper issues and analysis of stochastic systems. However, this book restricts analysis to a level of operational probability in order to make the concepts, analysis, and results accessible to more people.

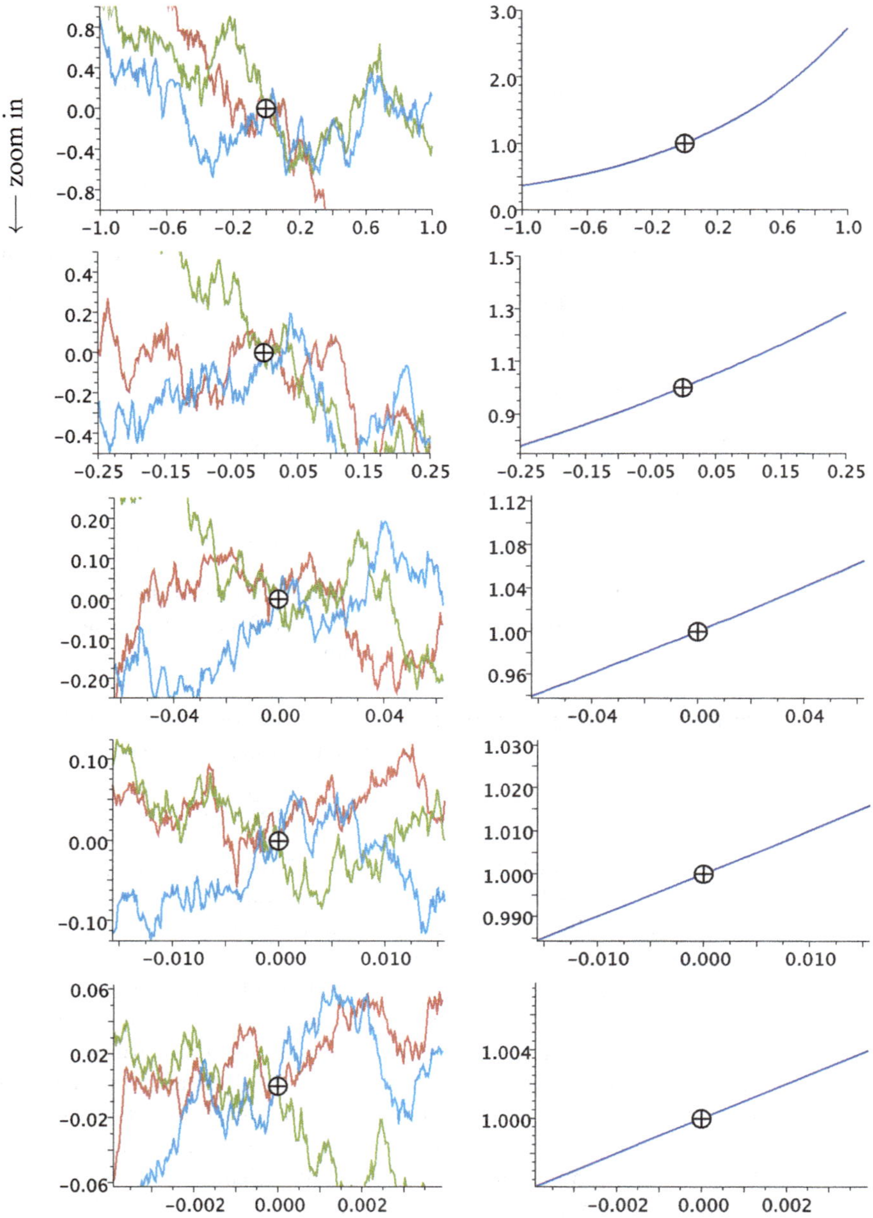

Figure 20.3. *Left, from top to bottom, zoom into three realizations of a Wiener process showing the never-ending level of detail and jaggedness; right, zoom into the relatively boring smooth exponential* e^t. *The horizontal axis is time* t.

20.1.2 ▪ Stochastic functions build on the Wiener process

Given the Wiener process, we may construct many sorts of stochastic functions by algebraic combination. Figures 20.4–20.7 plot realizations of four different stochastic functions built from the Wiener process.

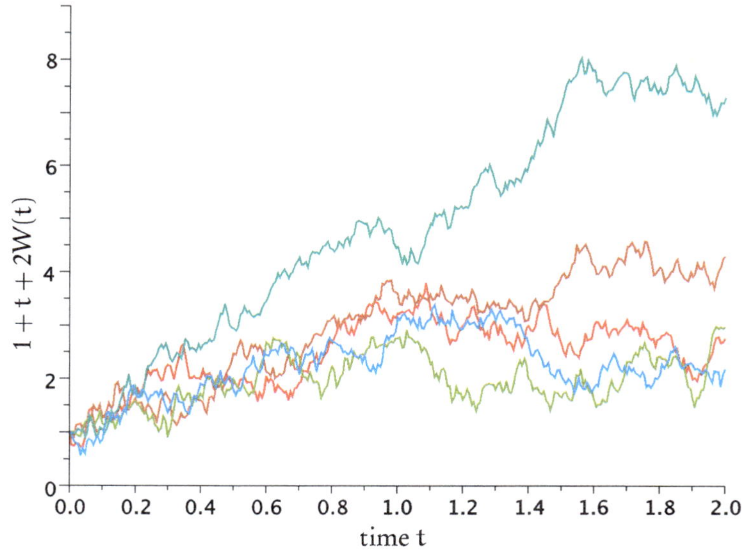

Figure 20.4. *Five realizations of* $f(t) = 1 + t + 2W(t)$.

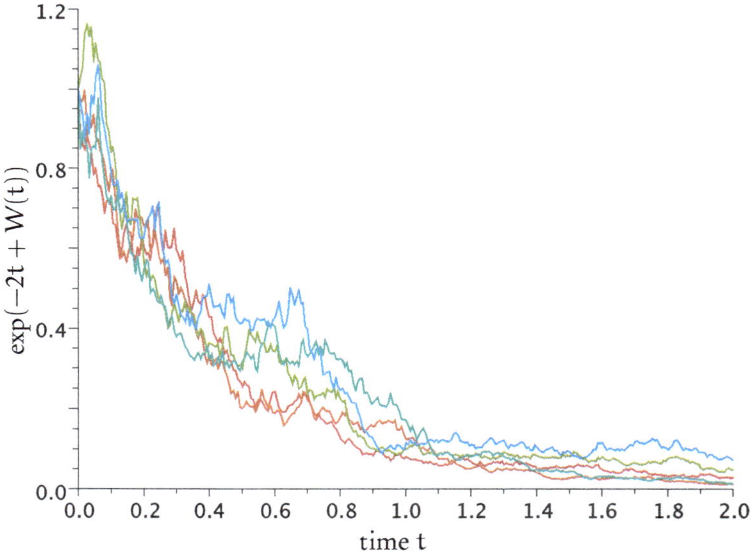

Figure 20.5. *Five realizations of* $f(t) = \exp(-2t + W(t))$.

- Figure 20.4 shows a scaled Wiener process simply added to a linear function.
- Figure 20.5 shows a stochastic decay: the $\exp(-2t)$ decay is affected by stochastic fluctuations; although the decay ultimately wins, the fluctuations cause large excursions.
- Figure 20.6 shows a stochastic growth: the stochastic fluctuations have a large effect on the $\exp(t)$ growth; although the growth ultimately wins, the fluctuations cause large excursions amplified by the exponential function.

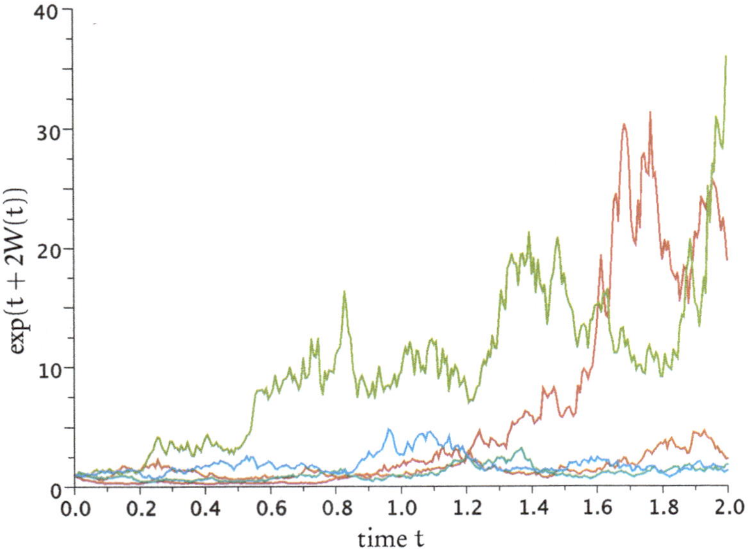

Figure 20.6. *Five realizations of* $f(t) = \exp(t + 2W(t))$.

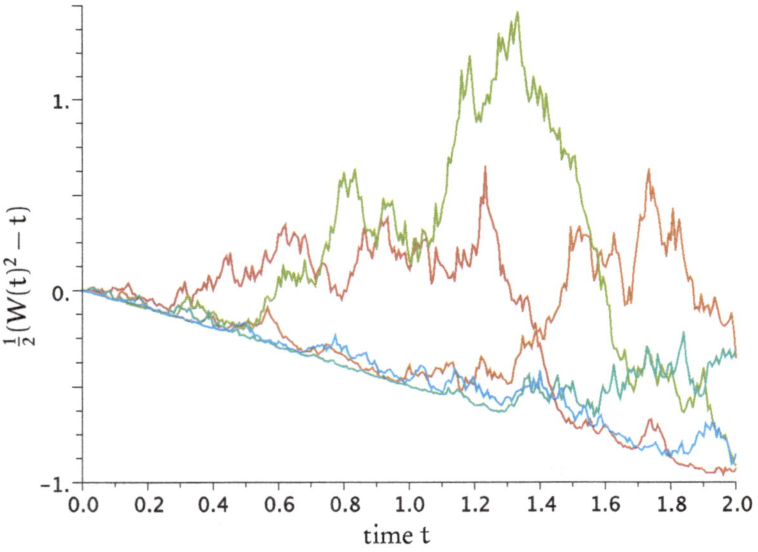

Figure 20.7. *Five realizations of* $f(t) = \frac{1}{2}(W(t)^2 - t)$.

- Figure 20.7 shows a function with mean zero: in the function $(W(t)^2 - t)/2$ the linear $-t$ decrease is balanced by the occasional, but persistently occurring, large excursions caused by the $W(t)^2$ term. In some (Itô) sense this function $(W(t)^2 - t)/2 = \int W\,dW$.

These four figures show just some of the combinations and interplay between noise and deterministic functions.

20.2 ▪ Herein choose the Stratonovich interpretation, not Itô

To deal with noise mathematically (and numerically) we need a valid calculus of noise. We need algebraic manipulation to work with noise. However, with stochastic noise we are working on an extreme edge of valid algebra and calculus. One false manipulation and our algebra falls into nonsense. We must be careful.

Some formal rules of thumb to avoid gross error in differential equations are the following.

- Despite a Wiener process $W(t)$ being nondifferentiable, we do allow expressions linear in their first derivative. Think of $\dot{W}$ as a random white noise (with a flat power spectrum)—as in section 20.1 a white noise integrates to the Wiener process.
- However, we are not allowed to have products or other nonlinear functions of white noise—expressions must be, at worst, linear in $\dot{W}$.
- Neither do we allow any further differentiation of a Wiener process: the white noise $\dot{W}$ is bad enough, but two or more derivatives are unbearable.

However, we allow integration of a white noise $\dot{W}$, as integration is a smoothing process—integration leads to the sort of stochastic functions plotted in Figures 20.4–20.7. Nonetheless, even integration has a nasty surprise.

Example 20.2. In terms of a Wiener process $W(t)$ and its derivative the white noise $\dot{W}$, some formally acceptable expressions are

$$\tfrac{1}{2}(W^2 - t), \qquad W\dot{W}, \qquad \exp(tW),$$
$$1 + W + \dot{W}\sin W, \qquad W = \int \dot{W}\,dt, \qquad \int_0^\infty e^{-s}\dot{W}(t-s)\,ds\,.$$

Conversely, some unbearable invalid expressions are

$$\dot{W}^2 - t, \qquad \exp(t\dot{W}), \qquad (t + \dot{W})(1 + \dot{W}),$$
$$\frac{d}{dt}\left[\dot{W}\sin W\right], \qquad 3 - 2W^2 + \ddot{W}, \qquad \int_0^\infty e^{-s}\ddot{W}(t-s)\,ds\,.$$

Now, what is the surprise with integrals of white noise? The issue is that although we accept such integrals, the answer to an integral depends upon how we define integration. Such sensitivity reflects our teetering on the edge of the abyss of mathematical nonsense. In stochastic calculus there are two common definitions of integration: the Itô and the Stratonovich. The *entire* stochastic calculus and our algebraic manipulations depend upon which definition we choose.

Let us look at the two definitions for the example of the integral $\int W\dot{W}\,dt$. Because of the "cancellation" of the differentials dt, we write this integral as $\int W\,dW$. Indeed, this rewriting is why we allow expressions *linear* in white noise $\dot{W}$, as such expressions are formally equivalent to strictly valid integrals; for example, we interpret the formal differential equation $\dot{x} = 1 + W + \dot{W}\sin W$ as the strictly valid integral equation $x = \int(1 + W)\,dt + \int \sin W\,dW$. But what such integrals actually mean depends upon the definition of the integral $\int \cdot\,dW$, as we now see for the example $\int W\,dW$.

Example 20.3 (basic Itô integral). We deduce $I(T) = \int_0^T W(t)dW$ for a Wiener process $W(t)$. The Riemann definitions of classic deterministic integrals use sums of rectangles; Itô integrals similarly use the height of the rectangles from the *left* function value. Divide the integration interval $[0, T]$ into n subintervals, of equal time step $h = T/n$. Thus let times $t_j = jh$ for $j = 0, \ldots, n$, and let $W_j = W(t_j)$ be the Wiener process at these times. Define the Itô approximation to the integral

$$I(T) = \int_0^T W(t)\, dW \quad \text{as} \quad I_n = \sum_{j=0}^{n-1} W_j \Delta W_j\,.$$

That is, the Itô integral is approximately the sum of the integrand evaluated at the beginning of a time subinterval times the change in W over the subinterval, $\Delta W_j = W_{j+1} - W_j$.

To evaluate the sum I_n observe that

$$\begin{aligned}\Delta(\tfrac{1}{2}W_j^2) &= \tfrac{1}{2}(W_{j+1}^2 - W_j^2) = \tfrac{1}{2}(W_{j+1} + W_j)(W_{j+1} - W_j) \\ &= \tfrac{1}{2}(2W_j + \Delta W_j)\Delta W_j = W_j \Delta W_j + \tfrac{1}{2}(\Delta W_j)^2.\end{aligned}$$

Thus the Itô sum becomes

$$I_n = \sum_{j=0}^{n-1} \left[\Delta(\tfrac{1}{2}W_j^2) - \tfrac{1}{2}(\Delta W_j)^2\right] = \tfrac{1}{2}W_n^2 - \tfrac{1}{2}W_0^2 - \tfrac{1}{2}\sum_{j=0}^{n-1}(\Delta W_j)^2.$$

Now $W_n = W(T)$ and $W_0 = W(0) = 0$, but what is the sum $y := \sum_{j=0}^{n-1}(\Delta W_j)^2$? We show that it is almost certainly just the time T. First, the expectation

$$E(y) = \sum_{j=0}^{n-1} E\left(\Delta W_j^2\right) = \sum_{j=0}^{n-1} \mathrm{Var}(\Delta W_j) = \sum_{j=0}^{n-1} h = nh = T\,,$$

as $\Delta W_j \sim N(0, h)$ and the ΔW_j are independent of each other. Second, the variance of y is almost surely negligible:

$$\begin{aligned}\mathrm{Var}(y) &= \sum_{j=0}^{n-1} \mathrm{Var}\left(\Delta W_j^2\right) = \sum_{j=0}^{n-1} E\left[(\Delta W_j^2 - h)^2\right] \\ &= \sum_{j=0}^{n-1} \left\{E\left[\Delta W_j^4\right] - 2h\,E\left[\Delta W_j^2\right] + h^2\right\} \\ &= \sum_{j=0}^{n-1} \left\{3h^2 - 2h^2 + h^2\right\} \\ &= \sum_{j=0}^{n-1} 2h^2 = 2nh^2 = 2Th \to 0 \quad \text{as} \quad h \to 0\,.\end{aligned}$$

When the number n of subintervals becomes infinite, $h \to 0$, then the variance of y becomes zero. Thus almost surely $y \to T$. Hence, the Itô integral

$$I(T) := \lim_{n\to\infty} I_n = \tfrac{1}{2}W(T)^2 - \tfrac{1}{2}T\,.$$

The surprise is that if the normal rules of calculus were to be followed we would write $\int W\, dW = \frac{1}{2}W^2$, but here the Itô integral is instead $\int W\, dW = \frac{1}{2}W^2 - \frac{1}{2}t$. The normal rules of calculus are invalid!

The most far reaching outcome of this example is that in the Itô calculus a term such as ΔW^2 has a macroscopic deterministic effect: when accumulated over many microscopic time steps, ΔW^2 is effectively Δt. Written in terms of infinitesimals, somehow dW^2 is effectively dt (Higham, 2001; Roberts, 2009a, e.g.). In contrast, in the usual calculus we consistently neglect such squares of microscale increments. Because of the effective "$dW^2 = dt$," Itô stochastic calculus has extra complications. Nonetheless, with care one uses ordinary calculus for many purposes by employing the alternative Stratonovich interpretation.

Example 20.4 (basic Stratonovich integral). Conversely, define the Stratonovich approximation to an integral using the trapezoidal rule; that is, approximate the integrand by the average of the beginning value and the ending value, that is, $\frac{1}{2}(W_{j+1} + W_j)$. Hence, here define the integral $J(T) = \int_0^T W(t)dW$ via its approximations[243]

$$\begin{aligned} J_n &:= \sum_{j=0}^{n-1} \tfrac{1}{2}(W_{j+1} + W_j)\Delta W_j \\ &= \sum_{j=0}^{n-1} \tfrac{1}{2}(W_{j+1} + W_j)(W_{j+1} - W_j) \\ &= \sum_{j=0}^{n-1} \tfrac{1}{2}(W_{j+1}^2 - W_j^2) \\ &= \sum_{j=0}^{n-1} \Delta(\tfrac{1}{2}W_j^2) = \tfrac{1}{2}W_n^2 - \tfrac{1}{2}W_0^2 = \tfrac{1}{2}W(T)^2. \end{aligned}$$

That is, this Stratonovich integral is the same as given by the normal rules of calculus.

Example 20.5 (basic differential equation). Consider the Itô differential equation $dX = X\,dW$ and the symbolically "equivalent" Stratonovich differential equation $\dot{X} = X\dot{W}$. Argue that the solutions are different, and hence these are not equivalent differential equations after all.

Solution:

- Let's first solve the Stratonovich differential equation $\dot{X} = X\dot{W}$ using classic calculus. This differential equation is separable, so write as $\frac{dX}{X} = \dot{W}\,dt$. Then integrate to $\log X = \int \dot{W}\,dt = W(t) + c$, where the integration constant is determined from the initial condition. Suppose the initial condition is $X(0) = X_0$; as by convention $W(0) = 0$, then the constant $c = \log X_0$. That is, $\log X = W(t) + \log X_0$, and taking the exponential of both sides gives the solution $X(t) = X_0 \exp W(t)$. What else could be the solution?

[243]To distinguish the two different sorts of stochastic integration, Itô and Stratonovich, many would write this integral as $\int_0^T W(t) \circ dW$, where the circle in $\circ\, dW$ denotes that the Stratonovich interpretation is being used. Because we generally use the Stratonovich interpretation, and in order to connect most closely with most engineers and physicists, I generally write such Stratonovich integrals as $\int W\dot{W}\,dt$. However, as here, sometimes the context is clear.

- Let's contrast this solution with the solution of the Itô differential equation $dX = X\,dW$. The Itô solution is to be the limit as time steps vanish in the discrete equations $\Delta X_j = X_j\,\Delta W_j$; that is, $\frac{\Delta X_j}{X_j} = \Delta W_j$. Now consider

$$\begin{aligned}\Delta \log X_j &= \log X_{j+1} - \log X_j = \log(X_{j+1}/X_j)\\ &= \log(1+\Delta X_j/X_j) = \log(1+\Delta W_j)\end{aligned}$$

from the rearrangement of the discrete equation. Now invoke the Taylor expansion of $\log(1+x)$ to deduce

$$\Delta \log X_j = \Delta W_j - \tfrac{1}{2}(\Delta W_j)^2 + \cdots .$$

Suppose we take n steps to reach time t for large n. Then

$$\begin{aligned}\log X_n - \log X_0 &= \sum_{j=0}^{n-1} \Delta \log X_j\\ &= \sum_{j=0}^{n-1} \Delta W_j - \tfrac{1}{2}\sum_{j=0}^{n-1}(\Delta W_j)^2 + \cdots\\ &= W_n - W_0 - \tfrac{1}{2}y + \cdots ,\end{aligned}$$

where $y = \sum_{j=0}^{n-1}(\Delta W_j)^2$. As shown in Example 20.3, as the number of time steps $n \to \infty$, almost surely $y \to t$, and the terms implicitly in the "$\cdots$" tend to zero. Since $X_n = X(t)$, $W_n = W(t)$, and $W_0 = 0$, we thus deduce that $\log X(t) - \log X_0 = W(t) - \tfrac{1}{2}t$. Taking the exponential of both sides gives the Itô solution $X(t) = X_0 \exp\left[-\tfrac{1}{2}t + W(t)\right]$.

These two different solutions for what appears at first sight to be the same differential equation prove that we must take care in stochastic calculus. We must be clear whether to invoke calculus in the Itô sense or the classic Stratonovich sense.

These two examples, integrated two different ways, surely indicate that we should adopt the Stratonovich integral, as it appears compatible with the normal calculus. Indeed for the bulk of Part VII we use the normal calculus for the formal analysis and modeling of stochastic systems. Hence, *in our conventional algebraic manipulations to model stochastic systems we implicitly use the Stratonovich interpretation of stochastic integration.*

However, be aware of three important reasons why many others insist on the Itô interpretation of stochastic integration. These reasons are why the rest of this chapter, but not other chapters in Part VII, mostly invoke the Itô interpretation.

- The theoretical support for stochastic integration is much stronger for the Itô interpretation.

- As section 20.3 discusses, simple numerical methods to solve SDEs converge to the Itô solution, not the Stratonovich (and nonsimple classic numerical methods often do not apply at all to stochastic systems).

  Indeed, Penland (2003) commented that "uncorrected higher-order Runge–Kutta schemes do not converge to either Itô or Stratonovich calculus."

- When continuous time is a convenient approximation to a physical process that actually occurs in discrete time steps, the Itô interpretation is usually correct, and correspondingly the Stratonovich incorrect, by causality.

For example, the Stratonovich interpretation of $\dot{I} = W\dot{W}$, via the sum of terms $\frac{1}{2}(W_{j+1} + W_j)\Delta W_j$, requires knowledge of the future value $W_{j+1} = W(t_{j+1})$ to step from time t_j and so may violate causality in the application.

Conversely, if the noise process W comes from some deterministic chaotic dynamics, then at any step the future value $W_{j+1} = W(t_{j+1})$ is deterministically predictable from the state W_j at time t_j, and so a Stratonovich interpretation would *not* violate causality.

20.2.1 ▪ Define and explore the Itô integral

In formal algebraic manipulation we allow multiplication by a factor of white noise $\dot{W}$ because in rigorous mathematics such multiplication corresponds to a well-defined integral, the Itô integral. Then section 20.3.2 shows how to transform stochastic differentials to and fro between the rigorous Itô form and the normal calculus of the Stratonovich form that most of this part employs.

This section *outlines* the definition of an Itô integral $\int_a^b f\,dW$ and its properties. Oksendal (1998, §3.1) and Kloeden and Platen (1992, §3.1) gave full details of a rigorous development of the integral.

Definition 20.2. *Use $\mathcal{V}$ to denote the class of stochastic functions that we may integrate (over the interval $[a, b]$). The class $\mathcal{V}$ is defined to be composed of functions $f(t,\omega)$ such that*

- *the expected square integral* $E\left[\int_a^b f(t,\omega)^2 dt\right] < \infty$*;*
- $f(t,\omega)$ *depends only upon the history of the Wiener process,* $W(s,\omega)$*, for* $s \leqslant t$ *— for example,* $f = W(t,\omega)^3$, $f = \sin W(t,\omega)$*, or* $f = \int_0^t W(s,\omega)ds$*, but not* $f = W(t+2,\omega)$.

The Itô integral for step functions Our first task is to define the Itô integral for step functions such as those drawn in Figure 20.8 and investigate the key properties of the Itô integral.

Definition 20.3. *Let the class of piecewise constant functions, called* **step functions***, be denoted by $\mathcal{S} \subset \mathcal{V}$. That is, for each step function $\phi(t,\omega) \in \mathcal{S}$, there exists a finite partition* $a = t_0 < t_1 < t_2 < \cdots < t_n = b$ *such that* $\phi(t,\omega) = \phi(t_j,\omega)$ *for* $t_j \leqslant t < t_{j+1}$.

Figure 20.8 plots four different members of $\mathcal{S}$ on the interval $[0, 1]$. Given any such partition we use ϕ_j or $\phi_j(\omega)$ to denote $\phi(t_j,\omega)$, just as we use W_j to denote $W(t_j,\omega)$.

Definition 20.4. *For any step function $\phi(t,\omega)$ define the* **Itô integral**

$$I(\omega) = \int_a^b \phi(t,\omega)dW(t,\omega) = \int_a^b \phi\,dW := \sum_{j=0}^{n-1} \phi_j \Delta W_j\,. \tag{20.3}$$

Three desirable properties immediately follow; the first two are the familiar linearity and union (proofs left as exercises), and the third is the nonanticipation property:

- *linearity*:

$$\int_a^b (\alpha\phi + \beta\psi)dW = \alpha\int_a^b \phi\,dW + \beta\int_a^b \psi\,dW; \tag{20.4}$$

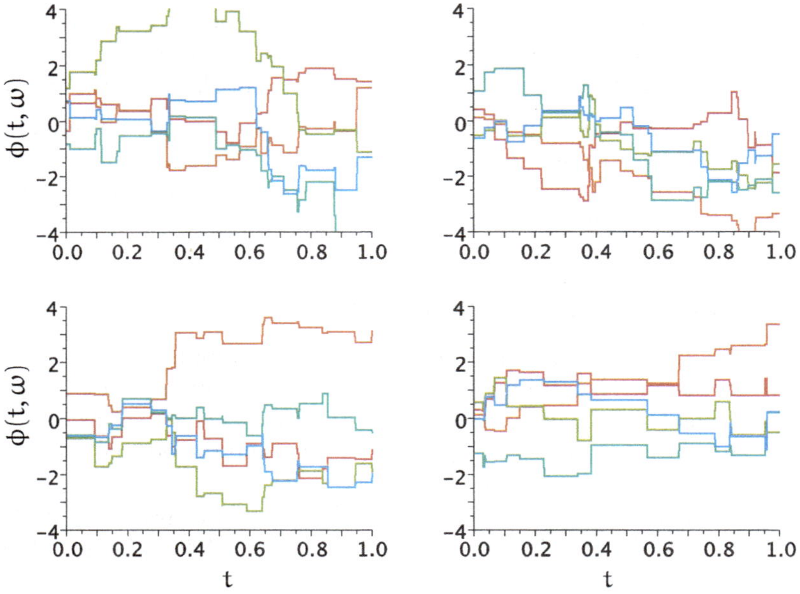

Figure 20.8. *Five realizations (different colors) for each of four different stochastic step functions* $\phi(t,\omega) \in \mathcal{S}$*: different step functions* $\phi(t,\omega)$ *have different partitions; different realizations of the one step function* $\phi(t,\omega)$ *have the same partition but different values distributed according to some probabilities.*

- *union*:
$$\int_a^b \phi\, dW + \int_b^c \phi\, dW = \int_a^c \phi\, dW; \tag{20.5}$$
- *history dependence*: the integral $\int_a^b \phi\, dW$ only depends upon the history of the Wiener process up to time b.

Two most useful properties of the Itô integral arise. Remarkably, irrespective of the integrand in an Itô integral, we always know the mean value of an Itô integral, namely zero! But even more remarkably, we also readily determine the variance, the spread, of an Itô integral. These two most important aspects of the stochastic integral $I(\omega) = \int_a^b \phi\, dW$, its mean and variance, can be determined *without* ever actually computing the Itô integral! The next two theorems establish these properties for step function integrands.

Theorem 20.5 (martingale property). *For the Itô integral* (20.3) *of a step function* ϕ*, its mean, average, or expected value is* always *zero:*

$$\mathrm{E}\left[\int_a^b \phi(t,\omega)\, dW(t,\omega)\right] = 0. \tag{20.6}$$

Proof. Recall the linearity of the expectation and that ϕ_j is independent of ΔW_j, as $\phi_j = \phi(t_j,\omega)$ depends only upon the earlier history of the Wiener process, whereas the increment ΔW_j is independent of the earlier history and on average is zero ($\mathrm{E}\left[\Delta W_j\right] = 0$).

Thus for the relevant partition, such as the one in Figure 20.8,

$$\begin{aligned}
\mathrm{E}\left[\int_a^b \phi\, dW\right] &= \mathrm{E}\left[\sum_{j=0}^{n-1} \phi_j \Delta W_j\right] && \text{by definition (20.4)}\\
&= \sum_{j=0}^{n-1} \mathrm{E}\left[\phi_j \Delta W_j\right] && \text{by linearity}\\
&= \sum_{j=0}^{n-1} \mathrm{E}\left[\phi_j\right] \underbrace{\mathrm{E}\left[\Delta W_j\right]}_{=0} && \text{by independence}\\
&= 0\,.
\end{aligned}$$

□

Theorem 20.6 (Itô isometry). *Consider the Itô integral* (20.3) *of a step function* $\phi(t,\omega)$. *Its variance is the integral of the expectation of the squared integrand:*[244]

$$\operatorname{Var}\left[\int_a^b \phi(t,\omega)\, dW(t,\omega)\right] = \mathrm{E}\left[\left(\int_a^b \phi\, dW\right)^2\right] = \int_a^b \mathrm{E}\left[\phi(t,\omega)^2\right] dt\,. \tag{20.7}$$

Proof. This proof invokes properties of the expectation, and that the zero mean increment ΔW_j is independent of any earlier stochastic quantity. For the relevant partition, as in Figure 20.8,

$$\begin{aligned}
\mathrm{E}\left[\left(\int_a^b \phi\, dW\right)^2\right] &= \mathrm{E}\left[\left(\sum_{j=0}^{n-1} \phi_j \Delta W_j\right)^2\right]\\
&= \mathrm{E}\left[\left(\sum_{j=0}^{n-1} \phi_j \Delta W_j\right)\left(\sum_{k=0}^{n-1} \phi_k \Delta W_k\right)\right]\\
&= \mathrm{E}\left[\sum_{j,k=0}^{n-1} \phi_j \Delta W_j \phi_k \Delta W_k\right]\\
&= \sum_{j,k=0}^{n-1} \mathrm{E}\left[\phi_j \Delta W_j \phi_k \Delta W_k\right].
\end{aligned}$$

Three cases arise within this double sum: $j < k$, $j > k$, and $j = k$.

- If $j < k$, then ΔW_k is independent of ϕ_j, ΔW_j, and ϕ_k—as the first depends only upon times $t > t_k$, whereas the last three depend only upon times $t \leqslant t_k$—so the expectation may be split,

$$\mathrm{E}\left[\phi_j \Delta W_j \phi_k \Delta W_k\right] = \mathrm{E}\left[\phi_j \Delta W_j \phi_k\right] \mathrm{E}\left[\Delta W_k\right] = \mathrm{E}\left[\phi_j \Delta W_j \phi_k\right] \times 0 = 0\,,$$

  and so all such terms vanish.

[244] As if the square can be taken inside the integration and $dW^2 = dt$.

- If $j > k$, then ΔW_j is independent of ϕ_k, ΔW_k, and ϕ_j, so the expectation may be split,

$$E\left[\phi_j \Delta W_j \phi_k \Delta W_k\right] = E\left[\phi_j \Delta W_k \phi_k\right] E\left[\Delta W_j\right] = E\left[\phi_j \Delta W_k \phi_k\right] \times 0 = 0,$$

and so all such terms also vanish.

- Last, if $j = k$, then

$$\begin{aligned}
E\left[\left(\int_a^b \phi\, dW\right)^2\right] &= \sum_{j=0}^{n-1} E\left[\phi_j^2 \Delta W_j{}^2\right] \\
&= \sum_{j=0}^{n-1} E\left[\phi_j^2\right] E\left[\Delta W_j{}^2\right] && \text{by independence} \\
&= \sum_{j=0}^{n-1} E\left[\phi_j^2\right] \Delta t_j && \text{by variance of } \Delta W_j \\
&= \int_a^b E\left[\phi^2\right] dt
\end{aligned}$$

by the definition of ordinary integration for piecewise constant integrands such as $E\left[\phi(t,\omega)^2\right]$. □

Example 20.6 (verify Itô isometry). Recall from Example 20.3 that the Itô integral $I(\omega) = \int_0^t W(s,\omega)dW(s,\omega) = \frac{1}{2}[W(t,\omega)^2 - t]$. Verify the Itô isometry for this integral.

From the algebraic expression for the integral,

$$\begin{aligned}
\text{Var}[I(\omega)] &= \text{Var}\left[\tfrac{1}{2}(W(t,\omega)^2 - t)\right] \\
&= E\left[\tfrac{1}{4}(W^2 - t)^2\right] \quad \text{as } E[I(\omega)] = 0 \\
&= \tfrac{1}{4} E\left[W^4 - 2W^2 t + t^2\right] \\
&= \tfrac{1}{4}\left(3t^2 - 2 \cdot t \cdot t + t^2\right) \\
&= \tfrac{1}{2} t^2,
\end{aligned}$$

whereas invoking the Itô isometry (20.7) gives

$$\begin{aligned}
\text{Var}[I(\omega)] &= E\left[\left(\int_0^t W(s,\omega)dW(s,\omega)\right)^2\right] \\
&= \int_0^t E\left[W(s,\omega)^2\right] ds \\
&= \int_0^t s\, ds \\
&= \tfrac{1}{2} t^2.
\end{aligned}$$

This equality between the two verifies the Itô isometry for this Itô integral.

Avoid complicated technicalities Comprehensive treatises on stochastic calculus, such as those by Oksendal (1998) and Kloeden and Platen (1992), then lead us through technical details to establish the above properties for general Itô integrals. One first establishes that step functions approximate sufficiently well any stochastic function in the class $\mathcal{V}$. Second, one defines the Itô integral to be the limit of the Itô integrals of sufficiently good approximating step functions. Last, one proves that the crucial properties of Itô integrals of step functions carry over through the limit to general Itô integrals. These details are not sufficiently relevant to us to discuss further.

The five main properties for general Itô integrals that thus follow from integrals of step functions are as follows:

- *linearity*:
$$\int_a^b (\alpha f + \beta g)\, dW = \alpha \int_a^b f\, dW + \beta \int_a^b g\, dW; \tag{20.8}$$
- *union*:
$$\int_a^b f\, dW + \int_b^c f\, dW = \int_a^c f\, dW; \tag{20.9}$$
- *history dependence*: the Itô integral $\int_a^b f(t,\omega)\, dW$ only depends upon the history of the Wiener process up to time b;
- *martingale property*: the mean, average, or expected value is *always* zero:
$$E\left[\int_a^b f(t,\omega)\, dW\right] = 0; \text{ and} \tag{20.10}$$
- *Itô isometry*: the variance is the ordinary integral of the expectation of the squared integrand:
$$\operatorname{Var}\left[\int_a^b f(t,\omega)\, dW\right] = E\left[\left(\int_a^b f(t,\omega)\, dW\right)^2\right] = \int_a^b E\left[f(t,\omega)^2\right] dt\,. \tag{20.11}$$

20.2.2 ▪ Basic Itô's formula

The fundamental rule of stochastic differentiation is Itô's formula, with the symbolic identity $dW^2 = dt$ at its heart. We present a basic version of Itô's formula and then the full version in section 20.2.3.[245]

Let $f(t,w)$ be a smooth function of two arguments—smooth means that it is differentiable at least several times and Taylor's theorem applies. Figures 20.9 and 20.10 show two example surfaces of two such smooth functions. Then letting $W(t)$ denote a Wiener process, the process $X(t) = f(t, W(t))$ is an Itô process—see the black curves wiggling across the surfaces in Figures 20.9 and 20.10. Note that $f(t,w)$ itself is smooth; the stochastic part of $X(t)$ comes only via the use of the Wiener process in the evaluation of $f(t, W(t))$. Now consider the change in X that occurs over some small change in time, Δt. The change ΔX occurs through the direct dependence upon t and indirectly

[245]Chapters 5 and 6 of the book by Stampfli and Goodman (2001) provide appropriate reading to supplement this section.

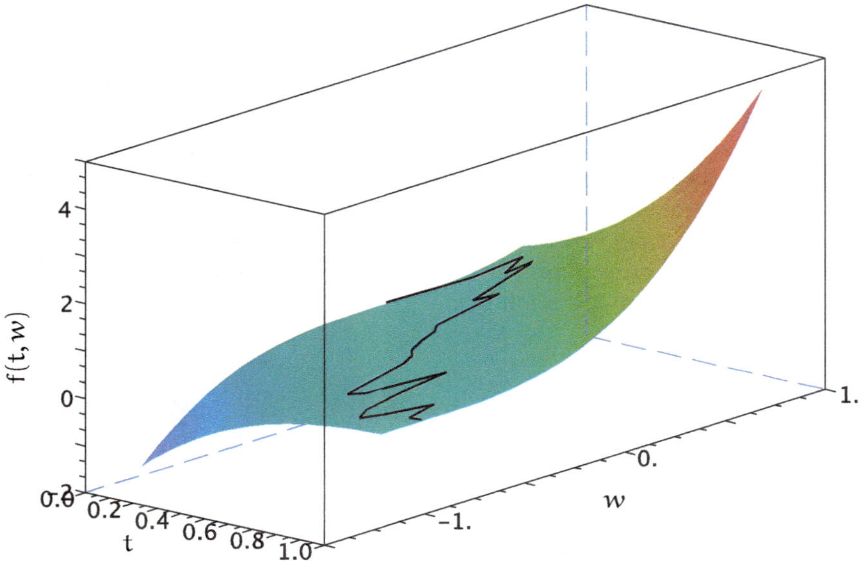

Figure 20.9. *The smooth surface* $f = (t + w)^3$ *with one realization of a Wiener process* $w = W(t)$ *evolving on the surface: the Itô process* $X = f(t, W(t))$ *is the height of the curve as a function of time and changes due to direct evolution in time and through evolution of* w.

through its dependence upon W. Writing $X(t+\Delta t) = X+\Delta X$ and $W(t+\Delta t) = W+\Delta W$, expand $f(t + \Delta t, W + \Delta W)$, using a multivariable Taylor series,

$$\begin{aligned} X + \Delta X &= f(t + \Delta t, W + \Delta W) \\ &= f(t, W) + \frac{\partial f}{\partial t}\Delta t + \frac{\partial f}{\partial w}\Delta W \\ &\quad + \tfrac{1}{2}\frac{\partial^2 f}{\partial t^2}\Delta t^2 + \frac{\partial^2 f}{\partial t \partial w}\Delta t\, \Delta W + \tfrac{1}{2}\frac{\partial^2 f}{\partial w^2}\Delta W^2 + \cdots \\ &\approx X + \frac{\partial f}{\partial t}\Delta t + \frac{\partial f}{\partial w}\Delta W + \tfrac{1}{2}\frac{\partial^2 f}{\partial w^2}\Delta t, \end{aligned}$$

as by the earlier rules for differentials $\Delta t^2 = \Delta t \Delta W = \cdots \approx 0$ and $\Delta W^2 \approx \Delta t$. In the limit as $\Delta t \to 0$ the differences become differentials and thus

$$dX = \left(\frac{\partial f}{\partial t} + \tfrac{1}{2}\frac{\partial^2 f}{\partial w^2}\right) dt + \frac{\partial f}{\partial w}\, dW, \tag{20.12}$$

where the partial derivatives are evaluated at (t, W). This is the basic version of *Itô's formula* and gives the differential of an Itô process X that depends directly and smoothly upon a Wiener process.

Example 20.7. What is the differential of $X(t) = (t + W(t))^3$? Hence deduce an Itô SDE that $X(t)$ satisfies.

Solution: Here $f(t, w) = (t+w)^3$ (see Figure 20.9), so $f_t = f_w = 3(t+w)^2$ and $f_{ww} = 6(t + w)$; thus Itô's formula (20.12) gives

$$dX = [3(t + W)^2 + \tfrac{1}{2}6(t + W)]dt + 3(t + W)^2 dW.$$

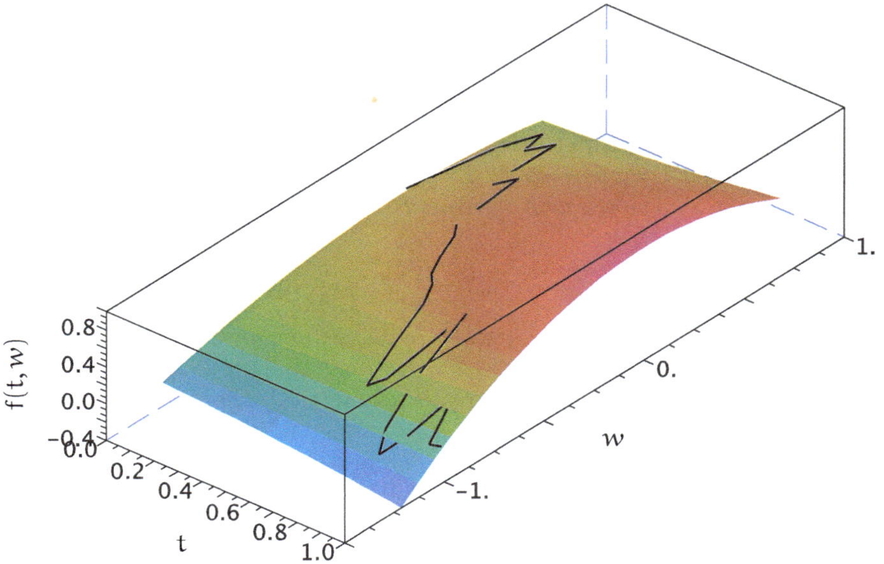

Figure 20.10. *The smooth surface* $f = 0.3\,e^t(1 - w^2)$ *with one realization of a Wiener process* $w = W(t)$ *evolving on the surface: the Itô process* $X = f(t, W(t))$ *is the height of the curve as a function of time and changes due to direct evolution in time and through evolution of* w.

Recognizing that $t + W = X^{1/3}$, we rewrite this as the SDE

$$dX = [3X^{2/3} + 3X^{1/3}]dt + 3X^{2/3}dW.$$

Either SDE is valid, but this last may be preferred as it has no explicit time dependence on the right-hand side.

Example 20.8 (revisit a basic SDE). Determine the differential of $X(t) = c\exp[at + bW(t)]$ and hence find the parameters a and b such that the process $X(t)$ satisfies the SDE $dX = X\,dW$ (compare with Example 20.5).

Solution: Here the function $f(t, w) = ce^{at+bw}$, whence $f_t = ace^{at+bw}$, $f_w = bce^{at+bw}$, and $f_{ww} = b^2ce^{at+bw}$. Thus Itô's formula (20.12) asserts that

$$\begin{aligned} dX &= \left(ace^{at+bW} + \tfrac{1}{2}b^2ce^{at+bW}\right)dt + bce^{at+bW}dW \\ &= ce^{at+bW}[(a + \tfrac{1}{2}b^2)dt + b\,dW]. \end{aligned}$$

Substituting $X = ce^{at+bW}$ on the right-hand side gives $dX = (a + \frac{1}{2}b^2)X\,dt + bX\,dW$. This SDE is the same as the specified SDE provided both $a + \frac{1}{2}b^2 = 0$ and $b = 1$. Thus $a = -\frac{1}{2}$. Hence, as in Example 20.5, the solution of the Itô SDE $dX = X\,dW$ is $X(t) = c\exp[-\frac{1}{2}t + W(t)]$.

20.2.3 ▪ The stochastic chain rule is Itô's lemma

The basic form (20.12) of Itô's formula rests upon the Taylor series of $f(t, w)$ and the rule that the only quadratic differential to retain is $dW^2 = dt$. The general form of Itô's formula, often referred to modestly as Itô's lemma, is more general in that it addresses the differential of a function of a stochastic process, rather than just a function of a Wiener process.

Lemma 20.7 (Itô's lemma). *Let* X *be an Itô process with drift* $\mu(t, X)$ *and volatility* $\sigma(t, X)$, *that is,* $dX = \mu\, dt + \sigma\, dW$. *For any smooth function* $f(t, x)$, *the function* $Y(t) := f(t, X(t))$ *is also a Itô process with differential*

$$dY = \frac{\partial f}{\partial t}dt + \frac{\partial f}{\partial x}dX + \tfrac{1}{2}\frac{\partial^2 f}{\partial x^2}dX^2, \tag{20.13}$$

$$\text{where in } dX^2 \text{ retain only } dW^2 \mapsto dt,$$

and the partial derivatives are evaluated at (t, X). *Equivalently, expanding expression* (20.13) *gives*

$$dY = \left(\frac{\partial f}{\partial t} + \mu\frac{\partial f}{\partial x} + \tfrac{1}{2}\sigma^2\frac{\partial^2 f}{\partial x^2}\right)dt + \sigma\frac{\partial f}{\partial x}dW. \tag{20.14}$$

Although formula (20.14) is more explicit, the formula (20.13) appears a more memorable form of Itô's formula.

Example 20.9. Example 20.7 shows that if $X(t) = (t + W(t))^3$ then $dX = (3X^{2/3} + 3X^{1/3})dt + 3X^{2/3}dW$. Hence deduce the simple $dY = dt + dW$ for $Y = X^{1/3}$.

Solution: Use Itô's formula with $f(t, x) = x^{1/3}$ for which $f_t = 0$, $f_x = \frac{1}{3}x^{-2/3}$, and $f_{xx} = -\frac{2}{9}x^{-5/3}$:

$$\begin{aligned}
dY &= f_x\, dX + \tfrac{1}{2}f_{xx}\, dX^2 \\
&= \tfrac{1}{3}X^{-2/3}\left[(3X^{2/3} + 3X^{1/3})dt + 3X^{2/3}dW\right] \\
&\quad + \tfrac{1}{2}(-\tfrac{2}{9})X^{-5/3}\left[(3X^{2/3} + 3X^{1/3})dt + 3X^{2/3}dW\right]^2 \\
&= (1 + X^{-1/3})dt + dW - \tfrac{1}{9}X^{-5/3}9X^{4/3}dW^2 \\
&= (1 + X^{-1/3})dt + dW - X^{-1/3}dt \\
&= dt + dW.
\end{aligned}$$

Example 20.10 (product rule). In classic calculus the product rule for differentiation is $d(fg) = f\, dg + g\, df$. The product rule for Itô calculus has an extra term. Let $X(t)$ and $Y(t)$ be stochastic processes with differentials

$$dX = \mu\, dt + \sigma\, dW \quad \text{and} \quad dY = \nu\, dt + \rho\, dW.$$

Using our quick symbolic rules we may deduce that

$$\begin{aligned}
d(XY) &= (X + dX)(Y + dY) - XY \\
&= X\, dY + Y\, dX + dX\, dY \\
&= X\, dY + Y\, dX + \sigma\rho\, dt
\end{aligned}$$

by retaining only the $dW^2 = dt$ term in the product $dX\,dY$. The product rule $d(XY) = X\,dY + Y\,dX + \sigma\rho\,dt$ is correct, but now deduce it using Itô's formula.

Solution: The difficulty is that the product XY is a function of two Itô processes, whereas Itô's formula (20.14) only applies to a function of one Itô process. To circumvent this difficulty, write $XY = \frac{1}{2}[Z^2 - X^2 - Y^2]$, where $Z(t) := X + Y$ is a new stochastic process. Because Z is simply the sum of X and Y, it has differential $dZ = (\mu+\nu)dt + (\sigma+\rho)dW$. Then use Itô's formula separately on each of the components on the right-hand side of $d(XY) = \frac{1}{2}[d(Z^2) - d(X^2) - d(Y^2)]$:

$$\begin{aligned}
d(X^2) &= 0\,dt + 2X\,dX + \tfrac{1}{2}2\,dX^2 \\
&= (2X\mu + \sigma^2)dt + 2X\sigma\,dW; \\
\text{similarly}\quad d(Y^2) &= (2Y\nu + \rho^2)dt + 2Y\rho\,dW \\
\text{and}\quad d(Z^2) &= \left[2Z(\mu+\nu) + (\sigma+\rho)^2\right]dt + 2Z(\sigma+\rho)dW.
\end{aligned}$$

Substitute these, with $Z = X + Y$, into $d(XY)$:

$$\begin{aligned}
d(XY) &= \tfrac{1}{2}\left[d(Z^2) - d(X^2) - d(Y^2)\right] \\
&= \tfrac{1}{2}\left[(2X\mu + 2Y\mu + 2X\nu + 2Y\nu + \sigma^2 + 2\sigma\rho + \rho^2)dt\right. \\
&\qquad + (2X\sigma + 2Y\sigma + 2X\rho + 2Y\rho)dW \\
&\qquad - (2X\mu + \sigma^2)dt - 2X\sigma\,dW - (2Y\nu + \rho^2)dt \\
&\qquad \left.- 2Y\rho\,dW\right] \\
&= X(\nu\,dt + \rho\,dW) + Y(\mu\,dt + \sigma\,dW) + \sigma\rho\,dt \\
&= X\,dY + Y\,dX + \sigma\rho\,dt,
\end{aligned}$$

in agreement with the quick symbolic rules.

The Itô formula (20.13)–(20.14) enables us to differentiate stochastic Itô processes and hence solve some SDEs. However, it is better to establish the integral form (20.15) of the Itô formula via the Itô integral.

Theorem 20.8. *Let* $X(t)$ *be an Itô process with drift* $\mu(t,X)$ *and volatility* $\sigma(t,X)$*; that is,* $dX = \mu\,dt + \sigma\,dW$. *For any smooth function* $f(t,x)$ *satisfying certain integral bounds, the process* $f(t,X(t))$ *is also a Itô process such that*

$$[f(t,X(t))]_a^b = \int_a^b \left(\frac{\partial f}{\partial t} + \mu\frac{\partial f}{\partial x} + \tfrac{1}{2}\sigma^2\frac{\partial^2 f}{\partial x^2}\right)dt + \int_a^b \sigma\frac{\partial f}{\partial x}\,dW \tag{20.15}$$

for all intervals $[a,b]$.

Proof. For simplicity here we consider only the case where the drift μ and volatility σ are step functions (piecewise constant) with a common partition $a = t_0 < t_1 < \cdots < t_n = b$. Let $h = \max_j \Delta t_j$ so that the limit $h \to 0$ means that the partition has vanishingly small steps.

For brevity let an overdot denote $\partial/\partial t$ and a prime denote $\partial/\partial x$ so that, for example, $\dot{f}_j = \partial f/\partial t|_{(t_j,X_j)}$. Then the left-hand side of the integral Itô formula (20.15) transforms as follows:

$$\begin{aligned}
[f]_a^b &= \sum_{j=0}^{n-1} \Delta f_j = \sum_{j=0}^{n-1} \left[f(t_{j+1}, X_{j+1}) - f(t_j, X_j)\right] \\
&= \sum_{j=0}^{n-1} \left[f(t_j + \Delta t_j, X_j + \Delta X_j) - f(t_j, X_j)\right] \\
&= \sum_{j=0}^{n-1} \left[\dot{f}_j \Delta t_j + f'_j \Delta X_j + \tfrac{1}{2}\ddot{f}_j \Delta t_j^2 + \dot{f}'_j \Delta t_j \Delta X_j + \tfrac{1}{2} f''_j \Delta X_j^2 + R_j\right] \qquad (20.16)
\end{aligned}$$

by expanding $f(t, x)$ in a Taylor series, as $f(t, x)$ is smooth, and where the residual term $R_j = \mathcal{O}(|\Delta t_j|^3 + |\Delta X_j|^3)$.

Recall that the drift μ and volatility σ are step functions and are thus constant on each time step in the partition, so $\Delta X_j = \mu_j \Delta t_j + \sigma_j \Delta W_j$ exactly. Then consider each term in turn on the right-hand side of (20.16).

- $\sum_{j=0}^{n-1} \dot{f}_j \Delta t_j \to \int_a^b \dot{f}\, dt$ as $h \to 0$ to give the first term in Itô's integral formula (20.15).
- Whereas two more terms are obtained from

$$\begin{aligned}
\sum_{j=0}^{n-1} f'_j \Delta X_j &= \sum_{j=0}^{n-1} f'_j \mu_j \Delta t_j + \sum_{j=0}^{n-1} f'_j \sigma_j \Delta W_j \quad \text{as step functions} \\
&\to \int_a^b f' \mu\, dt + \int_a^b f' \sigma\, dW \quad \text{as } h \to 0 .
\end{aligned}$$

- The last term in Itô's integral formula comes from the last of the quadratic terms in the sum, as we now show that the other quadratic terms are zero, starting with

$$\begin{aligned}
\left|\sum_{j=0}^{n-1} \tfrac{1}{2}\ddot{f}_j \Delta t_j{}^2\right| &\leqslant \sum_{j=0}^{n-1} \tfrac{1}{2}|\ddot{f}_j| \Delta t_j{}^2 \\
&\leqslant \tfrac{1}{2} h \sum_{j=0}^{n-1} |\ddot{f}_j| \Delta t_j \quad \text{as } h = \max_j \Delta t_j \\
&\to \tfrac{1}{2} h \int_a^b |\ddot{f}| dt \to 0 \quad \text{as } h \to 0,
\end{aligned}$$

provided this integral exists, and hence this term must contribute nothing to (20.15).

- The next quadratic term contributes nothing because

$$\begin{aligned}
\sum_{j=0}^{n-1} \dot{f}'_j \Delta t_j \Delta X_j &= \sum_{j=0}^{n-1} \dot{f}'_j \Delta t_j (\mu_j \Delta t_j + \sigma_j \Delta W_j) \\
&= \sum_{j=0}^{n-1} \mu_j \dot{f}'_j \Delta t_j^2 + \underbrace{\sum_{j=0}^{n-1} \sigma_j \dot{f}'_j \Delta t_j\, \Delta W_j}_{=Y \text{ say}} ,
\end{aligned}$$

and the first term here vanishes by the previous case, assuming $\int_a^b |\mu \dot{f}'| dt$ exists, whereas the second term, Y, is more delicate because it is stochastic. Consider the second moment

$$\begin{aligned}
E[Y^2] &= E\left[\left(\sum_{j=0}^{n-1}\sigma_j\dot f_j'\Delta t_j\,\Delta W_j\right)\left(\sum_{k=0}^{n-1}\sigma_k\dot f_k'\Delta t_k\,\Delta W_k\right)\right]\\
&= E\left[\sum_{j,k=0}^{n-1}\sigma_j\dot f_j'\Delta t_j\,\Delta W_j\,\sigma_k\dot f_k'\Delta t_k\,\Delta W_k\right]\\
&= \sum_{j,k=0}^{n-1}E\left[\sigma_j\dot f_j'\Delta W_j\,\sigma_k\dot f_k'\Delta W_k\right]\Delta t_j\Delta t_k\,.
\end{aligned}$$

All terms in this sum vanish except those for which $j = k$ because,[246] for example, if $j < k$ then the factor ΔW_k is independent of all other factors inside the expectation and so the expectation factors to

$$E\left[\sigma_j\dot f_j'\Delta W_j\,\sigma_k\dot f_k'\Delta W_k\right] = E\left[\sigma_j\dot f_j'\Delta W_j\,\sigma_k\dot f_k'\right]\underbrace{E\left[\Delta W_k\right]}_{=0},$$

and similarly for $j > k$, thus leaving only the terms $k = j$ in

$$\begin{aligned}
E[Y^2] &= \sum_{j=0}^{n-1}E\left[\sigma_j^2\dot f_j'^2\Delta W_j^2\right]\Delta t_j^2\\
&\quad\text{as } \Delta W_j \text{ is independent of earlier history}\\
&= \sum_{j=0}^{n-1}E\left[\sigma_j^2\dot f_j'^2\right]\underbrace{E\left[\Delta W_j^2\right]}_{=\Delta t_j}\Delta t_j^2\\
&\leqslant h^2\sum_{j=0}^{n-1}E\left[\sigma_j^2\dot f_j'^2\right]\Delta t_j \to h^2\int_a^b E\left[\sigma^2\dot f'^2\right]dt \to 0 \text{ as } h \to 0,
\end{aligned}$$

and thus almost surely the term $Y \to 0$ as $h \to 0$.

- Whereas the only significant quadratic term is

$$\begin{aligned}
&\sum_{j=0}^{n-1}\tfrac12 f_j''\Delta X_j^2 = \sum_{j=0}^{n-1}\tfrac12\left(\mu_j\Delta t_j + \sigma_j\Delta W_j\right)^2 f_j''\\
&= \sum_{j=0}^{n-1}\tfrac12\mu_j^2 f_j''\Delta t_j^2 + \sum_{j=0}^{n-1}\mu_j\sigma_j f_j''\Delta t_j\,\Delta W_j + \sum_{j=0}^{n-1}\tfrac12\sigma_j^2 f_j''\Delta W_j^2\,,
\end{aligned}$$

and of these three terms the first two vanish by almost identical arguments to the previous two cases. For the third term, set $c_j = \frac12\sigma_j^2 f_j''$ for brevity; then aim to establish $\sum_{j=0}^{n-1}c_j\Delta W_j^2 \to \int_a^b c\,dt$. We thus compare $\sum_{j=0}^{n-1}c_j\Delta W_j^2$ with

[246] This is exactly the same argument used to prove the Itô isometry (20.7), and we use it again later.

$\sum_{j=0}^{n-1} c_j \Delta t_j$ by showing the vanishing of

$$\begin{aligned}
& E\left[\left(\sum_{j=0}^{n-1} c_j \Delta W_j^2 - \sum_{j=0}^{n-1} c_j \Delta t_j\right)^2\right] = E\left[\left(\sum_{j=0}^{n-1} c_j (\Delta W_j^2 - \Delta t_j)\right)^2\right] \\
&= E\left[\left(\sum_{j=0}^{n-1} c_j (\Delta W_j^2 - \Delta t_j)\right)\left(\sum_{k=0}^{n-1} c_k (\Delta W_k^2 - \Delta t_k)\right)\right] \\
&= E\left[\sum_{j,k=0}^{n-1} c_j c_k (\Delta W_j^2 - \Delta t_j)(\Delta W_k^2 - \Delta t_k)\right] \\
&= \sum_{j,k=0}^{n-1} \underbrace{E\left[c_j c_k (\Delta W_j^2 - \Delta t_j)(\Delta W_k^2 - \Delta t_k)\right]}_{=0 \text{ unless } k=j \text{ by independence of increments}} \\
&= \sum_{j=0}^{n-1} E\left[c_j^2 (\Delta W_j^2 - \Delta t_j)^2\right] \\
&\quad \text{as } \Delta W_j^2 - \Delta t_j \text{ is independent of earlier history} \\
&= \sum_{j=0}^{n-1} E\left[c_j^2\right] E\left[(\Delta W_j^2 - \Delta t_j)^2\right] \\
&= \sum_{j=0}^{n-1} E\left[c_j^2\right] E\left[\Delta W_j^4 - 2\Delta W_j^2 \Delta t_j + \Delta t_j^2\right] \\
&= \sum_{j=0}^{n-1} E\left[c_j^2\right] (3\Delta t_j^2 - 2\Delta t_j \Delta t_j + \Delta t_j^2) \\
&= 2\sum_{j=0}^{n-1} E\left[c_j^2\right] \Delta t_j^2 \\
&\leqslant 2h \sum_{j=0}^{n-1} E\left[c_j^2\right] \Delta t_j \to 2h \int_a^b E\left[c^2\right] dt \to 0 \quad \text{as } h \to 0 .
\end{aligned}$$

Thus almost surely

$$\sum_{j=0}^{n-1} c_j \Delta W_j^2 \to \sum_{j=0}^{n-1} c_j \Delta t_j = \sum_{j=0}^{n-1} \tfrac{1}{2}\sigma_j^2 f_j'' \Delta t_j \to \int_a^b \tfrac{1}{2}\sigma^2 f'' dt$$

as $h \to 0$, to give the last term in Itô's integral formula (20.7).

- For the last task, Exercise 20.4 shows that cubic and higher-order terms vanish, $\sum_{j=0}^{n-1} R_j \to 0$ as $h \to 0$, leaving just the integral form (20.7) of the Itô formula.

Although this proof applies only to step functions, arbitrary reasonable drift μ and volatility σ may be effectively approximated by a sequence of such step functions. A more extensive proof then rigorously justifies the soundness of the limit of such a sequence of approximations (Oksendal, 1998; Kloeden and Platen, 1992, e.g.). □

20.3 ▪ Stochastic differential equations and their solution

Mostly we seek models of stochastic systems written as SDEs. Stochastic partial differential equations express the more complex spatiotemporal systems. This section sketchily introduces Itô and Stratonovich SDEs, a transformation between them, and a couple of useful basic numerical methods to simulate stochastic systems.

20.3.1 ▪ Multiplicative noise affects exponential growth

As a first nontrivial example of SDEs, let's explore how noise affects exponential growth. This example is one of the key differential equations in stochastic calculus, just as the solution to the deterministic system $\dot{x} = \alpha x$ is the important exponential function $x(t) = ce^{\alpha t}$. In a stochastic system the growth or decay is affected by chance events, and when the effect is also proportional to the current value, then a formal model is the SDE $\dot{X} = \alpha X + \beta X\dot{W}$. Let's explore what this sort of equation means in both a Stratonovich and an Itô sense.

In the Stratonovich interpretation of an SDE the ordinary rules of calculus apply—provided we do not do anything forbidden such as differentiate $\dot{W}$ any further, nor square $\dot{W}$. Thus classically solve $\dot{X} = \alpha X + \beta X\dot{W}$ as a separable differential equation:[247]

$$\begin{aligned} &\frac{dX}{X} = (\alpha + \beta\dot{W})dt \\ \implies &\int \frac{dX}{X} = \int (\alpha + \beta\dot{W})dt \\ \implies &\log X - \log X_0 = \alpha t + \beta W(t) \\ \implies &X = X_0 \exp[\alpha t + \beta W(t)]. \end{aligned}$$

This solution is classic exponential growth at rate α ameliorated by the noise. Figure 20.5 plots decaying realizations for the case $\alpha = -2$ and $\beta = 1$, whereas Figure 20.6 plots growing realizations for the case $\alpha = 1$ and $\beta = 2$. Although the Wiener process $W(t)$ makes a marked difference in the solutions as $W(t)$ grows roughly like $\beta\sqrt{t}$—see the figures—the linear term αt ultimately dominates to determine whether the realizations all decay to zero or all grow indefinitely.

Conversely, in the Itô interpretation of an SDE the ordinary rules of calculus do not apply, so we resort to basic arguments—at least for the moment. Recall that to distinguish between the Itô and the Stratonovich forms, we adopt the notation that an Itô SDE will be written in terms of differentials, as is usual in the field; that is, $dX = \alpha X\,dt + \beta X\,dW$ is the Itô SDE.

Recall that Example 20.8 uses Itô's formula (20.12) to deduce that the process $X(t) = c\exp[at + bW(t)]$ has differential $dX = (a + \frac{1}{2}b^2)X\,dt + bX\,dW$. This SDE will be the same as the specified Itô SDE provided $a + \frac{1}{2}b^2 = \alpha$ and $b = \beta$; hence set $b = \beta$ and $a = \alpha - \frac{1}{2}\beta^2$. That is, the solution to the Itô SDE $dX = \alpha X\,dt + \beta X\,dW$ is

$$X(t) = c\exp\left[(\alpha - \tfrac{1}{2}\beta^2)t + \beta W(t)\right].$$

[247]Most treatises on stochastic calculus would denote such a Stratonovich differential equation as $dX = \alpha X\,dt + \beta X \circ dW$, where the circle in $\circ dW$ denotes that the Stratonovich interpretation is to be used. But to relate most closely with the practice of engineers and physicists, this book adopts the classic mathematical notation $\dot{X} = \alpha X + \beta X\dot{W}$ for Stratonovich SDEs to reflect that classic calculus rules may be used—with care.

This solution is classic exponential growth, but in the Itô sense observe that the noise, $\beta \neq 0$, depresses the growth to rate $\alpha - \frac{1}{2}\beta^2$. Thus Figure 20.5 plots decaying realizations for the case $\alpha = -3/2$ and $\beta = 1$, whereas Figure 20.6 plots growing realizations for the case $\alpha = 3$ and $\beta = 2$.

20.3.2 ▪ Transform between Itô and Stratonovich

Observe in section 20.3.1 that the two interpretations have exactly the same solution provided the coefficient of the drift is modified: $\alpha_{\text{Itô}} = \alpha_{\text{Str}} + \frac{1}{2}\beta_{\text{Str}}^2$ and $\beta_{\text{Itô}} = \beta_{\text{Str}}$. This is an example of a very general relationship that empowers us to transform from the Stratonovich form of an SDE, which we use in algebraically constructing models, to the Itô form of an SDE, which we use in numerical solutions of SDEs, and vice versa.

Theorem 20.9 (single variable). *The solutions of the Stratonovich SDE* $\dot X = \mu^{Str}(t,X) + \sigma^{Str}(t,X)\dot W$ *and the Itô SDE* $dX = \mu^{Itô}(t,X)dt + \sigma^{Itô}(t,X)dW$ *are identical when the volatilities are the same,* $\sigma^{Itô}(t,X) = \sigma^{Str}(t,X)$*, while the drifts are related by* $\mu^{Itô} = \mu^{Str} + \frac{1}{2}\sigma\partial\sigma/\partial X$*, where* σ *is either* $\sigma^{Itô}$ *or* σ^{Strat}*, as they are equal.*

Example 20.11 (Ornstein–Uhlenbeck process). The Stratonovich SDE $\dot X = -\alpha X + \sigma\dot W$ for constant volatility σ has identical solutions to the Itô SDE $dX = -\alpha X\,dt + \sigma\,dW$ with unchanged coefficients.

This previous example illustrates the general result that if the volatility σ is constant, then the Itô and Stratonovich SDEs have precisely the same drift and volatility factors.

Example 20.12 (varying volatility). The Stratonovich SDE $\dot X = \sqrt{1+X^2}\dot W$ has identical solutions to the Itô SDE $dX = \frac{1}{2}X\,dt + \sqrt{1+X^2}\,dW$. This follows from $\sigma\partial\sigma/\partial X = \sqrt{1+X^2} \times X/\sqrt{1+X^2} = X$ for the given volatility $\sigma = \sqrt{1+X^2}$.

Conversely, the Itô SDE $dX = \sqrt{1+X^2}\,dW$ corresponds to the Stratonovich SDE $\dot X = -\frac{1}{2}X + \sqrt{1+X^2}\dot W$.

The first case may be solved analytically, as ordinary rules of calculus hold for Stratonovich SDEs: rearrange to $\dot X/\sqrt{1+X^2} = \dot W$; then integrate, via substitution $X = \sinh Y$, to obtain $Y = c + W(t)$, that is, $X = \sinh(c + W(t))$. The second case cannot be solved in this way.

In general, a stochastic system will have many evolving variables and potentially many independent noise sources. We need to transform between the Stratonovich version of a stochastic system and the Itô version. The following theorem generalizes Theorem 20.9 (Kloeden and Platen, 1992, §4.9).

Theorem 20.10 (multiple variables). *Consider the Stratonovich system of* n *SDEs* $\dot{\mathbf X} = \mu^{Str}(t,\mathbf X) + \sigma^{Str}(t,\mathbf X)\dot{\mathbf W}$ *and the system of* n *Itô SDEs Itô* SDE $d\mathbf X = \mu^{Itô}(t,\mathbf X)dt + \sigma^{Itô}(t,\mathbf X)d\mathbf W$*, where vector* $\mathbf W$ *represents* m *independent Wiener processes and the volatility matrices* σ *are* $n \times m$*. The solutions of the two forms of SDEs are identical when the volatility matrices are the same,* $\sigma_{i,j}^{Itô} = \sigma_{i,j}^{Str}$*, and the drifts relate by* $\mu_i^{Itô} = \mu_i^{Str} + \frac{1}{2}\sum_{j,k}\sigma_{j,k}\partial\sigma_{i,k}/\partial X_j$.

As before, if a volatility matrix is constant, then the Itô and Stratonovich expressions of the stochastic dynamics are identical.

Algorithm 20.2 *Basic Octave/MATLAB/Scilab code to generate Euler–Maruyama solutions of the Itô SDE* $dX = X\,dW$ *with* $X(0) = 1$ *in Example* 20.13.

```
n=1024; % number of time steps
t=linspace(0,2,n+1);
dw=sqrt(diff(t)).*randn(1,n);
x=cumprod([1 1+dw]);
plot(t,x)
```

20.3.3 ▪ The Euler–Maruyama scheme has low accuracy

We now turn to a couple of straight-forward computation schemes to numerically solve SDEs. Kloeden and Platen (1992) wrote the classic treatise on such computational schemes for SDEs. One of the most important messages to remember is that, of all the classic methods for deterministic ODEs, such as Runge–Kutta schemes, *trust none to apply* to noisy SDEs. The notable exception is the simple Euler scheme.

Consider general SDEs in the Itô form

$$dX = \mu(t,X)dt + \sigma(t,X)dW, \tag{20.17}$$

with drift μ and volatility σ. Why this form? Because the Doob–Meyer decomposition theorem assures us that a huge variety of continuous stochastic processes $X(t)$ have unique drift and volatility.

The explicit Euler–Maruyama scheme replaces the differentials in the SDE (20.17) with forward differences:

$$\Delta X_j = \mu(t_j, X_j)\Delta t_j + \sigma(t_j, X_j)\Delta W_j\,,$$

where the Wiener increment ΔW_j, as always, is random with normal distribution of mean zero and variance Δt_j.[248] Note the evaluation of the drift and volatility at the current time t_j to give the explicit scheme

$$X_{j+1} = X_j + \mu(t_j, X_j)\Delta t_j + \sigma(t_j, X_j)\Delta W_j\,. \tag{20.18}$$

Theory asserts that this explicit Euler–Maruyama scheme constructs approximations with error $\mathcal{O}(\sqrt{\Delta t})$ (Kloeden and Platen, 1992, e.g.).

Example 20.13. A simple example of the Euler–Maruyama scheme is to numerically solve $dX = X\,dW$ with $X(0) = 1$ (Example 20.5 gives the algebraic solution). Figure 20.11 plots one realization obtained numerically. The Euler–Maruyama scheme becomes $\Delta X_j = X_j\,\Delta W_j$, that is, $X_{j+1} = X_j + X_j\Delta W_j$.

The computation is $X_{j+1} = X_j(1 + \Delta W_j)$, so here the solution may be written $X_n = \prod_{j=0}^{n-1}(1+\Delta W_j)$. Algorithm 20.2 lists how we might code the computation. To observe the convergence in any given realization, as plotted in Figure 20.11, one

[248] Although for many practical purposes, and because of a central limit theorem, one could choose microscale increments $\Delta W_j = \pm\sqrt{\Delta t_j}$, each with probability $\frac{1}{2}$. The resulting macroscale simulations are effectively the same.

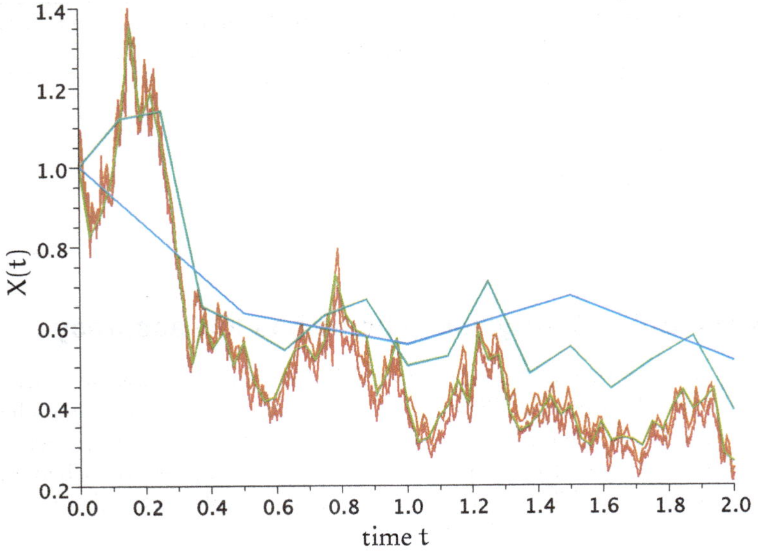

Figure 20.11. *A realization of the solution to* $dX = X\,dW$ *via the simple Euler–Maruyama scheme for five different time steps* $\Delta t = \frac{1}{2}, \frac{1}{8}, \frac{1}{32}, \frac{1}{128}, \frac{1}{512}$ *(colored blue to red, respectively). Very small time steps are needed to reasonably approximate the realization.*

Algorithm 20.3 *Basic Octave/MATLAB/Scilab code to generate Euler–Maruyama solutions of the Itô SDE* $dX = \frac{1}{2}X\,dt + \sqrt{1+X^2}\,dW$ *with initial condition* $X(0) = 0$ *in Example* 20.14.

```
m=5; % number of realizations
n=1024; % number of time steps
t=linspace(0,2,n+1);
dt=diff(t(1:2)); % constant time step
dw=sqrt(dt)*randn(m,n);
x=zeros(m,n+1); % dimension and initial condition
for j=1:n
x(:,j+1)=x(:,j)+0.5*x(:,j)*dt+sqrt(1+x(:,j).^2).*dw(:,j);
end
plot(t,x)
```

first generates the Wiener process on the finest time step and then subsamples to use the same realization of the Wiener process on coarser time steps.

Example 20.14. Generally we must use a for-loop to progress in time. Consider solving the Itô SDE $dX = \frac{1}{2}X\,dt + \sqrt{1+X^2}\,dW$ with initial condition $X(0) = 0$. The computation is then $X_{j+1} = X_j + \Delta X_j = X_j + \frac{1}{2}X_j\,\Delta t_j + \sqrt{1+X_j^2}\,\Delta W_j$. Algorithm 20.3 lists code that generates the five realizations plotted in Figure 20.12. Since we know, from Example 20.12, that algebraic solutions with this initial condition are $X = \sinh[W(t)]$, Figure 20.13 also plots the error in the numerical solution: observe that the error, at up to about 0.1 compared to solutions of size 2–4, is as a relative error about as large as the expected $\sqrt{\Delta t} \approx 10^{-3/2} \approx 0.03$.

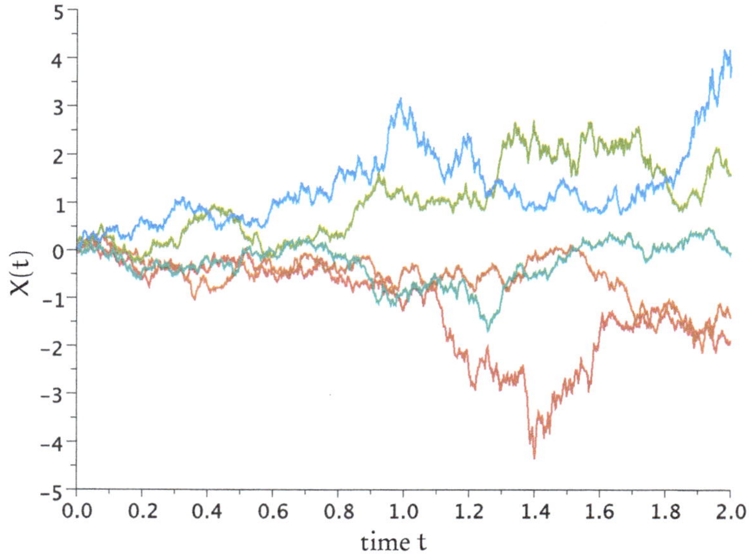

Figure 20.12. *Five realizations of the numerical Euler–Maruyama solution to the Itô* SDE $dX = \frac{1}{2}X\,dt + \sqrt{1+X^2}\,dW$ *with initial condition* $X(0) = 0$ *in Example* 20.14.

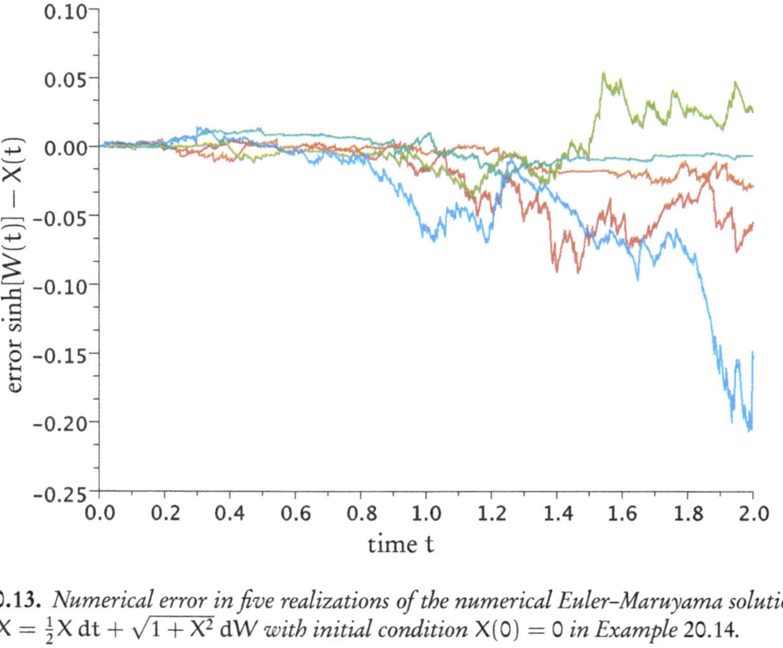

Figure 20.13. *Numerical error in five realizations of the numerical Euler–Maruyama solution to the Itô SDE* $dX = \frac{1}{2}X\,dt + \sqrt{1+X^2}\,dW$ *with initial condition* $X(0) = 0$ *in Example* 20.14.

20.4 ▪ Modify the improved Euler scheme to integrate SDEs

More than 20 years ago Kloeden and Platen (1992) described schemes for numerically integrating SDEs. Intervening research led to more-recent developments of useful Runge–Kutta-like methods for Itô SDEs by Rößler (2010, 2009) and for Stratonovich SDEs by Komori (2007c,a,b). These numerical integration schemes for SDEs are quite complicated

Algorithm 20.4 *Modified Runge–Kutta scheme to integrate the general Itô* SDE (20.19) *(set* $S_k = 0$ *to integrate Stratonovich SDEs).*

Given time step h, and given the value $X(t_k) = X_k$, estimate $X(t_{k+1})$ by X_{k+1} for time $t_{k+1} = t_k + h$ via

$$\begin{aligned} K_1 &= h\mu(t_k, X_k) + (\Delta W_k - S_k\sqrt{h})\sigma(t_k, X_k), \\ K_2 &= h\mu(t_{k+1}, X_k + K_1) + (\Delta W_k + S_k\sqrt{h})\sigma(t_{k+1}, X_k + K_1), \\ X_{k+1} &= X_k + \tfrac{1}{2}(K_1 + K_2), \end{aligned} \tag{20.20}$$

- where $\Delta W_k = \sqrt{h} Z_k$ for normal random $Z_k \sim N(0, 1)$, and
- where $S_k = \pm 1$, with each alternative chosen with probability 1/2.

The above describes only one time step. Repeat this time step $(t_m - t_0)/h$ times in order to integrate an SDE (20.19) from time $t = t_0$ to $t = t_m$.

and typically do not easily reduce to accurate deterministic schemes—although they are of good accuracy for SDEs. This section introduces a Runge–Kutta scheme for SDEs that does straightforwardly reduce to a well-known deterministic scheme—the variously called improved Euler, Heun, or Runge–Kutta 2 scheme.

Consider the vector stochastic process $X(t) \in \mathbb{R}^n$ that satisfies the general Itô SDE

$$dX = \mu(t, X)\, dt + \sigma(t, X)\, dW, \tag{20.19}$$

where drift μ and volatility σ are sufficiently smooth functions of their arguments. The noise is represented by the differential dW, which symbolically denotes infinitesimal increments of the random walk of a Wiener process $W(t, \omega)$. The symbolic form of the SDE (20.19) follows from the most basic approximation to an evolving system with noise that over a time step Δt_k the change in the dependent variable is

$$\Delta X_k \approx \mu(t_k, X_k)\Delta t_k + \sigma(t_k, X_k)\Delta W_k,$$

where $\Delta W_k = W(t_{k+1}, \omega) - W(t_k, \omega)$ symbolizes some random effect. This basic approximation has low accuracy and needs improving for practical applications.

Algorithm 20.4 lists the modified Runge–Kutta scheme for the general Itô SDE (20.19). The appeal of Algorithm 20.4 as an entry to stochastic integrators is its close connection to deterministic integration schemes. When the stochastic component vanishes, $\sigma = 0$, the integration step (20.20) is precisely the improved Euler, Heun, or Runge–Kutta 2 scheme that most engineers, scientists, and mathematicians learn in undergraduate coursework.

This connection has another useful consequence in application: for systems with small noise we expect that the integration error of the SDE is only a little worse than that of the deterministic system. Although section 20.4.4 proves the typical $\mathcal{O}(h)$ error of the stochastic Algorithm 20.4, as demonstrated in the examples of section 20.4.1, when the noise is small expect the error to be practically better than this order of error suggests.

Sections 20.4.2–20.4.4 also prove that Algorithm 20.4 integrates Stratonovich SDEs to $\mathcal{O}(h)$ provided one sets $S_k = 0$ throughout (instead of choosing ± 1).

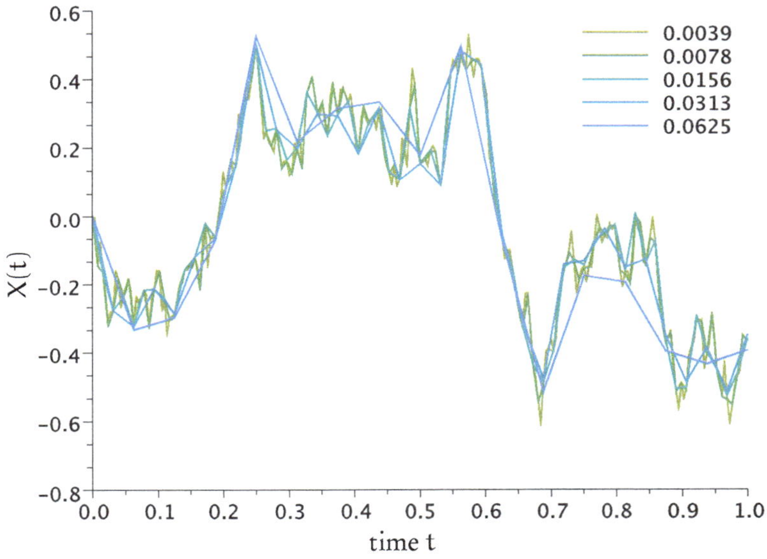

Figure 20.14. *As the time step is successively halved,* $n = 16, 32, 64, 128, 256$ *time steps over* $0 \leqslant t \leqslant 1$, *the numerical solutions of the SDE* (20.21) *via Algorithm* 20.4 *appear to converge.*

An outstanding research challenge is to generalize the integration step (20.20) to multiple noise sources.

20.4.1 ▪ Examples demonstrate that $\mathcal{O}(h)$ error is typical

This section applies Algorithm 20.4 to three example SDEs for which, for comparison, we know the algebraic solution by Kloeden and Platen (1992). Two of the examples exhibit errors $\mathcal{O}(h)$, as is typical, whereas the third exhibits a error $\mathcal{O}(h^2)$, which occurs for both deterministic ODEs and a special class of SDEs. These errors are "pathwise" errors, which means that for any one given realization ω of the noise process $W(t, \omega)$ we refer to the order of error as the time step $h \to 0$ for a fixed realization ω.

Example 20.15 (an autonomous SDE). Consider the autonomous SDE

$$dX = \left[\tfrac{1}{2}X + \sqrt{1 + X^2}\right] dt + \sqrt{1 + X^2}\, dW \quad \text{with} \quad X(0) = 0 \tag{20.21}$$

for some Wiener process $W(t, \omega)$. The SDE is not strictly autonomous because the noise dW introduces time dependence; we use the term "autonomous" to indicate that the drift a and volatility b are independent of time. For the SDE (20.21), Kloeden and Platen (1992) list the algebraic solution as $X(t, \omega) = \sinh\left[t + W(t, \omega)\right]$.

Such algebraic solutions are straightforwardly checked via the basic version of Itô's formula (20.12).

The proposed numerical Algorithm 20.4 was applied to integrate the SDE (20.21) from $t = 0$ to end time $t = 1$ with a time step of $h = 1/n$ for $n = 2^{15}, 2^{14}, \ldots, 2^4$ steps. For each of 700 realizations of the noise $W(t, \omega)$, the Wiener increments, $\Delta W \sim N(0, 2^{-15})$, were generated on the finest time step and subsequently aggregated to the corresponding increments for each realization on the coarser time steps. Figure 20.14 plots the predicted $X(t, \omega)$ obtained from the numerical

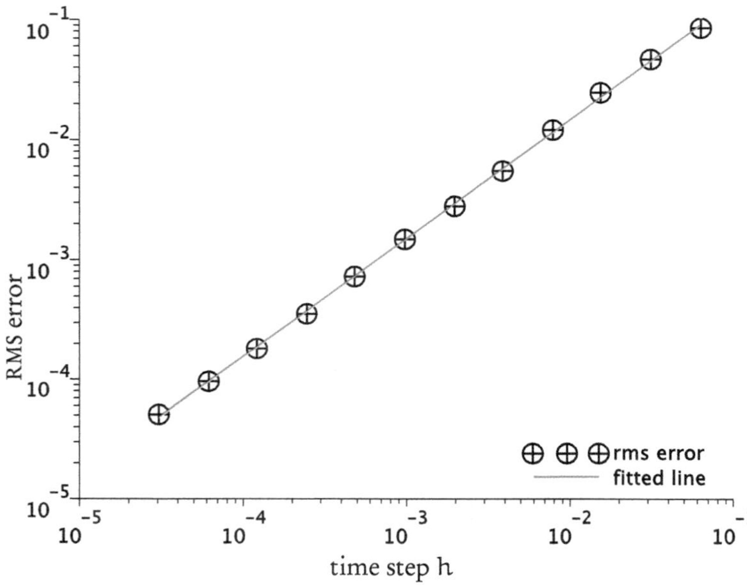

Figure 20.15. *Average over* 700 *realizations at each of* 12 *different step sizes for the SDE* (20.21)*: at* $t = 1$, *the RMS error in the predicted* $X(1, \omega)$ *decreases linearly in time step* h.

Algorithm 20.4 for just one realization ω using different time steps. The predictions do appear to converge to a well-defined stochastic process as the step size is refined.

For each size of time step, Figure 20.15 uses the algebraic solution to find the RMS error of the predicted $X(1, \omega)$, averaged over 700 realizations ω. This RMS error estimates the square root of the expectation $E[(X_m - X(1, \omega))^2]$. Figure 20.15 uses a log-log plot to show that the RMS error at time $t = 1$ decreases linearly with time step size h (over four orders of magnitude in time step). That is, empirically we see that Algorithm 20.4 has RMS error $\mathcal{O}(h)$.

Example 20.16 (a nonautonomous SDE). Consider the nonautonomous SDE

$$dX = \left[\frac{X}{1+t} - \frac{3}{2}X\left(1 - \frac{X^2}{(1+t)^2}\right)^2\right] dt + (1+t)\left(1 - \frac{X^2}{(1+t)^2}\right)^{3/2} dW, \tag{20.22}$$

with initial condition that $X(0) = 0$, for some Wiener process $W(t, \omega)$. Here both the drift a and the volatility b have explicit time dependence. Itô's formula (20.12) confirms that the algebraic solution to this SDE (20.22) is $X(t, \omega) = (1+t)W(t, \omega)/\sqrt{1 + W(t, \omega)^2}$.

To determine the order of error of Algorithm 20.4, the same approach was adopted here as described in Example 20.15. The slope of the log-log plot in Figure 20.16 shows that again the RMS error of the predicted $X(1, \omega)$ is $\mathcal{O}(h)$ for time step h over four orders of magnitude in h.

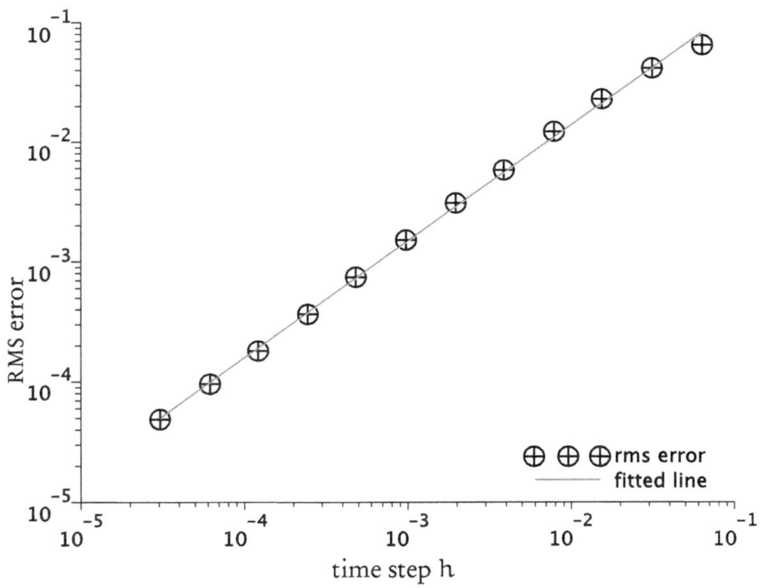

Figure 20.16. *Average over 700 realizations at each of 12 different step sizes for the SDE (20.22): at $t = 1$, the RMS error in the predicted $X(1, \omega)$ decreases linearly in time step h.*

Example 20.17 (second-order error). Consider the following SDE, which is linear in X:

$$dX = \left[\frac{2X}{1+t} + (1+t)^2\right] dt + (1+t)^2 dW, \quad \text{with } X(0) = 1, \tag{20.23}$$

for some Wiener process $W(t, \omega)$. Itô's formula (20.12) confirms that the algebraic solution to SDE (20.23) is $X(t, \omega) = (1+t)^2[1 + t + W(t, \omega)]$.

To determine the order of error of Algorithm 20.4, the same approach was adopted here as described in previous examples. The difference is that the slope of the log-log plot in Figure 20.17 shows that here the RMS error of the predicted $X(1, \omega)$ is $\mathcal{O}(h^2)$. There appear to be some SDEs for which the error of Algorithm 20.4 is quadratic in the time step h rather than linear.

20.4.2 ▪ Error for Itô integrals

This and the following two sections (sections 20.4.3–20.4.4) use stochastic integration to establish the general order of accuracy of the numerical integration scheme of Algorithm 20.4.

Proofs that numerical schemes do indeed approximate SDE solutions are often complex. The plan here is to elaborate three successively more complicated cases, with the aim of developing a feel for the analysis before it gets too complex. Lemma 20.11 first proves that the Runge–Kutta-like Algorithm 20.4 approximates the simplest Itô integrals $X = \int_a^b \sigma(t)\, dW$ to first order in the time step. Second, section 20.4.3 identifies a class of linear SDEs with additive noise when Algorithm 20.4 is of second order (such as Example 20.17). Third, section 20.4.4 proves the first-order global error of Algorithm 20.4 when applied to general SDEs. Those familiar with stochastic Itô integration could proceed directly to section 20.4.4.

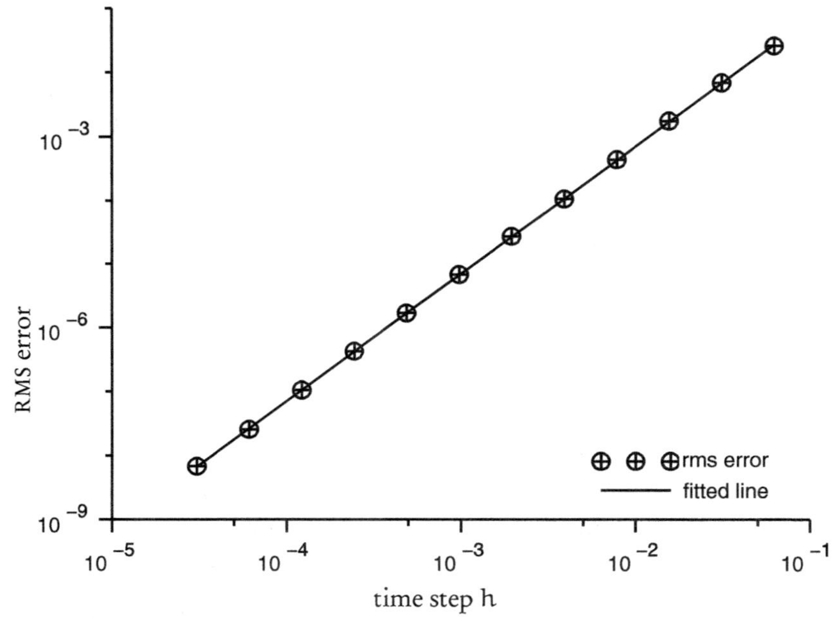

Figure 20.17. *Averaging over 700 realizations at each of 13 different step sizes for the linear SDE (20.23): at $t = 1$, the RMS error in the predicted $X(1,\omega)$ decreases quadratically, like h^2.*

One outcome of these three sections is to precisely identify the requisite properties of the choice of random signs S_j in the time step (20.20).

This first section establishes the order of error in computing the Itô integral $X = \int_a^b \sigma(t)\,dW$ if one were to invoke Algorithm 20.4 on the scalar SDE $dX = \sigma(t)\,dW$. Before proceeding, recall the two useful properties on the expectation and variance of Itô integrals: the martingale property (20.10) and Itô's isometry (20.11). These empower us to quantify errors in the integrals that approximate solutions of SDEs as in the following lemma.

Lemma 20.11. *The Runge–Kutta-like Algorithm 20.4 has $\mathcal{O}(h)$ global error when applied to $dX = \sigma(t)\,dW$ for twice differentiable functions $\sigma(t)$.*

Proof. Without loss of generality, start with the time step from $t_0 = 0$ to $t_1 = t_0 + h = h$. Applied to the very simple SDE $dX = \sigma(t)\,dW$, one step of the scheme (20.20) computes

$$K_1 = (\Delta W - S\sqrt{h})\sigma_0\,, \quad K_2 = (\Delta W + S\sqrt{h})\sigma_1$$

and then estimates the change in X as

$$\Delta\hat{X} = \tfrac{1}{2}(\sigma_0 + \sigma_1)\Delta W + \tfrac{1}{2}(\sigma_1 - \sigma_0)S\sqrt{h}\,, \tag{20.24}$$

where the integrand values $\sigma_0 = \sigma(0)$ and $\sigma_1 = \sigma(h)$. The classic polynomial approximation theorem (Kreyszig, 2011, p. 800, e.g.) relates the estimate (20.24) to the exact integral. Here write the integrand as the linear interpolant with remainder:

$$\sigma(t) = \tfrac{1}{2}(\sigma_1 + \sigma_0) + \tfrac{1}{h}(\sigma_1 - \sigma_0)(t - h/2) + \tfrac{1}{2}t(t-h)b''(\tau)$$

for some $0 \leqslant \tau(t) \leqslant h$. Then the exact change in $X(t)$ is

$$\Delta X = \int_0^h \sigma(t)\,dW = \tfrac{1}{2}(\sigma_1+\sigma_0)\Delta W + \tfrac{1}{h}(\sigma_1-\sigma_0)\int_0^h (t-h/2)\,dW + \tfrac{1}{2}\int_0^h t(t-h)b''(\tau)\,dW. \tag{20.25}$$

The error in one step of the scheme (20.20) is the difference between the increments (20.24) and (20.25). That is, the true integral increment $\Delta X = \Delta\hat{X} + \epsilon_0$, where the error

$$\epsilon_0 = \frac{\sigma_1-\sigma_0}{h}\left[-\tfrac{1}{2}Sh^{3/2} + \int_0^h (t-h/2)\,dW\right] + \tfrac{1}{2}\int_0^h t(t-h)b''(\tau)\,dW. \tag{20.26}$$

How big is this error? First, take expectations, invoke the martingale property (20.10) for the two stochastic integrals, and see that $E[\epsilon_0] = 0$ provided that $E[S] = 0$. *Thus the signs S must be chosen with mean zero.*

Second, compute the variance of the error ϵ_0 to see the size of the fluctuations in the error. Since the expectation $E[\epsilon_0] = 0$, the variance $\operatorname{Var}[\epsilon_0] = E[\epsilon_0^2]$. Look at various contributions in turn. The first term in the error (20.26) has variance $E[(Sh^{3/2})^2] = h^3\,E[S^2] = \mathcal{O}(h^3)$ provided that the signs S have bounded variance. Choosing the signs S independently of the noise W, there are then no correlations between the S terms and the other two terms. The second term in the error (20.26) has variance

$$\begin{aligned} E\left[\left(\int_0^h (t-h/2)\,dW\right)^2\right] &= \int_0^h (t-h/2)^2 dt \quad \text{by Itô isometry (20.11)} \\ &= \tfrac{1}{12}h^3 = \mathcal{O}(h^3). \end{aligned}$$

The third term in the error (20.26), by the Itô isometry (20.11), has variance

$$\begin{aligned} E\left[\left(\int_0^h t(t-h)b''(\tau)\,dW\right)^2\right] &= \int_0^h t^2(t-h)^2 b''(\tau)^2 dt \\ &\leqslant B_2^2 \int_0^h t^2(t-h)^2 dt = \frac{1}{30}B_2^2 h^5, \end{aligned} \tag{20.27}$$

provided that the second derivative is bounded, $|b''(t)| \leqslant B_2$. Last, the correlation between these previous two integrals is small as, by a slightly more general version of the Itô isometry (20.11),

$$\begin{aligned} &\left|E\left[\int_0^h (t-h/2)\,dW \int_0^h t(t-h)b''(\tau)\,dW\right]\right| \\ &= \left|\int_0^h (t-h/2)t(t-h)b''(\tau)\,dt\right| \\ &\leqslant B_2 \int_0^h |(t-h/2)t(t-h)|\,dt = \mathcal{O}(h^4). \end{aligned}$$

Hence the local one-step error is dominated by the first two contributions and has variance $\operatorname{Var}[\epsilon_0] = \mathcal{O}(h^3)$.

To estimate the global integral, $\int_a^b \sigma(t)\,dW$, we take $n = \mathcal{O}(1/h)$ time steps. With n steps the global error is the sum of n local errors: Algorithm 20.4 approximates the correct solution with global error $\epsilon = \sum_{j=0}^{n-1} \epsilon_j$. First, $E[\epsilon] = 0$ as $E[\epsilon_j] = 0$ for all time steps. Second, the errors on each time step are independent, since the increments dW and the signs S are independent; hence the variance

$$\operatorname{Var}[\epsilon] = \sum_{j=0}^{n-1} \operatorname{Var}[\epsilon_j] = n \operatorname{Var}[\epsilon_0] = \mathcal{O}(nh^3) = \mathcal{O}(h^2).$$

Thus, for the SDE $dX = \sigma(t)\,dW$, Algorithm 20.4 has global error $\mathcal{O}(h)$. □

20.4.3 ▪ Error for linear SDEs with additive noise

This second lemma addresses somewhat more general scalar SDEs. Not only does it serve as a stepping stone to a full theorem, but it also illustrates two other interesting properties. First, we identify a class of SDEs for which Algorithm 20.4 is second order accurate in the time step, as seen in Example 20.17. Second, the proof suggests that the sign S in the scheme (20.20) relates to substep properties of the noise W that are independent of the increment ΔW.

Lemma 20.12. *The Runge–Kutta-like Algorithm* 20.4 *has global error* $\mathcal{O}(h)$ *when applied to the additive noise, linear SDE* $dX = \mu(t)X\,dt + \sigma(t)\,dW$ *for functions* μ *and* σ *twice differentiable. Further, in the exact differential case when* $\mu\sigma = d\sigma/dt$, *the global error is* $\mathcal{O}(h^2)$.[249]

Proof. In this case, straightforward algebra shows that the first step in the scheme (20.20) predicts the change

$$\begin{aligned}\Delta X &= h\tfrac{1}{2}(\mu_0 + \mu_1)X_0 + \tfrac{1}{2}h^2\mu_0\mu_1 X_0 + \tfrac{1}{2}(\sigma_0 + \sigma_1)\Delta W \\ &\quad + \tfrac{1}{2}\mu_1\sigma_0 h(\Delta W - S\sqrt{h}) + \tfrac{1}{2}S\sqrt{h}\Delta\sigma,\end{aligned} \tag{20.28}$$

where the coefficient values $\mu_0 = \mu(0)$, $\mu_1 = \mu(h)$, $\sigma_0 = \sigma(0)$, and $\sigma_1 = \sigma(h)$. We compare this approximate change over the time step h with the true change using iterated integrals. For simplicity we also use subscripts to denote dependence upon "time" variables t, s, and r. We start by writing the SDE $dX = \mu_t X_t\,dt + \sigma_t\,dW$ as an integral over the first time step:

$$\begin{aligned}\Delta X &= \int_0^h \mu_t X_t\,dt + \int_0^h \sigma_t\,dW_t \\ &\quad \text{(substituting } X_t = X_0 + \textstyle\int_0^t \mu_s X_s\,ds + \int_0^t \sigma_s\,dW_s \text{ in the first integral)} \\ &= \int_0^h \mu_t \left[X_0 + \int_0^t \mu_s X_s\,ds + \int_0^t \sigma_s\,dW_s\right] dt + \int_0^h \sigma_t\,dW_t \\ &= X_0 \int_0^h \mu_t\,dt + \int_0^h \mu_t \int_0^t \mu_s X_s\,ds\,dt \\ &\quad + \int_0^h \mu_t \int_0^t \sigma_s\,dW_s\,dt + \int_0^h \sigma_t\,dW_t \\ &\quad \text{(substituting } X_s = X_0 + \textstyle\int_0^s \mu_r X_r\,dr + \int_0^s \sigma_r\,dW_r \text{ in the 2nd integral)}\end{aligned}$$

[249] A solution to the SDE is then $X = \sigma(t)W$.

$$
\begin{aligned}
&= X_0 \int_0^h \mu_t \, dt + \int_0^h \mu_t \int_0^t \mu_s \left[X_0 + \int_0^s \mu_r X_r \, dr + \int_0^s \sigma_r \, dW_r \right] ds \, dt \\
&\quad + \int_0^h \mu_t \int_0^t \sigma_s \, dW_s \, dt + \int_0^h \sigma_t \, dW_t \\
&= X_0 \int_0^h \mu_t \, dt + X_0 \int_0^h \mu_t \int_0^t \mu_s \, ds \, dt + \int_0^h \mu_t \int_0^t \mu_s \int_0^s \mu_r X_r \, dr \, ds \, dt \\
&\quad + \int_0^h \mu_t \int_0^t \mu_s \int_0^s \sigma_r \, dW_r \, ds \, dt + \int_0^h \mu_t \int_0^t \sigma_s \, dW_s \, dt + \int_0^h \sigma_t \, dW_t \,. \qquad (20.29)
\end{aligned}
$$

For the last part of the lemma in the case of higher-order error, we need to expand to this level of detail in six integrals. Of these six integrals, some significantly match the components of the numerical step (20.28) and some just contribute to the error. Recall that the proof of Lemma 20.11 identified that errors had both mean and variance. To cater for these two characteristics of errors, and with perhaps some abuse of notation, let's introduce the notation that $\mathcal{O}(h^p; h^q)$ denotes quantities with mean $\mathcal{O}(h^p)$ and variance $\mathcal{O}(h^q)$. For example, $\mathcal{O}(h^p; 0)$ classifies deterministic quantities $\mathcal{O}(h^p)$, whereas $\mathcal{O}(0; h^q)$ characterizes zero mean stochastic quantities of standard deviation scaling like $h^{q/2}$. The previous proof looked closely at the variances of error terms; here we simplify by focusing only upon their order of magnitude. In particular, let's show that the six integrals in (20.29) match the numerical step (20.28) to an error $\mathcal{O}(h^3; h^5)$.

Consider separately the integrals in (20.29).

- First, $X_0 \int_0^h \mu_t \, dt = X_0 h \frac{1}{2}(\mu_0 + \mu_1) + \mathcal{O}(h^3)$ by the classic trapezoidal rule. This matches the first component in the numerical (20.28).
- Second, using the linear interpolation $\mu_t = \mu_0 + \frac{\Delta\mu}{h} t + \mathcal{O}(t^2)$, where as usual $\Delta\mu = \mu_1 - \mu_0$, the double integral

$$
\begin{aligned}
\int_0^h \mu_t \int_0^t \mu_s \, ds \, dt &= \int_0^h \left(\mu_0 + \frac{\Delta\mu}{h} t\right)\left(\mu_0 t + \frac{\Delta\mu}{2h} t^2\right) + \mathcal{O}(t^3) \, dt \\
&= \int_0^h \mu_0^2 t + \mu_0 \frac{3\Delta\mu}{2h} t^2 + \mathcal{O}(t^3) \, dt \\
&= \tfrac{1}{2}\mu_0^2 h^2 + \mu_0 \frac{\Delta\mu}{2} h^2 + \mathcal{O}(h^4) \\
&= \tfrac{1}{2}\mu_0 \mu_1 h^2 + \mathcal{O}(h^4).
\end{aligned}
$$

  Multiplied by X_0, this double integral matches the second term in the numerical (20.28).
- Third, the triple integral

$$
\int_0^h \mu_t \int_0^t \mu_s \int_0^s \mu_r X_r \, dr \, ds \, dt = \mathcal{O}(h^3),
$$

  because, as seen in the previous two items, each ordinary integration over a time of $\mathcal{O}(h)$ multiplies the order of the term by a power of h.
- Fourth, look at the single stochastic integral in (20.29), the last term. From the proof of the previous lemma, equations (20.25) and (20.27) give

$$
\int_0^h \sigma_t \, dW_t = \tfrac{1}{2}(\sigma_1 + \sigma_0)\Delta W + \frac{\Delta\sigma}{h} \int_0^h \left(t - \tfrac{h}{2}\right) dW_t + \mathcal{O}(0; h^5). \qquad (20.30)
$$

The first term here matches the third term in the numerical (20.28). The second term on the right-hand side is an integral remainder that will be dealt with after the next two items.

- Fifth, change the order of integration in the double integral:

$$\begin{aligned}
&\int_0^h \mu_t \int_0^t \sigma_s \, dW_s \, dt \\
&= \int_0^h \sigma_s \int_s^h \mu_t \, dt \, dW_s \\
&= \int_0^h \sigma_s \int_s^h \mu_1 + \mathcal{O}(h-t) \, dt \, dW_s \\
&= \int_0^h \sigma_s \mu_1 (h-s) + \mathcal{O}((h-s)^2) \, dW_s \\
&= \int_0^h \sigma_0 \mu_1 (h-t) + \mathcal{O}(h^2) \, dW_t \\
&\quad \text{(by the martingale property (20.10) and Itô isometry (20.11))} \\
&= \int_0^h \sigma_0 \mu_1 (h-t) \, dW_t + \mathcal{O}(0; h^5) \\
&= \int_0^h \tfrac{h}{2} \sigma_0 \mu_1 + \sigma_0 \mu_1 (\tfrac{h}{2} - t) \, dW_t + \mathcal{O}(0; h^5) \\
&= \tfrac{1}{2} h \sigma_0 \mu_1 \Delta W - \sigma_0 \mu_1 \int_0^h (t - \tfrac{h}{2}) \, dW_t + \mathcal{O}(0; h^5).
\end{aligned}$$

The first term here matches the first part of the fourth term in the numerical (20.28). The second term on the right-hand side is an integral remainder that will be dealt with after the last item.

- Last, the triple integral

$$\int_0^h \mu_t \int_0^t \mu_s \int_0^s \sigma_r \, dW_r \, ds \, dt = \mathcal{O}(0; h^5),$$

because, as in the last item, changing the order of integration to do the stochastic integral last, the integral transforms to $\int_0^h \mathcal{O}(h^2) \, dW$, which by the martingale property (20.10) and Itô isometry property (20.11) is $\mathcal{O}(0; h^5)$.

Hence we now identify that the difference between the Runge–Kutta-like step (20.28) and the increment (20.29) in the true solution is the error

$$\begin{aligned}
\epsilon_0 = &-\tfrac{1}{2}\mu_1 \sigma_0 h^{3/2} S + \tfrac{1}{2} S \sqrt{h} \Delta\sigma + \sigma_0 \mu_1 \int_0^h (t - \tfrac{h}{2}) \, dW_t \\
&- \frac{\Delta\sigma}{h} \int_0^h (t - \tfrac{h}{2}) \, dW_t + \mathcal{O}(h^3) + \mathcal{O}(0; h^5) \\
= &\left[\tfrac{1}{2} S h^{3/2} - \int_0^h (t - \tfrac{h}{2}) \, dW_t \right] \left\{ -\mu_1 \sigma_0 + \frac{\Delta\sigma}{h} \right\} + \mathcal{O}(h^3; h^5).
\end{aligned} \tag{20.31}$$

Two cases arise corresponding to the main and the provisional parts of Lemma 20.12.

- In the general case, the factor in square brackets, $[\cdot]$, in (20.31) determines the order of error. Choosing the signs S randomly with mean zero, we have $Sh^{3/2} = \mathcal{O}(0; h^3)$. Recall that the integral $\int_0^h (t - \frac{h}{2})\, dW_t = \mathcal{O}(0; h^3)$ also. Thus the leading error is then $\mathcal{O}(h^3; h^3)$. This is the local one-step error. Summing over $\mathcal{O}(1/h)$ time steps gives that the global error is $\mathcal{O}(h^2; h^2)$. That is, the error due to the noise dominates, variance $\mathcal{O}(h^2)$, and is generally first order in h, as the standard deviation of the error is of order h.

  But as the noise decreases to zero, $b \to 0$, the factor in curly braces, $\{\cdot\}$, goes to zero. In this decrease, the order of error (20.31) transitions smoothly to the deterministic case of local error $\mathcal{O}(h^3)$ and hence global error $\mathcal{O}(h^2)$.

- The second case is when the factor in curly braces in (20.31) is small: this case occurs for the integrable case $\mu\sigma = d\sigma/dt$, as then the term in braces is $\mathcal{O}(h)$, so that the whole error (20.31) becomes $\mathcal{O}(h^3; h^5)$. Again this is the local one-step error. Summing over $\mathcal{O}(1/h)$ time steps gives that the global error is $\mathcal{O}(h^2; h^4)$. That is, in this case the error is of second order in time step h, through both the deterministic error and the variance of the stochastic errors. Figure 20.17 shows another case when the error is second order.

This concludes the proof. □

Interestingly, we would decrease the size of the factor in brackets in the error (20.31) by choosing the sign S to cancel as much as possible the integral $\int_0^h (t - \frac{h}{2})\, dW_t$. This substep integral is one characteristic of the substep structure of the noise and is independent of ΔW. If we knew this integral, then we could choose the sign S to cause some error cancellation; however, generally we do not know the substep integral. But this connection between the signs S and the integral $\int_0^h (t - \frac{h}{2})\, dW_t$ does suggest that the sign S relates to substep characteristics of the noise process W.

For example, if one uses Brownian bridges to successively refine the numerical approximations for smaller and smaller time steps, then it may be preferable to construct a Brownian bridge compatible with the signs S used on the immediately coarser step size.[250]

20.4.4 ▪ Global error for general SDEs

Section 20.4.3 established the order of error for a special class of linear SDEs. The procedure is to repeatedly substitute integral expressions for the unknown wherever it appears (analogous to Picard iteration). In section 20.4.3 each substitution increased the number of integrals in the expression by two. For general SDEs, this section employs the same procedure, but now the number of integrals doubles in each substitution. The rapid increase in the number of integrals is a major complication, so we only consider the integrals necessary to establish that the global error is $\mathcal{O}(h)$.

Further, the following theorem is also proved for vector SDEs in $\mathbb{R}^n$, whereas the previous two sections only considered special scalar SDEs.

---

[250]The Brownian bridge stochastically interpolates a Wiener process to half-steps in time if all one knows is the increment ΔW over a time step h. The Brownian bridge asserts that the change over half the time step, $h/2$, is $\frac{1}{2}\Delta W - \frac{1}{2}\sqrt{h}Z$ for some $Z \sim N(0,1)$; the change over the second half of the time step is correspondingly $\frac{1}{2}\Delta W + \frac{1}{2}\sqrt{h}Z$. Factoring out the half, these substeps are $\frac{1}{2}(\Delta W \mp Z\sqrt{h})$, which match the factors $(\Delta W \mp S\sqrt{h})$ used by the scheme (20.20): the discrete signs $S = \mp 1$ have mean zero and variance 1, just like the normally distributed Z of the Brownian bridge.

Theorem 20.13. *The Runge–Kutta-like numerical Algorithm 20.4 generally has global error $\mathcal{O}(h)$ when applied to the SDE (20.19) for sufficiently smooth drift and volatility functions $\mu(t,x)$ and $\sigma(t,x)$.*

Proof. The proof has two parts: the first is the well-known standard expansion of the solution of the general SDE (20.19) by iterated stochastic integrals leading to the Milstein scheme (Higham, 2001; Kloeden, 2001, e.g.); the second shows how the step (20.20) matches the integrals to the given order of error.

First look at the repeated integrals for one time step; without loss of generality, start with a time step from $t_0 = 0$ to $t_1 = t_0 + h = h$, as the analysis for all other time steps is identical with shifts in the times of evaluation and integration. The stochastic Taylor series analysis starts from the integral form of Itô formula (20.12): for a stochastic process $X(t)$ satisfying the general Itô SDE (20.19), for operators L_t^k, any smooth function $f(t,x)$ of the process satisfies

$$f(t,X_t) = f(0,X_0) + \int_0^t L_s^0 f(s,X_s)\,ds + \int_0^t L_s^1 f(s,X_s)\,dW_s\,, \tag{20.32}$$

$$\text{where}\quad L_s^0 = \left[\frac{\partial}{\partial t} + \mu_i\frac{\partial}{\partial x_i} + \frac{1}{2}\sigma_i\sigma_j\frac{\partial^2}{\partial x_i\partial x_j}\right]_{t=s}, \quad L_s^1 = \left[\sigma_i\frac{\partial}{\partial x_i}\right]_{t=s}.$$

For conciseness we use subscripts 0, t, s, and r to denote evaluation at these times, and similarly $f_t = f(t,X_t)$, and use subscripts i and j to denote components of a vector, with the Einstein summation convention for repeated indices. As you would expect, when stochastic effects are absent, volatility $\sigma = 0$, the integral formula (20.32) reduces, through the first two components of L_s^0, to an integral version of the well-known deterministic chain rule: $f(t,X_t) = f(0,X_0) + \int_0^t [\partial_t f(s,X_s) + \mu_i\partial_{x_i} f(s,X_s)]\,ds$. Now turn to the SDE (20.19) itself: it is a differential version of an integral equation, which over the first time step gives

$$\begin{aligned}
\Delta X = X(h,\omega) - X(0,\omega) &= \int_{t=0}^{t=h} dX \\
&= \int_0^h \mu(t,X_t)\,dt + \int_0^h \sigma(t,X_t)\,dW_t \\
&\quad\text{(applying the Itô formula (20.32) to both } \mu(t,X_t) \text{ and } \sigma(t,X_t)) \\
&= \int_0^h \left[\mu_0 + \int_0^t L_s^0\mu_s\,ds + \int_0^t L_s^1\mu_s\,dW_s\right] dt \\
&\quad + \int_0^h \left[\sigma_0 + \int_0^t L_s^0\sigma_s\,ds + \int_0^t L_s^1\sigma_s\,dW_s\right] dW_t \\
&\quad\text{(applying the Itô formula (20.32) to } L_s^1\sigma_s) \\
&= \int_0^h \mu_0\,dt + \int_0^h\int_0^t L_s^0\mu_s\,ds\,dt + \int_0^h\int_0^t L_s^1\mu_s\,dW_s\,dt \\
&\quad + \int_0^h \sigma_0\,dW_t + \int_0^h\int_0^t L_s^0\sigma_s\,ds\,dW_t \\
&\quad + \int_0^h\int_0^t \left[L_0^1\sigma_0 + \int_0^s L_r^0L_r^1\sigma_r\,dr + \int_0^s L_r^1L_r^1\sigma_r\,dW_r\right] dW_s\,dW_t \\
&\quad\text{(rearranging these eight integrals in order of magnitude)}
\end{aligned}$$

$$
\begin{aligned}
&= \mu_0 \int_0^h dt + \sigma_0 \int_0^h dW_t + L_0^1\sigma_0 \int_0^h \int_0^t dW_s\, dW_t \\
&\quad + \left[\int_0^h \int_0^t L_s^1 \mu_s\, dW_s\, dt + \int_0^h \int_0^t L_s^0 \sigma_s\, ds\, dW_t \right. \\
&\qquad \left. + \int_0^h \int_0^t \int_0^s L_r^1 L_r^1 \sigma_r\, dW_r\, dW_s\, dW_t \right] \\
&\quad + \left\{ \int_0^h \int_0^t L_s^0 \mu_s\, ds\, dt + \int_0^h \int_0^t \int_0^s L_r^0 L_r^1 \sigma_r\, dr\, dW_s\, dW_t \right\}.
\end{aligned} \tag{20.33}
$$

- Simplify the first line in this last expression (20.33) for $\Delta\mathbf{X}$ using the well-known integrals $\int_0^h dt = h$, $\int_0^h dW_t = \Delta W$, and $\int_0^h \int_0^t dW_s\, dW_t = \int_0^h W_t\, dW_t = \frac{1}{2}(\Delta W^2 - h)$ (Higham, 2001, eq. (3.6), e.g.). The last of these three integrals follows from applying Itô's formula applied to $F(t, W_t) = \frac{1}{2}W_t^2$ to deduce $dF = \frac{1}{2}\,dt + W_t\, dW_t$, and integrating a rearrangement gives $\int W_t\, dW_t = \int dF - \int \frac{1}{2}\,dt = \frac{1}{2}W_t^2 - \frac{1}{2}t$. Also simplify the first line by defining the matrix $\sigma_0' = \left[\partial\sigma_i/\partial x_j\right]_{t=0}$ so that $L_0^1\sigma_0 = \sigma_0'\sigma_0$.

- The three integrals in square brackets in expression (20.33) all have expectation zero and variance $\mathcal{O}(h^3)$. Recall that with two arguments, $\mathcal{O}(h^p; h^q)$ denotes quantities with mean $\mathcal{O}(h^p)$ and variance $\mathcal{O}(h^q)$. Thus these three integrals in square brackets are $\mathcal{O}(0; h^3)$.

- The two integrals in curly braces in expression (20.33) are all $\mathcal{O}(h^2)$ in magnitude and hence are $\mathcal{O}(h^2; h^4)$.

Combining all these leads to the well-established *Milstein scheme* for the change in $\mathbf{X}$ over one time step from t_0 to t_1 as

$$
\Delta\mathbf{X} = \mu_0 h + \sigma_0 \Delta W + \sigma_0'\sigma_0 \tfrac{1}{2}(\Delta W^2 - h) + \mathcal{O}(h^2; h^3). \tag{20.34}
$$

Second, we proceed to show that the time step (20.20) matches Milstein scheme (20.34). Note that $\mathbf{K}_1 = h\mu_0 + (\Delta W - S\sqrt{h})\sigma_0 = \mathcal{O}(h; h)$, so the product $\mathbf{K}_1\mathbf{K}_1 = \mathcal{O}(h; h^2)$, and so on. Hence, by Taylor series in the arguments of the smooth drift μ and volatility σ,

$$
\begin{aligned}
\mathbf{K}_2 &= h\left[\mu_0 + \mu_0'\mathbf{K}_1 + \mathcal{O}(h; h^2)\right] \\
&\quad + (\Delta W + S\sqrt{h})\left[\sigma_0 + h\dot{\sigma}_0 + \sigma_0'\mathbf{K}_1 + \tfrac{1}{2}\sigma_0''\mathbf{K}_1\mathbf{K}_1 + \mathcal{O}(h^2; h^3)\right],
\end{aligned}
$$

where $\sigma''\mathbf{K}_1\mathbf{K}_1$ denotes the tensorial double sum $\partial^2\sigma/\partial x_i \partial x_j K_{1i}K_{1j}$ and where the overdot denotes the partial derivative with respect to time, $\dot{\sigma} = \partial\sigma/\partial t$. Combining $\mathbf{K}_1$ and $\mathbf{K}_2$, the corresponding first step (20.20) predicts the change

$$
\begin{aligned}
\Delta\mathbf{X} &= \mu_0 h + \sigma_0\Delta W + \tfrac{1}{2}\sigma_0'\sigma_0(\Delta W^2 - S^2 h) \\
&\quad + \tfrac{1}{2}(\Delta W - S\sqrt{h})\left[h\mu_0'\sigma_0 + \tfrac{1}{2}(\Delta W^2 - S^2 h)\sigma_0''\sigma_0\sigma_0\right] \\
&\quad + \tfrac{1}{2}h(\Delta W + S\sqrt{h})(\dot{\sigma}_0 + \sigma_0'\mu_0) + \mathcal{O}(h^2; h^4).
\end{aligned} \tag{20.35}
$$

Provided $S^2 = 1 + \mathcal{O}(h; h)$, the first lines match to $\mathcal{O}(h^2; h^3)$: normally $S^2 = 1$ as specified in (20.20). Other terms detailed in (20.35) are $\mathcal{O}(0; h^3)$, provided $E(S) = \mathcal{O}(0; h)$, normally set to be zero, as specified in (20.20). Hence one step of Algorithm 20.4 matches the solution to $\mathcal{O}(h^2; h^3)$. The local error over one step of $\mathcal{O}(h^2; h^3)$ leads to, over $\mathcal{O}(1/h)$ steps, a global error of $\mathcal{O}(h; h^2)$. □

This proof confirms the order of error seen in the earlier examples. Further, because we can readily transform between Itô and Stratonovich SDEs, we now prove that a minor variation of the numerical scheme applies to Stratonovich SDEs.

Corollary 20.14 (Stratonovich SDEs). *The Runge–Kutta-like Algorithm* 20.4, *upon setting* $S = 0$, *has errors* $\mathcal{O}(h)$ *when the SDE* (20.19) *is to be interpreted in the Stratonovich sense, that is, as* $\dot{X} = \mu(t, X) + \sigma(t, X)\dot{W}$.

Proof. Interpreting the SDE (20.19) in the Stratonovich sense implies that solutions are the same as the solutions of the Itô SDE

$$dX = (\mu + \tfrac{1}{2}\sigma'\sigma)\, dt + \sigma\, dW.$$

Apply the time step (20.20) (with $S = \pm 1$ as appropriate to an Itô SDE), or the analysis of the previous proof, to this Itô SDE. Then, for example, the one-step change (20.35) becomes

$$\Delta X = (\mu_0 + \tfrac{1}{2}\sigma_0'\sigma_0)h + \sigma_0 \Delta W + \tfrac{1}{2}\sigma_0'\sigma_0(\Delta W^2 - h) + \mathcal{O}(h^2; h^3).$$

The component of the deterministic drift term that involves $\sigma_0'\sigma_0$ cancels, leaving, in terms of the coefficient functions of the Stratonovich SDE,

$$\Delta X = \mu_0 h + \sigma_0 \Delta W + \tfrac{1}{2}\sigma_0'\sigma_0 \Delta W^2 + \mathcal{O}(h^2; h^3). \tag{20.36}$$

Now apply the scheme (20.20) with $S = 0$ to the Stratonovich SDE: Taylor series expansions obtain the one-step numerical prediction as (20.35) upon setting $S = 0$. This one-step numerical prediction is the same as (20.36) to the same order of errors. Thus Algorithm 20.4 with $S = 0$ solves the Stratonovich interpretation of the SDE (20.19). □

20.5 ▪ The Fokker–Planck equation describes the probability density distribution

The stochastic calculus described so far concentrates on the properties of individual realizations of a stochastic process. Alternatively we may describe the statistics of stochastic processes: their mean and variance or, more generally, the probability distribution of the realizations. Recall that in Markov chains we typically discuss the probability density distribution over the states of the chain (Kao, 1997, Chap. 4, e.g.). In Markov chains, the transition matrix guides the evolution of the probabilities; for SDEs, the Fokker–Planck equation analogously describes the evolution of the probability density distribution.

Example 20.18 (Wiener processes diffuse). Consider a collection of realizations of a Wiener process $W(t)$; the five realizations shown in Figure 20.2 and the ten realizations of Figure 20.18 provide examples. As time increases, these realizations spread—they diffuse. Almost immediately from Definition 20.1 we know that at any time t a Wiener process is distributed as an $N(0, t)$ random variable. Thus its

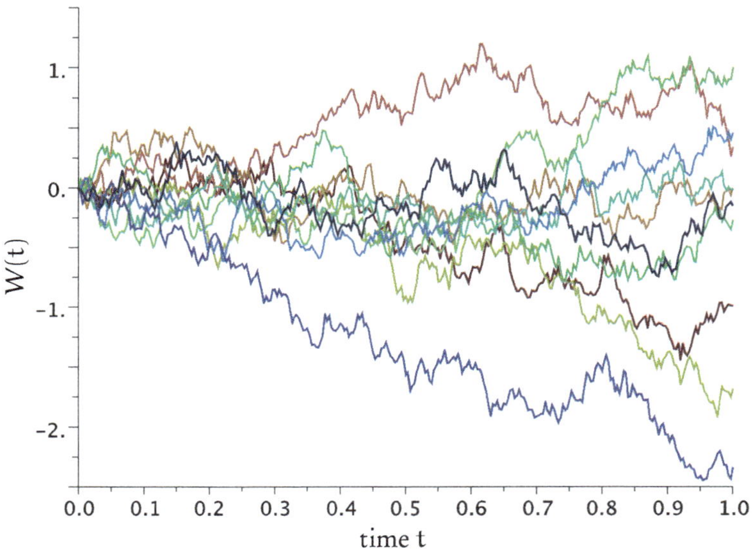

Figure 20.18. *Ten realizations of a Wiener process* $W(t)$.

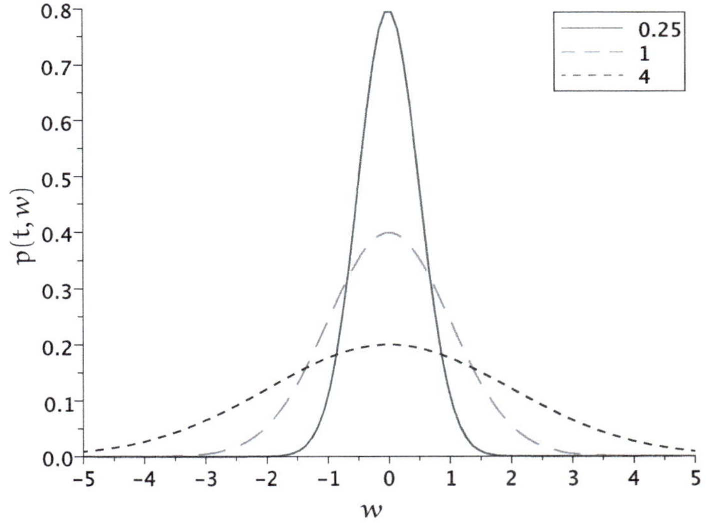

Figure 20.19. *The spreading Gaussian* PDF $p(t,w)$ *of a Wiener process at three times:* $t = 0.25$, $t = 1$, *and* $t = 4$.

probability distribution function (PDF) is the Gaussian normal distribution $p(t,w) = \exp(-w^2/2t)/\sqrt{2\pi t}$: Figure 20.19 plots this Gaussian and shows that it spreads in time corresponding to the spreading realizations. Recall that this Gaussian distribution $p(t,w)$ satisfies the PDE for diffusion (Kreyszig, 2011, §11.5–6, e.g.)

$$\frac{\partial p}{\partial t} = \tfrac{1}{2}\frac{\partial^2 p}{\partial w^2}\,.$$

This PDE, which governs the evolution of the PDF $p(t, w)$, is the simplest example of a Fokker–Planck equation describing the evolving distribution of a stochastic process.

Useful statistics are obtained from the PDF and sometimes show overall features missed by individual realizations of the SDE! Further, when we cannot solve the SDE, sometimes we can solve the Fokker–Planck equation and hence make at least partial analytic progress.

Example 20.19 (expect rare large fluctuations). The solution of the Itô SDE $dX = \alpha X\,dt + \beta X\,dW$ is the stochastic process $X(t) = \exp[(\alpha - \frac{1}{2}\beta^2)t + \beta W(t)]$. This process almost always decays to zero provided $\alpha - \frac{1}{2}\beta^2 < 0$ (even when the drift rate $\alpha > 0$). However, this remarkable statement misleads, as rare fluctuations are significant.

Determining the expected value of $X(t)$ from its PDF illustrates the significance of the rare fluctuations. First, determine that

$$E\left[e^{\beta W(t)}\right] = e^{\beta^2 t/2}, \tag{20.37}$$

using the PDF of the Wiener process. From the definition of expectation in terms of the PDF:

$$\begin{aligned}
E[e^{\beta W(t)}] &= \int_{-\infty}^{\infty} e^{\beta w} p(t, w)\,dw \\
&= \int_{-\infty}^{\infty} e^{\beta w} \frac{1}{\sqrt{2\pi t}} e^{-w^2/2t}\,dw \\
&= \int_{\infty}^{\infty} \frac{1}{\sqrt{2\pi t}} \exp\left[-\frac{w^2}{2t} + \beta w\right] dw \\
&= \int_{-\infty}^{\infty} \frac{1}{\sqrt{2\pi t}} \exp\left[-\frac{1}{2t}(w - t\beta)^2 + \frac{\beta^2 t}{2}\right] dw \quad \text{(upon completing the square)} \\
&= e^{\beta^2 t/2} \int_{-\infty}^{\infty} \frac{1}{\sqrt{2\pi t}} e^{-(w - t\beta)^2/(2t)}\,dw \\
&= e^{\beta^2 t/2} \int_{-\infty}^{\infty} \frac{1}{\sqrt{2\pi t}} e^{-u^2/(2t)}\,du \quad \text{(substituting } u = w - t\beta\text{)} \\
&= e^{\beta^2 t/2},
\end{aligned}$$

as the integral of the Gaussian is 1. This derivation confirms the growing expectation (20.37) of fluctuations.

Second, we conclude that the solutions to exponential Brownian motion, although almost always tending to zero for large enough β, nonetheless have a growing expectation whenever the coefficient $\alpha > 0$:

$$\begin{aligned}
E[X(t)] &= E\left[X_0 e^{(\alpha - \beta^2/2)t + \beta W(t)}\right] \\
&= E\left[X_0 e^{(\alpha - \beta^2/2)t} e^{\beta W(t)}\right] \\
&= X_0 e^{(\alpha - \beta^2/2)t}\, E\left[e^{\beta W(t)}\right] \\
&= X_0 e^{(\alpha - \beta^2/2)t} e^{\beta^2 t/2} \\
&= X_0 e^{\alpha t}.
\end{aligned}$$

For example, the almost surely decaying realizations $X(t) = e^{-t+2W(t)}$ of the SDE $dX = X\,dt + 2X\,dW$ (such that $X(0) = 1$) nonetheless have an exponentially growing expectation $E[X(t)] = e^t$. Very rare but very large fluctuations in $X(t)$ must occur so that the expectation grows despite almost all of the realizations decaying to zero.

The Fokker–Planck equation establishes a crucial transformation between the solution of SDEs, their distribution, and certain PDEs.

20.5.1 ▪ Probability density distributions evolve forward in time

We now discover how to describe the evolution of the probability density distribution function $p(t,x)$ for a general Itô process with some drift and volatility.

Analogue with Markov chains Consider a PDF $p(t,x)$. The PDF $p(t,x)$, being the probability of realizations being near x at time t, is analogous to the vector of probabilities $\mathbf{p}(t)$, where the jth component, $p_j(t)$, is the probability of being in a state j at time t discussed in Markov chains such as birth and death processes (Kao, 1997, §4.1, e.g.).

Theorem 20.15 (Fokker–Planck equation). *Consider the Itô process $X(t)$ with drift $\mu(t,X)$ and volatility $\sigma(t,X)$ and hence satisfying the SDE $dX = \mu\,dt + \sigma\,dW$. Then the PDF of the ensemble of realizations $p(t,x)$ satisfies the* **Fokker–Planck equation,** *alternatively called the* **Kolmogorov forward equation,**

$$\frac{\partial p}{\partial t} = -\frac{\partial}{\partial x}(\mu p) + \frac{\partial^2}{\partial x^2}\left(\tfrac{1}{2}\sigma^2 p\right). \tag{20.38}$$

Further, in n dimensions, for $\mathbf{X}(t) \in \mathbb{R}^n$, with m noises, Wiener processes $\mathbf{W}(t) \in \mathbb{R}^m$, the Fokker–Planck equation for the PDF $p(t,\mathbf{x})$ of the Itô SDE $d\mathbf{X} = \boldsymbol{\mu}\,dt + \sigma\,d\mathbf{W}$ for drift vector $\boldsymbol{\mu}(t,\mathbf{x}) \in \mathbb{R}^n$ and volatility matrix $\sigma(t,\mathbf{x}) \in \mathbb{R}^{n\times m}$ is

$$\frac{\partial p}{\partial t} = -\boldsymbol{\nabla}\cdot(\boldsymbol{\mu}p) + \boldsymbol{\nabla}^2 : \left(\tfrac{1}{2}\sigma\sigma^T p\right) = -\sum_{i=1}^{n}\frac{\partial(\mu_i p)}{\partial x_i} + \sum_{i,j=1}^{n}\frac{\partial^2(D_{ij}p)}{\partial x_i \partial x_j} \tag{20.39}$$

for symmetric diffusivity tensor $D_{ij}(t,\mathbf{x}) := \frac{1}{2}\sum_{k=1}^{m}\sigma_{ik}\sigma_{jk}$.

Stratonovich Fokker–Planck equation Beware: the Fokker–Planck equation for a Stratonovich SDE has a different algebraic form. The Fokker–Planck equation for the PDF $p(t,\mathbf{x})$ of the Stratonovich SDE $\dot{\mathbf{X}} = \boldsymbol{\mu} + \sigma\dot{\mathbf{W}}$ is

$$\frac{\partial p}{\partial t} = -\sum_{i=1}^{n}\frac{\partial(\mu_i p)}{\partial x_i} + \frac{1}{2}\sum_{k=1}^{m}\sum_{i,j=1}^{n}\frac{\partial}{\partial x_i}\left[\sigma_{ik}\frac{\partial}{\partial x_j}(\sigma_{jk}p)\right] \tag{20.40}$$

or, equivalently, upon defining $\boldsymbol{\sigma}_k(t,\mathbf{x}) \in \mathbb{R}^n$ to be the kth column of volatility matrix σ,

$$\frac{\partial p}{\partial t} = -\boldsymbol{\nabla}\cdot(\boldsymbol{\mu}p) + \frac{1}{2}\sum_{k=1}^{m}\boldsymbol{\nabla}\cdot[\boldsymbol{\sigma}_k\boldsymbol{\nabla}\cdot(\boldsymbol{\sigma}_k p)].$$

These Stratonovich Fokker–Planck equations follow straightforwardly from the Fokker–Planck equation (20.39) for the Itô SDE that corresponds to the given Stratonovich SDE, obtained via Theorem 20.10.

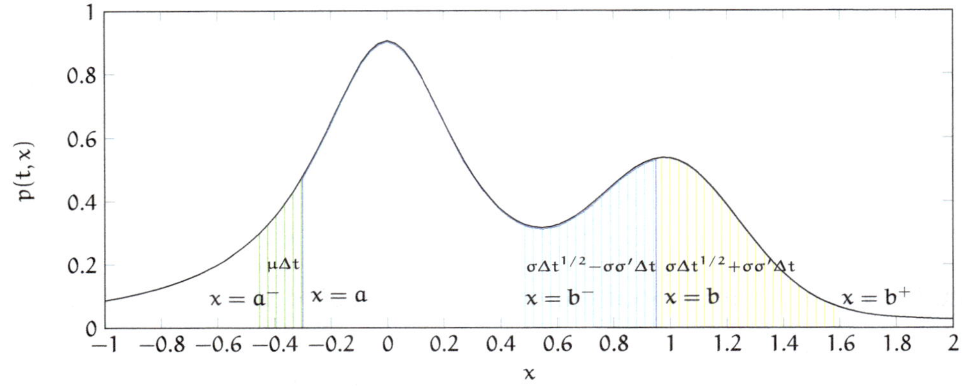

Figure 20.20. *The PDF* $p(t,x)$ *over an interval* $[a,b]$ *increases by drift into the interval from the left (green), by drift out of the interval to the right (not shown), and by random jumps into (yellow) or out of (cyan) the interval near each end but only shown on the right.*

Analogue The Fokker–Planck equation is analogous to the evolution of probability distributions in Markov chains: $p(t+1) = p(t)P$ for some transition matrix P. In the Fokker–Planck equation (20.38), the x-derivatives involving the drift and volatility operate exactly like the transition matrix P—they operate to dictate how probability distributions evolve in time.

Example 20.20 (Wiener process Fokker–Planck). Recall the Wiener process of Example 20.18. It satisfies the Itô SDE $dX = 0\,dt + 1\,dW$. The Fokker–Planck equation (20.38) for its probability distribution then must be

$$\frac{\partial p}{\partial t} = -\frac{\partial}{\partial x}(0p) + \frac{\partial^2}{\partial x^2}(\tfrac{1}{2}1^2 p) = \tfrac{1}{2}\frac{\partial^2 p}{\partial x^2},$$

namely the well-known basic diffusion equation.

Proof of Theorem 20.15. For simplicity we only derive the scalar Fokker–Planck equation (20.38).

Let's use conservation of probability in 1D (Roberts, 1994, e.g.). Consider any interval $[a,b]$ of the x-line such as illustrated in Figure 20.20. The total probability of random variable X being in this interval is $P(t) := \int_a^b p(t,x)\,dx$. Let's explore how this total probability changes with time.

Subsequent paragraphs argue that over the interval $[a,b]$ the total probability changes according to

$$\frac{dP}{dt} = (\mu p)|_{x=a} - (\mu p)|_{x=b} - \frac{\partial}{\partial x}\left(\tfrac{1}{2}\sigma^2 p\right)\Big|_{x=a} + \frac{\partial}{\partial x}\left(\tfrac{1}{2}\sigma^2 p\right)\Big|_{x=b}. \tag{20.41}$$

Using the fundamental theorem of calculus to write these differences as integrals of their derivative, this rate equation implies

$$\frac{dP}{dt} = \int_a^b \frac{\partial p}{\partial t}\,dx = -\int_a^b \frac{\partial}{\partial x}(\mu p)\,dx + \int_a^b \frac{\partial^2}{\partial x^2}\left(\tfrac{1}{2}\sigma^2 p\right)dx.$$

That is, gathering all terms all under the one integral,

$$\int_a^b \left[\frac{\partial p}{\partial t} + \frac{\partial}{\partial x}(\mu p) - \frac{\partial^2}{\partial x^2}\left(\tfrac{1}{2}\sigma^2 p\right)\right] dx = 0\,.$$

As is standard in conservation arguments (Roberts, 1994, e.g.), for this integral to be zero for all intervals $[a, b]$ requires the integrand to be everywhere zero in the domain of x. Here, setting the integrand to zero establishes the Fokker–Planck equation (20.38).

Now let's justify equation (20.41), which led to this Fokker–Planck equation. There are two mechanisms in (20.41) for the probability P to change: via the drift $\mu\, dt$ of the SDE, or via stochastic fluctuations $\sigma\, dW$. We quantify each in turn.

Choose a small time increment Δt. During such a small time increment the drift moves probability to the right a distance $\mu\Delta t$ in x (to the left when $\mu < 0$). Consequently, over a time Δt the drift moves a "slab" of probability of width $\mu\Delta t$, green in Figure 20.20, *into* the interval $[a, b]$ from the left, that is, an amount $\mu\Delta t\, p|_{x=a}$. Correspondingly, the drift moves a slab of probability of width $\mu\Delta t$, not shown in Figure 20.20, *out of* the interval $[a, b]$ to the right; that is, an amount $\mu\Delta t\, p|_{x=b}$ leaves the interval. Thus the change in total probability P from the drift over the time increment Δt is

$$\Delta P \approx \mu\Delta t\, p|_{x=a} - \mu\Delta t\, p|_{x=b}.$$

Dividing by Δt and letting $\Delta t \to 0$ gives the first two terms on the right-hand side of (20.41).

Stochastic fluctuations also move probability into and out of the interval $[a, b]$, but we need more care to justify the remaining two terms in (20.41). Over a small time increment Δt, the stochastic fluctuations move X by an amount $\sigma\Delta W$, where $\Delta W \sim N(0, \Delta t)$. The continuous distribution $N(0, \Delta t)$ is sufficiently well modeled on the microscale by simply choosing $\pm\sqrt{\Delta t}$, each with probability $\frac{1}{2}$; thus we set the stochastic "jumps" in X to be $\pm\sigma\sqrt{\Delta t}$, each with probability $\frac{1}{2}$. The difficulty is that the size of the jumps varies in x with the volatility $\sigma(t, x)$.

Let's analyze the stochastic jumps just at the right end of the interval $[a, b]$.

- Figure 20.20 also shows that a cyan slab of probability, and all those in this slab that jump to the right (half of them), will leave the interval $[a, b]$ across $x = b$.

- Conversely, Figure 20.20 shows that a yellow slab of probability, and all those in this slab that jump to the left (half of them), will enter the interval $[a, b]$ across $x = b$.

The end points of these slabs, $x = b^{\pm}$ shown in Figure 20.20, are determined by the varying size of the jumps as the solution $b^{\pm}$ of the algebraic equation $b = b^{\pm} \mp \sigma(t, b^{\pm})\sqrt{\Delta t}$. Rearrange this equation to $b^{\pm} = b \pm \sigma(t, b^{\pm})\sqrt{\Delta t}$. Solving approximately by iteration, a first approximation to these end points of the slabs is $b^{\pm} = b \pm \sigma(t, b)\sqrt{\Delta t} + \mathcal{O}(\Delta t)$. But the $\mathcal{O}(\Delta t)$ error is too big for our purposes, so we invoke another iteration to find

$$\begin{aligned} b^{\pm} &\approx b \pm \sigma\left(t, b \pm \sigma(t, b)\sqrt{\Delta t}\right)\sqrt{\Delta t} \\ &\approx b \pm \sigma(t, b)\sqrt{\Delta t} + \frac{\partial\sigma}{\partial x}(t, b)\sigma(t, b)\Delta t. \end{aligned}$$

This gives the widths of the slabs, namely $\sigma\sqrt{\Delta t} \pm \sigma\partial\sigma/\partial x\Delta t$, as illustrated in Figure 20.20. Because these slabs are relatively wide, namely $\mathcal{O}(\sqrt{\Delta t})$, we use the midpoint rule to estimate the area of probability in each slab, as colored in Figure 20.20: the areas are, where quantities are evaluated at $x = b$ except where specified in the first line,

$$\begin{aligned}\Delta P &\approx p\left(t, b \pm \tfrac{1}{2}\sigma\sqrt{\Delta t} + \tfrac{1}{2}\sigma\frac{\partial\sigma}{\partial x}\Delta t\right) \times \left[\sigma\sqrt{\Delta t} \pm \sigma\frac{\partial\sigma}{\partial x}\Delta t\right]\\ &\approx \left[p \pm \tfrac{1}{2}\sigma\sqrt{\Delta t}\frac{\partial p}{\partial x}\right] \times \left[\sigma\sqrt{\Delta t} \pm \sigma\frac{\partial\sigma}{\partial x}\Delta t\right]\\ &\approx p\sigma\sqrt{\Delta t} \pm \left(\tfrac{1}{2}\sigma^2\frac{\partial p}{\partial x} + \sigma\frac{\partial\sigma}{\partial x}p\right)\Delta t = p\sigma\sqrt{\Delta t} \pm \frac{\partial}{\partial x}\left(\tfrac{1}{2}\sigma^2 p\right)\Delta t\,.\end{aligned}$$

Recall that of these, only half jump across $x = b$, and also that half leave from the left and half enter from the right: the total stochastic contribution is then

$$\begin{aligned}\Delta P &\approx -\tfrac{1}{2}\left[p\sigma\sqrt{\Delta t} - \frac{\partial}{\partial x}\left(\tfrac{1}{2}\sigma^2 p\right)\Delta t\right] + \tfrac{1}{2}\left[p\sigma\sqrt{\Delta t} + \frac{\partial}{\partial x}\left(\tfrac{1}{2}\sigma^2 p\right)\Delta t\right]\\ &\approx \frac{\partial}{\partial x}\left(\tfrac{1}{2}\sigma^2 p\right)\Delta t\,.\end{aligned}$$

Dividing by Δt and letting $\Delta t \to 0$, this contribution gives the last term in equation (20.41). Corresponding arguments at the left end of the interval $[a, b]$ give the third term in equation (20.41), completing the stochastic effects. This concludes our justification of the scalar Fokker–Planck equation (20.38). □

20.5.2 ▪ Steady-state probability density

The next chapter uses the Fokker–Planck equation to explore multiscale stochastic systems in quasi-equilibrium. A first use of the Fokker–Planck equation is simply to predict equilibrium PDFs for some stochastic processes.

Example 20.21 (Ornstein–Uhlenbeck process). The so-called **Ornstein–Uhlenbeck process** combines deterministic exponential decay with an additive noise. The form of the SDE is generally

$$dX = -\alpha X\,dt + \sigma\,dW \tag{20.42}$$

for some constants α and σ measuring the rate of deterministic decay and the level of stochastic forcing, respectively. Figure 20.21 shows 10 realizations, and Figure 20.22 shows how the corresponding PDF, $p(t, x)$, evolves. Derive the steady-state PDF for this process, that is, the density of realizations in Figure 20.21 that emerges *after* the initial transients.

Solution: The Fokker–Planck equation (20.38) for the PDF is

$$\frac{\partial p}{\partial t} = -\frac{\partial}{\partial x}(-\alpha x p) + \frac{\partial^2}{\partial x^2}\left(\tfrac{1}{2}\sigma^2 p\right).$$

Obtain the differential equation for the steady-state PDF by setting $\partial p/\partial t = 0$. As the volatility σ is constant, this Fokker–Planck equation becomes

$$\begin{aligned}0 &= \frac{\partial}{\partial x}\left[\alpha x p + \tfrac{1}{2}\sigma^2\frac{\partial p}{\partial x}\right]\\ \implies \text{constant} &= \alpha x p + \tfrac{1}{2}\sigma^2\frac{\partial p}{\partial x}, \qquad \text{upon integrating in } x.\end{aligned}$$

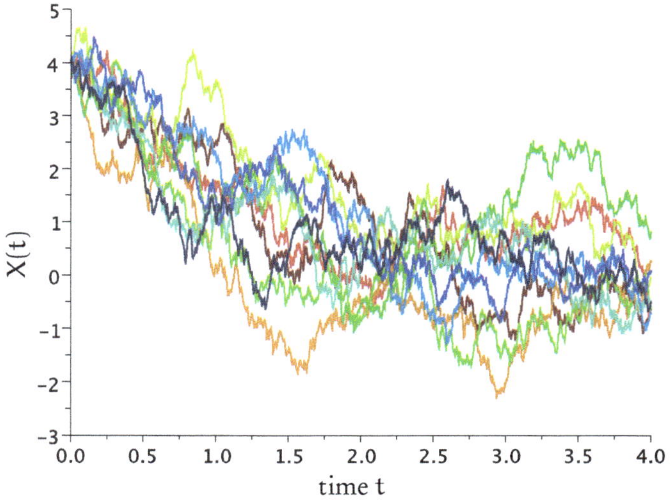

Figure 20.21. *Ten realizations of an Ornstein–Uhlenbeck process* $X(t)$ *with parameters* $\alpha = 1$ *and* $\sigma = \sqrt{2}$*, all with initial condition* $X(0) = 4$*. Each realization decays toward* $X = 0$ *with sizable fluctuations.*

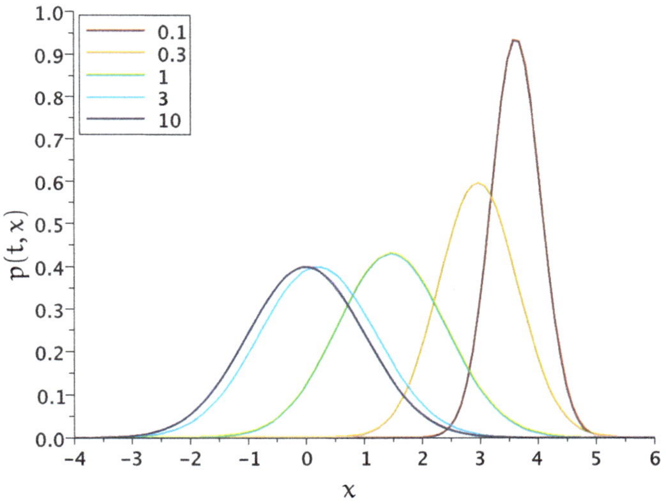

Figure 20.22. *Evolving PDF* $p(t, x)$ *for the Ornstein–Uhlenbeck process* $dX = -X\,dt + \sqrt{2}\,dW$ *with initial condition* $X(0) = 4$ *and plotted at times* $t = 0.1, 0.3, 1, 3,$ *and* 10*. These PDFs match the density of realizations plotted in Figure 20.21 at each corresponding time.*

The constant on the left-hand side is zero, as the PDF $p(t, x)$ and its derivatives must vanish for large enough x for the integral of p to be 1. Rearranging as a separable ODE (Kreyszig, 2011, §1.3–4, e.g.) leads to

$$\begin{aligned} \tfrac{1}{2}\sigma^2 \int \frac{dp}{p} &= \int -\alpha x\,dx \\ \implies \log p &= -\alpha x^2/\sigma^2 + \text{constant} \\ \implies p &= Ce^{-\alpha x^2/\sigma^2} \end{aligned}$$

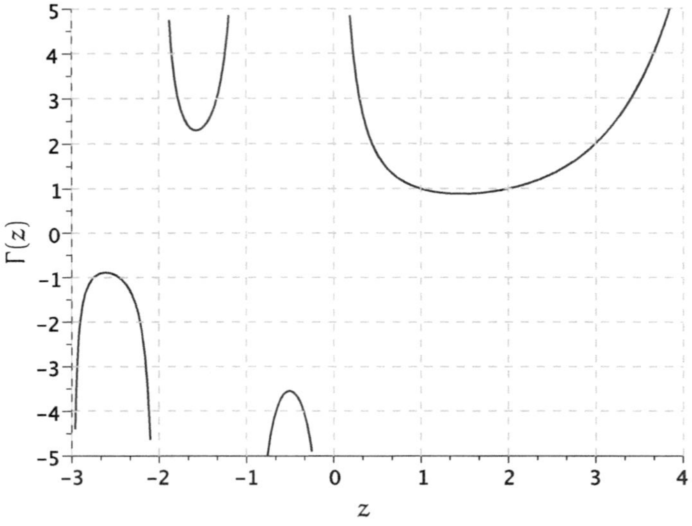

Figure 20.23. *The gamma function* (20.43) *for real argument z.*

for some integration constant C. This Gaussian distribution is centered on $x = 0$ and has width proportional to $\sigma/\sqrt{\alpha}$. The constant of proportionality is well known to be $C = \sqrt{\alpha/\pi}/\sigma$ to ensure that the total probability, the area under $p(x)$, is 1. The additive noise of the Ornstein–Uhlenbeck process has just "smeared out" the stable fixed point at $X = 0$. The width of the smearing is proportional to the strength of the noise, σ, and proportional to the inverse square root of the rate of attraction of the fixed point, $1/\sqrt{\alpha}$.

Analogue The steady-state probability distribution $p(x)$ is analogous to the steady-state distributions π found for discrete-state stochastic models. The difference here is the continuum of possible states. Just as you normalize steady-state distributions, so we normalize here: for a finite number of states $\sum_j \pi_j = 1$, here $\int p(x)\,dx = 1$.

Gamma function The next example invokes the *gamma function*, $\Gamma(z)$, which you possibly have already met in other study (Kreyszig, 2011, pp. A54–A55, e.g.):

$$\Gamma(z) := \int_0^\infty x^{z-1}e^{-x}\,dx, \tag{20.43}$$

as plotted in Figure 20.23 for real z. For example, $\Gamma(1) = \int_0^\infty x^0e^{-x}\,dx = 1$. Integration by parts shows that $\Gamma(z+1) = z\Gamma(z)$. Hence, for example, for integer n, $\Gamma(n+1) = n!$. Other special values are $\Gamma(\frac{1}{2}) = \sqrt{\pi}$ and $\Gamma(\frac{3}{2}) = \frac{1}{2}\sqrt{\pi}$.

Example 20.22 (a two-humped camel). Investigate the steady-state PDF of the SDE $dX = (3X - X^3)dt + X\,dW$ and relate it to the deterministic dynamics. Compare with the steady state of the SDE $dX = (3X - X^3)dt + 2X\,dW$, which has twice the volatility.

Solution: First, investigate the deterministic dynamics of $dX = (3X - X^3)dt$, as done in courses on ODEs (Kreyszig, 2011, §3.3–5, e.g.). This ODE has equilibria (fixed points) where $3X - X^3 = 0$, namely $X = 0$ and $X = \pm\sqrt{3}$: the fixed point at $X = 0$ is unstable, as the linearized dynamics is $dX = 3X\,dt$ with exponentially growing solutions, whereas the fixed points at $X = \pm\sqrt{3}$ are stable, as the local dynamics, say $X = \pm\sqrt{3} + Y(t)$, is $dY = -6Y\,dt$ with exponentially decaying solutions. Thus all deterministic trajectories evolve to one or other of the fixed points $X = \pm\sqrt{3}$.

Second, consider the steady-state PDF $p(x)$ of the SDE $dX = (3X - X^3)dt + X\,dW$. It satisfies the time-independent Fokker–Planck equation

$$0 = -\frac{\partial}{\partial x}\left[(3x - x^3)p\right] + \frac{\partial^2}{\partial x^2}\left[\tfrac{1}{2}x^2 p\right].$$

One integral with respect to x leads to

$$-(3x - x^3)p + \frac{\partial}{\partial x}\left[\tfrac{1}{2}x^2 p\right] = \text{constant},$$

but this constant has to be zero, as p and its derivatives must vanish for large enough x. Expanding, rearranging, and recognizing that the ODE is separable:

$$\begin{aligned}
\tfrac{1}{2}x^2\frac{\partial p}{\partial x} &= (2x - x^3)p \\
\Longrightarrow \int\frac{dp}{p} &= \int\frac{4x - 2x^3}{x^2}\,dx \\
\Longrightarrow \log p &= \int\frac{4}{x} - 2x\,dx \\
\Longrightarrow \log p &= 4\log|x| - x^2 + \text{constant} \\
\Longrightarrow p &= Cx^4 e^{-x^2}.
\end{aligned}$$

Thus the steady-state PDF, as shown in Figure 20.24, is zero near $x = 0$, increases away from $x = 0$ by the x^4 factor, but soon is brought back to zero by the rapid decay of the e^{-x^2} factor. The two humps of the probability distribution correspond to the two stable equilibria of the deterministic ODE.

Requiring that the area equal 1 under the PDF determines the integration constant C: using symmetry, integration over half the domain requires

$$\begin{aligned}
\tfrac{1}{2} &= \int_0^\infty Cx^4 e^{-x^2}\,dx \\
&= \frac{C}{2}\int_0^\infty u^{3/2}e^{-u}du \quad \text{upon substituting } u = x^2 \\
&= \frac{C}{2}\Gamma(5/2) = \frac{3\sqrt{\pi}}{8}A;
\end{aligned}$$

thus $C = 4/(3\sqrt{\pi})$.

Last, a perusal of the deterministic part of the SDE with stronger noise, $dX = (3X - X^3)dt + 2X\,dW$, again suggests that there should be two humps in the PDF near the two stable deterministic equilibria $X = \pm\sqrt{3}$. But it is not so. The steady solutions satisfy the corresponding Fokker–Planck equation

$$0 = -\frac{\partial}{\partial x}\left[(3x - x^3)p\right] + \frac{\partial^2}{\partial x^2}\left[2x^2 p\right].$$

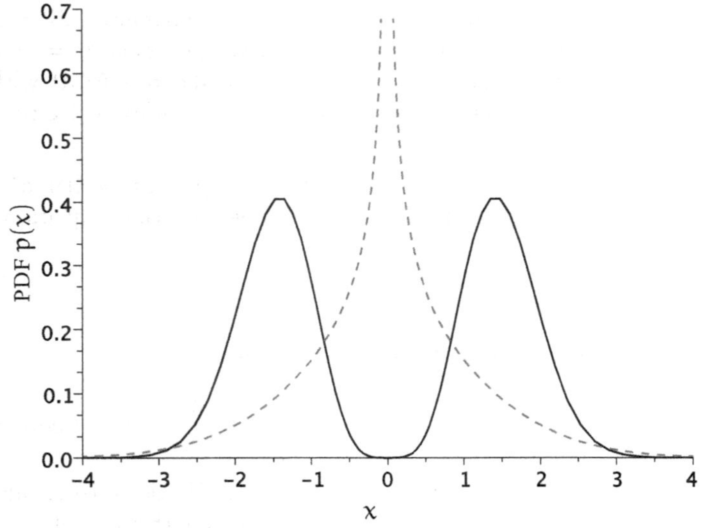

Figure 20.24. *Steady-state probability distributions for Example 20.22: solid,* $dX = (3X - X^3)\,dt + X\,dW$ *with lower volatility has two humps; dashed,* $dX = (3X - X^3)\,dt + 2X\,dW$ *with larger volatility peaks at the origin.*

Derive solutions via

$$\begin{aligned}
&-(3x - x^3)p + \frac{\partial}{\partial x}\left[2x^2 p\right] = 0 \\
\implies\ & 2x^2 \frac{\partial p}{\partial x} = (-x - x^3)p \\
\implies\ & \int \frac{dp}{p} = \int \frac{-x - x^3}{2x^2}\,dx \\
\implies\ & \log p = \int -\frac{1}{2x} - \tfrac{1}{2}x\,dx \\
\implies\ & \log p = -\tfrac{1}{2}\log|x| - \tfrac{1}{4}x^2 + \text{constant} \\
\implies\ & p = \frac{C}{\sqrt{|x|}} e^{-x^2/4}.
\end{aligned}$$

Figure 20.24 shows that doubling the level of the multiplicative noise, now $2X\,dW$, effectively stabilizes the equilibrium at the origin. The large spike of probability at the origin has finite area, as $1/\sqrt{|x|}$ is integrable. Requiring the total area under the PDF to be 1 implies

$$\begin{aligned}
\tfrac{1}{2} &= \int_0^\infty C x^{-1/2} e^{-x^2/4}\,dx \\
&= \frac{C}{\sqrt{2}} \int_0^\infty u^{-3/4} e^{-u}\,du \quad \text{upon substituting } u = x^2/4 \\
&= \frac{C}{\sqrt{2}} \Gamma(1/4)\,;
\end{aligned}$$

thus $C = 1/[\sqrt{2}\Gamma(1/4)] = 0.1950$.

20.6 ▪ Exercises

Exercise 20.1. Pick a normally distributed random variable $Z \sim N(0,1)$; then define $W(t) = Z\sqrt{t}$. Show that $W(t)$ is not a Wiener process.

Exercise 20.2 (linear combination of Wiener). Let $W(t)$ and $\tilde{W}(t)$ be independent Wiener processes and ρ a fixed number $0 < \rho < 1$. Show that the linear combination $x(t) = \rho W(t) + \sqrt{1-\rho^2}\tilde{W}(t)$ is a Wiener process.

Exercise 20.3 (linearity and union). Prove the linearity (20.4) of the Itô integral for step functions. Prove the union (20.5) of the Itô integral for step functions.

Exercise 20.4 (increments vanish). Argue that for exponents $p \geqslant 3$, almost surely the sum $\sum_{j=0}^{n-1} |\Delta W_j|^p \to 0$ as $h \to 0$, where $h = \max_j \Delta t_j$. Use that $E[|Z|^p]$ is finite for $Z \sim N(0,1)$.

Exercise 20.5 (Wiener moments). Consider the kth moment $E\left[W(t)^k\right]$ for a Wiener process $W(t)$. Apply Itô's formula to $d(W(t)^k)$ and invoke the martingale property of Itô integrals to deduce that

$$E\left[W(T)^k\right] = \tfrac{1}{2}k(k-1)\int_0^T E\left[W(t)^{k-2}\right] dt \quad \text{for } k \geqslant 2.$$

Hence determine $E\left[W(t)^2\right]$, $E\left[W(t)^4\right]$, and $E\left[W(t)^6\right]$.

Answer: $E\left[W(t)^2\right] = t$, $E\left[W(t)^4\right] = 3t^2$, and $E\left[W(t)^6\right] = 15t^3$.

Exercise 20.6 (verify integral properties). Consider $d(W^3 - 3tW)$ and hence find the Itô integral $I(\omega) = \int_0^T W(t,\omega)^2 - t\, dW(t,\omega)$, where $W(t,\omega)$ is a Wiener process. Hence verify the martingale property and Itô isometry for this Itô integral.

Answer: $I(\omega) = \frac{1}{3}W(T,\omega)^3 - TW(T,\omega)$ and $\operatorname{Var} I = \frac{2}{3}T^3$.

Exercise 20.7 (verify SDE solutions). Use Itô's formula (20.12) to confirm that the algebraic solutions given below satisfy the corresponding given SDE.

1. $dX = \frac{1}{2}(X-t)\,dt + (X-t-2)\,dW, X(0) = 3$; solution $X = 2 + t + \exp[W(t)]$.
2. $dX = X\,dW, X(0) = 1$; solution $X = \exp[W(t) - t/2]$.
3. $dX = -X(1-X^2)\,dt + (1-X^2)\,dW, X(0) = 0$; solution $X = \tanh[W(t)]$.
4. $dX = -X\,dt + e^{-t}dW, X(0) = 0$; solution $X = e^{-t}W(t)$.
5. $dX = -\frac{3}{2}X(1-X^2)^2dt + (1-X^2)^{3/2}dW$, $X(0) = 0$; solution $X = W(t)/\sqrt{1+W(t)^2}$.
6. $dX = -\left[\frac{1}{2}X + 2t\sqrt{1-X^2}\right]dt - \sqrt{1-X^2}dW, X(0) = 1$; solution $X = \cos(t^2 + W)$.

Exercise 20.8 (compare with numerics). Apply the scheme (20.20) to some of the SDEs of Exercise 20.7. Compare the numerical predictions, for different time step sizes, to the given algebraic solution. For which SDEs is the error $\mathcal{O}(h^2)$?

Exercise 20.9 (Hermite polynomials). Define $I_0(t) := 1$, $I_1(t) := W(t)$, $I_2(t) := W(t)^2 - t$, $I_3(t) := W(t)^3 - 3tW(t)$, $I_4(t) := W(t)^4 - 6tW(t)^2 + 3t^2$. Use Itô's formula to show that $dI_n = nI_{n-1}\,dW$. What is the analogue in classic calculus? Check some guesses with Itô's formula to determine corresponding I_5 and I_6.[251]

Exercise 20.10 (iterated integrals). Consider the scalar SDE $dX = X\,dW$. This SDE is shorthand for the Itô integral $X_t = X_0 + \int_0^t X_s\,dW_s$. Over a small time interval $\Delta t = h$, this integral gives $X_h = X_0 + \int_0^h X_t\,dW_t$. Use this as the start of an iteration to provide successively more accurate approximations to X_h: successive approximations are successive truncations of

$$X_h \approx X_0 + X_0\,\Delta W + X_0\left[\tfrac{1}{2}(\Delta W)^2 - \tfrac{1}{2}h\right] + X_0\left[\tfrac{1}{6}(\Delta W)^3 - \tfrac{1}{2}h\Delta W\right].$$

Determine the integral remainder for each of the approximations.

Exercise 20.11 (quadratic convergence). Adapt the proof of Lemma 20.12 to prove that in the specific case when the drift $\mu = \alpha(t) + \beta(t)X$ and the volatility, independent of x, satisfies $\dot{\sigma} = \beta\sigma$, then the scheme has local error $\mathcal{O}(h^3; h^5)$ and hence global error $\mathcal{O}(h^2; h^4)$, as seen in Figure 20.17.

Exercise 20.12 (variance growth). Consider the process $X(t) = X_0\exp[(\alpha - \frac{1}{2}\beta^2)t + \beta W(t)]$. Show that $\operatorname{Var}[X(t)] = X_0^2\left[e^{(2\alpha+\beta^2)t} - e^{2\alpha t}\right]$. Give an example of parameters α and β that illustrate that even when the expectation decays to zero, the variance may nonetheless grow in time.

Exercise 20.13 (estimate PDFs). Consider in turn each of the stochastic processes illustrated in Figures 20.25–20.28. Each figure shows 10 realizations of some stochastic process, a different stochastic process for each figure. Assuming each stochastic process is settling upon some steady-state PDF $p(x)$, sketch the PDF.

Exercise 20.14. Three out of four realizations plotted in Exercise 20.13 come from Itô processes in Exercise 20.15. Identify as best you can which Itô processes of Exercise 20.15 correspond to which three of the four figures.

[251]These $I_n(t)$ are closely related to the Hermite polynomials (Kreyszig 2011, pp. 246–7 or Abramowitz and Stegun 1965, Chap. 22).

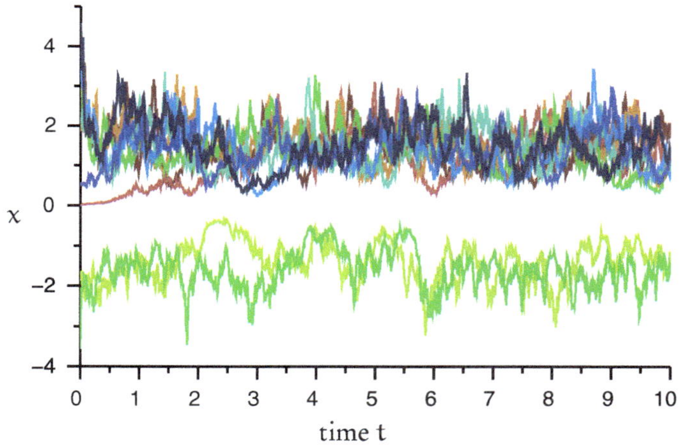

Figure 20.25. *Estimate the PDF from these realizations.*

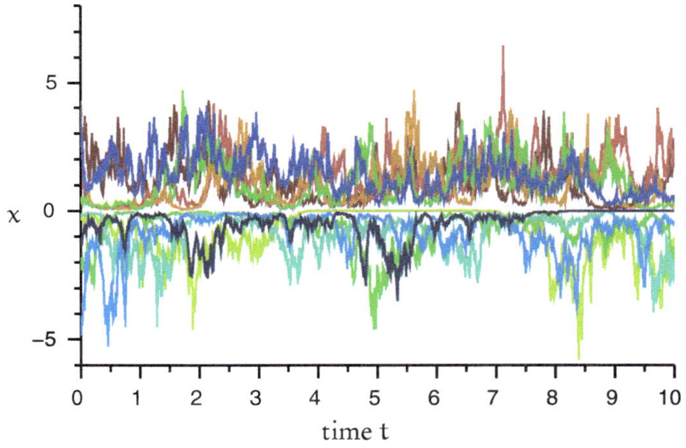

Figure 20.26. *Estimate the PDF from these realizations.*

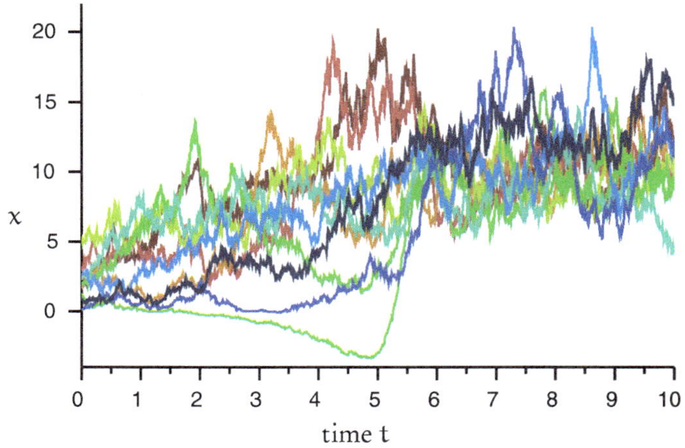

Figure 20.27. *Estimate the PDF from these realizations.*

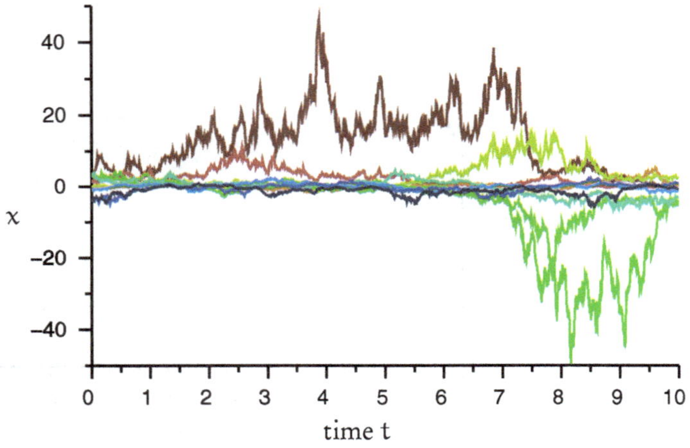

Figure 20.28. *Estimate the PDF from these realizations.*

Exercise 20.15. Use the Fokker–Planck equation (20.38) to determine the steady-state PDFs for the following Itô SDEs:

1. $dX = (2X - X^2)dt + X\,dW$ for $X(t) \geqslant 0$;
2. $dX = (3X - 2X^2)dt + 2X\,dW$ for $X(t) \geqslant 0$;
3. $dX = \beta\sqrt{1+X^2}\,dW$;
4. $dX = -X\,dt + \sqrt{1+X^2}\,dW$;
5. $dX = X\,dt + (1+X^2)\,dW$;
6. $dX = -X\,dt + \sqrt{2}(1+X^2)^{1/4}dW$;
7. $dX = -2X\,dt + \sqrt{1+X^2}\,dW$.

Answer: You should determine the constant of proportionality for the following:

1. $p \propto x^2 e^{-2x}$—the hump in probability near $x = 2$ corresponds to the deterministic fixed point.
2. $p \propto e^{-x}/\sqrt{x}$—here the deterministic equilibrium at $x = 3/2$ is washed out by the noise, which instead stabilizes the origin and generates an integrable peak in probability at $x = 0$.
3. $p \propto 1/(1+x^2)$—for small x the system spreads by a random walk, $dX \propto dW$, but the stabilizing effect of multiplicative noise for large x, where $dX \propto X\,dW$, stabilizes this spreading to generate this "Cauchy distribution," which has long tails (curiously parameter β has no influence on the steady-state distribution).
4. $p \propto 1/(1+x^2)^2$—by stabilizing the origin, the previous distribution has much smaller tails.
5. $p \propto \frac{1}{(1+x^2)^2} e^{-1/(1+x^2)}$—here the exponential factor in the solution does not affect the distribution very much, so the PDF is much like the previous one.
6. $p \propto \frac{1}{\sqrt{1+x^2}} e^{-\sqrt{1+x^2}}$—for small x this process is like an Ornstein–Uhlenbeck process, $dX \approx -X\,dt + \sqrt{2}\,dW$, but the PDF decays a little faster for large x because the noise increases like $\sqrt{X}$.
7. $p \propto 1/(1+x^2)^3$.

Exercise 20.16. Consider the class of Itô SDEs $dX = \mu(X)dt + \sigma(X)dW$. If a steady-state probability distribution exists, show that the steady-state PDF may be written as

$$p(x) = \frac{C}{D(x)} \exp\left[\int \frac{\mu(x)}{D(x)} dx\right],$$

where $D(x) = \frac{1}{2}\sigma(x)^2$ and C is a normalization constant.

Exercise 20.17 (Stratonovich Fokker–Planck equation). Derive the Fokker–Planck equation (20.40) for Stratonovich SDEs.

Chapter 21
Strong and weak models of stochastic dynamics

Contents

Modeling stochastic systems is notoriously delicate. Nonetheless, almost all of the necessary methods have been established in previous chapters:

- the techniques for modeling nonautonomous ODEs in Chapter 19 also empower strong modeling of stochastic dynamics and resolve many poorly understood issues,
- whereas the approach of Part III for constructing slowly varying solutions, applied to Fokker–Planck equations, derives weak models of stochastic systems.

Example 21.1 (basic modeling of a stochastic system). Consider the dynamics of $x(t)$ and $y(t)$ according to the coupled nonlinear ODEs (19.1):

$$\dot{x} = -xy \quad \text{and} \quad \dot{y} = -y + x^2 - 2y^2 + \sigma w(t),$$

but with the difference here that the forcing $w(t)$ denotes some white noise stochastic forcing of overall strength σ. Figure 21.1 plots some sample trajectories. Despite the noisy forcing, the trajectories do seem to be attracted to the vicinity of $y \approx x^2$. Surely we can form an approximate model that reflects such attraction.

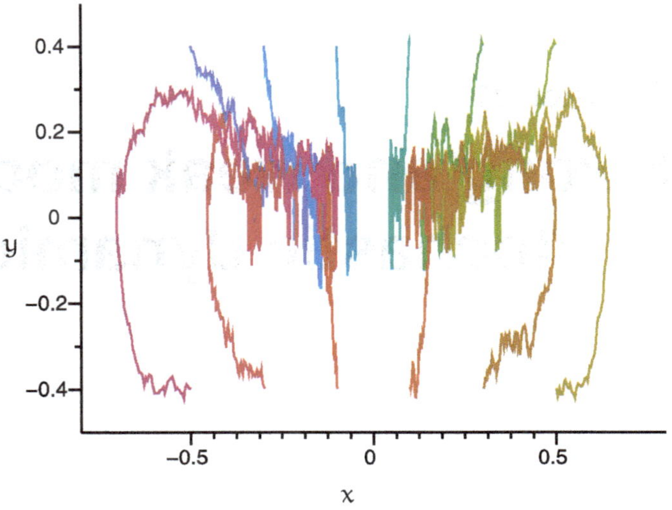

Figure 21.1. *Trajectories of the example system* (19.1) *from different initial conditions for white noise stochastic forcing* $w(t)$ *of strength* $\sigma = 0.1$ *(instead of the deterministic forcing of Figure* 19.1*). The trajectories are attracted to* $y \approx x^2 + \sigma e^{-t}\star w$.

Recall that section 19.1 analyzed the system (19.1) for arbitrary deterministic forcing $w(t)$. There we found that a time-dependent nonlinear coordinate transform to new variables (X, Y) separated the system (19.1) into a useful normal form, namely,

$$\begin{aligned} \dot{X} &= -X^3 - \sigma X w + 2\sigma^2 X w e^{-t}\star w + \mathbb{O}(4), \\ \dot{Y} &= -Y - 4\sigma Y w + \mathbb{O}(3). \end{aligned} \tag{21.1}$$

Crucially, the coordinate transform does not depend upon differentiating the forcing $w(t)$; the coordinate transformation (19.14a) and (19.11a) depends only upon integrals of the forcing $w(t)$ (via the convolutions). Consequently, the coordinate transform (19.14a) and (19.11a) still applies for white noise stochastic forcing $w(t)$. That is, the stochastic coordinate transforms, such as those illustrated by Figure 21.2, separate the fast and the slow dynamics of the stochastic version of the system (19.1).

Separating the stochastic system into the normal form (21.1) empowers modeling the stochastic dynamics. The Y SDE indicates that Y decays exponentially quickly for almost all realizations of the noise provided it is of small enough strength σ. The X SDE then indicates that the deterministic decay is buffeted by a multiplicative noise and affected by some nonlinear combination of the noise. Section 21.2.1 explores how such nonlinear noise combinations generate mean drift effects.

For reasons outlined in section 19.1.6, I do not expect a singular perturbation approach to resolve all of these effects.

A little stochastic history One of the first systematic efforts at reducing the dimensionality of stochastic systems was by Knobloch and Wiesenfeld (1983), who proposed analysis of the Fokker–Planck equations for the probability density function (PDF) of a

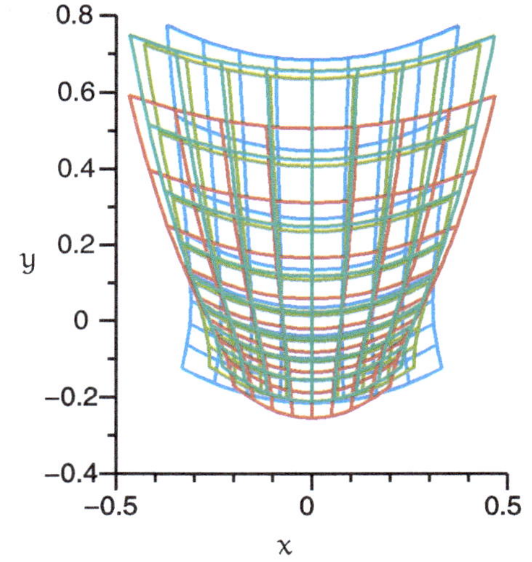

Figure 21.2. *Four different (colored) meshes represent four instants of the coordinate curves in the* xy*-plane of the time-dependent* (X, Y) *coordinate system* (19.14a) *and* (19.11a) *that simplifies the algebraic description of the dynamics of the system* (19.1) *for white noise stochastic forcing* $w(t)$ *of intensity* $\sigma = 0.3$ *(instead of the deterministic forcing of Figure* 19.2*).*

stochastic system; see section 21.2. Hutt et al. (2007, 2008) recently used such Fokker–Planck modeling in PDEs. Then Coullet et al. (1985) analyzed a noisy Hopf bifurcation from a center manifold view and identified a stochastic resonance that we discuss in section 21.1.3. Schöner and Haken (1986, 1987) perhaps initiated the direct construction of stochastic slow manifolds. Sri Namachchivaya and Leng (1990) and Sri Namachchivaya and Lin (1991) introduced how to view modeling via the beautiful stochastic normal-form argument, analogous to the previously given example and taken further in section 19.2 (albeit there for nonautonomous systems). Boxler (1989, 1991) proved some stochastic slow manifold theorems and gave some practical constructions. Arnold (2003), Arnold and Xu Kedai (1993), Arnold et al. (1996), and Arnold and Imkeller (1998) later developed such theory more extensively. Here we greatly simplify such theoretical support by invoking backward theorems that, in contrast to asserting the existence of a slow manifold model for a specific given stochastic system, assert that a nearby stochastic system has an obvious slow manifold model.

Subsequent to the above research, several approaches to modeling stochastic dynamics continue to be developed. Chao and Roberts (1996) connected the previous theory with formal methods to derive stochastic models more systematically. This empowered computer algebra to handle the multitude of noise interactions that occur in nonlinear systems (Roberts, 2006), later extended to the normal-form view (Roberts, 2008) with small noise as expounded here. Bensoussan and Flandoli (1995) and Schmalfuss and Schneider (2008) complementarily approached the issues using the view of inertial manifolds. Keller and Ochs (1999) constructed stochastic center manifolds numerically and found that the dynamics was more complicated than many believed. Many researchers have also approached the modeling from an averaging point of view, as introduced for nonautonomous systems in Chapter 18 (Berglund and Gentz, 2003; Khasminskii and Klebaner, 2003; Pavliotis and Stuart, 2008, e.g.). Mohammed et al. (2013) recently modeled bifurcation in stochastic partial differential equations (SPDEs) with some rigor.

21.1 ▪ Stochastic slow manifolds exist and usually emerge

Under suitable constraints, Boxler (1989) guaranteed the existence, emergence, and approximation of a stochastic center manifold in some finite neighborhood of the origin for a stochastic system in the form (the same form as (19.15))

$$\dot{x} = Ax + f(t, x, y), \quad \dot{y} = By + g(t, x, y). \tag{21.2}$$

This chapter explores the particular case of a stochastic slow manifold (SSM) because here we assume matrix A has only zero eigenvalues; it does not have complex eigenvalues.

The challenge that bedevils stochastic modeling is that suitable rare events will "push" a stochastic system so far off a stochastic slow manifold that the system may not return to the SSM, or if it does return, then relevant coherence with the past will be destroyed.

We sidestep this challenge by utilizing a backward approach. Here, as Example 21.1 introduced, the coordinate transform methodology of Chapter 19, and properties derived from the coordinate transforms, also apply to stochastic systems. The reason is that the construction of the coordinate transform only depends upon integrals of the time-dependent forcing, integrals that are well defined for stochastic forcing. Because Chapter 19 employed classic calculus, the stochastic analysis and interpretation here must be in the Stratonovich interpretation, not Itô (unless otherwise noted).

Proposition 21.1 (SSM existence and emergence). *Propositions* 19.3 *and* 19.4 *apply to Stratonovich stochastic systems in the form* (21.2) *under the conditions of section* 19.2.

That is, there exist stochastic systems asymptotically close to any such specified (21.2) that possesses an SSM through the origin.[252] Moreover, for as long as a solution of the close stochastic system stays within some specific finite domain about the origin, then each such solution tends exponentially quickly to a solution on the SSM. This backward result caters for the rare events that push solutions far away from the SSM because "far away" means that such solutions leave the finite domain of the proposition.

21.1.1 ▪ Noise transformations, anticipation, nondeterminism, and interactions clarified

Additive microscale noise appears as multiplicative macroscale noise Recall the system of Example 21.1. There the additive noise in the fast variable y feeds through the nonlinearity to appear as a multiplicative noise in the slow variable X: analysis deduces $\dot{X} \approx -X^3 - \sigma X w$ (in the Stratonovich sense). In general, by the construction of a normal-form coordinate transform (section 19.2), additive noise in the slow variables feeds directly into additive noise of the macroscale model, and we also expect nonlinear effects to feed additive noise through into multiplicative noise in the macroscale model.

Anticipation All who write down and then use macroscale models of time-dependent and/or stochastic dynamics are implicitly soothsayers. In writing down a macroscale model, modelers neglect the many details of any quickly decaying modes: they are ignored in the modeling as being insignificant. However, such a macroscale model is inevitably of some microscale system that involves many quickly decaying modes. Theorem 19.1 assures us that generally the neglect of such microscale modes requires us to know aspects of the near future of the ignored modes. This knowledge is essential in order to

[252]By the freedom inherent in the construction of the multivariate coordinate transform in section 19.2, there are infinitely many such systems asymptotically close to a specified system.

decouple the macroscale modes from the uninteresting details of the microscale modes. We require such knowledge to be sure of staying on the time-dependent slow manifold of the macroscale model by looking at least a little way into the future of the microscale dynamics. Nonetheless, Theorem 19.2 assures us that nonanticipative macroscale models do exist and may be accurate for all time.

Nonanticipation of slow manifolds A major theme of the discussion in section 19.2 about possible coordinate transformations is the anticipatory convolution integrals that necessarily arise in modeling nonautonomous dynamics. However, Theorem 19.2 assures us that such anticipation need never occur in the slow manifold dynamics itself. In the example of section 19.1 the slow X-evolution (19.14b) has no anticipatory convolutions. Further, upon setting the fast variable $Y = 0$, being the emergent slow manifold in the transformed variables, there are no anticipatory convolutions in the location, equations (19.11a) and (19.14a), of the slow manifold in the xy-plane. Similarly in the example of section 19.2.4, the slow manifold (19.30) with $Y = 0$ and the evolution (19.31) on the slow manifold have no anticipatory integrals.

The full coordinate transforms explored in section 19.2 resolve the details of the fast Y-dynamics, and when applied to stochastic dynamics establishes that a useful emergent slow manifold model exists. However, in application we are primarily interested in the slow manifold and the evolution thereon, $Y = 0$. Recall that section 19.3 discusses constructing just the slow manifold of nonautonomous dynamics and the simplifications that arise from not resolving transients in the fast variables. Thus section 19.3 also applies to stochastic dynamics, in the Stratonovich sense, and provides a more practical route to stochastic modeling.[253]

In contrast, Arnold and Xu Kedai (1993) and Arnold and Imkeller (1998) recorded anticipatory convolutions in the slow modes of their stochastic examples, respectively their (12) and (4.6). I contend that such anticipatory convolutions are undesirable in seeking to use a coordinate transform to support macroscale models.

The average SSM is not the deterministic slow manifold Many people assume the contrary: for example, Constable et al. (2013) go further and "examine the behaviour of a stochastic system in the SDE framework under the condition that its trajectories are confined to the slow manifold of the deterministic version of the system." Such an assumption may be a practical approximation in certain circumstances, but an example shows that the deterministic slow manifold will generally not be the average SSM.

Example 21.1 discussed how the toy ODE (19.1) with stochastic forcing could be modeled via the coordinate transform of section 19.1. Recall that the coordinate transform, namely (19.14a) and (19.11a), is

$$\begin{aligned} x &= X + XY + \tfrac{3}{2}XY^2 + \sigma\left[Xe^{-t}\star w + XY(3e^{-t}\star w - 2e^{+t}\star w)\right] \\ &\quad - \sigma^2 X(\tfrac{1}{2} + 2e^{-t}\star)(e^{-t}\star w)^2 + \mathcal{O}(4), \\ y &= Y + X^2 + 2Y^2 + \sigma\left[e^{-t}\star w + 4Ye^{-t}\star w\right] - 2\sigma^2 e^{-t}\star(e^{-t}\star w)^2 + \mathcal{O}(3). \end{aligned}$$

Since parameter σ measures the strength of the stochastic forcing $w(t)$, setting $\sigma = 0$ shows that a deterministic slow manifold is $y = x^2 + \mathcal{O}(3)$.[254] In general the SSM fluctuates about a mean location that is a little different from this deterministic slow manifold.

[253] At the time of writing, my web service constructs SSMs for any in a wide class of systems you may be interested in (Roberts, 2009e).

[254] Part of the appeal of this example system is that a deterministic slow manifold is precisely $y = x^2$.

From (19.14a) and (19.11a) with fast variable $Y = 0$, the SSM is

$$x = X + \sigma X e^{-t}\star w - \sigma^2 X(\tfrac{1}{2} + 2e^{-t}\star)(e^{-t}\star w)^2 + \mathcal{O}(4),$$
$$y = X^2 + \sigma e^{-t}\star w - 2\sigma^2 e^{-t}\star(e^{-t}\star w)^2 + \mathcal{O}(3).$$

Take expectations of these two equations: using the stochastic convolution properties (19.18d) and (19.18e), the expectations become

$$\mathsf{E}[x] = (1 - \tfrac{5}{4}\sigma^2)X + \mathcal{O}(4) \quad \text{and} \quad \mathsf{E}[y] = X^2 - \sigma^2 + \mathcal{O}(3). \tag{21.3}$$

Observe that $\mathsf{E}[y] \neq \mathsf{E}[x]^2$. Instead $\mathsf{E}[y] \approx (1 + \frac{5}{2}\sigma^2)\mathsf{E}[x]^2 - \sigma^2$ so that the average SSM is both displaced downward by σ^2 and a steeper parabola shape, with coefficient $(1 + \frac{5}{2}\sigma^2)$, than the deterministic slow manifold. Typically, such quadratic interaction of the noise processes deforms the average SSM away from the deterministic.

Multitude of microscale interactions Recall that section 19.3.3 comments on the combinatorial explosion of nonlinear interactions between forcing processes in nonautonomous systems. The arguments also apply to stochastic systems. Such a combinatorial explosion is enormously aggravated when modeling SPDEs, as then we typically want to postulate noise in an infinite number of spatial modes (Roberts, 2006, e.g.). The only practical approach appears to be to assume that the spectrum of noise decays quickly enough, as a function of mode number, so that a relatively low order truncation of the spectrum is sufficiently accurate for the macroscale model. But we need to resolve enough of the multitude of noises to get a reasonable estimate of the cumulative effect of the noise-noise nonlinear interactions.

A currently open research project would be to deduce the mean effect of noise-noise interactions without resolving the multitude of fluctuations in such noise-noise interactions.

Discretising SPDEs would have even more interactions Spatially discretizing SPDEs has many applications. Techniques could be based upon finite element or finite volume methods. As well as the multitude of noise interactions within each element or volume, one should perhaps account for the noise interactions between neighboring and near neighboring noises (Roberts, 2007).

Hopf bifurcations selectively filter noise Persistent oscillations are another vitally important class of dynamics. Hopf bifurcation is an example frequently considered, but many other cases occur, including wave propagation. The challenge with oscillations is how to consistently model the evolution of oscillations *over long time scales* when the oscillations are fast and in the presence of a force that fluctuates on some time scale. The challenge is exacerbated for stochastic forcing, where the fluctuations occur on all time scales and hence inevitably have some components that interact strongly with the oscillations and some components that do not.

Coullet et al. (1985) first constructed a normal form of Hopf bifurcations with noise. Subsequently, further research on stochastic bifurcations elucidated some fascinating fine structure. For example, Keller and Ochs (1999) explored the structure of the random limit cycle attractor using a stochastic version of the subdivision algorithm of Dellnitz and Hohmann (1997), whereas Arnold and Imkeller (1998) explored the structures using a normal-form approach very close to that used here. Roberts (2008) looked at the issues afresh to explore the characteristics of a long-term stochastic model of such stochastic oscillatory dynamics. Such considerations should underpin the multiscale modeling of general stochastic oscillations and waves. More research is needed.

21.1.2 ▪ Forecast from initial conditions

Suppose at time $t = 0$ we observe the state (x_0, y_0), or a distribution of such states. What forecast can we make with the SSM SDE (19.33)? That is, what forecasts can we make with the slow variable model $\dot{X} = AX + F(t, X)$.

Example 21.2 (basic stochastic system). Recall Example 21.1. Section 19.1 derived the time-dependent coordinate transform (19.11a) and (19.14a) to put the system into the normal form (21.1). We now want to use the model

$$\dot{X} \approx -X^3 - \sigma X w + 2\sigma^2 X w e^{-t}\star w \tag{21.4}$$

to make forecasts from some specified initial value (x_0, y_0).

Start by translating the given initial value (x_0, y_0) to the transformed variables. Revert the asymptotic expansion of the stochastic coordinate transform (19.14a) and (19.11a) to deduce

$$\begin{aligned} X &= x + x^3 - xy + \tfrac{3}{2}xy^2 + 2\sigma x y e^{+t}\star w \\ &\quad - 2\sigma^2 x(e^{+t}\star w)(e^{-t}\star w) + \mathcal{O}(4), && (21.5\text{a}) \\ Y &= y - x^2 - 2y^2 - \sigma e^{-t}\star w + \sigma^2(1 + e^{-t}\star)(e^{-t}\star w)^2 + \mathcal{O}(3). && (21.5\text{b}) \end{aligned}$$

Then the correct initial condition for the long-term dynamics on the SSM, governed by the SDE (21.4), is the X component of this reversion, (21.5a), evaluated at the observed state, namely

$$\begin{aligned} X(0) &= x_0 + x_0^3 - x_0 y_0 + \tfrac{3}{2}x_0 y_0^2 + 2\sigma x_0 y_0 e^{+t}\star w \\ &\quad - 2\sigma^2 x_0(e^{+t}\star w)(e^{-t}\star w) + \mathcal{O}(4). \end{aligned} \tag{21.6}$$

This is a projection of the observed initial state onto the SSM to provide an initial condition $X(0)$ for the slow mode. However, this projection involves convolutions requiring both memory of history ($t < 0$) and anticipation ($t > 0$) of the forcing. There are at least three interesting issues with computing this initial $X(0)$.

- First, suppose that at the initial instant we know neither the future nor the past, so the terms in the projection (21.6) involving the forcing w are unknown. Using the expectations (19.18d) and (19.18e), the projection $X(0)$ has known mean

  $$E[X(0)] = x_0 + x_0^3 - x_0 y_0 + \tfrac{3}{2}x_0 y_0^2 + \mathcal{O}(4),$$

  with known variance

  $$\operatorname{Var}[X(0)] \approx 2\sigma^2 x_0^2 y_0^2 + \sigma^4 x_0^2.$$

  That is, a deterministic given observed initial state (x_0, y_0) is mapped to a stochastic initial state for the evolution of the slow mode model on the SSM.

- Second, if this initial state $X(0)$ for the slow mode is to be used in a simulation to make forecasts, then we know the future of the forcing w, as the future values of forcing w are just those to be used in integrating the slow mode ODE (21.4). Thus we do soon know the anticipatory convolutions $e^{+t}\star w$ in (21.6), but not the memory convolution $e^{-t}\star w$. In this case of not knowing the history of the noise, the mean of the projection (21.6) is

$$E[X(0)] = x_0 + x_0^3 - x_0 y_0 + \tfrac{3}{2}x_0 y_0^2 + 2\sigma x_0 y_0 e^{+t}\star w + \mathcal{O}(4),$$

with variance $\operatorname{Var}[X(0)] \approx 2\sigma^4 x_0^2 (e^{+t}\star w)^2$.

- Last, if we made additional observations for times $t < 0$, then the additional information could determine the history of the forcing w and hence estimate the memory convolution $e^{-t}\star w$.

These three cases emphasize that the initial state $X(0)$ of the slow variable depends upon more than just the observed state (x_0, y_0) at an initial instant.

Now return to the general challenge. Suppose at the initial time $t = 0$ we observe the state $(\mathbf{x}_0, \mathbf{y}_0)$; what forecast can we make with the general slow manifold model (19.25a)? As introduced in Example 21.2, such forecasting requires we anticipate the future and remember history.

We relate the slow manifold model (19.25a) to the original state space via the stochastic near identity coordinate transform (19.16)—typically a multinomial coordinate transform. In principle we invert this transform evaluated at time $t = 0$, namely $\mathbf{x}_0 = \mathbf{x}(0, \mathbf{X}_0, \mathbf{Y}_0)$ and $\mathbf{y}_0 = \mathbf{y}(0, \mathbf{X}_0, \mathbf{Y}_0)$, to deduce initial values $\mathbf{X}_0 = \mathbf{X}(\mathbf{x}_0, \mathbf{y}_0)$ and $\mathbf{Y}_0 = \mathbf{Y}(\mathbf{x}_0, \mathbf{y}_0)$ for some functions. This inverse exists over the domain in which the coordinate transform is a diffeomorphism. Then, since the slow manifold SDE (19.25a) is independent of the fast variables $\mathbf{Y}$, the correct initial condition for the SDE (19.25a) is that $\mathbf{X}(0) = \mathbf{X}_0$. This initial condition empowers strong prediction of the stochastic solutions.

The devil is in the detail: according to (19.22) and (19.24), the coordinate transform generically involves convolutions both into the future and into the past, the history, albeit integrals over a fast time scale. Thus at the initial time $t = 0$ we must know both the history of the noise in $t < 0$ and anticipate the future of the noise. If either the future or the history of the noise is unavailable at the initial instant, then inevitably the initial condition for the slow manifold SDE (19.25a) will involve some distribution of possibilities, and hence any simulation with the slow manifold SDE (19.25a) will be a weak prediction.

21.1.3 ▪ Irreducible noise is its own SSM

Recall the evolution equations (19.25a) for the slow $\mathbf{X}$ in some nonautonomous/stochastic normal-form coordinates. As noted in Case 1 on page 599, the SDE (19.25a) generally contains nonlinear forcing such as $we^{-t}\star w$, in defiance of Principle 5 which aims to banish all fast processes from the slow model. Example systems, such as (19.31), (19.49), and (21.1), confirm the prevalence of such quadratic noise-noise interactions. This section establishes that over long times such irreducible noise-noise interactions could be replaced by an effective drift and an effective independent noise; for example, Chao and Roberts (1996) and Roberts (2006) established that $we^{-t}\star w$ in a slow SDE could be replaced with $\frac{1}{2} + \frac{1}{\sqrt{2}}\tilde{w}$ for some effectively new white noise forcing $\tilde{w}(t)$. Analogous replacement was also justified by Khasminskii (1966), as described by Sri Namachchivaya

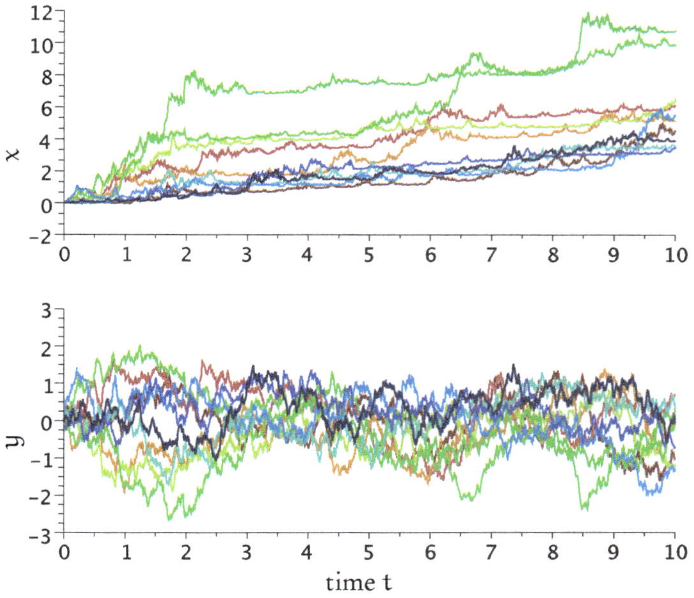

Figure 21.3. *Ten realizations of the SDE* (21.7) *show the mean drift and significant fluctuations in the noise-noise interaction process* $x(t)$ *driven via the Ornstein–Uhlenbeck process* y.

and Leng (1990). Importantly, such quadratic forcing, in effect, generates a mean deterministic drift term in the slow dynamics (Sri Namachchivaya and Leng, 1990; Sri Namachchivaya and Lin, 1991, e.g.). In applications such drifts can be vital.

The issue then is to determine the long-term dynamics of terms such as $we^{-t}\star w$. Surely we have the tools: we just have to ask what is its slow manifold evolution; in doing so the following argument finds that the term $we^{-t}\star w$ is irreducible.

Slow manifold model fails to make progress To analyze $we^{-t}\star w$ using our slow manifold approach we want to characterize $x(t)$ such that $\dot{x} = we^{-t}\star w$. This SDE for $x(t)$ is equivalent to the 2D stochastic system

$$\dot{x} = yw \quad \text{and} \quad \dot{y} = -y + w, \tag{21.7}$$

as the solution to the y SDE is the convolution $y = e^{-t}\star w$ (upon neglecting fast transients). Figure 21.3 shows how realizations of this system exhibit long-term drift in $x(t)$. The system (21.7) has an SSM, so let's extract it via a stochastic coordinate transform to new (X, Y) variables.

- Substituting $y = Y + \hat{y}(t, X, Y)$ such that $\dot{Y} = -Y + \hat{G}(t, X, Y)$ in the fast SDE of (21.7) leads straightforwardly to the homological equation

$$\hat{G} + \frac{\partial \hat{y}}{\partial t} + \hat{y} - Y\frac{\partial \hat{y}}{\partial Y} = w(t),$$

  with solution $\hat{G} = 0$ and $\hat{y} = e^{-t}\star w$. That is, $y = Y + e^{-t}\star w$ such that $\dot{Y} = -Y$ exactly satisfies the fast SDE (21.7).

- Substituting $x = X + \hat{x}(t, X, Y)$ such that $\dot{X} = F(t, X)$ into the slow SDE of (21.7) then leads straightforwardly to the homological equation

$$\hat{F} + \frac{\partial \hat{x}}{\partial t} - Y\frac{\partial \hat{x}}{\partial Y} = Yw + we^{-t}\star w.$$

  The Y terms lead to $\hat{x} = -Ye^{+t}\star w$. But as $we^{-t}\star w$ has unbounded integrals in general, then we must choose $\dot{X} = we^{-t}\star w$.[255] That is, $x = X - Ye^{+t}\star w$ such that $\dot{X} = we^{-t}\star w$ exactly satisfies the slow SDE (21.7).

Consequently, the slow manifold $Y = 0$ emerges exponentially quickly, and the evolution on the slow manifold is $\dot{X} = we^{-t}\star w$. However, this slow evolution is exactly what we started with. The quadratic noise-noise interaction $we^{-t}\star w$ is an irreducible kernel in slow manifold construction.

But in a slow model we cannot really abide the fast time convolutions in $we^{-t}\star w$. Here we make progress by reluctantly abandoning strong modeling. Let's permit ourselves to model, over long times, only the statistical effects of noise-noise interaction terms such as $we^{-t}\star w$.

Analyze statistics in Itô calculus Invoke Theorem 20.10 to translate the fundamental noise-noise interaction system (21.7) into its equivalent Itô SDE system

$$dx = \tfrac{1}{2}\,dt + y\,dW \quad \text{and} \quad dy = -y\,dt + dW, \tag{21.8}$$

where in a formal sense the Wiener process $W = \int w\,dt$. One of the beauties of such an Itô representation is that one can just take expectations and set the expectation $E(dW) = 0$ by the martingale property (20.10). Here, define $\bar{x} := E(x)$, and then the expectation of the slow x SDE in (21.8) immediately gives the long-term drift that $d\bar{x} = \frac{1}{2}dt$. That is, $\bar{x} = \frac{1}{2}t + \bar{x}(0)$ increases at an average rate of $\frac{1}{2}$.

The fluctuations around this drift are significant: let's derive the variance. To find the variance $\bar{v} := E[(x - \bar{x})^2]$, define the stochastic process $v(t) := [x(t) - \bar{x}(t)]^2$. Invoke Itô's formula (20.14) to find the differential

$$\begin{aligned}
dv &= 2(x - \bar{x})(dx - d\bar{x}) + \tfrac{1}{2}2(dx - d\bar{x})^2 \\
&= 2(x - \bar{x})(\tfrac{1}{2}dt + y\,dW - \tfrac{1}{2}dt) + (\tfrac{1}{2}dt + y\,dW - \tfrac{1}{2}dt)^2 \\
&= 2(x - \bar{x})y\,dW + y^2 dW^2 \\
&= 2(x - \bar{x})y\,dW + y^2 dt\,.
\end{aligned}$$

Take expectations, recalling that in Itô calculus the increment dW at any time t is independent of processes x and y at that time, to argue that the variance satisfies

$$\begin{aligned}
d\bar{v} = E(dv) &= 2\,E[(x - \bar{x})y\,dW] + E[y^2 dt] \quad \text{(from above)} \\
&= 2\,E[(x - \bar{x})y]\,E[dW] + E[y^2]dt \quad \text{(by independence)} \\
&= E[y^2]dt \quad (\text{as } E(dW) = 0).
\end{aligned}$$

The process y, by the SDE (21.8), is a well-known Ornstein–Uhlenbeck process from which we straightforwardly derive the expectation $\overline{y^2} := E(y^2)$. Invoke Itô's

[255] There is potential for splitting $we^{-t}\star w$ into the parts that would cause unbounded effects and those that would not, and then assigning them separately and respectively to the slow evolution and the slow coordinate. However, I am not aware of anyone who has yet researched how to do so and the consequences.

formula (20.14) again to find the differential

$$\begin{aligned} d(y^2) &= 2y\,dy + \tfrac{1}{2}2\,dy^2 \\ &= 2y(-y\,dt + dW) + (-y\,dt + dW)^2 \\ &= -2y^2dt + 2y\,dW + dW^2 \\ &= (-2y^2 + 1)dt + 2y\,dW. \end{aligned}$$

Take expectations: $d\overline{y^2} = (-2\overline{y^2} + 1)dt$ with general solution $\overline{y^2} = \frac{1}{2} + ce^{-2t}$. Use this in the differential for the variance to deduce that

$$d\bar{v} = E[y^2]dt = (\tfrac{1}{2} + ce^{-2t})dt \to \tfrac{1}{2}dt,$$

upon neglecting rapid transients. That is, the process x has fluctuations about the mean $\bar{x}$, where the variance of the fluctuations grows at rate $\frac{1}{2}$; Figure 21.3 (upper) is consistent with such growth in fluctuations. Such growth in the variance may be modeled by an increment in noise of $d\tilde{W}/\sqrt{2}$ for some Wiener process $\tilde{W}$.

That is, we effectively model the process x by the SDE $dx = \frac{1}{2}dt + \frac{1}{\sqrt{2}}d\tilde{W}$ to account for its drift and fluctuations. Recall that in the Stratonovich calculus used for our analysis of dynamics we defined, via the system (21.7), that $\dot{x} = we^{-t}\star w$. Hence the argument here asserts that, in effect, $\dot{x} = \frac{1}{2} + \frac{1}{\sqrt{2}}\tilde{w}$ for some white noise $\tilde{w} = d\tilde{W}/dt$. In conclusion, in our slow manifold modeling we may justifiably replace irreducible noise-noise interaction terms $we^{-t}\star w$ by $\frac{1}{2} + \frac{1}{\sqrt{2}}\tilde{w}$.

But such replacement needs some care. Consequently section 21.2 addresses the modeling of probability density functions where we find, for example, that the above replacement emerges but with algebraically decaying transients.

21.2 ▪ The Fokker–Planck equation guides weak models

Recall that the Fokker–Planck equation (20.39) governs the PDF of stochastic processes. Instead of forming models of stochastic dynamics using the SSM of an SDE, this section addresses modeling by constructing the slow manifold of the corresponding Fokker–Planck equation. Haken (1983) and Knobloch and Wiesenfeld (1983) initiated this approach. The slow manifold evolution is a Fokker–Planck equation of the slow variables in the system. In turn, the slow Fokker–Planck equation provides a weak model for the original dynamics: "weak" because it only describes the distribution of solutions, without necessarily maintaining fidelity with realizations.

Our first example, in section 21.2.1, predicts the long-term dynamics of the irreducible noise-noise interaction of section 21.1.3 (as first established by Chao and Roberts (1996, §4)).

The approach taken invokes the techniques of modeling slowly varying spatial systems developed in Part III. Indeed all the dispersion problems addressed in Chapter 7, instead of being viewed as the concentration of indefinite numbers of particles, could equally well be viewed as the modeling of the PDF of just one particle undergoing a random walk with space-dependent drift. One difference in the detail is that in Chapter 7 the random walker is laterally restricted by the walls of the pipe or channel, whereas here the random walker is laterally restricted by the attractiveness of the locale $y \approx 0$.

Indeed, in one view, most macroscale continuum models are Fokker–Planck equations for the density of some particles or their properties. For example, the Navier–Stokes equations of fluid dynamics may be interpreted as Fokker–Planck equations for the PDF of the molecular location (density) and the molecular velocity (momentum density).

Another difference between Chapter 7 and the analysis of Fokker–Planck equations here is that here most Fokker–Planck equations have significant dependence upon the slow variables $\mathbf{x}$. This dependence arises through the slow variables almost always appearing in the right-hand sides of the corresponding SDEs. Thus here the evolution of the pattern of the PDF is strongly affected by the heterogeneity in $\mathbf{x}$. Our analysis here is an example of heterogeneous pattern formation and evolution.

21.2.1 ▪ Irreducible noise generates drift and fluctuations

Recall the irreducible noise-noise interaction system (21.7), $\dot{x} = yw$ and $\dot{y} = -y + w$, with realizations plotted in Figure 21.3. Its Itô form (21.8) is $dx = \frac{1}{2}\,dt + y\,dW$ and $dy = -y\,dt + dW$. This section discovers its long-term properties using the corresponding Fokker–Planck equation.

The Fokker–Planck equation (20.39) is phrased in terms of the drift and volatility of the Itô SDE (21.8):

$$\mu = (\tfrac{1}{2}, -y) \quad \text{and} \quad \sigma = (y, 1).$$

Thus the Fokker–Planck equation governing the PDF $p(t, x, y)$ for the irreducible system (21.7), equivalently (21.8), is

$$\frac{\partial p}{\partial t} = -\frac{1}{2}\frac{\partial p}{\partial x} + \frac{\partial (yp)}{\partial y} + \frac{1}{2}y^2\frac{\partial^2 p}{\partial x^2} + \frac{\partial^2 (yp)}{\partial x \partial y} + \frac{1}{2}\frac{\partial^2 p}{\partial y^2}. \tag{21.9}$$

What do we do with this PDE? Recall the realizations of the process plotted in Figure 21.3: the vertical scale for the xt plot is much larger than the vertical scale for the yt plot, even over this relatively short time span; the disparity in scales between x and y grows more with time, as seen in Figure 21.4. Over long times the realizations spread out over large scales in x while still being compact in y. Thus we proceed to seek solutions $p(t, x, y)$ of the Fokker–Planck equation (21.9) that are slowly varying in x.

Let's adapt the approach of Part III for modeling slowly varying spatial dynamics. Two crucial ingredients are the marginal PDF $P(t, x) := \int_{-\infty}^{\infty} p(t, x, y)\,dy$ and the Gaussian $G(y) := e^{-y^2}/\sqrt{\pi}$. This section aims to derive that the PDF adopts a near product form, $P(t, x)G(y)$, as illustrated for large times in Figure 21.4. More precisely, we find that the emergent slow manifold PDF is

$$p(t, x, y) \approx PG(y) + \frac{\partial P}{\partial x}(\tfrac{1}{2} - y^2)G(y), \quad \text{where } \frac{\partial P}{\partial t} \approx -\frac{1}{2}\frac{\partial P}{\partial x} + \frac{1}{4}\frac{\partial^2 P}{\partial x^2}. \tag{21.10}$$

Then we interpret the PDE for the marginal P as the Fokker–Planck equation for an SDE: here it would be the Itô SDE $dX = \frac{1}{2}\,dt + \frac{1}{\sqrt{2}}dW$, from which we would predict a mean drift rate of $\frac{1}{2}$ and a volatility rate of $1/\sqrt{2}$, as seen in the realizations of Figure 21.3.

To derive the model (21.10) of such slowly varying PDFs, let's rearrange the Fokker–Planck equation (21.9) as

$$\frac{\partial p}{\partial t} = \underbrace{\frac{\partial (yp)}{\partial y} + \frac{1}{2}\frac{\partial^2 p}{\partial y^2}}_{\text{linear } \mathcal{L}p} \underbrace{- \frac{1}{2}\frac{\partial p}{\partial x} + \frac{\partial^2 (yp)}{\partial x \partial y} + \frac{1}{2}y^2\frac{\partial^2 p}{\partial x^2}}_{\text{nonlinear } f(p)}. \tag{21.11}$$

Here all terms involving the small derivatives $\partial/\partial x$ are grouped into the perturbing nonlinear function $f(p)$, whereas the remaining large terms are grouped into the linear operator $\mathcal{L}p$. Following the methodology of section 7.3, it is the properties of the linear

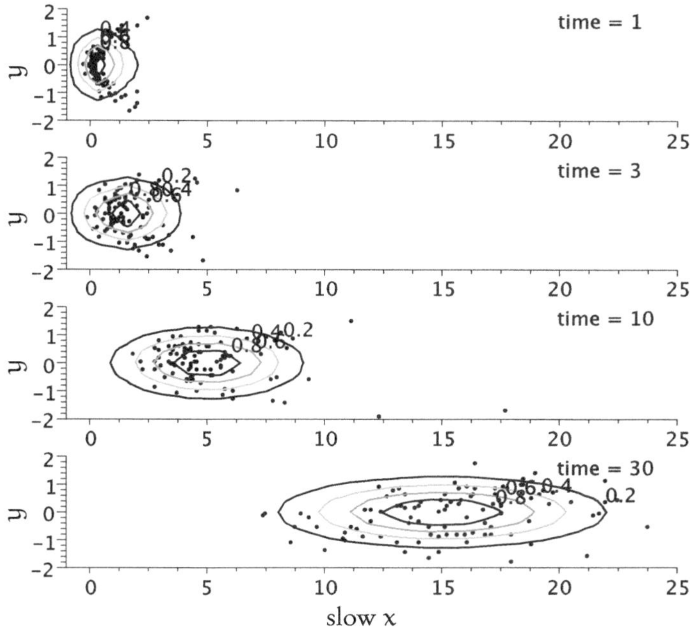

Figure 21.4. *One hundred realizations of the SDE (21.7) plotted as dots at four different times. The plots show the realizations approach a PDF that is compact in y and slowly varying in x. For comparison are contours of an approximate PDF from the Fokker–Planck equation.*

operator $\mathcal{L}p$ that determine the existence and emergence of a slow manifold model for the PDF p.

Linear spectrum implies a slow manifold As the Fokker–Planck equation (21.11) is linear in the PDF p, the linear analysis applies to all amplitudes of solutions. Seeking solutions to the linear PDE, $\partial p/\partial t = \mathcal{L}p$, we substitute $p = e^{\lambda t}G(y)v(y)$ for Gaussian $G = e^{-y^2}/\sqrt{\pi}$ and deduce the eigenproblem $v'' - 2yv' - 2\lambda v = 0$. With the boundary condition that $v(y)$ must not grow faster than polynomial as $y \to \pm\infty$, the eigenfunctions of this equation are the well-known Hermite polynomials, $v = H_n(y)$, for the eigenvalues $\lambda_n = -n$ for $n = 0, 1, 2, \ldots$. Thus the linear dynamics has one zero eigenvalue, corresponding to eigenfunction $G(y)H_0(y) = G(y)$ as the Hermite polynomial $H_0 = 1$, and the remaining eigenvalues are negative, $\lambda_n \leqslant -1$ for $n \geqslant 1$. Thus assuming small perturbation by the x-derivatives, the Fokker–Planck equation (21.11) has a slow manifold that should emerge from transients roughly like e^{-t}; for example, the non-Gaussian structures apparent for times $t \leqslant 3$ in Figure 21.4 have decayed for times $t \geqslant 10$.

A first nontrivial approximation The initial approximation to the slow manifold is the subspace of the zero eigenvalue. Here, the initial approximation to the slow manifold is thus $p \approx P(t,x)G(y)$ for the marginal PDF $P(t,x)$. Such a product-form solution is very common in applied probability. But we need corrections to account for the slow evolution of structures in x. Let's find the first nontrivial correction (21.10) by substituting that the PDF $p = PG(y) + \hat{p}(P,y)$ such that $\partial P/\partial t = \hat{g}(P)$ and where x-derivatives of P are small. The Fokker–Planck equation (21.11) then becomes, using subscripts on P

to denote partial derivatives,

$$G\hat{g} + \frac{\partial \hat{p}}{\partial P}\hat{g} = \mathcal{L}\hat{p} - \tfrac{1}{2}GP_x - \tfrac{1}{2}\frac{\partial \hat{p}}{\partial P}P_x + (1-2y^2)GP_x + \frac{\partial(yP_x\partial\hat{p}/\partial P)}{\partial y} + \tfrac{1}{2}y^2GP_{xx} + \tfrac{1}{2}y^2\frac{\partial(\partial\hat{p}/\partial P\, P_x)}{\partial x}.$$

As in section 7.3, simplify by omitting products of small terms to obtain the homological equation

$$G\hat{g} - \mathcal{L}\hat{p} \approx -\tfrac{1}{2}GP_x + (1-2y^2)GP_x + \tfrac{1}{2}y^2GP_{xx}.$$

Neglect the last term, in P_{xx}, as it is unreliable in this first iteration. To invoke the solvability condition for this homological equation, recall that the linear operator, $\mathcal{L}p = \partial/\partial y(yp + \tfrac{1}{2}\partial p/\partial y)$, is an exact differential, so its integral over all y must be zero. Take the integral over all y of the homological equation to deduce $\hat{g} \approx -\tfrac{1}{2}P_x$, since $\int_{-\infty}^{\infty} G\,dy = 1$, and since $1-2y^2 \propto H_2(y)$ is orthogonal on the domain. Consequently the update $\hat{p}$ to the PDF must satisfy

$$-\mathcal{L}\hat{p} \approx (1-2y^2)GP_x.$$

Again $1-2y^2 \propto H_2(y)$ so that $(1-2y^2)G(y)$ is an eigenfunction of $\mathcal{L}$ proportional to the eigenvalue $\lambda_2 = -2$ and so the PDF must be $-1/\lambda_2 = 1/2$ of this term; that is, $\hat{p} \approx (\tfrac{1}{2} - y^2)G(y)P_x$. Putting these together we find that the first nontrivial approximation to the slow manifold of the Fokker–Planck equation (21.9) is

$$p \approx G(y)P + (\tfrac{1}{2} - y^2)G(y)P_x, \quad \text{where } \frac{\partial P}{\partial t} \approx -\tfrac{1}{2}P_x,$$

as consistent with the previous claim (21.10).

We could perform further hand computations to find the next term in (21.10). However, computer algebra is more fun.

Computer algebra finds further approximations Algorithm 7.1 from section 7.3 is readily adapted to our purposes here. After modifying this algorithm, Algorithm 21.1 generates the following higher-order approximation to the slow manifold of the Fokker–Planck equation (21.9) and the evolution thereon:

$$p = G(y)P + (\tfrac{1}{2} - y^2)G(y)\frac{\partial P}{\partial x} + (-\tfrac{1}{4} - \tfrac{1}{4}y^2 + \tfrac{1}{2}y^4)G(y)\frac{\partial^2 P}{\partial x^2} + (\tfrac{1}{4} + \tfrac{1}{8}y^2 - \tfrac{1}{6}y^6)G(y)\frac{\partial^3 P}{\partial x^3} + \mathcal{O}(\partial_x^4), \tag{21.12a}$$

$$\frac{\partial P}{\partial t} = -\tfrac{1}{2}\frac{\partial P}{\partial x} + \tfrac{1}{4}\frac{\partial^2 P}{\partial x^2} - \tfrac{1}{4}\frac{\partial^3 P}{\partial x^3} + \mathcal{O}(\partial_x^4). \tag{21.12b}$$

Algorithm 21.1 works as follows.

- The key to the algorithm is the iteration at the end. The iteration refines approximations to the slow manifold based upon the current residual of the Fokker–Planck equation (21.9).
- The iteration terminates only when the residual is zero to the error specified by the neglect of high-order derivatives in line 14.

Algorithm 21.1 *Iteration finds the slow manifold of the Fokker–Planck equation* (21.9) *for the* PDF *of the irreducible noise-noise interactions* (21.7).

```
1  % Gaussian exp(-y^2)/sqrt(pi)
2  depend gg,y; let df(gg,y)=>-2*y*gg;
3  % operators: zero mean is inbuilt in linv
4  operator inty; linear inty;
5  let { inty(gg,y) => 1
6      , inty(y^~~q*gg,y)=>(q-1)/2*inty(y^(q-2)*gg,y) };
7  operator linv; linear linv;
8  let linv(y^~~q*gg,y)=>-(y^q-inty(y^q*gg,y))/q*gg
9      +(q-1)/2*linv(y^(q-2)*gg,y);
10 % slow manifold depends upon pp, and linear approx
11 depend pp,x,t; let df(pp,t)=>g;
12 p:=pp*gg;  g:=0;
13 % iteration, retaining terms up to some order in d/dx
14 let df(pp,x,~p) => 0 when numberp(p) and p>3;
15 for iter:=1:99 do begin
16     res:=df(p,t)+1/2*df(p,x)-df(y*p,y)
17         -y^2/2*df(p,x,2)-df(y*p,x,y)-1/2*df(p,y,2);
18     g:=g+(gd:=-inty(res,y));
19     p:=p+linv(res+gd*gg,y);
20     if res=0 then write iter:=1000000+iter;
21 end;
22 resamp:=inty(p,y)-pp;
```

- The iteration starts from the product-form approximation established in lines 11–12, using the Gaussian $G(y) := \exp(-y^2)/\sqrt{\pi}$ defined by its differential property on line 2.

- Then the operator `inty` on lines 4–6 evaluates $\int_{-\infty}^{\infty} \cdots dy$ using integration by parts to write

$$\int_{-\infty}^{\infty} y^q G(y)\,dy = \int_{-\infty}^{\infty} -\tfrac{1}{2}y^{q-1}\,dG(y) = \frac{q-1}{2}\int_{-\infty}^{\infty} y^{q-2}G(y)\,dy$$

  recursively until the basic $\int_{-\infty}^{\infty} G(y)\,dy = 1$.

- Analogously, via lines 7–9, the operator `linv` evaluates $\mathcal{L}^{-1}(y^q G)$ recursively from the observation that

$$\mathcal{L}(y^q G) = -qy^q G + \tfrac{1}{2}q(q-1)y^{q-2}G\,.$$

  Dividing by q, applying $\mathcal{L}^{-1}$ to this equation, and rearranging gives the coded transform that

$$\mathcal{L}^{-1}(y^q G) = -\frac{1}{q}y^q G + \tfrac{1}{2}(q-1)\mathcal{L}^{-1}(y^{q-2}G).$$

Executing Algorithm 21.1 derives the slow manifold PDF (21.12a). The PDF is approximately the product form $G(y)P(t,x)$ but with modifications due to supposed slow variations in x through derivatives $\partial P/\partial x$ and $\partial^2 P/\partial x^2$. At first sight this looks eminently reasonable. However, there is one problem: (21.12a) predicts that one may have ridiculous negative probabilities. Such negativity may occur for larger y due to, for example, $(\frac{1}{2} - y^2)G(y)$ terms. An argument is that these negative effects are negligible due to the rapid decay of the mollifying Gaussian $G(y)$ and the small gradients in x. This argument implies that this nonuniform convergence in y has no effect on the evolution on the slow manifold governed by the PDE (21.12b).

Approach to normality in the Fokker–Planck equation Recall the discussion of the first approximation (21.10). There the slow manifold $p \approx G(y)P$ such that the marginal PDF $P(t, x)$ evolves according to the advection-diffusion PDE $P_t \approx -\frac{1}{2}P_x + \frac{1}{4}P_{xx}$: such an advection-diffusion PDE predicts positive drift and diffusive spreading, as noted previously. The new slow manifold PDE (21.12b) refines this advection-diffusion model by including effects from the third derivative P_{xxx} of the marginal, and higher-order effects from any higher-order analysis. The slow manifold PDE (21.12b) is known as a Kramers–Moyal expansion.

The interest in Kramers–Moyal expansions lies in the following contrast. On the one hand, we want to model microscopic processes on macroscales by an SDE, such as the SDE $\dot{x} = \frac{1}{2} + \tilde{w}/\sqrt{2}$. But on the other hand, although such SDEs have a PDF that satisfies an advection-diffusion Fokker–Planck equation, accurate modeling of the microscale processes, as in (21.12b), generally finds indefinitely high-order PDEs.

Indeed, about 50 years ago, Pawula (1967) proved that a Kramers–Moyal expansion in derivatives, such as (21.12b), either truncates as an advection-diffusion PDE or else continues indefinitely to infinite order. Pawula (1967) concluded "it is logically inconsistent to retain more than two terms in the Kramers–Moyal expansion unless all of the terms are retained" (Tutkun and Mydlarski, 2004, p. 5). But conversely, building on Part III, recent theory (Roberts, 2013) suggests that a finite truncation of a Kramers–Moyal expansion should have a quantifiable error that is small in useful domains of application.

The field of shear dispersion of material along a channel or pipe addressed the analogous issue. Recall that the concentration field in dispersion corresponds to the PDF for an SDE problem. Chatwin (1970) discussed the approach to normality of the concentration/PDF because when the concentration field is the normal Gaussian it satisfies the commonly used advection-diffusion PDE, which in turn corresponds to a Kramers–Moyal expansion truncated to second-order derivatives. Chatwin (1970) showed that the normal Gaussian eventually emerges, but it does so with algebraically decaying transients. That is, higher-order terms in a Kramers–Moyal expansion reflect such algebraically decaying transients. Then the usual and context-dependent modeling subjective decision must be made: do we need to resolve such transients in the problem domain under consideration or not? The answer informs us where to truncate a Kramers–Moyal expansion such as (21.12b) in our slow manifold modeling of Fokker–Planck equations.

21.2.2 ▪ Nonlinearity bends the Gaussian PDF

For nonlinear stochastic dynamical systems the roughly Gaussian shape of the PDFs should be bent. This section introduces the issue via one of our favorite example systems, whose deterministic slow manifold is the exact parabola $y = x^2$. Modifications of the techniques of section 21.2.1 are necessary.

Knobloch and Wiesenfeld (1983) also bent the PDF and argued that a product-form PDF still applied to describe the shape of the slow manifold. This section finds that modifications to the basic product form are necessary to account for gradients in the marginal PDF $P(t, x)$.

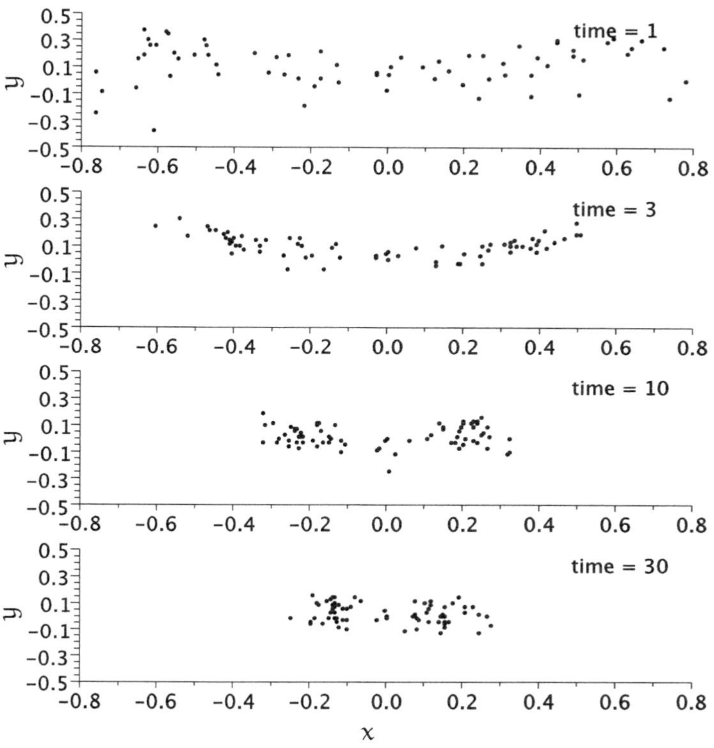

Figure 21.5. *One hundred realizations of stochastic system* (19.1) *with noise strength* $\sigma = 0.1$. *The plots of locations* (x, y) *at four times show that solutions rapidly become scattered near to the deterministic slow manifold* $y = x^2$ *and then slowly evolve into two clumps.*

Analogues of our bending modifications here to the PDF are the modeling of shear dispersion in varying and twisting pipes by Mercer and Roberts (1994) and Roberts (2004). Crucially the bending variations in the slow variable x have to be in some sense slowly varying in x.

Recall Example 21.1, which considers the dynamics of $x(t)$ and $y(t)$ according to the coupled nonlinear ODEs (19.1):

$$\dot{x} = -xy \quad \text{and} \quad \dot{y} = -y + x^2 - 2y^2 + \sigma w(t),$$

where noise $w(t)$ is the formal derivative, dW/dt, of a Wiener process $W(t)$. Figure 21.1 plots some trajectories showing that solutions are attracted to stochastic fluctuations near the deterministic slow manifold of $y = x^2$. Figure 21.5 plots 100 realizations of solutions to the system (19.1) at four successive times. These plots indicate the evolving PDF: initially the PDF concentrates over the slow manifold $y \approx x^2$, but over long times it clumps into two stochastically induced concentrations. Our aim is to model this evolving PDF as suggested by Figure 21.5.

To achieve this aim we construct the slow manifold of the Fokker–Planck equation for the system (19.1). But first recall that the shape of the PDF is dominated by the balance between the decay of y and the forcing by the noise w: here it is $\dot{y} \approx -y + \sigma w$, whereas in section 21.2.1 it is $\dot{y} = -y + w$. To most directly use the techniques of section 21.2.1 we scale the variables so that these two have matching dominant balance: here change to capital letter variables by the rescaling $(X, Y) := (x, y)/\sigma$; that is, for small noise

Algorithm 21.2 *Iteration finds the slow manifold of the Fokker–Planck equation* (21.14) *for the* PDF *of the nonlinear system* (21.14). *Simply modify Algorithm* 21.1 *by changing the asymptotic truncation and the computation of the residual of the Fokker–Planck equation.*

```
...
% let df(pp,x,~p) => 0 when numberp(p) and p>3;
let sigma^3=>0;
...
res:=df(p,t)+df(-sigma*x*y*p,x)
    +df(-y*p+sigma*(x^2-2*y^2)*p,y)
    -1/2*df(p,y,2);
...
```

strength σ this scaling zooms into the origin as (X,Y) of size 1 corresponds to (x,y) of small size σ. The dynamical system (19.1) is equivalently

$$\dot{X} = -\sigma XY \quad \text{and} \quad \dot{Y} = -Y + \sigma(X^2 - 2Y^2) + w(t). \tag{21.13}$$

We explore the PDF of this system.

The Fokker–Planck equation (20.40) for the PDF $p(t,X,Y)$ of the system (21.13) is straightforwardly[256]

$$\frac{\partial p}{\partial t} = -\frac{\partial}{\partial X}[-\sigma XYp] - \frac{\partial}{\partial Y}\left[-Yp + \sigma(X^2 - 2Y^2)p\right] + \tfrac{1}{2}\frac{\partial^2 p}{\partial Y^2}\,. \tag{21.14}$$

One significant difference between this and the example of section 21.2.1 is that here the Fokker–Planck equation depends upon the slow variable X. Mercer and Roberts (1990, 1994), argued that modeling slowly varying solutions for systems with such slowly varying coefficients is still valid.

Consequently, let's construct the slow manifold model of the Fokker–Planck equation (21.14). Given the computer algebra of Algorithm 21.1, and that the dominant balance of the Fokker–Planck equation (21.14) is the same as that for Algorithm 21.1, the quickest construction of a slow manifold here is to appropriately modify Algorithm 21.1. Algorithm 21.2 lists the two modifications necessary:

- first change the computation of the residual to that appropriate to the Fokker–Planck equation (21.14), and
- due to the nonlinearity in the SDE (21.13) it is not enough to truncate the asymptotic expansion in orders of derivatives, but it is sufficient to truncate in powers of the nonlinearity, which for system (21.13) are conveniently counted by orders of σ.

Executing Algorithms 21.1 and 21.2 gives that the slow manifold PDF is

$$p = GP + \sigma(Y - \tfrac{4}{3}Y^3 + 2X^2Y)GP + \sigma XYG\frac{\partial P}{\partial X} + \mathcal{O}(\sigma^2) \tag{21.15a}$$

such that the marginal PDF $P(t,X)$ evolves according to

$$\begin{aligned}\frac{\partial P}{\partial t} &= \sigma^2\left[(-\tfrac{1}{2} + 3X^2)P + (\tfrac{1}{2}X + X^3)\frac{\partial P}{\partial X} + \tfrac{1}{2}X^2\frac{\partial^2 P}{\partial X^2}\right] + \mathcal{O}(\sigma^3)\\ &= \sigma^2\frac{\partial}{\partial X}\left[-(X - X^3)P + \tfrac{1}{2}X\frac{\partial(XP)}{\partial X}\right] + \mathcal{O}(\sigma^3).\end{aligned} \tag{21.15b}$$

[256] As the noise in (21.13) is additive there is no difference between the Itô and Stratonovich interpretations.

This PDE neatly explains the clumps emerging over long times in Figure 21.5. Using (20.40), we interpret the PDE (21.15b) as the Fokker–Planck equation of the Stratonovich SDE

$$\dot{X} = \sigma^2(X - X^3) + \sigma X w'(t)$$

for some multiplicative white noise w'. Deterministically, this SDE predicts stable equilibria at $X = \pm 1$, which the noise would spread into the clumps. Recalling that the SDE is in scaled variables, in the original variables of Figure 21.5 the clumps would be about equilibria at $x = \pm\sigma$ as in Figure 21.5.

But wait—stochastic dynamics are not quite so simple The asymptotic expansion (21.15a) leading to the Fokker–Planck equation (21.15b) is not uniformly valid, so our further deductions are dubious.

To see the lack of uniform validity, let's temporarily return to unscaled coordinates. Recall that the slow manifold (21.15a) is in variables $(X, Y) = (x, y)/\sigma$. Reverting the slow manifold (21.15a) to the original variables, it becomes

$$p = G\left(\frac{y}{\sigma}\right) P + \left(y - \tfrac{4}{3}y^3/\sigma^2 + 2x^2y/\sigma^2\right) G\left(\frac{y}{\sigma}\right) P + \frac{xy}{\sigma} G\left(\frac{y}{\sigma}\right) \frac{\partial P}{\partial x} + \mathcal{O}(\sigma^2).$$

Divisions by σ indicate undesirable large rates of change in the corresponding terms as the noise strength $\sigma \to 0$. Nonetheless, factors y/σ are acceptable because of the rapid decay in $G(y/\sigma) = \exp(-y^2/\sigma^2)$. But, for example, the term $(x^2y/\sigma^2)G$ is not normally acceptable because the extra division by σ implies nonuniform validity in x. Many more such apparently nonuniform terms occur at higher order.

We repair such nonuniformity by bending space. The bending must somehow correspond to the mean shape of the PDF seen in Figure 21.5, especially for early times, and must surely reflect the deterministic slow manifold $y = x^2$ when noise strength $\sigma \to 0$. Returning to scaled coordinates, let's suppose the Gaussian G, instead of being $\exp(-Y^2)/\sqrt{\pi}$, is instead $G(\mathcal{Y}) = \exp(-\mathcal{Y}^2)/\sqrt{\pi}$, where the shifted coordinate $\mathcal{Y} = Y - h(X)$ for asymptotically small $h(X)$. For example, we expect that here $\mathcal{Y} \approx Y - \sigma X^2$ so that the maximum of the Gaussian is near the deterministic slow manifold $Y - \sigma X^2 = 0$. How do we change Algorithms 21.1 and 21.2?

- We must correct the computation of the residual of the Fokker–Planck equation (21.14). Correct residuals follow via the chain rule for derivatives: we recognize that the Gaussian G also depends upon X and code the chain rule as in Algorithm 21.3.
- The vertical shift is to be small, so we start the iteration with $\mathcal{Y} := Y$, as coded.
- Since $\mathcal{Y} \approx Y$, the existing `linv` is correct enough to use the residuals to guide corrections exactly as before, except that we need to explicitly ignore its deficiencies by mapping `linv(gg,y)=>0`.
- The vertical integration operator, `inty`, is used to compute residuals of the amplitude and the $\mathcal{Y}$-definition, so it must be correct. Via integration by parts,

$$\begin{aligned}
&\int_{-\infty}^{\infty} Y^q G(\mathcal{Y})\, dY \\
&= \int_{-\infty}^{\infty} -\tfrac{1}{2}Y^{q-1}\left[-2(Y-h)G(\mathcal{Y})\right] dY + h\int_{-\infty}^{\infty} Y^{q-1}G(\mathcal{Y})\, dY \\
&= \frac{q-1}{2}\int_{-\infty}^{\infty} Y^{q-2}G(\mathcal{Y})\, dY + (Y-\mathcal{Y})\int_{-\infty}^{\infty} Y^{q-1}G(\mathcal{Y})\, dY,
\end{aligned}$$

as coded in Algorithm 21.3.

- Then we might choose the vertical coordinate $\mathcal{Y}$ so that $\int_{-\infty}^{\infty} \mathcal{Y} p \, dY = 0$; that is, the mean location over Y of the PDF p is to be at $\mathcal{Y} = 0$. But this is infeasible without generating divisions by the marginal PDF P, so instead we just require this constraint on that part of the PDF p in the marginal PDF P but not its X-derivatives. Algorithm 21.3 thus computes the residual of this integral and then updates $\mathcal{Y}$ and the PDF p accordingly.

- These updates generally change the amplitude of the PDF p, so to restore its definition, that $P = \int_{-\infty}^{\infty} p \, dY$, Algorithm 21.3 computes its residual and updates the PDF.

- As usual, terminate the iteration when all residuals are zero to the specified order of error.

Executing Algorithm 21.3 gives that the slow manifold PDF is

$$p = GP + \sigma(2Y - \tfrac{4}{3}Y^3)GP + \sigma XYG\frac{\partial P}{\partial X} + \mathcal{O}(\sigma^2), \tag{21.16}$$

which eliminates the somewhat embarrassing X^2YGP term in the previous PDF (21.15a). The reason is that now the Gaussian $G = \exp(-\mathcal{Y}^2)$ is shifted in the vertical via

$$\mathcal{Y} = Y + \sigma(\tfrac{1}{2} - X^2) + \mathcal{O}(\sigma^3). \tag{21.17}$$

Algorithm 21.3 also derives that the marginal PDF P satisfies exactly the same PDE, namely the Fokker–Planck equation (21.15b). Thus the deductions from the Fokker–Planck equation (21.15b) do hold, such as the prediction of two stochastic-induced equilibria about scaled coordinate $X = \pm 1$, as seen in the simulations about unscaled $x = \pm\sigma$ in Figure 21.5.

This analysis also brings out, from (21.17), that the mean vertical location of the PDF p is $Y \approx \sigma(X^2 - \frac{1}{2})$, which is the perturbed deterministic slow manifold, as in unscaled coordinates it is $y \approx x^2 - \frac{1}{2}\sigma^2$. This information is hidden in the expression (21.15a) for the PDF.

Higher-order modeling of the PDF One beauty of our computer algebra is that we easily proceed to higher-order expansions. Who can resist? Change Algorithm 21.3 to neglect $\mathcal{O}(\sigma^5)$ and execute to find that the vertical coordinate

$$\mathcal{Y} = Y + \sigma(\tfrac{1}{2} - X^2) + \sigma^3(3 + 2X^2) + \mathcal{O}(\sigma^5),$$

which in unscaled variables indicates that the PDF slow manifold is centered about

$$y \approx (1 - 2\sigma^2)x^2 - \tfrac{1}{2}\sigma^2 - 3\sigma^4.$$

When the noise magnitude σ is zero, this expression recovers the exact deterministic slow manifold $y = x^2$.

The evolution of the PDF is parametrized by the marginal PDF $P(t, X)$. As determined by Algorithm 21.3, the marginal P satisfies the PDE

Algorithm 21.3 *Iteration finds the slow manifold of the Fokker–Planck equation* (21.14) *for the* PDF *of the nonlinear system* (21.14). *Modify Algorithms 21.1 and 21.2 by changing the dependence of the Gaussian, adapting the vertical integration, ignoring deficiencies in the update rule, ensuring that the center of the Gaussian is at the mean, and ensuring that the amplitude is correct.*

```
% Gaussian exp[-Y^2]/sqrt(pi)
depend gg,y,x;
let df(gg,~z)=>-gg*2*yy*df(yy,~z);
% operators: zero mean is inbuilt in linv
operator inty; linear inty;
let { inty(gg,y) => 1
    , inty(y^~~q*gg,y)=>(q-1)/2*inty(y^(q-2)*gg,y)
                          +(y-yy)*inty(y^(q-1)*gg,y) };
operator linv; linear linv;
let linv(y^~~q*gg,y)=>-(y^q-inty(y^q*gg,y))/q*gg
                        +(q-1)/2*linv(y^(q-2)*gg,y);
let linv(gg,y)=>0; % ignore deficiencies in update
% slow manifold depends upon pp, and linear approx
depend pp,x,t; let df(pp,t)=>g;
p:=pp*gg;  g:=0; yy:=y;
% iteration, retaining terms to order in noise
let sigma^3=>0;
for iter:=1:99 do begin
    res:=df(p,t)+df(sigma*(-x*y)*p,x)
        +df(-y*p+sigma*(x^2-2*y^2)*p,y)
        -1/2*df(p,y,2);
    g:=g+(gd:=-inty(res,y));
    p:=p+linv(res+gd*gg,y);
    yy:=yy-(resyy:=inty(coeffn(p,pp,1)*yy,y));
    p:=p-2*pp*gg*y*resyy;
    p:=p-gg*(resamp:=inty(p,y)-pp);
    write lengthress:=map(length(~r),{res,resyy,resamp});
    if {res,resyy,resamp}={0,0,0}
    then write iter:=1000000+iter;
end;
```

$$\begin{aligned}\frac{\partial P}{\partial t} &= \sigma^2\left[(-\tfrac{1}{2}+3X^2)P+(\tfrac{1}{2}X+X^3)\frac{\partial P}{\partial X}+\tfrac{1}{2}X^2\frac{\partial^2 P}{\partial X^2}\right]\\ &\quad+\sigma^4\left[-3(1+3X^2)P+3X(1-5X^2)\frac{\partial P}{\partial X}-3X^2(1-X^2)\frac{\partial^2 P}{\partial X^2}-2X^3\frac{\partial^3 P}{\partial X^3}\right]\\ &\quad+\mathcal{O}(\sigma^5)\\ &= \frac{\partial}{\partial X}\left\{\left[-\sigma^2(\tfrac{3}{2}X-X^3)-\sigma^4(21X-9X^3)\right]P\right\}\\ &\quad+\frac{\partial^2}{\partial X^2}\left\{\left[\sigma^2\tfrac{1}{2}X^2+\sigma^4(15X^2-3X^4)\right]P\right\}\\ &\quad+\frac{\partial^3}{\partial X^3}\left\{-\sigma^4 2X^3 P\right\}+\mathcal{O}(\sigma^5). \end{aligned} \tag{21.18}$$

This is the first three terms of a Kramers–Moyal expansion for the marginal PDF. The end of section 21.2.1 discussed how the third- and higher-order derivative terms, such as that in (21.18), affect the interpretation of such an expansion as a Fokker–Planck equation of a stochastic process.[257]

21.2.3 ▪ Weak models of nonlinear SDE systems

This section establishes a moderately general algorithm to construct the slow subspace Kramers–Moyal expansion of the Fokker–Planck equation of slow-fast stochastic systems. Beware: for relative simplicity this section presents an algorithm that involves an expansion that is not uniformly valid. Postprocessing appears to recover uniform validity (see Exercise 21.8) and hence tentatively justifies this approach. Further research is needed.

Consider the class of autonomous Stratonovich SDEs of the form

$$\dot{x} = f(x, y, \sigma) + \sigma\Sigma_x(x, y, \sigma)\dot{W}, \tag{21.19a}$$
$$\dot{y} = g(x, y, \sigma) + \sigma\Sigma_y(x, y, \sigma)\dot{W}, \tag{21.19b}$$

where slow variables $x \in \mathbb{R}^m$, fast variables $y \in \mathbb{R}^n$, and noises $W \in \mathbb{R}^o$ (independent Wiener processes) so that $\Sigma_x \in \mathbb{R}^{m\times o}$, $\Sigma_y \in \mathbb{R}^{n\times o}$, and the overall strength of the noise is σ. Assume that the linearizations of f and g have eigenvalues of zero and negative real part, respectively. Then section 21.1 establishes that the slow x-dynamics generally emerge from the fast y decay, albeit stochastically fluctuating. Our aim is to analyze the Fokker–Planck equation of the PDF $p(t, x, y)$ for any SDE system in the class (21.19) in a regime of small noise strength σ. The outcome is a slow subspace model, in the form of a Fokker–Planck equation, for the marginal PDF $P(x, t) := \int p(t, x, y)\, dy$. Such a PDE model for the marginal would subsequently be interpreted as the Fokker–Planck equation for an SDE, which itself would be a slow manifold, *weak* model of the SDE (21.19).

For the particular algorithm to work, the SDE system (21.19) needs to be in a restricted class:

- the origin, $(x, y, \sigma) = (0, 0, 0)$, is an equilibrium;
- the right-hand sides are multinomial, or a good enough multivariable Taylor approximation;
- the multinomials must be of low enough order for the coded method of undetermined coefficients;
- the Jacobian A of f at the origin must be zero, $f(0, 0, 0) = 0$;
- the Jacobian B of g at the origin must be nonsingular, and all eigenvalues have negative real part;
- the noise in fast y must not vanish at the origin, specifically $\Sigma_y(0, 0, 0)$ must be invertible.

Example 21.3 (Levy area contraction). To illustrate the algorithm and results, consider first the interesting example of Pavliotis and Stuart (2008, §11.6.7) (used in

[257] It is not apparent in the Kramers–Moyal expansion (21.18), but the slow manifold of the PDF has an apparent nonuniformity in a term $\sigma^4 X^6 P G$ of relatively high order. Whether this is a concern needs further research.

section 19.2.4 as an example of avoiding homogenization): the variables $(x_1, x_2, x_3, y_1, y_2)$ evolve according to the coupled nonlinear SDEs

$$\dot{x}_1 = \sigma y_1\,, \quad \dot{x}_2 = \sigma y_2\,, \quad \dot{x}_3 = x_1 y_2 - x_2 y_1\,, \tag{21.20a}$$
$$\dot{y}_1 = -y_1 - y_2 + \sigma \dot{W}_1\,, \quad \dot{y}_2 = +y_1 - y_2 + \sigma \dot{W}_2\,. \tag{21.20b}$$

We form an approximate model that describes the cross-sectional $\mathbf{y}$ structure of the PDF, a structure moderated by the slowly evolving lateral $\mathbf{x}$ structure.

Scaling the SDE systems ensures that small noise is a regular perturbation. Scale $\mathbf{x} = \sigma \mathbf{X}$ and $\mathbf{y} = \sigma \mathbf{Y}$ so that the system (21.20) becomes

$$\dot{X}_1 = \sigma Y_1\,, \quad \dot{X}_2 = \sigma Y_2\,, \quad \dot{X}_3 = \sigma(X_1 Y_2 - X_2 Y_1), \tag{21.21a}$$
$$\dot{Y}_1 = -Y_1 - Y_2 + \dot{W}_1\,, \quad \dot{Y}_2 = +Y_1 - Y_2 + \dot{W}_2\,. \tag{21.21b}$$

Then in the regime of negligible noise, $\sigma \approx 0$, the slow variables $\mathbf{X}$ are frozen, leaving the $\mathbf{Y}$ decay to balance the forcing noise.

To capture this balance and analyze the perturbative effects of nonnegligible noise we find the slow subspace of the Fokker–Planck equation for the SDE system (21.21). Since the noise terms in (21.21) have constant coefficient, the Itô (20.39) and Stratonovich (20.40) Fokker–Planck equations are the same: the PDF $p(t, \mathbf{X}, \mathbf{Y})$ satisfies

$$\begin{aligned}\frac{\partial p}{\partial t} &= -\sigma\left[Y_1 \frac{\partial p}{\partial X_1} + Y_2 \frac{\partial p}{\partial X_2} + (X_1 Y_2 - X_2 Y_1)\frac{\partial p}{\partial X_3}\right] \\ &\quad + (Y_1 + Y_2)\frac{\partial p}{\partial Y_1} + (-Y_1 + y_2)\frac{\partial p}{\partial Y_2} + \tfrac{1}{2}\frac{\partial^2 p}{\partial Y_1^2} + \tfrac{1}{2}\frac{\partial^2 p}{\partial Y_2^2}\,.\end{aligned}$$

For small noise, the first line of this PDE implies that the dependence of the PDF p upon $\mathbf{X}$ is slow, whereas the second line describes the dominant balance between decay in $\mathbf{Y}$ and the mixing by the noise.

Hence the PDF of the example SDE (21.21) is to leading approximation a Gaussian distribution $p \approx A \exp\left[-Y_1^2 - Y_2^2\right] P$ in terms of the marginal PDF $P(t, \mathbf{X})$ and normalization constant $A = 1/\pi$, written $p \approx A e^S P$ for "action" $S = -Y_1^2 - Y_2^2$ (some people know this as an Eikonal approximation). The first correction to this PDF is

$$\begin{aligned}p = A e^S P + \sigma A e^S &\left\{ \tfrac{1}{2}\left[-\frac{\partial P}{\partial X_1} + \frac{\partial P}{\partial X_2} + (X_1 + X_2)\frac{\partial P}{\partial X_3}\right] Y_1 \right. \\ &\left. + \tfrac{1}{2}\left[-\frac{\partial P}{\partial X_1} - \frac{\partial P}{\partial X_2} + (-X_1 + X_2)\frac{\partial P}{\partial X_3}\right] Y_2 \right\} + \mathcal{O}(\sigma^2).\end{aligned}$$

Algorithms 21.4–21.8 construct this slow subspace PDF to errors $\mathcal{O}(\sigma^3)$.

Example 21.4, below, completes this example by reporting and interpreting the PDE governing the evolution of the marginal PDF $P(t, \mathbf{X})$ on the constructed slow subspace.

Computer algebra constructs the slow subspace model

1. The first step for Algorithm 21.4 is for us to specify the SDE to analyze in the general form (21.19). For example, here code the SDE system (21.20). Initially `xj` denotes x_j and `yj` denotes y_j.

Algorithm 21.4 *First part of Reduce code to construct weak, slow manifold models of general SDEs* (21.19) *via their Fokker–Planck equation: in the first four lines specify the particular SDE, which here, for example, is system* (21.20)*; in the second, check dimensions; in the third, scale slow and fast variables to* (21.22)*; in the fourth, define a useful procedure; continued in Algorithm* 21.5.

```
% 1st specify SDE with mat()
f:=tp mat((sigma*y1,sigma*y2,x1*y2-x2*y1));
sigmax:=tp mat((0,0,0),(0,0,0));
g:=tp mat((-y1-y2,y1-y2));
sigmay:=tp mat((1,0),(0,1));

% 2nd get and check dimensions of the specified system.
m:=part(length(f),1);
n:=part(length(g),1);
o:=part(length(sigmax),2);
if {length(f),length(sigmax),length(g),length(sigmay)}
   neq {{m,1},{m,o},{n,1},{n,o}}
then rederr "Dimension mismatch error";

% 3rd, scale variables to avoid singular perturbation
scaling:=append(
    for i:=1:m collect (mkid(x,i)=sigma*mkid(x,i)),
    for j:=1:n collect (mkid(y,j)=sigma*mkid(y,j)));
f:=sub(scaling,f)/sigma;
g:=sub(scaling,g)/sigma;
sigmax:=sub(scaling,sigmax);
sigmay:=sub(scaling,sigmay);

% 4th, use to get coefficients of multinomial expressions
procedure coefs(exps,vars);
    if vars={} then exps else coefs(
        foreach z in exps join coeff(z,first vars),
        rest vars);
```

2. Second, Algorithm 21.4 checks the dimensionality of the specified matrix and vector functions.

3. We focus on the regime of relatively small noise strength σ. For small noise the Fokker–Planck equation becomes a singular perturbation problem for the PDF. To avoid the singular and make regular, scale the variables: $x = \sigma \mathbf{X}$ and $y = \sigma \mathbf{Y}$. Then the governing SDEs (21.19) become

$$\dot{\mathbf{X}} = \frac{1}{\sigma} f(\sigma \mathbf{X}, \sigma \mathbf{Y}, \sigma) + \Sigma_x(\sigma \mathbf{X}, \sigma \mathbf{Y}, \sigma)\dot{W}, \tag{21.22a}$$

$$\dot{\mathbf{Y}} = \frac{1}{\sigma} g(\sigma \mathbf{X}, \sigma \mathbf{Y}, \sigma) + \Sigma_y(\sigma \mathbf{X}, \sigma \mathbf{Y}, \sigma)\dot{W}. \tag{21.22b}$$

In general and for negligible σ, this scaling reduces the SDEs (21.22) to a slow random walk, $\dot{\mathbf{X}} \approx \Sigma_x(0)\dot{W}$, together with a type of Ornstein–Uhlenbeck balance in fast

Algorithm 21.5 *Continuing Algorithm 21.4: define the marginal* P *and its derivatives; initially the PDF is Gaussian; iterate.*

```
% 5th, define marginal PDF and useful d/dx abbreviations
operator pp; depend pp,t;
for i:=1:m do depend pp,mkid(x,i);
let df(pp,t) => gg;
let for i:=1:m join { df(pp,mkid(x,i))=>pp(i)
    , df(pp(~j),mkid(x,i))=>pp(max(i,j),min(i,j)) };
let { df(pp(~i),t)=>df(gg,mkid(x,i))
    , df(pp(~i,~j),t)=>df(gg,mkid(x,i),mkid(x,j)) };

% 6th, the PDF:=A.exp(S(Y))+... for this aexp
operator aexp;
for j:=1:n do depend aexp,mkid(y,j);
let df(aexp,~y)=>aexp*df(ss,y);
indvars:=for j:=1:n collect mkid(y,j)$
p:=pp*aexp; gg:=0;

% 7th, iteratively construct slow subspace
for iter:=1:3 do begin   itern:=iter$
  if iter=2 then <<clear sigma; indvars:=sigma.indvars>>;
  for i:=iter:iter do let sigma^i=>0;
  in "sim1fpe3.red"$ % code of next algorithm
  in "sim1fpe4.red"$ % code of last algorithm
end;
```

variables, $\dot{Y} \approx BY + \Sigma_y(0)\dot{W}$. Algorithm 21.4 then implements this scaling so that thereafter `xj` denotes X_j and `yj` denotes Y_j.

4. Algorithm 21.4 defines a recursive procedure to obtain coefficients of multinomial expressions.

5. The amplitude of the slowly varying model is the marginal PDF $P(\mathbf{X}, t)$, denoted by `pp`. So Algorithm 21.5 defines the dependence of P and stores its time derivative in `gg`. The slow subspace involves **X**-derivatives of the marginal, so useful abbreviations for derivatives of the marginal P are to be `pp(i)` $= \partial P/\partial X_i$ and `pp(i,j)` $= \partial^2 P/\partial X_i \partial X_j$. By commutativity of differentiation, their time derivatives translate to **X**-derivatives of the evolution.

6. We assume the noise distributions will be Gaussian-like in the fast variables **Y**, so Algorithm 21.5 introduces the operator `aexp` to denote $A\exp(S)$ for some unspecified amplitude A, where the action, S, denoted `ss`, is some quadratic form in **Y** that we find later. Structures are to be expressed in the fast variables **Y** and collectively stored in the set `indvars`. Later, `sigma` is included in this set to help check that all is proceeding well.

   The initial approximation invokes the Gaussian, of arbitrary amplitude, so set the initial probability density $p = P(X, t)A\exp(S)$. To leading order the marginal PDF does not evolve, so set $\partial P/\partial t = gg := 0$.

Algorithm 21.6 *First part of the loop code for Algorithm 21.5; Algorithm 21.7 lists the second part: in the first iteration form a general quadratic for the action,* $S(\mathbf{Y})$*; in subsequent iterations form a general sixth-order multinomial for the* PDF $p(t, \mathbf{X}, \mathbf{Y})$ *and collect the as yet unknown coefficients into the set* `vars`.

```
  if itern=1 then begin operator c;
    ss:=for i:=1:n sum for j:=1:i sum
      c(i,j)*mkid(y,i)*mkid(y,j)$
    vars:=for i:=1:n join for j:=1:i collect c(i,j)$
  end else begin
    gg:=gg+sigma^(itern-1)*gd$
    p:=p+sigma^(itern-1)*aexp*( 0*c+
      for i:=1:n sum mkid(y,i)*(c(i)+
      for j:=1:i sum mkid(y,j)*(c(i,j)+
      for k:=1:j sum mkid(y,k)*(c(i,j,k)+
      for l:=1:k sum mkid(y,l)*(c(i,j,k,l)+
      for m:=1:l sum mkid(y,m)*(c(i,j,k,l,m)+
      for n:=1:m sum mkid(y,n)*(c(i,j,k,l,m,n))
      ))))))$
    vars:=gd.(
      for i:=1:n join c(i).(
      for j:=1:i join c(i,j).(
      for k:=1:j join c(i,j,k).(
      for l:=1:k join c(i,j,k,l).(
      for m:=1:l join c(i,j,k,l,m).(
      for n:=1:m collect c(i,j,k,l,m,n)
      ))))))$
  end;
```

7. Algorithm 21.5 performs three iterations to find successive terms in the Kramers–Moyal expansion of the evolution on the slow subspace model of the Fokker-Planck equation for the SDEs (21.22). We use σ to measure smallness, including the smallness of derivatives with respect to the slowly varying dependence upon X_i. In each successive iteration Algorithm 21.5 increases the order of error to $\mathcal{O}(\sigma^{\text{iteration}})$ to construct the slow manifold in tightly controlled steps. Set the order of error in a trivial for-loop because it needs the "instant evaluation" property of a loop index.

8. Within in each iteration we update the PDF p and the Kramers–Moyal expansion of the evolution, $\partial P/\partial t$. In the first iteration, we seek the action $S(\mathbf{Y})$ in the leading Ornstein–Uhlenbeck balance between the fast decay of $\mathbf{Y}$ and its forcing by noise to form a Gaussian-like distribution $p \approx A\exp[S(\mathbf{Y})]P$ for quadratic $S(\mathbf{Y})$. So Algorithm 21.6 forms the as yet unknown quadratic $S(\mathbf{Y}) = \sum_{i=1}^{n}\sum_{j=1}^{i} c_{ij}Y_iY_j$ and then collects the as yet unknown coefficients into the set of variables `vars`.

9. In the second and third iterations, we seek updates to the PDF in the form $p := p + \sigma^k A e^S \sum_{|\alpha|\leqslant 6} c_\alpha \mathbf{Y}^\alpha$ for multi-index of exponents $\boldsymbol{\alpha} = (\alpha_1, \ldots, \alpha_n)$ and for appropriate power of smallness σ^k. So Algorithm 21.6 forms and adds the as yet unknown sixth-order multinomial in $\mathbf{Y}$ and then collects the as yet unknown coefficients into the set of variables `vars`. Truncating the sum to sixth order is a

Algorithm 21.7 *Second part of the loop code for Algorithm 21.5, continuing the first part in Algorithm 21.6: compute the Stratonovich diffusive term and then the residual of the Fokker–Planck equation; equate coefficients of the independent variables and solve for unknowns; substitute the solution into the general form.*

```
dif:=for k:=1:o sum begin
    q:=(for i:=1:m sum df(sigmax(i,k)*p,mkid(x,i)))
      +(for j:=1:n sum df(sigmay(j,k)*p,mkid(y,j)));
return (for i:=1:m sum df(sigmax(i,k)*q,mkid(x,i)))
      +(for j:=1:n sum df(sigmay(j,k)*q,mkid(y,j))) end;
fpe:=-df(p,t) +1/2*dif
    -(for i:=1:m sum df(f(i,1)*p,mkid(x,i)))
    -(for j:=1:n sum df(g(j,1)*p,mkid(y,j)));

eqns:=coefs({fpe/aexp},indvars);
soln:=solve(eqns,vars);
if soln={} then rederr "Solution empty.";
write ss:=sub(part(soln,1),ss);
write p :=sub(part(soln,1),p );
write gg:=sub(part(soln,1),gg);
```

somewhat arbitrary limitation but is done to contain computational time. The coefficients c_α are to be functions of the marginal PDF P, its slow $\mathbf{X}$-derivatives, and $\mathbf{X}$ itself. Similarly, the as yet unknown update `gd` to the evolution $\partial P/\partial t$ also depends upon $\mathbf{X}$, P, ∇P, and so on.

10. In each iteration, Algorithm 21.7 computes the residual of the Fokker–Planck equation for the scaled version (21.22) of the specified system of SDEs (21.19). Here the SDEs are interpreted in the Stratonovich sense. From the Stratonovich Fokker–Planck equation (20.40), Algorithm 21.7 codes the diffusion term in the form $\sum_{k=1}^{o} \nabla \cdot [\sigma_k \nabla \cdot (\sigma_k p)]$ using two successive applications of the divergence operator $\nabla \cdot (\sigma_k \cdot)$.

    Then Algorithm 21.7 computes the residual of the Fokker–Planck equation (20.40) by adding this stochastic diffusion to the Liouville equation for the system (21.22).

11. Given the residual of the Fokker–Planck equation, Algorithm 21.7 equates coefficients of the multinomial with the function `coefs` and then solves these equations using the inbuilt `solve` function. If the solution is empty, then the code abandons the algorithm: an empty solution typically indicates that either the $\mathbf{X}$-dynamics are not slow (try multiplying some terms in the specified SDE by small σ) or a higher-order multinomial in $\mathbf{Y}$ is needed.

12. Given a solution for the multinomial, substitute the now known coefficients into the various parts of the solution PDF and evolution.

13. Upon finishing the iteration, Algorithm 21.8 postprocesses. Optionally we may write the slow subspace and the evolution thereon in terms of unscaled variables, or not, as we choose. The unscaling uses $\mathbf{X} = x/\sigma$, $\mathbf{Y} = y/\sigma$, and $\partial/\partial X_j = \sigma\partial/\partial x_j$.

Algorithm 21.8 *Upon completing the constructing loop of Algorithm 21.5, postprocess: optionally unscale the variables; finally, integrate the Kramers–Moyal expansion once to get the flux and again to determine the diffusion matrix and Itô drift.*

```
clear sigma^3;
% optionally undo the scaling of the variables
if 0 then begin scaling:=append(
    for i:=1:m join (mkid(x,i)=mkid(x,i)/sigma)
        .(pp(i)=pp(i)*sigma)
        .(for j:=1:i collect (pp(i,j)=pp(i,j)*sigma^2)),
    for j:=1:n collect (mkid(y,j)=mkid(y,j)/sigma));
    p:=sub(scaling,p); gg:=sub(scaling,gg)*sigma;
end;

% finally, flux of slow subspace evolution
operator ee; factor ee;
for i:=1:m do depend mkid(x,i),x;
operator intx; linear intx;
let { intx(pp(~i)*~~a,x) => ee^i*pp*a
      -intx(pp*df(a,mkid(x,i)),x)
    , intx(pp(~i,~j)*~~a,x) =>
      +ee^i*pp(j)*a/2-intx(pp(j)*df(a,mkid(x,i)),x)/2
      +ee^j*pp(i)*a/2-intx(pp(i)*df(a,mkid(x,j)),x)/2 };
ppflux:=coeff(intx(-gg,x),ee);

ItoDriftDiff:=
for i:=1:m collect coeff(intx(-2*part(ppflux,i+1),x),ee);
```

14. Finally, Algorithm 21.8 computes the flux vector F for the evolution of the marginal probability, $\partial P/\partial t + \nabla\cdot\mathsf{F} = 0$. The operator `intx` performs integration by parts, using various powers of `ee` to label various vector components. An error in the evolution could be flagged by the first element of the set of components, the zeroth power of `ee`, being nonzero.

    One determines a corresponding SDE for the Kramers–Moyal expansion more easily from this flux vector F than from the corresponding Fokker–Planck equation. Algorithm 21.8 codes a further integration by parts of the flux components to determine the diffusion matrix and the Itô drift (actually minus twice the drift is inside `intx(.,x)` operators).

Example 21.4 (Levy area contraction continued). In the example SDE system (21.21), the algorithm finds the slow evolution on the slow subspace to be $\partial P/\partial t = -\nabla\cdot\mathsf{F}$ for the flux

$$\mathsf{F} = \frac{\sigma^2}{4}\begin{bmatrix} -P_{X_1} + X_2 P_{X_3} \\ -P_{X_2} - X_1 P_{X_3} \\ +2P + X_2 P_{X_1} - X_1 P_{X_2} - (X_1^2 + X_2^2) P_{X_3} \end{bmatrix} + \mathcal{O}(\sigma^3). \tag{21.23}$$

We could choose to write down a corresponding Itô or Stratonovich SDE corresponding to this flux: Algorithm 21.8 easily finds that the Itô form of the flux (21.23) is

$$\mathsf{F} = \begin{bmatrix} 0 \\ 0 \\ \sigma^2/2 \end{bmatrix} P - \tfrac{1}{2}\nabla\cdot\{DP\}, \quad \text{where } D = \frac{\sigma^2}{2}\begin{bmatrix} 1 & 0 & -X_2 \\ 0 & 1 & X_1 \\ -X_2 & X_1 & X_1^2+X_2^2 \end{bmatrix}$$

is the symmetric diffusion matrix. A little guesswork finds that this diffusion matrix factors as

$$D = \Sigma_X\Sigma_X^T, \quad \text{where} \quad \Sigma_X = \frac{\sigma}{2}\begin{bmatrix} 1 & -1 \\ 1 & 1 \\ X_1 - X_2 & X_1 + X_2 \end{bmatrix}$$

is a volatility matrix. Remembering the drift term in the flux (21.23), this Kramers–Moyal expansion of the slow evolution of the $\mathbf{X}$ variables is the Fokker–Planck equation of the Itô SDE

$$\begin{aligned} dX_1 &= \tfrac{1}{2}\sigma(dW_1' - dW_2'), \quad dX_2 = \tfrac{1}{2}\sigma(dW_1' + dW_2'), \\ dX_3 &= \tfrac{1}{2}\sigma^2 dt + \tfrac{1}{2}\sigma\left[(X_1 - X_2)dW_1' + (X_1 + X_2)dW_2'\right]. \end{aligned}$$

Because of the form of the volatility matrix Σ_X, Theorem 20.10 asserts that the corresponding Stratonovich SDE system has the same algebraic form. Let's also return to unscaled variables, $\mathbf{x} = \sigma\mathbf{X}$, by multiplying the above SDEs by σ, namely

$$\begin{aligned} \dot{x}_1 &= \tfrac{1}{2}\sigma^2(\dot{W}_1' - \dot{W}_2'), \quad \dot{x}_2 = \tfrac{1}{2}\sigma^2(\dot{W}_1' + \dot{W}_2'), \\ \dot{x}_3 &= \tfrac{1}{2}\sigma^3 + \tfrac{1}{2}\sigma\left[(x_1 - x_2)\dot{W}_1' + (x_1 + x_2)\dot{W}_2'\right]. \end{aligned}$$

This is a weak model of the Levy area contraction system because we do not know the relationship between the noise processes W_j' here and the original noise processes W_j of the original SDE system (21.20). Nonetheless, from the weak model here we see that the noise processes that affect the fast $\mathbf{y}$ variables in the SDE (21.20) feed into the slow x_1 and x_2, to then interact and drive a mean drift of the slow x_3. The weak model here quantifies the drift as $\dot{x}_3 \approx \tfrac{1}{2}\sigma^3$.

Does one have to scale the original variables? We do not necessarily have to scale the variables. Algorithms 21.4–21.8 are reasonably general. In order to work their magic on a reasonably wide range of problems, the algorithms have to be fairly brutal with their assumptions. Vary them as suits your particular problem.

Example 21.5. Algorithms 21.4–21.8 also analyze one of our favorite SDEs: $\dot{x} = -xy$ and $\dot{y} = -y + x^2 - 2y^2 + \sigma w(t)$. Section 21.2.2 analyzed this system in detail. Here, just change the specification of the SDE in Algorithm 21.4 to

```
f:=tp mat((-x1*y1)); sigmax:=tp mat((0));
g:=tp mat((-y1+x1^2-2*y1^2)); sigmay:=tp mat((1));
```

Executing the whole algorithm leads to the Kramers–Moyal expansion of the marginal PDF $P(t, X)$ for scaled variable $X = x/\sigma$ of

$$\frac{\partial P}{\partial t} = -\frac{\partial}{\partial X}\left[\sigma^2(\tfrac{3}{2}X - X^3)P\right] + \tfrac{1}{2}\frac{\partial^2}{\partial X^2}\left[\sigma^2 X^2 P\right] + \mathcal{O}(\sigma^3).$$

This Kramers–Moyal expansion is the Fokker–Planck equation of the Itô SDE $dX = \sigma^2(\frac{3}{2}X - X^3)dt + \sigma X\,dW'$. Unscaling the slow variable, and invoking Theorem 20.9 to rewrite in Stratonovich form, gives the SDE $\dot{x} = (\sigma^2 x - x^3) + \sigma x \dot{W}'$. This is the same weak model derived in section 21.2.2.

21.3 ▪ Established direct stochastic center manifold theory

This book generally develops backward theorems to justify using center manifolds for modeling: namely, there exists a system with marvelous properties that is close to the system specified (section 19.3.1). To appreciate the relative simplicity of this approach and to connect with other approaches, this section summarizes extant direct stochastic center manifold (SCM) theory (Roberts, 2006, §A); namely, does a given system possess an emergent SCM? This section summarizes only parts of the SCM theory of Boxler (1989). As Arnold (2003, p. 317) comments, the theory is "inevitably very technical."[258]

The approach to SCMs typically addresses a formal SDE in d dimensions: $\dot{\mathbf{x}} = \mathbf{F}(\omega, \mathbf{x})$, where ω parametrizes all the different realizations of the noise. However, stochastic theory is almost always better expressed in terms of integrals (section 20.2.1). Hence, here the theory is phrased in terms of the formal integral of the SDE starting from a generic initial condition, namely, $\mathbf{x} = \mathbf{x}_0$ at $t = 0$. Thus the stochastic ensemble trajectory/orbit $\mathbf{x} = \boldsymbol{\varphi}(t, \omega, \mathbf{x}_0)$, is defined to represent a stochastic/random dynamical system (RDS) (Arnold, 2003, §1.1).

Definition 21.2 (RDS (Boxler, 1989, Def. 2.1)). *Let $(\Omega, \mathcal{F}, P)$ be a probability space, and let ϑ_t be a group of bimeasurable measure-preserving bijections on Ω. A map $\boldsymbol{\varphi} : \mathbb{R} \times \Omega \times \mathbb{R}^d \to \mathbb{R}^d$, $(t, \omega, \mathbf{x}) \to \boldsymbol{\varphi}(t, \omega, \mathbf{x})$, is called a* **random dynamical system** *of C^k-diffeomorphisms ($k \geqslant 1$) on $\mathbb{R}^d$ over the flow $\{\vartheta_t : t \in \mathbb{R}\}$ if the following properties are satisfied:*

1. *$\boldsymbol{\varphi}(t, \cdot, \mathbf{x})$ is $\mathcal{F}, \mathcal{B}(\mathbb{R}^d)$-measurable for any $t \in \mathbb{R}$, $\mathbf{x} \in \mathbb{R}^d$.*

There is a ϑ_t-invariant set $\Omega_0 \subset \Omega$, $\Pr(\Omega_0) = 1$, such that for all $\omega \in \Omega_0$:

2. *$\boldsymbol{\varphi}(t, \omega, \mathbf{x})$ is continuous in $(t, \mathbf{x}) \in \mathbb{R} \times \mathbb{R}^d$.*

3. *$\boldsymbol{\varphi}(t, \omega, \cdot)$ is a C^k-diffeomorphism for any $t \in \mathbb{R}$. Herein, $C^{k,1}$ means that $D^k\boldsymbol{\varphi}(t, \omega, \cdot)$ satisfies a global Lipschitz condition.*

4. *$\boldsymbol{\varphi}(t+s, \omega, \mathbf{x}) = \boldsymbol{\varphi}(s, \vartheta_t\omega, \cdot) \circ \boldsymbol{\varphi}(t, \omega, \mathbf{x})$ for any $t, s \in \mathbb{R}$, $\mathbf{x} \in \mathbb{R}^d$ (cocycle property).*

This last cocycle property generalizes the semigroup property of integrals of deterministic differential equations. The shift in time ϑ_t of the realization ω accommodates the change in origin of time for the evolution over a time t from time 0, to the evolution over a time s from time t.

The Lipschitz condition of Property 3 of Definition 21.2 is one major difference between the approaches. In the direct theory, the dynamical system has to be Lipschitz in order to be assured that the center manifold is attractive despite very rare large fluctuations. In our backward theory, section 19.3.1 for example, the system need not be

[258]The theory by Aulbach and Wanner (1996, 2000) for nonautonomous deterministic systems of infinite dimensions with potentially infinite-dimensional center manifolds only requires nonautonomous effects to be measurable. Thus it may be straightforward to extend their theory to stochastic systems.

Lipschitz because we are empowered to make statements valid for as long as the system stays within an appropriate finite domain.

The stochastic theory is considerably simplified by transforming to a moving coordinate system attached to the orbit $\varphi(t,\omega,x)$. Without loss of generality the random dynamical system φ is hereafter assumed to have an equilibrium point at the origin $x=0$ (Boxler, 1989, Prop. 4.1). This is a common assumption.

Based about the dynamics near the origin we partition the dynamics into slow center modes, exponentially decaying modes, and exponentially growing modes. Oseledec's multiplicative ergodic theorem (Osceledec et al., 2012, e.g.) asserts that there is a decomposition of the state space $\mathbb{R}^d$ into a finite number of Oseledec spaces (analogous to eigenspaces), with each Oseledec space characterized by a different (real) Lyapunov exponent λ_i obtained from the linearized dynamics, and these Oseledec spaces collectively span the state space $\mathbb{R}^d$. The upper and lower bounds $-\beta = \max_{\lambda_i<0}\lambda_i$ and $\gamma=\min_{\lambda_i>0}\lambda_i$ are crucial in characterizing the time scale separation between the center, decaying, and growing modes, as analogously used in section 11.2 for deterministic autonomous systems. Then $E_c(\omega)$, $E_s(\omega)$, and $E_u(\omega)$ denote the spaces spanned by those Oseledec spaces respectively corresponding to zero, negative, and positive Lyapunov exponent. Subscripts c, s, and u denote projections onto the subspaces $E_c(\omega)$, $E_s(\omega)$, and $E_u(\omega)$. For example, we write $x=(x_c,x_s,x_u)$. The SCM is a nonlinear counterpart of the Oseledec space $E_c(\omega)$.

Now distinguish the linear and nonlinear parts of the RDS by the decomposition $\varphi(t,\omega,x)=T\varphi(t,\omega,0)x+\Phi(t,\omega,x)$. The direct theory only applies to RDSs with a nonlinear part that has sufficiently small derivatives, namely $\|D^j\Phi(t,\omega,x)\|<\epsilon_j(\omega)$ for $j=0,\dots,k$, and a Lipschitz condition on $D^k\Phi(t,\omega,x)$. Denote this class of RDSs by $NL_{\epsilon_0,\dots,\epsilon_k}$. The following existence theorem then applies.

Theorem 21.3 (existence (Boxler, 1989, Thm. 5.1)). *There is a constant* L_0 *such that for any* L, $0<L\leqslant L_0$, *and any* α, $0<\alpha<\frac{1}{4}\min(\beta,\gamma)$, *there are constants* $\epsilon_0(L)\geqslant\cdots\geqslant\epsilon_k(L)>0$, *which depend on* α, *and for any random dynamical system of class* $NL_{\epsilon_0(L),\dots,\epsilon_k(L)}$, *which consists of* $C^{k,1}$*-diffeomorphisms, there exists a* **global stochastic** $C^{k,1}$**-center manifold** *for* P, *where almost all* $\omega\in\Omega$. *It may be written in the form*

$$M(\omega)=\{(x_s,h_s(\omega,x_s),h_u(\omega,x_s))\mid x_s\in E_c(\omega)\}$$

with a function $h\in A_k(L)$.

The space $A_k(L)$ mentioned in this theorem denotes strictly nonlinear stochastic functions, $h(\omega,0)=0$ and $D^1h(\omega,0)=0$, which are smooth enough, that is, for all $x,\tilde{x}\in E_c(\omega)$ $|D^jh(\omega,x)|\leqslant\frac{1}{2}L$ for all $j=0,\dots,k$, and the Lipschitz condition $|D^kh(\omega,x)-D^kh(\omega,\tilde{x})|\leqslant\frac{1}{2}L|x-\tilde{x}|$. That is, the SCM $M(\omega)$ is written as a graph over the Oseledec space $E_c(\omega)$.

Such a *global* existence theorem is too restricted in practice. Much more useful is the existence of a *local* center manifold based upon the linearization at an equilibrium. When the deterministic part of the linearization has eigenvalues that are all zero (the imaginary part is zero), then we call the center manifold by the more informative name of a slow manifold.

Theorem 21.4 (local existence (Boxler, 1989, Thm. 6.1)). *Let* φ *be a random dynamical system on* $\mathbb{R}^d$ *that consists of* C^k*-diffeomorphisms (*$k\geqslant 2$*) and satisfies the assumptions of Definition* 21.2. *If one of the Lyapunov exponents of the linearization vanishes, then there will be a local stochastic* C^{k-1} *center manifold for* φ.

For modeling, we require the dynamics to settle onto a low-dimensional manifold so that the dynamics is described by the evolution of the relatively few variables parametrizing the manifold. SCM theory assures this when there are no growing modes, that is, when $E_u(\omega) = \emptyset$. The theory also asserts that the evolution of the orbits approaching the SCM is the same, to an exponentially decaying difference, as that of those orbits on the SCM. The following theorem guarantees the emergence of the dynamics on the SCM $M(\omega)$.

Theorem 21.5 (relevance (Boxler, 1989, Thm. 7.1)). *Let $M(\omega)$ be the SCM of Theorem 21.3. If all Lyapunov exponents are $\leqslant 0$, then there is a map $c_s : \mathbb{R}^+ \to (0,\infty)$ satisfying $\lim_{t\to\infty} \frac{1}{t} \log c_s(t) < 0$ such that for any initial values $\mathbf{x}_c \in E_c(\omega)$ and $\mathbf{x}_s \in E_s(\omega)$,*

$$|\varphi(t,\omega,\mathbf{x}_c,\mathbf{x}_s) - h[\vartheta_t\omega, \varphi(t,\omega,\mathbf{x}_c,\mathbf{x}_s)]|^s_{\vartheta_t\omega} \leqslant c_s(t)|\mathbf{x}_s - \mathbf{h}_s(\omega,\mathbf{x}_c)|^s_\omega$$

for any $t \geqslant 0$ almost surely.

Thus the difference between orbits on and off $M(\omega)$ is bounded by the initial difference and an essentially exponentially decaying function $c_s(t)$. Consequently, orbits on $M(\omega)$ model the dynamics from a wide variety of initial conditions.

Last, we may construct SCMs by approximately satisfying the governing SDEs. The following theorem asserts that the error in the SCM is of the same order as the residual of the governing equations of the random dynamical system. Consider that $\mathbf{x} = \eta(\omega,\mathbf{x}_c)$ approximates the actual SCM $\mathbf{x} = \mathbf{h}(\omega,\mathbf{x}_c)$, and let the operator V compute the residual of the governing equations, so that $V\eta$ is the residual for the approximation η. Then the following theorem determines the order of errors.

Theorem 21.6 (approximation (Boxler, 1989, Thm. 8.1)). *For a given $\eta \in A_k(L)$ assume $(V\eta)(\omega,\mathbf{x}_c) = \mathcal{O}(\|\mathbf{x}_c\|^q)$ for some $q > 1$, almost all $\omega \in \Omega$, and all $\mathbf{x}_c \in E_c(\omega)$ sufficiently small. Then*

$$\|\mathbf{h}(\omega,\mathbf{x}_c) - \eta(\omega,\mathbf{x}_c)\| = \mathcal{O}(\|\mathbf{x}_c\|^q)\,.$$

Finally, the theory of stochastic normal forms by Arnold (2003, §8.4) gives another and very appealing view of center manifold reduction. This direct theory of Arnold (2003) complements the backward theory we invoked in section 21.1. Corresponding to normal forms for deterministic systems, such a normal form of stochastic dynamics empowers us to simultaneously extract the SCM, derive the stochastic evolution thereon, and show emergence.

21.4 ▪ Exercises

Exercise 21.1 (example initial conditions). Recall the example of section 19.2.4. Recall that in the example the system (19.27) for the original variables $(x,\bar{x},x_3,y,\bar{y})$ is transformed to a system in $(X,\bar{X},X_3,Y,\bar{Y})$ expressed in equations (19.28), (19.29), and (19.31). Imagine we know initial values $(x_0,x_{3,0},y_0)$ and correspondingly for the complex conjugate variables. We need initial conditions, values, for $X(0)$ and $X_3(0)$ in order to make forecasts with the slow manifold dynamics (19.29) and (19.31).

- First, deduce that the corresponding initial $Y(0) = y_0 - e^{\beta t}\star w$.
- Second, deduce that $X(0) = x_0 - \frac{\epsilon}{\beta} y_0$. As for autonomous systems in general, $X_0 \neq x_0$.

- Third, rearrange the interesting slow variable transform and deduce

$$X_3(0) = x_{3,0} - \cdots - \frac{\epsilon^2\alpha}{|\beta|^2}\big(- \tfrac{1}{2}Y_0\bar{Y}_0 + Y_0 e^{-\beta t}\star e^{\bar{\beta} t}\star \bar{w} + \bar{Y}_0 e^{-\bar{\beta} t}\star e^{\beta t}\star w\big).$$

This expression involves convolutions $e^{\pm\beta t}\star$ and $e^{\pm\bar{\beta} t}\star$ that require knowledge of the future and the past of the forcing.

We cannot accurately forecast with the slow model without such anticipation and history.

Exercise 21.2 (noise-noise statistics I). Redo the analysis of section 21.1.3 to characterize the quadratic noise-noise interaction $w_1 e^{-t}\star w_2$ between *independent* white noises. Argue that $w_1 e^{-t}\star w_2$ may be replaced in a slow manifold model by simply $\tilde{w}/\sqrt{2}$ for some new white noise.

Exercise 21.3 (noise-noise statistics II). Via the Fokker–Planck equation, redo the analysis of section 21.2.1 to model the evolution of the PDF of the quadratic noise-noise interaction $w_1 e^{-t}\star w_2$ between *independent* white noises. Hence argue, as in exercise 21.2, that $w_1 e^{-t}\star w_2$ may be replaced in a slow manifold model by simply $\tilde{w}/\sqrt{2}$ for some white noise.

Exercise 21.4 (delayed noise interaction I). In many nonlinear problems, noise-noise interactions involve more history convolutions, such as $we^{-t}\star e^{-t}\star w$ (Roberts, 2006, e.g.).

- Argue that to model the Stratonovich interpretation of $we^{-t}\star e^{-t}\star w$ we need to understand $x(t)$ in the Stratonovich stochastic system

$$\dot{x} = zw\,, \quad \dot{y} = -y + w\,, \quad \dot{z} = -z + y\,. \tag{21.24}$$

- Derive that the Fokker–Planck equation for the system (21.24) for the PDF $p(t, x, y, z)$ is

$$\frac{\partial p}{\partial t} = \frac{\partial}{\partial y}[yp] + \frac{\partial}{\partial z}\big[(-y+z)p\big] + \tfrac{1}{2}z^2\frac{\partial^2 p}{\partial x^2} + z\frac{\partial^2 p}{\partial x \partial y} + \tfrac{1}{2}\frac{\partial^2 p}{\partial y^2}.$$

- Assuming x-derivatives are negligible, check that the PDFs $p \propto G(y,z) := \frac{2}{\pi}\exp(-2y^2+4yz-4z^2)$ are equilibria of this Fokker–Planck equation. Sketch contours of the PDFs G.
- Treating the marginal PDF $P(t,x)$ as a slowly varying function of t and x, check that the PDF

$$p = GP - 2z^2 G\frac{\partial P}{\partial x} + (2z^4 + \tfrac{5}{4}z^2 - \tfrac{3}{4}yz + \tfrac{1}{8}y^2)G\frac{\partial^2 P}{\partial x^2} + \mathcal{O}(\partial_x^3)$$

satisfies the Fokker–Planck equation provided that the marginal PDF obeys the diffusion equation $\partial P/\partial t = \frac{1}{8}\partial^2 P/\partial x^2 + \mathcal{O}(\partial_x^3)$.
- Hence argue that the long-term evolution of system (21.24) is modeled by $\dot{x} \approx \frac{1}{2}\tilde{w}$. Describe what this implies for the appearance of the convolution $we^{-t}\star e^{-t}\star w$ in a stochastic model.

Exercise 21.5 (delayed noise interaction II). Repeat Exercise 21.4 for the history convolution $w_1 e^{-t} \star e^{-t} \star w_2$, where w_1 and w_2 are independent Stratonovich white noise processes. Find that although the intermediate details are modified, the conclusion is the same.

Exercise 21.6 (bifurcation PDF). Explore the PDF of the stochastic bifurcation as parameter a crosses zero in the Stratonovich system

$$\dot{x} = ax - xy \quad \text{and} \quad \dot{y} = -y + x^2 + \sigma w(t)$$

for white noise w (Schöner and Haken 1986, §5; Chao and Roberts 1996, §2).

1. Scale the variables with the noise strength σ; then modify Algorithm 21.3 accordingly, and execute to construct the slow manifold to errors $\mathcal{O}(\sigma^3, a^3)$.
2. Deduce that the noise shifts the slow manifold slightly down relative to the deterministic case.
3. Find that the steady-state marginal PDF, if it exists, is of the form $P \propto |x/\sigma|^{2a-\sigma^2} e^{-(1-2a)x^2}$, approximately. Hence deduce that the PDF exists for parameter $a > -\sigma^2/2$ and is double peaked only for parameter $a > \sigma^2/2$, provided parameter a is small enough.
4. Interpret the PDE for the marginal $P(t, x)$ as a Fokker–Planck equation for a Stratonovich SDE. Similarly, what is a corresponding Itô SDE?

Exercise 21.7 (weak models via PDFs). Consider the following Stratonovich SDE systems. For each, and for a suitable parameter regime, apply Algorithms 21.4–21.8 to derive the slow subspace PDF and corresponding Kramers–Moyal expansion. Interpret the Kramers–Moyal expansion as the Fokker–Planck equation of a weak model for the original SDE system.

1. $\dot{x} = ax - xy$ and $\dot{y} = -y + x^2 + \sigma w(t)$ for small bifurcation parameter a (Schöner and Haken 1986, §5; Chao and Roberts 1996, §2). Compare with Exercise 21.6, which uses Algorithm 21.3.
2. Pavliotis and Stuart (2008) parametrized the "twist" in the fast variables by a parameter a, so model

$$\dot{x}_1 = \sigma y_1, \quad \dot{x}_2 = \sigma y_2, \quad \dot{x}_3 = x_1 y_2 - x_2 y_1,$$
$$\dot{y}_1 = -y_1 - a y_2 + \sigma \dot{W}_1, \quad \dot{y}_2 = +a y_1 - y_2 + \sigma \dot{W}_2.$$

3. Section 21.2.1 analyzed a basic irreducible noise-noise interaction: here model the next step in a hierarchy of irreducible noise-noise interactions (Roberts, 2006, §4.2):

$$\dot{x}_1 = \sigma y_1 \dot{W}, \qquad \dot{y}_1 = -\mu_1 y_1 + \sigma \dot{W},$$
$$\dot{x}_2 = \sigma y_2 \dot{W}, \qquad \dot{y}_2 = -\mu_2 y_2 + y_1.$$

What long-term drift appears in the weak Itô model?

4. The complementary problem to the previous hierarchy is when the interacting noise processes in the slow and fast variables are independent (Roberts, 2006, §5), so model

$$\dot{x}_1 = \sigma y_1 \dot{W}_1, \qquad \dot{y}_1 = -\mu_1 y_1 + \sigma \dot{W}_2,$$
$$\dot{x}_2 = \sigma y_2 \dot{W}_1, \qquad \dot{y}_2 = -\mu_2 y_2 + y_1 .$$

What long-term drift appears in the weak Itô model?

5. Weakly model the multiplicative triad system of Majda et al. (2002):

$$\dot{x}_1 = b_1 x_2 y, \quad \dot{x}_2 = b_2 x_1 y, \quad \dot{y} = -y + b_3 x_1 x_2 + \sigma \dot{W}.$$

6. Weakly model the additive triad system of Majda et al. (2002):

$$\dot{x} = b_1 y_1 y_2, \quad \dot{y}_1 = -y_1 + b_2 x y_2 + \sigma \dot{W}_1, \quad \dot{y}_2 = -y_2 + b_3 x y_1 + \sigma \dot{W}_2.$$

Notice that the multiple microscale noise sources show up as just one apparent noise in the macroscale weak model.

Exercise 21.8 (slow manifold from the Fokker–Planck equation). Consider Algorithms 21.4–21.8 for constructing the slow subspace of the Fokker–Planck equation for any given system of fast-slow SDEs. Strictly, the construction is nonuniform in slow variables $\mathbf{X}$. The challenge of this exercise is to show that postprocessing can, at least partially, achieve uniform validity. This is achieved by centering the Gaussian near the slow manifold of the SDEs, as in section 21.2.2. Show that changing the variables in the Gaussian from the fast variables $\mathbf{Y}$ to new variables $\mathcal{Y} = \mathbf{Y} - \mathbf{h}(\mathbf{X})$ for small $\mathbf{h}$ allows the shifted Gaussian $A \exp[S(\mathcal{Y})]P$ to match and eliminate some nonuniformities. Code postprocessing steps into Algorithms 21.4–21.8 that do this matching and elimination by requiring the mean $\int_{-\infty}^{\infty} \cdots \mathcal{Y}\, d\mathcal{Y} = \mathbf{0}$ for all PDF terms linear in the marginal PDF P. Check with section 21.2.2.

Part VII Summary

- Chapter 18 views the well-known method of averaging as a first approximation to a time-dependent coordinate transform that separates microscale fluctuations from macroscale mean dynamics.

  The analysis finds a couple of generic features that are little appreciated (section 18.3): the macroscale slow variable cannot be identical to the microscale slow variable, and the coordinate transform necessarily depends upon the future of the microscale fluctuations, albeit only a fast time scale ahead.

- In modeling nonautonomous systems, five principles guide the subjective decisions in constructing an illuminating coordinate transform: avoid unbounded terms, decouple fast processes from the slow, the "true" fast modes being zero is to form the exact slow manifold, minimize the number of terms in the slow system, and avoid fast time integrals in the slow evolution.

- Theorem 19.1 establishes that coordinate transforms can be constructed to satisfy these principles, in example systems (sections 19.1 and 19.2.4) and in general (section 19.2). The construction generally involves memory and anticipatory convolutions of micro–time scale fluctuations through solving homological equations (19.21) and (19.22) of the form

$$\hat{H} + \frac{\partial \hat{h}}{\partial t} - s\beta_j \hat{h} + \sum_{\ell=1}^{n} \beta_\ell Y_\ell \frac{\partial \hat{h}}{\partial Y_\ell} = \text{Res}\,, \quad s \in \{0, 1\},$$

  for corrections $\hat{H}$ and $\hat{h}$.

  However, on the slow manifold there need not be any anticipatory convolutions, and the evolution need not have memory convolutions in linear effects (Theorem 19.2).

- Section 19.3.1 uses the time-dependent normal-form coordinate transform to establish some backward propositions on existence and emergence. First, there exist asymptotically close systems that possess a nonautonomous slow manifold in a finite domain (Proposition 19.3). Second, for as long as the system stays within such a finite domain, the system approaches exponentially quickly a solution on the nonautonomous slow manifold (Proposition 19.4); that is, the slow manifold emerges.

- Section 19.3.3 shows how to use computer algebra to construct a slow manifold model of nonautonomous dynamics. Computer algebra is highly desirable in order to account for the potentially huge number of microscale nonlinear interactions of the various fluctuating modes. The approach is developed for the specific case of the system (19.42) but may be readily adapted to a variety of systems. The key challenge is to be able to code a separation of fluctuating components into secular and nonsecular components.

- As a special class of nonautonomous systems, stochastic differential equations (SDEs) represent dynamical systems with noise or uncertain effects.
- For modeling algebra and analysis we generally use the Stratonovich interpretation of stochastic calculus, not the Itô interpretation that is so useful in stochastic theory. To differentiate between the two interpretations in an SDE with drift μ and volatility σ,
  - the SDE $\dot X = \mu(X,t) + \sigma(X,t)\dot W$ or $\dot X = \mu(X,t) + \sigma(X,t)w$ denotes the Stratonovich interpretation,[259] whereas
  - the SDE $dX = \mu(X,t)dt + \sigma(X,t)dW$ denotes the Itô interpretation.

  To transform between the two, for a single variable and a single noise, the drift $\mu_{\text{Itô}} = \mu_{\text{Strat}} + \frac{1}{2}\sigma_{\text{Strat}}\sigma'_{\text{Strat}}$, while the volatility is unchanged, $\sigma_{\text{Itô}} = \sigma_{\text{Strat}}$.
- Itô's Lemma 20.7 gives the stochastic chain rule we need in some stochastic analysis: let $dX = \mu\, dt + \sigma\, dW$; then the process $Y(t) = f(t, X(t))$ has Itô differential

$$dY = \left(\frac{\partial f}{\partial t} + \mu\frac{\partial f}{\partial x} + \tfrac{1}{2}\sigma^2\frac{\partial^2 f}{\partial x^2}\right) dt + \sigma\frac{\partial f}{\partial x} dW.$$

- A good basic computational scheme for numerically integrating Itô SDEs is the Runge–Kutta-like Algorithm 20.4 (set $S_k = 0$ to integrate Stratonovich SDEs).

  One of the appealing features of Algorithm 20.4 is that it reduces, for small noise, to a well-known scheme for deterministic ODEs. Consequently, we expect the global error $\mathcal{O}(\|a\|h^2 + \|b\|h)$ for some norms of the drift and volatility. Such more general expressions of the error should be useful in multiscale simulations where the strength of the noise depends upon the macroscale time step, such as in the modeling of a stochastic Hopf bifurcation (Roberts, 2008, §5.4.2).
- The Fokker–Planck equation describes the evolution of the probability density function (PDF) $p(t,x)$ of a stochastic process. The PDF of the Itô SDE $d\mathbf{X} = \boldsymbol{\mu}\, dt + \sigma\, d\mathbf{W}$ satisfies the PDE (20.39),

$$\frac{\partial p}{\partial t} = -\sum_{i=1}^{n}\frac{\partial(\mu_i p)}{\partial x_i} + \sum_{i,j=1}^{n}\frac{\partial^2(D_{ij}p)}{\partial x_i\partial x_j},$$

  for symmetric diffusivity tensor $D_{ij} := \frac{1}{2}\sum_{k=1}^{m}\sigma_{ik}\sigma_{jk}$, whereas the PDF of the Stratonovich SDE $\dot{\mathbf{X}} = \boldsymbol{\mu} + \sigma\dot{\mathbf{W}}$ satisfies the PDE (20.40):

$$\frac{\partial p}{\partial t} = -\sum_{i=1}^{n}\frac{\partial(\mu_i p)}{\partial x_i} + \frac{1}{2}\sum_{k=1}^{m}\sum_{i,j=1}^{n}\frac{\partial}{\partial x_i}\left[\sigma_{ik}\frac{\partial}{\partial x_j}(\sigma_{jk}p)\right].$$

[259] Although many, especially mathematicians, use "$\circ\, dW$" to denote the Stratonovich interpretation.

- The time-dependent coordinate transform approach of Chapter 19 illuminates the modeling of slow-fast stochastic systems. The main reason is that the corresponding coordinate transforms for stochastic systems involve only the noise processes and their integrals (never their derivatives). Proposition 21.1 consequently appeals to Propositions 19.3 and 19.4 to assure us of the backward theorem that there are (many) stochastic systems arbitrarily asymptotically close to that specified that have emergent stochastic slow manifolds (SSMs).

  One difference with extant forward theory (section 21.3) is that forward theory is bedeviled by rare events that take the stochastic system outside the domain of validity, whereas the backward theory is easily phrased as "for as long as the system stays in the domain."

- The coordinate transform approach illuminates many aspects of modeling stochastic systems (section 21.1.1). Additive noises are transformed into multiplicative noise effects by nonlinearities. Although the decaying stable modes generally must anticipate the noise, the SSM itself and the evolution thereon need no anticipation. Due to the fast time scale noise, the slow variables are not quite what you might want to define. Although many people assume that the average SSM is the deterministic slow manifold, this is not quite correct.

  The stochastic coordinate transform also indicates that one of three cases occurs when using initial conditions to make predictions with a stochastic macroscale model (section 21.1.2): when the future and the history of the noise are known, when the future is known, or when neither is known.

- But in modeling nonlinear stochastic systems, irreducible noise-noise interactions, with fast time convolutions, necessarily occur in the supposed slow model (section 21.1.3): terms such as $we^{-\beta t}\star w$. Both analysis of the mean and variance of such terms, and the analytical modeling of the corresponding Fokker–Planck equation (section 21.2.1), indicate that such terms could be replaced by $\frac{1}{2}+\frac{1}{\sqrt{2\beta}}\tilde{w}(t)$ for some new independent white noise $\tilde{w}$ (Roberts, 2006, §4.6, §5). Further, an irreducible term $w_1e^{-\beta t}\star w_2$ for independent w_j could be replaced by just $\frac{1}{\sqrt{2\beta}}\tilde{w}(t)$. Such replacements then form a weak model.

- The slow manifold weak modeling of Fokker–Planck equations generalizes to a wide variety of stochastic systems (sections 21.2.2 and 21.2.3). We seek PDFs of the system that are approximately in product form—typically the product of a Gaussian with a marginal PDF. The marginal is then assumed to have slowly varying derivatives with respect to the slow variables so that the shear dispersion modeling of Chapter 7 applies. Given computer algebra algorithms may be readily adapted in order to similarly model any in a wide range of stochastic systems.

Bibliography

Abraham, R. H., and Shaw, C. D. (1983), *Dynamics—the geometry of behaviour. Part 1: Periodic Behaviour*, Aerial Press. (Cited on pp. 226, 227)

Abraham, R. H., and Shaw, C. D. (1988), *Dynamics—the geometry of behaviour. Part 4: Bifurcation behaviour*, Aeriel Press. (Cited on p. 227)

Abramowitz, M., and Stegun, I. A., eds (1965), *Handbook of mathematical functions*, Dover. (Cited on pp. 247, 680)

Abrams, D. M., and Strogatz, S. H. (2006), "Chimera states in a ring of nonlocally coupled oscillators," *Int. J. Bifurcation Chaos* **16**(1), 21–37. (Cited on p. 523)

Alekseenko, S. V., Markovich, D. M., and Shtork, S. I. (1996), "Wave flow of rivulets on the outer surface of an inclined cylinder," *Phys. Fluids* **8**, 3288–3299. (Cited on p. 295)

Alfimov, G. L., Kevrekidis, P. G., Konotop, V. V., and Salerno, M. (2002), "Wannier functions analysis of the nonlinear Schrödinger equation with a periodic potential," *Phys. Rev. E* **66**(046608), 1–6. (Cited on p. 501)

Arneodo, A., Coullet, P. H., Spiegel, E. A., and Tresser, C. (1985), "Asymptotic chaos," *Physica D* **14**, 327–347. (Cited on p. 136)

Arnold, L. (2003), *Random dynamical systems*, Springer Monographs in Mathematics, Springer. (Cited on pp. 586, 590, 591, 594, 628, 687, 714, 716)

Arnold, L., and Crauel, H. (1991), "Random dynamical systems," *in* L. Arnold, H. Crauel and J.-P. Eckmann, eds., *Lyapunov exponents*, Vol. 1486 of Lecture Notes in Mathematics, Springer, Berlin/Heidelberg, pp. 1–22. (Cited on p. 626)

Arnold, L., and Imkeller, P. (1998), "Normal forms for stochastic differential equations," *Probab. Theory Relat. Fields* **110**, 559–588. doi:10.1007/s004400050159. (Cited on pp. 178, 586, 591, 687, 689, 690)

Arnold, L., Sri Namachchivaya, N., and Schenk-Hoppé, K. R. (1996), "Toward an understanding of stochastic Hopf bifurcation: A case study," *Intl. J. Bifurcation & Chaos* **6**, 1947–1975. (Cited on p. 687)

Arnold, L., and Xu Kedai (1993), "Simultaneous normal form and center manifold reduction for random differential equations," *in* C. Perello, C. Simo, and J. Sola-Morales, eds., *Equadiff-91*, World Scientific, pp. 68–80. (Cited on pp. 687, 689)

Aulbach, B., and Wanner, T. (1996), "Integral manifolds for Caratheodory type differential equations in Banach spaces," *in* B. Aulbach and F. Colonius, eds., *Six lectures on dynamical systems*, World Scientific, Singapore, pp. 45–119. (Cited on pp. 151, 152, 250, 296, 488, 714)

Aulbach, B., and Wanner, T. (1999), "Invariant foliations for Caratheodory type differential equations in Banach spaces," *in* V. Lakshmikantham and A. A. Martynyuk, eds., *Advances of stability theory at the end of XX century*, Gordon & Breach Publishers. `http://citeseerx.ist.psu.edu/viewdoc/download?doi=10.1.1.45.5229&rep=rep1&type=pdf`. (Cited on pp. 151, 152)

Aulbach, B., and Wanner, T. (2000), "The Hartman–Grobman theorem for Caratheodory-type differential equations in Banach spaces," *Nonlinear Analysis* **40**, 91–104. doi:10.1016/S0362-546X(00)85006-3. (Cited on pp. 151, 152, 250, 296, 337, 488, 585, 593, 594, 600, 608, 714)

Balasuriya, S., and Padberg-Gehle, K. (2013), "Controlling the unsteady analogue of saddle stagnation points," *SIAM J. Appl. Math.* **73**(2), 1038–1057. (Cited on p. 603)

Batchelor, G. K. (1979), *An introduction to fluid dynamics*, Cambridge University Press. (Cited on pp. 233, 238, 244, 273, 412)

Baxter, M., and Rennie, A. (1996), *Financial calculus: An introduction to derivative pricing*, Cambridge University Press. (Cited on p. 572)

Bellman, R. (1955), "Perturbation methods applied to nonlinear dynamics," *J. Appl. Mech.* **22**, 500. (Cited on p. 308)

Bender, C. M., and Orszag, S. A. (1981), *Advanced mathematical methods for scientists and engineers*, McGraw–Hill. (Cited on pp. 6, 14, 31, 34, 45, 52, 54, 57, 58, 59, 61, 65, 93, 96, 97, 135, 464)

Bensoussan, A., and Flandoli, F. (1995), "Stochastic inertial manifold," *Stochastics and Stochastics Rep.* **53**, 13–39. (Cited on pp. 600, 687)

Bensoussan, A., Lions, J. L., and Papanicolaou, G. (1978), Asymptotic analysis for periodic structures, Vol. 5 of Studies in Mathematics and Its Applications. (Cited on p. 96)

Berglund, N., and Gentz, B. (2003), "Geometric singular perturbation theory for stochastic differential equations," *J. Diff. Equations* **191**, 1–54. doi:10.1016/S0022-0396(03)00020-2. (Cited on pp. 361, 368, 687)

Berry, M. V., and Shukla, P. (2011), "Slow manifold and Hannay angle in the spinning top," *European J. Phys.* **32**, 115. (Cited on pp. 164, 434)

Birch, D. A., and Young, W. R. (2006), "A master equation for a spatial population model with pair interactions," *Theoret. Population Biol.* **70**, 26–42. (Cited on p. 267)

Blanes, S., Casas, F., Oteo, J. A., and Ros, J. (2009), "The Magnus expansion and some of its applications," *Phys. Reports* **470**(5–6), 151–238. (Cited on p. 96)

Boffetta, G., Cencini, M., Falconi, M., and Vulpiani, A. (2002), "Predictability: A way to characterize complexity," *Phys. Reports* **356**, 367–474. (Cited on p. 526)

Bohl, P. (1904), "Über die Bewegung eines mechanischen Systems in der Nähe einer Gleichgewichtslage," *J. Reine und Angew. Math.* **127**, 179–276. (Cited on p. 151)

Boivin, N., Pierre, C., and Shaw, S. W. (1996), Non-linear normal modes, invariance, and modal dynamics approximations of a linear beam connected to a non-linear spring, Technical report. (Cited on p. 535)

Bonckaert, P., De Maesschalck, P., Doan, T. S., and Siegmund, S. (2014), Partial linearization for planar nonautonomous differential equations, Technical report, `http://www.researchgate.net/publication/259359409`. (Cited on pp. 618, 619)

Boxler, P. (1989), "A stochastic version of the center manifold theorem," *Probab. Th. Rel. Fields* **83**, 509–545. (Cited on pp. 594, 687, 688, 714, 715, 716)

Boxler, P. (1991), "How to construct stochastic center manifolds on the level of vector fields," *in* L. Arnold, H. Crauel and J.-P. Eckmann, eds., *Lyapunov Exponents*, Vol. 1486 of Lecture Notes in Mathematics, Springer Berlin/Heidelberg, pp. 141–158. (Cited on pp. 617, 687)

Brown, D. L., Bell, J., Estep, D., Gropp, W., Hendrickson, B., Keller-McNulty, S., Keyes, D., Oden, J. T., Petzold, L., and Wright, M. (2008), Applied mathematics at the U.S. Department of Energy: Past, present and a view to the future, Technical report, A Report by an Independent Panel from the Applied Mathematics Research Community, `http://science.energy.gov/~/media/ascr/pdf/program-documents/docs/Brown_report_may_08.pdf`. (Cited on p. x)

Bunder, J. E., and Roberts, A. J. (2013), Numerical integration of ordinary differential equations with rapidly oscillatory factors, Technical report, `http://arxiv.org/abs/1305.5022`. (Cited on p. 590)

Bykov, V., Goldfarb, I., Gol'dshtein, V., and Maas, U. (2006), "On a modified version of ILDM approach: Asymptotic analysis based on integral manifolds," *IMA J. Appl. Math.* **71**, 359–382. (Cited on p. 153)

Bykov, V., and Gol'dshtein, V. (2013), "Fast and slow invariant manifolds in chemical kinetics," *Comput. Math. Appl.* **65**, 1502–1515. (Cited on pp. 161, 368)

Camassa, R., Lin, Z., and McLaughlin, R. M. (2010), "The exact evolution of the scalar variance in pipe and channel flow," *Comm. Math. Sci.* **8**(2), 601–626. (Cited on p. 257)

Campbell, S. A. (2009), "Calculating center manifolds for delay differential equations using maple," *in* B. Balachandran et al., eds., *Delay differential equations: Recent advances and new directions*, Springer, Chap. 8. (Cited on pp. 105, 551)

Capiński, M. J., and Roldán, P. (2012), "Existence of a center manifold in a practical domain around L_1 in the restricted three-body problem," *SIAM J. Appl. Dyn. Systems* **11**, 285–318. (Cited on p. 125)

Carr, J. (1981), *Applications of centre manifold theory*, Vol. 35 of Applied Mathematical Sciences, Springer–Verlag. (Cited on pp. 113, 124, 128, 151, 197, 361)

Chang, H. C. (1994), "Wave evolution on a falling film," *Ann. Rev. Fluid Mech.* **26**, 103–136. doi:10.1146/annurev.fl.26.010194.000535. (Cited on pp. 283, 295)

Chang, H.-C., Demekhin, E., and Saprikin, S. S. (2002), "Noise-driven wave transitions on a vertically falling film," *J. Fluid Mech.* **462**, 255–283. doi:10.1017/S0022112002008856. (Cited on p. 295)

Chao, X., and Roberts, A. J. (1996), "On the low-dimensional modelling of Stratonovich stochastic differential equations," *Physica A* **225**, 62–80. doi:10.1016/0378-4371(95)00387-8. (Cited on pp. 398, 593, 687, 692, 695, 718)

Chapman, C. J., Childress, S., and Proctor, M. R. E. (1980), "Long wavelength thermal convection between non-conducting boundaries," *Earth & Planet. Sci. Lett.* **51**, 342–369. (Cited on p. 286)

Chapman, C. J., and Proctor, M. R. E. (1980), "Nonlinear Rayleigh–Benard convection between poorly conducting boundaries," *J. Fluid Mech.* **101**, 759–782. (Cited on pp. 286, 289)

Chatwin, P. C. (1970), "The approach to normality of the concentration distribution of a solute in a solvent flowing along a straight pipe," *J. Fluid Mech* **43**, 321–352. (Cited on pp. 254, 700)

Chicone, C. (2006), *Ordinary differential equations with applications*, Vol. 34 of Texts in Applied Mathematics, Springer. (Cited on pp. 157, 205, 337, 359, 372, 605)

Chicone, C., and Latushkin, Y. (1997), "Center manifolds for infinite dimensional nonautonomous differential equations," *J. Differential Equations* **141**, 356–399. `http://www.ingentaconnect.com/content/ap/de/1997/00000141/00000002/art03343`. (Cited on pp. 594, 608)

Chikwendu, S. C., and Ojiakor, G. U. (1985), "Slow-zone model for longitudinal dispersion in two-dimensional shear flows," *J. Fluid Mech.* **152**, 15–38. (Cited on p. 465)

Chow, S.-N., Drachman, B., and Wang, D. (1990), "Computation of normal forms," *J. Comput. Appl. Math.* **29**(2), 129–143. (Cited on p. 326)

Chueshov, I. D., and Scheutzow, M. (2001), "Inertial manifolds and forms for stochastically perturbed retarded semilinear parabolic equations," *J. Dynamics Differential Equations* **13**(2), 355–380. (Cited on pp. 126, 511)

Condon, M., Deano, A. A., and Iserles, A. (2010), "On systems of differential equations with extrinsic oscillation," *Discrete Cont. Dynamical Systems* **28**, 1345–1367. (Cited on p. 590)

Constable, G. W. A., McKane, A. J., and Rogers, T. (2013), "Stochastic dynamics on slow manifolds," *J. Phys. A: Math. Theoret.* **46**(29), 295002. (Cited on pp. 206, 689)

Cotter, C. J., and Reich, S. (2006), "Semigeostrophic particle motion and exponentially accurate normal forms," *Multiscale Model. Simul.* **5**(2), 476–496, `http://epubs.siam.org/doi/abs/10.1137/05064326x`. (Cited on p. 360)

Coullet, P. H., Elphick, C., and Tirapegui, E. (1985), "Normal form of a Hopf bifurcation with noise," *Phys. Lett.* **111A**(6), 277–282. (Cited on pp. 687, 690)

Coullet, P. H., and Spiegel, E. A. (1983), "Amplitude equations for systems with competing instabilities," *SIAM J. Appl. Math.* **43**, 776–821. (Cited on pp. 112, 169)

Cox, S. M., and Roberts, A. J. (1991), "Centre manifolds of forced dynamical systems," *J. Austral. Math. Soc. B* **32**, 401–436. (Cited on pp. 398, 399)

Cox, S. M., and Roberts, A. J. (1994), Initialization and the quasi-geostrophic slow manifold, Technical report, `http://arXiv.org/abs/nlin.CD/0303011`. (Cited on p. 429)

Cox, S. M., and Roberts, A. J. (1995), "Initial conditions for models of dynamical systems," *Physica D* **85**, 126–141. (Cited on pp. 370, 379, 382, 383, 384, 406, 585, 599)

Cross, M. C., and Hohenberg, P. C. (1993), "Pattern formation outside of equilibrium," *Rev. Mod. Phys.* **65**(3), 851–1112. doi:10.1103/RevModPhys.65.851. (Cited on pp. 211, 289, 292, 313, 622)

Cross, M. C., Tesauro, G., and Greenside, H. S. (1986), "Wavenumber selection and persistent dynamics in models of convection," *Physica D* **23**, 12–18. (Cited on pp. 266, 289, 292, 313)

de Carvalho Braga, D., Dias, F. S., and Mello, L. F. (2010), "On the stability of the equilibria of the Rikitake system," *Phys. Lett.* **374**, 4316–4320. (Cited on p. 556)

Dean, R. G., and Dalrymple, R. A. (1991), *Water wave mechanics for engineers and scientists*, World Scientific. (Cited on pp. 223, 228)

Dellnitz, M., Froyland, G., Horenkamp, C., Padberg-Gehle, K., and Gupta, A. S. (2009), "Seasonal variability of the subpolar gyres in the Southern Ocean: A numerical investigation based on transfer operators," *Nonlin. Processes Geophys.* **16**, 655–664. (Cited on p. 185)

Dellnitz, M., and Hohmann, A. (1997), "A subdivision algorithm for the computation of unstable manifolds and global attractors," *Numer. Math.* **75**, 293–317. doi:10.1007/s002110050240. (Cited on p. 690)

Domb, C., and Sykes, M. F. (1957), "On the susceptibility of a ferromagnetic above the Curie point," *Proc. Roy. Soc. Lond. A* **240**, 214–228. (Cited on pp. 195, 310)

Duan, J., Lu, K., and Schmalfuss, B. (2003), "Invariant manifolds for stochastic partial differential equations," *Ann. Probab.* **31**, 2109–2135. doi:10.1214/aop/1068646380. (Cited on p. 626)

Eisenberg, B. (2012), Life's solutions are complex fluids. A mathematical challenge, Technical report, `http://arxiv.org/abs/1207.4737`. (Cited on p. 224)

Elphick, C., Tirapeugi, E., Brachet, M. E., Coullet, P., and Iooss, G. (1987), "A simple global characterization for normal forms of singular vector fields," *Physica D* **29**, 95–127. doi:10.1016/0167-2789(87)90049-2. (Cited on pp. 379, 585)

Erneux, T. (2009), *Applied delay differential equations*, Vol. 3 of Surveys and Tutorials in the Applied Mathematical Sciences, Springer. (Cited on pp. 545, 551, 563)

Farrell, B. F., and Ioannou, P. J. (2009), "A stochastic structural stability theory model of the drift wave–zonal flow system," *Physics of Plasmas* **16**(11), 112903. (Cited on p. 557)

Fateman, R. (2003), "Comparing the speed of programs for sparse polynomial multiplication," *ACM SIGSAM Bulletin* **37**(1), 4–15. (Cited on p. 19)

Fenichel, N. (1974), "Asymptotic stability with rate conditions," *Indiana Univ. Math J.* **23**(2), 1109–1137. (Cited on pp. 126, 379)

Ferguson, J. (2004), A brief survey of the history of the calculus of variations and its applications, Technical report, http://arxiv.org/abs/arXiv:math/0402357. (Cited on p. 373)

Feynman, R. P. (1998), *Six easy pieces: The fundamentals of physics explained*, Penguin. (Cited on p. 359)

Fiedler, B., Liebscher, S., and Alexander, J. C. (2000), "Generic Hopf bifurcation from lines of equilibria without parameters: I. Theory," *J. Differential Equations* **167**, 16–35. (Cited on p. 159)

Foias, C., Jolly, M. S., Kevrekidis, I. G., Sell, G. R., and Titi, E. S. (1988), "On the computation of inertial manifolds," *Phys. Lett. A* **131**, 433–436. doi:10.1016/0375-9601(88)90295-2. (Cited on pp. 126, 511)

Foias, C., Jolly, M. S., Kevrekidis, I. G., and Titi, E. S. (1991), "Dissipativity of numerical schemes," *Nonlinearity* **4**, 591–613. (Cited on p. 487)

Fornberg, B. (1973), "On the instability of the leap-frog and Crank–Nicolson approximations of a nonlinear partial differential equation," *Math. Comput.* **27**, 45–57. (Cited on p. 487)

Froyland, G., and Padberg, K. (2009), "Almost-invariant sets and invariant manifolds—Connecting probabilistic and geometric descriptions of coherent structures in flows," *Physica D* **238**(16), 1507–1523. (Cited on p. 603)

Froyland, G., Padberg, K., England, M. H., and Treguier, A. M. (2007), "Detection of coherent oceanic structures via transfer operators," *Phys. Rev. Lett.* **98**(224503). (Cited on p. 185)

Gallay, T. (1993), "A center-stable manifold theorem for differential equations in Banach spaces," *Commun. Math. Phys.* **152**, 249–268. (Cited on p. 250)

Gear, C. W., Li, J., and Kevrekidis, I. G. (2003), "The gap-tooth method in particle simulations," *Phys. Lett. A* **316**, 190–195. doi:10.1016/j.physleta.2003.07.004. (Cited on p. 484)

Geigenmüller, U., Titulaer, U. M., and Felderhof, B. U. (1983), "Systematic elimination of fast variables in linear systems," *Physica A* **119**, 41–52. (Cited on pp. 377, 383)

Giddings, J. C. (1986), "Crossflow gradients in thin channels for separation by hyperlayer FFF, SPLITT cells, elutrition, and related methods," *Separation Sci. & Tech.* **21**, 831–843. (Cited on p. 262)

Grad, H. (1963), "Asymptotic theory of the Boltzmann equation," *Phys. Fluids* **6**, 147–181. (Cited on pp. 377, 383)

Grcar, J. F. (2011), "John von Neumann's analysis of Gaussian elimination and the origins of modern numerical analysis," *SIAM Review* **53**(4), 607–682. (Cited on pp. 152, 359, 423, 441, 604)

Greenside, H. S., and Coughran, W. M. (1984), "Nonlinear pattern-formation near the onset of Rayleigh–Benard convection," *Phys. Rev. A* **30**, 398–428. (Cited on pp. 211, 289, 292, 313, 622)

Guckenheimer, J. (1975), "Isochrons and phaseless sets," *J. Math. Biol.* **1**, 259–273. (Cited on pp. 377, 378, 379)

Guckenheimer, J., and Holmes, P. (1983), *Nonlinear oscillations, dynamical systems, and bifurcations of vector fields*, Springer–Verlag. (Cited on p. 178)

Guckenheimer, J., Johnson, T., and Meerkamp, P. (2012), Rigorous enclosures of a slow manifold, Technical report, `http://arxiv.org/abs/1201.1948`. (Cited on p. 125)

Haake, F., and Lewenstein, M. (1983), "Adiabatic drag and initial slip in random processes," *Phys. Rev. A* **28**, 3060–3612. (Cited on pp. 377, 384)

Haberman, R. (1987), *Elementary applied partial differential equations: With Fourier series and boundary value problems*, 2nd edition, Prentice–Hall. (Cited on p. 171)

Haken, H. (1983), *Synergetics: An introduction*, Springer, Berlin. (Cited on pp. 133, 695)

Haken, H. (1996), "Slaving principle revisited," *Physica D* **97**, 95–103. (Cited on p. 133)

Haragus, M., and Iooss, G. (2011), *Local bifurcations, center manifolds, and normal forms in infinite-dimensional dynamical systems*, Springer. doi:10.1007/978-0-85729-112-7. (Cited on pp. 121, 152, 178, 488, 548)

Hasselmann, K. (1976), "Stochastic climate models, Part I. Theory," *Tellus* **28**(6), 473–485. (Cited on p. 572)

Higham, D. J. (2001), "An algorithmic introduction to numerical simulation of stochastic differential equations," *SIAM Review* **43**(3), 525–546. (Cited on pp. 637, 666, 667)

Hinch, E. J. (1991), *Perturbation methods*, Cambridge Texts in Applied Mathematics, Cambridge University Press. (Cited on pp. 6, 14, 52, 63, 93)

Hutt, A. (2008), "Additive noise may change the stability of nonlinear systems," *Europhys. Lett.* **84**, 34003. (Cited on pp. 266, 289, 292, 313)

Hutt, A., Lefebvre, J., and Longtin, A. (2012), "Delay stabilizes stochastic systems near a non-oscillatory instability," *Europhys. Lett.* **98**(2), 20004. (Cited on p. 620)

Hutt, A., Longtin, A., and Schimansky-Geier, L. (2007), "Additive global noise delays Turing bifurcations," *Phys. Rev. Lett.* **98**, 230601. (Cited on p. 687)

Hutt, A., Longtin, A., and Schimansky-Geier, L. (2008), "Additive noise-induced Turing transitions in spatial systems with application to neural fields and the Swift–Hohenberg equation," *Physica D* **237**, 755–773. (Cited on p. 687)

Ion, A.-V. (2007), "An example of Bautin-type bifurcation in a delay differential equation," *J. Math. Anal. Appl.* **329**, 777–789. (Cited on pp. 561, 562)

Ion, A. V., and Georgescu, R. M. (2013), "Bautin bifurcation in a delay differential equation modeling leukemia," *Nonlinear Anal., Theory, Methods Appl.* **82**, 142–157. (Cited on p. 559)

Iooss, G., and Adelmeyer, M. (1992), *Topics in bifurcation theory*, World Scientific. (Cited on pp. 124, 127)

Iooss, G., and Lombardi, E. (2010), "Approximate invariant manifolds up to exponentially small terms," *J. Differential Equations* **248**(6), 1410–1431. (Cited on p. 360)

Iserles, A. (2004), "On the numerical quadrature of highly-oscillating integrals I: Fourier transforms," *IMA J. Numer. Anal.* **24**(3), 365–391. (Cited on p. 590)

Jolly, M. S., and Rosa, R. (2005), "Computation of non-smooth local center manifolds," *IMA J. Numer. Anal.* **25**, 698–725. (Cited on pp. 152, 153)

Jones, C. K. R. T., Kaper, T. J., and Kopell, N. (1996), "Tracking invariant manifolds up to exponentially small errors," *SIAM J. Math. Anal.* **27**, 558–577. (Cited on p. 360)

Just, W., Kantz, H., Rodenbeck, C., and Helm, M. (2001), "Stochastic modelling: Replacing fast degrees of freedom by noise," *J. Phys. A: Math. Gen.* **34**, 3199–3213. (Cited on p. 586)

Kan, X., Duan, J., Kevrekidis, I. G., and Roberts, A. J. (2013), "Simulating stochastic inertial manifolds by a backward-forward approach," *SIAM J. Appl. Dyn. Systems* **12**(1), 487–514. `http://arxiv.org/abs/1206.4954`. (Cited on p. 153)

Kao, E. P. C. (1997), *An introduction to stochastic processes*, Duxbury Press. (Cited on pp. 668, 671)

Keller, H., and Ochs, G. (1999), "Numerical approximation of random attractors," *in* H. Crauel and M. Gundlach, eds., *Stochastic dynamics*, Springer. `http://www.math.uni-bremen.de/Math-Net/preprints/ohne-metadaten/Report431.ps.gz`. (Cited on pp. 687, 690)

Kelley, A. (1967a), "On the Liapunov subcenter manifold," *J. Math. Anal. Appl.* **18**, 472–478. (Cited on p. 535)

Kelley, A. (1967b), "The stable, center-stable, center, center-unstable and unstable manifolds," *J. Differential Equations* **3**, 546–570. doi:10.1016/0022-0396(67)90016-2. (Cited on p. 151)

Kevrekidis, I. G., and Samaey, G. (2009), "Equation-free multiscale computation: Algorithms and applications," *Ann. Rev. Phys. Chem.* **60**, 321—44. (Cited on pp. 153, 187)

Khasminskii, R., and Klebaner, F. (2003), "A note on averaging and homogenization," *Stochastics and Dynamics* **3**(1), 113–120. (Cited on p. 687)

Kheshgi, H. S. (1989), "Profile equations for film flows at moderate Reynolds numbers," *AIChE Journal* **35**, 1719–1727. doi:10.1002/aic.690351017. (Cited on p. 295)

Kim, K., Mueller, C., and Sowers, R. B. (2010), A stochastic moving boundary value problem, Technical report, University of Illinois at Urbana-Champaign. (Cited on p. 209)

Kirkinis, E. (2012), "The renormalization group: A perturbation method for the graduate curriculum," *SIAM Review* **54**(2), 374–388. (Cited on pp. 74, 77)

Kirkinis, E., and O'Malley Jr., R. E. (2011), Asymptotic transformations for singularly perturbed ordinary and partial differential equations, Technical report, Department of Applied Mathematics, University of Washington. (Cited on p. 104)

Kitavtsev, G., Recke, L., and Wagner, B. (2011), "Centre manifold reduction approach for the lubrication equation," *Nonlinearity* **24**(8), 2347–2369. (Cited on p. 284)

Kloeden, P. E. (2001), A brief overview of numerical methods for stochastic differential equations, Technical report, Fachberiech Mathematik, Johann Wolfgang Goethe Universitat. (Cited on p. 666)

Kloeden, P. E., and Platen, E. (1992), *Numerical solution of stochastic differential equations*, Vol. 23 of Applications of Mathematics, Springer–Verlag. (Cited on pp. 590, 639, 643, 650, 652, 653, 655, 657)

Knobloch, E., and Wiesenfeld, K. A. (1983), "Bifurcations in fluctuating systems: The center-manifold approach," *J. Stat. Phys.* **33**, 611–637. (Cited on pp. 686, 695, 700)

Knobloch, H.-W., and Aulbach, B. (1982), "The role of center manifolds in ordinary differential equations," *in* M. Gregus, ed., *Proceedings of the fifth Czechoslovak conference on differential equations and their applications*, BSB B.G. Teubner Verlagsgesellschaft, Leipzig, pp. 179–189. (Cited on p. 151)

Kobayasi, K., and Takagi, S. (2003), On local center unstable manifolds, Technical report, `http://home.att.ne.jp/air/satorut/math/p1-002.pdf`. (Cited on p. 361)

Komori, Y. (2007a), "Multi-colored rooted tree analysis of the weak order conditions of a stochastic Runge–Kutta family," *Appl. Numer. Math.* **57**(2), 147–165. (Cited on p. 655)

Komori, Y. (2007b), "Weak order stochastic Runge–Kutta methods for commutative stochastic differential equations," *J. Comput. Appl. Math.* **203**(1), 57–79. (Cited on p. 655)

Komori, Y. (2007c), "Weak second-order stochastic Runge–Kutta methods for non-commutative stochastic differential equations," *J. Comput. Appl. Math.* **206**(1), 158—173. (Cited on p. 655)

Krauskopf, B., Osinga, H. M., Doedel, E. J., Henderson, M. E., Guckenheimer, J., Vladimirsky, A., Dellnitz, M., and Junge, O. (2005), "A survey of methods for computing (un)stable manifolds of vector fields," *Internat. J. Bifurcation & Chaos Appl. Sci. Engrg.* **15**(3), 763–791. (Cited on p. 152)

Krauskopf, B., Osinga, H. M., Doedel, E. J., Henderson, M. E., Guckenheimer, J., Vladimirsky, A., Dellnitz, M., and Junge, O. (2006), "A survey of methods for computing (un)stable manifolds of vector fields," *in* E. J. Doedel, G. Domokos and I. G. Kevrekidis, eds., *Modelling and computation in dynamical systems*, Vol. 13 of World Scientific series on Nonlinear Science, Series B, World Scientific, Chap. 4, pp. 67–96. (Cited on p. 326)

Kreyszig, E. (2011), *Advanced engineering mathematics*, 10th edition, Wiley. (Cited on pp. 45, 47, 59, 98, 114, 261, 262, 452, 454, 660, 669, 675, 676, 680)

Kuznetsov, Y. A. (1995), *Elements of applied bifurcation theory*, Vol. 112 of Applied Mathematical Sciences, Springer–Verlag. (Cited on pp. 328, 329)

Lamarque, C.-H., Touzé, C., and Thomas, O. (2012), "An upper bound for validity limits of asymptotic analytical approaches based on normal form theory," *Nonlinear Dynamics* **70**, 1931–1949. doi:10.1007/s11071-012-0584-y. (Cited on pp. 125, 535)

Laminie, J., Pascal, F., and Temam, R. (1993), "Implementation of finite element nonlinear Galerkin methods using hierarchical bases," *Comput. Mechanics* **11**, 384–407. (Cited on p. 211)

Lebedev, L. P., and Cloud, M. J. (2003), *The calculus of variations and functional analysis with optimal control and applications in mechanics*, World Scientific. (Cited on p. 373)

Lebiedz, D., and Rehberg, M. (2013), A numerical slow manifold approach to model reduction for optimal control of multiple time scale ODE, Technical report, `http://arxiv.org/abs/1302.1759v1`. (Cited on p. 620)

Leith, C. E. (1980), "Nonlinear normal mode initialisation and quasi-geostrophic theory," *J. Atmos. Sci.* **37**, 958–968. (Cited on p. 535)

Li, Z. (1999), Modelling shallow turbulent fluid dynamics and thin 3D fluid flows, PhD thesis, University of Southern Queensland. (Cited on p. 141)

Litzén, A. (1993), "Separation speed, retention, and dispersion in asymmetrical flow field-flow fractionation as functions of channel dimensions and flow rates," *Anal. Chem.* **65**, 461–470. (Cited on p. 263)

Lorenz, E., and Krishnamurty (1987), "On the non-existence of a slow manifold," *J. Atmos. Sci.* **44**, 2940–2950. (Cited on pp. 423, 429)

Lorenz, E. N. (1986), "On the existence of a slow manifold," *J. Atmos. Sci.* **43**, 1547–1557. (Cited on pp. 123, 129, 377, 400, 415, 416, 417, 418, 423, 429, 525)

Lorenz, E. N. (1996), "Predictability—A problem partly solved, *in* T. Palmer and R. Hagedorn, eds., *Predictability of weather and climate*, Cambridge University Press, pp. 40–58. (Cited on pp. 526, 527, 555, 556, 557)

Lorenz, E. N., and Emanuel, K. A. (1998), "Optimal sites for supplementary weather observations: simulation with a small model," *J Atmos. Sci.* **55**, 399–414. (Cited on p. 390)

Lu, K., and Schmalfuss, B. (to appear), Invariant manifolds for infinite dimensional random dynamical systems, World Scientific Review, World Scientific, chapter 10, pp. 315–345. (Cited on p. 626)

Maas, U., and Pope, S. B. (1992), "Simplifying chemical kinetics: Intrinsic low-dimensional manifolds in composition space," *Combustion and Flame* **88**, 239–264. (Cited on p. 153)

MacKenzie, T. (2005), Create accurate numerical models of complex spatio-temporal dynamical systems with holistic discretisation, PhD thesis, University of Southern Queensland. (Cited on pp. 446, 486, 505)

MacKenzie, T., and Roberts, A. J. (2003), Holistic discretisation of shear dispersion in a two-dimensional channel, *in* K. Burrage and R. B. Sidje, eds., *Proceedings of 10th computational techniques and applications conference* CTAC-2001, Vol. 44, pp. C512–C530. `http://journal.austms.org.au/ojs/index.php/ANZIAMJ/article/view/694`. (Cited on pp. 446, 505)

Majda, A., Timofeyev, I., and Vanden-Eijnden, E. (2002), "A priori tests of a stochastic mode reduction strategy," *Physica D* **170**, 206–252. (Cited on p. 719)

Mcphaden, M. J., and Ripa, P. (1990), "Wave-mean flow interactions in the equatorial ocean," *Ann. Rev. Fluid Mech.* **20**, 167–205. (Cited on p. 554)

Mercer, G. N., and Roberts, A. J. (1990), "A centre manifold description of contaminant dispersion in channels with varying flow properties," *SIAM J. Appl. Math.* **50**, 1547–1565. `http://epubs.siam.org/doi/abs/10.1137/0150091`. (Cited on pp. 247, 249, 310, 702)

Mercer, G. N., and Roberts, A. J. (1994), "A complete model of shear dispersion in pipes," *Jap. J. Indust. Appl. Math.* **11**, 499–521. (Cited on pp. 243, 255, 377, 701, 702)

Mitchell, S. L., and Myers, T. G. (2010), "Application of standard and refined heat balance integral methods to one-dimensional Stefan problems," *SIAM Review* **52**(1), 57–86. (Cited on pp. 147, 148)

Mocek, W. T., Rudnicki, R., and Voit, E. O. (2005), "Approximation of delays in biochemical systems," *Math. Biosic.* **198**(2), 190–216. (Cited on pp. 551, 558, 559, 560)

Mohammed, W. W., Blomker, D., and Klepel, K. (2013), Multi-scale analysis of SPDEs with degenerate additive noise, Technical report, Institut fur Mathematik, Universitat Augsburg. (Cited on p. 687)

Monahan, A. H., and Culina, J. (2011), "Stochastic averaging of idealized climate models," *J. Climate* **24**(12), 3068–3088. (Cited on pp. 583, 618)

Moon, S. J., Ghanem, R., and Kevrekidis, I. G. (2005), Coarse-graining the dynamics of coupled oscillators, Technical report, `http://arXiv.org/abs/nlin.AO/0509022`. (Cited on p. 523)

Muncaster, R. G. (1983), "Invariant manifolds in mechanics II: Zero-dimensional elastic bodies with directors," *Arch. Rat. Mech. Anal.* **84**, 375–392. (Cited on p. 411)

Murdock, J. (2003), *Normal forms and unfoldings for local dynamical systems*, Springer Monographs in Mathematics, Springer. (Cited on pp. 195, 199, 357, 358, 363, 380, 628)

National Physical Lab (1961), *Modern computing methods*, Vol. 16 of Notes on Applied Science, 2nd edition, Her Majesty's Stationery Office, London. (Cited on p. 481)

Nayfeh, A. H. (2008), "Order reduction of retarded nonlinear systems—The method of multiple scales versus center-manifold reduction," *Nonlinear Dynamics* **51**(4), 483–500. (Cited on pp. 564, 565, 567)

Nguyen, L. T., and Balakotaiah, V. (2000), "Modeling and experimental studies of wave evolution on free falling viscous films," *Phys. Fluids* **12**, 2236–2256. (Cited on p. 295)

Oksendal, B. K. (1998), *Stochastic differential equations: An introduction with applications*, Springer. (Cited on pp. 639, 643, 650)

Oliynyk, T. (2010), "Cosmological post-Newtonian expansions to arbitrary order," *Commun. Math. Phys.* **295**, 431–463. `http://dx.doi.org/10.1007/s00220-009-0931-0`. (Cited on p. 413)

O'Malley, Jr., R. E., and Kirkinis, E. (2010), "A combined renormalization group-multiple scale method for singularly perturbed problems," *Studies Appl. Math.* **124**, 383–410. (Cited on pp. 97, 98, 166)

Oron, A., Davis, S. H., and Bankoff, S. G. (1997), "Long-scale evolution of thin liquid films," *Rev. Mod. Phys.* **69**, 931–980. http://link.aps.org/abstract/RMP/v69/p931. (Cited on p. 284)

Osceledec, V., Izhikevich, E. M., Denninger, T., Orbeck, N., Meiss, J., and Arnold, L. (2012), Oseledets theorem, Technical report, Scholarpedia http://www.scholarpedia.org/article/Oseledets_theorem. (Cited on p. 715)

Pavliotis, G. A., and Stuart, A. M. (2008), *Multiscale methods: Averaging and homogenization*, Vol. 53 of Texts in Applied Mathematics, Springer. (Cited on pp. 368, 577, 601, 602, 687, 706, 718)

Pawula, R. F. (1967), "Approximation of the linear Boltzmann equation by the Fokker-Planck equation," *Phys. Rev.* **162**, 186–188. (Cited on p. 700)

Penland, C. (2003), "A stochastic approach to nonlinear dynamics: A review (extended version of the article – 'Noise out of chaos and why it won't go away')," *Bulletin of the American Meteorological Society* **84**(7), 925–925. (Cited on pp. 628, 638)

Potzsche, C., and Rasmussen, M. (2006), "Taylor approximation of integral manifolds," *J. Dynamics and Differential Equations* **18**, 427–460. (Cited on pp. 152, 250, 296, 346, 488, 594)

Potzsche, C., and Rasmussen, M. (2008), "Computation of integral manifolds for Carathéodory differential equations," *IMA J. Numer. Anal.* **30**(2), 401–430. (Cited on p. 594)

Prokopiou, T., Cheng, M., and Chang, H. C. (1991), "Long waves on inclined films at high Reynolds number," *J. Fluid Mech.* **222**, 665–691. (Cited on p. 295)

Rabier, F., Klinker, E., Courtier, P., and Hollingsworth, A. (1996), "Sensitivity of forecast errors to initial conditions," *Quart. J. Roy. Meteorol. Soc.* **122**, 121–150. (Cited on p. 390)

Rand, R. H., and Armbruster, D. (1987), *Perturbation methods, bifurcation theory and computer algebra*, Vol. 65 of Applied Mathematical Sciences, Springer–Verlag. (Cited on pp. 71, 176)

Renson, L., Deliege, G., and Kerschen, G. (2012), Finite element computation of nonlinear normal modes of nonconservative systems, *in Proceedings of the ISMA 2012 conference*. (Cited on p. 562)

Rhines, P. (1979), "Geostrophic turbulence," *Ann. Rev. Fluid Mech.* **11**, 401–441. (Cited on p. 413)

Riaza, R. (2012), "Manifolds of equilibria and bifurcations without parameters in memristive circuits," *SIAM J. Appl. Math.* **72**(3), 877–896. (Cited on p. 159)

Ribe, N. M. (2001), "Bending and stretching of thin viscous sheets," *J. Fluid Mech.* **433**, 135–160. (Cited on p. 304)

Roberts, A. J. (1985), "An introduction to the technique of reconstitution," *SIAM J. Math. Anal.* **16**, 1243–1257. http://epubs.siam.org/doi/pdf/10.1137/0516089. (Cited on p. 167)

Roberts, A. J. (1988), "The application of centre manifold theory to the evolution of systems which vary slowly in space," *J. Austral. Math. Soc. B* **29**, 480–500. doi:10.1017/S0334270000005968. (Cited on pp. 258, 303)

Roberts, A. J. (1989a), "Appropriate initial conditions for asymptotic descriptions of the long term evolution of dynamical systems," *J. Austral. Math. Soc. B* **31**, 48–75. (Cited on pp. 380, 389, 391, 398, 425)

Roberts, A. J. (1989b), "The utility of an invariant manifold description of the evolution of a dynamical system," *SIAM J. Math. Anal.* **20**, 1447–1458. http://epubs.siam.org/doi/abs/10.1137/0520094. (Cited on pp. 366, 511)

Roberts, A. J. (1992a), "Boundary conditions for approximate differential equations," *J. Austral. Math. Soc. B* **34**, 54–80. doi:10.1017/S0334270000007384. (Cited on p. 343)

Roberts, A. J. (1992b), "Planform evolution in convection—An embedded centre manifold," *J. Austral. Math. Soc. B* **34**, 174–198. (Cited on p. 267)

Roberts, A. J. (1993), "The invariant manifold of beam deformations. Part 1: The simple circular rod," *J. Elas.* **30**, 1–54. (Cited on p. 411)

Roberts, A. J. (1994), *A one-dimensional introduction to continuum mechanics*, World Scientific. (Cited on pp. 223, 228, 233, 672, 673)

Roberts, A. J. (1996), "Low-dimensional models of thin film fluid dynamics," *Phys. Lett. A* **212**, 63–72. doi:10.1016/0375-9601(96)00040-0. (Cited on pp. 283, 295, 297)

Roberts, A. J. (1997), "Low-dimensional modelling of dynamics via computer algebra," *Computer Phys. Comm.* **100**, 215–230. doi:10.1016/S0010-4655(96)00162-2. (Cited on pp. 388, 587)

Roberts, A. J. (2000), "Computer algebra derives correct initial conditions for low-dimensional dynamical models," *Comput. Phys. Comm.* **126**(3), 187–206. (Cited on pp. 388, 392, 394, 402, 403)

Roberts, A. J. (2001), "Holistic discretisation ensures fidelity to Burgers' equation," *Appl. Numer. Modelling* **37**, 371–396. doi:10.1016/S0168-9274(00)00053-2. (Cited on p. 485)

Roberts, A. J. (2003a), "Derive boundary conditions for holistic discretisations of Burgers' equation," *in* K. Burrage and R. B. Sidje, eds., *Proceedings of 10th computational techniques and applications conference* CTAC-2001, Vol. 44, pp. C664–C686. http://journal.austms.org.au/ojs/index.php/ANZIAMJ/article/view/701. (Cited on pp. 489, 490)

Roberts, A. J. (2003b), "A holistic finite difference approach models linear dynamics consistently," *Math. Comput.* **72**, 247–262. (Cited on p. 493)

Roberts, A. J. (2004), "Shear dispersion along circular pipes is affected by bends, but the torsion of the pipe is negligible," *SIAM J. Appl. Dyn. Systems* **3**, 433–462. http://epubs.siam.org/doi/abs/10.1137/030600886. (Cited on pp. 240, 243, 701)

Roberts, A. J. (2005), Computer algebra resolves a multitude of microscale interactions to model stochastic partial differential equations, Technical report, http://eprints.usq.edu.au/archive/00001242/. (Cited on p. 607)

Roberts, A. J. (2006), "Resolving the multitude of microscale interactions accurately models stochastic partial differential equations," *LMS J. Comput. Math.* **9**, 193–221. http://journals.cambridge.org/article_S146115700000125x. (Cited on pp. 607, 687, 690, 692, 714, 717, 718, 719, 723)

Roberts, A. J. (2007), Resolve subgrid microscale interactions to discretise stochastic partial differential equations, Technical report, http://arxiv.org/abs/math.DS/0601702. (Cited on p. 690)

Roberts, A. J. (2008), "Normal form transforms separate slow and fast modes in stochastic dynamical systems," *Physica A* **387**, 12–38. doi:10.1016/j.physa.2007.08.023. (Cited on pp. 593, 594, 600, 601, 608, 687, 690, 722)

Roberts, A. J. (2009a), *Elementary calculus of financial mathematics*, Vol. 15 of Mathematical Modeling and Computation, SIAM, Philadelphia. Appeared twice (two printings) in SIAM's top-25 best sellers for the year ending April 2010. (Cited on pp. 572, 637)

Roberts, A. J. (2009b), Holistic discretisation of dynamical partial differential equations, Technical report, http://www.maths.adelaide.edu.au/anthony.roberts/holistic1.html. (Cited on pp. 490, 500, 502)

Roberts, A. J. (2009c), "Model dynamics across multiple length and time scales on a spatial multigrid," *Multiscale Model. Simul.* **7**(4), 1525–1548. (Cited on pp. viii, 484)

Roberts, A. J. (2009d), Normal form of stochastic or deterministic multiscale differential equations, Technical report, http://www.maths.adelaide.edu.au/anthony.roberts/sdenf.php. Last revised June 2013. (Cited on pp. 347, 355, 586, 593, 603)

Roberts, A. J. (2009e), Slow manifold of stochastic or deterministic multiscale differential equations, Technical report, http://www.maths.adelaide.edu.au/anthony.roberts/sdesm.php. Last revised February 2012. (Cited on pp. 607, 689)

Roberts, A. J. (2012), Construct centre manifolds of ordinary or delay differential equations (autonomous), Technical report, http://www.maths.adelaide.edu.au/anthony.roberts/gencm.php. (Cited on pp. 100, 326, 541, 552)

Roberts, A. J. (2013), Macroscale, slowly varying, models emerge from the microscale dynamics in long thin domains, Technical report, http://arxiv.org/abs/1310.1541. (Cited on pp. 250, 257, 258, 275, 279, 284, 289, 303, 316, 384, 700)

Roberts, A. J., and Kevrekidis, I. G. (2007), "General tooth boundary conditions for equation free modeling," *SIAM J. Sci. Comput.* **29**(4), 1495–1510. (Cited on pp. 188, 194, 212, 484, 485)

Roberts, A. J., and Li, Z. (2006), "An accurate and comprehensive model of thin fluid flows with inertia on curved substrates," *J. Fluid Mech.* **553**, 33–73. doi:10.1017/S0022112006008640. (Cited on pp. x, 312)

Roberts, A. J., MacKenzie, T., and Bunder, J. (2014), "A dynamical systems approach to simulating macroscale spatial dynamics in multiple dimensions," *J. Engineering Mathematics* **86**(1), 175–207. (Cited on pp. 192, 446, 478, 486, 487, 505)

Robinson, J. C. (1996), "The asymptotic completeness of inertial manifolds," *Nonlinearity* **9**, 1325–1340. `http://www.iop.org/EJ/abstract/0951-7715/9/5/013`. (Cited on p. 126)

Romanovski, V. G., Mencinger, M., and Ferčec, B. (2013), "Investigation of center manifolds of three-dimensional systems using computer algebra," *Programming and Computer Software* **39**(2), 67–73. (Cited on pp. 525, 553)

Rosencrans, S. (1997), "Taylor dispersion in curved channels," *SIAM J. Appl. Math.* **57**, 1216–1241. (Cited on pp. 96, 243)

Rößler, A. (2009), "Second order Runge–Kutta methods for Itô stochastic differential equations," *SIAM J. Numer. Anal.* **47**(3), 1713–1738. (Cited on p. 655)

Rößler, A. (2010), "Runge–Kutta methods for the strong approximation of solutions of stochastic differential equations," *SIAM J. Numer. Anal.* **48**(3), 922–952. (Cited on p. 655)

Roy, R. V., Roberts, A. J., and Simpson, M. E. (2002), "A lubrication model of coating flows over a curved substrate in space," *J. Fluid Mech.* **454**, 235–261. doi:10.1017/S0022112001007133. (Cited on p. 377)

Ruyer-Quil, C., and Manneville, P. (2000), "Improved modeling of flows down inclined planes," *Eur. Phys. J. B* **15**, 357–369. `http://www.springerlink.com/link.asp?id=9ffp55pnhqkxe0lk`. (Cited on p. 295)

Samaey, G., Roose, D., and Kevrekidis, I. G. (2005), "The gap-tooth scheme for homogenization problems," *Multiscale Model. Simul.* **4**, 278–306. doi:10.1137/030602046. (Cited on p. 484)

Sanders, J. A., Verhulst, F., and Murdock, J. (2007), *Averaging methods in nonlinear dynamical systems*, Vol. 59, 2nd edition, Applied Math. Sciences. (Cited on p. 577)

Schmalfuss, B., and Schneider, K. R. (2008), "Invariant manifolds for random dynamical systems with slow and fast variables," *J. Dynamics and Differential Equations* **20**(1), 133–164. (Cited on p. 687)

Schöner, G., and Haken, H. (1986), "The slaving principle for Stratonovich stochastic differential equations," *Z. Phys. B—Condensed Matter* **63**, 493–504. (Cited on pp. 687, 718)

Schöner, G., and Haken, H. (1987), "A systematic elimination procedure for Ito stochastic differential equations and the adiabatic approximation," *Z. Phys. B—Condensed Matter* **68**, 89–103. (Cited on p. 687)

Schure, M. R., Barman, B. N., and Giddings, J. G. (1989), "Deconvolution of nonequilibrium band broadening effects for accurate particle size distributions by sedimentation field-flow fractionation," *Anal. Chem.* **61**, 2735–2743. (Cited on p. 262)

Schwartz, I., and Smith, H. (1983), "Infinite subharmonic bifurcations in an SEIR epidemic model," *J. Math. Biol.* **18**, 233–253. (Cited on p. 207)

Shaw, S. W. (1994), "An invariant manifold approach to nonlinear normal modes of oscillation," *J. Nonlinear Sci.* **4**, 419–448. (Cited on p. 414)

Shaw, S. W., and Pierre, C. (1993), "Normal modes for non-linear vibratory systems," *J. Sound Vibration* **164**(1), 85–124. (Cited on p. 414)

Shaw, S. W., and Pierre, C. (1994), "Normal modes of vibration for nonlinear continuous systems," *J. Sound Vibration* **169**(3), 319–347. (Cited on p. 414)

Shilnikov, L. P., Shilnikov, A. L., Turaev, D. V., and Chua, L. O. (1998), *Methods of qualitative theory in nonlinear dynamics. Part I*, Vol. 4 of World Scientific Series on Nonlinear Science, Series A, World Scientific. (Cited on p. 124)

Sijbrand, J. (1985), "Properties of center manifolds," *Trans. Amer. Math. Soc.* **289**, 431–469. (Cited on pp. 413, 414, 442)

Smith, R. (1987), "Diffusion in shear flow made easy: The Taylor limit," *J. Fluid Mech.* **175**, 201. (Cited on p. 256)

Sri Namachchivaya, N., and Leng, G. (1990), "Equivalence of stochastic averaging and stochastic normal forms," *J. Appl. Mech.* **57**, 1011–1017. (Cited on pp. 687, 692, 693)

Sri Namachchivaya, N., and Lin, Y. K. (1991), "Method of stochastic normal forms," *Internat. J. Nonlinear Mechanics* **26**, 931–943. (Cited on pp. 687, 693)

Stampfli, J., and Goodman, V. (2001), *The mathematics of finance: Modeling and hedging*, Brooks/Cole. (Cited on pp. 572, 643)

Stoleriu, I. (2012), Periodic orbits of a pair of coupled oscillators near resonance, Technical report, University of Iasi. (Cited on pp. 103, 555)

Stone, E., and Askari, A. (2002), "Nonlinear models of chatter in drilling processes," *Dynamical Systems* **17**(1), 65–85. (Cited on p. 105)

Summers, D. M. (2005), "Eddy diffusion in the sea: Reinterpreting an early experiment," *Proc. R. Soc. A* **461**, 1811–1827. (Cited on p. 222)

Sun, J., Bollt, E. M., and Nishikawa, T. (2009), Constructing generalized synchronization manifolds by manifold equation, Technical report, `http://arXiv.org/abs/0804.4180v2`. (Cited on p. 523)

Suslov, S. A., and Roberts, A. J. (1998), "Similarity, attraction and initial conditions in an example of nonlinear diffusion," *J. Austral. Math. Soc. B* **40**(E), E1–E26. `http://journal.austms.org.au/ojs/index.php/ANZIAMJ/article/view/453`. (Cited on p. 149)

Suslov, S. A., and Roberts, A. J. (1999), "Advection-dispersion in symmetric field-flow fractionation channels," *J. Math. Chem.* **26**, 27–46. doi:10.1023/A:1019169408365. (Cited on pp. 213, 262)

Suslov, S. A., and Roberts, A. J. (2000), "Modelling of sample dynamics in rectangular asymmetrical flow field-flow fractionation channels," *Anal. Chem.* **72**(18), 4331–4345. (Cited on pp. 213, 243)

Swift, J., and Hohenberg, P. C. (1977), "Hydrodynamic fluctuations at the convective instability," *Phys. Rev. A* **15**, 319–328. (Cited on pp. 266, 289, 292, 313)

Takens, F. (1974), "Singularities of vector fields," *Publ. Math. Inst. Hautes Etudes Sci.* **43**, 47–100. (Cited on pp. 4, 65, 70)

Takens, F., and Vanderbauwhede, A. (2010), "Local invariant manifolds and normal forms," *in* F. T. Henk Broer and B. Hasselblatt, eds, *Handbook of dynamical systems*, Vol. 3 of Handbook of Dynamical Systems, Elsevier Science, Chap. 3, pp. 89–124. (Cited on p. 178)

Takeshita, D., and Feres, R. (2010), "Higher order approximation of isochrons," *Nonlinearity* **23**(6), 1303–1323. (Cited on p. 96)

Taylor, G. I. (1953), "Dispersion of soluble matter in solvent flowing slowly through a tube," *Proc. Roy. Soc. Lond. A* **219**, 186–203. (Cited on pp. 243, 383)

Taylor, G. I. (1954), "Conditions under which dispersion of a solute in a stream of solvent can be used to measure molecular diffusion," *Proc. Roy. Soc. Lond. A* **225**, 473–477. (Cited on pp. 243, 255)

Temam, R. (1990), "Inertial manifolds," *Math. Intelligencer* **12**, 68–74. (Cited on pp. 126, 511)

Toledo, Y., and Agnon, Y. (2010), "A scalar form of the complementary mild-slope equation," *J. Fluid Mech.* **656**, 407–416. (Cited on pp. 515, 552)

Tutkun, M., and Mydlarski, L. (2004), "Markovian properties of passive scalar increments in grid-generated turbulence," *New J. Phys.* **6**. doi:10.1088/1367-2630/6/1/049. (Cited on p. 700)

Valorani, M., Goussis, D. A., Creta, F., and Najm, H. N. (2005), "Higher order corrections in the approximation of low-dimensional manifolds and the construction of simplified problems with the CSP method," *J. Comput. Phys.* **209**, 754–786. (Cited on pp. 153, 161)

Van Dyke, M. (1964), "Higher approximations in boundary-layer theory. Part 3. Parabola in uniform stream," *J. Fluid Mech.* **19**, 145–159. (Cited on pp. 308, 311, 464)

van Dyke, M. (1982), *An album of fluid motion*, Parabolic Press. (Cited on pp. 225, 226, 237)

van Dyke, M. (1984), "Computer-extended series," *Ann. Rev. Fluid Mech.* **16**, 287–310. (Cited on pp. 309, 310)

van Kampen, N. G. (1985), "Elimination of fast variables," *Phys. Reports* **124**, 69–160. (Cited on pp. 126, 163, 411, 419, 420)

Vanderbauwhede, A., and Iooss, G. (1988), "Center manifold theory in infinite dimensions," *Dynamics Reported* **1**, 125–163. (Cited on pp. 488, 548)

Verdugo, A., and Rand, R. (2008), "Center manifold analysis of a DDE model of gene expression," *Commun. Nonlinear Sci. Numer. Simulation* **13**, 1112—1120. (Cited on p. 545)

Verhulst, F. (1989), *Nonlinear differential equations and dynamical systems*, Springer-Verlag. (Cited on p. 577)

Verhulst, F. (2005), *Methods and applications of singular perturbations: Boundary layers and multiple timescales*, Vol. 50 of Texts in Applied Maths, Springer. (Cited on pp. 125, 127, 160, 368, 575, 577)

Wahlund, K.-G., and Giddings, J. G. (1987), "Properties of an asymmetrical flow field-flow fractionation channel having one permeable wall," *Anal. Chem.* **59**, 1332–1339. (Cited on p. 263)

Watt, S. D., and Roberts, A. J. (1995), "The accurate dynamic modelling of contaminant dispersion in channels," *SIAM J. Appl. Math.* **55**(4), 1016–1038. `http://epubs.siam.org/doi/abs/10.1137/S0036139993257971`. (Cited on p. 256)

Watt, S. D., and Roberts, A. J. (1996), "The construction of zonal models of dispersion in channels via matching centre manifolds," *J. Austral. Math. Soc. B* **38**, 101–125. (Cited on p. 256)

Wayne, C. E. (1997), "Invariant manifolds for parabolic partial differential equations on unbounded domains," *Arch. Rat. Mech. Anal.* **138**(3), 279–306. (Cited on p. 123)

Wiggins, S. (1990), *Introduction to applied nonlinear dynamical systems and chaos*, Springer–Verlag. (Cited on p. 515)

Winfree, A. (1974), "Patterns of phase compromise in biological cycles," *J. Math. Biol.* **1**, 73–95. (Cited on pp. 377, 378)

Wright, E. S. (2012), "Macrotransport in nonlinear, reactive, shear flows," *Commun. Pure Appl. Anal.* **11**(5), 2125–2146. (Cited on p. 243)

Yablonsky, G. S., Mareels, I. M. Y., and Lazman, M. (2003), "The principle of critical simplification in chemical kinetics," *Chem. Eng. Sci.* **58**, 4833–4842. (Cited on pp. 161, 371)

Yamaguchi, I., Ogawa, Y., Jimbo, Y., Nakao, H., and Kotani, K. (2011), "Reduction theories elucidate the origins of complex biological rhythms generated by interacting delay-induced oscillations," *PLoS ONE* **6**(11), e26497. (Cited on p. 545)

Young, W. R., Roberts, A. J., and Stuhne, G. (2001), "Reproductive pair correlations and the clustering of organisms," *Nature* **412**, 328–331. 19th July. (Cited on p. 267)

Index